2022 中非纺织服装国际论坛论文集

邱夷平　主编

東華大學出版社
·上海·

图书在版编目(CIP)数据

2022中非纺织服装国际论坛论文集 / 邱夷平主编
. --上海：东华大学出版社，2023.2
ISBN 978-7-5669-2108-6

Ⅰ.①2… Ⅱ.①邱… Ⅲ.①纺织工业-中国、非洲-国际学术会议-文集②服装工业-中国、非洲-国际学术会议-文集 Ⅳ.①TS1-53②TS941-53

中国版本图书馆CIP数据核字(2022)第162318号

2022中非纺织服装国际论坛论文集
Proceedings of 2021 Sino-Africa International Symposium on Textiles and Apparel
邱夷平　主编
责任编辑：符　芬　特约编辑：刘凡茜子　郑连刚　张　昆　魏笑笑　白志青　陈　伟

出 版 发 行：东华大学出版社(上海市延安西路1882号,邮政编码：200051)
本 社 网 址：dhupress.dhu.edu.cn
天猫旗舰店：http://dhdx.tmall.com
营 销 中 心：021-62193056 62373056 62379558
印　　刷：上海龙腾印务有限公司
开　　本：889 mm×1194 mm 1/16　印　张：37.75
字　　数：1 480千字
版　　次：2023年2月第1版　印　次：2023年2月第1次印刷
书　　号：ISBN 978-7-5669-2108-6　定　价：468.00元

Organization

Chairperson:

QIU Yiping (China) NZILA Charles (Kenya)

Advisory Committee of 2022 SAISTA

Chairperson:

YU Jianyong President of Donghua University

Honorary Advisor:

KIPROP Ambrose Acting Deputy Vice Chancellor of Moi University

QING Fengling Vice President of Donghua University

Academic Committee of 2022 SAISTA

Chairperson:

YU Jianyong Donghua University

Committee Members: (Alphabetically Sorted by Last Name)

DING Xuemei Donghua University

GU Bohong Donghua University

LI Jun Donghua University

OKECH Chris Moi University

QING Fengling Donghua University

WANG Zhidong Donghua University

XU Weilin Wuhan Textile University

ZHAO Lijun Donghua University

Organizing Committee of 2022 SAISTA: (Alphabetically Sorted by Last Name)

Chairperson:

GU Bohong (China)

Vice-Chairperson:

OKECH Chris (Kenya) ZHAO Mingwei (China)

Secretaries:

FU Fen (China) LIU Fanxizi (China) ZHAO Yuan (China)

Preface

Sino-Africa International Symposium on Textiles and Apparel and Sino-Africa Symposium on Cultural Exchange have been held for the eighth time since 2015 and it has been a great success in exchange of textiles and apparel research, education and culture between African and Chinese universities, research institutes, business, government, and international agencies. It focuses on not only progress of textile and apparel technologies, but also the environmentally friendly social and economic development of the world for a Community of Shared Future for Mankind. This indicates that the scope of the symposium has been expended from a specific domain of textiles and apparel to more generalized and global interests for a better future of mankind. A total of 125 research papers have been selected for the proceeding, almost doubling the number in the previous one. The authors come from 15 universities in 6 countries including Kenya, Ethiopia, Sudan, Uganda, Hungary, and China. These papers cover a wide range of topics such as protective clothing, textile engineering, nano technology, textile materials, wearable technologies, apparel design and technology, textile management and marketing, textile history and development trend, as well as Sino-Africa culture exchanges.

This symposium provides a valuable opportunity to review the latest progress in the collaboration and exchange in the textiles and apparel education and research, culture and social economic development between China and Africa. It will also shed a light on development of textile business and strategies for our battle against the social economic difficulties and problems in the future.

Professor YU Jianyong
Academician of Chinese Academy of Engineering
President, Donghua University

Contents

Silkworm Cocoon Defect Detection Based on K-YOLOX Algorithm

XING Bo[1,2], **WANG Jun**[3*]

[1] Department of Technical Textiles, College of Textiles, Donghua University, Shanghai *201620*, China

[2] University Lille, ENSAIT, ULR *2461* -GEMTEX-Génie et Matériaux Textiles, F-*59000* Lille, France

[3] *Key Laboratory of Textile Science & Technology, Ministry of Education, Donghua University, Shanghai 201620, China*

[*] *Corresponding author's email*: junwang@ dhu. edu. cn

Abstract: To address the current problem of low automatic cocoon inspection, a cocoon defect detection algorithm (K-YOLOX) is proposed based on the YOLOX target detection algorithm, combined with cocoon characteristics. The traditional YOLOX algorithm is lightened and improved by using the K-means algorithm for clustering, which is used to obtain an anchor that matches the cocoon size; the loss function's composition ratio is changed, and the GIoU border loss function is added to improve cocoon classification and localization accuracy. According to the test findings, the average detection accuracy of the K-YOLOX algorithm for cocoon photos is 94. 2%, and the complete evaluation index *F*1 is 92. 5%, an improvement of 10. 3% in accuracy and 14. 2% in *F*1 over YOLOX.

Keywords: Defect detection; Silkworm cocoon; YOLO

1 Introduction

With a lengthy history of the silk industry, China is the world's origin of the silk industry and the first country to find and use silk[1]. Following the collection of fresh cocoons, silk reeling enterprises often dry them first before converting them to raw silk through a series of processes such as mixing, stripping, filtering, boiling, and spinning. Cocoons can be separated into normal and defective cocoons throughout the selection procedure based on their use value in silk spinning manufacture. Normal cocoons are those that are rid of flaws in terms of color, thickness, and fold. Defective cocoons are those that have major issues that prevent or make it impossible to produce regular raw silk cocoons.

Currently, cocoon sorting relies on the cocoon selector to identify the completion of manual sorting. However, because manual sorting is time-consuming and labor-intensive, and the quality of identification is heavily reliant on the cocoon selector's subjective consciousness, researchers at domestically and overseas have begun to investigate how to solve this problem using image recognition technology in recent years. Chen Hao et al.[2] proposed an auxiliary detection method for cocoon selection, using MATLAB to graphically process and numerically calculate cocoon images and accurately calibrate them with a camera to finally derive the surface defect area of cocoons. Prasobhkumar et al.[3] proposed a cocoon excellent quality rating system by statistically measuring the RGB color threshold of yellow spot cocoons with the specific parameters of normal cocoon ellipse aspect ratio. Yang Guangyou et al.[4] proposed the idea of a total neural network application for cocoon detection, integrating BP and RBF neural networks, and the accuracy of cocoon species detection reached 87. 8%.

The consistency of domestic and foreign scholars' research on normal and defective cocoons is lesser, which cannot meet the requirements of cocoon selection accuracy in the scheduled production process; the existing methods in the literature for cocoon detection cannot meet the needs of simultaneous multiple targets and cannot be applied to the aforementioned research on cocoon defect detection algorithm.

The target detection algorithms based on deep learning that have emerged in recent years provide new ideas for cocoon defect detection, of which the representative algorithms are R-CNN[5] series and YOLO[6] series. The R-CNN series algorithm is a target detection algorithm based on region generation. Although the detection accuracy is high, the detection time is long and does not meet the criteria for real-time cocoon detection in the production line. The YOLO series algorithm is a target detection algorithm based on the regression method, with a multi-scale detection mechanism, fast recognition speed, and accuracy, can also basically reach the level of the R-CNN series algorithm.

In this paper, we use defective and normal cocoons as research objects, and we use the K-means[7] clustering algorithm to obtain the anchor size suitable for cocoons; furthermore, we use the GIoU[8] function in the loss function of YOLOX and adjust the classification ratio to build the K-YOLOX (K-mean-YOLOX) network structure based on the basic shape of cocoons and recognition needs. Experiments demonstrate that the new algorithm not only achieves high accuracy in cocoon defect detection, but also satisfies the need for multi-target identification.

2 Experiments

2.1 Construction of cocoon image dataset

The image collection equipment was used to acquire 500 photographs of cocoons, and the cocoon data was augmented by scaling, flipping, and panning for cocoons in various postures and forms, resulting in a total of 10 000 images. In batches, all photos were randomly separated into training and test sets.

The data set was labeled using Labelimg software, which automatically generated the corresponding x, y, w, and h values based on the location of the labeled cocoons in each image, recorded the specific cocoon type information, generated the cocoon image labeling file in XML format, and stored it in a standard VOC data set format after participating in the cocoon image acquisition.

2.2 K-mean clustering anchor

Because there are so many distinct things to identify, varied sizes of anchors must be built ahead of time to optimize prediction matching. The standard YOLOX method is pre-trained using 20 items of various sizes from the VOC dataset as examples, resulting in a variety of anchor sizes. However, because cocoon defect detection does not need a substantial size difference between the anchors, the cocoon defect detection anchor must be reset.

The K-mean algorithm clusters the dataset by selecting K objects at random as cluster centers, dividing each of them to the nearest cluster center by calculating the distance between the rest of the objects and the cluster center, and repeating this process until the cluster center no longer changes after the next division.

2.3 K-YOLOX detection network

This research builds the K-YOLOX detection network in terms of resetting the loss function for the cocoon's own features to increase the accuracy of cocoon defect identification based on the YOLOX network.

The conventional YOLOX algorithm's typical loss function consists of three-loss functions: coordinate, confidence, and classification. In YOLOX, We primarily increase cocoon flaw detection accuracy by substituting the coordinate loss function and altering the overall percentage of classification loss in the standard loss function. Fig. 1 shows the improved network structure.

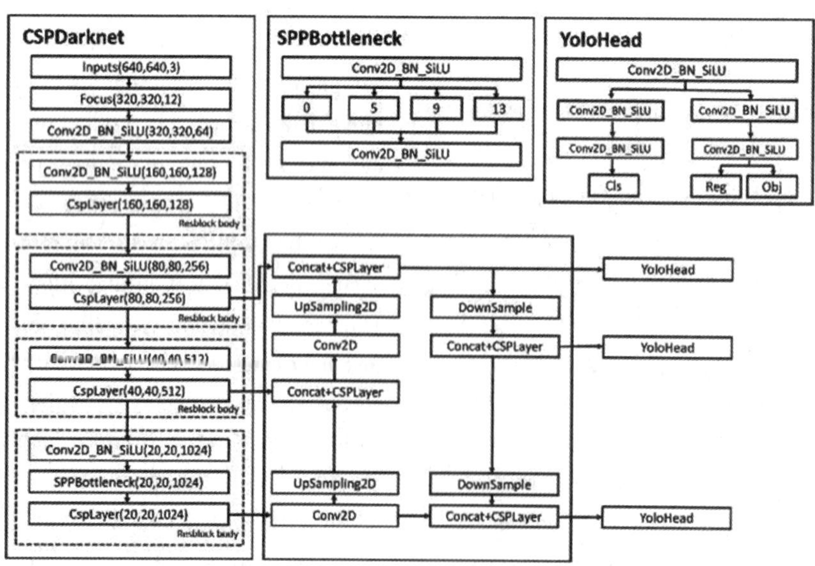

Fig. 1 Improved network structure

The mean square error function is utilized in the classic YOLOX technique for border coordinate regression (MSE). The MSE function has the drawback of treating the center point coordinates, width, and height of cocoon defect detection as distinct variables when there is a relationship among them, and it will enhance the instability of the border regression coordinates. As a result, this paper introduces the GIoU function to replace the mean square error function. The GIoU function is an optimization function of IoU, and its duty is to precisely determine the degree of overlap between the predicted frame of cocoon flaws and the real frame. L_{GIoU} loss function specific formula is derived as follows.

$$IoU = \frac{A \cap B}{A \cap B}$$

$$GIoU = IoU - \frac{C - (A \cup B)}{C}$$

$$L_{GIoU} = 1 - GIoU$$

For the non-overlapping section between A and B, the L_{GIoU} loss function is likewise included in the attention range. As a result, the L_{GIoU} loss function is chosen as the standard loss function in order to improve the accuracy of the coordinates of the faulty cocoon prediction framework.

When the recognition object is an object with low differentiation of surface features, such as the defect region on the cocoon, the standard loss function is ineffective. Thus this paper proposes the idea of strengthening the classification weight parameter to increase the total percentage of classification loss in the standard loss function for cocoon defect detection. The redesigned loss function formula is as follows.

$$\text{Loss} = \varphi_{co} \sum_{i=0}^{s^2} \sum_{j=0}^{B} I_{ij}^{obj} L_{GIoU} + \sum_{i=0}^{s^2} \sum_{j=0}^{B} I_{ij}^{obj} (C_i^j - \hat{C}_i^j)^2 + \varphi_{no} \sum_{i=0}^{s^2} \sum_{j=0}^{B} I_{ij}^{obj} (C_i^j - \hat{C}_i^j)^2 + 10 \sum_{i=0}^{s^2} I_i^{no} \sum_{c \in classes}^{s^2} [P_i^j(c) - \hat{P}_i^j(c)]^2$$

where s is the number of grids in the output layer of the network, B is the number of cocoon identification a priori frames, φ_{co} is the coordinate loss weight of the cocoon prediction frame, φ_{no} is the confidence loss function for not containing cocoons in the prediction frame, I_{ij}^{obj} is the specific number of cocoons in the ith grid in the cocoon image, P_i^j is the prediction probability when the jth prior frame of the ith grid identifies the defective cocoons, and $\hat{P}_i^j$ is the true probability that the jth prior frame of the ith grid identifies the specific class of cocoon.

2.4 Evaluation indicators

To evaluate the quality of cocoon defect identification, precision P, recall R, mean average accuracy MAP, and $F1$ score were used as evaluation indices. The calculation equations are as follows.

$$P = T_P/(T_P + F_P)$$

$$R = T_P/(T_P + T_N)$$

$$F1 = 2PR/(P + R)$$

$$MAP = \frac{1}{C} \sum_{K=1}^{C} J(P, R, K)$$

where T_P is the number of correct judgments in the prediction, F_P is the number of incorrect judgments, T_N is the number of missed detections, C is the number of cocoon defective species, K is the species serial number, and J is the area function of the P-R curve and the coordinate axis.

3 Results and Discussion

3.1 Desizing effect on fabric tensile properties

The improvement effect of K-YOLOX cocoon detection algorithm is tested by extracting images of cocoons with different species distributions and different number distributions in the defective cocoon dataset. The performance tests were performed by constructing three cocoon defect detection algorithms: the original YOLOX algorithm (method 1), the YOLOX algorithm based on K-means clustering (method 2), and the YOLOX algorithm based on K-means clustering with adjusted loss function (method 3). The evaluation indices are mainly P, MAP, and $F1$ score, and the test results are shown in Tab. 1.

Tab. 1 Improvement effect test

Method	P(%)	MAP(%)	$F1$(%)
Method 1	86.8	85.4	80.9
Method 2	90.6	91.9	93.4
Method 3	95.3	95.7	95.1

According to the data in Tab. 1, it can be obtained that the K-means clustering method improves the mean accuracy of cocoon defect detection by 6.5% and increases the $F1$ score. By comparing the $F1$ score and MAP between method 2 and method 3, it can be concluded that resetting the loss function can effectively improve the recognition accuracy of cocoon defect detection.

The visual inspection results are shown in Fig. 2. The K-YOLOX algorithm can precisely locate and mark defective cocoons with defective markings.

The K-YOLOX cocoon defect detection algorithm proposed in this paper further adds K-means clustering and resets the loss function on the basis of YOLOX, and according to the test results, the $F1$ and MAP of K-YOLOX reach 95. 1% and 95. 7%, respectively. The K-YOLOX algorithm can basically meet the technical requirements for intelligent detection of cocoon selection for the simultaneous identification of multiple samples in the cocoon selection process.

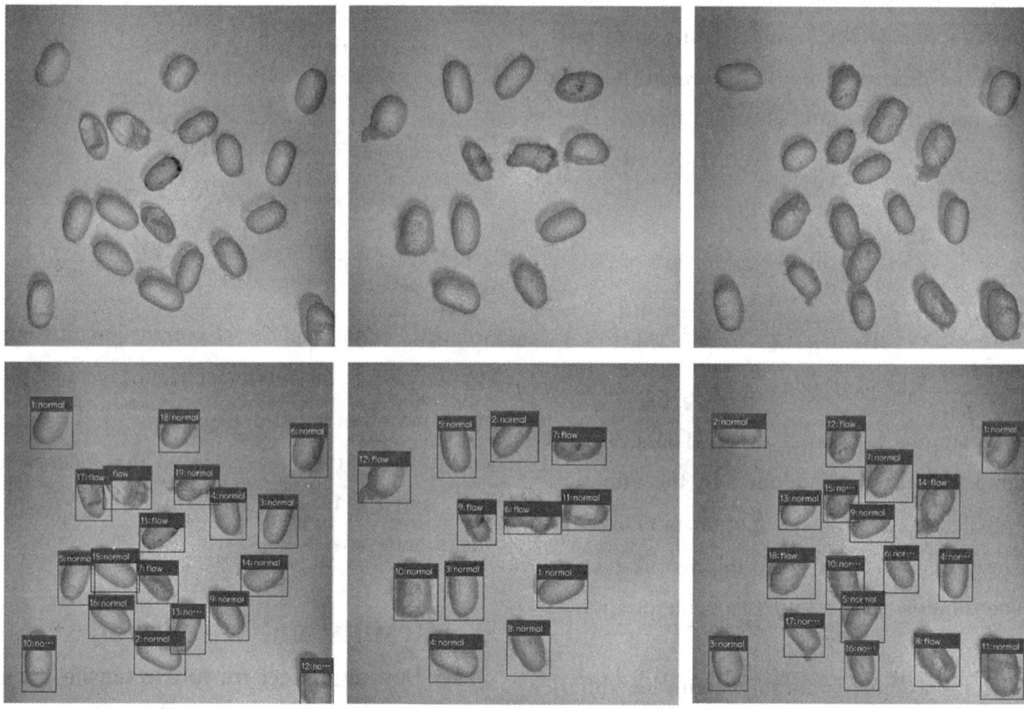

Fig. 2 Visual inspection effect

4 Conclusions

To solve the technical problem of fast and accurate identification of defective cocoons in the cocoon selection process, this paper proposes a cocoon defect detection algorithm (K-YOLOX). Taking defective cocoons and normal cocoons as the research object, the K-mean method is used to cluster the cocoon data set, so as to obtain a priori frames suitable for cocoons, which provides technical support for cocoon detection and localization; the classification accuracy of cocoons is improved by adjusting the classification ratio in the loss function and replacing the localization loss function. Comparative experiments are conducted to verify the correctness of the improved behavior by comparing the algorithm before and after the improvement, and the evaluation indices such as the average recognition accuracy and $F1$ score are used to compare YOLOX with K-YOLOX. From the experimental results, it can be obtained that the K-YOLOX algorithm has high accuracy, and the average accuracy and $F1$ score reach 95. 7% and 95. 1% respectively, which can meet the cocoon defect detection needs.

References

[1] THORLEY J. The silk trade between China and the Roman Empire at its height[J]. Greece & Rome, 1971, 18(1): 71-80.

[2] HAO C, ZHENG Y, XIA L, et al. Study on auxiliary testing method for mulberry silkworm cocoon sorting based on MATLAB[J]. Journal of Silk, 2016, 53(3): 32-36.

[3] PRASOBHKUMAR P, FRANCIS CGORTHI S S. Automated quality assessment of cocoons using a smart camera based system[J]. Engineering in Agriculture, Environment and Food, 2018, 11(4): 202-210.

[4] YANG G Y, JIANG J Y, ZHOU G Z, et al. Application of neural network ensemble on the non-destructive testing

of pod[J]. Silk, 2008, 5:28-30.

[5] REN S, HE K, GIRSHICK R, et al. Faster r-cnn: Towards real-time object detection with region proposal networks[J]. Advances in Neural Information Processing Systems, 2015.

[6] LEE Y-HKIM Y. Comparison of CNN and YOLO for object detection[J]. Journal of the Semiconductor & Display technology, 2020, 19(1): 85-92.

[7] HARTIGAN J AWONG M A. Algorithm AS 136: A k-means clustering algorithm[J]. Journal of the Royal Statistical Society. Series C (Applied Statistics), 1979, 28(1): 100-108.

[8] CHENGMING ZRONGGANG X. YOLOX target detection algorithm combining GIOU and focal loss[J]. Computer Engineering and Applications, 2020, 56(24): 214-222.

Investigation of Polymer Nano-Yarns by Direct Electrospinning

SHAO Qingqing[1,2], **DU Zhaoqun**[1,2], **YU Weidong**[1,2*]

[1] *Key Laboratory of Textile Science & Technology, Ministry of Education, Donghua University, Shanghai 201620, China*

[2] *Department of Textile Materials and Design, College of Textiles, Donghua University, Shanghai 201620, China*

* *Corresponding author's email*: wdyu@ dhu. edu. cn

Abstract: Yarns are key building blocks to construct complicated fibrous structures for diverse applications. In this paper, we report a simple but efficient and versatile method to prepare nano-yarns directly through an electrospinning process. An auxiliary electrode was used to induce the spinneret tips with relatively short collection distances during the electrostatic spinning process. The twisting structure of nanofibers is a very important factor in obtaining polymeric nano-yarns from jets of fiber bundles by this electrospinning method, which may have an important impact on the tensile properties of the yarn.

Keywords: Polymer material; Nanofiber; Nano-yarn; Electrospinning

1 Introduction

Micro-nano products have a very high volumetric specific surface area, about 1 000 times less than that of a human hair, as well as small size, interfacial effects, quantum size effects, and quantum tunneling effects, which can be widely used in nano-catalysis, tissue scaffolds, protective clothing, filtration, and electronic devices[1]. Methods to fabricate micro-nano fiber products include phase separation techniques[2], microfluidics[3], 3D printing[4], and electrospinning[5], which has been regarded as the most effective and versatile technology for producing nanofibers with controlled fiber morphology, dimensions and functional components from various polymeric materials[6-8]. However, most electrospun fibers are produced in the form of randomly oriented nonwoven fiber mats. The relatively low mechanical strength and the difficulty in tailoring the fibrous structure have restricted their applications[9-12].

To solve these problems, the researchers tried to improve the spinning device to make the fiber assembly into a fiber bundle with a specific orientation or to be bundled and twisted to obtain a yarn with a fineness of micrometers and certain mechanical properties. For example, the nanofiber bundles are prepared by placing the electrospun nanofibers in a highly oriented arrangement utilizing different receiving devices[13-17]. Twisting increases the direct connection, and cohesion of the nanofibers promotes the friction between them, and thus effectively improves the yarn's uniformity. So, the mechanical properties of the yarns are strong. More recently, there are different methods of adding a twist to nanofiber bundles to become nano-yarns during electrospinning that have been reported, which can be divided into core twisting[18-20] and non-core twisting[21-22]. However, issues remain with these methods in controlling the twist and yarn dimensional spinning stability. In the case of two oppositely placed syringes to produce a "tapered" hollow nanofiber network on the middle funnel collector, and continuously pulled out the yarn from the "tapered" apex[23]. The twist of the yarn could pass through the funnel, which currently used electrospinning method[24].

Although existing electrospinning techniques can prepare different polymer nano-yarns, these techniques offer an indirect way, usually with special spinning, receiving, and guiding devices. In this study, we have demonstrated a simple, but versatile and efficient process to directly spin nano-yarns with a simple receiving device. Polyacrylonitrile (PAN) was used as a model polymer to elucidate the process of nano-yarn preparation.

2 Experiments

2.1 Materials

Powdered polyacrylonitrile (PAN) and *N*, *N*-dimethylformamide (DMF) were obtained from Sigma-Aldrich. Powdered PAN was dissolved in DMF at 80 ℃ to form a solution containing 17% PAN, which had a viscosity of 1712 cP and a conductivity of 118 μS/cm.

2.2 Electrospinning of nano-yarns

Polymer nano-yarns were collected directly using an auxiliary electrode as shown in Fig. 1. The polymer solution was filled in a syringe. The spinneret was connected to a positive high voltage power supply of +20 kV. The auxiliary electrode was a copper electrode with a diameter of 2.5 mm and a length of 30 cm, which was grounded. The distance from the nozzle to the collector used in this work was kept at 5 cm. A metal needle tip with an inner diameter of 0.71 mm was installed on the syringe containing the polymer solution. The flow rate of the fluid in the syringe can be controlled by a micro-pump. During electrospinning, the environmental temperature was (25±2) ℃ and the relative humidity was maintained at 55%±5%.

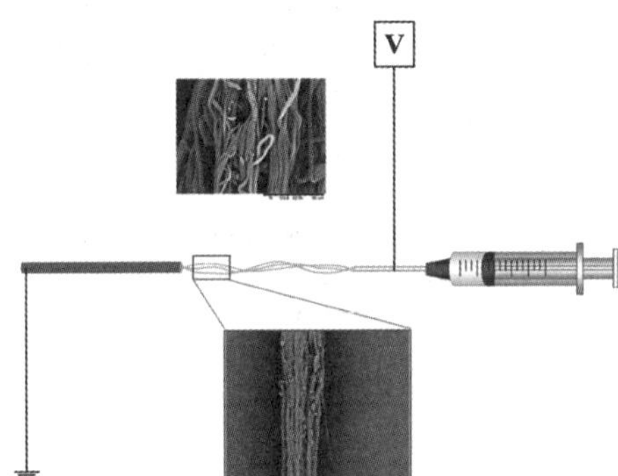

Fig. 1 The schematic diagram illustrating the experimental setup for polymer nano-yarns

2.3 Characterizations

The morphology of the prepared nano-yarns was observed under a scanning electron microscope (SEM) (TM3000). The twist angle of the yarn was measured based on the SEM images using image analysis software (Image J). That is, the higher the twist angle, the higher the twist of the yarn, and the better the mechanical and morphological stability of the yarn. The tensile properties were measured with the electronic single fiber tester (LLY-06 EDC), at a gauge length of 10 mm, a crosshead speed of 10 mm/min, and the preload tension of 0.1 cN.

$$Stress = \frac{Breakingforce}{\pi \times Radius \times Radius}$$

$$Strain = \frac{Breakinglength - Gaugelength}{Gaugelength} \times 100\%$$

3 Results and Discussion

The main difference between our electrospinning device and the conventional electrospinning apparatus is that an auxiliary electrode is used as a receiving device. In the presence of an electrostatic field, the positively charged fibers electrospun from the spinneret could be spontaneously attracted to the grounded collector and formed into yarns by self-generated electric field perturbation. All these changes in the electrospinning setup in this work were beneficial to make the self-bundling more broadly applicable and easier to adopt and use.

The representative SEM image of the self-forming polymer nano-yarns prepared by this direct electrospinning method is shown in Fig. 2. It can be seen that the electrospun fiber aggregates are arranged into bundles of fibers in a certain direction, rather than disordered nonwoven membrane. The magnified image of the nano-yarns in Fig. 2 (a) is shown in Fig. 2 (b), and it clearly shows that there are a large number of beads in the yarn. One of the reasons may be that the solvent does not have time to evaporate due to the short receiving distance. The twisting structure could be observed in Fig. 2 (c), which may be beneficial in increasing the strength of the polymer nano-yarns. Using Image J to measure the twist angle of the yarn to characterize the twist level of the nano-yarns, the twist angle of the yarn was about 17.60°, which indicated that the adhesive force between the fibers of nano-yarns was poor, and the morphological stability and mechanical properties of the yarn were weak. The nanofibers in polymer nano-yarns presented two main types, namely straight fibers and non-straight fibers, which included bent and knotted ones in Fig. 2 (d). During the movement of electrospun nanofibers toward the auxiliary electrode, most of the fibers would gradually straighten and align along the direction of the auxiliary electrode due to the electric field force. However, some fibers could not be completely straightened during the orientation process, forming bent or knotted fibers. Although knowledge of the mechanism is limited, this likely involved the receiving distance and electric field.

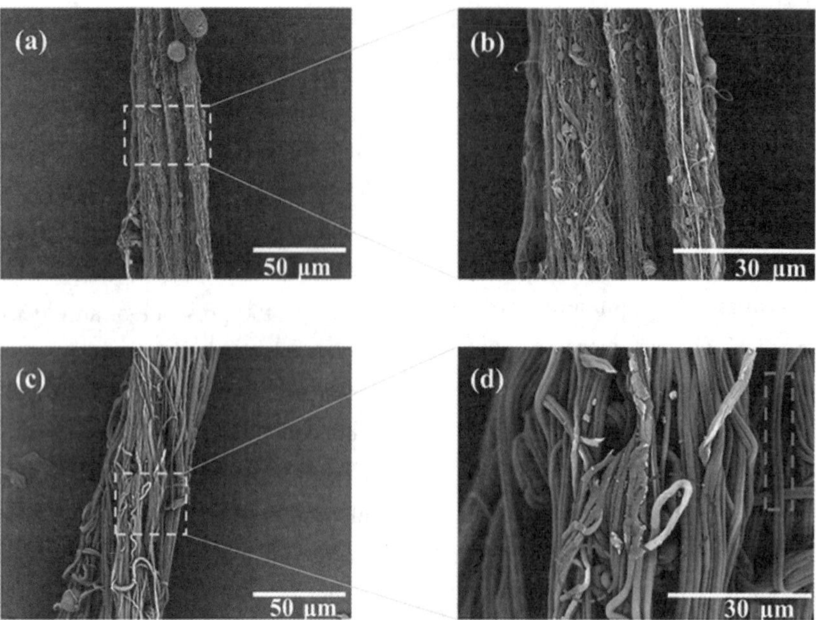

Fig. 2 SEM images of (a) the fiber bundle and (b) larger magnification of panel; (c) the nano-yarns and (d) larger magnification of panel

To further analyze the influence of the electric field during yarn formation, COMSOL Multiphysics software was used to simulate the electrostatic field during the yarn preparation. For the convenience of the calculation, the whole device was simplified into a two-dimensional model with specific parameters of the injector (0.71 mm × 15 mm), grounded rod (2.5 mm × 30 mm), and steady-state simulation to obtain the relevant field strength and potential distribution. It can be seen (Fig. 3) that the positively charged jet generated during electrospinning is attracted by the grounded rod, concentrating the charged jet between the ground rod and the syringe needle tip, which is consistent with a good distribution of the fiber aggregates between the two. However, due to various disturbances and perturbations of the electric field by the instrumental parts, the reason for the twisting of the fiber aggregates cannot be explained clearly, which can be explained by further studies.

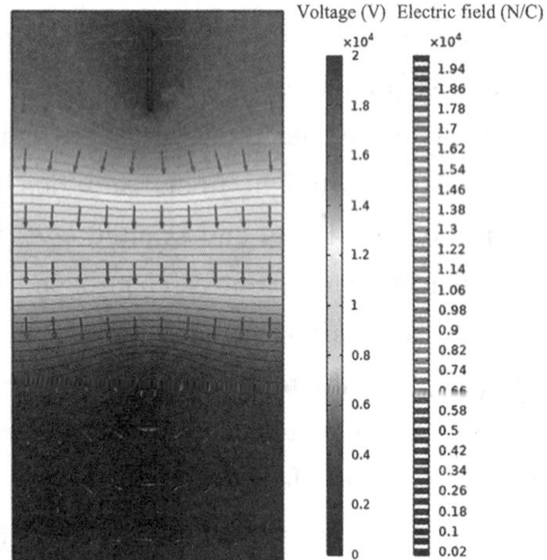

Fig. 3 The calculated static electric field distribution between spinnerets and receiving electrodes

Tab. 1 Mechanical properties of the nano-yarns and fiber nonwoven strip (width 2 mm)

Materials	Stress (MPa)	Strain (%)	Initial modulus (MPa)
Nano-yarn	18.78	55	852
Fiber nonwoven strip	12	110	750

The nano-yarn showed higher tensile strength than that of the fiber nonwoven strip (Tab. 1). The yarn tensile strength and elongation at break were 18.78 MPa and 55%, respectively, while the tensile elongation at break of the fiber nonwoven strip is 12 MPa, and the elongation at break is 110%. The author proposed that the twisted fibers in the yarn were beneficial to improve the mechanical properties. In addition, these yarns could be further processed into knitted fabrics and other fiber structures.

4 Conclusions

We have demonstrated a direct electrospinning process for making polymer nano-yarns, using a special setup consisting of an auxiliary electrode. The process showed excellent ability to produce nano-yarns, which have better mechanical properties than membranes. Since a grounded electrode was employed in yarn production, the device could also be used to produce and combine two different nanofibers into a single yarn with two syringes. Nanofiber yarns produced by this method may be useful for the construction of complicated nanofibrous assemblies for various applications, especially in biomedical equipment, catalysis, reinforcement, and fiber-based microelectronic.

References

[1] HUANG Z M, ZHANG Y Z, KOTAKI M, et al. A review on polymer nanofibers by electrospinning and their applications in nanocomposites [J]. Composites Science and Technology, 2003, 63(15): 2223-2253.

[2] ZHAO J, HAN W, CHEN H, et al. Preparation, structure and crystallinity of chitosan nano-fibers by a solid-liquid phase separation technique [J]. Carbohydrate Polymers, 2011, 83(4): 1541-1546.

[3] DU X Y, LI Q, WU G, et al. Multifunctional micro/nanoscale fibers based on microfluidic spinning technology [J]. Advanced Materials, 2019, 31(52): 1903733.

[4] LIASHENKO I, ROSELL-LLOMPART J, CABOT A. Ultrafast 3D printing with submicrometer features using electrostatic jet deflection[J]. Nature Communications, 2020, 11(1): 1-9.

[5] HERRICKS T E, KIM S H, KIM J, et al. Direct fabrication of enzyme-carrying polymer nanofibers by electrospinning[J]. Journal of Materials Chemistry, 2005, 15(31): 3241-3245.

[6] FRENOT A, CHRONAKIS I S. Polymer nanofibers assembled by electrospinning [J]. Current Opinion in Colloid & Interface Science, 2003, 8(1): 64-75.

[7] LI D, WANG Y, XIA Y. Electrospinning of polymeric and ceramic nanofibers as uniaxially aligned arrays [J]. Nano Letters, 2003, 3(8): 1167-1171.

[8] DERSCH R, GRAESER M, GREINER A, et al. Electrospinning of nanofibres: Towards new techniques, functions, and applications[J]. Australian Journal of Chemistry, 2007, 60(10): 719-728.

[9] NARAGHI M, CHASIOTIS I, KAHN H, et al. Novel method for mechanical characterization of polymeric nanofibers[J]. Review of Scientific Instruments, 2007, 78(8): 085108.

[10] CROISIER F, DUWEZ A S, JÉRÔME C, et al. Mechanical testing of electrospun PCL fibers[J]. Acta Biomaterialia, 2012, 8(1): 218-224.

[11] NEUGIRG B R, BURGARD M, GREINER A, et al. Tensile versus AFM testing of electrospun PVA nanofibers: bridging the gap from microscale to nanoscale[J]. Journal of Polymer Science Part B: Polymer Physics, 2016, 54(23): 2418-2424.

[12] WU J, WANG N, ZHAO Y, et al. Electrospinning of multilevel structured functional micro-/nanofibers and their applications[J]. Journal of Materials Chemistry A, 2013, 1(25): 7290-7305.

[13] TEO W, ERAMAKRISHNA S. Electrospun fibre bundle made of aligned nanofibres over two fixed points [J]. Nanotechnology, 2005, 16(9): 1878-1884.

[14] HANSEN B J, LIU Y, YANG R, et al. Hybrid nanogenerator for concurrently harvesting biomechanical and biochemical energy [J]. ACS Nano, 2010, 4(7): 3647-3652.

[15] LIU S L, LONG Y Z, ZHANG Z H, et al. Assembly of oriented ultrafine polymer fibers by centrifugal electrospinning[J]. Journal of Nanomaterials, 2013, 2013: 713275.

[16] SENTHILRAM T, MARY L A, VENUGOPAL J R, et al. Self crimped and aligned fibers [J]. Materials Today, 2011, 14(5): 226-229.

[17] YOUSEFZADEH M, LATIFI M, TEO W E, et al. Producing continuous twisted yarn from well-aligned nanofibers by water vortex [J]. Polymer Engineering & Science, 2011, 51(2): 323-329.

[18] PENG H K, WU M M, WANG Y T, et al. Enhancing piezoelectricity of poly (vinylidene fluoride) nano-wrapped yarns with an innovative yarn electrospinning technique[J]. Polymer International, 2021, 70(6): 851-859.

[19] SU C I, LAI T C, LU C H, et al. Yarn formation of nanofibers prepared using electrospinning[J]. Fibers and Polymers, 2013, 14(4): 542-549.

[20] BUSOLO T, SZEWCZYK P K, NAIR M, et al. Triboelectric yarns with electrospun functional polymer coatings for highly durable and washable smart textile applications[J]. ACS Applied Materials & Interfaces, 2021, 13(14): 16876-16886.

[21] PERSANO L, DAGDEVIREN C, SU Y, et al. High performance piezoelectric devices based on aligned arrays of nanofibers of poly(vinylidenefluoride-co-trifluoroethylene)[J]. Nature Communications, 2013, 4(1): 1633.

[22] CHANG G, LI A, XU X, et al. Twisted polymer micro/nanofiber yarns prepared by a direct fabrication[J]. Industrial & Engineering Chemistry Research, 2016, 55:

[23] ALI U, ZHOU Y, WANG X, et al. Direct electrospinning of highly twisted, continuous nanofiber yarns [J]. The Journal of the Textile Institute, 2012, 103(1): 80-88.

[24] LIAO X, DULLE M, DESOUZAESILVA J M, et al. High strength in combination with high toughness in robust and sustainable polymeric materials [J]. Science, 2019, 366(6471): 1376.

Preparation and Testing of CNT/WPU Coated Conductive Yarn

CAI Mingkun[1], **CHEN Wei**[1], **LIU Wei**[2], **QIU Yiping**[1], **XU Fujun**[1*]

[1] *College of Textiles, Donghua University, Shanghai 201620, China*

[2] *College of Fashion Technology, Shanghai University of Engineering Science, Shanghai 201620, China*

[*] *Corresponding author's email*: fjxu@ dhu. edu. cn

Abstract: CNT/WPU-coated DTY (CNT/WPU-DTY) with high electrical conductivity was prepared by coating the fluffy, porous draw texturing yarn (DTY) with carbon nanotube/waterborne polyurethane (CNT/WPU) conductive ink, and then curing it under stretched conditions. The morphological structure of the conductive yarn was characterized by SEM, and the conductivity of the conductive yarn was measured with the two-probe method. The results showed that the addition of WPU could effectively improve the interfacial binding performance of CNTs and polyester. In addition, because of the porous structure of DTY, the conductivity of CNT/WPU-DTY could reach 1 494. 27 S/m.

Keywords: Carbon nanotube (CNT); Waterborne polyurethane (WPU); Draw texturing yarn (DTY); Coat; Conductive yarn

1 Introduction

Due to their large length-to-diameter ratio, excellent mechanical properties, and conductive thermal conductivity, carbon nanotubes (CNTs) have been widely used in the preparation of conductive yarns[1]. Organic solvent spinning and wet spinning were common methods for preparing conductive polymer/CNT conductive yarns with high conductivity. Yet the mechanical properties of conductive yarns prepared by wet spinning are usually poor. To further improve the performance of CNT-based conductive yarns, coating CNTs onto the surface of the yarn substrate was used to prepare the conductive yarns with better mechanical properties[2].

However, because of the poor interfacial bonding performance of CNTs and chemical fibers with high strength and good wear resistance such as polyethylene terephthalate (PET) and polypropylene (PP), which limits the development and application of conductive yarns in a certain extent, and the low loading capacity of the coated CNT conductive yarn leads to the poor performance. To this end, adding adhesives to CNT dispersion followed by coating it onto yarns with porous structure is an effective method to overcome the above challenges. Among the adhesives, waterborne polyurethanes are widely used due to their good interfacial adhesion with flexible substrates such as PET and polyamide (PA), as well as their low toxicity and environmental friendliness[3].

In this study, we demonstrate an efficient and simple way to prepare a CNT/WPU coated conductive yarn by coating draw texturing yarn (DTY) with conductive coatings made of CNTs and WPU, followed by curing yarn at 10% stretching tension. The resulting CNT/WPU-DTY conductive yarn shows great improvement in the interfacial bonding properties between CNT and DTY and has a better yarn structure compared with CNT-DTY. In addition, compared with other conductive yarns coated with carbon nanotubes, CNT/WPU-DTY conductive yarn showed better electrical conductivity.

2 Experiments

2.1 Materials

CNT dispersion used in this research was provided by the Chengdu Institute of Organic Chemistry, Chinese Academy of Sciences, and prepared by ultrasound and dispersion (CNT: 10%). Scanning electron microscope (SEM) image (Fig. 1, Hitachi TM3000 at 1.0 VK) depicted the microstructure of carbon nanotube in CNT dispersion.

WPU was provided by LiNa plasticizer, silicone defoamer was provided by Shanghai Yuanye Biotechnology Co., Ltd., and polyester false twist textured yarn (DTY) was provided by Zhejiang Jinxia New Material Technology Co., Ltd.

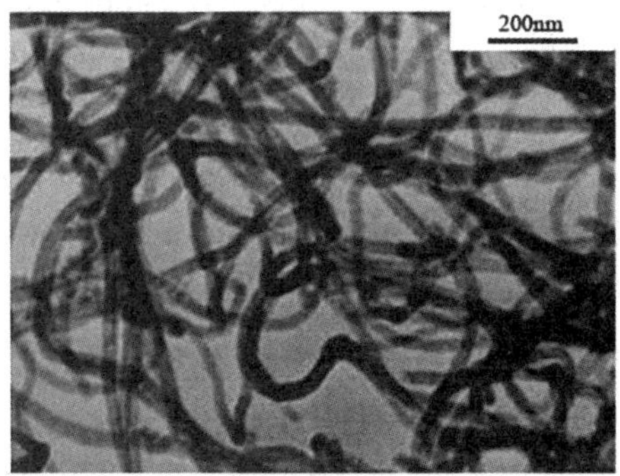

Fig. 1 SEM image of CNT dispersion

2.2 Preparation of CNT/WPU-DTY conductive yarn

Firstly, the DTY yarn was immersed in CNT dispersion (10%) and CNT/WPU (PU: 3%) conductive coating at 25 ℃ for 15 min. Before the DTY yarn was infiltrated, the CNT/WPU conductive coating should be stirred continuously at 800 r/min in the magnetic stirrer at 25 ℃ for 60 min. The treated yarn was taken out. Secondly, the DTY yarn was straightened vertically until the droplets on the surface of the yarn were completely drained, and then suspended and solidified in the oven under the condition of elastic tension of 10%. Finally, repeat the above steps four times to obtain CNT-DTY and CNT/WPU-DTY conductive yarns. A schematic diagram of the preparation process is shown in Fig. 2.

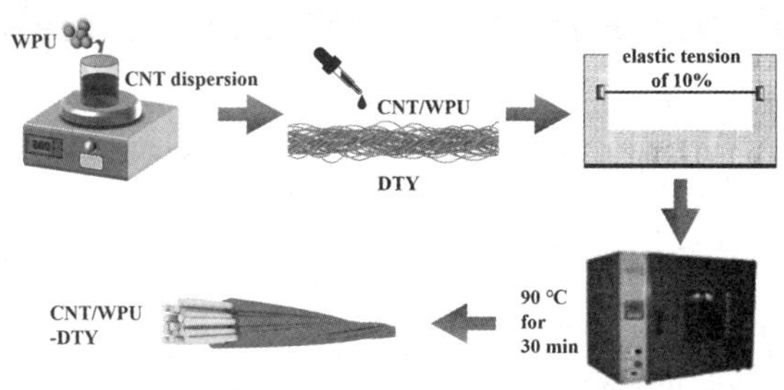

Fig. 2 Schematic diagram of samples preparation

2.3 Electrical properties

The resistance of CNT-DTY and CNT/WPU-DTY conductive yarns was measured by a standard two-probe method on an Agilent 34405A digital multi-meter. The length of the sample is 1 cm, copper wire was used as an electrode, and liquid silver glue was used for bonding, curing in the oven at 60 ℃ for about 1 h, until the silver glue was completely dry, and the distance between the electrodes is 10 mm. The sample diagram is shown in Fig. 3.

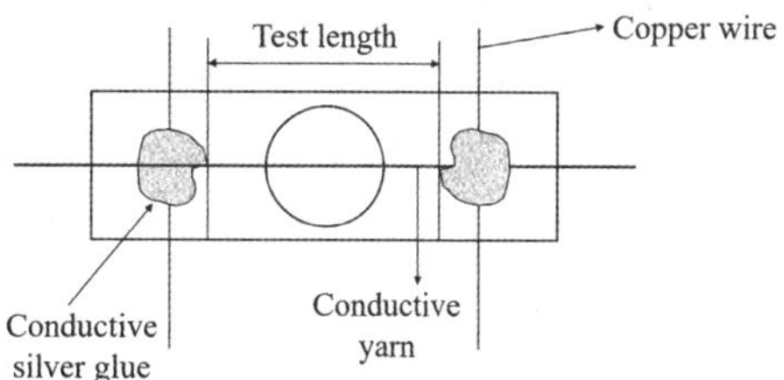

Fig. 3 An image of the specimen for electrical performance testing

At least 5 samples were measured under the same conditions, and then samples were photographed with a polarizing microscope. The diameter of the sample was measured by software, and the average diameter over 10 samples used as the yarn diameter. The electrical conductivity of the yarns was calculated.

$$\sigma = \frac{L}{RS} \tag{1}$$

where S, L, R, and σ are the area, length, volume resistance, and electrical conductivity of the specimen, respectively.

An SEM (Hitachi S-4800, Hitachi, Japan) was used to analyze the apparent morphology of CNT-DTY and CNT/WPU-DTY conductive yarns.

3 Results and Discussion

3.1 Electrical properties

As shown in Fig. 4 (a), with the increase in coating times, the conductivity of CNT-DTY increased at first and then decreased. The conductivity of CNT-DTY increased to maximum at the secondary coating and reached 1 956.11 S/m. While the yarn diameter decreased at first and then increased. This indicated that the elastic tension made the yarn structure compact, which helped to improve the electrical conductivity of the yarn. The decrease in the electrical conductivity of the CNT-DTY in the next may be due to the poor interface performance between CNT and DTY, which caused a large number of CNT to fall off during the preparation process. The electrical conductivity of CNT/WPU-DTY always showed an upward trend, and the rising rate gradually slowed down. After the fourth coating, the electrical conductivity of CNT/WPU-DTY increased to 1 494.27 S/m. The yarn diameter first decreased, then remained basically unchanged, and is always smaller than the diameter of CNT-DTY. This indicated that the addition of WPU improved the interface performance between CNT and DTY. With the increase in coating times, the load of CNT increased continuously and made the yarn structure more compact, which helped to improve the electrical conductivity of the yarn.

Fig. 4 (b) compared the conductivity of CNT/WPU-DTY with that of CNT-based conductive yarn prepared by coating in some literatures, and the results showed that CNTs/WPU-DTY had excellent electrical conductivity in contrast[4-8].

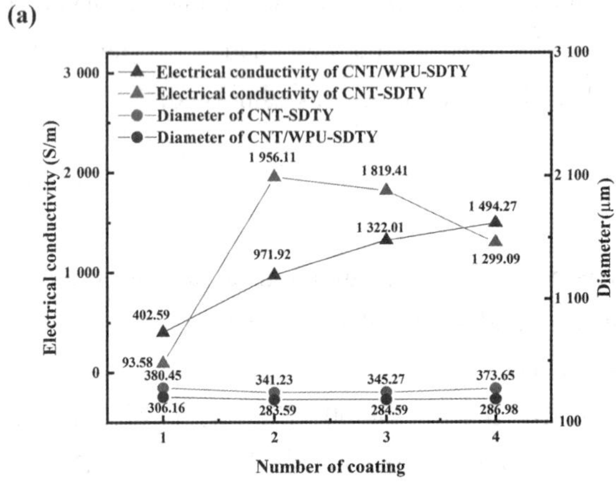

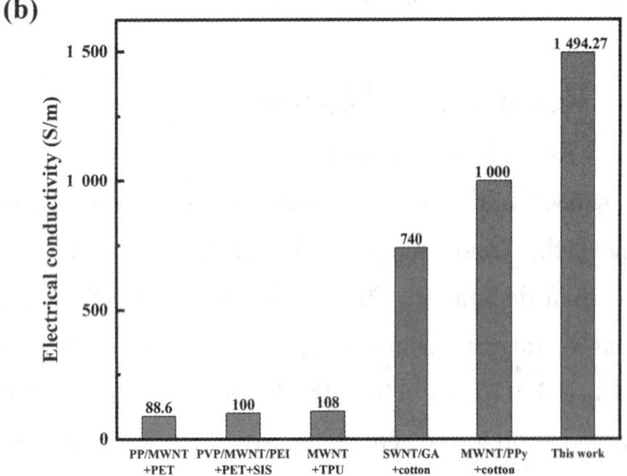

Fig. 4 (a) A graph of the electrical conductivity and yarn diameter of CNT-DTY and CNT/WPU-DTY as a function of coating time; (b) comparison of electrical conductivity of CNT/WPU-DTY and other CNT coated conductive yarns

3.2 Morphological and structural analysis

WPU has the alternating structure of soft and hard segments, and there are ester bonds, ether bonds, and other polar groups on the molecular chain, which can form hydrogen bonds with the polar groups on the surface of PET, PA, and other polar flexible substrates. The results indicates that the conductive coating containing WPU can make the carbon nanotubes completely adhere to the DTY yarn, and will not fall off after many times of coating. And based on the additional 10% tension during curing, the structure of the yarn was more compact so that the conductive yarn had higher electrical conductivity.

As shown in Fig. 5, after one coating, the CNT load of CNT/WPU-DTY was significantly higher than that of CNT-DTY, the internal structure of the yarn was more compact, and the arrangement of fibers was more regular. At the same time, CNT on CNT-DTY was only coated on the surface of each fiber, while a thin protective shell was formed on the outside of CNT/WPU-DTY. It fully indicated that the addition of WPU could effectively improve the interface properties between CNT and DTY.

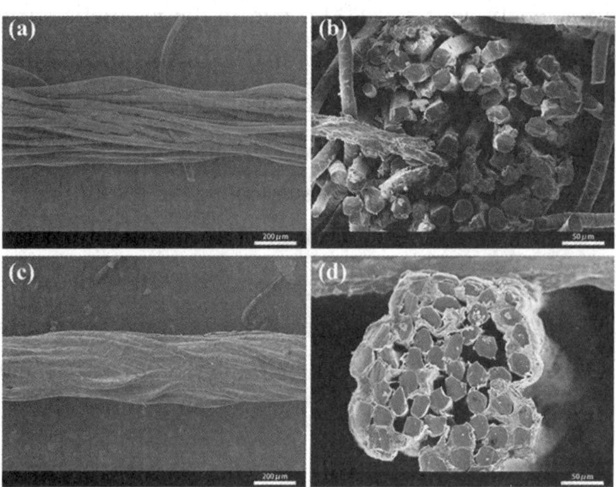

Fig. 5 SEM images of primary coated conductive yarn: (a) surface of CNT-DTY; (b) cross-section of CNT-DTY; (c) surface of CNT/WPU-DTY; (d) cross-section of CNT/WPU-DTY

After several times of coating, as shown in Fig. 6, the structure of CNT-DTY was still loose and the outer layer of CNT-DTY cannot form a carbon shell protective layer. While with the increase of coating times, the load of CNT inside CNT/WPU-DTY pores increased, the carbon protective layer became thicker, the yarn surface was more uniform, and the structure was still very tight.

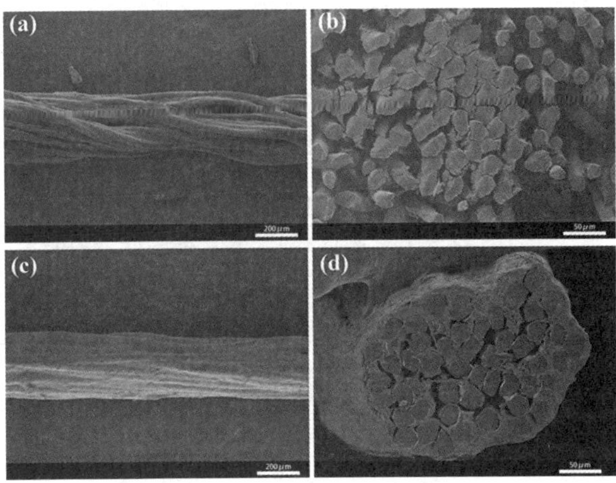

Fig. 6 SEM images of four times coated conductive yarns: (a) surface of CNT-DTY; (b) cross-section of CNT-DTY; (c) surface of CNT/WPU-DTY; (d) cross-section of CNT/WPU-DTY

4 Conclusions

In summary, CNT/WPU-DTY was fabricated by impregnating the coating and curing under elastic loading. This is a simple and cost-effective method to fabricate CNT-coated conductive yarns. The addition of WPU improved the interfacial properties of CNTs to DTY, elastic loading led to a tighter yarn structure, and the porous structure of DTY led to a higher CNT loading of the yarn, resulting in a yarn with high conductivity of 1 494.27 S/m.

References

[1] SHAHIDI S, MOAZZENCHI B. Carbon nanotube and its applications in textile industry — a review[J]. Journal of the Textile Institute, 2018, 109(12): 1653-1666.

[2] LIN Z I, LOU C Y, PAN Y J, et al. Conductive fabrics made of polypropylene/multi-walled carbon nanotube coated polyester yarns: mechanical properties and electromagnetic interference shielding effectiveness[J]. Composites Science and Technology, 2017, 141: 74-82.

[3] SHAO J Z, WANG C L, ZHOU J L, et al. Waterproof and moisture permeable coating of polyester fabrics using a novel waterborne polyurethane agent modified with silk fibroin and polyvinylpyrrolidone[J]. The Journal of the Textile Institute, 2017, 108(5): 864-869.

[4] LIN J H, LIN Z I, PAN Y J, et al. Manufacturing techniques and property evaluations of conductive composite yarns coated with polypropylene and multi-walled carbon nanotubes[J]. Composites Part A: Applied Science and Manufacturing, 2016, 84: 354-363.

[5] RYAN J D, ANJA L, HOFMANN A I, et al. All-organic textile thermoelectrics with carbon-nanotube-coated n-type yarns[J]. ACS Applied Energy Materials, 2018, 1(6): 2934-2941.

[6] LI Y H, ZHOU B, ZHENG G Q, et al. Continuously prepared highly conductive and stretchable swnt/mwnt synergistically composited electrospun thermoplastic polyurethane yarns for wearable sensing[J]. Journal of Materials Chemistry C, 2018, 6(9): 2258-2269.

[7] ZHU S, WANG M, QIANG Z, et al. Multi-functional and highly conductive textiles with ultra-high durability through "green" fabrication process[J]. Chemical Engineering Journal, 2021, 406: 127-140.

[8] LIMA R M A P, ALCARAZ-ESPINOZA J J, DA SILVA F A G, et al. Multifunctional wearable electronic textiles using cotton fibers with polypyrrole and carbon nanotubes[J]. ACS Applied Materials & Interfaces, 2018, 10(16): 13783-13795.

UHMWPE Fiber-Carbon Fiber/Epoxy Hybrid 3D Braided Composites with Bilateral Structure for Outstanding Electromagnetic Interference Shielding

SHEN Mengdi[1], QUAN Zhenzhen[1*], GU Qiujin[1], ZHANG Chengming[1], QIN Xiaohong[1]

[1] *College of Textiles, Donghua University, Shanghai 201620, China*

[*] *Corresponding author's email*: quan@ dhu. edu. cn

Abstract: Increasing electromagnetic pollution has driven enormous endeavors in electromagnetic wave absorption materials, but it remains a big challenge to prepare versatile composites to address the issue. Herein, a versatile bilateral structure of ultra-high molecular weight polyethlene(UHMWPE) fiber-carbon fiber/epoxy hybrid three-dimensional (3D) braided composites was designed via bilateral four step rectangular braiding and vacuum-assisted resin transfer molding (VARTM) method. As-prepared UHMWPE fiber-carbon fiber/epoxy hybrid 3D braided composites present excellent electromagnetic wave absorption as well as impact-resistant performance. The maximum total SE (SET) value can reach 22. 42 dB at 8. 2 GHz, and the composites become absorption-dominated electromagnetic shielding material at the frequency range from 8. 2 GHz to 12. 4 GHz. These remarkable properties were originated from the great impedance matching and dipole polarization along with conductivity loss. Considering the novel UHMWPE fiber-carbon fiber/epoxy hybrid 3D braided composites possessing a unique combination of outstanding electromagnetic interference shielding performance and 3D braided high structural integrity, these multifunctional features would be highly promising for the next-generation structural-functional integrated materials.

Keywords: Electromagnetic interference(EMI) shielding effect; Three-dimensional hybrid braiding; Carbon fiber; UHMWPE fiber; Composites

1 Introduction

With the development of 5G, various electronic devices have been widely used to improve people's living standards, but also bring serious EMI and radiation. These miniaturized devices generate electromagnetic waves that can adversely affect the performance of electronic systems. Some medical implants or devices like hearing aids, insulin pumps, and cardiac pacemakers are prone to malfunction in alternating electric fields. In addition, if human is exposed to electromagnetic radiation for a long time, it can have harmful effects on human health, such as causing nausea, headaches, eye problems, and even cancer[1]. Thus there is a great demand for EMI shielding materials to block or reduce the propagation of electromagnetic energy in space[2].

It is a great way to shield the electronic device from EMI by enclosing it with metallic materials like gold, silver, aluminum, and copper[3]. Although these metals provide adequate shielding against EMI, they have high densities and poor corrosion resistance, and are easy to cause secondary electromagnetic pollution due to high reflectivity, which limits their application in modern electronic devices[3-4]. Such limitations have resulted in the desire of lighter and more stable materials for EMI shielding applications and recent attentions have been focused on developing polymeric EMI shielding materials[5]. In general, there are two main approaches to enhance the electrical conductivity of polymeric materials. One is coating them with conductive materials[6]. The other is to disperse conductive materials in polymer matrix to make composites, and many porous conductive polymer composites (CPCs) are usually achieved in this way[5, 7]. For instance, Ameli et al. [8] reported polypropylene/carbon fiber composite foams by using dissolved pressurized nitrogen gas, which achieved a shielding effectiveness (SE) of around 25 dB with a volume fraction of 7. 5%. Yang et al. [9] fabricated a novel carbon nanotube-polystyrene foam composite with carbon nanotubes (CNTs) (mass fraction of 7%), achieving an excellent SE of 33. 1 dB ·

cm^3/g over a frequency range of 8.2–12.4 GHz.

In spite of polymeric materials' light weight, flexibility, easy processability, corrosion resistance, and excellent EMI shielding performance, their poor mechanical properties prevent them from being used as load-bearing parts in aviation, aerospace, automotive, shipbuilding, and other fields. Three-dimensional (3D) braided prefabricated parts are produced in an integral manner, resulting in great anti-delamination property[10]. In addition, 3D braided composites have high specific modulus and strength, impact resistance, and excellent fatigue resistance, gradually becoming the main application materials in the aerospace field[11]. Hence, it is a great way to endure 3D braided composites with EMI shielding performance, so that 3D braided composites can be used as load-bearing structural members with excellent EMI shielding performance, such as aircraft skins and shielding room shells. Wang et al.[12] prepared nickel-plated carbon fiber 3D braided prefabricated parts and then cured with epoxy resin. It was reported that the impact strength of the composite was 170 kJ/m^2, and the SE values were 52 dB to 65 dB over a frequency range of 30–1 200 MHz, indicating that the composites possessed not only excellent mechanical properties but also good electromagnetic shielding performance.

In this paper, a robust ultra-high molecular weight polyethylene (UHMWPE) fiber/carbon fiber (CF) 3D braided composite cured with epoxy resin was presented for EMI shielding applications. The UHMWPE fiber was chosen from four different functional fibers as the surface layer on account of its great impedance matching, and carbon fiber was made into the back layer, achieving great electromagnetic energy absorption performance. The UHMWPE fiber/CF braided composite is a kind of bilateral hybrid composite, herein, UHMWPE fiber/CF as top & substrate layers, which helps to achieve the desirable EMI "reflection (transmittance)-absorption" mechanism. As for the substrate layer, CFs form the braided skeleton, building efficient conductive networks and synergistically contributing to the electrical conductivity and EMI shielding performance effectively.

2 Experiments

2.1 Materials

The J98-1000D UHMWPE fiber, the 1 000D PBO and the 12K GF were obtained from Yueshun New Material company. Kevlar 29–3 000 d fiber was purchased from DuPont company. The T300B-3K carbon fiber was purchased from Toray Inc. (Japan). JC-02A epoxy resin and JC-02B hardener were both provided by Changshu Jiafa.

2.2 Preparation of 3D 4-directional braided composite

Five types of functional fibers (carbon fiber, UHMWPE fiber, Kevlar fiber, glass fiber, and PBO fiber) were used in this study, and they were separately made into 3D 4-directional braided performs by 4-step braiding while the 3D 4-directional UHMWPE fiber-CF hybrid braided preforms were prepared by bilateral four step rectangular braiding.

All performs were cured using vacuum-assisted resin transfer molding (VARTM) method. At the first stage, epoxy resin was degassed for 2 h at 60 ℃. Then the defoamed mixture obtained previously was injected into the 3D braided preforms by the VARTM method. When a length of continuous one-meter-long bubble-free resin appears at the outlet end, it indicates that the composite fabric has been completely saturated by fluid[13]. Therefore, it should be sealed, and then it was heated at 90, 110, 130 ℃ for 2, 1, and 4 h, respectively. Eventually, the composite fabric was totally cooled down at room temperature. All specimens were cut into rectangles of 10.16 mm by 22.86 mm. Figure 1 shows the morphology of five different single-fiber braided composites as well as UHMWPE fiber-CF hybrid braided composites.

2.3 EMI performance measurement

For EMI SE testing, the composites were cut into cubic specimens with the dimension of 10.16 mm×22.85 mm×5.70 mm. The EMI SE measurements were conducted using a vector network analyzer (Keysight PAN-L N523B) by a waveguide method in the frequency range of 8.2–12.5 GHz (X band). As a result, the parameters including S_{11}, S_{12}, S_{21}, and S_{22} were obtained to calculate the absorbed power A, reflected power R, and transmitted power T. In addition, microwave absorption SEA, microwave reflection SER, and total value of EMI SE SET were calculated from the scattering parameters (S_{11} and S_{21}) expressed as Eqs. (1)-(6)[13]. The simulation results were obtained using CST Studio Student version for the same frequency range and sam-

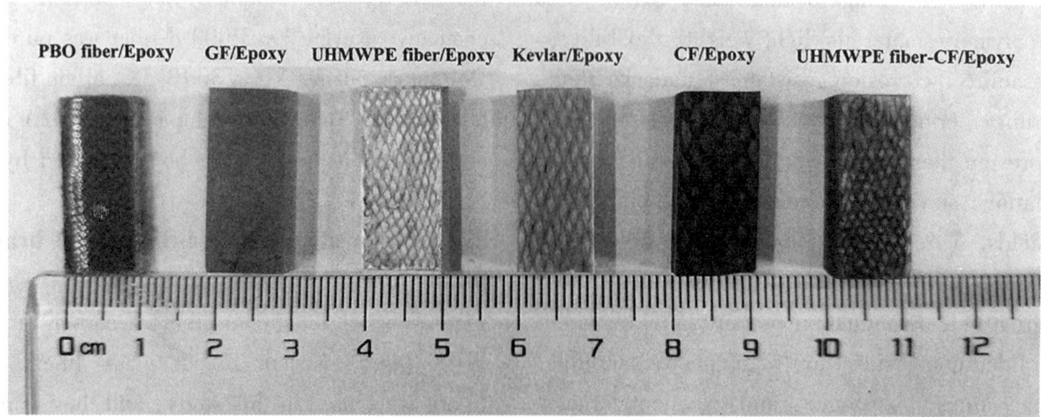

Fig. 1 Optical photographs of the PBO fiber/epoxy, GF/epoxy, UHMWPE fiber/epoxy, Kevlar/epoxy, CF/epoxy and UHMWPE fiber-CF/epoxy

ple dimensions. And the electrical conductivity and magnetic permeability were obtained from experiments, used as input data for these simulations.

$$R = | S_{11} |^2 \quad (1)$$

$$T = | S_{21} |^2 \quad (2)$$

$$A = 1 - R - T \quad (3)$$

$$SE_R = -10\lg(1 - R) \quad (4)$$

$$SE_A = -10\lg \frac{T}{1 - R} \quad (5)$$

$$SE_T = SE_A + SE_R + SE_M \quad (6)$$

3 Results and Discussion

3.1 Material selection and structural design

According to microwave transmission line theory, there are two principles to follow to achieve great absorption performance. (1) the surface impedance of absorbing materials is close to the air impedance, reducing the reflection of incident electromagnetic waves on the surface of the material; (2) the inner layer of absorbing material achieves efficient loss absorption of incoming electromagnetic waves, reducing the reflection of electromagnetic waves on the bottom surface.

Commonly, there are two types of impedance matching structures. One is the impedance gradient type, in which the impedance increases sequentially along the electromagnetic wave incidence direction, and the other is the impedance turnover type, in which a matching layer with lower impedance is inserted behind the first layer along the electromagnetic wave incidence direction. In this paper, the impedance tapering type structure is chosen.

When a plane electromagnetic wave with the frequency of f is uniformly and vertically incident on the medium from free space, the value of the reflection loss of electromagnetic wave energy (reflection loss, RL) (dB) by the medium can be calculated by the following equation.

$$R = 20\lg \left| \frac{Z_{in} - Z_0}{Z_{in} + Z_0} \right| \quad (7)$$

where Z_0(377 Ω) is the wave impedance of electromagnetic waves in free space, and Z_{in} is the wave impedance at the air-material interface. It is apparent that the absolute value of reflection loss will be infinitely large when Z_{in} close to Z_0. Therefore, Z is defined as normalized impedance in order to reflect the relationship between Z_0 and Z_{in} intuitively.

$$Z = \frac{Z_{in}}{Z_0} \quad (8)$$

Notably, the material would achieve a good impedance matching performance, when Z_0 is infinitely close to Z_{in} and the quotient of the two, Z is approaching 1.

Based on the transmission line theory:

$$Z_{in} = Z_0 \sqrt{\frac{\mu_r}{\varepsilon_r}} \tanh\left(j \frac{2\pi fd}{c} \sqrt{\mu_r \varepsilon_r} \right) \quad (9)$$

Consequently, the impedance matching can also be expressed as in general:

$$\frac{\varepsilon''}{\varepsilon'} = \frac{\mu''}{\mu'} \quad (10)$$

In other words, the ratio of the imaginary part to the

real part of the permittivity is equal to the ratio of that of the permeability.

It was seen from Fig. 2 (a) that the order of the absolute value of the difference between μ''/μ' and $\varepsilon''/\varepsilon'$ from small to large is UHMWPE fiber/epoxy, PBO fiber/epoxy, GF/epoxy and Kevlar/epoxy braided composites. Therefore, UHMWPE fiber was chosen as the material of the surface layer of the hybrid preform in order to reduce the reflection of incident electromagnetic waves on the surface, so that more electromagnetic waves can enter the interior of the material to be absorbed sufficiently. It can be seen from Fig. 2 (b) that the shielding reflection loss of UHMWPE is the smallest, indicating that most of the electromagnetic waves enter the material, which is consistent with the results of Fig. 2 (a). Therefore, UHMWPE was chosen as the surface layer of the hybrid braided composite.

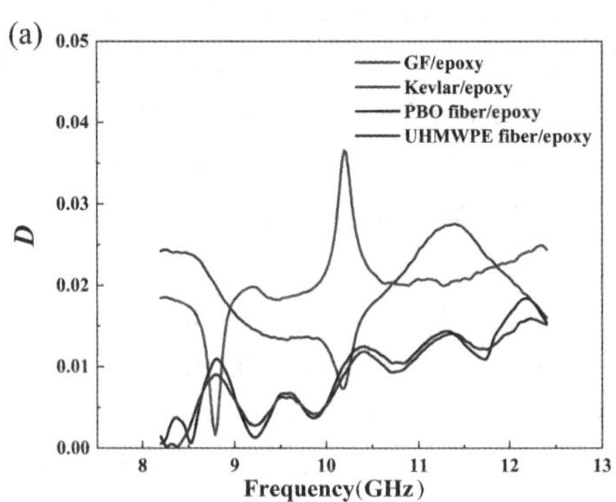

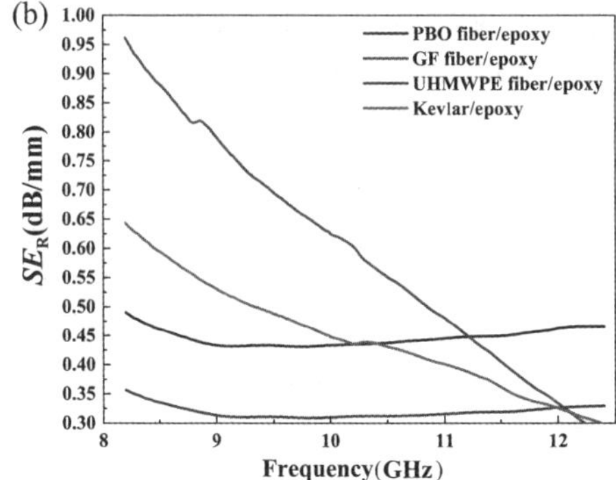

Fig. 2 (a) *D* of the composites; (b) *SE* refection part per thickness of the composites at X band (8.2–12.4 GHz) (D represents the absolute value of the difference between $\frac{\mu''}{\mu'}$ and $\frac{\varepsilon''}{\varepsilon'}$)

3.2 EMI Performance of the UHMWPE fiber/CF hybrid braided composite

From Fig. 3 (a), it can be seen that the highest total shielding effectiveness of the UHMWPE fiber/CF hybrid braided composite can reach 22.42 dB at 8.2 GHz. From Fig. 3 (b), it is found that the material is an absorption-oriented material in the frequency range of 8.2–12.4 GHz, which means that most of the incident electromagnetic waves are absorbed. The higher the frequency, the shorter the wavelength. As a result, more energy was absorbed.

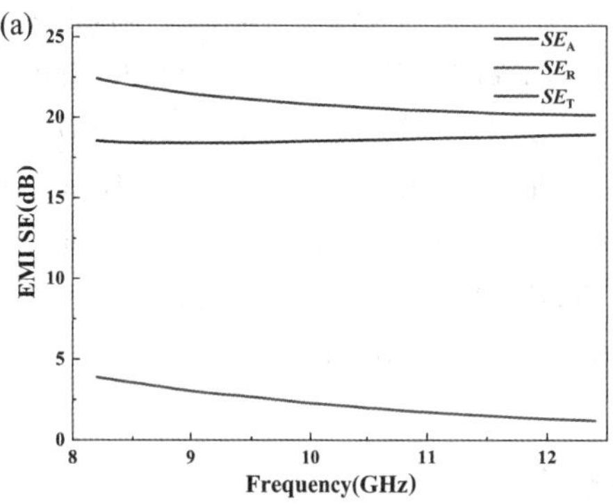

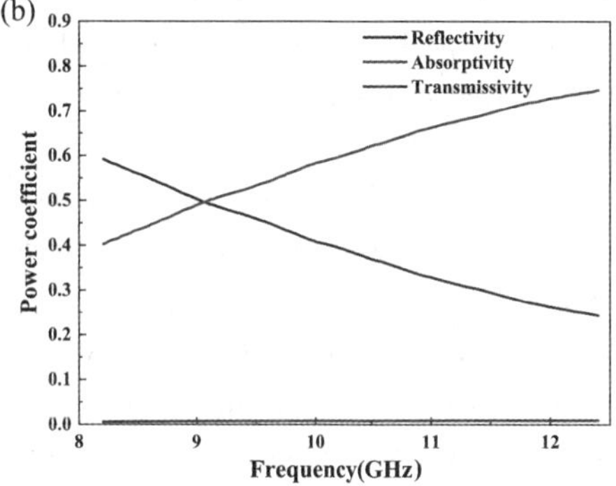

Fig. 3 (a) EMI SE values; (b) power coefficient versus frequency (8.2 – 12.4 GHz) of the UHMWPE fiber/CF hybrid braided composite

4 Conclusions

This study has designed a brand new UHMWPE fiber/CF hybrid 3D braided composites. The surface layer of the adapted fabric can form a good impedance match with the air layer, allowing more electromagnetic waves to enter the interior of the material and thus be absorbed sufficiently. Besides, the back layer made of carbon fiber, which possesses good electrical conductivity so that can consume more electromagnetic energy. The maximum *SET* can reach 22.42 dB at 8.2 GHz, and the composites become absorption-dominated electromagnetic shielding material at the frequency range from 9.0 GHz to 12.4 GHz. These remarkable properties were originated from the great impedance matching and dipole polarization along with conductivity loss.

Acknowledgements

This work was supported by the Fundamental Research Funds for the Central Universities (Nos. 2232021G-02 and 2232020D-15), the Open Fund of Shanghai Key Laboratory of Lightweight Composite (2232020A4-01), the Young Elite Scientists Sponsorship Program by CAST, and the Fund of National Engineering and Research Center for Commercial Aircraft Manufacturing (F-CXKT-0301C, COMAC-SFGS-2021-667) to Prof. Zhenzhen Quan, and the Fundamental Research Funds for the Central Universities and Graduate Student Innovation Fund of Donghua University (CUSF-DH-D-2019059) to Ms. Qiujin Gu.

References

[1] PANAGOPOULOS D J, CHROUSOS G P. Shielding methods and products against man-made electromagnetic fields: protection versus risk[J]. The Science of the Total Environment, 2019, 667: 255–262.

[2] GUPTA S, TAI N H. Carbon materials and their composites for electromagnetic interference shielding effectiveness in X-band[J]. Carbon, 2019, 152: 159–187.

[3] WANASINGHE D, ASLANI F. A review on recent advancement of electromagnetic interference shielding novel metallic materials and processes[J]. Composites Part B: Engineering, 2019, 176.

[4] GEETHA S. EMI shielding: methods and materials-a review[J]. Journal of Applied Polymer Science, 2009, 112(4): 2073–2086.

[5] SINGH A K. A review of porous lightweight composite materials for electromagnetic interference shielding[J]. Composites Part B: Engineering, 2018, 149: 188–197.

[6] PANIGRAHI R, SRIVASTAVA S K. Trapping of microwave radiation in hollow polypyrrole microsphere through enhanced internal reflection: a novel approach[J]. Science Reports, 2015, 5: 7638.

[7] CHEN Q. Design of 3D lightweight Ti_3C_2T MXene porous film with graded holes for efficient electromagnetic interference shielding performance[J]. Ceramics International, 2022, 48(10): 14578–14586.

[8] AMELI A, JUNG P U, PARK C B. Electrical properties and electromagnetic interference shielding effectiveness of polypropylene/carbon fiber composite foams[J]. Carbon, 2013, 60: 379–391.

[9] KO F K. 3D textile reinforcements in composite materials[J]. Composites Science and Engineering, 1999: 9–42.

[10] GU Q. Structural modeling and mechanical characterizing of three-dimensional four-step braided composites: a review[J]. Composite Structures, 2019, 207: 119–128.

[11] RUI W. Research on surface metallization of continuous carbon fiber and the electromagnetic shielding property of its composite[D]. Tianjin: Tianjin University, 2012.

[12] SHIN P S. Manufacturing and qualitative properties of glass fiber/epoxy composite boards with added air bubbles for airborne and solid-borne sound insulation[J]. Composites Science and Technology, 2020, 194.

[13] WANG M. Fabrication, mechanisms and perspectives of conductive polymer composites with multiple interfaces for electromagnetic interference shielding: a review[J]. Carbon: an International Journal Sponsored by the American Carbon Society, 2021, 177: 377–402.

Research on Clean Sizing-Dyeing Technology of Denim Warp Yarns

XU Yajing[1], PAN Yongzhen[1], GUO Lamei[2*]

[1] *College of Textiles, Donghua University, Shanghai 201620, China*

[2] *Key Laboratory of Textile Science and Technology, Ministry of Education, Donghua University, Shanghai 201620, China*

[*] *Corresponding author's email*: lmguo@ dhu. edu. cn

Abstract: With the national's requirements for clean production of enterprises, the environmentally friendly denim warp sizing-dyeing process has gradually become a hot research topic in the field of denim textile clothing. Therefore, we studied the new "integration of sizing and dyeing" process, and researched the optimal colouring paste components, so as to reduce energy consumption and wastewater discharge, and make the yarn obtain better sizing and dyeing performance. The results show that the concentration of phthalocyanine blue coating is 140 g/L, the adhesive ratio of 601C to 410A is 1 : 1, the total adhesive concentration is 160 g/L, and the concentration of cross-linking agent is 2. 5%, the colouring paste uptake rate is maintained at 12%–13%; under the premise of strength increasing, the elongation also increases; the cotton staining fastness reaches grade 4, and the fading fastness reaches grade 4–5, dry rubbing fastness is 4, and wet rubbing fastness is 3. In general, it can meet the weaving needs of denim warp yarns.

Keywords: "Integration of sizing and dyeing" process; Environmentally friendly; Sizing-dyeing performance

1 Introduction

From the inception of denim clothing, it has quickly swept the world with its unique advantages of fashion, comfort, wear resistance and universality and has endured to this day[1]. In the production process of denim, the sizing and dyeing process of denim warp yarns is unique. Traditionally, the sizing-dyeing machine is used to complete this process, then through the weaving process to get denim products[2]. In the sizing process, the starch added with polyvinyl alcohol (PVA) is often used for sizing yarns. This traditional sizing agent make the yarn must be desizing, which not only brings about environmental pollution because of the desizing wastewater discharge, but also causes the waste of resources and energy[3]. In the dyeing process, denim yarns are mainly dyed with indigo, which is an insoluble dye. Indigo needs to be reduced and oxidized before it eventually fixates on the fiber. Indigo will produce a great deal of printing and dyeing wastewater in the reduction process. In addition, it is necessary to go through several times of padding and oxidation to make the yarn get darker color, and finally increase the dyeing process and the energy consumption[4].

With the requirements for enterprises to carry out clean and green production and the proposal of ecological textiles, all links of the textile industry need to be environmentally friendly, including reducing the discharge of dyeing wastewater and the energy consumption of desizing[5]. While the traditional sizing and dyeing processes can no longer meet the sustainable development requirements of the country[6-7]. Therefore, seeking a new type of sizing and dyeing technology and realizing cleanness production of denim warp yarns have gradually become a hot research topic in the field of denim textile clothing.

This study focuses on the "integration of sizing and dyeing" process of pure cotton denim warp yarns. This technology is based on phthalocyanine blue coating and a new type of water-soluble polyurethane (WPU) adhesive, which combines yarn sizing and dyeing into one step. By using the optimal colouring paste components, combined with dyeing-yarn with pigments technology and non desizing technology, we strive to achieve the effect of energy saving and emission reduction while ensuring the quality of sizing and dyeing yarns.

2 Experiments

2.1 Materials

Materials: 36. 4 tex cotton yarn; phthalocyanine blue coating; WPU 601C and 401A adhesives (obtained from Anhui Hefei Huayue New Material Co. , Ltd.); non-formaldehyde cross-linking agent.

2.2 Clean sizing-dyeing technology with phthalocyanine blue coating

Colouring paste components:

phthalocyanine blue coating	140 g/L
601C and 401A adhesive ratio	x
total adhesive concentration	y g/L
cross-linking concentration	z g/L

Clean sizing-dyeing process: single yarn →padding colouring paste (1-dip-1-nip, dyeing at room temperature) → drying at 50 ℃ → curing at 160 ℃ for 3 min → single yarn that meets the requirements of weaving.

2.3 Testing on colouring paste uptake rate

Take thirty yarns 10 cm, dry them in an oven at 110 ℃ to constant weight, and use an electronic balance to weigh them. The original yarns are weighted as w_1, the sizing-dyeing yarns are weighted as w_2, and the sample size is 3. Take the average value and calculate according to Eq. (1).

$$Colouring\ paste\ uptake\ rate = \frac{w_2 - w_1}{w_1} \times 100 \quad (1)$$

2.4 Tensile testing on yarns

According to GB/T 3916—2013, the breaking strength and elongation of yarns were tested. The breaking strength of the original yarn shall be recorded as P_1, and that of the yarn after sizing-dyeing shall be recorded as P_2. The elongation at break of the original yarn is recorded as ε_1, and that of the yarn after sizing-dyeing is recorded as ε_2. The sample size is 30. Take the average value and calculate according to Eqs. (2) and (3).

$$Increasing\ strength\ rate = \frac{P_2 - P_1}{P_1} \times 100\% \quad (2)$$

$$Reducing\ elongation\ rate = \frac{\varepsilon_2 - \varepsilon_1}{\varepsilon_1} \times 100\% \quad (3)$$

2.5 Color fastness testing on yarns

According to GB/T 3921—2008 and GB/T 3920—2008, the colour fastness of yarns was tested.

3 Results and Discussion

3.1 The principle of "integration of sizing and dyeing"

This study used WPU adhesives, which is not water-soluble after film forming. It can bind pigment and sizing fiber when processing denim warp yarns. Its process is shown in Fig. 1.

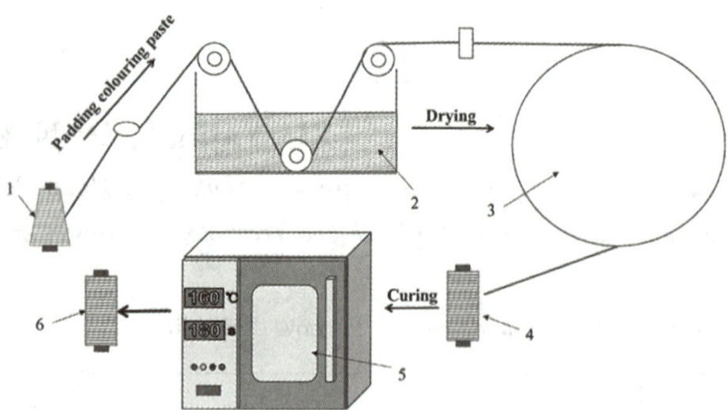

1-original yarns; 2-dye vat; 3-heat cylinder; 4-dried yarns; 5-oven; 6-finished dyed yarns.

Fig. 1 Sizing-dyeing process

After the yarns is padded with "pigment colouring paste", through the drying and curing process, the moisture on the yarns volatilizes. The adhesive molecules and cross-linking agent molecules are close to each other, and cross-linking reaction of active groups developes between fiber and adhesives, and between adhesives macromolecule, finally forming a 3D network structure. This structure not only wraps the pigment particles inside, but also firmly adheres to the yarns, achieving the purpose of yarn sizing-dyeing.

3.2 Adhesive ratio effect on sizing-dyeing performance of yarn

In the sizing-dyeing process, the total concentration of adhesive is 120 g/L, and the concentration of cross-linking is 1.5%. The effect of 601C and 401A adhesive ratio on sizing-dyeing performance of denim warp yarns was studied.

Since the yarns show poor color fastness when 601C or 401A adhesives are used alone, these two adhesives are mixed. The sizing and dyeing performance of yarns are shown in Fig. 2 and Tab. 1. When 601C or 401A is used alone, the adhesive molecules and the pigment particles lack a strong complex effect, so that the pigment particles are repelled by adhesive molecules during the film-forming process[8]. As a result, the increasing strength rate of yarns is not good and the dyeing performance is worse. When these two adhesives are mixed, the breaking strength and elongation of the yarns are increased, and the color fastness shows a trend of greatly increasing. This is because both of them are anionic resin materials, the intermolecular forces such as van der Waals force and hydrogen bond between coating and adhesive are greatly improved after mixed use, and their performance is fully exerted. From Tab. 1, we know that with the increase of the amount of 601C adhesive, more color paste coating on the yarn, the colouring paste uptake rate of the yarn increases, the adhesive film thickens, the yarn becomes brittle and hard, and the increasing strength rate increases while the reducing elongation rate decreases. Continuing to increase the amount of 601C adhesive, the viscosity of the color paste increases, which limits the penetration of coating particles into the fiber, the colouring paste uptake rate of the yarn has a downward trend, and the color fastness of the yarn is also reduced. When the ratio of adhesive is 1 : 1, the color fastness of yarn is the best, and the data of sizing yarn performance is the most appropriate.

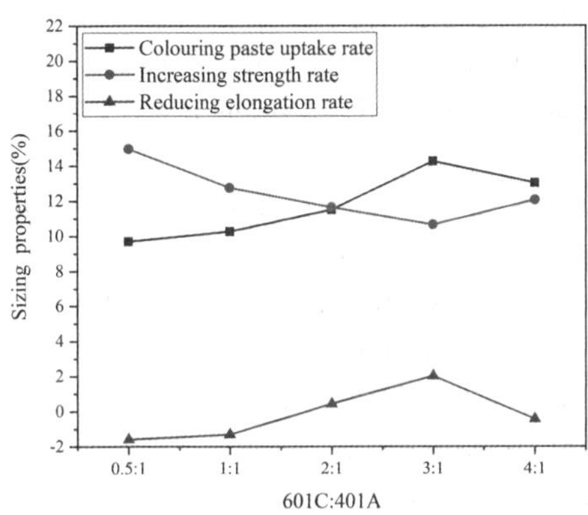

Fig. 2 The effect of adhesive ratio on sizing performance of denim warp yarns

Tab. 1 The effect of adhesive ratio on dyeing performance of denim warp yarns

601C : 401A	Staining fastness (grade) Cotton	Colour fastness to fading (grade)	Colour fastness to rubbing(grade) Dry	Wet
0.5 : 1	3-4	4	3	1-2
1 : 1	3-4	4	3	1-2
2 : 1	3-4	3-4	3	1-2
3 : 1	3-4	3-4	4	2
4 : 1	3	2	3-4	1

3.3 Total adhesive concentration effect on sizing-dyeing performance of yarn

Select the good test results in sections 3.2, and keep the cross-linking agent concentration at 1.5%. Research the effect of total adhesive concentration on sizing-dyeing performance of denim warp yarns.

The effect of total adhesive concentration on sizing performance of denim warp yarns are shown in Fig. 3. The colouring paste uptake rate and increasing strength rate of yarns increase with the increase of the total adhesive concentration. Due to the microcosmic phase-separate structure formed by the soft and hard segments of WPU, the elongation at break of yarns has a small increase. When the total adhesive concentration increases to 160 g/L, the elongation at break of the yarn is the same as that of the original yarn. If the total adhesive concentration continues to increase, the elongation at break of the yarn will decrease due to its hard feel.

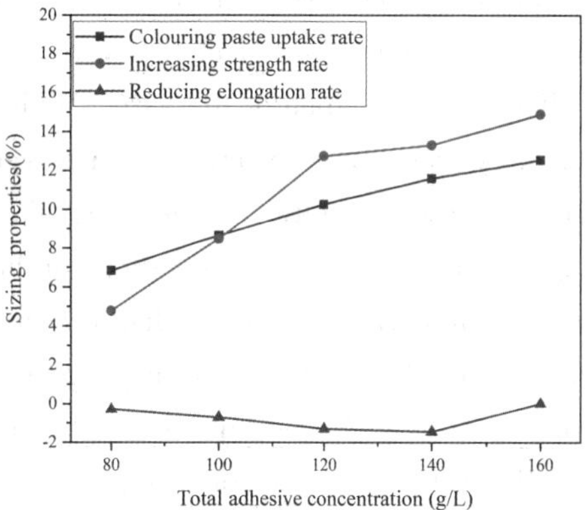

Fig. 3 The effect of total adhesive concentration on sizing performance of denim warp yarns

Tab. 2 The effect of total adhesive concentration on dyeing performance of denim warp yarns

Total adhesive concentration (g/L)	Staining fastness (grade) Cotton	Colour fastness to fading (grade)	Colour fastness to rubbing(grade) Dry	Wet
80	2	3	2	1
100	3	3	2-3	1
120	3-4	4	3	1-2
140	3-4	4-5	3	1-2
160	4	4-5	3	1-2

Pigment dyeing relies on adhesive to coating or inlaying the pigment particles in the adhesive film, less amount of adhesive cannot completely coat pigment particles, therefore, the amount of adhesive directly affects the colour fastness of pigment dyeing[9]. It can be seen from Tab. 2 that with the increase of the total adhesive concentration, the color fastness of the yarn gradually increases, and the adhesive gradually coveres the coating particles completely. When the total adhesive concentration increases to 160 g/L, the color fastness reaches the maximum. In order to make the yarn achieve better sizing-dyeing performance and not cause the waste, the appropriate adhesive concentration is 160 g/L.

3.4 Cross-linking agent concentration effect on sizing-dyeing performance of yarn

Select the test results in sections 3.2 and 3.3, and research the effect of cross-linking agent concentration on sizing-dyeing performance of denim warp yarns.

The cross-linking agent can improve the water resistance and mechanical performance of the WPU, so that the adhesive film wrapped on the yarn has better strength and elongation. It can be seen from Fig. 4 that the concentration of cross-linking agent will not affect the colour paste content coated on the yarn, but only slightly affect the elongation of the yarn. With the increase of the concentration of cross-linking agent, the elongation increases slightly. In general, the cross-linking agent concentration has little effect on the sizing performance. From Tab. 3 we know that low cross-linking agent concentration can improve the colour fastness of the yarn, while high concentration will cause the yarn to feel hard. Thus its concentration should be strictly controlled, and 2.5% is good enough.

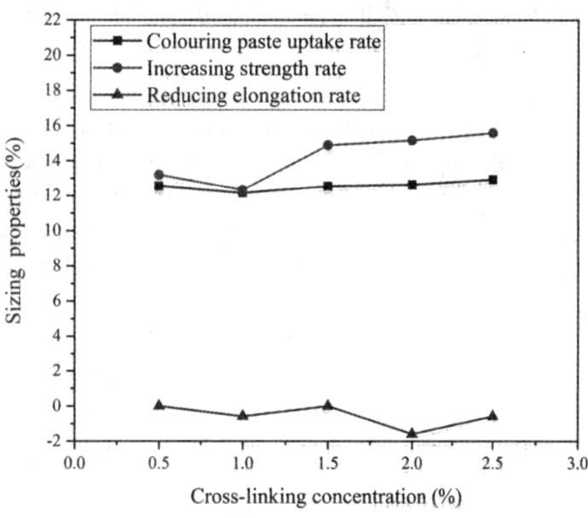

Fig. 4 The effect of cross-linking agent concentration on sizing performance of denim warp yarns

Tab. 3 The effect of cross-linking agent concentration on dyeing performance of denim warp yarns

Cross-linking concentration (%)	Staining fastness (grade) Cotton	Colour fastness to fading (grade)	Colour fastness to rubbing (grade) Dry	Wet
0.5	2-3	2-3	3	1
1	3	3	3	1-2
1.5	4	4-5	3	1-2
2	4	4-5	3-4	2-3
2.5	4-5	4-5	4	3

4 Conclusions

The results show that the concentration of phthalocyanine blue coating is 140 g/L, the adhesive ratio of 601C to 410A is 1 : 1, the total adhesive concentration is 160 g/L, and the concentration of cross-linking agent is 2.5%. The colouring paste uptake rate is maintained at 12%-13%; under the premise of strength increasing, the elongation also increases; the cotton staining fastness reaches grade 4, the fading fastness reaches grade 4-5, dry rubbing fastness is grade 4, and wet rubbing fastness is grade 3. In general, it can meet the weaving needs of denim warp yarns.

References

[1] XU Z L, CHENG S X, ZHANG Y H. Analysis of the technical innovation and development trend of China's denim and denim yarn[J]. China Textile Leader, 2020(5): 50-56.

[2] ZHAO Y M. Denim production technology research and product development[J]. Textile Reports, 2021, 40(8): 18-19,43.

[3] ZHANG Z J. Research on sizing technology of denim special size[J]. 2012: 346-348.

[4] LI R Q, WANG W, WEI B J, et al. Innovative indigo dyeing of denim yarn [J]. China Dyeing & Printing, 2019, 45(23): 26-28,38.

[5] YANG J. Sizing process practice of "non desizing finishing" [J]. Chemical Fiber & Textile Technology, 2020, 49(11): 31-33.

[6] LIU T T. Comparison of water pollutant discharge data of different caliber in printing and dyeing industry[J]. Textile Dyeing and Finishing Journal, 2021, 43(1): 1-2, 14.

[7] YANG J Y, WANG L X. SWOT analysis and research on denim clothing industry in Guangdong province[J]. Light Industry Science and Technology, 2020, 36(11): 71-73.

[8] QI D M, XU J, CHEN Z J, et al. Dispersion of pigment particles in polyacrylate latex film containing phthalocyanine blue [J]. CIESC Journal, 2013, 64(5): 1876-1883.

[9] PENG S N. Innovative technology for pigment dyeing of cotton yarn [D]. Zhengzhou: Zhongyuan university, 2017.

VR-Oriented Personalized Head and Face Texture Generation Technology of Dressed Human Body

CHEN Jinwen[1], WANG Xin[1], LUO Weihao[1], MEI Chennan[1], WEI Jingyan[1], ZHONG Yueqi[1,2*]

[1] *College of Textiles, Donghua University, Shanghai 201620, China*

[2] *Key Laboratory of Textile Science & Technology, Ministry of Education, Donghua University, Shanghai 201620, China*

* *Corresponding author's email*: zhyq@dhu.edu.cn

Abstract: In order to enhance the user experience in virtual reality, the human body wearing the same dress can have various customized user faces. By inputting user's single frontal face image, DECA neural network is used to generate the corresponding 3D head model with initial texture. After that, the initial texture is corrected by illumination, and the skin color of the main area and edge of the face is extracted by a given mask. Furthermore, the facial texture is in-painting twice to obtain a complete face texture. Finally, neural style transferring is employed to tone the skin texture by the face texture through PAMA neural network. Experimental results show that the facial texture restored by our proposed method can preserve the user's facial information with natural texture transition between face and body. The customized figure of user can be a suitable avatar in various virtual reality applications.

Keywords: Face texture; Illumination correction; Texture in-painting; Neural style transfer; Virtual reality

1 Introduction

In recent years, the concept of the metaverse, powered by virtual reality, augmented reality, and mixed reality technologies, has flourished. In particular, virtual avatars based on user customization have become more and more widely used in virtual reality. From the perspective of technical composition, the realization of the 3D-dressed human body has been relatively common[1-2]. However, it is still a proposition worthy of in-depth exploration to be able to assign the user's head and facial features to the dressed human body and apply it in virtual reality and even metaverse scenes. Because of this, this paper proposes to reconstruct the 3D head model with only a single frontal photo of the user's face through a minimum amount of input and unify the texture and skin tone of the face, neck, and body parts through texture fusion and restoration technology. The whole process does not require manual intervention, and can automatically realize a customized human body with user characteristics, which can be used in various virtual reality application scenarios.

In the aspect of face reconstruction based on a single image, Aaron et al. [3] proposed a method of directly reconstructing a 3D face from a single 2D image by using a convolutional neural network (CNN). This method can also process any facial image downloaded from the network, that is, any facial image with arbitrary posture, facial expression, and occlusion, but the accuracy of the reconstructed 3D face is not high; DECA[4] reconstructed a 3D head and face model from a single face image, but texture mapping was limited to the face, lack of completion and mapping of overall textures.

In the aspect of customizing the human body based on head and facial features, Li et al. [5] proposed a virtual fitting method based on model and user photos, which included three steps skin color replacement, head replacement, and body shape adjustment. Although the reconstruction speed is fast, the user interaction based on 2D virtual reconstruction is not enough, and the dressing effect cannot be observed in all directions during fitting. The 3D virtual dressed human body based on head reconstruction is realized, but the texture mapping is only processed for the face, and there are obvious skin color stitching traces between the face and the body edge. Styleme and Virtisse reconstruct a roughly

3D human body by manually inputting various human parameters, but the user's skin color, hairstyle, and face are replaced by templates, and the user's real experience still need to be improved.

It can be seen that the current technological progress has made it possible to infer the user's 3D facial features from a single photo through deep learning technology using a template face or head model. But the texture of the resulting face, which is often far from the textured skin tone of the neck and body parts, does not blend naturally. In order to solve this problem, this paper proposes a technique of head and face texture fusion and restoration so that the generated head and face model can seamlessly connect with the dressed human body.

2 Experiments

2.1 Illumination correction for facial textures

Through a series of loss functions such as reconstruction loss, feature point mapping loss, eye closure loss, illumination loss, identity loss, and shape consistency loss, the DECA network forces the deep neural network to pay attention to the overall and detailed features of the head and face during the inference process. Therefore, the geometric model of the obtained 3D face has high accuracy. However, the network lacks corresponding optimization objects for texture mapping in local space and skin color transformation in color space, so the initial texture of the obtained face needs to be processed according to the user's neck and body texture skin color.

In virtual reality application scenarios, the textures of the dressed human body are usually divided into clothing textures, body textures, and facial textures. As shown in Fig. 1, the facial texture in the head and neck texture of the original dressed human body is deleted, and the obtained head and neck template texture is U_m. The initial texture of the DECA face obtained according to the user's input is U_0 (the resolution is 256×256), the work of this paper aims to obtain the natural fusion of the two head and face texture U_{re} through a series of image processing, and map the texture to the 3D head and face model for rendering, which can obtain a good visual rendering effect.

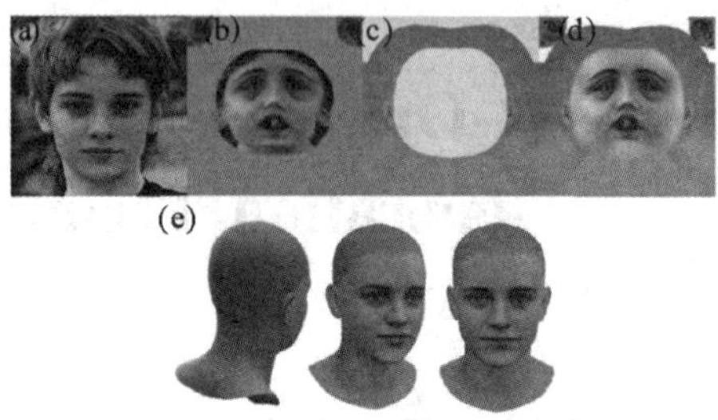

Fig. 1 Flow chart of facial texture fusion and repair: (a) user input image; (b) initial texture U_0; (c) head and neck template texture U_m; (d) texture generated in this paper U_{re}; (e) texture mapping rendering results

In most cases, the frontal face photos input by the user will be affected by ambient light and natural light, resulting in inconsistent skin brightness on the left and right sides. If texture mapping is performed directly, the color difference between the left and right faces will be large, which is not conducive to subsequent rendering applications. In order to solve this problem, on the basis of retaining the original facial information of the user, this paper adopts the illumination correction algorithm[6], which uses the distribution characteristics of the illumination components to adjust the parameters of the 2D gamma function, reduces the brightness value of the image in the area with excessive illumination, and improves the brightness of the image. According to the brightness value of the image in the dark area, the adaptive correction processing of the unevenly illuminated texture image is finally realized, so that the brightness difference between the two sides of the skin is reduced (Fig. 2).

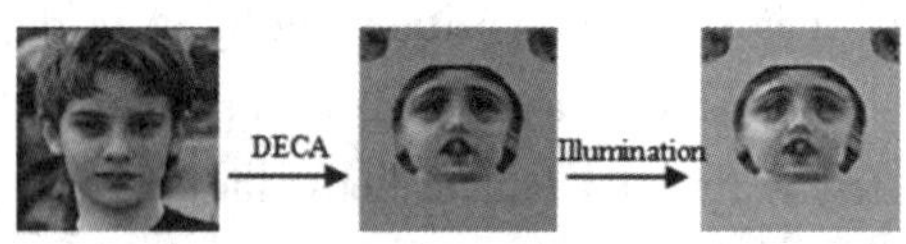

Fig. 2 Flow chart of illumination correction

2.2 Automatic restoration of facial texture

The initial texture generated by the DECA network only has the facial texture in the central area, and the texture in the side face is incomplete (Fig. 1). In order to obtain a complete facial texture, it is necessary to remove the hair occlusion and background in the texture image U received after the illumination mentioned above correction, extract an effective facial texture U_{tex}, and then perform texture completion and repair.

As shown in Fig. 3, firstly, a binarized face mask U_{mask}

is obtained according to the skin color and contour of the face in the texture U (that is, the pixel value of the face and eye area is 0, and the pixel value of the rest area is 1). Then U and U_{mask} are added element by element to get the facial texture U_{tex}. On this basis, to make the color transition between the face texture edge and the neck texture natural, firstly calculate the RGB average value of the U_{tex} edge pixels as the skin color fill value, and obtain a solid color image I_m. Then mix and add I_m and the head and neck template texture U_m to get the head and neck texture U_{head} after skin color conversion, as shown in Fig. 4.

$$qI_m + (1 - q)U_m = U_{head}$$

where q is the weight coefficient.

Fig. 3 Flow chart of facial texture extraction

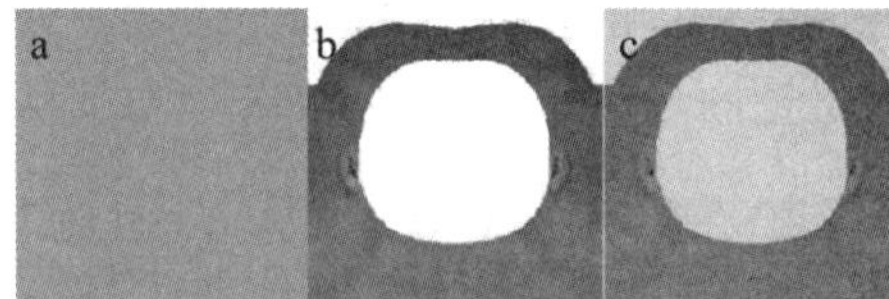

Fig. 4 (a) Solid color image I_m; (b) head and neck template texture U_m; (c) head and neck texture U_{head}

It is not difficult to find that if the textures U_{tex} and U_{head} representing the user's face are directly superimposed, there will be a blank area U_{blank} between them (Fig. 5). In order to make this area as natural as possible to blend with U_{head} and U_{tex} after coloring, it is regarded as an image in-painting task (image in-painting), and the texture in it is completed by a two-step in-painting method using a deep neural network.

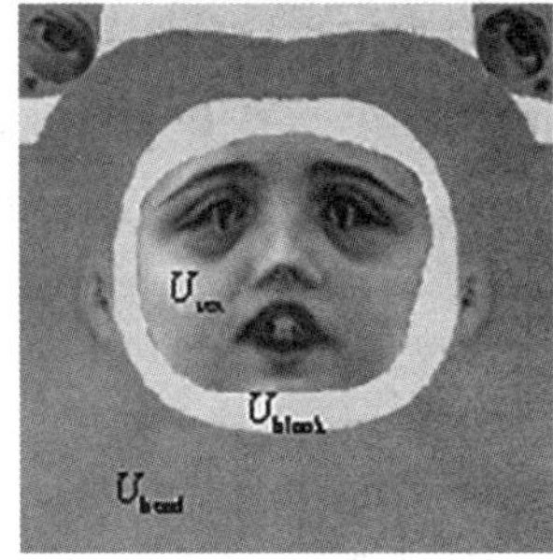

Fig. 5 Incomplete texture map

As shown in Fig. 6, in order to complement the face texture, first enlarge the 256×256 U_{tex} to 286×286, and intercept the 256×256 area from the center to get the texture U_{sym}; then make the background of U_{tex} transparent and U_{sym} image Superimpose, and use the repair method based on the image surrounding relationship[7] for the edge, and obtain the texture U_a that is initially completed. Then the background transparency of U_a is superimposed on the image of U_{head}, and the blank area is repaired using the method based on fast marching[8]. Then the eye texture above the U_0 texture is extracted and transparentized using a mask and superimposed with the repaired texture. The completed texture U_{re} can be obtained.

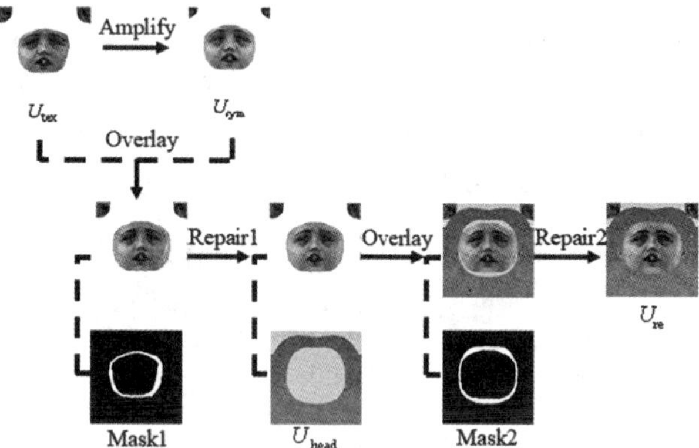

Fig. 6 Flow chart of texture completion

2.3 Style transfer of body textures

After the above processing, the texture of the user's head and face can be naturally integrated into the texture U_m of the head and neck of the original dressed human body. In order to further improve the sense of realism, as shown in Fig. 7, the head and face texture at this time is used as the style image, and the original body texture is used as the content image. Results when

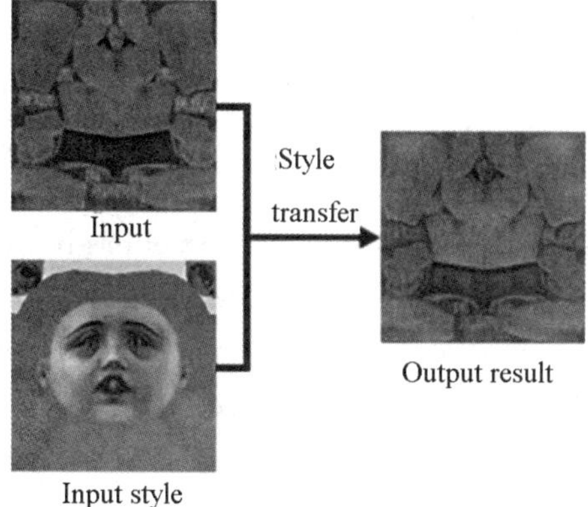

Fig. 7 Display diagram of body texture style migration

differences in facial skin tones are minimized. Since the core of this task is to preserve the content of body texture as much as possible while constraining the regional consistency and skin tone style quality before and after the transfer, this paper adopts the PAMA network[9] to solve it.

3 Results and Discussion

For the texture inpainting task, when only the inpainting method based on the image surrounding relationship is used, the edge fusion effect of facial texture U_{tex} and head and neck texture U_{head} are poor, and there are artifacts, as shown in Fig. 8(a). When only the fast-moving-based repair method is used, although the edges of the facial texture U_{tex} and the head and neck texture U_{head} have an excellent fusion effect, the repaired regional texture comes from the edge diffusion of the head and neck texture, as shown in Fig. 8(b). Only when two ways are combined, the effect is the best, as shown in Fig. 8(c).

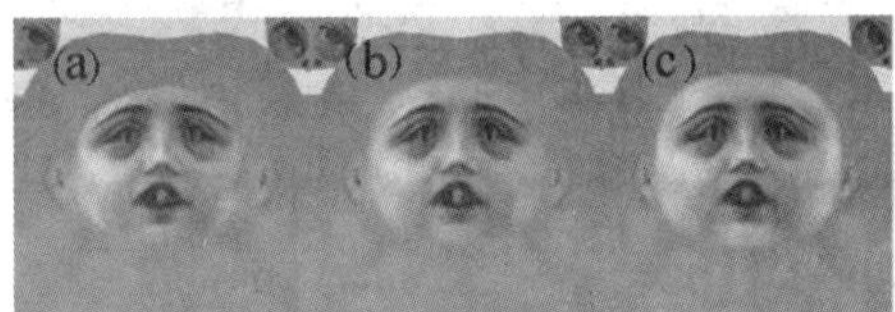

Fig. 8 (a) Only using the results of the image surrounding relation assisted inpainting method; (b) use only the repair method results based on fast travel; (c) the results of using two repair method U_{re}

Figure 9 shows the results of texture restoration for different users using the method in this paper. It can be seen that after the original texture is restored and completed, the texture of the user's head and face model is the same as the neck skin color in the original 3D dressed human model. Similarly, the boundary fusion is natural, the user's original facial information is preserved, and the whole process does not require user intervention. It shows that the method proposed in this paper is reasonable and practical.

After the textured head and face model has been generated, it can be mounted on a 3D rigged human body. This process mainly involves the affine transformation of the head, posture calibration, Boolean (add) operation for fusion with the body, etc. Since these contents are standard technologies in virtual fitting, they do not need to repeat here.

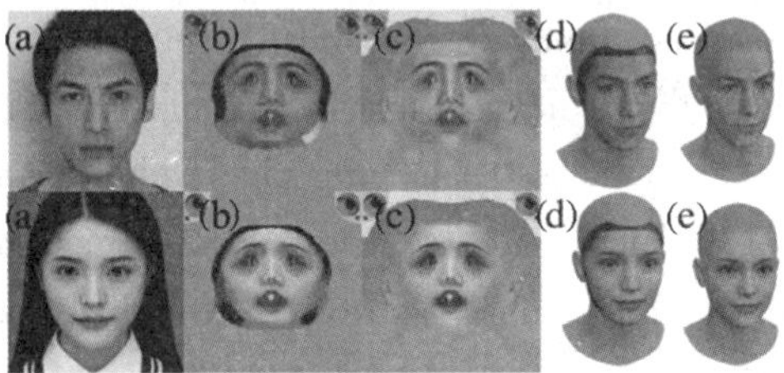

Fig. 9 Texture restoration results show. (a) the user's frontal view; (b) initial texture map; (c) the results of this paper; (d) the initial texture rendering result; (e) rendering results of this paper

As shown in Fig. 10, the 3D head and face model generated from a single user photo (Fig. 10 (a)) is replaced with the dressed human body (Fig. 10(b)) to obtain the user's dressed human body (Fig. 10(c)), and the user's head and face model is fused with the dressed human body. Figure 11 shows the rendering results of the dressed human bodies of different users in various virtual scenes. It can see that the model can present natural light and shadow effects in different indoor and outdoor lighting environments, which proves that the method in this paper is suitable for VR scene practicability.

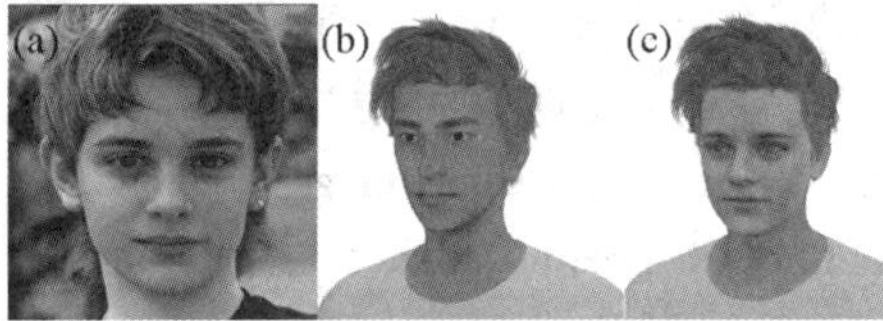

Fig. 10 Virtual reality application effect display: (a) the user's frontal view; (b) wearing the original human body; (c) the user dresses the human body

Fig. 11 VR rendering effect display

4 Conclusions

Starting from the application of virtual reality, this paper studies the generation of personalized head and face textures of dressed human bodies and proposes a head and face texture fusion and restoration technology. By inputting a single frontal face image, complete head and face textures and corresponding body textures can be generated, which solves the problem of the natural fusion of head and face textures and body texture and skin color from different sources. They can meet the requirements of virtual fitting in various VR environment application.

Acknowledgements

This work was supported by Shanghai Natural Science Foundation (No. 21ZR1403000).

References

[1] ALLDIECK T, ZANFIR M, SMINCHISESCU C. Photorealistic Monocular 3D Reconstruction of Humans Wearing Clothing[J]. ArXiv e-prints, 2022-4-19.

[2] ALEXANDER O, ROGERS M, LAMBETH W, et al. The Digital Emily project: photoreal facial modeling and animation[C]//ACM SIGGRAPH 2009 Courses, 2009.

[3] JACKSON A S, BULAT A, ARGYRIOU V, et al. Large Pose 3D Face Reconstruction from a Single Image via Direct Volumetric CNN Regression[C]//International Conference on Computer Vision. IEEE Computer Society, 2017.

[4] FENG Y, FENG H, Black M J, et al. Learning an Animatable Detailed 3D Face Model from In-The-Wild Images[J]. ACM Transactions on Graphics, 2020, 40(4): 88.

[5] LI J, ZHANG M M, PAN Z G, et al. Virtual try-on by replacing the person in image[J]. Journal of Computer-Aided Design & Computer Graphics, 2015, 27(9): 1694-1700.

[6] LIU Z C, Wang D W, LIU Y, et al. Adaptive adjustment algorithm for non-uniform illumination images based on 2D gamma function[J]. Transactions of Beijing Institute of Technology, 2016, 36(2): 191-196,214.

[7] ZENG Y, LIN Z, LU H, et al. CR-Fill: Generative Image Inpainting with Auxiliary Contexutal Reconstruction[J]. Arxiv: 2011.12836, 2020-11-25.

[8] TELEA A. An image inpainting technique based on the fast-marching method [J]. Joumal of Graphics Tools, 2004, 9(1): 25-36.

[9] XUAN L, ZHEN H, LINKANG Y, et al. Consistent arbitrary style transfer[J]. Arxiv: 2201. 02233, 2022-01-06.

3D Virtual Fitting Network Based on Attention Mechanism and Transformer

YUAN Tiantian[1], WANG Xin[1], LUO Weihao[1], MEI Chennan[1], WEI Jingyan[1], ZHONG Yueqi[1, 2*]

[1] *College of Textiles, Donghua University, Shanghai 201620, China*

[2] *Key Laboratory of Textile Science & Technology, Ministry of Education, Donghua University, Shanghai 201620, China*

* *Corresponding author's email*: zhyq@ dhu. edu. cn

Abstract: Aiming at the problems of blurred edges of 3D human models, unexpected garment deformation and artifacts in 3D virtual fitting networks, attention mechanism and visual transformer structure are introduced into the deep neural network. By fusing garment deformation information and depth estimation information, 3D virtual fitting can be realized. Since the attention mechanism can guide the network to pay attention to the image details, it can realize the purpose of constraining the deformed garment to conform to the human body posture, and effectively deal with the edge adhesion of the 3D human model. The quantitative evaluation results show that the structural similarity is improved by 0. 015 7, and the peak signal-to-noise ratio is improved by 0. 113 2; the absolute relative error of the depth estimation value is reduced by 0. 037, and the squared relative error is reduced by 0. 014. Both quantitative and qualitative analyses show that this method can achieve more accurate prediction of 3D virtual fitting.

Keywords: Virtual fitting; Transformer; Attention mechanism; Depth estimation; 3D reconstruction

1 Introduction

At present, the virtual fitting technologies based on deep learning mainly use 2D images or 3D reconstruction to simulate how the target garment will look on the consumer. Among them, the image based 2D virtual fitting technologies mainly uses thin plate spline (TPS) to deform the garment and fuse it with the human body to achieve 2D virtual fitting. For example, the classic VITON[1] network uses a coarse-to-fine network to extract the texture information of the garment, and controls the deformation of the garment through TPS transformation to realize virtual fitting. Building on this, CP-VTON[2] added a geometric matching module and a mask maximization loss function to further constrain the distortion of garment deformation. The later method introduces the GAN loss function[3] to optimize the deformation and conformity of the garment. However, 2D virtual fitting technology still has drawbacks, which cannot truly represent the body shape information of and the side wearing effect of the garment.

Human reconstruction is very vital for 3D virtual fitting. Common algorithms include parametric, non-parametric, and hybrid methods. The originator of the parametric method is SMPL[4], which drives the standard human template through template parameters to generate 3D human models with different body shapes and poses. Non-parametric methods utilize implicit functions, such as the network represented by Pifu[5], which can generate 3D clothed human bodies with rich textures. Through an implicit function that discriminates the inside and outside of the position, the inside and outside probability field on a continuous space is predicted, and the iso-surface is extracted from it to obtain the human body surface. Geo-Pifu[6] designs a structure-sensitive network to learn implicit voxel features, reducing the generation of artifacts and distortions. The hybrid method combines the advantages of both of the above. For example, ARCH[7] combines the pose information of the 3D human body model under the framework of implicit surface function and adds a normal prediction network to enhance the 3D reconstruction details.

The attention mechanism allows the model to give greater computational weight to high values in attention map while ignoring low values in attention map by generating an attention map. Many related works, such as SeNet[8], DANet[9], CBAM[10], Transformer[11], have shown that using attention mechanism can guide the network to focus on image details to improve performance.

The advancement of the above technologies makes it possible to find a solution that requires the least amount of input but produces the best output. Such methods generate 3D dressing effects from a single image of a person and an image of a single item of garment to be tried on. For example, M3D-VTON[12] inputs a single image, and can obtain 3D try-on results by fusing garment deformation information with human depth information. However, the edges of the 3D human model generated by the network have adhesions and artifacts, and the garment deformation appears distorted. One of the reasons for this problem is that the network loses the details of the image in the computing process. In order to improve the model's computation of details with the images, this paper proposes a 3D virtual fitting network T3D-VTON based on the attention mechanism. T3D-VTON inputs the user image I_{user} and target garment image I_{sku}. Then the 3D model is obtained end-to-end for the user trying on the garment Ω_{sku}.

2 Structure of T3D-VTON

2.1 Overall network architecture

The overall architecture of the T3D-VTON network is shown in Fig. 1. The network consists of three main modules: 2D garment deformation module, depth estimation module and 3D fusion module. The inputs are the user image I_{user} and the target garment image I_{sku}. The output is the 3D model of the user trying on the garment Ω_{sku}.

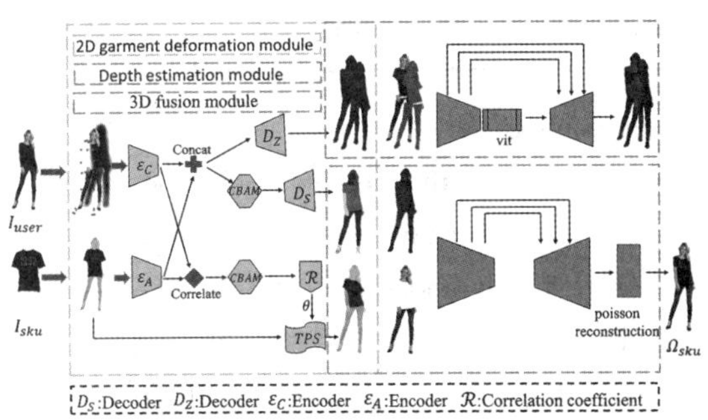

Fig. 1 Overall network structure of T3D-VTON

2.2 2D garment deformation module

In order to make the deformed garment more in line with the human posture, this paper adds a convolutional attention module (including channel attention and spatial attention) before the TPS transformation to focus on important features and reduce the interference of irrelevant information. As shown in Fig. 2, the convolutional attention module includes two feature optimization mechanisms, channel attention and spatial attention.

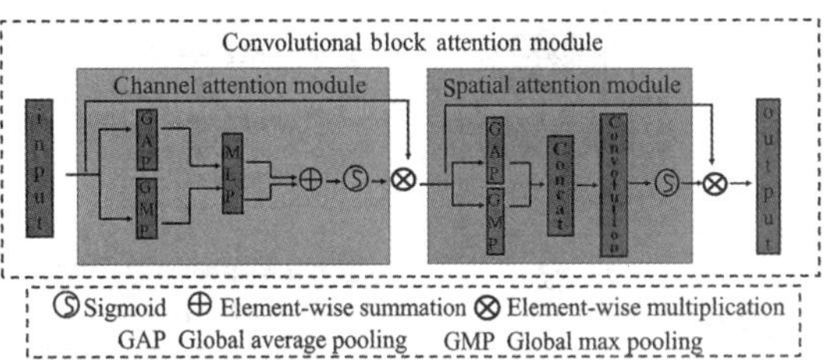

Fig. 2 Convolutional block attention module

2.3 Depth estimation module

Depth estimation is crucial for reasoning about 3D information contained in 2D images. The inputs to this module include the given front-to-back depth maps D_i, deformed garment C_w, non-target garment body parts I_p, and shadow information I_g. To address the problems of blurred edges and unclear outlines of characters in the M3D-VTON network, this paper proposes a depth estimation network combining convolutional neural network and Transformer for multi-scale feature fusion, as shown in Fig. 3.

First, the encoder-decoder architecture is designed with Resnet50 as the backbone network. Among them, the encoder extracts spatial and semantic information at four different resolutions at multiple scales (1/2, 1/4, 1/8, and 1/16 of the input image respectively). Then, the Transformer structure is added. Each Transformer layer mainly includes layer normalization (layer norm), multi-head attention (MHSA) and MLP. The extracted features are further processed by taking advantage of the Transformer's ability to take into account global information as well as locally focused information. Finally, the decoder performs up-samples the extracted features and combines the multi-scale features extracted by the encoder to achieve upper and lower-level feature fusion to improve the utilization of contextual information, and fuses the semantic and spatial information extracted by the encoder to obtain more accurate depth information.

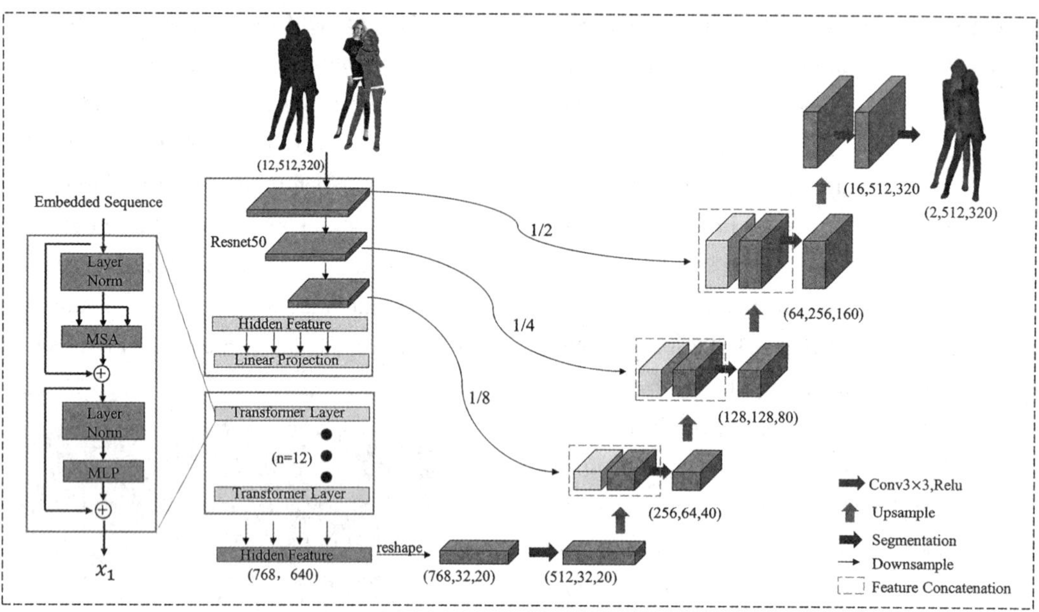

Fig. 3 Depth estimation network

2.4 3D fusion module

The 3D fusion module of T3D-VTON adopts the same U-net network as M3D-VTON to fuse 2D information (such as deformed garment and human body segmentation map) and 3D information (such as depth maps from front and back view).

3 Experimental Results and Discussion

3.1 Experimental dataset

The experimental dataset used in this paper was MPV-3D women's garment dataset, which contained 6 566 pairs of image groups. The images were divided into target garment, person images, and front and back depth maps. The resolution of the image was 512×320. According to the ratio of 6 : 1, 5 632 images were used for training and 934 images were used for testing.

3.2 Experimental setup

All experiments in this paper were conducted on a PC with NVIDIA GeForce RTX1080ti dual graphics card (12 G video memory), and were implemented by PyTorch programming.

During training, the Adam optimizer was used and supplemented by learning rate decay. The initial learning rate was set to 0.000 1. After every 50th epoch, the learning rate was multiplied by 0.98 to decay, for a total of 100 training rounds.

3.3 Discussion of results

This paper analyses the specific effects of the aforementioned improvements from the effect of garment deformation and the depth value error of depth estimation.

3.3.1 Discussion of garment deformation effect

The effect of garment deformation is analyzed quantitatively and qualitatively by means of 2D virtual fitting effect on the intermediate output.

Quantitative evaluation:

Structural similarity (SSIM)[13] was used to evaluate the similarity between the 2D tested try-on image and the real image, and peak signal to noise ratio (PSNR)[14] was used to evaluate the image quality. As can be seen from Tab. 1, the addition of the convolutional attention module increases SSIM by 0.015 7, and the PSNR by 0.113 2. The results show that the method in this paper produces higher quality images that retain many original details.

Tab. 1 The quantitative evaluation of 2D virtual try-on

Network name	SSIM	PSNR
M3D-VTON	0.9218	20.4205
T3D-VTON	0.9375	20.5337

Qualitative comparison:

As shown in Fig. 4, from left to right, the target garment, the original human body image, the 2D fitting effect of M3D-VTON and the 2D fitting effect image of T3D-VTON are represented respectively.

The virtual fitting effect produced by M3D-VTON has blurred garment on the shoulders, and even exposes a large area of human skin color. The shape of the short sleeves after trying on is closer to the geometric area of the original wear, while T3D-VTON can well retain the geometric shape of short sleeves, reducing the phenomenon of long sleeves becoming short sleeves and short sleeves becoming sleeveless. The deformation of the neckline is more natural. The above results show that the convolutional attention module introduced by T3D-VTON proposed in this paper can effectively regulate the fusion of garment deformation and target characters when dealing with complex textures, while retaining the pattern effect of garment textures, such as logos and text on garment.

(a)Target garment (b)Clothed human (c)M3D-VTON (d)T3D-VTON

Fig. 4 The visual comparison of 2D virtual try-on

3.3.2 Discussion of depth estimation results

As previously mentioned, depth estimation mainly affects the 3D fitting results, and its quantitative and qualitative comparative analysis is as follows.

Quantitative evaluation:

Absolute relative error (AbsRel) and square relative error (SqRel) are used to evaluate the effect of the depth estimation network.

As shown in Tab. 2, compared with M3D-VTON, the T3D-VTON network proposed in this paper reduces AbsRel by 0.037 and SqRel by 0.014. It shows that the depth value obtained by T3D-VTON is more in line with the true value, because the transformer and multi-scale feature fusion network can effectively fuse semantic information and spatial information, thereby improving the accuracy of depth estimation.

Tab. 2 The quantitative evaluation of depth estimation

Network name	AbsRel	SqRel
M3D-VTON	7.223	0.375
T3D-VTON	7.186	0.361

Qualitative comparison:

Fig. 5 shows the qualitative comparison results of the optimized networks (from left to right, the target garment, the original body image, and the frontal and side images of the output of the two networks). As shown in Fig. 5, the human body presents a clearer contour edge, while eliminating the common adhesion problem at the junction of the human arm and abdomen. When the knees of the human body are close, the T3D-VTON proposed in this paper can effectively eliminate the adhesion between the knees and obtain a 3D human body model with a clearer outline. It is proved that the depth information obtained by the network is more in line with the shape of the human body. In other words, under the guidance of depth information and combined with 2D fitting images, the final 3D human body model can better restore human body surface details and texture information on the premise of ensuring the restoration of human body contours.

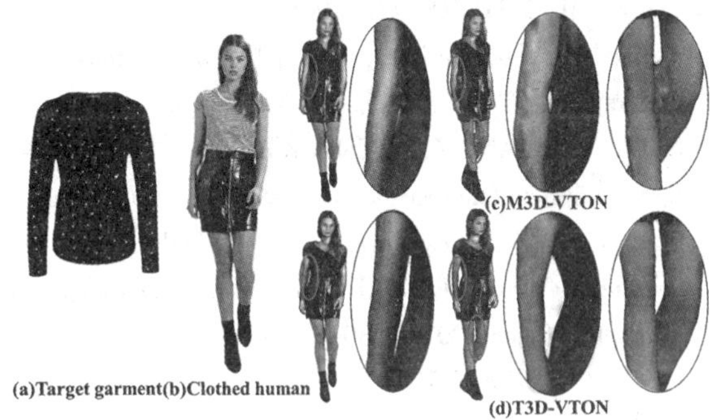

Fig. 5 The visual comparison of 3D virtual fitting

4 Conclusions

This paper optimized M3D-VTON and proposed T3D-VTON, a 3D virtual fitting network based on attention mechanism and Transformer. The experimental results indicate that the added convolutional attention module skillfully guides the generation of TPS deformation parameters, constrains the degree of garment deformation, and makes the generated segmentation map more accurate. This ensures the reasonable integration of deformed garment and human body shape and posture. The use of Transformer can more accurately predict the front and back depth information of the human body. By integrating the segmentation map information, deformed garment information and depth estimation information, the results of 3D virtual fitting would be predicted more accurately, which can provide a useful reference for 3D clothed human reconstruction and virtual fitting based on a single image.

Acknowledgements

This work was supported by Shanghai Natural Science Foundation (No. 21ZR1403000).

References

[1] HAN X T, WU Z X, WU Z, et al. Viton: an image-based virtual try-on network[C]//Salt Lake City, UT, USA: Proceedings of the IEEE Conference on Computer Vision and Pattern Recognition. 2018: 7543-7552.

[2] WANG B C, ZHENG H B, LIANG X D, et al. Toward characteristic-preserving image-based virtual try-on network[C]//Munich, Germany: Proceedings of the European Conference on Computer Vision (ECCV). 2018: 589-604.

[3] GOODFELLOW I J, POUGET-ABADIE J, MIRZA M, et al. Generative adversarial nets[J]. Advances in Neural Information Processing Systems, 2014, 27: 2672-2680.

[4] LOPER M, MAHMOOD N, ROMERO J, et al. SMPL: a skinned multi-person linear model[J]. ACM Transactions on Graphics (TOG), 2015, 34(6): 1-16.

[5] SAITO S, HUANG Z, NATSUME R, et al. Pifu: pixel-aligned implicit function for high-resolution clothed human digitization[C]//Los Angeles CA, United States: Proceedings of the IEEE/CVF International Conference on Computer Vision. 2019: 2304-2314.

[6] HE T, COLLOMOSSE J, JIN H L, et al. Geo-pifu: geometry and pixel aligned implicit functions for single-view human reconstruction[J]. Advances in Neural Information Processing Systems, 2020, 33: 9276-9287.

[7] HUANG Z, XU Y L, LASSNER C, et al. ARCH: animatable reconstruction of clothed humans[C]//Seattle, WA, USA: Proceedings of the IEEE/CVF Conference on Computer Vision and Pattern Recognition. 2020: 3093-3102.

[8] HU J, SHEN L, SUN G. Squeeze-and-excitation networks[C]//Salt Lake City, UT, USA: Proceedings of the IEEE Conference on Computer Vision and Pattern Recognition. 2018: 7132-7141.

[9] FU J, LIU J, TIAN H, et al. Dual attention network for scene segmentation[C]//Los Angeles CA, United States: Proceedings of the IEEE/CVF Conference on Computer Vision and Pattern Recognition. 2019: 3146-3154.

[10] WOO S, PARK J, LEE J Y, et al. Cbam: convolutional block attention module[C]//Munich, Germany: Proceedings of the European Conference on Computer Vision (ECCV). 2018: 3-19.

[11] VASWANI A, SHAZEER N, PARMAR N, et al. Attention is all you need[C]//Long Beach, CA, USA: Advances in Neural Information Processing Systems. 2017: 5998-6008.

[12] ZHAO F W, XIE Z Y, KAMPFFMEYER M, et al. M3d-vton: a monocular-to-3d virtual try-on network[C]//Nashville, TN, USA: Proceedings of the IEEE/CVF International Conference on Computer Vision. 2021: 13239-13249.

[13] WANG Z, BOVIK A C, SHEIKH H R, et al. Image quality assessment: from error visibility to structural similarity[J]. IEEE Transactions on Image Processing, 2004, 13(4): 600-612.

[14] HUYNH-THU Q, GHANNBARI M. Scope of validity of PSNR in image/video quality assessment[J]. Electronics Letters, 2008, 44(13): 800-801.

Green Fabrication of Washable and Long-Lasting Fragrant Nanocapsules/Cotton Fabrics without Adhesive via Atomization Spray

FAN Jiaxuan, ZHAO Yi *

Engineering Research Center of Technical Textiles, Ministry of Education, Donghua University, Shanghai 201620, China

* *Corresponding author's email*: zhaoyi@ dhu. edu. cn

Abstract: Control-released aromatic nanocapsules/cotton fabrics with good wash durability and long-lasting fragrance release capability have been fabricated by spraying with high efficiency and low consumption. The aromatic nanocapsules with lavender essence as the core materials and epichlorohydrin modified cyclodextrin as the wall materials were sprayed after atomization onto the cotton fabric. Under alkaline conditions, the ring-opening reaction occurred with the hydroxyl group on the cotton fiber, so as to realize the adhesive free, uniform and firm loading of aromatic nanocapsules on the surface of the cotton fabrics. The fragrances of the treated fabrics could release until 90 d, and the residue rate of the essential oil content was up to 78.6% after 6 cycles of washing, which was 4 times of that of the aromatic fabric prepared by traditional padding. Atomized spraying method not only endows cotton fabrics with fragrance-releasing properties, highly improves the utilization rate of aromatic nanocapsules, but also effectively reduces the cost of wastewater treatment during after-finishing processes, which is potentially more suitable for large-scale production of functional textiles with high efficiency, less pollution and low energy consumption.

Keywords: Aromatic nanocapsule; Cotton fabrics; Atomization spray; Environmentally friendly

1 Introduction

The high pressure and fast pace of life make a large number of modern people suffer from anxiety, depression and other problems. Compared with the addiction and dependence of drug therapy, aromatherapy with a long history has gradually entered the public view. Aromatic textiles were developed by combining the essential oils of plants with textiles, which can help sleeping[1], and anti-anxiety[2], thus providing a new idea for the application and promotion of aromatherapy. The traditional preparation method of aromatic textiles is to fix the aromatic essential oil directly on the textiles through immersing[3], coating[4], and microwave[5], but aromatic fabrics prepared by this method failure to achieve the effect of long-lasting release of fragrances, which seriously hinders the further development of aromatic textiles. Microcapsule[6] technology can wrap aromatic essential oils in the wall material and then prolong the service life of aromatic textiles. Aromatic textiles prepared by loading aromatic microcapsules onto textiles using adhesives are facing elimination due to adverse effects on the human body and the environment during use. To overcome the above problems, the direct load technology of aromatic micro/nanocapsules on textiles without adhesives comes into being. Ma et al.[7] prepared three-dimensional superabsorbent nonwovens with liquid-triggered fragrance release via green, fast and scalable multi effect hot-air anchoring method which physically crosslinks the joint thermo-bonding fibers and anchores fragrance microcapsules/super absorbent polymer (SAP) onto adjacent thermo-bonding fibers simultaneously. The loading of aromatic nanocapsules by chemical grafting method mostly relies on the modification of wall materials of nanocapsule. Among many wall materials, cyclodextrin stands out due to its advantages of low cost and strong packing capacity.

Epichlorohydrin modified cyclodextrin is one kind of modified cyclodextrin. In alkaline environment, the epoxy group can react with the hydroxyl of cotton fiber to form a firm covalent bond as shown in Fig. 1 (a), and improve the combination fastness between nanocapsules and fibers. Traditional padding processes will expose the aromatic microcapsules to alkaline environment in advance or pollute the aromatic microcapsule solution, resulting in the aggregation of aromatic nanocapsules. In response to this question, our previous work[8] proposed to rely on the inside-outside diffusion and penetration throughout fibers, and immersed the lower end of the cotton fabric in an alkaline solution to complete the in situ immobilization of the aromatic nanocapsules on the cotton fabric. However, it was hard to be applied to large scale and high efficiency factory production.

Herein, the cotton fabrics were padded in alkali solution, and then the aromatic nanocapsules with epichlorohydrin modified cyclodextrin as the wall materials were loaded on the cotton fabric via atomization spray, so as to creating a strong bond between the two through the ring-opening reaction. The preparation process in practical production is shown in Fig. 1 (b). This preparation method was more adapted to the large-scale production with high efficiency, less pollution, and low energy consumption, providing solutions for the industrialization of aromatic textiles with long-lasting release of aroma and washing durability.

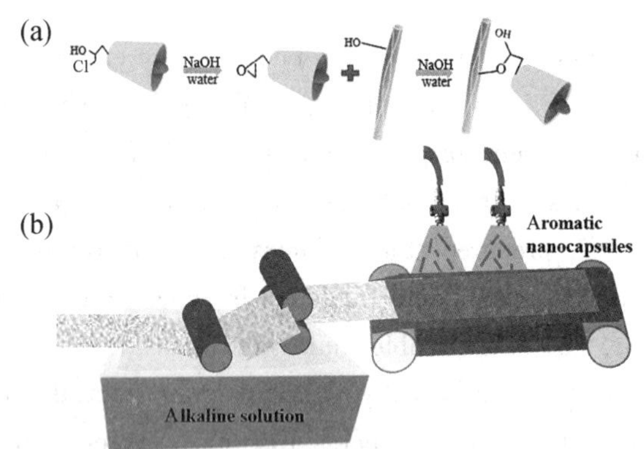

Fig. 1 (a) Schematic diagram of the reaction between aromatic nanocapsules and cotton fabrics; (b) flowchart of aromatic fabrics industrial preparation

2 Experiments

2.1 Materials

Well-prepared 100% cotton woven fabrics were used in all experiments. Sodium hydroxide (NaOH) was obtained from HUSHI. Ethanol (≥99.7%) was purchased from HUSHI. Aromatic nanocapsules with the core material of lavender essence and the wall material of epichlorohydrin modified cyclodextrin were provided by Shanghai Institute of Technology. Deionized water was used in all experiments. All chemicals were used as received without further purification.

2.2 Immobilization of aromatic nanocapsules on cotton fabrics

Cotton fabrics were cleaned by alternate ultrasonic irradiation in ethanol and deionized water for 10 min to remove the dirt and dust. The cleaned cotton fabrics (round cloth with a radius of 4 cm, approximately 0.8 g each) were immersed in different concentrations of sodium hydroxide solution (0%, 1.0%, 1.5% and 2.0%) for 1 min. A spray gun is used to spray 10% (v/v) aromatic nanocapsules solution onto the cotton fabric impregnated with alkaline solution. Maintained the spray holder 25 cm away from the fabric throughout the preparation process. Control samples were prepared by traditional padding process with 1.5% sodium hydroxide solution (sample was denoted as 1.5*). After the ring-opening reaction, the nanocapsules which did not react on cotton fiber were washed with deionized water and the prepared fabrics were air-dried for 24 h before characterization.

2.3 Characterization

The particle size distribution of the provided aromatic nanocapsules was analyzed by a dynamic/static laser light scatterer (BI-200SM, America). The zeta potential of aromatic nanocapsules was measured using a nano particle size and zeta potential analyzer (Nanotrac Wave II, America). The morphologies and microstructures of blank and treated fabrics were investigated through a field emission scanning electron microscope (FESEM) (SU8010, China). The surface element compositions of treated cotton fabric were analyzed by energy dispersive spectroscopy (EDS). The aromatic retentions of aromatic fabrics were evaluated by gas chromatography mass spectrometry (QP2010Ultra, America).

2.4 Characterization of long-lasting fragrance and wash durability of aromatic fabrics

The long-lasting fragrance performance of aromatic fabrics and control fabrics that were kept at room temperature for 1, 30 and 90 days was measured by gas ultrafast gas chromatographic electron nose (Heracles II, France). The wash durability of the aromatic textiles was tested after washing by automatic shrinkage testing machine (Y089D, China), using ISO 6330-2012 3N as a reference standard for different times and evaluated the aromatic retentions of aromatic fabrics by gas chromatography mass spectrometry (QP2010Ultra, America).

3 Results and Discussion

3.1 Characterization of aromatic nanocapsules

The aromatic nanocapsules with 95% encapsulation efficiency were composed of epichlorohydrin modified β-cyclodextrin as the wall materials and lavender essence as the core materials. Fig. 2 (a) showed the photo of an aromatic nanocapsules solution. It could be seen that the solution was clear, indicating that the aromatic nanocapsules could be evenly dispersed in water solution. As shown in Fig. 2 (b), the average particle size of aromatic microcapsules was 10.1 nm. The zeta potential of the aromatic nanocapsules solution was −80 mV. The zeta potential can be used to predict the stability of particle aggregation in colloids, and the boundary of dispersion stability of nanoparticles in aqueous phase is generally considered as ± 30 mV. When the absolute value of zeta potential is greater than 30 mV, nanoparticles can be considered as stable dispersion. Therefore, it was shown that the solubility of β-cyclodextrin modified by epichlorohydrin in water solution was greatly improved.

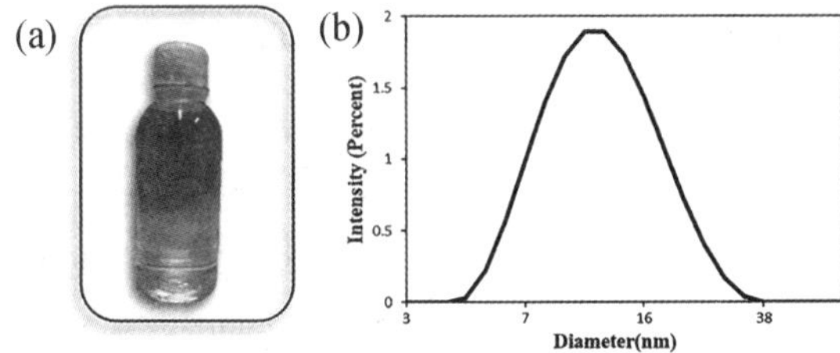

Fig. 2 (a) Digital image of lavender essence aromatic nanocapsules solution; (b) particle size distribution diagram of the aromatic nanocapsules

3.2 Immobilization of aromatic nanocapsules on cotton fabrics

The surface morphology of untreated cotton fabric and aromatic cotton fabric prepared by spraying at different NaOH concentrations was shown in Figs. 4 (a)−4(d). It could be seen from Fig. 4 (a) that the surface of untreated cotton fabric was relatively smooth, while some small particles existed on the surface of treated cotton fabric, indicating that the wall material of aromatic nanocapsules had ring-opening reaction with cotton fiber in alkaline solution. With the increase of NaOH concentration, the loading of aromatic nanocapsules increased first, then decreased, and reached the highest value at 1.5% NaOH. When NaOH concentration reached 2.0%, the aromatic nanocapsules appeared to agglomerate, indicating that alkali concentration had a great influence on the agglomeration of nanoparticles, as shown in Figs. 4 (b)−4(d). Compared with the aromatic cotton fabric prepared by spraying methods shown in Fig. 4 (c), the surface of the aromatic cotton fabric prepared by padding showed the agglomeration of aromatic nanocapsules in Fig. 4 (e), which resulted in a decrease in load uniformity. The results showed that aromatic cotton fabric prepared by spraying could effectively reduce the agglomeration of aromatic nanocapsules.

Once immersing into the alkali solution, the aromatic nanocapsules rapidly underwent ring-opening grafting reaction to form the small agglomerated particles. When contacting with the cotton fabrics, the agglomerated nanoparticles grafted with the hydroxyl groups on the cotton fabrics, thereby forming the aromatic nanocapsules with a certain degree of agglomeration evenly dispersed on the surface of the cotton fabrics. The chemical reaction formula of this process was shown in Fig. 3 (a). Spraying aromatic nanocapsules on the surface of cotton fabric, the ring-opening reaction occurred rapidly at the interface after the two contacted, which could realize the firm adhesion of aromatic nanocapsules and reduce the possibility of the ring-opening reaction between the nanocapsules. The chemical reaction formula of this process was shown in Fig. 3 (b). Atomization spray method produced droplets with a particle size of 3−10 μm. The spraying was uniform and delicate

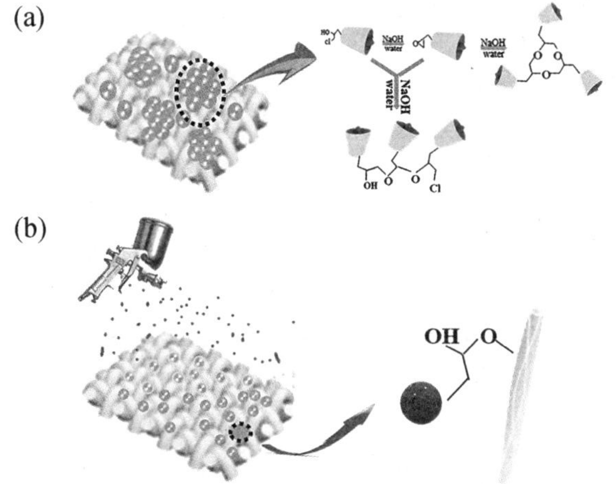

Fig. 3 Schematic diagram of (a) the agglomeration of aromatic nanocapsules on cotton fabric by padding and (b) the uniform load on cotton fabric via atomization spray

with a wide range, so that aromatic nanocapsules could be evenly dispersed on the surface of cotton fabric.

In order to exclude the possibility that small particles on the surface of treated cotton fabric were NaOH crystals, aromatic fabrics prepared under 1.5% NaOH were tested by EDS. As shown in Fig. 4 (f), only C, O, and Al existed on the surface of aromatic cotton fabric without Na. The results showed that the small particles on the surface of the treated cotton fabric were indeed aromatic nanocapsules. The change of essence content of aromatic fabrics with different NaOH concentrations was shown in Fig. 4 (g). When NaOH concentration was 1.5%, aromatic cotton fabric prepared by spraying had the highest essential oils content. Compared with the control sample prepared by padding, the content of essential oil increased by about 59%, demonstrating that spraying method can effectively improve the loading of aroma nanocapsules on aromatic cotton fabric.

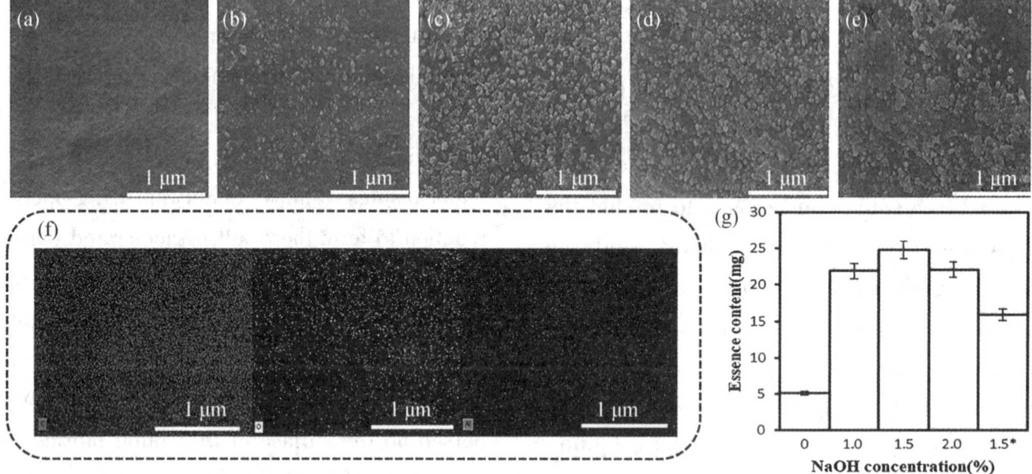

Fig. 4 SEM images of aromatic fabrics prepared under different alkali concentration conditions: (a)-(d) 0%, 1.0%, 1.5% and 2.0% NaOH by spraying and (e) by traditional padding; (f) EDS spectra of aromatic fabrics prepared under 1.5% NaOH; (g) essence content diagram of aromatic fabrics per gram under 0%, 1.0%, 1.5%, and 2.0% NaOH by spraying and under 1.5% NaOH by padding

3.3 Characterization of long-lasting fragrance and wash durability of aromatic fabrics

The height of the peak in the radar diagram represents the intensity of the aroma. Figs. 5 (a)−5(b) respectively represented the polar aroma intensity radar maps of aromatic fabrics perfumed by the aromatic nanocapsules after 1 d and 90 d of staying in natural condition. As shown in Fig. 5 (a), in the beginning, the aromatic fabrics treated with 1.5% NaOH got the highest peak, that was because it had the most aromatic nanocapsules load, and the aromatic fabrics prepared by traditional padding came second. However, the padding process would cause the rupture of aromatic nanocapsules, resulting in the rapid volatilization of essential oil. Therefore, after 90 d of storage, the aroma intensity of the aromatic fabrics prepared by traditional padding was significantly weakened. The results demonstrated that the spray loading method could ensure the encapsulation property of nanocapsules and obtain the aromatic cotton fabric with long-lasting fragrance release capacity.

The changes of the essence content of the aromatic fabrics prepared by different NaOH concentrations and processes after different washing cycles were shown in Fig. 5 (c). For the samples did not immersed in alkaline solution, the essence content of cotton fabric decreased significantly after washing because the epoxy group on the wall material of aromatic nanocapsules could not form chemical bond with cotton fiber. Via atomization spray, the aromatic fabrics treated with 1.5% NaOH after 6 cycles of washing had the highest content of essential oil and only decreased by 21.4%, showing good washing durability. Due to the large amount of aggregation of aromatic nanocapsules on the surface of the cotton fabric, the reaction sites with the cotton fiber decreased. The combination fastness between the two was insufficient, so it was very easy to fall off after repeated cycles of washing. Under the condition of the same concentration of NaOH, the residual amount of essence oil prepared by spraying could be about 4 times higher than that prepared by padding. As can be seen from Figs.

5 (d)-5(f), the number and distribution of aromatic nanocapsules on the surface of the fibers did not change significantly before and after 3 or 6 cycles of washing, which indicated that the process of spraying leaded to good washing durability of the aromatic fabrics.

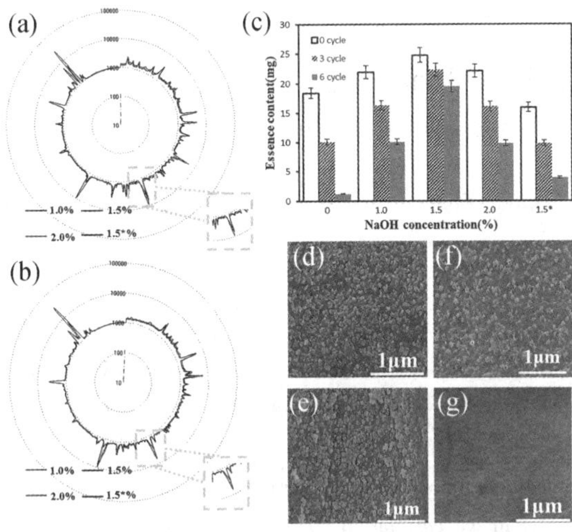

Fig. 5 Polar and nonpolar aroma intensity radar maps of (a) and (b) aromatic nanocapsules perfumed aromatic fabrics after 1 d and 90 d of staying in natural condition; (c) essence content diagram of aromatic fabrics per gram under 0-2.0% NaOH by spraying and under 1.5% NaOH by traditional padding after 1-6 washing cycles. SEM images of aromatic fabrics after 0 cycle (d), 3 cycles (e), 6 cycles (f) of washing and untreated cotton fabric (g)

4 Conclusions

The aromatic textiles with high washing durability and long-lasting fragrance were prepared by atomization spray method. Compared with the aromatic cotton fabric prepared by the traditional padding, the load uniformity and load volumes of aromatic nanocapsules greatly improved, and the content of essential oil increased by about 59%. The aroma could still be released after 90 d at room temperature. After 6 times of washing, the essential oil content decreased by only 21.4%, which was about 4 times higher than the residual essence of the aromatic cotton fabric prepared by padding. It provides a new solution for industrial production of adhesive-free aromatic cotton fabric with control-released and washing durability.

Acknowledgments

This research was supported by Capacity building project of local universities Science and Technology Commission of Shanghai Municipality (No. 19090503500).

References

[1] KIM K, BU Y, JEONG S, et al. Memory-enhancing effect of a supercritical carbon dioxide fluid extract of the needles of Abies koreana on scopolamine-induced amnesia in mice[J]. Bioscience Biotechnology and Biochemistry, 2006, 70(8): 1821-1826.

[2] KUMAR D, BHAT Z A, SHAH M Y. Anti-anxiety activity of successive extracts of angelica archangelica line on the elevated t-maze and forced swimming tests in rats[J]. Journal of Traditional Chinese Medicine, 2012, 32(3): 423-429.

[3] SHARKAWY A, FERNANDES I P, BARREIRO M F, et al. Aroma-loaded microcapsules with antibacterial activity for eco-friendly textile application: synthesis, characterization, release, and green grafting[J]. Industrial & Engineering Chemistry Research, 2017, 56(19): 5516-5526.

[4] XU X, SONG K, XING B, et al. Thermal-tenacity-enhanced and biodegradable textile sizes from cellulose nanocrystals reinforced soy protein for effective yarn coating[J]. Industrial Crops and Products, 2019, 140(111701).

[5] GHAYEMPOUR S, MORTAZAVI S M. Microwave curing for applying polymeric nanocapsules containing essential oils on cotton fabric to produce antimicrobial and fragrant textiles[J]. Cellulose, 2015, 22(6): 4065-4075.

[6] MAMUSA M, SOFRONIOU C, RESTA C, et al. Tuning the encapsulation of simple fragrances with an amphiphilic graft copolymer[J]. ACS Applied Materials & Interfaces, 2020, 12(25): 28808-28818.

[7] MA J, ZHANG N, CHENG Y, et al. Green fabrication of multifunctional three-dimensional superabsorbent nonwovens with thermo-bonding fibers[J]. Advanced Fiber Materials, 2022, 4(2): 293-304.

[8] MA J, XU W, KOU X, et al. Green fabrication of control-released, washable, and nonadhesives aromatic-nanocapsules/cotton fabrics via electrostatic-adsorption/in situ immobilization[J]. ACS Sustainable Chemistry & Engineering, 2020, 8(40): 15258-15267.

Axial Impact Properties of 3D Braided Composite Multi-Way and Single Circular Tubes

WANG Qihang[1], **GAO Xinyu**[2], **SUN Baozhong**[1,2*]

[1] *Shanghai Collaborative Innovation Center of High Performance Fibers and Composites, College of Textiles, Donghua University, Shanghai 201620, China*

[2] *Shanghai Frontier Science Research Center for Modem Textiles, College of Textiles, Donghua University, Shanghai 201620, China*

* *Corresponding author's email*: sunbz@ dhu. edu. cn

Abstract: Tubular carbon fiber reinforced polymers (CFRPs) have great application potential as main load-bearing thin-walled structures, on account of their prominent energy absorption capacity and lightweight characteristics. The 3D braided multi-axialtubes based on four-step 1×1 circular weaving can avoid the poor integrity caused by suture or splicing. This study introduced the special preform braiding method and curing molding process, the mechanical properties of multi-way and single tubes under low-velocity impact (LVI) were also compared via digital image correlation (DIC) technique. The results showed that the boundary condition affected the failure mode, and the single tube with free-end was more resistant to impact than the multi-way tube constrained by the clamps. However, the catastrophic damage at the junction of the multi-way tube could avoid the failure of the connected part, so single tube was more suitable for use as joints. According to the strain cloud diagram, the surface strain of the two samples represented opposite trends in x and y directions. There also existed structural effect on the strain distribution of the single tube, where the strain at the interweaving point was larger than that at the float because of the weak strength of the matrix.

Keywords: Tubular CFRPS; DIC; Boundary conditions

1 Introduction

As structural composites with complex geometries and special-shaped sections, tubular carbon fiber reinforced polymers (CFRPs) have received widespread attention for the advantages of excellent anti-corrosion properties, high specific strength, and stiffness to weight ratio[1]. Applied in the fields of the automobile industry, aerospace, and transportation, they can reduce the mass of the frame and achieve the purpose of energy-saving and carbon emission reduction[2]. At the same time, because of their eminent damage resistance and damage tolerance, they are expected to serve as the main load-bearing component instead of metallic material[3]. However, the current forming method for 3D braided composites with two or more axial directions is to manufacture the individual components of the preform separately and then connect them via stitching or co-curing techniques. The disadvantage of this method is that only the suture fiber and the matrix are stressed at the junction leading to the poor integrity of the multi-way tube.

The existing researches on the mechanical performance of tubular CFRPS are extensive and focus particularly on the high-speed axial impact[4], quasi-static[5] or residual impact compression[6], tension[7], torsion[8], bending[9], and fatigue[10] conditions. Whereas, there are few studies concerned about the dynamic damage evolution of composite tubes under low-velocity impact (LVI) load, let alone multi-way circular tubes. Composites will undergo low-speed impact events such as tool drop, debris, and bird strike in manufacture and service[11]. A violent energy conversion occurs at the moment of the impact. One part of the kinetic energy is converted into elastic deformation of composites[12]. And the other part is absorbed by the material through the failure modes such as tension, compression, and shear damage, resulting in irreversible destruction and potential safety risks.

This research overcame the difficulty in the forming process of 3D braided composite multi-way tube and used digital image correlation (DIC) technology to explore the axial compression performance under LVI load. The difference in mechanical response between multi-way and single tubes was also compared to find the dual effects of boundary conditions and braiding structure.

2 Materials and Methods

2.1 Specimen preparation

Single and multi-way tubular preforms of the same inner diameter (20 mm) and wall thickness (1.25 mm) were produced on the braiding machine. The basic principle was followed by the four-step 1×1 circular weaving technique aided with the manual beating-up operation as shown in Fig. 1. In the first step, each carrier of the main yarn first moved to the adjacent position in the circumferential direction according to the law of "outer layer-clockwise, and inner layer-counterclockwise". Subsequently, the carrier moved one step radially as indicated by the arrow. The movement directions of the following two steps were opposite accordingly. So far one braiding cycle was completed, and a pitch length of 5 mm was formed on the surface of the fabric. It is worth noting that the number of carbon tows required for a multi-way tube was twice that of a single tube (48, 2×16 yarn array).

At the end of the manufacture of the first branch, the other half of the yarn was added manually to participate in the braiding according to the movement trend and interweaving rule at the intersection region to ensure that the structure of each branch tube was consistent.

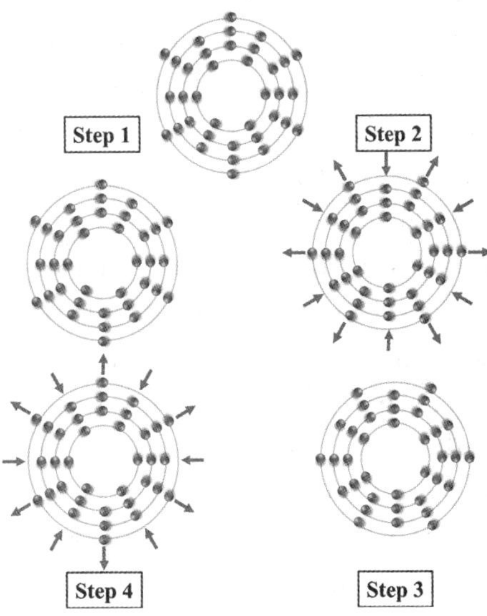

Fig. 1 Schematic diagram of four-step 1×1 circular weaving technique

As depicted in Fig. 2 (a), the preform was placed in an oven for 30 min to release the prestress. Then the pre-proportioned (100 : 85) and defoamed resin/curing agent mixture was injected into the mold from two conduits and flowed out from one outlet to ensure full impregnation. After curing at high temperature for 7 h, it was cooled to room temperature, demolded and polished, and finally, the composite tube was obtained as shown in Fig. 2 (b). The sample parameters were summarized in Tab. 1.

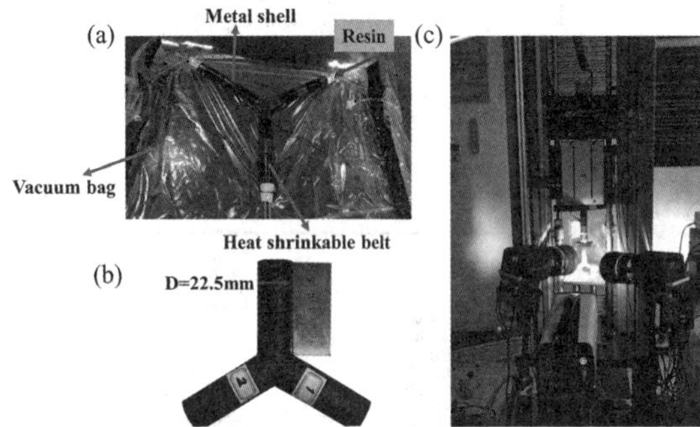

Fig. 2 (a) Curing process; (b) multi-way tube sample; (c) drop weight test system

Tab. 1 Sample parameters

Sample ID	Braiding angle (°)	Mass (g)	Tube length (mm)
Multi-way tube	60	19.65	6
Single tube	60	7.03	6

2.2 Low-velocity impact (LVI) test

The LVI test was carried out on drop weight system (Instron Dynatup 9250HV), and the real-time mechanical response was recorded by the photoelectric sensing device as shown in Fig. 2 (c). The impact energy was determined to be 30 J by pre-experiment and the drop height of the punch (7.255 kg) was 0.42 m. The high-speed camera (i-Speed 716) was equipped in front of the specimen, with brightness enhanced by two spotlights. The pictures were taken every 20 μs, and the surface strain distribution diagrams were obtained by processing the speckle images in MATLAB software.

3 Results and Discussion

3.1 Axial impact performance

The axial impact mechanical response curves of the two kinds of tubes were plotted in Fig. 3, and the initial slope of the load-deflection curve (the interval from the zero point to the first load undulate point) represented the initial compressive stiffness. Four different damage propagation stages were labeled by feature points A, B, C, and D, respectively.

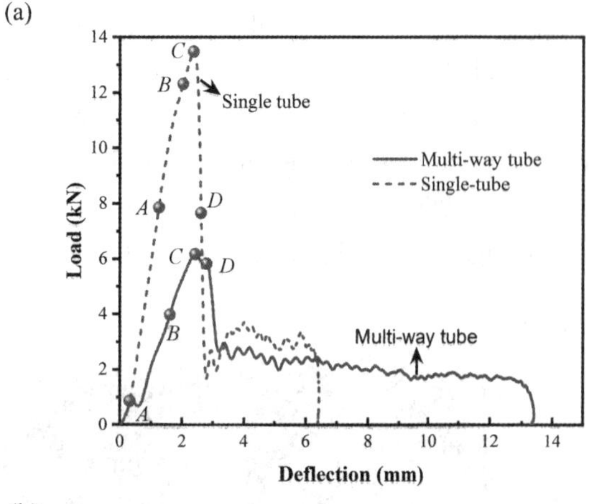

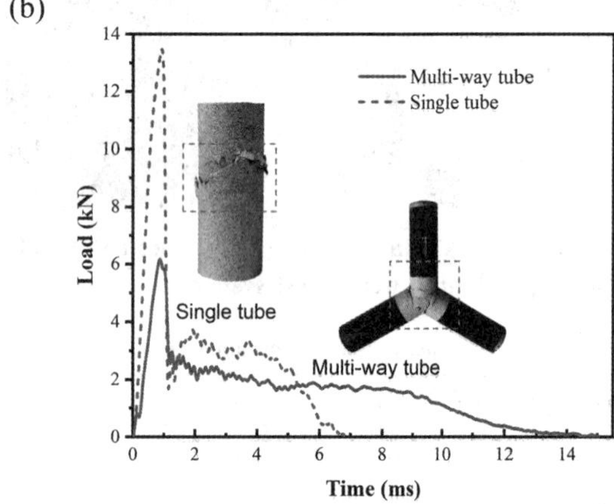

Fig. 3 (a) Load-deflection curve; (b) load-time curve of tubes

It can be seen from Fig. 3 that both specimens exhibited distinct brittle failure at 30 J impact energy. Before the peak load, the load of the single tube increased linearly, while occasionally fluctuating for the multi-way tube, indicating that internal damage such as matrix cracking and interface debonding happened. Finally, the damage of the single tube was concentrated in the middle part, forming an obvious compression shear band. However, for the multi-directional braided tube, it converged on the triangle region, and the branch part only played the role of transmitting the load with no injury.

The comparison of the impact resistance capacity of the two samples was listed in Fig. 4. The initial compression stiffness, peak load, and specific energy absorption (SEA) of the single tube were 3.06, 2.18, and 2.74 times higher than those of the multi-way tube, respectively. It can be inferred that boundary conditions combined with structural effects have a great influence on the final failure mode. Under free-end conditions, the fibers and resin within the single tube shared the load together, which absorbed most of the energy, causing catastrophic failure. Although the impact resistance of the multi-way tube with both ends constrained was weakened, the damage was localized and concentrated at the joint, avoiding the failure of other components connected to it.

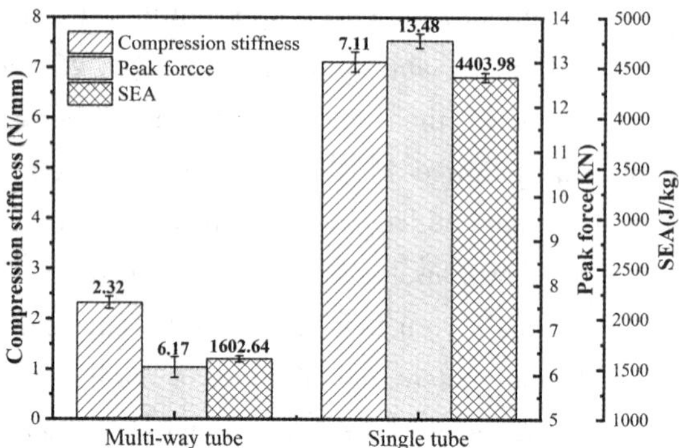

Fig. 4 Comparison of mechanical performance of tubes

3.2 Progressive surface strain analysis

The surface strain contours of the two samples at four feature points were summarized in Fig. 5, and the large deformation region was selected as the area of interest (AOI). It could be seen from Fig. 5 that the surface deformation of single tube was manifested as buckling expansion. It was mainly subjected to tensile load in the x-direction (perpendicular to the impact direction) and was compressed in the y-direction. The crack propagated from the back to the front, and with the occurrence of fiber buckling, a V-shaped interfacial shear crack was formed. The strain distribution at the triangle area of the multi-way tube was opposite. The obvious tensile deformation occurred in the middle part, and the junction region was severely compressed. It was speculated that the dominant factor was the boundary condition.

The average strain curve extracted from single tube surface was plotted in Fig. 6 (a). It was found that when index = 49 (0.98 ms), E_{xx} and E_{yy} dropped suddenly, representing the failure formation, and the change slowed down after point D. More interestingly, the strain distribution on the surface of single tube showed the characteristics related to the braided structure as illustrated in Fig. 6 (b). Along the selected path L0, the interweaving points and edges of the fiber bundles (corresponding to the troughs of E_{yy}) displayed more serious compressive stress concentration, while the

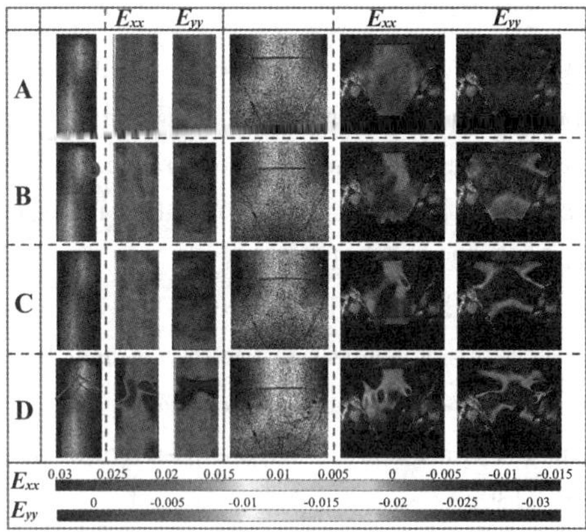

Fig. 5　Surface strain contours of tubes at feature points

strain value of the outer fiber floats was smaller (corresponding to the wave crest). This is because the fiber interweaving point is the resin-enriched area. The damage is prone to occur because the load-bearing capacity of the matrix is weaker than that of the fiber[13].

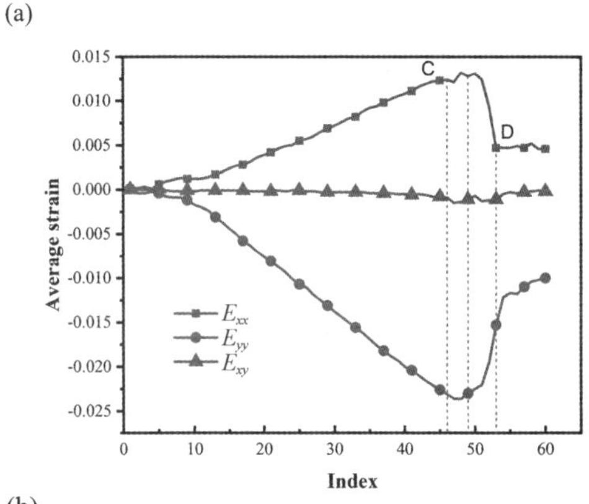

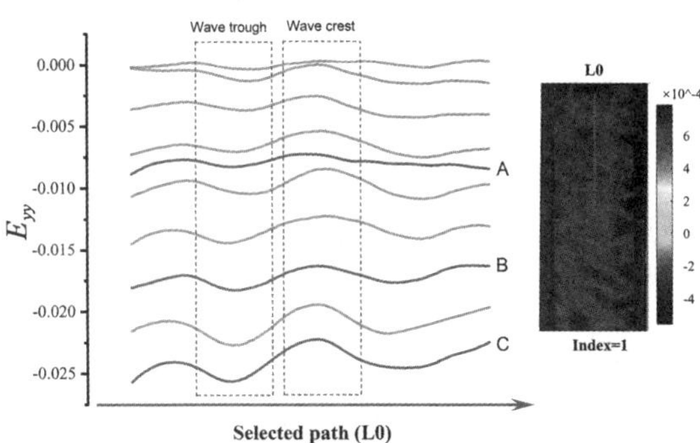

Fig. 6　(a) Average strain curve (E_{xx}, E_{yy}, and E_{xy}); (b) strain distribution along the selected path (L0) on the surface of single tube

4　Conclusions

In this study, the forming method of the 3D braided multi-way tube was explored and the mechanical performance was compared with the single tube under the same impact conditions. The boundary condition was found to have a great influence on the failure mode. The single tube exhibited buckling expansion and produced V-shaped shear bands. The damage of the multi-way tube was localized so as to avoid the failure of the connected parts. The strain distribution on the surface of the single tube was also related to the braided structure, where the interweaving point bore larger load because of the weak strength of the matrix.

Acknowledgments

The authors acknowledge the financial supports from the Chang Jiang Scholars Program, National Science Foundation of China (No. 51875099) and the Fundamental Research Funds for the Central Universities (No. 2232022D-12).

References

[1] SHI L, WU Z, CHENG X, et al. Transverse impact response of hybrid biaxial/uniaxial braided composite tubes [J]. Engineering Structures, 2021, 244: 112816.

[2] CHAI Y, WANG Y, YOUSAF Z, et al. Following the effect of braid architecture on performance and damage of carbon fibre/epoxy composite tubes during torsional straining[J]. Composites Science and Technology, 2020, 200: 108451.

[3] CHEN Y, YE L, ESCOBEDO-DIAZ J P, ZHANG Y X. Effect of initiator geometry on energy absorption of CFRP tubes under dynamic crushing[J]. International Journal of Crashworthiness, 2021, 26(5): 526-536.

[4] WU X, ZHANG Q, ZHANG W, et al. Axial compressive deformation and damage of four-step 3-D circular braided composite tubes under various strain rates[J]. Journal of the Textile Institute, 2016, 107(12): 1584-1600.

[5] SHANG Z F, MA J Y, YOU Z, et al. Lateral indentation of a reinforced braided tube with tunable stiffness[J]. Thin-Walled Structures, 2020, 149: 106608.

[6] LABAN O, GOWID S, MAHDI E, et al. Experimental investigation and artificial intelligence-based modeling of the residual impact damage effect on the crashworthiness of braided carbon/Kevlar tubes[J]. Composite Structures, 2020, 243: 112247.

[7] ZHAO G, WANG J, HAO W, et al. Numerical study on the tensile behavior of 3D four directional cylindrical braided composite shafts[J]. Applied Composite Materials, 2018, 25(5): 1103-1114.

[8] CHAI Y, WANG Y, YOUSAF Z, et al. Damage evolution in braided composite tubes under torsion studied by in-situ X-ray computed tomography[J]. Composites Science and

Technology, 2020, 188: 107976.

[9] WANG Y, CHAI Y, SOUTIS C, et al. Evolution of kink bands in a notched unidirectional carbon fibre-epoxy composite under four-point bending[J]. Composites Science and Technology, 2019, 172: 143-152.

[10] TATE J S, KELKAR A D, WHITCOMB J D. Effect of braid angle on fatigue performance of biaxial braided composites[J]. International Journal of Fatigue, 2006, 28(10): 1239-1247.

[11] WANG M, CAO M, WANG H, et al. Drop-weight impact behaviors of 3D angle interlock woven composites after thermal oxidative aging[J]. Composite Structures, 2017, 166: 239-255.

[12] CAO M, GU B H, SUN B Z. Low-velocity impact and residual flexural behaviors of 2.5-D woven composite under accelerated thermal ageing: experiment and numerical modelling[J]. International Journal of Damage Mechanics, 2020, 29(3): 413-434.

[13] GU Y H, ZHANG D T, ZHANG W, et al. Torsion damage mechanisms analysis of two-dimensional braided composite tubes with digital image correction and X-ray micro-computed tomography[J]. Composite Structures, 2021, 256: 113020.

Knitted Arch-Structure Triboelectric Nanogenerator for Energy Harvesting and Human Motion Monitoring

JIN Xiaomeng[1], **LIU Yanping**[1*]

[1]*Department of Knitting Technology, College of Textiles, Donghua University, Shanghai 201620, China*

* *Corresponding author's email*: liuyp@ dhu. edu. cn

Abstract: Wearable triboelectric nanogenerator (TENG) has attracted lots of attention in recent years because of the ability to continuously power and being comfortable to wear. Several studies based on textile TENG for improving wearability have been reported. However, the comfort of these TENGs is far less than that of the full-textile TENG. Herein, we present a knitted arch-structure TENG composed of two kinds of yarns. One yarn is a double-covered yarn that presents an arched structure and acts as an electrode (core yarn) and a negative friction material (covered yarn) at the same time. The other yarn is knitted as the base of the fabric, acting as the positive triboelectric material and providing elasticity to the fabric. The special arch structure supplies the space for the TENG to contact under pressure and separate by itself. Therefore, the single-electrode TENG can convert mechanical energy to electrical energy under pressure. The current and voltage of the TENG (25 cm^2 effective area) reached 1. 56 μA and 67. 03 V at 30 N pressure. The TENG combined with clothing can realize self-powered monitoring of human motion as a wearable sensor. Furthermore, the TENG is stretchable, flexible, lightweight, soft, breathable, durable, washable, and easy to produce with advanced knitting technology. The TENG is projected to be the ideal solution for human motion monitoring wearables due to these properties.

Keywords: Knitted; TENG; Wearable; Energy harvesting; Human motion monitoring

1 Introduction

Wearable electronic devices have received much attention due to people's desire for an improved quality of life[1]. In 2012, a strategy for providing continuous electricity for wearable devices was proposed, making the continuous power supply for wearable devices possible[2]. Triboelectric nanogenerators (TENGs) have been developed for energy harvesting and human motion monitoring combined with clothing[3-7]. Arched structures can provide natural contact separation space for TENGs, and various TENGs with arched structures have been reported[8-13]. For example, an arch-shaped TENG composed of films whose energy conversion efficiency is as high as 10%–39%, utilizes the contact electrification between a polymer thin film (PDMS) and a metal thin foil (Al)[8]. However, the comfort and wearability of the TENG are important for people to wear. A wearable TENG utilizing a textile composed of pile-embroidered fibers (rayon and polyester) and Ni-coated polyester fabrics exhibits a high output power[9]. However, the coating and embroidering process affect the durability and washability of the TENG and limit mass production.

Herein, we report a knitted arch-structure TENG, which has both excellent energy harvesting ability and the advantage of textiles, including durability, breathability, washability, and softness. A crucial part of the TENG is a double-covered yarn made of a core yarn (Ag-plated nylon) and two outer yarns (polypropylene). The double-covered yarn forms the arches on the based fabric composed of the nylon-covered spandex yarn by the knitting technique. Polypropylene (PP) multifilament and nylon 66 multifilament were selected as negative and positive triboelectric materials respectively, and Ag-plated nylon yarn was selected as conductive materials. The TENG can contact under pressure and separate by itself due to the arched structure. The output power TENG can reach 67. 03 V and 1. 56 μA at a compression force of 30 N and a frequency of 3 Hz when the effective area is 25 cm^2. In addition to this, the TENG can achieve the monitoring of human motion by attaching the textile under the arms.

The electric output is significantly different under different intensities of exercise.

2 Experiments

2.1 Materials

Polypropylene multifilament (240 D/24 F) was bought from Zhuji Huachuang Chemical Fiber Co., Ltd., Zhuji, Zhejiang, China. Ag-plated nylon6 multifilament (100 D/36 F) was bought from Qingdao Tianyin Textile Technologies Co., Ltd., Qingdao, Shandong, China. Nylon-covered spandex yarn (70/70 D) was bought from Haining Kaiwei Textile Co., Ltd., Haining, Zhejiang, China.

2.2 Preparation of double-covered yarn

The double-covered yarn was manufactured on a coating machine (HKV141A (II)-32 spindles). Firstly, take Ag-plated nylon6 multifilament as the core yarn, and two strands of 240 D polypropylene multifilament as the inner covered yarn to wrap the core yarn with a Z twist. Secondly, take two strands of 240 D polypropylene multifilament as the outer covered yarn with Z twist to wrap the covered yarn that finished the inner covering. Then the double-covered yarn with different twist directions of inner and outer layers is produced, which has a good covered effect.

2.3 Fabrication of the TENG

The TENG is designed to form the arched structure, which is consisted of two parts corresponding to two kinds of yarns: aches and substrate fabric. As depicted in Fig. 1, the arches are formed by double-covered yarns (Y2) and the substrate fabric is knitted by the nylon-covered spandex yarn (Y1 and Y3-8).

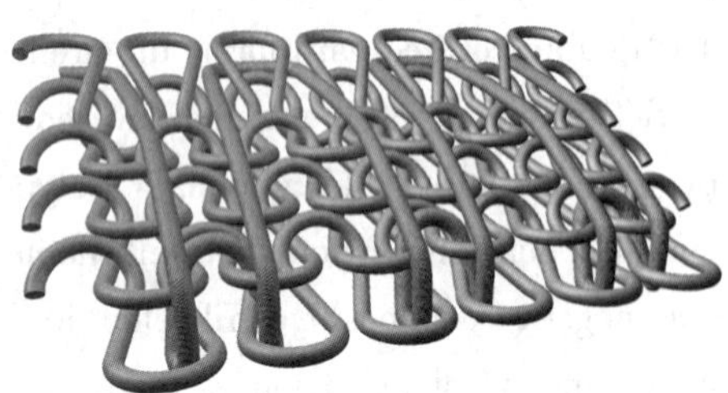

Fig. 1 The 3D structure of the TENG

Figure 2 (a) shows the knitted sequence diagram and the key states of loops during the knitting process. Through the cooperation of the stitches on the front and rear needle beds, the loops formed by the double-covered yarn at the Y2 row are hung continuously for 4 rows, and one loop forms two vertical arches on the front of the fabric. The double-covered yarn is knit on the front needle bed and tucked on the back, as shown in Fig. 2(d). Figures 2 (c) and 2(b) show the loops are hung for the first and the fourth rows. The machine knitted TENG is fabricated on a computerized flat knitting machine (Stoll CMS ADF 530-32 W E7.2, Reutlingen, Germany).

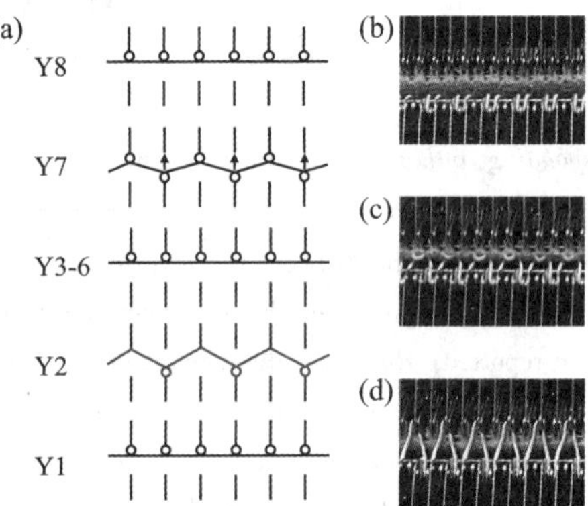

Fig. 2 (a) The knitted sequence diagram; Y6 (b), Y3 (c), and Y2 (d) state of loops on the needle beds during the knitted process

2.4 Characterization

The enlarged section of the fabric was captured with an optical microscope (ANDONSTAR AD207, China). The section of the double-covered yarn and the section and surface of the nylon-covered spandex yarn were captured with SEM (JSM-IT300, Japan). A linear motor (LinMot P01-37X120F/300X380-HP, USA) was applied to provide a periodic reciprocating pressure for t-TENG. An electrometer (Keithley 6517B, USA) was used to test the short-circuit current and transferred charge of the t-TENG.

3 Results and Discussion

3.1 Fabrication of the arch-structure TENG

The wearable arch-structure TENG is knitted directly on a computerized flat knitting machine. The straight arched structure forms by the smart design of the knitted structure and the different elasticities of yarns. The double-covered yarn as the electrode and friction material and the nylon-covered spandex yarn as another friction material are both used for knitting the TENG. As seen in Fig. 3 (a), the surface image of the nylon-covered spandex yarn shows that polyester filament occupies the surface of the yarn, because of the shrinkage characteristic of spandex. The cross-sectional and surface (Fig. 3 (b)) SEM images of the double-covered yarn suggest that the core yarns are well covered. Therefore, the friction material can contact completely due to the yarn's character and separate by itself due to the arched structure. The TENG can be integrated into the middle of the fabric (Fig. 3 (c)) with the use of knitting technology. The middle TENG is 5 cm×5 cm in size and contains 832 arches. As Fig. 3 (d) shows, each loop knitted with the double-covered yarn forms two axisymmetric arches. The shape of the arch is shown in Fig. 3 (e), and it can be seen that the top arc is hung by the elastic yarn to form an arch.

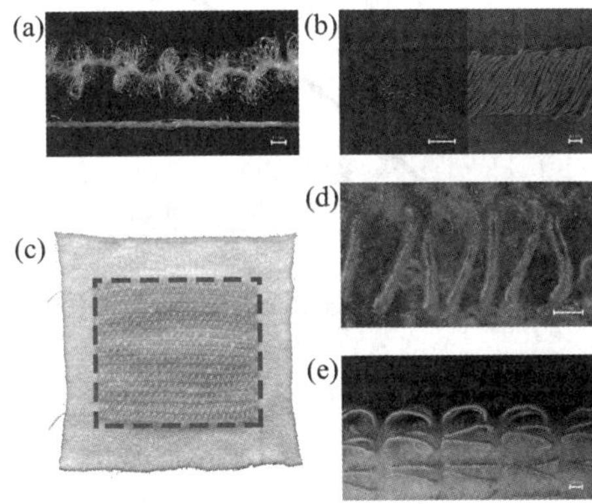

Fig. 3 (a) Microscopic image of the surface of the nylon-covered spandex yarn; (b) SEM images of the section and surface of the double-covered yarn; (c) the front photo of the t-TENG with a size of 5 cm× 5 cm; (d) microscopic image of the arched structure; (e) the section of the arched structure

3.2 Principle and performance of the TENG

The TENG realizes energy harvesting through the motions of contact and separation. The arches are excellent structure for the TENG to recover from deformation. As shown in Fig. 4, the arches composed of the double-covered yarn and the substrate fabric composed of the nylon-covered spandex yarn are kept apart in the initial state. When the TENG is pressed by a compression force, the two surfaces will touch and rub. Nylon tends to lose electrons, while polypropylene, the outer layer of the double-covered yarn is inclined to get electrons. Therefore, equal and opposite charges accumulate on the two surfaces. When the force removes, the arch structure begins to recover and the two surfaces begin to separate. The electric potential between two layers increases as the distance between the two layers increases. Electrons will flow from the conductive core yarn to the earth to balance the electric potential difference. Similarly, when the fabric is pressed again, the electric potential difference between the two triboelectric surfaces reduces and generates a reversed current.

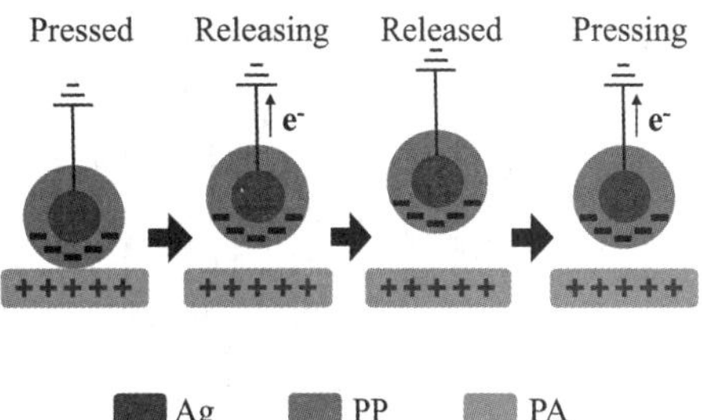

Fig. 4 Working principle of the TENG

The arch-structure TENG is as comfortable to wear as a normal textile and has a high electrical output performance. According to Figs. 5 (a)-5(c), the electrical output of the TENG (25 cm^2 effective area) reaches 1.56 μA (current), 67.03 V (voltage), and 25.63 nC (charge transfer) under pressure of 30 N at 3 Hz. Besides, we test the electrical output at three different frequencies (1 Hz, 2 Hz, and 3 Hz). As the frequency increases, the output of the TENG increases. This is because with the increase of the compression frequency, the relative movement speed between the double-covered yarn and the nylon-covered spandex yarn increases, the contact and separation speed between the two materials accelerates, and the amount of electron flow per unit time increases. Furthermore, the TENG can still output a considerable electrical signal at a lower frequency (1 Hz).

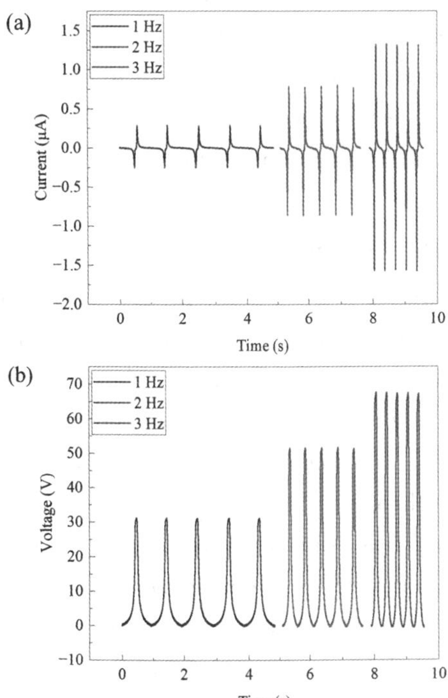

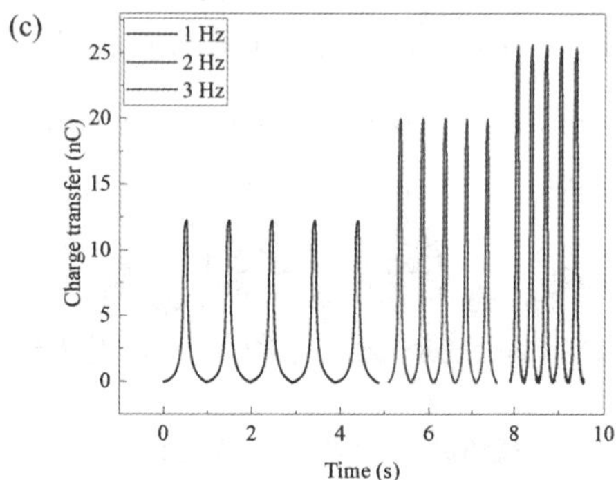

Fig. 5 The output performance of the TENG: (a) current; (b) voltage; (c) charge transfer of the TENG with compression frequencies at 1 Hz, 2 Hz, and 3 Hz

3.3 The application of the TENG

To verify the power supply performance of TENG, we use TENG to power the LEDs. As shown in Fig. 6(a), The TENG lights 54 LEDs under pressure of 30 N at 3 Hz. As a result, TENG can provide sustainable energy for wearable lighting systems. When the TENG is triggered, the generated electricity cannot be used directly by most wearable electronic devices. We design a rectifier circuit (Fig. 6(b)) to charge a 3.3 μF capacitor and then power the wearable electronic devices. As depicted in Fig. 6(c), the voltage of the capacitor reaches 6 V after 650 of charging. Besides, the TENG has the potential of monitoring human motion. We integrate the TENG on both sides of the garment, and the sleeves of the garment are made of nylon. The electrical signal is generated by the pressure and friction created between the arms and body. As shown in Fig. 6(d), in different degrees of motion, the electrical signal output of the TENG is significantly different.

4 Conclusions

In summary, a knitted arch-structure TENG was designed and applicated in energy harvesting and human motion monitoring. The arches are the ideal structure for contacting and separating, which can be knitted directly with continuous yarns thanks to the developed knitting technology. The TENG is stretchable, breathable, light, soft, durable, washable, and comfortable to wear because of its all-textile composition. Besides, the current and voltage of the TENG reach about 1.56 μA and 67.03 V, and 54 LEDs were lit. The TENG is integrated into the garment and applied to motion sensors, which can monitor the intensity of exercise. The TENG can be used for energy harvesting and show promising applications in wearable sensors.

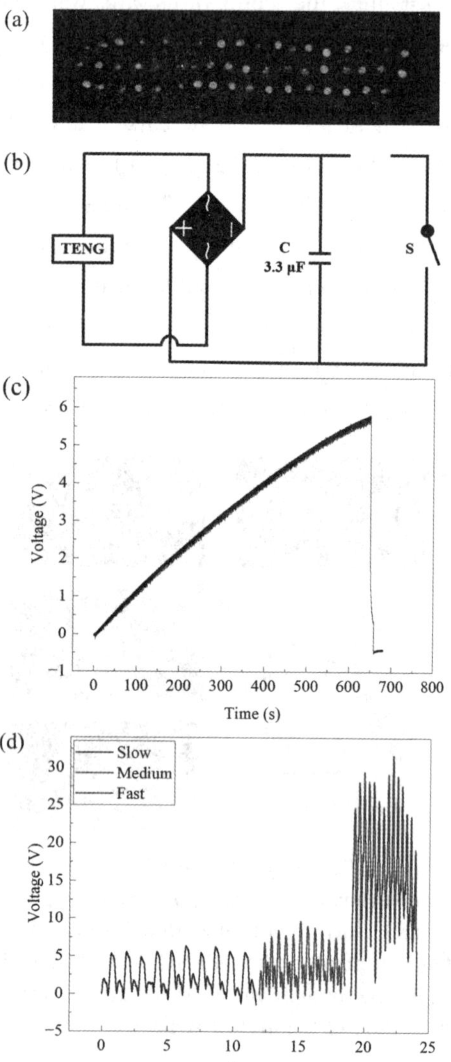

Fig. 6 The application of TENG. (a) The TENG lit 54 LEDs; (b) the rectifier circuit for charging the capacitor; (c) the voltage curve of capacitor charging and discharging; (d) the voltage of the TENG under different exercise frequencies

References

[1] XU Z, HAI T D, DAN L W, et al. Wearable multi-functional sensing technology for healthcare smart detection [J]. Micromachines, 2022, 13(2): 254.

[2] FAN F, TIAN Z, WANG Z. Flexible triboelectric generator[J]. Nano Energy, 2012, 1(2): 328-334.

[3] YE C, DONG S J, REN J. Ultrastable and high-performance silk energy harvesting textiles[J]. Nano-Micro Letters, 2020, 12: 12.

[4] LIU R, LI M P. A textile-based triboelectric nanogenerator for long jump monitoring[J]. Materials Technology,

2022, 2: 1-8.
[5] LI Y Y, ZHANG Y H, YI J, et al. Large-scale fabrication of core shell triboelectric braided fibers and power textiles for energy harvesting and plantar pressure monitoring[J]. EcoMat, 2022, 4: e12191.
[6] KIM J, KIM W J, JANG G, et al. 1D stretchable block copolymer yarn-based energy harvesters via BaTiO3/ polydimethylsiloxane composite-carbon conductive ink[J]. Advanced Energy Materials, 2020, 10(6): 1903217.
[7] ZHANG Z X, HE T Y, ZHU M L, et al. Deep learning-enabled triboelectric smart socks for IoT-based gait analysis and VR applications[J]. Flexible Electronics, 2020, 4: 29.
[8] WANG S H, LIN L, WANG Z L. Nanoscale triboelectric-effect-enabled energy conversion for sustainably powering portable electronics[J]. Nano Letters, 2012, 12: 6339-6346.
[9] PYOL S, KIM M O, KWON D S, et al. All-textile wearable triboelectric nanogenerator using pile-embroidered fibers for enhancing output power[J]. Smart Materials and Structures, 2020, 29(5): 055026.
[10] CHEN X, HE J, SONG L L, et al. Flexible one-structure arched triboelectric nanogenerator based on common electrode for high efficiency energy harvesting and self-powered motion sensing[J]. AIP Advances, 2018, 8: 045022.
[11] SHI X X, ZHANG S D, GONG S Q. A self-powered and arch-structured triboelectric nanogenerator for portable electronics and human-machine communication[J]. Journal of Materials Chemistry A, 2020, 2: 8997.
[12] ZHAO Z F, PU X, DU C H, et al. Freestanding flag-type triboelectric nanogenerator for harvesting high-altitude wind energy from arbitrary directions[J]. ACS Applied Nano Materials, 2016, 10: 1780-1787.
[13] DONG S S, XU F, SHENG Y L, et al. Seamlessly knitted stretchable comfortable textile triboelectric nanogenerators for E-textile power sources[J]. Nano Energy, 2020, 78: 105327.

Clothing Sewing Pattern Reconstruction from Sketch based on Deep Learning

WU Liuliu[1], **WANG Xin**[1], **MEI Chennan**[1], **WEI Jingyan**[1], **ZHONG Yueqi**[1,2*]

[1] *College of Textiles, Donghua University, Shanghai 201620, China*

[2] *Key Laboratory of Textile Science & Technology, Ministry of Education, Donghua University, Shanghai 201620, China*

[*] *Corresponding author's email*: zhyq@ dhu. edu. cn

Abstract: The traditional production of a clothing sewing pattern is cumbersome since designers cannot obtain the pattern through sketch unless they master computer-aided design (CAD) skills. In this paper, a method is proposed to rapidly convert a clothing sketch to a sewing pattern based on deep learning technology. A dataset of 5 classes of clothing sketches, including the front piece of top, the back piece of top, the front piece of pants, the back piece of pants, and the sleeves, is created to support training and validation. The backbone of the network is Densenet 121, which takes the sketch as input and predicts parameters of the line and Bézier curve. These predictions can be further converted into other commonly-used formats like drawing exchange format (DXF). The experiments find that this method can accurately regress the parameters for different class sketches with mean squared error (MSE) as 5.205 7 average. The research results show that based on deep learning, an end-to-end method is raised to efficiently resolve the problem of transition from sketch to clothing sewing pattern, enabling fast production of parameterized sewing pattern after sketching.

Keywords: Deep learning; Parameterize sketch; Sewing pattern; Bézier curve; Computer-aided design

1 Introduction

The production of clothing sewing patterns is an integral part of clothing production. Fashion designers can obtain the final clothing pattern through hand-drawn sketches or with the help of computer-aided design (CAD) software. Hand-drawing can help designers freely implement their inspiration for fashion patterns without concerning about computer skills. However, the handing-drawing pattern is not suitable for massive production unless they are converted into parametric sewing patterns. Currently, no method can directly realize the transformation from sketch to the parametric sewing pattern.

Presently, the following methods can be used to generate parameterized clothing patterns.

Dimension driven pattern making uses a set of parameters like dimension values and constraints of geometric figures to obtain the final figure. Jiang[1] uses AutoLISP by inputting key parameters with commands, which calls the AutoCAD dialog box to draw straight lines and curves, realizing the parametric and intelligent generation of clothing patterns. Chen[2] uses body parameters (waist circumference, hip circumference, etc.) to drive the generation of clothing patterns, which builds the geometric relationship and constraint between the body parameters and each line segment of the clothing pattern. Liu[3] proposed a novel method to establish a mapping relationship between the fashion pattern and the sewing clothing pieces, realizing the rapid generation from the fashion pattern to the clothing sewing pattern.

Although these methods perform well, their process relies on manually input parameters. When we cannot obtain relevant body parameters, such methods lose their advantages.

In this article, we propose a new method to solve the problem. First, we build a dataset of clothing sketches. Specific parameters generate the sketches so that they can be learned through neural networks about the mapping. The parameters predicted by the neural network can be directly used to generate industrial sewing patterns with DXF format, which is conducive to the development of a secondary design. Therefore, this method can realize the rapid transformation from clothing sketches to parametric clothing sewing patterns, improving the efficiency of garment pattern making.

2 Methods

The general scheme to generate parametric clothing patterns from sketches comprises building datasets and training a neural network. Each sample in the dataset is built following two steps: (1) separate the sketch into straight and Bézier curved segments; (2) specific parameter expression of each sketch. The workflow of the neural network will be discussed in section 2.3.

2.1 Sketch of drawing principle

To parameterize a sketch, the critical point of drawing a sketch contains parametric straight and curved segments. As shown in Fig. 1, the top front piece contains five straight segments (AB, CD, EF, FG, and GH) and three curved segments (BC, DE, and HA). The front piece of the pants each contains one straight segment (MN) and seven curved segments (IJ, JK, KL, LM, NO, OP, and PI). The sleeve contains three straight segments (RS, ST, and TU) and two curved segments (QR and QU). The back piece of the top and pants have the same structure as the front, which shows in the fine line in Fig. 1.

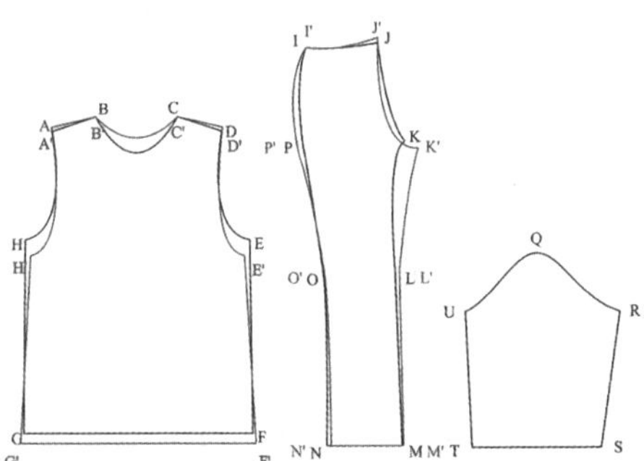

Fig. 1 Five categories of clothing sketches, the thick line represents the front piece, the fine line represents the back piece

A straight segment is drawn by connecting the start and end points of the line segment. The curve segment is drawn by Bézier curve parametric equation. The specific parameter equation[4] is shown in Eq. (1), where n is the order of the Bezier curve, τ is the curve point generation parameter, $\tau \in [0, 1]$, and P_j is the j control point of the Bezier curve.

$$Bézier(n, \tau) = \sum_{j=0}^{n} \frac{n!}{j!(n-j)!} \tau^{i}(1-\tau)^{n-i} P_j \tag{1}$$

In this paper, the third-order Bézier curve is selected, so there are four control points, which are the two endpoints of the curve P_{start} and P_{end}, and control points P_{e1} and P_{e2}. This paper calculates P_{e1} and P_{e2} through three points P_{start}, P_{mid} and P_{end} on the curve. Assuming that the three points are on the same circle, as shown in Fig. 2, draw a tangent to the circle through P_{mid}, and let P_{e1}, P_{e2} and P_{mid} be on the same tangent, according to De Casteljau Algorithm[5], combined with Eqs. (2) and (3), we can get P_{e1} and P_{e2}.

$$\begin{cases} d_1 = \| P_{start} - P_{mid} \| \\ d_2 = \| P_{end} - P_{mid} \| \\ t = \dfrac{d_1}{d_1 + d_2} \end{cases} \tag{2}$$

$$\begin{cases} P_{e1} = P_{mid} + t \times d \\ P_{e2} = P_{mid} - (1 - t) \times d \end{cases} \tag{3}$$

The value of d represents the distance between P_{e1} and P_{e2}, which can be a random value. However, in order to facilitate the calculation and make the four control points on the same straight line possibly, we set the value of d to one-third of the distance between P_{start} and P_{end}.

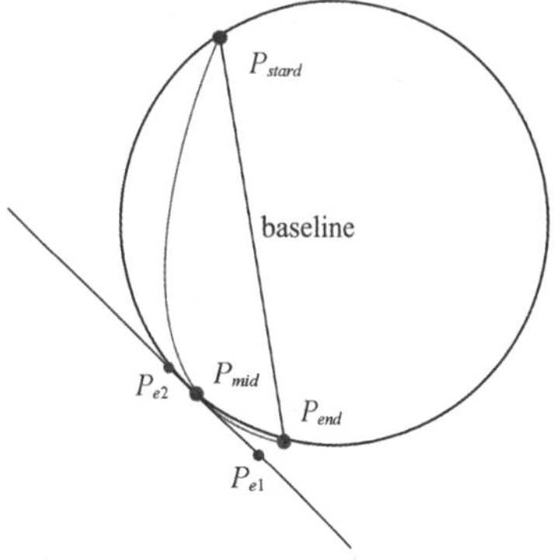

Fig. 2 P_{start}, P_{mid}, and P_{end} are on the same circle, P_{e1}, P_{e2}, and P_{mid} are on the same tangent, and the fine line represents the Bézier curve generated by P_{start}, P_{end}, P_{e1}, P_{e2}.

2.2 Specific parameter expression:

The parameters are composed of the coordinates of each key point for generating the garment pattern, such as the vertex coordinates and the control point coordinates of the curve. The unit of the parameters is pixels. The specific ratio refers to the actual clothes, pants, and sleeves length. The ratio of the actual clothes size to the image pixels is 1 cm to 10 px. The specific coordinates of each line segment of the front piece of the top in Fig. 1 are shown in Tab. 1.

Tab. 1 Specific the coordinates of each line segment of the front piece of top

Segment	Clothing parameters	Coordinates (px)
Line AB	Shoulder width (L_0) Shoulder angle (α)	$P_A(P_{A_x}, P_{A_y})$ $P_B(P_{A_x} + L_0 \times \cos(\alpha \times \pi/180), P_{A_y} - L_0 \times \sin(\alpha \times \pi/180))$
Curve BC	Collar width (L_1) Collar height (H_3)	$P_{B_{mid}}(P_{B_x} + L_1 * 0.5, P_{B_y} + H_3)$ $P_{B_{e1}}(P_{B_{mid_x}} + t \times d, P_{B_{mid_y}} + t \times d)$ $P_{B_{e2}}(P_{B_{mid_x}} - (1 - t) \times d, P_{B_{mid_y}} - (1 - t) \times d)$ $P_C(P_{B_x} + L_1, P_{B_y})$ The midpoint between BC: $P_m((P_{B_x} + P_{C_x}) \times 0.5, (P_{B_y} + P_{C_y}) \times 0.5)$
Line CD	Shoulder width (L_0)	$P_D(P_{C_x} + L_0 \times \cos(\alpha \times \pi/180), P_{C_y} - L_0 \times \sin(\alpha \times \pi/180))$
Curve DE	Cuff width (L_2) Cuff height (H_1)	$P_{D_{mid}}(P_{D_x} - 15.0, P_{D_y} + 5.0)$ $P_{D_{e1}}(P_{D_{mid_x}} + t \times d, P_{D_{mid_y}} + t \times d)$ $P_{D_{e2}}(P_{D_{mid_x}} - (1 - t) \times d, P_{D_{mid_y}} - (1 - t) \times d)$ $P_E(P_{D_x} + L_2, P_{D_y} + H_1)$
Line EF	Clothes length (H_2) Hem width (L_3)	$P_F(P_{m_x} + L_3 \times 0.5, P_{m_y} + H_2)$
Line FG	Hem width (L_3)	$P_G(P_{m_x} - L_3 \times 0.5, P_{m_y} + H_2)$
Line GH	Clothes length (H_2)	$P_H(2 \times P_{m_x} - P_{D_x}, P_{D_y})$
Curve HA	Cuff width (L_2) Cuff height (H_1)	$P_{H_{mid}}(P_{A_x} + 50.0, P_{D_y} + 5.0)$ $P_{H_{e1}}(P_{H_{mid_x}} + t \times d, P_{H_{mid_y}} + t \times d)$ $P_{H_{e2}}(P_{H_{mid_x}} - (1 - t) \times d, P_{H_{mid_y}} - (1 - t) \times d)$ $P_A(P_{A_x}, P_{A_y})$

The production methods of the pants sketches and the sleeve sketches are similar.

2.3 Neural network

In this paper, the workflow of the neural network is shown in Fig. 3. The sketch is first converted into a 224×224×3 vector as input and sent to the neural network for training. The corresponding key point parameters $\varepsilon = \{\varepsilon_1, \varepsilon_2, \varepsilon_3, \cdots, \varepsilon_n\}$ are output. For different types of clothing sketches, we use the same neural network structure but with different weights and biases. The number of parameters of different types of sketches differs, where the top has 36 parameters as output, the pant has 48 parameters, and the sleeve has 20 parameters.

We found that compared with relatively lightweight networks such as mobilenetv 2 and resnet 50, the more layers of the network, the more accurate the prediction. Therefore, we choose the pre-trained Densenet 121[6] as the model, modifying the linear of the last layer and changing the output features to the corresponding number of parameters n.

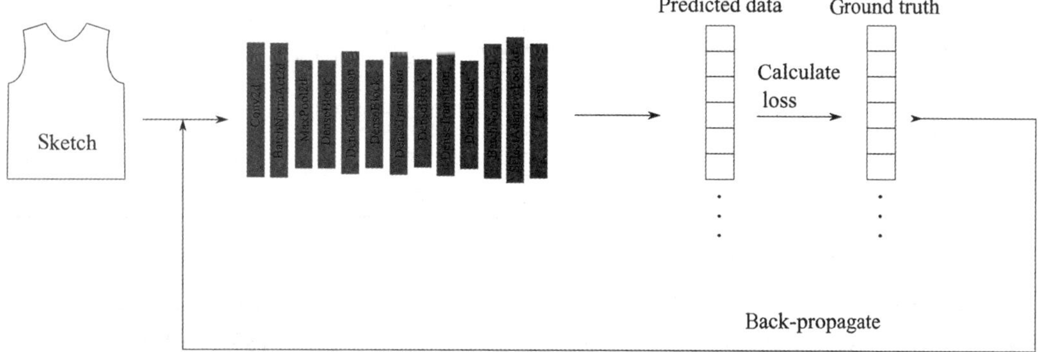

Fig. 3 The workflow of the transformation from sketch to crucial parameters, where neural network learns the mapping between the pair of sketch and data by back-propagate through calculating the bias between prediction and ground truth

3 Experimental Results and Discussion

3.1 Establishment of the dataset

The dataset in this paper comprises pairs of clothing sketches and critical parameters of the sketch. The sketches include five types: front pieces of tops, back pieces of tops, front pieces of pants, back pieces of pants, and sleeves. Considering the richness of the dataset, we modify the parameters of clothing parts; each cloth piece has four different clothing styles, as shown in Fig. 4. In the end, we got 5k sketches for each clothing category.

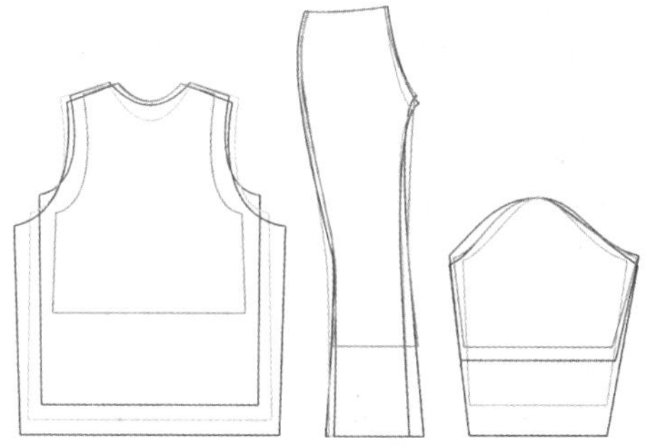

Fig. 4 Four styles of clothing sketches of each category

We use the same model architecture for different clothing types but train separately to get different model parameters. We train 80% content of the dataset and use the remaining 20% for testing.

3.2 Experimental setup

All experiments in this paper were conducted on a PC with NVIDIA GeForce RTX 3090 dual graphics card (24 G video memory) and were implemented by PyTorch programming. The experiment uses the SGD optimizer to make the training converge fast, where the learning rate is 0.001, the momentum is 0.9, and the weight decay is 0.000 1.

3.3 Training

The critical parameter ε actually corresponds to the vital point coordinates of the clothing pattern map. Its value is unevenly distributed, which is likely to cause overfitting phenomena. Therefore, it is necessary to normalize the parameters to make them satisfy $\varepsilon \sim N_n(0, 1)$. In order to learn the parameters of the network, we choose mean square error (MSE) to measure the difference between the ground truth and the network output. MSE is an ordinary loss function for this kind of regression task learning. However, MSE treats each parameter equally, so the critical point parameters cannot be learned entirely and effectively. For the sketch, the critical point parameters of the pant, shirt, and sleeve corners have more distinct characteristics than other parameters. Therefore, we added a weight mask W_i to the original loss function:

$$Loss = \sum \| \varepsilon_i - \tilde{\varepsilon}_i \|^2 \cdot W_i \qquad (4)$$

where ε_i is the model output, $\tilde{\varepsilon}_i$ is the ground truth, the weight of clothing corners is set to 10, and the others are set to 1.

3.4 Discussions

We use MSE, mean absolute error (MAE), R-square (R^2), and root mean square error (RMSE) to judge the accuracy of model predictions. The specific results are shown in Tab. 2. The MSE, MAE, and RMSE are relatively low, and R^2 is close to 1. The average MSE is 5.205 7, indicating that the prediction accuracy of the model is high.

Tab. 2 The accuracy of model predictions

Error	Front top	Back top	Front pant	Back pant	sleeve
MSE	5.795 1	10.908 1	2.185 3	3.599 8	3.540 4
MAE	1.571 9	1.927 2	0.822 7	0.893 5	1.075 1
R2	0.998 6	0.998 0	0.998 3	0.998 6	0.998 6
RMSE	2.260 4	2.847 5	1.345 3	1.481 4	1.732 9

The parameters obtained from the output can be used to draw the corresponding parameterized pattern using python's cairosvg library. As shown in Fig. 5, compared with the ground-truth image (thick line), our prediction effect (fine line) is almost the same. Meanwhile, the parameters obtained can be drawn using the ezdxf library of python to obtain the corresponding industrial parameterized sewing pattern with drawing exchange format (DXF).

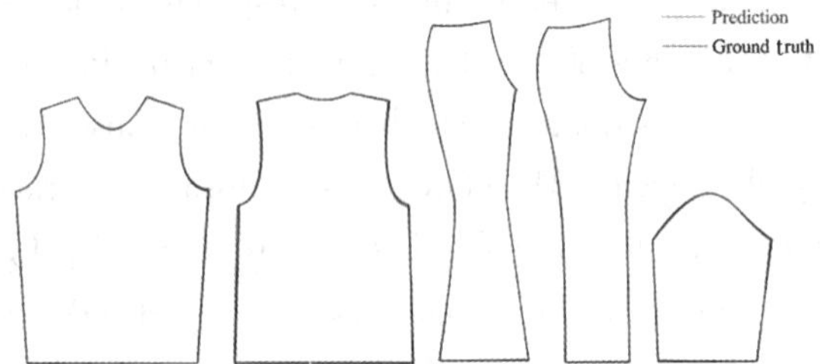

Fig. 5 Comparison between prediction and ground truth

3.5 Limitations

This paper also has some problems that can be solved in the future, for example parameterizing higher complexity of clothing sketches or hand paintings in actual production, including more design details.

4 Conclusions

We propose to obtain the mapping between clothing sketches and key parameters through a deep neural network, which can realize the rapid transformation from clothing sketches to parametric clothing sewing patterns. The prediction of parameters for different class sketches with mean squared error (MSE) was 5.205 7 on average. The MSE of the front pants is the smallest, which is 2.185 3, while the results of predicting the back piece of the top work worse, which is 10.908 1.

Acknowledgments

This work was supported by Shanghai Natural Science Foundation (No. 21ZR1403000).

References

[1] JIANG X, LIN D. Application of autoLISP language to pattern parameterized intelligent design [J]. Journal of Beijing Institute of Clothing Technology, 2007, 27(1): 41-45.

[2] BAO C, MIAO Y, GU B, et al. 3D interactive garment parametric pattern-making and linkage editing based on constrained contour lines [J]. International Journal of Clothing Science and Technology, 2021, 33(5): 696-723.

[3] LIU K, ZENG X, TAO X, et al. Associate design of fashion sketch and pattern [J]. IEEE Access, 2019, 7: 48830-48837.

[4] ZHANG J. C-Bézier curves and surfaces [J]. Graphical Models and Image Processing, 1999, 61(1): 2-15.

[5] BOEHM W, MÜLLER A. On de Casteljau's algorithm [J]. Computer Aided Geometric Design, 1999, 16(7): 587-605.

[6] IANDOLA F, MOSKEWICZ M, KARAYEV S, et al. Densenet: implementing efficient convnet descriptor pyramids [J]. arXiv preprint, 2014, 1: 1-11.

Insoluble Electrospun Porous Membrane for Protective Clothing

LIU Xiang[1], DU Zhaoqun[1], YU Weidong[1*]

[1] *College of Textiles, Donghua University, Shanghai 201620, China*

[*] *Corresponding author's email*: wdyu@ dhu. edu. cn

Abstract: In this work, a mixture of polyacrylonitrile (PAN) and polyamic acid (PAA) was used as raw materials to prepare an insoluble electrospun porous fiber membrane by electrospinning and heat treatment technology. The process optimization of the porous membrane was carried out by adjusting the mass ratio of PAN to PAA. The mechanical properties and air permeability of the prepared porous membranes were tested, and their surface morphology, chemical structure and thermal stability were characterized. The results show that the PAN/PAA porous membrane undergoes an imidization reaction during the heat treatment to become a polyacrylonitrile-polyimide (PAN-PI) fibrous membrane, which is not dissolved by any solvent. The breaking stress and elongation of the membrane are greatly improved and good breathability is maintained. This work provides a promising candidate for the inner layer material of protective clothing.

Keywords: Electrospinning; Protective clothing; Insoluble; Heat treatment

1 Introduction

Each area of the protective clothing field has its own unique hazards and threats that must be understood before ensembles and equipment can be developed to protect the individual at risk[1]. The basic design idea of protective clothing is to shield potential hazards from the human body in a targeted manner. Today a variety of new functional protective fabrics have been developed, many of which are multifunctional[2-3]. But it should be noted that, although the effect of many fabrics appears to be perfect, the actual application environment is not fully considered. For example, non-porous film or coated chemical protective clothing is usually completely isolated from outside air and liquid, but it is difficult to balance its lightness and comfort. Protective clothing for firefighters requires fabrics with excellent breathability and moisture permeability, as well as lightness and comfort. Heavy fabrics or non-porous coated fabrics are difficult to meet the requirements of lightness and comfort, while ordinary electrospun porous membranes are difficult to meet their high temperature resistance requirements[4].

In this work, insoluble and high temperature resistant electrospun porous membranes were successfully fabricated by step-by-step electrospinning and heat treatment technology. First, electrospinning was performed after mixing polyacrylonitrile (PAN) and polyamic acid (PAA) to obtain an electrospun porous membrane with uniform fiber diameter. Then, the chemical structure of the porous membrane was changed from PAN/PAA to PAN-PI through a heat treatment process. Benefiting from the stable chemical structure of PAN-PI, the porous membrane is insoluble and exhibits excellent thermal stability, tensile stress and elongation at break. It is expected that the successful synthesis of these porous membranes could satisfy the requirements of protection as well as comfort, thereby providing a version to fabricate high-performance protective clothing.

2 Experiments

2.1 Materials

PAN (Mw = 85 000) was provided from Shanghai Chemical Fibers Institute. PAA (15%) were supplied by Changzhou Furun Special Plastic Materials Co., Ltd. N-Methyl pyrrolidone (NMP) and N, N-dimethylformamide (DMF) were purchased from China National Medicines Co. Ltd.

2.2 Fabric treatment

PAN solution with a concentration of 12% was prepared by dissolving PAN powder in DMF, and the solutions were stirred for 10 h at room temperature. Then the PAN and PAA solutions were mixed uniformly at the mass ratios of 3 : 0, 3 : 1, 3 : 2 and 3 : 4, respectively, and electrospinned to obtain four kinds of PAN/PAA porous membranes, which were named PAN/PAA-0, PAN/PAA - 1, PAN/PAA - 2, and PAN/PAA - 4. During electrospinning, an voltage of 25 kV was applied to a collecting distance of 15 cm between the needle tip and the collector, the flow rate of the solution was 1.2 mL/h controlled by a digital syringe pump, and the relative humidity and temperature were 45%- 50% and 26-29 ℃, respectively. Finally, PAN/PAA porous membranes were placed in an oven at a temperature of 250 ℃ for 15 min to obtain PAN-PI porous membranes, which were named PAN-PI - 0, PAN-PI-1, PAN-PI-2, and PAN-PI-4.

2.3 Membrane characterization and performance experiments

The morphology of samples was observed by the scanning electron microscope (SEM, DXS-10ACKT). The Fourier transform infrared (FTIR) spectra of samples were recorded on a Fourier transform infrared spectrometer (FTIR, Spectrum Two). Thermogravimetric analysis (TGA) was performed with a thermogravimetric tester (TGA 4000) under a nitrogen atmosphere of 30 ℃ to 800 ℃ with a heating rate of 20 ℃/min. The air permeability was evaluated via the air permeability tester (YG 461 E). Mechanical property was tested via the electronic fiber strength tester (LLY-06).

3 Results and Discussion

3.1 Surface morphology

The surface morphologies of the porous membranes with different PAN/PAA ratios before and after heat treatment were characterized by SEM. It could be seen from Fig. 1 (a) that after the pure PAN was treated at 250 ℃, the fiber morphology changed from straight to twisted, and multiple fibers were adhered together. The fiber membrane after adding PAA was adhered after heat treatment, and the degree of adhesion increased with the increase of PAA content. In PAN-PI-1, the fibers only adhered at the fiber junction points. Where multiple fibers in PAN-PI-2 and PAN-PI-4 intersect or are close together, fusion occurs between fibers, which looks like two or more fibers that are close together are fused into one. Therefore, it is easy for people to have a visual misjudgment, that is, a fiber is split into two. The diameters of PAN-PI-0, PAN-PI-1, PAN-PI-2 and PAN-PI-4 were measured to be 234.5, 484, 491 and 599.5 nm, respectively.

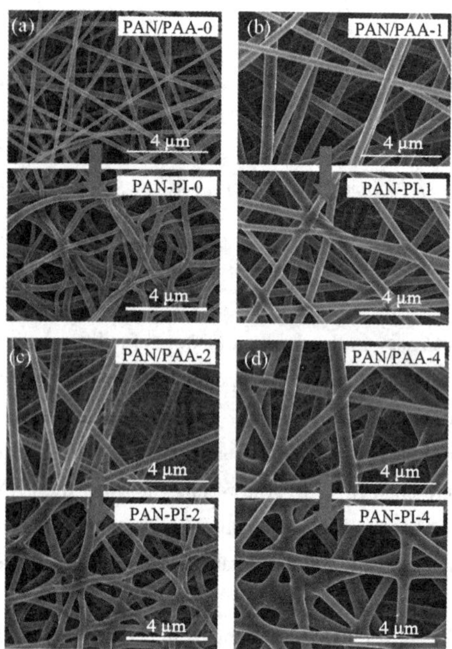

Fig. 1 SEM of PAN/PAA and PAN-PI porous membranes

3.2 Mechanical properties and air permeability

The typical stress-strain curves of the PAN-PI membranes are shown in Fig. 2 (a). In PAN/PAA-0 without PAA component, the breaking strength after heat treatment is 6.52 MPa, the breaking elongation is 6.40%, and the elastic modulus is 1.66 MPa. It exhibits high strength and modulus, but is brittle and easily breaks after bending. The breaking strengths of PAN-PI-1, PAN-PI-2 and PAN-PI-4 were 7.94, 8.39 and 8.74 MPa, while the breaking elongations reached 17.53%, 27.86% and 31.06%, respectively. The elastic moduli are 0.98, 0.92 and 0.81 MPa, respectively. Figure 2 shows that after the introduction of PAA into the spinning solution system, its elastic modulus decreases with the increase of the proportion of PAA in the system, but the breaking strength, the breaking elongation and breaking work show an increasing trend. The excellent mechanical properties of the fiber membrane ensure that it can be used as a protective clothing material for a long time.

In order to compare the air permeability of porous membranes with different PAN/PAA ratios, the permeability tests under different pressure differences were carried out on the porous membranes with four ratios. It can be seen from Fig. 2(b) that PAN-PI-0 has better air per-

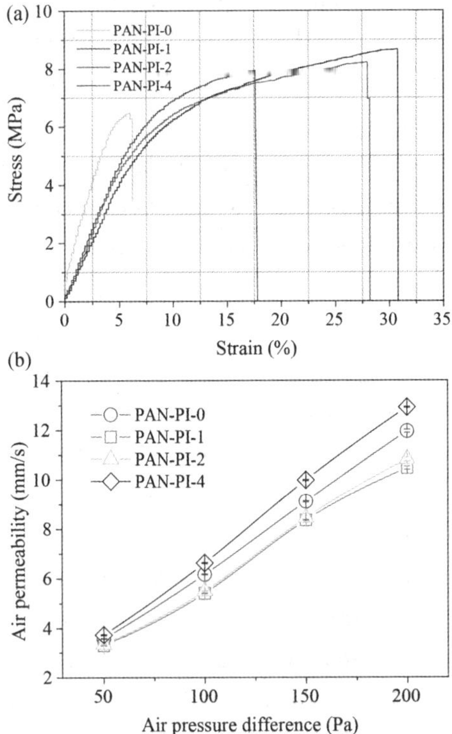

Fig. 2 (a) The typical stress-strain curves and (b) air permeability of PAN-PI porous membranes

meability. When the air pressure difference is 50 and 200 Pa, the air permeability is 3.57 and 11.95 mm/s, respectively. With the increase of the PAA ratio, the air permeability of the porous membrane has a decreasing trend. When PAA was added to the porous film, the air permeability of the heat-treated PAN-PI-1, PAN-PI-2 and PAN-PI-4 at a pressure difference of 50 Pa were 3.29, 3.33 and 3.73 mm/s, respectively. When the difference increases to 200 Pa, the air permeability increases to 10.46, 10.81 and 12.91 mm/s, respectively. This change should be related to the average pore size of the porous membrane. According to relevant studies, a smaller average pore size leads to a decrease in the air permeability of the porous membrane.

3.3 Chemical structure

The FTIR of the membranes before and after heat treatment was compared.

As shown in Fig. 3(a), the stretch vibration peak of —C=O in the amide group in PAA at 1 658 cm^{-1} is called the amide I band of PAA[5]. 1 545 cm^{-1} is the N—C stretching vibration peak in the amide group and the —NH variable angle vibration peak, which is called the amide II band. The peak at 1 605 cm^{-1} is empirically considered to be the ring stretching vibration peak of the benzene ring in PAA. The strong peak at 1 500 cm^{-1} is the stretching vibration peak of the para-substituted benzene ring. The doublet at 1 243 cm^{-1} is the stretching vibration peak of the aromatic ether $—C_6H_6—O—C_6H_6—$. It can be clearly seen from Fig. 3 that with the increase of PAA content, the vibration intensity of PAA characteristic peak is higher, while the PAN characteristic peak is opposite.

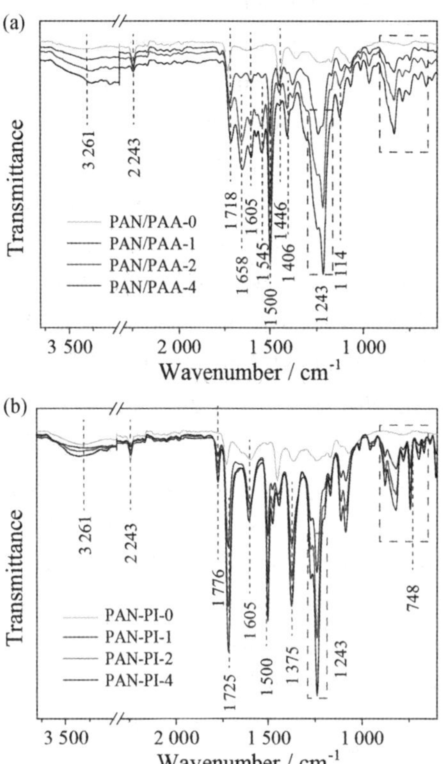

Fig. 3 FTIR of (a) PAN/PAA and (b) PAN-PI membranes

As shown in Fig. 3(b), after heat treatment, the peak at 1 776 cm^{-1} is two —C=O symmetrical stretching vibration peaks on the PI imine ring, which is called the imide I band. The strong peak at 1 725 cm^{-1} is the two asymmetric stretching vibration peaks of —C=O on the imine ring, which is called the imide II band. The strong peak at 1 375 cm^{-1} is the stretching vibration peak of C—N in the CONCO group on PI, which is called the imide III band. The strong peak at 748 cm^{-1} is the variable angle vibration peak of —C=O in the imine ring, which is called the imide IV band. The appearance of four bands indicates the conversion of PAA to PI.

3.4 Thermal stability

Figure 4 shows TG and DTA curves of porous membranes. The thermal weight loss of PAN/PAA-0 mainly occurs in two stages. The first stage is 250–300 ℃, and its weight loss is mainly caused by cyclodehydrogenation. The second stage is 330 – 500 ℃, and the weight loss at this stage is caused by the removal of small molecules during PAN dehydrogenation and the elimination of non-carbon elements during pre-oxidation to form a stable cyclized structure. While PAN-PI – 0 was heat treated at 250 ℃, the thermal weight loss mainly occurred in the range of 330 – 500 ℃. PAN/PAA – 2 showed three stages of mass. The first stage degradation occurred in the range of 175–260 ℃, and the maximum weight loss was at 230 ℃. Its weight loss was earlier than that of PAN/PAA–0 due to the removal of residual solvent and water during the imidization process. And there is no obvious weight loss at 260 – 300 ℃, which may be because a large number of —COOH groups catalyze the cyclization of PAN below 250 ℃ after the introduction of PAA. The second-stage degradation also occurred in the range of 330–500 ℃, and the residual amount at 500 ℃ was 68.9%. The weight loss at this stage was caused by the degradation of PAN[6]. The final stage occurs at 520 – 670 ℃, mainly due to the degradation of PI[7]. That is, the C_{Ph}—N bond on the main chain and the C—N and C—C bonds on the imide ring are broken and decomposed into small molecules such as CO, CO_2, aromatic hydrocarbons, and imide. Unlike PAN/PAA, the weight loss curves of PAN-PI–2 after heat treatment only appeared at 330–500 ℃ and 520–670 ℃. It was proved that pre-oxidation and imidization reactions occurred during the heat treatment of the fiber membrane.

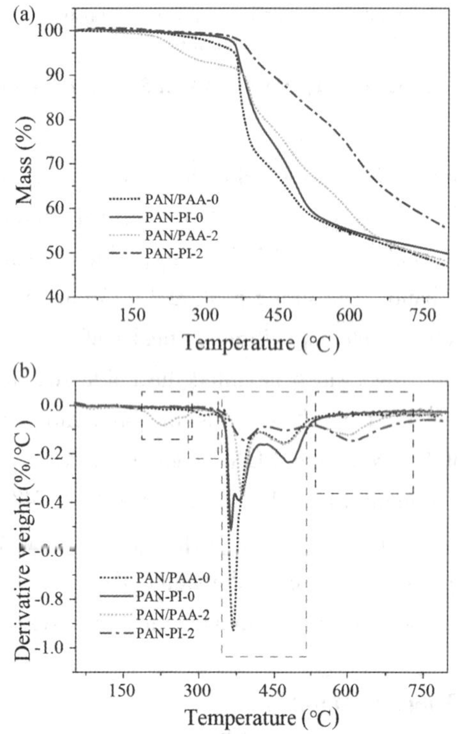

Fig. 4 (a) TG and (b) DTA of PAN/PAA and (b) PAN-PI porous membranes

3.5 Insolubility and reaction mechanism

DMF and NMP were selected as solvents for PAN and PAA, respectively, to prepare spinning solutions in the electrospinning process. To verify its insolubility, after immersing the PAN/PAA porous membranes in NMP and DMF, respectively, the porous membranes swelled and became transparent immediately after being immersed in the solvent, and then dissolved within seconds. However, the state of the fibrous membrane changed with time after immersing the PAN-PI porous membrane in NMP and DMF, respectively, and the color of the fibrous membrane became darker in both solvents[8], indicating that it had been penetrated by the solvent. Until the end, after 20 h soaking, the porous membrane did not become transparent in the two solvents, indicating that the fibrous membrane did not dissolve or swell. This indicates that after heat treatment, the chemical structure of the porous film is changed from PAN/PAA to PAN-PI, and the effect of resisting the dissolution of the solvent is achieved.

According to the previous analysis, the reaction mechanism for the conversion of PAN/PAA to PAN-PI is shown in Fig. 5. In the PAN/PAA hybrid system, the acidic component —COOH group on the PAA molecular chain acts as an initiator to accelerate the cyclization

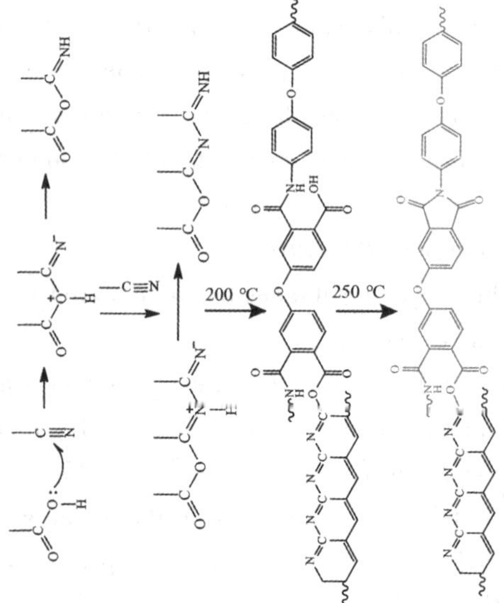

Fig. 5 Cross-linking forming mechanism of PAN-PI

of the PAN component in the system, and at the same time, the PAA begins to undergo imidization. As time goes on, the PAA component completes imidization to form PI. Finally, a stable and insolubility PAN-PI porous membrane was obtained.

4 Conclusions

In this work, the insoluble electrospun porous membranes were prepared by electrospinning and heat treatment techniques. The process was optimized by adjusting the ratio of PAN/PAA, SEM showed that the fiber diameter and pore size in the porous film were uniform, and the fiber contact points were fused. The tensile stress and elongation of the prepared porous membrane increased with the increase of the PAA ratio, and the highest value could reach 8.74 MPa and 31.06%, and the air permeability reached 12.91 mm/s. FTIR analysis showed that the chemical structure of the porous membrane changed after heat treatment, and the TG test showed that the onset temperature of thermal decomposition of the porous membrane was as high as 330 ℃. Solvent dissolution tests showed that the porous membranes acquired insolubility. Finally, the chemical reaction mechanism of the porous membrane is given. It can be seen that the prepared porous membrane can be a promising candidate for the inner layer material of heat- or chemical-resistant protective clothing.

References

[1] YU W D. Textiles Materials[M]. 2nd Edition. Beijing: China Textile & Apparel Press, 2018.

[2] LIU X, CHEN J L, SHAO Q Q, et al. Robust, flame-retardant and colorful superamphiphobic aramid fabrics for extreme conditions[J]. SCIENCE CHINA Tech-nological Sciences, 2021, 64: 1765-1774.

[3] LIU X, SHAO Q Q, CAO J Q, et al. Superamphiphobic and flame-resistant cotton fabrics for protective clothing[J]. Cellulose, 2022, 29(1): 619-632.

[4] GU J, GU H, CAO J, et al. Robust hydrophobic polyurethane fibrous membranes with tunable porous structure for waterproof and breathable application[J]. Applied Surface Science, 2018, 439(1): 589-597.

[5] ISIDA H, WELLINGHOFF S T, BAER E, et al. Spectroscopic studies of poly[*N*, *N'*-bis (phen-oxyphenyl) pyromellit imide]. 1. Structures of the polyimide and three model compounds[J]. Macromolecules, 1980, 13(4): 826-834.

[6] KO T, LIN C, TING H. Structural changes and molecular motion of polyacrylonitrile fibers during pyrolysis [J]. Journal of Applied Polymer Science, 1989, 37: 553-566.

[7] MIAO Y E, ZHU G N, HOU H, et al. Electrospun polyimide nanofiber-based nonwoven separators for lithium-ion batteries[J]. Journal of Power Sources, 2013, 226: 82-86.

[8] GRASSIE N, HAY J N. Thermal coloration and insolubilization in polyacrylonitrile[J]. Journal of Polymer Sciencs, 1962, 56: 189-202.

Design of Drag Reducing Cycling Clothing Based on Wind Tunnel Tests

JIANG Wei[1], ZHANG Jie[2], QIU Jun[2], MA Guoqiang[2], LIU Lifang[1]*

[1] *Department of Technical Textiles, College of Textiles, Donghua University, Shanghai 201620, China*

[2] *Shanghai Research Institute of Sports Science, Shanghai 200030, China*

* *Corresponding author's email*: lifangliu@ dhu. edu. cn

Abstract: Aiming at the different speeds of the athletes' body parts during cycling, the cycling suit is divided into four parts, and a design scheme of the drag reducing cycling suit is proposed through the combination of the splicing between different fabrics. Through wind tunnel test, the fabric with the best drag reduction effect under different wind speeds is obtained; compared with the smooth fabric, the fabric with dimpled structure and grooved structure has a better drag reduction effect. According to the speed distribution of each part of the body, the best fabric is matched in different parts, and then the design scheme of drag reduction cycling suit is completed.

Keywords: Warp knitting fabric; Cycling clothing; Wind tunnel test; Drag reduction; Non-smooth surface

1 Introduction

Cycling is one of the most influential sports with the largest number of participants. It has been found that the performance of cyclist is not only related to the athletes' own ability, but also influenced by many objective factors, one of which is wind drag[1]. A twofold increase in riding speed results in a fourfold increase in corresponding resistance, and in order for athletes to gain a competitive advantage, ways must be found to overcome this resistance. The performance of cyclists is closely related to posture, equipment and clothing[2]. Owning clothes with excellent drag reduction performance has gradually become the key to win the race, so it is very important to design a set of bicycle clothes with good drag reduction performance.

At present, people mainly improve the drag reduction performance of cycling clothes from two aspects: the seams of clothes and the degree of fitting with the human body[3]. Even if an athlete rides at a constant speed, the speed of each part of his body is different. Therefore, matching the fabric with the best drag reduction performance according to the speed of various parts of the human body can provide a new idea for the design of drag reduction bicycle clothing.

Most sportswear uses knitted fabrics, among which warp knitted fabrics have better dimensional stability and breathability, and can weave complex structures, so they are widely used in professional sportswear. In this paper, 20 kinds of warp knitted fabrics are selected to test their drag reduction effect through wind tunnel tests, and the fabrics with the best drag reduction effect at different speeds are obtained.

2 Experiments

2.1 Materials

The surface structure of samples is shown in Fig. 1. These samples can be divided into three types according to the surface structure. The first one was in smooth surface, marked as No. 1–No. 2, the second is the dimpled surface, marked as No. 3–No. 10, and the last one is the grooved surface, marked as No. 11–No. 20.

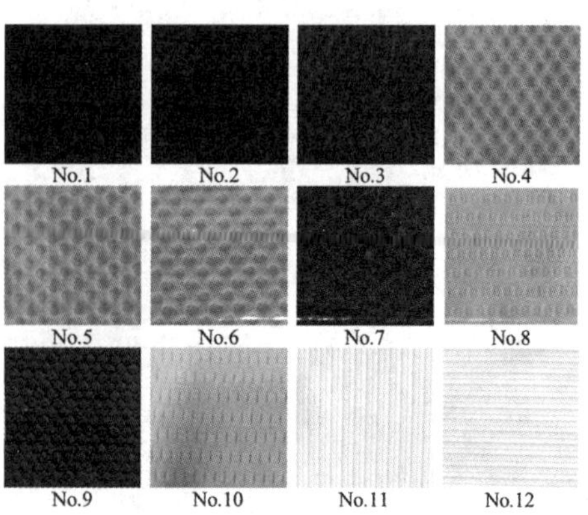

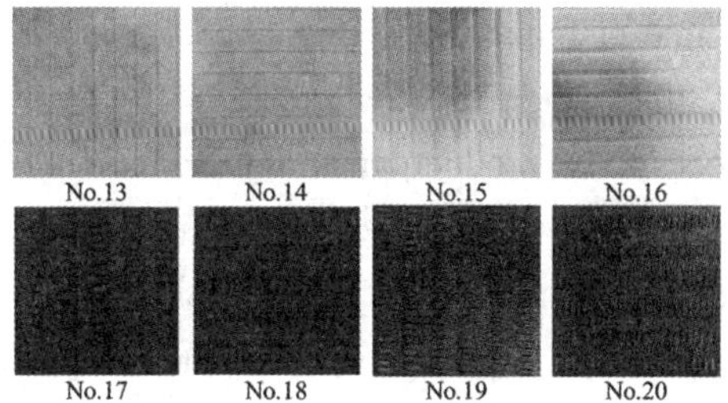

Fig. 1 Surface structure of fabric

2.2 Wind tunnel test

The sprint speed of an athlete during a short distance cycling race is about 12.5–18.1 m/s[4], and the wind speed experienced by each part of the athlete's body is different even at a constant cycling speed as found by numerical simulation. In order to study the wind velocity experienced by each part of the athlete's body while riding, the velocity distribution of the athlete riding at a speed of 18 m/s was simulated, as shown in Fig. 2. From Fig. 2, it can be seen that the highest velocity is 22–26 m/s at the inner and outer sides of the arms; the velocity at the chest is around 14 m/s; the velocity at the back is around 16 m/s; the velocity at thigh is around 18–20 m/s. Therefore, the wind tunnel test is carried out at the speed of 12–26 m/s.

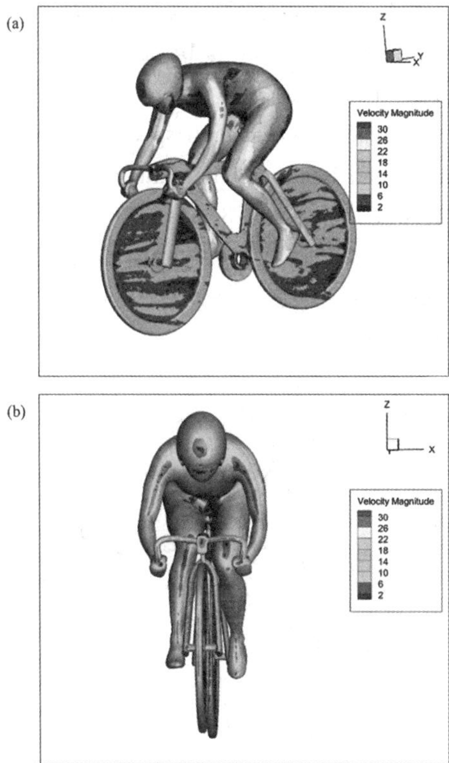

Fig. 2 Wind velocity diagram of various parts of the body during riding

Because of the similarity of aerodynamic properties between the cylinder model and the human model, this experiment covered the cylinder with fabric to study the aerodynamic properties of the fabric[5]. The wind speed was controlled by a computer during the test, and the wind speed was stabilized to obtain the drag of six components through a force balance as well as an electronic pressure scanning valve[6-7]. The device of wind tunnel test is shown in Fig. 3. The wind drag coefficient and drag reduction rate of the samples were calculated according to the following equation.

$$C_d = \frac{2F}{\rho S V^2} \tag{1}$$

$$R = \frac{C_{d1} - C_{d2}}{C_{d1}} \times 100\% \tag{2}$$

where: C_d is the wind drag coefficient; F is the drag, N; ρ is the air density, kg/m^3; S is the windward area of the object, m^2; V is the relative motion velocity of the object and air, m/s; R is the drag reduction rate, %; C_{d1} is the drag coefficient of nude cylinder; C_{d2} is the drag coefficient of the cylinder wrapping fabrics.

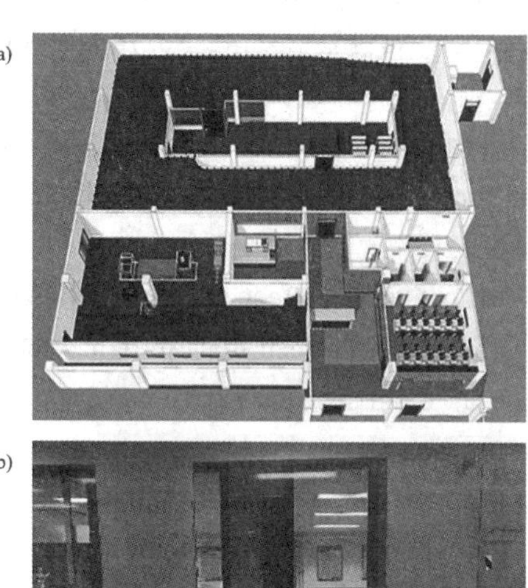
(a)

(b)

Fig. 3 Wind tunnel test setup

3 Results and Discussion

3.1 Wind tunnel test results and discussion

Figure 4 shows the variation of drag coefficient with speed. The smaller the drag coefficient is, the better the drag reduction effect is. From Fig. 4, the change in drag coefficient is very small as the speed increases when the model without any coating. However, with the increase in speed, the drag coefficient of the sample coated with fabric changes obviously, and it is easy to find that the drag coefficient of the fabric with different morphologies is significantly different.

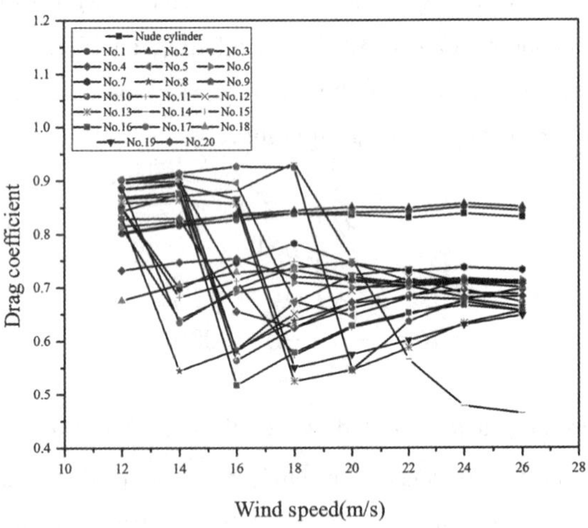

Fig. 4 Variation of drag coefficient with speed

It can be seen from Fig. 4 that the samples with the best drag reduction effect when the wind speed is 12, 14, 16, 18, 20, 22, 24 and 26 m/s are No. 20, No. 8, No. 3. No. 13, No. 13, No. 14, No. 14 and No. 14, respectively.

Next, we analyze the variation of drag coefficient with speed of three kinds of fabrics with different surface structures.

3.2 Drag reduction effect of smooth fabric

From Fig. 5, it can be seen that No. 1 has the effect of reducing drag when speed is 12–16 m/s, the best effect of reducing drag is obtained when speed is 12 m/s, and the rate of reducing drag is 1. 58%. No. 2 has the effect of drag reduction when speed is 12–14 m/s, the best effect of drag reduction is obtained when speed is 12 m/s, and the drag reduction rate is 1. 44%. Overall, the drag reduces of the smooth fabric at low speed and increases at high speed, but the drag reduction effect is poor compared with the non-smooth surface fabric.

3.3 Drag reduction effect of fabric with dimpled structure

From Fig. 6, it can be seen that the drag reduction rate of the fabrics with dimpled structure changes with the change of wind speed. The largest drag of No. 3 and No. 10 is obtained when speed is 14 m/s, the best drag reduction effect is obtained when speed is 16 m/s, and then the drag reduction rate starts to decrease to a stable state. The largest resistance of No. 4 is obtained when speed is 14 m/s, the best drag reduction effect is obtained when speed is 18 m/s, and the drag reduction rate is 25. 39%. The drag of No. 5 is the largest when speed is 14 m/s and reaches the optimal reduction effect when speed is 20 m/s, and the drag reduction rate is 22. 64%. The drags of No. 6, No. 7 and No. 8 increase when speed is 12 m/s and reach the optimal reduction effect when speed is 14 m/s, and then the reduction rate decreases with the increase of speed and reaches the steady state at high speed. No. 9 increases drag when speed is 12–18 m/s and reaches the maximum drag when speed is 18 m/s. The drag is the largest when speed is 12–18 m/s, the best effect is obtained when speed is 20 m/s, and the drag reduction rate is 34. 69%.

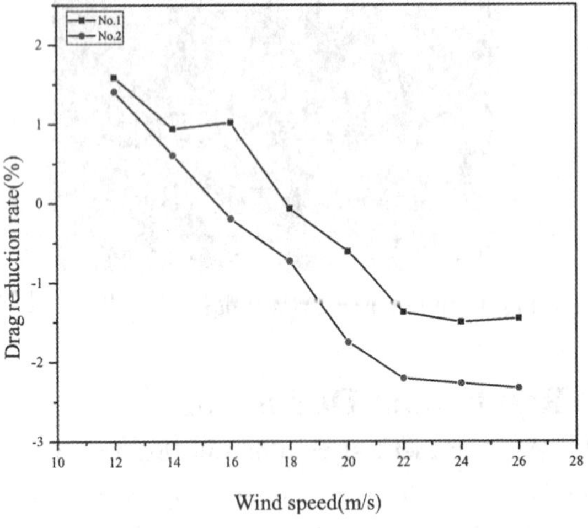

Fig. 5 Variation of drag reduction rate with speed of smooth fabrics

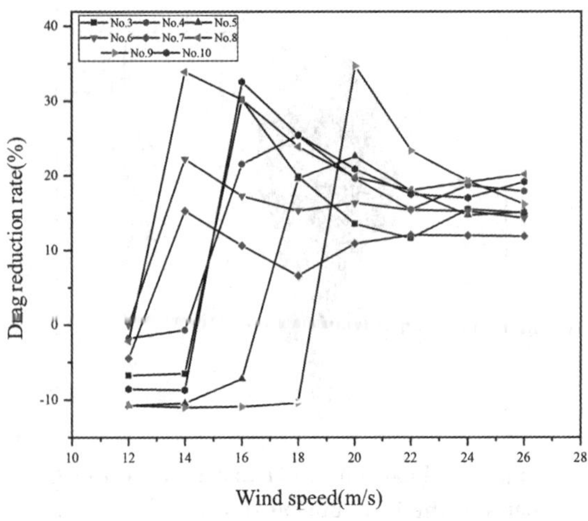

Fig. 6 Variation of drag reduction rate with speed of dimpled fabrics

3.4 Drag reduction effect of with grooved structure

As can be seen from Fig. 7, the drag of NO. 11 reaches the maximum when speed is 12 m/s, and starts to have the effect of drag reduction after the wind speed increases. After that, the drag reduction effect decreases and becomes stable. No. 13 and No. 15 both reach the maximum drag value when speed is 14 m/s, and the best drag reduction effect is obtained when speed is 18 m/s. After that, the drag reduction effect decreases and becomes stable with the increase of speed. No. 14 has the maximum drag when speed is 18 m/s, after that, it starts to have the drag reduction effect, and the drag reduction effect increases with the increase of speed. The best drag reduction effect is obtained when speed is 26 m/s, and the drag reduction rate is 44.18%. The best effect of drag reduction of No. 17 is obtained when speed is 14 m/s, the drag reduction rate is 23.05%, and the speed increases the drag reduction effect decreases and tends to stabilize. No. 18 can reduce drag at all speeds, the best drag reduction effect is obtained when speed is 12 m/s, and the drag reduction rate is 17.04%. The drag of No. 19 increases when speed is 12-16 m/s, the drag reduces when speed is 18-26 m/s, the best effect of drag reduction is obtained when speed is 18 m/s, and the drag reduction rate is 34.32%. No. 20 have the effect of reducing resistance at all speeds, the best effect of drag reduction is obtained when speed is 26 m/s, and the drag reduction rate is 15.10%.

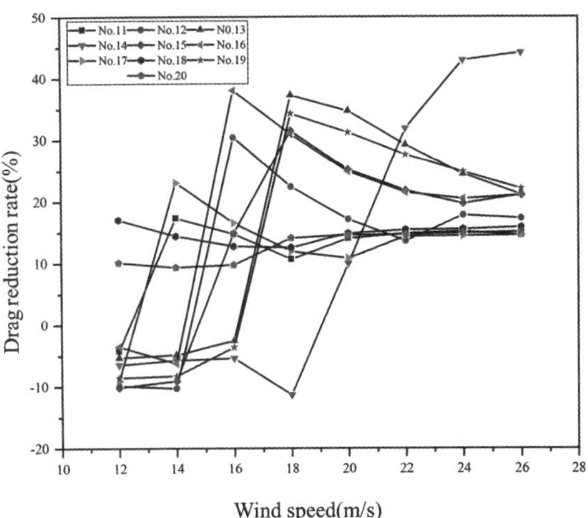

Fig. 7 Variation of drag reduction rate with speed of grooved fabrics

3.5 Design of drag reducing cycling suit

According to the wind tunnel test results, the fabrics with the best drag reduction effect at different speeds are obtained. The fabrics are matched according to the speed distribution of various parts of the human body, and the design scheme of bicycle clothes is completed through splicing. The design drawing is shown in Fig. 8.

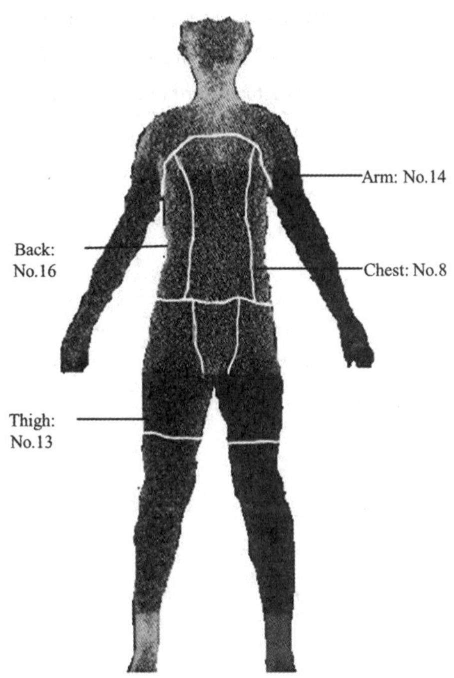

Fig. 8 Cycling clothing design diagram

4 Conclusions

The aerodynamic performance of bicycle clothing fabric was tested and analyzed by using the wind tunnel test device. By calculating the resistance coefficient and resistance reduction rate, the fabric with the best drag reduction effect under different wind speeds was obtained. Compared with the smooth fabric, the fabric with dimpled structure and grooved structure has better drag reduction effect. According to the velocity distribution of different parts of the human body, the fabric with the best drag reduction effect is matched, and then through the combination of different fabrics, the design scheme of drag reduction bicycle clothing is put forward.

References

[1] MALIZIA F, BLOCKEN B. Bicycle aerodynamics: history, state-of-the-art and future perspectives[J]. Journal of Wind Engineering and Industrial Aerodynamics, 2020, 200: 104-134.

[2] CHOWDHURY H, ALAM F. An experimental study on

aerodynamic performance of time trial bicycle helmets[J]. Sports Engineering, 2014, 17(3): 165-170.

[3] YAN Y, NIE Y, WU J, et al. Study on aerodynamic frictional drag on the surface of flexible fabric[J]. Textile Research Journal, 2016, 87(11): 1326-1334.

[4] LI X. A study indicator system of performance results influential factors for Olympic short-distance track cyclist [D]. Beijing: Beijing Sport University, 2018.

[5] TROYNIKOV O, BARDAL LM, OGGIANO L, et al. Influence of fabric structural attributes on their aerodynamic behavior[J]. Journal of Engineered Fibers and Fabrics, 2013, 8(3): 112-124.

[6] JAFARI M, HOU F, ABELKEFI A. Wind-induced vibration of structural cables[J]. Nonlinear Dynamics, 2020, 100(1): 351-421.

[7] DENG Z Q, TANG H J, LI Y L, et al. Experimental research on the wind-induced vibration of stay cables with lighting fixtures[J]. Journal of Vibration and Shock, 2020, 39(6): 44-50.

3D Geometrical Simulation of Woven Fabrics Based on TexGen

XU Zhilong[1], WANG Xi[2], ZHANG Tianshu[3], LI Qiao[1,*]

[1] *Key Laboratory of Textile Science & Technology, Ministry of Education, College of Textiles, Donghua University, Shanghai 201620, China*

[2] *College of Information Science and Technology, Donghua University, Shanghai 201620, China*

[3] *College of Science, Donghua University, Shanghai 201620, China*

[*] *Corresponding author's email*: qiaoli@ dhu. edu. cn

Abstract: This study presents an operational procedure based on the software TexGen, allowing the creation of 3D geometric models of woven fabrics. Easy and visual modeling can be accomplished by observing the relevant parameters of the fabric and inputting them in the software. Meanwhile, the established model can be adjusted with parameters to achieve easily modifying the model, by assigning material properties and meshing in the software, which can be imported to the finite element environment for finite element analysis at the later stage to predict its mechanical properties.

Keywords: Numerical simulation; TexGen; Woven fabric; Mechanical properties

1 Introduction

The structure of a fabric plays a determinative role in its mechanical and physical properties. Fabric structure and mechanical properties underlie the design, manufacture, testing and use of textiles. Therefore, an accurate geometrical description of the fabric structure is essential for modeling, whereby accurate modeling contributes to the prediction of property changes when the fabric receives external nonlinear loads. Nowadays, a significant part of the design of fabric structures comes from experience and conjecture, and idealized assumptions, such as the Pierce model and its derivatives, are often used for the design of yarn trajectories and yarn cross-sections.

Presently, numerous researchers have applied finite element analysis (FEA) to the field of fabrics. Gasser[1] and Boisse[2] assumed the yarn in the fabric as a kind of orthogonal subject as well as analyzed the yarn as a soft beam unit, which described the effect of yarn interaction, offering a new method for the description of fabric deformation mechanism. However, with this method, the fabric needs to be modeled three dimensionally and requires more accurate capture of the fabric structure, high accuracy of the model established, as well as subsequent modification of the model parameters according to the experimental requirements, which is undoubtedly a huge task, while the current fabric simulation software, such as textile CAD, still resides at the level of 2D simulation of the fabric, which is obviously different from the 3D structure of the fabric. Therefore, it is necessary to explore an efficient, fast and easy way to build 3D geometrical models of fabrics.

TexGen[3], a fabric geometrical modeling software developed by the University of Nottingham, provides accurate 3D modeling of yarns by setting a variety of input parameters such as yarn density, spacing, cross-sectional shape, etc., to simulate the structure of most yarns with the ability to locally change the cross-sectional shape of the yarn for each yarn, and to separately set parameters[4] such as thread density, Poisson's ratio, Young's modulus, etc. Parameters such as linear density, Poisson's ratio and Young's modulus can be set separately to achieve further accurate simulation of the fabric. The 3D model of yarn established by TexGen can directly generate mesh and other properties, which can be subsequently imported into the finite element environment for simulation and analysis[5], providing new ideas for predicting fabric mechanical properties and developing fabric applications.

Tab. 1 Measurement data

Fabric code	Yarn width(mm) (warp/weft)	Yarn height(mm) (warp/weft)	Yarn spacing(mm) (warp/weft)	Cross section	Fabric thickness (mm)
A	0.16/0.18	0.11/0.09	0.18/0.25	Ellipse	0.19

2 Experiments

2.1 Materials

A plain fabric composed of nylon fibers was selected as shown in Fig. 1.

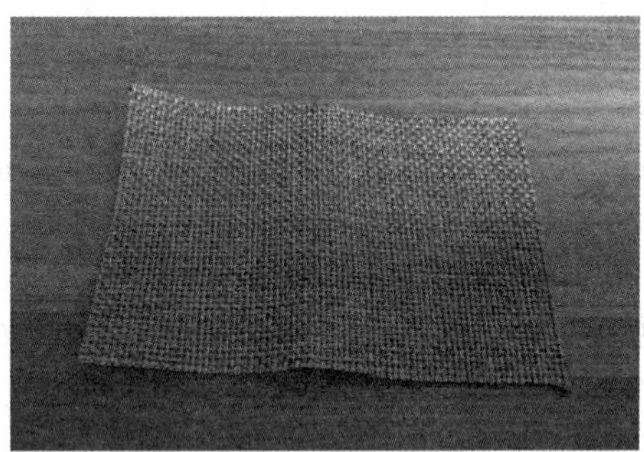

Fig. 1 Experimental sample

2.2 Determination of structural parameters

In this section, the selected fabric was observed through a microscope for accurate description of the geometric structure of the fabric.

First, reference points A/B/C/D were marked on the three woven fabrics used in the experiment for subsequent measurements, as shown in Fig. 2. The microstructure of this sample was observed by the microscope, and the distance between the reference points was measured, with A-B as the distance between two adjacent warp yarns and C-D as the distance between two adjacent weft yarns. Next, the sample fabric was sliced, and its cross-section was observed through a microscope to obtain data such as cross-sectional shape, yarn diameter, yarn width, yarn height, etc. The final measured data is shown in Tab. 1.

Fig. 2 Reference point marking

2.3 Construction of 3D models of fabrics with weave wizard

The geometric modeling of the fabric has a decisive influence on the subsequent textile simulation analysis, and the geometry of the model affects the accuracy of the final simulated results. This section introduces two methods to construct a 3D model of the fabric, by using TexGen to generate the geometry of the fabric through a comprehensive description based on the position of the yarn centerline, as well as the shape and size of the cross section.

Based on the data obtained from Tab. 1, in the weave wizard window of TexGen software in Fig. 3 (a), set the number of warp yarns, number of weft yarns, yarn spacing, yarn width, cross-sectional shape and other parameters, and set the yarn weaving mode in the weave pattern window in Fig. 3 (b).

2.4 Construction of 3D models of fabrics with modeler

Although using weave wizard to create fabric models is fast and easy, the modeler can be called for fabric model building in order to further improve the control of the model and facilitate subsequent modification of parameters to debug the model.

The yarn cross section is the two-dimensional shape of the vertical cross section of the yarn, which can be set as ellipse, polygon, hybrid, etc. in TexGen. As an example, the ellipse shape can be adjusted by modifying the parameters such as ellipse eccentricity, and then the cross-section shape can be modified.

In the Assign Properties window, individual properties can be assigned to each yarn, including yarn linear density, fiber diameter, Young's modulus, Poisson's ratio and other indicators, in preparation for subsequent export of the model to the finite element environment.

2.5 Model export

After setting the parameters according to the above, the fabric model created can be set in TexGen in the model output format, such as finite element file containing yarn information, surface mesh output and other formats. Figure 4 (a) shows the 3D structure diagram of the fabric under the microstructure, and the 3D model structure diagram obtained by modeling with TexGen. Figure 4 (b) shows a 3D cross-sectional view of the fabric. The selected finite element format will be output as a. tg3 format file, which can be used for future finite element analysis in ABAQUS environment as shown in Fig. 5.

When exporting the model, you can choose the type of finite element meshing, such as using C3D8R mesh sweeping method to mesh the model, in order to avoid the highly distorted elements affecting the subsequent analysis. In addition, the same mesh should be generated for the boundary on the repeated cells. Set the constraints of periodic boundary conditions, and define the contact area, etc.

3 Results and Discussion

The method proposed in this study successfully establishes a 3D model of woven fabric by TexGen and provides a new way of thinking for the performance prediction of fabric models by setting the parameters, which can be easily modified later, and by assigning the corresponding material properties to the model and dividing a high-precision mesh on the surface of the model, which can then be used to export the model to ABAQUS for finite element analysis.

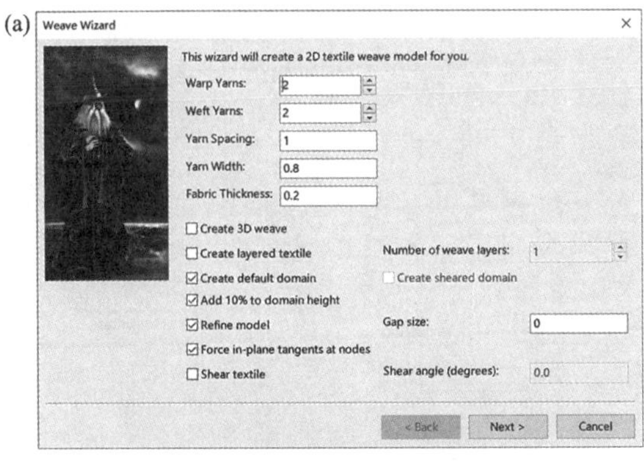

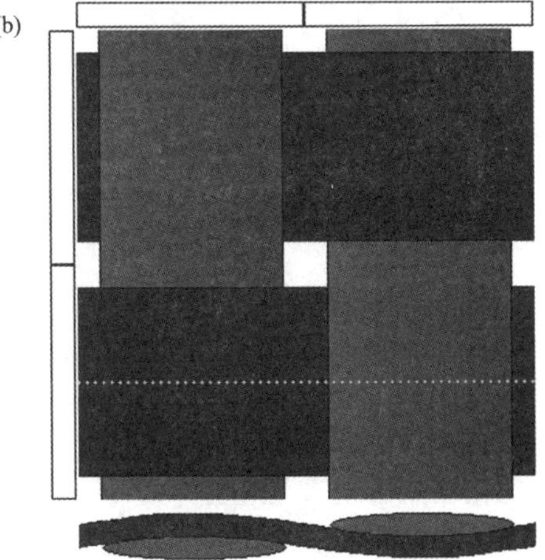

Fig. 3 (a) Weave wizard window; (b) weave pattern window

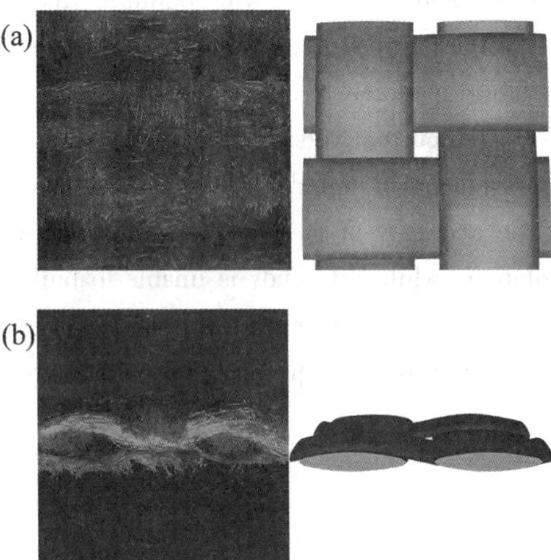

Fig. 4 (a) Fabric structure diagram (left) and model 3D model diagram (right); (b) fabric cross sections (left) and 3D model cross sections (right)

Fig. 5 Fabric model in FEM environment

4 Conclusions

In this study, the fabric model simulation is explored, presenting the process of visualizing 3D fabric modeling by using TexGen software, combining observation data, setting different parameters in the corresponding session, as well as setting relevant constraints to optimize the model, in order to achieve the purpose of accurate, convenient and visual modeling, with the possibility of modifying the parameters according to needs, offering the maximum flexibility of fabric performance.

At the same time, various parameters can be set in the

output model, and the created model can be exported to external analysis software for fabric performance analysis in the form of various data. In the future, it will be possible to modify the relevant conditions in the ABAQUS environment for the analysis of the mechanical properties of the fabric, and to predict and verify the fabric model, which will optimize the design process of the fabric structure and have a certain impact on the fabric in practical applications.

However, the current version of TexGen shows potential for improvement in the definition of boundary conditions, the accuracy of meshing, and the algorithm for interpolation, while this study is unable to build overly complex models, and more attempts may be needed in the optimization of subsequent model parameters.

References

[1] BOISSE P, BUET K, GASSER A, et al. Meso/macro-mechanical behavior of textile reinforcements for thin composites[J]. Composites Science and Technology, 2001, 61: 395-401.

[2] BADEL P, VIDAL E, BOISSE P. Computational determination of in-plane shear mechanical behavior of textile composite reinforcements[J]. Computational Material Science, 2007, 40: 439-448.

[3] TexGen from: http://www.texgen.sourceforge.net.

[4] HUA L, MIKE C, ANDREW C, et al. A finite element approach to the modelling of fabric mechanics and its application to virtual fabric design and testing[J]. The Journal of The Textile Institute, 2012, 103(10): 1063-1076

[5] LIN H, ZENG X, SHERBURN M, et al. Automated geometric modelling of textile structures[J]. Textile Research Journal. 2012, 82(16): 1689-1702.

Effect of Pile Height on Mechanical Properties of Three-Dimensional Hollow Fabric Composites

LENG Xiaoxiao[1,2], SHAO Huiqi[1,2], CHEN Nanliang[1,2], JIANG Jinhua[1,2*]

[1] *Engineering Research Center of Textile Technical, Education Ministry, Shanghai 201600, China*

[2] *College of Textiles, Donghua University, Shanghai 201620, China*

[*] *Corresponding author's email*: jiangjinhua@ dhu. edu. cn

Abstract: As a relatively novel material in three-dimensional woven composites, three-dimensional hollow fabric composites have the advantages of high strength, wear resistance, impact resistance, delamination resistance and so on. With the application of three-dimensional hollow fabric composites more and more widely, their mechanical properties have attracted the attention of researchers at home and abroad. In this paper, the bending and compression properties of three-dimensional hollow fabric composites are tested and analyzed from the height of the core pile, and the test process phenomenon and failure mode are expounded. The results show that the bending load and compressive properties of three-dimensional hollow fabric composites decrease significantly with the increase of core pile height.

Keywords: Three-dimensional hollow fabric composites; Core pile height; Bending performance; Flattening performance

1 Introduction

Three-dimensional hollow fabric composites are new types of three-dimensional woven composites. Their microstructure is composed of surface warp, weft and two surface layers connected to form Z-directional fibers in the core[1]. Due to the advantages of high strength, high modulus, delamination resistance and impact resistance, the composites are widely used in aerospace, shipbuilding, rail transit and other fields[2]. Although traditional sandwich composites, such as honeycomb sandwich and foam sandwich composites, have been widely used, there are still many limitations. This is because the traditional sandwich is bonded by adhesive, which is prone to delamination during loading. At present, existing studies have shown that the mechanical properties of integral hollow sandwich composite are equivalent to those of honeycomb sandwich, but it overcomes the shortcomings of low interlayer strength of traditional sandwich structure materials and has excellent overall performance[3-5]. In this paper, from the height of the pile, the three-dimensional hollow fabric composites with different structural parameters are systematically compared. And the relevant performance parameters and failure laws of the material are obtained, which set up a foundation for further research and development of three-dimensional hollow fabric composites.

2 Experiments

2.1 Raw materials and equipment

Three-dimensional spacer fabrics: fiber raw material is warp and weft 132×132 tex alkali-free glass fiber yarn, and pile warp 200 tex alkali-free glass fiber yarn, and fabric spacing height is 8 mm, 10 mm, and 15 mm.

Resin matrix: WBH-120 was selected as heat-soluble epoxy resin and its supporting curing agent, which was prepared according to 3 : 1 mass ratio and fully stirred.

Main equipment: electronic precision balance, triangle digital angle meter, vacuum assistant equipment and consumables, MTS electronic universal testing machine, electric heating blower dryer, scanning electron microscope, etc.

2.2 Combined forming process

The first step in the preparation of three-dimensional hollow fabric composites is the selection and preparation of resin matrix. The resin used in this paper is WBH-120 and its supporting curing agent. With good physical properties and operation processing, it can be used for carbon fiber, glass fiber and other prepreg materials manufacturing, dedicated to wind power carbon fiber girder and other industrial composite materials. In this paper, some optimizations are carried out based on the vacuum bag method, as shown in the Fig. 1. The three-dimensional hollow fabric is cut into a certain size, and a layer of stripping layer and high permeability medium are added on the fabric. Then the whole package was wrapped and processed with a vacuum bag to check its air tightness. Finally, with the help of negative pressure, the prepared resin was sucked into the vacuum bag, so that it fully penetrated into the fabric and infiltrated it. The impregnated fabric was placed between two aluminum plates and placed in an oven. The fabric was kept at 50 ℃ for 300 min and then heated to 80 ℃ for 480 min to obtain a three-dimensional hollow fabric composite.

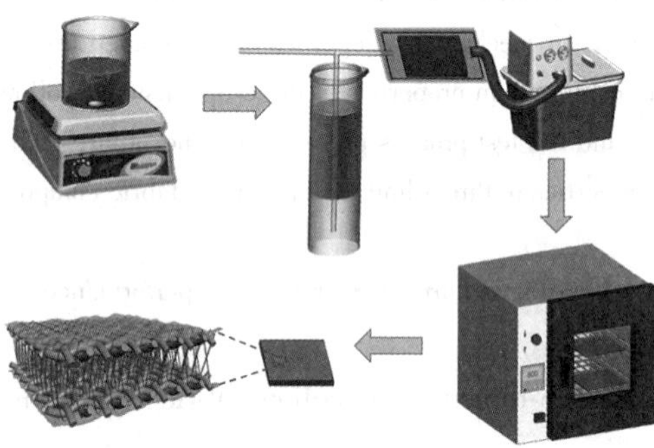

Fig. 1 Composite molding process of three-dimensional hollow fabric

According to the above preparation process, three-dimensional hollow fabric composites with different specifications and different process parameters are obtained. Fabrication of three-dimensional hollow fabric composites with different pile height using 8 mm, 10 mm and 15 mm three-dimensional hollow fabrics is shown in Tab. 1.

Tab. 1 Pile height of three-dimensional hollow fabric composite

Sample number	SW8-1	SW10-1	SW15-1
Height of fabric pile (mm)	8	10	15

2.3 Compression testing

Reference standard GB/T 1453—2005. The sample size is 30 mm×30 mm. The average value of five parallel samples in each group was taken, and the loading speed of the instrument was 2 mm/min.

2.4 Three-point bending testing

Referring to GB/T 1446—2005 and GB/T 1456—2005, the sample preparation and three-point bending test of the three-dimensional hollow fabric composites were carried out. The specimen is 140 mm long and 30 mm wide. The bending test was carried out on a MTS electronic universal testing machine with a loading speed of 10 mm/min and a span of 100 mm to obtain the relevant indicators.

3 Results and Discussion

3.1 Effect of sample pile height on material compression performance

As shown in Fig. 2, with the increase of the pile height, the compressive strength of the composites shows a significant downward trend.

When the height increases from 8 mm to 10 mm, the compressive strength decreases from 1.4 MPa to 0.96 MPa. When the height increases from 10 mm to 15 mm, the compressive strength decreases from 0.96 MPa to 0.42 MPa. The analysis shows that with the increase of pile height, the compressive strength and compressive strength decrease greatly, which is mainly related to the fabric structure. When the pile height is small, the '8' -shaped yarn of three-dimensional hollow fabric is closely bonded, and the hollow void of the fabric is filled after filling the resin. At this time, the core material forms a solid rod, so the flat compressive strength of the composite material with small core height is larger. With the increase of pile height, the core material is easy to lose stability, so the strength decreases rapidly. In this paper, the formula and curve of compressive strength are obtained by data fitting. The formula model can better fit the compressive strength of three-dimensional hollow fabric composites with different pile heights. Based on this, three-dimensional hollow fabric composites with other pile heights can also be deduced.

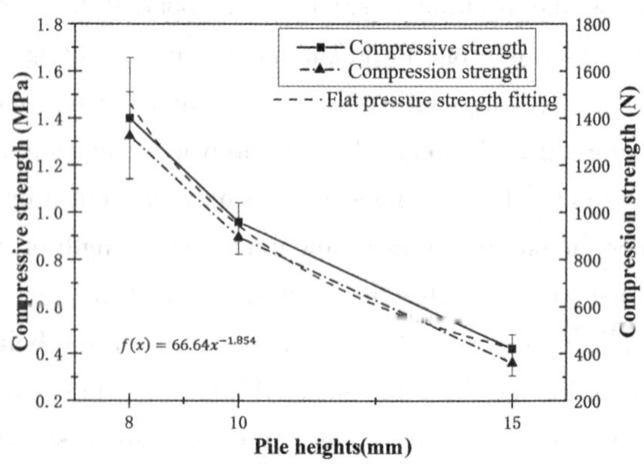

Fig. 2 Compression properties of three-dimensional hollow composites with different pile heights

3.2 Effect of sample pile height on material bending performance

As shown in Fig. 3, with the increase of spacer height, the bending properties of three-dimensional hollow fabric composites show a significant downward trend. The analysis shows that when the pile height increases to 10 mm, the strength decreases greatly. This is mainly determined by the fabric structure. The fabric core is '8' shaped structure. When the height is small, the two yarns are tightly bonded, and the resin will fill two gaps and become a coarse rod. When the height of the core material reaches more than 10 mm, the core material is a slender bending rod, and the instability strength decreases significantly.

In addition, when the composite material is subjected to bending loading, the fibers above the central layer are compressed, and the fibers below are tensioned. In theory, the normal stress on the cross section is linearly distributed along the beam height. The bending load formula of three-dimensional hollow fabric composites can be obtained by formula derivation. σ_f is bending load (or bending stress when deflection is 1.5 times of sample thickness), MPa; p is the failure load (or the maximum load, or the load when the deflection is 1.5 times the sample thickness), N; l is span, mm; h is the sample thickness, mm; b is sample width, mm. It can be seen that the relationship between specimen thickness and bending load is quadratic. When the surface thickness is constant, the higher the pile height is, the worse the bending load is.

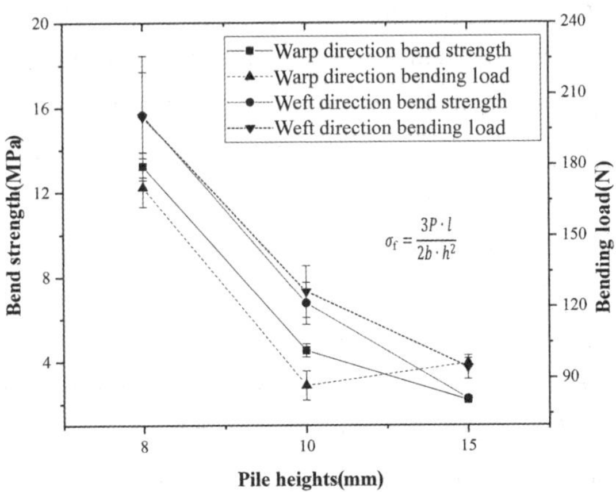

Fig. 3 Bending load of three-dimensional hollow fabric composites with different pile heights

4 Conclusions

Through the compression test and three-point bending test, this paper explores the influence of pile height on three-dimensional hollow fabric composites, and draws the following conclusions:

With the increase of pile height, the compressive strength of three-dimensional hollow fabric composites decreased significantly. This is related to the fabric structure. When the pile height is small, the core material is a coarse solid rod, and the flat compressive strength of the composite material is large. With the increase of column height, the core material is prone to instability, so the strength decreases rapidly.

With the increase of pile height, the bending load of three-dimensional hollow fabric composite decreases significantly. The relationship between sample thickness and bending load is quadratic, when the surface thickness is constant. The higher the pile height is, the worse the bending load is.

Acknowledgements

This work was sponsored by the Shanghai Natural Science Foundation of Shanghai Municipal Science and Technology Commission (No. 20ZR1400600, 20ZR1401600), and Fundamental Research Funds for the Central Universities (No. 2232020D－09, 2232020G-06, 2232021G-06).

References

[1] VERPOEST I, WEVERS M. 2.5 D-and 3D-fabrics for delamination resistant composite laminates and sandwich structures[J]. SAMPE Journal, 1989, 25(3): 51-56.

[2] HOSUR M V, ABDULAH M, JEELANI S. Manufacturing and low-velocity impact characterization of foam filled 3D integrated core sandwich composites with hybrid face sheets[J]. Composites Structures, 2004, 69(2): 167-181.

[3] FAN H, WEI Y, ZHOU Q. Experimental research of compressive responses of multi-layered woven textile sandwich panels under quasi-static loading[J]. Composites Part B: Engineering, 2011, 42(5): 1151-1156.

[4] FAN H, ZHAO L, CHEN H, et al. Ductile deformation mechanisms and designing instructions for integrated woven textile sandwich composites[J]. Composites Science & Technology, 2012, 72(12): 1338-1343.

[5] SAFARI H, KAREVAN M, NAHVI H. Mechanical characterization of natural nano-structured zeolite/polyurethane filled 3D woven glass fiber composite sandwich panels[J]. Polymer Testing, 2018, 67: 284-294.

Novel Phosphorus-Nitrogen Flame Retardant and Its Application in Cotton Fabric

KANG Gaowei[1], CHEN Hao[1], WU Dequn[1,2*]

[1] *Key Laboratory of Textile Science & Technology, Ministry of Education, Donghua University, Shanghai 201620, China*

[2] *Innovation Center for Textile Science and Technology, Donghua University, Shanghai 201620, China*

* *Corresponding author's email*: dqwu@ dhu. edu. cn

Abstract: The poor thermal properties of cotton fabric limit its deep application in various fields. In this paper, a new phosphorus-nitrogen flame retardant (TPU) was synthesized based on tromethamine (THAM), dicyandiamide (DCA), phosphoric anhydride (PA) and urea. Fourier transform infrared spectroscopy (FTIR) test results show the successful synthesis of TPU. The flame retardant is grafted onto the cotton fabric to prepare flame-retardant cotton fabric. The modified cotton fabrics have been characterized by FTIR, a scanning electron microscope (SEM), UL-94 vertical burning tests, limit oxygen index (LOI) tests and other technical means. The results showed that the TPU was successfully grafted onto the cotton fabric. And when the grafting amount of flame retardant is 15% of cotton fabric, the LOI of flame-retardant cotton fabric could reach 30%. These results indicated that TPU grafted onto cotton fabric as a flame retardant could achieve a great flame retardant effect, which had great potential for improving the fire safety of cotton fabric.

Keywords: Phosphorus-nitrogen flame retardant; Durable flame retardant; Cotton fabric; Flammability

1 Introduction

Fire accident is a kind of disaster with great harm and frequent occurrence in today's society. Many of these fires are caused by textiles, because most of the textiles we use in our daily life are flammable. Therefore, fire-retardant textiles are very important to our life. Among them, because of their renewable, wide source, large output and other excellent characteristics, cotton products are favored by various fields, such as textile clothing, home textiles, etc. However, cotton fiber is an extremely flammable high molecular material[1-2] (decomposition temperature: 210-230 ℃, limit oxygen index value: 18%). In the process of combustion, the flame propagation speed is fast, while the heat is high, and the smoke is large. It is easy to ignite other materials, and then cause and spread fire[3]. This feature limits its in depth application in more fields.

According to the types of flame retardant elements, flame retardants can be roughly divided into halogen flame retardants, phosphorus flame retardants, nitrogen flame retardants and silicon flame retardants[4-5]. Halogen flame retardant research started earliest. It can achieve high flame retardant efficiency with a small amount content[6]. However, because of environmental pollution and personal injury, it has been banned to be used. Phosphorous flame retardants, as the preferred alternative to traditional halogen flame retardants, can play a flame retardant effect in both condensed and gas phases. It has become one of the most rapidly developing flame retardants on account of the low toxicity, smoke generation and corrosion[7]. The compatibility is well between nitrogen flame retardant and polymer. It releases a large number of non-flammable gas or refractory gas along with low toxicity and low smoke during the combustion process, such as N_2, NO_2, NH_3 and H_2O. N_2 can capture free radicals, playing a gas phase flame retardant effect[8]. In addition to giving the properties of low combustion rate, heat release rate and melt drop resistance to polymer materials, the unique Si—O bond of silicone flame retardants can increase the flexibility of molecular chain and improve the processing performance, mechanical properties and heat resistance of polymer materials. In recent years, many studies have focused on synergistic flame retardants containing various flame retardants elements for achieving excellent flame retardant effect with a small amount of flame retardant in practical application.

On this basis, we successfully synthesized a new phospho-nitrous flame retardant (TPU) from tromethamine, dicyandiamide, phosphorus pentoxide and urea. It is grafted onto cotton fabric, and the flame-retardant properties and mechanism have been preliminary studied.

2 Experiments

2.1 Materials

Tromethamine (THAM) was supplied by Gaipu Chemical Co., Ltd. (Shanghai, China). Dicyandiamide (DCA) was purchased from Shanghai Xushuo Biotechnology Co., Ltd. (Shanghai, China). Hydrochloric acid was supplied by Sinopharm Chemical Reagent Co., Ltd. (Shanghai, China). Phosphorus pentoxide was purchased from Shanghai Yien Chemical Technology Co., Ltd. (Shanghai, China). Urea was supplied by Shanghai Titan Technology Co., Ltd. (Shanghai, China). Absolute ethanol was purchased from Shanghai Gaoxin Chemical Glass Instrument Co., Ltd. (Shanghai, China). All of them are analytical grade without further purification before use. Cotton fabric (245 g/m^2) was provided by Xiajin Renhe Textile Technology Co., Ltd. (Dezhou, China).

2.2 Preparation of flame retardants

TPU was synthesized by one-pot method. The dicyandiamide (0.92 g, 0.011 mol) was dissolved in the deionized water and the pH value was adjusted by hydrochloric acid. Tromethamine (1.21 g, 0.01 mol) was added into the three-mouth flask when the temperature was 65 ℃ and stayed for 3 h. And then the phosphorus oxide was gradually added to it. The reaction mixture was mixed for 2 h at 80 ℃. Whereafter, urea (2.0 g, 0.033 mol) was added and the temperature was raised to 130 ℃. Finally, the reaction mixture was concentrated via a rotary evaporator. The synthesis route of TPU was shown in the Fig. 1.

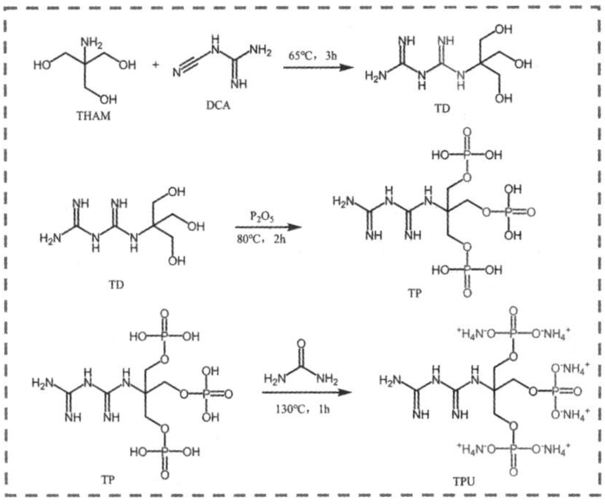

Fig. 1 Synthetic route of TPU

2.3 Preparation of flame retardant cotton fabric

Flame retardant solutions of various concentrations were prepared by dissolving TPU in deionized water. Then the dicyandiamide whose weight was 5% of the weight of solution was added as the catalyst. Cotton fabrics were soaked in TPU solution of different concentrations at 70 ℃ with a bath ratio of 1 : 30 in constant temperature water bath for 1 h. After soaking. The liquid carrying rate of the fabric reached 200%. Finally, the treated cotton fabrics were reacted at 170 ℃ for 5 min. The grafting rate of modified cotton fabrics were 5%±0.2%, 10%±0.2% and 15%±0.2%, respectively. The grafting rate (A) was calculated by weighting method of Eq. (1) after pre-humidification at chamber of constant temperature and humidity (21 ± 1) ℃, 60%±2% for 48 h.

$$A = \frac{W_f - W_i}{W_i} \times 100\% \qquad (2\text{-}1)$$

In this equation, W_f is the weight of the grafted cotton fabrics and W_i is the weight of the pure cotton fabrics.

2.4 Spectroscopic analyses

Fourier transform infrared (FTIR) spectra of the TPU over the wavenumber ranging from 4 000 cm^{-1} to 400 cm^{-1} were achieved at room temperature with PerkinElmer FTIR Spectra 1 000.

2.5 Morphological observation

Scanning electron microscopy (SEM) images were obtained by using Japan JEOL TM-3000 instrument to observe the morphology of the cotton fabrics before and after burning. All the samples were sputter-coated with a conductive gold layer before SEM testing.

2.6 Vertical combustion test

Vertical burning test (VBT) was performed in a UL-94 vertical flame chamber with specimen dimensions of 300 mm×89 mm×3 mm according to the GB/T 5455—2014 standard. There were five specimens in each experiment.

2.7 Limit oxygen index test

Limiting oxygen index (LOI) test was performed by using Qingdao ZR－1 according to the GB/T 5454—1997 standard with specimen dimensions of 150 mm× 58 mm×3 mm. There were fifteen specimens in each experiment.

2.8 Mechanical tests

To clarify the effect of modification on tensile property of fabrics, tensile test of fabrics was conducted on electronic universal tensile testing machine at the tensile rate of 100 mm/min with the gauge length of 100 mm. The effective width of specimens was kept to 50 mm. Five specimens were tested for each experiment.

3 Results and Discussion

3.1 Infrared spectra and analysis of TPU and modified cotton fabrics

The FTIR spectra of flame retardant and TPU-treated cotton fabrics were shown in Fig. 2. As shown in Fig. 2 (a), the absorption peak at 1 596 cm^{-1} was attributed to C ═N. The peaks at 1 269 cm^{-1} and 1 064 cm^{-1} were attributed to P ═O and P—O—C, and the peak at 931 cm^{-1} was assigned to the P—O. These results of infrared analysis showed that the synthesis of TPU was successful. In the Fig. 2 (b), compared with the original cotton, there was a new peak at 1 638 cm^{-1} in modified cotton fabrics, which was attributed to C ═N, and the peaks at 1 225 cm^{-1}, 998 cm^{-1} and 929 cm^{-1} were attributed to P ═O, P—O—C and P—O, respectively, indicating that the flame retardant TPU was successfully grafted onto the cotton fabric.

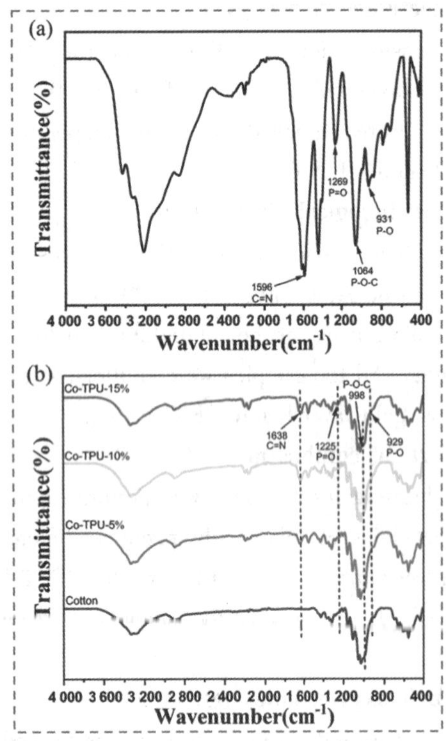

Fig. 2 (a) FTIR spectra of TPU and (b) FTIR spectra of control cotton fabrics and modified cotton fabrics

3.2 Morphology of modified cotton fabrics

To further compare the difference in surface morphology between untreated cotton fabrics and TPU-treated cotton fabrics, the samples were characterized by SEM in Fig. 3. As shown in Fig. 3, it can be clearly noticed that the surface morphology of the untreated cotton fabrics was relatively smooth and retained their complete natural bending. In contrast, there were obvious particles on the surface of the modified cotton fabrics, and the particles increased with the increase of the amount of flame retardant, indicating that TPU was successfully grafted onto the fiber surface.

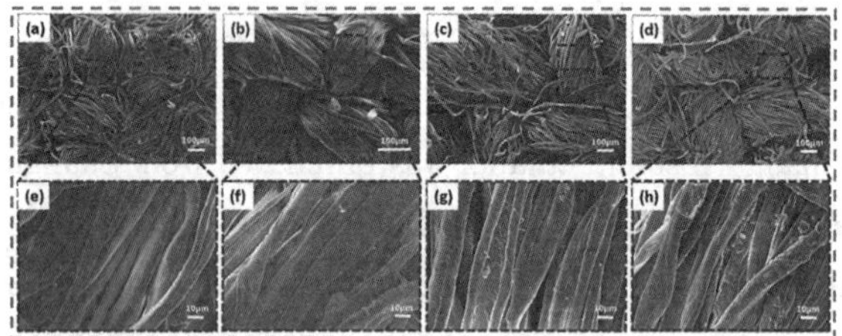

Fig. 3 SEM images of control cotton fabrics (a) and modified cotton fabrics: (b) Co-TPU-5%; (c) Co-TPU-10%; (d) Co-TPU-15%; (e), (f), (g) and (h) are the magnifications of (a), (b), (c) and (d), respectively

3.3 Evaluation of flame retardancy of TPU-treated cotton fabrics

The results of LOI were shown in Tab. 1. Cotton fabrics were easy to ignite at a low LOI value (18%). When 5%, 10% and 15% of TPU was grafted to cotton fabrics, it could dramatically increase the LOI value to 23%, 27% and 30%, respectively. These results revealed that the abilities of ignition and combustion of cotton fabric both decreased with the increasing of TPU.

The vertical combustion tests of pure cotton fabrics and modified cotton fabrics were conducted by an UL-94 instrument to analyze the flame-retardant property. The samples were ignited for 10 s by methane gas, and the height of the nozzle from the sample was 10 mm. As shown in Tab. 1, the combustion of the pure cotton fabrics was rapid and complete after ignition with no char left. The modified cotton fabric could self-extinguish after removing the fire. And with the increase of TPU, the self-extinguishing speed will be faster. The sample of Co-TPU-15%, in particular, did not burn at all. It greatly improved the thermal stability of cotton fabric. These phenomena illustrated the inhibition of TPU on the combustion performance of cotton fabrics directly.

Tab. 1 UL-94 vertical burning tests and limit oxygen index of modified cotton fabrics.

Sample	LOI (%)	Vertical burning test		
		After-flame time(s)	After-grow time(s)	Char length (mm)
Cotton	18	N. R	N. R	T. D
Co-TPU-5%	23	20	0	230
Co-TPU-10%	27	5	0	75
Co-TPU-15%	30	0	0	50

The carbon residue of cotton fabric before and after modification was tested by SEM to study its flame retardant mechanism. It can be seen from Fig. 4 that the raw cotton fabric presents a loose and disordered state after combustion, which was conducive to the diffusion of oxygen and combustible gas and further combustion of the material. In contrast, the modified cotton fabric could remain intact after combustion. This is because when the modified cotton fabric is burned, a layer of dense carbon layer will be formed on the fiber surface under the action of flame retardant to protect the fiber. At the same time, there were some bubbles on the surface of the carbon residue of the modified cotton fabric, which indicated that the self-extinguishing behavior was occurred during the combustion process and no further combustion was occurred, so the bubbles did not burst.

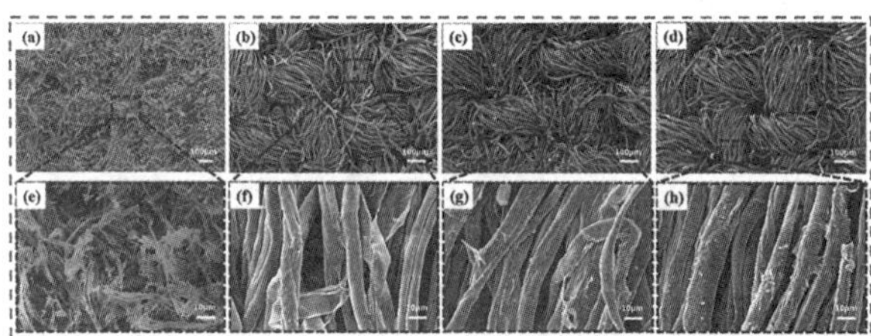

Fig. 4 SEM images of control cotton fabrics char residues (a) and modified cotton fabrics char residues: (b) Co-TPU-5%, (c) Co-TPU-10% and (d) Co-TPU-15%. (e), (f), (g) and (h) are the magnifications of (a), (b), (c) and (d), respectively

3.4 Mechanical property of modified cotton fabrics

In order to investigate the degree of damage of TPU-treated cotton fabrics, the tensile test results of untreated cotton fabrics and TPU-treated cotton fabrics with different weight increments were shown in Fig. 5. After treatment, the tensile strength of samples in warp and weft directions were reduced slightly. It's pretty obvious that the tensile strength of warp and weft gradually decreased with the increase of TPU. And when the weight increment was achieved 15%, the tensile strength in warp direction was decreased by 12.2%, while that in weft direction was decreased by 12.5% via calculating from Fig. 5. The strength loss of the treated cotton fabrics may be due to the high temperature and the formation of acidic substances during grafting. But these reductions will not affect the application of cotton fabrics in clothing, interior decoration and other fields.

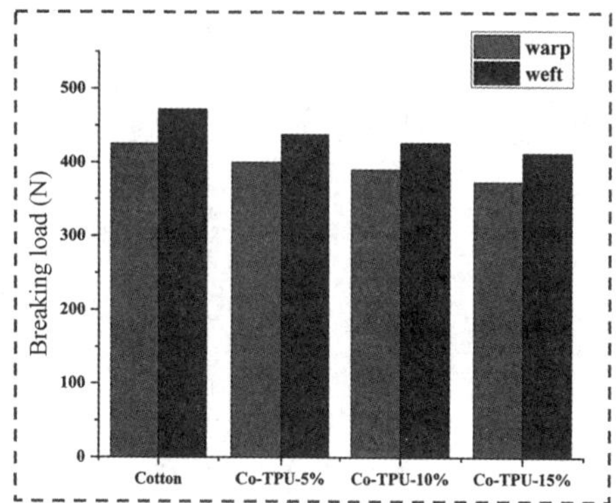

Fig. 5 Breaking load of the control cotton fabrics and the modified cotton fabrics

4 Conclusions

In this paper, a novel phosphorus-nitrogen flame retardants was successfully fabricated and added onto cotton fabric by grafting method to improve its flame retardancy. TPU had both condensed phase flame retardancy mechanism and gas phase flame retardancy mechanism. It was decomposed to form phosphoric acid during combustion, and phosphoric acid could promote carbon formation on fiber surface. At the same time, it generated N_2, NO_2, NH_3 and other nonflammable gases, which could dilute the concentration of oxygen and combustible gases. As a result, the LOI of the modified cotton fabric increased from 18% to 30%. Although the strength of warp and weft yarn decreases, it will not affect the application of cotton fabric in various fields.

References

[1] ZHANG J J, CHEN B, LIU J, et al. Multifunctional antimicrobial and flame retardant cotton fabrics modified with a novel N, N-di(ethyl phosphate) biguanide[J]. Cellulose, 2020, 27: 7255-7269.

[2] LI N, CHEN P P, LIU D N, et al. Novel P/Si based nanoparticles for durable flame retardant application on cotton[J]. Cellulose, 2022, 29(3): 2063-2076.

[3] LAZAR S T, KOLIBABA T J, GRUNLAN J C, et al. Flame-retardant surface treatments[J]. Nature Reviews Materials, 2020, 5(4): 259-275.

[4] LI N, KANG G W, LIU H, et al. Hybrid nanoparticles based on novel Schiff Base for durable flame retardant and antibacterial properties[J]. Composites Part B: Engineering, 2022, 238: 109905.

[5] LUO X L, LI Z Y, SHEN J Y, et al. A facile strategy to achieve efficient flame-retardant cotton fabric with durable and restorable fire resistance[J]. Chemical Engineering Journal, 2022, 430: 132854.

[6] LI N, KANG G W, LIU H, et al. Fabrication of eco-friendly and efficient flame retardant modified cellulose with antibacterial property[J]. Journal of Colloid and Interface Science, 2022, 618: 462-474.

[7] LI P, WANG B, LIU Y Y, et al. Fully bio-based coating from chitosan and phytate for fire-safety and antibacterial cotton fabrics[J]. Carbohydrate Polymers, 2022, 237: 116173.

[8] FAN S, PENG B, YUAN R C, et al. A novel Schiff base-containing branched polysiloxane as a self-crosslinking flame retardant for PA6 with low heat release and excellent anti-dripping performance[J]. Composites Part B: Engineering, 2020, 183: 107684.

Synthesis of Gel-Spun Poly (Vinyl Alcohol) Fibers Reinforced with High Content Lignin

SUN Xiaorui[1,2], **LU Chunhong**[1,2*]

[1] *Shanghai Frontiers Science Center of Modern Textiles, Donghua University, Shanghai 201620, China*

[2] *College of Textiles, Donghua University, Shanghai 201620, China*

* *Corresponding author's email*: chlu@ dhu. edu. cn

Abstract: Lignin is a biological renewable polymer and it has been used as an environmental filler in composite fibers. However, lignin aggregation occurred at high lignin content, which significantly hindered the further enhancement of fiber performance. Esterified hydrophilic lignin obtained from the esterification of organosolv lignin (OL) and D-gluconic acid has good interaction with poly (vinyl alcohol) (PVA) matrix in gel-spun fibers at low content. In order to explore the effects of this modified lignin (ML) on the performance of high concent lignin/PVA composite fibers, quality gel-spun PVA fibers were prepared with the incorporation of 30% lignin. Considering the discrepancy in the reinforcement effect of ML and OL on PVA, in this work, we explored the properties of gel spinning fibers reinforced with two different fillers (30% OL/PVA and 30% ML/PVA). 30% ML/PVA fibers exhibited optimal mechanical properties. Compared with 30% OL/PVA fibers, the tenacity of 30% ML/PVA fibers increased from 5. 50 cN/dtex to 6. 60 cN/dtex, while the average specific modulus increased from 134. 62 cN/dtex to 192. 31 cN/dtex, and the toughness increased from 17. 20 cN/dtex to 19. 03 J/g. This work demonstrates the possibility of employing bio-based high content lignin filler for the reinforcement of polymeric high-performance fibers to achieve better sustainability with no sacrifice of fiber properties.

Keywords: PVA fibers; Lignin; High content; Gel spinning; Mechanical properties

1 Introduction

High performance fibers (i. e., aramid, poly acrylonitrile (PAN), polyethylene terephthalate (PET), polyvinyl alcohol (PVA), polyethylene (PE) and polypropylene (PP), etc.) play an important role in biomedicine, construction industry, geotextiles, aerospace, and electronics fields. Among them, PVA shows great potential as a biodegradable material in biomedical wound dressings[1], packaging films[2], and construction fields[3]. Due to the strong inter- and intra-molecular forces within the PVA molecular chain, it is difficult to achieve near-parallel orientation and complete crystallization of the molecular chain by multifold stretching as in the case of polyethylene fibers. The strength and modulus of currently developed high performance PVA fibers are still a long way from the theoretical values (tensile strength of 27 GPa, Young's modulus of 255 GPa)[4-5]. Therefore, how to further improve the mechanical properties of PVA fibers has become an ongoing challenge for researchers.

Next to cellulose, lignin is the second most abundant organic polymer in nature. Due to its properties such as biodegradability and low cost, lignin has been filled into different polymers to strengthen the polymer matrix. Lignin/PVA fibers are mainly prepared by wet spinning and gel spinning techniques[6]. It was found that lignin at a low doping concentration can form a good interaction with PVA and improve the crystallinity of the fiber, thus greatly improving the mechanical properties of the fiber[7]. However, the intercross-linked polyaryl structure of lignin leads to its high hydrophobicity. This also results in the fact that it is not very compatible with highly polar PVA fibers, which largely affects its application in reinforcing polar materials such as PVA. Lin et al. [8] successfully modified lignin with D-gluconic acid for the first time and prepared ML-reinforced PVA fibers. The results showed that the introduction of 5% ML from different reaction durations significantly enhanced the elongation and crystallinity of PVA fibers, yielding high-quality fibers. In order to further investigate the effect of ML on the properties of composite fibers with high lignin content, ML_{48} prepared by our previous work[8] was used as raw material to prepare the spinning dope at 30 wt. % lignin concentration in this work. The introduction of ML in 30 wt. % lignin/PVA fibers significantly improved the mechanical properties of PVA, reasulting in quality fibers with an optimal tenacity of 6. 60 cN/dtex and specific modulus of 192. 31 cN/dtex. This study will possibly expand the application of lignin and low-cost bio-based high-performance fibers in industry.

2 Experiments

2.1 Materials

PVA (Mw = 146 000–186 000, 99% hydrolysis, Sigma-Aldrich Co. , Ltd. , USA) and D-gluconic acid solution (49% – 53%, Sigma-Aldrich Co. , Ltd. , USA) were used as received. Organosolv softwood lignin (OL, 90%, Hubei Yunmei Co. , Ltd. , China) was purified before use. Dimethyl sulfoxide (DMSO, 99%), methanol (99. 5%) and acetone (99.5%) were obtained from Shanghai Titan Scientific Co. , Ltd. , China. Water was deionized before use.

2.2 Preparation of spinning dopes

OL was dispersed in acetone to remove low-molecular-weight fractions with vacuum filtration. Then, lignin was washed by deionized (DI) water for several times, and dried in a vacuum oven at 65 ℃ overnight before further usage. Modified lignin (ML) with reaction time of 48 h was synthesized according to the method described by our previous work[8] since it effectively enhanced the mechanical performance of PVA gel-spun fibers at low content of 5%. The steps for the preparation of spinning dopes are as follows. 10 g of PVA powder was dissolved in 100 mL of DMSO at 85 ℃ for 4 h until complete dissolution. 3 g of lignin powders were added to PVA dopes to yield 30% OL/PVA and 30% ML/PVA solution mixtures, respectively. All prepared spinning dopes were maintained at 65 ℃ before spinning.

2.3 Gel spinning

The gel spinning process is illustrated in Fig. 1. Each spinning dope (50 mL) was loaded into a stainless-steel syringe that was maintained at 65 ℃. Spinning dopes were extruded through a syringe needle (0. 7 mm inner diameter). The air gap between the syringe tip and the coagulation bath was 3–5 mm. The as-spun gel fibers were injected into acetone/methanol (85/15, v/v) coagulation solvents at 0–5 ℃ and then collected by a rotating take-up winder. The collected as-spun fibers were air drawn at room temperature (Fig. 1 (b)) before the immersion in the 85/15 acetone/methanol coagulation solvents for aging at – 20 ℃ for 24 h (Fig. 1 (c)). Finally, gel fibers went through silicone oil baths for four stages at high temperature (100 – 220 ℃) to achieve fully drawn fibers (Fig. 1 (d)).

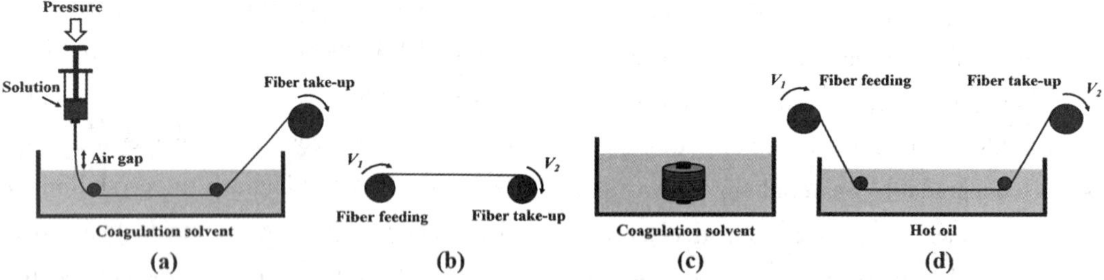

Fig. 1 Schematic illustration of gel spinning process with four steps: (a) as-spun gel fiber formation; (b) air drawing (V_2>V_1); (c) gel-aging and (d) multiple-stage fiber drawing (V_2>V_1)

2.4 Mechanical testing

The mechanical properties of each composite fiber were obtained by the XQ-1C testing system according to the ASTM D3379. In detail, the gauge length was set as 20 mm, and strain rate was 15 mm/min. The sample size for each type of fiber was at least 5.

2.5 Image analysis

To better investigate morphologies of the fiber fracture tips after the measurement of mechanical performance, each type of fiber was sputter coated with gold, and later imaged by a field emission scanning electron microscope (FESEM, S–4800, Hitachi, Japan). Scanning electron microscope (SEM) images were acquired at an accelerating voltage of 15 kV.

3 Results and Discussion

3.1 Effect of high content lignin fillers on fiber mechanical properties

The influence of high content lignin mixture on the mechanical performance of gel-spun PVA fibers will be elucidated in this section. In Fig. 2, the representative tensile stress-strain curves of 30% OL/PVA and 30% ML/PVA fibers are shown. It is obvious that all fibers exhibit approximately linear stress-strain curves, implying their elastic behavior.

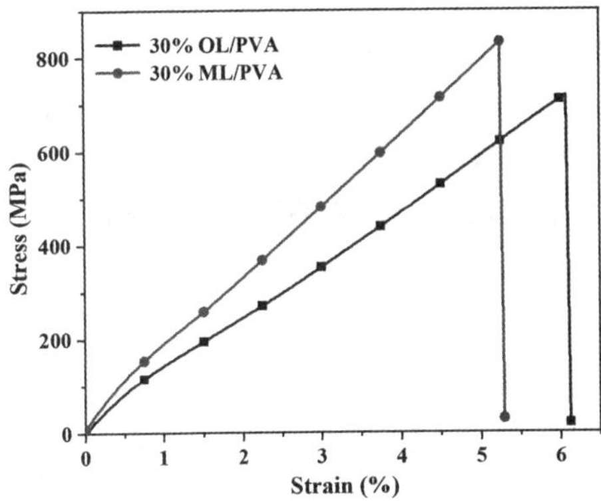

Fig. 2 Representative stress-strain curves of 30% OL/PVA and 30% ML/PVA fibers

In Fig. 3, the average tenacity and specific modulus of 30% OL/PVA fiber are 5.50 cN/dtex and 134.62 cN/dtex, respectively. With the filler changed from OL to ML, the mechanical properties of the composite fiber are significantly changed. In detail, the tenacity and specific modulus values of 30% ML/PVA composite fiber are 6.60 cN/dtex and 192.31 cN/dtex, respectively. Modulus is a reflection of the interatomic bonding force. In 30% ML/PVA fiber, more modified lignin molecules are involved in the interaction with PVA and form stronger intermolecular forces with it, resulting in less intra- and intermolecular forces in PVA.

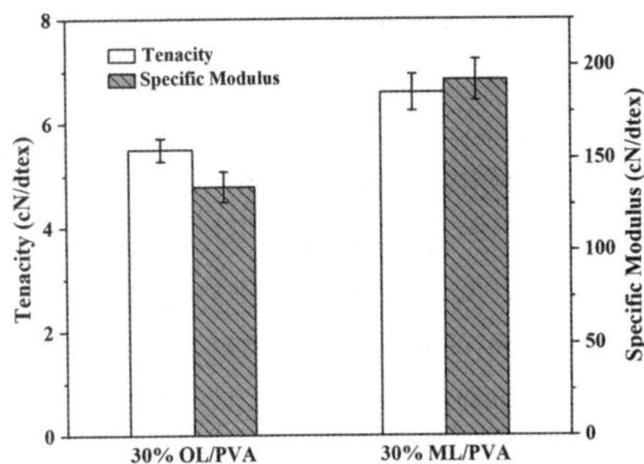

Fig. 3 Tenacity and specific modulus of 30% OL/PVA and 30% ML/PVA fibers

The effect of lignin modification on the toughness and elongation at the break of PVA fibers is demonstrated in Fig. 4. The elongation at break values of 30% OL/PVA and 30% ML/PVA fibers are 5.8% and 5.2%, respectively. The fiber elongation at break had a slightly decreasing trend, but still remained at a good level. It is worth noting that the toughness and tenacity of the composite fibers show a consistent trend of change. The toughness of 30% OL/PVA fiber is 17.20 J/g. With the incorporation of ML in PVA fibers, fiber toughness reaches an optimal value of 19.03 J/g for 30% ML/PVA fiber. Toughness reflects the energy transfer and dissipation under stress, which is related to the strength and plastic deformation. It is suggested that ML can reduce the brittleness of fibers with high lignin content and absorb and transmit load more efficiently than fibers containing OL, which may be the result of stronger interfacial bonding between ML and PVA.

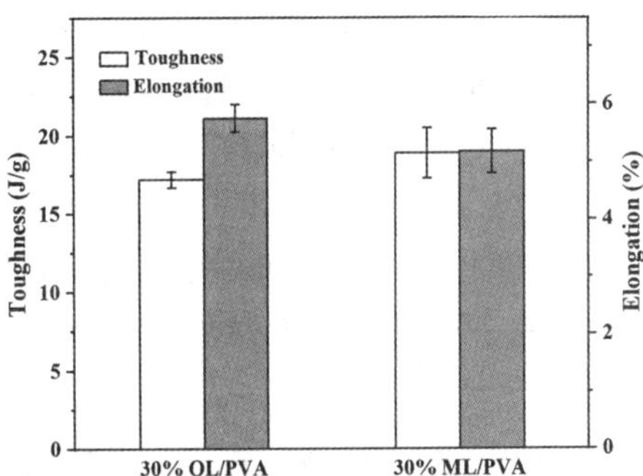

Fig. 4 Toughness and elongation of 30% OL/PVA and 30% ML/PVA fibers

3.2 Effect of high content lignin fillers on fiber macrostructure

Macrostructure of fiber fracture tips after mechanical testing has been demonstrated by SEM images (Fig. 5). The smooth surface and dense structure of the fracture tip of 30% OL/PVA fibers (Fig. 5 (a_1), 5 (a_2)) indicate that 30% unmodified lignin is well dispersed in PVA fibers, which is consistent with the homogeneous dispersion of the 30% OL/PVA spinning dope (Fig. 6 (a)). The fracture tips of the 30% ML/PVA fibers are characterized by significant plasticization (Fig. 5 (b_1), 5(b_2)), and the polymer matrix deforms plastically under stress, which indicating that the fibers have excellent toughness.

From the optical images of spinning dopes in Fig. 6, it can be observed that modified lignin has been over-solubilized in the spinning dope due to the enhanced hy-

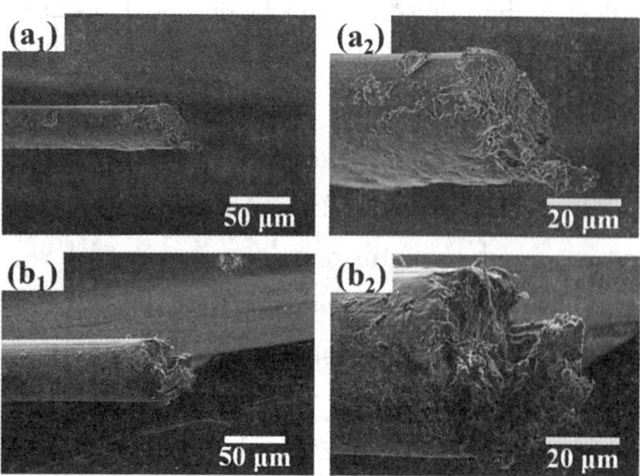

Fig. 5 SEM images of fracture tips of (a) 30% OL/PVA and (b) 30% ML/PVA fibers at (1) low and (2) high magnification

drogen bonding within the modified lignin molecule, and was dispersed as particles in the spinning dopes. In addition to the agglomeration of undissolved lignin within the fiber, some of the dissolved lignin may also precipitate out with solvent exchange in the coagulation bath and volatilization of the solvent in the hot oil drawing stages. The inhomogeneous dispersion of lignin in the polymer matrix greatly reduces the homogeneity of the fiber structure. Thus, the addition of modified lignin made the structure of 30% lignin/PVA fibers more inhomogeneous, but the plasticity characteristics also became more pronounced.

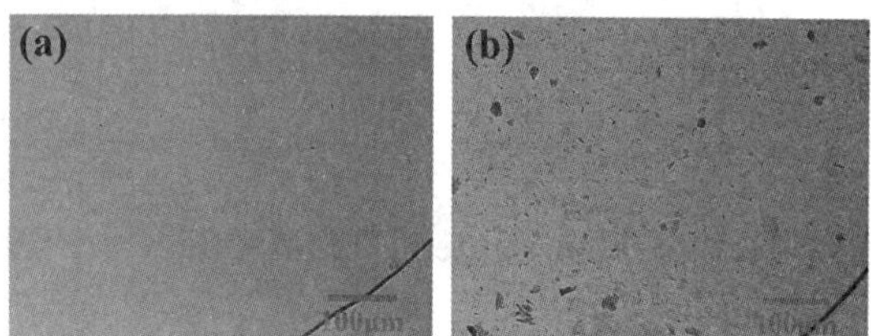

Fig. 6 Optical micrograph of (a) 30% OL/PVA and (b) 30% ML/PVA spinning dopes

4 Conclusions

In this work, we successfully fabricated 30% lignin/PVA gel-spun fibers with different fillers and investigated the effect of high content lignin fillers on their mechanical properties and fiber structures. The mechanical properties of 30% ML/PVA fiber are significantly improved compared with 30% OL/PVA fiber due to the strong interaction between ML and the PVA matrix. It had an optimal tenacity of 6.60 cN/dtex, specific modulus of 192.31 cN/dtex, and toughness of 19.03 J/g, respectively. These results strongly suggest that high content ML can significantly strengthen PVA fibers. However, there is still much room for improvement in the compatibility of lignin with PVA substrates. It is of great importance to develop an effective method that reduces the particle aggregation of lignin and improves the compatibility of lignin with different polymer matrices. This will better exploit the advantages of lignin as a bio-based resource and its high value application in the field of polymer reinforcement.

Acknowledgements

The project was funded by Shanghai Frontier Science Research Center for Modern Textiles, Donghua University. The authors are also deeply grateful for the financial support from the National Natural Science Foundation of China (No. 51903033) and Shanghai Sailing Program, China (No. 19YF1400800).

References

[1] TAVAKOLI J, TANG Y. Honey/PVA hybrid wound dressings with controlled release of antibiotics: structural, physico-mechanical and in-vitro biomedical studies[J]. Materials Science and Engineering: C, 2017, 77: 318-325.

[2] SARWAR M S, NIAZI M B K, JAHAN Z, et al. Preparation and characterization of PVA/nanocellulose/Ag nanocomposite films for antimicrobial food packaging[J]. Carbohydrate Polymers, 2018, 184: 453-464.

[3] MENG Y, ZHAO X, YE L. Construction of dual orientation crystalline structure in poly(vinyl alcohol)/graphene oxide nano-composite hydrogels and reinforcing mechanism[J]. Industrial & Engineering Chemistry Research, 2019, 58(25): 10908-10921.

[4] CHA W I, HYON S H, IKADA Y. Gel spinning of poly(vinyl alcohol) from dimethyl sulfoxide/water mixture[J]. Journal of Polymer Science Part B: Polymer Physics, 1994, 32(2): 297-304.

[5] TANIGAMI T, NAKASHIMA Y, MURASE K, et al. High strength and high modulus poly(vinyl alcohol) by the gel ageing method[J]. Journal of Materials Science Letters, 1995, 30(20): 5110-5120.

[6] CHENG Y, LIN J, ZHENG Y, et al. High-performance gel-spun poly(vinyl alcohol) fibers reinforced by organosolv lignin-graft-poly(acrylic acid)[J]. Industrial & Engineering Chemistry Research, 2022, 61(17): 6037-6051.

[7] LU C, BLACKWELL C, REN Q, et al. Effect of the coagulation bath on the structure and mechanical properties of gel-spun lignin/poly(vinyl alcohol) fibers[J]. ACS Sus-

tainable Chemistry & Engineering, 2017, 5(4): 2949-2959.

[8] LIN J, CHENG Y, LI Z, et al. Synthesis of modified lignin as an antiplasticizer for strengthening poly(vinyl alcohol)-lignin interactions toward quality gel-spun fibers[J]. ACS Applied Polymer Materials, 2022, 4(3): 1595-1607.

Preparation and Application of Fructus Cnidii Oil/β-CD Microcapsules

WANG Haizheng[1,2], **GUO Lamei**[1,2], **GAO Jing**[1,2*], **WANG Lu**[1,2]

[1] *Key Laboratory of Biomedical Textile Materials & Technology in Textile Industry, Shanghai 201620, China*

[2] *College of Textiles, Donghua University, Shanghai 201620, China*

* *Corresponding author's email*: gao2001jing@ dhu. edu. cn

Abstract: The essential oils show amazing properties in various domains. They also present the disadvantage of chemical instability in the presence of temperature, oxygen, and light. In order to improve the long-term effectiveness of Fructus Cnidii essential oil, a Fructus Cnidii oil/β-cyclodextrin microcapsule was prepared using the aqueous solution stirring method in this study. The morphology, particle size, thermal stability and sustained release effect of the Fructus Cnidii essential oil/β-cyclodextrin microcapsules were characterized, respectively. The results showed that the microcapsules had a regular external shape and uniform sizes, with an average particle of 3.09 μm, and the use of β-CD prolonged the release time with excellent thermal stability. They can continuously release the active ingredients of the oil at normal temperature in 180 d. In order to further examine the application effect of the Fructus Cnidii essential oil/β-cyclodextrin microcapsules, they were loaded on knitted fabrics by the soaking method. Air permeability of the finished fabric decreased slightly, but it could still meet the wearability of the fabrics well. The microcapsules prepared in this study can find applications in domains as allergy prevention, medicine, and other related products.

Keywords: Functional fabric; Microcapsule; β-CD; Fructus Cnidii oil; Chemical instability

1 Introduction

Fructus Cnidii essential oil possesses various medicinal properties including allergy prevention, muscle relaxation spasm, prevention of fungal infection, etc. However, the use of Fructus Cnidii oil has been limited due to its low solubility in water and easy oxidation in air.

Microencapsulation is a technique to package solids, liquids or gaseous substances within polymer coatings, and the contents would release under specific conditions[1]. The packaged materials are normally called core materials. Since the core-shell structure of microcapsules, they were found having great advantages in protecting the active agents from the environment and controlling release of core materials.

Cyclodextrin (CD) is the cyclic oligosaccharide composed of glucopyranose units linked by α-1,4 glycosidic bond, containing six (α-CD), seven (β-CD), or eight (γ-CD) glucopyranose, respectively[2]. The presence of hollow cavities in the molecules of cyclodextrin provides a channel to attract some kinds of gases, liquids or solids through van der Waals force, which contributes to the formation of inclusion complexes. Higher stability and lower price are found in β-CD, leading to the extensive application in industry[3]. In addition, it has been proved that β-CD could be used as wall material of microcapsules to enhance the physicochemical properties of core materials and offer a protection to the active ingredients of inclusion[4].

In this work, a Fructus Cnidii oil/β-cyclodextrin microcapsule was prepared using the aqueous solution stirring method and characterized to confirm the formation of composite inclusion. And the microcapsules were coated on fabrics to propose the practical bases for its textile's application.

2 Experiments

2.1 Materials

β-CD was purchased from MACKLIN. Fructus Cnidii essential oil was offered by Chinese Fragrant Plants Co., Ltd. (Jilin, China). Waterborne polyurethane (WPU) with a mass fraction concentration of 60% in water was supplied by MACKLIN. Sodium dodecyl benzene sulfonate (SDBS) was obtained from ALADDIN. Well-prepared knitted fabrics were purchased from Jihua Trading Co., Ltd. (Nanping, China).

2.2 Preparation of microcapsules

Fructus Cnidii essential oil was firstly added to an aqueous solution of β-CD (4%). The mass ratio of oil to β-CD was 1 : 6. The mixtures were kept under stirring at 50 ℃ for 4 h. After being refrigerated at 2 ℃ for 12 h, the mixtures were filtered. Finally, the powders were placed into a vacuum oven at 50 ℃ for 6 h.

2.3 Fabric finishing process

Weigh a certain amount of microcapsules (40 g/L) and SDBS (2 g/L), dissolve them in ultrapure water, and stir them adequately. Thereafter, 100 g/L WPU was added into the mixtures. Fabrics were dipped into the mixtures at a bath ratio (the ratio of ultrapure water weight to fabric weight) of 20 : 1. Maceration was conducted at 45 ℃ for 90 min. Afterwards, the finished fabrics were oven-dried at 80 ℃.

2.4 Morphology observation

The morphology of microcapsules and finished fabrics were observed using scanning electron microscope (SEM, SU 8200). The particle size of microcapsules was analyzed by Zetasizer Nano ZS 90 (Malvern, UK).

2.5 Fourier transform infrared (FTIR) analysis

FTIR spectra of β-CD, Fructus Cnidii essential oil, and microcapsules were obtained using Thermo Nicolet NEXUS 670 FTIR (wave numbers 400–4 000 cm^{-1}) at room temperature.

2.6 Thermal stability

Thermal stability was carried out with a Perkin Elmer-TGA 8000 at the heating range from 50 ℃ to 600 ℃ at the rate of 10 ℃/min.

2.7 Content of essential oil in microcapsules and release test

A certain amount of microcapsules were added to 50 mL ethanol. After sonication for 4 h, the solution was centrifuged and the precipitation was removed. The microcapsules were placed in an oven at 37 ℃ with samples extracted at regular intervals. Fructus Cnidii essential oil was extracted following the method above. The absorbance of the supernatant at 308 nm was measured using the UV-visible spectrophotometer (Metash X-6, Shanghai). Each experiment was triplicated. Then the oil release at different times and the content of oil in microcapsules were obtained according to the standard curve.

2.8 Air permeability of the fabrics

Air permeability tests of the untreated and finished fabrics were performed using an air permeability tester (YG461E) according to GB/T 5453—1997. Data analysis was performed with Origin software (version 2022).

3 Results and Discussion

3.1 SEM and particle size of the microcapsules

The morphology of β-CD and microcapsules is presented in Figs. 1 (a) and 1 (b), respectively. β-CD molecules appeared as discrete particles with irregular dimensional structures, which had concordant results with Fauziah[5]. Nevertheless, the microcapsules indicated

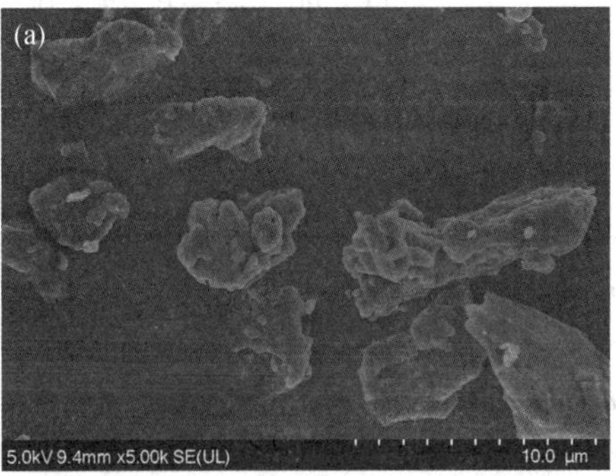

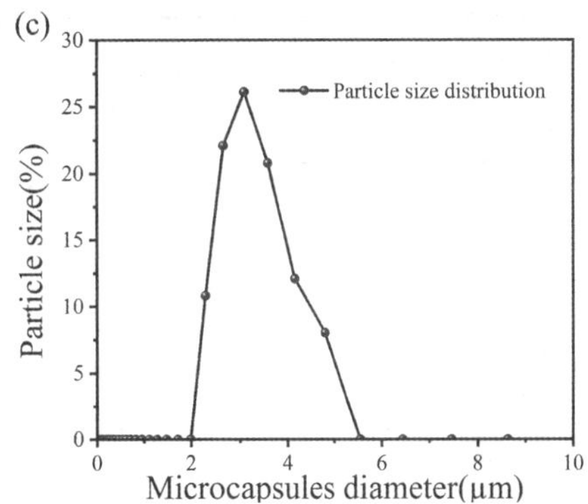

Fig. 1 SEM of (a) β-CD and (b) microcapsules; (c) particle size of the microcapsules

a sheet-like structure with smooth surface[6]. The different shapes of β-CD and microcapsules suggested the formation of inclusion complex between β-CD and Fructus Cnidii essential oil.

The particle size was shown in Fig. 1 (c). The size of the microcapsules ranged from 2 μm to 5.56 μm, and the particles of the size at 3.09 μm accounted for 26.1%, following a normal distribution.

3.2 FTIR and TG

Infrared spectra of Fructus Cnidii essential oil, β-CD and microcapsules were shown in Fig. 2 (a). The spectrum of β-CD had C—O stretching at 1 160 cm^{-1}, the absorption peak at 1 030 cm^{-1} showed C—O—C stretching vibrations, and the band at 3 410 cm^{-1} represented the vibration of stretching of —OH groups[6]. The stretching vibration of unsaturated C=C was observed at 3 008.55 cm^{-1} on the spectrum of Fructus Cnidii essential oil. The symmetric stretching vibration of methyl occurred at 2 854.41 cm^{-1}. There were also present the same peaks for the microcapsules, with a much lower intensity peak, which indicated the synthesis of microcapsules. The bands of microcapsules and β-CD were similar, but the intensity of the specific peaks changed substantially. Additionally, a significant broadening of 3 410 cm^{-1} spectra was observed for the microcapsules, which presented the hydrogen bonds form between β-CD and Fructus Cnidii essential oil. Formation of microcapsules may be thus evidenced.

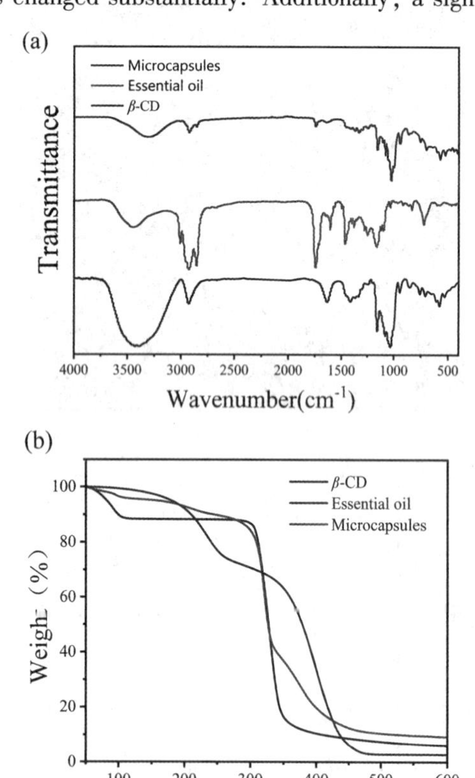

Fig. 2 (a) FT-IR spectra and (b) thermogravimetry curves of β-CD, essential oil, and microcapsules

The results of thermogravimetry were shown in Fig. 2 (b). At 100 ℃, the first stage of weight loss of β-cyclodextrin occurred as a result of water, and the major loss occurred at 300 ℃[4]. The degradation temperature of β-CD in microcapsules moved from 300 ℃ to 275 ℃. And the significant difference in TG curves between β-CD and microcapsules revealed the interaction between the guest molecules and β-CD[2].

3.3 Encapsulation efficiency and release property

The encapsulation yield was an important index for assessing microcapsules embedding. And the encapsulation rate of Fructus Cnidii essential oil microcapsules was 32.09% when the mass ratio of oil to β-CD was 1 : 6. The efficiency value was similar to the result reported by Abarca et al.[7].

The release profile of Fructus Cnidii essential oil was shown in Fig. 3. And the release data were fitted to the Ritger-Peppas equation to describe the release mechanism. The release exponent n, which is 0.345 7, lower than 0.450 0, represents that the release of oil from microcapsules corresponds to Fickian diffusion mechanism[8]. Additionally, the results of microcapsules after 180 d shows that there is 3.73% oil remain, which indicates the long-term release of microcapsules.

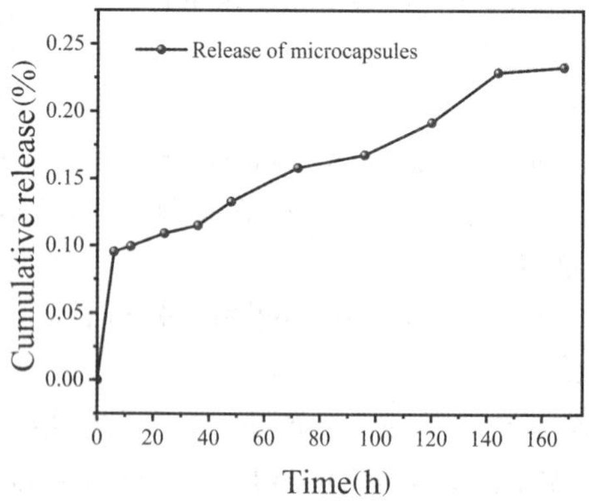

Fig. 3 Cumulative release profile of Fructus Cnidii essential oil at 37 ℃

3.3 Characterization of finished fabric

Figs. 4 (a) and 4(b) showed the SEM images of the fabrics before and after coating treatment. As shown in Fig. 4 (b), the surface of treated knitted fabric was uniformly covered by the finishing agent, which contained adhesive and microcapsules. Results of air permeability of the the treated fabric presented 687.47 mm/s, 44.2% off compared to the untreated fabric (1 233.62 mm/s). Yet, it could still meet the wearability of the fabrics well.

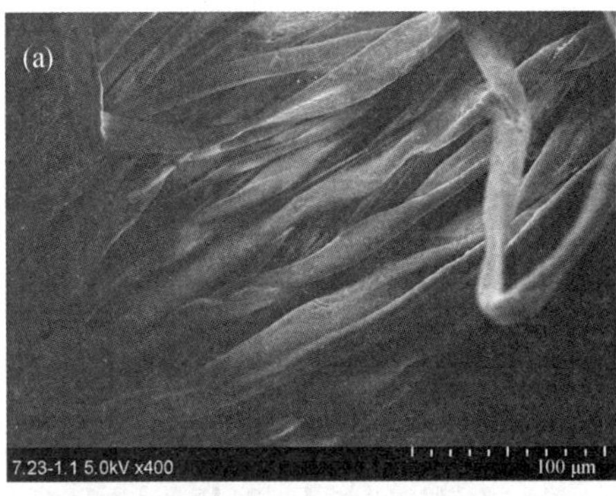

Fig. 4 SEM of (a) original and (b) treated fabric

Results above proved that the synthesized microcapsules could be used as a fabric finishing agent to endow textiles with various functions, without destroying basic properties of fabrics.

4 Conclusions

The results of this study clearly demonstrate that β-cyclodextrin can form stable inclusion complexes with Fructus Cnidii essential oil. FTIR and TGA observation have conformed the formation of microcapsules. Moreover, the release study suggests that β-CD provides the protection for Fructus Cnidii essential oil, which supports 180 d of deliver. These findings suggest that β-CD can be used to improve the oxidative stability and long-term effectiveness of essential oils. In addition, further studies on characterization of the finished fabric show that the prepared microcapsules can be applicated to fabricate multifunctional fabrics. This work is believed to promote the application of microcapsules in textile industry, such as allergy prevention textiles, sleep promotion textiles, and other domains.

References

[1] LI L, AU W, HUA T, et al. Improvement in antibacterial activity of moxa oil containing gelatin-arabic gum microcapsules[J]. Textile Research Journal, 2013, 83(12): 1236-1241.

[2] ABBEHAUSEN C, FORMIGA A, SABADINI E, et al. A beta-cyclodextrin/siloxane hybrid polymer: synthesis, characterization and inclusion complexes[J]. Journal of the Brazilian Chemical Society 2010, 21(10): 1867-1876.

[3] ZHANG Y, ZHANG H, WANG F, et al. Preparation and properties of ginger essential oil beta-cyclodextrin/chitosan inclusion complexes[J]. Coatings, 2018, 8(9): 305-317.

[4] MUNHUWEYI K, CALEB O J, VAN REENEN A J, et al. Physical and antifungal properties of β-cyclodextrin microcapsules and nanofibre films containing cinnamon and oregano essential oils[J]. LWT, 2018, 87(1): 413-422.

[5] FAUZIAH C I, ZAIBUNNISA A H, OSMAN H, et al. Thermal analysis and surface morphology study of cholesterol: β-cyclodextrin inclusion complex[J]. Advanced Materials Research, 2013, 812(9): 221-225.

[6] ZHANG W, LI X, YU T, et al. Preparation, physicochemical characterization and release behavior of the inclusion complex of trans-anethole and β-cyclodextrin[J]. Food Research International, 2015, 74(8): 55-62.

[7] ABARCA R L, RODRíGUEZ F J, GUARDA A, et al. Characterization of beta-cyclodextrin inclusion complexes containing an essential oil component[J]. Food Chemistry, 2016, 196(7): 968-975.

[8] IRFAN S A, RAZALI R, KUSHAARI K, et al. A review of mathematical modeling and simulation of controlled-release fertilizers[J]. Journal of Controlled Release, 2018, 271(3): 45-54.

Polyamide Copper Plating Based on $CaBr_2$ Modification

SHI Haixia[1], **WANG Siyan**[1], **LIU Ruohua**[1*]

Department of Technical Textiles, College of Textiles, Donghua University, Shanghai 201620, China

* *Corresponding author's email*: liuruohua@dhu.edu.cn.

Abstract: In recent years, polyamide copper-coated fabric has received extensive attention from researchers. Due to the smooth surface of polyamide, it is necessary to roughen the polyamide. In addition, in the conventional process, the catalyst layer is generally prepared by the sodium borohydride method which causes serious environmental pollution. This paper presents a new process of polyamide copper-plated fabric. Based on the high sensitivity of silver bromide to ultraviolet light, calcium bromide is used as a roughening reagent and the catalytic layer is prepared by environmentally friendly ultraviolet light reduction to obtain polyamide copper-coated fabric. The appropriate calcium treatment time is determined after several experiments such as fabric strength tests, adhesion tests, and scanning electron microscope (SEM) analysis tests. Also, the new process is compared with the traditional process in terms of the copper plating quality. The results of these experiments indicate that the suitable treatment experimental time is 30 s. The average resistance of our new copper-plated fabric is 15 Ω, compared to 5 Ω for the conventional method. The data shows that our new method is not far behind the classical method, but has high research value.

Keywords: Calcium ion; Roughening; Fabric strength; Electroless copper plating

1 Introduction

Polyamide copper-plated fabric is the perfect use of electroless copper plating technology to make polyamide (PA) fabric have conductive properties[1]. Electroless copper plating technology is a way of depositing copper ions on the substrate material with catalytic properties. Electroless copper plating of non-conductive materials such as textiles, requires the use of a catalyst that is coated onto the surface of the substrate. This catalyst initiates a redox reaction which allows copper to be reduced from ions in the electrolyte at a regular rate with formaldehyde as a reducing agent, resulting in a uniform metallic coating. The deposited copper coating then catalyzes subsequent copper plating, and hence the process is often described as auto-catalytic. Obtaining a uniform copper deposit is heavily dependent upon the efficacy of the catalyst. Silver is the latest new popular catalyst. In the traditional electroless plating process, sodium borohydride ($NaBH_4$) reduction is often used to deposit a thin layer of silver. However, $NaBH_4$ is toxic and will cause great harm to the environment. Thus, it is necessary to develop an environmentally friendly polyamide copper plating process.

In this paper, calcium bromide ($CaBr_2$) was utilized as a roughening reagent for polyamide in the pretreatment process of polyamide. On the one hand, the surface of polyamide is too smooth to attach particles. It is the calcium ion that can roughen the surface of polyamide[2-3]. On the other hand, the bromide ion is easily combined with the silver ion to form silver bromide which is sensitive to ultraviolet (UV) light[4]. In this paper, the silver layer was obtained on the surface of PA by ultraviolet light, not $NaBH_4$. Additionally, roughening treatment of calcium ions will destroy the tissue structure of the fibers and affect the original properties[5-6]. In particular, the tensile breaking load will be greatly reduced. In the treatment process, high concentration and quick time are commonly used to achieve the ideal effect that the fiber surface is rough and the internal structure is not damaged. In this paper, we would like to study the effect of new copper-plated fabric technology. The appropriate roughening time is explored through fabric tensile experiments and iron powder adhesion experiments. After the copper plating is completed, the advantages and disadvantages of the method are analyzed in comparison with the traditional sodium borohydride method.

2 Experiments

2.1 Materials

PA fabrics were purchased from Wuxi Jinkun Weaving Co., Ltd. (Wuxi, China)

Calcium bromide, anhydrous ethanol, iron powder, silver nitrate, ammonia water, silver nitrate, sodium hydroxide, copper sulfate pentahydrate, EDTA, potassium ferrocyanide, 2,2′-bipyridine, formaldehyde, and potassium sodium tartrate were sourced from Sinopharm Chemical Reagent Co. Ltd. and lab-made deionized water.

2.2 Experimental methods

2.2.1 Fiber degreasing

The PA fabric was put into a 10 g/L sodium dodecylbenzene sulfonate solution, and the solution was placed in an ultrasonic cleaning instrument and shaken for 30 min. The PA fabric was taken out and cleaned with deionized water for three times, then put into the oven, and baked to constant weight.

2.2.2 Roughening

Put a batch of PA samples into the calcium bromide solution for roughening. The roughening time was 5 min, 4 min, 3 min, 2 min, 1 min, 50 s, 40 s, 30 s, 20 s, and 10 s, respectively.

2.2.3 Sensitization

One group of PA samples was immersed in 2 g/L silver ammonia solution for 10 min, and then placed in a UV light box for 20 min irradiation; another group of samples was placed in 25 mL of 20 g/L sodium borohydride solution and taken out after 5 min[7].

2.2.4 Copper plating

The copper plating solution was prepared by mixing sodium hydroxide, copper sulfate pentahydrate, EDTA, potassium ferrocyanide, 2, 2′-bipyridine, formaldehyde, potassium, and sodium tartrate[8]. Put the samples after the above series of treatments into the copper plating solution for 30 min.

2.3 Testing

2.3.1 Tensile testing on the fabric

The tensile load of the roughened polyamide fabric was tested by the fabric load tester. The fabric samples were cut uniformly by the strip method with the same experimental parameters for each test. Five samples were taken for each test to obtain an average value. After the test, the load loss rate of the fabric could be expressed by the load loss rate. The load loss rate can be obtained by Eq. (1).

$$w_0 = \frac{N_0 - N_1}{N_0} \times 100\% \tag{1}$$

where w_0 is the load loss rate(%); N_0 is the load of the fabric before roughening (N); N_1 is the load after roughening (N).

2.3.2 Iron powder adhesion test

After roughening, the surface of polyamide had more areas to combine with other particles. The roughening effect could be directly reflected by the adhesion degree of iron powders. The specific operation method was: after the roughening treatment, spread the iron powders evenly on the polyamide fabric, and then pressed the fabric with a heavy object for 15 min. After pressing, put the fabric into a constant temperature oven to dry, and then washed with deionized water 200 mL for 10 min. The iron powders that did not adhere to the fabric were washed and weighed m_2. Then put the fabric into the ultrasonic cleaner to clean for 10 min, to wash away the iron powder that had been adhered to but not firmly adhered to the fabric, and weighed m_3. The weight of the fabric before roughening is m_1.

The adhesion rate and the weight loss rate of iron powders could be obtained by Eqs. (2) and (3).

$$w_1 = \frac{m_2 - m_1}{m_1} \times 100\% \tag{2}$$

$$w_2 = \frac{m_2 - m_3}{m_2} \times 100\% \tag{3}$$

where w_1 is the adhesion rate(%); w_2 is the weight loss rate(%); m_1 is the initial weight (g); m_2 is the weight after washing (g); m_3 is the weight after the ultrasound (g).

2.3.3 SEM test

The samples obtained after roughening could be used to prepare crystal spray samples. Then, the microstructure of the polyamide surface was observed by SEM.

2.3.4 Conductivity

An electric meter was used to measure the resistance on the left and right diagonals of the fabric after copper plating under different conditions. Measured both the front and the back, and took the average value.

3 Results and Discussion

3.1 Roughening effect on PA fabric tensile properties

Since it was an exploratory experiment, we first conducted experiments with a roughening time of 1–5 min. The fabric tensile test can reflect the effect of roughening time on the polyamide load. Therefore, the purpose of these experiments went to look for a suitable roughening time when the PA fabric load loss rate did not exceed 10% through the tensile test.

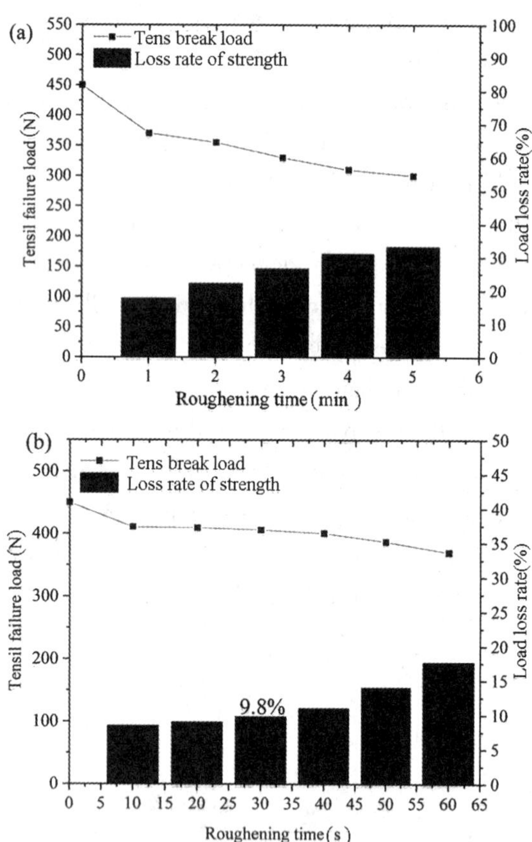

Fig. 1 Curves of polyamide load with the changing roughening time: (a) 1–5 min; (b) 0–60 s

It could be seen from Fig. 1 (a) that with the increase of time, the load of PA fabric gradually decreased, and the load loss rate increased. It was clear that with the extension of time, the speed of this change was gradually accelerating. From the data analyses, when the roughening time was 1 min, the load loss rate was 14.9%. To ensure that the PA fabric has excellent mechanical properties, we hope to control the load loss rate of the fabric after roughening to less than 10%. Therefore, it was necessary to continue to shorten the roughening time. The change in the load of PA fabric after roughening by calcium bromide for 0–60 s was depicted in Fig. 1 (b). It can be seen that the load loss rate is mostly small within 10–60 s. Moreover, the load loss rate reached 9.8% at 30 s, lower than 10%.

3.2 Roughening effect on iron powder adhesion

The critical purpose of roughening is to make PA fabrics obtain excellent adhesion properties. Through the previous fabric tensile experiments, a reasonable roughening time has been found. However, it is still necessary to conduct experiments to determine whether PA fabric obtains excellent adhesion properties. In this experiment, the roughening effect which was intuitively reflected by the adhesion of iron powder was illustrated in Fig. 2.

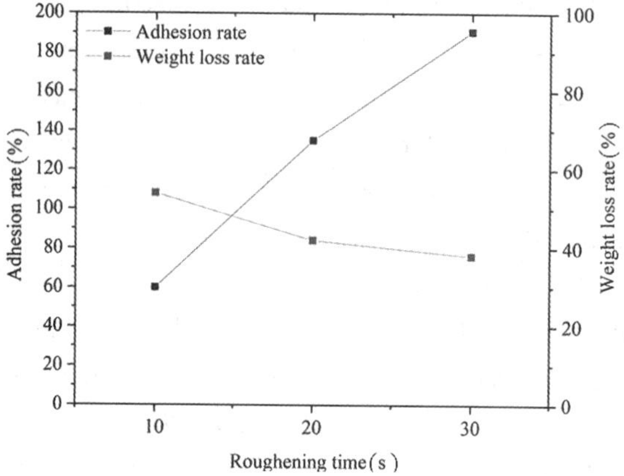

Fig. 2 Curves of PA fabric adhesion property

As shown in Fig. 2, with the increase in roughening time, the adhesion rate of iron powder gradually increased and the weight loss rate decreased. It was demonstrated that the longer the roughening time was, the stronger the ability of PA fabric to absorb particles. Under the condition that the load complies with the requirement, when the roughening time was 30 s, the adhesion rate of iron powder was the highest and the weight loss rate was the lowest. Therefore, we selected the roughening time as 30 s.

3.3 Analysis of SEM results

In this experiment, the roughening effect of PA fabrics was observed by an SEM. Micromorphology of PA fabric surface was displayed in Fig. 3. Obviously, the surface of the unroughened PA fabric was smooth and flat. After being roughened by $CaBr_2$, the PA fabric became bumpy, facilitating more particles' adhesion. With the increase in roughening time, the roughening effect on the PA fabric surface became stronger and stronger.

When the roughening time was 30 s, it could be observed that the roughening effect of polyamide was the clearest. It was consistent with the previous conclusion that the iron powder adhered the most in 30 s.

After undertaking the data of these experiments, we concluded that the reasonable roughening time was 30 s.

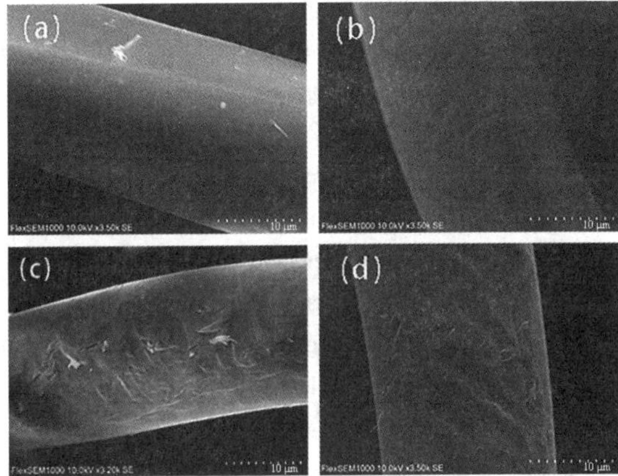

Fig. 3 SEM microscopic picture of polyamide: (a) 0 s; (b) 10 s; (c) 20 s; (d) 30 s

3.4 Copper plating effect comparison

This experiment mainly compared our new PA copper-coated fabric with the PA copper-coated fabric prepared by the traditional process. In addition, by comparing the electrical conductivity, we analyzed the advantages and disadvantages of our new fabric.

As shown in Fig. 4, all three methods can obtain copper-plated fabrics. A represents no sensitization treatment; B is our new method for obtaining silver catalyst by ultraviolet light reduction of silver bromide; C is the traditional method of reducing silver catalyst by $NaBH_4$. The resistance of copper-coated fabrics without the sensitization treatment was 285 Ω. Our new method was 15 Ω, and the resistance of the traditional sodium borohydride method was 5 Ω. Copper-coated fabrics can also be obtained without the sensitization process. The reason is that copper has an autocatalytic effect, but the quality of copper plating that only depends on the autocatalytic effect of copper will be poor. Therefore, without the sensitization process, the conductivity is poor[6]. It can be seen from the comparison that although our newly copper-plating process is not as good as the traditional process, it can significantly improve the conductivity of the copper-plated fabric.

In addition, the copper plating process, which takes calcium bromide as the coarsening agent and uses ultra-

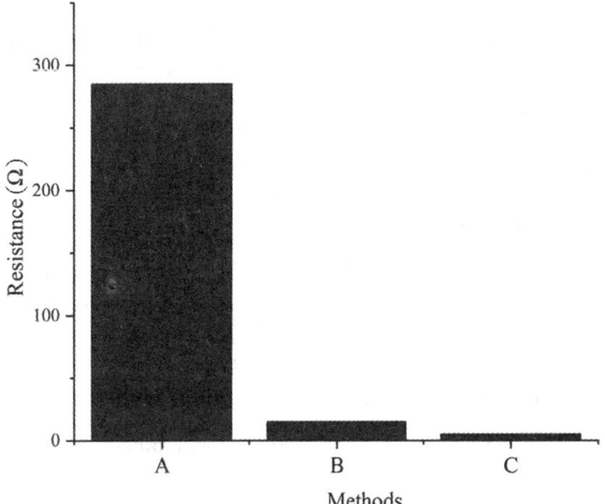

Fig. 4 Comparison of different reduction methods (A: no treatment; B: UV; C: $NaBH_4$)

violet light to reduce the catalytic layer of silver bromide, is a new environmentally friendly copper plating process route, which can greatly reduce environmental pollution. It meets the need for the green development of industry and provides a new idea for the green development of electroless plating technology.

4 Conclusions

According to the sensitivity of silver bromide to ultraviolet light, we propose a new process route of polyamide copper plating: calcium bromide is used as a roughening agent in the pretreatment process, which has a variety of applications. Firstly, calcium ion roughening enhanced the interfacial binding ability of the polyamide surface. In addition, bromine ion combines with silver ion to form silver bromide. And silver bromide is easily reduced by UV light. It is a physical method that does not require the addition of any chemical reagents. Through many experiments, we get the ideal time of calcium roughening 30 s. Compared with the traditional copper plating process, it is concluded that our method can also significantly improve the electrical conductivity of the fabric, which is not different from the traditional copper plating process. At the same time, our method provides a relatively novel and green environmental protection copper-plated fabric production route.

Acknowledgment

This research was funded by the National Natural Science Foundation of China (No. 51603039), the Fundamental Research Funds for the Central Universities, the Key Laboratory of Textile Science and Technology (Donghua University), Ministry of Education, and the Initial Research Funds for Young Teachers of Donghua University.

References

[1] XIONG L L, LI X M, WANG T, et al. Electromagnetic shielding fabric of nylon 66 composite plating[J]. Surface Technology, 2020, 49(1): 180-186.

[2] YONG K. Progress in modification of PA[C]//Limas, Peru: 2018 6th ICPESM International Conference on Educational Research, Leisure, Sport and Tourism (EIST 2018), 2018-12-27.

[3] GLEISSNER C, BIERMAIER C, BECHTOLD T, et al. Complexation-mediated surface modification of polyamide-66 textile to enhance electroless copper deposition[J]. Materials Chemistry and Physics, 2022, 288.

[4] WANG J Q, ZHANG X Z, SUN L P, et al. Laser-induced deposition of silver nanoparticles films and microstructures onto glass substrates[J]. Chinese Journal of Lasers, 2011, 38: 180-184.

[5] GYUNG G K, JOUNG A K, JI H K, et al. Metallization of the polymer through a novel surface modification applying a photocatalytic reaction[J]. Surface & Coatings Technology, 2006, 201(6): 3761-3766.

[6] ALI A, AHETI V, VIK M, et al. Copper electroless plating of cotton fabrics after surface activation with deposition of silver and copper nanoparticles[J]. Surface and Coatings Technology, 2006, 201: 3761-3766.

[7] HUANG R. The research of sulfhydryl/silver catalytic treatment of the PA6 before electroless copper plating[D]. Shanghai: Donghua University, 2013.

[8] LIU Y X. Study on preparation of conductive material by polyphenol surface modification on fabric[D]. Shanghai: Donghua University, 2021.

Machine Learning Empowered Gesture Recognition Glove

LUO Jun[1], QIAN Yuze[1], ZHUANG Qinliang[1*], ZHANG Kun[1*]

[1] *Key Laboratory of Textile Science & Technology (Ministry of Education), College of Textiles, Donghua University, Shanghai 201620, China*

* *Corresponding author's email*: kun.zhang@dhu.edu.cn

Abstract: Recently, gesture recognition technology has attracted increasing attention, because it provides another means of information exchange in some special occasions, especially for auditory impaired individuals. At present, the fusion of sensor signals and artificial intelligence algorithms is the mainstream trend of gesture recognition technology. Therefore, the machine learning empowered gesture recognition glove was designed and a flexible strain sensor with sandwich structure, which has high sensitivity and good cycle stability, was fabricated. After the sensors are configured in the knitted gloves, the smart gloves can respond to different gestures. And according to the representation characteristics and recognition targets of sampled signal data, we explore a segmented processing method of dynamic gesture recognition based on Logit Adaboost algorithm. After classification training, the recognition accuracy of smart gloves can reach 97%.

Keywords: Gesture recognition; Flexible sensor; Machine learning; Logit Adaboost algorithm

1 Introduction

As a method of nonverbal communication[1], standardized gestures play an important role in the transmission of message on specific occasions, such as communication between auditory impaired individuals[2] and information interaction in AR/VR environments[3] or military field[4]. However, without long-term systematic learning, it is difficult for common people to understand the meaning of gestures[5]. At present, gesture recognition technology offers a feasible solution to this communication barrier, which can effectively judge the specific movements of the hand.

Machine vision based image taking and processing has received extensive attention as a means of gesture recognition due to simple equipment requirements[6]. However, it is troubled by a series of problems in the process of application, including complex operation mechanism and variable lighting condition[7]. Accelerometers and gyroscopes are widely used in commercial gesture recognition devices, which can track the direction of the hand movement in three-dimensional space[8]. But, rigid materials and excessive volume limit the application scenarios of these devices, which bring inconvenience to users. Benefiting from the rapid advancement of flexible electronics and artificial intelligence, wearable flexible data gloves based on flexible sensors have been developed. Common flexible sensors for gesture recognition include piezoresistive sensors[9], capacitive sensors[10], piezoelectric sensors[11], triboelectric sensors[12], and EMG sensor arrays[13]. Through the electromechanical performance of flexible sensors, the deformation degree of the hand can be reflected correspondingly.

Furthermore, the classification of sensor signals and the extraction of gesture features are the key factors to determine the accuracy of gesture recognition. It generally requires intelligent algorithms to train sensor signals and generate appropriate classifiers[14]. However, due to the differences in gesture behavior patterns of individuals, the generalization ability of the trained classifiers is low if the sample capacity is limited[15]. A common approach is to increase the number of people participating in algorithm training, which is time- and energy-consuming[16]. Therefore, there is a new idea to remove the irrelevant factors of sensor signals as much as possible through special designs.

In this study, we report the machine-learning empowered gesture recognition glove. Flexible piezoresistive strain sensors with high sensitivity and cycle stability were configured in knitted gloves. Then, in order to enhance the generalization ability of gesture recognition model, a multi segment classification model of dynamic gesture recognition based on Logit AdaBoost algorithm was explored. Static signals collected at intervals of gesture changes are usually considered invalid signals because of unobvious characteristics and uncertain duration. So the model distinguished static signals from effective dynamic signals due to finger movements. And subsequence training and classification were only related to dynamic signals, avoiding misjudgment caused by static signals.

2 Experiments

2.1 Materials

Reduced graphene oxide (r-GO) was supplied by Qingdao university, China. Polydimethylsiloxane(PDMS) was purchased from Microflu Co. Ltd (Changzhou, China). *N*, *N*-Dimethylformamide (DMF, 99.5%) was bought from Aladdin Co. Ltd(Shanghai, China).

2.2 Fabrication of flexible strain sensor

The r-GO dispersion (6%) was prepared by two steps. The reduced graphene oxide powders were firstly dispersed in *N*, *N*-dimethylformamide by mechanical shearing for 1 h at 3 500 r/min, and secondly, it was necessary to remove air bubbles of the solution using ultrasonic for 5 min. Then, the solution was poured onto the PDMS substrate (the volume ratio of PDMS matrix to curing agent was 10 : 1 and the curing condition was 70 ℃ for 1 h) and heated at 50 ℃ for 3 h using a hot plate. After the solvent evaporates completely, there was a composite film containing the r-GO layer and PDMS substrate. Finally, two copper wires were adhered to both sides of the r-GO layer with silver paste, which were used as the electrodes of the strain sensor. And PDMS was poured and cured on the top of the composite film, resulting in a sandwich structure. The size of the final sample is about 35 mm×4 mm×1 mm.

2.3 Characterization

A scanning electron microscope (SEM, S-4800) was used to characterize the surface morphology of the r-GO film. In order to test the sensing performance of the strain sensors, the flexible strain sensors were fixed between the metallic clamps of a stretching machine (ZQ-CI701G, 1 000 N), and a digital source-meter (Keithley 2400) was used to measure the resistance of the sensors continuously in the process of stretching. The stretching speed was maintained at 30 mm/min.

2.4 Preparation of data gloves and data acquisition

The configuration of strain sensors in data gloves is shown in Fig. 1. To monitor the bending degree of fingers, the sensors were attached to the corresponding finger joint position of knitted gloves. And the resistance signal acquisition of sensors was realized by the data acquisition card (Ni-6211). In the detection circuit, sensors were connected in parallel, and divider resistors were connected in series with the sensor on each branch. Based on the differential mode of the data acquisition card, the voltage on both sides of the sensors were collected completely and displayed in a laptop program designed with LabVIEW. The resistance of the sensor can be converted by Ohm's law.

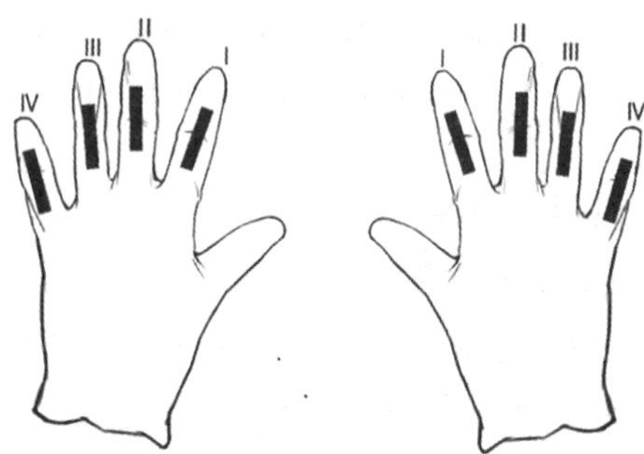

Fig. 1 Configuration of flexible sensor in data gloves

2.5 Design of machine-learning algorithm

A flow diagram of the machine-learning process is shown in Fig. 2. During the training process, the change of voltage due to predetermined gestures was firstly merged into a matrix as the input of mRMR algorithm to discard useless information and retain relevant signals, which improves the generalization ability of the model and reduces the complexity of the algorithm. Subsequently, the multi-segment classification model was applied to classify gesture by using the extracted features with Logit AdaBoost algorithm. The classification process included two steps. First, static signals and dynamic signals were effectively distinguished. And then, combined with the sample characteristics and distribution, the motion trajectory of the finger was further subdivided. Finally, 10-fold cross-validation method was used to evaluate the model. It can make full use of the samples in the training process, which avoids over fitting.

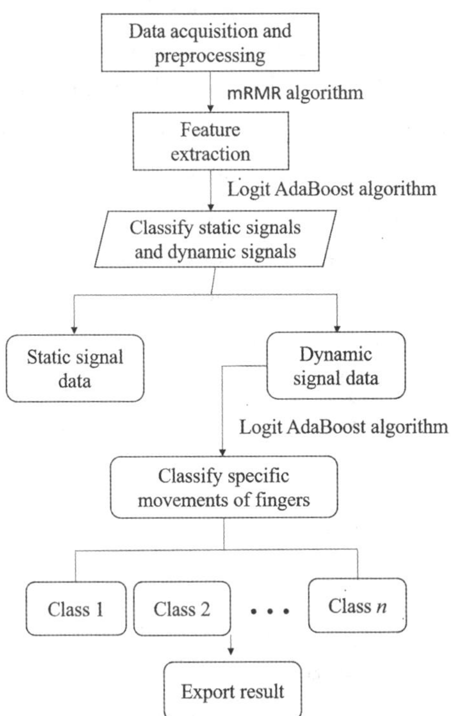

Fig. 2 The process of a multi-segment classification model

3 Results and Discussion

3.1 Morphology characterization

In the r-GO layer, the r-GO is uniformly distributed, as shown in Figs. 3 (a) and 3(b), which is the basis to ensure the stable electrical performance of the sensor. At the same time, based on a sandwich structure composed of a r-GO layer and two PDMS layers, the overlapping state of graphene sheets can change during the deformation of the sensor and recover under the constraint of elastic material (PDMS), resulting in the electromechanical response of sensors.

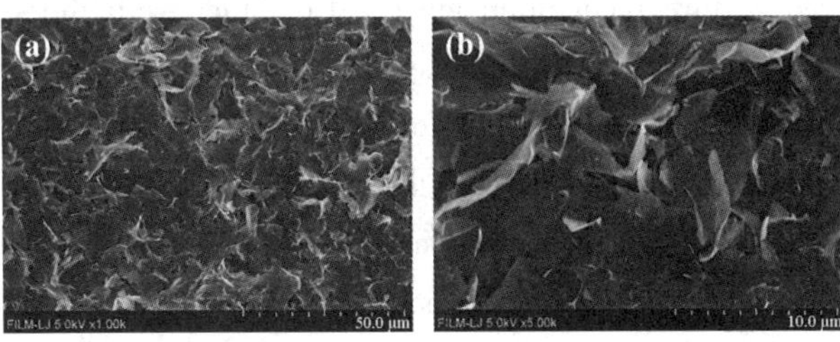

Fig. 3 SEM images of the r-GO film with magnifications of (a) 1 000 and (b) 5 000

3.2 Piezoresistive sensing response

Figure 4 (a) shows the relationship between the sensor resistance and applied strains, from which the sensitivity of the strain sensor can be obtained. The sensitivity of the strain sensor can be calculated by the following equations.

$$GF = \frac{\Delta R/R_0}{\varepsilon} \tag{1}$$

$$\Delta R = R - R_0 \tag{2}$$

where ε is the strain of the strain sensor, R represents the resistance of sensor after stretching, R_0 is the resistance of sensors at $\varepsilon = 0\%$. Fig. 4 (a) shows that the strain sensor provides high sensitivity for detecting hand movements (GF－2. 69 at strains below 12%, GF－5. 05 at the strain range from 12% to 36%). Furthermore, the strain sensors also show a stability response over a large number of stretch-release cycles (above 1 500 cycles), as shown in Fig. 4 (b).

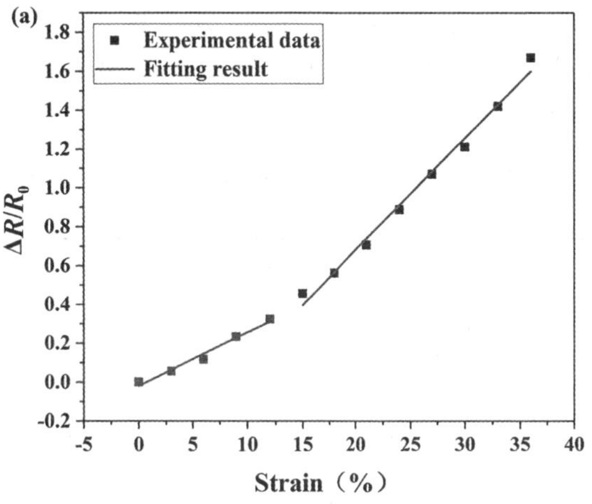

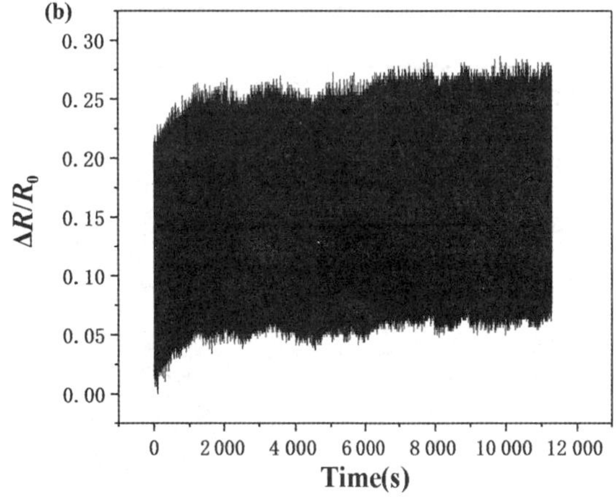

Fig. 4 (a) The relationship between the sensor resistance and the applied strain; (b) sensor stability under 1 500 stretch- release cycles at 10% strain

3.3 Gesture recognition modeling

In the first stage, dynamic signals are separated from all samples for subsequent training. Positive predictive value (PPV) is an index to evaluate the level of correct classification. The lower PPV of dynamic signals is, the more sample capacity is lost in the second stage, resulting in unobvious and incomplete features as gesture output. And false discovery rate (FDR) represents the probability of misclassification, which affects the interference degree of data directly. Tab. 1 shows the classification results of static signals and dynamic signals. High PPV level (96.9%) and low FDR level (3.1%) illustrate that effective gesture features are completely preserved.

The ROC curve after 10-fold cross-validation is shown in Fig. 5. Area under curve (AUC) is a statistical concept. It represents the probability that the predicted value of positive cases is greater than the predicted value of negative cases. Assuming that the dynamic signal is positive, the AUC value of the model in the first stage is 0.99 and the accuracy reaches 98.4%, indicating that the model can eliminate invalid static signals further.

Tab. 1 Evaluation of classification model to distinguish static signals and dynamic signals

Index	Static signals	Dynamic signals
True positive rate	99.3%	94.0%
False positive rate	0.7%	6.0%
Positive predictive value	98.7%	96.9%
False discovery rate	1.3%	3.1%

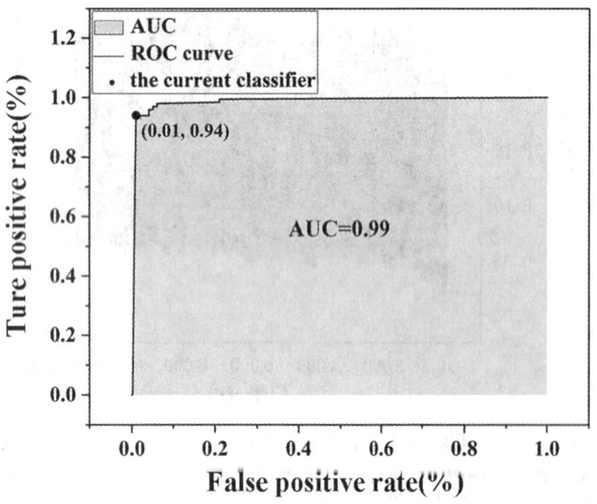

Fig. 5 ROC curve of classification model to distinguish static signals and dynamic signals

Based on defined gestures, the movement of the hand is divided into the bending and straightening of fingers. Tab. 2 shows the accuracy of classification remains at a high level. And Fig. 6 illustrates that the AUC value of the model is 0.97, which only has subtle decrease compared with the first stage. The good classification results of two stages indicates that the multi-segment classification model has the generalization ability for gesture recognition.

Tab. 2 Evaluation of classification model to judge finger motion trajectory

Index	Bend fingers	Straighten fingers
True positive rate	95.0%	95.0%
False positive rate	5.0%	5.0%
Positive predictive value	96.6%	92.7%
False discovery rate	3.4%	7.3%

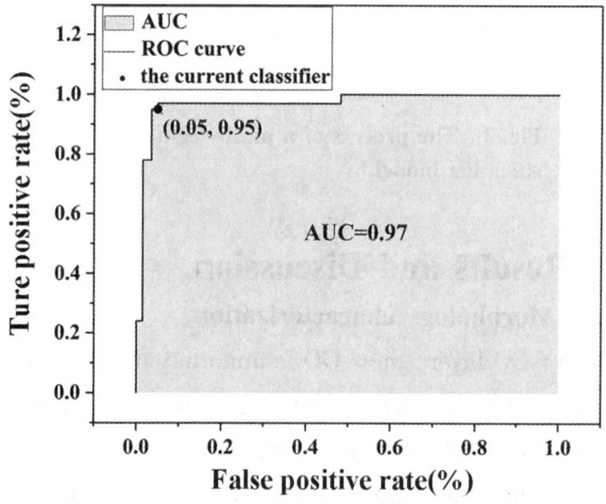

Fig. 6 ROC curve of classification model to judge finger motion trajectory

4 Conclusions

This article reported a type of machine-learning empowered gesture recognition gloves. Combined with the electrical characteristics of flexible strain sensors and the signal processing ability of machine learning algorithms, data gloves judged the specific actions of hands accurately, which showed many commendable features such as low weight and good comfortableness. At the same time, we paid attention to the interference of invalid static signals in the training and classification process. The model successfully extracted effective dynamic signals from all signals through multi-segment classification based on Logit AdaBoost algorithm. It is of significance to enhance the generalization ability of the recognition model and increase the accuracy of gesture recognition. However, not all irrelevant factors have been excluded for this model, which requires further research.

References

[1] JESUS G R, CARLOS M T, ACCAYMO T, et al. Perspective and evolution of gesture recognition for sign language: a review[J]. Sensors, 2020, 20: 35-71.

[2] PAN W, ZHANG X, YE Z. Attention-based sign language recognition network utilizing keyframe sampling and skeletal features[J]. IEEE Access, 2020, 8: 215592-215602.

[3] TSIRONI E, BARROS P, WEBER C, et al. An analysis of convolutional long short-term memory recurrent neural networks for gesture recognition[J]. Neurocomputing, 2017, 268: 76-86.

[4] SAGGIO G, CAVRINI F, PINTO C A. Recognition of arm-and-hand visual signals by means of SVM to increase aircraft security[J]. Studies in Computational Intelligence. 2017, 669: 444-461.

[5] XUE Q, LI X, WANG D, et al. Deep forest-based monocular visual sign language recognition[J]. Applied Sciences, 2019, 9(9): 19-45.

[6] WEICHER F, BACHMANN D, RUDAK B, et al. Analysis of the Accuracy and Robustness of the Leap Motion Controller[J]. Sensors, 2013, 13: 6380-6393.

[7] AMEEN S, VADERA S. A convolutional neural network to classify American Sign Language fingerspelling from depth and colour images[J]. Expert Systems, 2017, 34(3): 1-8.

[8] SHUKOR A Z, MISKON M F, JAMALUDDIN M H, et al. A new data glove approach for Malaysian sign language detection[J]. Procedia Computer Science, 2015, 76: 60-67.

[9] LEE B G, LEE S M. Smart wearable hand device for sign language interpretation system with sensors fusion[J]. IEEE Sensors Journal, 2018, 18: 1224-1232.

[10] ABHISHE K S, QUBELEY L C F, HO D. Glove-based hand gesture recognition sign language translator using capacitive touch sensor[C]//2016 IEEE International Conference on Electron Devices and Solid-State Circuits (EDSSC). IEEE, 2016: 334-337.

[11] LECUN Y, BENGIO Y, HINTON G. Deep learning[J]. Nature, 2015, 521: 436-444.

[12] ZHOU Z, CHEN K, LI X, et al. Sign-to-speech translation using machine-learning-assisted stretchable sensor arrays[J]. Nature Electronics, 2020, 3(9): 571-578.

[13] WU J, SUN L, JAFARI R. A wearable system for recognizing American sign language in real-time using IMU and surface EMG sensors[J]. IEEE Journal of Biomedical and Health Informatics, 2016, 20: 1281-1290.

[14] RIVERA-ACOSTA M, ORTEGA-CISNERO S, RIVERA J, et al. American sign language alphabet recognition using a neuromorphic sensor and an artifcial neural network[J]. Sensors, 2017, 17: 21-76.

[15] CAI Y D, FENG K Y, LU W C, et al. Using LogitBoost classifier to predict protein structural classes[J]. Journal of Theoretical Biology, 2006, 238(1): 172-176.

[16] HO C Y F, LING B W K, DENG D X, et al. Tachycardias classification via the generalized mean frequency and generalized frequency variance of electrocardiograms[J]. Circuits Systems and Signal Processing, 2022, 41(2): 1207-1222.

Hygroscopic and Heat-Generating Properties of Hygroscopic Exothermic Knitted Fabric with Moisture-Diffused Conduction

ZHANG Limin[1,2], CHENG Longdi[1*], ZHANG Ruiyun[1], XUE Wenliang[1], MA Wanwan[1], HAN Xueqing[1]

[1] *Key Laboratory of Textile Science & Technology, Ministry of Education, College of Textiles, Donghua University, Shanghai 201620, China*
[2] *College of Textiles and Clothing, Xinjiang University, Urumqi 830046, China*

[*] *Corresponding author's email*: ldch@ dhu. edu. cn

Abstract: This paper aims to study the moisture absorption and heating behavior of the hygroscopic exothermic knitted fabric exploring the mode of moisture-diffused conduction by changing the path of moisture transmission. In the experiment, the treated fabrics were compared with the untreated one, and it was verified that the moisture-diffused conduction was suitable for delaying and uniforming the moisture absorption and heating of the fabric. The interaction between fiber and water in the inner layer of the fabric after composite finishing is enhanced, which can quickly disperse moisture, and reduce the difference between moisture conduction velocity along the fabric thickness and perpendicular to the thickness. Thus, larger-area absorption and transmission of moisture in the middle layer of the fabric becomes possible, which further endows a larger-area and uniform heating in the outer layer of the fabric. Finally, it is possible to realize the overall heating delay of the fabric in a continuous high-humidity of the micro-environment and heating evenly with the moisture increasing until high-humidity. These results demonstrate that the moisture-diffused conduction in the fabric can provide an effective option to solve the problem of rapid, local and non-uniform heating of the fabric after the moisture absorption.

Keywords: Hygroscopic exothermic fabric; Moisture-diffused conduction; Uniform and slow moisture absorption and heat generation

1 Introduction

Human body feels comfortable in the microclimate, the area between the human body and the inner layer of the clothing, with the temperature of (32±1) ℃, the relative humidity of (50±10)%, and the air velocity of (0.25±0.15) m/s. In recent years, the development of science and technology has transformed people's heat preservation and chilliness resistance of garment from traditional heat insulation to active heat generation[1]. A number of studies focused on attaching the hygroscopic exothermic property to the textile materials by using the functional fibers such as EKS, Softwarm, Thermogear, coffee carbon acrylic[2-5] or finishing techniques[6-7]; improving the dyeing performance of hygroscopic exothermic products[8-9]; exploring test methods and devices on hygroscopic exothermic characteristics[10-11]; manufacturing one-way transport hygroscopic exothermic draperies based on different fibers, yarns and fabrics combined with different structures[12]. Moreover, in order to explain the process of heat and moisture transfer in porous materials more objectively, scholars have defined the CLO value[13], moisture permeability index im[14], heat, and moisture transfer resistance[15]. The researches on the mechanism of heat and moisture transfer have also been expanded from one-dimensional to three-dimensional[16-17], which shows the process of heat and moisture transfer in fabrics more comprehensively.

According to the classic model of two-dimensional coupled heat and moisture transfer[18], it is demonstrated that fast permeability, concentrate transmission and uneven distribution of moisture are bound to give the sequels on rapid heating, local heating and uneven heating. However, human sweating is divided into three stages: little sweat precipitation, mass sweat perspiration, and sweat evaporation. The first two stages, if added with the rapid moisture absorption and heating of the fabric, will not bring a comfortable experience to the human body. In the third stage, the sweat evaporation takes away the heat of the human body, resulting in the wet and cold feeling of the skin. The sudden drop in temperature after the moisture absorption and heating of the fabric will also aggravate the discomfort of the human body. Therefore, according to human needs, appropriately delaying, slowing, and homogenizing the moisture absorption and heat generation of the fabric will help to meet the comfort requirements of clothing with moisture absorption and heat generation after sweating, which is of great significance.

It can be seen from Fig. 1 that, changing the capillary wicking effect in any dimension (x or y) will further change and control the moisture transmission in the fabric. Then, the process above may eventually make the fabric absorb the moisture slowly, and heat slowly and uniformly.

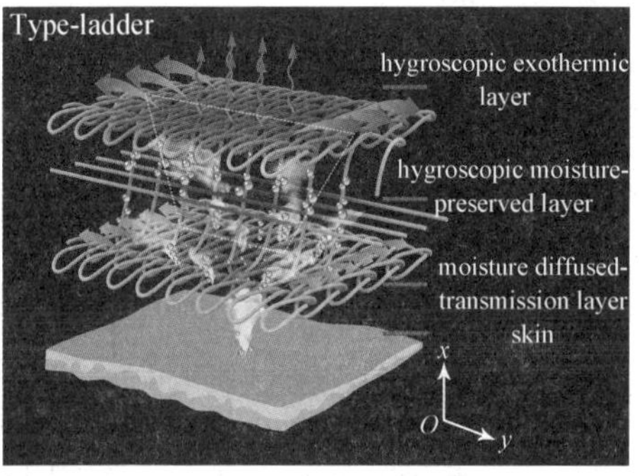

Fig. 1 Mechanism of "type-ladder"

In this study, the main objective is to propose a type of hygroscopic heating fabric with slow and uniform heating after moisture absorption by changing the capillary wicking effect in x and y directions according to the diffused-transmission mechanism named "type-ladder" shown in Fig. 1. The internal mechanism from experiments to regulate the hygroscopic exothermic process by controlling the moisture transmission is also discussed.

2 Experiments

2.1 Materials

The three-layer knitted fabric was prepared shown in Tab. 1 and Fig. 1.

2.2 Fabric treatment

Step 1: single padding finishing. The finishing solution was prepared by deionized water and HP moisture-wicking finishing agent (40 g/L, bath ratio 1 : 30). The fabric was put into the finishing solution at 25 ℃ for 30 min (padding rate 80%), pre-baked at 80 ℃ for 3 min and heated at 150 ℃ for 50 s, washed and dried, which was named "treated fabric".

Step 2: padding-printing composite finishing. The printing paste was prepared by TF-629C finishing agent (concentration 4%), deionized water, and sodium alginate, which was printed to the outer layer of the treated fabric. Then the printed fabric was pre-baked at 80 ℃ for 10 min, baked at 150 ℃ for 5 min, washed, and dried, which was named "multi-treated fabric".

The fabric based on diffused-transmission mechanism of "type-ladder" contains three layers: diffused moisture transmission on the inner layer, diffused moisture absorption and storable moisture transmission on the middle layer, and uniform hygroscopic exothermic function on the outer layer.

Tab. 1 The materials of the fabric

Type	Inner layer	Middle layer		Outer layer
		Connect yarn	Weft insertion	
Three-layer quilted fabric	75D/72F cooldry and cotton (100)	75D/36F PET	32^S cotton	32^S comfortable & hot fiber/viscose (40/60)

2.3 Scanning electron microscopy (SEM)

The surface topography of the materials in inner and middle layers of the fabric was obtained from SEM using SEM (DXS-10ACKT, Shanghai, China) at 10 kV, 20 ℃, and relative humidity (RH) 65%.

2.4 Contact angle test

The contact angle between water and the surface of materials was tested using the contact angle surface performance tester (OCA15EC, Germany).

2.5 Hygroscopic and exothermic properties test

2.5.1 Curve of moisture absorption and heating

According to the standard of GB/T 29866—2013, ΔT_{max} is the maximum temperature rising in 30 min. In addition, $\Delta \overline{T}$ is the average temperature rising within 30 min. ΔT_{max} and $\Delta \overline{T}$, as well as the curve of moisture absorption and heating, are obtained in a constant temperature and humidity climate chamber at (20.0 ± 0.5) ℃ and RH of (90±3)%.

2.5.2 IR camera test

The real-time IR thermal images on the outer layer of

the fabrics and the RH of the microclimate environment were monitored every 10 s with the RH of the microclimate environment rising from (65±3)% to (90±3)% and keeping in (90±3)%.

2.6 Moisture management test

2.6.1 Liquid moisture management test

According to the standard of AATCC 195—2020, liquid moisture transmission properties of fabrics were tested by the liquid moisture management system (M290, SDL Atlas, USA) at (20±2) ℃ and RH of (65±3)%.

2.6.2 Wicking height test

According to the standard of FZ/T 01071—2008, wicking height of the fabrics was obtained using the capillary effect tester (YG871-Ⅱ, Wenzhou, China) at (20±2) ℃ and RH of (65±3)%.

3 Results and Discussion

3.1 Surface topography

As shown in Fig. 2 (c), the longitudinal surface of the yarn in the inner layer of the fabric was covered by a layer of film formed by the finishing agent, as well as surface of the polyester in the middle layer of the fabric shown in Fig. 2 (d). The results above indicated that the finishing agent had been attached to the surface of polyester fibers after padding finishing. In addition, due to the unique groove structure of the polyester in the inner layer of the fabric, the coating of the finishing agent was more obvious.

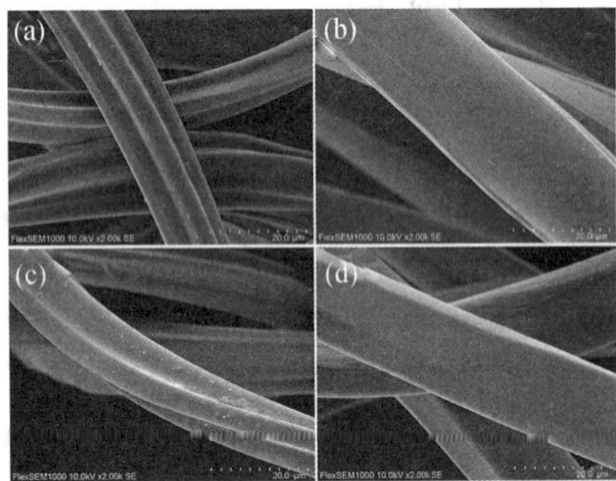

Fig. 2 (a) Untreated cooldry (polyester in the inner layer of the fabric); (b) untreated polyester in the middle layer of the fabric; (c) treated cooldry; (d) treated polyester

3.2 Contact angle

As shown in Figs. 3 (d)-3(f), the contact angle was reduced after finishing, indicating that the padding finishing had a certain degree of hydrophilic modification on the polyester, which was beneficial to realize the rapid moisture diffusion along the y-axis, large area moisture diffusion in the inner layer of the fabric, more uniform moisture distribution, and a long-term moisture storage function in the middle layer of the fabric.

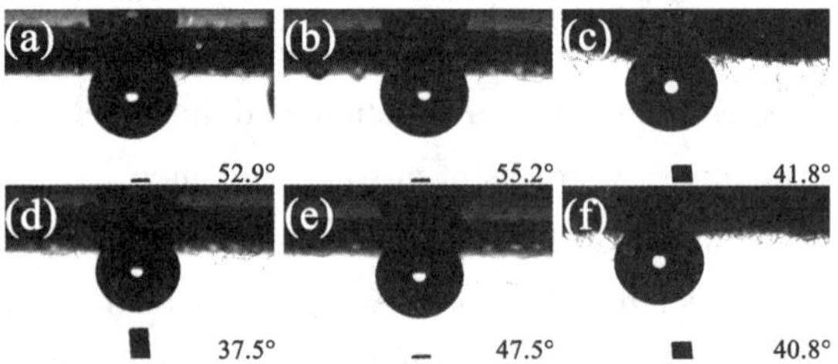

Fig. 3 (a) Untreated cooldry (inner layer); (b) untreated polyester (middle layer); (c) untreated surface of fabric inner layer; (d) treated cooldry; (e) treated polyester; (f) treated surface of fabric inner layer

3.3 Hygroscopic and exothermic properties

3.3.1 Curve of moisture absorption and heating

As shown in Fig. 4 (a), the curve of heating-up of the treated fabric and multi-treated fabric showed an obvious heating delay in the classic moisture absorption and heating stage. Combined with Fig. 4 (c), the delay degree of heating time was gradually aggravated with the heating process, showing a rise of 1-5 ℃ delaying by 17 s, 24 s, 30 s, 30 s, and 64 s after single padding and 27 s, 34 s, 40 s, 56 s, and 104 s after composite finishing, respectively. The results above show that the diffusion of moisture along the y direction is enhanced, as shown in Fig. 1 (b). Once the moisture touches the

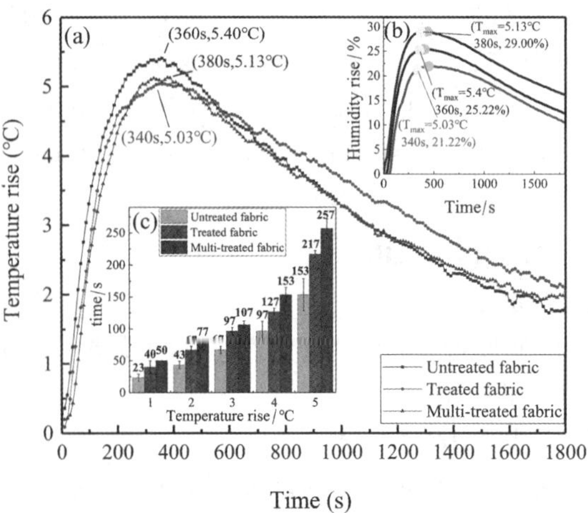

Fig. 4 (a) The curve of heating-up; (b) the curve of humidity-up; (c) the degree of retarded heating-up

inner surface of the fabric, it will be adsorbed and diffused, which increases the diffusion area and reduces the moisture distribution density. Then, it increases the combination of water-fiber at the same moisture content, which is conducive to avoiding the moisture accumulation from the inner layer to the middle layer of the fabric and slowing down the transmission of moisture along the x direction as shown in Fig. 1 (b).

3.3.2 IR thermal image

As shown in Fig. 5 (a), when RH of microclimate rises from 65% to 90% and maintains 90%, the temperature rise, heating duration (90 s), heating area, and heating uniformity at different times are inferior to the finished fabrics as shown in Figs. 5 (b) and (c). On the contrary, the fabrics as shown in Figs. 5 (b) and 5 (c) can continue to heat for a long time after moisture absorption (200 s and 240 s), the heating area significantly increases, and the heating uniformity is significantly improved. The results show that, the large-area moisture diffusion along the y axis, the large-area output and transmission along the x axis, a long-time moisture storage in the middle layer of the fabric can trigger a long-time, large-area, and uniform heating after moisture absorption.

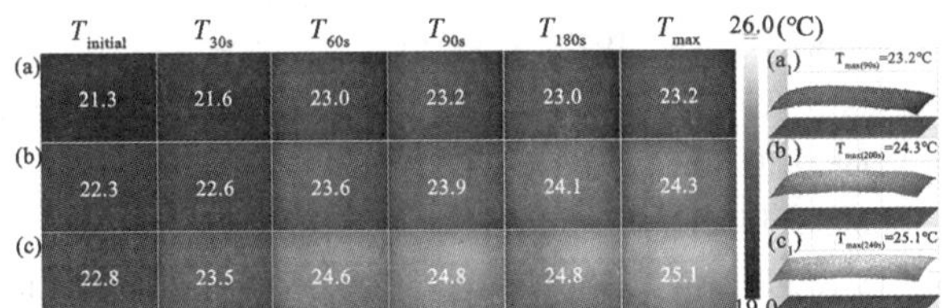

Fig. 5 (a), (a_1) Infrared thermal images of untreated fabric at different times and 3D-IR images; (b), (b_1) infrared thermal images of treated fabric at different times and 3D-IR images; (c), (c_1) infrared thermal images of multi-treated fabric at different times and 3D-IR images

3.4 Moisture management

As shown in Fig. 6, Fig. 7 and Tab. 2, compared to the untreated fabric, the moisture transmission in treated and multi-treated fabrics presented almost the same trend, indicating that higher moisture diffusion rate and larger diffusion area in the inner layer of the fabric with further moisture diffusion and storage in the middle layer of the fabric would make the slow and uniform heating come true.

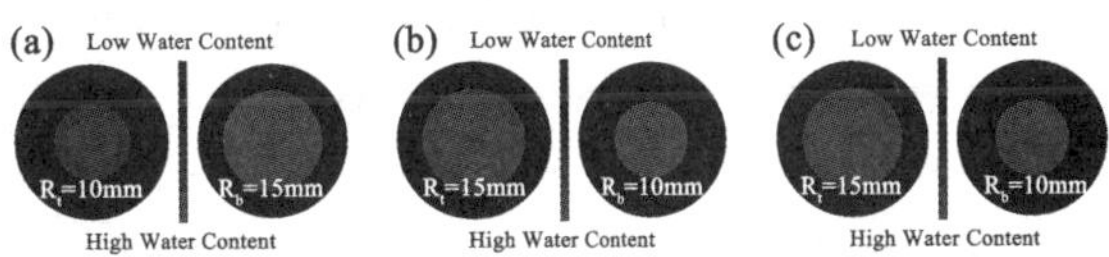

Fig. 6 (a) Liquid moisture management properties of untreated fabric; (b) liquid moisture management properties of treated fabric; (c) liquid moisture management properties of multi-treated fabric

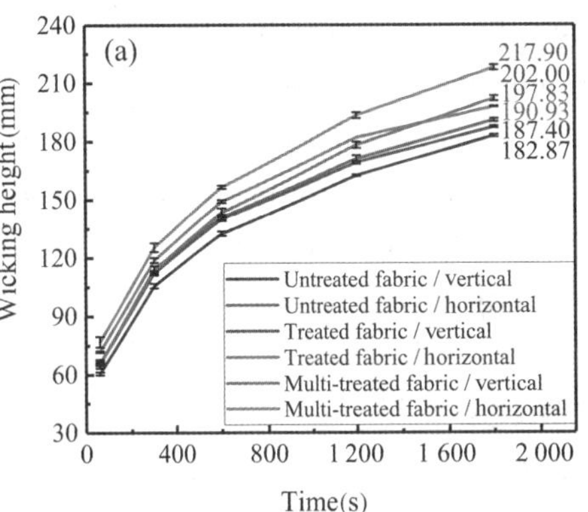

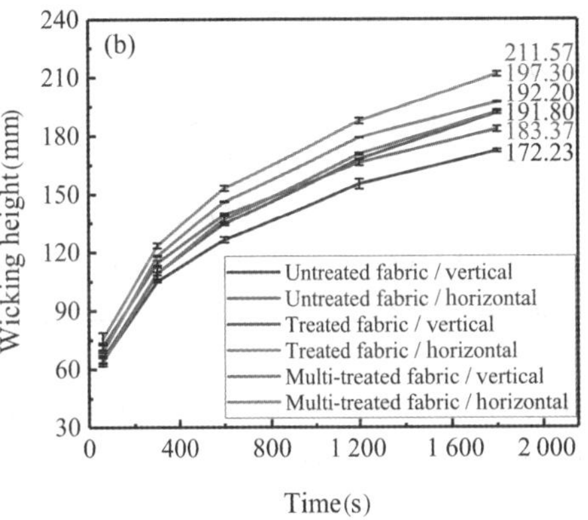

Fig. 7 (a) Wicking height on inner layer of the fabric; (b) wicking height on outer layer of the fabric

Tab. 2 Liquid moisture management properties

Test metrics	Untreated fabric	Treated fabric	Multi-treated fabric
Wetting time top/s	8.33	3.93	4.12
Top spreading speed/mm · s^{-1}	1.11	2.53	2.42
Wetting time bottom/s	6.83	5.24	5.90
Bottom spreading speed/mm · s^{-1}	1.81	1.42	1.32

4 Conclusions

This study has focused on the problems of excessive heating, local heating, and uneven heating caused by the rapid one-way transport of the moisture, small-area diffusion of the moisture, and accumulated conduction of moisture in hygroscopic exothermic fabrics, proposed the "type-ladder" diffused-transmission mechanism, and prepared the corresponding fabrics according to that.

Through the property evaluation of the hygroscopic exothermic fabric, it is confirmed that the "type-ladder" diffused-transmission of moisture has advantages in delaying heating and promoting the overall uniform heating by changing the transmission form of moisture in the fabric, that is, accelerating the diffusion of moisture up along the y direction, increasing the diffusion area, and strengthening the role between water and fiber. The fabric after finishing can delay the heating by more than 50% in the continuous high humidity environment, and can continue to heat evenly for 240 s in the gradually rising humidity environment.

References

[1] ZHANG Q, WANG Z, LIU C, et al. Study on the preparation and performance of heat-storing textiles[J]. Progress in Textile Science & Technology, 2019(8): 20-23.

[2] CAO J, MENG J G, LI L. Studies on properties of EKS heat-generating fiber and yarn[C]. Applied Mechanics and Materials, 2012: 1216-1220.

[3] JIE L, MENG J G, LIU X. The analysis on the wearing properties of softwarm heat-generating fiber knitted fabric and the comprehensive evaluation[J]. Knitting Industries, 2013(6): 28-31.

[4] CAO Q L, WANG L, SHEN Y. Functional properties of coffee carbon acrylic blended knitted fabric[J]. Knitting Industries, 2015(10): 17-19.

[5] CUI Y, GAO S Y, ZHANG R Y, et al. Study on the moisture absorption and thermal properties of hygroscopic exothermic fibers and related interactions with water molecules[J]. Polymers, 2020, 12(1).

[6] LI M Y, ZHAO T. Hygroscopic exothermic of nylon fabric by surface modification[J]. Dyeing and Finishing, 2019, 45(23): 6-10,25.

[7] LIU Z Y, LIU Y J, WANG X H. Research progress of heating materials and their applications in the clothing field [J]. Technical Textiles, 2019, 37(8): 1-11.

[8] WANG J, XU D T, YU J K. An analysis on the dyeing property of the moisture-absorbing and heat-emitting eks blended fabric[J]. Knitting Industries, 2012(4): 32-33.

[9] WU Y J, QI L. Research progress in hygroscopic and exothermic material[J]. New Chemical Materials, 2012, 40(1): 47-48,83.

[10] YUAN Z L, LI F X. Test method for performance of moisture adsorption and heat release of textiles[J]. China Textile Leader, 2011(8): 105-106.

[11] SHAN X L, LIU N. Study of testing method of textile's moisture-absorption and heat-generating property[J]. Knitting Industries, 2015(3): 67-69.

[12] YOO S, BARKER R L. Moisture management properties of heat-resistant workwear fabrics-effects of hydrophilic finishes and hygroscopic fiber blends[J]. Textile Research Journal, 2004, 74(11): 995-1000.

[13] GAGGE A P, BURTON A C, BAZETT H C. A practical system of units for the description of the heat exchange of man with his environment[J]. Science, 1941, 94 (2445): 428-430.

[14] WOODCOCK A H. Moisture transfer in textile systems, Part I[J]. Textile Research Journal, 1962, 32(8): 628-633.

[15] MECHEELS J, UMBACH K. The psychrometric range of clothing systems[J]. Clothing Comfort. Michigan, Ann Arbor Science, 1977: 133-151.

[16] LI F, LI Y, WANG Y. A 3D finite element thermal model for clothed human body[J]. Journal of Fiber Bioengineering and Informatics, 2013, 6(2): 149-160.

[17] LI F, WANG Y, LI Y. A transient 3-D thermal model for clothed human body considering more real geometry[J]. JCP, 2013, 8(3): 676-684.

[18] ZHONG M, WANG R M, LI Y, et al. Analysis and implement of simulation system of functional textile product[J]. Computing Technology and Automation, 2005, 24(4): 96-98.

Influence of Preform Structure on Bending Properties of Needle Punched Carbon Fiber Composites

CHEN Xi[1], **CHEN Nanliang**[1*], **JIANG Jinhua**[1], **SHAO Huiqi**[1]

[1] *Engineering Research Center of Technical, Education Ministry, College of Textiles, Donghua University, Shanghai 201600, China*

[*] *Corresponding author's email*: nlch@ dhu. edu. cn

Abstract: Carbon fiber composites have been widely used in aircraft structural parts, power generation blades, lightweight automotive parts, and other specific industrial products, due to their excellent mechanical and thermal properties. In this study, we used woven carbon fabric/mesh needle-punched fabrics as reinforcement and epoxy resin as matrix, to prepare three types of needle-punched composites with different structures. The effect of the carbon fabric structure on the flexural properties of composites was studied by the three-point bending test, and the failure behavior and the damage mechanism of the material during the test were analyzed and discussed. The results showed that among the composites reinforced by unidirectional non-weft fabrics, plain weave fabrics and twill weave fabrics respectively, the twill weave fabrics reinforced composite had the highest bending load. The bending failure characteristics of the three materials are brittle fracture, and the main failure mode is the bottom V-shaped cracking failure.

Keywords: Carbon fabric structure; Needle-punched carbon fiber composite; Bending performance

1 Introduction

Because of their low density, high strength, corrosion resistance, and strong designability, carbon fiber composites are widely used in aerospace, automobile industry, military, and civilian fields[1-2]. Needle-punched carbon fiber composites refer to three-dimensional (3D) reinforced materials with special structures obtained by needle-punched technology. Using the regular movement of the barbed felting needle, the in-plane fibers of the carbon fiber fabric/mesh material are introduced into the thickness direction. Generating vertical fiber clusters, so that the carbon fiber fabric and the mesh layer are tightly combined, to obtain 3D structural materials with certain strength both in-plane and between layers.

In recent years, a large number of scholars have deeply explored the factors affecting the mechanical properties of needle-punched carbon fiber composites. For example, Qi et al.[3] used a qualitative analysis method to fully consider the influence of needle depth, needle arrangement, and interlayer density on preforms, and successfully established a functional relationship between the stiffness properties of the composites. Jiang et al.[4] compared the bending properties of 3D needle-punched composites with two fiber architectures at room and elevated temperatures. The experimental results proved that the material exhibited obvious linear elastic and brittle fracture characteristics at room temperature, while the curves at elevated temperature showed non-linear and long plasticity plateau. Zhang et al.[5] proposed a line-line simplified model for predicting the effective elastic properties of 3D needle-punched C/C composites, which fitted well with experimental data and provided guidance for product performance design. From the domestic and foreign literature which have been published, we can find that the research on needle-punched composites mainly focuses on how the needle parameters affect the mechanical properties of the material, and how to establish a micro-structure model to effectively predict their performance. However, there are few studies on the prefabricated fiber structure composed of carbon fiber composites. Therefore, in this paper, we used unidirectional non-weft fabric (N), plain weave fabric (P), twill weave fabric (T) and a certain density of carbon fiber mesh for needle punching, combined with vacuum-assisted resin transfer molding (VARTM), to prepare different needle-punched carbon fiber preforms, and to explore the influence of the preform structure on the bending properties of composite materials, as well as the damage mechanism.

2 Experiments

2.1 Materials

The raw material woven fabrics tested in this paper are T700 3k plain/twill weave carbon fabric with an area density of 200 g/m^2, and T700 12k unidirectional non-weft carbon fabric with an area density of 200 g/m^2. The nonwoven mesh is made of short carbon fiber, and the area density is 30 g/m^2.

2.2 Preparation process

In the experiment of this paper, a layer of carbon fiber fabric and a layer of carbon fiber mesh are defined as a needling unit layer, and the unit layer is relay needle punched by the laminated needling process to obtain a carbon fiber preform. The needling process is shown in Fig. 1. The needling density is 13 punch/cm^2, and the needling depth is 8 mm. After the preform is made, VARTM process is used. The VARTM process has significant advantages such as low cost, high efficiency, and environmental friendliness in the preparation of large-scale composite components. In recent years, it has been widely used in the manufacture of ship hulls, wind turbine blades, and various aviation parts[6]. The specific process flow is shown in Fig. 2.

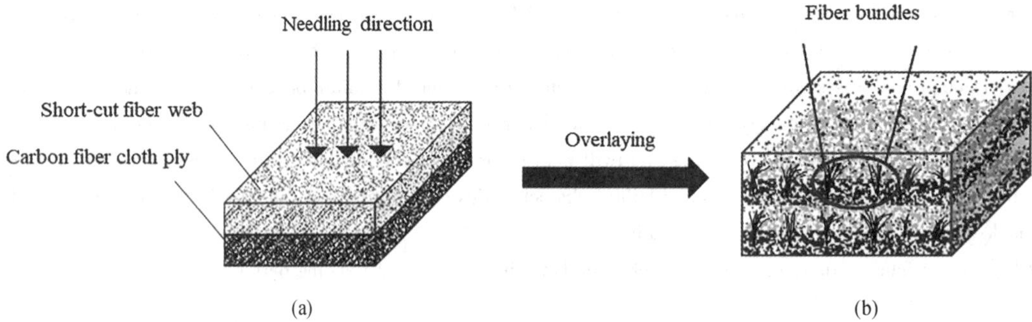

Fig. 1 Needle-punched process flow chart: (a) needle punching; (b) needle punch preform

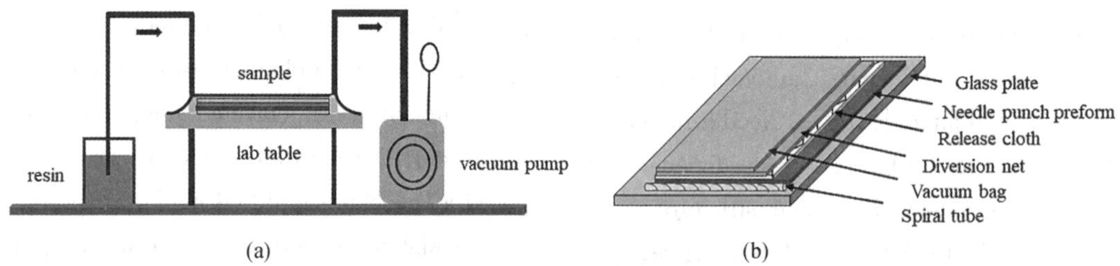

Fig. 2 VARTM process flow chart: (a) VARTM process flow diagram; (b) sample laying order

2.3 Test Method

The three-point bending test is used based on GB/T 1447—2005. The size of the sample is 80 mm×15 mm×4 mm, the span is 64 mm, and the loading rate is 5 mm/min. The bending strength is calculated as follows:

$$\sigma_F = \frac{3Pl}{2bh^2} \tag{1}$$

Where P is the maximum load of failure (N), l is the span (mm), and b and h are the width (mm) and the thickness (mm), respectively. There are no less than 5 valid samples for all tests, and the test result is the average value of the valid samples.

3 Results and Discussion

3.1 Bending strength

Three-point bending tests were carried out on needle-punched carbon fiber composite prepared from carbon fabrics with different structures, and the bending strength and load-displacement curves are shown in Figs. 3 and 4, respectively.

As shown in Fig. 3, after continuous needle punching of carbon fabric and carbon fiber mesh with three different structures, the obtained reinforcement (preform) shows different bending properties. Among them, the bending strength of the T group material reaches the highest 195.094 MPa, which is 39.21% and 19.55% higher

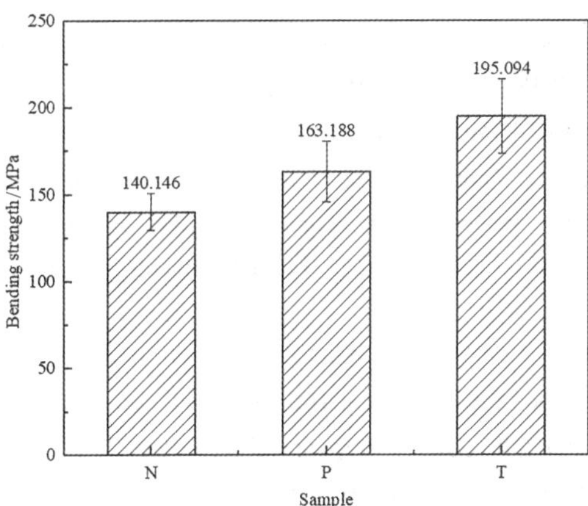

Fig. 3 Bending strength of composite with different carbon fabric structures

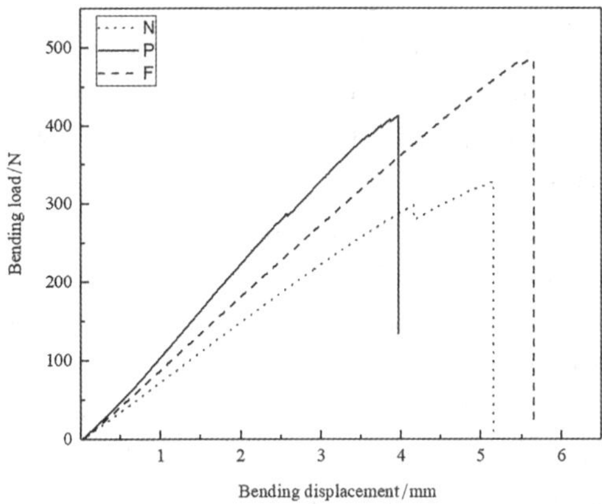

Fig. 4 Load-displacement curves of composite with different carbon fabric structures

than that of N and P groups, respectively. Under the continuous action of the needle, some fibers in the carbon fabric are punctured, and the overall mechanical properties of the material are damaged. However, the needle with a special structure during the needling process will continuously hook up fibers or fiber clusters and bring them to the thickness direction of the material, thereby increasing the interconnection between fibers and enhancing the interlayer performance of the material. In this experiment, under the same needle-punched process parameters, the T group had better bending properties than the N group and the P group. This depends on the twill weave carbon fabric with fewer warp and weft yarn interlacing points, long floating lines, small pores between yarns, large holding force, and tighter arrangement, which makes it possible for fibers to slip and spread during the needle punching process. The fiber content increases and the porosity decreases, thus endows the twill needle-punched carbon fiber composite with better bending properties.

3.2 Bending fracture mechanism

Carbon fiber reinforced composites are prone to damage under the continuous action of external loads. According to the researches, the common damage forms of layered composite materials are fiber breakage, fiber pullout, matrix cracking, fiber-matrix debonding, interlaminar damage, etc.[7]. Figure 4 is the load-displacement curve three different carbon fabric structure preforms. It can be seen that the bending failure of three needle-punched carbon fiber composites shows brittle fracture characteristics. With the increase of load, the internal damage of the material becomes more and more serious, and the bending load and the bending displacement always maintain a linear relationship. Figure 5 describes the general process and specific damage form of the sample bending fracture during the test. After reaching the maximum fracture stress of the material, the crack rapidly expands inside the material, resulting in regional fracture failure of the lower end fiber and the matrix of the sample, which is manifested as a sudden drop in the load-displacement curve. When materials of N and P groups are damaged, the samples are bent into a "V" shape as a whole, and individual samples are directly broken from the middle. The fracture surface is relatively flat, with a small amount of fibers pulled out, and fiber cracks can be clearly seen. This shows that the bending damage of the N and P groups is mostly fiber breakage and fiber pullout, which may be accompanied by part of the matrix cracking. The fracture surface of the samples in the T group is slightly uneven compared with that in the N and P groups. There are many grooves formed after the fibers are pulled out and the matrix is cracked. In the Z direction, layer-to-layer separation can also be observed, so the failure of the T group is a synergistic effect of multiple damages.

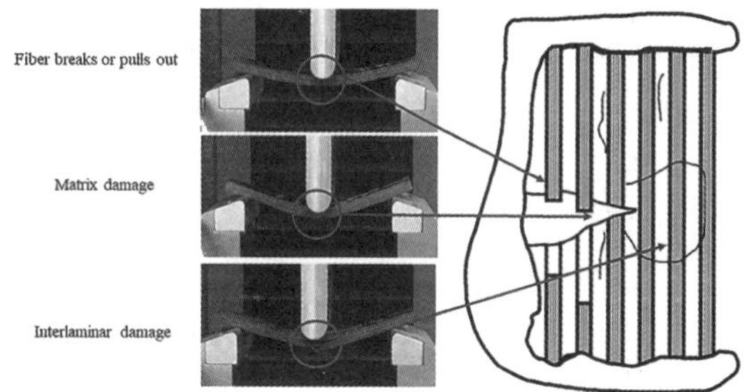

Fig. 5 Bending fracture and damage patterns of carbon fiber preforms

4 Conclusions

The research in this paper focuses on the differences in the bending properties of carbon fiber reinforced composites prepared from three groups of carbon fabrics with different micro-structures.

After testing, it was found that the bending strength of the T group was 39.21% higher than that of the N group. The bending failure modes of the three materials are also slightly different. N and P groups are mainly due to fiber fracture and pullout under load, accompanied by a small amount of matrix cracking. The failure of T group materials is the combined effect of various damage forms, including fiber pullout, matrix cracking, and interlaminar damage.

Acknowledgements

This work was sponsored by the Shanghai Natural Science Foundation of Shanghai Municipal Science and Technology Commission (No. 20ZR1401600, 20ZR1400600), and Fundamental Research Funds for the Central Universities (No. 2232020D-09, 2232020G-06, 2232021G-06).

References

[1] LIU Y F, LIN X L, LI Z L, et al. Design and application of carbon fiber composite material in end box of rail transit vehicles[J]. IOP Conference Series: Materials Science and Engineering, 2019, 612(3): 032200.

[2] EVDOKIMOV A A, DONETSKLL K I, SIDORINA A I, et al. Manufacture of three-dimensional reinforced fabric preforms for making aviation products in russia and abroad — a review[J]. Fibre Chemistry, 2019, 51(2): 105-108.

[3] QI Y, FANG G, WANG Z, et al. An improved analytical method for calculating stiffness of 3D needled composites with different needle-punched processes[J]. Composite Structures, 2020, 237: 111938.

[4] JIANG N, LI D S, YAO Q Q, et al. Influence of temperature on the bending properties and failure mechanism of 3D needle-punched carbon/epoxy composites[J]. Fibers and Polymers, 2017, 18(2): 313-321.

[5] ZHANG J, AN P. Modeling quasi-3D needle-punched C/C composites using a linear simplification representative volume element model[J]. AIP Advances, 2019, 9(3): 035344.

[6] GENKI M, HITOSHI N, TAKAHIRO M, et al. Mechanical behavior of CFRP on steel surface molded and bonded by vacuum-assisted resin transfer molding technology[J]. SN Applied Sciences, 2019, 1(6): 1-13.

[7] ZHANG J, LUO R, XIANG Q, et al. Compressive fracture behavior of 3D needle-punched carbon/carbon composites[J]. Materials Science and Engineering: A, 2011, 528(15): 5002-5006.

Systematically Investigating Influence Factors of Tensile Tear Properties for Airship Envelope Materials

ZHANG Yue[1,2], **ZHANG Chenglong**[1,2], **LI Wei**[1,2], **JIANG Jinhua**[1,2*]

[1] *Engineering Research Center of Textile Technical, Ministry of Education, Donghua University, Shanghai 201620, China*;

[2] *College of Textiles, Donghua University, Shanghai 201620, China*;

[*] *Corresponding author's email*: jiangjinhua@dhu.edu.cn

Abstract: The stratosphere airship, because of its economic advantages of high efficiency and long service time, is widely used in military and civil fields. However, the envelope materials are susceptible to tear damage due to tiny flaws or stress concentration in preparation, transportation, and the working process of the flight, which may lead to envelope overall failure. Therefore, in this paper, the tensile tearing properties of a kind of airship envelope materials are studied. The influence of slit angle and other factors on tensile tearing performance and failure mechanism under single edge slit is investigated in detail. The results show that the envelope material exhibits the best tensile tearing performance when the single edge slit angle is 30°. And the projected transverse length of the slit affects the tensile tearing strength of the envelope material a lot. Additionally, DIC results also confirm that the existence of the slit will form stress concentration, which is the main cause of envelope material tearing.

Keywords: Stratospheric airship; Envelope material; Single edge slitted crack; Digital image correlation technology

1 Introduction

The stratospheric airship is a type of lighter-than-air, low-speed near space aircraft that operates in near-space at an altitude of 20 km to 100 km above the ground, and is widely used in resource exploration, information transmission, and other fields because of its long residence time, stable work, low energy consumption, high resolution, and good safety performance[1-5]. Due to the harsh service environment, as the main structure of the airship, its envelope material needs to meet the requirements of light weight, high strength, good mechanical properties, and low cost. Therefore, the envelope materials are generally made of a film-fabric laminated structure[6]. Its main structure includes bearing layer, weather-proof layer, helium barrier layer, etc. The functional layers are bonded to each other to form the overall structure.

This paper is based on a new high-performance envelope material to study the change of envelope material tensile tearing performance in the presence of a slit. The effects of slit length and slit angle on the tearing properties of the envelope material with a single edge slit are investigated under a single variable, and the tensile tearing process is analyzed, and the effects of the slit on the mechanical properties of the flexible laminated composites and the extension behavior of the slit are combined with digital image correlation techniques. The test results would provide data to support the study of the tensile tear performance of stratospheric airship envelope materials in the presence of single edge slit.

2 Experiments

2.1 Materials

The experiment is based on a new airship envelope material with a Vectran base fabric as the load-bearing layer and a high melting point TPU film as the outer layer, whose main functions are weather-proof and helium barrier. Figure 1 shows a schematic diagram of the structure of the envelope material. Among these, the Vectran base fabric is a plain weave fabric with a yarn fineness of 400 D. The warp and weft yarn densities of the Vectran base fabric are both 36 yarns/(3 cm), with a thickness of 0.157 mm and an area density of 122.2 g/mm^2. The TPU adhesive has a melting point of 130–140 ℃ and a thickness of 0.025 mm.

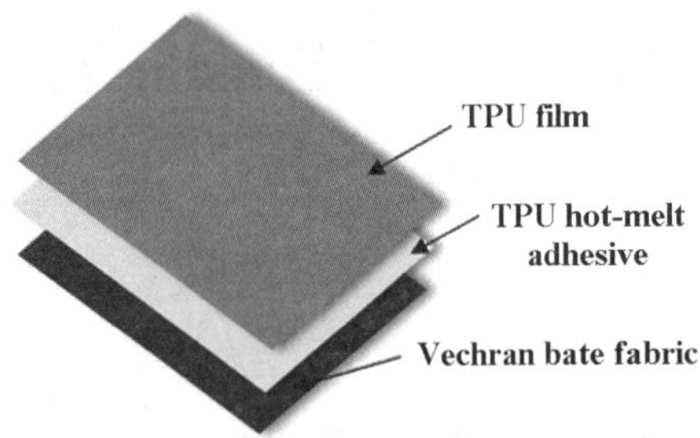

Fig. 1 Schematic diagram of stratospheric airship envelope material structure

2.2 Equipment and environment

The MTS E42.503 microcomputer-controlled electronic universal testing machine was used to perform tensile tear test on the envelope material, and the range of the universal testing machine was 5 000 N. The test environment standard refered to ISO 139:2005, and the test was conducted at a controlled temperature of (20±2) ℃, a relative humidity of (65±4)%, and a standard atmospheric pressure.

2.3 Mechanical testing methods

The tensile tear performance of the envelope material was studied and analyzed with reference to Japanese Industrial Standard JIS L-1096. The specimens for tensile tearing test were made along the warp direction. The length of the specimen is 300 mm, the width is 50 mm, the clamped distance of the fixture at both ends of the specimen is 50 mm, the initial spacing is 200 mm, and the stretching speed is set to be 20 mm/min. In order to investigate the effect of a single side slit on the tearing properties of the masking material, a slit was produced on the side of the specimen with a hobby knife based on the tensile strip size. The specimen with the slit is shown in Fig. 2. To ensure the reliability of the test data, three valid data should be obtained for each set of tests. The angle of the slit perpendicular to the warp direction is 0° and the length of the slit is expressed in terms of n (the number of cut yarns). The numbers of the specimens with different seam angles and seam lengths are shown in Tab. 1.

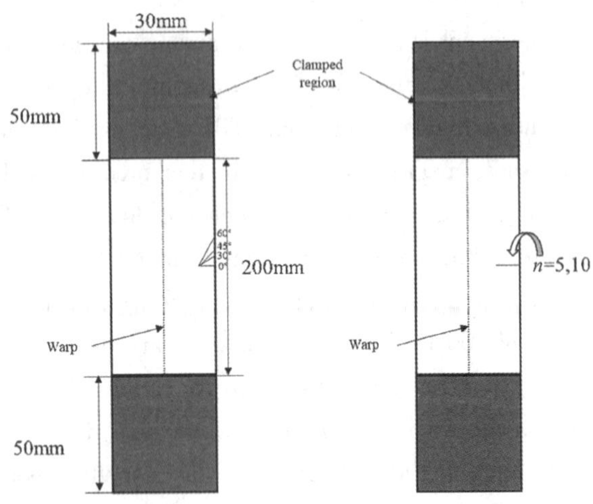

Fig. 2 Specimen dimensions

Tab. 1 Specimen numbering for different slit angles and lengths

Parameter	0°	30°	45°	60°
n=5	s5-0	/	s5-45	/
n=10	s10-0	s10-30	s10-45	s10-60

3 Results and Discussion

3.1 Specimen tearing process

The test shows that the tensile tearing process of the specimen containing the slit is roughly divided into four stages: slit open, slit expansion, yarn fracture, and final destroy, as shown in Fig. 3. Take s10-60 as an example, firstly, as the loading force increases, the stretched yarn is taut and straightened, the material on both sides of the slit gradually opens and curls outward, and the tip of the slit appears convex along the stretching direction. Then the load continues to increase. The angle no longer increases, the tearing sound begins to appear, and the outer covering layer peels off the bearing layer fabric and is damaged by tearing. The stretched yarn is exposed but not fractures. Then, as the first fracture sound appeared, the first yarn fractured, accompanied by continuous fracture sound, and the cutting tip continued to expand inward. The tear triangle expanded, and the remaining yarns fractured in turn, till all yarns completely fractured.

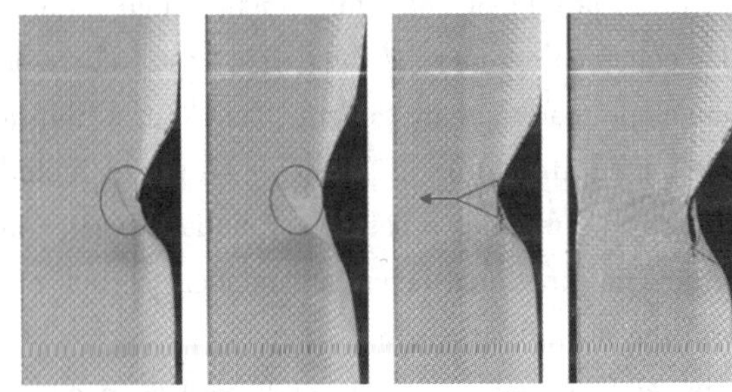

Fig. 3 Single edge slit specimen tensile tearing process

3.2 Effect of slit length

In order to study the effect of slit length on the tensile tear performance of the envelope material, specimens with slit lengths n of 5 and 10 were selected for analysis, and the tensile tearing curves and tearing strengths of the specimens with two slit lengths were compared when the slit angle was 0° and 45°. Figure 4 (a) shows the displacement-stress curves of specimens for two slit lengths when the slit angle is 0°. It can be seen that the curves correspond to different stages of the tensile tearing process. At the beginning of the test, the slit opened and its tip expanded inward. The outer covering film of the material was destroyed first. The tearing strength increased continuously, and the curve continued to rise. The curve peaked for the first time when the first yarn broke. Subsequently, the remaining tensile yarns continued to break, and the tensile curve fluctuated frequently until the last yarn broke. Then the stress dropped to 0 N/mm and the material failed. In contrast, when n is 5, there are more yarns under tension, the tearing of the cover layer will precede the specimen with n 10, and the first yarn will break faster, so the specimen reaches its stress peak sooner.

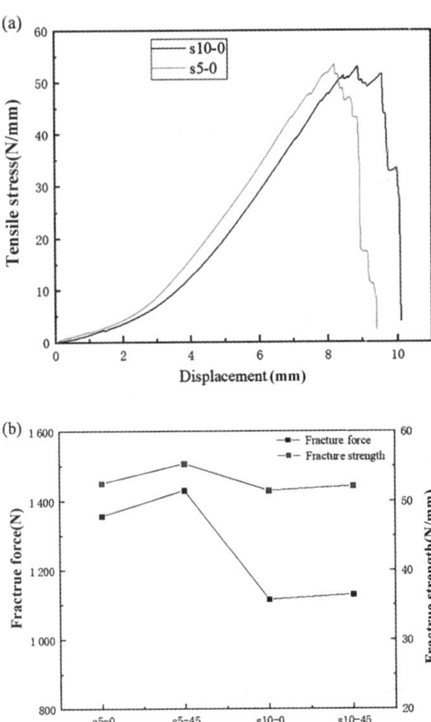

Fig. 4 (a) Tensile curves of two specimens at slit angle of 0°; (b) comparison of tearing performance of different specimens

Figure 4 (b) shows the tear force and strength at 0° and 45° for slit lengths of 5 and 10. When the slit angle is constant, the tear strength of the material decreases as the slit length increases. The maximum tearing stress of specimen s10−0 decreases by 2% with respect to s5−0 when the slit angle is 0°, and by 5.8% when the slit angle is 45°. This is because when the slit length increases, fewer yarns are stretched and the bearing width decreases, so the required force to pull off the specimen decreases. Yarns break when the slit angle is changed.

3.3 Effect of slit angle

To study the effect of slit angle on the tearing performance of the envelope materials, tests were conducted on specimens with slit angles of 0°, 30°, 45°, and 60°. Figure 5 shows the stretching process of the specimens at different slit angles when the slit length is 10, and the curves are shifted by 0.8% in order to make the tensile process of the specimens clear. The curves show that shapes and trends of the tensile curves are similar for specimens with different slit angles. However, the curves of the specimens with a slit angle of 0° and 45° are more consistent in shape at the yarn fracture stage, because at 0°, only the warp yarn is pulled, and at 45°, both the warp and weft yarns are stressed, while the bearing layer of the material is plain woven fabric, and the stresses in the warp and weft directions are more consistent.

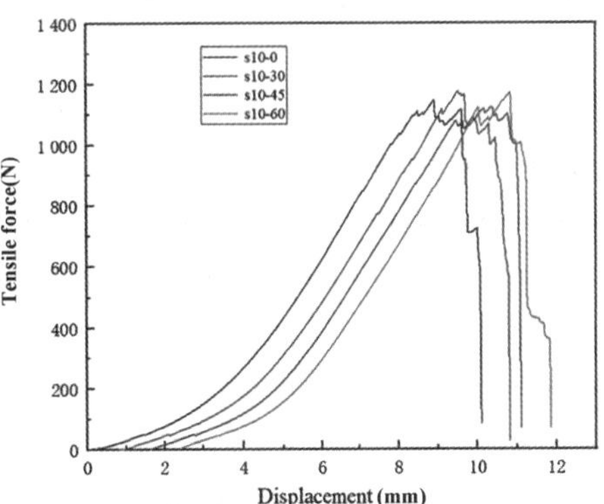

Fig. 5 Loading curves of specimens with different slit angles

As shown in Fig. 6, the maximum tear stress value of the specimen at a slit angle of 30° is greater than that at other angles. The overall tearing stress is higher because the warp yarns are more stressed than the weft yarns and the warp yarn tension is slightly higher than the weft yarns under the cover layer. However, the tearing stress is only 5.58% higher than that at 0°. This indicates that changing the slit angle has less effect on the maximum tearing stress of the specimen when the projection length of the slit in the direction perpendicular to the tensile direction is equal[7]. According to the experimental results in the previous section, the effect of changing the slit angle on the tear strength of the specimen is greater for n of 5 than that for n of 10, which indicates that changing the slit angle has a greater effect on the tearing properties of the specimen when the slit length is small.

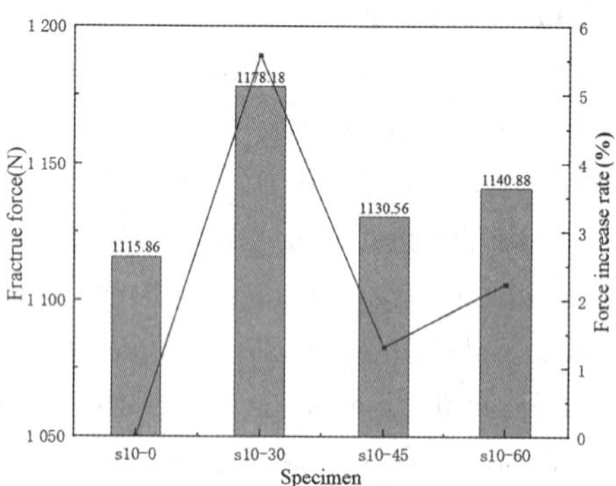

Fig. 6 Tensile tearing results of specimens with different slit angles

3.4 Digital image correlation

To further study the expansion of the specimen slit in the test, images of the displacement field in the test were acquired using digital image correlation techniques. From Fig. 7, it can be found that the stress concentration region at the tip of the slit has the largest displacement during the fracture process, and even the position of some points cannot be captured. As the stretching proceeds, the tangential seam expands along the direction perpendicular to the tensile direction, rather than extending along the direction of the tangential seam. In this process, the region near the tangential seam stretches more uniformly, and the upper and lower sides are elongated by the force simultaneously.

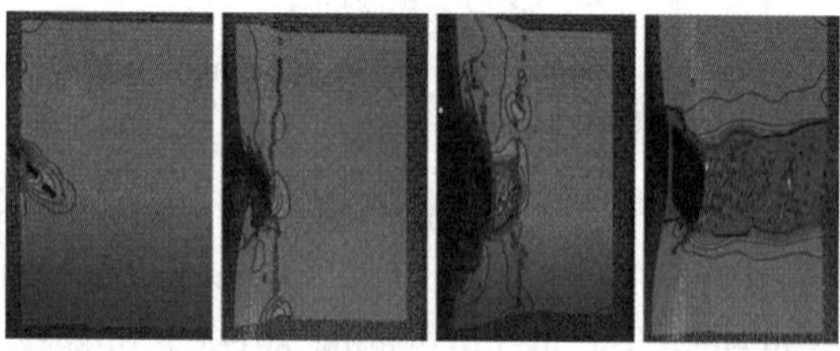

Fig. 7 Single edge slit specimen tensile tearing process

4 Conclusions

This paper discusses the tearing properties of a new stratospheric airship envelope material, the results are as follows.

The failure process of single edge slit specimens is roughly divided into four stages: slit open, slit expansion, yarn fracture, and final destroy. When the slit angle is constant, the increase of the slit length causes the tear strength of the material decrease, and the difference is more significant when the slit exists at a certain angle. When the slit length is constant and the slit angle is 30°, the maximum tear stress is greater than that at other angles. When the slit length is shorter, the effect of changing the slit angle on the lower performance of the specimen will be greater.

Acknowledgements

This work was financially supported by the Shanghai Natural Science Foundation of Shanghai Municipal Science and Technology Commission (No. 20ZR1400600), the Fundamental Research Funds for the Central Universities (Nos. 2232021G - 06, 2232020A4 - 09, 2232020D-09).

References

[1] TAN H F, LIU Y Y, et al. Research progress and requirement analysis of envelope materials for near space airship[J]. Acta Materiae Compositae Sinica, 2012, 29(6): 1-8.

[2] LIN L, IGOR P. A review of airship structural research and development[J]. Progress in Aerospace Sciences, 2009. 45(4): 83-96.

[3] LIU J H, LIU Z H, JI X M, et al. Research status of stratoshpheric airship envelope material[J]. Information Recording Materials, 2016, 17(2): 1-5.

[4] ZHAO D, LIU D X, SUN K W, et al. Research status, technical difficulties and development trend of stratospheric airship[J]. Acta Aeronautica et Astronautica Sinica, 2016, 37(1): 45-56.

[5] XU Y M, ZHU W Y, LI J, et al. Improvement of endurance performance for high-altitude solar-powered airships: a review[J]. Acta Astronautica, 2020, 167: 245-259.

[6] KANG W, SUH Y, WOO K, et al. Mechanical property characterization of film-fabric laminate for stratospheric airship envelope[J]. Composite Structures, 2006, 75(1/2/34): 151-155.

[7] WANG F X. Research on airship envelope material of tensile and tearing properties[D]. Shanghai: Shanghai Jiao Tong University, 2019.

Preparation and Properties of Fiber Foaming Coated Aramid Needle Felt Filter

CHEN Qi[1], CHEN Miaomiao[1], WANG Hong[1*]

[1]*Department of Nonwovens Materials and Engineering, College of Textiles, Donghua University, Shanghai 201620, China*

* *Corresponding author's email*: wanghong@ dhu. edu. cn

Abstract: In this study, Betaine is selected as the foaming agent, and glass fibers with different diameters are added to finish the coating of the aramid needle felt filter material, so as to obtain an industrial filter material with excellent performance. It has been found that the smaller the diameter of the short fibers, the better the performance of the filter material. By testing the surface morphology, air permeability, pore size, and filtration performance of the aramid needle felt before and after coating, it has been found that the air permeability and pore diameter of the aramid needled felt treated with short fiber coating decrease significantly. A layer of evenly distributed short fibers is obviously attached to the surface of the needle felt, and the filtration performance of the filter material is significantly improved.

Keywords: Short fiber; Needle felt; Foam coating; Filter material; Performance

1 Introduction

Air pollution mainly includes inhalable particulate matter (PM_{10}), fine particulate matter ($PM_{2.5}$), and exhaust gas[1]. These pollutants not only cause environmental pollution, but also cause serious damage to human health. Therefore, the treatment of high temperature flue gas is imminent. Aramid filter material has the advantages of light weight, high modulus, high strength, high temperature resistance, acid and alkali resistance, insulation, good aging resistance, and long service life[2], and is widely used in the field of high temperature flue gas filtration[3]. However, the filtration accuracy of pure aramid needle felt is often insufficient in practical applications due to its large pore size. How to improve the filtration performance of aramid needle felt for high temperature dust removal is a current hot research topic.

Through post-finishing, such as dipping, padding, film coating and coating, the method of processing the aramid needle felt to improve the filtration performance is favored by researchers. Among them, the coating finishing process is relatively simple and convenient, and is widely used in the field of textile processing. Foam coating is a common coating method, which refers to a process in which a foaming agent is added to the finishing fluid to form a certain quality of foam, and then the foam is uniformly applied to the surface of the base fabric[4-6]. It has the advantages of low cost, uniform coating, easy control of coating amount, and little influence on the air permeability of the base fabric. At present, high-efficiency filtration is an important research direction of aramid needle felt. If a high-efficiency filtration aramid needle felt filter can be obtained by finishing with foam coating, the effect of aramid needle felt in the field of high temperature flue gas filtration will be improved. The application is of great significance[7-9].

In this study, Betaine is selected as the foaming agent, and the water-based foam is obtained by high-speed stirring. The short glass fibers with different diameters are uniformly dispersed in the water-based foam, and then the foam is uniformly coated on the aramid needle felt. Vacuum filtration is performed to form a coating composed of short fibers on the surface of the aramid needle felt to prepare an aramid filter with a fiber foam coating, and the air permeability and filtration performance of the material are tested.

2 Experiments

2.1 Materials

Aramid needle felt filter (585. 5 g/m^2) was provided by Yanpai Filtration Technology Co. Ltd. (Zhejiang, China). Glass fiber powder (average particle size of

15 μm) was purchased from Zhuzhong Technology Co. Ltd. (Henan, China). Glass staple fibers (average length of 3 mm, and average diameter of 8 μm) were purchased from Gaoze Refractory Co. Ltd. (Henan, China). Glass fiber paper (average diameter of 2 μm) was purchased from Zhenpu Medical Co. Ltd. (Shanghai, China). Betaine (TEGO Betain F50) was purchased from Hongte Chemical Co. Ltd. (Guangzhou, China). Polytetra fluoroethylene (PTFE) emulsion (type TE3893, solid content of 60%) was purchased from DuPont Co. Ltd., USA.

2.2 Pretreatment of fiberglass paper

The glass fiber paper was calcined at high temperature in a muffle furnace at 400 ℃ for 24 h, so as to remove the binder on the surface of the glass fiber paper and enhance its dispersion uniformity in water. The glass staple fiber with a diameter of 2 μm was obtained.

2.3 Preparation of filter

Add a certain mass of short glass fibers into the container, and at the same time add a certain mass of deionized water and a foaming agent to the container. Then the suspension was stirred with a stirrer. The speed was fixed at 5 000 r/min and the stirring time was 5 min to ensure that the volume of foam was stable. The aramid needle felt filter was cut into a circular sample with a diameter of 10 cm and placed in a vacuum suction filter bottle. The foam was evenly poured onto the needle felt to form a foam coating with a thickness of 6 cm. At the same time, vacuum filtration was carried out to make the short fibers evenly distributed on the surface of the aramid needle felt after the foam burst and vacuum drainage. Then, the samples were immersed in the PTFE emulsion for 10 min, and then dried at 250 ℃, so that the short fibers could better adhere to the surface of the aramid needle felt.

The sample number and foaming liquid parameters of the fiber foam coating aramid needle felt filter material are shown in Tab. 1, and the number of the aramid needle felt filter material without coating finishing is 0#.

Tab. 1 Sample number and foaming liquid parameters of fiber foam coating aramid needle-punched felt filter material

Sample serial number	Short fiber diameter(μm)	Short fiber content(%)	Betaine content(%)
1#	15	0.5	2
2#	8	0.5	2
3#	2	0.5	2

2.4 Observation of surface structure

The surface structure of the filter material was observed by the FlexSEM - 1000 scanning electron microscope (SEM). First, the sample was made into a sample suitable for the size of the instrument sample stage, and the sample was adhered to the sample stage with conductive adhesive. Then the gold spraying treatment was carried out. Finally, the surface structure of the filter material was observed.

2.5 Air permeability test

Using the YG461E automatic air permeability tester to test the air permeability of the prepared samples. The area of the presser foot was 20 cm^2. The dynamic pressure difference was 200 Pa, and the diameter of the nozzle was automatically adjusted by the instrument.

2.6 Aperture test

The pore size and distribution of the samples were tested by the CFP - 1100AI aperture analyzer (PMI, USA), and the sample size was 25 mm×25 mm. In this experiment, the performance of the sample was mainly characterized by the pore size distribution histogram and the most distributed pore size.

2.7 Filtration efficiency and resistance test

The TSI8130 automatic filter material instrument was used to characterize the filtration efficiency and filtration resistance of filter materials. At a certain air flow rate, particles with a specified concentration and particle size distribution were used to penetrate the filter material. The concentration of particles before and after filtration and the change of pressure drop were measured, and the instantaneous pressure drop and instantaneous penetration rate were calculated.

3 Results and Discussion

3.1 Surface structure analysis of filter material before and after coating

The SEM images of the aramid needle felt filter material coated with glass fibers with different diameters are given in Fig. 1. It can be seen from the Fig. 1 (a) that the entanglement between the aramid fibers can be clearly seen on the surface of the uncoated aramid needle felt, and the pores between the fibers are relatively large. As shown in Fig. 1 (b), after the aramid needle felt filter material is coated with 15 μm diameter glass fibers, a layer of particles is obviously attached to the surface of the aramid fibers, adhesion occurs between adjacent fibers, and the pores between fibers become smaller. In Fig. 1 (c), the surface of the filter material has been covered by glass fibers with a diameter of 8 μm and PTFE particles, and the aramid fiber can not be observed on the surface, and the glass fibers are filled with PTFE particles. It can be seen from Fig. 1 (d) that PTFE particles and glass fibers with a diameter of 2 μm form a dense film on the surface of the aramid needle felt. The glass fibers are evenly distributed and adhered to each other, and the gaps between the fibers are small.

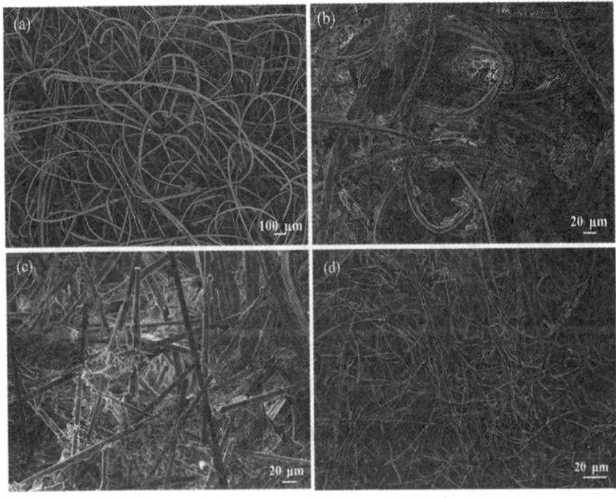

Fig. 1 SEM images of (a) 0# sample: uncoated aramid needle felt; (b) 1# sample: 15 μm glass fiber coating; (c) 2# sample: 8 μm glass fiber coating; (d) 3# sample: 2 μm glass fiber coating

3.2 Air permeability analysis of filter material before and after coating

The change of air permeability of the aramid needle felt filter material before and after coating is shown in Fig. 2. It can be seen that the air permeability of the aramid needle felt filter material after the short fiber foam finishing is lower than that of the uncoated aramid needle felt filter material. The smaller the diameter of the glass fibers in the foam is the more the air permeability decreases, especially the filter material coated with glass fibers with a diameter of 2 μm, and the air permeability decreased by 81.3%. This is because the foam coating is coated on the surface of the aramid needle felt. After the foam is ruptured by vacuum filtration, the short fibers and the PTFE emulsion form a film on the surface of the filter material, which greatly reduces the pores between the fibers and reduces the air permeability of the filter material.

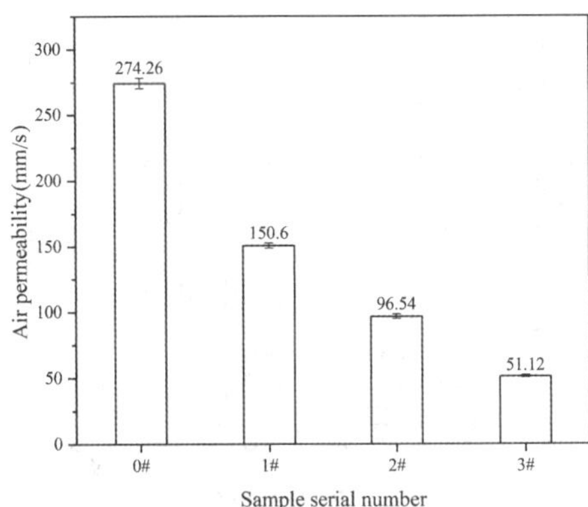

Fig. 2 Air permeability of aramid needle felt filter material before and after coating

3.3 Pore size analysis of filter material before and after coating

The pore size distribution of the aramid needle felt filter material coated with glass fibers with different diameters is given in Fig. 3. The pore size distribution of the uncoated aramid needle felt filter material is relatively dispersed, the pore size range is 10–40 μm, and the maximum pore size reaches 60.59 μm. Most of the pore size distribution is 10–25 μm. The pore size is relatively large. As a filter material, it can only filter larger soot particles, and the filtering performance is poor.

The pore size of the aramid needle felt filter material coated with 15 μm glass fibers is mainly distributed between 0 and 30 μm, the maximum pore size is 48.50 μm, and the minimum pore size is 0.42 μm. Compared with the uncoated filter material, the pore size is greatly reduced, which can filter smaller particles. The pore size distribution of the filter material coated with 8 μm glass fibers is very concentrated, most of the pore sizes are less than 50 μm, and the minimum pore size is 3.40 μm. But compared with the 1# sample, the filtration performance is not greatly improved. The pore size of the filter material coated with 2 μm glass fibers is mainly distributed between 0 and 5 μm, and the pore size is greatly reduced. The minimum pore size is 0.43 μm, and the average pore size reaches 2.79 μm, which can effectively filter out particles with a diameter of less than 5 μm and achieve high efficiency filter.

3.4 Analysis of filtration performance of filter material before and after coating

It can be seen from Fig. 4 that the filtration performance of the uncoated aramid needle felt filter material is

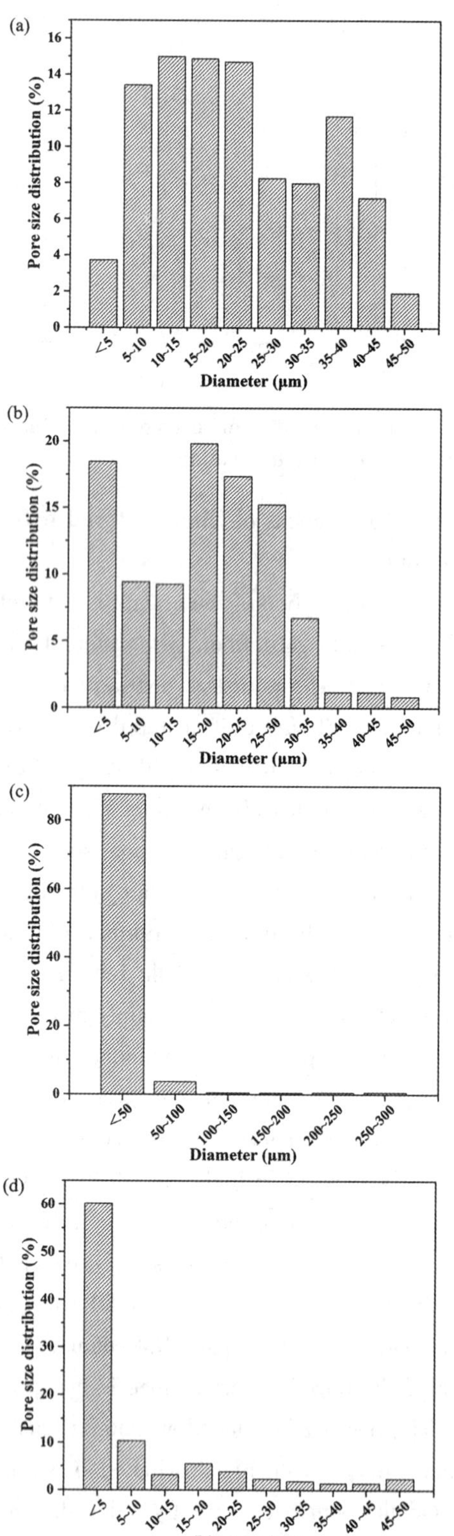

Fig. 3 Pore size distribution diagram of (a) 0# sample; (b) 1# sample; (c) 2# sample; (d) 3# sample

poor, the filtration efficiency is only 53. 34%, and the filtration resistance is 12. 26 Pa. After finishing with glass fibers coating with different diameters, the filtration efficiency of the filter material has been improved on the whole. The filter material coated with 2 μm glass fibers has the most excellent improvement effect, and its filtration efficiency is as high as 81. 96%, which is 54% higher than that of the uncoated filter material. The filtration performance of the aramid needled felt has been greatly improved. At the same time, due to the coating on the surface of the needle felt, its filtration resistance also greatly increases. However, it can be seen from the air permeability test that the overall filter material still has a relatively large air permeability, so the filtration performance of this filter material is very excellent.

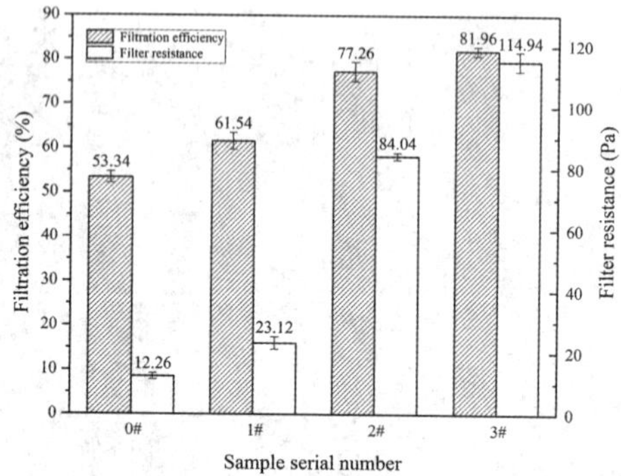

Fig. 4 Filtration performance of aramid filter material before and after coating

4 Conclusions

The aramid needle felt is coated with glass fibers with different diameters after foaming, and the changes in the surface structure, air permeability, pore size, and filtration performance of the aramid needle felt before and after foam coating have been investigated. The results show that the performance of the aramid needle felt after coating is improved in all aspects, especially the aramid needle felt filter material coated with glass fibers with a diameter of 2 μm. A dense film is attached to the surface of the filter material, and the glass fibers are evenly dispersed, so that the average pore size of the filter material reaches 2. 79 μm, and the pore size is significantly reduced, resulting in a large filtration resistance, but the air permeability is still good, and the filtration efficiency is greatly improved 53. 34% significantly increased to 81. 96%, which meets the requirements of high-efficiency filtration of aramid needle felt.

References

[1] ZHANG L. Analysis on the status quo and prevention

measures of atmospheric environmental pollution in my country [J]. Shandong Industrial Technology, 2018, (13): 208.

[2] KATO T, MAGARIO A. The wear of aramid fiber reinforced brake pads: the role of aramid fibers[J]. Tribology Transactions, 2008, 37 (3): 559-565.

[3] WANG Z L, LI Y P, ZHU W, et al. Discussion on the progress of aramid fiber research and the importance of establishing relevant standards [J]. China Quality and Standards Guide, 2019(1): 65-68.

[4] SU C, LI Y, CAO H, et al. Novel PTFE hollow fiber membrane fabricated by emulsion electrospinning and sintering for membrane distillation[J]. Journal of Membrane Science and Engineering Ethics, 2019, 583: 200-208.

[5] KIM J U, PARK B H, LEE M H. Critical parameters to determine mean bubble size of generated foams from a foam generator[J]. Journal of Applied Polymer Science, 2013, 130(3): 2062-2067.

[6] PARK B H, LEE M H, KIM S B, et al. Preparation and characterization of porous composite filter medium by polytetrafluoroethylene foam coating [J]. Air & Waste Management Association, 2010, 60(2): 137-141.

[7] HU J, LI Y, LI C, et al. Mechanical properties of fluorinated ethylene propylene (FEP) foils in use for neutrino detector project[J]. Polymer Testing, 2017, 59: 362-370.

[8] FANG D, WANG Q, LI H, et al. Mortality effects assessment of ambient $PM_{2.5}$ pollution in the 74 leading cities of China[J]. Science of the Total Environment, 2016, 569/570: 1545-1552.

[9] WANG H, ZHAO J, ZHU Y, et al. The fabrication, nano/micro-structure, heat-and wear-resistance of the superhydrophobic PPS/PTFE composite coatings[J]. Journal of Colloid and Interface Science, 2013, 402: 253-261.

Design and Development of Biodegradable 3D Nanofibrous Oil-Absorbing Sponge

SHA Qiyu[1*], LIU Ruohua[1]

[1] *Department of Technical Textiles, College of Textiles, Donghua University, Shanghai 201620, China*

* *Corresponding author's email*: qiyusha4645@ 163. com

Abstract: The application of environmentally friendly oil-absorbing materials in oil pollution treatment is receiving increasing attention. In this study, a fully biodegradable 3D fluffy porous structure nanofiber sponge was prepared by using polycaprolactone (PCL) as the main material, and Pluronic F127 as the foaming agent, followed by sodium borohydride ($NaBH_4$) gas expansion. The effects of Pluronic F127 and $NaBH_4$ content on the formability and conformability of the 3D nanofiber sponge were explored. It was shown that the expansion height of the sponge became larger as the concentration of Pluronic F127 and $NaBH_4$ increased, but too high a concentration made the sponge structure lose and soft and did not contribute significantly to its oil absorption and oil retention effect. The best overall effect of the 3D nanofiber sponge is achieved when the concentrations of Pluronic F127 and $NaBH_4$ are 4% and 10%, respectively. Compared with the nanofiber membrane, using a nanofiber sponge body as oil-absorbent material can maintain high adsorption capacity without reducing the oil retention rate, which provides a new way to develop oil-absorbent materials.

Keywords: Electrostatic spinning; Sponge; Oil-absorbent properties; Polycaprolactone (PCL); Pluronic F127

1 Introduction

Polycaprolactone (PCL) has received much attention because of its biodegradability, environmental friendliness, and excellent physical and mechanical properties. PCL is selected as the substrate and prepared into the membrane by electrostatic spinning, with additives and 3D treatment to enhance the membrane's oil-water separation performance and adsorption space[1-2]. Pluronic F127 is one such additive that can improve composite membranes' hydrophilic and adsorption capacity and promote the molding of 3D structures[3-4]. Traditional oil-absorbing materials mostly use sponge structure; through compound modification and other processing, three-dimensional structure, adsorption, and oil-water separation performance of the new sponge material are more excellent[5-7]. It has a higher adsorption capacity and is more environmentally friendly than conventional oil-absorbing materials[8]. Currently, the most direct and effective way is to modify it by coating and to deposit low surface energy polymeric materials on its surface[9].

In this study, electrostatic spinning fiber is selected as the research direction, together with chemical modification and $NaBH_4$ swelling treatment. A degradable PCL/Pluronic F127 composite sponge body with a 3D structure is prepared, and its oil adsorption capacity and oil absorption and retention rate are evaluated.

2 Experiments

2.1 Materials

Pluronic F127 (analytical grade) and polycaprolactone (Mw = 50 kDa) were supplied by Beijing Chemical Reagent Factory, China. Hexafluoroisopropanol (H811026) was provided by Sinopharm Chemical Reagent Co. Ltd., China. Sodium borohydride, analytical grade, was supplied by Sinopharm Chemical Reagent Co. Ltd., China. Engine oil, soybean oil, and silicone oil were provided by Sinopharm Chemical Reagent Co. Ltd., China.

2.2 Preparation of PCL/Pluronic F127 nanofiber sponge

The concentration of PCL was kept constant at 10% at room temperature, and the mixture of PCL/Pluronic F127 with different ratios was prepared by varying the percentage of Pluronic F127 in PCL (0%, 2%, 4%, and 6%) using hexafluoroisopropanol (HFIP) as the solvent. PCL/Pluronic F127 nanofiber membranes were prepared using the electrostatic spinning technique. The nanofiber membranes of 15 cm×15 cm made at 4% concentration of Pluronic F127 were swollen with different concentrations of $NaBH_4$ (6%, 8%, 10%, and 12%) for 30 min to obtain sponge bodies, and the swelling effect of sponge bodies was observed.

2.3 Oil absorption capacity test of PCL/Pluronic F127 nanofiber sponge

To clarify the magnitude of the oil adsorption capacity of the membranes with different Pluronic F127 concentrations and the sponges swelled with different concentrations of $NaBH_4$. Three oil adsorption experiments were conducted, in which the prepared fibrous membrane and sponge were immersed in a surface dish containing engine oil, soybean oil, and silicone oil. The oil absorption rate was calculated:

$$q = (m_1 - m_0)/m_0 \tag{1}$$

where q is the adsorption capacity of the fibrous membrane and sponge ($g \cdot g^{-1}$), and m_0(m_1) is the mass of the fibrous membrane and sponge before (after) adsorption (g). Oil absorption experiments were conducted three times for all three oils, and the average value was taken.

2.4 Oil retention capacity test of PCL/Pluronic F127 nanofiber sponge

The excellent oil retention helps avoid secondary contamination due to oil droplet leakage after oil absorption. After the oil absorption experiment, each sample group analyzed the oil retention effect. The membranes and sponges were left in the filter after oil absorption, weighed separately after 24 h, and the oil retention rate was calculated:

$$w = m_2/m_1 \tag{2}$$

where w is the oil retention rate after 24 h, and m_1 (m_2) is the mass of the fibrous membrane and sponge after oil adsorption (24 h of resting) (g).

3 Results and Discussion

3.1 Sponge physical picture analysis

The height of sponge expansion for different Pluronic F127 concentrations is shown in Fig. 1.

The swelling effect of the spongy body is the best at a Pluronic F127 concentration of 4%, and the layers are well defined without being limp.

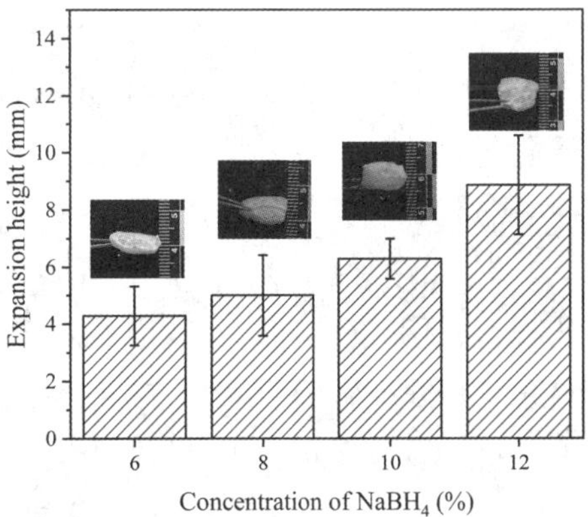

Fig. 1 Height of sponge expansion for different Pluronic F127 concentrations

The height of sponge expansion for different $NaBH_4$ concentrations is shown in Fig. 2.

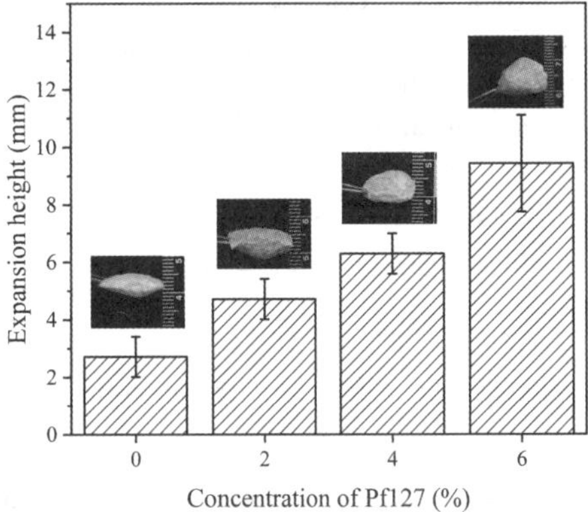

Fig. 2 Height of sponge expansion for different $NaBH_4$ concentrations

It is shown that: the swelling height is too high, and the spongy body is too loose when the concentration of $NaBH_4$ is 12%; the swelling produced by 6% $NaBH_4$ is insufficient, and the layers are poorly defined. The final comparison concluded that when 10% $NaBH_4$ is added, the expansion height and thickness of the spongy body are moderate, and the expansion effect is good.

3.2 Scanning electron microscope (SEM) analysis

The SEM images of the nanofiber membrane and the sponge are shown in Fig. 3.

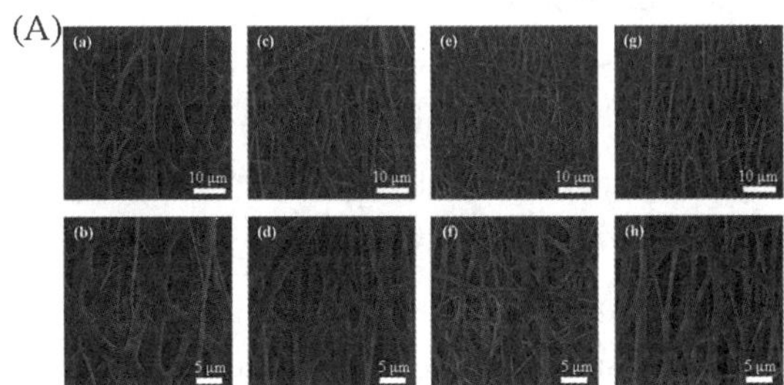

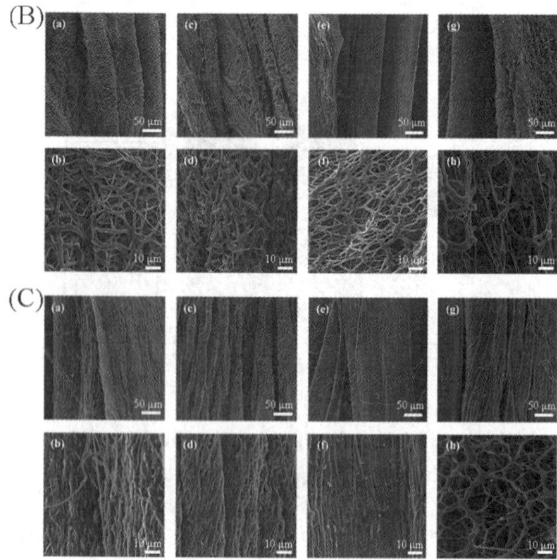

Fig. 3 (A) SEM images of nanofibrous membranes with different Pluronic F127 concentrations, 0%: (a), (b); 2%: (c), (d); 4%: (e), (f); 6%: (g), (h); (B) SEM images of nanofiberbre sponges with different Pluronic F127 concentrations, 0%: (a), (b); 2%: (c), (d); 4%: (e), (f); 6%: (g), (h); (C) SEM image of nanofiber sponge with different $NaBH_4$ concentrations, 0%: (a), (b); 2%: (c), (d); 4%: (e), (f); 6%: (g), (h)

The morphological surface of the electrostatic spun nanofiber membrane is smooth and flat, with some adhesion; the fiber diameter decreases, and the pores between the fibers increase as the concentration of Pluronic F127 is added. The sponges have an excellent 3D structure, obvious lamellar structure, and a clear mesh structure between fibers; with a higher concentration of Pluronic F127, the reaction in aqueous $NaBH_4$ solution is more rapid, and the swelling effect is more obvious than before. As the concentration of $NaBH_4$ increases, the swelling effect of the nanofiber membrane is enhanced, the swelling height of the sponge becomes larger, the lamellar structure is more obvious, the pores between the fibers become larger, and the adsorption is stronger.

3.3 Infrared spectral analysis

The FTIR spectra of three different nanofiber membranes are shown in Fig. 4.

The characteristic peaks of the PCL/Pluronic F127 nanofiber membrane contain all the characteristics of PCL and Pluronic F127.

A comparison of the IR spectra of the membrane and the sponge is shown in Fig. 5.

The nanofiber sponge is identical to the membrane, indicating that the expansion does not alter the chemical structure of the sponge during the swelling treatment, and there is no $NaBH_4$ residue present.

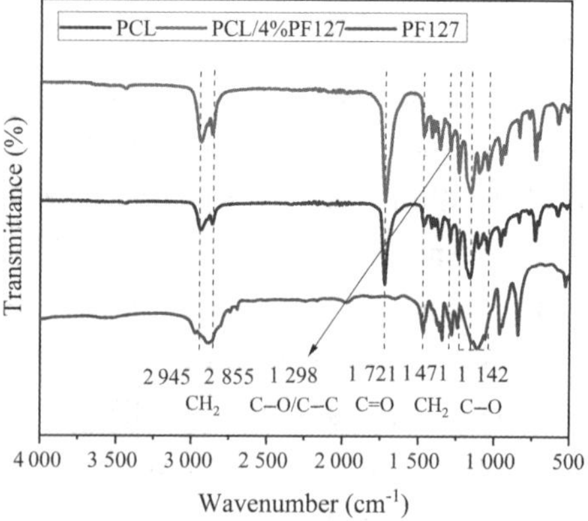

Fig. 4 Infrared spectra of PCL, PCL/Pluronic F127, and Pluronic F127

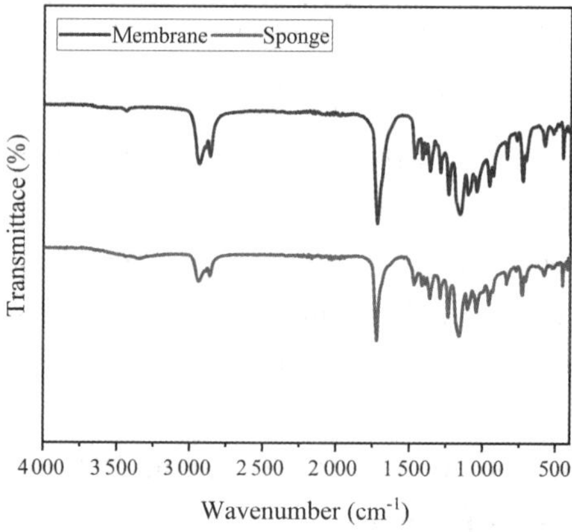

Fig. 5 Comparative infrared spectra of PCL/4% Pluronic F127 nanofiber membranes and sponges

3.4 Oil absorption ability test

The oil absorption of the membranes and sponges is shown in Figs. 6 and 7.

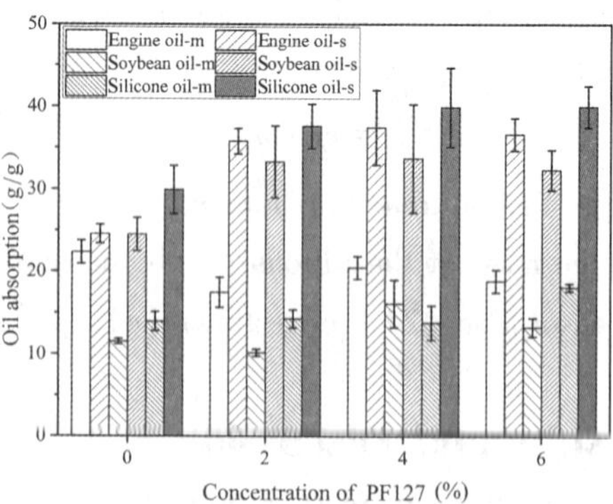

Fig. 6 Comparison of oil absorption by nanofiber membranes and sponges at different Pluronic F127 concentrations

The oil absorption capacity of the sponge is 2–3 times that of the membrane. When the concentration of Pluronic F127 is 4%, the combined oil absorption of the two is optimal, where the oil absorption of the sponge on silicone oil is up to 39.91 $g \cdot g^{-1}$.

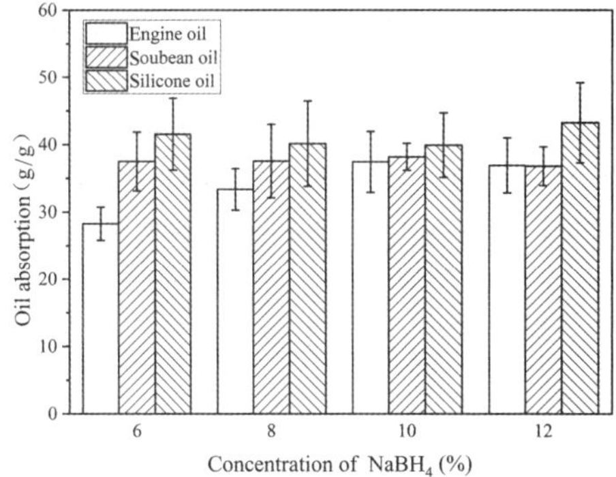

Fig. 7 Oil absorption of nanofiber sponges swollen with different $NaBH_4$ concentrations

The oil absorption increases as the concentration of $NaBH_4$ increases. When the concentration is too high, the sponge structure is too loose, which is not conducive to moulding and affects its oil absorption and energy absorption. Thus it is best to choose 10%.

3.5 Oil retention ability test

The oil retention rates are shown in Figs. 8 and 9.

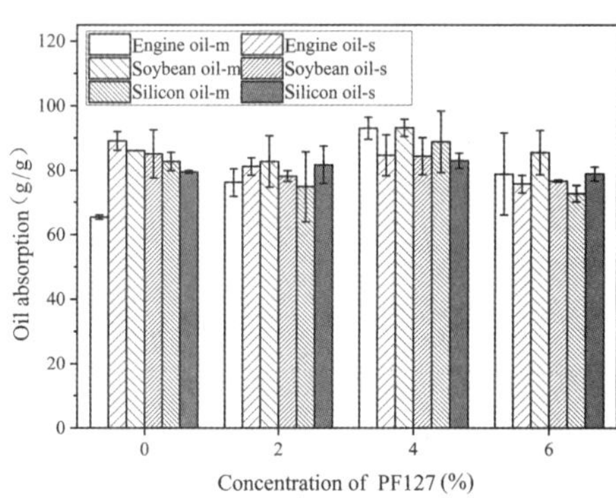

Fig. 8 Comparison of oil retention rates of different Pluronic F127 concentration nanofiber membranes and sponges

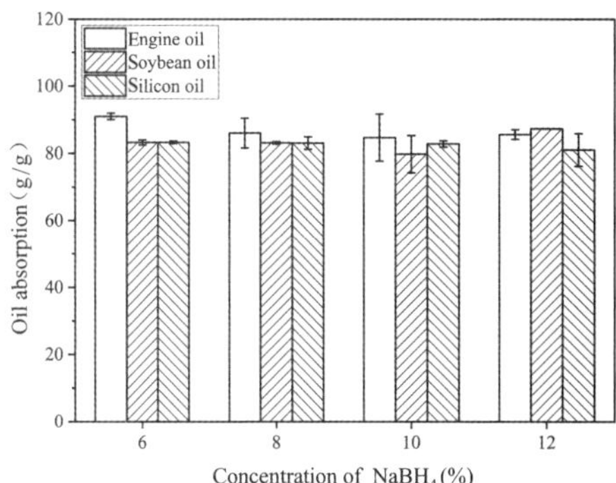

Fig. 9 Oil retention of nanofiber sponges swollen with different $NaBH_4$ concentrations

At different concentrations of Pluronic F127, the oil retention rates of the membrane and sponge were close to each other, and the oil retention effect of both was slightly better at 4% concentration. Therefore, after the fiber membrane expands into a sponge, the oil retention rate fluctuates less while the oil absorption capacity increases.

It can be seen that the oil retention effect of the sponges does not vary much, and the mesh structure of the sponges with different heights has a more negligible effect on the retention of the absorbed oil, which indicates that the process of swelling the PCL/Pluronic F127 nanofiber membrane into sponges does not reduce its oil retention performance.

4 Conclusions

In this study, 3D fluffy sponges were obtained by $NaBH_4$ gas expansion treatment of nanofiber membranes prepared from PCL and Pluronic F127. It was shown that: (1) the swelling height of the 3D nanofiber sponge increased with the increase of Pluronic F127 and $NaBH_4$ concentrations; (2) when the concentrations of both were 4% and 10%, the sponge had the best swelling effect and had a 3D mesh structure, obvious nanofiber layer and suitable size of fiber pores; (3) the adsorption capacity of the sponge reached 35–50 $g \cdot g^{-1}$, which is 2–3 times of the membrane, and the oil retention effect is slightly increased, so the 3D nanofiber sponge can maintain the high adsorption capacity without making the oil retention effect worse.

References

[1] ZHANG G H. Preparation of novel polycaprolactone-based superhydrophobic membranes and their oil-water separation performance[D]. Jilin: Jilin University, 2019.

[2] ZHANG X X. Preparation and performance of bionic polycaprolactone-based high-efficiency oil-water separation membranes[D]. Jilin: Jilin University, 2020.

[3] LI J. Preparation of anti-pollution polyethersulfone ultrafiltration membrane and its application in oil-water separation[D]. Nanjing: Southeast University, 2017.

[4] ZHAO W, SU Y, LI C, et al. Fabrication of antifouling polyethersulfone ultrafiltration membranes using Pluronic F127 as both surface modifier and pore-forming agent[J]. Journal of Membrane Science, 2008, 318(1/2): 405–412.

[5] LI B B. Preparation and performance of lipophilic and hy-

drophobic sponge and membrane-based oil-water separation materials[D]. Shanghai: Shanghai University, 2016.
[6] XIE J. Preparation of lipophilic hydrophobic sponges and textile fabrics and their properties[D]. Gung Zhou: South China University of Technology, 2020.
[7] ARUNAGIRI V, et al. Facile fabrication of eco-friendly polycaprolactone (PCL)/Poly-D, L-Lactic acid (PDLLA) modified melamine sorbent for oil-spill cleaning and water/oil (W/O) emulsion separation[J]. Separation and Purification Technology, 2021, 259: 118081.
[8] MELBIAH J S B, NITHYA D, MOHAN D. Surface modification of polyacrylonitrile ultrafiltration membranes using amphiphilic Pluronic F127/$CaCO_3$ nanoparticles for oil/water emulsion separation[J]. Colloids and Surfaces A: Physicochemical and Engineering Aspects, 2017, 516: 147-160.
[9] WU W J. Preparation of high-performance oil-absorbing materials and their oil-water separation performance[D]. Qingdao: Qingdao University of Science and Technology, 2014.

Self-Floating Solar Evaporation Device Constructed from Carbon Nanotube-Modified Cotton Fabric for Seawater Desalination

JIAN Ying[1], ZHUANG Qinliang[1*], GUO Jiansheng[1]

[1]Department of Technical Textiles, College of Textiles, Donghua University, Shanghai 201620, China

* Corresponding author's email: qlzhuang @ dhu. edu. cn

Abstract: The earth's surface is covered by a large amount of water, but the freshwater resources that can be directly used by humans are very scarce. In the existing freshwater resources, the shortage of freshwater resources is a common problem faced by human beings. Seawater desalination is a green and promising way to solve the scarcity of freshwater resources. Cotton fabric has good hygroscopic properties, and the voids between the fibers in the fabric are conducive to the passage of water molecules during evaporation, which is a very good base material for photothermal evaporation. In this paper, carbon nanotubes and sodium alginate are mixed uniformly and attached to the surface of cotton fabric by a spraying method to prepare photothermal materials. Through testing, it is found that the carbon nanotube-coated cotton fabric has excellent photothermal properties, and the surface temperature can reach 56. 7 ℃ within 3 min under the illumination power of 1 kW · m^{-2}. Through the contact angle test, it is proved that the cotton fabric has good hydrophilic properties. A self-floating evaporation device with excellent air mobility was built, and the evaporation test was carried out under the conditions of illumination power of 1 kW · m^{-2} and brine concentration of 3. 5 wt%. The results show that the evaporation device has excellent evaporation performance, and the evaporation rate can reach 2. 00 kg · m^{-2} · h^{-1}. It can be considered to be used for desalination to solve the shortage of fresh water resources.

Keywords: Cotton fabric; Solar evaporation; Desalination; Photothermal

1 Introduction

Freshwater resources play a vital role in the survival and development of human beings. The demand for freshwater resources is already a major difficulty faced by human beings, and the uneven distribution caused by factors such as geographical location aggravates the development of the problem[1]. 71% of the earth's surface is covered by water, but fresh water only accounts for 2. 5% of all water resources[2]. Some data show that by 2025, 60% of the world's total population will face extreme water use problems[3], which deserves attention. To solve this problem, scientists have developed techniques to distill seawater into fresh water[4]. Solar energy is a green and clean energy source, and it is an inexhaustible renewable energy source. Using it as an energy supply for evaporators is a field worthy of research[5].

Primitive solar evaporation is desalination of seawater by irradiating the water body itself with solar energy to obtain distilled water. Water directly acts as the main body of solar energy to absorb light and convert the absorbed solar energy into heat energy. Due to the low absorption rate of water in the wavelength range of solar energy[6], and the serious heat loss in the water body, the utilization efficiency of solar energy is low, and the efficiency of photothermal evaporation is low[7]. In recent years, people have continuously improved the solar water evaporation system (SSG)[8]. The evaporator structure can be divided into three parts: base material, light-to-heat conversion material, and heat insulation device. Photothermal materials generally include metal materials, carbon-based materials, such as Cu^{2+}, carbon black (CB), carbon nanotubes (CNTs), and polypyrrole (PPy). The thermal insulation device is generally designed by foam.

Here, cotton fabric is used as the base material, CNTs are used as the photothermal material, multi-wall carbon nanotubes (MWCNTs) and sodium alginate (SA) are evenly mixed, and the carbon material is sprayed onto the cotton fabric by a high-pressure spray gun to modify the cotton fabric to prepare photothermal material with excellent photothermal performance. An evaporation device with excellent air mobility is built independently.

2 Experiments

2.1 Materials

Plain woven cotton fabrics (560 g/m^2) were obtained from Shaanxi Xinxing Textile Co. Ltd., China. MWCNT was provided by Chinese Academy of Sciences Chengdu Organic Chemistry Co. Ltd., China. SA, sodium hydroxide (NaOH) and sodium dodecyl benzene sulfonate (SDBS) were provided by Sinopharm Chemical Reagent Co. Ltd., China. Anhydrous ethanol (C_2H_6O) was obtained from Shanghai Titan Technology Co. Ltd., China.

2.2 Fabric treatment

Simple pretreatment of cotton fabrics (30 mm × 12 mm) to remove surface impurities. In order to remove the organic impurities on the cotton fabric, soak the cotton fabric in absolute ethanol for ultrasonic cleaning for 15 min, then ultrasonically clean it with deionized water for 25 min, and rinse it with deionized water for several times to remove the impurities on the cotton fabric. The cotton fabric was then placed in a mixed solution of NaOH (30 mg/mL) and SDBS (3 mg/mL), and heated in a water bath at 60 ℃ for 1 h to remove impurities. The treated cotton fabric was washed several times with deionized water and dried in a blast oven at 65 ℃.

2.3 Composite manufacture

First configure the mixed solution of MWCNT (15 mg/mL) and SA (3 mg/mL), then stir the mixed solution for 30 min (600 r/min) until the solution is mixed evenly, and clean it in an ultrasonic cleaner (power of 50 W, and frequency of 80 kHz) in ultrasonic dispersion for 45 min. Finally a homogeneous mixed dispersion of MWCNT and SA (MWCNT & SA) was obtained, as shown in Fig. 1 (a). The obtained MWCNT & SA were loaded into the high-pressure atomization spray gun silo, the treated cotton fabric was placed under a mold with a radius of 30 mm. The cotton fabric was uniformly sprayed with a spray gun, and then cut to obtain a length of 280 mm and a width of 40 mm. Cotton fabric sprayed with MWCNT (CFSM) is shown in Fig. 1 (b).

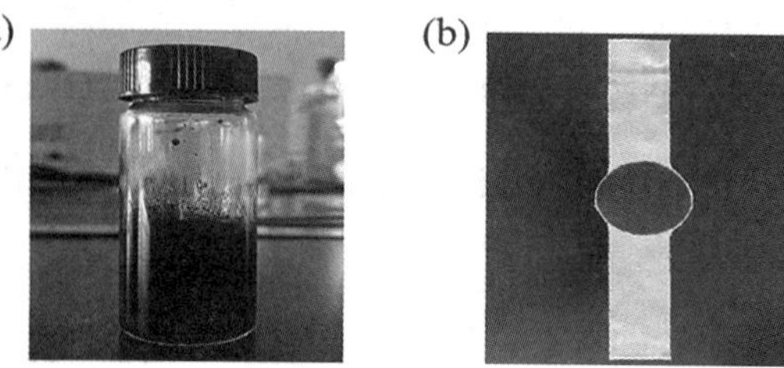

Fig. 1 (a) MWCNT & SA; (b) CF8M

2.4 Photothermal and hydrophilic performance testing of composites

The photothermal properties of CFSM were tested, and the comparison sample was the untreated cotton fabric. An infrared camera was used to test the surface temperature of the fabric during photothermal evaporation. By photographing the temperature of the material surface during actual evaporation, the maximum temperature and arrival time that CFSM and the comparison sample could reach were analyzed and compared, so as to determine the photothermal performance of CFSM. To further understand the hydrophilic properties of CFSM, contact angle tests were performed on CFSM.

2.5 Seawater evaporation testing of composites

A photothermal evaporation device was built, and the seawater evaporation test was carried out under the condition of a photothermal power of 1 kW · m^{-2}. The seawater was configured with NaCl (3.5 wt%). The evaporation rate was calculated by using the balance to record the change of the balance mass within 1 h, and the data was recorded every 5 min. Compare the evaporation of blank cotton fabric and CFSM to compare the seawater evaporation capacity of CFSM. The actual situation of seawater evaporation is shown in Fig. 2.

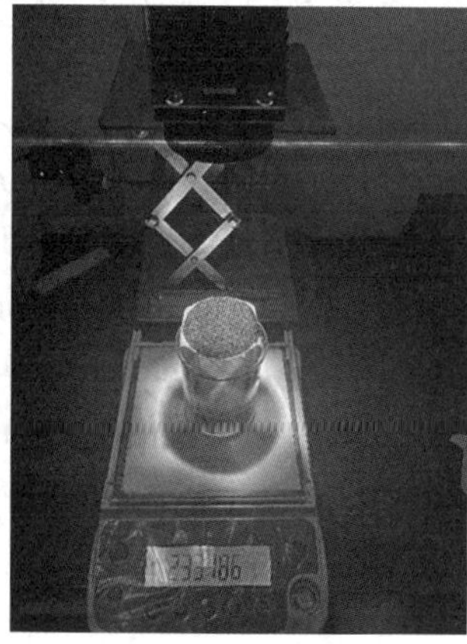

Fig. 2 Photothermal evaporation experiment

3 Results and Discussion

3.1 Analysis of MWCNT attachment

As shown in Fig. 3, it can be clearly seen that the cotton fibers are arranged in a crisscross pattern. The particles on the fibers are the MWCNT particles attached by spraying, and a large number of particles are attached to the cotton fibers. It can be seen that the particle distribution is uniform and sufficient. This distribution can ensure that the fabric has the ability to absorb light and heat, and has good hydrophilicity to ensure the water transmission capacity of the fabric. It can be seen that the preparation of CFSM is successful.

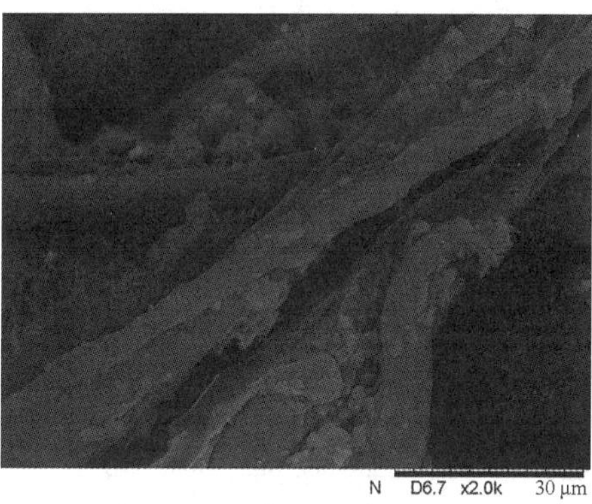

Fig. 3 SEM of cotton fabric sprayed with MWCNT

3.2 Photothermal capacity and hydrophilic performance analysis

Compared with the cotton fabrics without MWCNT treatment, CFSM showed great photothermal properties. The initial temperature of the untreated cotton fabric was 20.3 ℃ in the absence of light, the temperature rose to 29.6 ℃ in 60 s, and rose to 30.5 ℃ in 120 s, and then the temperature was basically maintained at about 30.5 ℃ large changes (shown in Fig. 4 (a)). Under the same conditions, the temperature of CFSM increased to 51.7 ℃ in 60 s, and temperature could reach 56.2 ℃ in 180 s, and the temperature kept rising with the increase of time (shown in Fig. 4 (b)). It can be seen that CFSM has excellent photothermal properties.

Photothermal evaporation requires substrate materials with good hydrophilic properties to ensure continuous water transport during the evaporation process. In order to examine the hydrophilic properties of MWCNT-treated cotton fabrics, the fabrics were analyzed by the contact angle test. It can be seen from Fig. 5 that the fabric has good hydrophilic properties. It is proved that the treatment of MWCNT does not really affect the hydrophilic properties of CFSM, which is beneficial to the photothermal evaporation.

(a)

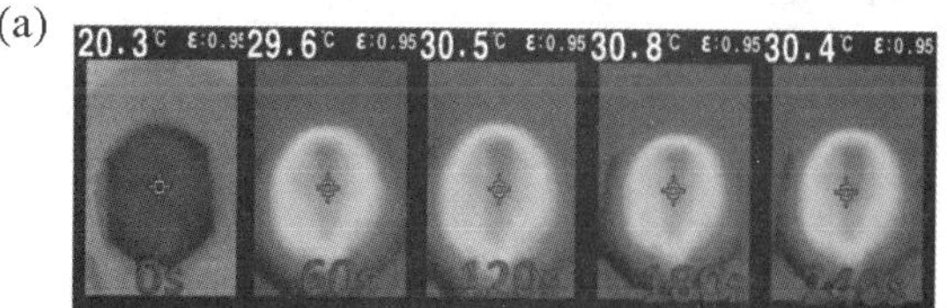

(b)

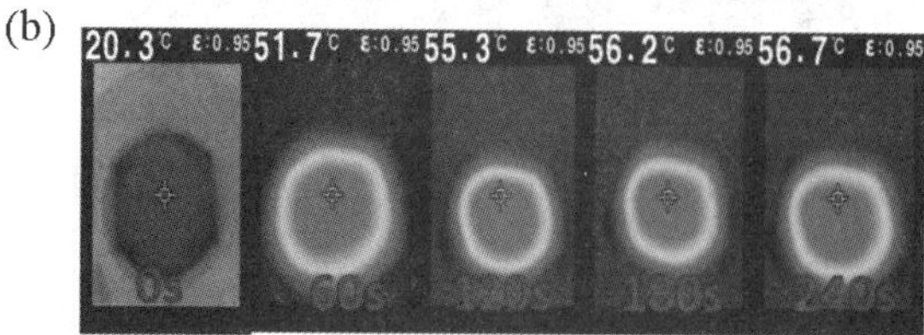

Fig. 4 (a) Temperature change of pure cotton cloth under the condition of the illumination power of 1 kW · m^{-2}; (b) temperature variation of CFSM under the condition of the illumination power of 1 kW · m^{-2}

Fig. 5 Surface contact angle testing of CFSM

3.3 Analysis of seawater evaporation capacity

As time goes on, the water evaporation effect of the cotton fabric is poor, the mass loss of seawater is less. There is no very significant evaporation performance, and the brine mass is only 2.617 g in the 1 h evaporation process. However, CFSM showed a good evaporation effect, and the quality of brine decreased significantly. The quality of brine decreased by 5.648 g in 1 h, which was significantly higher than the evaporation effect of untreated cotton fabrics. It can be seen that the evaporation effect of the untreated cotton fabric is relatively insignificant every 5 min, while the results of CFSM change very significantly (Fig. 6). By calculating the evaporation rate, it can be obtained that the evaporation rate of the untreated cotton fabric is 0.926 kg · m^{-2} · h^{-1}, while the rate of the CFSM is 2.00 kg · m^{-2} · h^{-1}, which is significantly higher than that of the blank cotton fabric, and also higher than that of many other the article has a high evaporation rate.

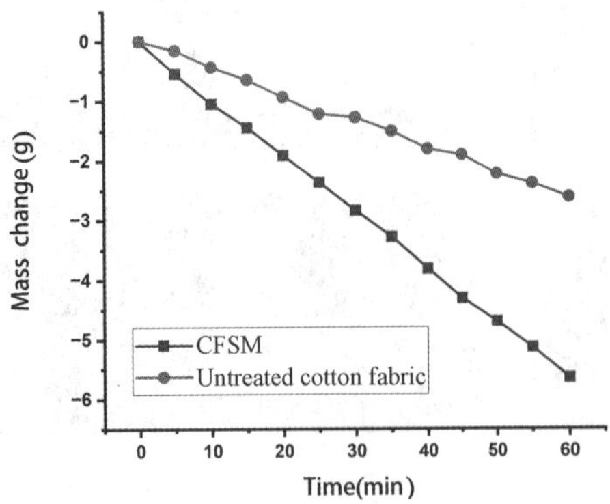

Fig. 6 Evaporation rates of CFSM and cotton fabrics in actual seawater

Evaporation rate $\Delta M(\mathrm{kg} \cdot \mathrm{m}^{-2} \cdot \mathrm{h}^{-1})$ calculation formula:

$$\Delta M = m/St$$

where m (kg) is the mass loss during time t (h) and S (m^2) is the surface area of the evaporator. The test environment temperature is 25 ℃, and the relative humidity is 50%-60%.

4 Conclusions

In this paper, the spraying of MWCNTs on the surface of cotton fabrics was studied, the photothermal evaporation device was built independently, and the photothermal properties, hydrophilic properties, and seawater desalination properties of CFSM were studied. Through experiments, it is found that CFSM has good photothermal properties, and the surface temperature can reach 56.2 ℃ within 120 s. By the contact angle test, it is proved that CFSM has good hydrophilic properties. Through the seawater evaporation test, it is found that the evaporation rate of CFSM is as high as 2.00 $\mathrm{kg} \cdot \mathrm{m}^{-2} \cdot \mathrm{h}^{-1}$, which has excellent evaporation performance and is of great significance for seawater evaporation.

References

[1] OKI T, KANAE S, MUSIAKE K. Global hydrological cycle and world water resources[J]. Membrane, 2003, 28(5): 206-214.

[2] POSTEL S L, DAILY G C, EHRLICH P R. Human appropriation of renewable fresh water[J]. Science, 1996, 271(5250): 785-788.

[3] LIU Y, LOU J, NI M, et al. Bioinspired bifunctional membrane for efficient clean water generation[J]. ACS Applied Materials & Interfaces, 2016, 8(1): 772-779.

[4] KULLAB A, MARTIN A. Membrane distillation and applications for water purification in thermal cogeneration plants[J]. Separation & Purification Technology, 2011, 76(3): 231-237.

[5] SHARSHIR S W, PENG G, YANG N, et al. A continuous desalination system using humidification dehumidification and a solar still with an evacuated solar water heater[J]. Applied Thermal Engineering, 2016, 104: 734-742.

[6] WANG Y, ZHANG L, WANG P. Self-floating carbon nanotube membrane on macroporous silica substrate for highly efficient solar-driven interfacial water evaporation[J]. ACS Sustainable Chemistry & Engineering, 2016, 4(3): 1223-1230.

[7] LIU Y, CHEN J, GUO D, et al. Floatable self-cleaning and carbon-black-based superhydrophobic gauze for the solar evaporation enhancement at the air-water interface[J]. ACS Applied Materials & Interfaces, 2015, 7(24): 13645-13652.

[8] HUANG Z, LI S, CUI X, et al. A broadband aggregation independent plasmonic absorber for highly efficient solar steam generation[J]. Journal of Materials Chemistry A, 2020, 8(21): 10742-10746.

Design and Optimization of a Four-Bar Beating Up Mechanism for Rapier Loom

HUANG Tao[1], CHEN Nanliang[1*], JIANG Jinhua[1], SHAO Huiqi[1], JIANG Fei[2], CHEN Yi[2]

[1] *Engineering Research Center of Textile Technical, Education Ministry, College of Textiles, Donghua University, Shanghai 201600, China*

[2] *Changzhou Saijia Machinery Co., Ltd., Changzhou 213132, China*;

[*] *Corresponding author's email*: nlch@dhu.edu.cn

Abstract: Four-bar beating up mechanism is widely used in looms because of its simple structure and low manufacturing cost. At present, single objective optimization is used in loom optimization, but the multi-objective optimization is seldom studied. In this paper, the mathematical simulation software is used to optimize the beating force and reed swing angle of the beating up mechanism. After the design variables are determined, the multi-objective optimization is carried out by using the ideal point method. The simulation results show that the maximum swing angle increases by a quarter and the beating force is nearly doubled by adjusting the size of the mechanism. The efficiency of the loom is greatly improved, which provides a new idea for the design and optimization of the beating up mechanism.

Keywords: Optimization design; Four bar beating up mechanism; Motion simulation

1 Introduction

As one of the five major mechanisms of looms, beating up mechanism is mainly used to push the new weft into the shed. The general beating up mechanism can be divided into two categories: connecting rod beating up and conjugate cam beating up. Among them, four-bar beating up mechanism and six-bar beating up mechanism[1] are commonly used. The conjugate cam beating up mechanism inserts weft through the curve design of the cam, in which the reed achieves relative static and completes the weft insertion action[2-3]. The six-bar linkage mechanism is to realize the weft insertion when the main shaft stops for a long time and the main shaft is at the rear center position. Four-bar linkage is widely used in air-jet loom, water jet loom and rapier loom because of its simple structure and flexible processing. However, it is difficult to insert weft without static time[4].

In a previous study, Yang[5] and other people have deeply studied the kinematic process of the four-bar beating up mechanism. Later, with the iteration of modern analysis software, the research gradually entered the simulation stage. Qiu[6] and others studied its dynamic load. Xi[7] and others reduced the vibration suffered by the mechanism during re-operation by changing the rotational inertia of the counterweight on the connecting rod. Fang[8] introduced a parametric modelling and simulation analysis method for a planar four-link mechanism using ADAMS as the software platform by using the secondary development and development language provided by ADAMS. However, most of the past studies are single-objective optimization methods, and there are fewer multi-objective optimization methods for its dynamics.

In the actual production process, the structural characteristics of the four-link mechanism itself are limited, so there are often many optimization points that need to be taken into account. This paper proposes the use of the ideal point method, a multi-objective optimization method, to optimize the dimensional parameters of the mechanism, providing a new way of thinking about the optimal design of weaving machines.

2 Method and Model

2.1 Principle and structure of four-bar beating up mechanism

The schematic diagram of the four-bar beating mechanism is shown in Fig. 1. In Fig. 1, AB rod is the crank, BC rod is the handle, CD rod is the rocker, and AD is the frame. When the crank AB rotates under the drive of the motor, the crank drives the rocker to and fro through the hand BC. In the process of the reed swinging back and forth, when the weft is introduced, the reed on the reed will drive the weft into the weaving mouth, and the reed will move to the front dead center position.

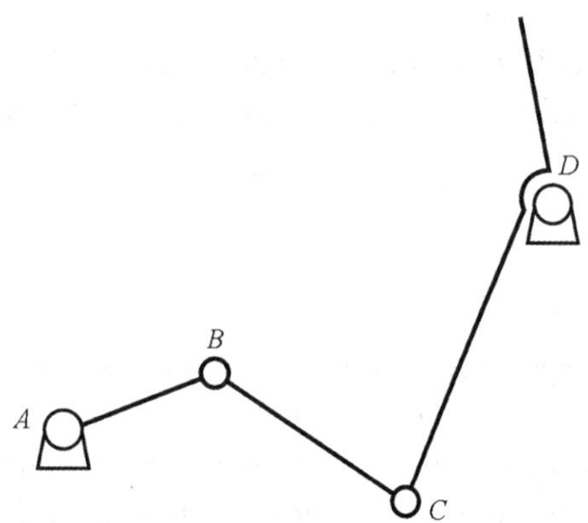

Fig. 1 Sketch of four-bar beating up mechanism

2.2 Motion analysis of four-bar beating up mechanism

The coordinate system is established with A as the coordinate origin and AD as the x-axis. The length of each rod is $AB = L_1$, $BC = L_2$, $BC = L_3$, $CD = L_4$, and the dead center positions of front and rear are $C1$ and $C2$, as shown in Fig. 2. The angle between each pole and the horizontal is θ_1, θ_2, and θ_3. Closed graph vector expression can be established from closed graph ABCD:

$$\overrightarrow{l_1} + \overrightarrow{l_2} = \overrightarrow{l_3} + \overrightarrow{l_4} \tag{1}$$

According to Euler's formula, it can be expressed in the complex form.

$$l_1 e_1^{i\theta_1} + l_2 e^{i\theta_2} = l_3 e_3^{i\theta_3} + l_4 \tag{2}$$

The swing angle of reed can be obtained by geometric relation β.

$$\begin{cases} \beta_1 = \arccos \dfrac{l_4^2 + l_3^2 - (l_2 + l_1)^2}{2 l_4 l_3} \\ \beta_2 = \arccos \dfrac{l_4^2 + l_3^2 - (l_2 - l_1)^2}{2 l_4 l_3} \\ \beta = \beta_1 - \beta_2 \end{cases} \tag{3}$$

The acceleration α_3 of the rocker arm can be obtained from Eq. 4 by taking the time t quadratic derivation of Eq. 2.

$$\begin{bmatrix} -l_2 \sin \theta_2 & l_3 \sin \theta_3 \\ l_2 \cos \theta_2 & -l_3 \cos \theta_3 \end{bmatrix} \begin{bmatrix} \alpha_2 \\ \alpha_3 \end{bmatrix} + \begin{bmatrix} -\varpi_2 l_2 \cos \theta_2 & \varpi_3 l_3 \cos \theta_3 \\ -\varpi_2 l_2 \sin \theta_2 & \varpi_3 l_3 \sin \theta_3 \end{bmatrix} \begin{bmatrix} \varpi_2 \\ \varpi_3 \end{bmatrix} = \varpi_1 \begin{bmatrix} \varpi_1 l_1 \cos \theta_1 \\ \varpi_1 l_1 \sin \theta_1 \end{bmatrix} \tag{4}$$

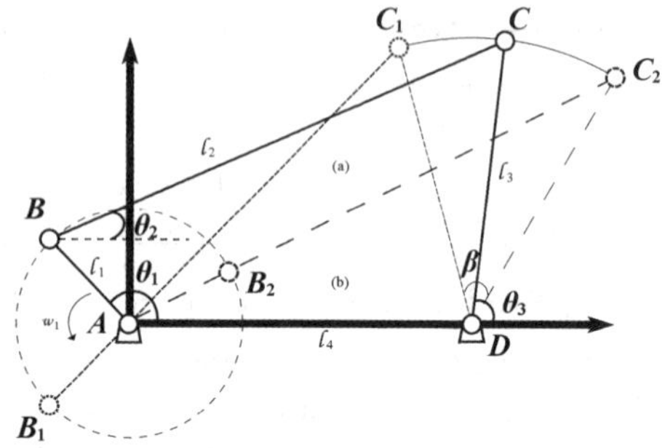

Fig. 2 Diagram of the movement of the four-link weft beater

Taking a four-bar loom as an example, $L_1 = 70$ mm, $L_2 = 105$ mm, $L_3 = 301$ mm, $L_4 = 315$ mm, and vehicle speed $n = 150$ r/min. Matlab simulation program is compiled to solve the problem. The angular position, angular velocity and angular acceleration of rocker 3 change as shown in Fig. 3 when the spindle moves once. The reed swings back and forth between 147.2° and 174.0° with a maximum swing angle of 26.8° and a maximum angular velocity of 5 780°/s, which appears in front of the back dead center. Tangential acceleration provides sufficient beating up force for reed. When move to the front dead center position, the angular acceleration is −1 176°/s², the angular velocity is about 0° and the inertial beating up force is sufficient.

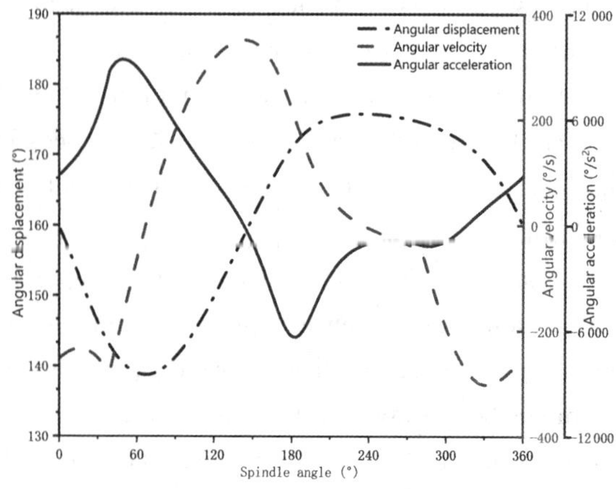

Fig. 3 Reed motion curve of optimized four-bar beating up mechanism

2.3 Multivariable and multi objective optimization model

2.3.1 Determination of objective function and design variables

In the actual beating up process, the beating up mechanism should meet the following requirements: (1) reed should have enough swing angle to ensure sufficient weft insertion time; (2) the beating up mechanism can provide enough beating force to ensure the good formability of the fabric; (3) the overall size of the mechanism is small which is easy to process and weave. The overall structure is compact, and the force is uniform so that the floor area and cost are reduced.

From the above three conditions, three objective functions can be established.

$$\min: f_1(x) = l_1 + l_2 + l_3 + l_4 \quad (5)$$

$$\max: f_2(x) = \beta \quad (6)$$

$$\max: f_3(x) = \alpha_3 \quad (7)$$

According to Eqs. (2) and (3), the total swing angle and maximum beating up force of sley are related to the four-bar size (L_1, L_2, L_3, L_4). Therefore, the design variables of this problem are $x=[a, b, c]^T$, where $a=L_1$, $b=L_2$, $c=L_3$, and $d=L_4$=constant.

2.3.2 Determination of constraints

AB is driven by the motor to rotate. ABCD is a crank connecting rod mechanism, so the length of four sides should meet the following conditions:

(1) AB is the shortest bar.

(2) The shortest bar+the longest bar≤the sum of the other two bars

$$g_1(x) = \min(a, b, c) = a \quad (8)$$

$$g_2(x) = a - b - c - d + 2(b, c, d)_{\max} \leqslant 0 \quad (9)$$

In order to ensure the transmission efficiency, the smaller the transmission angle is, the greater the effective component force of the driven rod movement is, and the more favorable for the mechanism transmission. The minimum transmission angle of the four-bar linkage appears in one of the two positions where the active crank and the frame are collinear. In general, the transmission angle should be greater than 40° in mechanical design. $\Gamma_{\min}=[\gamma_1, \gamma_2]>40$. In other words:

$$g_3(x) = \frac{b^2 + c^2 - (d + a)^2}{2bc} + 0.76 \leqslant 0 \quad (10)$$

$$g_4(x) = 0.76 - \frac{b^2 + c^2 - (d - a)^2}{2bc} \leqslant 0 \quad (11)$$

2.3.3 Build an evaluation function

This paper is a multi-objective function optimization, so the ideal point method is used to solve the problem.

$$f'(x) = \sqrt{(f_1(x) - ff_1(x))^2 + (f_2(x) - ff_2(x))^2 + (f_3(x) - ff_3(x))^2} \quad (12)$$

In view of the difficulty of actual machining, the precision $k=1$, considering the actual old machine transformation, the optimization range of a, b, c is 50% lower than the original basis. After programming with brute force method, the optimized rod length is: $L_1=95$ mm, $L_2=122$ mm, $L_3=300$ mm, and $L_4=315$ mm.

3 Results and Discussion

With the development of science and technology and the improvement of CAE technology in recent years, Adams, a virtual product development software for mechanical design, provides a new way for mechanical system dynamics, statics and kinematics analysis. Based on the Lagrange equation method, Adams calculates multi-body dynamics, and simulates the product with computer to obtain various performance data.

According to the results of program optimization, the following model is established in ADAMS software. According to the actual transmission relationship, various constraints are added. The completed virtual prototype is shown in Fig. 4. Because the actual size of the four-bar mechanism cannot be determined in space, it is necessary to carry out virtual assembly first and then simulate it.

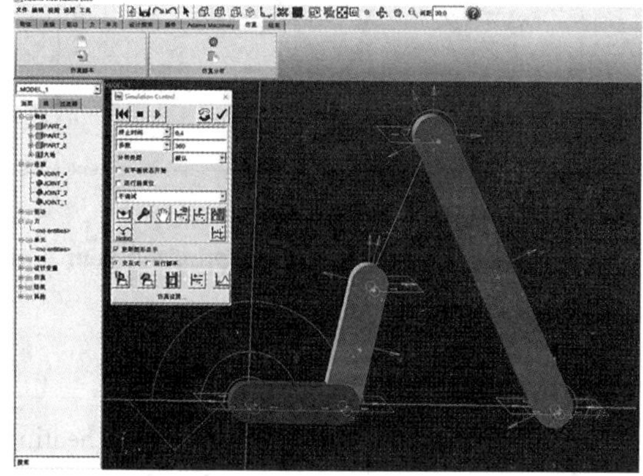

Fig. 4 Simulation model of four-bar beating up mechanism

Finally, the output of simulation and measurement curve is processed to obtain the image shown in Fig. 5. By adjusting the length of crank and rocker, the motion range of sley angle is from 139° to 176°, the maximum swing angle is 37° and the overall swing angle increases by a quarter compared with 26° before optimization, which gives the weft insertion mechanism enough time to weave wide fabric, and improves the problem that the relative time of back dead center time of four-bar mechanism is short. In terms of angular acceleration, the maximum negative angular acceleration is −6 357°/s^2, which is beneficial to the inertial beating up of the loom. However, at the front dead center position, it is about − 8 600°/s^2, and the speed changes rapidly, which may cause large vibration of the loom. The maximum positive angular acceleration is 9 412°/s^2, which is located near the back dead center. Compared with that before optimization, the maximum positive angular acceleration is nearly doubled, and the beating up force greatly increases. The optimized mechanism makes full use of the quick return characteristic of the four-bar linkage mechanism, which can make the fabric keep good formability in theory. However, the angular acceleration of the rocker fluctuates greatly. In the process of the spindle rotation, the rapid replacement of the angular acceleration may cause the vibration of the loom and the rapid loss of the rocker.

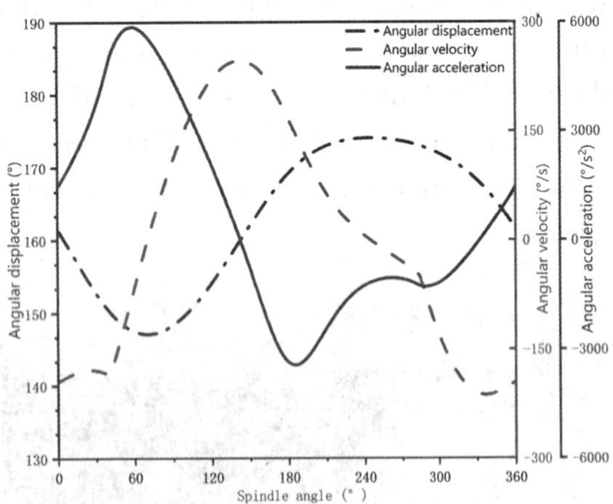

Fig. 5 Reed motion curve of optimized four bar beating up mechanism

4 Conclusions

In view of the weaving defects of the four-bar beating up mechanism, a new beating up mechanism design and optimization idea is provided on the basis of the traditional beating up mechanism. ADAMS software is used to carry out parametric modeling and accurate kinematic analysis of the optimized mechanism, which is verified with the traditional analysis results. After optimization, the total stroke of reed increases greatly, the weft insertion time is prolonged and the weaving width increases. The inertia beating up force of reed increases, and better optimization effect is achieved.

Acknowledgements

This work was sponsored by the Shanghai Natural Science Foundation of Shanghai Municipal Science and Technology Commission (Nos. 20ZR1401600 and 20ZR1400600), and Fundamental Research Funds for the Central Universities (Nos. 2232020D − 09, 2232020G−06, and 2232021G−06).

References

[1] ZHU Z. The design and study of four-Bar linkage beating-up mechanism for air-Jet weaving machine[J]. Journal of Textile Research, 1986(5): 9−16

[2] WEI Z. Flexible dynamic analysis of high-speed conjugate cam beating-up mechanism[J]. Journal of Mechanical Engineering, 2017, 53(3): 81.

[3] LIN W, WANG Y, ZHAO X, et al. Establishment and application of kinematics equations of beating-up motion based on Hermite polynomials[J]. Journal of Textile Research, 2011, 32(3): 127−132.

[4] WANG Y, HUI S. Computer aided analysis of loom beating-up mechanisms[J]. Textile Research Journal, 1998, 68(9): 630−634.

[5] YANG T A. Acceleration analysis of spatial four-link mechanisms [J]. Journal of Engineering for Industry, 1966, 88(3): 296.

[6] QIU H, WANG Y. Calculation and simulation of dynamic load for four-bar beating-up mechanism at both sides[J]. Journal of Textile Research, 2014, 35(9): 144−148+154.

[7] XI F, SINATRA R. Effect of dynamic balancing on four-bar linkage vibrations[J]. Mechanism and Machine Theory, 1997, 32(6): 715−728.

[8] FANG X. Research on simulation experiment platform of plane four-bar linkage based on ADAMS[J]. Journal of Mechanical Transmission, 2012, 36(10): 25−27.

SWOT Analysis of the E-Commerce Development for Personal Protective Textile Products in Ethiopia

XIE Yilin[1], Wu Xiaojun[3], TESSEMA Shimelis Adugna[4], TEGAW Esubalew Dessie[4], DING Xuemei[1, 2*]

[1] *College of Fashion and Design, Donghua University, Shanghai 201620, China*

[2] *Key Laboratory of Clothing Design & Technology (Donghua University), Ministry of Education, Shanghai 201620, China*

[3] *International Cultural Exchange School, Donghua University, Shanghai 201620, China*

[4] *Bahir Dar University, Bahir Dar, Ethiopia*

* *Corresponding author's email*: fddingxm@ dhu. edu. cn

Abstract: Electronic commerce (e-commerce) is valued and welcomed by African countries. But due to the overall backwardness of African countries in many aspects, restrictions and obstacles have been caused to the development of e-commerce. In the current study, the development status of e-commerce for personal protective textile products in Ethiopia was explored by SWOT analysis. The conclusion provided that the weak infrastructure, poor information and communications technology (ICT), imperfect e-payment system, low logistics efficiency, the lack of a legal structure and people's poor awareness of personal protection during the pandemic are challenges that blocking this booming market. Suggestions on promoting the production of personal protective textile products in Ethiopia through e-commerce were put forward. The study provides new ideas for measures to deal with the pandemic in Africa.

Keywords: Personal protective textile products; E-commerce; SWOT analysis; Ethiopia; Pandemic

1 Introduction

Under the outbreak and the global spread of COVID-19, the complete industrial chain for the production of personal protective textiles has enabled some countries to respond quickly to control the domestic pandemic with sufficient supply of protective masks. According to relevant papers and industry reports[1-4], the production process of disposable medical masks is shown in Fig. 1.

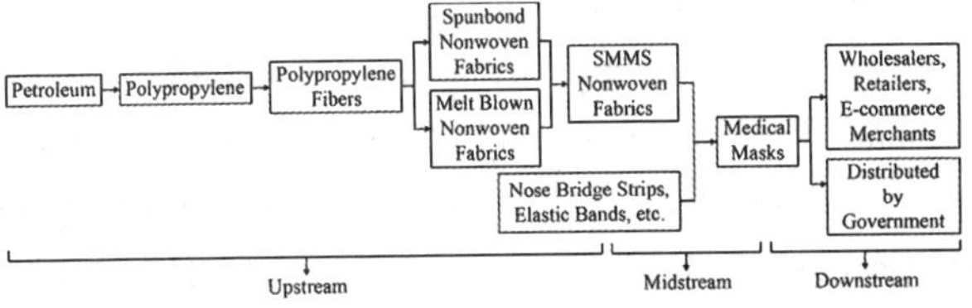

Fig. 1 Production process of disposable medical mask

With the development of Internet technology, electronic commerce (e-commerce), as a new form of economic development, is developing rapidly. The role of e-commerce in the production process of disposable medical masks can be divided into two categories: at the upstream and midstream, it mainly promotes information interaction to achieve an efficient connection between supply and demand of production materials, so that promotes the production of the end products; at the downstream, it promotes information interaction between supply and demand of end products, so that assists in the distribution by the government, the sale of the merchants and the purchase by the customers. During the pandemic, e-commerce can also reduce people's offline contact and curb the spread of the pandemic through convenient online communication.

E-commerce is valued and welcomed by many African countries such as Ethiopia, Kenya and Nigeria. However, due to the overall backwardness of African countries in manufacturing, education and business environment, many restrictions and obstacles have been caused to the development of e-commerce. In the current study, the development status of e-commerce for personal protective textile products in Ethiopia was explored by SWOT analysis. The analysis will cover the aspects including the recent market of e-commerce, the e-commerce platforms being used by the people, the status of the infrastructure, logistic, e-payment and the legal structure and people's awareness of personal protection during the pandemic in the country. Suggestions on promoting the production of personal protective textile products in Ethiopia through e-commerce were put forward, providing new ideas for measures to deal with the pandemic in Africa.

2 Development of Personal Protective Textiles Industry in Ethiopia

According to the data provided by Prof. Tessema and Dr. Tegaw, the Ethiopian textile experts in our research team, there are 28 manufacturers for personal protective textile products approved by the Ethiopian Food and Drug Administration (EFDA), including 16 producing medical masks and 5 producing cover-all-gown. There are also 303 importers of personal protective textile products, of which 258 import medical masks, 131 import cover-all-gown, and 100 import shoe covers. In April, 2020, the United States Agency for International Development (USAID) Digital Health Activity (DHA) team created systems to support the EFDA efforts to accelerate the availability of critical personal protective items needed to combat COVID-19[5]. The DHA team applied digital solutions to shorten the amount of time it takes manufacturers to register critical supplies from a minimum of three days to one. Also, companies in Hawassa industrial park shifted their business to produce face masks as part of efforts to halt spread of COVID-19[6].

In Ethiopia, the limited amount of material for personal protective textiles products and limited number of manufactures have become the problem during the pandemic. Therefore, to meet the emergent demand, mobilization of them to reach fast production is required. What e-commerce can bring to the industry is the fast information interaction between supply and demand. That will certainly shorten the production lead time and provide a better measure to deal with the pandemic.

3 SWOT Analysis of E-Commerce Environment for Personal Protective Textiles Products in Ethiopia

3.1 Strengths

E-commerce is in its infancy in Ethiopia, but it's developing rapidly. The COVID-19 pandemic has resulted in increasing number of Ethiopians using e-commerce services, including merchandizes purchasing and payment of water and electricity, which is no longer the exclusive services for Ethiopian technicians[7].

Moreover, along with the popularity of social media, such as Facebook and Telegram, the basic environment of e-commerce business has shown in Ethiopia. Today, there are three electronic payment platforms in the country, namely, Amole, Hello Cash and M-BIRR, through which users can access their bank accounts and conduct transactions[8].

Tab. 1 shows some e-commerce platforms using by Ethiopians. Among them, there are medical mask products on HellooMarket, Qefira and Jumia.

Tab. 1 The e-commerce platforms using by Ethiopians

	E-Commerce Platform	Address
1	EthioSuQ	https://www.EthioSuQ.com
2	Helloo-Market	https://www.helloomarket.com
3	Qefira	https://www.qefira.com
4	Jumia	https://www.jumia.com.et
5	Sheba Shopping	https://www.ShebaShopping.com
6	Guzomart	https://www.Guzomart.com
7	Shegar	https://www.sheger.net
8	Kikuu Ethiopia	https://www.kikuu.com
9	Kemmis	https://www.kemisd.com/
10	Shola Mart	https://www.Sholamart.com

3.2 Weaknesses

The present condition of e-commerce in Ethiopia is that small businesses are not developing due to a lack of infrastructure, financial capacity, and labor, as well as the usage of ICT to benefit businesses[9]. Also, the lack of information technology training for employees to improve their ability to utilize the internet is also a key problem[10]. According to the data presented by International Telecommunication Union and The World Bank, the value for mobile cellular subscriptions (per 100 people) in Ethiopia was 37.4 as of 2019 and for fixed broadband subscriptions (per 100 people) was 0.18 as of 2020. Moreover, Ethiopian banks use debit cards and automated teller machines (ATMs) but have not begun to issue credit cards. Most Ethiopians do not have credit cards and internet connections are slow, expensive and unreliable[11].

3.3 Opportunities

Over the 15 years until 2019, Ethiopia's economy had been amongst the fastest growing in the world at an average of 9.5 % per year[12]. At present, the goal of the country is to reach the lower middle-income level by 2025 through sustained economic growth[12]. That creates a good environment for e-commerce to grow.

What's more, since the finance, banking and telecommunications sectors of the country have been largely controlled by the government, the local market has not been monopolized by large foreign enterprises, so the e-commerce market is basically left to local enterprises in Ethiopia[7]. Abraham said that the government is preparing a draft national law to govern e-commerce, which will promote the capitalization of Ethiopia's digital economy, accelerate the digitization process, and increase the application of e-commerce[7].

In this year, Ethiopian Airlines Group will ramp up investment in cargo services and infrastructure and is looking to become a logistics hub for Africa's growing e-commerce market[13], which is obviously another opportunity for the development of the supply process of personal protective textiles products through e-commerce.

3.4 Threats

Ethiopia's status as a landlocked developing nation is a hindrance to its e-commerce development. According to the study[14], foreign logistics providers are not welcome, but private sector operators cannot effectively participate in e-commerce, even if they want to, because of a combination of crucial factors-inability to raise capital, inability to access land, confusion over customs bonded regulations and inability to set up private customs bonded warehouses. For the import of personal protective equipment, the complicated customs formalities in the country often discourage importers, who must submit a large number of documents before they can go through legal formalities with the customs and tax administration and other legal institutions of the country. These formalities are a heavy burden on not only importers, but small and medium-sized enterprises.

More importantly, the country lack of a legislative structure of e-commerce[11]. The consumer rights protection is normally thought of when the e-commerce is in relation to business-to-consumer transactions. Relevant national policies for consumer protection that encourage good practices should also be applicable to e-commerce in areas including information disclosure, contract terms, secure payment mechanisms, consumer privacy and data security and dispute resolution and redress. The trade competition and consumer protection proclamation of Ethiopia has a section on consumer protection issues and rights. But the reading of the provisions of the proclamation does not have an overt provision that covers the aspects of e-commerce consumers.

According to the survey of scholars, many Ethiopians do not have enough medical protection knowledge. In Dessie Referral Hospital, 13.6% of the medical staff does not have the professional knowledge of mask wearing[15]. In the police health facilities of Addis Ababa, 24% of the workers know nothing about medical masks[16]. These cases indirectly limit the domestic demand for masks and the diversified development of mask trading methods.

4 Development Strategy of E-Commerce for Personal Protective Textiles Products in Ethiopia

The obstacles to the development of e-commerce in Ethiopia basically include the following aspects: weak infrastructure, poor ICT, imperfect e-payment system, low logistics efficiency, the lack of a legal structure and people's poor awareness of personal protection during the pandemic. Each aspect needs to cope with and to improve.

The survival and promotion of e-commerce cannot be separated from perfect ICT. It is necessary to accelerate the development of the foundation of network communication infrastructure, the establishment and improvement of a complete mobile network system, the expansion of the mobile network user group, and the enhancement of the popularity of the Internet in Ethiopia.

The continuous progress and development of e-commerce is also directly related to the development of third-party platforms. It is necessary to improve the e-payment system and develop online payment platforms, improve the security, stability, reliability and convenience of it, and further improve the efficiency of e-payment.

Most importantly, the government and related departments need to increase policy and financial support for e-commerce and encourage PPE enterprises to develop this way of doing business. Meanwhile, they need to support more on the logistics industry, improve the construction of the infrastructure, learn from foreign advanced logistics management experience and establish an innovative logistics management method in coincidence with domestic conditions.

High quality and high-level talents are the necessary conditions for industrial development. Therefore, Ethiopia should actively introduce talents in the development of e-commerce through the cooperation with relevant local universities. This not only increases the number of talents in enterprises, but also provides employment opportunities for local college graduates. In addition, enterprises should strengthen the training of internal employees, and improve the comprehensive skills of them, so as to improve the overall competitiveness of enterprises.

Normally, it takes a long time for people to accept a new technology and integrate it into their daily life. The government and relevant departments should cooperate with enterprises and universities to popularize the knowledge of e-commerce to the public, especially its benefits during the pandemic. At the same time, the public's awareness of COVID-19 and the importance of personal protection need to improve, and the COVID-19 prevention guidelines also need to update.

Over all, personal protective textile products are actually a kind of product that is easy to enter the e-commerce industry. Customers only need to pay attention to the certification and price of them. Its material and size are basically the same. So, when customers purchase them online, they have not much concern about returning. The government should improve the market supervision of personal protective textile products and strictly punish those who do not meet the certification standards. At the same time, the government should also strengthen the legislation of e-commerce, establish and improve relevant legal systems and improve the supervision, so that ensure the e-commerce merchants carry out business activities reasonably within the scope of laws and enable the e-commerce industry to grow legally.

5 Conclusions

This study identified the strengths, weaknesses, opportunities and threats of the e-commerce for personal protective textile products in Ethiopia.

It is concluded that the future of e-commerce for personal protective textile products in Ethiopia is bright, though the weak infrastructure, poor ICT, imperfect e-payment system, low logistics efficiency, the lack of a legal structure and people's poor awareness of personal protection during the pandemic are challenges that blocking this booming market.

The government and related departments along with the relative enterprises and universities need to work closely together to manage all these problems to create a better environment for the e-commerce. It is necessary to accelerate the development of the ICT system and logistic system, develop secure, stable and reliable online payment platforms, increase policy and financial support for e-commerce, improve relevant legal systems, improve the supervision, introduce talents, improve people's awareness of personal protection during pandemic and widely introduce the advantage of e-commerce to the public.

Acknowledgements

The funding supports to this research are provided by the Construction of Cooperation Platform for International Science and Technology Organizations under the "the Belt and Road" Initiative of China Association for Science and Technology through 2021ZZGJB071527 and Shanghai Science and Technology Committee through project 21640770300. We would also like to thank the reviewers of the paper.

References

[1] YAN B Y. Discussion on the production process of SMS nonwovens[J]. Nonwoven Fabric, 2006(2): 13-17.

[2] ZHOU C, LIU C, JIN X Y. Study on the performance of external electret of medical protective mask meltblown filter material[J]. Nonwoven Fabric, 2010, 18(6): 38-41.

[3] LIU C. Effect of electret treatment on filtration performance of melt blown air filter material[J]. Industrial Textiles, 2013, 10(8): 34-37.

[4] TAO Y L, CHEN X, XIANG K J. Application of polypropylene in medical mask[J]. Rubber and Plastic Technology and Equipment, 2020, 46(8): 33-36.

[5] JSI. Accelerating Ethiopia's production of critical COVID-19 supplies[Z]. 2020.

[6] Embassy of Ethiopia Washington DC. Companies in Hawassa industrial park begin face mask production[Z]. 2020.

[7] Haowang Observation. Ethiopia's e-commerce market has a rapid momentum, local e-commerce platforms have emerged one after another[Z]. 2020.

[8] HelloCash. Amole, HelloCash, M-BIRR urges Ethiopian to use e-payments[Z]. 2020.

[9] SHAIKH M A. Ethiopian banker's perception of electronic banking in Ethiopia — a case of Adama city[J]. Internatonal Journal of Scientific and Research Publication. 2014, 4(9): 1-7.

[10] YILMA K M, ABRAHA H H. The internet and regulatory responses in Ethiopia: telecoms, cybercrimes, privacy, e-commerce, and the new media[J]. Mizan Law Review. 2015, 9(1): 108-153.

[11] International Trade Administration. Ethiopia — country commercial guide[Z]. 2022.

[12] SOHU Marketing Observation. Masreza Andaru: Ethiopia is expected to become a middle — income country in 2025[Z]. 2018.

[13] GEBRE, S. Ethiopia plans to build African e-commerce logistics hub[Z]. 2022.

[14] HURIA A. Facilitating trade and logistics for e-commerce: Building blocks, challenges, and ways forward[J]. World Bank, 2019.

[15] GEZIE H, LETA E, ADMASU F, et al. Health care workers knowledge, attitude and practice towards hospital-acquired infection prevention at Dessie referral hospital, northeast Ethiopia[J]. Clinical Journal of Nursing Care and Practice, 2019(3): 59-63.

[16] TADESSE T, TESFAYE T, ALEMU T, et al. Healthcare worker's knowledge, attitude, and practice of proper face mask utilization, and associated factors in police health facilities of Addis Ababa, Ethiopia[J]. Journal of Multidisciplinary Healthcare, 2020(13): 1203.

Preparation of Aromatic Pillow Nonwovens via Ultrasonic Cavitation and Hot-Air Anchoring of Fragrance Microcapsules

XIA Yichang[1], **ZHAO Yi**[1 *]

[1] *Engineering Research Center of Technical Textiles, Ministry of Education, Donghua University, Shanghai 201620, China*

[*] *Corresponding author's email*: zhaoyi@ dhu. edu. cn

Abstract: Nowadays, sleep problems have become a common "epidemic disease". In order to help people improve sleep qualities in a non-medicine way, fragrant textiles with sleep aids have attracted increasing attention. However, the preparation of fragrant textiles often has some drawbacks, such as high requirements for technologies, not eco-friendly finishing process and poor fragrance effect. This work studies a green method of preparing long-term aromatic nonwovens via ultrasonic cavitation and hot-air anchoring of lavender polyurea microcapsules (LPMs) onto the hollow polyester through-air thermal bonded nonwovens (H-TA) without extra additives. The influence of pore size on add-on of microcapsules and fragrance release effect have also been studied. The results showed that the fragrance microcapsules were well loaded and evenly anchored on nonwovens through the thermal bonding, air permeability of which reached 2 626. 7 mm/s. LPMs@ H-TA showed a good lavender aroma release effect in the flash GC E-nose test, indicating potential application in health care and sleep aid.

Keywords: Aroma finishing; Through-air thermal bonded nonwovens; Environmental protection; Microcapsules

1 Introduction

With the continuous improvement on people's living standards, the demand for health has gradually increased, and "sleeping soundly" has become one of people's pursuits for a better life. Adequate sleep is an important physiological process for human body to maintain metabolic balance and regulate body health[1]. However, sleep problems become a common disease nowadays due to the intensely competitive and fast-paced contemporary. "2020 Chinese Internet Netizens Sleep White Paper" shows that about 56% of Chinese netizens have sleep quality problems, and more and more people are willing to pay for high-quality sleep, with natural essential oils and bedding among the most popular sleep aids.

Through-air thermal bonded nonwovens (TAs) are a kind of heat-adhesive nonwovens (calender bonded, through-air thermal bonded), which are reinforced by hot air penetrating the fiber web to melt the low-melting thermoplastic fiber components to generate bonding points. TAs have a good three-dimensional mesh porous spatial structure, and the liquid can penetrate or infiltrate the fabric through the tortuous pores. Moreover, TA can be used as high-quality textile filler materials due to its soft, fluffy, warm, transparent and elastic properties[2-3].

Lavender essential oil has the excellent functions of soothing spirits and helping sleep, antibacterial and antioxidant. However, the poor stability of natural plant essential oils leads to the disadvantages of short effective action time, large concentration changes within the release cycle and poor processability when used directly. Microencapsulation is a common method to improve the stability of essential oil and broaden its application field[4-5].

The development and application of aromatic microcapsules in textiles not only enhance the grade of fabrics, but also endows textiles with new connotations. The aroma finishing process of loading microcapsules onto fabrics by coating printing, padding and direct dipping is widely used because of its simple process. Li et al. [6] used the extract of geranium as the core, prepared aromatic microcapsules by the interfacial polymerization method, and prepared a scented health care textile that could release fragrance intelligently in the presence of sweat by dry coating. However, in order to improve the affinity between microcapsules and fabrics, the aroma finishing needs to add chemical binders to increase its adhesion on the surface of the base fabric, which is contrary to the concept of green and sustainable development, and is a problem that needs to be studied and improved[7].

In this study, the uniform loading and anchoring of microcapsules in fabrics was accomplished without the use of chemical auxiliaries. The advantages of ultrasonic cavitation and hot-air anchoring technology for aromatic functionalization of H-TA were analyzed. The relationship between material pore size and application performance and aroma finishing result was comparatively studied.

2 Experiments

2.1 Materials

Lavender polyurea microcapsules (LPMs) in this research were provided by Shanghai Research Institute of Fragrance & Flavor Industry (SRIFFI, Shanghai, China), and prepared by interfacial polymerization. Fig. 1 (DXS-10ACKT at 10.0 kV) depicted microcapsule microtopography. Hydrophobic ES fibers (Polyester-Propylene side by side, the melting point of the cortex is 127 ℃ 2.00 De, 38 mm) were obtained from Sichuan Huvis Corporation (Sichuan, China). Hollow polyester fibers (HPET, 6.67 dtex, 64 mm) were provided by Sinopec Yizheng Chemical Fibre Co. Ltd. (Jiangsu, China).

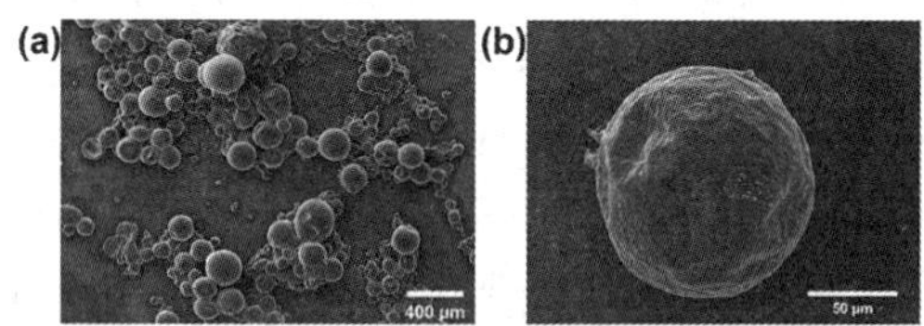

Fig. 1 Scanning electron microscopy (SEM) images of LPMs (a) entirety, (b) single

2.2 Preparation of H-TA substrate

Weigh 200 g of ES fiber and hollow polyester (HPET) fiber in a certain ratio (ES : HPET = 3 : 7, 4 : 6, 5 : 5, 6 : 4, 7 : 3). After pre-wetting, opening and mixing the fiber mass in turn, use BG218A-100 single-cylinder double-doffer carding machine (Changshu Weicheng Nonwoven Equipment Co. Ltd., Jiangsu, China) to carry out two messy carding. Cut the output fiber web into 30 cm×30 cm, and use R-3 type automatic self-setting dryer (Xiamen Rapid Precision Machinery Co. Ltd., Fujian, China) to conduct through-air thermal reinforcement at 150 ℃ for 2 min to obtain HPET-TATBNs substrate.

2.3 Aromatic functional finishing

At room temperature, deionized water (DW) is used to prepare 100 mL of LPMs dispersion with several concentration gradients (1%, 3%, 5%). The H-TA were immersed in the dispersion liquid, followed by ultrasonic cavitation treatment for 10 min with F-040S ultrasonic oscillator (Shenzhen Fuyang Technology Group Co. Ltd., Shenzhen, China), then taken out with tweezers and placed flat in an oven for drying at 60 ℃ for 4 h. Under the condition of 130 ℃, the self-setting dryer was used to carry out through-air thermal treatment for 20 s on the dried fragrant fabric to complete the anchoring of microcapsules in the H-TA.

2.4 Aroma finishing effect evaluation

A flash GC E-nose was used to test the release strength of LPMs@ H-TA. The loading uniformity and anchoring effect of LPMs on H-TA were observed and analyzed by an optical microscope (OM) and SEM. Zeta-potential and nanoparticle size analyzer were used to evaluate the effect of ultrasonic cavitation technology on promoting dispersion of LPMs in DW. Since both ES fibers and HPET fibers have good hydrophobicity, the loading weight gain rate (WGR) and utilization rate (UR) of LPMs on TA can be calculated using a simple weighing method, respectively:

$$WGR(\%) = \frac{G_1 - G_0}{G_0} \times 100\% \tag{1}$$

$$UR(\%) = \frac{G_1 - G_0}{g} \times 100\% \tag{2}$$

Where G_0 is the weights of the fabric, G_1 is the weight of the fabric after loading LPMs and drying, g is the total added mass of LPMs.

2.5 Application performance testing

According to the GB/T 5453—1997 standard test method, the YG461H automatic air permeability tester was used for air permeability test to evaluate the application performance of LPMs@ H-TA. The pore size distribution of H-TA has been measured by the bubble point method on a CFP-1100A capillary flow porometer, and the infiltrating fluid is Silwick reagent. The basic relationship for the flow porosity calculation is based on:

$$D = \frac{4\gamma\cos\theta}{p} \quad (3)$$

Where D, γ, θ, p are the pore diameter, surface tension of liquid, contact angle, and pressure difference, respectively.

3 Results and Discussion

3.1 Assistive technology effect on aroma finishing result

The H-TA prepared in this experiment is shown in Fig. 2 (a). Using HPET fiber with greater rigidity and hollow structure as the fabric skeleton can improve the light texture of the product and meet people's requirements for fluffy and soft pillow products. The microscopic topography of H-TA in Fig. 2 (b) shows the good bonding between fibers after thermoplastic setting, and the fabric has a good three-dimensional porous spatial structure, which is favorable for capturing microcapsules during impregnation. By comparing the microstructures of LPMs@ H-TA before and after hot-air anchoring, it was easy to find that before treatment (Fig. 2 (c)), the LPMs particles were restricted by the fiber structure and were simply bound in the porous structure of the fabric, which had a risk of falling. After hot-air anchoring (Fig. 2 (d)), the LPMs particles were bonded with the low-melting thermoplastic PE to form obvious bonding points, thus being stably anchored in the fabric structure, which was consistent with our previous research results[8]. Therefore, the use of hot-air anchoring assisted finishing can give full play to the role of ES fiber and reduce the risk of aromatic functional particles falling during storage and use.

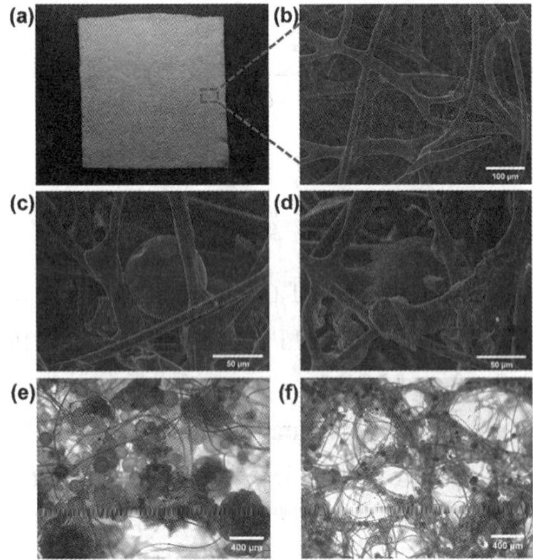

Fig. 2 The surface morphologies of LPMs @ H-TA (a) macroscopic digital imaging and (b) microscopic SEM image; SEM images of the LPMs (c) before and (d) after the hot-air anchoring; OM images of H-TA after (e) direct impregnation and (f) ultrasonic cavitation aromatization (3 wt% LPMs)

Since ultrasonic wave will produce a large number of small bubbles when acting on the liquid, with the movement, growth and rupture of the small bubbles will generate high temperature, high pressure and shock wave on the surrounding liquid environment, so it can inhibit the agglomeration of tiny particles in the liquid and promote their continuous movement[9]. It is the reason why the assisted preparation by ultrasonic cavitation technology can obtain better microcapsule loading effect compared with direct immersion, as well as the results of OM (Figs. 2 (e)-2(f)). In the process of direct immersion, most LPMs are loaded on the surface of the fabric by natural sedimentation. Therefore, most of the loaded LPMs are large-size LPMs that are prone to settlement. Moreover, polyurea microcapsules were easily aggregated due to agglomeration and clogged on the surface of the fabric (Fig. 2 (e)). In contrast, ultrasonic cavitation can promote the movement of LPMs before they were captured by H-TA, while small-sized LPMs are easier to be captured because of more vigorous than large-sized ones, and can be uniformly and densely distributed inside the fabric (Fig. 2 (f)). It should be noted that the LPMs inside H-TA space can fully contact with fiber structure, which will facilitate the bonding of LPMs in the hot-air anchoring stage, so ultrasonic cavitation and hot-air anchoring are more suitable. In addition, the particle size distributions of LPMs after mechanical stirring and ultrasonic cavitation are shown in Figs. 3 (a)-3(b), respectively. After stirring by rotor, there were two obvious peaks in the histogram, one near 100 μm and the other near 1 000 μm, while there was only one obvious peak near 100 μm after ultrasonic cavitation. As shown in Fig. 1 (a), the particle size of LPMs is less than 400 μm, so the peaks larger than 400 μm were caused by the agglomeration of microcapsules, which further confirm that ultrasonic cavitation treatment can improve the dispersion of LPMs in aqueous solution significantly.

In the aroma finishing experiments with different LPMs concentrations, the D- and V-LPMs@ H-TA prepared by direct immersion method and ultrasonic cavitation method, respectively, showed little overall difference in WGR (Fig. 3 (c)). Only at high concentrations of LPMs, the WGR of D-LPMs@ H-TA was slightly higher than that of V-LPMs@ H-TA, presumably because LPMs were prone to agglomerative sedimentation at high concentrations, while the crowded conditions inside the dispersion restricted the movement of particles. However, large-sized LPMs aggregates tend to block on the fabric surface and cannot enter the internal space of H-TA, which is not conducive to the subsequent bonding and fixing of LPMs in the three-dimensional space of H-TA by hot-air anchoring. In addition, both D-and V-LPMs@ H-TA have significant aroma release intensity peaks (Fig. 4), indicating that lavender essential oil was successfully loaded, and polyurea microcapsules

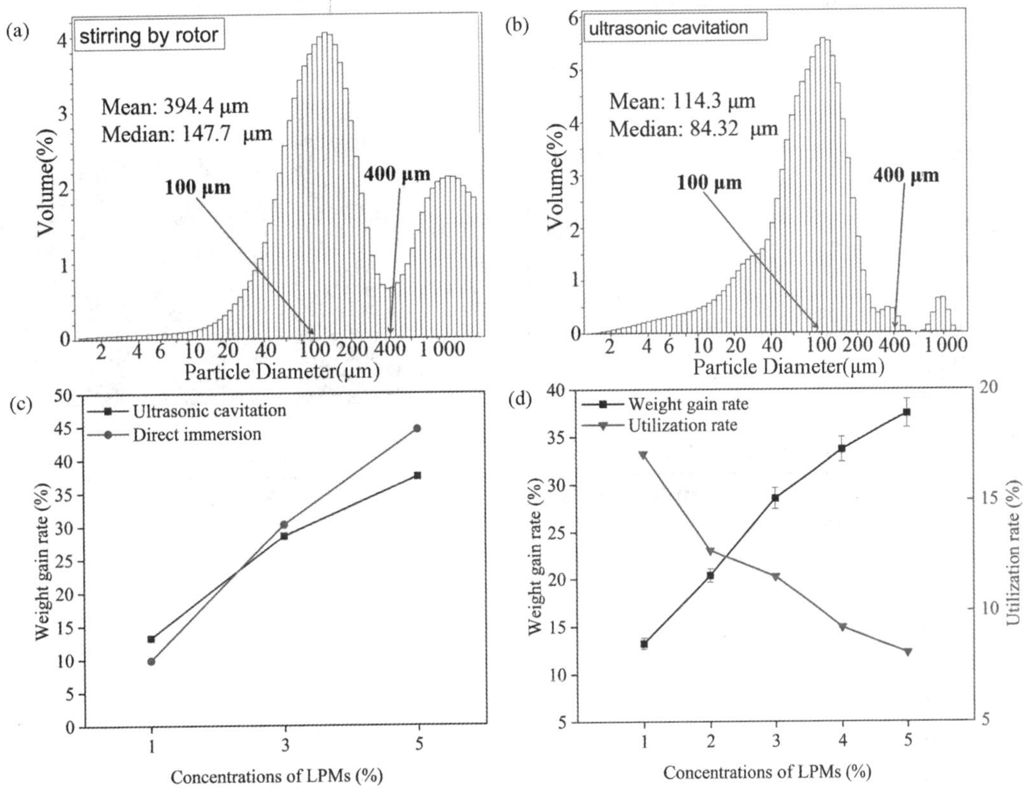

Fig. 3 Particle size distribution of 3 wt% LPMs after (a) stirring and (b) ultrasonic cavitation; (c) WGR of D-and V-LPMs@ H-TA as a function of LPMs concentration gradient; (d) WGR and UR of V-LPMs@ H-TA as a function of LPMs concentration gradient

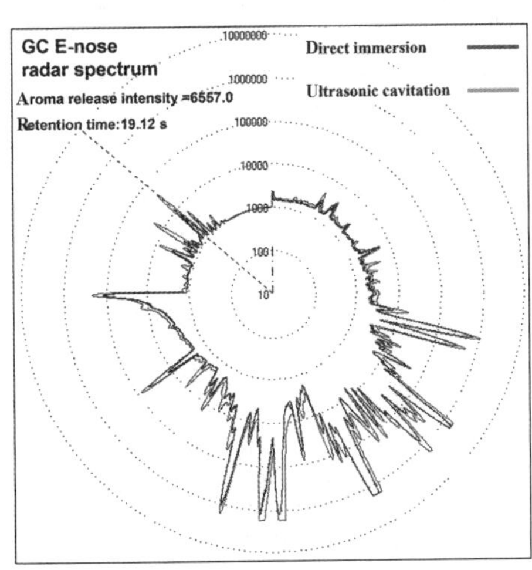

Fig. 4 Flash GC E-nose spectra of D-and V-LPMs@ H-TA

have good thermal stability, without premature rupture due to the additional thermal environment provided by ultrasonic cavitation and hot-air anchoring. Interestingly, V-LPMs@ H-TA have more obvious aroma release performance, which may be because the coating ability of small-sized polyurea microcapsules on lavender essential oil is better than that of large-sized one, while V-LPMs @ H-TA has a large number of small-size LPMs. Fig. 3 (d) shows that the WGR of V-LPMs@ H-TA is positively correlated with the LPMs concentration, while the UR is opposite, which indicates that the higher the concentration of LPMs dispersion uses in single-time fabric aroma finishing, the more LPMs will be wasted. Therefore, the contradiction between the desired aroma release intensity and economic cost should be considered comprehensively in actual production.

3.2 Applied performance analysis

The variation of the average pore size of H-TA with the ratio of HPET is shown in Fig. 5 (a), and they are positively correlated. Since HPET has high rigidity and high flexural strength, it can support the formation of large pores during processing. Under the same preparation process, the product with larger pore size has less resistance to flowing air and greater air permeability. Therefore, the permeability of the H-TA is consistent with the change of pore size, and positively correlated with HPET content (Fig. 5 (b)), up to 2 626.7 mm/s. Pillow products with good air permeability will bring users a more comfortable sleeping experience. Among them, the pore size distribution histogram of TA is shown in Fig. 5 (c). Considering that the particle size of LPMs is distributed in the range of 1 – 400 μm (Fig. 3 (b)), the agglomeration problem can be ignored

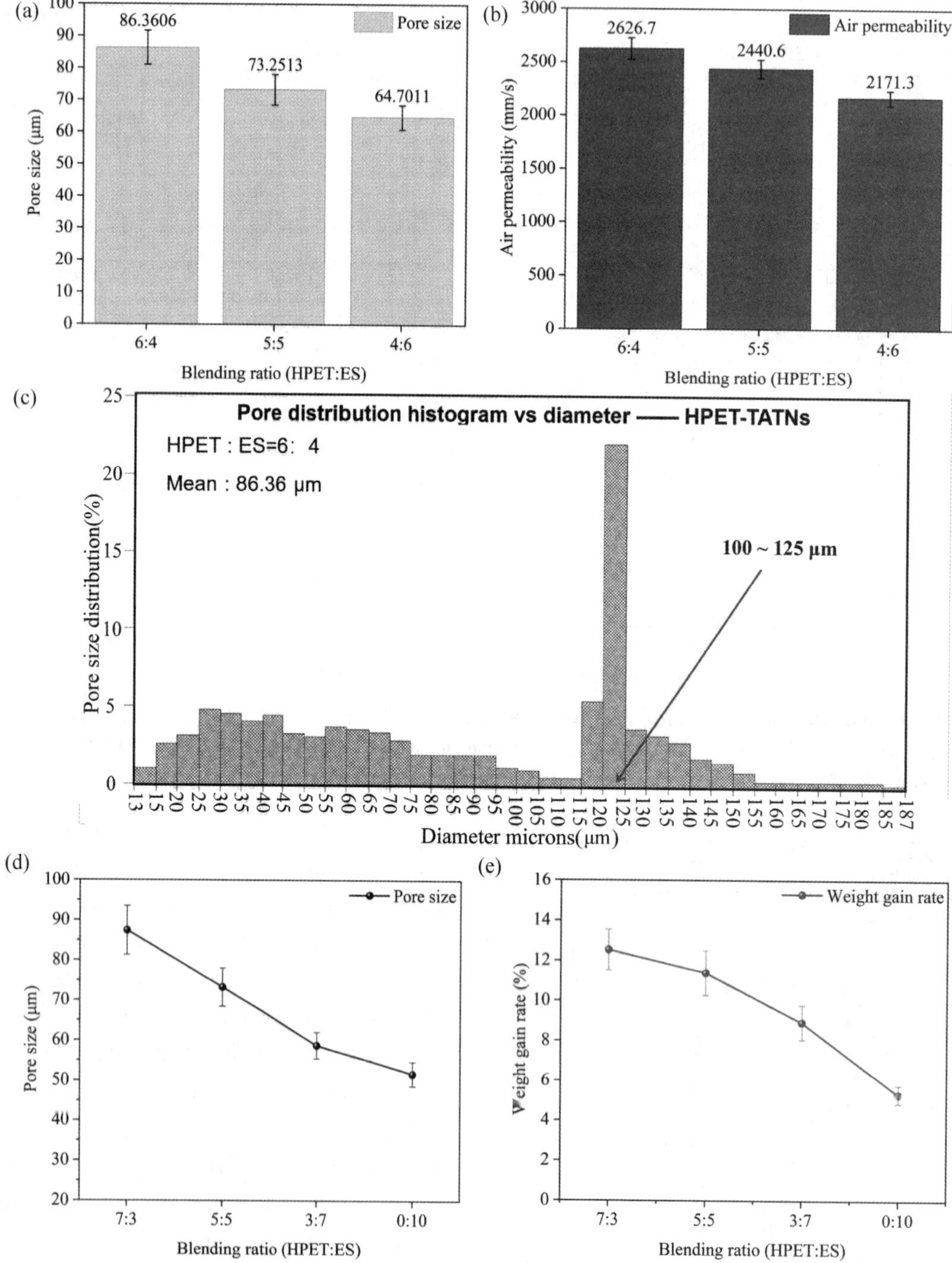

Fig. 5 Diagram of pore size (a), (d); air permeability (b) and WGA (e) changes with HPET : ES ratio of H-TA; (c) typical pore size distribution histogram of H-TA (HPET : ES=6 : 4)

after ultrasonic cavitation. So most of the LPMs can enter the H-TA by motion or flow with the dispersion, which is consistent with the phenomenon observed in Fig. 3 (f), indicating compatibility between TA and LPMs. When the change gradient of fiber ratios was further expanded, and pure ES fabric was used as the control group. As shown in Figs. 5(d)-5(e), with the increase of HPET content, the average pore size distribution and WGR of the fabric also increases gradually, indicating that in the process of ultrasonic cavitation assisted aromatization, the larger pore size is also more conducive to the entry of LPMs, so as to obtain a better aroma finishing effect.

4 Conclusions

In this study, a nonwoven material for aroma slow-release pillow with the potential of sleep aid was prepared by ultrasonic cavitation method and hot-air anchoring method, which was a green, eco-friendly and short-process aromatic fabric preparation technic. Compared with the direct immersion aroma finishing process, the ultrasonic cavitation method has obvious advantages in uniform loading of microcapsule particles, and the hot-air anchoring method makes full use of the bonding properties of ES fiber thermoplastic components to achieve stable anchoring of loaded microcapsules. Furthermore, by analysing the influence of the fiber ratio of HPET and ES on the application performance and aroma-carrying capacity of the H-TA, it is found that the fabric with larger pore size has more excellent air permeability and loading capacity of microcapsules.

References

[1] VLEESHOUWERS J, KNARDAHL S, CHRISTENSEN J O. Effects of psychosocial work factors on number of pain sites: the role of sleep quality as mediator[J]. BMC Musculoskeletal Disorders, 2019, 20(1): 581-589.

[2] ZAKRIYA G M, RAMAKRISHNAN G G, ABINAYA D, et al. Design and development of winter over coat using Jute and hollow conjugated polyester non-woven flexible composite[J]. Journal of Industrial Textiles, 2018, 47(5): 781-797.

[3] MIAO D, HUANG Z, WANG X, et al. Continuous, spontaneous, and directional water transport in the trilayered fibrous membranes for functional moisture wicking textiles[J]. Small, 2018, 14(32): 1801527.

[4] XIAO Z, LIU M, NIU Y, et al. Lavender fragrance sol-gel encapsulated in ORMOSIL nanospheres[J]. Flavour and Fragrance Journal, 2019, 34(1): 21-27.

[5] KFOURY M, AUEZOVA L, GREIGE H, et al. Encapsulation in cyclodextrins to widen the applications of essential oils[J]. Environmental Chemistry Letters, 2019, 17(1): 129-143.

[6] LI M, WANG R, LIU X. Preparation and slow release mechanism of geranium wilfordii maxim microcapsules loaded composite fabrics[J]. Journal of Textile Research, 2018, 39(2): 138-143.

[7] MA J, XU W, KOU X, et al. Green fabrication of control-released, washable, and nonadhesives aromatic-nanocapsules/cotton fabrics via electrostatic-adsorption/in situ immobilization[J]. ACS Sustainable Chemistry & Engineering, 2020, 8 (40): 15258-15267.

[8] MA J, ZHANG N, CHENG Y, et al. Green fabrication of multifunctional three-dimensional superabsorbent nonwovens with thermo-bonding fibers[J]. Advanced Fiber Materials, 2022, 4(2): 293-304.

[9] BAI L, YAN J, ZENG Z, et al. Cavitation in thin liquid layer: a review[J]. Ultrasonics Sonochemistry, 2020, 66: 105092

Deformation and Simulation of One-Step Molding for Solvent Responsive Composite Glass Fabric/Hydrogel Bilayer Actuator

ZHANG Xintian[1,2], HU Xuefeng[1], NI Jiaqian[4], WANG Jilong[3*], SHEN Hua[1], XU Fujun[1,2]

[1] *Key Laboratory of Textile Science & Technology, Ministry of Education, Donghua University, Shanghai 201620, China*
[2] *Department of Textile Engineering, College of Textiles, Donghua University, Shanghai 201620, China*
[3] *College Textile and Garment, Shaoxing University, Shaoxing 312000, China*
[4] *Zhejiang Zhongda Testing Technology Services Co., Ltd., Shaoxing 312030, China*

[*] *Corresponding author's email*: 2022000018@usx.edu.cn

Abstract: Hydrogel actuators are widely used in the field of soft robots due to their good biocompatibility and multi-stimulus responsiveness. Hydrogel actuators can also be compounded with other materials to achieve more complex deformation effects and better mechanical properties. However, the complex preparation method and weak interface connection need to be further improved. In this paper, we use the one-step molding method to prepare the solvent responsive glass fabric/polyacrylamide/calcium alginate (GF/PAAm/CA) hydrogel bilayer actuator by using the glass fabric reinforced bilayer hydrogel actuator interface and mechanical strength. The actuator can produce reversible bending strain in deionized (DI) water and organic solvents. It can maintain good deformation performance with repeated deformation for more than 20 times. Besides, the bending angle can reach 720° or even greater, it can also use various tailoring and designing properties of hydrogel and fabric to achieve various deformation effects. The finite element simulation (FEM) is used to simulate the deformation of the actuator, which effectively saves the experimental cost.

Keywords: Hydrogel actuator; Glass fabric; Bilayer; FEM

1 Introduction

Soft robots have large degrees of freedom, continuous deformation ability and good environmental adaptability. It can replace the clumsy rigid robots in some occasions[1]. Soft robots are mostly made of flexible materials[2], which makes the driving and control design more complex[3]. At present, the main driving methods of the soft robots include shape memory alloy drive[4], fluid drive[5], electroactive polymer drive[6], and response hydrogel drive[2,7]. The responsive hydrogel is a polymer with three-dimensional net structure. The smart hydrogel can perceive the subtle stimulation of external energy fields such as temperature[8], light[9], chemistry[10] or magnetism[11], and exhibit the characteristics of water absorption and release, accompanied by macroscopic reversible volume transformation. The above soft driving materials have some problems, such as too small driving force, simple driving deformation or poor mechanical properties. In order to solve these problems, usually it is compounded with other flexible materials. Fabric is a kind of commonly used flexible material. There are many studies on the composite of fabric and hydrogel. When the hydrogel is combined with the fabric to form an asymmetric structure, complex deformation can be realized. At present, the anisotropic structure of hydrogel actuators is mainly divided into bilayer structure[12], gradient structure[13], patterned structure[14] and orientation structure[15]. Therefore, when the fabric and hydrogel are compounded layer by layer, the bilayer structure can be formed. There is a large expansion and contraction gap between the two layers, resulting in uneven internal stress distribution. However, the bilayer structure has the disadvantages of fracture and separation between two layers, thus affecting the deformation effect and service life of the actuator. The fabric has many crevices, which is conducive to the penetration of hydrogel and the formation of good physical entanglement. It can better solve the problem.

Gong et al.[16] combined the hydrogel and the glass fabric (GF), resulting in the synergistic improvement of the mechanical properties of the actuator. Glass fiber can also form hydrogen bonds with hydrogel to further improve the interface performance. In order to better realize the prediction and design of the actuating effect, the FEM can be used to build models and simulate.

In this article, we use polyacrylamide/calcium alginate (PAAm/CA) hydrogel as the active layer, which can expand in DI water and contract in organic solution. Using GF and hydrogel composite as the passive layer, great deformation can be achieved by good softness of fabric. There are great prospects for soft actuators development. FEM is used to simulate the deformation effect, which makes the deformation of the actuator accurate, controllable and predictable, greatly saves experimental time and resources.

2 Experiments

2.1 Materials

The sodium alginate (SA) and calcium chloride ($CaCl_2$, 500 g ≥ 96.0%, granular) were purchased from Shanghai TITAN Technology Co., Ltd. Acrylamide (AAm, ≥ 98.0%), acetone and ethanol (95%) were purchased from Sinopharm Chemical Reagent Co., Ltd. N, N_0-Methylenebis acrylamide (MBAA, 99%) and ammonium persulfate (APS, ≥ 98.0%) were purchased from Beijing Inokai Technology Co., Ltd. TEMED (99%) was purchased from Shanghai Aladdin Biochemical Technology Co., Ltd.

2.2 Fabrication of actuator

First, SA and AAm were completely dissolved in DI water, and then crosslinking agent was added to react with initiator. After complete dissolution and elimination of bubble, low-temperature accelerator was added in ice bath conditions. The solution was poured into the mold. Then the glass fiber fabric was laid on the solution. The solution was poured on the surface of the fabric, which was sealed with a glass plate, and then it was placed at room temperature for 24 h. After removing the mold, the sample was soaked in $CaCl_2$ solution for 15 min in order to promote the formation of alginate microcrystal and entanglement network. Therefore, a double network hydrogel with good mechanical properties was formed.

2.3 Characterization

Placed the tailored actuator in DI water and acetone solution and observed its bending deformation morphology. Using bending angle θ indicates the deformation of the actuator. Hu bbard et al. defined θ as the angle formed between two tangents at the end of the hydrogel actuator[17]. Equation (1) is used to calculate the induced strain in polymer hydrogels and it can define small and large bends.

$$S = \frac{\theta}{L} d \qquad (1)$$

Where L is the length, d is the thickness and S is the strain of the actuator.

The angle of bending toward the fabric layer is positive and bending toward the hydrogel layer is negative. The surface morphology of the GF hydrogel actuator was observed by an scanning electron microscope (SEM), which verified that the hydrogel could penetrate into the GF to form a good connection and interface. In addition, using the tailor ability of the actuator, the actuator can be cut into different shapes to achieve different deformation effects.

3 Results and Discussion

3.1 Preparation and SEM structural characterization of actuator

As mentioned above, the intervention of fabric well solved the problem of weak interface connection. Fig. 1 (a) and Fig. 1 (b) are the SEM images of GF that the warp and weft density of the fabric is small, and the yarn is composed of many fibers with many gaps. This is conducive to the penetration of hydrogel to form phys-

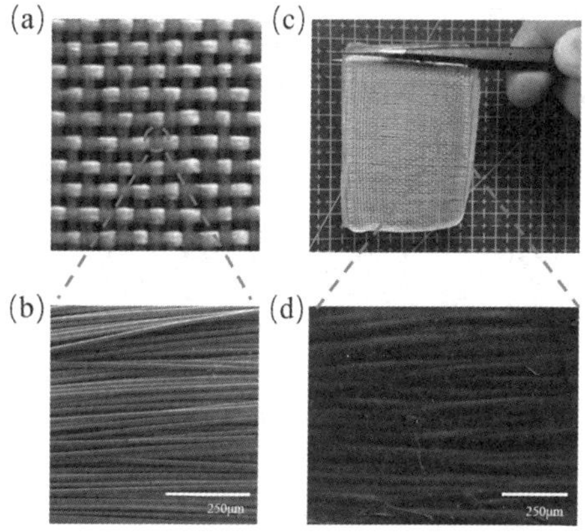

Fig. 1 (a) SEM of plain weave GF; (b) SEM of GF not soaked by hydrogel; (c) GF compounded with hydrogel; (d) SEM of GF soaked by hydrogel

ical entanglement. Figs. 1(c) and 1(d) are the physical picture and SEM of the composite of hydrogel and GF, which forms a close combination. The hydrogel can penetrate into the fabric well, which increases the chance of entanglement and forms a smooth surface. In addition, the main component of GF is SiO_2, which forms hydrogen bonds with acrylamide in hydrogel, thus greatly improving the mechanical properties and cycle times of the bilayer hydrogel actuator.

3.2 Deformation performance of actuator

The bilayer actuator prepared in this paper belongs to asymmetric structure. One layer is hydrogel impregnated GF, and the other layer is hydrogel. The CA/PAAm hydrogel layer can absorb water, expand in DI water and release water and contract in organic solvents. Although there are a few hydrogels in fabric, the fabric has limitations. Therefore, the fabric layer barely absorb water and do not deform in organic solvents, resulting in the bend of actuator to the fabric layer in DI water, and the bend in the opposite direction in organic solution.

In this experiment, the actuators were cropped into long strips of 60 mm×10 mm×3.8 mm and put into DI water and acetone solution, respectively. Fig. 2 (a) shows that the undeformed actuator was put into acetone solution. With the increase of time, the actuator gradually bent towards the fabric layer, and the deformation was faster in the initial stage. The corresponding strain was calculated according to Eq. (1), which was consistent with the angle change. Fig. 2 (b) showed the change of angle and strain when the sample was put into DI water, and the change trend of the value was consistent with the change when the sample was put into acetone. The absolute value of bending angle can reach 360° no matter which layer bends.

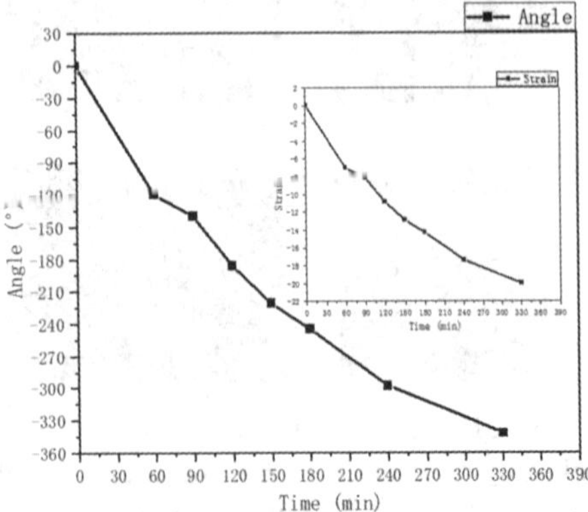

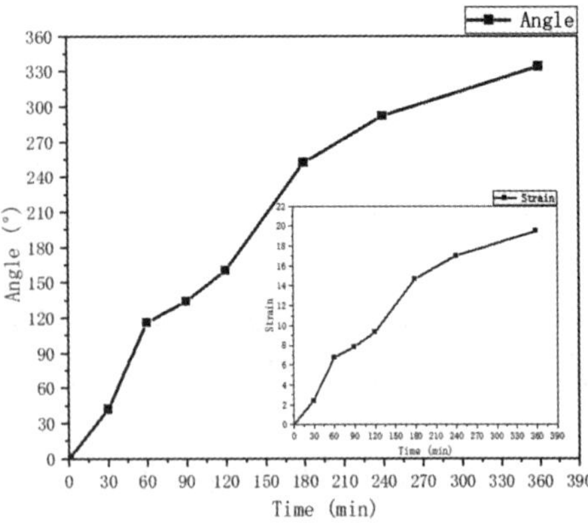

Fig. 2 (a) Variation of actuator bending angle and strain with time in acetone; (b) Variation of actuator bending angle and strain with time in DI water

3.3 Cycle experiment

In addition to meeting the requirements of large deformation, the actuator also needs to have good recycling performance. In this experiment, the performance of the actuator was verified by multiple cycle experiments. Fig. 3 (a) is a line chart of actuator angle and strain change after three cycles experiments. The change trend of the three cycles was roughly the same. Fig. 3 (b) shows the angle and strain values of the ac-

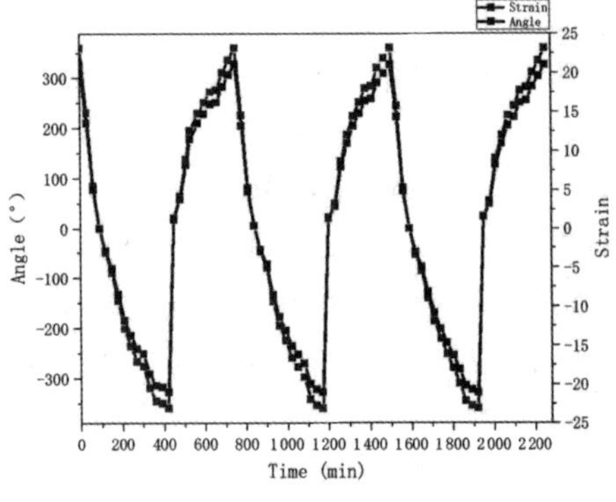

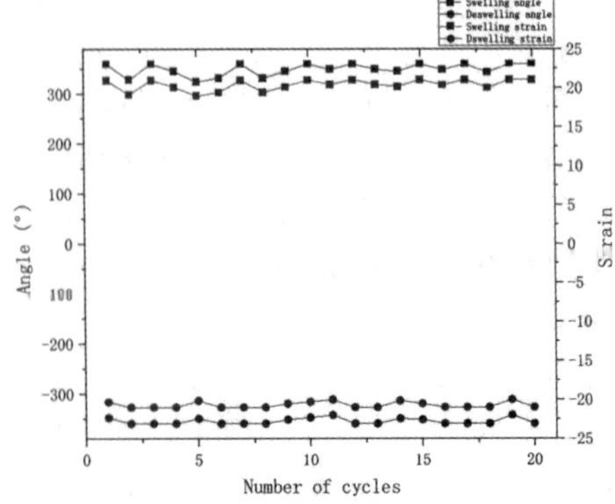

Fig. 3 Angle change of three cycles experiment and multiple cycles experiment: (a) the angle and strain of the actuator in three cycles change with time (measured every 30 min); (b) actuator 20 cycles angle and strain final state change

tuator after being put into DI water and acetone solution respectively, which complete deformation and was repeated 20 times. It showed that the actuator still maintained a good bending shape in multiple cycles.

3.4 Finite element simulation of complex actuator deformation

In order to better predict the deformation of the actuator, FEM was used to simulate the actuator and the actuator with special shapes. In this paper, according to the deformation effect of the above simple strip actuator, a hexagonal gripper was further prepared. The length and width of each actuator of hexagonal gripper were 6 mm and 1 mm, and the angle between each gripper was 60°. Compared with a single actuator, the functions of the hexagonal gripper were more diverse and could cooperate with each other. From Figs. 4 (a)−4(e), the gripper could realize "downward crawling" in DI water, after putting it in acetone solution, the hexagonal gripper gradually curled in the opposite direction until "upward fishing" was realized, and this process can be carried out circularly. FEM of complex gripper was also consistent with the experimental deformation, which further determined the feasibility and

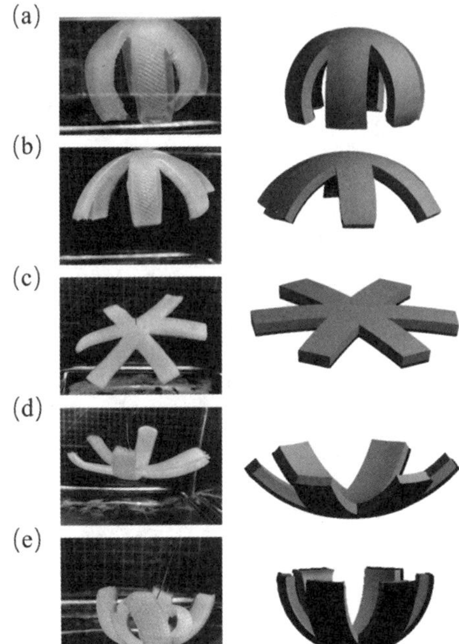

Fig. 4 (a) Actual and simulated diagram of hexagonal gripper under gravity when fully expanded in DI water; (b)-(d) The picture of the actual and simulated initial hexagonal gripper placed in acetone solution every 30 min; (e) Actual and simulated diagram of complete upward deformation of hexagonal gripper in acetone solution under the action of gravity

effectiveness of this simulation method. The center point of the gripper was fixed during FEM simulation, which was the same as the actual deformation. From the section, FEM can accurately simulate this complex deformation effect. Later, if improved, more functions can be achieved.

4 Conclusions

In this paper, the solvent responsive GF/PAAm/CA actuator with bilayer asymmetric structure was prepared by a one-step molding method. The interface performance of the actuator was improved and the service life was prolonged by the intervention of the GF. The actuator has simple structure and good performance. It can realize bending, grasping and other movements. It can also realize more complex deformation effects by changing the fabric structure, hydrogel pattern and actuator shape, and has good cycle repeatability. This provides new insights for the design and manufacture of intelligent actuator for bionic applications. In addition, the FEM method can better predict deformation and save experimental time, and promote the development of software actuators and soft robots.

References

[1] ZHI W L, VINCENT M N, YA D Y. Light-powered soft steam engines for self-adaptive oscillation and biomimetic swimming[J]. Science Robotics, 2021, 6(61).

[2] TASKISHIMA Y, YOSHIDA K, KHOSLA A, et al. Fully 3D-printed hydrogel actuator for jellyfish soft robots[J]. ECS Journal of Solid State Science and Technology, 2021, 10(3).

[3] NG C S X, TAN M W M, XU C Y, et al. Locomotion of miniature soft robots[J]. Advanced Materials, 2021, 33(19): 24.

[4] KIM C, KIM G, LEE Y, et al. Shape memory alloy actuator-embedded smart clothes for ankle assistance[J]. Smart Materials and Structures, 2020, 29(5): 9.

[5] WEHNER M, TRYBY R L, FITZGERALD D J, et al. An integrated design and fabrication strategy for entirely soft, autonomous robots[J]. Nature, 2016, 536(7617): 451.

[6] KHALDI A, MAZIZ A, ALICI G, et al. Bottom-up microfabrication process for individually controlled conjugated polymer actuators[J]. Sensors and Actuators B-Chemical, 2016, 230: 818−824.

[7] LI M, WANG X, DONG B, et al. In-air fast response and high speed jumping and rolling of a light-driven hydrogel actuator[J]. Nature Communications, 2020, 11(1).

[8] D'ERAMO L, CHOLLET B, LEMAN M, et al. Microfluidic actuators based on temperature-responsive hydrogels[J]. Microsystems & Nanoengineering, 2018, 4.

[9] REHOR I, MASLEN C, MOERMAN P G, et al. Photoresponsive Hydrogel Microcrawlers Exploit Friction Hysteresis to Crawl by Reciprocal Actuation[J]. Soft Robotics, 2021, 8(1): 10–18.

[10] DONG Z Q, CAO Y, YUAN Q J, et al. Redox- and glucose-induced shape-memory polymers[J]. Macromolecular Rapid Communications, 2013, 34(10): 867–872.

[11] TANG J, TONG Z, XIA Y, et al. Super tough magnetic hydrogels for remotely triggered shape morphing[J]. Journal of Materials Chemistry B, 2018, 6(18): 2713–2722.

[12] WANG J L, ZHANG X T, WANG A, et al. The synthesis and simulations of solvent-responsive bilayer hydrogel[J]. Polymer, 2020, 204.

[13] TAN Y, WANG D, XU H, et al. A fast, reversible, and robust gradient nanocomposite hydrogel actuator with water-promoted thermal response[J]. Macromolecular Rapid Communications, 2018, 39(8).

[14] LI T, WANG J, ZHANG L, et al. "Freezing", morphing, and folding of stretchy tough hydrogels[J]. Journal of Materials Chemistry B, 2017, 5(29): 5726–5732.

[15] QIN H, ZHANG T, LI N, et al. Anisotropic and self-healing hydrogels with multi-responsive actuating capability[J]. Nature Communications, 2019, 10.

[16] LE X X, ZHANG Y C, LU W, et al. A novel anisotropic hydrogel with integrated self-deformation and controllable shape memory effect[J]. Macromolecular Rapid Communications, 2018, 39(9).

[17] HUBBARD A M, CUI W, HUANG Y, et al. Hydrogel/elastomer laminates bonded via fabric interphases for stimuli-responsive actuators[J]. Matter, 2019, 1(3): 674–689.

Design and Knitting of a Fully Formed Glove by Elastic Yarns

WU Chubin[1], **WANG Xi**[2], **ZHANG Tianshu**[3], **LI Qiao**[1*]

[1] *Key Laboratory of Textile Science & Technology, Ministry of Education, College of Textiles, Donghua University, Shanghai 201620, China*

[2] *College of Information Science and Technology, Donghua University, Shanghai 201620, China*

[3] *College of Science, Donghua University, Shanghai 201620, China*

[*] *Corresponding author's email*: qiaoli@ dhu. edu. cn

Abstract: This paper introduces the method of knitting a fully formed glove by elastic yarns on a handmade flat knitting machine. In order to realize the integral forming of the knitted glove, it takes the length direction of the fingers in the glove as the weft case in the knitting process. The glove was firstly divided into three different parts, that is, hand, palm, as well as wrist. The different sizes of each part were then determined according to an industrial standard. By assuming the structure of the different parts as U shape or ring geometries, the stitches of different parts were calculated and adjusted according to the final size, the gauge of the knitting machine, as well as the mechanical properties of the elastic yarns. After knitting, heat setting was finally used to determine the final size of the knitted glove.

Keywords: Knitting technology; Fully formed glove; Elastic yarn; Heat setting

1 Introduction

In recent years, fully fashioned knitting technology has become the future trend in the knitting industry. As a common accessory, knitted gloves can be made with fully fashioned knitting technology[1-2]. They can be worn directly after getting off the machine, and the subjects feel soft and comfortable. The seamless forming technology of the product is realized by using the techniques of drawing and releasing needles, moving circle and resting knitting on the flat machine. This production mode greatly shortens the manufacturing process, reduces the material waste, and improves the comfort and fit of the product. However, the integrated technology requires high requirements on equipment, process and product design[3-4].

Fully formed knitted gloves are usually produced on a special flat machine. In 2010, Braga et al. [5] presented a technical solution of producing knitted gloves on an electronic flat knitting machine with a regular two needle-bed (CMS330TCE7, Stoll Co. Ltd., Germany) based on split technique. In 2018, Man Cuiping et al. [6] introduced the design and knitting method of fully formed knitted gloves based on LXC-352SCV computer-controlled flat knitting machine (Long Xing Co. Ltd., Jiangsu, China) and solved the problems encountered in the production process, which proved that the domestic computerized flat knitting machine was feasible for producing fully formed knitted gloves. This paper realizes the knitting of fully formed gloves on ordinary flat knitting machine by using the technology of cylinder knitting and shifting circle knitting.

2 Design of Knitted Fully Formed Gloves

2.1 Structure of glove

The structure of the fully formed knitted glove can be divided into three parts: finger, palm and wrist, as shown in Fig. 1[6]. For five fingers, it requires to determine the width and length of each finger, that is, the width of thumb (1), index finger (2), middle finger (3), ring finger (4), as well as little finger (5); the length of thumb (6), index finger (7), middle finger (8), ring finger (9), as well as little finger (10). For the palm, it demands to determine the length (11, 12 and 13) and width (14). For the wrist, it needs to determine the width (15).

The size of the glove is designed according to an indus-

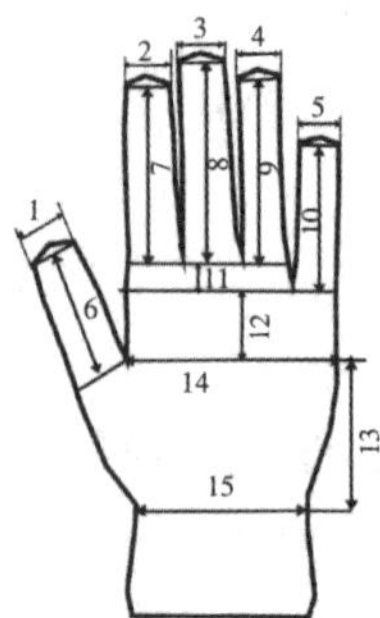

Fig. 1 Structure of the fully formed knitted glove
1-5: width of thumb, index finger, middle finger, ring finger and little finger; 6-10: length of thumb, index finger, middle finger, ring finger and little finger; 11: the length from the carpal metacarpal joint of ring finger to the carpal metacarpal joint of little finger; 12: the length of the line from the wrist and metacarpal joint of little finger to the jaws; 13: the length from the tiger line to the wrist stripe; 14: width of the hand; 15: width of the wrist.

try standard of FZ/T73047-2013[7] for knitted gloves. Take a male subject with 180/90 as an example to design and make fully formed knitted gloves, in which "180" and "90" respectively represent the length and width of the glove, and the unit is "mm". The length or width of each part of the glove is shown in Tab. 1.

Tab. 1 Width or length of each part of the glove

Parts	Width or length (mm)
1	27
2	25
3	25
4	24
5	21
6	57
7	62
8	77
9	62
10	61
11	13
12	37
13	32
14	101
15	78

2.2 Design of machine process drawing

As the glove presents a tubular structure as a whole, U-shaped and ring structures are selected to make the glove. Tubular knitting forms a tube by knitting one yarn on the two needle-beds alternately, only passing the yarn from one needle bed to the other at the two selvages[8]. Because the U-shaped structure is open at one end, it is used on the palm, wrist and four fingers (index, middle, ring and little fingers). The ring structure is closed at both ends, so it is used in the position of the thumb. The knitting diagrams of these two structures are shown in Fig. 2.

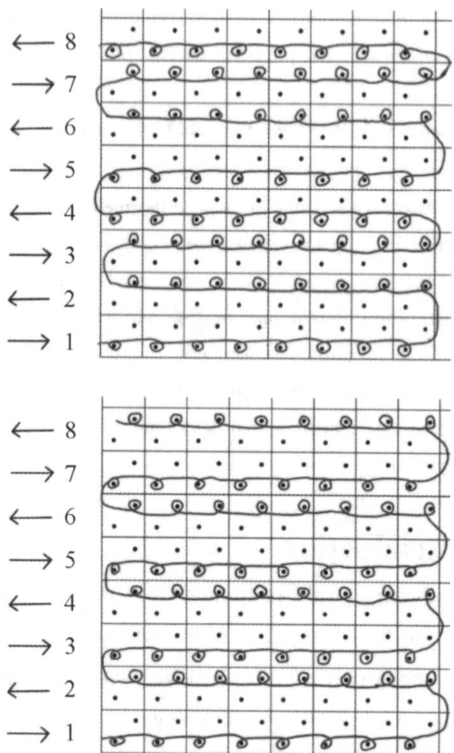

Fig. 2 (a) Knitting diagram of U-shaped structure; (b) knitting diagram of ring structure

It is necessary to calculate the density of the sample before knitting the glove, so as to determine the number of stitches and rotations in each part of the glove. The fabric sample with tubular structure is shown in Fig. 3. After measurement and calculation, the density of the fabric sample is 35 stitch/5 cm in the course direction

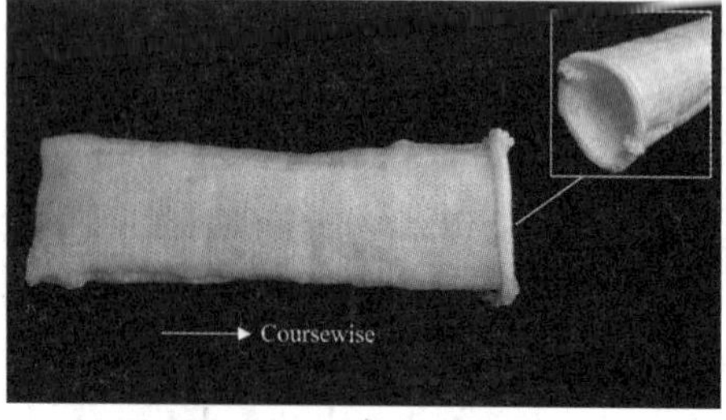

Fig. 3 Tubular fabric sample for calculation of the density

and 133 row/5 cm in the wale case. In the course of calculation, each part could be assumed as a tubular fabric. Since the fingers are not completely cylindrical, the circumference of the tube is approximately twice the width of the finger. According to the length and width of the five fingers, palm and wrist, as well as the density of the fabric sample in both course and wale directions, the number of stitches and rotations in the main parts of the glove can be calculated, as shown in Tab. 2.

Tab. 2 The number of stitches and rotations in each part of the glove

Gloves parts	Stitches	Rotations
Thumb	28	75
Index finger	44	33
Middle finger	54	33
Ring finger	44	32
Little finger	43	28
Palm	72	132
Wrist	22	107

Therefore, the knitting of integrated gloves can be realized only by using U-shape structure and ring structure. The machine diagram of the knitted gloves is shown in Fig. 4, where the thick number indicates the number of stitches. The fine number indicates the number of rows, and "4-1-45" means one stitch removed every four rows, with a total of 45 stitches removed, while "4+1-23" means one stitch added every four rows, for a total of 23 stitches added.

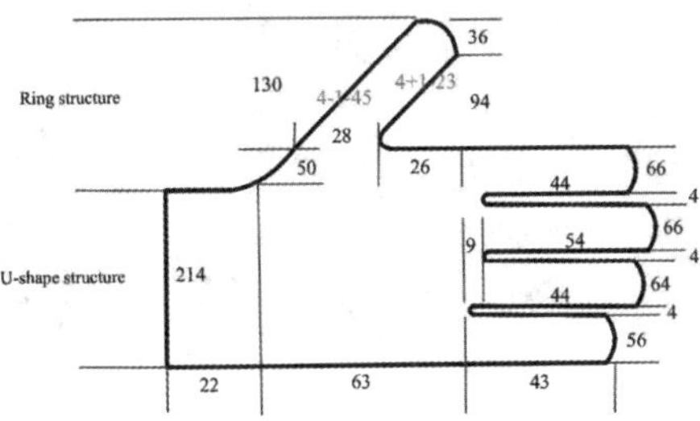

Fig. 4 Machine drawing of knitted full-form glove

3 Knitting Process of Fully Formed Glove

3.1 Equipment and raw materials

In the paper, the yunnei KH-868 (main machine) and KR-838 (secondary machine) flat knitting machine are used for knitting. The main technical parameters of the flat knitting machine are shown in Tab. 3, and the type and specification of the yarn used are shown in Tab. 4.

Tab. 3 Main specifications of yunnei KH-868 (main machine) and KR-838 (secondary machine) flat knitting machine

Product name	Five-needle machine (main machine)	Five-needle machine (secondary machine)
Model	KH-868	KR-838
Size	103 cm×20 cm	103 cm×13 cm
Gauge	5	
Applicable yarns	Wool thread Cashmere thread	
Braided tissue	Plain stitch, set circle, jacquard, local jacquard, relief, add yarn, add yarn set circle	Rib, unit treasure, double dollar treasure, ripple

Tab. 4 The type and specification of the yarn

Yarn name	Linear density
Spandex core-spun yarn	560, 100, 100 D

3.2 Parameters of the knitting process

The glove is knitted from the wrist, palm and little finger and then from the bottom to the top, and the little finger, ring finger, middle finger, index finger and thumb in turn. After each finger has been knitted, it is necessary to subtract the stitch to finish the knitting of the finger, and then add the stitch to the next finger. Each finger needs to be knitted by adding and subtracting stitches before all four fingers are finished. Therefore, in order to realize the purpose of subtracting multiple stitches and adding multiple stitches, this paper adopts the way of moving the loop stitch and adding the needle. The knitting diagrams of the two kinds of adding and subtracting needles are shown in Fig. 5.

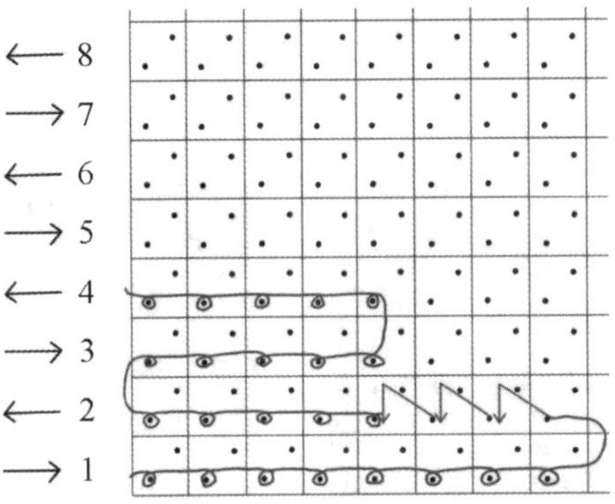

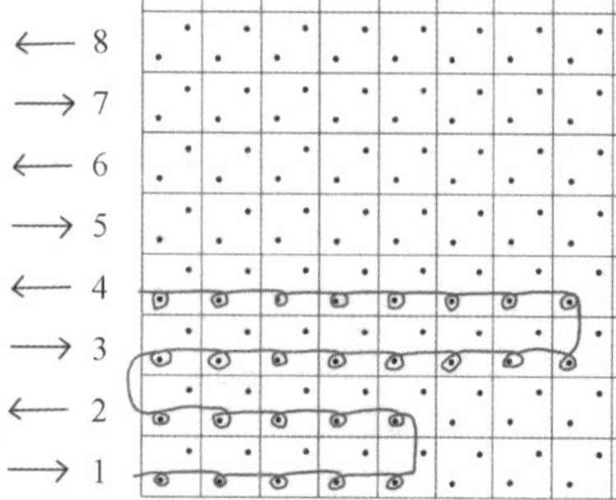

Fig. 5 The knitting diagram of (a) subtracting needle and (b) adding needle

At the same time, in order to ensure the shape of the thumb during the knitting process, it is necessary to keep the needle subtract and add on both sides of the thumb, as shown in Fig. 6. This way could ensure that the edge of the thumb is at an angle.

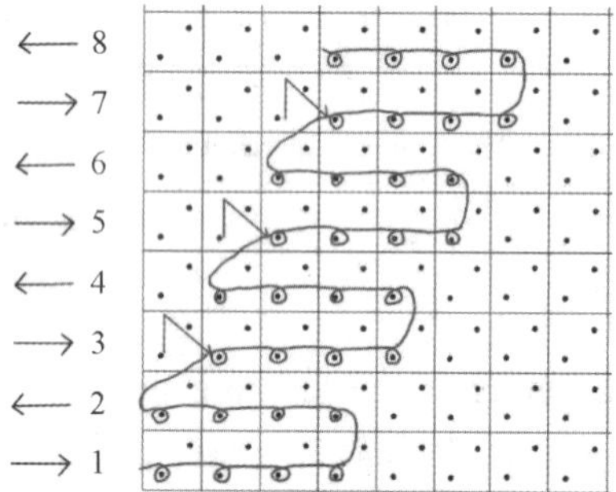

Fig. 6 The knitting diagram of subtracting and adding needles on both sides of the thumb

3.3 Knitting process

Fig. 7 shows the entire knitting process of the glove. The specific process was as follows. The density was firstly adjusted on a wheel, as shown in Figs. 7(a) and 7(b). The knitting mode was secondly adjusted to U-shaped knitting, as shown in Figs. 7(c) and 7(d). The fingers were thirdly knitted by adding and subtracting needles, as shown in Figs. 7(e) and 7(f). The knitting mode was then adjusted to ring knitting after reaching the rotations of the wrist, as shown in Fig. 7(c). Finally, the glove was removed after reaching the required rotations.

(a)

(b)

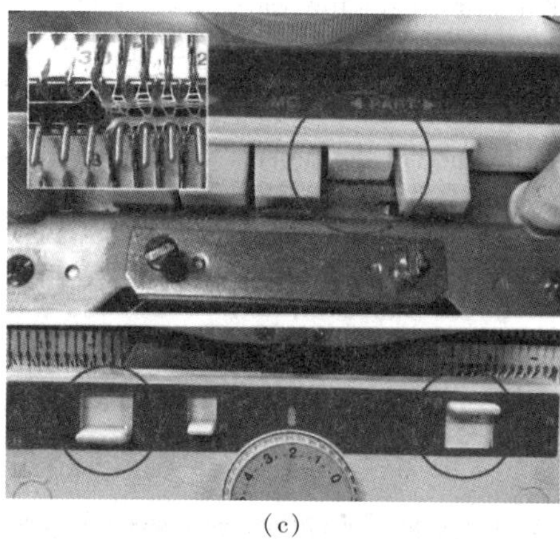

(c)

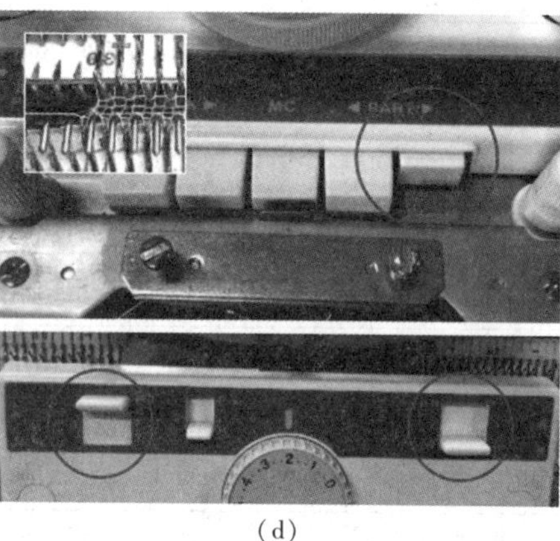

(d)

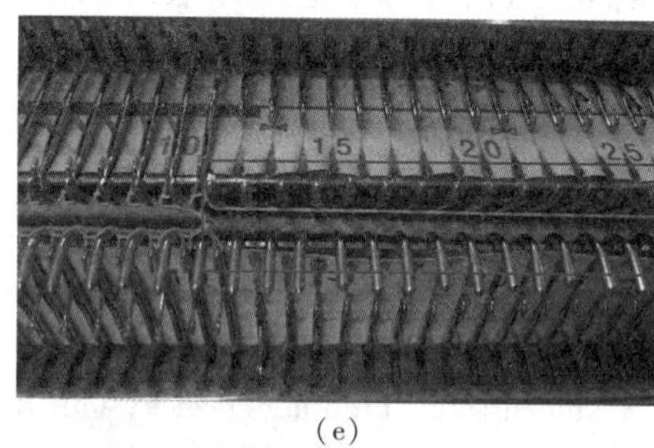

(e)

(f)

Fig. 7 (a) The density wheel of the main machine; (b) the density wheel of the secondary machine; (c) the position of the button when the yarn is on the needle bed above; (d) the position of the button when the yarn is on the needle bed below; (e) the position of the yarn after subtracting stitches; (f) the position of the yarn after adding stitches

3.4 Post treatment

The gloves made by the flat knitting machine cannot be used directly after they are removed from the machine (Fig. 8). In order to make the glove have more stable structure and elastic recovery, it is necessary to heat set. The specific process is as follows. First, CAD software is used to design the size and shape of the mold, as shown in Fig. 9 (a), and then laser cutting machine is used to cut the mold (Fig. 9 (b)) on the steel plate. Finally, the glove is covered on the mold (Fig. 9 (c)) and heat set with an iron. The glove after treatment is shown in the Fig. 10.

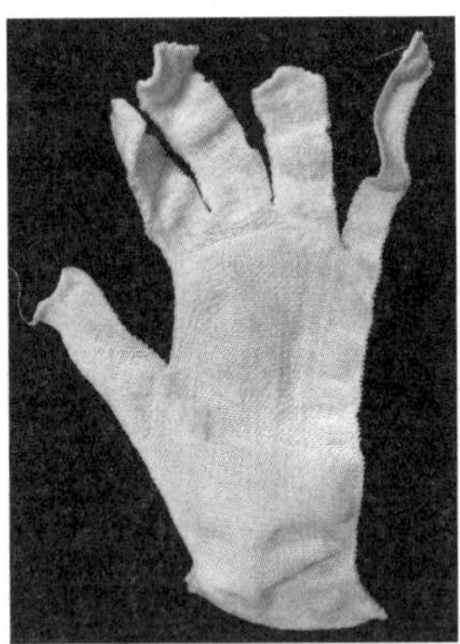

Fig. 8 The glove removed from the machine

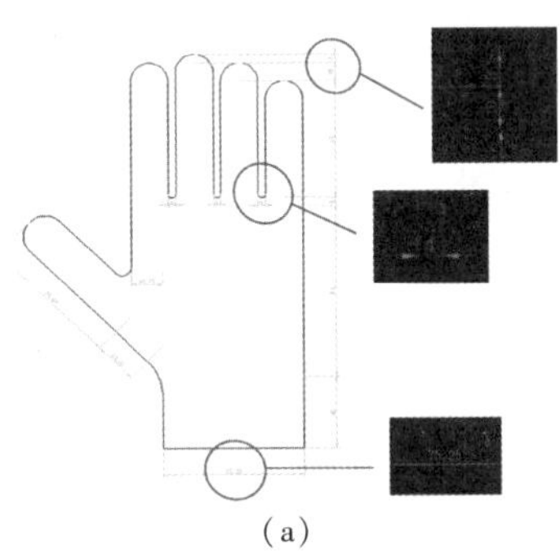

(a)

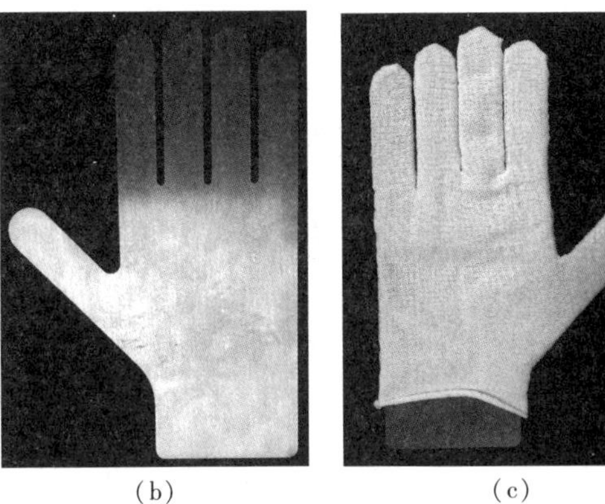

(b) (c)

Fig. 9 (a) The shape and size of the glove mold designed by CAD software; (b) the glove mold; (c) the glove is covered on the mold

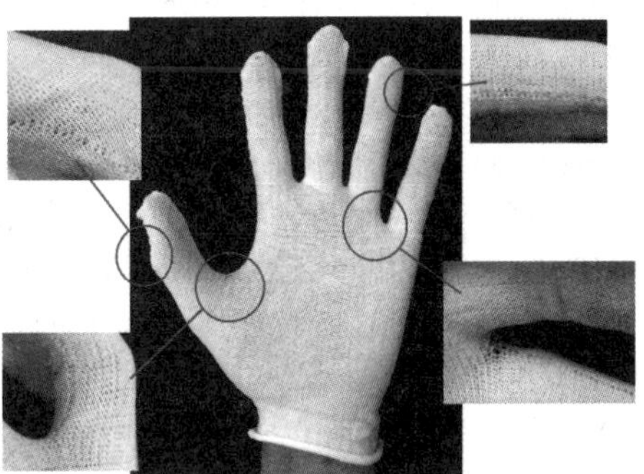

Fig. 10 The details of the glove

4 Conclusions

This paper discusses the forming principle of gloves on yunnei KH-868 (main machine) and KR-838 (secondary machine) flat knitting machines from three aspects of structure, pattern design and knitting process. Compared with the common method, a different knitting method is proposed in this paper, which is the weft direction of the glove is following the finger. At the same time, the integrated design of knitting gloves is made use of the characteristics of ring structure and U-shaped structure. Each finger is woven by the operation of subtracting and adding needles. Finally, the fully formed knitting gloves are realized.

Acknowledgements

This research was funded by the National Natural Science Foundation of China (Nos. 12002085 and 51603039), sponsored by the Shanghai Pujiang Program, and supported by the Fundamental Research Funds for the Central Universities, the Key Laboratory of Textile Science and Technology (Donghua University), Ministry of Education, as well as the Initial Research Funds for Young Teachers of Donghua University.

References

[1] WANG M, CONG H L, JIANG G M, et al. Whole garment knitting process on four-bed computerized flat knitting machine[J]. Journal of Textile Research, 2017, 38(4): 61-67.

[2] PENG J J, JIANG G M, CONG H L, et al. Structure and knitting principle of whole garment[J]. Journal of Textile Research, 2017, 38(11): 48-55.

[3] YUAN Y, XU X L, DING C M. Knitwear technology[J].

Knitting Industry, 2007(11): 23-26.

[4] ZHANG C D, LONG H R. Research on three-dimensional knitted fabric weaving technology on computerized flat knitting machine[J]. Shandong Textile Science and Technology, 2003, 44(1): 15-17.

[5] BLAGA M, PENCIUC M, DAN D, et al. Production of knitting gloves by flat machine[J]. International Textile Review, 2010, 38(8): 47-48,50.

[6] MAN C P, LONG H R. Design and preparation of fully-fashioned knitted gloves[J]. International Journal of Advanced Manufacturing Technology, 2018(4): 17-20.

[7] FZ/T 73047-2013. Knitted civil gloves[S]. Ministry of Industry and Information Technology of China. 2013: 1-12.

[8] CHEN X G. Advances in 3D textiles[M]. 1st Edition. Elsevier: Woodhead Publishing Limited in Association with The Textile Institute, 2015.

Study on the Recycling of LiBr-$NaHSO_3$ System in the Extraction of Wool Keratin

YANG Zhengqiang[1,2], LIU Fang[1,2], YAN Xiong[1,2*], YU Jinlin[2,3], LI Qiao[1,2]

[1] *Key Laboratory of Textile Science & Technology, Ministry of Education, Donghua University, Shanghai 201620, China*

[2] *College of Textiles, Donghua University, Shanghai 201620, China*

[3] *Jiangsu Danmao Textile Co., LTD., Danyang 212351, China*

[*] *Corresponding author's email*: yan_xig@yeah.net

Abstract: To reuse waste wool resources, this paper adopts LiBr-$NaHSO_3$ dissolution system to extract keratin from waste wools, and explores the recycling and treatment process of waste liquid generated after extraction. When the density of the first and second recovery after concentration and $BaCl_2$ treatment is 1.47 g/mL and 1.51 g/mL, respectively, dissolution rate of the wool, extraction rate of keratin and yield are close to their values in the original dissolution system. It provides an environmentally friendly and economical method for extracting keratin from waste wools.

Keywords: Wool dissolution; Keratin; LiBr; Recycling

1 Introduction

Keratin is one of the most abundant non edible proteins in the world. Every year, it produces a large number of non-woven short hairs, poultry feathers and human hair from the textile industry. However, these keratin wastes have not been fully utilized. At present, only some feathers are used as fillers for warm clothing and animal food, while the rest are incinerated and landfilled as waste, which not only wastes resources, but also seriously affects the protection of the natural ecological environment[1]. With the emergence of ecological and environmental problems and the accumulation of waste, people pay more and more attentions to the research of some naturally renewable polymers. Among them, the unique characteristics of keratin based materials are applied to biomedicine[2], keratin regenerated fiber materials[3], shape memory materials[4] and other fields.

The key step of extracting keratin from wool fiber rich in keratin is the dissolution and degradation of wool fiber. At present, the common methods of extracting keratin are mainly reduction method[5], oxidation method[6], acid-base hydrolysis method[7], microwave radiation[8], steam flash explosion[9], superheated water method[10], ionic liquid method[11], low eutectic solvent assisted extraction of keratin[12], enzyme degradation method[13] and microbial degradation method[14]. However, many methods still have a lot of disadvantages, such as expensive chemical reagents, great environmental hazards, threats to human health, etc., which are not conducive to large-scale use in the industry[15]. Therefore, the goal of this study is to develop an economically feasible, ecofriendly, resource efficient and minimally destructive method for extracting wool keratin.

In the traditional reduction methods, it is necessary to combine high-dose urea to extract keratin with complete structure[16]. However, these high doses of urea will produce a large amount of urea containing wastewater in the process of keratin purification, which will bring a series of environmental problems such as water eutrophication, acid rain formation and so on. Therefore, high-dose urea has brought many obstacles to the large-scale industrial extraction of waste wool keratin. LiBr-$NaHSO_3$ dissolution system is proposed in this paper. Among them, $NaHSO_3$ is a reducing agent. Compared with other similar common reducing agents, it has low toxicity, low price, low irritation, mild reaction conditions and is suitable for industrial production. LiBr, as a swelling agent, promotes the solid-liquid phase transition of keratin in solution[4]. In this system, LiBr does not react with other substances, and the use of LiBr is large. If it can be recycled, the cost can be controlled, thereby bringing economic benefits and utilization of resource. This study mainly explores and improves the recycling process of LiBr in the process of extracting wool keratin from LiBr-$NaHSO_3$ system.

2 Experiments

2.1 Materials

Experimental materials: wool (Jiangsu Danyang Woolen Textile Co., LTD.); interception of dialysis bags with molecular weight of 8 000-14 000; $NaHCO_3$(AR, Collins); $NaHSO_3$ (AR, RHAWN); LiBr (AR, RHAWN); $BaCl_2$(AR, Hushi); HCl (AR, Hushi).

Instruments: magnetic agitator; table centrifuge; freeze dryer; electronic analytical balance; oven; scanning electron microscope.

2.2 Recovery and treatment of dialysis liquid waste

The experimental steps are shown in Fig. 1. Firstly, the wool was immersed in $NaHCO_3$ solution (0.1 mol/L) with a bath ratio of 1 : 20 for 3 h. After that, it was rinsed with deionized water to neutral, and then dried and cut into the fibers with length of 2-5 mm. Secondly, the wool with the mass of 1.25 g was put into 25 mL LiBr solution (7 mol/L) and stirred at room temperature until fully dissolved. Thirdly, at 90 ℃, $NaHSO_3$ of the same quality as wool was added and stirred for 12 h. After the reaction, centrifuged them at 5 500 r/min for 45 min. Fourthly, to obtain the concentrated solution of dialysis recovery fluid (recorded as solution A), the supernatant was taken for dialysis for 72 h. The deionized water was changed every 4 h, and collected it after the first two dialysis to heat and concentrate. Fifthly, the 1.25 g wool was added to 25 mL solution A, and then, the above experimental process was repeated to recover the concentrated solution (recorded as solution B). Sixthly, the 10% $BaCl_2$ solution was prepared to titrate and precipitate solution A. And then, the solid and liquid of the suspension was separated by centrifugation. It was not until the solution no longer produces precipitation that the above operations were stopped repeating. And then, a clear solution (recorded as solution A_0) was obtained. It was not until the solution density of recovery liquid were 1.42 g/mL (recorded as solution A_1), 1.45 g/mL (recorded as solution A_2)

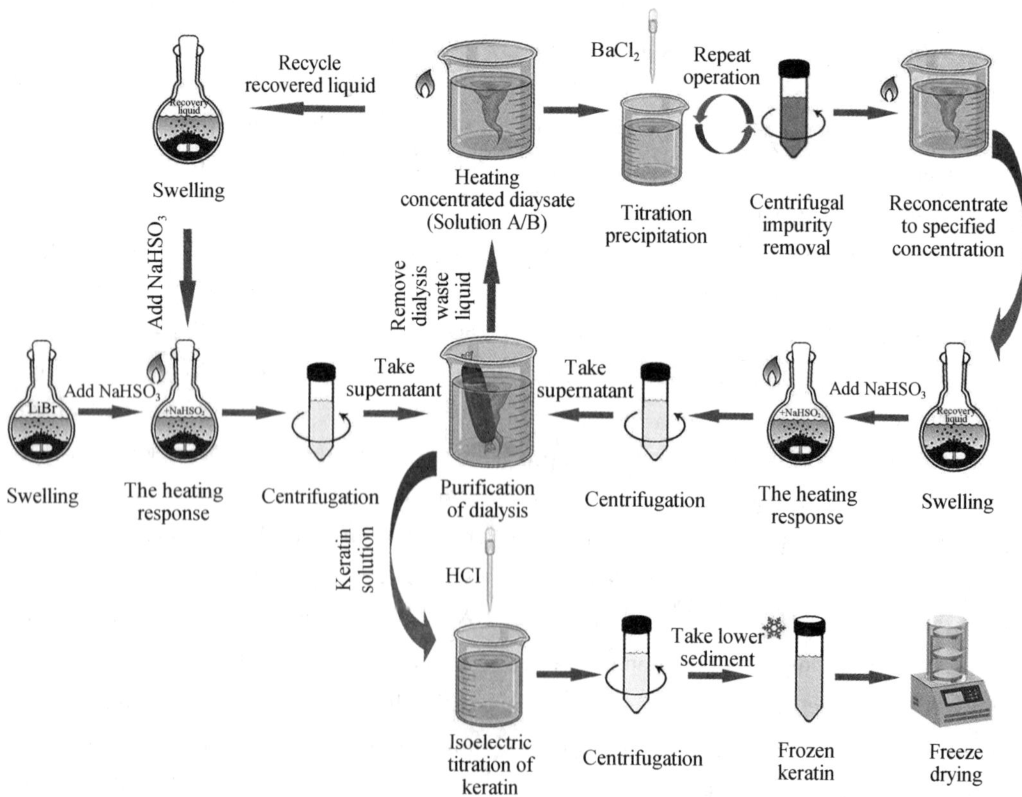

Fig. 1 Technological process of recovery and treatment of dialysis liquid waste

and 1.47 g/mL (recorded as solution A_3) that the treated solution A_0 was stopped heating and concentrating. Similarly, it was not until the solution density of recovery liquid were 1.47 g/mL (recorded as solution B_1), 1.49 g/mL (recorded as solution B_2) and 1.51 g/mL (recorded as solution B_3) that the treated solution B_0 was stopped titrating, precipitating and concentrating. 1.25 g wool was added to 25 mL the above solutions A_1, A_2, A_3, B_1, B_2 and B_3, respectively. And then, the above dissolution, dialysis and other experimental processes were repeated to compare the results.

Finally, the keratin solution obtained by wool dissolution dialysis with the original solution (recorded as solution Z), solutions A, B, A_1, A_2, A_3, B_1, B_2 and B_3 was titrated to the pH of 4.6 by hydrochloric acid solution, and then it was not until keratin precipitates completely that solution was stopped leaving to stand. The white precipitate obtained by centrifugation was freeze-dried to obtain white keratin powder.

2.3 Characterization

Thoroughly wash, dry and weigh the residue after wool dissolution reaction to obtain the mass of the residue.

The formula of wool solubility is:

$$W_1(\%) = \frac{M_0 - M_1}{M_0} \times 100\% \qquad (1)$$

where W_1 is the solubility of wool; M_0 is the mass of wool in the initial system (g); M_1 is the mass of residue (g).

The extraction rate of keratin is:

$$W_2(\%) = \frac{M_2}{M_0 - M_1} \times 100\% \qquad (2)$$

where W_2 is the extraction rate of wool keratin; M_2 is the mass of wool keratin obtained by freeze-drying (g).

The yield of keratin is:

$$W(\%) = W_1 \times W_2 \times 100\% \qquad (3)$$

where W is the yield of wool keratin.

3 Results and Discussion

After repeated recycling, the concentration of LiBr decreased with the accumulation of impurity ions in the solution. Through measurement and calculation, the density of LiBr solution (7 mol/L) was 1.42 g/mL at normal temperature and pressure. The density of the first recovered solution was 1.42 g/mL, and the density of the second recovered solution was 1.38 g/mL. Therefore, to reach the same level as the original LiBr solution, the density of the recovered solution after concentration should be greater than that of the previous LiBr solution. In order to explore an optimized process, wool was dissolved with different densities of recovery solution under different recovery times, and dissolution rate of the wool, extraction rate of keratin and yield of each group of recovery solution were compared.

Fig. 2 shows that the solubility and keratin yield of wool increase with the increment of density, while the extraction rate of wool keratin remains at about 80%. When the density of the first and second recovered solution was 1.47 g/mL and 1.51 g/mL, respectively, dissolution rate of the wool, extraction rate of the keratin and yield are equivalent to the original solution dissolution levels. This is mainly because the plasma containing SO_3^{2-} and SO_4^{2-} in the dialysis liquid waste has a great impact on the LiBr-$NaHSO_3$ dissolution system, while the Cl^- ion introduced after removal of $BaCl_2$ impurity has little impact on the dissolution system and the relative quality is low. When the recovered solution was concentrated to a certain density, the concentration of LiBr was close to that of the original LiBr solution (7 mol/L). At this time, LiBr in the recovered solution can fully induce the solid-liquid reversible transformation of crystalline keratin in water, achieving a relatively ideal effect. At the same time, LiBr mainly acted on the hydrogen bond between wool molecules to achieve the purpose of reducing agent dissolving wool,

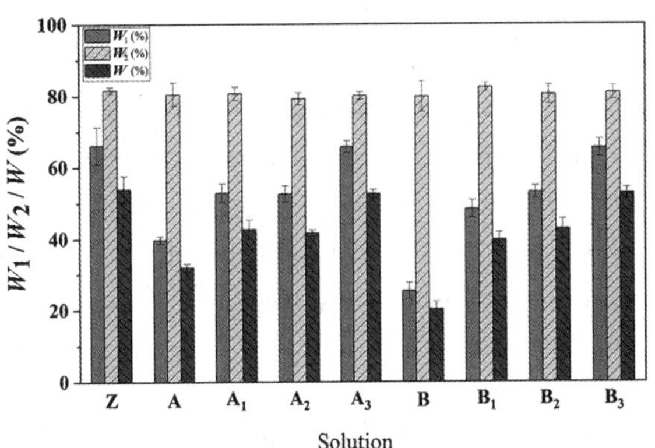

Fig. 2 Effect of different densities of recycled solution on wool dissolution and keratin extraction

and would not cause serious damage to the macromolecular chain of keratin, so the extraction rate of keratin was basically stable at the same level.

From the comparison in Fig. 3, the wool structure had completely disappeared after the reaction of the original solution, and the residue after the reaction of solution A and solution B had obvious fibrous residue. After $BaCl_2$ treatment, the wool dissolved more completely with the increase of density. This phenomenon showed that the concentration of LiBr in the solution played a great role in the swelling effect of wool. When the concentration of LiBr solution was small, the solution rate and keratin yield of wool were low. When the concentration of LiBr solution increased to about 7 mol/L, the wool dissolved well.

Fig. 3 SEM images of wool residues after reaction under different treatment conditions: (a) residue after the reaction of the original solution; (b) residue after the reaction of solution A; (c) residue after the reaction of solution B; (d) residue after the reaction of solution A_1; (e) residue after the reaction of solution A_2; (f) residue after the reaction of solution A_3; (g) residue after the reaction of solution B_1; (h) residue after the reaction of solution B_2; (i) residue after the reaction of solution B_3

4 Conclusions

The recovery and treatment process of dialysis liquid waste was improved and a comparative discussion and analysis of dissolution rate of the wool, extraction rate of the keratin, keratin yield and morphology of the residue after wool dissolution was made. The following conclusions are obtained:

(1) The experimental results showed that LiBr in wool dissolution system did not participate in the reaction, and the recovery and reuse of dialysis liquid waste was feasible.

(2) The wool dissolution rate and keratin yield of the recovered solution after concentration were not high when the wool was dissolved again. However, titration and concentration of the dialysis liquid waste through $BaCl_2$ could better remove the ions in the liquid waste that had a great impact on the dissolution system, so that the wool dissolution effect was improved.

(3) When the density of the recovered solution reached 1.47 g/mL and 1.51 g/mL respectively, the solubility and keratin yield obtained by dissolving wool were close to the level of keratin extracted from the original solution, achieving a better recovery treatment effect. It

provides a relatively environmental and economic method to reduce the pressure on the environment caused by the accumulation of waste wool.

References

[1] YAN R R, GONG J S, SU C, et al. Preparation and applications of keratin biomaterials from natural keratin wastes [J]. Applied Microbiology and Biotechnology, 2022, 106(7): 2349-2366.

[2] FEROZ S, MUHAMMAD N, RATNAYAKE J, et al. Keratin — based materials for biomedical applications [J]. Bioactive Materials, 2020, 5(3): 496-509.

[3] SHAVANDI A, SILVA T H, BEKHIT A A, et al. Keratin: dissolution, extraction and biomedical application [J]. Biomaterials Science, 2017, 5(9): 1699-1735.

[4] CERA L, GONZALEZ G M, LIU Q H, et al. A bioinspired and hierarchically structured shape-memory material [J]. Nature Materials, 2021, 20(2): 242-249.

[5] SU C, GONG J S, QIN J F, et al. Glutathione enables full utilization of wool wastes for keratin production and wastewater decolorization [J]. Journal of Cleaner Production, 2020, 270: 122092.

[6] RAJABINEJAD H, PATRUCCO A, CARINGELLA R, et al. Preparation of keratin-based microcapsules for encapsulation of hydrophilic molecules [J]. Ultrasonics Sonochemistry, 2018, 40: 527-532.

[7] FOUNTOULAKIS M, LAHM H W. Hydrolysis and amino acid composition analysis of proteins [J]. Journal of Chromatography A, 1998, 826(2): 109-134.

[8] CHEN J Y, DING S Y, JI Y M, et al. Microwave-enhanced hydrolysis of poultry feather to produce amino acid [J]. Chemical Engineering and Processing-Process Intensification, 2015, 87: 104-109.

[9] ZHANG Y Q, ZHAO W, YANG R J. Steam flash explosion assisted dissolution of keratin from feathers [J]. ACS Sustainable Chemistry & Engineering, 2015, 3(9): 2036-2042.

[10] BERTINI F, CANETTI M, PATRUCCO A, et al. Wool keratin-polypropylene composites: properties and thermal degradation [J]. Polymer Degradation and Stability, 2013, 98(5): 980-987.

[11] ZHENG S S, NIE Y, ZHANG S J, et al. Highly efficient dissolution of wool keratin by dimethylphosphate ionic liquids [J]. ACS Sustainable Chemistry & Engineering, 2015, 3(11): 2925-2932.

[12] WANG D Y, YANG X H, TANG R C, et al. Extraction of keratin from rabbit hair by a deep eutectic solvent and its characterization [J]. Polymers, 2018, 10(9): 993.

[13] MARTINEZ J P D, CAI G Q, NACHTSCHATT M, et al. Challenges and opportunities in identifying and characterising keratinases for value-added peptide production [J]. Catalysts, 2020, 10(2): 184.

[14] KORNIŁŁOWICZ-KOWALSKA T, BOHACZ J. Biodegradation of keratin waste: theory and practical aspects [J]. Waste Management, 2011, 31(8): 1689-1701.

[15] ZHANG N, WU Z R, JIANG Z, et al. pH mediated l-cysteine aqueous solution for wool reduction and urea-free keratin extraction [J]. Journal of Polymers and the Environment, 2022, 30(7): 2714-2726.

[16] RAMYA K R, THANGAM R, MADHAN B. Comparative analysis of the chemical treatments used in keratin extraction from red sheep's hair and the cell viability evaluations of this keratin for tissue engineering applications [J]. Process Biochemistry, 2020, 90: 223-232.

Preparation and Properties of Boron Nitride Loaded Regenerated Cooling Fiber

LU Yanping[1], LI Zhijiang[1,2], ZHANG Peihua[1*]

[1] *College of Textiles, Donghua University, Shanghai 201620, China*

[2] *College of Mechanical and Electrical Engineering, Tarim University, Alar 843300, China*

* *Corresponding author's email*: phzh@dhu.edu.cn

Abstract: The development of hygroscopic and cool textiles has become an inevitable trend in the development of textiles used in summer and under extreme environmental conditions. To obtain regenerated cooling fibers with excellent thermal conductivity and cooling properties, in this work, the spinning dope was prepared by dissolving the flax fiber recovered from the factory in a green and efficient ionic liquid 1-butyl-3-methylimidazolium chloride. The boron nitride (BN) nanoparticles are then added to the spinning solution. Finally, regenerated cooling fiber was prepared using the wet spinning method. To study the effect of different addition amounts of boron nitride on the chemical structure, surface morphology, mechanical properties, thermal stability and thermal conductivity of regenerated fibers and conduct characterization analysis. The results show that the tensile breaking strength of the regenerated cooling fiber is 121.68 MPa and the elongation is 13.37 % when the ionic liquid and organic solvent compound system is used to dissolve the flax fiber and the mass fraction of BN is 60 wt%. The fabric made of BN with the addition of 60 wt% has a thermal conductivity of 0.357 51 W/(m · K) which can effectively dissipate and conduct the heat. Therefore, it can be used as a raw material for cool wear fabrics in summer, giving the fabric a cool feeling.

Keywords: Boron nitride; Flax fiber; Wet spinning; Regenerated cooling fiber; Performance

1 Introduction

With global warming and frequent high temperatures in summer, people put forward higher requirements for the functionality of clothing textiles. The development of cool and comfortable summer clothing fabrics can make people feel happy and healthy, and then improve work and study efficiently[1]. For natural cool fibers, flax fiber has a strong cooling feeling compared with other fibers due to its molecular structure, a large number of hydrophilic groups[2], and its special surface structure[3].

Boron nitride (BN) has excellent heat resistance, thermal stability, chemical stability, insulating properties and thermal conductivity[4-6]. It has a wide range of applications in various aspects. For example, Wu et al.[7] prepared a CNF (cellulose nanofiber)/BNNS (boron nitride nanosheets) thermally conductive composite film by simple vacuum-assisted filtration. When the content of BN was 70%, the thermal conductivity of the film was as high as 30.25 W/(m · K). The multilayer films prepared by Song et al.[8] used a vacuum-assisted self-assembly method with CNF as the matrix and BN and graphene as fillers, exhibiting high thermal conductivity and electrical insulation. Kwon et al.[9] used ball milling to process BN and prepared boron nitride nanosheets and polyvinyl alcohol thermally conductive composite films by electrospinning technology. It is superior to the common BN polyvinyl alcohol composite film in thermal conductivity, glass transition temperature, heat resistance index and other comprehensive properties.

In this study, recycled flax fiber is used as raw material, and ionic liquid is to dissolve cellulose, and an appropriate amount of BN nanoparticles are added. Therefore, a cool-feeling fiber with excellent thermal conductivity is prepared by spinning process. The ionic liquid is used as the dissolving system, which has the characteristics of good solubility, recyclability, and environmental protection. The addition of BN has excellent thermal conductivity and the channel on the surface of the regenerated fiber forms a synergistic effect.

2 Experimental

2.1 Materials

Flax fiber (Recover degummed) was obtained from Hangzhou Shanglu silk Co., Ltd. Ionic liquid ([BMIM]Ac, AR) was purchased from Shanghai Fujie Chemical Co., Ltd. The organic solvent (DMSO, AR) was purchased from Tianjin kemio Chemical Reagent Co., Ltd. Boron nitride (BN, Diameter = 1 000 nm, density = 2.29 g · cm^{-3}) was made by Shanghai xiangtian nano material Co., Ltd.

2.2 Configuration of spinning solution

Cut the recovered flax fiber into pieces, and dissolve the cellulose in the ionic liquid according to the mass ratio of flax fiber to ionic liquid of 1 : 20 in a constant temperature water bath of 80 ℃, the rotating speed of the magnetic stirrer is 600 r/min, and the solution is stirred for 2 h. Then, DMSO with a mass ratio of ionic liquid to organic solvent of 1 : 3 was added to the above solution and stirred at constant temperature for 3 h. Moreover, add BN with mass fractions of 0%、20%、40%、60% and 80% of the mass of flax fiber to configure spinning solutions with different BN contents.

2.3 Preparation of regenerated cooling fiber

Use a syringe to extract the above-prepared spinning stock solution. Place it on a syringe pump and use a spinning machine for spinning. Using a 21 G needle squeeze it into the coagulation bath of deionized water at a speed of 30 mm/h and the coagulation time is 15 min. Then, it was wound on a bobbin and dried in a drying oven at 40 ℃ for 12 h to obtain regenerated cooling fiber.

2.4 Characterization

The micro morphology of the fiber was observed by scanning electron microscope (SEM, S-4800, Hitachi, Japan). The crystallinity of the fiber was measured by X-ray diffractometer (XRD, D/max-2550VB/PC, Rigku Corporation, Japan). The thermal conductivity of spinning dope and fabric were measured by DRE-multifunctional rapid thermal conductivity tester (Xiangtan Xiangyi Instrument Co., Ltd.). The surface temperature of fabrics and skins was measured by Tasi K-type thermocouple thermometer TA612C (Suzhou Tasi Electronics Co., Ltd.). According to GB/T 14337—2008 test method for dimensional tensile properties of chemical fiber staple fibers, the breaking strength and elongation at break of fibers were measured by LLY-06 EDC electronic single fiber strength instrument. The chemical structure of the fiber was measured by infrared spectrometer (Nicolet 6700, Thermo Fisher, American). The thermal stability of regenerated cooling fiber was tested by thermogravimetric analyzer (TGA-4000, Thermo Fisher, American).

3 Results and Discussion

3.1 Structural characterization of regenerated cooling fiber

The surface morphology of regenerated cooling fibers with different BN content is characterized respectively. The SEM results of the longitudinal section are shown in Figs. 1(a)-1(e). The surface morphology of the fiber can be seen in the cross section of the fiber. Interestingly, with the increase of the BN content, the surface morphology of the fiber gradually becomes rough and grooved. The diameters of regenerated fibers with different amounts of BN are shown in Fig. 1(f). In the spinning process, the same spinning speed and needle diameter are used, but the diameter of the regenerated fiber will vary in the content of the organic solvent in

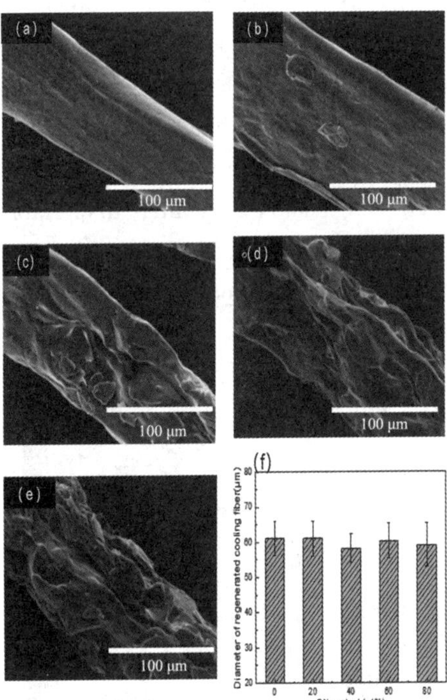

Fig. 1 SEM image of Regenerated cooling fiber and Diameter diagram of regenerated cooling fiber: (a) 0% BN; (b) BN 20%; (c) BN 40%; (d) BN 60%; (e) BN 80%

the fiber due to the difference in the content of BN, and then the degree of diffusion in the coagulation bath will be different.

3.2 Thermo-conductive properties

Adding BN to the spinning solution can improve the thermal conductivity of the spinning solution, thereby improving the thermal conductivity of the regenerated fiber. The thermal conductivity of the regenerated fiber is the key performance that determines the heat dissipation capacity and cooling effect of the textile. Measuring the thermal conductivity of spinning dope and fabric (Fig. 2 (a)). With the increase of BN content, the thermal conductivity of spinning stock solution and fabric increases. When the addition amount of BN is 60 wt%, the thermal conductivity of the spinning solution is 0.316 13 W/(m · K), and the thermal conductivity of the fabric reaches 0.357 59 W/(m · K), which is 2.34 times that of the fabric without BN fiber. Therefore, the addition of BN can indeed improve the thermal conductivity of the fiber and meet the cooling performance requirements of regenerated fiber.

Fig. 2 (b) shows the temperature data of the fabric woven from regenerated cooling fiber through a thermocouple test. The exposed skin surface is easily taken away by the wind, and the heat temperature is 32.36 ℃, which is much lower than the temperature after covering the fabric. The temperature of the measured skin surface decreases gradually with the BN content. Compared with the fabric without BN, the temperature of the regenerated cooling fiber fabric with 60 wt% BN decreased by 1.39 ℃, and the regenerated cooling fiber has the superior cooling power.

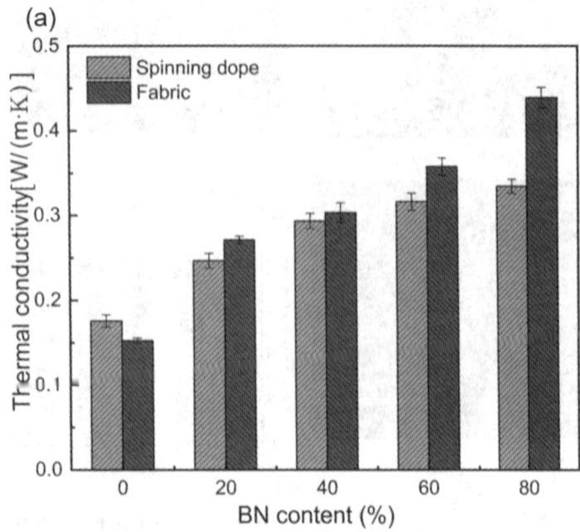

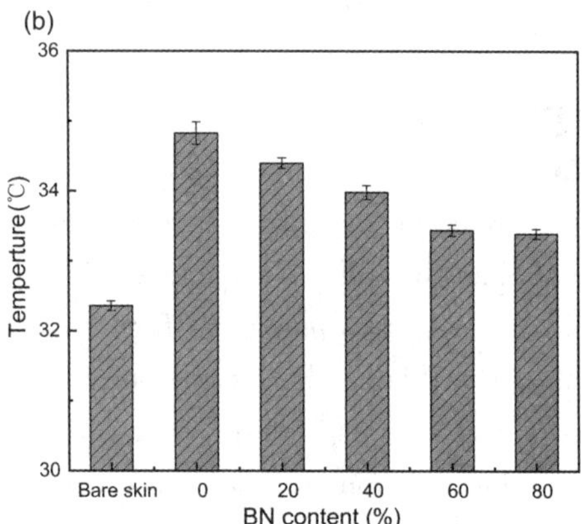

Fig. 2 (a) Spinning dope and fabric thermal conductivity; (b) comparison of skin temperature covered by different fabrics

3.3 Mechanical properties

Mechanical property is significant for practical applications of our regenerated fiber and they are illustrated in Fig. 3. With the increase of BN content, the tensile breaking strength and elongation at break of regenerated fiber decreased gradually. Notably, when the content of BN was 60 wt%, the tensile breaking strength and elongation at the break of the regenerated fiber were 121.68 MPa and 13.37%. However, when the content

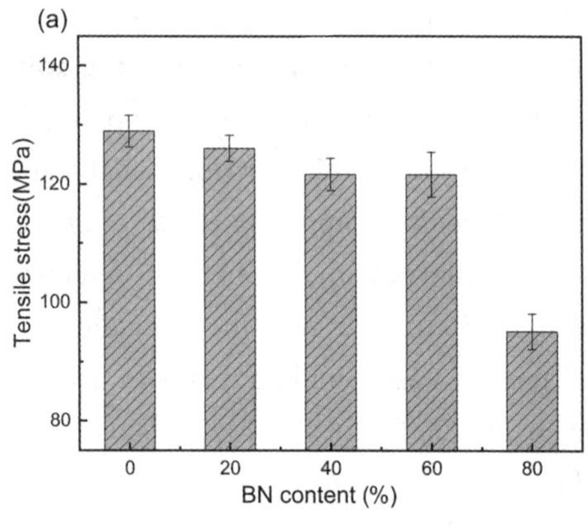

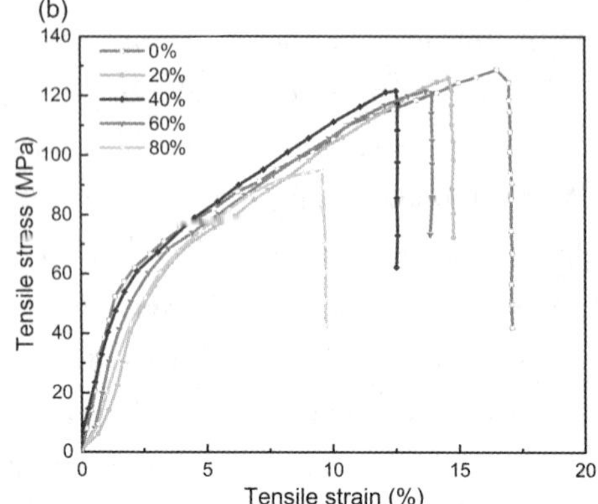

Fig. 3 Mechanical properties of regenerated cooling fiber

of BN was 80%, the tensile breaking strength and elongation at the break of the regenerated fiber decreased sharply to 95.18 Mpa and 9.59%. Because the directional structure of the regenerated fiber is destroyed by BN at a high BN load.

3.4 Chemical structure and thermal stability

Fig. 4 (a) shows the XRD spectrum of regenerated fibers with different BN addition. When cellulose dissolves in ionic liquid to form newly regenerated fibers, the molecular chains rearrange to form a cellulose type II structure. Therefore, broad absorption peaks appear and the diffraction peak is 28° − 22°. Characteristic peaks appear near 26.7°, 41.6°, 50.2°, 55.5°, 75.9°and 82.1°, corresponding to hexagonal boron nitride (002), (100), (102), (004), (104) and (112) crystal planes respectively. With the increase of BN content, the higher the intensity of absorption peak and the better the crystallinity.

The infrared spectra of the regenerated cooling fibers with BN contents of 0%, 20%, 40%, 60% and 80%, respectively, are shown in Fig. 4 (b). A wide absorption peak appears at about 3 400 cm^{-1}, which is the result of the stretching vibration of all O—H bonds in the added cellulose. 2 900 cm^{-1} is the symmetric and asymmetric stretching vibration of C—H bond, and 1 100 cm^{-1} is the stretching vibration peak of C—O bond in cellulose. In the infrared spectrum, there are two obvious absorption peaks at 1 380 cm^{-1} and 780 cm^{-1}, which correspond to the stretching vibration peak of the B—N bond of BN and the out-of-plane bending vibration of the B—N—B bond, respectively. After the addition of BN, the regenerated cooling fibers did not generate new chemical bonds. Adding different amounts of BN would affect the projected intensity and lead to differences in the height of the absorption peak.

The thermal stability of regenerated fibers with different BN content is tested by thermogravimetric analysis and the results are shown at Figs. 4(c)−4(d). The pyrolysis process of regenerated fibers can be mainly divided into two stages: The first stage is the loss of water molecules, and the overall moisture content is very small, so the quality remains almost the same. The second weight loss occurred at 190 ℃ to 370 ℃. At this stage, cellulose is rapidly decomposed, and the mass loss is very large. It can be seen from the DTG curve that when the addition amount of BN is 0%, the mass

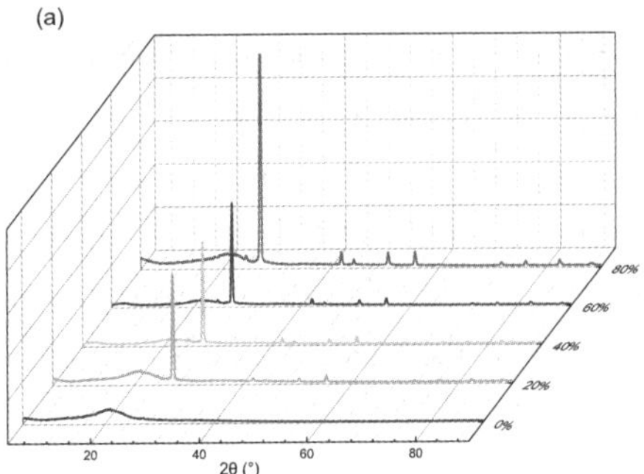

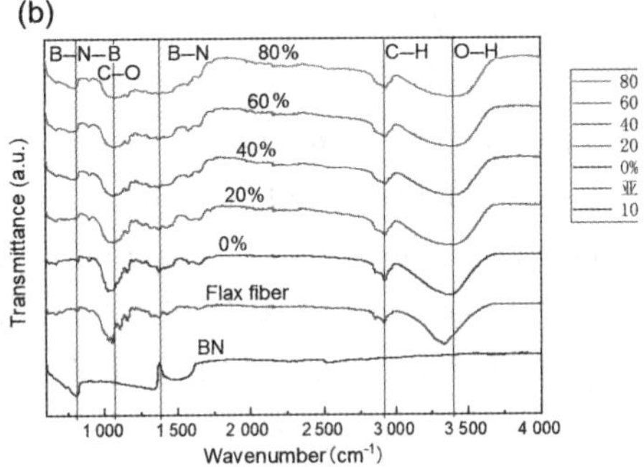

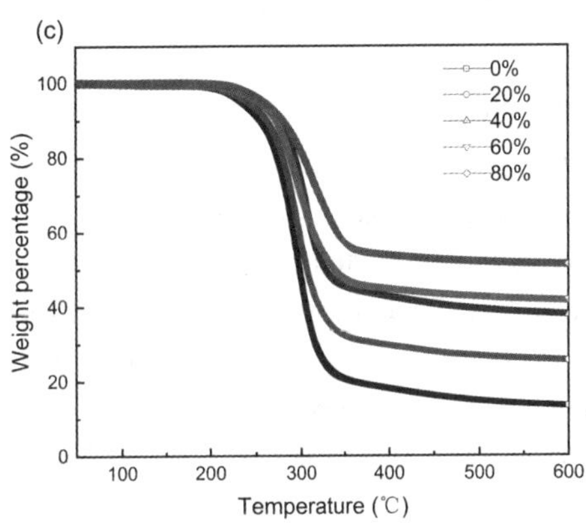

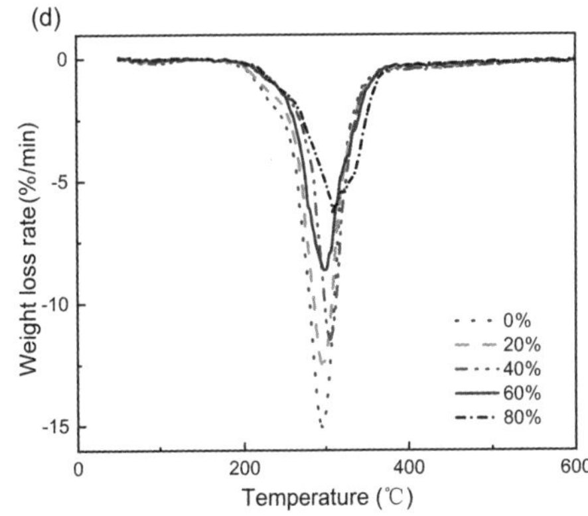

Fig. 4 (a) XRD pattern of regenerated cooling fiber; (b) infrared spectra of regenerated cooling fiber; (c) TG curve of regenerated cooling fibers; (d) DTG curve of regenerated cooling fiber

loss rate of the regenerated fiber is 15.05 %/min, and the thermal decomposition temperature is 295.66 ℃. When the addition amount of BN is 80%, the mass loss rate of the regenerated fiber is 6.32 %/min, and the thermal decomposition temperature is 308.81 ℃. The results show that with the increase of BN content, the mass loss rate of regenerated fiber decreases and the thermal decomposition temperature increases. This is because with the increase of BN content, the crystalline area of regenerated fiber increases, while the structure is more stable and the thermal stability of regenerated fiber is improved.

4 Conclusions

In summary, inspired by the stack effect that is usually applied in compound material processing for filler, we have successfully prepared multifunctional regenerated cooling fiber. The regenerated cooling fiber can have good mechanical properties (121.68 MPa) and thermal conductivity (0.357 5 W/(m · K)) when the BN content is 60%, which is enough to meet the requirements of cool fabric materials in summer.

References

[1] LI D, ZHANG L. The Influences of material and structure on the UV protection of summer fabrics[J]. Advanced Materials Research, 2011, 13(27): 298-298.

[2] PAK S Y, KIM, D L, SONG Y S, et al. Micro-macroscopic coupled modeling for the prediction of synergistic improvement on the thermal conductivity of boron nitride and multi-walled carbon nanotube reinforced composites[J]. Composites Part A, 2021, 148: 14-28.

[3] BOLSHAKOV B O, GALIAK R F, SMYSLOV A M. The structure and properties of composite compacts made of high-alloy powder containing boron nitride additives[J]. Inorganic Materials: Applied Research, 2021, 12(2): 278-298.

[4] YONG Z, TANG P, ZHANG M X, et al. Temperature effects on the dielectric properties and breakdown performance of h-BN/epoxy composites[J]. Materials, 2019, 12(24): 61-68.

[5] WU K, FANG J C, MA J R. Achieving a collapsible, strong, and highly thermally conductive film based on oriented functionalized boron nitride nanosheets and cellulose nanofiber[J]. Mater. Interfaces, 2017, 35: 4-9.

[6] CHIN S K, GUJAR P D. Significance of moisture management for high performance textile fabric[J]. International Journal of Innovative Research in Science, Engineering and Technology, 2013, 3: 814-819.

[7] SONG N, PAN D H, LIANG X F. Structural design of multilayer thermally conductive nanofibrillated cellulose hybrid film with electrically insulating and antistatic properties[J]. Mater. Chem. C, 2018, (26).

[8] KWON O H, HA T, KIM D G. Anisotropy-driven high thermal conductivity in stretchable poly (vinylalcohl)/hexagonal boron nitride nanohybrid films[J]. Appl Mater Interfaces, 2018, (10).

[9] YANG X, GUO Y, HAN Y, et al. Significant improvement of thermal conductivities for BNNS/PVA composite films via electrospinning followed by hot-pressing technology[J]. Composites Part B, 2019, 175.

Design and Performance of Textile Circularly Polarized Antennas Based on Different Substrates

LI Mengdie[1], LI Erhong[1], YAO Lan[1]*, QIU Yiping[1]

[1] College of Textiles, Donghua University, Shanghai 201620, China

* Corresponding author's email: yaolan@ dhu. edu. cn

Abstract: In the area of military personal protection as well as medical and life monitoring, the circularly polarized antennas are increasingly used due to the good capability of anti-interference. This type of antenna based on textile materials is more suitable for functioning in the smart wearable or on-body wireless communication systems as textile itself has good flexibility and is easy to be conformal to the human body. Different textile substrates with different thicknesses and dielectric properties would lead to different antenna properties. In this study, to investigate the effect of textile substrates on the antenna performance, four types of woven fabrics were chosen. They are polyester felt, polyester woven fabric, polyester/spandex blended fabric and "ke ti la" polyester woven fabric. The simulation work has been done to obtain the optimum size parameters of the antennas. And after simulation, the antenna prototypes were fabricated by using the screen printing method. The measured antenna results showed that the return losses had good agreement with the simulated results and S_{11} value were lower than −10 dB, indicating that good impedances were achieved. Additionally, the axial ratios of the antennas met the requirement of the value lower than 6 dB, showing the validity of the circularly polarized antennas.

Keywords: Circularly polarized antennas; Textile substrates; Screen printing; Electromagnetic performance

1 Introduction

The electromagnetic waves of the linearly polarized antennas are characterized by high requirements for the direction, easy to attenuate in polarization deflection, and easy to form multipath effects. The circularly polarized wave forms a reverse deflection after encountering obstacles, which has the advantages of rain and fog proof, strong anti-interference ability and nearly no performance impact by carrier movement[1-2]. In recent years, the satellite navigation systems of various countries have developed rapidly, such as the Navy Navigation Satellite System (NNSS) of the United States, the GALILEO system of the European Union, and the Beidou positioning system of China, which has accelerated the application of circularly polarized antennas in communications, radar, military and other fields[3-4]. Textile circularly polarized antenna is a new type in the circular polarized antenna, as researchers apply textile materials as the substrate of the circularly polarized antenna, which can not only achieve the original function, but also be conformal with the human body and easy to be used in the on-body environment[5-10].

In this study, the influence of different textile substrate materials on the antenna performance was investigated. Four different textile materials were used as the substrates of the antennas, and the textile circularly polarized microstrip antennas were designed and fabricated. Finally, the performance of the antennas was tested and compared.

2 Experiments

2.1 Selection of textile substrate materials

In previous studies on textile antennas, the selection of substrate materials includes washed cotton, denim, polyester felt, polyacrylic felt, Lycra, corduroy, etc. In the choice of substrate materials, due to the requirements of the antenna substrate itself, it was necessary to take into account the stability of dielectric properties, therefore in this study four commercially available textile materials were used as the substrates. They are polyester felt (Pf), polyester woven fabric (Pw), polyester/spandex blended fabric (Ps) and "ke ti la" polyester woven fabric (Pc). Among them, the polyester/spandex blended fabric contains 2% spandex, and the rest of the woven fabrics were composed of 100% polyester. Due to the low moisture regain of polyester itself, the moisture regains of the four textile materials were low and the dielectric properties were relatively stable.

2.2 Antenna design and simulation optimization

The schematic diagram of the circularly polarized microstrip antenna is shown in Fig. 1. In the electromagnetic simulation software HFSS, the radiation element, dielectric substrate, reference site, coaxial feed core and air box were built respectively to obtain the complete models of the circularly polarized microstrip antennas.

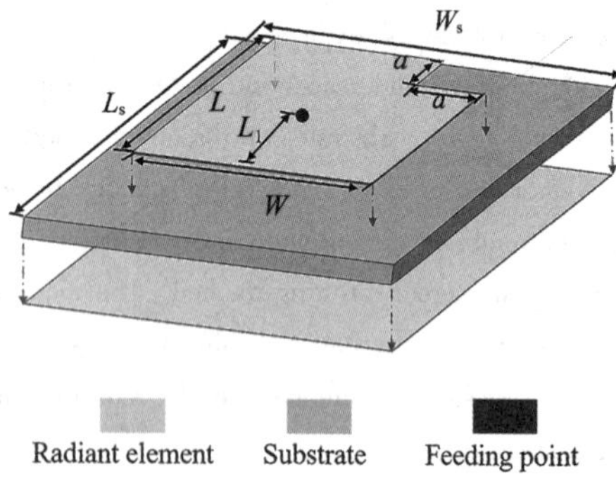

Fig. 1 Schematic diagram of circular polarized microstrip antenna structure

The length (L_s) and width (W_s) of the antenna substrates were 80 mm. The radiation element length (L), the radiation element width (W), the feed point position (L_1), and the tangent angle width (a) were optimized on the simulation software. The target axial ratio was 3 dB, and the resonant frequency was 3.50 GHz. The final optimized parameters are shown in Tab. 1. Tab. 2 shows the specific parameters of the antenna simulation performance. The relative bandwidths, return losses and gains of the antennas met the working requirements, and the axial ratios were less than 3 dB, showing the circular polarization performance was excellent.

Tab. 1 Dimensional parameters of textile circular polarization microstrip antennas

Textile substrate material	L(mm)	W(mm)	L_1(mm)	a(mm)
Pf	32.4	31.8	8.0	6.0
Pw	25.9	25.0	10.0	6.0
Ps	27.5	25.5	13.0	7.0
Pc	28.8	27.6	9.5	5.9

Tab. 2 Specific parameters of antenna simulation performance

Textile substrate material	Relative bandwidth(%)	Return loss(dB)	Gain (dB)	Axial ratio (dB)
Pf	8.25	-21.13	9.41	1.88
Pw	12.43	-23.81	6.39	0.47
Ps	16.90	-46.95	4.57	2.24
Pc	9.96	-23.01	7.43	1.87

2.3 Fabrication and antenna performance test

The designed textile antennas were fabricated by the method of screen printing. Before printing, the screen plate and the textile substrates were stabilized and the conductive silver paste was placed on the side of the screen plate. Then the conductive silver paste was evenly scraped with a scraper, and the printed conductive silver paste was attached to the substrates surface to form radiation elements. The whole system was then dried and cured in the vacuum heating box. Finally, SMA connectors were installed on the antennas and the antenna prototypes were shown in Fig. 2.

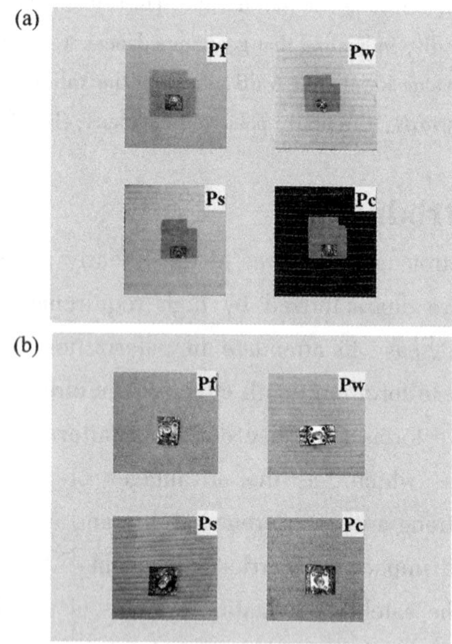

Fig. 2 Prototype of the antenna: (a) front of the antenna; (b) back of the antenna

In this study, the return losses of the textile antennas were tested using the vector network analyzer KC901V. The radiation patterns and axial ratios of textile antenna samples were measured by the vector network analyzer Agilent N5225A in the microwave anechoic chamber.

3 Results and Discussion

3.1 Return loss

The simulated and measured return loss curves for the

four circularly polarized microstrip antennas with polyester felt, polyester woven fabric, polyester/spandex blended fabric and"ke ti la" polyester woven fabric as the substrates are shown in Fig. 3.

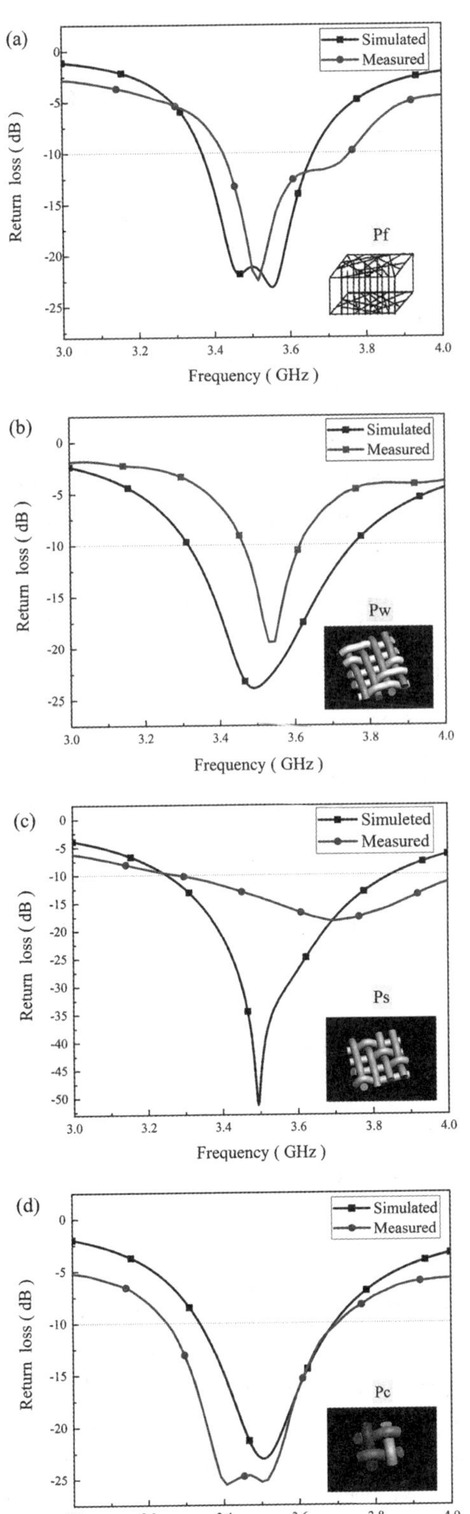

Fig. 3 Return loss curves of antennas: (a) polyester felt; (b) polyester woven fabric; (c) polyester/spandex blended fabric; (d) "ke ti la" polyester woven fabric

As can be seen from Fig. 3, under the measured conditions, the resonant frequency of these four different textile circularly polarized antennas had a small resonant frequency shift from the designed curves and all the return losses were less than −10 dB showing meeting the working requirements. When the measured values of the circularly polarized microstrip antennas corresponding to the four textile substrate materials were compared, it is noticed that when the polyester felt and the "ke ti la" polyester woven fabric were used as the substrates, the resonant frequencies of antennas were closer to the target frequency of 3.50 GHz and the bandwidths were wider, which was highly consistent with the simulation results, indicating excellent performance for the above two substrate antennas. When the polyester woven fabric and polyester/spandex blended fabric were used as the substrates, the resonance frequencies showed a shift to the higher frequency, and the shift was more severe when the polyester/spandex blended fabric was used as the substrate, showing these two types of fabric structures had greater influence.

3.2 Radiation pattern

The radiation pattern is a far-field superposition of electric currents, and its shape and effcct are closely related to the structure of the radiation element itself. The measured radiation patterns of the circularly polarized microstrip antennas with the substrate of polyester felt, polyester/spandex blended fabric and "ke ti la" polyester woven fabric are shown in Fig. 4.

As can be seen in Fig. 4, the radiation patterns of the four textile circularly polarized microstrip antennas were larger in the front direction and smaller in the back direction. When polyester felt was used as the substrate, the radiation patterns of the two main planes of *XOZ* and *YOZ* were similar in shape, the measured gain was as high as 10.05 dB when the angle was near 0° showing good radiation performance. When polyester woven fabric, polyester/spandex blended fabric and "ke ti la" polyester woven fabric were used as the substrates, the maximum measured gains of the circularly polarized microstrip antennas were 5.89 dB, 7.58 dB, and 7.46 dB respectively. Therefore, the gains of the above four textile circularly polarized microstrip antennas met the operating requirements.

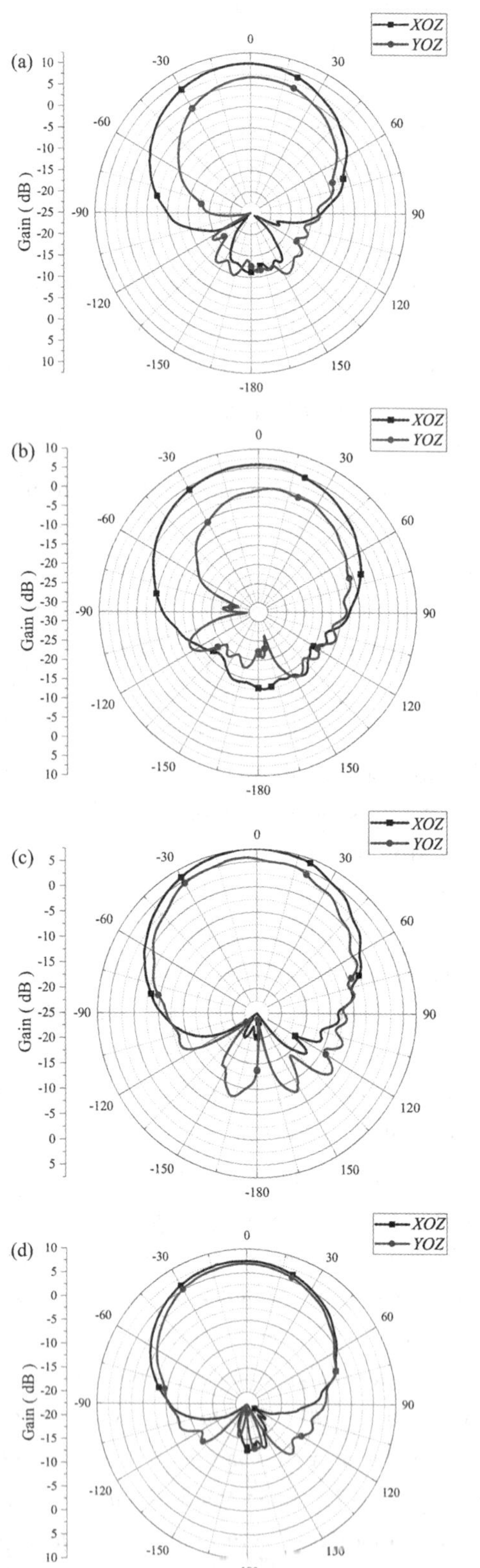

Fig. 4 Measured radiation patterns of the antennas: (a) polyester felt; (b) polyester woven fabric; (c) polyester/spandex blended fabric; (d) "ke ti la" polyester woven fabric

3.3 Axial ratio

The measured axis ratios of the four circularly polarized microstrip antennas with polyester felt, polyester woven fabric, polyester/spandex blended fabric and "ke ti la" polyester woven fabric as the substrates were 3.21 dB, 5.98 dB, 1.69 dB and 0.52 dB at their resonant frequencies, which were less than 6 dB, meeting the circularly polarized performance. Among them, the axial ratio of the "ke ti la" polyester woven fabric was the lowest implying the circular polarization of this antenna was the best.

4 Conclusions

In this study, four kinds of circularly polarized microstrip antennas with different textile substrates were designed, and the dimensional parameters of the antennas were optimized by simulation. The antenna prototypes were fabricated by the screen printing method and the electromagnetic performance of the antennas were tested. The results showed that the return losses, gains and axis ratios of the four textile circularly polarized antennas met the working requirements and realized the circular polarization characteristics. Polyester felt and "ke ti la" polyester woven fabric antennas had better return losses, gains and axis ratios. The better performance was due to the stable dielectric properties of pure polyester composition, felt surface and less deformation of "ke ti la" polyester woven fabric surface yarn. In addition, good contact with the feeding connector was helpful to get the proper performance.

Acknowlegement

The authors would like to thank Tong Guolin, Shanghai Jiaotong University, Shanghai, China, for his help in far-field measurement.

References

[1] MAZZINGHI A, ALBANI M, FRENI A. Double-spiral linearly polarized RLSA[J]. IEEE Transactions on Antennas and Propagation, 2014, 62(9): 4900-4903.

[2] ABADI S M A M H, BEHDAD N. Wideband linear-to-circular polarization converters based on miniaturized-element frequency selective surfaces[J]. IEEE Transactions on Antennas and Propagation, 2016, 64(2): 525-534.

[3] LUMEI L, JIANXING L, BIN H, et al. A compact circularly polarized microstrip antenna with bandwidth enhancement[J]. Prog Electromagn Res Lett (USA), 2016, 61: 85-89.

[4] SREEMATHY R, HAKE S, SULAKHE S, et al. Slit loaded textile microstrip antennas[J]. IETE Journal of Research, 2020(12): 1-9.

[5] JOLER M, BOLJKOVAC M. A sleeve-badge circularly polarized textile antenna[J]. IEEE Transactions on Antennas and Propagation, 2018, 66(3): 1576-1579.

[6] MAO C X, VITAL D, WERNER D H, et al. Dual-polarized embroidered textile armband antenna array with omnidirectional radiation for on-/off-body wearable applications[J]. IEEE Transactions on Antennas and Propagation, 2020, 68(4): 2575-2584.

[7] ZHOU L, FANG S, JIA X. Dual-band and dual-polarised circular patch textile antenna for on-/off-body WBAN applications[J]. IET Microw Antennas Propag, 2020, 14(7): 643-648.

[8] KUMAR S, NANDAN D, SRIVASTAVA K, et al. Wideband circularly polarized textile MIMO antenna for wearable applications[J]. IEEE Access, 2021, 9: 108601-108613.

[9] SAHA P, MITRA D, PARUI S K. A frequency and polarization agile disc monopole wearable antenna for medical applications[J]. Radioengineering, 2020, 29(1): 74-80.

[10] MAO C, WERNER P L, WERNER D H, et al. Dual-polarized armband embroidered textile antenna for on-/off-body wearable applications[C]//2019 IEEE International Symposium on Antennas and Propagation and USNC-URSI Radio Science Meeting, Atlanta, 2019-07-12.

Effect of Plasma Treatment on the Mechanical Properties of Hemp/Jute Fiber Mat Reinforced Polylactic Acid Composites

REN Danyao[1], YANG Jiahua[1], YAO Lan[1*], QIU Yiping[1]

[1] *College of Textiles, Donghua University, Shanghai 201620, China*

[*] *Corresponding author's email*: yaolan@ dhu. edu. cn

Abstract: With the enhancement of people's awareness of environmental protection and improvement of life quality, the natural fiber reinforced composite, as a new type of environmentally friendly material, is increasingly studied and applied in industry. In this study, the hemp/jute needle-punched fiber mat reinforced polylactic acid (PLA) composites were prepared to investigate the effect of fiber mat preparation process and plasma treatment technology on the properties of the composites. Hemp and jute fibers were blended, carded and needle-punched into fiber mat reinforcements, which were then treated by plasma and the effect were detected by a scanning electron microscope, contact angle etc. The above reinforcements and polylactic acid were consolidated to form composites by the hot pressing process. The results showed that the mechanical properties of the composites differed according to the fiber blend mass ratio, and the tensile and bending strengths increased with the increase of hemp fiber content. Meanwhile, plasma treatment improved the interfacial adhesion between fibers and PLA, which had a positive effect on the mechanical properties of composites. These results would provide enough significance for investigating the fabrication process and performance characteristics of natural fiber reinforced composites.

Keywords: Hemp/jute needle-punched fiber mat; Polylactic acid; Composite; Plasma; Mechanical properties

1 Introduction

Nowadays, synthetic fiber composites have been widely used in production and life. However, due to their non-biodegradability causing some pollution to the environment, people are looking for their alternatives. Natural fibers are attracting attention as an ideal substitute for synthetic fibers because of their good mechanical properties, biodegradability and lower price[1]. Natural fibers are closely related to people's lives, not only in the daily necessities of life, but also have a wide range of applications in packaging, plastics and other areas. The performance of their composites can meet the requirements of these fields. Moreover, it can also effectively reduce costs and pollution to the environment, which is in accord with the national environmental protection and energy-saving development concept. Ultimately, it can realize the sustainable development of economy and environment, and has an invaluable role for future development[2-3].

Natural fibers as reinforcement for composite materials need to have high cellulose content and low microfiber angle. Flax, hemp, jute and ramie possess this property and have good tensile strength values[4-6]. Among them, the combination of hemp and jute fibers has given rise to research on their use as reinforcement in polylactic acid (PLA) resin[7]. However, some problems existed when hemp and jute fiber were applied to composites, the typical of which is the poor interfacial adhesion between the fiber and resin, as the fibers have high moisture regain, and the resin is hydrophobic. Therefore, in order to improve the interfacial adhesion of the natural fibers and the resin, various treatments of the fibers are required[8].

Traditionally, physical methods can be used to improve the interfacial adhesion capability. The adhesion ability of the fiber and resin interface is influenced by the changes in fiber structure and surface properties after physical modification treatment, which improves the mechanical properties of composites. In addition, chemical treatments can remove impurities and some non-cellulosic substances from the fiber surface, and the microscopic morphology of the fiber surface changes, which improves the adhesion between the fiber and the resin[9-10].

In this study, the fabrication process and performance characteristics of hemp/jute reinforced fiber composites were investigated. Hemp and jute fiber were needle-punched into fiber mat reinforcements, which were then treated by plasma and the effect were detected by a scanning electron microscope (SEM), contact angle etc. The above reinforcements and polylactic acid were consolidated to form composites by the hot pressing process and the mechanical properties were evaluated for the composites. This would provide enough significance for investigating the fabrication process and performance characteristics of natural fiber reinforced composites.

2 Experiments

2.1 Materials

The jute fiber used in the experiment was provided by Nanyuan New Material Co. Ltd. (Hunan, China). Hemp fiber was provided by Shangde Textile Co. Ltd. (Zhejiang China) and its moisture regain was 12%. PLA (Nature Works® 4032D) was provided as pellets by Suzhiyuan Suhua Co., Ltd. (Guangdong, China). The purity specification of anhydrous ethanol used was analytical reagent.

2.2 Fiber mat preparation

The jute and hemp fibers were cut to a length of 5 - 8 cm after the moisture treatment. 20 g of mixed hemp/jute fibers with fiber mass ratios of 0 : 100, 100 : 0, 25 : 75, 50 : 50, and 75 : 25 (noted as 0/100 J/H, 100/0 J/H, 25/75 J/H, 50/50 J/H, and 75/25 J/H, respectively) were weighed on an analytical electronic balance (Jingke Co. Ltd., Shanghai, China) and fed into a small carding machine for carding into a mat. Each sample was carded on average 2-3 times to obtain a uniformly mixed hemp/jute blend. Then the fluffy fiber mats were reinforced by needling treatment to produce needle mats with high mechanical properties. The final needle mats were cut into 115 mm×115 mm size.

2.3 Plasma treatment

The prepared hemp/jute needled mats were soaked in anhydrous ethanol for 20 min for pretreatment. After drying, the fiber mats were treated with a plasma processor (PSPT - 2000C, Nanjing Voltech Electronic Technology Co. Ltd., Jiangsu, China) with a power setting of 400 W. The handle was moved at a uniform speed so that the ceramic rods could be discharged uniformly for 30 s (Each fiber mat after plasma was recorded as D-0/100 J/H, D-100/0 J/H, D-25/75 J/H, D-50/50 J/H, D-75/25 J/H, respectively).

The surface micromorphology of the fiber mats before and after the plasma treatment was tested using scanning electron microscopy (FlexSEM1000, Japan). The hydrophilicity and hydrophobicity of the mats were tested by optical contact angle measurement using bubble method and suspension drop method, respectively. The changes in chemical elemental composition of hemp/jute needled mats were also tested by X-ray photoelectron spectroscopy (XPS).

2.4 Composite manufacturing

The fiber mats were alternately stacked with PLA films to obtain samples with a fiber/resin mass ratio of 2 : 5. The temperature of the upper and lower plates of the compression molder (XLB-400×400×2, Qicai Hydraulic machinery Co. Ltd., Shanghai, China) was set to 185 ℃ and the pressure was set to 2 MPa for preheating. After that, the stacked PLA film and fiber mats were quickly put into the template, then pressed on the preheated metal plate, and put into the hot pressing between the upper and lower plates of the compression molder in parallel for 3 min. After the completion of the first hot pressing, it was held and rested for 1 min. Then the second hot pressing was carried out for 2 min. Finally, the metal plate was removed and a heavy weight was placed on the upper metal plate to achieve the effect of pressurization. The composite samples were removed from the mold after natural cooling at room temperature.

2.5 Mechanical testing of composites

In the tensile performance test, the samples were cut into 100 mm × 10 mm size according to ASTM D3039 and put into a constant temperature and humidity environment (temperature 20 ℃, humidity 60%) for 24 h. Tensile test was conducted along the direction of the reinforcement until the samples fractured by electronic universal testing machine (INSTRON4302, America). Each group of valid samples were required to be at least 5.

The bending performance test was conducted according to ASTM D3039, and the sample size was 130 mm× 10 mm with a span of 50 mm and a loading rate of 2 mm/min. The bending test used the three-point bending method with electronic universal testing machine for testing, each group of effective samples for 5.

3 Results and Discussion

3.1 Effects of plasma on the fiber surface

The microscopic morphology of hemp and jute fiber before and after plasma treatment is shown in Fig. 1. The microscopic morphology of the untreated fiber surface is shown in Fig. 1 (a) and Fig. 1 (c). It is noticed that the fiber surface is smooth, with few impurities and no fiber shedding. The microscopic morphology of the treated fiber surface is shown in Fig. 1 (b) and Fig. 1 (d), which show that the fiber surface is obviously etched, the surface is rough and the fiber surface material is partially shed.

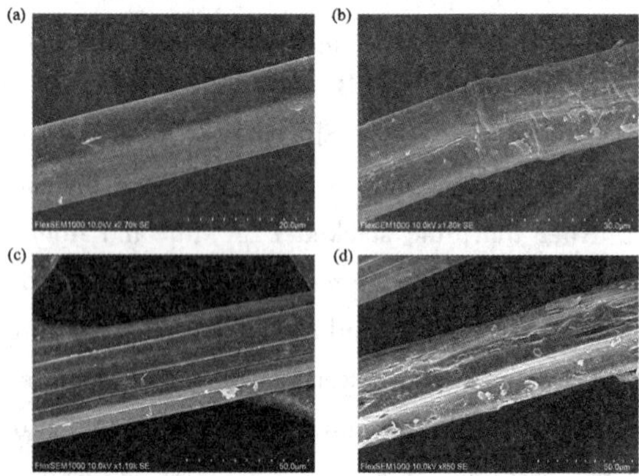

Fig. 1 SEM of hemp and jute fiber surface before and after plasma treatment: (a) untreated hemp; (b) treated hemp; (c) untreated jute; (d) treated jute

The analysis of surface wettability of hemp/jute needle fiber mats before and after plasma treatment is shown in Fig. 2. After plasma treatment, the contact angle of hemp/jute fibers became obviously larger, and both changed from hydrophilic to hydrophobic state.

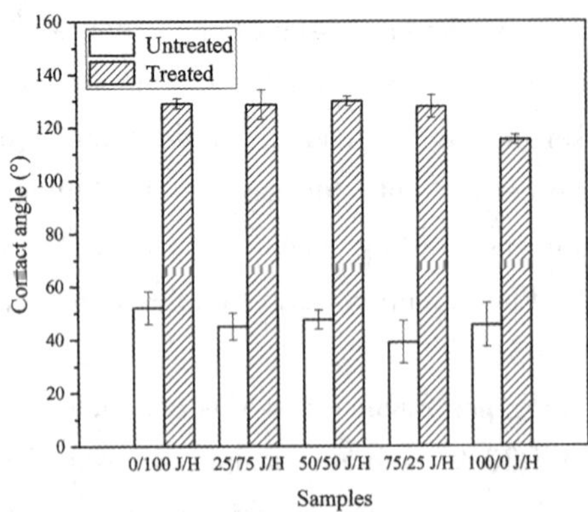

Fig. 2 Fiber contact angle before and after plasma treatment

The surface element content of fiber mat before and after plasma treatment are shown in Tab. 1. After plasma treatment, the trends of C, O and N content of the mats with different fiber mass ratios were the same. The element content of C and N decreased, and the element content of O and O/C ratio increased.

Tab. 1 The surface element content of fiber mat before and after plasma treatment.

Samples	Content (At%)			Ratio
	C	O	N	O/C
0/100 J/H	72.71	26.74	0.55	0.37
D-0/100 J/H	63.53	36.00	0.47	0.57
25/75 J/H	68.59	31.00	0.41	0.45
D-25/75 J/H	60.51	38.86	0.63	0.64
50/50 J/H	69.39	30.81	0.80	0.45
D-50/50 J/H	61.28	38.72	0.00	0.63
75/25 J/H	66.66	33.04	0.30	0.49
D-75/25 J/H	60.23	39.40	0.37	0.65
100/0 J/H	60.40	38.68	0.93	0.64
D-100/0 J/H	58.12	41.88	0.00	0.72

These changes were due to the fact that plasma treatment could effectively remove various organic substances such as non-cellulose from the fiber surface, resulting in a rougher fiber surface. At the same time, free radicals were generated on the fiber surface, which grafted with ethanol molecules and their ionization products to produce hydrophobic groups on the fiber surface. It made the bonding between fiber and resin interface more tightly, so that the mechanical properties of the composite were able to be improved.

3.2 Tensile properties of the composites

Fig. 3 (a) and Fig. 3 (b) show the results of the tensile properties of the untreated and plasma-treated composites with different fiber mass ratios. It can be seen that the tensile properties of the composites gradually decreased with the increase of jute fiber content in the blended fibers. This was due to the fact that hemp fiber had a higher cellulose content than jute fiber, and the tensile properties of hemp fiber were better than jute fiber. Compared to the untreated composites, the composites treated by plasma increased the tensile strength by 1.2%, 5.5%, 1.4%, 3.1%, and 11.7%, respectively. This was due to the increase in fiber surface roughness and the change from hydrophilic to hydrophobic state after etching of the fiber surface by plasma.

The interfacial adhesion between the fiber and resin was improved and the fiber tensile properties were enhanced.

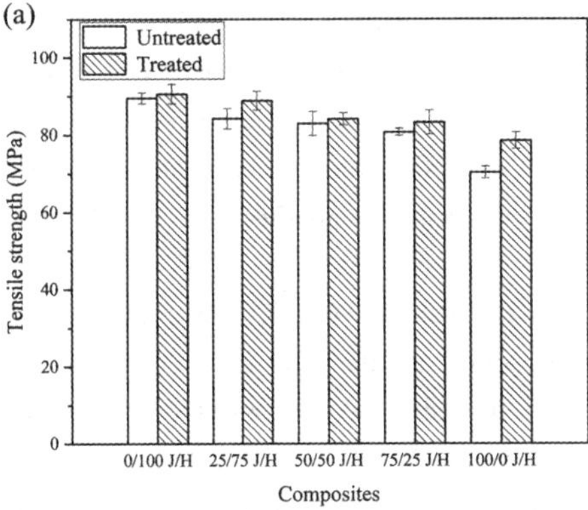

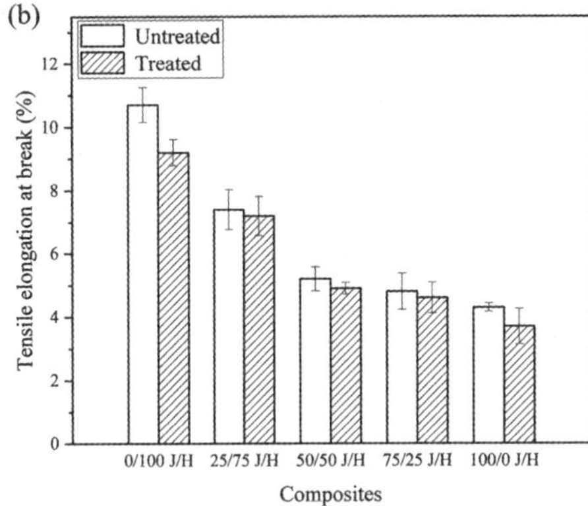

Fig. 3 (a) Tensile strength of different composites; (b) tensile elongation at break of different composites

3.3 Bending properties of the composites

The bending test results are shown in Fig. 4 (a) and Fig. 4 (b). The bending properties of the composites gradually decreased as the jute fiber content in the blended fibers was increased. Similarly, this was due to the fact that hemp fibers contain higher cellulose content and higher fiber strength than jute fibers. Therefore, the composite had higher bending strength when the content of hemp fiber was higher in the blended fibers.

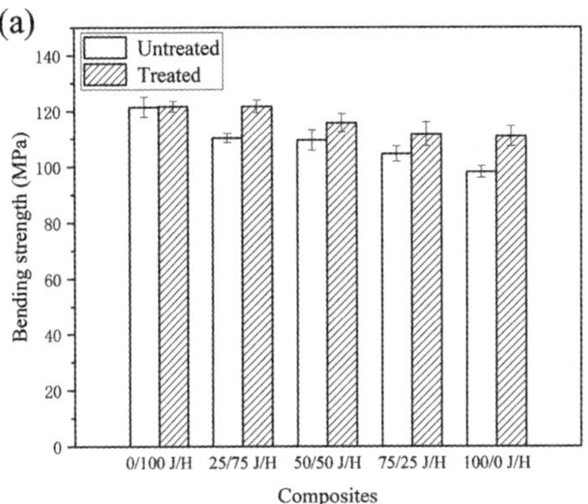

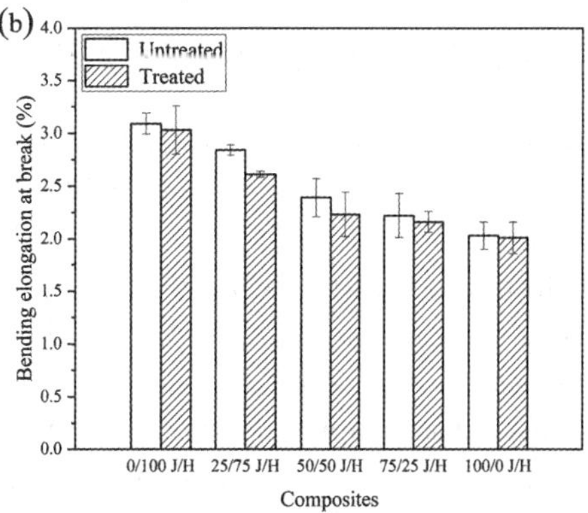

Fig. 4 (a) Bending strength of different composites; (b) bending elongation at break of different composites

Compared to the untreated composites, the bending strength increased by 0.2%, 10.3%, 5.8%, 6.8%, and 13.3%, but the bending elongation at break of the plasma-treated composites decreased. The reason may lies in after plasma treatment, the fiber surface was changed to some extent. The decreased mechanical properties of the fibers themselves led to their more brittle fracture.

4 Conclusions

In this study, a series of work was carried out to investigate the effect of plasma treatment on the mechanical properties of hemp/jute fiber mat reinforced PLA composites. After the plasma treatment, physical etching occurred on the surface of the fiber mats, and the non-cellulose material on the fiber surface was removed. At the same time, hydrophobic groups were generated on the fiber surface, which changed the fiber mats from hydrophilic to hydrophobic state. The interfacial adhesion between the fiber and resin was enhanced, resulting in an increase in the tensile and bending strength of the composite.

In the future, the combing direction of fiber mats and the choice of plasma treatment parameters can be further explored in order to achieve fiber-reinforced composites with better mechanical properties.

References

[1] DAS P P, CHAUDHARY V. Moving towards the era of bio fibre based polymer composites[J]. Cleaner Engineering and Technology, 2021, 4: 100182.

[2] MOHIT H, SELVAN V A M. A comprehensive review on surface modification, structure interface and bonding

mechanism of plant cellulose fiber reinforced polymer-based composites [J]. Composite Interfaces, 2018, 25 (5/6/7): 629-667.

[3] FARUK O, BLEDZKI A K, FINK H P, et al. Biocomposites reinforced with natural fibers: 2000—2010 [J]. Progress in Polymer Science, 2012, 37(11).

[4] MANRAL A, AHMAD F, CHAUDHARY V. Static and dynamic mechanical properties of PLA bio-composite with hybrid reinforcement of flax and jute [J]. Materials Today: Proceedings, 2019, 25(2020): 577-580.

[5] EJAZ M, AZAD M M, UR A, et al. Mechanical and biodegradable properties of jute/flax reinforced PLA composites[J]. Fibers and Polymers, 2020, 21: 2635-2641.

[6] MOTRU S H V V, ADITHYAKRISHNA V H, BHARATH J, et al. Development and evaluation of mechanical properties of biodegradable PLA/flax fiber green composite laminates[J]. Materials Today: Proceedings, 2020, 24: 641-649.

[7] MAHESH V, JOLADARASHI S, KULKARNI S M. Physio-mechanical and wear properties of novel jute reinforced natural rubber based flexible composite[J]. Materials Research Express, 2019, 6(5): 055503.

[8] BEHERA S, GAUTAM R K, MOHAN S, et al. Hemp fiber surface modification: its effect on mechanical and tribological properties of hemp fiber reinforced epoxy composites[J]. Polymer Composites, 2021, 42 (10): 5223-5236.

[9] SUNNY T, PICKERING K L, LIM S H. Alkali treatment of hemp fibres for the production of aligned hemp fibre mats for composite reinforcement [J]. Cellulose, 2020, 27(5): 2569-8252.

[10] CRUZ J, FANGUEIRO. Surface modification of natural fibers: a review [J]. Procedia Engineering, 2016, 155: 285-288.

Fabrication of Polypyrrole Conductive Cotton Fabric and Its Use on Textile Antennas

WANG Nan[1], YAO Lan[1*], QIU Yiping[1]

[1] *College of Textiles, Donghua University, Shanghai 201620, China*

* *Corresponding author's email*: yaolan@ dhu. edu. cn

Abstract: In the wireless communication area, the development of wearable antennas has increasingly become deep and extensive. Textile wearable antennas have received more concern due to their ease of conformability with the human body. In this study, three kinds of conductive cotton fabrics with different covering factors were prepared, and the conductive element was polypyrrole (PPy), which was obtained by the pyrrole (Py) as the monomer and ferric chloride as the oxidizing agent. Using the above conductive cotton fabrics as the radiation patches of textile microstrip antennas, the textile microstrip antennas were fabricated and the antenna performance was tested. The results showed that the prepared conductive cotton fabrics had good conductivity and the relatively good agreement was achieved between the measured return losses of all three antenna samples and the simulation results. And their optimum resonant frequencies were consistent with the design frequencies, proving the effectiveness of fabrication of conductive cotton fabric and textile antennas.

Keywords: Polypyrrole; Conductive cotton fabric; Radiation patch; Textile microstrip antenna; Return loss

1 Introduction

In recent years, wearable antennas have received extensive attention due to their advantages in lightweight, low-cost, flexible, and portable wireless communication and sensing[1]. Textile antenna refer to wearable antennas, in which the fabric materials are used as the dielectric substrates. Because of their soft, conformal, good integratability and other characteristics, they have been widely used in medical, health monitoring and other fields[2-5]. Furthermore, textile antennas play a vital role in body centric communications such as personal area network and body area network (BAN)[6]. With the development of high-performance textiles, textile fabrics have expanded from traditional clothing and home decoration fields to military, civil aviation and other fields. By conducting conductive treatment on textiles to make them have excellent conductivity, they can not only effectively broaden the application field of textile materials, but also adapt to the current development trend of flexible electronic materials, providing ideal flexible conductive materials for the development of flexible electronic devices[7-10].

In this study, we would like to investigate the antenna performance of textile microstrip antennas with conductive cotton fabrics with different covering factors as the radiating patches. The in-situ polymerization method was applied, where the pyrrole and ferric chloride were used as the monomer and the oxidizing agent, and finally three kinds of PPy conductive cotton fabrics with different covering factors were prepared. The conductive properties of the PPy conductive cotton fabrics were measured and compared, and the textile microstrip antennas with PPy conductive cotton fabrics as radiation patches were fabricated, and the performance of the antennas was tested and analyzed.

2 Experiments

2.1 Materials

The three kinds of cotton fabrics with different covering factors were common plain cotton fabrics. Ferric chloride ($FeCl_3$, CP, ≥97.0%), acetone (C_3H_6O, CP, ≥99.0%) were purchased from Sinopharm Chemical Reagent Co., Ltd. Pyrrole (C_4H_5N, 99%) was purchased from Titan Technology Co., Ltd. (Shanghai, China). Deionized water was made in the laboratory. Wool/polyester fancy suiting fabrics were provided by Danmao Textile Co., Ltd. (Jiangsu, China) with a thickness of 2.34 mm, a relative permittivity of 1.66, and a loss tangent of 0.015 8.

2.2 Fabrication of PPy conductive cotton fabrics

First, the cotton fabric samples were soaked in acetone solution for 20 min to remove impurities on their surface. Then the cotton fabrics were taken out from the acetone solution and were washed with deionized water. Then the cleaned cotton fabrics were dried in the oven at 60 ℃ and taken out. Then, the treated cotton fabrics were soaked in pyrrole solution (0.50 mol/L) for 1 h, and then ferric chloride solution (0.25 mol/L) was slowly added dropwise. After 2 h of reaction, the obtained PPy conductive cotton fabrics were put into the deionized water and rinsed for several times until the solution was free of black PPy. Finally, the conductive cotton fabrics were dried in the oven at 60 ℃ to the constant weight and taken out.

2.3 Fabrication of textile microstrip antennas

The Schematic diagram of a coaxial-fed microstrip antenna is shown in Fig. 1. Three kinds of PPy conductive cotton fabrics with different covering factors with a

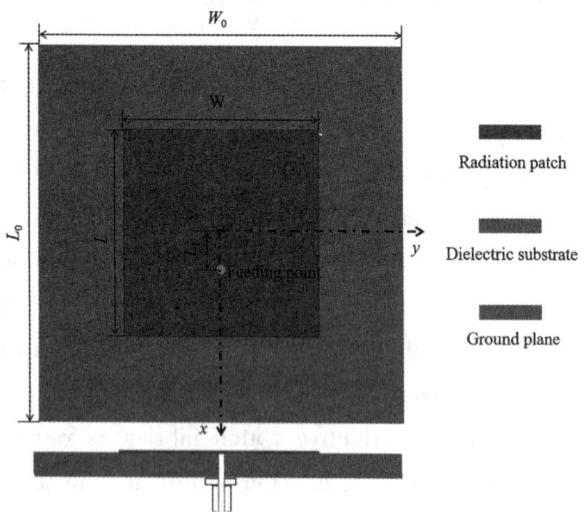

Fig. 1 Schematic diagram of a coaxial-fed microstrip antenna

cutting length (L) of 44.5 mm and a width (W) of 44.0 mm were used as the radiation patches of the textile microstrip antennas. The wool/polyester fancy suiting fabric and copper foil with a cutting length (L_0) of 80.0 mm and a width (W_0) of 80.0 mm were used as the dielectric substrate and ground plane of the textile microstrip antennas, respectively. The radiation patch (PPy conductive cotton fabric) and the upper surface of the dielectric substrate (wool/polyester fancy suiting fabric) were adhered together by using conductive glue, and the copper foil was also adhered on the back of the dielectric substrate (wool/polyester fancy suiting fabric). Then SMA connector was installed at the feeding point which length (L_1) from the origin of the coordinates was 9.0 mm. Thus the fabrication of the textile microstrip antennas was completed. The physical images of three different types of the after-prepared textile microstrip antennas are shown in Fig. 2 and Fig. 3.

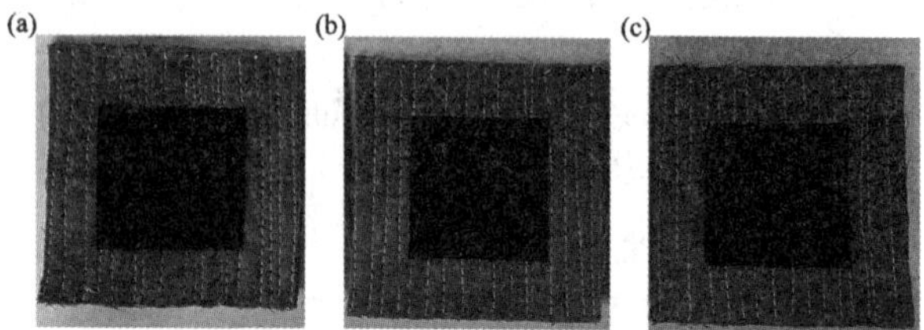

Fig. 2 The front sides of the textile microstrip antennas: (a) antenna 1; (b) antenna 2; (c) antenna 3

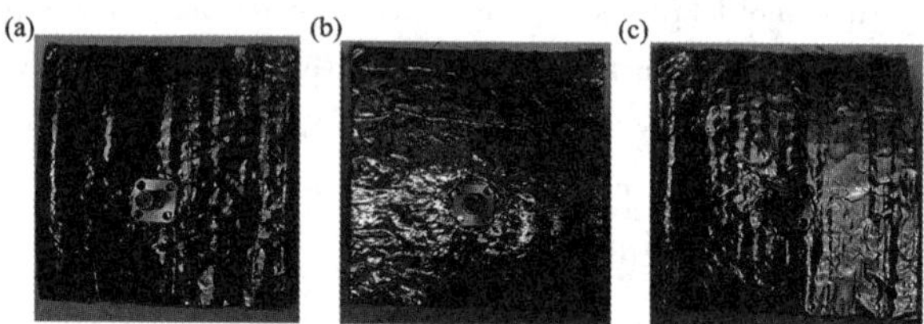

Fig. 3 The back sides of the textile microstrip antennas: (a) antenna 1; (b) antenna 2; (c) antenna 3

2.4 Performance testing of conductive cotton fabrics

The samples of untreated cotton fabrics and the samples of PPy conductive cotton fabrics were weighed using an FC/104 analytical electronic balance (Jingke Co., Ltd., Shanghai, China), and their increased weight percentages were calculated. The resistance values of the prepared 4 cm×4 cm PPy conductive cotton fabrics were tested using DT9205A digital multimeter (Sinaike Electronics Co., Ltd., Shenzhen, China). The red electric pointer and the black electric pointer were placed on both ends of the same side of the PPy conductive cotton fabrics, and the appropriate gear was selected to measure the surface resistance of the conductive fabrics.

2.5 Performance testing of textile microstrip antennas

The KC901V vector network analyzer was used to test the return loss of the textile microstrip antennas. First, the instrument was calibrated. Then the S_{11} mode was selected on the analyzer interface. 2.45 GHz was set as the center frequency, and the frequency sweeping test was performed in the 2.0–3.0 GHz frequency range.

3 Results and Discussion

3.1 Conductivity analysis of conductive cotton fabrics

The Structure diagrams of three kinds of cotton fabrics with different cover factors are shown in Fig. 4. It can be seen that the covering factors of the three kinds of cotton fabric samples was according to the following sequence: Fabric 1<Fabric 2<Fabric 3. The weight increase and surface resistance of the prepared PPy conductive cotton fabrics are shown in Tab. 1. The loadings and resistance values of the PPy conductive cotton fabrics prepared from the three kinds of cotton fabric samples were significantly different. When the above parameters were considered, it can be found that the conductive Fabric 1 with sparsely arranged warp and weft yarns had the highest loading, and the resistance values in warp and weft directions were lowest. The conductive Fabric 3 with closely arranged warp and weft yarns had the lowest loading, and the resistance values in warp and weft directions were highest. With the increase of the loading amount of the PPy conductive cotton fabrics, both the resistance values in warp and weft directions decreased. The conductive Fabric 1 with the highest loading had a weight increase of 23.15% compared with the untreated cotton fabric, and its resistance values were lowest with the resistance values in warp and weft directions of 642 Ω and 651 Ω, respectively. It was speculated that due to the different covering factors of the three kinds of cotton fabric samples, the loadings and resistance values of the three kinds of PPy conductive cotton fabrics were quite different. The warp and weft yarns of Fabric 1 were sparsely arranged, and the covering factor of fabric was low. So during the in-situ polymerization of PPy, the generated PPy substances were more deposited on the fabric. Therefore, in the prepared PPy conductive cotton fabrics, the conductive Fabric 1 had the highest loading, the lowest resistance values, and excellent conductivity. The warp and weft yarns of the conductive Fabric 3 were closely arranged, and the covering factor of fabric was high. So the generated PPy substances were less deposited on the fabric. Therefore, the conductive Fabric 3 had the lowest loading and the highest resistance values, and the conductivity was lower than that of the conductive Fabric 1 and Fabric 2.

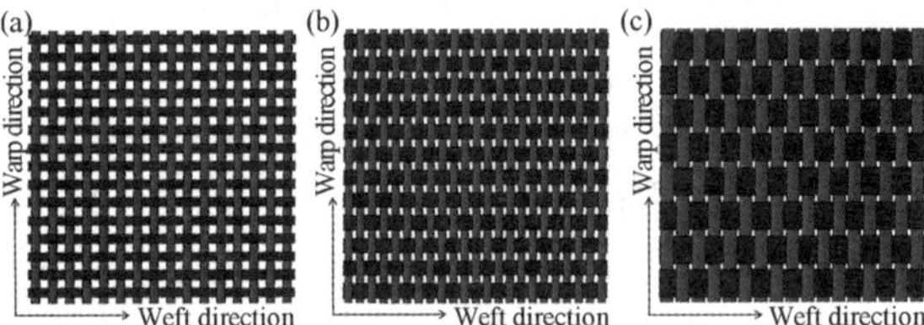

Fig. 4 Structure diagrams of three kinds of cotton fabrics: (a) fabric 1; (b) fabric 2; (c) fabric 3

Tab. 1 Specific parameters of PPy conductive cotton fabrics

Samples	Warp density (end/10 cm)	Weft density (end/10 cm)	Weight increase(%)	Resistance in warp direction (Ω)	Resistance in weft direction (Ω)
Fabric 1	388	348	23.15	642	651
Fabric 2	548	280	14.49	687	937
Fabric 3	400	184	5.18	1 211	1 690

In addition, it was found through testing that the resistance values in warp and weft directions for three kinds of PPy conductive cotton fabrics were inconsistent, and the resistance values in warp direction were all lower than the resistance values in weft direction. The higher warp density than weft density caused the cotton fabrics to be more closely arranged in warp direction, and the PPy had better connectivity in warp direction, forming a smooth conductive path. So the resistance values of fabrics in warp direction were lower than the resistance values in weft direction. Furthermore, it was speculated that due to the close arrangement of the warp yarns, it was equivalent to connecting multiple resistors in parallel, thus reducing the resistance values of the conductive cotton fabrics in warp direction.

3.2 Antenna performance analysis of textile microstrip antennas

The measured and simulated results of the return losses of the fabricated textile microstrip antennas are shown in Fig. 5. It can be seen that the three different types of the fabricated textile microstrip antennas had the resonant frequencies of 2.47 GHz, 2.45 GHz, 2.46 GHz, and the corresponding return losses were −33.41 dB, −27.38 dB, −29.53 dB, respectively. The above measured results had good agreement with the simulated results, in which the return loss was −27.57 dB at the resonant frequency of 2.45 GHz, basically meeting the antenna design requirements.

The resonant frequencies and bandwidths of the fabricated three different types of textile microstrip antennas are shown in Tab. 2. It can be seen that the resonant frequencies and bandwidths of the textile microstrip antennas fabricated by three different types of radiating patches were different. The resonant frequency of Antenna 2 was consistent with the simulation value, and the resonant frequency of Antenna 1 was 0.02 GHz higher than the simulation value; while the resonant frequency of Antenna 3 was 0.01 GHz higher than the simulation value. The resonant frequency of Antenna 2 was consistent with the design frequency, indicating that the processing size of the radiation patch was consistent with the design. It also indicated that the warp and weft yarns of cotton Fabric 2 were arranged moderately with the fabric surface being flat, and had little influence on the conductive radiation patch. Fabric 1 had a relatively sparse yarn arrangement, which led to a high unevenness on the surface of the fabric, affecting the size of the antenna radiation patch. And the surface flatness of the cotton Fabric 3 was also poor due to the large difference in the fineness of the warp and weft yarns, and the surface unevenness was high, so their resonant frequencies deviated slightly from the design frequencies.

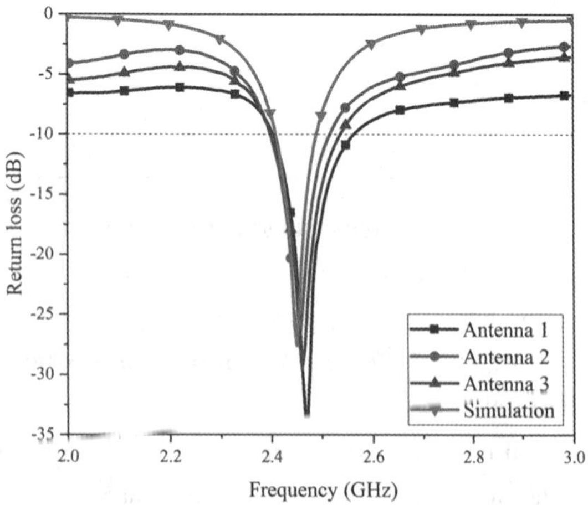

Fig. 5 Comparison of the measured and simulated results of return loss of three kinds of textile microstrip antennas

In terms of bandwidth, as the microstrip antennas themselves are narrow-band antennas, special measures need to be taken to widen the bandwidth, such as changing the shape of the patch or slotting the patch. At present, the simple conductive radiation patch is attached to the fabric, and the bandwidth does not exceed 10%. Therefore, the three kinds of textile antennas are currently narrow-band antennas. In the future, effective methods can be taken to widen the bandwidth and optimize the antenna performance.

Tab. 2 Resonant frequency and bandwidth of textile microstrip antennas

Samples	Radiation patch types	Resonant frequency (GHz)	Bandwidth (%)
Antenna 1	Fabric 1	2.47	6.15
Antenna 2	Fabric 2	2.45	4.45
Antenna 3	Fabric 3	2.46	4.88

4 Conclusions

In this study, the antenna performance of the fabricated textile microstrip antennas by using PPy conductive cotton fabrics with different covering factors as the radiating patches was investigated. The results showed that the fabricated PPy conductive cotton fabrics with pyrrole as the monomer and ferric chloride as the oxidizing agent had good conductivity. The maximum weight increase of the PPy conductive cotton fabrics was 23.15%, and the minimum resistance values in warp and weft directions were 642 Ω and 651 Ω, respectively. Using PPy conductive cotton fabric as the radiation patch, the wool/polyester fancy suiting fabric as the dielectric substrate, and the copper foil as the ground plane, the textile microstrip antennas were fabricated. The measured results of the return loss of the three kinds of antennas were in good agreement with the simulated results. The measured return losses were −33.41 dB, −27.38 dB, and −29.53 dB, respectively. And the return loss characteristics were excellent. The optimal resonant frequency of the fabricated antennas was 2.45 GHz, which was consistent with the design frequency, and the maximum bandwidth was 6.15%.

References

[1] PARACHA K N, RAHIM S K A, SOH P J, et al. Wearable antennas: a review of materials, structures, and innovative features for autonomous communication and sensing

[J]. IEEE Access, 2019, 7: 56694-56712.

[2] YADAV A, KUMAR SINGH V, KUMAR BHOI A, et al. Wireless body area networks: UWB wearable textile antenna for telemedicine and mobile health systems[J]. Micromachines, 2020, 11(6): 558.

[3] PADMANABHARAJU M, MADHAV B T P, KISHORE D S P, et al. Conductive fabric material based compact novel wideband textile antenna for wireless medical applications [J]. Materials Research Express, 2019, 6(8): 086327.

[4] THANGASELVI E, MEENA ALIAS JEYANTHI K. Implementation of flexible denim nickel copper rip stop textile antenna for medical application[J]. Cluster Computing, 2019, 22(1): 635-645.

[5] SINGH H, KANAUJIA B K, KUMAR A, et al. Dual-resonance ultra-miniaturised textile antenna for ISM/wearable applications [J]. International Journal of Electronics, 2021: 1-21.

[6] SREEMATHY R, HAKE S, SULAKHE S, et al. Slit loaded textile microstrip antennas[J]. IETE Journal of Research, 2020: 1-9.

[7] DAS P S, PARK J Y. A flexible touch sensor based on conductive elastomer for biopotential monitoring applications [J]. Biomedical Signal Processing and Control, 2017, 33: 72-82.

[8] HUANG W, LI J, ZHAO S, et al. Highly electrically conductive and stretchable copper nanowires-based composite for flexible and printable electronics[J]. Composites Science and Technology, 2017, 146: 169-176.

[9] UZUN S, HAN M, STROBEL C J, et al. Highly conductive and scalable Ti3C2Tx-coated fabrics for efficient electromagnetic interference shielding [J]. Carbon, 2021, 174: 382-389.

[10] ZHANG Y, REN H, CHEN H, et al. Cotton fabrics decorated with conductive graphene nanosheet inks for flexible wearable heaters and strain sensors [J]. ACS Applied Nano Materials, 2021, 4(9): 9709-9720.

A Multicomponent Braided Yarn for Absorbable Ventricular Septal Defect Occluder Better Suited to Cardiac Tissue

ZHANG Yunyue[1], LI Yiming[3], LI Chaojing[1,2], LING Jing[1,2], WANG Lu[1,2], WANG Fujun[1,2*]

[1] *Department of Technical Textiles, College of Textiles, Donghua University, Shanghai 201620, China*

[2] *Key Laboratory of Textile Science & Technology, Ministry of Education, Donghua University, Shanghai 201620, China*

[3] *Xinhua Hospital Affiliated to Shanghai Jiao Tong University School of Medicine, Shanghai 200092, China*

[*] *Corresponding author's email*: wfj@ dhu. edu. cn

Abstract: Ventricular septal defect is a common congenital heart disease. Among the existing treatment methods, minimally invasive implanted occluder is an effective method. Although existing commercial non-degradable nickel-titanium alloy occluders can meet basic occlusion needs, prolonged presence of the occluder may cause various complications. Through continuous efforts, many results have been achieved, but existing fully absorbable occluder bodies generally consist of a single component. The concentrated degradation of this single-component occluder after implantation in the body will produce a large amount of acidic degradation products in a similar time, which may cause a strong inflammatory response. Combining different materials may be a good way to avoid short-term concentrated degradation. At the same time, the endothelialization process is generally completed within 3 m after the implantation of the occluder. In order to better adapt to the cell growth time and avoid centralized degradation, studies have shown that three materials with different degradation times, PPDO (3–6 m), PGA (60–90 d), PGLA (70–110 d), braided into yarns can achieve a better effect of reducing inflammation. At the same time, materials with mechanical properties closer to cardiac tissue are more conducive to tissue growth. In this study, PPDO, PGA, and PGLA were selected, and the braiding height was changed, in order to obtain braided yarns with mechanical properties closer to the heart tissue. By adjusting the braid height, we made two kinds of yarns with different pitches. We tested the tensile and flexural properties of both yarns. The results show that the lower the braiding height and the smaller the pitch, the lower the mechanical properties but the better the bending properties. For subsequent braiding of future multicomponent occluders, yarns with moduli closer to cardiac tissue and with better flexural properties may be a better choice.
Keywords: Occluder; Braided yarn; Absorbable; Multi-components; Mechanical properties

1 Introduction

A ventricular septal defect (VSD) refers to an underdevelopment of the ventricular septum during the embryonic period, resulting in abnormal communication and a left-to-right shunt at the level of the ventricle. According to the "Report on the Prevention and Treatment of Birth Defects in China" issued by the Ministry of Health in 2012[1], there are more than 10 000 new cases of ventricular septal defect in China every year, causing a huge economic burden. There are many ways to treat ventricular septal defect, among which the method[2] with less damage and better curative effect is the intervention of VSD occluder. Existing commercial occluders are mostly made of nickel-titanium alloy materials[3]. Nitinol is a material with shape memory property, superelasticity and biocompatibility. Nitinol occluder can meet the basic needs of occlusion, but it will be gradually corroded when placed in the body for a long time, releasing toxic nickel ions[4], causing side effects. It is clear that long-term foreign matter in the body is not good for patient. Therefore, it is of great significance to develop absorbable occluder.

The absorbable occluders that have been studied include[5] Double Umbrella occluder, Chinese Lantern occluder, PCL-PLGA/Collagen occluder, etc. These researches have played a role in promoting the development of the absorbable occluder. But there is still room for improvement in existing absorbable occluders. After implantation, the absorbable occluder will cause an inflammatory response. The main materials of many existing absorbable occluders are all made of a single material, and the degradation of a single material is similar, which may lead to its acidic substances. Released in large quantities at the same time, causing a severe inflammatory response is not conducive to tissue growth. Using a combination of different materials may be a way to avoid this. For the occluder, the endothelialization process after implantation is generally completed within 3 M[6], so the mechanical properties of the material should be maintained for a period of 3 M, and then the self-degradation is completed, so that the defect site is completely covered by autologous tissue.

Studies[7] have shown that three materials with different degradation times, Poly (p-dioxanone) (PPDO, 3-6 M), Polygl-ycolide acid (PGA, 60-90 d), Poly (DL-glyco)-lide-co-lactide (PGLA, 70-110 d), can be used to relieve inflammation after braiding into yarns. At the same time, studies have[8] shown that materials with mechanical properties closer to cardiac tissue are conducive to cell growth. Therefore, we selected the same kind of material and obtained different braiding yarns by adjusting the braiding height. Then, the prepared samples were subjected to tensile test and bending test, and the best yarn that better fit the mechanical properties of cardiac tissue was obtained. Preparing for future multi-component ventricular septal defect occluders.

2 Experiments

2.1 Materials

PPDO monofilaments (USP Size 7-0, Monosorb®, d = 0.090 mm ± 0.001 mm), PGA multifilaments (USP Size 7-0, Trisorb® II, d = 0.060 mm ± 0.004 mm), and PGLA multifilaments (USP Size 7-0, Neosorb®, d = 0.060 mm ± 0.003 mm) were purchased from Samyang Biopharmaceuticals Corporation (Korea).

2.2 Basic properties of materials

The appearances of all materials were observed by a biological microscope (ECLIPSE E200, Shanghai Nikon Instrument Co., Ltd., China). Diameters of all materials were measured by a benchtop thickness gauge (Shanghai Liuling Instrument Factory, China). The tensile properties (breaking strength, stiffness, modulus) were determined using a HY-940FS testing system instrument (Shanghai Heng Yu Instrument Co., Ltd., China). Referring to the existing standards, we conducted tensile tests with a gauge length of 130 mm at a loading speed of 300 mm/min.

2.3 Design and preparation of braided yarn

Similar to existing research, in order to obtain the gradient degraded yarn, two PPDO, one PGA and one PGLA were used for 4-bobbin diamond braid with an 8-bobbin braiding machine (manufactured by Donghua University, China). The arrangement of bobbins on the braiding machine is shown in Fig. 1. In order to obtain braided yarns that are closer to the mechanical properties of cardiac tissue, we chose 0.06 mm multifilaments. By adjusting the braiding height, two kinds of multi-component degradable yarns with different pitches were prepared (the braiding heights were 9 cm and 8 cm, respectively).

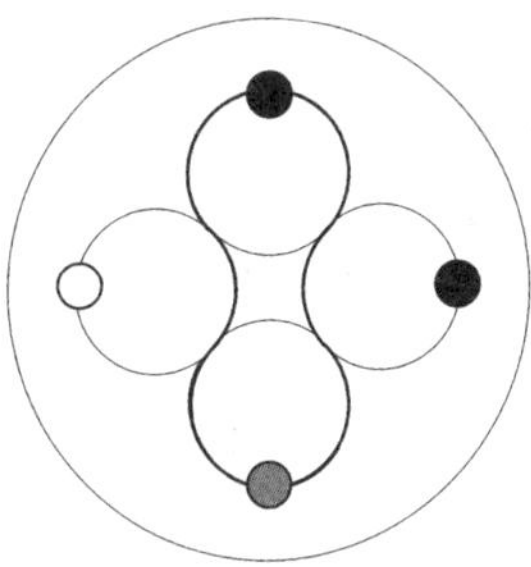

Fig. 1 The arrangement of bobbins on the braiding machine (black for PPDO, grey for PGA, white for PGLA)

2.4 Surface morphology of braided yarns

The braided yarns were examined by a stereo-microscope (SMZ745T, Shanghai Nikon Instrument Co., Ltd., China). Pitches and braiding angles were measured by MB-Ruler software. Diameters were measured by a benchtop thickness gauge. The braided height yarns of 9 cm are multi-component degradable loose yarns (MDL), and the yarns of 8 cm are multi-component degradable tight yarns (MDT).

2.4 Tensile test of braided yarns

Tensile test of braided yarns was conducted on a HY-940FS testing system instrument at a loading speed of 300 mm/min with a gauge length of 130 mm. At least 10 samples were tested (5 samples for each).

2.5 Bending test of braided yarns

Took 20 braided yarns of 100 mm length and fixed them between two cardboards at 5 mm intervals.

KES multi-purpose bending tester (KES-FB2S, Kato Tech Co., Ltd., Japan) was used to test the bending properties of MDL and MDT. After the test, the bending curve and bending hysteresis curve of the sample were obtained.

All materials were stored at −20 ℃ in freezer when not tested or used.

3 Results and Discussion

3.1 Basic properties of raw materials

PGA and PGLA multi-filaments are braided from multiple filaments, and the whole is uniform. The surface of PPDO monofilament is smooth, and the surface has no obvious defects. Fig. 2 shows the typical stress-strain curves of the three materials. It can be seen that the breaking strength of the multifilament is better than that of the monofilament, while the elongation at break is lower than that of the monofilament, which may be due to the difference in material types and preparation processes.

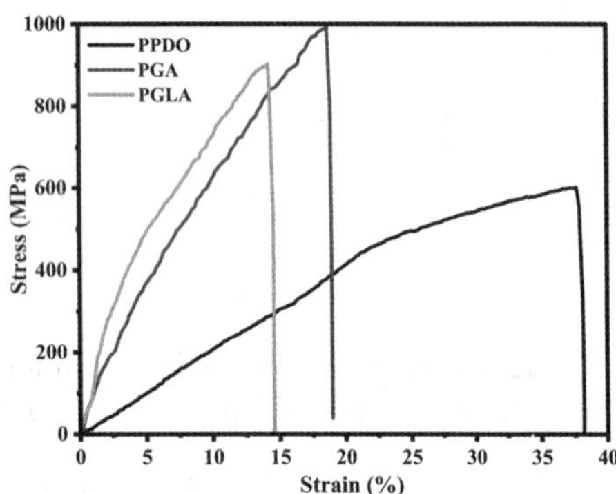

Fig. 2 Typical stress-strain curves of PPDO, PGA and PGLA

3.2 The effect of braiding height on braided yarn tensile properties

As shown in Fig. 3, in the braided sample, it can be clearly seen that MDT is more tightly braided between filaments than MDL. Combining the measurement results in Tab. 1, it is clear that after the braiding height decreases, the pitch decreases, the diameter increases, and the braiding angle also increases. It can be seen that the stress-strain curve in Fig. 4, the breaking strength of MDT is lower than that of MDL, and the Young's modulus and elongation at break of MDT are also lower than those of MDL. The reason for this phenomenon may be due to the increased braid between the yarns due to the tighter winding between the braided yarns. It is known that the Young's modulus of cardiac tissue is 22 kPa to 500 kPa[9], and the modulus of both materials are higher than this value. There are also studies believe that materials closer to the mechanical properties[8] of cardiac tissue will be beneficial to tissue growth. Therefore, although the mechanical properties of MDT are lower than that of MDL, its mechanical properties are closer to those of cardiac tissue. In addition, the curve of MDL presents a three-segment type, which may be caused by the looseness of the three braided filaments, which leads to the rupture of the three materials in sequence after being stretched by force. MDL may not be suitable for the preparation of subsequent occluders due to its looseness. At the same time, MDT, which have closer mechanical properties to the heart tissue may be more conducive to cell growth. Therefore, MDT may be more suitable for subsequent preparation than MDL.

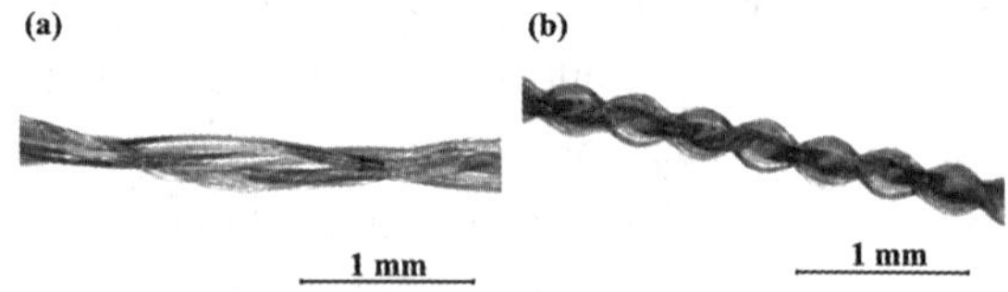

Fig. 3 Stereomicroscopic images of (a) MDL and (b) MDT

Tab. 1 Structural parameters of braided yarn

	MDL	MDT
Diameter (mm)	0.19±0.01	0.26±0.01
Braiding angle (°)	10.80±2.51	36.56±0.74
Pitch (mm)	2.92±0.12	0.74±0.04

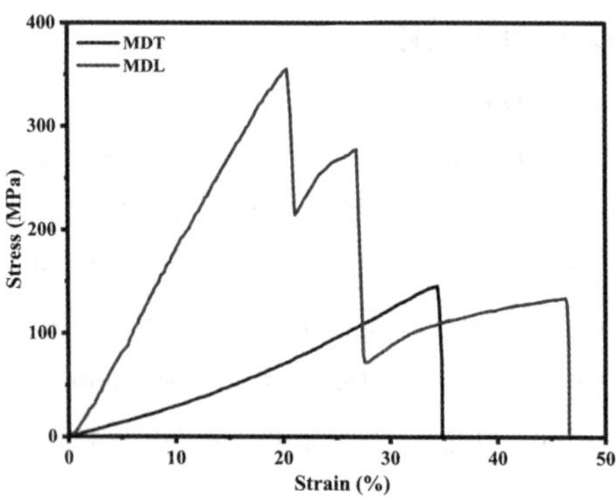

Fig. 4 Typical tensile stress-strain curves of MDL and MDT

3.3 The effect of braiding height on braided yarn flexural properties

One end of the braided yarn was clamped and fixed, and the other end of the braided yarn was clamped by the instrument to bend back and forth at a constant speed along the fixed track. The relationship between bending moment and curvature per unit width is obtained, and its schematic diagram is shown in Fig. 5. From this, the flexural rigidity and bending hysteresis of the yarns can be calculated, which are shown in Tab. 3.

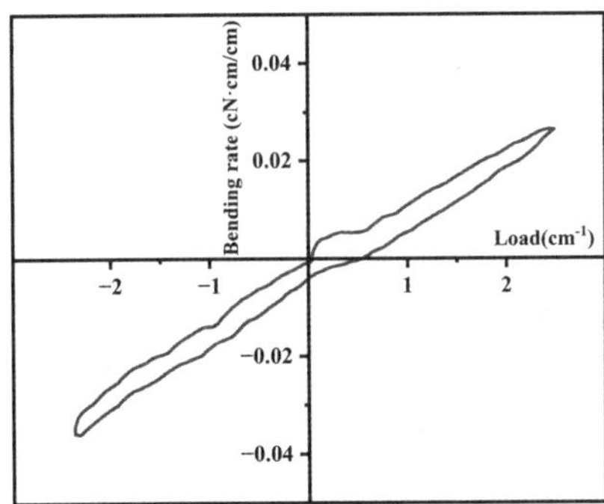

Fig. 5 Schematic flexural curves of yarns

Tab. 3 Flexural properties of braided yarn

	MDL	MDT
Flexural rigidity (cN · cm^2/cm)	0. 02	0. 05
Bending hysteresis (cN · cm/cm)	0. 02	0. 08

Because of its higher braiding height than MDT, MDL is loosely braided between filaments, there are larger gaps in braided filaments, and the friction between filaments is smaller. The friction between the filaments[10] affects bending performance, so the flexural rigidity and bending hysteresis of MDL are less than half of those of MDT. At the same time, since the braided yarns are braided from monofilament and multi-filament, the flexural rigidity and bending hysteresis of them are both small. Considering that the subsequent braided yarns will be used to braid the occluder, softer yarns are easier to braid. From this aspect, both yarns can be used for braiding. However, the final occluder stent also needs to have a certain ability to resist bending deformation. Therefore, between the two, MDT with better bending resistance is preferred as the yarns of the subsequent braided occluder.

4 Conclusions

In this study, multi-component biodegradable braided yarns with certain mechanical properties and bending resistance that can be gradually degraded were prepared. The braided yarns were made for subsequent preparation of VSD occluders. Braided yarns have lower strength than filaments performance due to friction and torsion between filaments. As for the two braided yarns, MDT with lower braiding height is tighter, and the tensile strength has decreased, but its mechanical properties are more suitable with cardiac tissue. The flexural performance and flexural hysteresis of MDT with lower braid height are significantly higher than that of MDL. In order to make the prepared occluder have better bending resistance and mechanical properties suited to cardiac tissue, MDT should be used as the basic material for the subsequent preparation of multi-component occluders.

References

[1] The Ministry of Health. The report on the prevention and treatment of birth defects in China (2012) [J]. China Pharmacy, 2012(39): 3693.

[2] BAUMGARTNER H, DE BACKER J, BABU-NARAYAN S V, et al. 2020 ESC guidelines for the management of adult congenital heart disease[J]. Eur Heart J, 2021, 42(6): 563-645.

[3] LI B, XIE Z, WANG Q, et al. Biodegradable polymeric occluder for closure of atrial septal defect with interventional treatment of cardiovascular disease[J]. Biomaterials, 2021, 274.

[4] ZHANG Z X, FU B F, ZHANG D Y, et al. Safety and efficacy of nano lamellar tin coatings on nitinol atrial septal defect occluders in vivo[J]. Materials Science and Engineering: C, 2013, 33(3): 1355-1360.

[5] LIN C, LIU L, LIU Y, et al. Recent developments in next-generation occlusion devices[J]. Acta Biomaterialia, 2021, 128: 100-119.

[6] SODERBERG B, VASKELYTE L, SIEVERT K, et al. Prospective single center first in human (FIH) clinical trial to evaluate the safety and effectiveness of a septal occluder with bioresorbable framework in patients with clinically significant atrial septum defect (ASD) or patent foramen ovale (PFO)[J]. Journal of the American College of Cardiology, 2016, 68(18): 334-334.

[7] WU S. Design and preparation of multi-component gradient degradation fully absorbable ventricular septal defect occluder[M]. Shanghai: Donghua University, 2021.

[8] ZHAO L M, WANG L, ZHANG W Q, et al. Promotion of right ventricular outflow tract reconstruction using a novel cardiac patch incorporated with hypoxia-pretreated urine-derived stem cells[J]. Bioactive Materials, 2022, 14: 206-218.

[9] WU Y K, YU J. The role of tissue engineering in cellular therapies for myocardial infarction: a review[J]. Journal of Materials Chemistry B, 2015, 3(31): 6401-6410.

[10] CHATTOPADHYAY R, TYAGI G K, GOYAL A. Studies of the hybrid effect in mechanical properties of tencel blended ring-, rotor- and air-jet spun yarns[J]. The Journal of The Textile Institute, 2013, 104(3): 339-349.

The Improvement of the Opaque Performance of High-Count Lightweight Pure Cotton Fabrics

MA Qingsong[1], ZHOU ZHiquan[1], ZHANG Yanru[1], WANG Ni[1]*

[1] *Key Laboratory of Textile Science & Technology of Ministry of Education, College of Textiles, Donghua University, Shanghai 201620, China*

* *Corresponding author's email*: wangni@ dhu. edu. cn

Abstract: Technological progress and the improvement of living standards have made cotton fabrics tend to develop in high count and thin. However, fabrics with both light weight and light-colored are prone to the problem of insufficient shielding, so how to improve the opacity of cotton fabrics has received widespread attentions. The paper aims to improve the opaque property of the pure cotton fabrics by the application of a finishing agent CM-5000. The finishing agent was first padded onto pure cotton fabrics and the properties of fabrics such as thickness, opaque performance and moisture permeability, before and after padding were compared. The results indicated that: at a bath ratio of 1 : 30 and a CL-328A finishing agent dosage of 10 g/L, a satisfactory rolling rate can be obtained by controlling the padding pressure and rotation speed to 0. 1 MPa and 6. 0 m/min, respectively. After finishing, the thickness of the fabric increased, and the greater concentration of the finishing agent, the thicker the fabric and the better the opaque performance. Both the whiteness and the reflectivity of the fabric increased after the padding treatment, and the opaque characterized by shielding rate increased, while the moisture permeability decreased, and as the concentration of finishing agent increased, the more the moisture permeability decreased.

Keywords: Opaque; Cotton fabric; Padding; Shielding rate

1 Introduction

In summer, people choose thin, UV-proof clothing to achieve the demand for physical cooling, but "thinness" and "transparency" often go hand in hand. With the demand for comfort in textiles, clothing is becoming thinner and thinner, and as the increasing demand for personalized and multiple functions, the visual shielding property of fabrics has received considerable attentions.

The main object of the visual shielding or opaque property of textiles is "light". With the in-depth study of the optical model of fabrics, Yang[1] proposed the four-component model, the interaction of incident light with the textiles could be divided into four parts: reflection, absorption, scattering, and transmission. While reflected light is made of regular reflection and scattered reflection, and transmission could be further divided into the direct transmission and diffuse transmission, where the sum of scattered reflected light and diffuse transmission light is scattered light.

The ways to enhance the visual shielding of textiles includes:

(1) Improving fiber and yarn structure. Liu[2] produced opaque PET filaments by adding various weight percentages of TiO_2. Yanagihara et al. [3] proposed a composite DTY yarn with a core-sheath structure. The skin layer has microporous forming agents as well as other polymers. The core layer is a yarn composed of inorganic TiO_2 particles of 3. 0%- 12. 0% by weight.

(2) Improving the fabric structure to change the thickness and tightness of the fabric to improve the visual-shielding property is a relatively easy technique to achieve. Esfandiar et al. [4] studied the application of hollow glass microspheres and TiO_2 nanoparticles in developing flame retardant, UV resistant, heat-insulating, and sound absorbing textiles. Typical examples included "Rikado" by Shikishima and "Refiruka" by Fuji Spun in Japan[5].

In this study, a high-count pure cotton lightweight fabric was treated with CM-5000 opaque finishing agent (Suzhou Weiming Chemical Industry Co. Ltd., China) by padding to improve the visual-shielding property.

2 Experiments

2.1 Materials and instruments

White twill cotton fabric (124.81 g/m^2) were obtained from Lu Thai Textile (Shandong, China). CM-5000 and CL-328A, was supplied by Weiming Chemical Industry (Jiangsu, China). $CaCl_2$, was provided by Kuling Fine Chemical (Shanghai, China).

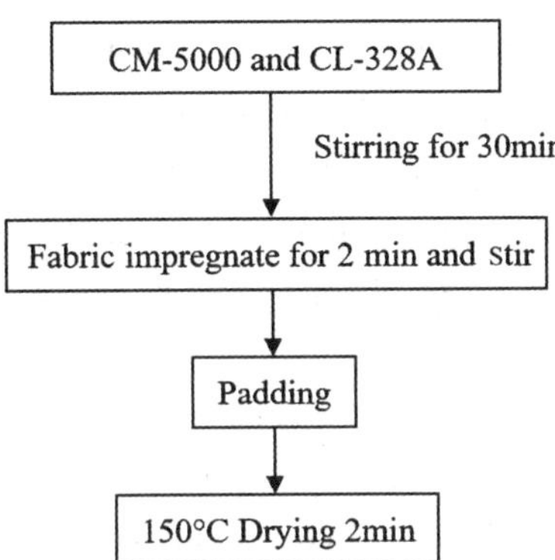

Fig. 1 Padding process

The finishing agent solution was prepared as the processed shown in Fig. 1 and the solvent volume was calculated according to the solution mass with a bath ratio of 1 : 30. Firstly, CM-5000 and CL-328A finishing agent were added to the beaker with an appropriate amount of water. Then the finishing agent was stirred for 30 min coated on the fabric by dipping rolling method. In detail, the fabric was added to the completed finishing agent and impregnated for 2 min with stirring and the fabric was put into the rolling roller and padded. Finally, the fabric was dried in the constant temperature dryer at 150 ℃ for 2 min.

Fabrics were weighted before and after the padding process, and the data was recorded as G_{0i} and G_i. According to the pre-experiment results, a good visual shielding effect could be obtained with the rolling rate of 70%-90%. The rolling rate (r) is given below as Eq. (1):

$$r = \frac{G_i - G_{0i}}{G_{0i}} \tag{1}$$

2.2 Thickness and moisture permeability

The thickness and the moisture permeability were measured according to GB/T 3820—1997 and GB/T 12704.1—2009 respectively.

2.3 Optical analysis

The Datacolor 850 spectrometer (Datacolor Trading Co. Ltd. Shanghai, China) was used to characterize the whiteness, shielding and reflectivity of the fabric samples before and after finishing. After calibration, the fabric was folded into 4 layers and tested 3 times in different parts to take the average value.

3 Results and Discussion

3.1 Rolling rate summary of the fabric

A satisfactory rolling rate (70% - 90%) can be obtained by controlling the padding pressure and rotation speed to 0.1 MPa and 6.0 m/min. The more water removed, the less liquid the fabric carried and the more efficient the mill rolled the liquid[6]. This interval of rolling rate not only met the process requirements, but also reduced the waste of resources and energy. And the large bath ratio used was a guide for practical production[7].

In order to analyze the relationship between the properties of the fabric and padding, the pad index K was proposed as the comprehensive effect of both the finishing agent concentration and the rolling rate to characterize the performance change after finishing for Tab. 1.

Tab. 1 Pad index of the fabric

Fabric	Concentration (g/L)	Rolling rate (%)	Pad index
1#	20	75.56	15.112
2#		88.87	17.774
3#		93.21	18.642
4#	30	68.03	13.606
5#		84.44	16.888
6#	40	74.70	14.940
7#		79.92	15.984
8#		82.06	16.412

The pad index was defined as Eq. (2).

$$K = r \cdot C \tag{2}$$

where: K is the pad index; r is the rolling rate, %; C is the finishing agent concentration, g/L.

3.2 Thickness of the fabric

The thickness results of all fabrics after padding were shown in Fig. 2. It could be found that the thickness of the fabric was significantly increased after finishing. As the concentration of the finishing agent increased, the fabric thickness showed a rising trend, with the highest thickness of 0.252 mm. Shi et al.[8] pointed out that fabric thickness is the main factor in analyzing the factors affecting the opaque performance of fabrics. After the fabric is padded, the finishing agent adheres to the fabric to increase the thickness of the fabric. The light reflection of the fabric is enhanced and the transmission is reduced, improving the visual-shielding property of the fabric.

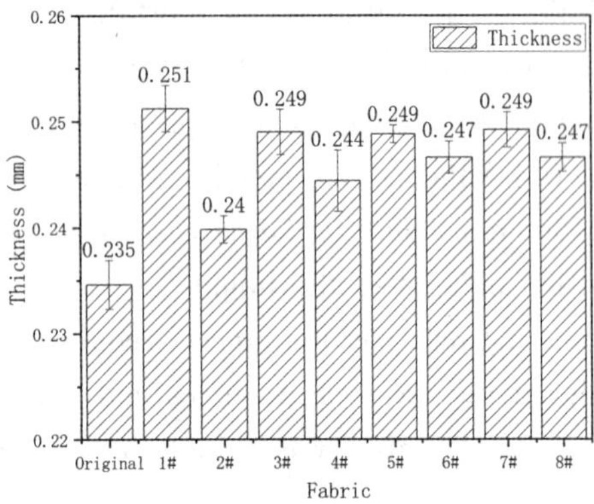

Fig. 2 Thickness of the fabric

3.3 Whiteness of the fabric

From Fig. 3, it was found that the whiteness of the fabrics increased after padding, and with the increase of finishing agent concentration, the whiteness of the fabric showed an increasing trend. It was also found that when the finishing agent CM-5000 was 20 g/L, the pad index increased from 15.112 to 18.642, and its whiteness became higher. In the high concentration of finishing agent CM-5000, the rise of the whiteness became less. However, when 40 g/L finishing agent CM-5000 was added, the whiteness of 8# fabric with a rolling index of 16.412 increased by 6.65% compared to that of the original fabric. This indicated that at high concentrations, the finishing agent improved fabric whiteness significantly. Zhang et al. [9] also pointed out a significant correlation between whiteness and opaque

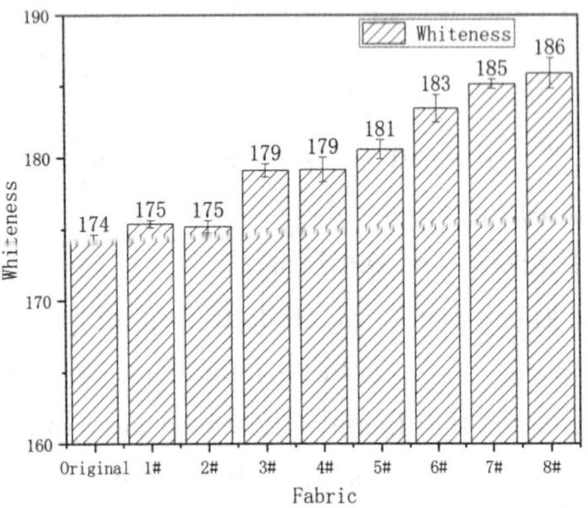

Fig. 3 Whiteness of the fabric

when the fabric was white or light-colored.

3.4 Shielding of the fabric

Visible light shielding of fabrics, also known as visual shielding, refers to the degree of fabric obstruction to the human eye[10]. Based on what was derived in the analysis of Wang et al. [11] for a white paper, the paper opacity (shielding rate) reflects the ability to resist light projection. The Datacolor spectrometer contains an integrating sphere in which the light is diffusely reflected several times to calculate the total energy emitted by the light source. Thus, the shielding rate or opacity of the substance is measured. The visual shielding effect of the fabric can be characterized by the shielding rate. Fig. 4 showed that the shielding rate of the fabrics after finishing was on the rise. Among them, fabric 4# had a low rolling rate, resulting in the minimum padding index and the lowest shielding rate improvement. At a low concentration of 20 g/L CM-5000, such as fabric 1#-3#, the enhancement for shielding rate was low. However, under the high concentration of 40 g/L CM-5000, such as fabric 6#-8#, the enhancement for shielding rate was improved more, among which the shielding rate of fabric 8# was improved the best with an enhancement of 3.86% compared with that of the original fabric.

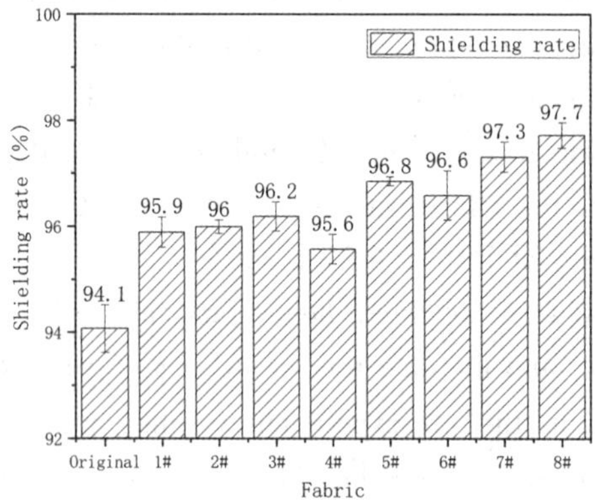

Fig. 4 Shielding rate of the fabric

3.5 Reflectivity of the fabric

From Fig. 5 (a), it was found that all fabrics peaked at wavelengths between 450 nm and 470 nm. When the wavelength was 380-460 nm, the reflectivity was on an upward trend, and while the wavelength was between 460-560 nm, the reflectivity was on a downward trend. The reflectivity in 450-500 nm wavelength was increased compared with that of the original.

From the study of LIN[12], it was found that the human

eye is most sensitive to yellow-green light at a wavelength of 550 nm under normal illumination. For this purpose, the 550-700 nm reflectance images were intercepted, as shown in Fig. 5 (b). When the wavelength was greater than 550 nm, with the increase of wavelength, the reflectivity of the fabric also rose, and with the increase of finishing agent CM-5000 concentration, the peak reflectivity of the fabric increased. For fabric 4#, due to its smaller padding index, the reflectivity was lower than that of the other fabrics. LIU[13] pointed out that reflectance on white fabric was closely related to the visual shielding by human eyes. From Fig. 5 (b), it can be found that when using a high concentration of 40 g/L finishing agent CM-5000 with the largest padding index, its reflectivity is greater, then the opaque effect is better.

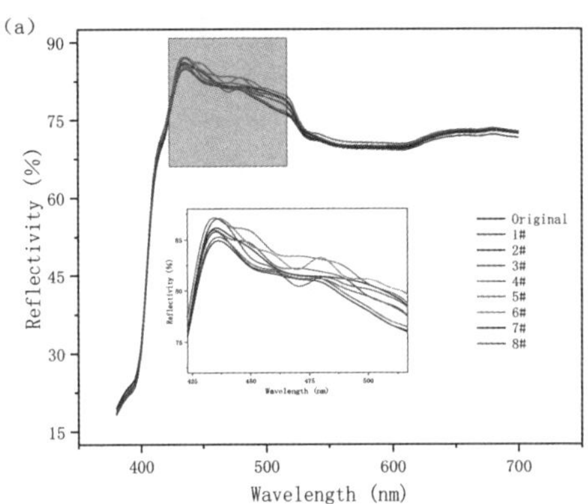

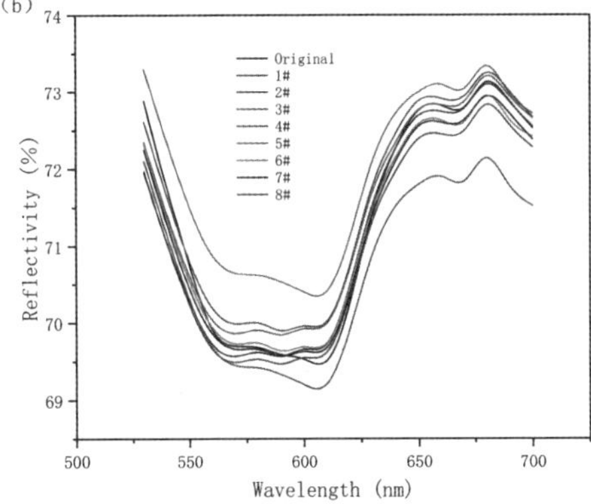

Fig. 5 Reflectivity of the fabric: (a) 380-700 nm; (b) 530-700 nm

3.6 Moisture permeability of the fabric

From Fig. 6, it can be found that as the concentration of finishing agent CM-5000 increased, the moisture permeability decreased. For fabric 4#, its padding index was smaller, and its moisture permeability decreased less compared to the fabrics with low concentration of finishing agent CM-5000. When 40 g/L finishing agent was applied and the index of padding was the largest, the moisture permeability decreased the most, as about 17.56% compared with the original fabric. The reason was that after padding, the finishing agent in the fabric surface formed a thin film or fill the original fabric voids. When the thickness of the fabric increases after finishing, the transmission pathway of the fiber became longer, which increased the opportunity for contact between the fabrics and increased the blockage of water molecules in the pathway (wet resistance).

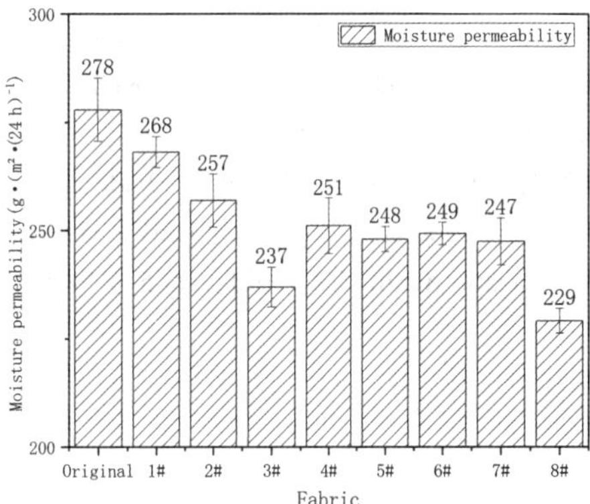

Fig. 6 Moisture permeability of the fabric

4 Conclusions

This study has focused on the improving of the opaque property of the high-count lightweight pure cotton fabric. Under the conditions of bath ratio 1:30, 70%-90% rolling rate can be obtained after adjusting the pressure and speed of the press roll to 0.1 MPa and 6.0 m/min, respectively. The fabric thickness was elevated after finishing. The increase of whiteness and shielding rate was used to characterize the opaque performance after padding. The reflective effect of the fabric was improved under the effect of high concentration of finishing agent. However, the moisture permeability of cotton fabrics showed a decreasing trend, and the moisture permeability decreased more with the increase of CM-5000 concentration.

References

[1] YANG H Y. Study on the model for fabric optical behaviors and its applications[D]. Shanghai: Donghua University, 2004

[2] LIU Z L, YANG Y, WANG Y W, et al. Preparation and properties of opaque polyethylene Terephthalate/TiO2 filaments[J]. Materials Science-Medziagotyra, 2021, 27

(3): 43-44.

[3] YANAGIHARA M, NAKAJIMA T. Composite false-twisted yarn excellent in transparency preventing property and method for producing the same [P]. Japan, JP2007239146. 2007-09-20.

[4] ESFANDIAR P, MARYAM N, SIMA K, et al. Functional cotton fabric using hollow glass microspheres: focus on thermal insulation, flame retardancy, UV-protection and acoustic performance[J]. Progress in Organic Coatings, 2020, 141: 105553.

[5] ZHOU Q. Research status of anti-UV textiles[J]. Progress in Textile Science & Technology, 2019(10): 6-8, 43.

[6] HOU Z Y. Effect of padder pick up on energy consumption stenter setting machine[J]. Textile Dyeing and Finishing Journal, 2015, 37(4): 50-52.

[7] FAN J, MIAO J H, AN Y, et al. Padding/non-aqueous medium fixation process of cotton fabrics with reactive dyes [J]. Textile Auxiliaries, 2021, 38(5): 38-42.

[8] SHI M W, ZHANG Y. Study of the visual masking synthetic fiber and its knitted fabric[J]. Knitting Industries, 2010(10): 6-8.

[9] ZHANG N, WANG N, et al. Comparative study on evaluating index for visual shielding property of textiles[J]. Journal of Textile Research, 2015, 36(9): 38-43.

[10] JING Q Q, SHEN L P, SHI Y. Current research status of visible light shielding based fabrics[J]. Synthetic Fiber in China, 2019, 48(6): 34-38.

[11] WANG X, ZHANG M Y, LUO Y. An introduction to the optical properties of paper[J]. Heilongjiang Pulp & Paper, 2012, 40(3): 37-39+46.

[12] LIN L H, SHEN R H, et al. Test methods of visual shielding property of fabrics[J]. Textile Auxiliaries, 2021, 38(8): 55-60.

[13] LIU S Y, YAN C H. Application of reflectivity tester in test of fabric visual shielding performance[J]. Advanced Textile Technology, 2021, 29(4): 63-68.

Numerical Simulation of Liquid-Solid Two-Phase Conductive Mixture Under Different Models

ZHANG Hao[1], LI Liqing[1], ZHANG Tianshu[2], LI Qiao[1]*

[1] *Key Laboratory of Textile Science & Technology, Ministry of Education, College of Textiles, Donghua University, Shanghai 201620, China*

[2] *College of Science, Donghua University, Shanghai 201620, China*

* *Corresponding author's email*: qiaoli@ dhu. edu. cn

Abstract: The mixing of fluid materials and solid conductive materials and the distribution of solid materials in the fluid are important parts in the preparation of conductive mixture in the fabric sensors. In order to achieve a uniform distribution of the solid conductive particles in the liquid matrix thus improving the sensing performance of the fabric sensors, this paper used the Euler two-fluid model and k-ε turbulence model to simulate and calculate the liquid-solid two-phase mixing model with the help of computational fluid dynamics (CFD) numerical method. The volume fraction of solid particles under different parameters was analyzed by changing the size ratio of the model and the shape of the stirring magneton, and the influence of different specifications on the characteristics of liquid-solid two-phase mixing and the distribution of conductive particles was explored. On the basis of the results, it can provide a reference for further optimization of the distributions of conductive particles in the liquid matrix.

Keywords: Multiphase flow; Viscosity; CFD; Simulation

1 Introduction

Fabric temperature sensor is composed of flexible substrate and conductive mixture, in which the quality of conductive mixture greatly affects the performance of sensor. The distribution of solid granular conductive materials in liquid polymer is a big problem in the process of preparing conductive mixture. For the problem of liquid-solid two-phase mixing, researchers use computational fluid dynamics (CFD) to simulate and analyze related cases. Kasat et al. [1] used Euler multiphase flow model to explore the influence of suspension properties on liquid phase mixing in stirred vessel. Micale et al. [2] used the k-ε turbulence equation in CFD to simulate concentration distribution of the particles in two-phase stirring based on the three-dimensional flow field.

In this research, based on the process of preparing polydimethylsiloxane-platinum conductive mixed materials for fabric sensors, CFD was used to simulate the mixing of liquid-solid two-phase materials in the stirring process, and the two-phase mixing characteristics and the distribution of conductive particles in the mixture with different specifications of container and stirring devices were analyzed.

2 Experiments

2.1 Geometric modeling and meshing

In this simulation, a simplified beaker and magneton model is used to describe the process of magnetic stirring. The beaker and the magneton used in the experiment have various specifications. In order to explore the influence of different parameters on the mixing characteristics and particle distribution, this experiment simulates different ratios of beaker width to the maximum length of the magneton and the shape and size of the magneton. The specific specifications are shown in Tab. 1.

Tab. 1 Specification parameters

Model	Breaker length(mm)	Breaker height(mm)	Magneton length(mm)	Magneton height(mm)
5-A7	20	10	15	7
10-C6	28	10	21	6
5-C6	20	10	18	6
10-A9	30	12	21	9

The established model was divided into grids, and some

areas were selected to be meshed. The grid type was set to triangular grid in unstructured grid with the grid size of 0.2 mm. The quality of the checked grid meets the requirements of simulation analysis. Fig. 1 is the grid diagram of 5-C6 model.

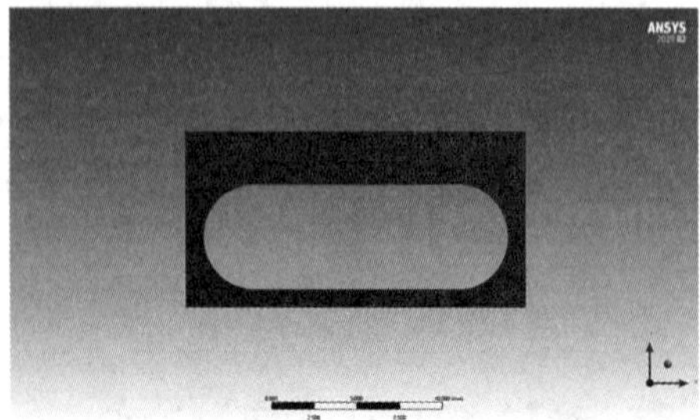

Fig. 1 Mesh diagram of 5-C6 model

2.2 Mathematical model

In this simulation, Eulerian two-fluid model was used to analyze the distribution of liquid-solid two-phase materials after stirring. In this model, the liquid phase is considered as a continuous medium and the solid phase is considered as a quasi-fluid. The two coexist and permeate each other at any position[3]. The Euler model is fast in calculating the system with complex computational domain and uncertain number of particle phases. When Euler model is used, it is difficult to determine the viscosity of solid particle materials. The viscosity of fluid refers to the internal friction force (shear stress) required by the fluid to produce a unit velocity gradient, that is, the resistance of the fluid to the flow[4]. For solid materials, the shear viscosity of solid includes three parts[5]: collision, dynamics and friction, and the friction viscosity will only exist when solid materials are piled up. The specific expression is as follows.

$$\mu_s = \mu_{s,col} + \mu_{s,kin} + \mu_{s,fr} \tag{1}$$

When defining the second phase of particles, granular viscosity, granular bulk viscosity and frictional viscosity should be defined in particle attributes. Constant value, Syamlal-Obrien equation and Gidaspow equation are generally used to express the collision viscosity and dynamic viscosity of particles. In this simulation, Gidaspow equation is used to describe collision viscosity and dynamic viscosity[6], as shown in Eqs. (2) and (3) respectively.

$$\mu_{s,col} = \frac{4}{5}\alpha_s\rho_s d_s g_{0,ss}(1+e_{ss})\left(\frac{\theta_s}{\pi}\right)^{1/2} \tag{2}$$

$$\mu_{s,kin} = \frac{10\rho_s d_s\sqrt{\theta_s\pi}}{96\alpha_s(1+e_{ss})g_{0,ss}}\left[1+\frac{4}{5}g_{0,ss}\alpha_s(1+e_{ss})\right]^2 \tag{3}$$

In low-shear dense flow, the secondary volume fraction of solid phase approaches the stacking limit, and the stress is mainly caused by the friction between particles. Constant value and Schaeffer equations are generally used to express the frictional viscosity generated by the accumulation of granular materials when defining the frictional viscosity. The expression of Schaeffer equation[7] is as follows.

$$\mu_{s,fr} = \frac{p_{friction}\sin\varphi}{2\sqrt{I_{2D}}} \tag{4}$$

Granular bulk viscosity expresses the resistance of particles to compression and expansion. By default, the granular bulk viscosity of particles is generally set to 0, which can also be described by Lun equation[8] or user-defined function. In this experiment, Lun equation is used to describe the volume viscosity of solids, and the expression is as follows:

$$\lambda_s = \frac{4}{3}\alpha_s^2\rho_s d_s g_{0,ss}(1+e_{ss})\left(\frac{\theta_s}{\pi}\right)^{1/2} \tag{5}$$

In the above formula, the meanings and units represented by each symbol are shown in Tab. 2.

Tab. 2 Meaning and unit of symbol

Symbol	Meaning	Unit
μ_s	Granular shear viscosity	Pa · s
$\mu_{s,col}$	Granular collision viscosity	Pa · s
$\mu_{s,kin}$	Granular kinetic viscosity	Pa · s
$\mu_{s,fr}$	Granular friction viscosity	Pa · s
λ_s	Granular bulk viscosity	Pa · s
α_s	Solid volume fraction	—
ρ_s	Density of solid phase	kg/m^3
d_s	Particle diameter	mm
$g_{0,ss}$	Radial distribution function	—
e_{ss}	Coefficient of particle collision recovery	—
Θ_s	Particle temperature	m^2/s^2
φ	Internal friction angle	—
I_{2D}	Second invariant of partial stress tensor	—

2.3 Simulation system and method

The two-phase system in this study is PDMS-Pt, in

which the particle size of Pt is 30 μm and the density is 21 450 kg/m^3; PDMS has a density of 1 000 kg/m^3 and a viscosity of 0.001 Pa · s. A region with a height of 2 mm is divided at the bottom of the model, and the volume fraction of solid particles is set at 30% in this region. Set the calculation time step as 0.001 s, the number of steps is 12 500, and the maximum number of iterations is 20.

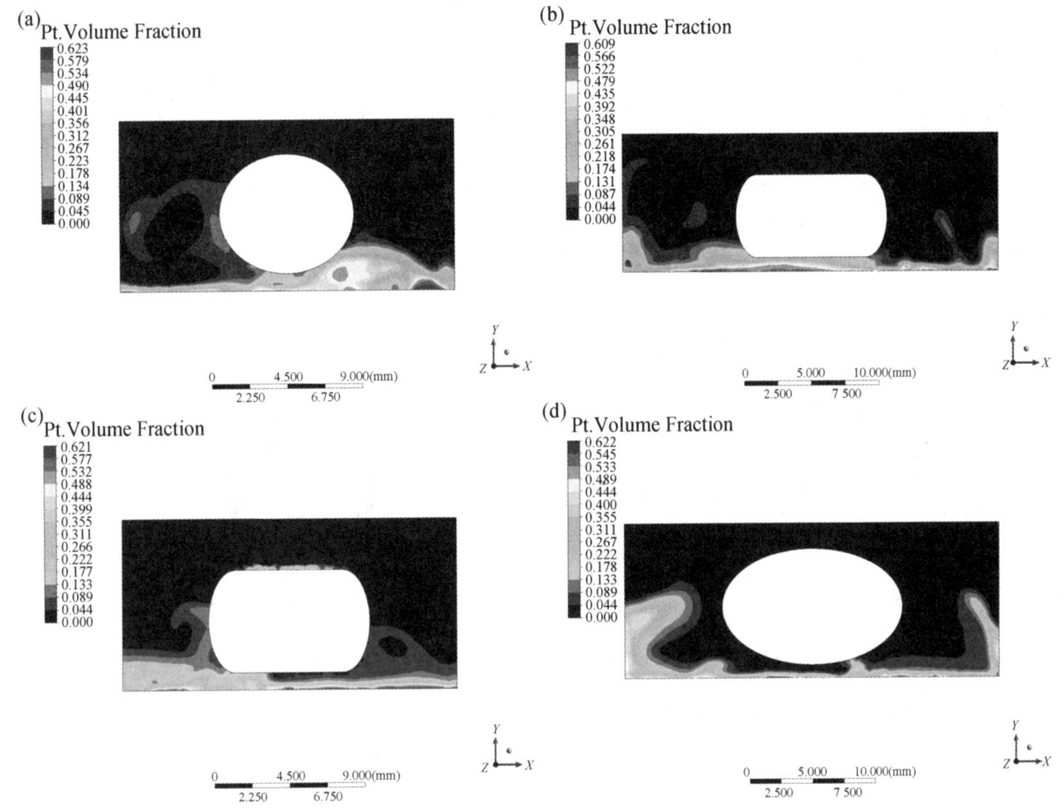

Fig. 2 Nephogram of Pt volume fraction in different models: (a) 5-A7; (b) 10-C6; (c) 5-C6; (d) 10-A9

3 Results and Discussion

According to the simulation, the cloud map of Pt's volume fraction is shown in Fig. 2, in which (a), (b), (c) and (d) are the calculated results of four models: 5-A7, 10-C6, 5-C6 and 10-A9, respectively. From the cloud image, it can be seen that Pt is mainly concentrated in the bottom of the four models, and the dispersion effect is not ideal. The influence of model size on the distribution of Pt is not obvious.

In order to further analyze the influence of model size on Pt distribution, the volume fraction of Pt in four groups of models is analyzed. From the bottom of the beaker to the lowest edge of the magneton, the distance is 1 mm, so there are three straight lines along the X axis every 0.25 mm, with L1, L2 and L3 from bottom to top, respectively. The example is shown in Fig. 3.

Fig. 4 shows the volume fraction of solid particles along L1, L2 and L3 in different models, where (a), (b), (c) and (d) respectively show the volume fraction of Pt on three straight lines in four models of 5-A7, 10-C6, 5-C6 and 10-A9. As the properties of liquid-solid two-phase materials are quite different, and the density of solid-phase granular materials is high, which is greatly influenced by gravity during stirring, the volume fraction of Pt gradually decreases from L1 to L3. Moreover, a certain centrifugal force will be generated in the stirring process, and the solid particles will accumulate and distribute to both sides under its action, resulting in the phenomenon that the volume fraction of both sides along the straight line is larger than that in the middle.

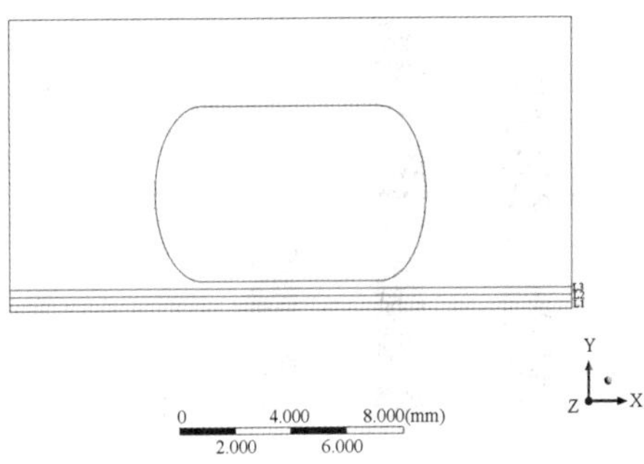

Fig. 3 Schematic diagram of straight-line position

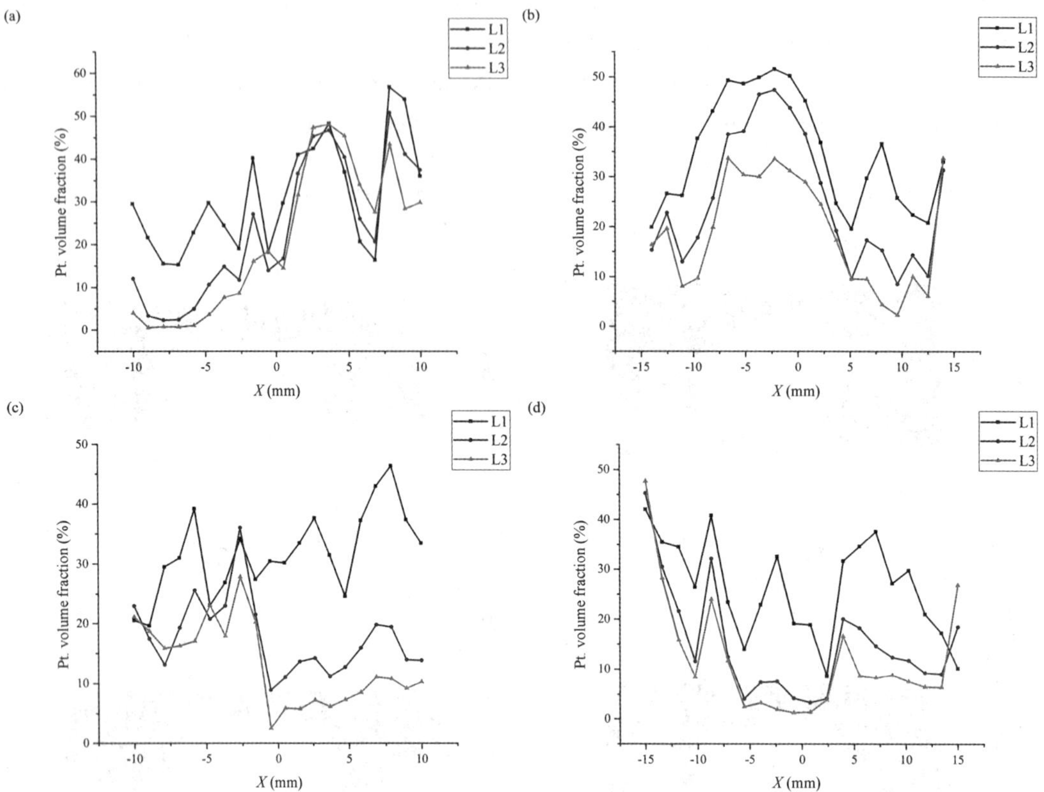

Fig. 4 Changes of Pt volume fraction in L1, L2 and L3 of each model: (a)5-A7; (b)10-C6; (c) 5-C6; (d)10-A9

Fig. 5 shows the mean and standard deviation of the volume fraction of solid particles in each model. From Fig. 5, it can be seen that the average value of Pt volume fraction in each model is less than 30% of the set value, indicating that there is a phenomenon that Pt volume fraction is larger in some places and smaller in some places in the four models. Compared with the other three models, the standard deviation of particle volume fraction in the 5 − C6 model is about 10. 336%, which is smaller than the other three models. This shows that although the Pt volume fraction in this model does not reach the target value, the difference of Pt volume fraction is small and the distribution is relatively uniform.

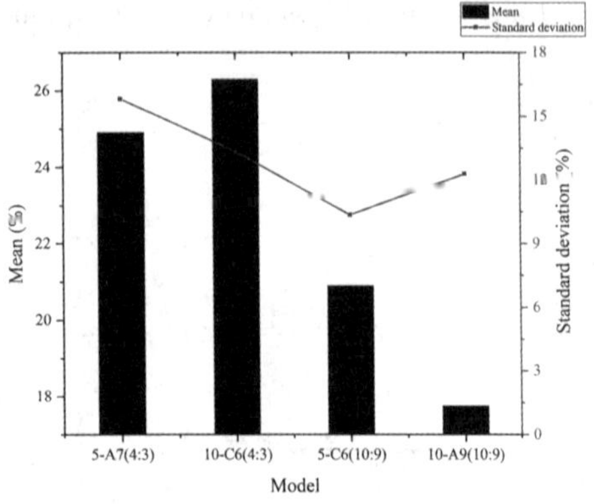

Fig. 5 Average and standard deviation of Pt volume fraction in each model

4 Conclusions

In the paper, Eulerian multiphase flow model and k-ε turbulence model were used to simulate the mixing process of conductive mixture, and the effects of different specifications of containers and stirring equipment on the mixing characteristics of two phases and the distribution of solid particles of conductive materials were analyzed. The analysis shows that the conductive particles in 5 − C6 model are distributed relatively evenly, and the mixing effect is better than other models. This simulation provides a certain reference for selecting the specifications of stirring vessel and equipment when preparing conductive mixture, but further research is needed to obtain better parameters and specifications of stirring equipment.

Acknowledgements

This research was funded by the National Natural Science Foundation of China (No. 51603039), and supported by the Fundamental Research Funds for the Central Universities, the Key Laboratory of Textile Science and Technology (Donghua University), Ministry of Education, as well as the Initial Research Funds for

Young Teachers of Donghua University.

References

[1] KASAT G R, KHOPKAR A R, RANADE V V, et al. CFD simulation of liquid-phase mixing in solid-liquid stirred reactor[J]. Chemical Engineering Science, 2008, 63(15) : 3877-3885.

[2] MICALE G, MONTANTE G, GRISAFI F, et al. CFD simulation of particle distribution in stirred vessels [J]. Chemical Engineering Research and Design, 2000, 78 (3) : 435-444.

[3] BARRUE H, BERTRAND J, CRISTOL B, et al. Eulerian simulation of dense solid-liquid suspension in multi-stage stirred vessel[J]. Journal of Chemical Engineering of Japan, 2001, 34(5) : 585-594.

[4] MARTIN J E, ANDERSON R A. Chain model of electrorheology [J]. Journal of Chemical Physics, 1996, 104 (12) : 4814-4827.

[5] WADNERKAR D, TADE M O, PAREEK V K, et al. CFD simulation of solid-liquid stirred tanks for low to dense solid loading systems[J]. Particuology, 2016, 29: 16-33.

[6] GIDASPOW D, BEZBURUAH R, DING J. Hydrodynamics of circulating fluidized beds: kinetic theory approach [J]. Engineering, 1991: 1-5.

[7] SCHAEFFER D G. Instability in the evolution-equations describing incompressible antigranulocytes flow[J]. Journal of Differential Equations, 1987, 66(1) : 19-50.

[8] LUN C K K, SAVAGE S B. A simple kinetic theory for granular flow of rough, inelastic, spherical particles [J]. Transactions of the ASME Journal of Applied Mechanics, 1987, 54(1) : 47-53.

Investigation on 3D Printing of Continuous Fiber Reinforced Thermoplastic Composite (CFRTPC)

LIU Cheng[1], QUAN Zhenzhen[1*], GU Qiujin[1], QIN Xiaohong[1]

[1] *College of Textiles, Donghua University, Shanghai 201620, China*

* *Corresponding author's email*: quan@dhu.edu.cn

Abstract: When applying the desktop FFF 3D printer and its commonly used process to print continuous fiber reinforced thermoplastic composite (CFRTPC), the fiber bundle tends to break and debond. To solve the problem, this study has investigated 3D printing technology of CFRTPC and performed fluid simulation. This paper has investigated the influence of the nozzle diameter and printing speed on the 3D printing process of CFRTPC. It has been found that the faster the printing speed and the smaller the nozzle diameter are, the more serious the phenomenon is. When the nozzle diameter becomes greater and the printing speed becomes faster, the deflection at the turn becomes more serious. The numerical simulation of fluid in the nozzle has revealed that in the contraction section and the exit section, the resin speeds up, while the local pressure decreases significantly. The pressure drop brings insufficient feeding force, making the extrusion of fiber fall behind that of the resin. Therefore, at the turn of the path, the fiber bundle will be pulled from the deposition wire which has not been completely cured under the viscous force of the resin. Based on the above conclusions, the optimization method of 3D printing process of CFRTPC is prospected.

Keywords: 3D printing; Fiber reinforced thermoplastic composites; In-situ composite printing; Fluent simulation

1 Introduction

At present, the increasing shortage of non-renewable natural resources, such as oil, makes the lightweight manufacturing of aerospace aircraft, automobile and other fields become the general trend of its development. Fiber reinforced composites have many advantages such as light weight, high specific strength, high specific modulus and strong designability, which have been required in aerospace, automobile and rail transportation[1]. Continuous fiber reinforced thermoplastic composite (CFRTPC) makes use of continuous high performance fiber and can adjust the orientation and distribution of fiber along the stress direction to improve the capability to carry load, which brings high strength weight ratio, high stiffness weight ratio, good stability as well as excellent characteristics of load carrying and transferring, making CFRTPC a good choice of producing strong and lightweight products[2].

Currently, the 3D printing process of CFRTPC can be roughly divided into pre-impregnation and in-situ composition. The former requires special pre-impregnation equipment and process[3], which increases the manufacturing cost, while the latter can be directly achieved by ordinary desktop printers. In printing practice, 3D printing CFRTPC using the same process as printing pure resin will lead to fiber deposition failure, pulling of deposited fibers and fiber damage. In order to obtain suitable process parameters for stably and successfully printing CFRTPC and to find out the causes of the above phenomenon, this work conducted a series of printing experiments based on variable nozzle diameter and printing speed and studied the flow of resin in the nozzle using the Fluent simulation tool.

2 Experiments

2.1 Materials and 3D printer

Polylactic Acid (PLA) filament with the density of 1.24 g/cm^3 and printing temperature of 190 - 210 ℃ was provided by Shenzhen Kuaizao Technology Co., Ltd. (Guangdong, China). Kevlar fiber bundles (1.44 g/cm^3, 200 D) were provided by Dongguan Shengmao Special Weaving Technology Co., Ltd. (Guangdong, China).

Snapmaker A350T 3D printer was provided by Shenzhen Kuaizao Technology Co. , Ltd. (Guangdong, China). The equipment is equipped with one single feeding port for feeding resin. The research group made an oblique hole below the wire feeding gear for fiber to enter and extrude for compositing with resin.

2.2 Preparation before printing

The model was drawn in Solidworks and then was imported into Repetier Host software and finally was sliced for nozzle diameter of 0.8 mm, 1.0 mm, 1.5 mm, 2.0 mm and the printing speed of 1 mm/s, 2 mm/s, 3 mm/s, 4 mm/s and 5 mm/s to obtain 20 Gcode files. The printing temperature was set uniformly to 210 ℃, and the printing height was about half of the nozzle diameter.

2.3 Printing

Before printing and when the nozzle was changed, the equipment should be leveled. The fiber bundle was fed into the side hole, and the resin is fed vertically through the wire feeding gear for printing. During the printing process, the printing height was adjusted dynamically according to the deposition of the material, and the adjustment amount is about ±0.05 mm. The fiber deposition failure, pulling of deposited fibers, fiber damage mainly occurs at the turn of the printing path. Therefore, this experiment studied the printing situation of CFRTC at different printing speeds and nozzle diameters by printing several 90° turning points.

3 Simulation

In the in-situ 3D printing of CFRTC, the resin matrix is actively fed at a certain speed by feeding gear and the fiber is transported by the viscous force between them. The flow of the resin is key to the printing process. Therefore, the study of the resin flow helps to understand the 3D printing process of CFRTC.

The theoretical study on the flow of resin in the nozzle started from Phan et al. [4], who studied the flow and heat transfer mechanism of resin. Recently, finite element simulation tools becomes popular, the researchers have used fluid simulation software to study the resin flow in the hot end, getting to know the melting process of resin, the recirculation zone between the resin and wall, the pressure distribution, the velocity distribution, the temperature distribution, the viscosity distribution, the local shear rate, extrusion force and so on[5-7], the results of which can guide the optimization of process parameters, the improvement of printer and the design of printer parts.

This paper aims to study the influence of printing speed and nozzle diameter on resin flow. Ansys Workbench 2019 R2 of Ansys Company was used for simulation. Model Designer was used for modeling. Meshing was used for Meshing, and Fluent was used as a solver. Cfd Post was used for post-processing. Origin 2018 was used to draw diagrams after exporting post-processing data.

3.1 Performance parameters of materials

For PLA, the densities in molten state and solid state are 1 070 kg/m^3 and 1 250 kg/m^3 respectively[8], the liquidus temperature is 428 K[4], and the solid phase temperature is 40 - 50 ℃ higher than T_g[9], taking 373 K, melting (100% crystallization) enthalpy is 93 600 J/kg[10], specific heat capacity is 1 800 J/(kg · K)[11], thermal conductivity is 0.20 W/(m · K)[4]. For air, the default parameter in Fluent is used, that's to say, the air is set as incompressible ideal gas with specific heat of 1 006.43 J/(kg · K) and heat conduction coefficient of 0.024 2 W/(m · K)[12].

3.2 Preprocessing and solving set up

The domain inside the nozzle was modeled and divided into a mixed grids dominated by quadrilateral grids with the size of 0.02 mm. The total number of grids was 79 332 and the average grid quality was 0.963 94.

The models used for solving the problem are: multiphase flow volume of fluid (VOF) model; energy equation; laminar flow model; radiation model (S2S); melting and solidification model, the solution method is PISO, considering gravity.

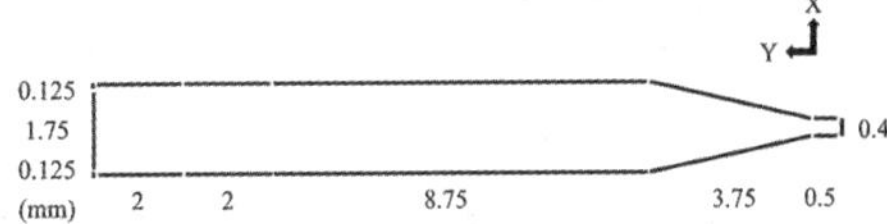

Fig. 1 Nozzle fluid domain model and boundary conditions

4 Results and Discussion

4.1 Results of printing experiments

The prints are shown in Tab. 1, where v stands for printing speed and D stands for nozzle diameter.

It can be seen that at lower printing speeds, such as 1 mm/s, nozzles with different diameters can print stably and successfully. When the speed increases, the fiber bundle begins to be partially pulled off or damaged. This phenomenon is more significant for the nozzle with small diameters, and the debonding fiber is longer, which may be caused by the small amount of extruding resin, that is, the surface fiber of the deposited wire is not consolidated by the resin. For cases with large nozzle diameters, in addition to the above rules, there is also a deviation of the fiber bundle from the midline of the path at the turn. This is because the steering of the fiber bundle during deposition is mainly controlled by the edge of the nozzle. The larger the diameter of the nozzle is, the larger the free movement space of the fiber bundle is, the easier it is to deviate from the midline.

Tab. 1 Prints at different printing speeds and nozzle diameters

v(mm/s) / D(mm)	1	2	3	4	5
0.8	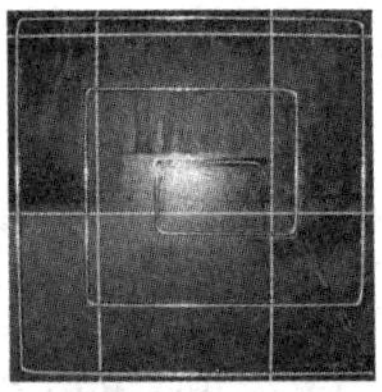	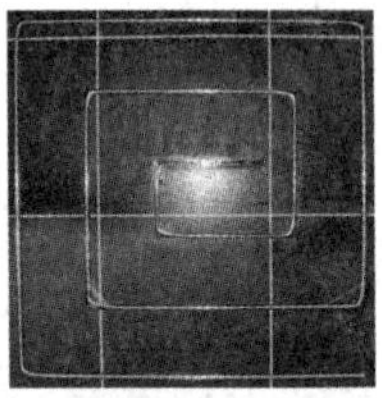	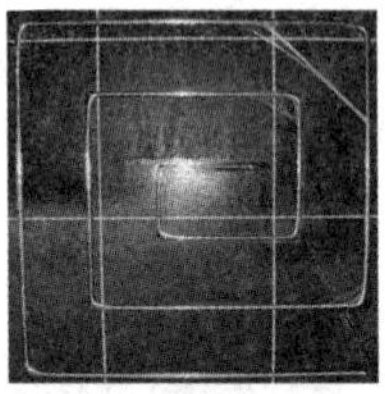	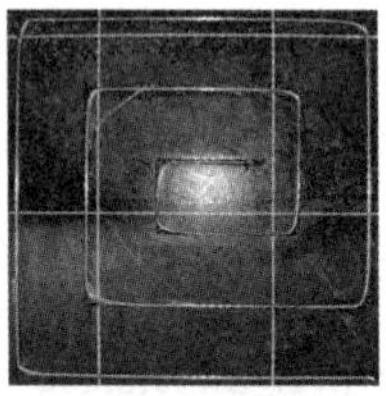	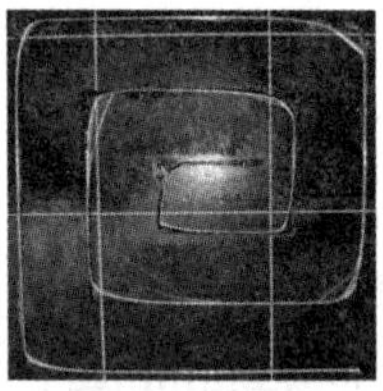
1.0	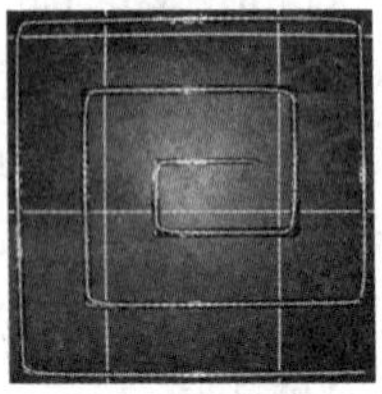	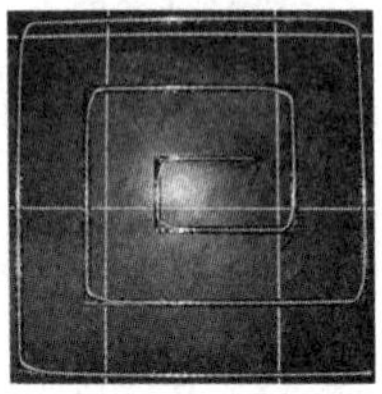	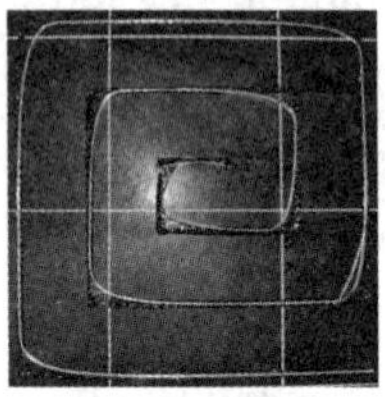	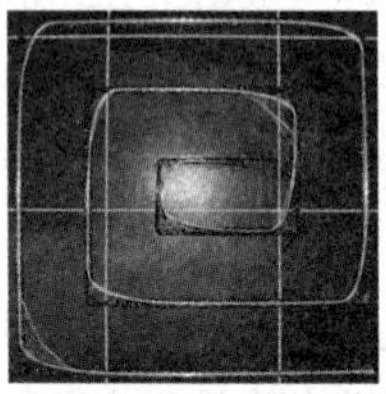	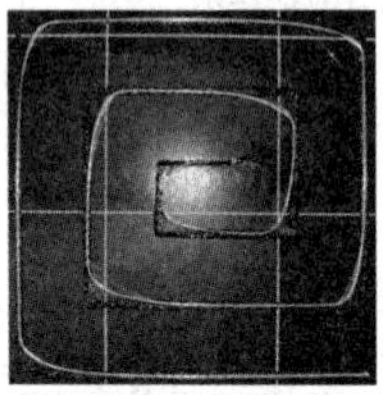
1.5	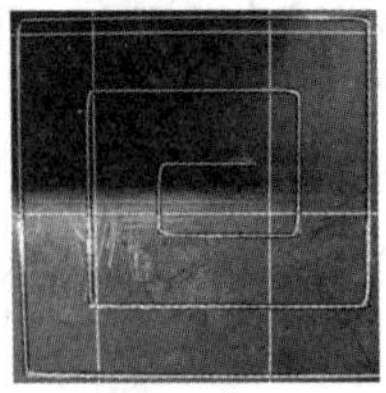	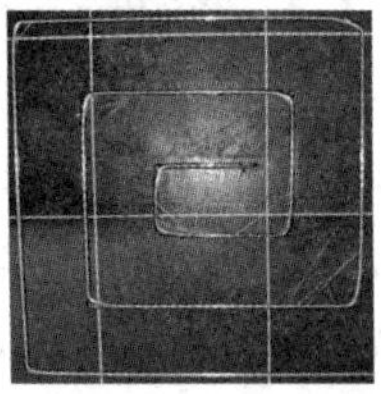	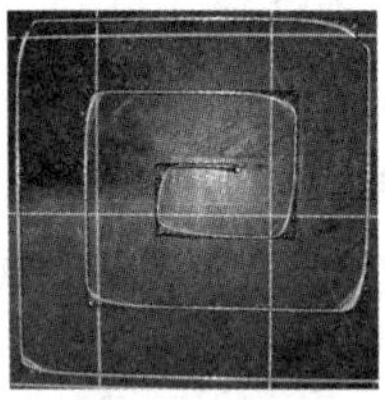	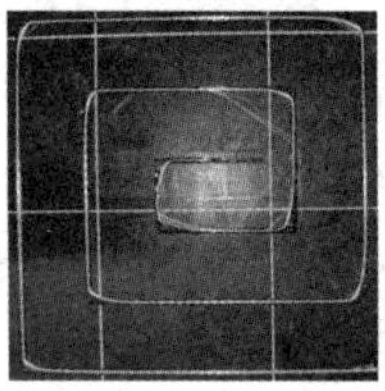	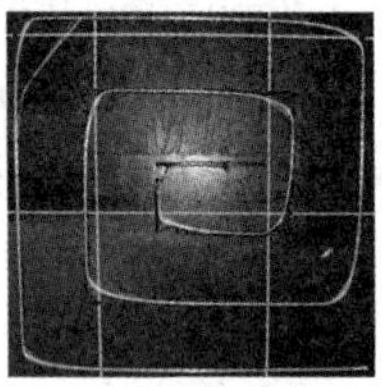
2.0	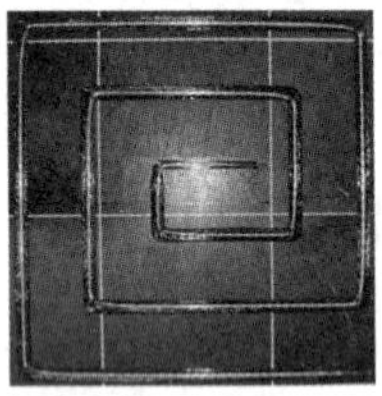	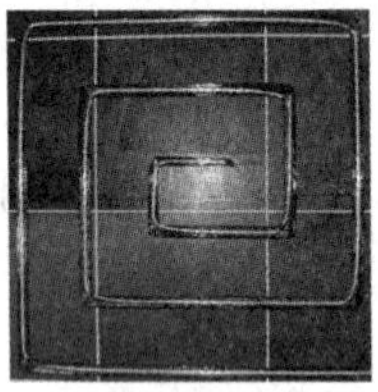	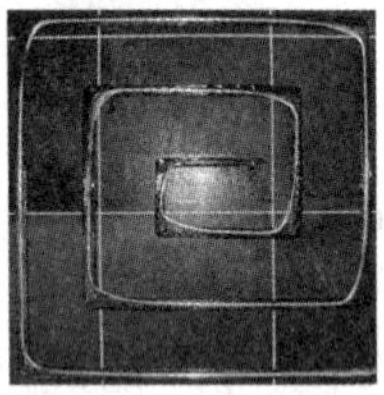	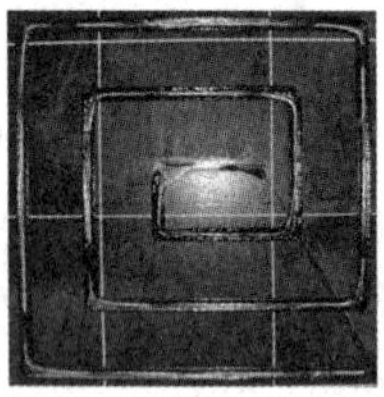	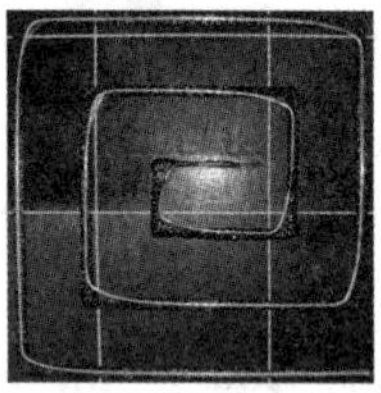

4.2 Results of simulation

At the nozzle diameter of 0.4 mm and the printing speed of 1 mm/s, 2 mm/s, 3 mm/s, 4 mm/s and 5 mm/s, the ways the pressure and the ratio of printing speed to feeding speed changes with the height of the nozzle are shown in Fig. 2.

From Fig. 2 (a), it can be seen that from the upper inlet to the outlet below, the pressure decreased, especially in the outlet segment with the vertical height of 0.5 mm. As the printing speed increases, the pressure at the inlet increases almost linearly, resulting in a larger pressure drop at the vertical outlet. As can be seen from Fig. 2 (b), the velocity remains relatively stable in the vertical section, but increases sharply in the contraction section, and increases slowly in the vertical outlet section. The outlet section velocity is about 5 times of the feeding speed. For different feeding speeds, the curves of the ratio of outlet velocity to feeding velocity changing with vertical height almost overlap.

At the printing speed of 1 mm/s and the nozzle diameter of 0.4 mm, 0.6 mm, 0.8 mm, 1.0 mm, the ways the pressure and the outlet speed changes with the height of the nozzle are shown in Fig. 3.

It can be seen from Fig. 3 (a) that the pressure variation trends of different nozzle diameters are similar. The larger the nozzle diameter is, the smaller the initial pressure in the inlet section and the smaller the pressure drop. The feeding speed affects the initial inlet pressure more than the nozzle diameter. It can be seen from Fig. 3 (b) that for different nozzle diameters, the acceleration stage of resin mainly occurs at the contraction section. The outlet velocity decreases with increasing nozzle diameter.

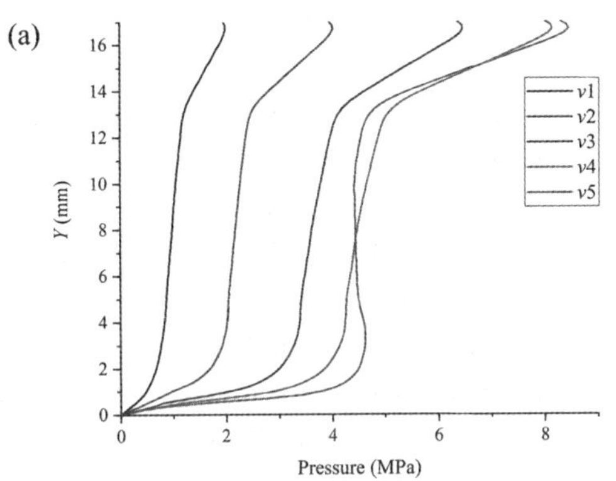

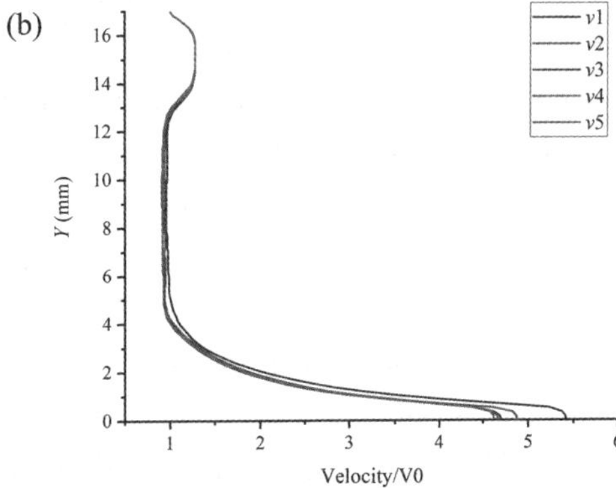

Fig. 2 (a) Pressure and (b) ratio of outlet velocity to feeding speed vary with nozzle height at different printing speeds

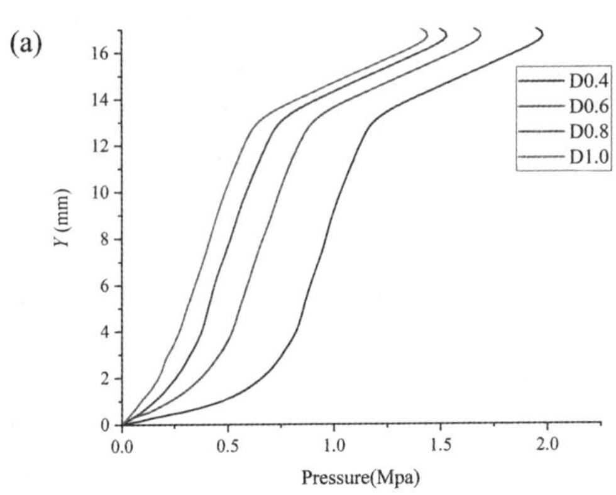

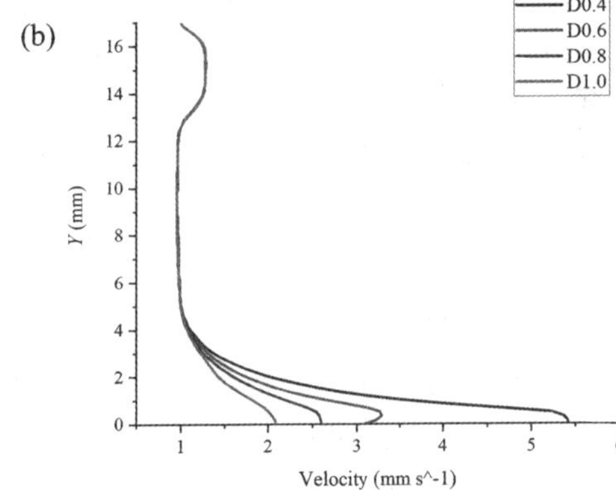

Fig. 3 (a) Pressure and (b) outlet velocity vary with nozzle height under different nozzle diameters

4.3 Comprehensive analysis

The simulation results show that the resin flow at the nozzle exit is accelerated flow with dropping pressure. In the 3D printing of CFRTPC, the resin is actively fed through the wire feeding gear and serves as the fiber conveying medium. When the pressure at the outlet end drops, the extrusion force of the fiber bundle decreases, while the resin flows out at an accelerated speed. There is an obvious speed difference between them. Thus, the synchronous extrusion of the resin and fiber bundle depends largely on the consolidation force of the deposited part to the fiber bundle. When using a nozzle with a small diameter, the amount of resin extruded is less, coupled with the incomplete impregnation of the resin onto the fiber bundle, the consolidation of the fiber bundle is limited[13], fiber bundle break and fiber debonding are more likely to occur. As the printing speed increases, the exit speed increases by the same proportion, and the pressure in the nozzle decreases more, the speed difference between the fiber bundle and the resin becomes larger correspondingly, so the fiber is more likely to debond and being broken. When the nozzle diameter is larger, the resin extruded is more, the exit speed increase and the pressure drop are smaller, the deposition and consolidation of the fiber bundle are relatively smooth, but there is path deviation at the turn.

5 Conclusions

This paper has conducted a series of printing experiments of continuous aramid fiber reinforced poly (lactic acid) (PLA), based on varying nozzle diameters and printing speeds. It was found that when using nozzles with larger diameters and smaller printing speed, the print process can go smoothly, and when applying nozzles with relatively smaller diameters and larger printing speed, the fiber bundle is prone to separate from the resin and be damaged. This paper also has carried on the simulation research of the resin flow inside the nozzle. It is shown that, the flow speeds up at the exit where the pressure drops, and when the printing speed is faster, the diameter of nozzle is smaller, the above phenomenon is more obvious, which causes non-simultaneous extrusion of the resin and the fiber, requiring stronger consolidation strength provided by the deposited part to achieve synchronization extrusion. When the consolidation force of deposited wire to fiber bundle is insufficient, the viscous force of resin to monofilaments or the friction force of nozzle edge to monofilaments is too large, resulting in the debonding and damage of fiber. Based on the above conclusions, three technological optimization methods can be adopted for 3D printing of CFRTPC: using large diameter nozzles or pre-impregnated fiber bundles; decreasing the printing speed at the turn when planning the printing path; accelerating the cooling of the deposited part.

Acknowledgements

This work was supported by the Fundamental Research Funds for the Central Universities (Nos. 2232021G-02 and 2232020D-15), the Open Fund of Shanghai Key Laboratory of Lightweight Composite (No. 2232020A4-01), the Young Elite Scientists Sponsorship Program by CAST, and the Fund of National Engineering and Research Center for Commercial Aircraft Manufacturing (Nos. F-CXKT-0301C, and COMAC-SFGS-2021-667) to Prof. Zhenzhen Quan, and the Fundamental Research Funds for the Central Universities and Graduate Student Innovation Fund of Donghua University (No. CUSF-DH-D-2019059) to Ms. Qiujin Gu.

References

[1] LI J, DURANDET Y, HUANG X, et al. Additively manufactured fiber-reinforced composites: a review of mechanical behavior and opportunities[J]. Journal of Materials Science and Technology, 2022, 119: 219-244.

[2] GOH G D, DIKSHIT V, NAGALINGAM A P, et al. Characterization of mechanical properties and fracture mode of additively manufactured carbon fiber and glass fiber reinforced thermoplastics[J]. Materials & Design, 2018, 137.

[3] JIA M Y, LI C X, XUE P, et al. Research on the melt impregnation of continuous carbon fiber reinforced nylon 66 composites[J]. IOP Conference Series: Materials Science and Engineering, 2016, 137(1).

[4] PHAN D D, SWAIN Z R, MACKAY M E. Rheological and heat transfer effects in fused filament fabrication[J]. Journal of Rheology, 2018, 62(5): 1097-1107.

[5] UFODIKE C O, NZEBUKA G C. Investigation of thermal evolution and fluid flow in the hot-end of a material extrusion 3D Printer using melting model[J]. Additive Manufacturing, 2022, 49: 102502.

[6] KATTINGER J, EBINGER T, KURZ R, et al. Numerical simulation of the complex flow during material extrusion in fused filament fabrication[J]. Additive Manufacturing, 2022, 49: 102476.

[7] PHAN D D, HORNER J S, SWAIN Z R, et al. Computational fluid dynamics simulation of the melting process in the fused filament fabrication additive manufacturing technique[J]. Additive Manufacturing, 2020, 33: 101161.

[8] SERDECZNY M P, COMMINAL R, PEDERSEN D B, et al. Experimental and analytical study of the polymer melt flow through the hot-end in material extrusion additive manufacturing[J]. Additive Manufacturing, 2020, 32: 100997.

[9] COLÓN QUINTANA J L, HIEMER S, GRANDA DUARTE N, et al. Implementation of shear thinning behavior in the fused filament fabrication melting model: analytical solution and experimental validation[J]. Additive Manufacturing, 2021, 37: 101687.

[10] PYDA M, BOPP R C, WUNDERLICH B. Heat capacity of poly(lactic acid)[J]. The Journal of Chemical Thermodynamics, 2004, 36(9): 731-742.

[11] RUDOLPH N, CHEN J, DICK T. Understanding the temperature field in fused filament fabrication for enhanced mechanical part performance[J]. AIP Conference Proceedings, 2019, 2055: 140003.

[12] Ansys Fluent 2019 R2. Fluent User's Guide[M]. 1st Edition Pittsburgh: ANSYS Inc., 2019.

[13] TENGFEI L, XIAOYONG T, YAYUAN Z, et al. High-pressure interfacial impregnation by micro-screw in-situ extrusion for 3D printed continuous carbon fiber reinforced nylon composites[J]. Composites Part A: Applied Science and Manufacturing, 2020, 130(C).

Effect of Structural Parameters of Wovens on the Performance of Flexible Keyboard Fabrics

LIU Saidi[1], ZHUANG Qinliang[1*]

[1] *Department of Textile Engineering, College of Textiles, Donghua University, Shanghai 201620, China*

[*] *Corresponding author's email*: qlzhuang@ dhu. edu. cn

Abstract: With the development of intelligent textiles for wearable pieces of equipment, the flexible keyboard is getting attention. The original textile-based keyboard usually consists of 3−5 individual fabrics. They are laminated together to form a flexible keyboard, which is cumbersome and inefficient. This method of forming is only feasible on a laboratory scale, greatly limiting the popularity and application of fabric keyboards. In this paper, the fabrication of keyboard fabric with three-dimensional structure is designed and described, which can serve to simplify the weaving process and improve its functional stability. And the fabricating method has shown a great potential for mass production. In this study, the effect of structural parameters of the fabric, i. e., supporting width, supporting height and cavity width, on the fabric's compression resilience and the support strength of the support part and the contact pressure of the fabric are investigated in detail. The results show that the three structural parameters have significant effects on all three performance indicators, and it has been found that optimum structural parameters exist for the best performance of keyboards fabric.

Keywords: Flexible keyboard; Spacer fabrics; Three-dimensional fabrics; Compression resilience

1 Introduction

The textile-based keyboard has a series of advantages such as light, soft and foldable. The original flexible keyboard was made of several layers of independently woven fabric (mainly including upper and lower conduction and insulation layers) laminated together, which had non-negligible defects in the manufacturing process and functional stability. In seeking a more efficient manufacturing process and more stable performance, attempts have been made to apply special fabric structures to fabric keyboards. Spacer fabric is a special three-dimensional fabric, when its surface layer is under pressure, the middle support part can play a role in dispersing the load and reducing stress concentration so that the fabric can resist a certain degree of pressure and present a compression resilience[1-3]. Comparing the structure of fabric keyboard and spacer fabric, the upper and lower layers of the spacer fabric can be used as the upper and lower conductive layers of the fabric keyboard, while the supporting part is used as the insulating layer[4-5], as shown in Fig. 1. In this paper, we will briefly introduce the design method of spacer fabric applied to flexible keyboard and weave the specimens. The effect of structural parameters on the performance of the fabric keyboard will be investigated by testing specimens of different sizes and the optimum structural parameters for the flexible keyboard will be obtained.

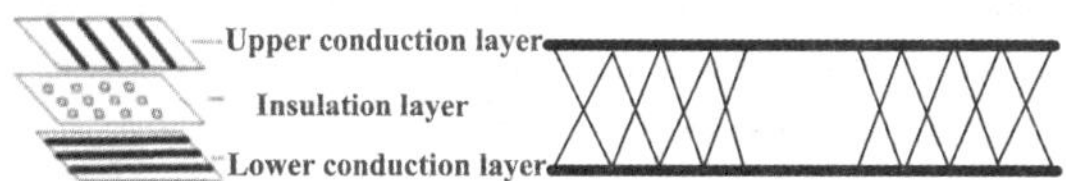

Fig. 1 Comparison of original fabric keyboard functional layers and spacer fabric structure

2 Experiments

2.1 Fabrication of the spacer fabric for flexible keyboard

The pattern design of the spacer fabric includes the design of the face layer and the intermediate layer. In this paper, we choose plain weave as the surface layer, and the "X" shaped crossed yarn is selected as the support form, which can reduce the weaving difficulty and simplifies the weaving process, thus increasing production efficiency[6]. The structural parameters, i. e. supporting width, supporting height and cavity width, are designed according to the above selection of surface pattern and support form. The meaning of each parameter is referred to the following Fig. 2, and the design of each part of the fabric and the levels of each parameter selected is shown in Tab. 1.

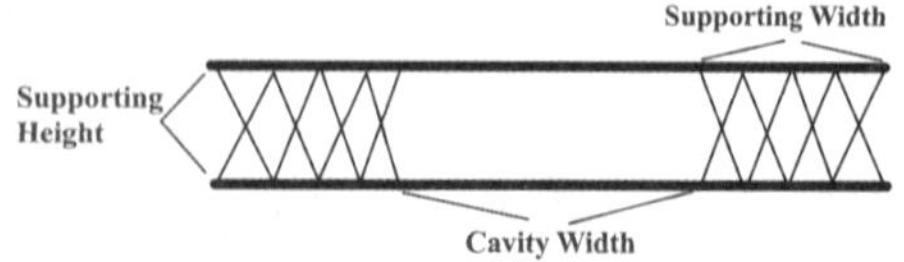

Fig. 2 Meaning of fabric structure parameters

Tab. 1 Structural parameters and levels of woven specimens

Surface layer	Supporting form	Supporting width (mm)	Supporting height (mm)	Cavity width (mm)
Plain weave	"X" shape	0.5	2	8
		1	4	11
		2	6	14
		3	8	17

The design of the spacer fabric has to consider both the face layer and the middle layer, which is difficult to design with the traditional design method. It is necessary to draw the weft profile of the fabric first (such as Fig. 3), and the regular of each warp yarn is observed before drawing the pattern draft[7-8].

Fig. 3 The weft profile of the fabric (circles and numbers represent weft yarns and weft yarn order, lines marked with Arabic numerals are top warp yarns, and lines marked with Roman numerals represent spacer yarns)

Observe the interweaving rules between warp and weft yarns in Fig. 3, and draw the pattern draft on the point

Fig. 4 Pattern draft of the fabric with 3 mm supporting width and 11 mm cavity width (Arabic numerals indicate warp yarns, and Roman numerals indicate spacer yarns)

paper. Fig. 4 shows the pattern draft of the fabric with 3 mm supporting width and 11 mm cavity width.

After comparing several yarn materials, the polyester yarn with linear density of 29.2 tex was selected. Double warp beam feeding by weight control and weaving with the required supporting height under the action of the height setter. The weaving technology is shown in Tab. 2, and the woven specimens are shown in Fig. 5 below.

Tab. 2 Weaving processing parameters

Cloth width (cm)	Warp density (end/10 cm)	Warp numbers (end)	Reed size	Number of ends per dent
5	540	540	62	8

Fig. 5 Finished specimens

2.2 Compression resilience test of fabrics

In this paper, three structural parameters are selected, and each parameter is selected at four levels. In order to improve efficiency and save materials, the parameters and levels were designed orthogonally and 16 specimens were finally selected for testing.

The experiment tested three indicators, namely resilience recovery rate, support strength and contact pressure. The resilience of elasticity reflects the compressive resilience of the fabric. Support strength reflects the maximum strength that the support part and the cavity part can withstand. These two indicators are tested by applying the presser foot of the instrument to a cavity and both sides of the support part. The test of contact pressure is to apply the presser foot to the middle part of a cavity, reflecting the pressure required for contact between the top and bottom. All indicators are tested by the compression instrument equipped with LLY-06D compression test system, manufactured by Laizhou Electronic Instruments Co. Ltd. (Shandong, China). After debugging and calibrating the instrument, the specimen was tested using the constant distance compression system of the tester. Set the presser foot up and down the speed to 15 mm/min, contact tension and starting force value to 0. 1 cN. Ten test points were randomly selected on each specimen and compressed into contact with the upper and lower layers.

3 Results and Discussion

3.1 Effect of structural parameters on compression resilience of fabrics

Orthogonal analysis of the resilience recovery data was performed. The results showed that all three structural parameters had significant effects on the resilience recovery performance of the fabrics.

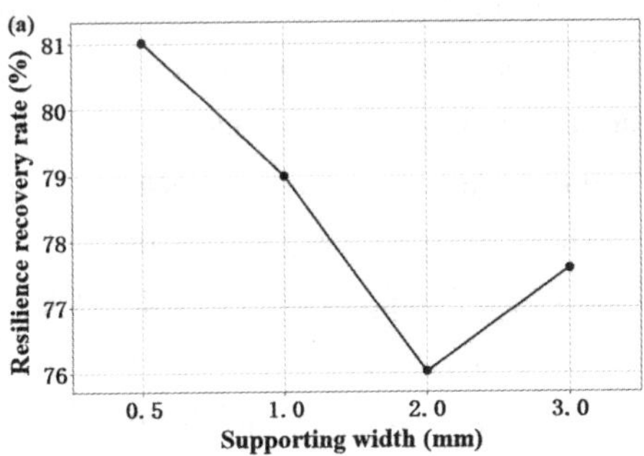

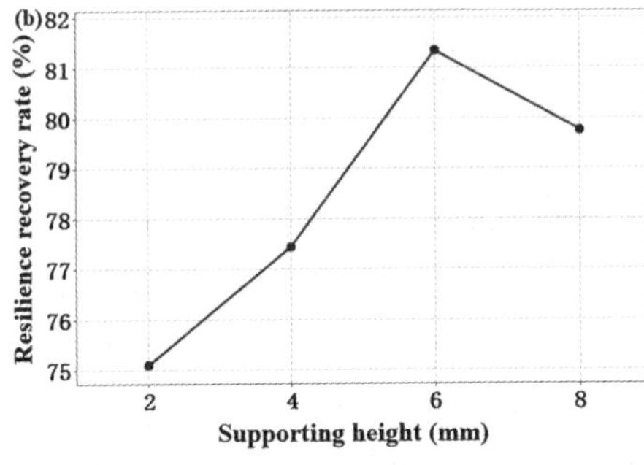

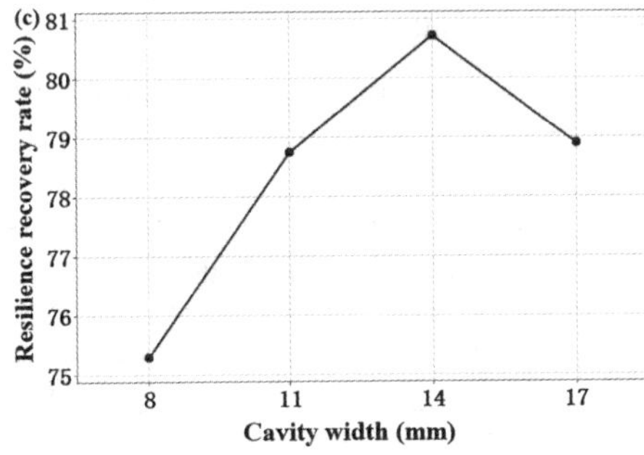

Fig. 6 (a) Mean value of resilience recovery rate at different supporting widths; (b) mean value of resilience recovery rate at different supporting heights; (c) mean value of resilience recovery rate at different cavity widths

It can be seen from Fig. 6 (a) that the resilience recovery rate is highest at the supporting width of 0. 5 mm, and the resilience recovery rate is lower than this level when the width of the support increases. This is because an increase in supporting width may cause friction between the spacer yarns and hinder resilience recovery[9]. From Fig. 6 (b), it can be seen that the resilience recovery rate tends to increase and then decrease as the supporting height increases, and the fabric shows the best compression recovery performance when the supporting height is 6 mm. When the height is too high, the support yarn tilts to one side under pressure. Even if the pressure is withdrawn, the resulting tilt is not easily recovered, so the resilience recovery rate decreases. From Fig. 6 (c), it can be seen that the cavity width increases and the resilience recovery increases first and then decreases, and the best compression resilience performance is achieved when the width is 14 mm, which indicates that the cavity part can share the pressure when it is properly enlarged. However, the cavity part collapse because the role of the support part becomes limited.

3.2 Effect of structural parameters on supporting load of fabrics

The results of the orthogonal analysis show that all three structural parameters had a significant effect on the support strength.

As can be seen from Fig. 7 (a), the support load increases with the increase of the supporting width, and the support load can reach more than 700 cN when the supporting width reaches 3 mm, it shows that the support part brings stronger support ability. Fig. 7 (b) shows that the support load is enhanced with the in-

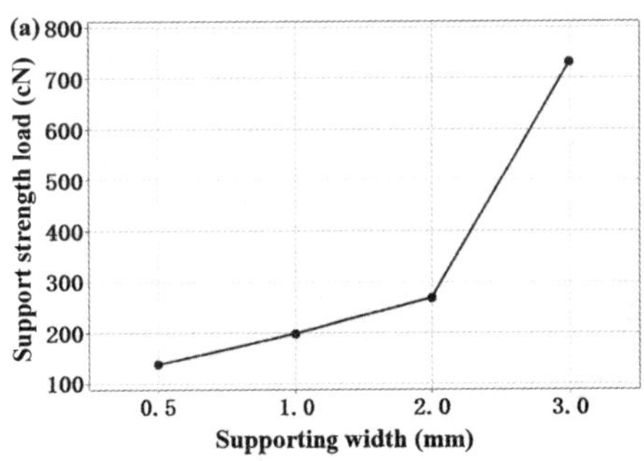

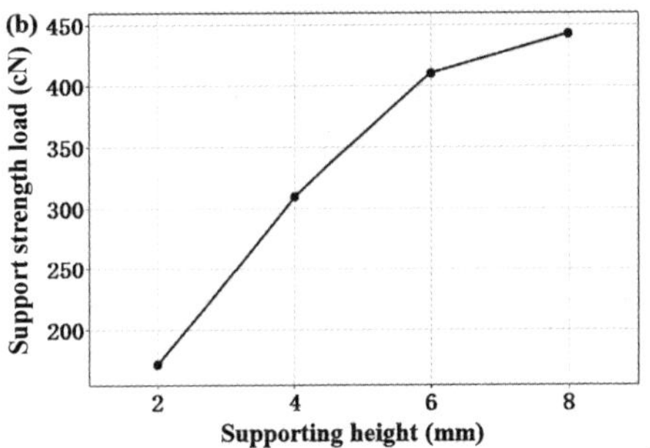

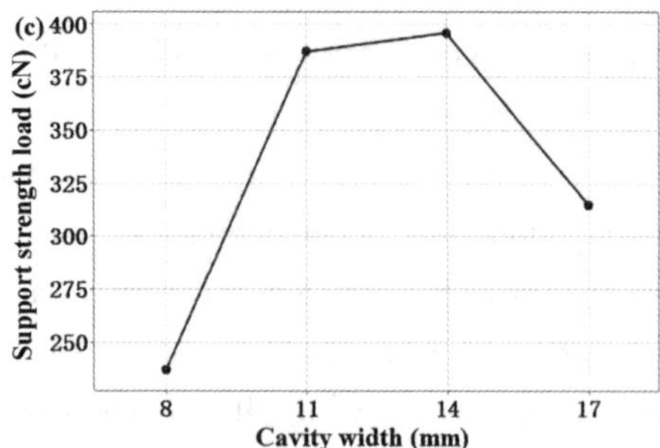

Fig. 7 (a) Mean value of support strength at different supporting widths; (b) mean value of support strength at different supporting heights; (c) mean value of support strength at different cavity widths

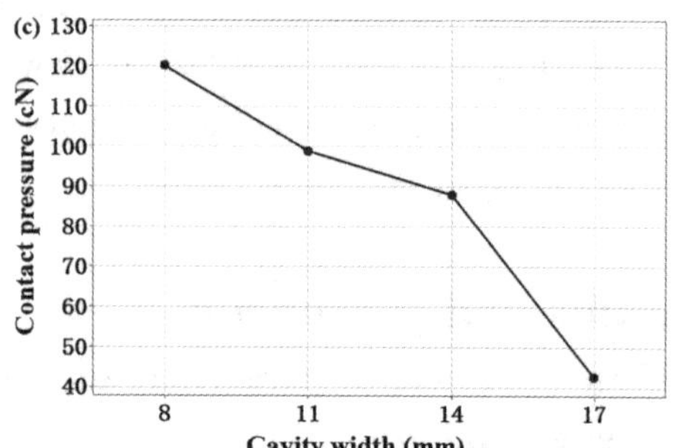

Fig. 8 (a) Mean value of contact pressure at different supporting widths; (b) mean value of contact pressure at different supporting heights; (c) mean value of contact pressure at different cavity widths

crease of the supporting height. This means that the increase in supporting height can act as a cushion against pressure. As can be seen in Fig. 7 (c), the support load tends to increase and then decrease as the cavity width increases, reaching the strongest at 14 mm. This indicates that the cavity width can play the role of pressure sharing, but too wide will limit the support capacity of the support part.

3.3 Effect of structural parameters on contact pressure of fabrics

After studying the effect of the three parameters on contact pressure, it was found that all three parameters had a significant effect on contact pressure.

It is clear from Fig. 8 (a) that the contact pressure increases with the increase of the supporting width, indicating that the increase of the width helps to enhance the support strength. As can be seen in Fig. 8 (b), as the supporting height increases, the contact pressure shows a trend of increasing and then decreasing, and the contact pressure reaches the maximum at a height of 6 mm. This means that the increase in supporting height cushions the pressure, but if the height is too high the yarn in the support section will tilt so that the support capacity decreases. From Fig. 8 (c), we can see the contact pressure decreases as the cavity width increases, and the decrease becomes larger as the cavity width increases to 17 mm. This indicates that the increase in cavity width makes the support part limited in its ability to support the cavity part.

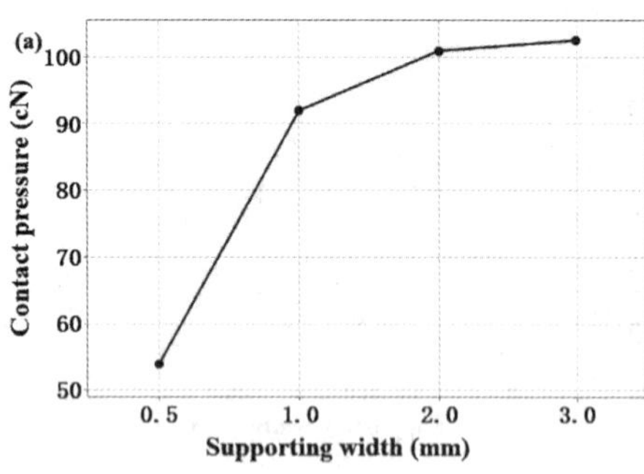

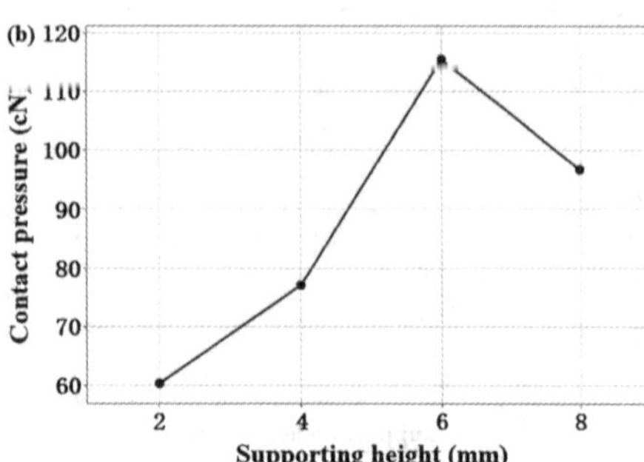

3.4 Optimization of structural parameters of fabric

Since the experiment involves three indicators, in order to compare the importance of each indicator and the performance under each parameter, the ranking scoring method is used to determine the priority and level of the parameters. The results of the extreme difference analysis of the queueing scoring method obtained are shown in Tab. 3.

Tab. 3 The results of the extreme difference analysis of the queueing scoring method

	A	*B*	*C*
K_1	54.33	50.89	64.76
K_2	62.62	61.24	51.69
K_3	57.32	68.62	65.19
K_4	76.17	69.70	55.92
k_1	13.58	12.72	16.19
k_2	15.65	15.31	12.92
k_3	14.33	17.16	16.30
k_4	19.04	17.42	13.98
R	5.46	4.70	3.38

($K_1 \sim K_4$: sum of the scores at the levels of parameters 1, 2, 3, and 4; $k_1 \sim k_4$: mean of the composite scores; A: supporting width; B: supporting height; C: Cavity width)

From the R values in Tab. 3, it can be concluded that the order of factors affecting the overall performance of the fabric are supporting width, supporting height, and cavity width. And the optimum structural parameters scheme is $A_4B_4C_3$, which means the supporting width is 3 mm, supporting height is 8 mm, and cavity width is 14 mm.

Based on the optimal structural parameters obtained from the above analysis to weave the specimens, the resilient response rate was measured to be 89. 14%, while the support strength was 1215. 56 cN, and the contact pressure was 140. 8 cN. It can be seen that all the properties were significantly improved.

4 Conclusions

This paper investigates the effects of different structural parameters on compressive resilience, support strength and contact pressure of the fabric by conducting a compressive test on the fabric. The results showed that all three structural parameters had significant effects on measured indexes. An increase in supporting width results in increased friction between the yarns, so the resilience recovery rate decreases. But the support strength and contact pressure increase due to the increased support capacity provided by the support part. When the supporting height increases, the support strength increases. However, too high a height causes the yarn to tilt, which reduces the support capacity. Therefore, the resilience recovery rate and the contact pressure both show an increase and then a decrease, and the increase of the support strength becomes slower. A moderate increase in the width of the cavity causes an increase in the elastic recovery rate and support strength. But too wide will cause a decreasing trend due to the limited support capacity of the support part. The contact pressure decreases as the cavity width increases. This paper finally analyzed the combined effect of three parameters to obtain the optimal structural parameters, i. e. , 4 mm supporting width, 8 mm supporting height and 14 mm cavity width. And it was measured that the fabric properties under the optimal structural parameters were all significantly improved.

References

[1] GU P. Fabric Weaves and Structures[M]. 2nd Edition. Shanghai: Donghua University Press, 2009.

[2] CHEN M Y, SUN R J, ZHANG C Q, et al. Pressure reduction property of warp-knitted spacer fabric[J]. Journal of Textile Research, 2019, 40(7): 58-63.

[3] DELAVARI K, DABIRYAN H. Mathematical and numerical simulation of geometry and mechanical behavior of sandwich composites reinforced with 1×1-Rib-Gaiting weft-knitted spacer fabric: compressional behavior[J]. Composite Structures, 2021, 268.

[4] HUANG X M, TANG H, ZHANG C C, et al. Integral structure and weaving of keyboard fabric[J]. Journal of Textile Research, 2010, 31(8): 45-48.

[5] YIP J, NG S-P. Study of three-dimensional spacer fabrics: Physical and mechanical properties[J]. Journal of Materials Processing Technology, 2008, 206: 359-364.

[6] HUANG J B, ZHU C Y, ZHANG H X, et al. Design of three-dimensional spacer fabrics based on rapier looms [J]. Journal of Textile Research, 2021, 42(6): 166-170.

[7] ZHONG Z L, SUN H, SHI R X, et al. Design and weaving of three-layer woven spacer fabric[J]. Shanghai Textile Science & Technology, 2020, 48(5): 46-49.

[8] LYU L H, SHI Y L, ZHANG X F, et al. Design method of three sections three-dimensional spacer woven fabric [J]. Cotton Textile Technology, 2017, 45(4): 76-79.

[9] WANG D, LIU G, LU D D, et al. Study on compression properties of spacer fabric based on viscoelastic model[J]. Journal of Textile Science and Engineering, 2022, 39(2): 1-6.

Effect of Heat Setting on Fabric Density and Elasticity of Knitted Fabrics

YANG Xiuyi[1], **YANG Xudong**[1*], **LI Qiao**[1]

[1] *Key Laboratory of Textile Science & Technology, Ministry of Education, College of Textiles, Donghua University, Shanghai 201620, China*

[*] *Corresponding author's email*: xdyang@ dhu. edu. cn

Abstract: Heat setting is a process to make the fabric more dimensionally stable. Most of the fabric will adopt the heat setting process to improve its physical characteristics or mechanical performance. In the heat-setting process, temperature, time, and tension of the fabric are main parameters. This paper firstly selected a pure cotton fabric and an elastic fabric containing polyurethane yarns in the knitted structure. Heat setting was then performed on the fabrics to identify the fabric density and fabric dimension after being stretched. The parameter of time was finally compared to optimize the physical properties of the knitted fabric.

Keywords: Heat setting; Fabric parameters; Knitted fabrics.

1 Introduction

Heat-setting process usually refers to a process that makes the fabric structure and shape more stable under the action of certain temperature and external force[1-3]. The main factors affecting heat setting include temperature, time, tension, etc.[4-5]. The heat setting of the fabric is mainly divided into four stages. The first stage is the heating stage, in which the surface of the fabric is heated to the setting temperature. The second stage is the thermal equilibrium stage, in which the heat energy penetrates into the fiber, so that the fiber surface and interior reach the same set temperature. Third stage for molecular adjustment phase, the fibers are under stress, when reaching the design temperature, the weaker secondary bond in the fiber structure is damaged, and fiber rearranges the molecular chain. The fourth stage for cooling stage, fabric is rapidly cooling, so the shape of the fabric is in accordance with fixed fiber molecular state of the new arrangement[6-7]. In the process of heat-setting process, fiber structure was readjusted, and the original molecular orientation and crystallinity of the dimensional stability increases. The shape is more stable. Physical and mechanical properties are improved. The density of the fabric has also changed[8], and fabric density is one of the main factors to decide the fabric structure. Fabric elasticity is an important mechanical property of woven fabrics. Therefore, this paper focuses on the effect of thermal setting on fabric density and fabric elasticity.

2 Experiments

2.1 Materials

Cotton plain woven fabric. Spandex elastic plain woven fabric (3% spandex, 97% polyester). Cloth mirror (9005A). Household small iron (YD18A07J-12).

2.2 Fabric treatment

Cotton fabric and spandex fabric were cut into 21 square samples of the same size of 10 cm×10 cm, and then the samples of each fabric were divided into two groups. One group was used for the experiments to measure fabric density, while the other group was used for the experiments to detect fabric elasticity.

First, the first group of samples were taken out and arranged neatly, and the density was measured and recorded with a cloth mirror. The iron was heated to 210 ℃, and the samples were placed under the heated iron for 1 min, 5 min, and 10 min, respectively. Then the samples after heat setting were taken out and the density was measured with the cloth mirror.

Take out the second group of samples, then heat the iron to 210 ℃, and place the 9 samples under the heated iron for 1 min, 5 min, and 10 min, respectively. Then, the samples after heat setting were taken out, and the other three samples were not treated with any treatment. Then, the 12 samples were stretched to 60% of the original and kept for 15 min before loosening, and the width and length of the samples were measured.

3 Results and Discussion

3.1 Fabric density

3.1.1 Cotton fabric

According to Tab. 1 and Fig. 1, with the increase of time, under the condition of constant setting temperature, the fabric density in the wale and course directions of cotton fabric is improved. And it may be due to the structural characteristics of plain fabric, the growth of fabric density in the wale case is not very obvious, and the growth ratio of the fabric density in the course direction is greater than that of in the wale cases.

Tab. 1 Effect of time of heat setting on density of cotton fabric

Time(min)	0		1		5		10	
Density	Wale (/5 cm)	Course (/5 cm)	Wale (/5 cm)	Course (/5 cm)	Wale (/5 cm)	Course (/5 cm)	Wale (/5 cm)	Course (/5 cm)
1	30	40	30	43	31	46	33	50
2	30	40	31	42	31	45	34	51
3	29	39	30	43	32	48	34	52
Average	30	40	30	43	31	46	43	51
CV(%)	0.019	0.014	0.019	0.013	0.018	0.032	0.017	0.019

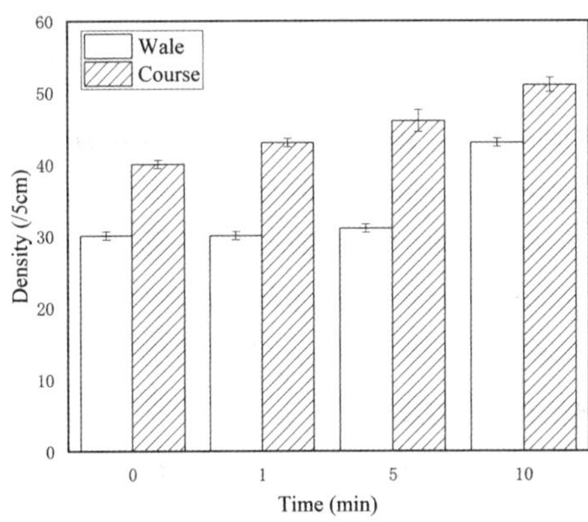

Fig. 1 Effect of heat setting time on density of cotton fabric

3.1.2 Spandex fabric

According to Tab. 2 and Fig. 2, the same as the cotton fabric, with the increase of time, in the shape under the condition of constant temperature, fabric density in the wale and course directions of the elastic fabric was also improved. The fabric density in the course case increased more greatly than that in the wale case. Since glass transition temperature of the polyester and spandex fiber is lower than the cotton fiber, the influence of heat setting is more obvious on the elastic fabric than that of cotton fiber. Therefore, the density change of spandex fabric is more obvious than that of cotton fabric.

Tab. 2 Effect of time of heat setting on density of spandex fabric

Time(min)	0		1		5		10	
Density	Wale (/5 cm)	Course (/5 cm)	Wale (/5 cm)	Course (/5 cm)	Wale (/5 cm)	Course (/5 cm)	Wale (/5 cm)	Course (/5 cm)
1	35	71	35	71	37	74	41	82
2	35	70	36	72	37	73	43	81
3	35	70	35	72	36	75	41	81
Average	35	70	35	72	37	74	42	81
CV(%)	0	0.008	0.016	0.008	0.015	0.0313	0.027	0.007

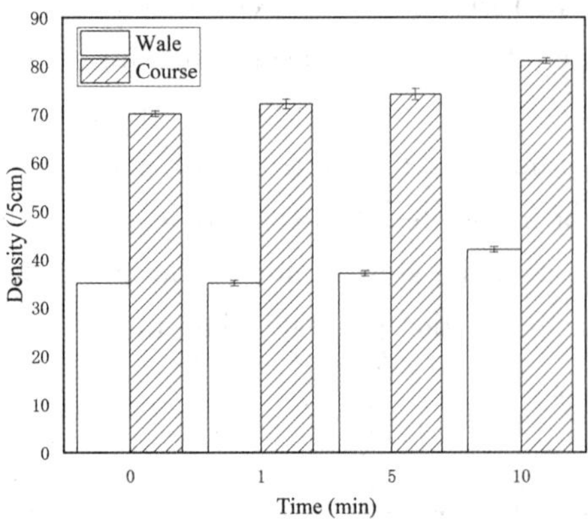

Fig. 2 Effect of tine of heat setting on density of spandex fabric

3.2 Elastic property

3.2.1 Cotton fabric

After being stretched for 15 min, Tab. 3 and Fig. 3 show that compared with the fabric without heat setting, the elasticity changes very little after heat setting of the fabric. When the setting time is 5 min and 10 min, the elasticity of the fabric after heat setting did not change obviously. When the setting time is 5 min and 10 min, it may be that the setting time is too long, which leads to the brittleness of the fabric, and the elasticity of the fabric decreases slightly.

Tab. 3 Effect of time of heat setting on dimensional size of cotton fabric

Time(min)	0		1		5		10	
Dimension	Width (cm)	Length (cm)	Width (cm)	Length (cm)	Width (cm)	Length (cm)	Width (cm)	Length (cm)
First	14.5	14.9	14.6	14.8	14.8	15.1	15.2	15.5
Second	14.7	15.0	14.7	14.7	14.9	15.2	15.0	15.3
Third	14.7	15.1	14.6	14.9	14.7	15.2	15.1	15.4
Average	14.6	15.0	14.6	14.8	14.8	15.2	15.1	15.4
CV(%)	0.008	0.007	0.004	0.007	0.007	0.004	0.007	0.006

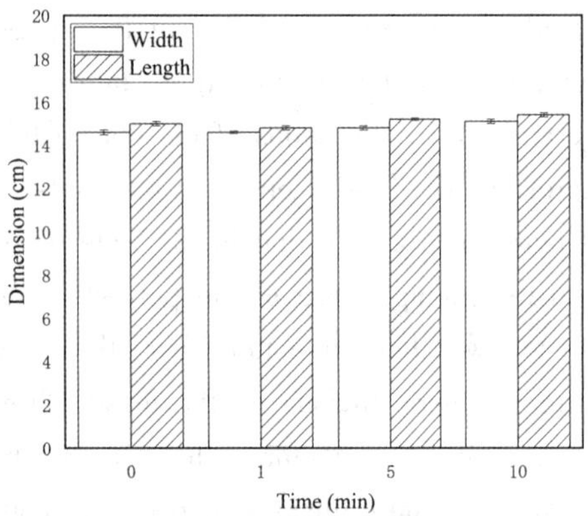

Fig. 3 Effect of time of heat setting on dimensional size of cotton fabric

3.2.2 Spandex fabric

Tab. 4 and Fig. 4 show that, similar to cotton fabric, the elasticity of spandex fabric after heat setting gradually decreases with the increase of heat setting time, and the change is also insignificant.

4 Conclusions

Through the experiment and analysis, it is shown that heat setting not only makes the fabric structure more stable, but also makes the fabric "heat shrink", which increases the density of the fabric. At the same thermal setting temperature, with the increment of the setting

Tab. 4 Effect of time of heat setting on dimensional size of spandex fabric

Time(min)	0		1		5		10	
Dimension	Width (cm)	Length (cm)	Width (cm)	Length (cm)	Width (cm)	Length (cm)	Width (cm)	Length (cm)
First	11.2	12.5	11.5	12.4	12.0	12.8	12.5	13.5
Second	11.5	12.4	11.6	12.5	11.8	12.5	13.0	13.4
Third	11.3	12.5	11.4	12.6	12.1	12.4	12.5	13.7
Average	11.3	12.5	11.5	12.5	12.0	12.6	12.7	13.5
CV(%)	0.013	0.005	0.007	0.008	0.013	0.017	0.022	0.011

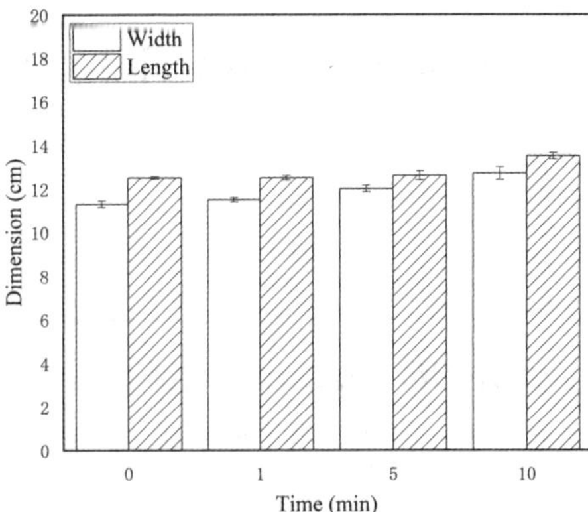

Fig. 4 Effect of time of heat setting on dimensional size of spandex fabric

time, the fabric density increases. In addition, by comparing the recovery size of the fabric without heat-setting process and after heat setting, it can be concluded that the elasticity of the fabric after heat setting is slightly lower than that of the fabric without heat setting.

Acknowledgement

This research was funded by the National Natural Science Foundation of China (No. 51603039), supported by the Fundamental Research Funds for the Central Universities, the Key Laboratory of Textile Science and Technology (Donghua University), Ministry of Education, and the Initial Research Funds for Young Teachers of Donghua University.

References

[1] ZHOU L, GAO L L, JIN F J. Process parameters optimization of heat setting process based on multi-objective genetic algorithm[J]. Journal of Clothing Research, 2012, 11(4): 453-457.

[2] ZENG L. Principle and practice of fabric heat-setting (2) [J]. Textile Dyeing and Finishing Journal, 2012, 34(1): 5-9.

[3] ZENG L. Principle and practice of fabric heat-setting (1) [J]. Textile Dyeing and Finishing Journal, 2011, 33(12): 1-6.

[4] SCHELLENBERGER G. Control of heat setting[J]. Textile Technology Oversea, 1981(22): 6-10.

[5] FENG Y, SHEN W, CHAO C X, et al. Discussion on heat setting technology of polyester-cotton knitted fabric before dyeing[J]. Knitting Industries, 2009(1): 59-61.

[6] SU H, HU Y C. Study and design of extra-wide width contraction setting machine[J]. Shanghai Textile Science & Technology, 2012, 40(6): 15-16,31.

[7] CUI S. 33 kinds of finishing processes make the fabric more beautiful[J]. Textile Science Research, 2019(2): 71-73.

[8] JIANG Y. Study on the Construction and properties of soybean knitted fabric[D]. Shanghai: Donghua university, 2004.

Extraction of Crack Parameters of a Fabric Strain Sensor Based on MATLAB

ZHAO Yuqing[1], YANG Xudong[1*], WANG Xi[2], ZHANG Tianshu[3], LI Qiao[1]

[1] *Key Laboratory of Textile Science & Technology, Ministry of Education, College of Textiles, Donghua University, Shanghai 201620, China*

[2] *College of Information Science and Technology, Donghua University, Shanghai 201620, China*

[3] *College of Science, Donghua University, Shanghai 201620, China*

* *Corresponding author's email*: xdyang@ dhu. edu. cn

Abstract: Knitted strain sensor has low Young's modulus similar to human skin. The strain measurement range is more than 30%, and the fatigue life is more than 100 000 times. SEM observation shows that surface coating of the fabric strain sensor occurs cracks in the repeated stretching process, which governs the sensing mechanism and sensing performance of the sensor. This paper is mainly based on the quantitative analysis of the cracks produced by the sensor in the tensile process. Firstly, the SEM images with cracks are preprocessed by PHOTOSHOP software. Then, the crack and coating are segmented and impurities are removed from the image by MATLAB software. Finally, the length, width, number, area and proportion of the cracks in the whole image are calculated, so as to facilitate the subsequent study of the sensing mechanism of the sensor.

Keywords: Fabric strain sensor; MATLAB; Crack/Coating; Characteristic parameters

1 Introduction

At present, fabric-based flexible strain sensors can be roughly divided into two types: weaving or knitting conductive yarns into fabrics or directly coating conductive materials on the fabrics[1]. Coating method is to attach metal particles, carbon-based materials or polymer conductive composites to the fabric surface to form conductive fabrics by deposition or printing, which is the most common preparation method of flexible fabric sensor at present. The flexible fabric sensor prepared by coating method has the advantages of high sensitivity, good linearity and wide measurement range[2]. However, the conductive fabric with a coating will produce cracks in the process of stretching, and the resistive change of the sensor is mainly determined by the conductive coating on the surface of the fabric. The production and growth of crack on conductive coating, and the opening and closing of cracks not only determine the connected state of conductive pathway, but also affect the sensing mechanism of the sensor[3-4].

Scanning electron microscopy (SEM) is used to observe and record the surface crack changes of the sensor under tensile condition, and the obtained SEM images are quantitatively analyzed. Quantitative analysis of the SEM images mainly uses digital image processing techniques. It is simple and effective to apply Photoshop to quantitative study of SEM images, but Photoshop is mainly a software for image art processing. It has few functions that can be used for quantitative analysis of images, and the data analysis method is relatively simple[5]. The image processing function of MATLAB is powerful, and the format, read, write and display of image files are very convenient. It supports various types of image files and has strong data processing ability[6]. Therefore, the microstructure SEM image could be processed by PHOTOSHOP software, and further processed by MATLAB. The combination of the two types of the software would obtain satisfactory image processing results.

2 Research Object

A knitted fabric strain sensor is taken as the research object, which is a plain knitted fabric with carbon nanoparticles/silicone elastomer/silicone oil (CNP/SE/SO) conductive coating, and the size is 10 mm × 10 mm × 0. 55 mm. The fabric strain sensor has good sensing performance. The gauge factor was found as good as 3. 5 throughout the working range of 0%–60%[7]. In the range of 30% working strain, the fatigue life is more than 100 000 times[8].

The resistance change of conductive fabric is not only related to the properties of the material itself, but also must consider the resistivity change of the sensor during the tensile process. The change of fabric resistivity is related to the surface morphology of conductive materials[3]. The sensor resistance change is mainly determined by the conductive coating on the fabric surface, and the coating on the sensor surface will produce a series of microcracks in the process of stretching. The change of the crack on the conductive coating determines the connected state of the conductive path and affects the resistance change of the sensor. SEM is used to observe and record the surface change of the sensor under tensile condition.

3 SEM Image Processing of Sensor Surface

3.1 SEM observation in situ

Microscopic experiments are carried out under a scanning electron microscope with an in-situ loading device, which mainly observed the change of surface morphology of the sensor under large strain. With the extension, the SEM images of the sensor surface morphology at different loading times can be obtained. When the conductive fabric is stretched along the longitudinal direction to 60%, the crack distribution on the surface of the sensor is shown in Fig. 1. It can be observed through the electron microscope that the conductive fabric will produce a series of cracks after stretching. The cracks are distributed along the transverse direction (the vertical direction of the applied force), and the cracks are closely arranged. It is worth noting that cracks are generated in places with holes. The width and area of cracks in places with holes are wider than those without holes. Multiple cracks near the holes are connected to a large crack. It can be observed from Fig. 1 that a series of cracks are generated on the sensor surface along the vertical direction of the applied pressure. With the progress of tension, the microcracks continue to expand, and the length, width and number of microcracks are also changing. When the force is unloaded, the open cracks are closed with the recovery of tension, and the resistance recorded in the test is also changing with the progress of tension.

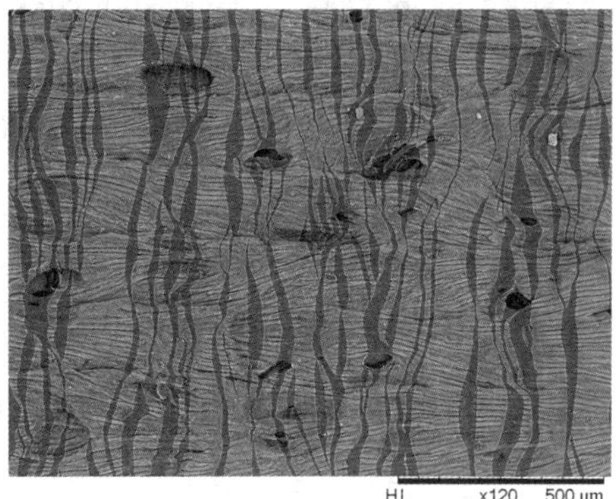

Fig. 1 The SEM image of sensor surface

Knitted fabric plays a supporting role in the sensor, and the change of sensor resistance is mainly determined by the conductive composite coating on the surface of knitted fabric. As the stretch progresses, the coating on the sensor surface produces a series of cracks, and the appearance of the cracks on the coating changes the resistance of the composite material. In other words, the morphological change of the microcracks leads to a change in the resistivity of the sensor, which makes the resistance of the sensor increase. In order to obtain more effective information about the surface morphology of the conductive fabric from the SEM image, the image is processed by commercial software PHOTOSHOP and MATLAB.

3.2 SEM image processing

On the basis of qualitative research on the micro-stretching process of the sensor surface by in-situ tensile scanning electron microscope, the image is binarized, morphologically calculated, and noise removed based on digital image analysis technology. It is convenient for subsequent quantitative analysis of the image. PHOTOSHOP software is used to preprocess the image, which is convenient for further segmentation of cracks and coatings by MATLAB software. First, the boundary of the crack is extracted. Then, the interior of the crack is filled with white and the coating part is filled with black. Fig. 2 shows the image after filling the color, but there are still pixels in the image except the coating and crack.

Next, the image processing is continued by using the MATLAB software. Image binarization is to determine

Fig. 2 PHOTOSHOP image pre-processing

the relationship between the gray value of each pixel in the gray image and the threshold by setting the gray value threshold. The gray value of pixels larger than the threshold is set to 255, and the gray value of pixels smaller than the threshold is set to 0. The image is transformed into black and white images with only intensity value of 0 and intensity value of 255[9]. Through graythresh and im2bw instructions of MATLAB, the crack and coating are segmented by automatically selecting the appropriate threshold. The pixel gray value of the crack is 255, and the pixel gray value of the coating is 0. After binarization, cracks and coatings have been obviously separated. The white part represents cracks, and the black part represents coatings. However, there are still some small white clutters in the coating. Using the bwareaopen instruction in MATLAB[10], the bwareopen function can delete small area objects to achieve the purpose of removing noise. After repeated tests, when the threshold value is taken to 10, the white impurities in the coating are removed and the crack morphology is relatively clear, as shown in Fig. 3.

Fig. 3 MATLAB image processing

4 Extraction of Characteristic Parameters

To investigate the effect of the microcracks on the sensor sensing behavior, the parameters of micro-cracks in Fig. 3 are extracted by means of image processing function in MATLAB. In order to better describe the characteristics of cracks, five characteristic parameters (crack length, crack width, crack number, crack area and crack rate) are selected as the quantitative characteristics of images in this paper.

4.1 Length, width and number of cracks

The crack length and width in the image are automatically calculated and output by MajorAxisLengh and MinorAxisLengh instructions in the Regionprops function in MATLAB[10]. However, before using the Regionprops function, the bwlabel function must be called to label the connected region[11-12]. After labeling, the connected and non-connected regions can be distinguished. The connected regions are framed by red lines, and the number of connected regions is denoted as the number of cracks. The annotated image is shown in Fig. 4. After the MATLAB operation, there are 28 number of cracks, whose length and width are shown in Fig. 5. The average length of the cracks is 557.26 pixels, and the average width is 57.65 pixels.

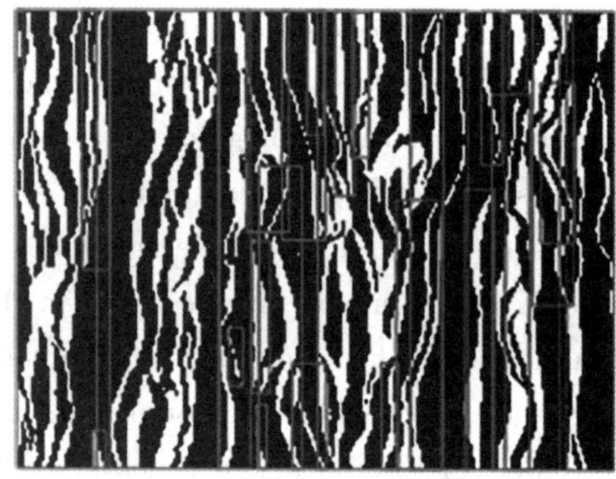

Fig. 4 Fracture target figure

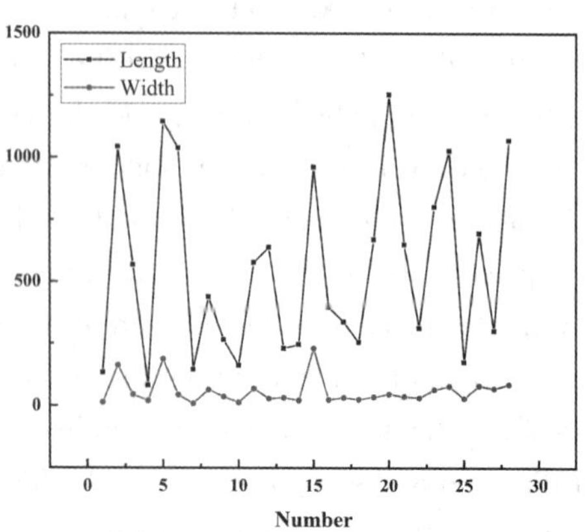

Fig. 5 Length, width and number of cracks (pixel values)

4.2 Crack area and crack rate

The area of the crack is the number of white pixels in the picture, that is, the crack area at the pixel scale. The number of white pixels that can be found through the find function in the MATLAB is 383 914, and the number of pixels in the whole picture is 1 220 175. In order to study the development law of cracks more conveniently, after calculating the area of cracks, the crack rate at pixel scale is further calculated to reflect the relative size of cracks. Crack Rate (CR)[13] refers to the ratio of the area (pixel) of the crack to the area (pixel) of the whole image, and the calculated fracture rate is 31.46%.

The expression of crack rate is

$$CR = \frac{A_0}{A} \times 100\%$$

where A_0 is the crack area (pixel); A is the area (pixel) of the entire image.

4.3 Calibration pixel

Since the calculated crack characteristic parameters are represented by the number of pixels, the actual image should be calibrated according to the resolution of the camera. The actual length of each pixel in the image (conversion coefficient μ) is calculated. Then the actual length, actual width and actual area of the crack can be calculated by multiplying the number of pixels calculated by the conversion coefficient μ[14-15]. The spatial resolution of the SEM image after noise removal is 1 275 × 957 pixels, that is, the image occupies 1 275 pixels in the width direction. According to the scale in Fig. 1, the conversion coefficient μ is calculated as equal to 1.145 μm. After calculation, the average crack length is 638.062 7 μm; the average crack width is 66.009 25 μm; the actual area of cracks is 0.5 mm^2.

5 Conclusions

In this experiment, the SEM image is processed by PHOTOSHOP and MATLAB software. The crack area in the image is located and marked. The effective information of crack length, width, number and area on the surface morphology of the sensor is obtained from the SEM image, which can be used to quantitatively analyze the crack of the sensor more conveniently and quickly. However, this paper only analyzes the cracks under a single strain. In the subsequent study, the characteristic parameters of cracks under different strains will be statistically analyzed. Based on the quantitative relationship between the change of microcracks and strain, the relationship between resistance and strain will be further established. The influence of crack parameters on the change of resistance under different strains will be analyzed, and the internal mechanism of microcracks affecting the sensing behavior of fabric sensors will be further analyzed.

Acknowledgement

This research was funded by the National Natural Science Foundation of China (No. 12002085, 51603039), sponsored by the Shanghai Pujiang Program, supported by the Fundamental Research Funds for the Central Universities, the Key Laboratory of Textile Science and Technology (Donghua University), Ministry of Education, and the Initial Research Funds for Young Teachers of Donghua University.

References

[1] WANG S, LIU W, LIU X X. Research progress of flexible fabric strain sensor [J]. Transducer and Microsystem Technologies, 2017, 36(12): 1-3,9.

[2] SUN J Q, YU X K, WANG K Y. Research status and development of flexible fabric sensor[J]. Journal of Functional Materials and Devices, 2020, 26(1): 16-23.

[3] WANG J P. Study of sensing behavior and its mechanisms of conductive fibers/fabric[D]. Xi'an: Northwestern Polytechnical University, 2014.

[4] LI T. Fabrication and characterization of electro-mechnical hystersis of polypyrrole-coated yarn strain sensors [D]. Shanghai: Donghua University, 2020.

[5] MAO L T, XUE R, AN Q L. Quantitative analysis on SEM image of microstructure with MATLAB[J]. Journal of Chinese Electron Microscopy Society, 2004(5): 579-583.

[6] LIU X H, XU L, XIAO C J, et al. Identifying the crack of silicon solar cells based on Matlab image processing[J]. Journal of Shanghai Jiao Tong University, 2010, 44(7): 926-930.

[7] WANG X, LI Q, TAO X M. Sensing mechanism of a carbon nanocomposite-printed fabric as a strain sensor[J]. Composites Part A: Applied Science and Manufacturing, 2021, 144: 106350.

[8] YI W J. Fabric strain sensor integrated with CNPECs for repeated large deformation[D]. Hong Kong: Hong Kong

Polytechnic University, 2012.
[9] LIU J H, MENG X, DUAN P J, et al. Study on calculation method of concrete crack width based on MATLAB [J]. Materials Reports, 2022, 36(6): 21010082-21010086.
[10] ZHANG H M. Calculation of porosity and fracture parameters based on rock micro image [D]. Jingzhou: Yangtze University, 2021.
[11] KANG S Y, NIE W. Research on road crack recognition based on MATLAB [J]. Computer Knowledge and Technology, 2020, 16(30): 191-193.
[12] JIN M J, HUANG Z, HAN Z Q, et al. Study on bridge crack width detection based on Matlab image processing method [J]. Journal of China & Foreign Highway, 2017, 37(5): 120-123.
[13] JIAN W X, YU J F, REN J, et al. The statistical method of rock surface crack based on Matlab [J]. Safety and Environmental Engineering, 2015, 22(5): 128-133.
[14] DU J M, DUAN W H. Research on automatic identification technology of bridge crack based on Matlab [J]. Construction & Design for Project, 2019(1): 186-188.
[15] YAO X, XU C P, SHU A Q, et al. Sci-tech journal editors should know the application of scale bars in micrographs [J]. Acta Editologica, 2014, 26(4): 338-340.

Self-Powered Electro-Responsive PVDF/PEDOT Electrospun Nanofiber Membranes for Drug Delivery System

CHEN Min[1,2], BIAN Minyue[2], LI Chaojing[1,2], WANG Fujun[1,2*], WANG Lu[1,2]

[1] *Key Laboratory of Textile Science & Technology, Ministry of Education, Donghua University, Shanghai 201620, China*

[2] *Department of Technical Textiles, College of Textiles, Donghua University, Shanghai 201620, China*

* *Corresponding author's email*: wangfujun@ dhu. edu. cn

Abstract: Electro-responsive drug delivery system is one of the most popular stimuli-responsive drug delivery systems. However, most designs need external electrical stimulation, which limits their application. Piezoelectric materials could transform external force into electricity, which means the drug delivery system could provide electricity by itself. Thus, we utilize electrospinning technology to fabricate polyvinylidene difluoride (PVDF) fiber membranes with a high β phase as the electro-responsive part, transferring ultrasonic into electricity. Conductive polymer Poly(3,4-ethylenedioxythiophene) nanoparticles (PEDOT NPs) were mixed into the spinning solution, which would vibrate to destroy the fiber integrity when electro-response is applied, thus the drug could release rapidly. Hydrophilic Rhodamine B (RhB) and hydrophobic Curcumin (CUR) were chosen as drug models. The SEM images showed that the morphology was improved because of the PEDOT NPs. FTIR spectrums showed that the content of β phase has significantly increased due to the PEDOT NPs. The drug release of the PVDF-PEDOT-RhB fiber membrane was 8.33% higher than that of the control group for single response. It can be preliminarily proved that the membrane has the controlled drug release effect with the addition of PEDOT NPs.

Keywords: Electro-responsive; PVDF; PEDOT NPs; Conductive polymer; Drug delivery system

1 Introduction

Electro-responsive drug delivery system is one of the most widely used smart stimuli responsive drug delivery systems in recent years[1-3]. For instance, biocompatible electro conductive hydrogels have been designed to mimic the redox switching and electrical properties of inherently conducting polymers (CPs), facilitating drug release[4]. CPs, such as polypyrrole and polythiophene and their derivatives, behave as linear actuators when they suffer electro stimulation[5]. The actuation principle of CP films is based on the motion of ions in and out of the film during oxidation and reduction[6]. In that case, it could transform electrochemical energy into linear mechanical energy. Poly (3, 4-ethylenedioxythiophene) (PEDOT) is one of the most widely used CPs, and also is widely used in biomedical fields because of its biocompatibility[7-9].

Polyvinylidene difluoride (PVDF) is one of the most popular piezoelectric polymers because of its β phase[10]. Electrospinning is the mainstream preparation process, which not only provides high voltage electric field polarization, but also provides mechanical stretching[11]. Since the current electrical responsive controlled drug delivery system has the shortcoming of external power supply. In this study, PVDF was chosen as the power, transforming ultrasonic to electricity[12], which controlled the vibration of PEDOT NPs dispersing in the fiber. Hydrophilic Rhodamine B (RhB) and hydrophobic Curcumin (CUR) were chosen as drug models to explore whether the wettability of the drug would impact the result.

2 Experiments

2.1 Materials

PVDF(M_w = 4 kDa) was purchased from Macklin Biochemical Co., Ltd. (Shanghai, China). Rhodamine B (RhB) and 3, 4-Ethylene-dioxythiophene (EDOT) were purchased from Shanghai Titan Science Co., Ltd. (Shanghai, China). Curcumin was purchased from Rhawn Co., Ltd. (Shanghai, China). Sodium dodecyl benzene sulfonate (DBSA) was purchased from Shanghai Aladdin Biochemical Technology Co., Ltd.

(Shanghai, China). Ammonium persulfate (APS), sodium chloride, sodium dihydrogen phosphate, potassium dihydrogen phosphate, acetone and N, N-Dimethylformamide were purchased from Sinopharm Chemical Reagent Co., Ltd. (Shanghai, China).

2.2 Synthesis of the PEDOT NPs

Firstly, 0.016 3 g DBSA was added to 4.5 mL of deionized water and the solution was stirred for 1 h at 750 r/min at room temperature. This was followed by the addition of 23.6 mg of EDOT monomer and, again, was allowed to stir for 1 h at 750 r/min at room temperature. Then, 91.2 mg of APS dissolved in 0.5 mL of deionized water was added to the mixture. The reaction was maintained in agitation at 40 ℃ overnight, protected from light with aluminum foil. The side products and unreacted chemicals were removed by a sequence of 3 centrifugations at 11 000 r/min for 40 min at 4 ℃. The PEDOT NPs obtained were dried in a vacuum oven for later use[13].

2.3 Electrospinning of PVDF-based composite fiber membranes

The composite fiber membranes were fabricated using an electrospinning machine (Tong Li Weina Technology Co., Ltd., Guangdong, China). Firstly, polymer solution with a concentration of 18 wt% was prepared by dissolving PVDF powder in a mixture of DMF and acetone (3 : 7, v/v), followed by dispersing PEDOT NPs, RhB and CUR. The specific amount of additions to each substance is detailed in Tab. 1.

Tab. 1 The content of fiber membranes

Sample name	PVDF (g)	PEDOT (mg)	RhB (mg)	CUR (mg)
PVDF	1.8	—	—	—
PVDF-PEDOT	1.8	10.8	—	—
PVDF-RhB	1.8	—	1.1	—
PVDF-RhB-PEDOT	1.8	10.8	1.1	—
PVDF-CUR	1.8	—	—	1.1
PVDF-CUR-PEDOT	1.8	10.8	—	1.1

2.4 Characterization of composite PVDF-based fiber membranes

Surface morphologies of fiber membranes were observed by scanning electron microscopy SU-8010 (Hitachi Co., Ltd., Tokyo, Japan). An attenuated total reflectance Fourier transform infrared spectroscopy Nicolet 6700 (Thermo Fisher Scientific Co., Ltd., MA, America) was applied to investigate the β phase of PVDF. The piezoelectric properties of the fiber membranes are measured by a multimeter (Owon Co., Ltd., Fujian, China).

2.5 Drug release studies

PVDF-PEDOT-RhB fiber membrane and control sample PVDF-RhB fiber membrane was added to 4 mL of PBS (pH=7.4) and placed into an ultrasonic cleaner immediately. Then they were sonicated at maximum ultrasonic power (720 W) and 40 kHz for 15 min, 3 mL of the extract was taken to measure the absorbance (optical density value) at 554 nm wavelength with a UV-Vis spectrophotometer (Shanghai Metash Instruments Co., Ltd., Shanghai, China). The drug release of the RhB-loaded PVDF fiber membrane was calculated according to the calibration curve of RhB finally.

3 Results and Discussion

3.1 Characterization of the PEDOT NPs

Fig. 1 (a) shows the morphology of the PEDOT NPs, demonstrating that PEDOT NPs were successfully synthesized. The effective diameter is 126.78 nm±27.51 nm. The Raman spectrum of PEDOT NPs is displayed in Fig. 1 (b). The peaks at 1 371 cm^{-1} and 1 502 cm^{-1}

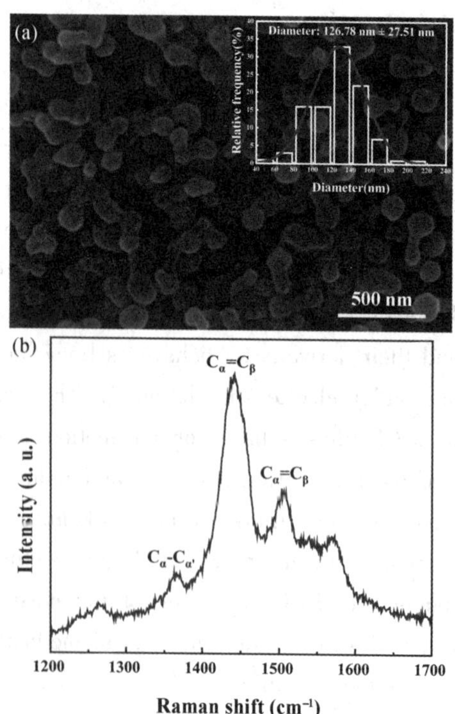

Fig. 1 (a) SEM image of PEDOT NPs. Inset: effective diameter histogram derived from SEM measurements; (b) Raman spectrum of PEDOT NPs

refer to $C_\alpha - C_{\alpha'}$ inter-ring stretching vibrations and asymmetric $C_\alpha = C_\beta$ stretching, respectively. The peak of $C_\alpha = C_\beta$ stretching vibration mode at 1 424 cm^{-1} experiences a blue shift toward 1 440 cm^{-1}, indicating that the PEDOT NPs are p-doped[13, 14].

3.2 Structural analysis of the PVDF-based composite fiber membranes

The morphology of the PVDF-based composite fiber membranes is shown in Fig. 2. Because of the hydrophilicity of RhB, the conductivity of the spinning solution is improved so that the morphology of the fiber is smooth and homogeneous [Fig. 2 (a)]. Adversely, beads occur due to the hydrophobic CUR[Fig. 2 (b)]. With the addition of PEDOT NPs, fibers seem to be smoother and have fewer beads because of PEDOT NPs [Fig. 3 (d) and (e)]. This phenomenon proves that PEDOT NPs could improve the conductivity of the solution, enabling a better stretching of fibers under the electric field.

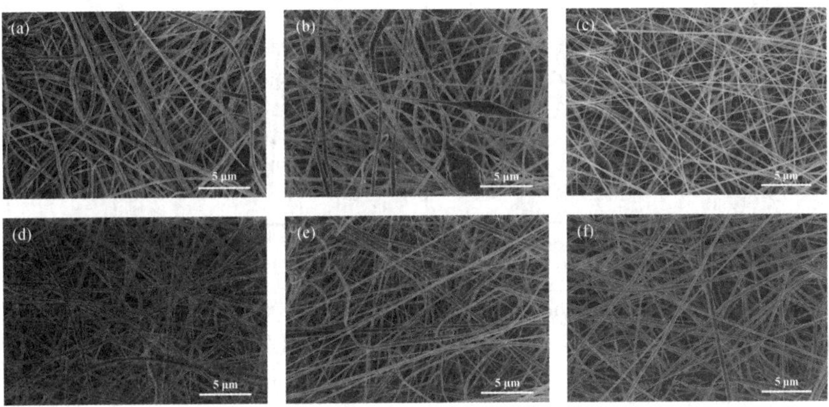

Fig. 2 SEM images of PVDF-based composite fiber membranes: (a) PVDF-RhB; (b) PVDF-CUR; (c) PVDF; (d) PVDF-RhB-PEDOT; (e) PVDF-CUR-PEDOT; (f) PVDF-PEDOT

Fig. 3 shows the FTIR result of PVDF-based composite fiber membranes. The β-phase of PVDF is evaluated from the following formula:

$$F(\beta) = \frac{A_\beta}{1.26A_\alpha + A_\beta}$$

where, A_α and A_β are the absorption intensity for β phase (840 cm^{-1}) and α phase (763 cm^{-1}), respectively[15].

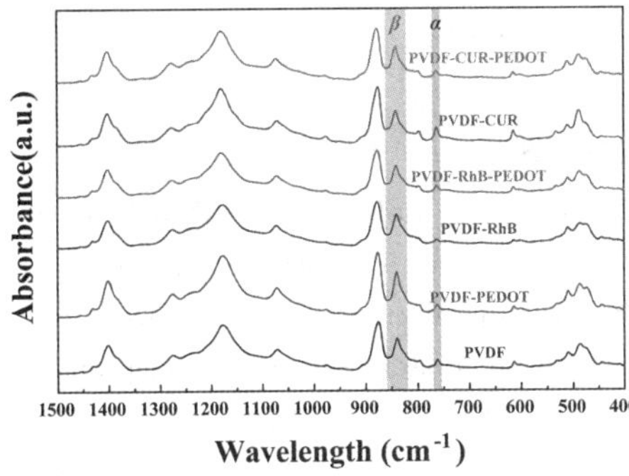

Fig. 3 FTIR spectrums of PVDF-based composite fiber membranes

The content of the β phase of PVDF is about 68.80%, due to the electrospinning process. Both RhB and PEDOT are helpful to the polarization of PVDF, in which β phase content is 75.09% and 78.33%, respectively. Meanwhile, CUR decreases the β phase content by 60.59% because of the beads formation of the fiber[16]. The β phase of PVDF-RhB-PEDOT and PVDF-CUR-PEDOT results in contrary results, which are 69.93% and 72.73% respectively. But they are still higher than the control.

3.3 Piezoelectric performance of the PVDF-based composite fiber membranes

As shown in Fig. 4 (a), the piezoelectric performance of the PVDF-based composite fiber membranes is measured. Figs. 4 (b) to 4 (d) show that PEDOT NPs doesn't seem to enhance the performance tremendously. According to Shao et al. [10], short circuit may occur when the thickness of the fiber membrane is lower than 20 μm. Another potential reason is that the accuracy of the multimeter cannot meet the need of requirements. In despite of these problems, the voltage generated by the piezoelectric response of PVDF can directly act on the PEDOT NPs, resulting in vibration and controlled drug release.

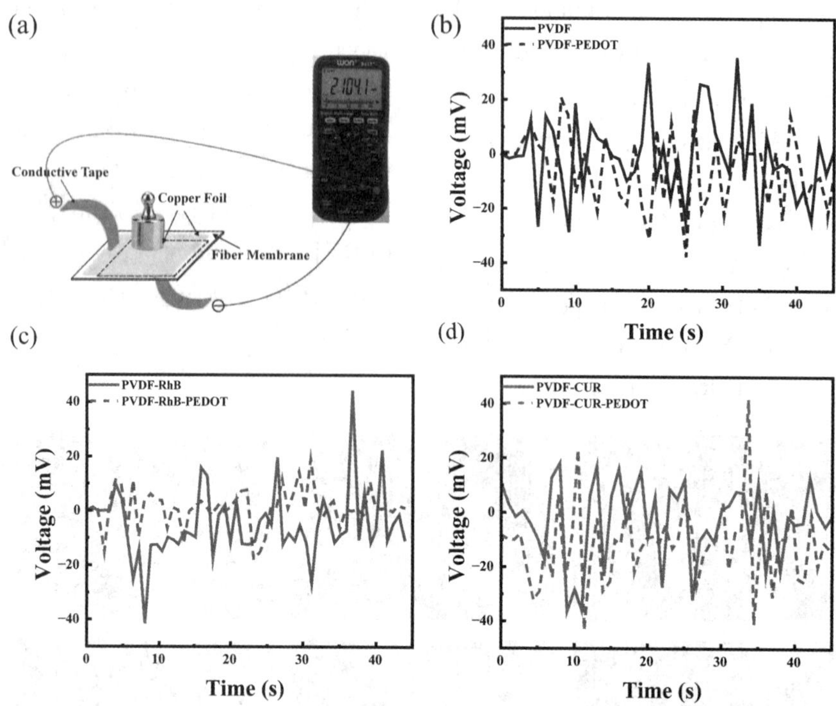

Fig. 4 (a) Scheme of fiber membrane piezoelectric test method; piezoelectric performance of (b) PVDF and PVDF-PEDOT; (c) PVDF-RhB and PVDF-RhB-PEDOT; (d) PVDF-CUR and PVDF-CUR-PEDOT

3.4 In vitro stimulate drug release

Since PVDF could transform ultrasonic wave into electricity, ultrasonic was chosen as the energizing signal. However, ultrasonic itself can also accelerate the drug release from fiber membranes. In that case, drug loaded PVDF fiber membrane was chosen as the control sample and RhB was tested at first. From the calibration curve of RhB, the relationship between absorbance and the concentration of Rhodamine B is $y = 0.027\,65x - 0.012\,32$, $R^2 = 0.998\,3$. Tab. 2. shows the result of single ultrasonic response. With the help of PEDOT NPs, the hydrophilic drug model RhB is released better than that without PEDOT NPs. According to Anna et al.[13], we predict that the CUR group will have a similar result.

Tab. 2 The amount of drug release for single stimulation

Sample	Optical density (OD)	Drug release (μg/mL)
PVDF-RhB	0.001	0.48
PVDF- RhB-PEDOT	0.002	0.52

4 Conclusions

In this study, PEDOT NPs have been synthesized successfully with an effective diameter of 126.78 nm ± 27.51 nm. In order to make the drug delivery system self-powered, PVDF with the piezoelectric effect is chosen as the base material. SEM images show that PEDOT NPs could improve the spinnability of fiber and the FTIR spectrums show that with the help of electrospinning process, the content of β phase has been significantly improved, which is related to piezoelectric properties. The in-vitro drug release shows that PEDOT NPs could increase the release amount of RhB. Overall, the result indicates that this kind of ultrasonic-responsive piezoelectric drug delivery system has a prosperous prospect for targeted drug release, such as ultrasound tumor therapy.

References

[1] SÁNCHEZ K R, LEDEZMA-ESPINOZA A, SÁNCHEZ-KOPPER A, et al. Polysaccharide κ-carrageenan as doping agent in conductive coatings for electrochemical controlled release of dexamethasone at therapeutic doses[J]. Molecules, 2020, 25(9): 2139.

[2] WOEPPEL K M, ZHENG X S, SCHULTE Z M, et al. Nanoparticle doped pedot for enhanced electrode coatings and drug delivery[J]. Advanced Healthcare Materials, 2019, 8(21): 1900622.

[3] GELMI A, SCHUTT C E. Stimuli-responsive biomaterials: Scaffolds for stem cell control[J]. Advanced Health-

care Materials, 2021, 10(1): 2001125.

[4] LAVRADOR P, ESTEVES M R, GASPAR V M, et al. Stimuli-responsive nanocomposite hydrogels for biomedical applications[J]. Advanced Functional Materials, 2021, 31(8): 2005941.

[5] WU J G, CHEN J H, LIU K T, et al. Engineering antifouling conducting polymers for modern biomedical applications[J]. ACS Applied Materials and Interfaces, 2019, 11(24): 21294-21307.

[6] PUIGGALÍ-JOU A, DEL VALLE L J, ALEMÁN C. Drug delivery systems based on intrinsically conducting polymers [J]. Journal of Controlled Release, 2019, 309: 244-264.

[7] PARADEE N, SIRIVAT A. Electrically controlled release of benzoic acid from poly(3,4- ethylenedioxythiophene)/ alginate matrix: effect of conductive poly(3,4-ethylenedioxythiophene) morphology[J]. Journal of Physical Chemistry B, 2014, 118(31): 9263-9271.

[8] CHEN C, CHEN X, ZHANG H, et al. Electrically-responsive core-shell hybrid microfibers for controlled drug release and cell culture[J]. Acta Biomaterialia, 2017, 55: 434-442.

[9] PUIGGALĹ-JOU A, DEL VALLE L J, ALEMÁN C. Encapsulation and storage of therapeutic fibrin-homing peptides using conducting polymer nanoparticles for programmed release by electrical stimulation[J]. ACS Biomaterials Science and Engineering, 2020, 6(4): 2135-2145.

[10] SHAO H, FANG J, WANG H, et al. Effect of electrospinning parameters and polymer concentrations on mechanical-to-electrical energy conversion of randomly-oriented electrospun poly(vinylidene fluoride) nanofiber mats [J]. RSC Advances, 2015, 5(19): 14345-14350.

[11] JIYONG H, YUANYUAN G, HELE Z, et al. Effect of electrospinning parameters on piezoelectric properties of electrospun PVDF nanofibrous mats under cyclic compression[J]. The Journal of the Textile Institute, 2018, 109(7): 843-850.

[12] WAN X, ZHANG X, LIU Z, et al. Noninvasive manipulation of cell adhesion for cell harvesting with piezoelectric composite film[J]. Applied Materials Today, 2021, 25: 101218.

[13] PUIGGALÍ-JOU A, CEJUDO A, DEL VALLE L J, et al. Smart drug delivery from electrospun fibers through electroresponsive polymeric nanoparticles[J]. ACS Applied Bio Materials, 2018, 1(5): 1594-1605.

[14] JANGRA M, DAM S, THAKUR A, et al. Modulating the conductivity of free-standing, flexible composite films of poly(vinylidene fluoride)/poly(3, 4-ethyelenedioxythiophene): poly(styrelenesulfonate)[J]. Thin Solid Films, 2021, 737: 138918.

[15] GREGORIO R, CESTARI M. Effect of crystallization temperature on the crystalline phase content and morphology of poly(vinylidene fluoride)[J]. Journal of Polymer Science Part B: Polymer Physics, 1994, 32(5): 859-870.

[16] ČERNOHORSKY P, PISARENKO T, PAPEŽ N, et al. Structure tuning and electrical properties of mixed PVDF and nylon nanofibers[J]. Materials, 2021, 14(20): 6096.

Design and Electromagnetic Properties of a Novel Knitted Dipole Antenna

KUANG Ye[1,2*], XUE Rui[1], CHEN Yanru[1], LOU Lin[2,3*], ZHENG Chenhua[1], YAN Yanhong[1]

[1] *Clothing Engineering Research Center of Zhejiang Province, Zhejiang Sci-Tech University, Hangzhou 310018, China*

[2] *Key Laboratory of Advanced Textile Materials and Manufacturing Technology (Zhejiang Sci-Tech University), Ministry of Education, Hangzhou 310018, China*

[3] *Zhejiang Provincial Key Laboratory of Fiber Materials and Manufacturing Technology, Zhejiang Sci-Tech University, Hangzhou 310018, China*

* *Corresponding author's email*: y.kuang@zstu.edu.cn, loulin@zstu.edu.cn

Abstract: Flexible antenna plays an important role in smart wearable systems. Among many types of flexible antennas, textile antennas show advantages in terms of comfort, price and feasibility. In order to seamless embed textile antennas into garments while improving the productivity of textile antennas, an intarsia knitting method was used to fabricate dipole antenna in this paper. The arms and substrate of the knitted antenna were composed of silver-plated yarn and polyester yarn respectively, the prototype was small and easy to integrate on the surface of garments. The measured S_{11} of the knitted antenna at operating frequency 2.45 GHz was −16.07 dB and the radiation pattern of the antenna had a near omnidirectional feature, indicating the feasibility of the knitted antenna.

Keywords: Smart textiles; Textile antennas; intarsia knitting; electromagnetic properties

1 Introduction

Internet of Things (IoT) is changing the lives of people with the booming development of network technology. Among the many technologies supporting IoT, the establishment of body centric communication system is especially essential. Body area network is a set of networks centered on human body used for real-time physiological indicators monitoring, which provides the technical foundation for wearable devices potentially used in medical, military, sports and entertainment fields[1-2]. In order to decrease the disturbance of wearable devices to body movements, the wearable devices should achieve the requirements of flexible, lightweight, and comfortable. In addition, signal transmission between the wearable devices and terminals should be wireless. Antenna is an indispensable component in wearable devices, which is used to transmit and receive electromagnetic signals. Wearable antenna is usually made of flexible materials to improve comfortability. Textile material is one of the most preferred materials for making flexible antennas in many flexible materials because of its reasonable price and easy to be integrated to garments. The most common textile antenna fabrication methods are printing and embroidering. The conductive materials of antenna radiation element fabrication are conductive ink and conductive yarn for printing and embroidering methods respectively. Traditional fabrics such as polyester or cotton are used in antenna substrate for both printing and embroidering methods[3-6]. However, the radiation element formed with conductive ink by using printing is easy to deform after friction and stretching. Moreover, the performance of printed embroidered antenna is susceptible to moisture absorption properties of the substrate. In previous study, a woven antenna was proposed to enhance durability and reduce the probability of delamination between radiation element and substrate. Nevertheless, the effect of substrate on antenna performance could not be eliminated by the weaving method[7].

In order to reduce the influence of substrate on antenna performance and improve the integrity of textile antenna, a novel knitted dipole antenna is proposed in this paper. Intarsia knitting technique was used to from the antenna and the radiation element and substrate were made of silver-plated nylon yarn and polyester yarn. Furthermore, essential antenna electromagnetic properties such as return loss and radiation pattern were measured.

2 Experiments

2.1 Antenna design and fabrication

The configuration of knitted dipole antenna is shown in Fig. 1, the antenna consisted of two symmetrical rectangular arms with length of L and width of W, the gap width between the two arms was g. A substrate with length of L_0 and width of W_0 was used to fix the structure of knitted dipole antenna. The design operation frequency of the antenna was 2.45 GHz, and the size parameters of the knitted dipole antenna were listed in Tab. 1.

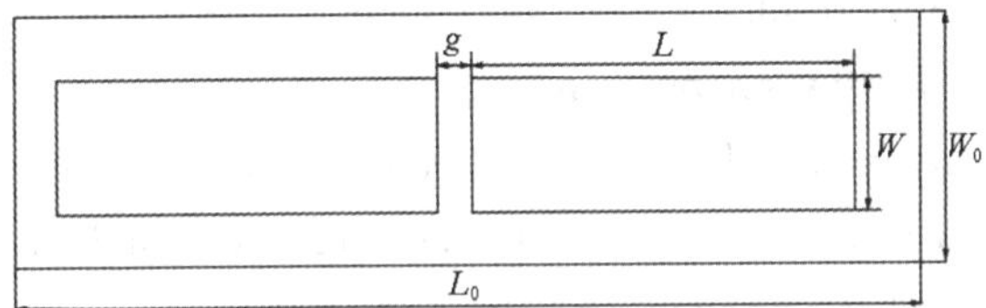

Fig. 1 Configuration of the knitted dipole antenna

Tab. 1 Design size parameters of the knitted dipole antenna (mm)

L_0	W_0	L	W	g
70	25	25.5	6	2

The material of antenna substrate was polyester yarn with thickness of 40 s/2, the dielectric constant and loss tangent of the yarn was 1.9 and 4.5×10^{-3}[8]. The material of antenna arm was silver-plated nylon yarn. In this study, four types of silver-plated nylon yarns with a thickness of 100 D, 200 D, 300 D, 400 D and conductivity of 4 Ω/cm, 3 Ω/cm, 2 Ω/cm and 1 Ω/cm were used. Both antenna substrate and antenna arms were knitted plain structures with interwoven loops at the edges of both contours to form a single piece of fabric. The antenna was fed by a coaxial connector, the two arms were connected to the inner conductor and outer conductor respectively. A piece of copper wire was placed between the antenna arm and coaxial connector conductor for connection reinforcement, since soldering flexible textile antenna arms and rigid metal coaxial connector together was difficult. Even then, the antenna prototype was easy to integrate into garments as shown in Fig. 2.

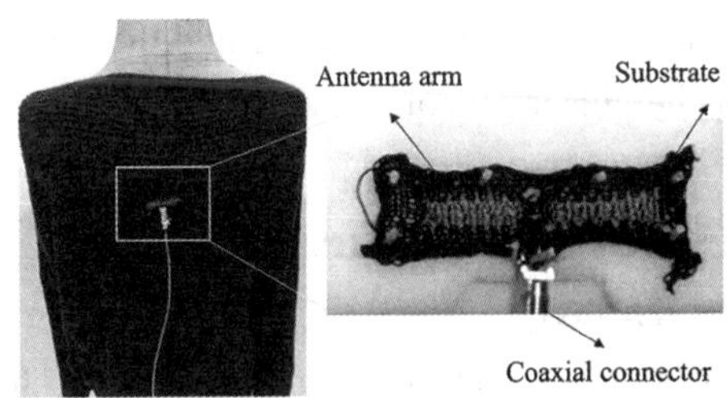

Fig. 2 Photos of the knitted dipole antenna prototype

2.2 Antenna performance tests

Return loss and radiation pattern are the two most essential performances to characterize an antenna. The return loss was measured by using KC901V vector network analyzer and expressed as parameter S_{11}. The expressions for return loss (RL, dB) and S_{11}(dB) are shown as follows:

$$RL = 10\log_{10}\left(\frac{P_i}{P_r}\right) = -S_{11} \tag{1}$$

where P_i and P_r are the incident and reflected powers. The radiation pattern was measured by using signal generator and spectrum analyzer. As shown in Fig. 3, a horn antenna was connected to the signal generator for signal transmission, the knitted dipole antenna prototype was connected to the spectrum analyzer for signal receiving, a cystosepiment was placed behind the knitted dipole antenna as a support. The radiation pattern characterizes the far-field radiation energy of an antenna in three-dimensional space, the two-dimensional radiation pattern could be obtained by rotating the measured antenna sample from 0° to 360°. The E-plane pattern and the H-plane pattern are the most representative two-dimensional antenna radiation patterns, which parallel to antenna electric field and magnetic field respectively. In test condition, the center of the horn antenna was aligned with the center of the antenna, the distance

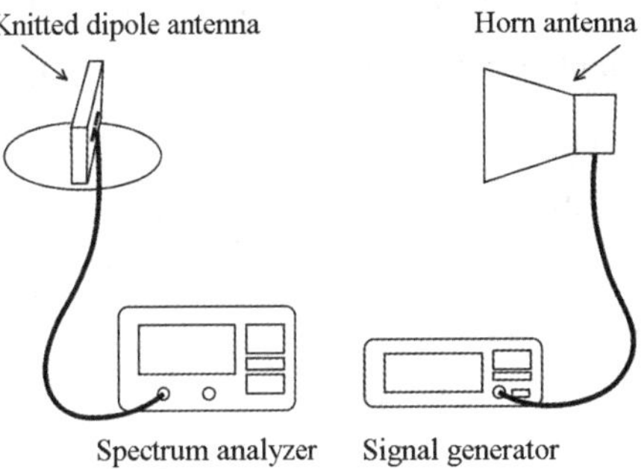

Fig. 3 Schematic diagram of the radiation pattern measurement set-up

between the two antennas was 2 m, and the transmitting power of the signal generator was 30 dB.

3 Results and Discussion

3.1 Return loss

The measured resonant frequencies and S_{11} for four types of antenna samples with conductive yarn thickness of 100 D to 400 D are listed in Tab. 2. Among the four samples, the resonant frequencies of antenna fabricated with 300 D and 400 D conductive yarns were 2.46 GHz, which were the most approximate to the operating frequency of 2.45 GHz. Furthermore, the S_{11} of antenna fabricated with 300 D conductive yarn was −13.65 dB at 2.45 GHz and −13.71 dB at antenna resonant frequency (shown in Fig. 4 (a)), which meets the requirement of −10 dB or less for S_{11} in antenna operating frequency. Therefore, the antenna made of 300 D conductive yarn was further discussed.

Tab. 2 Measured return loss for antenna samples with different conductive yarn thickness.

Yarn thickness (D)	Resonant frequency (GHz)	Minimum S_{11}(dB)
100	2.49	-11.17
200	2.40	-6.26
300	2.46	-13.71
400	2.46	-8.28

As shown in Fig. 4 (a), the knitted dipole antenna had two resonant frequencies of 1.91 GHz and 2.46 GHz from 1.5 GHz to 3.5 GHz, which was different from the simulated result. It was found that the redundant resonant frequency was due to the excessively long copper wires between the antenna arms and coaxial connector for connection. In order to eliminate the redundant resonant frequency, the copper wires were shortened from 5.7 mm to 1.6 mm. The return loss of the antenna with shortened copper wire is shown in Fig. 4 (b),

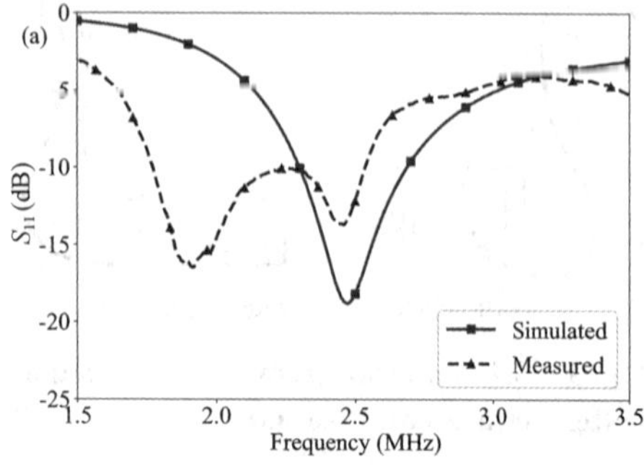

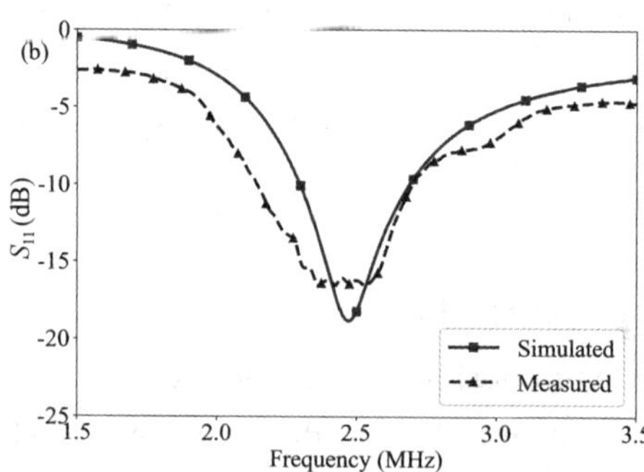

Fig. 4 Return loss of the knitted dipole antenna fabricated with 300 D conductive yarn: (a) before reducing the length of wire between the antenna arms and coaxial connector; (b) after reducing the length of wire between the antenna arms and coaxial connector

the measured S_{11} at operating frequency was −16.07 dB implying better impedance matching performance, moreover, the measured return loss agreed well with the simulated result for modified antenna.

3.2 Radiation pattern

The antenna had near-omnidirectional radiation characteristics in accordance with both simulated and measured radiation patterns as shown in Fig. 5. The simulated and measured H-plane radiation patterns of the an-

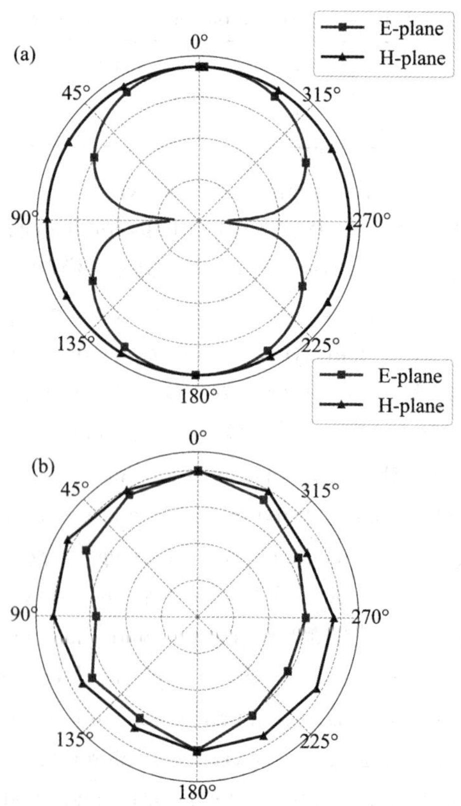

Fig. 5 Radiation patterns of the knitted dipole antenna fabricated with 300 D conductive yarn: (a) simulated results; (b) measured results

tenna were approximately circular, indicating that the radiation intensity of the antenna was similar at all angles in the plane parallel to the magnetic field. In comparison, discrepancies could be seen between the simulated and measured E-plane radiation pattern. The radiation intensity at 0° and 180° were higher and the radiation intensity at 90° and 270° were lower for the simulated E-plane radiation pattern, while the radiation intensity was similar at all angles for measured E-plane radiation pattern. The distortion of the measured results may be because the conductive material used for antenna arm composing was a discontinuous knitted structure.

4 Conclusions

In this study, a knitted dipole antenna operating at the frequency of 2.45 GHz was proposed and the antenna performance was measured. The measured return loss of the antenna agreed well with the simulated result and the measured S_{11} at operating frequency was −16.07 dB. Moreover, the antenna had near-omnidirectional radiation pattern that matches the typical radiation characteristics of a dipole antenna. This study provides the fundamental basis for the design and performance optimization of knitted dipole antennas.

Acknowledgements

This work was supported by the Science Foundation of Zhejiang Sci-Tech University (ZSTU) (Nos. 19072466-Y and 17072156-Y); Zhejiang Sci-Tech University (ZSTU) Outstanding Youth Grant Program; National Natural Science Foundation of China (Nos. 50803010 and 52003245), Zhejiang Provincial Natural Science Foundation of China (No. LQ18E030007).

References

[1] SHI J, LIU S, ZHANG L, et al. Smart textile-integrated microelectronic systems for wearable applications[J]. Advanced Materials, 2020, 32(5): 1901958.

[2] SAYEM A S M, TEAY S H, SHAHARIAR H, et al. Review on smart electro-clothing Systems (SeCSs)[J]. Sensors, 2020, 20(3): 1-23.

[3] EL GHARBI M, FERNANDEZ-GARCIA R, GIL I. Textile antenna-sensor for in vitro diagnostics of diabetes[J]. Electronics, 2021, 10(13): 1-9.

[4] KAPETANAKIS T N, PAVEC M, IOANNIDOU M P, et al. Embroidered bow-tie wearable antenna for the 868 and 915 MHz ISM bands[J]. Electronics, 2021, 10(16): 1-14.

[5] LUO C, GIL I, FERNANDEZ-GARCIA R. Wearable textile UHF-RFID sensors: a systematic review[J]. Materials, 2020, 13(15): 3292.

[6] LABIANO I I, ARSLAN D, YENIGUN E O, et al. Screen printing carbon nanotubes textiles antennas for smart wearables[J]. Sensors, 2021, 21(14): 1-28.

[7] KUANG Y, YAO L, YU S H, et al. Design and electromagnetic properties of a conformal ultra wideband antenna integrated in three-dimensional woven fabrics[J]. Polymers, 2018, 10(8): 861.

[8] OUYANG Y, CHAPPELL W J. High frequency properties of electro-textiles for wearable antenna applications[J]. IEEE Transactions on Antennas and Propagation, 2008, 56(2): 381-389.

3D Reconstruction of Fabric Drape Based on Drape Parameters

HU Xuhui[1], JI Feng[1*], ZHANG Ruiyun[1], XU Tengfei[1], MA Zhenyi[1], XING Hao[1]

[1] *Key Laboratory of Textile Science & Technology, Ministry of Education, Donghua University, Shanghai 201620, China*

* *Corresponding author's email*: jifeng@ dhu. edu. cn

Abstract: Drape is one of the most important properties describing the appearance of apparel fabrics both in the real and virtual world. In the virtual world, information about fabric drape is transferred through internet in the form of digital data. The speed of transmission largely depends on the form of data. Hence information of fabric drape can be transferred much faster in the form of digits than images and videos, especially when batch transmissions are needed. In this paper, a method of 3D reconstruction of fabric drape was proposed based on drape indices. This paper proposed a method that used MatLab language and fabric drape indices to rebuild fabric drape shape. First of all, by using the fabric drape indices including wave number, wave peak and wave valley, the edge contour curve of fabric drape shape was simulated. Then with the contour algorithm, the 3D drape of the fabric was rebuilt. Finally, the intuitional 3D reconstruction effect of fabric draping was obtained. This study provided a new method for the fabric drape shape reconstruction. This method helped to implement the fastest transmission of drape information in large quantities through internet. Meanwhile the characteristic drape shape of fabrics could be reconstructed accurately.

Keywords: Fabric; Drape indexes; Drape shape; 3D reconstruction

1 Introduction

Drape of fabric refers to the drape form and drape degree of fabric under the action of gravity. It is one of the most important factors affecting the visual form, style and aesthetic appearance of garment fabric, and also is an important index to describe the fabric performance[1].

With the development of Internet technology, how to realize the rapid transmission of fabric drape information and obtain the 3D drape form of fabric is a problem that many scholars are trying to solve. Yin[2] used a periodic sine function to simulate the 2D curve of the fabric drape. The limitation of using a periodic sine function to simulate the drape curve of the fabric is that because the fabric has various anisotropy, the drape map is not completely symmetrical. Chen et al.[3] developed a 3D fabric drape shape tester and put forward the uniformity and aesthetic coefficient. The 3D drape image of fabric was reconstructed by Bessel surface mathematical model combining drape coefficient and 3D drape coefficient. Li et al.[4] used a finite element method to reconstruct the drape form of woven fabric, and the simulation process was complex. Therefore, this paper proposed a method to reconstruct fabric drape shape based on the fabric performance index, which is helpful to realize the rapid transmission of fabric drape information and provided a new method for 3D reconstruction of fabric drape shape.

2 Fabric Drape Performance Indices

Drape coefficient is the most commonly used index, which reflects the drape degree of fabric under self-weight.

$$F = \frac{A_1 - A_2}{A_1 - A_0} \times 100\% \tag{1}$$

where A_0 is the area of the disk; A_1 is the original area of the fabric; A_2 is the projected area of fabric overhang; F is the drape coefficient, and the better the drape of the fabric, the smaller the F value.

Ripple number (n) refers to the number of ripples in the drape form of the fabric. The fabric with a better drape has more and even ripple count.

Wave peak angle (α) refers to the angle between two wave peaks.

Wave peak (R) refers to the distance from the peak to

the fabric center.

Wave valley (r) refers to the distance from the trough of the wave to the fabric center as shown in Fig. 1.

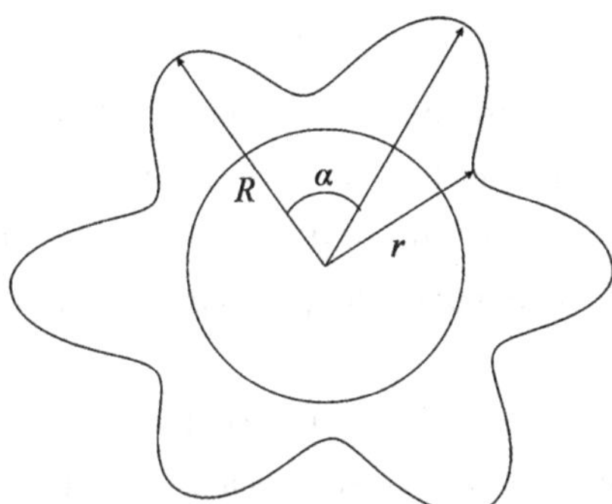

Fig. 1 Schematic diagram of wave peak, wave valley and wave peak angle

According to the above definition, the average value of 3 groups of corresponding indices and 3 groups of indices representing the unevenness of fabric drape form can be obtained.

1) Average wave peak angle and variation coefficient of wave peak angle:

$$\bar{\alpha} = \frac{1}{n}\sum_{i=1}^{n} \alpha_i \tag{2}$$

$$CV_\alpha = \frac{1}{\bar{\alpha}}\sqrt{\frac{1}{n-1}\sum_{i=1}^{n}(\alpha_i - \bar{\alpha})^2} \times 100 \tag{3}$$

where $i=1, 2, \cdots, n$

2) Average wave peak and variation coefficient of wave peak:

$$R = \frac{1}{n}\sum_{i=1}^{n} R_i \tag{4}$$

$$CV_R = \frac{1}{\bar{R}}\sqrt{\frac{1}{n-1}\sum_{i=1}^{n}(R_i - \bar{R})^2} \times 100\% \tag{5}$$

where $i=1, 2, \cdots, n$

3) Average wave valley and variation coefficient of wave valley:

$$\bar{r} = \frac{1}{n}\sum_{i=1}^{n} r_i \tag{6}$$

$$CV_r = \frac{1}{\bar{r}}\sqrt{\frac{1}{n-1}\sum_{i=1}^{n}(r_i - \bar{r})^2} \times 100\% \tag{7}$$

where $i=1, 2, \cdots, n$

Among the above indices, the variation coefficient of wave peak angle reflects the asymmetry degree of fabric drape projected edge contour. Wave peak variation coefficient and wave valley variation coefficient reflect the variation degree of peak value and depth in the fabric ripple unfolding diagram[4-5].

3 Simulation of Fabric Drape Pattern

3.1 Based on the drape performance indicators of 2D shape reconstruction

For the simulation of the 2D drape shape of fabric, some foreign scholars proposed the periodic sine function to simulate it, and its function model was as follows[6]:

$$r = \alpha + b \times \sin(n\theta) \tag{8}$$

where r represents the radius of fabric overhanging contour; α represents the average radius of fabric overhanging contour; b is the height difference between the crest and the trough; n represents the number of ripples.

The drape profile obtained by this model is symmetrical, which is ideal for the simulation of isotropic fabric with a good drape and uniform drape morphology. However, the model has the shortcoming that fabrics are anisotropic in reality. Because of this problem, this paper adopted the simulation of the drape shape based on the drape performance index, and calculated the corresponding sine function of each wave peak and wave valley for fitting, and the simulation effect was reasonable.

3.1.1 Reconstruction of wave expansion diagram

In order to realize the 3D fabric draping shape reconstruction based on draping performance parameters, the ripple expansion diagram of the fabric draping shape should be obtained first[7], as shown in Fig. 2.

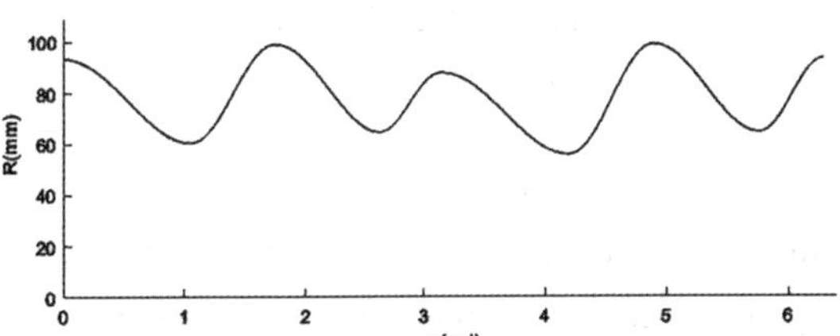

Fig. 2 Wave expansion diagram (R and α respectively represent the sag radius (mm) and the arc value (rad) corresponding to the polar angle)

As long as the coordinates of the corresponding wave peaks and wave valleys in the ripple expansion graph are obtained, the whole ripple curve can be reconstructed by sine function fitting of each key point.

As shown in Fig. 2, by taking any wave peak as the starting point of ripple expansion diagram, and combining it with the definition of drape performance index, the main value methods of the following key points can be obtained[8].

1) Obtain the horizontal coordinates of each key point.

Horizontal coordinate of wave peak: from the known ripple number (n), the average wave peaks angle is

$$\bar{\alpha} = \frac{360°}{n} \tag{9}$$

The horizontal coordinates of each wave peak can be obtained from the known variation coefficient of wave peak angle (CV_α):

$$\begin{cases} \alpha_1 = 0° \\ \alpha_{i+1} = \alpha_i + [\bar{\alpha} \pm \bar{\alpha} \times CV_\alpha] \\ \alpha_{n+1} = 360° \end{cases} \tag{10}$$

where $i = 1, 2, \cdots, n$

Horizontal coordinate of wave valley: in the ideal case, the trough is halfway between two adjacent wave peaks. However, due to the existence of the variation coefficient of the wave peak angle, the position of the trough point will also change:

$$v_i = \frac{(\alpha_i + \alpha_{i+1})}{2} \pm \bar{\alpha} \times CV_\alpha \tag{11}$$

where $i = 1, 2, \cdots, n$

2) Obtain the ordinate of each key point.

Ordinate of wave peak: From the known average wave peak ($\bar{R}$) and variation coefficient of wave peak (CV_R):

$$\begin{cases} R_i = \bar{R} \pm \bar{R} \times CV_R \\ R_{n+1} = R_1 \end{cases} \tag{12}$$

where $i = 1, 2, \cdots, n$

Ordinate of wave valley: From the known average wave valley ($\bar{r}$) and variation coefficient of wave peak (CV_r):

$$r_i = \bar{r} \pm \bar{r} \times CV_r \tag{13}$$

where $i = 1, 2, \cdots, n$

The above is the main value algorithm of key points, and the error range of the corresponding average value and variation coefficient of key points should be considered in MatLab programming.

3.1.2 Fitting of corrugated curves

As the sine function model:

$$y = A \times \sin(\omega x + \psi) + B \tag{14}$$

where A is determined by the peak; B represents the position relation of the ripple diagram on the Y-axis; ω determines the period, so it can be solved by the period formula; ψ determines the position of the waveform and the X-axis.

Through the above function model and the algorithm, we can get $2n$ group of wave peaks and wave valley coordinates, through each group of peaks and troughs. We can find out the corresponding expression of each sine function, and then through Matlab programming, fitting the corrugated curve of the fabric. The ripple curve under the rectangular coordinate system was obtained, but it is necessary to transform the rectangular coordinate into the polar coordinate, so that the 2D draping profile of the fabric can be obtained, as shown in Fig. 3.

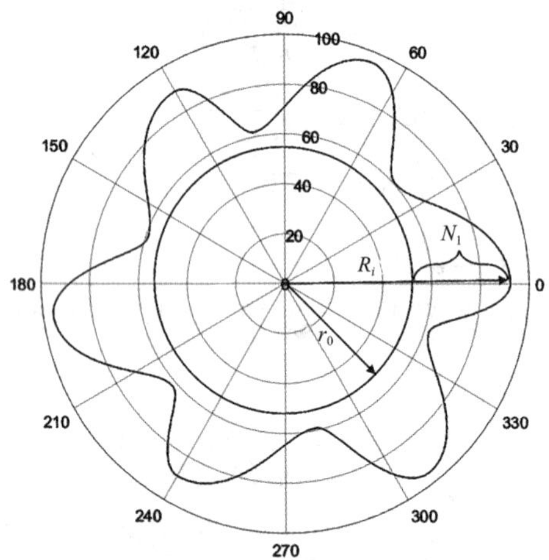

Fig. 3 The simulation figure of fabric drape contour

3.2 Reconstruct fabric 3D drape

3.2.1 Concentric circle algorithm for fabric drape simulation

Taking the center point of the disk as the center of the circle and the radius of the disk (r_0) as the starting point, extending outward, the fabric overhanging contour area is equally divided into N parts, as shown in Fig. 3.

Define the abscissa of wave peak and wave valley as array Q and the ordinate as array R, then the average distance of each concentric circle is:

$$\Delta R = \frac{R_i - r_0}{N} \tag{15}$$

where $i=1, 2, \cdots, N+1$

R_i represents the fabric drape radius, r_0 represents the disk radius, and N represents the number of portions of the drape area divided equally.

Then the radius corresponding to the j concentric circle is:

$$R_j = r_0 + \Delta R \times (j - 1) \tag{16}$$

where $j=1, 2, \cdots, N+1$

In the three-dimensional rectangular coordinate system with disk as the reference plane and Z-axis in the vertical direction, the corresponding height of R_j can be obtained, as shown in Fig. 4.

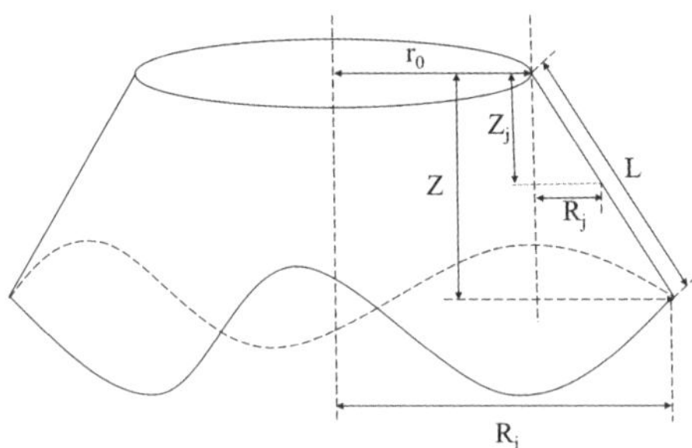

Fig. 4 Schematic diagram of Z_j

The height (Z_j) of the corresponding Z-axis direction can be obtained from the similar triangle relationship:

$$Z = \sqrt{L^2 - (R_i - r_0)^2} \tag{17}$$

$$Z_j = \frac{Z \times \Delta R \times (j - 1)}{R_i - r_0} \tag{18}$$

where $i=1, 2, \cdots, 2n+1$;

$j=1, 2, \cdots, N+1$

In summary, the three-dimensional coordinates of any point in the rectangular coordinate system are obtained:

$$\begin{cases} x_j = R_j \times \cos(Q_i) \\ y_j = R_j \times \sin(Q_i) \\ Z_j = \dfrac{Z \times \Delta R \times (j - 1)}{R_i - r_0} \end{cases} \tag{19}$$

where $i=1, 2, \cdots, 2n+1$;

$j=1, 2, \cdots, N+1$

Using the above algorithm, the fabric draping shape effect based on the concentric circle algorithm can be drawn, as shown in Fig. 5.

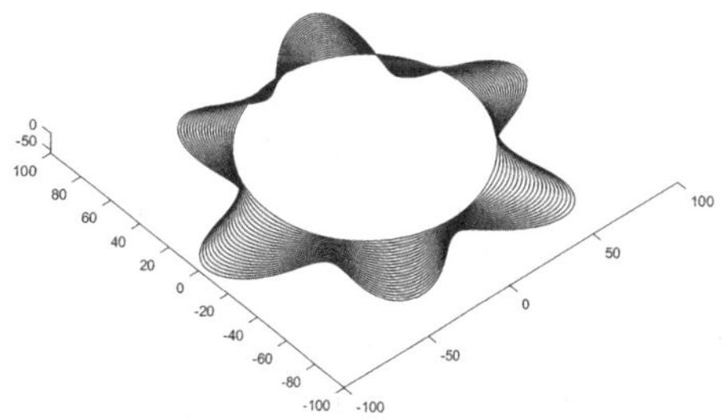

Fig. 5 Simulation diagram of fabric 3D drape shape concentric circle algorithm

3.2.2 Drawing of 3D draping shape of fabric

Through the above algorithm, the 3D surface can be drawn by using the mesh (x, y, z) function in Matlab language[9]. Using Matlab and the above algorithm, the relevant program is written, and the 3D fabric drape shape reconstruction effect is remarkable. Tab. 1 shows the drape index of sample A and B. The following is the comparison between the actual effect diagram, simulated concentric circles and the 3D reconstruction diagram of several fabrics, as shown in Fig. 6.

Tab. 1 The drape index of sample A and B

Sample	n	CV_α (%)	$\bar{R}$ (mm)	CV_R (%)	$\bar{r}$ (mm)	CV_r (%)
A	5	9.35	94.747	3.983	61.43	3.293
B	6	13.01	94.336	3.607	62.45	2.353

In Tab. 1, n represents the ripple number; CV_α represents the variation coefficient of wave peak angle; $\bar{R}$ represents the average wave peak; CV_R represents the variation coefficient of wave peak; $\bar{r}$ represents the average wave valley; CV_r represents the variation coefficient of wave valley.

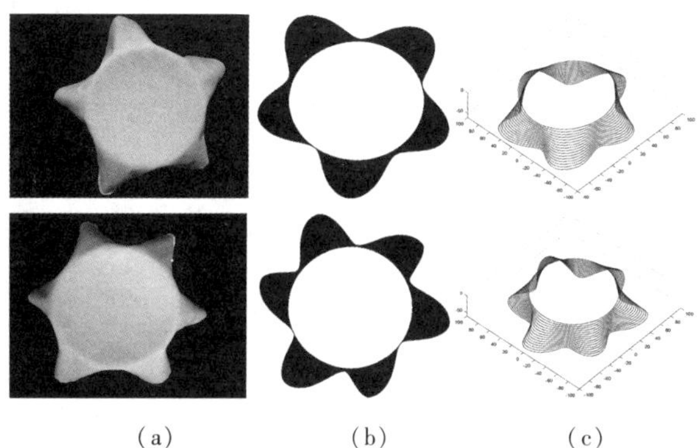

(a) (b) (c)

Fig. 6 Comparison diagram of the actual and simulated drape of sample A and B: (a) the actual overhang image of the sample; (b) the prone view of 3D reconstruction of drape contour; (c) the prone view of fabric drape 3D reconstruction graphics

4 Conclusions

Using the fabric drape performance indices to reconstruct the 3D drape morphology, not only the simulation effect is ideal, but also the digital transmission of fabric drape performance indices is realized from the image information transmission mode of fabric drape, which greatly accelerates the transmission speed and conversion efficiency of information.

References

[1] MEI Z H. Development of fabric unidirectional drape test system[D]. Hangzhou: Zhejiang University of Technology, 2016.

[2] YIN H Y. Fabric drape performance evaluation and simulation research [D]. Hangzhou: Zhejiang University of Technology, 2008.

[3] CHEN M, ZHOU H, YANG L J, et al. Fabric 3D drape shape test indicators and 3D reconstruction[J]. Journal of Textile Research, 2008(9): 51-55.

[4] LI C Y, ZHANG K T, XU J F. Numerical simulation of fabric drape buckling[J]. Journal of Tianjin University, 2007(1): 46-50.

[5] QI H Q. Development of fabric drape test and 3D prediction system[D]. Hangzhou: Zhejiang University of Technology, 2009.

[6] LO W. A study of fabric anisotropy [D]. Hong Kong: Hong Kong Polytechnic University, 2001.

[7] ZHOU L L, QI H Q, ZHOU H. Fabric drape morphology test and evaluation index research [J]. Modern Textile Technology, 2010, 18(3): 46-50.

[8] SHEN Y, QI H Q. Simulation of fabric draping morphology[J]. Journal of Textile Research, 2010(10): 6.

[9] SHEN X Y. Mathematical graphics analysis based on MATLAB [J]. Journal of Heihe University, 2018, 9(7): 2.

Fabrication of Hydrogel/Nanofibers Wound Dressings with High Water Absorption and Antibacterial Properties

LI Mengjuan[1], **NIU Liting**[1], **LI Liqing**[1*], **WU Dequn**[1,2]

[1] *Key Laboratory of Textile Science & Technology, Ministry of Education, Donghua University, Shanghai 201620, China*

[2] *Innovation Center for Textile Science and Technology, Donghua University, Shanghai 201620, China*

* *Corresponding author's email*: liliqing@ dhu. edu. cn

Abstract: The development of hydrogel wound dressings with antibacterial activity, water absorption capacity and certain mechanical strength is an urgent need for biomaterial science. The focus of this work is the development of a new wound dressing that can be hydrogel/nanofibers with high water absorption and antibacterial properties. Owing to the formation of catechol-catechol adducts, the hydrogel not only exhibited a high swelling ratio of up to 18.85 times but also possessed suitable adhesiveness. The morphology of the hydrogels was confirmed to be a porous and interconnected network structure by scanning electron microscopy (SEM). By blending quaternary ammonium salt antibacterial agent in polycaprolactone (PCL), the obtained nanofiber membrane is then compounded with hydrogel to prepare the hydrogel/nanofibers wound dressing. It is worth noting that the hydrogel/nanofibers wound dressing not only exhibits a wide range of antibacterial activities against *S. aureus* and *E. coli*, but also has good mechanical properties. The results show that the hydrogels/nanofibers wound dressings provide new prospects for the field of medical dressings.

Keywords: Hydrogel; Water absorption; Antibacterial nanofibers; Wound dressing

1 Introduction

The skin is the largest organ of the human body, consisting of epidermis, dermis and subcutaneous tissue, and is the first barrier to foreign invasion[1]. While wound healing is a complex physiological mechanism involving the interaction of various local growth factors (GFs) including cells, proteases and extracellular matrix (ECM) components. And this process involves four overlapping and independent stages[2], including hemostasis, inflammation, proliferation, and maturation (remodeling) stages. An ideal wound dressing should not only absorb wound exudate in time, but also maintain a relatively moist microenvironment. In addition, it needs to be able to prevent the growth of bacteria, as well as the possibility of blocking the contact of the wound with the external environment.

Compared with other wound dressing materials, hydrogel is an insoluble hydrophilic polymer with high water content. The three-dimensional (3D) network structure of hydrogels facilitates the absorption and retention of water. Therefore, the hydrogel can maintain a long-term moist environment for the wound, which is conducive to promoting wound healing and relieving wound discomfort[3]. And nanofibers have become widely accepted materials in the field of human health care due to their excellent properties such as softness, biodegradability, and air permeability[4]. Therefore, hydrogel/nanofibers wound dressings have been widely used as soft functional materials to provide moist pads for wound healing. However, due to the insufficient antibacterial properties and poor mechanical and bioadhesive properties of current hydrogel wound dressings, the clinical application of hydrogel/nanofibers wound dressings needs to be improved[5].

Here, we fabricated the hydrogel/nanofibers wound dressings with excellent water absorption and good antibacterial properties. The hydrogels were prepared in situ by the Schiff base reaction between the aldehyde group of oxidized carboxymethyl cellulose-dopamine hydrochloride (OCMC-DA) and the amino group of carboxymethyl chitosan (CMCS). The hydrogels have suitable adhesion due to the formation of catechol-catechol adducts. For antibacterial nanofibers, CTAB and PCL were used for electrospinning, which endowed the nanofibers with excellent antibacterial properties. Therefore, the hydrogel/nanofibers wound dressings integrates the good water absorption and suitable adhesion of hydrogels and the remarkable antibacterial ability of antibacterial nanofibers, which provides a new prospect for the field of medical dressings.

2 Experiments

2.1 Materials

Carboxymethyl cellulose (CMC), sodium periodate, and PCL were purchased from Sigma-Aldrich and used without further purification. Dopamine hydrochloride (DA), carboxymethyl chitosan (CMCS) and *N*, *N*-dimethylformamide (DMF) were purchased from Aladdin Industrial Inc. Hexadecyl trimethyl ammonium bromide (CTAB) was purchased from Sinopharm Chemical Reagent Co., Ltd. (Shanghai, China).

2.2 Synthesis of hydrogels

Oxidized carboxymethyl cellulose (OCMC) was synthesized according to the method reported in previous literature[6]. 10.0 g of carboxymethyl cellulose (CMC) and 10.0 g of sodium periodate ($NaIO_4$) were dissolved in distilled water (DW) and stirred for 4 h at 37 ℃. Then, 8.2 mL of ethylene glycol was added to the above solution and stirred for another 2 h. Finally, the mixture was dialyzed against DW for 48 h, and lyophilized to obtain OCMC. OCMC-DA was successfully synthesized by the following method[7]. 3.0 g of OCMC was dissolved in DW (80 mL, pH = 7.0) under N_2 protection, and then 3.0 g of dopamine hydrochloride (DA) was added. After 5 h, dialysis against DW for 48 h and final lyophilization were performed to give the product OCMC-DA. The hydrogel prepared by Schiff base reaction with a concentration of 5% OCMC and CMCS in a volume ratio (1 : 1) was named GEL-1. The hydrogel prepared by Schiff base reaction with a concentration of 5% OCMC-DA and CMCS in a volume ratio (1 : 1) was named GEL-2.

2.3 Preparation of modified PCL nanofibers and hydrogels/nanofibers wound dressings

Electrospinning solutions were prepared by dissolving 1.2 g PCL and 0.02 g, 0.04 g, and 0.06 g of CTAB in 10 mL of DMF, respectively. In an environment with a temperature of 20-25 ℃ and a humidity of 45%-50%. The distance from the needle to the receiver was set to 16 cm, the spinning voltage was set to 12 kV, and the flow rate of the spinning solution was 0.48 mL/h. Nanofiber membranes with different contents of antibacterial agents were prepared. They were named PCL-1, PCL-2, and PCL-3, respectively. It was compounded with GEL - 2 to prepare hydrogel/nanofibers wound dressings with different antibacterial agent contents, which were named H/NF-1, H/NF-2 and H/NF-3, respectively. The formula of hydrogel/nanofibers wound dressings is detailed in Tab. 1.

Tab. 1 The recipe for hydrogel/nanofibers wound dressings

Sample	CMCS (mL)	OCMC-DA (mL)	CTAB (g)	PCL (g)
H/NF-1	1	1	0.02	1.2
H/NF-2	1	1	0.04	1.2
H/NF-3	1	1	0.06	1.2

2.4 Spectroscopic analyses

Fourier transform infrared (FTIR) spectra over the wavenumber ranging from 4 000 cm^{-1} to 400 cm^{-1} were achieved at room temperature with PerkinElmer FTIR Spectra 1 000.

2.5 ^1H NMR spectrum

A nuclear magnetic resonance instrument (Mercury VX-300) was used to scan the hydrogen spectrum, and observe the structure of the monomer and the modified monomer. The deuterated reagent was D_2O.

2.6 Morphological observation

The structure of hydrogel/nanofibers wound dressings of the hydrogels and samples were observed by SEM (Hitachi, TM3000) after being coated with gold. Typically, the hydrogels should be freeze-dried before observation.

2.7 Swelling properties

The freeze-dried hydrogels were immersed in an excess of deionized water until swelling equilibrium was achieved. The initial weight of the freeze-dried hydrogel was recorded as w_0, and the wet weight of the hydrogel was recorded as w_t after being immersed every certain time. The swelling ratio was calculated by the following equation:

$$\text{Swelling ratio}(\%) = \frac{w_t - w_0}{w_0} \times 100\% \quad (1)$$

2.8 Antimicrobial activity property

The antibacterial zone and colony count method were used to determine antimicrobial activity of the treated nanofibers against the bacterium *E. coli* and *S. aureus* according to GB/T 20944. 3—2008. The antibacterial rate was calculated by the following equation:

$$Y = \frac{W_t - Q_t}{W_t} \times 100\% \qquad (2)$$

where W_t is the number of bacterial colonies in the control sample; Q_t is the number of colonies of bacteria treated with antibacterial nanofibers.

3 Results and Discussion

3.1 Synthesis and characterization of hydrogels

As shown in Figs. 1 (a) and 1(b), compared with CMC, the characteristic absorption peak at 1 730 cm^{-1} in OCMC is assigned to the symmetrical vibration of the aldehyde group, indicating the successful synthesis of OCMC. Compared with OCMC, the characteristic absorption peaks at 1 530 cm^{-1}, 1 448 cm^{-1} and 1 281 cm^{-1} in OCMC-DA can be assigned to C=N[8], aromatic C—C[8] and C—N[8], which indicates that DA has been incorporated into the OCMC polymer chain via the Schiff base reaction. As shown in Fig. 1 (c), In ^{1}H NMR images, the chemical shifts at $5.4\text{–}4.3\times10^{-6}$ in both OCMC and OCMC-DA showed new distinct broad peaks after sodium periodate oxidation, ascribing to the formation of hemiacetal groups. Some of the hydroxyl groups on the carboxymethyl cellulose chain are oxidized to aldehyde groups. The remaining hydroxyl groups react with adjacent aldehyde groups to form a cyclic hemiacetal structure. This structure is more stable than the free aldehyde. Meanwhile, in the polymer OCMC-DA, the chemical shift of 2.8×10^{-6} was assigned to the hydrogen in the methylene group. The chemical shift of 6.6×10^{-6} corresponds to the hydrogen in the aromatic ring in DA. The successful introduction of DA was thus confirmed.

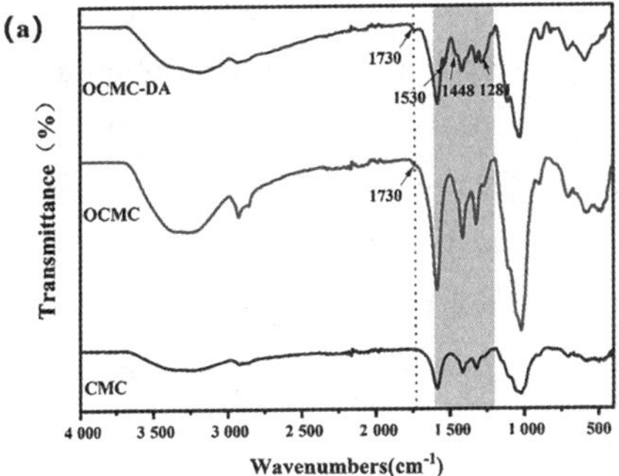

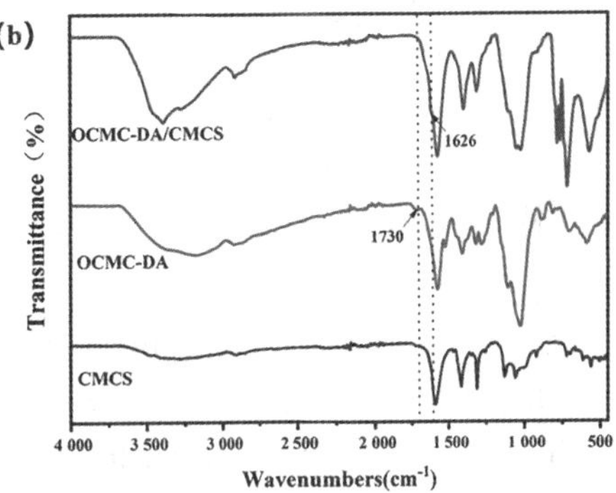

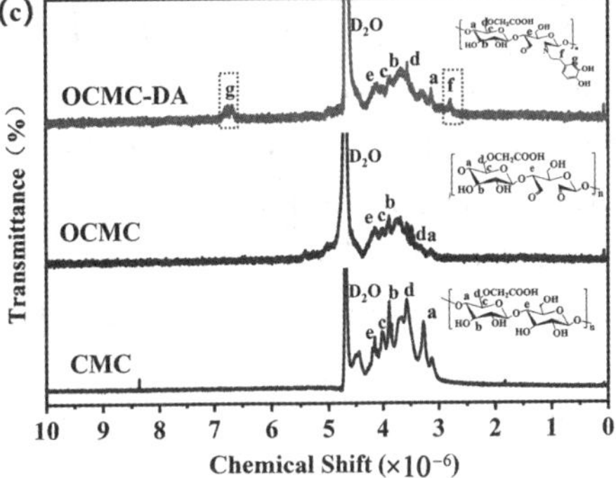

Fig. 1 (a) FTIR spectra of CMC, OCMC, and OCMC-DA; (b) FTIR spectra of CMCS, OCMC-DA, and OCMC-DA/CMCS; (c) ^{1}H NMR spectra of CMC, OCMC, and OCMC-DA

3.2 Structure of hydrogel/nanofibers wound dressings and hydrogels

SEM analysis was used to study the surface morphology of hydrogels and hydrogel/nanofibers wound dressings. In Fig. 2 (a), the SEM image shows that all injectable hydrogels have interconnected porous three-dimensional structures, and after grafting DA, GEL-2 has a larger pore size due to its small crosslinking density. H/NF-1 shows inter-cross-linked hole-like structures and nanofibers. In Fig. 2 (b), floccules appeared on the

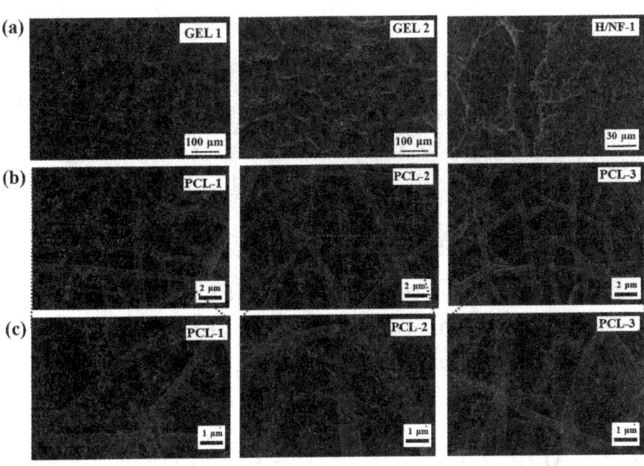

Fig. 2 (a) SEM micrographs of GEL-1, GEL-2, and H/NF-1; (b) SEM micrographs of PCL-1, PCL-2, and PCL-3; (c) SEM magnified micrographs of PCL-1, PCL-2, and PCL-3

surface of PCL nanofibers blended with CTAB, and CTAB was seen on the surface of a single modified nanofiber by further observation of the SEM magnified image.

3.3 Performance of hydrogels

It can be observed from Fig. 3 (a) that all hydrogel samples can achieve swelling equilibrium within 60 s, and the hydrogel samples show rapid swelling in the first 30 s, which may be the key factor to improving the wound environment by rapid aspiration of hydrogels. The maximum swelling ratio of GEL-1 can reach 13.57 times, and the maximum swelling ratio of GEL-2 can reach 18.85 times. After grafting DA, the swelling of GEL-2 was slightly increased due to the small cross-linking density, which corresponds to SEM result. Compared with GEL-1, GEL-2 has a larger pore size. As shown in Fig. 3 (b), the GEL-2 is able to adhere to the knuckles of the fingers, indicating that the hydrogel has excellent adhesion properties. And the adhesion mechanism of the hydrogel is shown in Fig. 3 (c). The phenolic hydroxyl groups in DA and the amino groups and hydroxyl groups in human tissues can quickly establish hydrogen bonds at the interface.

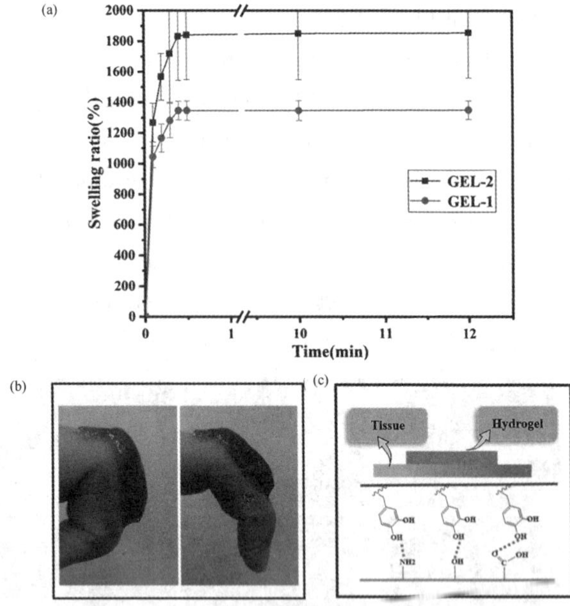

Fig. 3 (a) The swelling ratio of GEL-1 and GEL-2; (b) adhesive properties of GEL-2; (c) schematic illustration of the adhesive mechanism between hydrogels and substrates

3.4 Antibacterial properties of hydrogels and hydrogel/nanofibers wound dressings

E. coli and *S. aureus* were used as the test strains, the antibacterial properties of hydrogels samples and hydrogel/nanofibers wound dressings were detected by the method of counting the colonies. In Fig. 4 (a), the number of colonies on the agar plate after treatment with the antibacterial samples (H/NF-1, H/NF-2, H/NF-3) was less than that of the blank control group. In Fig. 4 (b), the antibacterial rates of GEL-2, H/NF-1, H/NF-2, H/NF-3 against *S. aureus* were 43.33%, 97.13%, 97.28%, 99.06%, respectively. In Fig. 4 (c), the antibacterial rates of GEL-2, H/NF-1, H/NF-2, H/NF-3 against *E. coli* were 28.57%, 97.29%, 98.48%, 99.28%, respectively. The results show that the hydrogel/nanofibers wound dressings have excellent antibacterial properties and meet the performance requirements of medical wound dressings.

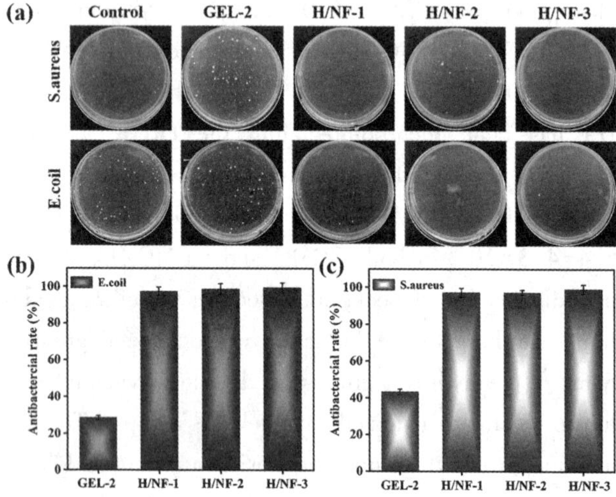

Fig. 4 (a) The pictures of agar plates treated with hydrogels samples and hydrogel/nanofibers wound dressings samples; (b)-(c) the antibacterial rate of hydrogels samples and hydrogel/nanofibers wound dressing samples

4 Conclusions

In conclusion, the design of the hydrogel/nanofibers wound dressings combine the functions of hydrogel and nanofibers with excellent water absorption and good antibacterial properties. The obtained hydrogels not only exhibited a high swelling ratio of up to 18.85 times, but also possessed suitable adhesion. FTIR and ^1H NMR demonstrated the successful oxidation of OCMC, the successful grafting of OCMC-DA and the state-of-the-art synthesis of the hydrogel. SEM shows that the hydrogels have interconnected porous three-dimensional structures and the successful preparation of antibacterial nanofibers. Studies have shown that the prepared hydrogel/nanofibers wound dressings have good antibacterial properties, which can reduce the chance of wound infection. Taken together, it can be considered that the hydrogel/nanofibers wound dressings would open a new path for wound healing.

References

[1] LIU Y, ZHOU S, GAO Y, et al. Electrospun nanofibers as a wound dressing for treating diabetic foot ulcer[J]. Asian Journal of Pharmaceutical Sciences, 2019, 14(2): 130-143.

[2] SHEN W, HE P, XIAO C, et al. From antimicrobial peptides to antimicrobial poly(alpha-amino acid)s[J]. Advance Health Material, 2018, 7(20): e1800354.

[3] NAHA P C, LIU Y, HWANG G, et al. Dextran-coated iron oxide nanoparticles as biomimetic catalysts for localized and pH-activated biofilm disruption[J]. American Chemical Society Nano, 2019, 13(5): 4960-4971.

[4] LI S, DONG S, XU W, et al. Antibacterial hydrogels [J]. Advance Science, 2018, 5(5): 1700527.

[5] CAO L, CAO B, LU C, et al. An injectable hydrogel formed by in situ cross-linking of glycol chitosan and multi-benzaldehyde functionalized PEG analogues for cartilage tissue engineering[J]. Journal of Materials Chemistry B, 2015, 3(7): 1268-1280.

[6] HU W, ZHANG Z, ZHU L, et al. Combination of polypropylene mesh and in situ injectable mussel-inspired hydrogel in laparoscopic hernia repair for preventing post-surgical adhesions in the piglet model[J]. ACS Biomaterial Science & Engineering, 2020, 6(3): 1735-1743.

[7] HU C, LONG L, CAO J, et al. Dual-crosslinked mussel-inspired smart hydrogels with enhanced antibacterial and angiogenic properties for chronic infected diabetic wound treatment via pH-responsive quick cargo release [J]. Chemical Engineering Journal, 2021, 411.

[8] ZHONG Y, XIAO H, SEIDI F, et al. Natural polymer-based antimicrobial hydrogels without synthetic antibiotics as wound dressings[J]. Biomacromolecules, 2020, 21(8): 2983-3006.

Fabrication of Hygroscopic Cotton/Nanofiber Composite Yarn for Atmosphere Water Harvesting

LIU Han[1], Li Ailin[1], WANG liming[1], QIN Xiaohong[1], YU Jianyong[1,2*]

[1] *Key Laboratory of Textile Science & Technology of Ministry of Education, College of Textiles, Donghua University, Shanghai 201620, China*

[2] *Innovation Center for Textile Science and Technology, Donghua University, Shanghai 201620, China*

* *Corresponding author's email*: yujy@ dhu. edu. cn

Abstract: Water is essential for human life. In order to alleviate the problem of water shortage, the way to efficiently obtain fresh water has attracted increasing attention, which contains different methods including desalination of seawater, wastewater treatment and other convention water treatment methods. Current research focuses on the development of atmosphere water collecting membrane materials, which have some limitations due to the size and durability. In this study, LiCl was used as a hygroscopic agent and CNT was used as a solar thermal agent to prepare nanofiber membranes with good hygroscopicity by electrospinning. The hygroscopicity can reach 94. 12% under the LiCl concentration of 20%. In addition, composite yarn was prepared by the combination of cotton fiber and nanofiber. Under the promotion of nano-structure, the hygroscopic performance of cotton yarn was improved from 10. 41% to 15. 34% under 90% humidity, which had the potential to be applied in the field of hygroscopic and water harvesting fabrics.

Keywords: LiCl; Electrospinning; Water harvesting; Composite yarn

1 Introduction

Water is an essential resource for human life. Although the earth is rich in water resources, the total freshwater resources only account for 2. 80% of the total water resources, and only 0. 34% of the freshwater can be directly used. Besides, weak economic development and poor infrastructure construction are also economic reasons for the inefficient use of water resources[1-3]. Therefore, in order to alleviate the problem of water shortage, how to obtain a large amount of fresh water efficiently has become the focus of scientists' research at present.

Fresh water is mainly distributed in liquid, solid and gaseous forms in glaciers, underground, surface and air. Water for human consumption is mostly purified from rivers and lakes, other conventional water treatment technologies including seawater desalination and sewage treatment are also the main sources. Seawater desalination technology includes reverse osmosis, multi-stage flash, freezing, steam distillation, electrodialysis and so on[2]. Wastewater treatment technologies include physical, chemical and biological methods. The above conventional water treatment technologies have solved the freshwater supply problem in most areas and are widely used nowadays. However, the above water treatment technologies are only suitable for areas with direct water sources, and there is still no effective method for remote areas with scarce water resources.

The water vapor content in the air is about 15 trillion liters, which is much larger than the surface marshes[2]. Therefore, collecting water from the air not only saves labor and energy, but also applies to a wide range of areas, and contains fewer impurities and bacteria than other water sources, which can simplify the complex water purification process to meet drinking water standards[1]. In terms of energy utilization, current air water harvesting technologies can be divided into three categories[2]. I) Passive water harvesting systems, which do not require additional energy input and can spontaneously collect water from the air, including rainwater harvesting, fog harvesting and passive surface cooling-induced condensation harvesting. The equipment is simple and the energy consumption is low. II) Active water collection system, which requires additional input of energy such as electric power, including surface cooling and membrane separation technology. This kind of system with large water production but the equipment is large and complex thus can cause high energy consumption. III) Solar driven water harvesting, the system absorbs and collects moisture in the air through a hygroscopic agent, and then collects fresh water through solar driven evaporation and desorption of moisture. The system has unlimited solar energy sources and no pollution problems and the system has simple equipment, low cost and high efficiency of water production. In recent years it is the most promising way for the development of air water harvesting.

Electrospinning technology is the most efficient way to prepare nanofibers. Nanofibers prepared by this technology have the characteristics of high porosity and large specific surface area, and are widely used in filtration, biomedicine, catalysis and other fields. In the field of photothermal driven moisture absorption and water harvesting, there have also been many studies using electrospinning to prepare composite membrane materials. Zhang et al. [9] prepared PAN@ MIL/LiCl hygroscopic layer by electrospinning technology, and combined it with PAN/CB photothermal layer to prepare hygroscopic pump for solar driven indoor continuous dehumidification. The results show that the nanofiber structure can significantly improve the adsorption-desorption rate of moisture.

In this study, cotton fiber and nanofiber were combined to prepare cotton/nanofiber (NF) composite yarn, in which the nanofiber was fabricated by polyacrylonitrile (PAN)/multi-wall carbon nanotubes (CNTs)/lithium chloride (LiCl). As the hygroscopic agent of yarn, LiCl could improve the hygroscopic performance of cotton yarn with the promotion of composite structure. And it has the potential to be applied in the field of air water collection. This study includes exploring the influence of the amount of LiCl on the hygroscopic properties of nanofibers, the preparation and morphology characterization of composite structure composite yarns, and the hygroscopic properties of composite yarns.

2 Experiments

2.1 Materials

Polyacrylonitrile (PAN) was provided by Shanghai Sinopharm Chemical Reagent Co., Ltd. *N*, *N*-dimethylformamide (DMF, AR ≥99.5%) was of analytical grade, purchased from Shanghai Lingfeng Chemical Reagent Co., Ltd. Anhydrous LiCl was purchased from Chengdu McCaxi Chemical Co., Ltd. Multi-wall carbon nanotubes (CNTs) were supplied by Aladdin Reagent (Shanghai) Co., Ltd.

2.2 Preparation of PAN/CNT @ LiCl fibrous membranes

An electronic balance (JA2003B, Shanghai Yueping Scientific Instrument Manufacturing Co., Ltd.) was used to weigh 10.5 g PAN powder, and then the PAN powder was added to 100 mL DMF. At the same time, anhydric LiCl with weights of 0 g, 0.525 g, 1.050 g, 1.575 g and 2.100 g were added. Then 5 wt% CNTs was added to the solution. Using digital display magnetic stirrers (MS-H280, Dalong Xingchuang Experimental Instrument Co. Ltd.) to stir for 12 h to form a uniform and stable solution, the spinning solution with PAN concentration of 10% and LiCl concentration of 0%, 5%, 10%, 15% and 20% was obtained. Continuous nanofibers were generated using a free surface electrospinning machine without a needle, which was used in the previously reported work in our group[8]. Collect the nanofibers on a uniformly rotating roller. The spinning voltage was set at 45 KV, the receiving distance was 25 cm, and the spinning time was 1 h. Finally, PAN/CNT @ LiCl nanofiber membrane with nanofiber structure was prepared.

2.3 Preparation of cotton/NF composite yarn

As described above, the spinning solution with PAN concentration of 10% and LiCl concentration of 15% was prepared. The cotton web carded by the carding machine was attached to the roller of the free surface electrospinning equipment. The spherical nozzle continuously generated multiple Taylor cones and became nanofibers which were then attached to the cotton mesh under electric field drafting. The conditions and parameters of electrospinning are as described above. The cotton web spun with nanofibers was peeled off, and then went through subsequent processes such as winding, drawing (DYJ305, Shandong Longmei Mining Machinery Co. Ltd.), roving (DHU C01, Donghua university) and ring spinning (DHU X-01, Donghua university). Then cotton/NF composite yarn was established.

2.4 Characterization

The morphology of nanofiber membranes and cotton/NF composite yarn were observed by scanning electron microscopy (SEM, DLXS-10ACKT, Hitachi Co. Ltd.).

Solar absorption was measured with a Fourier transformation infrared (FTIR) spectrometer (Spectrum Two, Perkin Elmer, Inc.) in the wavelength range of 300 mm to 2 500 nm. Fabric moisture permeability tester (Ningfang, YG 601H, China) was used as a constant temperature and humidity chamber to test the moisture absorption ability of samples.

The hygroscopic performance test is performed by controlling the ambient temperature at 25 ℃ and ambient humidity at 60%, 70%, 80%, and 90%. The nanofiber membranes were cut into circular samples with a diameter of 3 cm and placed in a vacuum oven to dry for 12 h at 120 ℃ before testing. After starting the test, place the fully dried sample in the setting temperature and humidity environment. The nanofiber membrane has a fast moisture absorption speed, so the weight is recorded every 2 min in the first 20 min, and then recorded every 5 min when the weight gradually gets stable. The weight of the nanofiber composite yarn was recorded every 5 min.

The hygroscopic capacity is calculated using the following formula:

$$C_{abs} = \frac{\Delta m}{m_0} \tag{1}$$

where m_0 is the initial mass and Δm is the hygroscopic capacity, which is also the mass change before and after hygroscopic absorption. The same samples were used to test the hygroscopic ability of the nanofiber membrane under different humidity to observe the enhancement of the hygroscopic ability of the nanofiber membrane by LiCl. The hygroscopic ability of pure cotton yarn and composite cotton yarn was also compared.

3 Results and Discussion

3.1 Morphology and structure of membrane and yarn

Figs. 1 (a)-1(b) shows the SEM images of PAN/CNT @ LiCl electrospun membrane. In Fig. 1 (a), clear nanofibers can be seen with good orientation. Some nanofibers are not completely stretched, and some adhesion and entanglement exist. Fig. 1 (b) shows that the diameter of the nanofibers is relatively uniform with a small amount of beads in it. The viscosity of electrospinning solution can cause a great influence on electrospinning effect. Because of the salt, the viscosity is reduced. In the meantime, the moisture absorption ability of LiCl is extremely strong and may absorb the moisture in the air, thus the solvent cannot completely evaporate during electrospinning process, leading to nanofiber adhesion phenomenon exists. Besides, LiCl and CNTs have strong electrical conductivity and cause high electrical conductivity of solution, resulting in more electrostatic residue between nanofibers and large electrostatic force, so nanofibers present a relatively fluffy state.

Fig. 1 Structure characterization: (a)-(b) SEM images of PAN/CNT @ LiCl nanofiber membrane at different magnifications; (c)-(d) SEM images of cotton/NF composite yarn at different magnifications

Figs. 1 (c)-1(d) shows the SEM images of the cotton/NF composite yarn with mosaic structure. Fig. 1 (c) shows the composite yarn with a certain twist, among which the waist shaped fibers are cotton fibers, and it can be observed that there are concentrated flake distributed nanofibers among the cotton fibers, indicating that the spinning method can make the nanofibers infiltrate into the cotton yarn. In Fig. 1 (d), a single nanofiber exists between two cotton fibers. The nanofiber is short in length, indicating that the nanofiber may be cut by the external force of the machine in the subsequent spinning process. The aggregation and dispersive distribution of nanofibers in the cotton yarn indicate that the spinning method can embed the nanofibers in the yarn, and the nanofibers and cotton fibers have good recombination.

Fig. 2 shows the FTIR spectra of PAN/CNT @ LiCl nanofiber membrane, cotton yarn and cotton/NF composite yarn. The nanofiber membrane has an absorption peak at 2 242 cm^{-1}, which is the characteristic absorption peak of —C≡N in PAN and can be clearly seen in PAN/CNT @ LiCl nanofiber membrane. Besides, a smaller absorption peak can be seen in composite yarn and cannot be seen in pure cotton yarn, indicating the existence of nanofiber in composite yarn. Both pure cotton yarn and composite yarn have a characteristic peak of 1 030 cm^{-1} as —C—O— bond, which exists in pure cotton yarn and composite yarn. The infrared image shows that there are a small number of nanofibers in the composite structure of cotton yarn, and the chemical structure of cotton/NF yarn has not changed significantly.

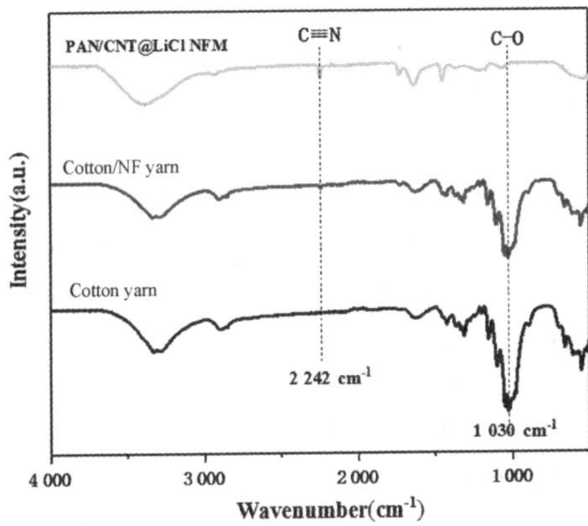

Fig. 2 Structure characterization: FTIR spectra of PAN/CNT@LiCl nanofiber membrane, cotton yarn and cotton/NF composite yarn

3.2 Moisture absorption performance of PAN/CNT@LiCl NFM

Fig. 3 shows the moisture absorption performance of PAN/CNT@LiCl NFM at 25 ℃ and 90% RH. As can be seen from Fig. 3, the nanofiber membrane gradually absorbed the moisture until get stable in 60 min. The PAN NFM has a stable hygroscopic capacity of 0.053 $g \cdot g^{-1}$. When the concentration of LiCl was 5%, 10%, 15% and 20%, the hygroscopicity of NFM was 0.240, 0.465, 0.743 and 0.941 $g \cdot g^{-1}$. The maximum concentration of LiCl was chosen to be 20%, because the amount of LiCl would have a great impact on the electrospinning effect, and it was difficult to spin successfully when the electrical conductivity of spinning solution was too high. As shown in Fig. 3, the addition of LiCl significantly improved the hygroscopicity of NFM. With the addition of LiCl, the hygroscopicity increased from 0.053 $g \cdot g^{-1}$ to 0.941 $g \cdot g^{-1}$. With the increase of LiCl concentration, the hygroscopicity of NFM is gradually enhanced. When the concentration of LiCl is 20%, the spinning effect is poor and the silk is unstable so the roller is difficult to collect nanofibers. In order to obtain the best hygroscopic ability and spinning effect, the concentration of LiCl was selected to be 15% in the subsequent preparation of composite yarn.

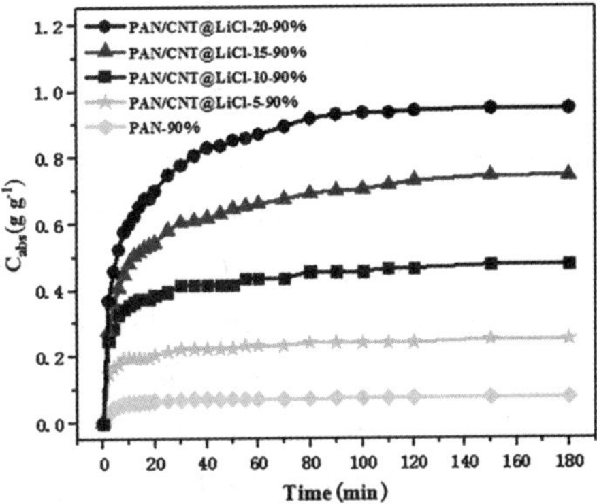

Fig. 3 Moisture absorption performance of PAN/CNT@LiCl nanofiber membrane with LiCl concentration of 0%, 5%, 10%, 15% and 20% at 25 ℃ and 90% RH

3.3 Moisture absorption performance of cotton/NF yarn

After selecting 15% as the subsequent operating concentration of LiCl, the subsequent composite electrospinning and spinning process were carried out, and the hygroscopic properties of cotton yarn, composite yarn and nanofiber film were compared. Under two humidity conditions of 60% and 90%, the hygroscopicity of the three samples is shown in Figs. 4 (a)−4(b). When

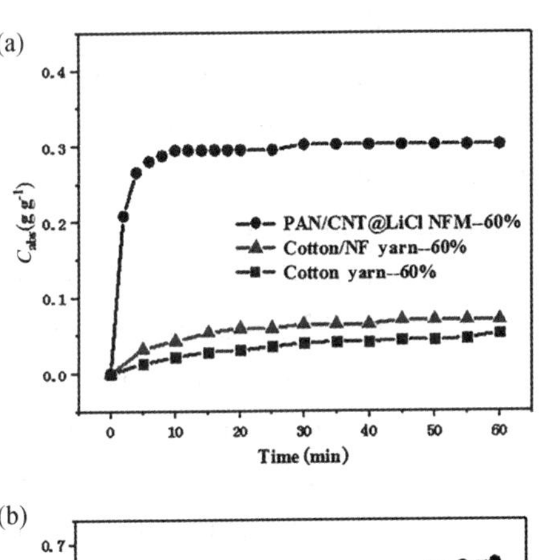

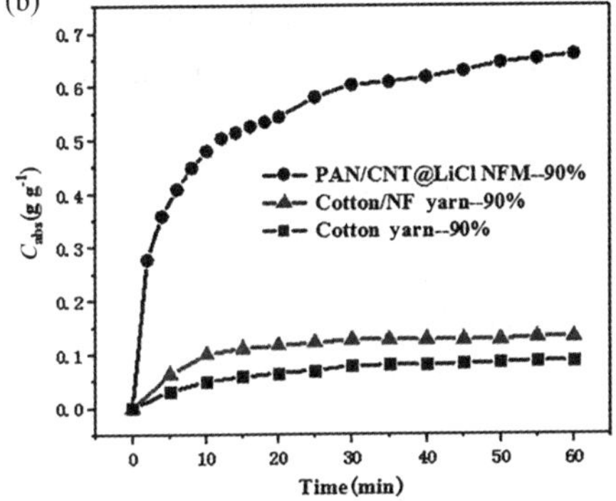

Fig. 4 Moisture absorption performance of PAN/CNT@LiCl nanofiber membrane with LiCl concentration of 15%, cotton yarn, cotton/NF composite yarn. The conditions are at (a) 25 ℃ and 60% RH; (b) 25 ℃ and 90% RH

the humidity is 90%, the hygroscopicity of cotton yarn and composite yarn is 0.010 4 $g \cdot g^{-1}$ and 0.153 $g \cdot g^{-1}$. When the humidity is 60%, the hygroscopic of cotton yarn and composite yarn are 0.054 $g \cdot g^{-1}$ and 0.075 $g \cdot g^{-1}$. With the increase in humidity, the hygroscopicity of both yarns showed an upward trend, and the hygroscopicity of composite yarn was significantly improved compared with cotton yarn. This is because the large porosity of nanofibers provides many microchannels for yarns, and the hygroscopic ability of yarns is further improved due to the existence of LiCl. However, compared with NFM, the hygroscopic capacity is inferior because the composite yarn contains less nanofibers.

4 Conclusions

In summary, a PAN/CNT@ LiCl nanofiber membrane has been successfully fabricated by electrospinning, and the enhancement of the hygroscopic ability of nanofiber membrane was investigated under the condition of 90% RH. Then the composite nanofiber yarn was prepared by combining electrospinning with traditional spinning. The improvement of the hygroscopic ability of composite yarn compared to cotton yarn was studied under humidity of 60% and 90%. It is concluded that LiCl can improve the hygroscopicity of nanofiber membrane from 0.053 $g \cdot g^{-1}$ to 0.941 $g \cdot g^{-1}$. With the concentration of LiCl increasing from 5% to 20%, the hygroscopicity gradually increased. The hygroscopic ability of nanofiber composite yarn is also improved from 0.010 4 $g \cdot g^{-1}$ to 0.153 $g \cdot g^{-1}$ at 90% RH. Besides, the addition of CNTs can make the nanofiber membrane have faster photothermal conversion efficiency and water evaporation rate under light, which is conducive to the desorption and collection of water. This method is expected to be used in the field of water harvesting yarn or clothing.

References

[1] KORKMAZ S, KARIPER I A. Fog harvesting against water shortage[J]. Environmental Chemistry Letters, 2020, 18(2): 361-375.

[2] ZHUANG S D, QI H S, WANG X Y, et al. Advances in solar-driven hygroscopic water harvesting [J]. Global Challenges, 2021, 5(1): 22.

[3] WANG B, ZHOU X C, GUO Z G, et al. Recent advances in atmosphere water harvesting: design principle, materials, devices, and applications[J]. Nano Today, 2021, 40.

[4] PAN S B, WANG J N, FANG Y N, et al. Preparation and characterization of electrospun PA6/CTS/LiCl nanofibrous membrane [J]. New Chemical Materials, 2019, 47(6): 93-96,101.

[5] LI A L, XIONG J, LIU Y, et al. A rapid-ab/desorption and portable photothermal MIL-101(Cr) nanofibrous composite membrane fabricated by spray-electrospinning for atmosphere water harvesting[J]. Energy & Environmental Materials, 2022, 8.

[6] JI J, WAGN J N, WAGN B, et al. Study on preparation of PAN/LiCl electrospun nanofiber membrane and its liquid desicant dehumidification performance [J]. New Chemical Materials, 2019, 47(12): 238-242.

[7] DU M, ZHAO Y, TIAN Y, et al. Electrospun multiscale structured membrane for efficient water collection and directional transport [J]. Small, 2016: 12(8): 1000-1005.

[8] YANG Y C, ZHAO Y J, QUAN Z Z, et al. An efficient hybrid strategy for composite yarns of micro-/nano-fibers[J]. Materials & Design, 2019, 184.

[9] ZHANG Y F, WU L, WANG X F, et al. Super hygroscopic nanofibrous membrane-based moisture pump for solar-driven indoor dehumidification[J]. Nature Communications, 2020, 11(1).

[10] XU J H, DU X X, XIN B J, et al. Moisture-Wicking and solar-heated coaxial fibers with a bark-like appearance for fabric comfort management[J]. ACS Applied Materials & Interfaces, 2021, 13(22): 26590-26600.

[11] CHEN J X, MAI J Z, WANG C, et al. Biomimetic aligned micro-/nanofibrous composite membranes with ultrafast water transport and evaporation for efficient indoor humidification[J]. ACS Applied Materials & Interfaces, 2022, 14(1): 1983-1993.

Patterned Woven Fabrics with Special Wettability for Efficient Fog Collection

ZHANG Ling[1], CHEN Xia[1*], LIU Hongling[1*]

[1] *College of Textiles, Key laboratory of Textile Science Technology, Ministry of education, Donghua University, Shanghai 201620, China.*

[*] *Corresponding author's email*: chenxia@ dhu. edu. cn, hlliu@ dhu. edu. cn

Abstract: Inspired by the phenomenon of fog collection on biological surfaces in nature, fog collection fabrics are attracting worldwide attention for their ability to capture and transfer fog droplets in a directional manner due to the differential distribution of surface wettability. We design and weave different patterned fog collection fabrics based on fabric structural design and material optimization. Environmentally friendly and fluorine-free superhydrophobic modified polyester yarn and hydrophilic viscose yarn are rationally combined to construct patterned woven fabrics in cord weave, honeycomb weave, pointed twill weave and square check weave. Scanning electron microscopes, optical microscopes, contact angle testers, and laboratory-made fog collectors were used to characterize the properties of the yarns and fabrics. The fog collection efficiency of the fabric with cord weave reached 1 983 mg · m^{-2} · $(6\ h)^{-1}$, which was equivalent to 79. 3 L water per day for 1 m^2 fabric. Fog collection fabrics show great promise for development by reducing external energy input, being clean and environmentally friendly, low cost, sustainable, and meeting human water needs. Special wettability patterned fabrics show great potential in the field of fog collection.

Keywords: Special wettability; Fog collection; Weave design; Woven fabric

1 Introduction

Water resources are essential for the survival of life on earth. However, the shortage of fresh water resources is not only a serious constraint to human daily production and life, but also poses a more critical challenge to all living organisms[1]. The unique water harvesting structures evolved by plants and animals in nature provide new inspiration for the human to design and prepare water harvesting devices[2-3]. Inspired by the patterned structure of the desert beetle's back to collect water, special wettability materials utilize the hydrophilic/hydrophobic properties of their surfaces to enable spontaneous water collection and transport, which reduces the additional energy that needs to be applied for liquid condensation and movement[4]. Wang et al. [5] created superhydrophobic fabric surfaces with photo-induced superhydrophilic bumps to accelerate the agglomeration of water droplets on them. Yu[6] constructed hybrid wettable water collecting fabric by interweaving hydrophilic and hydrophobic yarns. The water collection efficiency of the water-collecting fabric was 784. 3 mg · m^{-2} · $(6\ h)^{-1}$. Moazzam et al. [7] used a hydrophilic/hydrophobic patterned nonuniformly wettability fabric surface constructed by mask lithography with a water collection rate of 882. 4 mg · m^{-2} · $(6\ h)^{-1}$, which was equivalent to about 35 L of fresh water per m^2 for 1 d, showing significantly better water collection advantages than conventional hydrophilic/hydrophobic surfaces. However, this technology relies on mask plates and lithography machines, and the cost and process requirements are relatively high. Based on the fabric weave design method, we used the superhydrophobic yarns and hydrophilic yarns to prepare fog collection fabrics.

In this paper, green and fluorine-free superhydrophobic polyester yarns and viscose yarns with excellent hydrophilicity were selected as weaving materials. Rectangular, circular, triangular and square fabric patterns were used as the inspiration source to prepare cord weave, honeycomb weave, pointed twill weave, and square check weave fog collection fabrics. To explore the morphology of the fog collection fabric, the surface of the fabric was observed with a scanning electron microscope (SEM) and an optical microscope. The contact angle and fog water collection experiments were used to characterize the fog collection properties of the fabrics. The test results were expected to provide useful information for the preparation of efficient fog collection fabrics.

2 Experiments

2.1 Materials

Polyester yarn (16.4 tex) was provided by Star Textile Accessories Co., Ltd. (China). Viscose yarn (16.4 tex) was provided by Fujian Xinhuayuan Textile Co. Ltd. (China). *N*, *N*-ethylenebisacrylamide (MBAA), divinylbenzene (DVB), azodiisobutyronitrile (AIBN), vinyltrimethoxysilane (VTMS), polydimethylsiloxane (PDMS) and ethyl acetate were provided by Aladdin's Reagent Shanghai Co., Ltd. (China).

2.2 Preparation of superhydrophobic polyester yarn

DVB was synthesized into PDVB by the in situ polymerization method. AIBN, MBAA, VTMS and DVB were dissolved in the DMF solution for 60 min with an oscillator at 120 r/min. Then, the synthesis kettle was placed in a vacuum oven at a temperature of 100 ℃ for 24 h. The solution was dried and ground to obtain PDVB.

PDVB, PDMS and α-cyanoacrylate were fully dissolved in ethyl acetate solution in a certain proportion. The hydrophobic solution was obtained by ultrasonic shaking for 20 min. Then, the polyester yarns were then soaked in a hydrophobic solution for 10 min and dried to move the solution to ethyl acetate. Finally, polyester yarns were treated via the plasma technique in vacuum to obtain superhydrophobic polyester yarns.

2.3 Preparation of fog collection fabrics

SGA598 semi-automatic loom was used for weaving. The weaving process of fabric is divided into four steps: heald, reed, warping and weft beating. Two shuttles are needed in the weaving progress.

2.4 Wetting performance test

The contact angle of the fabric was measured by a contact angle tester.

2.5 Fog collection performance test

In order to verify the fog harvesting performance of the samples, a laboratory-made test system was constructed, as shown in Fig. 1. The prepared sample (8 cm×8 cm) was fixed on a holder in a plexiglass box (PMMA, length×width×height: 1 m×1 m×1 m), and its surface was perpendicular to the horizontal plane.

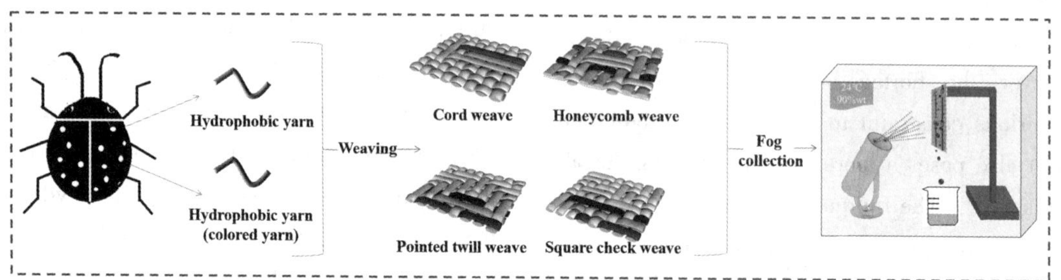

Fig. 1 Fog collection fabric preparation mechanism diagram

3 Results and Discussion

3.1 Surface morphology of fog collection fabric

Fig. 2 (a) shows the polyester fiber after superhydrophobic coating finishing. High resolution SEM shows a surface with multi-scale roughness with abundant mesopores/macropores. Meanwhile the polyester fibers are tightly covered by hydrophobic organic matter PDVB. These features are essential for superhydrophobicity and WCA is 159.7°. As shown in Fig. 2(b), the longitudinal microgrooves of viscose fibers make the specific surface area of viscose yarns large, hence it exhibits good hydrophilicity.

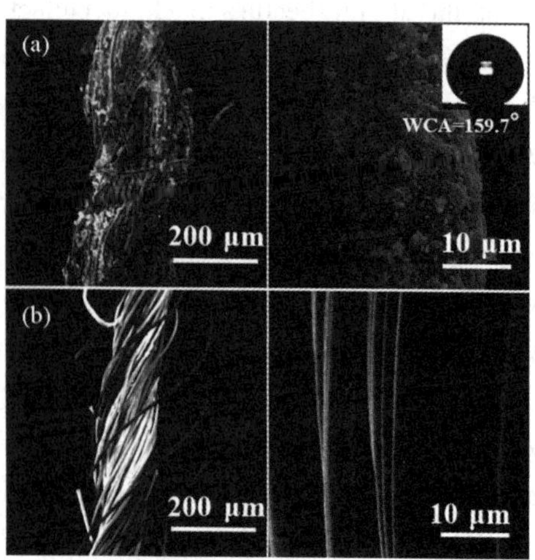

Fig. 2 SEM images of yarns: (a) superhydrophobic polyester yarn; (b) viscose yarn

The functionality of the fog collection fabric comes from the reasonable configuration of hydrophilic and hydrophobic areas on the fabric surface. In order to simulate the back structure of desert beetle, the patterned surface of fog collection fabric is designed with cord weave (1#), honeycomb weave (2#), and pointed twill weave (3#) and square check weave (4#), as shown in Fig. 3. Superhydrophobic polyester yarns are used as warp yarns, and viscose yarns and superhydrophobic polyester yarns are used as hydrophilic area weft yarns and hydrophobic area weft yarns respectively, forming a ratio of 25% of its surface hydrophilic/hydrophobic area, where the convex strip weave has a ratio of 28.1% due to its regularity.

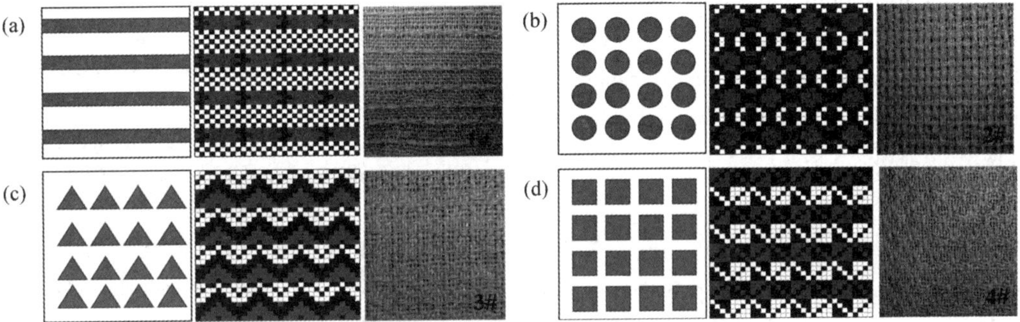

Fig. 3 Surface patterned design and optical microscope image of the fog collection fabric: (a) rectangle; (b) circle; (c) triangle; (d) square (the colored yarns are hydrophilic yarns)

3.2 Wettability of fog collection fabrics

The difference in wettability distribution on the fabric surface brings the driving force to transfer droplets spontaneously, therefore, the wettability of the fabric is the key influencing factor. Fig. 4 shows that the WCA of the hydrophobic area of the fabric is above 128°, and the WCA of the hydrophilic area is below 60°, due to a large difference in wettability between the two. Hydrophilic spots capture droplets from the air and form tiny bubble on the surface. When the bubble coalesces into droplets, the adhesion force at the hydrophilic area is not enough to support the droplets, which will converge to the hydrophobic area and flow to the collection device along the hydrophobic area. Under the synergistic effect of hydrophilic and hydrophobic properties of the fabric, the process of capturing droplets and collecting water is accomplished.

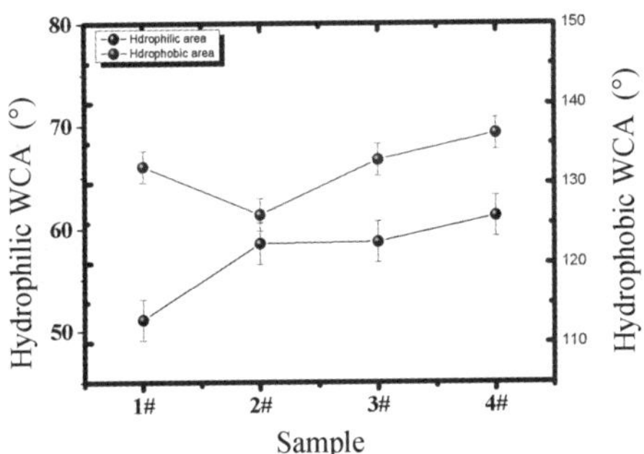

Fig. 4 WCA of hydrophilic and hydrophobic areas of fog collection fabric

3.3 Fog collection performance

Fog collection experiments were carried out by a device built independently in the laboratory. As shown in Fig. 5 (a), the fog collection fabric has been maintaining a relatively stable fog collection rate, and its 6-hour fog collection volume can reach more than 109 000 mg. As shown in Fig. 5 (b), the fog collection rate of 1 # fabric can reach 1 983.3 mg · (6 h)$^{-1}$ · cm^{-2}, which is equivalent to 79.3 L freshwater resources per day collected by 1 m^2 of 1# fabric.

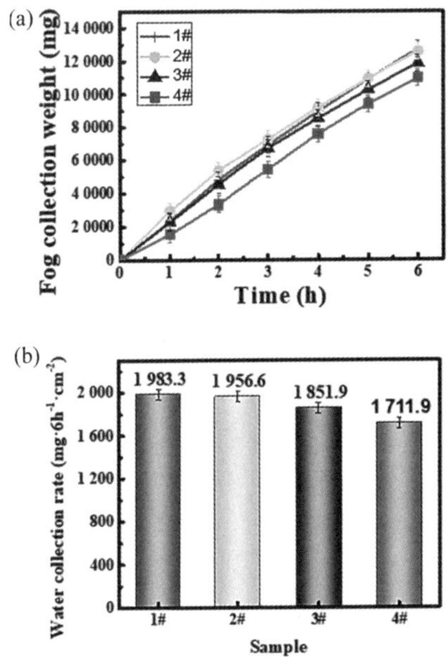

Fig. 5 Fog collection of different patterned fabrics: (a) fog collection weight; (b) 6 h fog collection efficiency

1# fabric is a cord weave, whose boundary between hydrophilic area and hydrophobic area is more clear and convenient for fog droplet capture. At the same time, the hydrophobic area of the fabric is continuous and uniform, which is conducive to the transfer of water droplets along the hydrophobic area to the collection device. In 4# fabric, the hydrophobic area at the weft yarn shared with the hydrophilic area has a liquid diffusion and penetration phenomenon. Because its back side is hydrophilic viscose yarn, making it easier for liquids to penetrate the fabric, and its fog collection efficiency is low.

4 Conclusions

Green fluorine-free superhydrophobic polyester yarns and hydrophilic viscose yarns were used as weaving materials. Combined with the fabric weave design, the surface of the fog collection fabric shows a regular hydrophilic/hydrophobic pattern. Through the fog collection experiments, it is proved that the fog collection efficiency of the four fabrics has reached more than 1 711.9 mg · $(6\ h)^{-1}$ · cm^{-2}, of which 1# fabric reaches 1 983.3 mg · $(6\ h)^{-1}$ · cm^{-2}, equivalent to 79.3 L clean water can be collected per square meter per day. The special wettability patterned fabric provides a new idea for the research of fog collection device, which has a broad application prospect in the field of fog collection.

References

[1] JURY W A, VAUX H J. The emerging global water crisis: managing scarcity and conflict between water users[J]. Advances in Agronomy, 2007: 1-76.

[2] ZHENG Y, BAI H, HUANG Z, et al. Directional water collection on wetted spider silk [J]. Nature, 2010, 463(7281): 640-643.

[3] GARROD R P, HARRIS L G, SCHOFIELD W, et al. Mimicking a Stenocara beetle's back for microcondensation using plasmachemical patterned superhydrophobic-superhydrophilic surfaces[J]. Langmuir, 2007, 23(2): 689-693.

[4] MOHD G, MAJID K, LONE S. Multiscale janus surface structure of trifolium leaf with atmospheric water harvesting and dual wettability features[J]. ACS Applied Materials & Interfaces, 2022, 14(3): 4690-4698.

[5] WANG Y, WANG X, LAI C, et al. Biomimetic water-collecting fabric with light-induced superhydrophilic bumps[J]. ACS Applied Materials & Interfaces, 2016, 8(5): 2950-2960.

[6] LIANG Y, GUO J. Research and fabrication of water collection device in textile-based fog[J]. Synthetic Fiber, 2017, 46(10): 29-33.

[7] MOAZZAM P, TAVASSOLI H, RAZMJOU A, et al. Mist harvesting using bioinspired polydopamine coating and microfabrication technology[J]. Desalination, 2018, 429, 111-118.

Evaluation of Performance of 2.5D Fabric-Based Sensor for Plantar Pressure Monitoring

ZHOU Wenna[1,2], TAO Qingyun[1,2], JIANG Ziqin[1,2], WANG Huanyu[1,2], HU Jiyong[1,2*], YANG Xudong[1,2]

[1] *Key Laboratory of Textile Science & Technology, Ministry of Education, Donghua University, Shanghai 201620, China*

[2] *College of Textiles, Donghua University, Shanghai 201620, China*

* *Corresponding author's email*: hujy@dhu.edu.cn

Abstract: With the emergence of an aging population society, the incidence of foot diseases such as diabetic foot continues to increase. Flexible textile-based pressure sensors have become increasingly important for monitoring plantar pressure and preventing foot diseases due to their softness, comfort, and biocompatibility. However, stability and durability under repeated pressure use remain a challenge for existing plantar pressure monitoring sensors. Moreover, most of the current textile-based pressure sensors are used to enhance the sensor performance by changing the electrode layer material, and few studies have been conducted on the fabric sensor structure itself. In this study, we designed an all-fabric piezoresistive pressure sensor based on a 2.5D bilayer fabric tissue structure to monitor plantar pressure changes in the form of a fabric sensing array. The study was first conducted by preparing fabric sensing elements of different proportions, evaluating their sensing performance parameters, and determining the optimal sensing element for the selected application. The results show that the unique structure makes the fabric piezoresistive sensing element with 10 : 6 ratio of conductive to common yarns exhibit high sensitivity ($GF=45.02\times10^{-3}(N/cm^2)^{-1}$), good linear response in the range of 0–1 000 N/cm², and good stable repeatability. Moreover, it shows good repeatability and detection performance in practical applications, meeting the performance requirements of daily plantar pressure monitoring. And it has the potential application to become the next generation of plantar pressure wearable monitoring device.

Keywords: All-fabric sensors; Sensing elements; Plantar pressure monitoring; Piezoresistive

1 Introduction

In recent years, the incidence of foot diseases such as diabetic foot has continued to increase as the population ages. Plantar pressure is significantly abnormal at the onset of diabetic foot disease. Plantar pressure distribution characteristics can reflect foot injuries or lesions, and the degree of diabetic foot ulcers. Monitoring plantar pressure can provide reasonable treatment plans and take effective protective measures[1]. Conventional plantar pressure collection systems include platform force measurement systems and pressure sensing walkways[2]. Due to the lack of portability, the application environment of these devices is limited to use in the laboratory and is not suitable for continuous wearable monitoring. They are also rigid, not biocompatible with human skin, and prone to deformation when subjected to repeated external pressure[3-5]. With the development of flexible electronic devices and wearable electronic devices in recent years, flexible textile pressure sensors based on electrical characteristics have been combined with traditional footwear to realize the acquisition, processing, and transmission of plantar pressure data. Textiles have become an ideal substrate for flexible wearable devices and are receiving more and more attention to monitoring plantar pressure and preventing foot diseases.

Flexible textile sensors are produced by weaving conductive fibers/yarns directly into the fabric using the basic electrical properties of the material on a conventional textile substrate. Among them, piezo-resistive fabric pressure sensors are widely used for plantar pressure measurement because of their high sensitivity, simple preparation process, and low cost. When it is stimulated by external pressure, the piezoresistive layer converts the applied pressure information into a resistive signal, and an electrical signal output proportional to the force change is obtained through the measurement circuit. The fabric structure and the choice of material play an important role in the construction of piezoresistive textile-based pressure sensors[6]. Generally speaking, textile-based sensor structures mostly use multilayer stacked structures, and most of the piezoresistive or electrode layers are coated to become conductive fabric substrates. For example, Lian et al.[7] prepared an all-fabric pressure sensor by inserting a cotton mesh matting plate between a double layer of silver nanowire-coated cotton fabric, which showed extremely high sensitivity due to the collective effect of the fiber/yarn/fabric multi-scale contact. Liu et al.[8] produced a textile pressure sensor using an interleaved structure containing Ni and CNT-coated cotton fabric. In addition, Ge et al.[9] fabricated a flexible pressure sensor using a 3D polyaniline sponge as a framework. However, such structures and coating treatments can lead to poor sensor stability and repeatability under long-term use or loss of sensor characteristics, such as sensitivity during the cleaning process. Nowadays, in the case of repeated use of pressure, it is still a challenge to maintain a balance between high stability and durability, and good sensitivity. Moreover, most of the studies nowadays focus on the change of the electrode or piezoresistive layer materials to improve the sensor performance, while only a few studies consider the design of the fabric structure itself. This study is based on the design of a multilayer fabric structure variation to achieve a balance among reliability and repeatability, low cost, sensitivity, and measurement range of the fabric sensor for plantar pressure monitoring.

In this study, an all-fabric piezoresistive pressure sensor with a 2.5D bilayer fabric structure was designed to monitor plantar pressure changes in the form of fabric sensing elements in a scattered pattern. The study prepared fabric sensing elements with different scales (different area sizes), evaluated their sensing parameters by the change of the resistance of the sensing elements with the applied pressure, and determined the best elements of the sensing array for the final application. It showed good stability and repeatability in the application, and provided an alternative for the next generation of plantar pressure wearable monitoring system devices.

2 Experiments

Based on the piezoresistive principle, 2.5D bilayer fabric was designed and weaved on a loom. Fabric sensing elements of different sizes were prepared. The influence of the sensing element size on the performance of the sensor was investigated, and the resistance-pressure change was used to analyze the performance of multiple unit points of specimens with different area sizes. Finally, the stability and repeatability of the sensing fabric were tested to evaluate the potential for application in the plantar pressure fabric sensing array.

2.1 Plantar force analysis

Plantar pressure distribution measurement is to obtain the posture state of a person. Before preparing the fabric sensing element, a certain understanding of the plantar pressure distribution area and the force load is required, which determines the placement of the fabric sensing element in the plantar force area and the measurement range. Existing studies have shown that the plantar pressure in the forefoot of diabetic foot patients is elevated, especially in the first toe, third, and fifth metatarsal regions. In this study, a scattered array design was adopted to obtain more accurate pressure information within the overall plantar contour through a few arrays unit points. In addition, some scholars have pointed out that the range of plantar pressure load size in different motion postures of humans is 0–96 N/cm^2, so the above-mentioned fabric pressure sensing element was evaluated on its detection capability in this range.

2.2 Preparation for the fabric of sensing element

To prepare the fabric pressure sensing elements under the above conditions, its structure was designed first. The structure is based on the 2.5D bilayer fabric. The structure is mainly divided into upper and lower layers, and the specific structure of each layer is shown in Fig. 1. The upper and lower layers are designed with a horizontal convex strip structure, which increases the cavity structure by increasing the linear density of the insulation yarn, and thus increases the resilience of the sensing element.

The conductive and insulated yarns are alternatively weaved into the sensing fabric. The ratio of different arrangement numbers of insulating yarns to conductive yarns determines the different area size of the sensing element, which in turn affects the sensing performance of the fabric pressure sensor array. Under the requirement for clear strip structure, three different yarn arrangement ratios (insulating yarn: conductive yarn) were set, i. e., sensing elements of different sizes were woven, and the specific parameters are listed in Tab. 1.

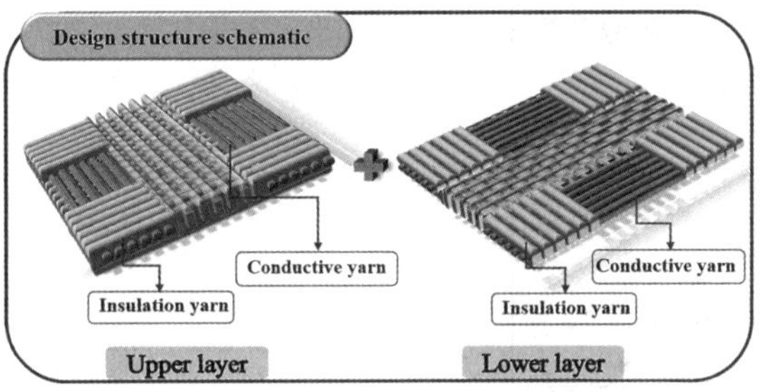

Fig. 1 Schematic of fabric cell structure for sensing elements

Tab. 1 Sensing elements on the machine parameters

Yarn number ratio	Warp yarn	Weft yarn
10 : 4	Polyester-cotton blended yarn (98 tex)	Polyester-cotton blended yarn (98 tex) + stainless steel yarn (356 tex)
10 : 6		
10 : 8		

2.3 Performance tests

2.3.1 Sensing uniformity of elements

For multiple sensing elements on different area size sensing array specimens, their resistance changes with pressure were compared, and the sensing uniformity among multiple sensing elements was evaluated. During testing, static pressure was applied at a random value in the range of 0−1 000 N/cm^2, and the resulting resistances were recorded before and after applied pressure. The sensing array specimens with three different ratios were prepared, and five sensing elements for each specimen were tested.

2.3.2 Sensitivity test

The sensitivity indicates the degree of response of the fabric sensing element to external pressure, and it is an important parameter for determining the performance of the sensing element. Under the applied pressure, the resistance of the sensing element changes, and then from the relationship between pressure and resistance change, the sensitivity was calculated. Five sensing elements from three specimens were tested under the pressure applied at a displacement rate of 30 pulse/s in the range of 0−1 000 N/cm^2, and the resistance was recorded. And then, a linear fit was performed for the resistance pressure, and the linearity was considered as the sensitivity of the flexible piezoresistive sensor.

2.3.3 Stability and repeatability tests

The stability and repeatability of fabric sensing arrays are an important prerequisite for plantar pressure applications, so it is important to calibrate the stability and repeatability performance of fabric sensing elements. The fabric sensing element is subjected to many loading and unloading pressure cycles to evaluate its performance monitoring plantar pressure. And it is expected that its output resistance value always remains the same as the initial state. Specifically, the stability and repeatability of the fabric sensing element with the optimal sensitivity and uniformity were evaluated under the pressure of 0−250 N/cm^2, cyclically loading and unloading the selected sensing element 1 000 times, and collecting resistance change data.

3 Results and Discussion

3.1 Sensing uniformity of elements

The resistance trends of multiple sensing elements with pressure are shown in Fig. 2. It can be seen that all three proportional sensing array specimens have a clear trend of resistance change with pressure (resistance decreases as pressure increases), and the measurement ranges all meet the range of foot pressure load size. It is noteworthy that the resistance changes at a high rate when the external pressure continuously increases by 200 N/cm^2 from 0 N/cm^2. And the measured 10 : 6 ratio specimens on the five sensing elements of the measured resistance change trend are highly consistent, indicating that the sensing array specimens under this area sensing uniformity is high.

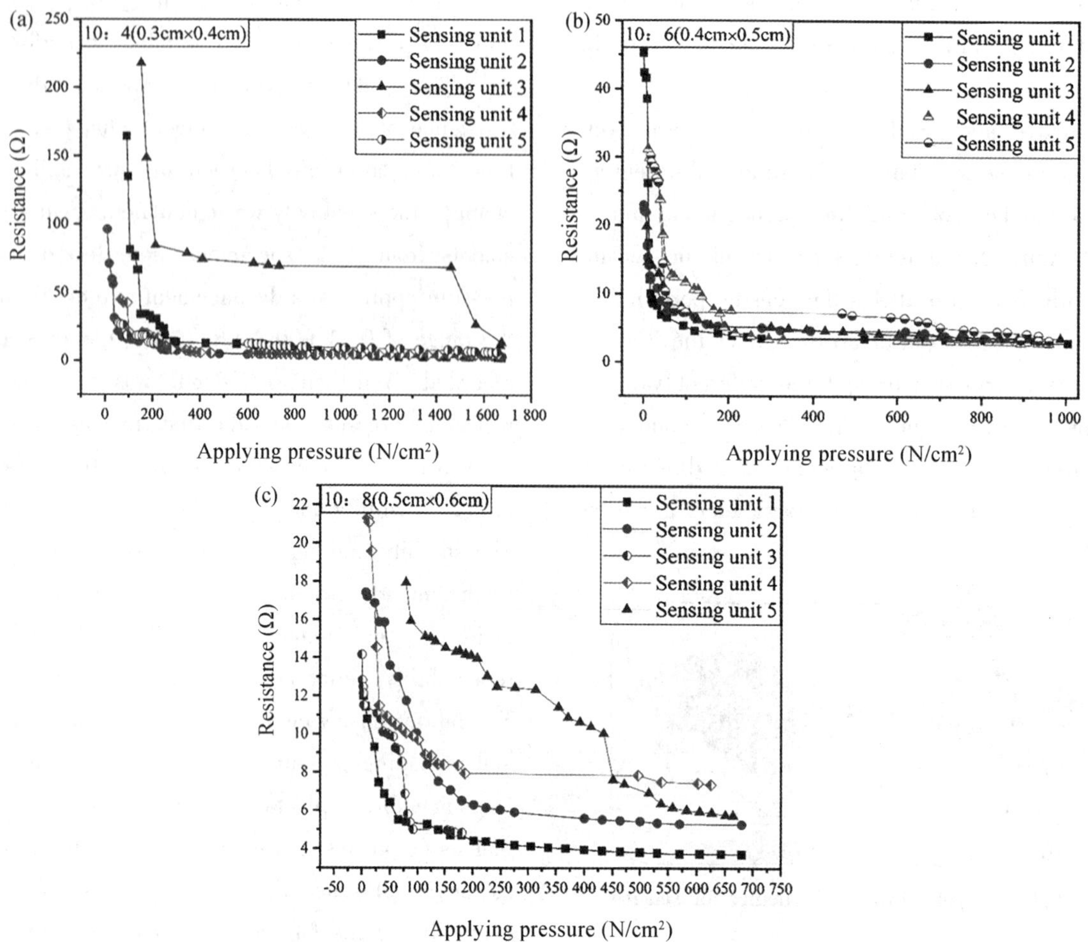

Fig. 2 Variation of resistance with pressure at different yarn number ratios: (a) 10 : 4; (b) 10 : 6; (c) 10 : 8

3.2 Sensitivity

For more accurate characterization, the relationship curve between the two is plotted using the pressure as the horizontal axis and the ratio of the change in resistance to the initial resistance as the vertical axis. According to the formula, the larger the *GF* value of the sensing element, the higher the sensitivity and the better the performance, i. e., the more obvious the change of resistance under smaller strain, provided that the actual application pressure range is satisfied. The closer R^2 is to 1, the better the fit and the better the linearity. The values are shown in Tab. 2.

$$GF = \frac{(R_1 - R_0)/R_0}{P} \tag{1}$$

R_0 and R_1 are the resistance values before and after the application of pressure, ΔR is the difference in the amount of resistance change, and P is the pressure after application.

Tab. 2 Sensitivity and linearity of different proportional sensing elements

Yarn number ratio	Sensitivity $(N/cm^2)^{-1}$	Linearity(R^2)
10 : 4	6.7×10^{-3}	0.971 21
10 : 6	45.02×10^{-3}	0.991 49
10 : 8	4.03×10^{-3}	0.986 39

As can be seen from Tab. 2 and Fig. 3, the 10 : 6 ratio specimen has the highest sensitivity ($45.02\times10^{-3}(N/cm^2)^{-1}$), and its relative resistance change value is in good agreement with the regression fitting curve value. The R^2 is above 0.99, which is due to the structure of the specimen providing more conductive yarn contact points with a moderate area. The conductive network paths increase with applied pressure on it, resulting in good sensing performance. Combined with the sensing element uniformity test, this scale specimen was determined to be the element specification for the final sensing array.

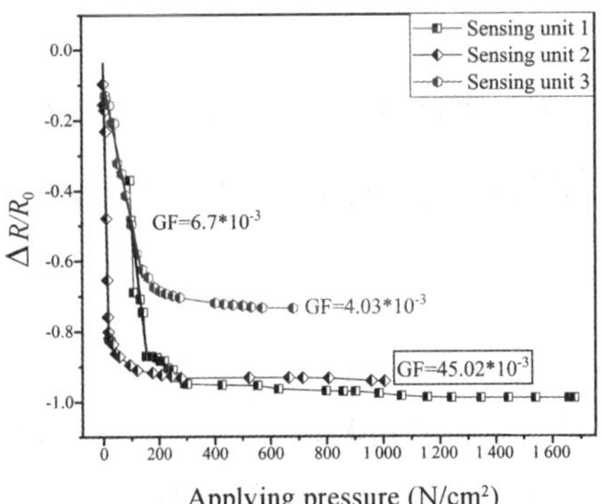

Fig. 3 Sensitivity curves for different scale area sensing elements

3.3 Stability repeatability

The results of the stability repeatability test on the 10 : 6 scale specimen are shown in Fig. 4. The inset shows the change of resistance values with time for the pre-load and post-load sections of the interception cycle. The results show that the response of the sensing element resistance is good after loading, and the amount of change of the sensing element resistance remains the same after more than 1 000 repetitions.

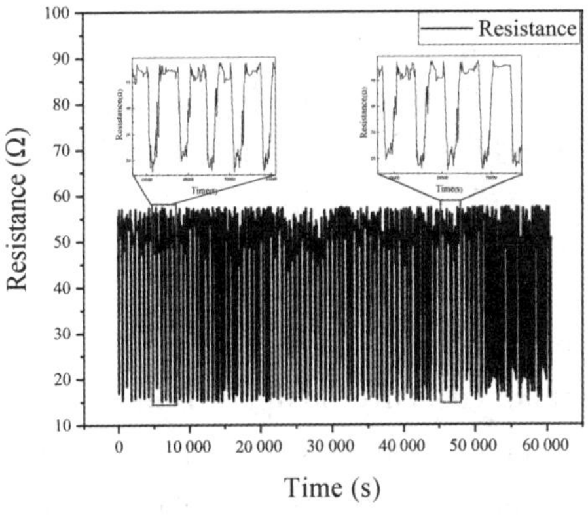

Fig. 4 Resistance response curve during pressure loading and unloading

4 Conclusions

In this paper, an all-fabric piezoresistive pressure sensor based on a 2.5D bilayer fabric tissue structure was designed, fabric sensing elements with different area sizes were prepared in the form of weaving, and their sensing performance parameters were evaluated and analyzed. The main conclusions obtained are as follows.

(1) The stability of the fabric pressure sensor can be improved by designing a 2.5D bilayer fabric monolithic structure.

(2) Three different ratios of sensing elements can be applied to plantar pressure monitoring, and the measurement ranges all meet the plantar pressure load range (0–96 N/cm^2).

(3) The 10 : 6 ratio specimen is selected as the final fabric sensing array preparation specification, which has high uniformity of each sensing element and superior sensitivity ($GF = 45.02 \times 10^{-3} (N/cm^2)^{-1}$).

(4) The 10 : 6 ratio specimen sensing element still has good and stable repeatability after 1 000 repetitions at pressures range of 0–250 N/cm^2, indicating that the sensing element application stability is better at this ratio, providing greater potential for plantar pressure monitoring.

References

[1] PIZARRO F, VILLAVICENCIO P, YUNGE D, et al. Easy-to-build textile pressure sensor[J]. Sensors, 2018, 18(4): 1190.

[2] WANG H, CHEN S, LIU J. Measuring system of a 3D force platform for plantar pressure distribution [C]. 2009 IEEE International Conference on Automation and Logistics, IEEE, 2009: 906–910.

[3] AQUEVEQUE P, OSORIO R, PASTENE F, et al. Capacitive sensors array for plantar pressure measurement insole fabricated with flexible PCB[C]. 2018 40th Annual International Conference of the IEEE Engineering in Medicine and Biology Society (EMBC), IEEE, 2018: 4393–4396.

[4] NGUELEU A M, BLANCHETTE A K, BOUYER L, et al. Design and accuracy of an instrumented insole using pressure sensors for step count[J]. Sensors, 2019, 19(5): 984.

[5] PAREDES-MADRID L, PALACIO C A, MATUTE A, et al. Underlying physics of conductive polymer composites and force sensing resistors (FSRs) under static loading conditions[J]. Sensors, 2017, 17(9): 2108.

[6] POSSANZINI L, TESSAROLO M, MAZZOCCHETTI L, et al. Impact of fabric properties on textile pressure sensors performance[J]. Sensors, 2019, 19(21): 4686.

[7] LIAN Y, YU H, WANG M, et al. Ultrasensitive wearable pressure sensors based on silver nanowire-coated fabrics[J]. Nanoscale Research Letters, 2020, 15(1): 1–8.

[8] LIU M, PU X, JIANG C, et al. Large area all textile pressure sensors for monitoring human motion and physiological signals[J]. Advanced Materials, 2017, 29(41): 1703700.

[9] GE G, CAI Y, DONG Q, et al. A flexible pressure sensor based on rGO/polyaniline wrapped sponge with tunable sensitivity for human motion detection [J]. Nanoscale, 2018, 10(21): 10033–10040.

Comparative Analysis of Yarn Formation Performance Between Conventional Ring Spinning and Additional Twisting Technology

AHALA Kalima[1,2], ZHANG Sudao[1,2], QIAN Jingfang[1,2], XUE Wenliang[1,2], HE Shanshan[1,2*]

[1] *Key Laboratory of Textile Science & Technology, Ministry of Education, Donghua University, Shanghai 201620, China*

[2] *College of Textiles, Donyhua University, Shanghai 201620, China*

[*] *Corresponding author's email*: heshanshan@ dhu. edu. cn

Abstract: Research on additional twists have been conducted to improve ring-spun yarn quality. The driving forces commonly used to attach more twist on fibers come from vortex air or contacting friction, and the devices that provide these forces have been named the twist adding devices. In this study, a twist adding device for air-driving and a twist adding device mechanical driving device are proposed and constructed successively on a ring-spun frame. The difference in structure and properties among conventional ring spun yarns (without the twist adding devices), and ring spun yarns with additional twists endowed by the air or mechanical friction have been discussed. The test results show that the performance and appearance of the yarn spun by the twist adding device are improved. The vortex air provided high-speed rotating airflow to actively wrap the outer fibers around the yarn body and form new twist, therefore its yarn strength and yarn breaking elongation increased by 16. 58% and 9. 89% separately, and yarn hairiness of 1 mm, 3 mm, and 5 mm reduced by 18. 13%, 52. 19% and 67. 13% respectively. The mechanical-driving device resulted in a tighter yarn structure through twisting. It led to an increase of 4. 12% in yarn strength, and an increment of 5% in yarn breaking elongation. The yarn hairiness of 1 mm, 3 mm, and 5 mm was reduced by 7. 58%, 16. 32% and 14. 81%, respectively. The results demonstrate the positive and effective effect of the twist adding device, and that the yarn structure and properties are different because of the different driving forces produced by the two devices.

Keywords: Ring spinning; Additional twist; Yarn properties; Yarn structure

1 Introduction

Spinning is the primary process of the entire industry chain. Its product quality and production efficiency of spinning are of great importance to the following manufacturing process. In the spinning industry, ring spinning technology is still the most used spinning method. For conventional ring spinning, it has a simple machine frame structure, and a wide range of fiber applications, including natural fibers, regenerative fibers, and synthetic fibers[1]. However, its twisting and winding mechanism not only limits the production efficiency but also brings yarn quality problems such as relatively more yarn hairiness[2]. Therefore, researchers have developed a variety of spinning methods based on conventional ring spinning technology, such as compact spinning, Siro spinning, Sirofil spinning, embedded composite spinning, and other technologies[3]. As textile machinery's automation level increases, systems such as single spindle broken-end detection, roving stop-feed, and spinning collective doffing will become accessories for the spinning frame. For example, the use of fully automated robotic ROBO spin technology significantly reduces yarn broken-end rates and labor costs, and greatly increases production efficiency[4]. The emergence of these technologies reduces the cost of labor and processing costs, and improves production efficiency to a certain extent for ring spinning. But at the same time, they may also lead to additional problems, such as extra high energy consumption.

Focusing on the ring spinning technology itself, production efficiency can be achieved by increasing the spindle speed. However, at present, the highest range of ring spinning spindle speed can only reach 20 000 – 25 000 r/min, and simply increasing the spindle speed results in a worse yarn quality[5]. The reason that limits its production speed is the contradiction between the traveler and the ring. The friction between the traveler and the ring generates much heat which soars with the relative speed between the ring and the traveler, and leads to rapid wear and tear on the ring. Therefore, the ring could only run at a speed lower than 40 m/s and this ceiling resistance to the development of the ring spinning. Researchers have made endeavors on extending the life of travelers and improved its operating speed by improving technologies such as size, material, surface roughness, and surface treatment[6]. At the same time, to fundamentally solve the problem of friction, the superconducting magnetic bearing twisting system (SMB) is developed. It consists of superconducting materials and permanent magnets. SMB can produce a yarn with equivalent quality to the conventional ring spinning yarn at a speed of 15 000 r/min[7]. However, this kind of technology does not match the technical level of general spinning enterprises and is not applied on a large scale because it is still in the stage of proof of concept. On the other hand, low-torque spinning technology has been widely used in recent years[8], greatly improving the problem of residual yarn torque. Adding a false twister between the front roller and the guide hook, changes the fiber tension and twist distribution in the spinning triangle, thus reducing the yarn twist. It has great application value. Therefore, only by fundamentally changing the twisting and winding mechanism of ring spinning can the ring spinning technology make a qualitative leap. Therefore, it is of great significance to further investigation, reform, and innovation in ring spinning technology[9]. Based on this background, the additional twisting technology is proposed. Currently, the main power sources used to drive the sliver to produce additional twists are mechanical force and airflow force[10-12]. In order to clarify the impact of both forms of drive on the yarn product, two devices are proposed and designed in this study and the yarns spun by these two spinning methods are compared with conventional ring spinning yarns.

2 Additional Twisting Ring Spinning Principle and Device

The additional twisting area is composed of a holding roller added to the area where the sliver passes through the drafting area and does not enter the twisting area, as shown in Fig. 1. In this way, it will not affect the drafting of the sliver, and it will be appropriately pre-twist. If the device is placed directly between the front roller and the guide hook, the movement of the guide hook up and down will cause a change in the guide angle, and the yarn does not get a stable twist, so the holding roller is added. The device has a simple structure and is convenient to manufacture and use. The two devices are:

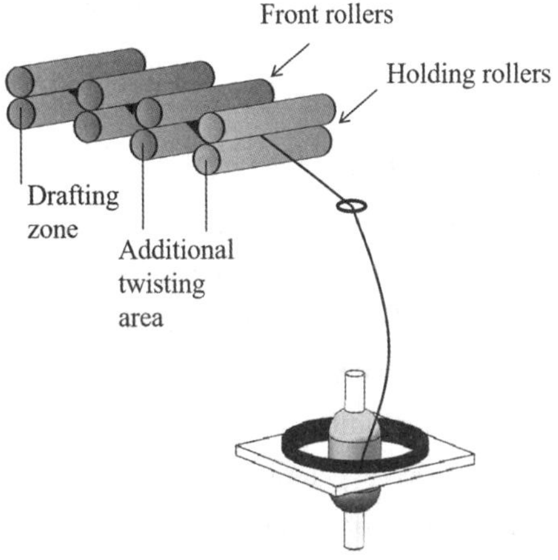

Fig. 1 Sketch of additional twisting area

(1) Air-driving twisting is the installation of an air-flow device in this area, which is designed by a combination of air-flow nozzles and a universal rod, whose function is to fix the air-flow nozzles and adjust their positions and angles. When air passes through, the fiber enters the cavity, the air near the inlet separates the fiber at the head end from the fiber body. And the fiber continues to move forward, the airflow causes the fiber at the head end to wrap around the sliver, finally outputs as from the export. This device is mainly used to wrap the fiber with the help of vortex air, thereby reducing yarn hairiness and improving yarn quality.

(2) Mechanically-driving twisting device is installed in the area of the disc twisting device, consisting of two twisting discs, a drive shaft, motor, external power supply, and universal rod bracket. Two discs are connected to the motor through the drive shaft, the motor is fixed by the universal rod, and universal rod and self-braced bracket can adjust the space position and angle of the discs, by changing the external power supply voltage and positive and negative connection to control the speed and direction of the discs. The two circular wheel disks operate with each other to squeeze fiber and produce a gathering effect on the sliver. Both ends of the sliver are in the state of being held, which is a false twisting. One end floating on the surface of the main

yarn is held, and the free end of the fiber is subjected to centrifugal force after passing through the circular disc twisting device. The holding end rotates with the sliver, and then the free end is involved in the sliver. The air-driving and mechanical-driving additional twist devices are based on the difference in the application of force, and the different driving forces acting on the sliver result in differences in the structural properties of the final yarn.

3 Experimental Design

To compare conventional ring spinning, air-driving additional twist yarn, and mechanical-driving additional twist yarn, the DHU-1X ring spinning machine was used in the spinning laboratory at the temperature of 20 ℃ and the humidity of 65%. Roving composition and spinning process parameters are shown in Tab. 1. To verify the effectiveness of the additional twisting device, the yarn was photographed with a scanning electron microscope (SEM), showing the longitudinal structure of the yarn. Secondly, the yarn strength properties and hairiness index are tested by the XL-1A yarn strength tester and the YG172A yarn hairiness tester. The strength properties of the three yarns are expressed in terms of breaking strength and breaking elongation.

Tab. 1 Parameter settings for experiments

Items	Parameter
Roving composition	70% cotton 30% polyester
Roving weight	4.5 g
Twist	500/600/700 twist/m
Twist direction	Z twist
Drafting ratio	13.8
Holding roller to front roller speed ratio	1.04
Air pressure	0.11 MPa
Additional twist degree	1587 α
Additional twist position	1/2 of additional twisting area

(**Continued**)

Note: α is defined as additional twisting efficiency, which is determined by factors such as slippage between the twisting disc and the fiber.

4 Results and Discussion

4.1 The structure of three yarns

Fig. 2 is SEM images of the yarn length of three kinds of spinning methods. The changes in yarn structure and hairiness on the surface of three kinds of yarns can be observed. Fig. 2 (a) shows an ordinary ring-spun fiber, which has an obvious tendency for inner and outer transfer, and its surface has more hairiness and disarray. The yarn spun with an additional twist device is shown in Figs. 2 (b) and 2(c). The surface hairiness of the yarn is significantly reduced. The airflow additional twisting yarn structure is loose, the surface is wrapped, whereas the yarn structure of the texture of the mechanical additional twisting device is relatively tight, and the surface hair feathers are less. Compared with the mechanical additional twisting yarn, the wrap of the outer fiber of the air-driving additional yarn is in an orderly direction. The experimental results show that the yarn spun under the additional twisting device has significantly less surface hairiness.

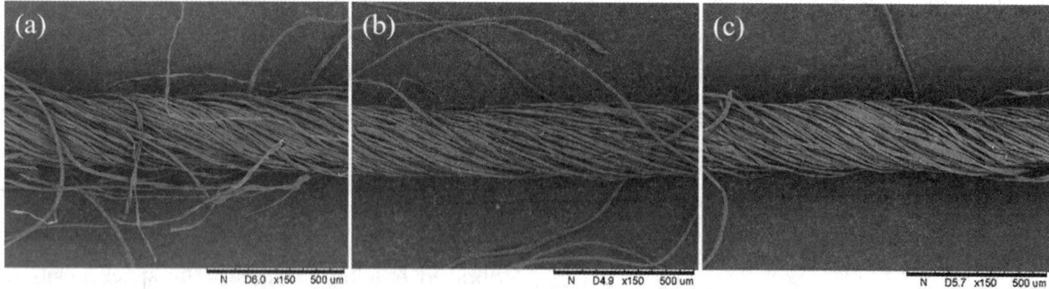

Fig. 2 SEM images of the (a) ring-spun; (b) air-driving twisted; (c) mechanically driving twisted yarns

4.2 Analysis of yarn performance

The hairiness test results of ordinary ring spinning yarn, air flow additional twist yarn, and mechanical additional twist yarn are shown in Fig. 3. Figs. 3(a), 3(b), and 3(c) are respectively 1 mm, 3 mm, and 5 mm hairiness of the yarn with and without devices at the three twist levels of 500, 600, and 700 T/m.

It can be seen that the hairiness of the yarn is reduced after the addition of the twist device. Because the device still has a certain aggregation effect on the sliver while wrapping the fiber at the free end, the twisting triangle area of the holding roller is reduced, so the yarn hairiness is reduced. The results show that the yarn hairiness of the yarn with a lower twist adding device is similar to that of the high twist yarn without adding device. The device cannot destroy the strength and hairiness of the yarn, and the production efficiency can be improved by reducing the twist setting, which lays a foundation to produce low-twist and high-performance yarn.

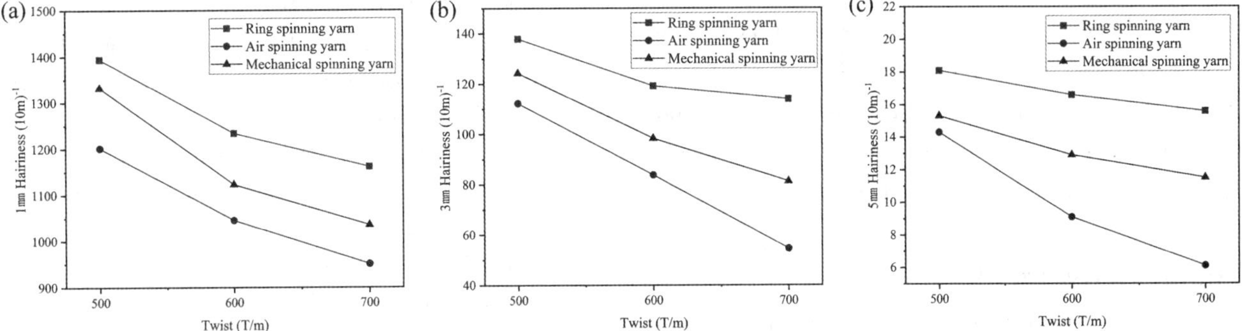

Fig. 3 Three yarns: (a) 1 mm; (b) 3 mm; (c) 5 mm hairiness performance

The load and elongation properties of the three yarns are shown in Fig. 4. The breaking load and elongation at the break of the yarns increased to a certain extent after adding twisting. The reason for air-driving twist yarn may be the wrapping action of air twist on the outer fiber of the main yarn, which makes the cohesion between the fibers stronger, and thus the load and elongation properties of the yarn are improved. In addition, when the twist is 600 twist/m, the breaking load of the yarn with the device is better than that without the device at 700 twist/m. When the twist is 500 twist/m, the breaking elongation of the yarn with the device is better than that without the device at 600 twist/m. These phenomena show that air-driving additional twisting device can reduce the setting twist to a certain extent and improve the production efficiency under the condition of ensuring the yarn load and elongation. The reason for the mechanical additional twist device is that its agglomeration effect on fibers is conducive to improving the parallel and straight arrangement of fibers, so the yarn load and elongation performance are improved. Calculations of the spinning speed show that the installation of the additional twisting device results in a 16.67% increase in productivity when the corresponding yarn twist is reduced from 700 twist/m to 600 twist/m, and a 20% increase in productivity when the twist is reduced from 600 d/m to 500 twist/m.

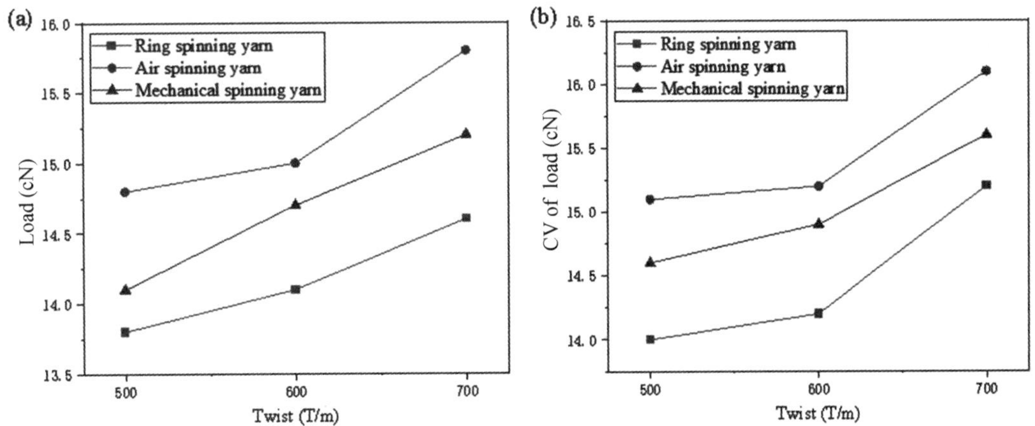

Fig. 4 (a) The load and (b) the breaking elongation of three yarns

5 Conclusions

In this paper, the structure and properties of three kinds of yarns were tested by adding a holding roller after the front roller to form an additional twisting area and adding air-driven and mechanical-driven twisting devices in this area. The results show that ring spun yarns with additional twisting have certain advantages over conventional ring spun yarns in terms of yarn hairiness and tensile properties, and the air-driving additional twisting device has a better effect. The reason is that the air twist device will be wrapping the free end of the fixed end of the fiber into the sliver, to increase the friction and cohesion among the fibers, so it has a more positive effect on the improvement of yarn properties. The effect of mechanical additional twist device on fiber aggregation is more conducive to the reduction of twisting triangle area, thus reducing hairiness.

Acknowledgement

This work was supported by the Applied Basic Research Programs of "Glory of Textiles" (No. J201807), the Shanghai Sailing Program (No. 21YF1401100), and the Fundamental Research Funds for the Central University (No. 2232021D-09).

References

[1] YIN R. Mathematical modeling and numerical simulation of nonlinearly elastic yarn in ring spinning[J]. Textile Research Journal, 2021, 91(3,4): 278-288.

[2] LI P, GUO M, SUN F, et al. Reducing yarn hairiness in ring spinning by an agent-aided system[J]. Textile Research Journal, 2019, 89(21,22): 4438-4451.

[3] XIA Z, XU W. A review of ring staple yarn spinning method development and its trend prediction[J]. Journal of Natural Fibers, 2013, 10: 62-81.

[4] SHI J, LIANG W, WANG H, et al. Recent advancements in cotton spinning machineries[J]. Cotton Science and Processing Technology, 2020: 165-190.

[5] KOLTE P, GULHANE S, PATIL J, et al. Effect of spindle speed of ring frame on yarn quality[J]. 2018, 12: 4-10.

[6] BEGUM H A, HELALI M M. A way to reduce friction between ring and traveler of a ring spinning frame[J]. Bangladesh Journal of Scientific and Industrial Research, 2013, 47(4): 421-426.

[7] HOSSAIN M, ABDKADER A, CHERIF C, et al. Innovative twisting mechanism based on superconducting technology in a ring-spinning system[J]. Textile Research Journal, 2014, 84(8): 871-880.

[8] XU B G, TAO X M. Techniques for torque modification of singles ring spun yarns[J]. Textile Research Journal, 2008, 78(10): 869-879.

[9] YIN R, LING Y L, FISHER R, et al. Viable approaches to increase the throughput of ring spinning: a critical review[J]. Journal of Cleaner Production, 2021, 323.

[10] YILMAZ D, USAL M R. Nozzle usage in spinning process and changes in yarn hairiness[C]. Hongkong: Fiber Society Spring 2011 Conference, 2011-5-23.

[11] FU J, YU W. The false twist compact drafting technique for cotton ring spinning and optimization design of its device[C]//ICAFPM 2009 — International Conference on Advanced Fiber and Polymer Materials, 2009-10-21.

[12] LYU J, CHENG L, XU B, et al. Numerical simulation of fiber motion in the condensing zone of lateral compact spinning with pneumatic groove[J]. Autex Research Journal, 2022, 22(1): 35-41.

Study on the Preparation and Performance of Breathable Down-Proof Electrospun Film Composite Cotton Fabrics

ZENG Zhen[1], WANG Rongwu[1*], ZENG Yongchun[1]

[1] Department of Textile Engineering, College of Textiles, Donghua University, Shanghai 201620, China

* Corresponding author's email: wrw@ dhu. edu. cn

Abstract: Maintaining good air permeability while satisfying the down-proof performance is still a major difficulty in down-proof products. In this study, the cotton fabrics were treated by electrospinning process, and the micro-morphology, down-proofness, air permeability and tensile properties of cotton fabrics and different composite fabrics were analyzed. The results indicated that with the increase in epoxy resin concentration, the sheet-like nodules between the fibers of the electrospun film became larger, and the pores became smaller. Compared with TPU composite fabrics, TPU-epoxy composite fabrics improve the down-proofness of the cotton fabrics down bag. At the same time, all composite fabrics maintain the air permeability of 30 mm/s or above, and the tensile breaking load increases compared with that of the cotton fabric itself, with a maximum increase of 18.5%. Cotton fabrics with an appropriate amount of TPU-epoxy resin concentration can improve their down-proof performance, have good air permeability, and have a certain enhancement effect on the tensile strength of the cotton fabrics.

Keywords: Electrospinning; Down-proof; Air permeability; Tensile strength

1 Introduction

Downs are fluffy, soft, and light, so the demand for down products in the market has been relatively strong. Consumers' requirements for down products are also getting higher and higher. The solution to the problem of down-proof has always been the biggest difficulty in down products[1].

At present, the commonly used methods to improve the down-proof performance of down products include using high-count and high-density chemical fiber fabrics as bile fabrics[2], adding quilt fabrics, and filling down between the two layers of fabrics, but the processed products have poor breathability and comfort. As well as coating finishing, calendering and baking treatment of quilt fabrics[3], but these processing will lead to a significant decrease in the softness and poor tear strength of the fabrics[4]. What is more, the composition of coating finishing agents is easy to produce the phenomenon of light yellowing[5] affecting the appearance of the product. There are polyester (PE) coated fabrics prepared by hot melt adhesive lamination process[6], but the air permeability is very poor, which affects the actual use of down products. And a polyfluorocarbon film-coated composite fabric has been proposed[7], which has great down-proof properties, but the film has poor air permeability, complex production process, high cost, and certain environmental problems.

In view of the balance of down-proof performance and air permeability of down products, soft, breathable, and natural skin-friendly cotton fabric is selected as the base fabric, and thermoplastic polyurethane (TPU) with excellent elasticity and wear resistance and epoxy resin are selected as electrospun film materials. The spinning solution is spun onto the cotton fabrics through the electrospinning process. Finally a composite fabric that is porous and breathable, and has the performance of down-proof is obtained.

2 Experiments

2.1 Materials

Twill cotton fabrics (60 S) were supplied by Jiangsu Canasin Textile Co., Ltd. (Jiangsu, China). Thermoplastic polyurethane (Wanthane® WHT-1190) was provided as pellets by Dongguan Jianzhang Plastic Co., Ltd. (Guangdong, China). Epoxy resin was obtained from Huachang Polymer Co., Ltd., East China University of Science and Technology (Shanghai, China).

Dimethyl formamide, analytical grade, was obtained from Shanghai Titan Technology Co. , Ltd. (Shanghai, China).

2.2 Preparation of spinning solution

A solution with a mass fraction of TPU of 13% and a solvent of dimethyl formamide was prepared. Then, five groups of mixed solutions with epoxy resin mass fractions of 8%, 10%, 12%, 14%, and 16%, TPU mass fractions of 13%, and dimethyl formamide as the solvent were prepared. The solution was stirred on a magnetic stirrer at room temperature for 8 h and left to stand for later use.

2.3 Composite manufacture

Prepare 30 pieces of 42 cm×42 cm cotton cloths and place them on a vertical drum take-up (Qingdao Nuokang Environmental Protection Technology Co. Ltd. , Qingdao, China). Take an appropriate amount of the prepared solution into a 10 mL syringe, place it on a syringe pump (LongerPump Co. , Ltd. , Hebei, China), adjust the spinning distance to 12 cm, and set the spinning speed, spinning voltage and the drum speed to be 1 mL/h, 16 kV and 320 r/min, respectively. After spinning for 1.2 h, air-dry the composite fabrics for 8 h for subsequent use. The temperature and relative humidity during spinning were 25 ℃ and 30%, respectively.

2.4 Morphological characterization

The desktop scanning electron microscope (SEM) model TM 3000 was used to observe the morphology at 3 000 times magnification, and the operating voltage of the test was 15 kV.

2.5 Down-proof testing on composites

According to the standard of GB/T 12705. 2—2009, prepare down bags with an effective size of 40 cm× 40 cm. The test was carried out by the same standard as above in the rotary box method down-proof instrument model YG(L)819D (Shaoxing Libixin Instrument Co. , Ltd. , Zhejiang, China). Each type of fabric was tested in three groups, the average value was taken as the final test result, and each fabric was tested at one time.

2.6 Air permeability testing on composites

The air permeability test of the corresponding fabrics was carried out on the automatic air permeability meter model YG461G (Ningbo Textile Instrument Factory, Zhejiang, China). The air permeability test is referred to EN ISO 9237. The pressure drop was 100 Pa, and the test head area was 20 cm^2. Different parts of different fabrics were tested 10 times, respectively. The average value was taken as the final test result.

2.7 Tensile testing on composites

The tensile tests were conducted on a multifunctional electronic fabric strength machine model YG026MB. The tests were done according to GB/T 3923. 1—2013. The effective width of the cloth sample was 50 mm, the gauge length was 200 mm, and the drawing speed was 100 mm/min. Five groups of each fabric were tested along the warp direction, and the average value was taken as the final test result.

All fabrics were conditioned at 20 ℃ and 65% relative humidity for 24 h before testing.

For the convenience of statistics, in the following, cotton fabrics, TPU composite fabrics, TPU-epoxy resin composite fabrics, TPU-epoxy resin-8% composite fabrics, TPU-epoxy resin-10% composite fabrics, TPU-epoxy resin-12% composite fabrics, TPU-epoxy resin-14% composite fabrics, TPU-Epoxy resin-16% composite fabrics were respectively named as Cotton, CT, CTE, CTE-1, CTE-2, CTE-3, CTE-4, CTE-5.

2.8 Statistical analysis

Data in tensile properties and air permeability of composites were analyzed using one-way analysis of variance and Fisher's pair-wise multiple analysis. *A* p-value was considered statistically significant if it was smaller than 0.05.

3 Results and Discussion

3.1 Morphological characterization

SEM images of the electrospun films with different epoxy resin concentrations are shown in Figs. 1 (a)-1(f). The electrospun films with epoxy resin have smaller voids than the film with TPU solely. When the epoxy resin concentration is 8%, the average void of fibers is the smallest, and there are lots of nodules between fibers. When the epoxy resin concentration is 10%, relatively good fiber uniformity is exhibited, and the nodules are relatively few and small. As the epoxy resin concentration increased to 16%, the nodules between the fibers became larger and more irregular, and the interfiber space gradually decreased.

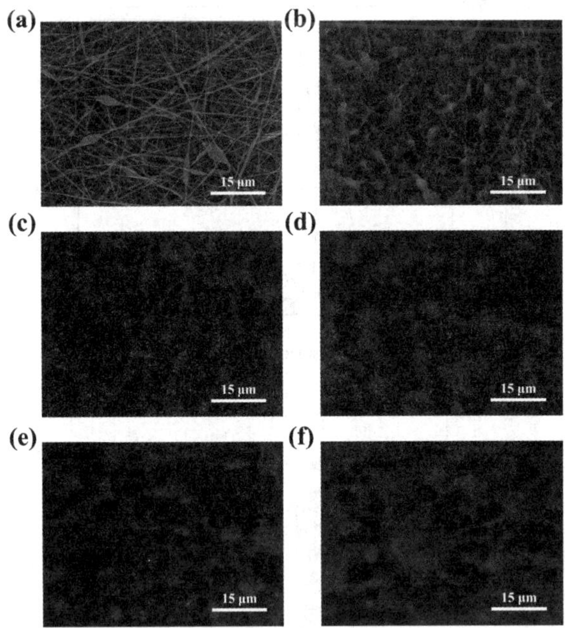

Fig. 1 (a)-(f) Represents SEM images of electrospun films of CT; CTE-1; CTE-2; CTE-3; CTE-4; CTE-5, respectively

3.2 Epoxy resin concentration effect on composite down-proof performance

The test results of down-proof performance of Cotton and CTE with different epoxy resin concentrations are shown in Tab. 1.

Tab. 1 Down-proof performance test on down feather package

Sample	Way of rotation		Nberum (piece)
	Forward rotation	Reverse rotation	
Cotton	≥200	≥200	≥400
CT	45	50	95
CTE-1	25	29	54
CTE-2	16	17	33
CTE-3	15	15	30
CTE-4	12	14	26
CTE-5	14	14	28

The down clothing standard GB/T 14272—2011 stipulates that the requirements for down-proof fabrics that are in direct contact with down are 5 or less for first-class products, 15 or less for first-class products, and 50 or less for qualified products[8]. It can be seen in Tab. 1 that after the forward rotation process of the down-proof test for the samples corresponding to Cotton, the number of drilling-out is far greater than 50, and Cotton does not have any down-proof properties. After Cotton is treated with TPU electrospun film, the average number of drilling-out is more than 50, which cannot meet the qualified standard. CTE-1 has an average number of drilling-out slightly greater than 50, and it cannot reach the qualified standard either. With the increase in epoxy resin concentration, the average number of drilling-out for each concentration of composite fabrics is less than 50, reaching the qualified standard. TPU electrospun film is thin and soft, and at the same time, it has poor bonding effect with Cotton, which leads to easy damage during the rotations. Compared with the former, CTE-1 has improved down-proofness, because the addition of epoxy resin increases the area of the stereotyped area and reduces the pores, which is beneficial to improve the down-proofness. However, due to the weak area between the fiber intersections in the nanomembrane, it is easy to be damaged during the working process, resulting in relatively poor down-proofness. With the increase in the epoxy resin concentration, the number of drilling-out first decreased and then increased. The reason is that the micropores of the membrane in the composite fabrics are small, which can only satisfy the movement of small molecules, and the weak points are relatively reduced. The hindering effect increases, and it is difficult for the down to drill out from the composite fabric. With the increase in resin concentration, the number of drilling-out for CTE-5 increased slightly, because there may be weak joints between the fibers in the stereotyped area and the non-stereotyped area, which led to a slight increase in the number of drill-outs.

3.3 Epoxy resin concentration effect on composite air permeability

The air permeability of Cotton and CTE with different concentrations is shown in Fig. 2. It can be seen from Fig. 2 that the TPU film has a weakening effect on the air permeability of the composite fabric. With the addition of epoxy resin, the air permeability of the composite fabric first decreases, then increases, and then continues to decrease. Among them, because the TPU film fiber and the cotton fiber overlap, the air permeable pore size is further reduced in space, and the gas flow rate is slow, so the addition of the TPU film will reduce the air permeability of the composite fabric. Due to the low viscosity of epoxy resin with a concentration of 8%, the electrostatic stretching effect in the spinning process is insufficient for its cohesive force. It is easy to form small droplets under the action of high voltage, which are closely dispersed in the electrospun membrane or in the fiber. The junction is formed by adhesion and overlapping, which leads to the reduction of the pore size between fibers and the further decrease of air permeability. As the concentration of epoxy resin increases, the viscosity of the solution increases, the cohesion of the solution and the electrostatic stretching effect reach a relative balance, the adhesion and agglomeration are less, and the air permeability of the fabric is relatively improved. When the epoxy resin concentration is greater than 12%, due to the increase in the viscosity of the solution, the cohesive force is greater than the electrostatic stretching, and the fiber macromolecules are more likely to agglomerate, resulting in a relatively reduced air permeability. Since the epoxy resin forms a stereotyped area after curing, if the molecular crystallinity of the material is higher, the barrier property of the material to gas will be stronger. And the diffusion of gas will be more difficult, so the addition of epoxy resin reduces the air permeability of the composite fabrics.

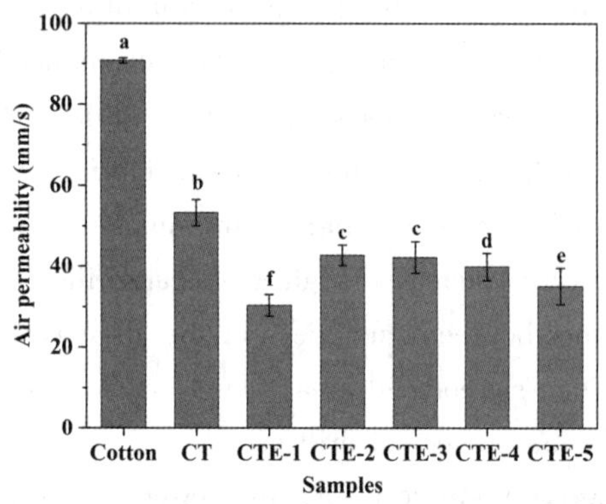

Fig. 2 Air permeability comparison among Cotton, CT, CTE-1, CTE-2, CTE-3, CTE-4, and CTE-5 (different letters labeled on top of the columns show the means with significant difference)

3.4 Epoxy resin concentration effect on composite tensile properties

Typical tensile load-displacement curves of the fabrics in the warp direction are given in Fig. 3 (a). Composite fabrics all presented a higher tensile load than Cotton, but the deformation decreases to varying degrees. The tensile failure load of the Cotton and different composite fabrics is shown in Fig. 3 (b). The tensile failure load of Cotton is 361.2 N, and the deformation is 60.8 mm. The tensile strength of CT is 372.9 N, and the deformation is 57.09 mm. The tensile failure load

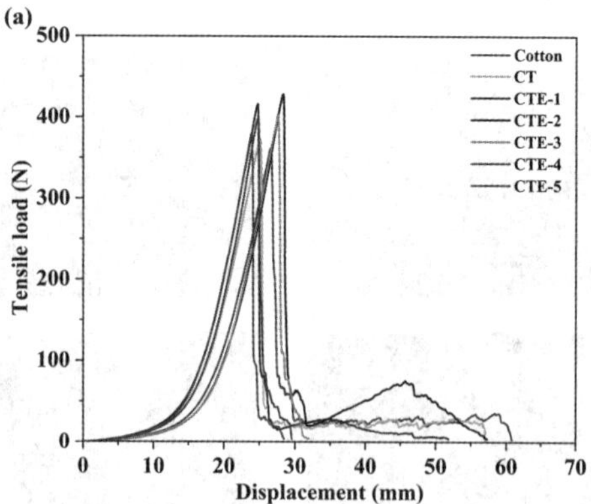

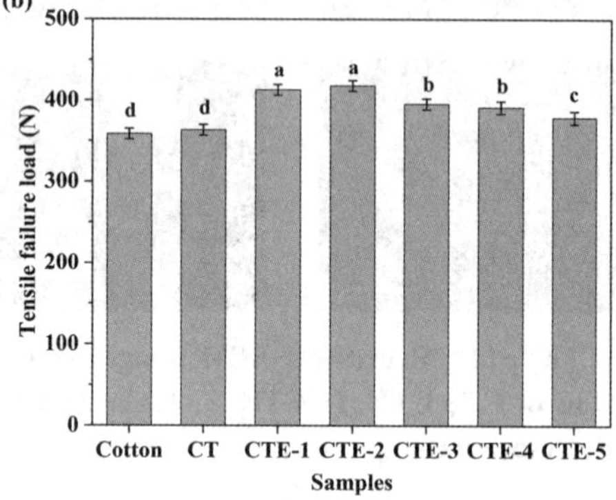

Fig. 3 (a) Typical tensile load-displacement curves of Cotton, CT, CTE-1, CTE-2, CTE-3, CTE-4, and CTE-5; (b) tensile failure load comparison among Cotton, CT, CTE-1, CTE-2, CTE-3, CTE-4, and CTE-5 (different letters labeled on top of the columns show the means with significant difference)

increases by 3.2%, and the deformation is basically unchanged compared with that of Cotton. It shows that the elongation at break of Cotton and TPU film is basically unaffected, because the macromolecular structure of TPU is composed of soft and hard segments. There are hydrogen bonds that enhance the interaction between the macromolecules of the polyurethane elastomer, thus giving the TPU higher strength. The soft segment is polyester with strong polarity, the molecular weight of the soft segment is large, and its structure has high regularity. When the TPU film is stretched, the soft segment is stretched until the TPU film is pulled apart. Therefore, after Cotton is compounded with TPU film, its tensile failure load will increase, while the deformation will remain basically unchanged. Compared with Cotton, the tensile failure load of composite fabrics contained with epoxy resin increased by 15.2%, 18.5%, 11.4%, 10.5% and 5.4%, respectively. This is because the crystallinity of epoxy resin molecules is higher than that of TPU molecules, and the hardness of TPU-epoxy resin film is higher than that of

TPU film, so the tensile strength is greater than that of TPU film. At the same time, it can be observed that with the increase in epoxy resin concentration, the tensile failure load of composite fabrics first increases and then decreases. When the epoxy resin concentration increases to 8%, due to the increase in the surface tension of the solution and the static electricity under the action of stretching, more dispersed droplets and fibers intersect on the base cloth. The presence of epoxy resin increases the tensile failure load and slightly reduces the deformation. As the epoxy resin concentration increases to 10%, the electrostatic stretching effect is balanced with the solution cohesion, and the TPU-epoxy fiber distribution is relatively uniform. It has good tensile failure load and increases the toughness of the fiber, which improves the tensile failure load. Elongation at break increases. As the epoxy resin concentration continues to rise, the solution viscosity increases, resulting in easier agglomeration of fiber macromolecules, reduced toughness, and poor ductility of the composite fabric, so the deformation is relatively reduced. At the same time, the increase in the curing area also enlarges the weak area with the fiber, which is an inducement of the decrease in the tensile failure load of composite fabrics.

4 Conclusions

This study mainly focused on the effect of epoxy resin concentration changes on the microstructure, down-proof, air permeability and tensile properties of Cotton and different composite fabrics. With the increase in the epoxy resin concentration, the sheet-like nodules between the fibers became larger and more numerous, resulting in the decrease of the interfiber space. Compared with CT, CTE has improved the down-proof of Cotton. The air permeability of CTE is lower than that of Cotton. Among them, the air permeability of CT is better than that of CTE, but the air permeability of the composite fabric remains above 30 mm/s. At the same time, the tensile failure loads of all composite fabrics have increased compared with Cotton. Among them, the tensile failure load of the CTE is greater than that of the CT, and the maximum tensile failure load is 416.1 N, which increases the tensile failure load of the Cotton by 18.5%. The above shows that the composite fabrics treated with electrospun film have certain down-proof properties and can well meet the air permeability requirements. Meanwhile, it can also improve the tensile failure load of Cotton, which is beneficial to the mechanical properties of the fabrics.

References

[1] LIU L M. Improvement of the problem of drilling out on light down jackets [J]. Jiangsu Textile, 2013 (8): 38-39.

[2] GUO K J, LIU Y M, DENG H W, et al. Production practice of fine and extra high-density a down-proof fabrics[J]. Cotton Textile Technology, 2014, 42(2): 71-74.

[3] YE M J, FENG L Q, CHEN W E, et al. Research on down-proof garment technology [J]. Shanghai Textile Technology, 2016, 44(12): 13-16.

[4] LUO S L, LIAO Y L, ZHANG Y Q. Discussion on the wearability of double-layer woven down-proof fabrics[J]. Advances in Textile Science and Technology, 2017(10): 40-42.

[5] WANG Z Q, LI C L, SUN R X, et al. Light yellowness of down duvet[J]. Journal of Donghua University (English Edition), 2017, 34(2): 216-220.

[6] WANG M R, LI C L, WANG Z Q, et al. The effect of lamination on the down-proof properties of cotton fabrics[J]. Anhui Engineering University Journal, 2018, 33(5): 42-45.

[7] ZHENG L, TIAN J, XIA Y H, LIN L, et al. Feasibility of the application of polyfluorocarbon nanofiber film in textile down-proof[J]. Modern Textile Technology, 2020, 28(3): 17-20.

[8] DOU M C, WU W Y, ZHU W J. Classification and comparison of test methods for down-proof of fabrics[J]. Wool Textile Technology, 2011, 39(12): 52-56.

Research on Lightness Performance Test Method of Luminous Fabrics with Polymetric Optical Fibers

HUANG Dan[1], MA Yanxue[1*], XUE Wenliang[1], WEI Mengyuan[2], HUANG Yu[2], HUANG Ruhua[3], LIN Jianfu[3]

[1] *College of Textiles, Donghua University, Shanghai 201620, China*

[2] *Shanghai Customs District P. R. China, Shanghai 200120, China*

[3] *Fujian Deliang Jacquard Weaving Technology Co Ltd, Quanzhou 350000, China*

* *Corresponding author's email*: yxma@ dhu. edu. cn

Abstract: As an important component of smart textiles, polymetric optical fiber (POF) luminous fabrics are increasingly used in fashion clothing, home decoration and other fields. However, there is no testing standard or specific method to measure the brightness of POF luminous fabrics. This paper presented a testing device and method to effectively measure the brightness of POF luminous fabrics, utilizing optical dark boxes, imaging type luminometer, and other devices in order to meet the testing requirements of POF luminous fabrics for luminous brightness parameters. Through testing the repeatability of various POF luminous fabric samples with different sizes, the best specimen size in this proposed test method was determined to obtain stable brightness test data. Establishment of the testing device and the validated test method could support manufacturers control of the quality of POF luminous fabrics, as well as market supervision and administration.

Keywords: Polymetric optical fiber; Luminous fabric; Brightness; Imaging luminance measurement device

1 Introduction

With the increasing demand for night consumption in cities, luminous fabric products appear more in cultural entertainment and commercial places, which are both decorative and fashionable[1]. According to the luminescent materials, luminous fabrics can be divided into three categories: fabrics made of photoluminescent materials, fabrics integrated with light-emitting diodes and luminous fabrics woven by polymetric optical fibers (POF)[2]. Compared with other types of luminous fabrics, POF luminous fabrics have the advantages of the controllable luminescent area, flexibility, light weight and changeable luminescent effects, which are widely used in wearable clothing and interior decoration such as stage clothing, festival clothing, shoes and curtain[3-4].

POFs have a good light transmission performance, which is used in luminous fabrics. When one end of POFs is illuminated by a light source, the fabric can present a luminous effect[5]. The brightness of POF luminous fabrics reflects the brightness of luminous effect perceived by human eyes, which is an important index to evaluate the luminous effect of products. At present, the main test methods for the brightness performance of luminescent materials at home and abroad are as follows: GB/T 24909—2010 "LED Lamps for Decorative Lighting"[6], SN/T 2558. 7—2014 "Testing Method of Import and Export of Functional Textiles-Part 7: Luminescent Properties of Luminous Fibers"[7], YD/T 1447—2013 "Plastic Optical Fiber for Communication"[8]. However, there are no relevant test methods and evaluation standards for POF luminous fabrics. The existing research on the brightness performance of POF fabrics mainly uses the point type luminance meter, and the measurement points are different for different sizes of POF fabrics. The wide application of POF luminous fabrics is limited because the test methods and evaluation requirements of luminous fabrics cannot be unified and standardized[9-12]. This paper proposes and builds a device and method to effectively measure the brightness of POF fabrics. By testing the brightness of different sizes of optical fiber woven fabrics, the appropriate sample size for this method can be explored, so as to obtain the test results with high stability and reliable data.

2 Experimental Design

2.1 Test device and principle

The test uses the image colorimetric method[13] to take a picture of the POF luminous fabric. Using a three-color filter to measure the brightness of the POF fabric, which can capture millions of brightness data information of the luminous area image on the surface of the fiber fabric in a few seconds. In this experiment, the CY- 1200C imaging luminance measurement device produced by Yande instrument Co. Ltd. was used to test the brightness.

The brightness of the POF luminous fabric is tested in an optical darkroom. The test device consists of a light source connecting part, a preparation plate, an imaging luminance measurement device and an optical dark box. The schematic diagram of the device is given in Fig. 1. The POF luminous fabric is placed on the prepared plate. The POF bundle at one end of the fabric is connected with the light source, so that the POF luminous fabric can emit light by controlling the light source. Brightness is measured by an imaging luminance measurement device at the top of the dark box. Considering the limitation of experimental space and the use conditions of the imaging luminance measurement device, the size of the optical dark box used in this paper is 80 cm×80 cm×100 cm, and the imaging luminance measurement device is 80 cm away from the POF fabric sample.

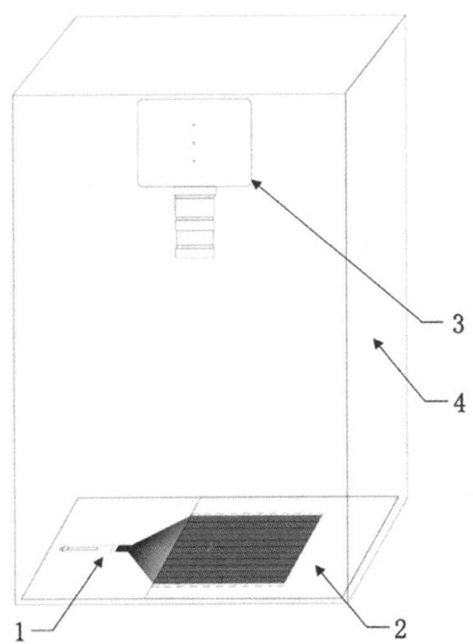

1—light source connecting part; 2—preparation plate; 3—imaging luminance measurement device; 4—optical dark box

Fig. 1 Schematic diagram of brightness test device

2.2 Specimen design and preparation

The samples prepared in this paper contain the cluster of POF fabric, light source lamp parts and power supply. As shown in Fig. 2 (a), the fiber bundle on one side of the POF fabric was formed after beam splitting treatment. The cross section of the fiber bundle was cut lengthwise. The fiber bundle was connected with the light source lamp in a head-to-head manner and fixed by a thermoplastic tube, so that the fiber fabric can pass through light normally and achieve maximum coupling efficiency. During the testing process, a double mask was used to fix the POF luminous fabric and determine the light-emitting area on the fabric surface. As shown in Fig. 2(b), the lower shade was used to fix the POF luminous fabric, and the upper shade was used to block the light at the light source and expose the fabric part that needs to be tested for brightness.

(a)

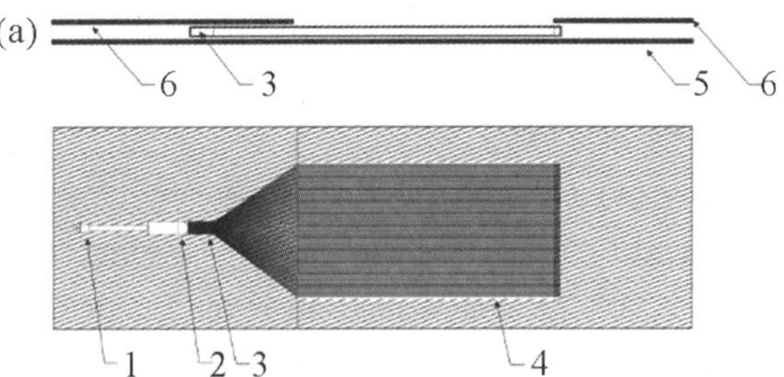

(b)

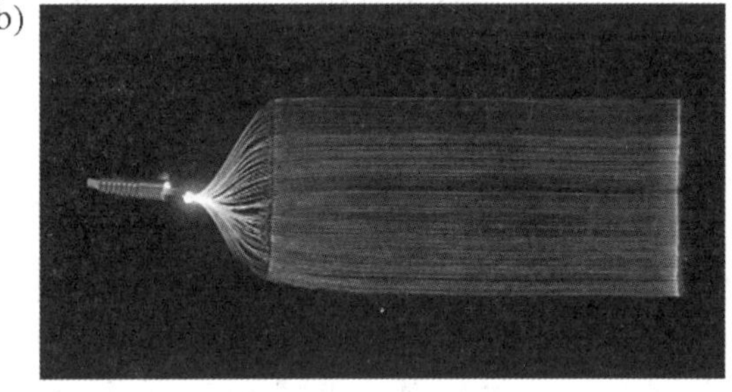

1—power supply; 2—light source lamp; 3—connection part; 4—sample test part; 5—lower shade; 6—upper shade

Fig. 2 (a) Schematic of specimen preparation; (b) Physical drawing of the sample

The sample size has a direct effect on the luminance test of POF luminous fabrics. The POF fabric width affects the cross-section diameter of the fiber bundle. When the cross-section diameter is too large, the fibers in the cluster center may slip and the coupling efficiency with the light source is low.

In order to reduce the impact of the sample preparation on the brightness performance test, six sizes of the same POF luminous fabrics were selected in the size range of the setup device and sampled for five times respectively. By exploring the brightness stability of samples of different sizes, the best sample size under the test method was obtained. Sample size specifications are shown in Tab. 1.

Tab. 1 Specimen preparation for 6 sizes

Specimen number	Specimen size(cm)
1#	12×6
2#	16×8
3#	20×10
4#	24×12
5#	28×14
6#	30×15

In addition, the experiment collects 10 samples of POF fabric with different luminous effects from different manufacturers after the sample size was determined. The surface brightness performance of the ten samples was tested three times to verify the validity and efficiency of the test method.

2.3 Test process

In the test process, the standard atmosphere for samples wetting and experiment is carried out according to GB/T 6529—2008 "Standard atmosphere for Textile humidification and Test"[14]. Other light sources need to avoid interference from the experiment during testing.

The following are the specific test procedures. Step 1: to ensure that it is parallel to the lens of the imaging luminance measurement device and within the size range of image capture, placing the sample plate fixed with the optical fiber fabric horizontally in the dark box. Step 2: turning on the power supply and wait for the POF fabric to light up and stabilize. Step 3: adjusting the focusing lens of the imaging luminance measurement device to make the image clear. The brightness image of the POF fabric sample can be generated in the matched computer software through one exposure. Step 4: framing the luminous area of the POF fabric in the image to obtain the surface brightness value of the POF luminous fabric, rounded to 0.001 cd/m^2. A screenshot of the test data is shown in Fig. 3.

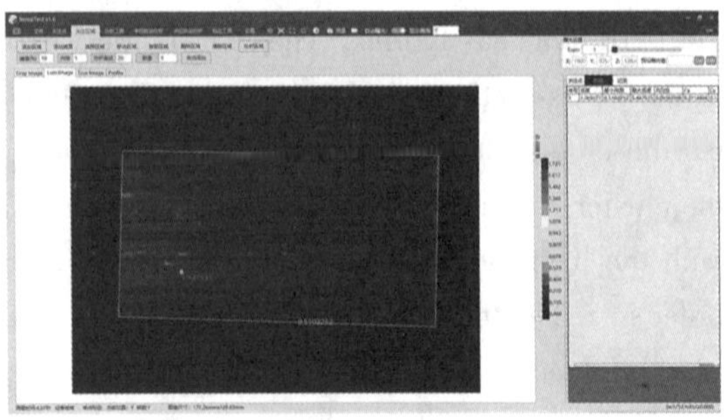

Fig. 3 Data screenshot of software test

3 Results and Discussion

3.1 Effect of sample size on brightness performance

It can be found that with the increase in sample size, the greater the optical attenuation of the POF fabric length direction, the smaller the average brightness of the sample. Meanwhile, the coefficient of variation of luminance is smaller as the sample size increases. And the volatility of the test results is less, and the test repetition is better. The test results are shown in Tab. 2.

Tab. 2 Brightness Performance of Six Specimen Sizes

Specimen number	1#	2#	3#	4#	5#	6#
Sample 1 brightness(cd · m^{-2})	0.619	0.596	0.551	0.435	0.386	0.336
Sample 2 brightness(cd · m^{-2})	0.977	0.727	0.644	0.445	0.396	0.353
Sample 3 brightness(cd · m^{-2})	0.746	0.612	0.531	0.463	0.373	0.349
Sample 4 brightness(cd · m^{-2})	0.666	0.648	0.474	0.505	0.403	0.355
Sample 5 brightness(cd · m^{-2})	0.540	0.506	0.543	0.458	0.384	0.345
Average brightness(cd · m^{-2})	0.710	0.618	0.548	0.461	0.388	0.348
Degree of dispersion	0.149	0.072	0.055	0.024	0.01	0.007
Coefficient of variation(%)	21.05	11.61	9.97	5.21	2.65	1.92

In order to compare the stability of the brightness test, the brightness values of five samples of each size were analyzed by line plot. As can be seen from Fig. 4, the

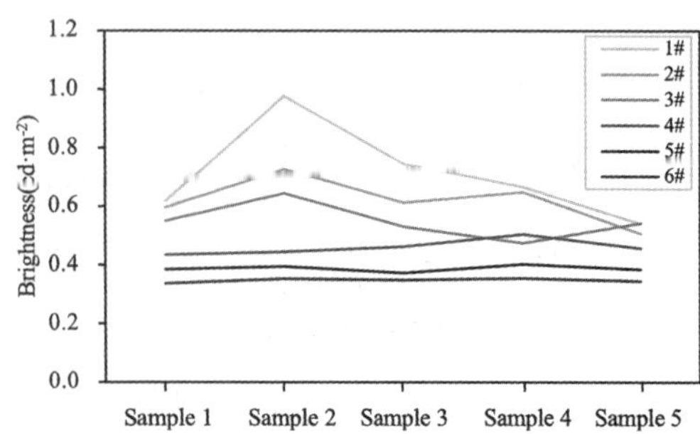

Fig. 4 Brightness comparison of samples with different sizes

test results of POF fabric with 6# are the most stable, and the dispersion degree reaches 0.007. Considering the convenience of manual fiber bundling and the coupling efficiency of light source, the sample size should not be too large. Therefore, the stability of test is best when the size specification is 30 cm×15 cm.

3.2 Validation of the testing method

After determining the sample size under the test method, 10 POF fabrics with 30 cm×15 cm size from different manufacturers and different luminescence effects were selected for the surface brightness performance test respectively. Each sample was tested 3 times to verify the effectiveness of the test method, and the test results are shown in Tab. 3.

Tab. 3 Brightness Performance Test of 10 Kinds of Optical Fiber Fabric

Sample number	Luminance average ($cd \cdot m^{-2}$)	Degree of dispersion	Coefficient of variation(%)
1#	0.419	0.001	0.00
2#	0.631	0.004	0.01
3#	1.817	0.002	0.00
4#	0.135	0.000	0.00
5#	0.253	0.000	0.00
6#	7.333	0.011	0.00
7#	0.637	0.001	0.00
8#	0.279	0.000	0.00
9#	1.441	0.000	0.00
10#	2.129	0.000	0.00

It can be seen from Tab. 3 that the coefficient of variation of all parameters is less than 0.01%, indicating that the repeatability of the three level test results is good. In addition, the brightness of different fiber fabrics varies greatly. The average brightness of each sample test was analyzed by a line chart. As can be seen from Fig. 5, the average brightness of 6# POF fabric is the highest, reaching 7.333 cd/m^2, while the average brightness of 8# POF fabric is the lowest, only

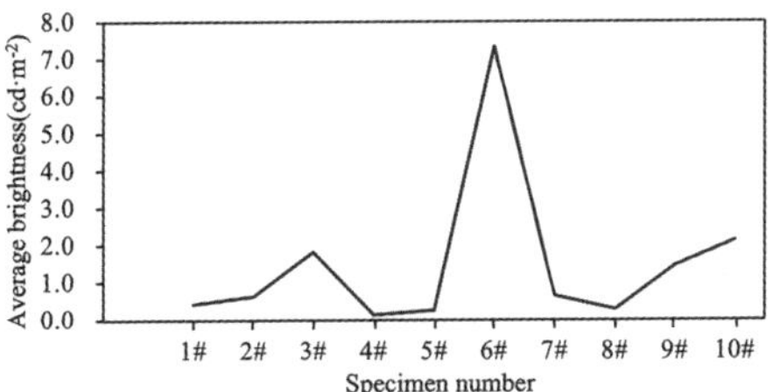

Fig. 5 Comparison of average brightness values of different samples

0.279 cd/m^2, with a difference of more than 26 times. It can be seen that the brightness performance of the POF fabric products on the market vary greatly. Meanwhile, the experimental method is relatively stable and suitable for the detection and evaluation of the brightness performance of POF luminous fabrics.

4 Conclusions

According to the luminescence characteristics of optical fiber fabric, this paper proposes a POF fabric brightness test method based on imaging luminance measurement device, and the corresponding test device is constructed. This paper analyzes the suitable sample size and verifies the stability and reliability of the method. The results show that the POF fabric surface brightness measurement method based on imaging luminance measurement device can provide accurate and reliable data on the basis of saving time and energy. And the stability of brightness test is best when the sample size is 30 cm×15 cm in this test device. The research content and results of this paper provide a new idea for brightness performance test of luminous fabrics, and provide technical support for the performance detection and evaluation standard of POF luminous fabric. Thus, it is conducive to the relevant production enterprises to control product quality, and help the market supervision department to carry out the relevant product testing.

References

[1] GONG Z, XIANG Z, OU Y X, et al. Wearable fiber optic technology based on smart textile: A review[J]. Materials, 2019, 12(20): 3311.

[2] LIN W J, MIAO X H. Application of optical fiber in luminescent fabric[J]. Journal of Textile Science, 2021, 42(7): 169-174.

[3] LV C. Development of polymer optical fiber luminescent fabric[D]. Tianjin: Tiangong University, 2013.

[4] YANG K, LI M Q, ZHANG C. Research progress of side-emitting optical fibers and fabrics[J]. Wool spinning technology, 2019, 47(7): 84-89.

[5] D. KONG, D. ZHANG, Y. YUAN, et al. Progress in research and application of plastic optical fiber[J]. Acta Photonica, 2019, 48(11): 71-85.

[6] WANG J, HUANG B, YANG B. Effect of weave structure on the side-emitting properties of polymer optical fiber jacquard fabrics[J]. Textile Research Journal, 2013, 83

(11): 1170-1180.

[7] LV C, ZHONG Z L, KUANG L Y, et al. Preparation and luminance test of polymer optical fiber luminescent fabric[J]. Journal of Tianjin Polytechnic University, 2013, 32(2): 42-45.

[8] CHEN Y Y, YANG B, JIN Z M. Fabrication of controllable luminescent fabric and its luminance characterization[J]. Journal of Textile Science, 2008(8): 38-41.

[9] HEE Y J, SEUNG C H, KUEN K H, et al. Influence of detailed structure and curvature of woven fabric on the luminescence effect of wearable optical fiber fabric[J]. Science of Emotion and Sensibility, 2018, 21(4): 55.

[10] MINAMISAWA R A, SANTOS L E R, PARADA M A, et al. Digital image analysis to standardize a photometric method in colorimetric quantification[J]. Instrumentation Science & Technology, 2007, 36(1): 97-104.

Application Properties of Bio-Based PVC Paste Resin in Automotive Interior Artificial Leather

ZHU Hui[1], GUO Jiaqi[2], XU Lanxin[2], MA Yao[1], GAO Jing[2]*

[1] *Benecke Changshun Auto Trim (Zhangjiagang) Co., Ltd, Zhangjiagang 215600, China*

[2] *College of Textiles, Donghua University, Shanghai 201620, China*

* *Corresponding author's email*: gao2001jing@dhu.edu.cn

Abstract: In order to transform and upgrade a green and sustainable automotive interior, bio-based PVC paste resin was utilized to create a new green PVC artificial leather material. In this work, the basic properties of the bio-based PVC paste resin were examined and the properties of the skin layer, physical foam layer and chemical foam layer prepared from bio-based PVC paste resin were studied to explore its performance in automotive interior artificial leather. The experimental results showed that the viscosity of fresh bio-based PVC skin plastisol, physical foaming plastisol and chemical foaming plastisol were 7 821 Pa · s, 11 320 Pa · s and 7 203 Pa · s respectively, which were similar to that of petroleum-based PVC plastisol. The odor ratings of the skin layer and physical foam layer industrial films made of bio-based PVC paste resin are 4.1 and 4.5, respectively, which were acceptable with the odor ratings between 4 and 5 just as petroleum-based PVC layer performed. The VOC test results indicated that the volatile content of the skin layer and physical foam layer made of bio-based PVC paste resin were 129 and 49, respectively. It proved that there was no obvious difference in volatile content between the above-mentioned films. The foaming ratios of bio-based PVC physical foaming at 190 ℃, 200 ℃ and 210 ℃ were 1.47, 1.40 and 1.24 respectively, and the chemical foaming ratios were 1.50, 2.82, and 3.55, which were slightly higher than the physical foaming ratios. The foaming of the side section of the industrial films also showed that the chemical foaming was fuller than physical foaming. In addition, the aging performance of bio-based and petroleum-based PVC coatings for automotive interiors was particularly similar. Therefore, bio-based PVC paste resin can be utilized to replace traditional petroleum-based PVC paste resin to apply in automotive interior synthetic leather materials.

Keywords: Automotive interior; PVC artificial leather; Bio base; Application performance

1 Introduction

In recent years, the automobile industry has developed vigorously. As a travel tool, automobiles bring convenience to people's lives. At the same time, the problems of high pollution and emissions caused by its industrial chain have also received more and more attention. PVC synthetic leather is an important part of automotive interiors, and its superior performance in all aspects is favored by the industry. The main raw material for PVC synthetic leather processing is PVC paste resin, one of the synthetic routes of which is that ethylene extracted from petroleum cracking is chlorinated to obtain vinyl chloride monomer, and then PVC paste resin is obtained by emulsion polymerization of vinyl chloride monomer. The traditional PVC paste resin industry is strongly dependent on non-renewable petroleum resources. In addition, petroleum cracking is a process with excessive energy consumption and emission causing irreversible damage to the environment. Therefore, in order to promote the green and sustainable development of the automobile industry and reduce the waste of petroleum resources, eco-friendly materials are demanded urgently. Today, bio-based ethylene has attracted extensive interests[1], which is obtained through the growth and metabolism of organisms using organic matter or biomass as raw materials[2].

The existing bio-based PVC paste resin is made by extracting ethylene from trees, then generating polyvinyl chloride through chlorination reaction. Finally, polyvinyl chloride is obtained after polymerization. Using this green and eco-friendly raw material instead of traditional petroleum-based PVC material will be expected to promote the sustainable development of the automotive industry. In order to verify whether bio-based PVC can meet the industrial requirements of mass production and utilization, the properties of bio-based PVC raw materials, such as plastisol viscosity, microscopic morphology of industrial films, odor and aging characteristics of finished products were evaluated and compared with the original petroleum-based products.

2 Experimental

2.1 Materials

Petroleum based PVC paste resin powder was obtained from Benecke Changshun Auto Trim (Zhangjiagang) Co., Ltd. Bio-based PVC paste resin powder was supplied by Benecke Changshun Auto Trim (Zhangjiagang) Co., Ltd.

2.2 Preparation of bio-based PVC industrial film and general PVC industrial film

The plastisol was prepared according to the specific formula, as shown in Tab. 1. The obtained plastisol was stirred at 1 500 r in a mixer for 10 min, and then baked at three temperatures of 190 ℃, 200 ℃, and 210 ℃ for 1.5 min to obtain a plasticized industrial film, respectively. The numbers of the prepared samples are as follows: 1# for petroleum-based PVC paste resin skin, 2# for bio-based PVC paste resin skin, 3# for petroleum-based PVC paste resin physical foaming, 4# for physical foaming of bio-based PVC paste resin, 5# for chemical foaming of petroleum-based PVC paste resin, and 6# for chemical foaming of bio-based PVC paste resin.

Tab. 1 Raw material formula

Raw material	1#	2#	3#	4#	5#	6#
Phthalate plasticizer-1	50	50			50	50
Phthalate plasticizer-2			45	45		
Calcium zinc stabilizer-1	1	1			1	1
Calcium zinc stabilizer-2			3.3	3.3		
Petroleum based PVC powder	50		50		50	
Bio-based PVC powder		50				50
ADC chemical blowing agent					2.68	2.68
Chemical foaming catalyst					0.3	0.3
Microsphere foaming agent			1	1		
Flame retardant			4	4	4	4
Filler					10	10

2.3 Characterization

Cut out 10 cm×10 cm samples to weigh and measure the thickness. Take the optical microscope photo of the side section to observe the foaming situation and calculate the foaming ratio by weight and thickness values. Prepare the plastisol and use viscometer to measure the viscosity at 23 ℃. Use an infrared moisture meter to measure the moisture content after 10 min treatment at 80 ℃.

Scraper fineness: the plastisol is prepared and applied on the scraper according to the ratio of PVC:DINP of 55 : 45.

Rising height: prepare the plastisol and use a vacuum pump to evacuate according to the ratio of PVC:DINP of 55 : 45. Measure the rising height of the graduated cylinder with a ruler.

Fogging degree: put the sample into a beaker, place it in a temperature-controlled box, and then cover it with a cooling plate. After constant temperature treatment, it is placed in a drying oven. Weigh the mass of the aluminum foil with atomized condensation and compare it with the mass of the aluminum foil without atomized condensation to obtain the fogging value.

Odor rating: use a special oven for odor to process the samples at 80 ℃±2 ℃ for 120 min±10 min. The treated samples were placed in glass bottles for manual evaluation.

VOC test: Use an oven to pretreat the sample, and put the sample into a polyvinylidene fluoride sampling bag in the sampling chamber, then replace the bag with gas, seal the bag and put it into the sampling chamber for 2 h at 65 ℃. The sorbent Tenax was used to detect VOCs, the DNPH was used to test aldehydes, and the mass spectrometer was used for qualitative and quantitative analysis of the detected substances.

Aging degree: place the samples at 120 ℃ for 1 d, 3 d, 7 d, 10 d and 14 d, respectively to observe samples the discoloration.

3 Results and Discussion

3.1 Micromorphology analysis

The micromorphology of petroleum-based and bio-based PVC is shown in Fig. 1. The secondary particles of bio-based PVC paste resin (agglomerates formed by primary particles) have rough surfaces, and its average particle size is significantly smaller than that of petroleum-based PVC paste resin. The average size of the primary particles of petroleum-based PVC paste resin (the smallest particles formed during the polymerization process) is smaller than the bio-based PVC paste resin. It is noteworthy that the molecules of petroleum-based PVC paste resin are more tridimensional and compact.

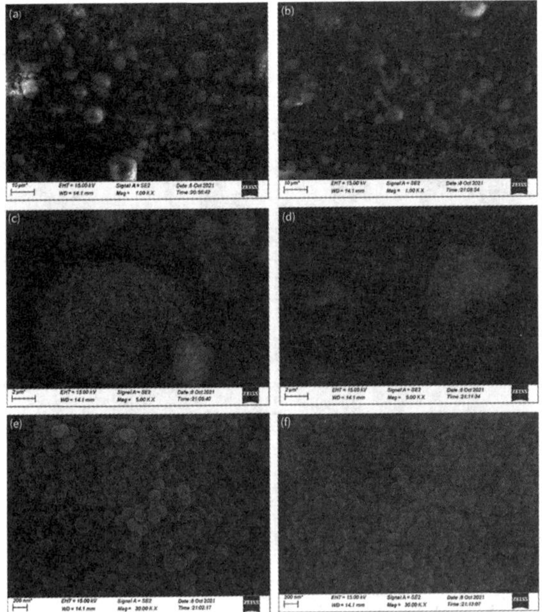

Fig. 1 Micro morphology of (a) petroleum-based PVC (1 000×), (b) bio-based PVC (1 000×); (c) petroleum-based PVC (5 000×); (d) bio-based PVC (5 000×); (e) petroleum-based PVC (30 000×); (f) bio-based PVC 30 000×

3.2 Raw material properties analysis

The performance test of raw materials mainly includes viscosity, water content, scraper fineness, rising height, fogging, etc. The experimental results are shown in Tab. 2. From the results, we can see that the properties of the bio-based PVC paste resin powder satisfy the standard of the PVC paste resin powder for automotive interiors.

Tab. 2 Test results of basic items

Test items	Petroleum-based PVC plastisol	Bio-based PVC plastisol
water content(%)	≤0.3	0.21
Scraper fineness(μm)	≤80	18
rising height(mm)	≤80	44
Fogging index(mg)	≤0.5	0.13
Shear rate $D=5\ s^{-1}$(Pa·s)	8–22	9 459
Shear rate $D=163\ s^{-1}$(Pa·s)	2–5	3 854

In Fig. 2, the viscosity of bio-based PVC plastisol approximates that of the original PVC plastisol. And the viscosity of physical foaming samples has a remarkable increase with the time up to 7 d. The initial viscosity of bio-based PVC paste resin is higher than that of petroleum-based paste resin due to more secondary particles but less loose primary particles in petroleum-based PVC paste resin making the particles arranged densely. However, the bio-based PVC has more primary particles and less secondary particles, which are easier to contact with the catalyst and bind the catalyst particles. That is why its viscosity is higher. In addition, the viscosity increase of PVC paste resin is faster after a few days. It may be explained that the secondary particles of bio-based PVC paste resin are smaller in size and dispersed more easily. On the contrary, the secondary particles of petroleum-based PVC paste resin are larger and disintegrate more slowly.

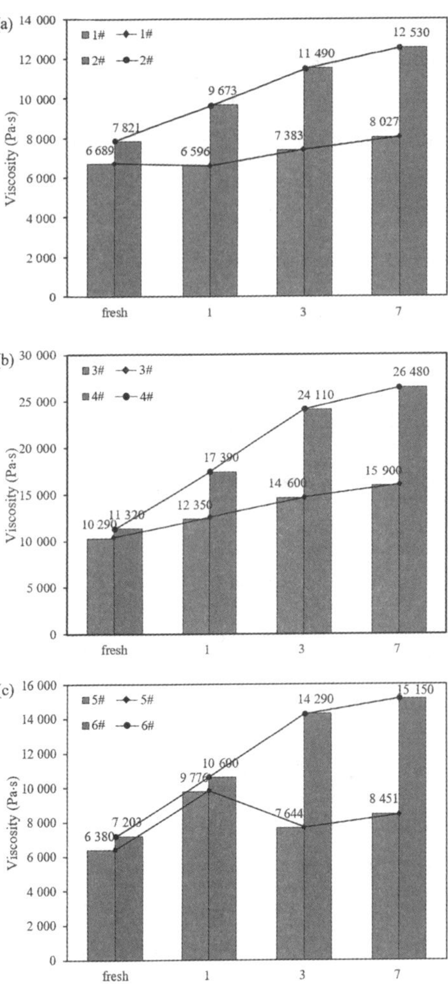

Fig. 2 Variation of plastisol viscosity of (a) sample1# and 2#; (b) sample 3# and 4#; (c) sample 5# and 6#

3.3 Odor

The results of odor evaluation of the industrial films prepared from bio-based PVC paste resin and petroleum-based PVC paste resin are shown in Tab. 3. The odor rating of 4 samples is all in the range of 4 to 5. There is no obvious difference in smell between the skin layer and foam layer industrial film made from above-mentioned PVC paste resin. It can be considered that they all satisfy the industrial production requirements.

Tab. 3 Odor rating

Sample	1#	2#	3#	4#
Odor rating	4. 3	4. 1	4. 4	4. 5

3.4 VOC

The VOC test results are shown in Tab. 4 and Tab. 5. Comparing the VOC test results of 1# with 2# sample, it can be considered that the difference in VOC content and the types and proportions of the substances contained between the bio-based PVC paste resin and the petroleum-based PVC paste resin skin industrial film are basically the same. Comparing the test results of 3# with 4# sample, it can be concluded that the VOC content of the bio-based PVC paste resin and the petroleum-based PVC paste resin foamed industrial film are almost the same, and the substances contained are similar, but the proportions are slightly different.

Tab. 4 VOC test results

Sample	1#	2#	3#	4#
VOC (ug/g)	108	129	46	49

Tab. 5 VOC content

Content	1#	2#
1-(2-Butoxyethoxy)ethanol	41. 5%	41. 9%
2-Ethylhexanol	13. 1%	10. 8%

3.5 Foaming

The side-cut foaming of the industrial film obtained by drying at 200 ℃ for 1. 5 min is visualized in Fig. 3, and the calculated foaming ratio is shown in Tab. 6. It can be seen that the bio-based PVC paste resin foamed approximately the same as the petroleum-based PVC paste resin at 190 ℃, 200 ℃ and 210 ℃. Among them, physical foaming is generally not as full as chemical foaming. The foaming ratio also proves that the foaming performance at three different temperatures is basically the same. In terms of foaming performance, bio-based PVC may have the potential to replace ordinary PVC.

Tab. 6 Foaming ratio

Foaming ratio	3#	4#	5#	6#
190 ℃	1. 43	1. 47	1. 49	1. 50
200 ℃	1. 40	1. 40	2. 83	2. 82
210 ℃	1. 24	1. 24	3. 55	3. 55

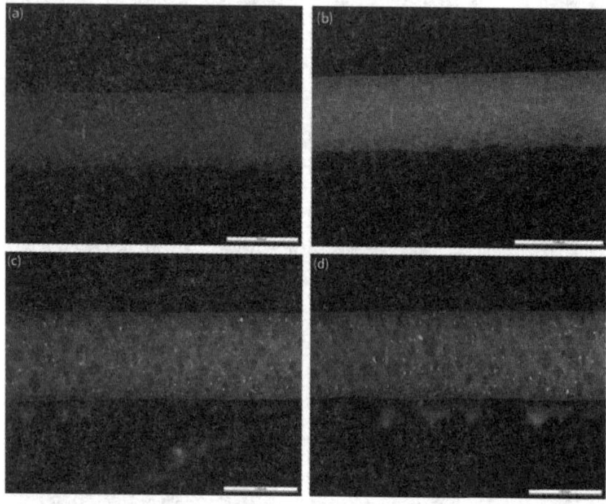

Fig. 3 Foaming of the side-cut surface of the industrial film obtained by drying at 200 ℃ for 1. 5 min: (a) 3#; (b) 4#; (c) 5# and (d) 6#

3.6 Aging

Fig. 4 shows the surface changes of industrial film samples after seven-day aging at 120 ℃. The skin industrial films of bio-based PVC paste resin and petroleum-based PVC paste resin changed from colorless and transparent to reddish-brown, and the physical foamed industrial film changed from white to brown after a certain time. The surface changes of bio-based and petroleum-based industrial films in 1, 3, 7, 10, and 14 d are roughly the same. It proves that the industrial films prepared with bio-based PVC paste resin have a well aging performance as industrial films prepared by petroleum-based PVC paste resin.

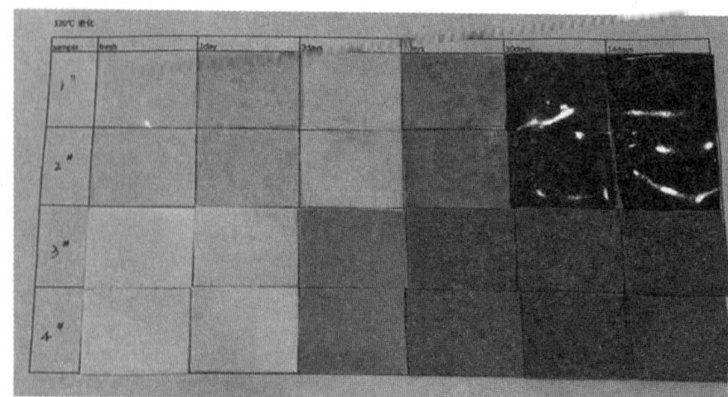

Fig. 4 Seven-d aging of industrial films

4 Conclusions

This study shows the different performances of bio-based PVC paste resin and petroleum-based PVC paste resin. For bio-based PVC paste resin, the performance of its raw material is acceptable for automotive interior. In addition, the odor, VOC content, foaming and aging performance of plastisol prepared by bio-based PVC paste resin as well as plastisol prepared by petroleum-based PVC paste resin. However, the viscosity of bio-based PVC plastisol is a little higher than petroleum-based PVC plastisol. This work has shown that the green bio-based PVC paste resin will be able to replace the petroleum-based PVC paste resin for the applicability of the automotive interior.

References

[1] HARO P, OLLERO P, TRIPPE F. Technoeconomic assessment of potential processes for bio-ethylene production[J]. Fuel Processing Technology, 2013, 114(4): 35-48.

[2] HU Y C, ZHAN N N, DOU C, et al. Selective dehydration of bio-ethanol to ethylene catalyzed by lanthanum-phosphorous-modified HZSM-5 Influence of the fusel[J]. Biotechnology Journal, 2010, 5(11): 1186-1191.

Mechanical Properties of a Novel Braided Ureteral Prototype Stent in Vitro

JIAN Wenping[1,2], LIN Jing[1,2], WANG Fujun[1,2*], WANG Lu[1,2*]

[1] *Key Laboratory of Textile Science and Technology, Ministry of Education, Donghua University, Shanghai 201620, China*

[2] *College of Textiles, Donghua University, Shanghai 201620, China*

* *Corresponding author's email*: wfj@ dhu. edu. cn; wanglu@ dhu. edu. cn

Abstract: Ureteral stents are commonly used in the treatment of ureteral obstruction to support the narrow ureteral cavity and maintain urine drainage. Due to the high mechanical strength of metal materials, metal ureteral stents are suitable for implantation in patients with ureteral obstruction caused by malignant tumors. Mechanical braiding technique as a controllable and efficient method has been used in the molding of polymer ureteral stents. However, mechanical braided metal ureteral stents have not been studied. In this study, two kinds of single-layer braided prototype stents (S1 and S2) and two kinds of double-layer braided prototype stents (D1 and D2) covered with a tight braided layer were prepared by automatic mechanical braided technology using medical nitinol wire. The radial compression and axial tensile properties of the four stents were tested. The results show that the covering layer of the double-layer stent significantly improves the radial compression performance, but does not affect the axial tensile performance, which is mainly determined by the inner layer with a smaller braiding angle. D2 stents have the highest radial compressive load 1 487. 93 cN±92. 13 cN), the best radial compressive elasticity (94. 95% ±2. 18%), the higher fracture load (90. 11 N±7. 83 N) and the lower elongation (20. 20%±3. 06%).

Keywords: Ureteral stents; Double-layer ureteral stents; Braid; Mechanical properties

1 Introduction

As an important component of urological medical devices, ureteral stents are used in the treatment of hydronephrosis and urethral stricture or obstruction, as well as enabling urine drainage and passive dilatation of the ureter[1].

The biomechanical performance of stents is one of the significant factors affecting therapeutic effects[2]. Moreover, the mechanical properties are largely influenced by the material and preparation of stents. Materials of ureteral stents are generally divided into Polyurethane and metal. Polyurethane stent (PUS) has weak radial support and is prone to lumen blockage. While the metal stent provides excellent radial support, maintains the lumen unobstructed for a long time, and is not prone to fracture failure[3]. Nitinol alloy is one of the materials for metal ureteral stents[4].

Mechanical braiding is one of the simple and important methods for preparing stent. By changing the density of braiding points and the diameter of the braiding wire, it is easy to regulate the performance of the stent. In addition, mechanical braiding realizes automatic production and greatly reduces the cost of the stents[5]. Braided polymer ureteral stent has been studied[6-7]. One of the advantages of braided stent implantation is that its radial support performance is not compromised by repeated contraction and relaxation of the urethra[8]. However, the braided metal ureteral stent has been poorly studied.

In this study, four kinds of braided metal prototype ureteral stents were prepared by mechanical braiding technology using medical nitinol (NiTi) wire[9], including two kinds of single-layer stents and two kinds of double-layer stents covered with braided layer. The structures of braided metal ureteral prototype stents were characterized. And in vitro mechanical properties such as radial compression and axial tensile properties were tested. Analyze comprehensively the influence of stents structure on the mechanical properties by comparing the properties of single-layer braided stents and double-layer braided stents.

2 Experiments

2.1 Materials

Three kinds of medical Nitinol wires with diameters of 0.08 mm, 0.065 mm and 0.04 mm were supplied by FJMT Co., Ltd. (Guangdong, China).

2.2 Design and preparation of ureteral stents

The design and preparation parameters are shown in Tab. 1. Mb－32 multi-functional automatic medical braiding machine in Biomedcail Textile Lab of DHU was used to braid the four kinds of ureteral prototype stents. The spindle speed was set 250 r/min. All stents were selected to braid in a regular braid configuration. The stainless steel mandrel diameter of 1.6 mm was selected to control the inner diameter of the stents. The single-layer stents were fabricated by braiding directly on mandrel, and the double-layer stents were fabricated by braiding an outer layer on the single-layer stents with mandrel.

Tab. 1 Design and preparation parameters

Number of braided stents	Preparation parameters (wire diameter/design PPI)	
	inner	outer
D1	0.08/40	0.04/110
D2	0.065/50	0.04/110
S1	0.08/40	—
S2	0.065/50	—

After braiding, the stents with mandrel were heat treated at 500 ℃ for 5 min in LT40/11/B410 muffle furnace (Nabertherm Co., Ltd., Lilienthal, Germany) and water quenching at 20 ℃ to eliminate residual stress and stable braiding structure.

2.3 Morphology and physical characteristics

TM3000 desktop scanning electron microscope (Hitachi Co., Ltd., Tokyo, Japan) was used to observe the morphology and structure of the stent. Image J software was used to measure the tube outer diameter, braiding angle, pitch, number of braid points per inch (PPI) according to the microscope image. And calculate the wall thickness.

2.4 Radial compression property

LLY－06D compression tester (Laizhou Electronic Instrument Co., Ltd., Shandong, China) was used to test the compression performance of stents. Referring to ISO 25539-2—2012, according to the characteristics of small diameter metal braided mesh stents, the plate compression method was selected to test the compression property of stents. The length of the sample was 40 mm and the diameter of the presser foot was 50 mm. The maximum compression distance was 50% of the inner diameter of the stent and the rate of compression process was 10 mm/min. When the presser foot reached the maximum compression distance, the compression strength was recorded. The presser foot stayed at the maximum compression distance for 10 s, and the compressive strength continued to decrease due to the stress relaxation of stents. Then the presser foot was unloaded at a rate of 10 mm/min until it left the surface of stents. Then waited for 10 s, presser foot was loaded for the second time and touched the surface of stents. The elastic recovery rate was calculated during the process.

2.5 Axial tensile property

YG (B) 026G－500 medical textiles multifunctional strength testing machine (Wenzhou Darong Textile Instrument Co., Ltd., Zhejiang, China) was used to test the tensile properties of every sample. Referring to YY/T 0872—2013, the tensile rate was set as 200 mm/min and the spacing was 50 mm. In order to simulate the internal environment as much as possible and ensure the tubular shape of the stents, short metal support bars with a diameter of 1.6 mm were inserted at both ends of the stents during the test. The fracture strength and elongation at the break of the stents were obtained.

Statistical analysis:

SPSS 24 statistical software was used for statistical analysis. All data obtained from braided stents were expressed as mean ± standard deviation. Students' *t*-tests were used for determining the statistical difference between samples. The data in the figures were marked by * for $p < 0.05$.

3 Results and Discussion

3.1 Morphology and physical characteristics

The SEM images in Fig. 1 show that the structure of braided prototype ureteral stents is stable.

The physical parameters of stents are listed in Tab. 2. The diameters of braided stents range from (1.83 ± 0.03) mm to (2.07±0.03) mm. The calculated wall thickness is equal to 2 times of the sum of the diameters of the participating wires. Due to the high rigidity of the

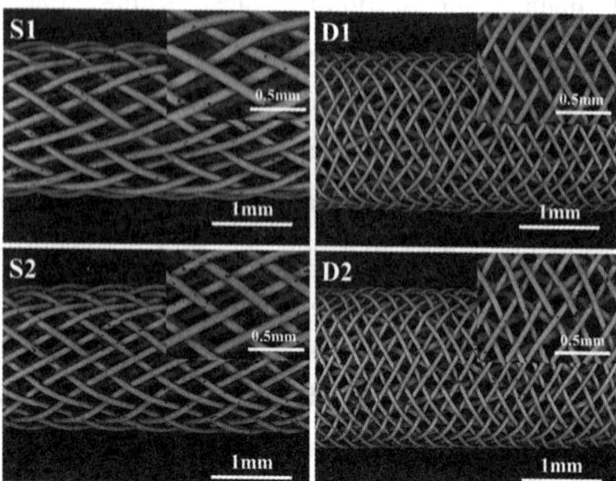

Fig. 1 Scanning electron microscope images of four kinds of prototype ureteral stents

wire, it is not easy to deposit along the circumferential direction and tends to be oriented along the axial direction. In addition, the friction between the wire and the mandrel is low, which makes it easier to slip in the axial direction. As a result, the actual PPI of group S stents is lower than the designed PPI. The wire spacing of group S stents is relatively uniform because of small dispersion of PPI. However, actual PPIs of group D stents are higher, especially D2 stents, which may be due to two reasons. The first is that wire of D2 outer layer is fine with low stiffness. The second is the influence of the braided inner of stents. Different from the smooth surface of the mandrel, the braided inner has buckling bumps at the interleaving points, and the NiTi wire of the outer slips at the interleaving point after deposition, resulting in uneven wire spacing. Group D stents have the same outers, but due to the influence of different inners, the pitch of D2 stents is smaller than that of D1 stents, and the actual outer structure is more compact. Braiding angle is a more intuitive and important structural parameter of braided stents, which has a corresponding relationship with PPI in Eq. (1).

$$n = 25.4 \times \frac{\frac{N}{2} \times \tan\alpha}{\pi D_m} \quad (1)$$

where n is PPI, α is braiding angle, N is number of bobbins, D_m is diameter of mandrel.

Tab. 2 Physical features of stents

	S1	S2	D1	D2
Tube outer diameter (mm)	1.92± 0.041	1.83± 0.027	2.07± 0.030	2.03± 0.047

(**Continued**)

	S1	S2	D1	D2
Wall thickness (mm)	0.16± 0.020	0.12± 0.013	0.23± 0.015	0.21± 0.024
Outer braiding angle(°)	28.8± 2.99	31.3± 1.91	64.09± 1.46	63.7± 1.19
Pitch (mm)	11.31± 0.322	8.80± 0.329	3.34± 0.179	3.08± 0.056
Actual PPI	35.6± 0.55	45.4± 0.79	119.5± 1.73	126.2± 3.96

Note: the braiding angles, pitches and the actual PPIs of D1 and D2 stents refer to the parameters of their braided outers, as their inners are the same as S1 and S2 stents.

3.2 Radial compression property

The compression behaviors of stents are shown in Fig. 2. The initial moduli of all stents are basically same and the initial compression resistances are similar. Slight stress relaxation is observed in group D stents when stents are subjected to compressive load for 10 s at 50% of the inner diameter. There are almost no irreversible compression deformations for four kinds of stents.

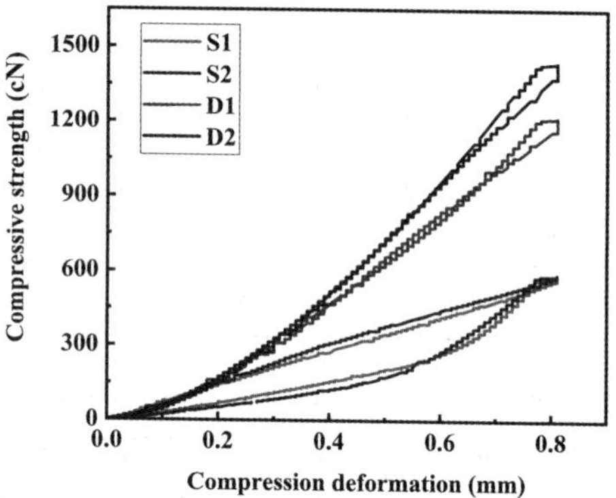

Fig. 2 Radial compression curves of stents

Radial compression load reflects the ability of the stents to resist external pressure and keep the lumen unobstructed. Fig. 3 shows the radial compression loads of stents. The covering of the tight outer significantly increases the radial compression load of group D stents. When the compression deformation reaches 50% of the inner, the radial compression loads of the two kinds of double-layer stents are 1 239.38 cN±38.83 cN and 1 487.93 cN ± 92.13 cN respectively, which are 1.4 times and 1.9 times higher than single-layer stents, and higher than those of PUS (1 270.70 cN). The elastic recovery rates of stents are shown in Fig. 4. The four kinds of braided stents all have good elastic recovery rates, and are over 90%. The elastic recovery

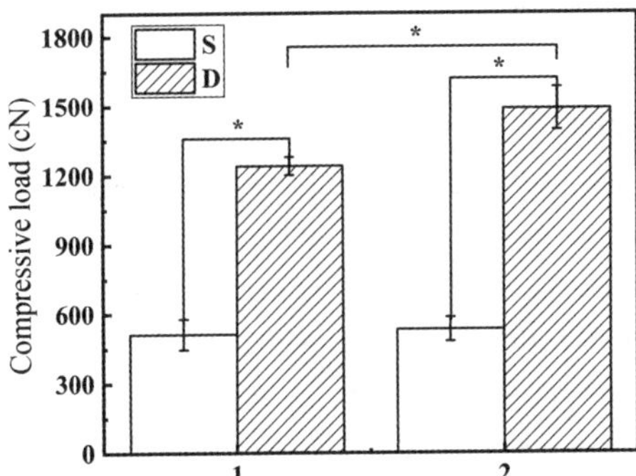

Fig. 3 Radial compression strengths of stents

rates of two kinds of group D stents are 94. 71% ± 2. 10% and 94. 95%±2. 18%, respectively, and slightly higher than those group S stents. It shows that the braided covering enhances the radial elasticity.

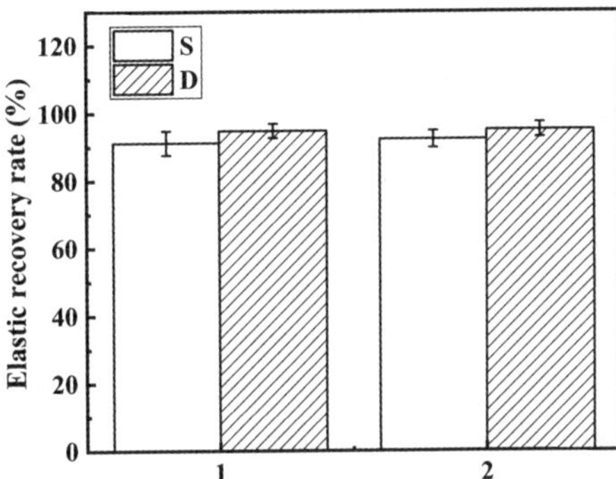

Fig. 4 Elastic recovery rates of stents

3.3 Axial tensile property

The typical tensile behaviors of the stents are shown in Fig. 5. NiTi braided stents have similar tensile behavior. In the initial tensile, the flexible interlace points in braid structure make the wires easy to orientate along the axial direction and the braiding angle is gradually reduced, so stents can be elongated under the action of a small force. The initial process can be calculated by Eq. 2. Once the wires are almost axial, the tensile behaviors of the stents are mainly affected by the tensile properties of the NiTi wire. The tensile fracture behaviors of the double-layer stents depend on the inner. According to Eq. 2, the elongation in the initial tensile stage of the outer is 128% ($1/\cos 64° - 1 \approx 128\%$). The outer is still in the initial stage when the inner fails. Therefore, D1 stents and S1 stents, D2 stents and S2 stents have more similar tensile behaviors. The braiding angle of S2 stents (31. 3°±1. 91°) is larger than that of S1 stents (28. 8°±2. 99°), so the initial stretching platform of S2 stents ($1/\cos 31.3° - 1 \approx 17\%$) is longer than that of S1 stents ($1/\cos 28.8° - 1 \approx 14\%$). In fact, the actual initial elongations of stents are less than the theoretical initial elongations, which are related to the stiffness of the wires, friction at interlacing points and other factors.

$$l_{imax} = \frac{1}{\cos\alpha} - 1 \tag{2}$$

where α is braid angle, l_{imax} is maximum elongation at initial stage.

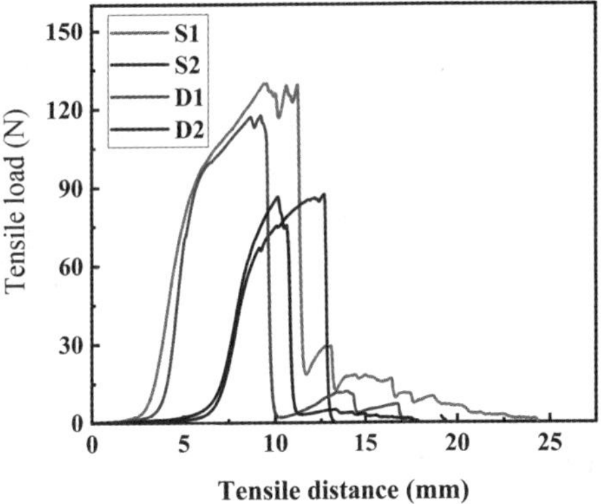

Fig. 5 Axial tensile typical curves of stents

Breaking load refers to the difficulty of fracture failure of stents during operation and using. Fig. 6 shows the axial breaking loads of stents. Because the fracture of stents mainly depends on the layer with a small braiding angle, there is no significant difference between the breaking loads of 1D stents and 1S stents, 2D stents and 2S stents. The breaking loads of group 1 stents are significantly higher than those of the corresponding group 2 stents, which may be due to the difference between 0. 08 mm NiTi wire and 0. 065 mm NiTi wire in tensile load.

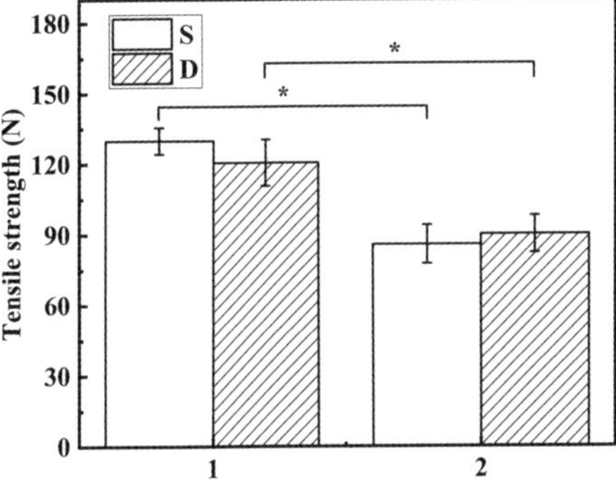

Fig. 6 Axial breaking loads of stents

Elongation reflects the operability of the stents, which is required to be as small as possible in the process of stent placement and removal. Fig. 7 shows the elongation of stents. The elongations of the four braided NiTi stents are much lower than that of PUS (286.69%). The elongations of the two kinds of double-layer stents are slightly lower than single-layer stents. There is no significant difference between the elongations of D1 and D2 stents (19.37%±2.51% and 20.20%±3.06%, respectively).

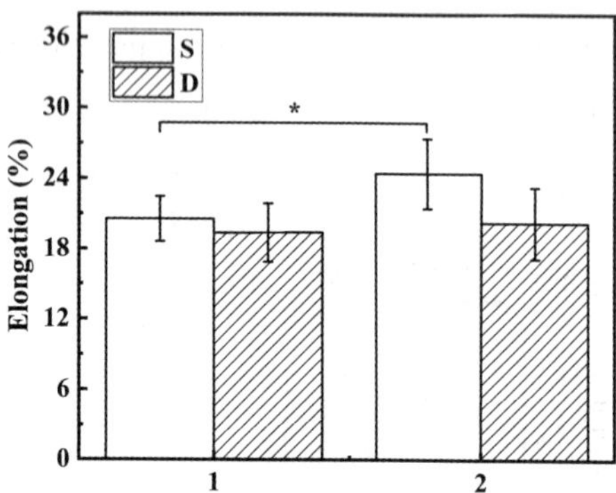

Fig. 7 Axial elongations of stents

4 Conclusions

In this study, four kinds of NiTi ureteral prototype stents were prepared by mechanical braided technology, including two kinds of single-layer braided stents (S1, S2) and two kinds of double-layer braided stents (D1, D2), and their mechanical properties were tested in vitro. The results show that the braided coverings significantly increase the radial compressive loads of the two single-layer stents by 1.4 and 1.9 times, respectively, but have no significant effect on the tensile properties. The compressive load (1 487.93 cN ± 92.13 cN), breaking strength (90.11 N ± 7.83 N) and elongation (20.20%±3.06%) of D2 stents are better than those of PUS, which have potential clinical application value.

References

[1] YIN K, DIVAKAR P, WEGST U. Freeze-casting porous chitosan ureteral stents for improved drainage[J]. Acta Biomaterialia, 2018, 84(15): 231-241.

[2] AMITAY-ROSEN T, NISSAN A, SHILO Y, et al. Failure of ureteral stents subject to extrinsic ureteral obstruction and stent occlusions[J]. International Urology and Nephrology, 2021, 53 (8): 1535-1541.

[3] ASAKAWA J, IGUCHI T, TAMADA S, et al. Treatment outcomes of ureteral stenting for malignant extrinsic ureteral obstruction: A comparison between polymeric and metallic stents[J]. Cancer Management and Research, 2018, 10: 2977-2982.

[4] PARK C, SHIN J H. Advances in ureteral stent design and materials[J]. Journal of the Korean Society of Radiology, 80(4): 631-642.

[5] REBELO R, VILA N, FANGUEIRO R, et al. Influence of design parameters on the mechanical behavior and porosity of braided fibrous stents[J]. Materials & Design, 2015, 86(5): 237-247.

[6] ZOU T, WANG L, LI W C, et al. A resorbable bicomponent braided ureteral stent with improved mechanical performance[J]. Journal of the Mechanical Behavior of Biomedical Materials, 2014, 38: 17-25.

[7] ZOU T. Preparation, structure and degradation behavior of a new "Fiber-film" resorbable ureteral stent[D]. Shanghai: Donghua University, 2015.

[8] CUI H P, ZHANG K, GAO C G, et al. Design and characterization of a novel braided biodegradable double-J ureteral stent [J]. Polymers for Advanced Technologies, 2022, 33 (8): 2519-2527.

[9] WANG L, JIAN W P, LIN J, et al. Ureteral stent: China Patent, 114588496A[P]. 2022-06-07.

Texture Display on Dynamically Draped Fabric Surface based on Particle-Spring Model

MA Zhenyi[1], ZHANG Ruiyun[1*], JI Feng[1], HU Xuhui[1], XING Hao[1], XU Tengfei[1]

[1] *Key Laboratory of Textile Science & Technology, Ministry of Education, Donghua University, Shanghai 201620, China*

* *Corresponding author's email*: ryzhang@ dhu. edu. cn

Abstract: The current virtual fabric texture research commonly focused on the static image display, or used the original image gloss light and shade to demonstrate the 3D effect. Most of these methods could not realize the dynamic display of the fabric texture. In this paper, we simulated the particle spring model of a virtual fabric with OpenGL based on the particle spring model, and mapped the relevant texture maps to the grids constructed by the particle spring model. Virtual fabrics could be seen as large grids of multiple small pieces of cloth. The vertices of the grids were regarded as particles, and the lines between the vertices were thought of as springs. The spring was composed of three aspects, namely the structural spring, the shear spring, and the bending spring. With OpenGL, the original image can be ideally mapped on a 2D texture. In order to accurately show the texture on fabric surfaces that were dynamically draped, we attempted to cut the original image into several color blocks, refined them into the same number of patches as the grids, and finally arranged the patches in the correct order. Through the arrangement and combination of color blocks, the texture image could be accurately mapped on the grids of particle spring model, and then the model presented a virtual fabric texture dynamic demonstration effect.

Keywords: Texture map; Virtual fabrics; Particle spring model

1 Particle Spring Model

The particle spring model was first put forward by Provot, the key point is to arrange fabric as several rules of particle collection. The fabric density was uniform, so the quality of the whole fabric was evenly on each particle. Between different particles were connected by a zero quality spring, spring and particle could show real physical deformation of fabrics[1]. The springs in the model were divided into three parts: structural springs, shear springs, and bending springs. Structural springs were used to fix the initial distance between the longitudinal and transverse particles to prevent excessive multi-displacement. Shear springs were used to prevent excessive deformation between oblique particles. Bend springs were used to connect two separated particles to prevent excessive deformation when the fabric was bent and deformed[2].

Therefore, depending on the connection between the three springs and particles, we could think of the model of particles and springs as a large mesh composed of several triangles. The nodes in the mesh were particles, and the lines between the nodes were springs. This was shown in Fig. 1.

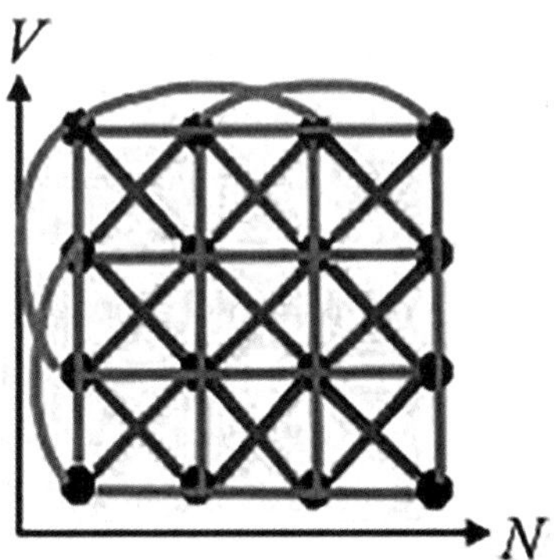

Fig. 1 Schematic of a particle spring model. Particle (black vertices), Structural springs (blue line), Shear springs (red line), Bending springs (green line)

2 Experimental Analysis

2.1 Analysis of forces

Throughout the particle spring model system, there were internal and external forces. The internal forces were spring force and damping force. The spring force was calculated by Hooke's law. Firstly, we gave an initial spring coefficient, and then calculated the type variable of the spring in real time to find the spring force. The role of the damping force was to cushion the excessive force of the spring, improve the overall stability of the fabric, and make the particle spring model truly simulate the dynamic effect of the fabric. The calculation of the damping force also required a given damping coefficient, which was determined by the difference in velocity between the particles. The external forces of the system were mainly gravity and wind, we only introduced the effect of gravity on the fabric and in this model. So we only needed to give the acceleration of gravity and the initial mass of the particle to get the corresponding gravity[3].

2.2 Method analysis

Most of the current academic research on fabric texture was mostly concentrated on 2D static pictures, or the use of a pseudo-3D effect to reflect the three-dimensional dynamic effect. These methods could not realistically simulate the physical characteristics of fabric texture. That is the reason we planned to map textures based on the particle spring model. When textured mapping a particle spring model, we discussed a variety of approaches, including overall or partial mapping of fabrics. When we mapped the entire original image to the fabric, there may be a situation where the local picture could not move steadily with the model, which would greatly reduce the three-dimensional dynamic effect of the particle spring model.

The establishment of the particle spring model provided a completely new idea and idea for the mapping of fabric textures. Eventually we decided to refine the overall model into small triangles and texture map them. As we textured each small piece of cloth in the fabric, the overall image moved synchronously with the movement of the particle spring model, thus achieving a dynamic display of the fabric texture. The comparison of the particle spring model before and after the map we made was shown in Fig. 2[4].

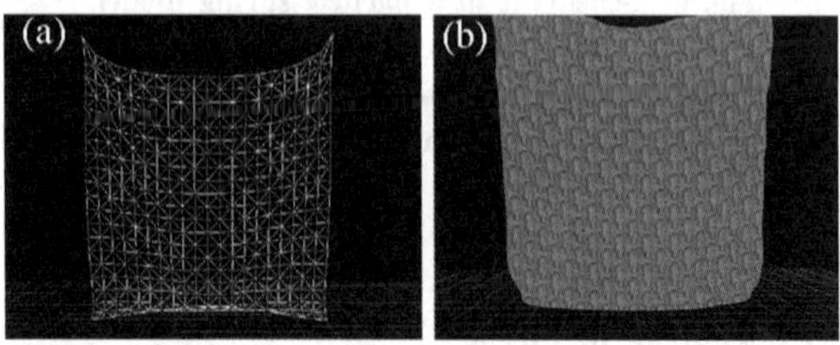

Fig. 2 Model map comparison: (a) before the map; (b) after the map

3 Texture Effect

3.1 Illumination

First of all, we compared the particle spring model with lighting added. In the case of lighting, the glossy effect of the fabric texture could bring a more realistic effect. Without light effect, the overall saturation of the fabric was higher, and the sense of realism was lower (Fig. 3). The specific effect was shown in Fig. 4[5-6].

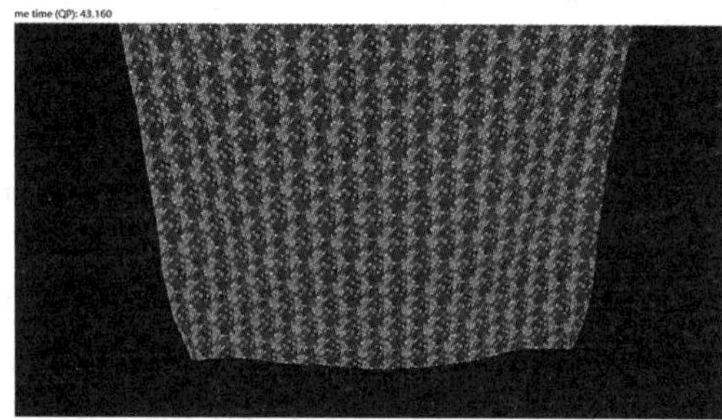

Fig. 3 No lighting effect

Fig. 4 Lighting effect

3.2 Patterns effect

Secondly, the experiment found that by mapping different pictures to the fabric, we selected different modification effects of the pictures, which could give the fabric different 3D effects.

3.2.1 Line patterns

As can be seen from Figs. 5 and 6, the more complex the pattern, fabrics on display looked and felt thicker, so the intuitive feeling was more realistic. Patterns of thick lines were visually more colorful than thin lines. However, there must be a certain limit to the complexity of the pattern, beyond this limit, the texture of the virtual fabric would become messy, which greatly reduced the three-dimensional effect of the fabric.

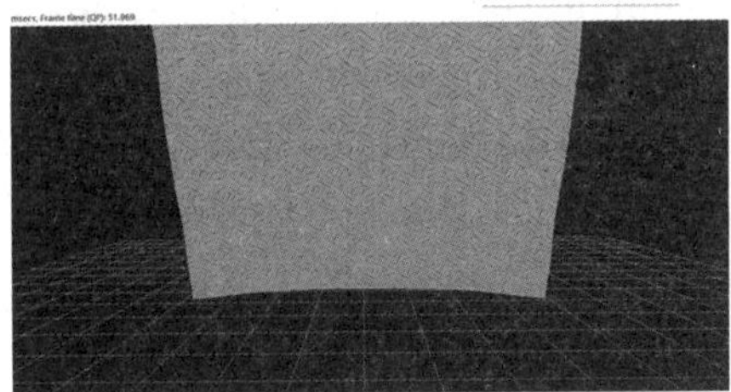

Fig. 5 Thin line pattern effect

Fig. 6 Thick line pattern effect

3.2.2 Mosaic effect

Taking Figs. 8 and 9 as an example, when the contemporary picture was transparent and concise, the particle spring model could not realistically simulate the texture of the fabric. After the mosaic processing of the white area of the picture, the three-dimensional dynamic effect of the model became more realistic and visual. The effect of this pattern mosaic was similar to increasing the convexity of the fabric texture, so that the fabric reflected a three-dimensional feeling with the help of the light and shade of the light.

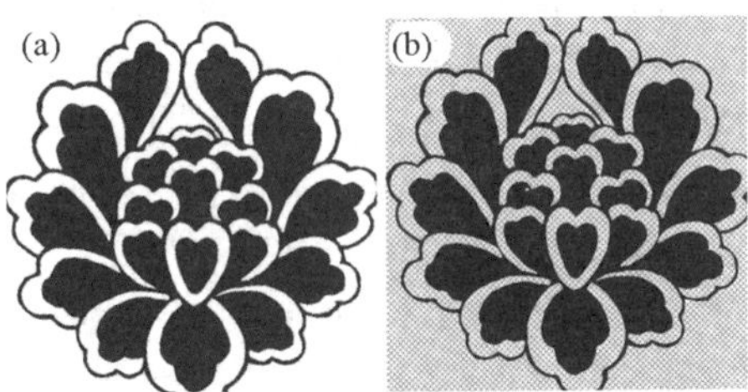

Fig. 7 Picture mosaic comparison: (a) no mosaics; (b) mosaics

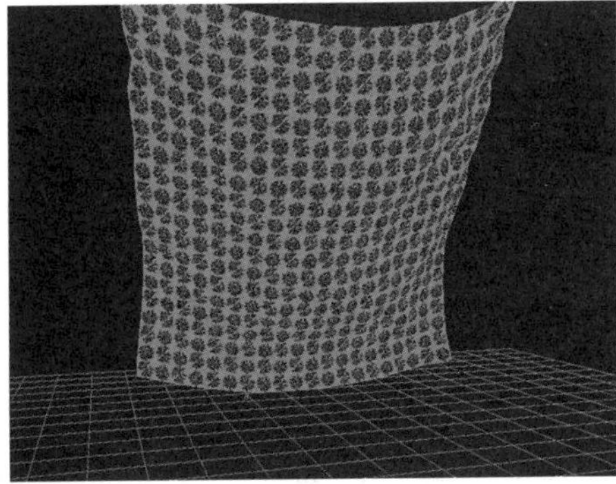

Fig. 8 No mosaic effect

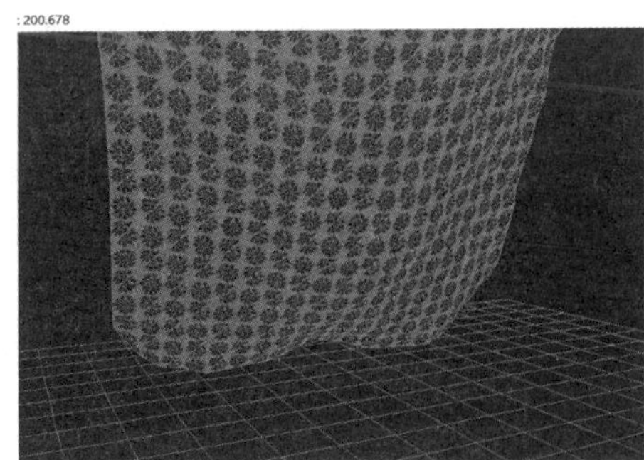

Fig. 9 Mosaic effect

3.2.3 Color vibrancy

Through the comparison of Figs. 10 and 11, we could see the different effects of color pictures and black and white pictures on the texture of the fabric. What is clear was that the color pattern could better grasp the details of fabric movement, while better showing the dynamic effect of the entire particle spring model. In addition, the choice of color also had a certain impact on the texture of the fabric, and warm colors were often more intuitive than cool colors.

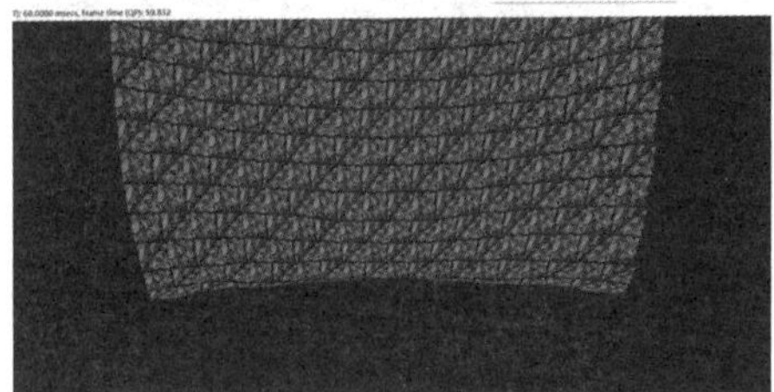

Fig. 10 Black and white picture effect

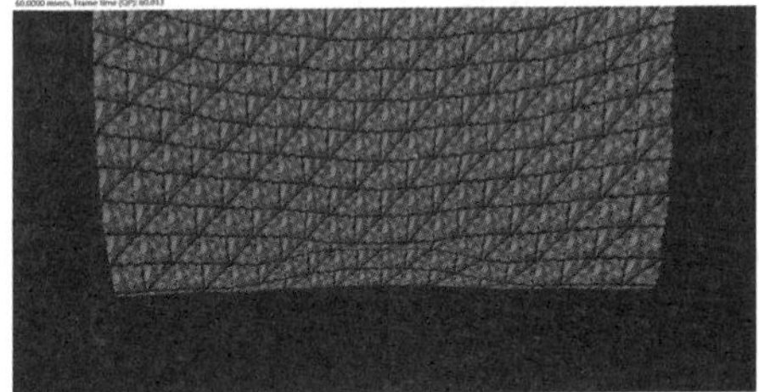

Fig. 11 Color picture effect

3.3 The number of pictures

Finally, we tried to compare the texture of a single image with the texture of four identical images (Fig. 12). On the one hand, the density of the texture increased, resulting in a decrease in the three-dimensional effect of the fabric. On the other hand, due to the performance of the GPU, the rendering effect of the model also had a decrease in frame rate. Therefore, we concluded that the number of meshes should be within a reasonable range, and too many or too few would affect the dynamic effect and visual presentation. This was shown in

Figs. 13 and 14.

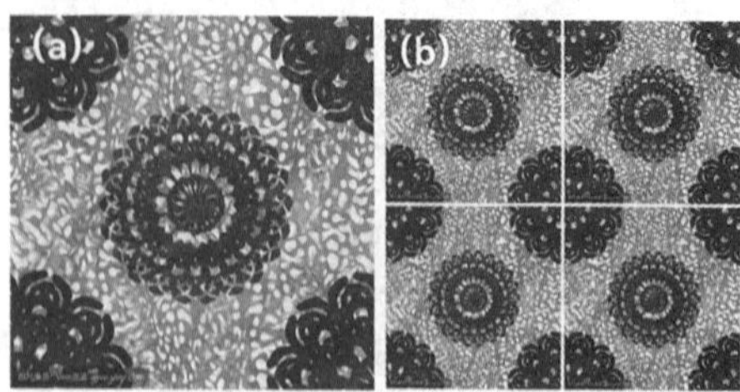

Fig. 12 Compare a single image with composite image: (a) single picture; (b) composite picture

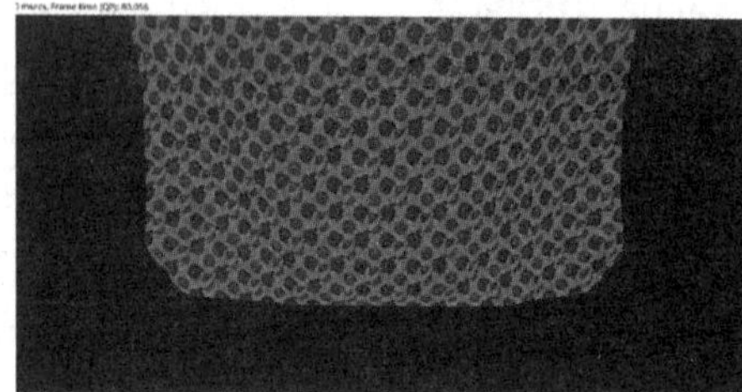

Fig. 13 Single picture effect

Fig. 14 Composite picture effect

4 Conclusions

In this paper, through the texture mapping of the particle spring model. With the helping of OpenGL's platform, the dynamic texture display of the virtual fabric was realized, which provided a choice for simulating the real fabric texture.

(1) Compared with the traditional mapping of the whole original picture to the particle spring model, the method of simultaneous mapping of several meshes introduced in this article could make the texture of a single small piece of cloth synchronized with the corresponding mesh for dynamic simulation, to achieve a more realistic image simulation effect.

(2) The introduction of lighting effects could effectively present the light and dark luster of the fabric texture, further making the texture effect of the particle spring model close to the realistic level, and improving the overall appearance of the texture.

(3) The correct selection of the picture had a role in promoting the overall look and feel of the fabric. During the operation of the virtual fabric, subtle differences and local movements could better please the viewer's eyeballs. We processed the complexity, mosaic and vividness of the pattern, which could improve the three-dimensionality of the fabric texture map to a certain extent, and presented the dynamic display of the real fabric texture.

(4) The number of meshes should be established in a reasonable range, the grid was added to improve the authenticity of the fabric texture, and exceeding the ideal threshold would reduce the display effect of the fabric texture.

References

[1] JI F, LI R Q, ZHANG R Y, Dynamic clothing simulation based on particle-spring model[J]. Acta Textile Sinica, 2004(6): 22-24.

[2] JI F, LI R, Simulation model of fabric based on particle-spring system[J]. Acta Textile Sinica, 2004(1): 42-44.

[3] GONG Y Y, LUO X N, Image upsampling based on motion-restricted spring proton model[J]. Acta Computer Sinica, 2011, 34(9): 1621-1628.

[4] XIAO Y Y, WANG Z N. Research and implementation of opengl texture mapping technology based on VC++[J]. Journal of Guizhou University (Natural Science Edition), 2008(2): 158-160.

[5] XIE Y M. Textile 3D scene simulation system based on OpenGL[D]. Hangzhou: Zhejiang University, 2013.

[6] ZHAO W W. Dynamic cloth simulation[D]. Shenyang: Liaoning Normal University, 2009.

Preparation and Performance of Nonwoven Needle-Punched Fabric Actuator Based on High Shrinkage Fiber

SHAO Dongyan[1], **FENG Wenxin**[1], **LIU Li**[1*]

[1] *Department of Technical Textiles, College of Textiles, Donghua University, Shanghai 201620, China*

* *Corresponding author's email*: liulill@ dhu. edu. cn

Abstract: The high-shrinkage fibers shrink above 60 ℃ and combine with cotton fibers, which hardly shrink when heated, to cause a change in the shape of the actuator by the significant volume change in the double-layered non-woven fabric when the joint interface is strong. The actuation effect of the high-shrinkage actuator is highly dependent on the temperature and the thickness ratio of the double layer. At 70 ℃, the thermal contraction of the high shrinkage fibers remains undamaged and the curvature of the actuator at this temperature is 125 m^{-1} with a 3 : 1 fiber ratio, with the most pronounced curling occurring. The actuator proposed in this paper has good mechanical properties and uses the needle-punching reinforcement technique to make the bilayer fibers tightly bonded, solving the delamination problem and making the actuator a candidate material for the study of artificial blood vessels, stents and other biomedical materials.

Keywords: High-shrinkage fibers; Cotton fibers; Needle-punched; Actuator; Tubular

1 Introduction

Intelligent actuators are mainly made of flexible materials, which have better flexibility and adaptability than traditional rigid robots. In the construction of an intelligent actuator based on a double-layer structure, the stimulus-responsive material is used as the active layer and the common material is used as the passive layer. Depending on different external stimuli (e. g. , pH[1-3] or temperature[4-6]), the actuator can be deformed in an edited shape.

At present, the most used are temperature-sensitive actuators and pH-sensitive actuators. The bilayer composites of the hydrophobic polymer polymethyl methacrylate (PMMA) and the temperature-sensitive polymer P (NIPAM) are reversible between flat and tubular[7]. However, most of the current research on actuators focuses on polymer actuators and relatively little research on fiber actuators. The preparation of tubular materials by the fiber connection method mainly includes three methods of knitting, three-dimensional machine weaving and non-woven. Non-woven artificial blood vessels can meet the requirement of a small diameter, which can facilitate the attachment of endothelial cells. Polyester materials have good performance in biocompatibility, physical strength, aging resistance and non-rejection, and are unanimously used at home and abroad as the preferred material for the manufacture of artificial blood vessels.

This topic is inspired by the temperature-sensitive polymer actuator, using the characteristics of high shrinkage polyester fiber heat shrinkage combined with cotton fiber. Non-woven needle punching technology is used to construct a double-layer actuator, the active layer fiber shrinks significantly when heated, while the passive layer fiber length is basically unchanged. The length difference of the double layer fiber enables the actuator to spontaneously form a tubular structure after heat treatment, which can be used in biomedical fields, such as minimally invasive surgery, drug delivery[8-9], artificial blood vessels and other in vivo applications.

2 Experiments

2.1 Materials

This subject selects high-shrinkage polyester fiber as the active layer material and absorbent cotton fiber as the passive layer material. The boiling water shrinkage rate of high-shrinkage fibers can reach 35% to 45%, and some can even reach 70%. Cotton fiber has certain dimensional stability, and the fiber basically does not change at high temperature.

2.2 Preparation of nonwoven needle-punched fabric actuator

Weigh the fiber with a total mass of 300 g in five groups of different proportions, the proportions are high shrinkage polyester fiber: cotton fiber = 3 : 1, 2 : 1, 1 : 1, 1 : 2, 1 : 3. The high-shrinkage fiber and the cotton fiber are respectively combed through the carding machine twice to obtain a high-shrinkage fiber web and a cotton fiber web with a certain thickness. After that, the two layers of fiber webs of the same group were superimposed together for two times of needling reinforcement. The needling frequency was set to 300 times/min for the first pass and 600 times/min for the second pass. Five groups of high-shrinkage fiber-cotton fiber composite double-layer needle-punched fabrics with different thickness ratios were obtained.

2.3 Characterization

Fourier infrared spectrometer (NEXUS－670), X-ray diffraction (XRD, Bruker D8 ADVANCE) were used for qualitative analysis of the material. A scanning electron microscope (SEM, DXS－10ACKT) was used to observe the microscopic morphology of the sample and measure the fiber diameter. Fiber specific resistance test instrument (XR－1A) to measure the specific resistance of the material. Digital thickness instrument (YG(B)141G) to measure the thickness of the sample. Electronic balance (AL104/ME204E) to measure the surface density of the sample. Electronic fabric strength instrument (HD026N) to measure the fracture of the sample strength and elongation. Sample heated in water bath. The sample size in the actuation performance study was 600 mm×300 mm.

The thermal shrinkage of the sample can be calculated using Eq. (1):

$$\text{Thermal shrinkage}(\%) = \frac{L_0 - L_1}{L_0} \times 100\% \quad (1)$$

where L_0 is the original length of the sample, L_1 is the length of the sample after heat treatment.

The curvature of the sample is calculated using Eq. (2):

$$K = \frac{1}{R} \quad (2)$$

where R is the radius of curvature of the sample.

3 Results and Discussion

3.1 Characterization of fibers

Infrared spectroscopy experiments and X-ray diffraction were carried out on the two fibers used in the experiment, and the two fibers used were determined to be high shrinkage polyester and cotton fibers. Fig. 1 (a) is the infrared spectrum of high shrinkage fibers, and an absorption peak is visible at 3 063.65 cm^{-1}. In the I region, the wavelength range of 3 100–300 cm^{-1} is C—H telescopic vibration on the aromatic ring. In the II zone, there is a very strong absorption peak at 1 713.97 cm^{-1}, which is in the wavelength range of the C═O group (1 900–1 600 cm^{-1}). And there are two strong absorption peaks in the range of 1 250 – 1 000 cm^{-1}, indicating that the compound contains C—O—C groups in the ester group, which is a high shrinkage polyester fiber. Fig. 1 (b) is a high shrink fiber XRD diagram. It has a wider peak type, showing a significant amorphous peak form. And its crystallinity is only 2.7%, which is in line with the characteristics of low crystallinity of high shrinkage polyester fiber. Fig. 1 (c) is an infrared spectral pattern of cotton fibers, 3 700–3 200 cm^{-1}, 3 000–2 800 cm^{-1} each has an absorption peak, 1 100–1 300 cm^{-1} can be seen between multiple absorption peaks, respectively, indicating the containing hydroxyl O—H, CH, CH_2 and C—

(a)

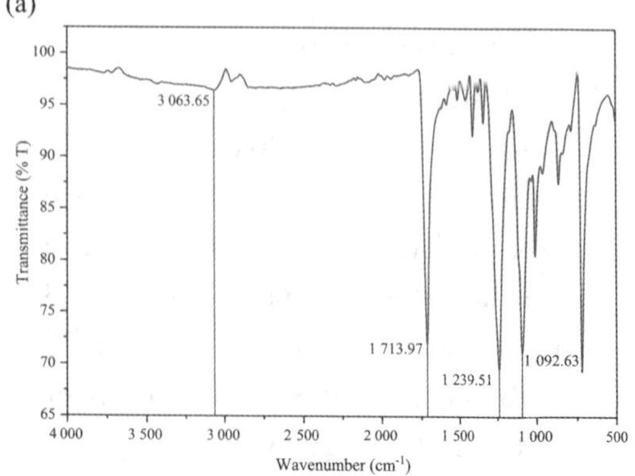

(b)

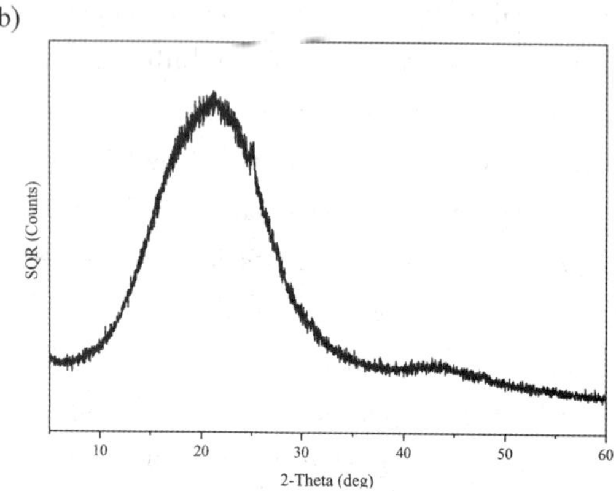

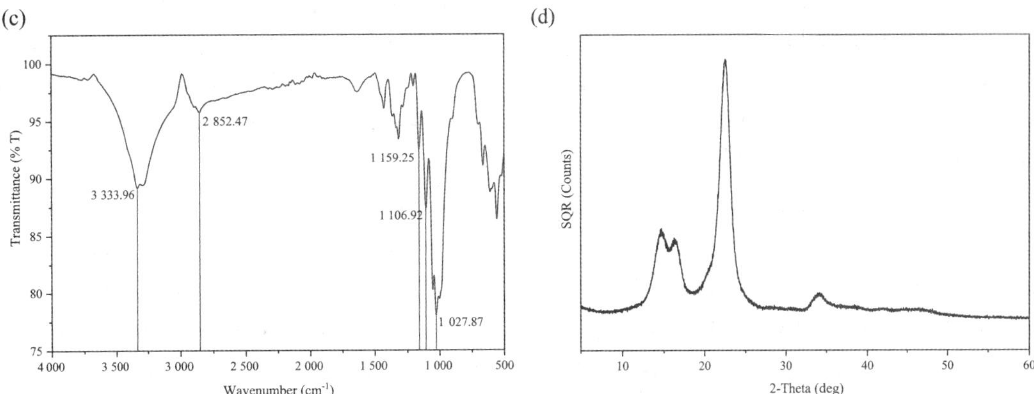

Fig. 1 (a) Infrared spectrum of high shrinkage PET; (b) XRD diagram of high shrinkage PET; (c) infrared spectrum of cotton fibers; (d) XRD diagram of cotton fibers

O—C groups, in line with the structural formula of cellulose fibers. The cotton fiber XRD shown in Fig. 1 (d) has sharp peaks at $2\theta = 14.57°$, $16.53°$, $22.62°$ and $33.73°$, and the $(1\bar{1}0)$, (110), (200) and (004) crystal surfaces of cellulose (standard card number 50-2241) coincide, indicating that the main component of the substance is cellulose, with a crystallinity of 69%.

From the SEM diagram of the two fibers in Figs. 2 (a) and 2(b), it can be observed that the surface of the high shrinkage fiber is relatively smooth, while the cotton fiber has a natural curved form. The diameter distribution of the two fibers in Figs. 2 (c) and 2 (d) was obtained by selecting 100 fibers from the SEM

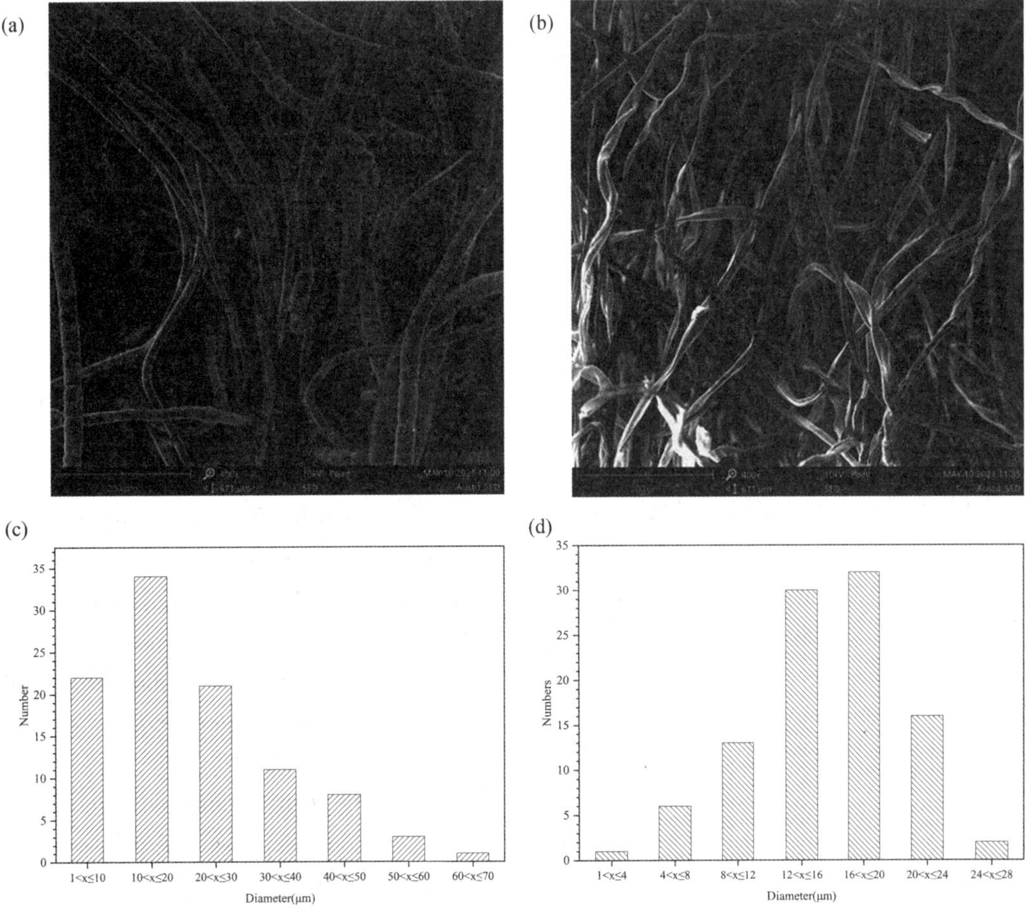

Fig. 2 (a) SEM image of high shrinkage PET; (b) SEM image of cotton fibers; (c) diameter distribution of high shrinkage PET; (d) diameter distribution of cotton fibers

plots of high shrink fibers and cotton fibers. It can be seen from the distribution map that the diameter of high shrinkage fibers is mainly distributed between 1 μm and 30 μm, and the diameter of cotton fibers is mostly distributed in the range of 12−20 μm. Through the test, the volume ratio resistance of high shrinkage polyester fiber is within the order of 10^{13}, while the volume specific resistance of cotton fiber is only $8.2\times10^{8}\ \Omega\cdot cm$, and the high shrinkage fiber needs to be treated with antistatic treatment to increase its processability. The fiber ratio resistance treated by the antistatic agent is reduced to less than $1\times10^{8}\ \Omega\cdot cm$, which greatly improves the electrostatic adhesion phenomenon during the carding process of high shrinkage fibers.

The heat shrinkage properties of the two experimental fibers are tested, and the heat shrinkage rate of the fibers at 50, 60, 70, 80 and 90 ℃ is tested. As can be seen from Fig. 3 (a), the cotton fibers maintain a shrinkage rate of less than 5% at each experimental temperature, the length of the cotton fibers does not change after heating, only because the fiber absorbs water and has a slight impact on the fiber length. The

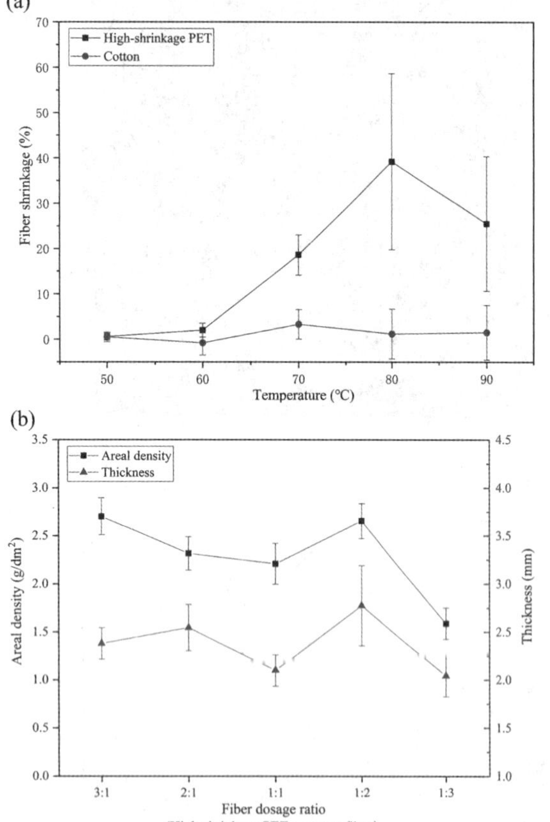

Fig. 3 (a) Shrinkage trend chart of high shrinkage PET and cotton fibers; (b) diagram of fiber dosage ratio-thickness and surface density

high shrinkage polyester fiber begins to show shrinkage performance at 60 ℃, and reaches the highest at about 80 ℃, and the fiber shrinkage rate can be close to 40%.

3.2 Properties of nonwoven needle-punched actuators

Superimposed high shrinkage fiber mesh and cotton fiber mesh, after needle punch reinforcement to obtain a double layer of nonwoven fabric, the basic performance of the test. Thickness and areal density were determined on five groups of samples with fiber ratios of 3 : 1, 2 : 1, 1 : 1, 1 : 2, and 1 : 3 (Fig. 3 (b)). The average thickness of the five groups of double-layer nonwovens is basically about 2.369 mm, and the average is density about 0.230 kg/m^2. Although the double-layer interface of nonwoven fabrics is tightly combined, the internal structure of the fabric is still fluffy. The MD and CD directions of the two single-layer and five double-layer samples were strongly tested separately to obtain tensile load-tensile elongation curves in both directions (Fig. 4). It can be seen from the experimental curve that the strength of pure cotton nonwovens is lower than that of high shrinkage fiber nonwovens in the two experimental directions of MD and CD, and the strength of the double-layer nonwovens is greater than that of single-layer nonwovens in the MD direction and CD direction. Whether it is a single layer or a double layer, the MD direction strength is generally better than the CD direction strength, because the fiber mesh combed by the carding machine has a certain orientation, and the fibers show a tendency to be arranged in the MD direction. In addition, it can be found that the strength of the double-layer material is affected by the fiber ratio, and the TENS strength of the CD direction is more obviously affected by the fiber ratio. It can be noted that with the increase of cotton fiber content, the strength of double-layer nonwovens shows a general downward trend, especially in the two groups of samples with more cotton fibers than high shrinkage fibers. There is no obvious correlation between elongation at break and fiber ratio.

3.3 Mechanism of nonwoven actuators

The actuation mechanism of the temperature-sensitive actuator mainly includes the stress release and thermal expansion difference mechanism. Actuators are usually made of a double layer of material, the size of the active layer changes when it is stimulated by the outside

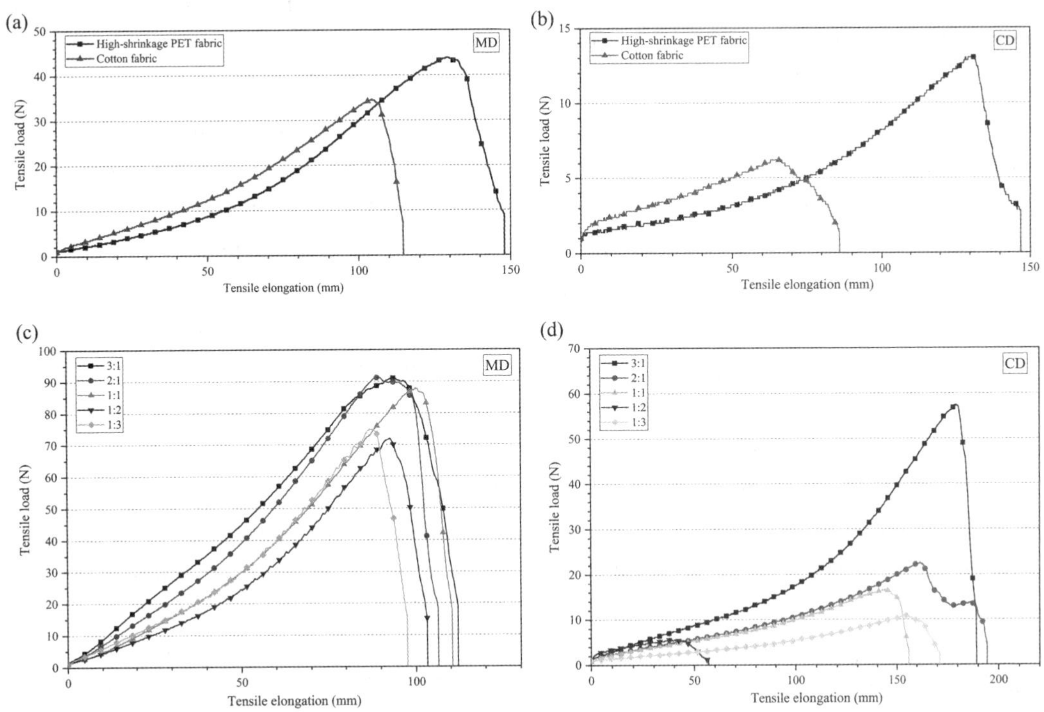

Fig. 4 (a) Tensile load-tensile elongation curve in MD direction of single-layer fabric; (b) tensile load-tensile elongation curve in CD direction of single-layer fabric; (c) tensile load-tensile elongation curve in MD direction of double-layer fabric; (d) tensile load-tensile elongation curve in CD direction of double-layer fabric

world, while the size of the passive layer hardly changes, so the double-layer material with a strong joint interface will cause the actuator shape change due to the size difference between the upper and lower layers. Using needle punch processing can strengthen the double-layer structure of high shrinkage fibers and cotton fibers, so that the joint interface is firm. When the double-layer actuator in reaches the temperature more than 60 ℃, the high shrinkage fiber layer shrinks significantly while the cotton fiber layer maintains the original size. The double layer produces obvious dimensional differences result in actuator curling.

3.4 The influence of fiber ratio and temperature on the actuation effect

Testing the actuation performance of the sample in hot water at different temperatures. It can be observed that after reaching a certain temperature, the double-layer specimen is actuated, resulting in a certain degree of curling. The curling state of the specimen at different temperatures is different, and the specimen with different thickness ratios have different degree of curling. Under experimental conditions of 70 ℃, the curling state of five kinds of double-layer materials with different thickness ratios is tested, as shown in Fig. 5. The thickness ratio from top to bottom is high shrinkage fiber: cotton fiber = 3 : 1, 2 : 1, 1 : 1, 1 : 2, 1 : 3. It can be observed that with the increase of high shrinkage fiber content, the curl of the double-layer material is more obvious. The specimen with a high shrinkage fiber content higher than the cotton fiber can form a complete tubular shape, while the specimen with a ratio of 1 : 1, 1 : 2, and 1 : 3 can only produce bending and cannot be closed into a tubular structure.

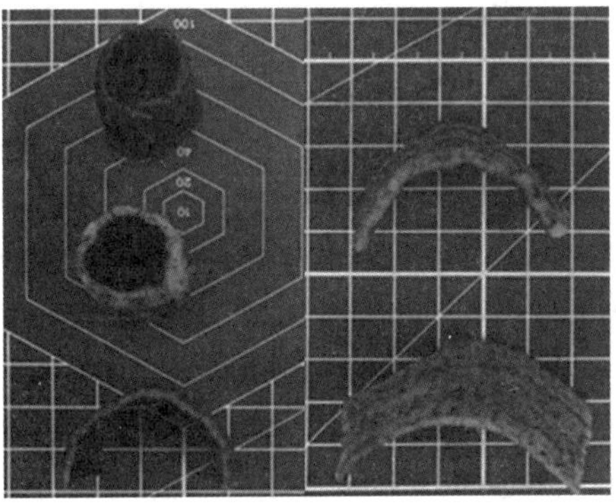

Fig. 5 Actuation of double-layer nonwoven fabric with different fiber ratios at 70 ℃

The degree of curling of the specimen after heat treatment is described by the curvature of the specimen. Measure the curvature of five sets of specimens of different proportions at 50, 60, 70, 80, and 90 ℃ to obtain the curvature-temperature curve of Fig. 6. Most high shrinkage actuators produce significant curl after 60 ℃, and the degree of curl further increases with the increase of temperature, and the actuator with high shrinkage fiber content has higher curvature. As can be seen from the figure, the fiber ratio is high shrink fiber: cotton fiber = 3 : 1, when the temperature is 90 ℃, the high shrink fiber actuator has the greatest degree of curl, and the curvature reaches 200 m^{-1}. However, at 90 ℃, the actuator curvature of the thickness ratio of 1 : 1 exceeds the actuator with a thickness ratio of 2 : 1, and the law has changed. And the actuator with a thickness ratio of 1 : 2 shows a downward trend in curvature after 70 ℃. The actuator actuation performance is the best at 70 ℃, and the actuator with a fiber ratio of 3 : 1 at this temperature has the largest curvature, reaching 125 m^{-1}, and the curling is the most obvious.

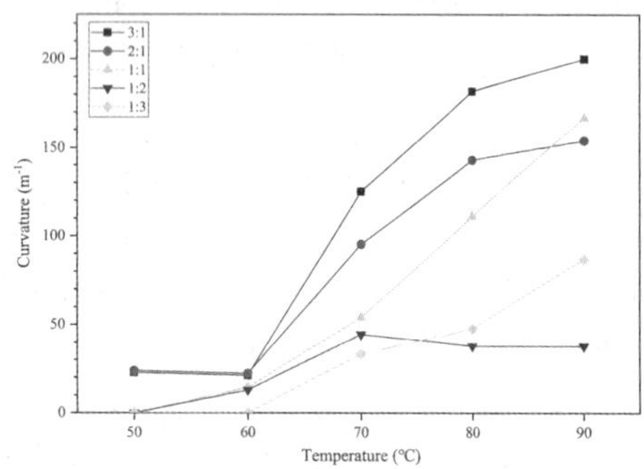

Fig. 6 Curvature-temperature curve of double layer nonwovens fabric

4 Conclusions

In this paper, non-woven fabrics with a double-layer structure are prepared by needle punching technology using the characteristics of high shrinkage fibers that shrink above 60 ℃, combined with cotton fibers that hardly produce shrinkage under heat. The obvious volume change of the double-layer nonwoven fabric with a firm joint interface when shrinking is caused to cause the shape change of the actuator. The actuation effect of the actuator can be obtained and related to the temperature and the thickness ratio of the double layer. In the temperature range that does not destroy the properties of the material, the higher the temperature and the greater the ratio of high shrinkage fibers to cotton fibers, the greater the degree of actuator curl. The experiment obtained that the heat shrinkage performance of the high shrinkage fiber at a temperature of 70 ℃ has not been destroyed, and the actuator curvature of the fiber ratio of 3 : 1 at this temperature is 125 m^{-1}, and the curl is the most obvious. Under the premise of basic activism, the actuator proposed in this paper has good mechanical properties at the same time, and uses needle punch reinforcement technology to make the double-layer fibers closely bonded, solve the delamination problem of the double-layer actuator.

References

[1] CHEN S, JIANG S, JIANG H. A review on conversion of crayfish-shell derivatives to functional materials and their environmental applications[J]. Journal of Bioresources and Bioproducts, 2020, 5(4): 238-247.

[2] HAN Z, WANG P, MAO G, et al. Dual pH-responsive hydrogel actuator for lipophilic drug delivery[J]. ACS Applied Materials & Interfaces, 2020, 12(10): 12010-12017.

[3] KWON S J, PARK J G, LEE S H. Morphological dynamics of swelling-induced surface patterns in metal-capped polymer bilayer[J]. Journal of Chemical Physics, 2005, 122(3): 031101.

[4] AGARWAL S, JIANG S, CHEN Y. Progress in the field of water- and/or temperature-triggered polymer actuators[J]. Macromolecular Materials and Engineering, 2019, 304(2): 1800548.

[5] LI J, MA Q, XU Y, et al. Highly bidirectional bendable actuator engineered by LCST-UCST bilayer hydrogel with enhanced interface[J]. ACS Applied Materials & Interfaces, 2020, 12(49): 55290-55298.

[6] LI L, ZHAO Q, WEN H, et al. Preparation and performance of galactomannan temperature-sensitive hydrogels[J]. Journal of Forestry Engineering, 2021, 6: 120-127.

[7] STOYCHEV G, REUTHER C, DIEZ S, et al. Controlled retention and release of biomolecular transport systems using shape-changing polymer bilayers[J]. Angewandte Chemie International Edition, 2016, 55(52): 16106-16109.

[8] HAMIDI M, AZADI A, RAFIEI P. Hydrogel nanoparticles in drug delivery[J]. Advanced Drug Delivery Reviews, 2008, 60(15): 1638-1649.

[9] LANGER R. New methods of drug delivery[J]. Science, 1990, 249(4976): 1527-1533.

Tensile and Flexural Properties of Glass Fiber Reinforced Polypropylene/Self-reinforced Polypropylene Interply Hybrid Composites

WANG Yan[1,2], **PAN Lijian**[1,2*], **ZHANG Shu**[3], **WU Jiamu**[3], **YUE Guangquan**[1,2]

[1] *College of Textiles, Donghua University, Shanghai 201620, China*

[2] *Center for Civil Aviation Composites, Donghua University, Shanghai 201620, China*

[3] *Suzhou Tianyi Composites Co., Ltd., Suzhou 215000, China*

* *Corresponding author's email*: xiaopanhrb@163.com

Abstract: Self-reinforced polypropylene (SRPP) composites have been widely used in many fields due to its excellent toughness. However, their less characterized in rigidity. Glass fiber reinforced polypropylene (GRPP) composites have excellent rigidity, but relatively insufficient toughness. The two composites were combined in a sandwich structure, three thickness ratios of 1 : 1, 1 : 2, and 1 : 3 were set, and the effects of different thickness ratios on the mechanical properties of the laminates were studied. The results show that although the ductility gradually deteriorates with the increase of GRPP core thickness ratio, the tensile and flexural rigidity properties of the laminates improve significantly. The optimum comprehensive properties of the laminates were obtained at the thickness ratio of 1 : 2, the tensile and flexural modulus were improved by 218% and 72%, and the specimens did not show obvious general GRPP-like fracture wreck during flexural test. Both the rigidity of GRPP and the toughness of SRPP have been largely preserved.

Keywords: Self-reinforced composites; Rigidity improve; Interply hybrid; Mechanical properties

1 Introduction

Glass fiber reinforced polymer (GFRP) occupies a major composites market position due to its strong rigidity and low cost. However, the toughness of GFRP is not ideal. It is fatiguing easy, and often shows fracture failure during loading[1-3].

As "All-PP" composites, SRPP greatly enlarges the toughness of polypropylene. The SRPP composites represented by "Curv" has a notched impact strength as high as 7 500 J/m, and still maintain good toughness at about −40 ℃[4]. However, SRPP has poor rigidity, and cannot undertake the comprehensive mechanical properties required by some mechanical structural components[5].

As all-thermoplastic material, SRPP is simple in secondary molding, and can be easily combined with rigid material for toughening[6]. In this study, GRPP and SRPP were laminated by vacuum assisted resin infusion (VARI) at thickness ratios of 1 : 1, 1 : 2 and 1 : 3. Through tensile and flexural experiments, investigated the rigidity and toughness mechanical properties of the interply hybrid composites.

2 Experiments

2.1 Materials

Fabric hot-pressed SRPP composites (thickness: 0.60–0.65 mm, 1.70–1.80 mm), polypropylene film (thickness: 0.1 mm) and glass fiber/polypropylene fiber blended fabric (plain weave, thickness: 0.65–0.75 mm) were obtained from Suzhou Tianyi Composites Co., Ltd (Jiangsu, China).

2.2 Preparation of GRPP/SRPP interplay hybrid Composites

Three kinds of hybrid structure parameters shown in Tab. 1 were set in this study. All interply hybrid composites were use in simple sandwich structure. Setting SRPP as the cover and GRPP as the core, make SRPP provide a certain pre-shrinkage to GRPP during cooling molding, and finally improve the ductility of the laminates[7-9].

Tab. 1 **Structural parameter setting of GRPP/SRPP interply hybrid composites**

Hybrid configuration	Thickness(mm)			Thickness ratio
	SRPP	GRPP	Laminates	
SRPP/GRPP/SRPP	0.65/0.65	1.30	2.60	1 : 1
	0.65/0.65	2.60	3.90	1 : 2
	0.65/0.65	3.90	5.20	1 : 3

VARI was used to prepare the laminates. GRPP needs to be prepared separately first. According to the structure that two layers of polypropylene films on both sides of each layer of glass fiber/polypropylene fiber blended fabric, lay at 0/90° to the required thickness. Use V750-30 intelligent oil-free vacuum pump (Taizhou Tengyuan Tool Co., Ltd., Zhejiang, China) evacuate to -100 KPa and keep the pressure constant, and then heat it to 195 ℃ in SMO-10 blower drying oven (Shanghai Yibo Industrial Co., Ltd., Shanghai, China) and keep the temperature for 1 hour. After cooling down, remove the vacuum pressure and demould to obtain GRPP.

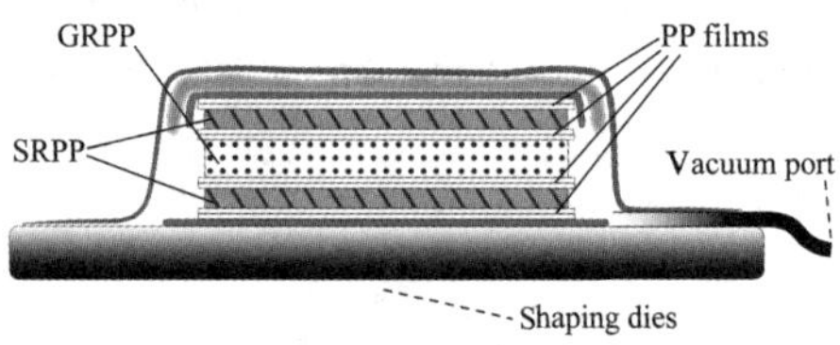

Fig. 1 Schematic diagram of vacuum assisted resin infusion for preparing interply hybrid composites

As the method shown in Fig. 1, GRPP, SRPP and polypropylene films were stacked according to the three structural parameters set, the stacking angle was 0/90°, vacuum was evacuated to -100 kPa and maintained the pressure, temperature was raised to 165 ℃ and kept for 0.5 h. The preparation procedure is the same as that of GRPP. Finally, three thickness ratios of GRPP/SRPP interply hybrid composites were obtained.

2.3 Tensile testing of laminates

All specimens' tensile properties were tested according to ASTM D 3039 using an ETM105D microcomputer-controlled electronic universal testing machine (Shenzhen Wance Testing Equipment Co., Ltd., Guangdong, China) under a 100 KN sensor. The specimen size was 250 mm×25 mm and the loading rate was 2 mm/min. The full-field strain of the specimen during test was collected using a FETECH 32-00 digital image correlated (DIC) full-field strain measurement system (AC-QTEC, Shanghai, China). Acquisition frequency 1 s/time, acquisition software: Vic-Snap 9, post-processing software: Vic-3D 9.

2.4 Flexural testing of laminates

All specimens' flexural properties were tested according to ASTM D 7264 using an ETM105D microcomputer-controlled electronic universal testing machine (Shenzhen Wance Testing Equipment Co., Ltd., Guangdong, China) under a 10 kN sensor. The test proceeded under three-point bending with a span-thickness ratio of 32 : 1. The specimen size was 160 mm×13 mm and the loading rate was 1 mm/min.

3 Results and Discussion

3.1 Tensile rigidity of laminates

The tensile strength and modulus of the two base composites SRPP, GRPP and the laminates with three thickness ratios are shown in Fig. 2. The tensile strength of three laminates improves with the increase of the GRPP core thickness ratio. The 1 : 1 laminate has a minimum strength of 86 MPa, which is less than 115 MPa for SRPP. The main reason is that the rigid carrier GRPP core in the laminate is too thin, making it more like a defect in the composites. Not only doesn't show its excellent tensile strength, but reduces the original tensile strength of SRPP. The 1 : 3 laminate has a maximum strength of 145 MPa, which is 26% higher than SRPP, but only 51% of that of GRPP.

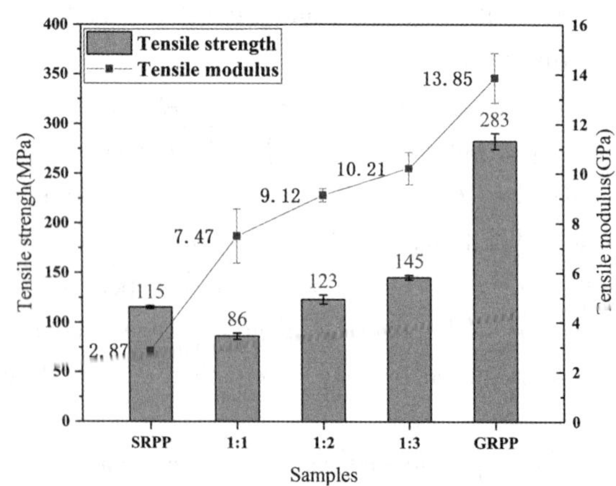

Fig. 2 Tensile strength and modulus of five composites

The tensile modulus of the five composites showed different trends. The tensile modulus of three laminates also improves with the increase of GRPP core thickness ratio, but it is higher than SRPP as a whole. The minimum and maximum modulus are 7.47 GPa at 1 : 1 and 10.21 GPa at 1 : 3, which increase by 160% and 256% compared with SRPP. Moreover, the retention rates of modulus at 1 : 2 and 1 : 3 are 66% and 74% relative to GRPP.

The tensile rigidity of the laminates will improve with the increase of the GRPP core thickness ratio. The tensile strain in the initial stage is small, and the strain rates of SRPP cover and GRPP core are relatively consistent, so the negative effect of SRPP on the laminates' modulus is weakened, resulting in higher improvement in modulus. In the later stage, GRPP core gradually approached its failure strain, the internal micro-damage increasing. The evenly distributed load on the cross-section began to transition from SRPP cover to GRPP core, resulting in a decrease in the overall strength. The strength of the laminates is not obviously improved.

3.2 Flexural rigidity of laminates

Fig. 3 shows the flexural strength and modulus of five composites. It should be emphasized that the three composites of SRPP, 1 : 1 and 1 : 2 did not show breakage failure during the whole procedure, so there is no "flexural failure strength" for them. The values shown in Fig. 3 are the maximum stress of three composites during the test, which are "pseudo-strength"

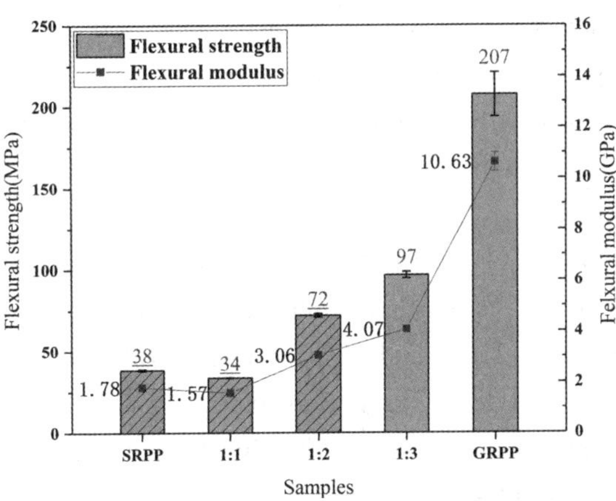

Fig. 3 Flexural strength and modulus of five composites

The flexural modulus trend of the five composites is consistent with the flexural strength, which all improves with the increase of the thickness ratio of GRPP core. The strength and modulus at 1 : 1 are both lower than those of SRPP. The reason is the same as the tensile test, because the GRPP core thickness ratio is too low, making it more like a defect in the laminate. The strength and modulus at 1 : 3 are the highest, which are 155% and 129% higher than SRPP. However, the retention rates are lower, 47% and 38% relative to GRPP, which is quite different from those of tensile. That is related to the loading direction. The tensile load is perpendicular to the cross section of the specimen, and there is a reasonable distribution and mutual transfer of the load between GRPP core and SRPP cover during loading. The flexural load is parallel to the cross section of the specimen. During the loading procedure, the SRPP cover and the GRPP core have low synergy, the strain rate has a large difference, and the load distribution is unreasonable, resulting in a low modulus.

3.3 Tensile and flexural toughness of laminates

Fig. 4 shows the typical tensile stress-strain curves of five composites. Different from the direct failure of two basic composites, three laminates still have a magnitude of load after the first failure, which rises firstly and then weakens until it reaches zero. The two stages correspond to the failure of GRPP core and SRPP cover in the laminates, respectively. Due to the sudden failure of the GRPP core, its load almost instantaneously transferred to the SRPP cover, causing a sharply load rise on the SRPP cover. Then the strain rate rises sharply, leading to the change in the tensile state. In the second stage, the SRPP cover does not exhibit high ductility, so the final strain of three laminates is similar to GRPP and much smaller than SRPP.

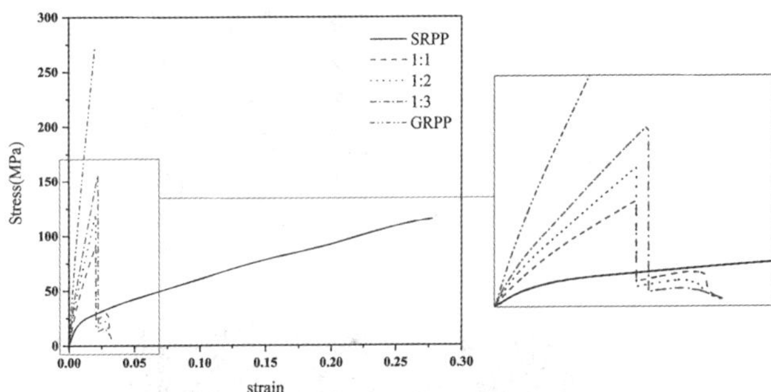

Fig. 4 Typical tensile stress-strain curves of five composites

Fig. 5 shows the typical flexural stress-strain curves of the five composites. The damage forms of the five composites are different. GRPP is a brittle failure of breakage type, while SRPP has almost no damage in the whole procedure. The three laminates all start to show small fluctuations in stress under the strain of about 0.023. The 1 : 1 laminate has only smooth fluctuation of stress in the whole procedure, no obvious damage occurs. The 1 : 2 and 1 : 3 laminates have several small amplitude stress dips and rises in the later stage of the fluctuation, which is accompanied by a crackling sound. However, only the 1 : 3 laminate showed a final breakage type failure similar to GRPP, and the 1 : 2 laminate did not show obvious failure.

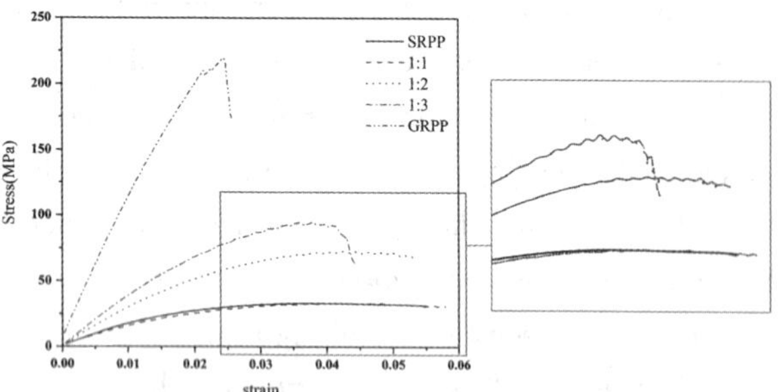

Fig. 5 Typical flexural stress-strain curves of five composites

Figs. 6 (a) and 6(b) show the primary damage strain of five composites and the macroscopic morphology of the samples 24 h after testing. Consistent with the information expressed in Fig. 5, only 1 : 2, 1 : 3, GRPP three composites have obvious damage, and only 1 : 3, GRPP two composites have ultimate failure. The primary damage strain of composites increases with the decrease of GRPP core thickness ratio.

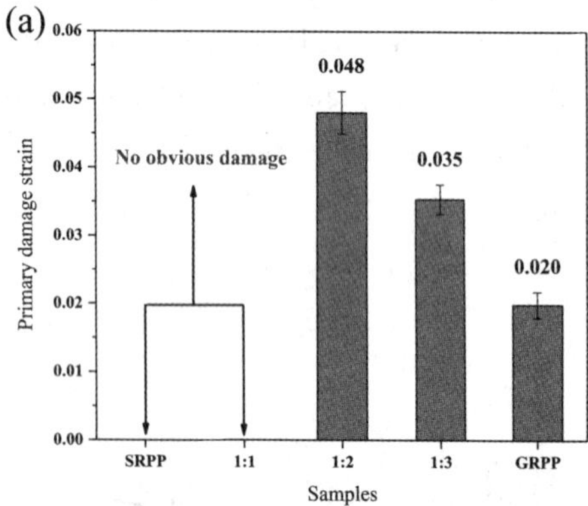

Fig. 6 (a) The primary damage strain of five composites during flexural test; (b) macroscopic appearance of five composites after flexural test for 24 h

Generally, with the decrease in GRPP core thickness ratio, the flexural toughness of the laminates gradually increases. This is related to the deforms way of the specimen during flexural loading. There is a central layer in the samples' section perpendicular to the loading direction, which is compressive deformation upward and gradually increases the amplitude, and downward is tensile deformation and gradually increases the amplitude. Therefore, the main flexural deformation in the laminates is borne by the SRPP covers upper and lower. As the GRPP core thickness ratio decrease, the deformation borne by the SRPP cover increases, and the overall toughness of the laminates increases.

4 Conclusions

This study has focused on the effect of thickness ratio on the rigidity and toughness characteristics of GRPP/SRPP interply hybrid composites under tensile and flexural. The tensile and flexural rigidity of laminates are more obvious with the increase of the GRPP core thickness ratio. The rigidity of 1 : 2 and 1 : 3 laminates are greatly improved compared with SRPP. Tensile strength improved by 7% and 26%, tensile modulus improved by 218% and 256%, flexural modulus improved by 72% and 129%, respectively.

The flexural toughness of laminates improves with the decrease of GRPP core thickness ratio, 1 : 1 and 1 : 2 laminates showed no failure, and the primary damage strain of 1 : 2 laminate improved by 140% compared with GRPP. There is no significant improvement in tensile toughness, and its relationship with the thickness ratio does not reflect an obvious mathematical connection.

The 1 : 2 laminate showed the best comprehensive properties. Its tensile modulus and flexural modulus increased by 218% and 72% compared with SRPP. Under flexural load, its performance is similar to SRPP with no obvious failure. The rigidity and toughness of the composites are overall well combined.

References

[1] YU G Y, LI X H, JIN S L, et al. Study on aging properties and life prediction of GMT-PP composites[J]. Guangzhou Chemical Industry, 2016, 44(10): 96-98.

[2] SONG Y C. All PP composites seize the GMT composites market[J]. Foreign Plastics, 2003(6): 28-40.

[3] SHEN Y K, ZHANG G Y, LIU Y. New trend of fiber reinforced thermoplastic composites—From GMT-LFT to All-PP[J]. Fiber Composites, 2004(2): 57-59.

[4] WARD I M, HINE P J. The science and technology of hot compaction[J]. Polymer, 2004, 45(5): 1413-27.

[5] REIS P N B, GORBATIKH L, IVENS J, et al. Viscoelastic behaviour of self-reinforced polypropylene composites under bending loads[C]//Belgrade, SERBIA: proceedings of the 22nd European Conference on Fracture (ECF) — Loading and Environmental Effects on Structural Integrity, 2018-2.

[6] SHAO S. An overview of the research progress of self-reinforced polypropylene composites[J]. Henan Science and Technology, 2014(19): 75-76.

[7] SELEZNEVA M, SWOLFS Y, KATALAGARIANAK-IS A, et al. The brittle-to-ductile transition in tensile and impact behavior of hybrid carbon fibre/self-reinforced polypropylene composites[J]. Composites Part A-Applied Science and Manufacturing, 2018, 109: 20-30.

[8] HASSANI F, MARTIN P J, FALZON B G. Progressive failure in interply hybrid composites of self-reinforced polypropylene and glass fibre[J]. Polymer, 2020, 195.

[9] TAKETA I, USTARROZ J, GORBATIKH L, et al. Interply hybrid composites with carbon fiber reinforced polypropylene and self-reinforced polypropylene[J]. Composites Part A-Applied Science and Manufacturing, 2010, 41(8): 927-932.

Study on the Effect of Silver Nitrate Solution on PA 66 Under High Temperature and High Pressure

WANG Siyan[1,2], **SHI Haixia**[1,2], **LIU Ruohua**[1,2*]

[1] *Key Laboratory of Textile Science & Technology, Ministry of Education, Donghua University, Shanghai 201620, China*

[2] *Department of Technical Textiles, College of Textiles, Donghua University, Shanghai 201620, China*

* *Corresponding author's email*: liuruohua@ dhu. edu. cn

Abstract: The chemical-plated fabric has good electrical conductivity, breathability and electromagnetic shielding effectiveness, and the texture is supple. It occupies an important position in anti-electromagnetic shielding materials. The premise of electroless plating is to form a catalytic layer on the surface of the substrate, but the high crystallinity and the low chemical mobility of the surface of polyamide 66 (PA 66) fibers are not conducive to the formation of a catalytic layer. In this paper, by using silver nitrate solution to treat PA 66 yarn under high temperature and high pressure conditions, Ag^+ is promoted to enter the fiber, destroy the macromolecular structure on the surface of PA 66 fiber, reduce its crystallinity, and facilitate the formation of a catalytic layer with excellent performance in the subsequent electroless plating process. At the same time, Ag^+ will have an impact on the properties of the fiber itself. Therefore, this topic discusses the effect of silver nitrate on PA 66 under high temperature and pressure conditions by characterizing tensile load, crystallinity and surface microstructure of PA 66 yarns, and determining the optimal treatment conditions.

Keywords: Polyamide 66; High temperature and high pressure; Silver nitrate solution

1 Introduction

Electroless plating is an excellent way of surface treatment. In the case of no external current, the use of the reductant reduces the catalytic layer so that the metal ions in the electroless plating solution reduce deposition on the fabric's surface, coating a continuous layer of metal with the surface of the fabric. The essence of electroless plating is the REDOX reaction[1-3]. The equation is as follows:

Cathode reaction: $M^{n+} + ne^- \longrightarrow M$

Anode reaction: $Red \longrightarrow Ox + ne^-$

Electroless plating has the advantages of uniform coating and stable performance and is widely applicable, including electroless copper, electroless silver, electroless nickel, Etc. Electroless plating fabrics have the wearability of fabrics and the conductivity and electromagnetic shielding properties of coated metals, so they have an excellent application prospect[4].

The premise of electroless plating is that the base materials' surface must have a catalytic activation layer. For electroless plating on the surface of PA 66 fabric, the surface of the PA 66 substrate has the characteristics of hydrophobicity, surface inertia, no wettability, and no catalytic activity. Therefore, electroless plating can not be carried out directly. Before electroless plating, an activated layer with catalytic activity should be prepared on the substrate surface. Metallic silver has excellent chemical activity and is very suitable for catalyst material[5-6].

The impregnation of PA 66 fiber with silver nitrate solution is an essential step in preparing the catalytic layer in electroless silver plating[7-8]. Under high temperature and high pressure conditions, pressure can help Ag^+ enter the fiber. In this article, high temperature and high pressure treatment will replace the conventional condition of silver nitrate impregnation so that Ag^+ can more fully enter the fiber to form a catalytic layer with better performance. However, the interaction between Ag^+ and PA 66 fiber will change the macromolecular structure inside the thread and affect the fiber strength. With the increase in temperature and prolonged processing time, this effect will be intensified. Consideration of silver ions into fiber interior needs certain pressure, the higher the pressure, the better the impregnation effect. The temperature determines the pressure, so the temperature should be greater than 100 ℃. Due to the synergistic effect of temperature and time, the research should focus on processing time. Therefore, it is necessary to combine the damage of silver nitrate solution on the load of fiber under high temperature and high pressure, and comprehensively consider the treatment time, treatment temperature and concentration of silver nitrate.

2 Experiments

2.1 Materials

70 D/24 F×2 PA 66 yarn and plain weave fabric made of PA 66 yarn of this specification.

2.2 Material treatment

(1) The PA 66 yarn was treated for 20 min, 40 min, 60 min, 80 min, 100 min, and 120 min under high temperature and high pressure conditions of 110 ℃ and 120 ℃, respectively, and then placed in an oven at 90 ℃ for drying.

(2) PA 66 yarn was treated with 0.01 g/L, 0.02 g/L, and 0.03 g/L silver nitrate solution at 110 ℃ for 20 min, 40 min, 60 min, and 80 min placed in an oven at 90 ℃ to dry.

(3) PA 66 yarn was treated with 0.05 g/L, 0.01 g/L, 0.015 g/L, 0.02 g/L, and 0.03 g/L silver nitrate solution for 20 min, 30 min, and 40 min at 110 ℃ respectively, and then dried in an oven at 90 ℃.

2.3 Tensile load testing on yarn

To determine the effect of treatment on the tensile properties of the yarn, the XL-1A yarn tensile strength tester of Shanghai New Fiber Instrument Company was used to test the tensile strength of the treated yarn. Set the pre-tension value of the instrument to 500 (mm/min) (cN) and the tensile speed to 500 (mm/min). Before testing, PA 66 yarns were conditioned at 20 ℃ and 65% relative humidity for 24 h. At least 20 samples were tested.

2.4 Crystallinity testing on yarn

After treatment, the crystallinity of the fibers was measured by an 18 kW rotation anode high power X-ray diffractometer. The change of crystallinity of the fibers characterized the effect of silver nitrate solution on PA 66 yarn.

2.5 Scanning electron microscopy (SEM) testing on fabric

A SEM is used to observed the surface microstructure of PA 66 fabric treated with silver nitrate solution to characterize the effect of silver nitrate solution on the surface of fabrics microstructure.

3 Results and Discussion

3.1 Effect of high temperature and high pressure treatment on tensile load

In the tensile load testing, it is found that the strength of the treated surface yarns differs significantly from that of the inner threads, and the load loss of the surface yarns is severe. Compared with the inner yarns, the hydrolyzed degree of the surface yarns is greater, and the load of the outer yarns is severely damaged. In addition, the outer yarns are in an environment with a higher concentration of silver ions, and the relative number of silver ions in contact with them is more. Hence, the damage to the crystalline structure is intense. Therefore, the experimental design should consider Cannikin's law. When determining the treatment conditions, the average tensile load of yarns should not be considered, and more attention should be paid to the average tensile load of the surface yarns. Instead of using cone yarn in production, use single yarn, Hank yarn, or handle fabric directly.

The average tensile load of surface yarns treated under high temperature and high pressure conditions is shown in Fig. 1. With the extension of the treatment time, the yarn load gradually decreased, and the yarn strength was incredibly lost after 100 min of processing, reaching 45.0%. At the same time, the yarn load loss is more severe at 120 ℃ under high temperature and high pressure conditions, up to 35.3% at 20 min. Furthermore, with the extension of treatment time, the load deteriorated further, and by the time of 80 min, the tensile load loss rate reached 86%. By analyzing the internal causes of strength loss, it can be seen that the macromolecules in PA 66 yarn hydrolyze at high temperature, the crystallinity is reduced, and the macromolecule chain is destroyed, resulting in load damage to the yarn. This effect becomes evident with the increase in temperature or time extension. Therefore, when determining the conditions for treating PA 66 with silver nitrate under high temperature and high pressure conditions, we should refer to the experimental results of high temperature and high-pressure treatment of yarn. In order to ensure the good mechanical properties of the silver-plated fabric, the initial treatment temperature is 110 ℃, and the time is 80 min. Further experiments were carried out to determine experimental parameters.

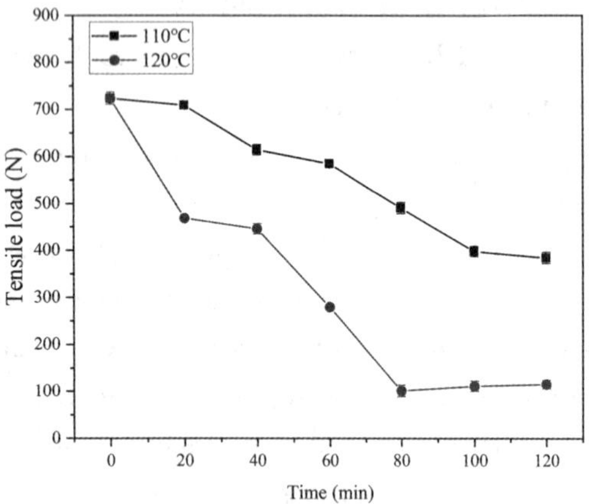

Fig. 1 Average tensile load of surface yarns treated at 110 ℃ and 120 ℃ under high temperature and high pressure conditions for different times

The average load of surface yarns treated with silver nitrate solution of different concentrations at 110 ℃ high temperature and high pressure conditions is shown in Fig. 2. It can be seen from the figure that the tensile load of silver nitrate solution at 110 ℃ under high temperature and pressure conditions is compared with that of a yarn treated at 110 ℃ high temperature and pressure at the same time. It can be seen that the silver nitrate solution has excellent damage to the yarn load. When 0.01 g/L silver nitrate solution is used to treat

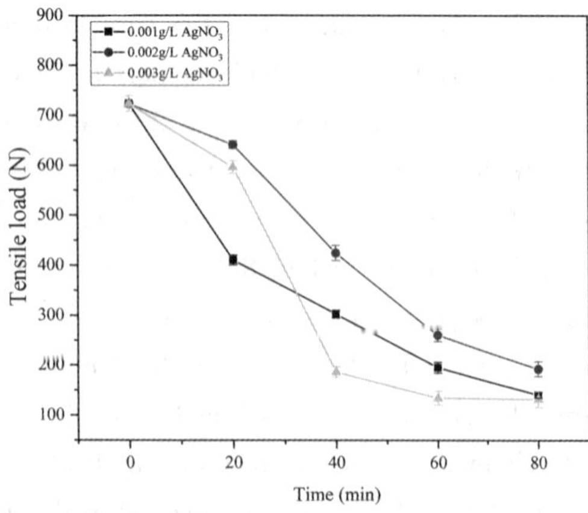

Fig. 2 Average load of surface yarns treated with silver nitrate solution of different concentrations for different times

yarn, the tensile load is seriously damaged. When the concentration of silver nitrate increased to 0.02 g/L, the yarn load increased compared with that of 0.01 g/L silver nitrate solution. In contrast, when the concentration of silver nitrate increased to 0.03 g/L, the yarn load decreased.

From the analysis of yarn's internal structure, the whole experiment process results from the comprehensive effect of temperature, water, and silver nitrate on the fiber. There are three main effects. High temperature promotes crystallization of PA 66 fiber. Under high-temperature conditions, the macromolecules in the fiber hydrolyze, and Ag^+ enters the fiber and interacts with it. After Ag^+ enters the fiber, it can destroy the original hydrogen bond in PA 66 fiber, form a coordination bond with the oxygen atom on the macromolecular chain, and change the microstructure of the PA 66 molecule.

When the yarn is treated with 0.01 g/L silver nitrate solution, the tensile load of the yarn is damaged under the comprehensive action of temperature, water, and silver nitrate on the fiber. When the concentration of silver nitrate solution increases to 0.02 g/L, Ag^+ enters the fiber and damages the hydrogen bond. Meanwhile, the orientation of macromolecules becomes better, and the simultaneous loading of macromolecular chains increases along the yarn axis. Therefore, the tensile load increased. When the solution concentration continued to rise to 0.03 g/L, Ag^+ destroyed the crystallization zone, hydrolysis became more severe, and the tensile load loss became severe.

At the same time, with the extension of treatment time, the tensile load decreases. When the treatment time reaches 40 min and above, the yarn treated with silver nitrate solution of different concentrations has reached significant damage. It indicates that within 40 min, each factor is enough to interact with PA 66 fiber, and Ag^+ can also enter the fiber, so the technological goal of silver nitrate treatment under high temperature and high pressure conditions has been achieved. Increasing the processing time will not make PA 66 yarn meet the technological conditions and product requirements of silver plating, so the time condition should be controlled within 40 min.

The surface tensile load of yarns treated with silver nitrate solution at different concentrations for different times at 110 ℃ under high temperature and high pressure conditions is shown in Fig. 3. As can be seen from Fig. 3: at 110 ℃ under high temperature and high pressure conditions, tensile load decreases gradually with the extension of treatment time. When the concentration of silver nitrate increases, surface tensile load first decreases, then increases and then decreases. The tensile load was damaged when treated with 0.005 g/L silver nitrate solution and continued to be damaged with the increase of the concentration of silver nitrate. When the concentration of silver nitrate increased to 0.01 g/L, the load increased, and when the concentration of silver nitrate continued to increase, the load began to decrease.

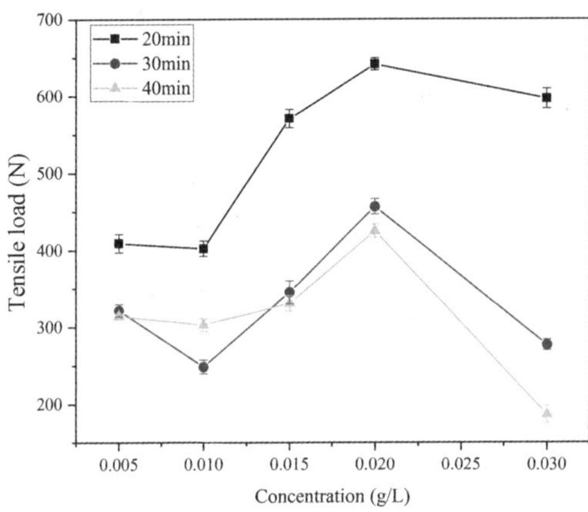

Fig. 3 Average load of surface yarns treated by silver nitrate with different solubilities for different time

When the concentration gradient of silver nitrate solution was reduced, the tensile load of yarn changed in a different trend. When the concentration of silver nitrate solution is 0.005 g/L, the yarn load is damaged by high temperature, hydrolysis, and Ag^+. When the concentration of silver nitrate solution increases to 0.01 g/L, the yarn load decreases compared with that of silver nitrate solution at 0.005 g/L. As the concentration of silver nitrate increases, its destruction of the crystallization zone is further strengthened, and macromolecules begin to slip out of the crystal zone. But they do not reach the degree of whole orientation arrangement. All functions weaken the connection between the zone. When the concentration of silver nitrate increases to 0.02 g/L, the macromolecular slip is sufficient, the orientation is more orderly, and the load increases. When the concentration of silver nitrate increases to 0.03 g/L, the tensile load decreases, which is consistent with the previous experimental conclusion. Ag^+ damage to the hydrogen bond and crystal structure in PA 66 fiber and more hydrolysis become the main factors affecting the tensile load, further reducing the tensile load. Combined with the above results of tensile load testing, the optimal treatment conditions are as follows. 110 ℃ at high temperature and high pressure conditions, the PA 66 yarn is treated with 0.02 g/L silver nitrate solution for 20 min.

3.2 Results of crystallinity testing

The crystallinity of fiber treated with silver nitrate solution of different concentrations for 20 min at 110 ℃ under high temperature and high pressure conditions is shown in Fig. 4. As can be seen from Fig. 4, when the PA 66 fiber was treated with 0.005 g/L silver nitrate solution at 110 ℃ under high temperature and high pressure conditions, its crystallinity increased compared with the control sample. As the concentration of silver nitrate solution increased to 0.01 g/L, the crystallinity of fiber continued to increase as the concentration of silver nitrate solution increased. It was found that the crystallinity decreased when the concentration of silver nitrate solution was 0.01 g/L, which also proved the above conclusion. Therefore, it was reasonable to speculate that when the concentration of silver nitrate solution continued to increase to 0.02 g/L, the crystallinity increased.

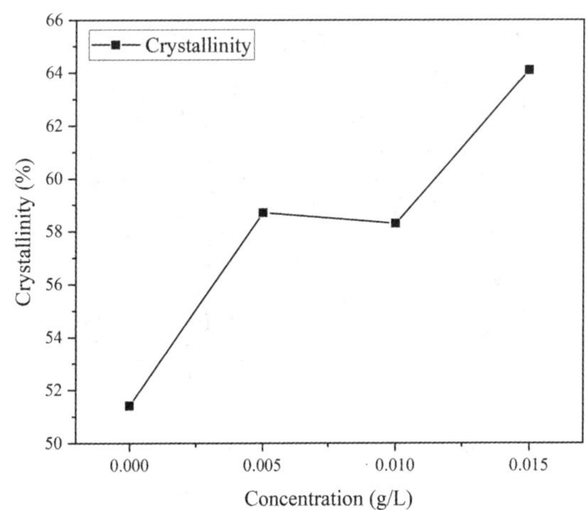

Fig. 4 Crystallinity of fiber treated with silver nitrate solution for 20 min at 110 ℃ under high temperature and high pressure condition

3.3 Results of SEM testing

The SEM photograph of the control sample is shown in Fig. 5 (a). The SEM photograph of the fabric treated with 0.02 g/L silver nitrate solution for 10 min is shown in Fig. 5 (b). The SEM photograph of the fabric treated with 0.02 g/L silver nitrate solution for 20 min in Fig. 5 (c). The surface of the control sample is smooth and uniform, while many attached small particles appear on the fiber's surface after treatment with 0.02 g/L silver nitrate solution for 10 min and 20 min. With the extension of treatment time, the number of particles increased, indicating that more Ag^+ interacted with the fiber and enters into the fiber interior or adheres to the fiber surface.

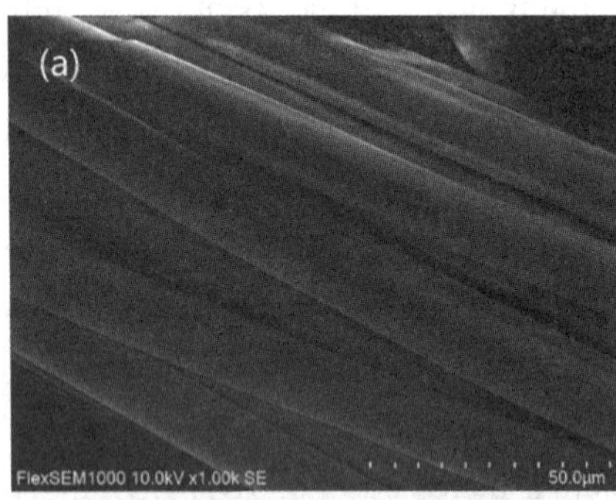

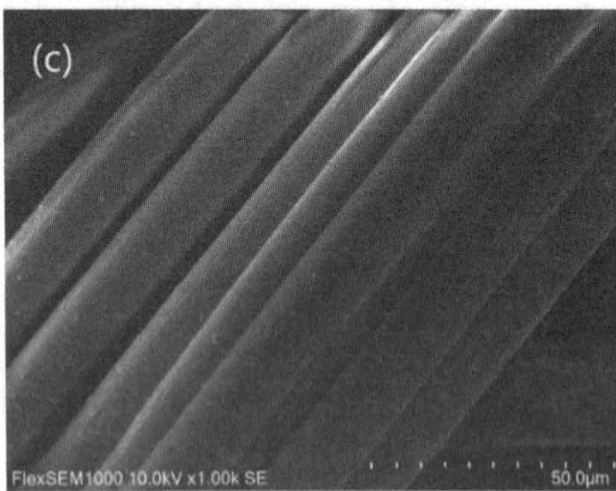

Fig. 5 (a) SEM photograph of the control sample; (b) SEM photograph of the fabric treated with 0.02 g/L silver nitrate solution for 10 min; (c) SEM photograph of the fabric treated with 0.02 g/L silver nitrate solution for 20 min

4 Conclusions

This paper discusses the effect of silver nitrate solution treatment on the entry of silver ions into the PA 66 fiber under high temperature and high pressure conditions, mainly through the tensile load, crystallinity, and surface morphology from the site characterization. After a series of tests, the tensile strength testing results show that the best treatment conditions are 110 ℃ with 0.02 g/L silver nitrate solution treatment of PA 66 fiber for 20 min. The experimental results of crystallinity and surface microstructure confirm this conclusion.

References

[1] JIANG X X, SHEN W. The fundamentals and practice of electroless plating [M]. 1st Edition. Beijing: National Defense Industry Press, 2000.

[2] MA Y H, WANG W, YU D, et al. Study on activation process of electroless plating on polyester fabric[J]. Technical Textiles, 2009, 27(4): 38-41.

[3] YANG Y Y. Electroless silver plating on the polyester fibers and polyamide fibers modified by 3-mercartopropyltriethoxysilane[D]. Shanghai: Donghua University, 2012.

[4] TAN G D. The preparation of composite plating electromagnetic shielding fabrics[D]. Shanghai: Shanghai University of Engineering Science, 2015.

[5] DING H T, XU Y. Research progress on electroless metal electromagnetic shielding fabric[J]. New Chemical Materials, 2022, 50(2): 11-14.

[6] GUO Y. Study on rapid electroless silver plating of polyamide 6[D]. Qingdao: Qingdao University, 2018.

[7] JIN D J, CAO B, MA F, et al. Preparation and pr operties study of electroless silver plated fabric[J]. Knitting Industries, 2020(10): 49-52.

[8] ZHANG M, ZHAO Y P, CAI Z Z. New progress in the preparation and application of silver-plated textiles[J]. Melliand China, 2020, 48(3): 10-12,14.

Predicting Fashion Compatibility Based on Visual Features and Text Representations

DONG Ruiqi[1], WAN Xianfu[1*], WANG Jun[2]

[1] *College of Textiles, Donghua University, Shanghai 201620, China*

[2] *Key Laboratory of Textile Science & Technology, Ministry of Education, Donghua University, Shanghai 201620, China*

* *Corresponding author's email*: wanxianfu@ dhu. edu. cn

Abstract: With people's pursuit of beauty, more and more attention has been paid to fashion compatibility. But most people are not experts in matching, and the computer recommendation system of fashion compatibility is a good choice. Up to now, deep learning becomes the research hotspot method in this area. However, it is still in the development stage, and there are some shortcomings in previous works. On the one hand, most visual features extraction models in previous works were not trained on special clothing dataset, resulting in the rough description of the visual features of clothing. On the other hand, for the textual information of clothing, the existing model has limited representation ability. And the semantic relationship among different textual information is not considered, resulting that the correlation between text representations is not fully analyzed. However, visual features and text representations are used as inputs to subsequent models, which seriously affect the final matching results. To solve these problems, some developments are introduced in this study. Firstly, a dataset in the clothing field was used to pre-train the visual features extraction model for the capture of more detailed visual features of clothing. Secondly, a text binary classification model was developed to analyze the clothing textual information, and the text representations obtained by trained model could restore the correlation among different textual information. Experiment results show that our method achieves an accuracy of 67% in fashion compatibility task on the IQON3000 dataset, which is 2. 2% higher than that of the previous methods.

Keywords: Fashion compatibility; Visual features; Text representations; Deep learning

1 Introduction

Clothing plays a significant role in people's lives, such as decorating individuals. Different outfits will give people different feelings, but not everyone is an expert in clothing matching, for these people, picking the suitable outfit is a huge challenge. Fortunately, some scholars have recently started to use deep neural networks to study fashion compatibility, trying to learn matching rules through a complex model, and then provide people with clothing matching suggestions.

Currently, there are two existing strategies: featured-based model learning and end-to-end model learning. The method of featured-based model learning is mainly divided into two steps. First, the clothing features are extracted through the pre-trained model, and then the clothing features are used as the input of the fashion compatibility model, training the model to learn complex clothing matching rules. Sagar[1] obtained the clothing visual features through AlexNet pre-trained on ImageNet, obtained the clothing textual features through Nwjc2vec, and used the visual features and textual features as the input of a fully connected network and a convolutional network, respectively. Liu[2] extracted clothing features by pre-trained CNNs and bag of style phrases, and then training matching model with BPR method. Song[3] extracted visual and textual features through convolutional network and bag of words model, and then processed clothing features through auto-encoder to obtain the final matching result. Zhang[4] extracted clothing features through ResNet, and carried out fashion compatibility according to the Hamming distance between the features. The end-to-end learning method takes clothing images and text directly as the input of the model, and the output is the fashion compatibility result. This method looks simpler, but the model here will be more complicated and difficult to train, and there is a requirement for the amount of clothing data. Chen[5] used an attention mechanism to process image information, and GRU processed textual information. Liu[6] used GAN to generate auxiliary templates to help complete the clothing matching task. Yang[7] combined convolutional network and attention mechanism to build a fashion compatibility model for clothing matching learning. Zhang[8] only used the clothing textual information to realize the clothing matching task through the LSTM model.

Although the above research has made some progress, fashion compatibility is non-trivial due to the following challenges.

(1) About visual features extraction, most of the existing deep learning models are not trained on professional clothing datasets, so the extracted clothing visual features are not precise enough.

(2) For clothing text description information, the existing model representation ability is limited, ignoring the semantic connection between texts, resulting in insufficient text representations.

To address the aforementioned challenges, we propose a new deep learning fashion compatibility model on the basis of GPBPR[9]. In particular, our model consists of two essential components. (1) A more refined model is trained on the clothing dataset to extract more precise visual features. (2) A text binary classification model is created to get more powerful text representations. Without changing other parts of the original model, experiments show that our method can effectively improve the effect of fashion compatibility.

2 Method

2.1 Model structure

The model structure is based on GPBPR[9]. As shown in Fig. 1, visual features and textual features are extracted from the top and bottom clothing images and their textual information respectively, and then these features are integrated to build a matching model.

The visual features are extracted by the ResNet[10] model. Different from previous studies, this paper trains ResNet model with a professional clothing dataset to extract more refined clothing visual features. Here is the general process: First, the top and bottom clothing images generate $V_i(V_j)$ through the ResNet model, and then obtain $V_{i'}(V_{j'})$ through the fully connected layer, where V denotes the visual features, i and j denote the top and bottom respectively. The specific training details of the model will be expanded in the second section.

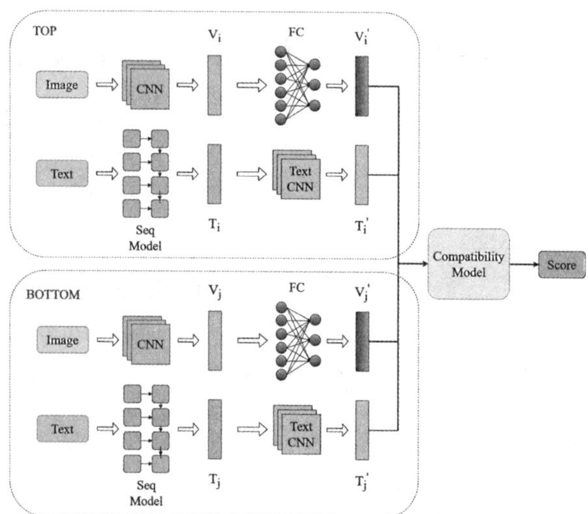

Fig. 1 Illustration of the proposed model, visual and text features extracted through neural network, are integrated with compatibility model

Text features are extracted by TextCNN[11]. Different from previous studies, we design a text binary classification model to make the text representations input into TextCNN more favorable. Here is the general process: The top and bottom clothing text description information generates $T_i(T_j)$ through the text binary classification model, and then obtains $T_{i'}(T_{j'})$ through TextCNN, where T denotes the Text features, i and j denote the top and bottom. The specific training details of the model will be expanded in the third section.

Regarding the matching model, we do not make any changes, which is the same as the original paper method. Specifically, the inner product of the vectors is used to express the distance between the feature vectors, and the visual features and text features of the top and bottom are respectively subjected to inner product operations, and the results of the visual and text inner products are summed, and the sum represents the matching score.

2.2 Visual feature extraction

Different from ImageNet[12], Deepfashion[13] is a professional dataset in the clothing field. Models trained on this dataset have more accurate visual feature representation capabilities. Regarding the specific work of training models, it is mainly explained by cleaning the dataset and selecting the model.

The first is cleaning the dataset. Deepfashion has problems such as complex data, wide categories, uneven distribution, and unclear labeling. We remove the categories that are too small and merge the same categories. After that, the revised dataset has hundreds of thousands of images with 12 categories.

The second is selecting a model. Since the models of the ResNet series have been used as benchmark models in the field of image classification, we choose the ResNet50 and ResNet101 models to train the Deepfashion respectively. By setting different hyperparameters, the result is that the accuracy of ResNet50 on the dataset is 72.7%, ResNet101 has an accuracy rate of 73.7% on the dataset. Since the difference in accuracy is not large, and the model used in the original paper is ResNet50, we choose the trained ResNet50 model as the visual feature extraction model.

2.3 Text feature extraction

In order to solve the problem of the poor representation ability of the text model, we designed a text binary classification model for the clothing matching task to judge whether the top and bottom are matched. Regarding the specific content of the binary classification model, it is mainly explained by the construction of classification tasks and model selection.

For the construction of the classification task, we used the IQON3000 dataset[9], there is a LikeCount item in original dataset, which expresses how much people like the outfit, and reflects whether the outfit is matched. We counted the LikeCounts of all the tops and bottoms in the dataset. Through data analysis and considering the ratio of positive and negative samples, we set the critical point of LikeCount as 40, that is, top and bottom with a LikeCount bigger than 40 is a positive sample, otherwise it is a negative sample.

Regarding selecting a model, we use the Bi-GRU[14], Bi-LSTM[15] models for training, as these two models are commonly used for text tasks. To do the classification task, the output of the last layer of models is taken out and mapped to the category space through a fully connected layer. By setting different hyperparameters, the result is that the accuracy of Bi-GRU on the dataset is 68.2%, Bi-LSTM has an accuracy rate of 68%. As the accuracy of Bi-GRU is better and this model has a simpler structure, we choose the trained Bi-GRU to acquire text representation.

3 Results and Discussion

We creat three models, GPBPR_V, GPBPR_T, and GPBPR_VT for the fashion compatibility task to evaluate the effectiveness of our proposed method. GPBPR_V uses the visual features extracted by the ResNet model pre-training on Deepfashion, GPBPR_T uses the text representations obtained by the text binary classification model, and GPBPR_VT uses both visual features and text representations. Except for the above-mentioned changes, the other model architectures and training methods of the three models are the same as GPBPR. The final results of different models are shown in Tab. 1.

Tab. 1 Performance comparison among different models in terms of area under the ROC curve

	Approach	AUC
Baseline	GPBPR	0.648
Proposed	GPBPR_V	0.657
	GPBPR-T	0.662
	GPBPR_VT	0.670

Through the analysis of the results, we can draw the following three conclusions:

(1) Both GPBPR_V and GPBPR_T are better than GPBPR, with the AUC reaching 65.7% and 66.2%, respectively, which are 0.9% and 1.4% higher than GPBPR. This means that both visual features extracted by model trained on professional datasets and text representations got through text binary classification model are beneficial to fashion compatibility task.

(2) The AUC of GPBPR_T is 0.5% higher than that of GPBPR_V, there are two reasons. On the one hand, the effect of image classification task is general; on the other hand, the visual features of the original paper are extracted by model pre-trained on ImageNet, but the text representations are obtained randomly, so changing the visual features does not improve the matching effect as much as changing the text representations.

(3) The AUC of GPBPR_VT is the best, reaching 67%, which is 2.2% higher than GPBPR. It shows that using visual features and text representations at the same time is the best, and also shows from the side that both methods are effective.

Of course, if the model structure is updated, the effect of fashion compatibility task will be further improved. It's just that the focus of this paper is to illustrate the importance of visual features and text representations, so the model will not be changed to achieve a more effective method of fashion compatibility.

4 Conclusions

This study extracts more precise clothing visual features by training models on professional clothing dataset, and designs a binary text classification model to obtain more effective text representations. The results of our designed experiments show that through these two changes, the effect of fashion compatibility task can be effectively improved. The results also indicate that the clothing features obtained in the early stage of the matching task have a great influence on the final results, and remind everyone not to ignore the precision of the initial representation of the features.

References

[1] SAGAR D, GARG J, KANSAL P, et al. Pai-bpr: personalized outfit recommendation scheme with attribute-wise interpretability[C]. New Delhi: 2020 IEEE Sixth International Conference on Multimedia Big Data. IEEE, 2020: 221-230.

[2] LIU J, SONG X, CHEN Z, et al. Neural fashion experts: I know how to make the complementary clothing matching [J]. Neurocomputing, 2019, 359(24): 249-263.

[3] SONG X, FENG F, LIU J, et al. Neurostylist: Neural compatibility modeling for clothing matching[C]. California: Proceedings of the 25th ACM International Conference on Multimedia. 2017: 753-761.

[4] ZHANG Z K. Clothing matching intelligent recommendation system based on deep learning[D]. Beijing: Beijing Institute of Fashion Technology, 2019.

[5] CHEN X, CHEN H, XU H, et al. Personalized fashion recommendation with visual explanations based on multimodal attention network: Towards visually explainable recommendation[C]. Pairs: the 42nd International ACM SIGIR Conference. 2019: 765-774.

[6] LIU J, SONG X, CHEN Z, et al. MGCM: Multi-modal generative compatibility modeling for clothing matching[J]. Neurocomputing, 2020, 414: 215-224.

[7] YANG X, SONG X, FENG F, et al. Attribute-wise explainable fashion compatibility modeling[J]. ACM Transactions on Multimedia Computi, Communications, and Applications (TOMM), 2021, 17(1): 1-21.

[8] ZHANG H, HUANG W, LIU L, et al. Learning to match clothing from textual feature-based compatible relationships[J]. IEEE Transactions on Industrial Informatics, 2019, 16(11): 6750-6759.

[9] SONG X, HAN X, LI Y, et al. GP-BPR: Personalized compatibility modeling for clothing matching[C]. Nice: Proceedings of the 27th ACM International Conference on Multimedia. 2019: 320-328.

[10] HE K, ZHANG X, REN S, et al. Deep residual learning for image recognition[C]. Amsterdam: Proceedings of the IEEE Conference on Computer Vision and Pattern Recognition. 2016: 770-778.

[11] CHEN Y. Convolutional neural network for sentence classification[D]. Waterloo: University of Waterloo, 2015.

[12] DENG J, DONG W, SOCHER R, et al. Imagenet: A large-scale hierarchical image database [C]. Miami: 2009 IEEE Conference on Computer Vision and Pattern Recognition. IEEE, 2009: 248-255.

[13] LIU Z, LUO P, QIU S, et al. Deepfashion: Powering robust clothes recognition and retrieval with rich annotations[C]. Amsterdam: Proceedings of the IEEE conference on computer vision and pattern recognition. 2016: 1096-1104.

[14] CHUNG J, GULCEHRE C, CHO K H, et al. Empirical evaluation of gated recurrent neural networks on sequence modeling[J]. Eprint Arxiv, 2014, 1412-1421.

[15] SHI X, CHEN Z, WANG H, et al. Convolutional LSTM network: a machine learning approach for precipitation nowcasting[J]. Advances in Neural Information Processing Systems, 2015, 28.

Evaluation of Luminance Property of Three-Layer Stretchable Optical Fiber Fabric

LIU Qiying[1,2], GE Lan[1,2*], XUE Wenliang[1,2], MA Yanxue[1,2], QIAN Jingfang[1,2]

[1] *Key Laboratory of Textile Science & Technology, Ministry of Education, Donghua University, Shanghai 201620, China*

[2] *College of Textiles, Donghua University, Shanghai 201620, China*

* *Corresponding author's email*: gelan@ dhu. edu. cn

Abstract: In recent years, as optical materials and textile technology have been continuously integrated, polymer fiber-optic fabrics with luminance properties have been developed. The luminance performance is one of the most important properties of optical fiber fabrics. However, the luminance property of dimensional and stretchable optical fiber fabric has not been tested. In this paper, the luminance performance of optical fiber fabrics was evaluated based on nine small samples of optical fiber fabrics with different weaving parameters and stretching properties. Using CY−1200C imaging brightness colorimeter, the luminance performance of optical fiber fabrics with different stretchable properties was tested to investigate the effect of varied weaving parameters on the luminance condition of optical fiber fabrics. We provide selection guidance for meeting the photometric requirements of three-dimensional stretchable luminescent materials for different applications and provide the corresponding weaving parameters.

Keywords: Optical fiber fabric; Luminance test; Three-layer fabric

1 Introduction

In recent years, optical fiber, as an excellent light-emitting material, has been used by more and more researchers for smart textile applications due to its novel and practical characteristics[1]. When the optical fibers are woven into the fabric, a macro bending phenomenon occurs and the incident light leaks out from the side of the optical fiber[2], which affects the luminance properties of the fabric. By influencing the degree of macro bending of the fiber in the fabric, the weaving parameters of the fabric can affect its optical properties[3]. Luminance performance is one of the most important properties of optical fiber fabrics, and the accuracy of its performance testing is crucial. Current researches mainly use luminance meters[4] or digital imaging chronometers[5] to test the luminance of optical fiber fabrics. The imaging colorimeter is capable of multi-point sampling and of fabrics to test different areas of luminance values, suitable for precise luminance testing, especially for stretchable optical fiber fabrics with dimensional surface areas.

In this paper, the CY−1200 imaging brightness colorimeter was used to test the luminance of optical fiber fabrics, which provides a suitable luminance performance testing method for stretchable optical fiber fabrics with dimensional surfaces. The orthogonal tests were designed to analyze the degree of influence of weft density, pattern size and weave structure on the luminance of stretchable optical fiber fabric, and the best combination of parameters designed for luminance was obtained, which provided theoretical parameters for the rational design of subsequent products.

2 Experiments

2.1 Materials

The samples used for testing in this paper are nine fabric samples with different stretchable properties woven with optical fiber and spandex yarns. The optical fibers used in the samples are plastic optical fibers obtained from Senwo Co., Ltd., and the specific parameters of performance are shown in Tab. 1.

Tab. 1 Optical fiber parameters

Yarn name	Yarn diameter (mm)	Material content		Color	Attenuation (dB/m)
Plastic optical fiber	0.25	PMMA	Fluorine resin	Transparent and colorless	≤0.20

2.2 Sample preparation

2.2.1 Weaving experimental design

To test the luminance performance of stretchable optical fiber fabrics and to investigate the effect of weaving parameters on the optical properties of stretchable optical fiber fabrics, the following orthogonal experiments were designed, and the specifications for the experiments are shown in Tab. 2.

The pattern size is the size of the repeatable unit; the weave structure indicates the structure of the surface layer and the bottom optical fiber layer. A three-factor and three-level orthogonal test was designed based on the parameters in Tab. 3.

Tab. 2 Orthogonal experimental parameter design

Level	Factors		
	Weft density (picks/cm)	Pattern size (cm)	Weave structure
1	39	4.5×4.5	Plain
2	33	2×4.5	Five-harness satin
3	27	6×4.5	Eight-harness satin

Tab. 3 Orthogonal design table

Orthogonal design table			
Sample number	Weft density (picks/cm)	Pattern size (cm)	Weave structure
1	39	4.5×4.5	Plain
2	39	2×4.5	Five-harness satin
3	39	6×4.5	Eight-harness satin
4	33	4.5×4.5	Five-harness satin
5	33	2×4.5	Eight-harness satin
6	33	6×4.5	Plain
7	27	4.5×4.5	Eight-harness satin
8	27	2×4.5	Plain
9	27	6×4.5	Five-harness satin

2.2.2 Bundling processing

The optical fiber floating yarns at both ends of the fabric are bundled with thermoplastic shrink tubing. The bunding part is then cut flat to be connected to the light source.

The distance between the test point and the light source is standardized as 5 cm to ensure that the subsequent luminance test results are not affected by the distance of the light source[6].

2.3 Testing of luminance properties

As the surface of the fabrics in this study is three-dimensional, the test surface of the brightness meter does not fit the fabric completely, so it is difficult to measure the accurate luminance value by using a brightness meter. An imaging colorimeter takes pictures of fabric for analysis, suitable for indicating uneven fabric. CY-1200 imaging brightness colorimeter was used in this paper to test the luminance of stretchable optical fiber fabrics.

The fabric was placed flat in the dark room and connected to the light source, and the fabric was put in the same area to take pictures to obtain the chromaticity map. The chromaticity map of the obtained specimen is analyzed in the software for its luminance value, and the method of multi-point sampling is used. Due to the different colors of warp yarns, the analysis is taken on the same color area of the fabric luminance value for comparison.

3 Results and Discussion

3.1 Luminance value test results

The underlying principle of luminous effect is the light loss of optical fiber in the process of transmission so that the light leaks from the surface of the fiber dispersion[7]. In this paper, three main weaving factors are varied to investigate their effect on the luminance performance of optical fiber fabric: fabric weft density, fabric weave structure and fabric pattern size.

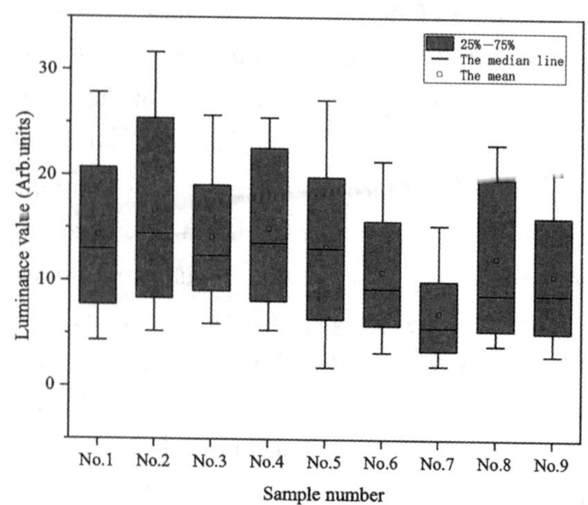

Fig. 1 The diagram of luminance values

Fig. 1 shows the diagram of the luminance values of nine fabrics. In terms of individual optical fiber fabric luminance values, the luminance values of each fabric fluctuate to a certain extent. The luminance of the point at the edge of the fabric closest to the fiber bundle is the highest, and the luminance becomes lower when the test point is farther away from the fiber bundle point.

Due to the fluctuation of the luminance value of the optical fiber fabric, the average luminance value was used as the optical performance index for subsequent comparison. The nine fabrics' mean luminance values show significant differences. The mean luminance values of the nine fabrics are shown in Fig. 1, and the top three fabrics with the highest mean luminance values are classified as sample 2, sample 4 and sample 1.

3.2 Chromatogram and micrograph analysis

Fig. 2 shows the chromaticity graphs of nine fabric samples by serial number, and Fig. 3 shows the microscope graphs of nine fabrics.

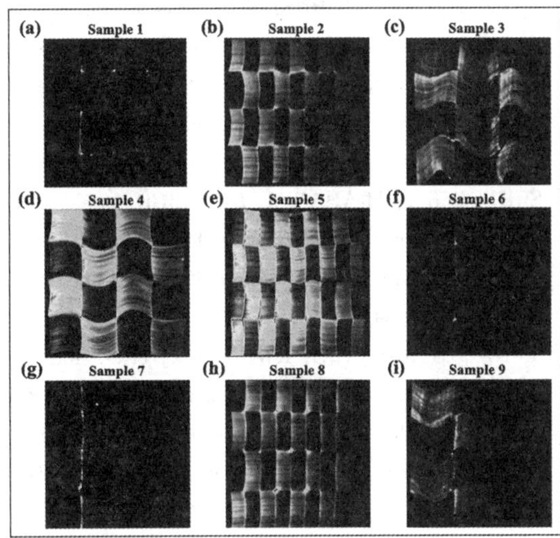

Fig. 2 Sample chromaticity graphs

As shown in Fig. 2, the luminance uniformity of sample 3, sample 4, sample 7 and sample 9 was poor. Observation of the fabric micrographs shows that optical fiber fabrics with different degrees of a stretch produce different degrees of bending of the fiber in the weft direction of the fabric, resulting in differences in the light leaking out from the side of the optical fiber at different locations. Observation of the chromaticity graphs also revealed that the luminance of the fabric gradually decreased along the weft direction. This is because the interweaving of the optical fiber and warp yarn will form a bend, which will make the light emitted from the surface of the optical fiber, the luminance gradually decreased.

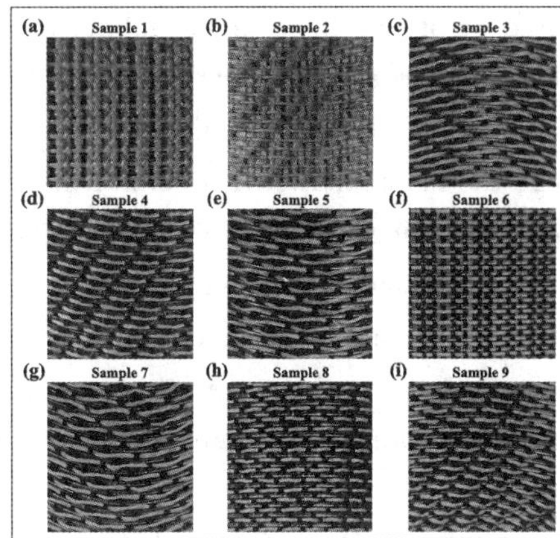

Fig. 3 Fabric microscope graphs

The luminance of sample 2, sample 4, and sample 5 decreases faster along the weft direction, which is due to the influence of the number of bends and the degree of bending, and over a certain degree light leakage in the optical fiber, resulting in a rapid decrease in luminance.

Tab. 4 Results of the orthogonal test

	Level	Weft density	Pattern size	Weave structure
K	1	65.024	53.362	55.271
	2	57.317	63.805	63.174
	3	45.675	51.049	49.571
Average of *K*	1	21.675	17.787	18.424
	2	19.106	21.268	21.058
	3	15.225	17.016	16.524
The best level		1	2	2
R		6.450	4.252	4.534

3.3 The degree of influence of weaving parameters on luminance

The results of the orthogonal tests in Tab. 4 were analyzed. The weft density represents the number of fibers in the fabric, and the luminance of the optical fibers is positively correlated with the number of fibers. However, when the weft density is too large, the fabric structure is more compact and difficult to be stretched, and the degree of bending of the optical fiber will be small. The increase in yarn weft density quantitatively increases the luminance of fabrics but reduces the bend degree of the optical fiber.

The impact of weave structure on the luminance value has two points, one is to affect the stretch properties of the fabric, the tighter the weave structure, the worse the stretchable properties of the fabric, and the less the bend degree of optical fiber in the fabric. Secondly, the number of interweaving points can affect the luminance of the fabric. The optical fiber forms bend at interweaving, the more interweaving points, the more the number of optical fiber bending, the more leakage from the fiber surface, and the higher the luminance. Besides, the interweaving points affect the weft coverage of fibers in the fabric, the more interweaving points, the fewer fiber floating lines, and the lower the weft coverage.

In this paper, the size of the repeatable unit can affect the stretchability of the fabrics, which will affect the bend degree of the optical fiber. Besides, different pattern sizes will affect the number of bends of the optical fiber. The smaller the size of the pattern the more the number of bends, but the bend degree of the optical fiber is less. The higher the number of bends increases the number of light leaking from the optical fiber surface and increases the luminance; the degree of bend affects the degree of light dispersion from the optical fiber surface.

The degree of influence of the 3 factors was further analyzed by the results of the orthogonal test. The optimal level of each factor can be derived from the k value. In this study, 39 picks/cm, 2 cm×4.5 cm and five-harness satin were the optimal levels of each factor. That is, the best luminance properties of stretchable fiber fabric can be obtained by using this scheme. Comparing the R values, it can be seen that the fabric weft density is the most significant factor affecting the luminance of the fabric, followed by the weave structure and finally the square size. It can be seen that in the subsequent design of stretchable optical fiber fabrics, priority can be given to higher weft density fabrics, while fabric weave structures and fabric patterns can be used as auxiliary factors to improve the stretchability and luminance of fabrics.

4 Conclusions

The number of yarns and the degree of bending of optical fibers in the fabric can affect the luminance properties of optical fiber fabrics. In this paper, nine types of stretchable optical fiber fabrics were produced with different levels of weaving parameters, and the effects of weave design, weft density and pattern size on the optical properties of stretchable optical fiber fabrics were studied.

In this study, the luminance of stretchable optical fiber fabrics was tested by using a CY-1200 imaging brightness colorimeter, and the chromaticity maps of nine small samples were analyzed by multi-point sampling. Studies and tests have shown that weft density, weave structure and pattern size have different degrees of influence on the optical properties of stretchable fiber optic fabrics. Stretchable optical fiber fabric luminance is positively correlated with weft density, but there is no significant linear correlation with pattern size and weave structure. Fabric pattern size and weave structure can affect the luminance of the fabric by affecting the degree of optical fiber bending in the fabric. The combination of fabric weaving parameters with the best luminance performance was finally obtained through the result analysis.

References

[1] BAI Z Q, TAN J, JOHNSTON C F, et al. Smart textiles: evaluation of optical fibres as embedded sensors for structure health monitoring of fibre reinforced composites[J]. International Journal of Clothing Science and Technology, 2015, 27(6): 870-894.

[2] SCHUSTER J, TRAHAN M, HEIDER D, et al. Influence of fabric ties on the performance of woven-in optical fibres [J]. Composites Part A: Applied Science and Manufacturing, 2003, 34(9): 855-861.

[3] YANG J H, CHO H S, KWAK H K, et al. Influence of detailed structure and curvature of woven fabric on the luminescence effect of wearable optical fiber fabric[J]. Korean Society for Emotion and Sensibility, 2018, 21(4): 55-62.

[4] WU R, JIN Z M, LU H J, et al. Development of optical fiber fabric with pattern color changeable[J]. Journal of Textile Research, 2010, 31(12): 43-46,52.

[5] LV C, ZHONG Z L, KUANG L Y, et al. Preparation and luminance test of polymer optical fiber light-emitting fabrics[J]. Journal of Tianjin Polytechnic University, 2013, 32(02): 42-45.

[6] ACHENBACH C, CODD J H. Computational studies of light acceptance and propagation in straight and curved multimodal active fibres[J]. Journal of Optics A: Pure and Applied Optics, 2003, 5(3): 239.

[7] MASUDA A, MURAKAMI T, HONDA K, et al. Optical properties of woven fabrics by plastic optical fiber[J]. Journal of Textile Engineering, 2006, 52(3): 93-97.

Comparison of Anaerobic Continuous Flow Degumming Effects of Ramie Under Different Pretreatment Methods

YANG Lishuang[1,2], **LI Yuling**[1,2*], **MA Yanxue**[1,2], **CHEN Xiaoguang**[3,4], **XIE Xuehui**[3,4], **WANG Xiaonuo**[1,2]

[1] *Key Laboratory of Textile & Technology (Donghua University), Ministry of Education, Shanghai 201620, China*

[2] *College of Textiles, Donghua University, Shanghai 201620, China*

[3] *School of Environmental Science &Engineering, Donghua University, Shanghai 201620, China*

[4] *National Environmental Protection Textile Industry Pollution Control Engineering Technology Center, Donghua University, Shanghai 201620, China*

* *Corresponding author's email*: lylu@ dhu. edu. cn

Abstract: Ramie fiber is an important raw material in the textile industry, but the raw ramie contains gum, which cannot be directly used for spinning. Thus it is necessary to remove the gum between the fibers. It is a need to pretreat raw ramie before degumming to reduce the burden of degumming and improve the uniformity of degumming ramie. The purpose of this study is to determine the pretreatment methord suitable for this dugumming system by comparing the degumming effect of boiling, warm water constant temperature soaking, warm water constant temperature soaking+soaking self-cirulation and cool water soaking of ramie. The results show that the fiber fineness and strength of raw ramie degummed by cold water soaking are 6. 30 dtex and 6. 84 cN/tex respectively, which is superior to the other three soaking methods. The pretreatment method is easy to operate and has good fiber quality, which is beneficial to promote the application of the system for ramie.

Keywords: Ramie; Anaerobic continuous flow biological degumming; Pretreatment; Boiling; Soaking

1 Introduction

Ramie is a perennial shrub of the Urticaceae. It has excellent strength, moisture absorption and anti-bacterial properties, and so on[1]. Ramie fiber is the longest and strongest natural fiber. These fibers are found in the bast fiber. The bast fiber contains other glial that cannot be directly used for spinning[2]. Therefore, the raw ramie needs to be degummed to remove the gum. Non-cellulosic colloid components mainly include pectin, hemicellulose, lignin, etc. [3]. At present, there are two main degumming methods for ramie. One is chemical degumming which seriously pollutes the environment. The second is biological degumming. The preparation of biological agents is separated from the degumming process, which is difficult to scale production[4]. In this paper, the anaerobic continuous flow biological degumming system (the system) was applied to degumming ramie[5-7]. Before formal ramie degumming, certain pretreatment is generally required to soften the stiff and dry raw ramie so as to facilitate cage loading and other operations. Meanwhile, it can loosen the bond between the raw ramie gum and fiber[8]. This reduces the burden of degumming and improves the uniformity of degumming ramie[9].

Previous studies used boiling pretreatment by the system to promote the dissolution of raw ramie components and realize the inactivation of miscellaneous bacteria carried on raw ramie[5]. But this method brought energy consumption and operational risks. This paper improves the pretreatment method and no longer adopts boiloing, but refers to the principle of fully automatic washing machine pre-impregnation, and realizes the softening of raw ramie and loosening colloid and fiber through soaking pretreatment for a certain period of time. In this study, the effects of different soaking treatments on the degumming effect of ramie were investigated and compared with the results of boiling pretreatment, which provides a basis for the selection of reasonable pretreatment and promotes the industrial application of the system.

2 Experiment Method

2.1 Raw materials

Ramie raw: Hunan Mingxing Hemp Co., Ltd. As shown in Fig. 1(a), there are many ramies bark, ramie stalk and other impurities on the surface of raw ramie, resulting in poor quality. Fig. 1(b) shows that multiple fibers are tightly adhered together under the action of gum.

Fig. 1 Physical view of raw ramie and longitudinal morphology electron microscope view

Bacteria: Anaerobic granular bacteria were obtained from the bioreactor of a sewage treatment project site.

2.2 Anaerobic continuous flow biological degumming

The system schematic diagram is shown in Fig. 2. It's composition, operation characteristics and function principle are detailed in reference[5-7].

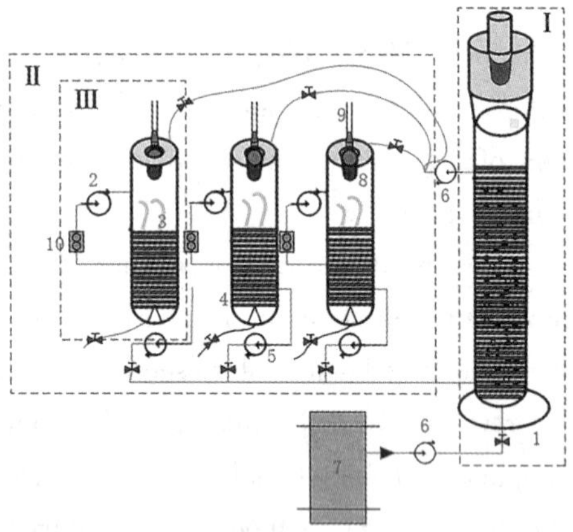

1—spiral symmetry stream anaerobic bio-reactor (SSSAB); 2—self circulating pump; 3—ramie fiber; 4—biological degumming tank; 5—external circulation pump; 6—water inlet pump; 7—refill tank; 8—water level balancer; 9—thermometer; 10—flowmeter; 11—heating belt; Ⅰ—anaerobic reactor subsystem; Ⅱ—external circulation subsystem; Ⅲ—internal circulation subsystem; Ⅳ—level balance subsystem.

Fig. 2 Schematic diagram of anaerobic continuous flow biological degumming system

2.3 Sample preparation

(1) Pretreatment

Method Ⅰ: raw ramie was boiled in an automatic cooking pot in accordance with the method in literature[13]. Basic parameters: preheat 0.5 h, boiling at 100 ℃, 1 h.

Method Ⅱ: the degumming tank of the system (Fig. 2) is to soak raw ramie at a constant temperature. Basic parameters: soaking time 24 h, soaking temperature 35 ℃.

Method Ⅲ: the degumming tank of this system (Fig. 2) is also used to soak raw ramie at a constant temperature, but the self-circulation system of the degumming tank is opened. Basic parameters: soaking time 24 h, soaking temperature 35 ℃, self-circulation rate 4.5 L/min.

Method Ⅳ: raw ramie is soaked in automatic cooking pot[5] in cold water. Basic parameters: soaking time 24 h, soaking temperature 15-20 ℃.

(2) Anaerobic continuous flow biological degumming

Degumming parameters were set according to reference[7].

(3) Post-processing

The degumming ramie taken out of the dugumming tank is cleaned by ERVP19209T automatic washing machine in accordance with standard washing procedures and dried naturally to obtain degummed ramie samples ①#, ②#, ③# and ④#.

2.4 Relevant indicators and detection methods

2.4.1 Relevant indicators

In order to select the appropriate pretreatment method, the following indexes were tested and analyzed on deglued ramie samples according to literature[5]. The specific contents are shown in Tab. 1.

Tab. 1 Fiber composition analysis and physical properties test content

Test indicators	The specific content
Composition analysis	Quantitative analysis of chemical composition
	Longitudinal form
Physical properties	Fineness
	Strength

2.4.2 Detection method

(1) Quantitative analysis of chemical components: according to the national standard GB 5889—1986 "Quantitative Analysis Method of Ramie Chemical Components".

(2) Longitudinal morphology observation of fiber:

TM3000 scanning electron microscopy.

(3) Fiber fineness test: YG231－10mm fiber cutter, JN-B－10－10mg precision torsion balance.

(4) Fiber strength test: LY－06E electronic fiber strength tester.

3 Results and Discussion

3.1 Quantitative analysis of chemical composition

The quantitative analysis results of the chemical composition of raw ramie and degummed ramie are shown in Tab. 2. The content of cellulose in literature[7] is 63.35%, while the content of cellulose in this paper is only 48.10%, and the content of non-cellulose is as high as 51.90%. The ramie used in this experiment is relatively mature.

The paper with different factors such as chemical composition and gel content of value to evaluate the difference between the degumming. Due to the differences of each factor is not consistent, therefore, this paper adopts the comprehensive evaluation method to assess the stand or fall of degumming. For the four data under each factor, scores of 1, 2, 3 and 4 are assigned from largest to the smallest (Tab. 3). After adding the score data under each factor, the influences of different pretreatment methods the on degumming effect are evaluated according to the total score, and the pretreatment method with a high total score is considered the best effect (Tab. 4).

Tab. 2 Quantitative analysis of chemical constituents of ramie raw and degummed ramie

	C(%)	W(%)	P(%)	H(%)	L(%)	A(%)	R(%)
Raw ramie	48.10	5.39	5.51	17.08	16.76	7.16	51.90
①#	70.52	1.83	2.10	12.19	10.20	3.16	29.48
②#	65.38	3.26	3.39	12.49	13.76	1.72	34.62
③#	66.09	3.24	1.60	13.47	13.84	1.76	33.91
④#	66.62	4.14	2.72	12.11	12.94	1.47	33.38

Tab. 3 Chemical composition and gel content factors

Score	W	P	H	L	A	R
1	④#	②#	③#	③#	①#	②#
2	②#	④#	②#	②#	③#	③#
3	③#	①#	①#	④#	②#	④#
4	①#	③#	④#	①#	④#	①#

Notes: C—cellulose; W—water soluble content; P—pectin; H—hemicellulose; L—lignin; A—ash; R—rubber content.

As can be seen from the score results in Tab. 4, the ①# and ④# degumming effects are better, ③# is the second and ②# is the worst. That is the cold water immersion can be comparable to the boiling degumming ramie sample.

Tab. 4 Comprehensive score under the factors of chemical composition and gel content

Sample No.	①#	②#	③#	④#
Total score	19	8	13	17

3.2 Analysis of fiber longitudinal morphology

The residual status of glue on the fiber surface can be seen in a more microscopic way, and the assigned value is also evaluated according to the residual amount of glue. The results are shown in Tab. 5. That is the degumming effect is the best after cold water immersion pretreatment.

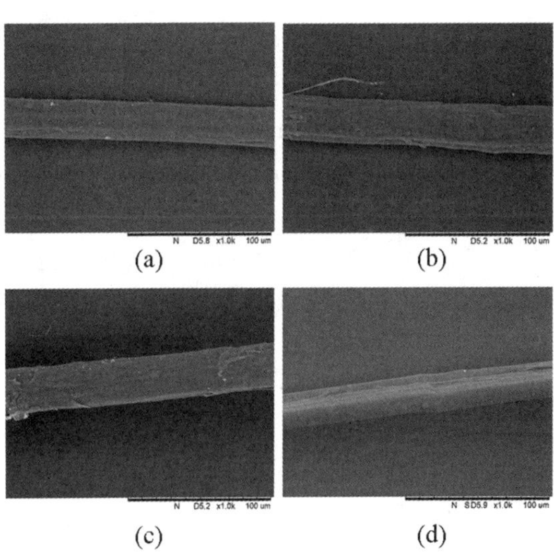

Fig. 3 Electron microscopy of longitudinal surface of degummed ramie

Tab. 5 Result of longitudinal form assignment

Score	1	2	3	4
Sample No.	③#	②#	①#	④#

3.3 Basic physical properties

Tab. 6 shows that the fineness of ④# is the smallest and the strength is the largest. The scoring results of fineness and strength are shown in Tab. 7. Thus the comprehensive score of ④# in Tab. 8 is also the highest. Combined with the previous evaluation results, it is considered that the fineness, strength and overall fiber quality of the degumming ramie pretreated by cold water immersion is the most suitable pretreatment method for this system.

Tab. 6 Fineness and strength of different dipping methods

Sample No.	Fineness (dtex)	Strength (cN/dtex)
①#	8.82	4.16
②#	7.81	4.55
③#	7.47	5.09
④#	6.3	6.84

Tab. 7 Scores under fineness and strength factors

Score	Fineness	Strength
1	①#	①#
2	②#	②#
3	③#	③#
4	④#	④#

Tab. 8 Overall score under fineness and strength factors

Sample No.	①#	②#	③#	④#
Total score	2	4	6	8

4 Conclusions

(1) Cold water immersion pretreatment is the most suitable pretreatment method for anaerobic continuous flow biological degumming system.

(2) No matter which pretreatment method is adopted, using anaerobic continuous flow biological degumming system can effectively remove the gelatin of raw ramie.

Acknowledgements

The authors would like to thank the Fundamental Research Funds for the Central Universities, China (No. 2232022G-01).

References

[1] CUI Y I, Manlan, et al. Research on the character and degumming process of different parts of ramie fiber[J]. Textile research journal, 2018, 88(17): 2013-2023.

[2] ZHANG Q, YAN S. Degumming of ramie bast fibers by Ca^{2+}-activated composite enzyme[J]. Journal of the Textile Institute, 2013, 104(1): 78-83.

[3] JIANG M. On the relationship between ramie fiber quality and textiles quality[J]. China Fiber Inspection, 2011 (11): 74-76.

[4] HAN H F, CUI Y D, CAI L B. Removing the resins of original hemp[J]. Cellulose Science and Technology, 2003(3): 50-54.

[5] LIU S H, LI Y L, CUI Y H, et al. Anaerobic microbial degumming effect of boiling water on cooked fresh ramie fresh ramie[J]. Chinese Science of Hemp Industry, 2020, 42(1): 31-37.

[6] LIU F, MA Y X, CHEN X G, et al. Study on process performance of ramie fiber anaerobic biological degumming system[J]. Textile Journals, 2020, 41(11): 98-103.

[7] TONG Y X, LI Y L, MA Y X, et al. Characteristics of anaerobic bio-degumming system for bast fibers[J]. Chinese Science of Hemp Industry, 2021, 43(1): 26-31+38.

[8] WANG Q, XUN X Y, CHEN W Y. Study of bio-chemical degumming process of african sisal fiber[J]. Journal of Zhejiang Textile and Garment Vocational Technical College, 2020,19(1): 1-6+13.

[9] MEI Z Q. Textile dictionaries[M]. 1st Edition. Beijing: China Textile & Apparel Publishing, 2007: 294.

Study on the Anti-Stabbing Mechanism of Semi-Rigid Stab-Proof Clothing

WANG Tianzeng[1], **YU Weidong**[1*]

[1] *College of Textiles, Donghua University, Shanghai 201620, China*

[*] *Corresponding author's email*: wdyu@ dhu. edu. cn

Abstract: Analyze the resistance change of the cutter in the process of piercing the resin sheet filled with artificial diamond particles, the shape and distribution of artificial diamond particles are important influence factors. The process of piercing the resin sheet makes the collision and displacement of diamond particles, but the collision between tool and diamond particles and the collision between diamond particles will also consume the energy of tool penetration. Based on this energy dissipation mechanism, a theoretical resin sheet model is put forward. The theoretical model improves the collision probability between tool and diamond particles and between diamond particles through the permutation and distribution of diamond particles, so as to optimize the stab resistance of resin sheet filled with artificial diamond particles, and reduce the thickness and quality of resin sheet without affecting the stab resistance of resin sheet.

Keywords: Theoretical resin sheet model; Distribution; Collision probability

1 Introduction

The stab-proof suit is a kind of personal protective equipment, which is suitable for anti-terrorism and anti-riot emergency treatment, and helps the human body to avoid the damage of nails, knives and other sharp objects. In the field of security protection, the research on the application of common stab-proof clothing and stab-proof mechanisms has matured due to the accumulation of quite excellent solutions in the cold weapon era. Since modern times, with the use of high-performance fibers such as aramid fibers, ultra-high molecular weight polyethylene fibers and other cut-proof materials, the research on stab-proof materials has shifted to flexible stab-proof clothing woven with ultra-high strength fibers or enhanced with resin coating. Compared with traditional rigid stab-proof clothing, modern high-performance fiber-based stab-proof clothing is light weight, flexible and has efficient performance against stabbing, cutting and slashing. The current types of stab-proof clothing can be divided into three categories according to the nature of the stab-proof material: rigid stab-proof clothing, semi-rigid stab-proof clothing, and flexible stab-proof clothing. Rigid stab-proof clothing is very effective in preventing stabbing, but it is heavier and too inflexible. Later researchers, in order to improve the comfort of the stab-proof layer as well as to reduce the weight, are now mostly considering the gradual change of the one-piece structural design to a discontinuous one[1]. Flexible stab-proof clothing is mainly based on fabrics woven with high performance filament yarns and processed by weaving, coating or impregnation. The current fabric made by Ai et al. [2] that can meet the requirements of Class A stab-proof standards has a surface density of 7. 85 kg/m^2. Semi-rigid anti-stab clothing mainly introduces rigid inorganic particles[3] and thermoplastic resin[4-5] into the fabric by coating or impregnation. Xia et al. [6] mixes boron carbide into epoxy resin and then coats aramid fabric to produce materials with significantly higher penetration resistance than untreated samples. The surface density of the semi-rigid anti-stab clothing made by Ma et al. [7] is 6. 39 kg/m^2, which meets the standard of Class A anti-stab clothing; The surface density of the composite resin sheet made by Yang[8] is 5. 82 kg/m^2, which is 8. 9% lower than the surface density of 6. 39 kg/m^2 made by Ma. The surface density of the resin sheet made by Chen[1] is 6. 67 kg/m^2. Liquid anti-stab clothing mostly uses shear thickening fluid to improve anti-stab performance. However, shear thickening fluid is often used to improve the performance of liquid stab proof suit, but if the stabbing speed of the tool is less than the critical shear speed of liquid stab proof suit, often cannot play a good stab proof effect[9-12]. Chen[1] found that the diamond particles in resin-based semi-hard puncture resistant clothing act on the tool in two forms: reverse cutting and hard blocking. Through the experimental verification, Yang[8] found that the barrier of diamond particles to the tool and the collision between diamond particles affect the anti-stabbing performance of the composite resin sheet. But they all did not establish the corresponding theoretical model.

This paper tries to put forward the theoretical model of semi-rigid stab-proof clothing and understand the mechanism of the semi-rigid stab-proof clothing based on resin. The exploration of such issues will help to solve the problem of the influence of the morphology of diamond particles, and their distribution in the resin sheet on the stab-proof performance of semi-rigid stab-proof clothing.

2 The Theoretical Model

2.1 Anti-stab properties of synthetic diamond-PI composite resin sheets

Chen Lifu[1] conducted dynamic anti-stabbing performance tests on pure resin sheets with a surface density of 2 600 g/m^2, and obtained that the maximum anti-stabbing force of resin sheets was 372 N. As a control, synthetic diamond-PI composite resin sheets with the surface density of 2 600 g/m^2 and 6 000 g/m^2 were prepared in this paper. After dynamic stab resistance test, the experimental results are shown in Tab. 1. The experimental results show that the resin sheet has better dynamic anti-stabbing performance at the same surface density, and the maximum anti-stabbing force is increased by 35%. At the same time, the structure of the composite resin sheet also significantly affects the anti-stabbing performance. Therefore, we will discuss the influence of resin sheet structure on the anti-stabbing performance.

Tab. 1 Dynamic penetration results of different resin sheets

Diamond particle size(μm)	Surface density (g/m^2)	Maximum stab resistance of tool(N)
300	2 600	504
300	6 000	1 260

2.2 Resistance analysis of diamond particles to cutting tools

After the cutting tool inserts into the resin sheet, the force cross section changes all the time, which is divided into three stages: the first stage is $0 \leqslant h \leqslant 3.98$ mm, and the cross-sectional area is denoted as S_1; The second stage is $3.98 < h \leqslant 11.8$ mm, and the cross-sectional area is denoted as S_2; The third stage is $11.8 < h \leqslant 27.99$ mm, and the cross-sectional area is denoted as S_3. The first and second stage tool front view is shown in Fig. 1 (a), the third stage tool front view is shown in Fig. 1 (d), the first stage force cross section is shown in Fig. 1 (b), and the second and third stage force cross section is shown in Fig. 1 (c). h is the depth of cutting tool penetration into resin sheet. X_1 and X_2 are the side lengths of the front view of the stressed part of the tool, L_1 and L_2 are the side lengths of the stressed section of the tool.

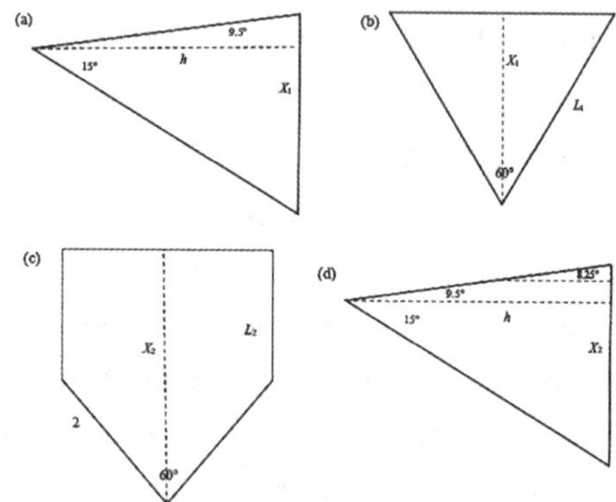

Fig. 1 (a) and (d) Front view of the force part of the tool; (b) and (c) cutting tool force section diagram

$$S_1 = 0.109h^2 \tag{1}$$

$$L_1 = \frac{2\sqrt{3}}{3}h(\tan 9.5° + \tan 15°) \tag{2}$$

$$S_2 = 0.87h - \sqrt{3} \tag{3}$$

$$L_2 = 0.435h - \sqrt{3} \tag{4}$$

$$S_3 = 0.826h - 1.21 \tag{5}$$

$$X_1 = 0.413h + 0.26 \tag{6}$$

The resistance of the tool when it collides with the resin sheet of diamond particles includes: the frictional resistance of the resin to the tool, the blocking and the cutting of the diamond particles to the tool. The initial energy of the tool W_0, the frictional resistance of the resin to the tool is related to the change in volume of the tool piercing into the resin sheet, noted as W_1. The blocking of the tool by the diamond particles and the cutting resistance, noted as W'. The energy of the tool after piercing into the resin sheet is noted as W.

$$W_0 = \frac{1}{2}mv^2 \tag{7}$$

$$W = W_0 - W' - W_1 \tag{8}$$

$$W' = \int_0^d F(h)\mathrm{d}h \tag{9}$$

The relationship between the resistance and the variation of the piercing depth d can be obtained by the pendulum type fabric anti-piercing. The cutting and chopping function combination measuring instrument developed by the group, and then find out the resistance of diamond particles to the tool.

2.3 Design of diamond particle-filled resin sheets

Since the frictional resistance of resin to the tool is low, this project considers increasing the resistance of diamond particles to the tool and thus achieving the anti-puncture effect. In addition to the direct collision with diamond particles, the collision between diamond particles and the cutting of diamond particles on the tool are also included in the process of piercing the resin sheet. The anti-puncture effect. The resin sheet design is as Fig. 2 (a) and Fig. 2 (b).

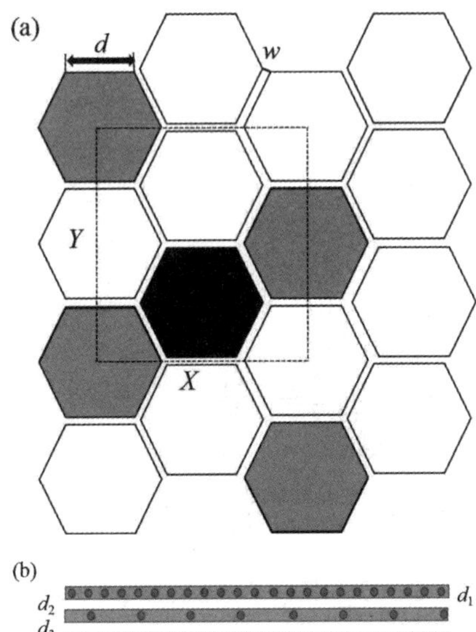

Fig. 2 (a) Distribution plan of diamond particles in resin sheet; (b) diamond particle distribution in resin sheet

d is the side length of the regular hexagonal surface of diamond particles, and w is the distance between two regular hexagonal planes. d_1 is the thickness of the granular layer, d_2 is the distance between the particles of the first layer and the second layer, and d_3 is the distance between the particles of the second layer and the third layer. The first layer of diamond particles is marked white; the second layer of diamond particles is marked blue; the third layer of diamond particles is marked black. According to the distribution as shown in Fig. 2 (a), in order to realize the successive collision of diamond particles between layers and increase the loss of energy for tool stabbing, it can ensure the collision of particles between layers at least twice, and the weight can be reduced by 9/16 at most.

$$X = 3d + \frac{4\sqrt{3}w}{3} \tag{10}$$

$$Y = 2\sqrt{3}d + 2w \tag{11}$$

$$P_1 = \frac{6\sqrt{3}d^2}{6\sqrt{3}d^2 + \frac{8\sqrt{3}w^2}{3} + 14wd} \tag{12}$$

X is the length of the basic unit in diamond particle distribution, and Y is the width. P_1 is the collision probability.

For a 6 cm×6 cm resin sheet, the fraction of volume of man-made diamond filling $V_f = P_1$. When the particles are distributed as shown in Fig. 2 (a) and Fig. 2 (b), the volume fraction of man-made diamond is $V_f = \frac{1}{4}P_1$, at this time the second layer of diamond particles and the first layer do not necessarily collide. Because the particle spacing is greater than the length of the opposite side of the man-made diamond particles $\sqrt{3}d$, when necessarily collide the volume fraction of the second layer of diamond particles is distributed as follows $\frac{1}{4}P_1 < V_f \leqslant P_1$. The volume fraction distribution of the third layer of diamond particles when the second and third layers of diamond particles necessarily collide is $\frac{1}{16}P_1 < V_f \leqslant P_1$.

3 The Theoretical Analysis

Theoretical analysis of the tool piercing process into the

resin sheet based on the design of inorganic diamond particles resin sheet.

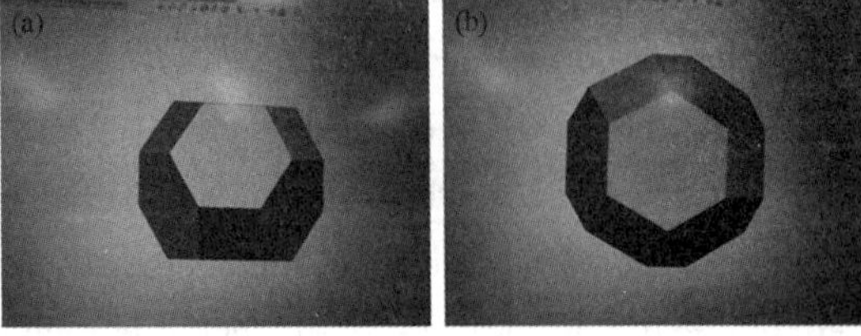

Fig. 3 (a) Side view of diamond particles; (b) top view of diamond particle

After 3D modeling of the decahedron, including eight regular hexagonal surfaces and six square faces, the angle between the regular hexagonal surfaces is measured to be 109. 47° and the angle between a regular hexagon face and a square face is 125. 26°.

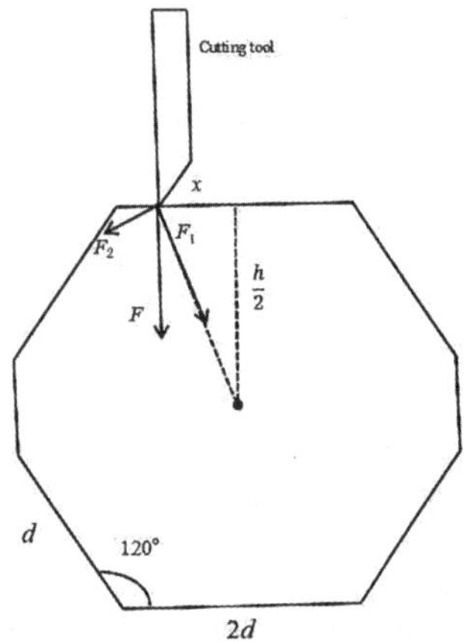

Fig. 4 Diagram of tool and diamond particles collision mechanics

When the tool collides with the horizontal hexagonal face of the diamond particle, the distance between the tool and the diamond particle is x, as shown in Fig. 4. The force F of the tool on the resin sheet is divided into F_1 and F_2, and $F_1>F_2$ because of $\theta_{max}=36.2°$. The analysis can be obtained if the particle can overcome the resistance to rotation, then rotation and movement occur simultaneously, divided into two cases: the particle only moves without rotation and the particle moves and rotation occur simultaneously.

When the diamond particle only moves without rotation, the particle moves downward for y and to the right for $y\times\tan\theta$. The maximum distance the diamond particle moves to the right is $\frac{3d}{2}$ and the maximum distance of downward movement is $\frac{\sqrt{3}d+d}{2}$ The distance between the first and second layer of particles d_2 cannot be greater than $\frac{\sqrt{3}d+d}{2}$. The distance between the first and second layer of particles d cannot be greater than. The tool piercing direction remains unchanged in the case of downward moving distance is $\frac{\sqrt{3}d+d}{2}$. In the case of 300 μm diamond particles, for example, $d=100$ μm, the cross-sectional area of the tool is 2.03×10^{-3} mm^2, much smaller than the area of the hexagon 34.64×10^3 mm^2. In the collision with the tool collision particles and the second layer of particles, the tool does not collide with the first layer of other particles, only by the tool collision particles with the first layer and the second layer of other particles, the maximum force particles will produce $\frac{3d}{2}$. The displacement, as shown in Fig. 5, yellow particles for the first layer in the collision with the tool particles. It is known that the first layer of particles and the second layer of particles collision probability of 100%, but only with the second layer of a particle collision. Only if $d_2+d_3+300<\frac{\sqrt{3}d+d}{2}$, the second layer of particles can collide with the third layer of particles. The tool continues to move down at this point with 25% probability of collision with a particle in the second layer and 6. 25% probability of collision with a particle in the third layer.

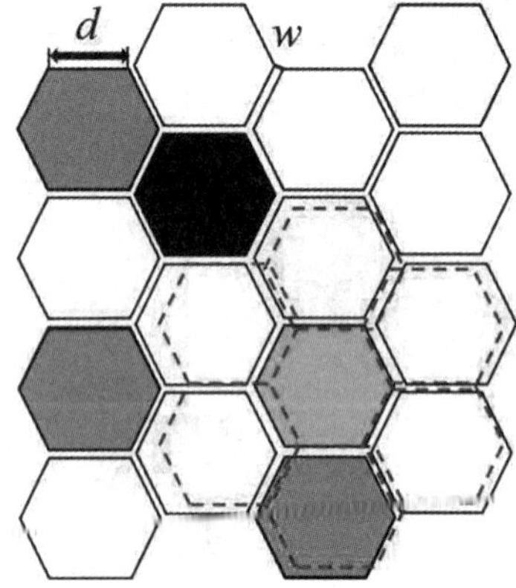

Fig. 5 Distribution of diamond particles after tool impact

Fig. 5 shows the movement diagram of diamond particles after the tool is pierced into the resin sheet. The red dashed particles represent the possible movement direction of the stressed particles.

When the movement and rotation of diamond particles occur at the same time, the stressed particles will collide with other particles in the first layer and be subject to resistance, and the particle rotation will be hindered. Because the particle rotation will make the downward moving distance smaller, less than $\frac{\sqrt{3}d + d}{2}$, and the spacing between the particle layers is too small, if the first and second layer of particles collide, then d_2 should also be reduced.

4 Conclusions

Therefore, we can consider increasing the density of the second and third layer of diamond particles appropriately, reducing the distance between the diamond particles to realize the collision between the intermediate particles in the process of cutting tool piercing. Finally, the resistance of diamond particles to the cutting tool can be increased.

This paper constructs and analyzes the theoretical model of resin sheet through dynamic stab-resistant experiment, preliminarily explores the influence of diamond particle distribution on stab-resistant performance of resin sheet. In the future, we will further study how to maximize the barrier effect of diamond particles on cutting tools, so as to improve the anti-stabbing effect of semi-hard anti-stabbing clothing.

References

[1] CHEN L F. Study on preparation and stab resistant mechanism of resin sheet filled by inorganic particle layer[D]. Shanghai: Donghua University, 2020.

[2] AI Q S, HU L, WU Z W, et al. Preliminary study on anti-puncture performance of aramid fiber weft free fabric[J]. Fiber Reinforced Plastics/Composites, 2019, 05: 106-109.

[3] ZHANG Z, LIU X Y, YU W D. Preparation of coating puncture-proof fabric and its mechanism of puncture-proof [J]. Journal of Textile, 2018, 39(3): 108-113,119.

[4] HOSUR M V, MAYO J B, WETZEL E, et al. Studies on the fabrication and stab resistance characterization of novel thermoplastic-kevlar composites[J]. Solid State Phenomena, 2008(136): 83-92.

[5] FIROUZI D, FOUCHER D A, BOUGHERARA H. Nylon-coated ultra high molecular weight polyethylene fabric for enhanced penetration resistance[J]. Journal of Applied Polymer Science, 2014(131): 1-9.

[6] XIA M M, QUAN Z Z, WANG X L, et al. Preparation and characterization of B4C particle coated composites for stab-resistance[J]. Composite Structure, 2019(228): 1-7.

[7] MA F F, XIE Z Q, WANG X H. Design and optimization of flexible anti-stab material formed by resin[J]. Industrial Textiles, 2014(5): 18-22.

[8] YANG C. Preparation and anti-stab mechanism of surface modified diamond particle composite resin sheet[D]. Shanghai: Donghua University, 2021.

[9] BROWN E, JAEGER H M. Shear thickening in concentrated suspensions: phenomenology, mechanisms and relations to jamming[J]. Reports on Progress in Physics, 2014, 77(4).

[10] KALMAN D P, WAGNER N J. Microstructure of shear-thickening concentrated suspensions determined by flow-USANS[J]. Rheologica Acta, 2009, 48(8): 897-908.

[11] CHENG X, MCCOY J H, ISRAELACHVILI J N, et al. Imaging the microscopic structure of shear thinning and thickening colloidal suspensions[J]. Science, 2011, 333(6047): 1276-1279.

[12] CHEN Q, HE Q Y, LIU M, et al. Mechanical properties and mechanism of shear thickening fluid[J]. Journal of Solid Mechanics, 2016, 37(6): 518-537.

Preparation Method of Large Folding Diameter Ratio and Variable Diameter Tubular Woven Fabric

WU Chenguang[1,2], CUI Yunhua[1,2*], MA Yanxue[1,2], LI Yuling[1,2], CHEN Lijun[3]

[1] *College of Textiles, Donghua University, Shanghai 201620, China*

[2] *Key Laboratory of Textile and Fabric Technology, Ministry of Education, Donghua University, Shanghai 201620, China*

[3] *Shanghai Bioman Medical Technology Co. Ltd.*

* *Corresponding author's email*: anniecui@ dhu. edu. cn

Abstract: Based on the design principle of equal filling density and the way of "flattening-weaving-restoring", the weaving factor of the balloon-shape tubular billet fabric was designed in this method. In the preparation process, the position of the fell is changed by the lifting and lowering technology of the front rest. On the basis, the reed-in numbers of special reed will be changed to achieve a large proportional change in fabric width. The results show that the mentioned design method of the loom set factor is reasonable and feasible, the preparation process is effective. And the prepared samples are dimensionally stable and well formed. The preparation process can provide certain method guidance for the wide application of large folding diameter (*FD*) ratio variable diameter tubular woven fabric in actual production.

Keywords: Tubular woven fabric; Balloon-shape; The lifting and lowering of the front rest; Large folding diameter ratio; Special reed; Reed-in number

1 Introduction

The balloon is formed by heating and inflating the tube, which is generally composed of high molecular polymers and blends such as polyester[1] in Fig. 1 (a). It has a wide range of applications, such as dilatation to treat duct stenosis, assisted stent implantation after compression and expansion, drug delivery[2-3], orthopedic treatment[4-5], and local laser chemotherapy[6]. In clinical use, the burst pressure requirements of some balloons are relatively high (>40 atm), but the balloons made of traditional polymer membrane materials cannot meet this requirement, and fibers must be added to improve the strength of the balloon.

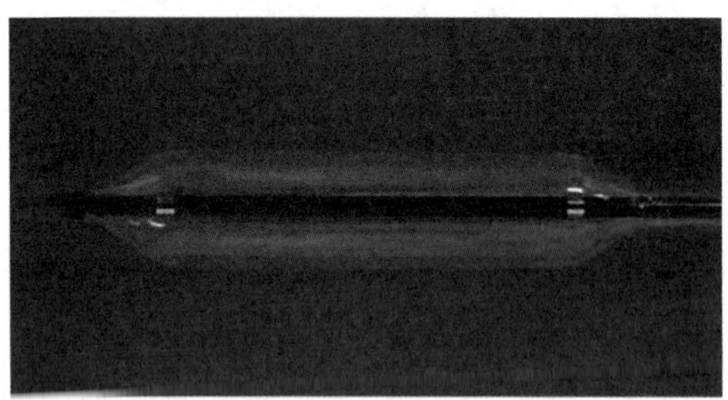

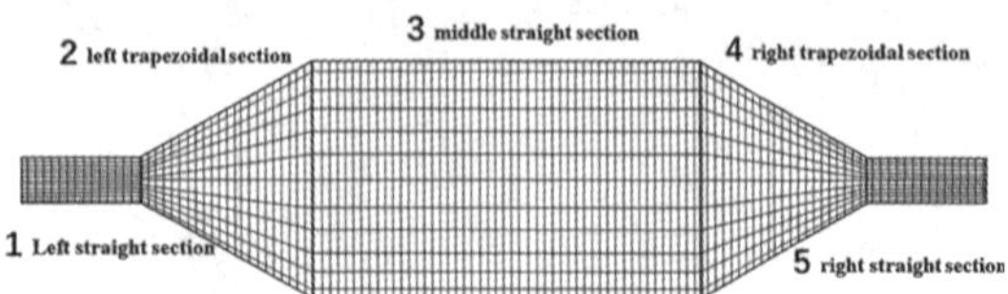

Fig. 1 (a) Traditional membrane balloon; (b) balloon with fiber reinforcement

At present, the fiber-reinforced structure of the fiber-reinforced balloon on the market is a filament winding structure, as shown in Fig. 1 (b). However, it was found that the burst pressure still could not reach the strength, so it was planned to use a one-time-forming balloon-shape woven fabric to strengthen the structure. In this paper, the design principle and preparation technology of balloon woven fabric will be researched.

2 Experimental Design

According to the balloon shape of Fig. 1(b), a balloon-shape woven fabric is designed as shown in Fig. 2(b). It is a tubular fabric with a drastic change in diameter, and its structure can be divided into five parts: left straight tube, left tapered tube, middle straight tube, right tapered tube and right straight tube.

In the actual preparation process of the tubular woven fabric, the three-dimensional design size is generally calculated and converted into two-dimensional data,

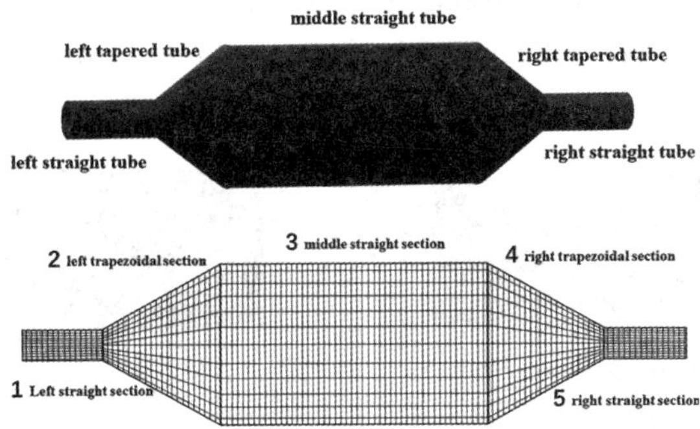

Fig. 2 (a) Balloon-shape woven fabric model; (b) tubular billet

and then the flat double-layer fabric is woven. Finally, it can be stretched and shaped after getting off the machine. This process involves the geometric transformation of the fabric structure[7]. That is to say, the three-dimensional balloon woven fabric must be converted into a two-dimensional woven fabric tubular billet in Fig. 2(b), including the corresponding left straight section, left trapezoidal section, middle straight section, right trapezoidal section, and right straight section.

2.1 Size design

The original shape and parameters of the balloon were provided by the cooperative company, as shown in Fig. 3 and Tab. 1.

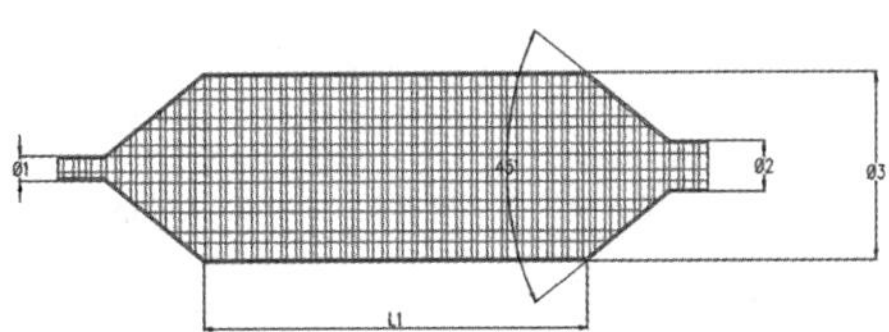

Fig. 3 Balloon woven fabric shape

This article will take the 8.0×100 mm sample as an example to describe the design of the size parameters and the preparation of the corresponding tubular billet fabric in detail.

Tab. 1 Size parameters

Spec(mm)	Ø1	Ø2	Ø3	L1
4.0×20	1.7±0.1	1.8±0.1	4.0±0.2	20±1
4.0×100	1.7±0.1	1.8±0.1	4.0±0.2	102±3
8.0×100	1.8±0.1	2.0±0.1	8.0±0.2	102±3
12.0×20	2.2±0.1	2.4±0.1	12.0±0.3	20±1
12.0×40	2.2±0.1	2.4±0.1	12.0±0.3	40±2

According to the specification requirements of 8.0×100 mm in Tab. 1, the basic size of the corresponding woven tubular billet is designed according to the formula on the fabric width (FD) $D=\pi d/2$ (where d is the diameter). Due to the tension fluctuation and other factors in the weaving process are closely related to its size, so its size does not set the precision value.

Tab. 2 Tubular billet size

Section	X	Diameter $_X$(mm)	Length $_X$(mm)
Left straight	1	≈3	10
Left trapezoidal	2	/	≈6.5
Middle straight	3	≈13	100
Right trapezoidal	4	/	≈6.5
Right straight	5	≈3	10

2.2 Process design

2.2.1 variable diameter

It can be seen from Tab. 2 that the *FD* ratio of the tubular billet is about 4.5. The width change methods of tubular woven fabrics are generally divided into two categories: warp density unchanged and warp density change[8]. In this study, in order to ensure the continuity and integrity of selvage, the method of warp density will be adopted.

The teeth of the traditional reed are evenly distributed, which means that the width of the fabric remains unchanged during the weaving process. In order to realize variable warp density weaving, a special reed is used in this paper, and the contact position between the fell and reed is changed in moving the front beam, so as to realize the change of the warp density of the fabric[9], as shown in Fig. 4.

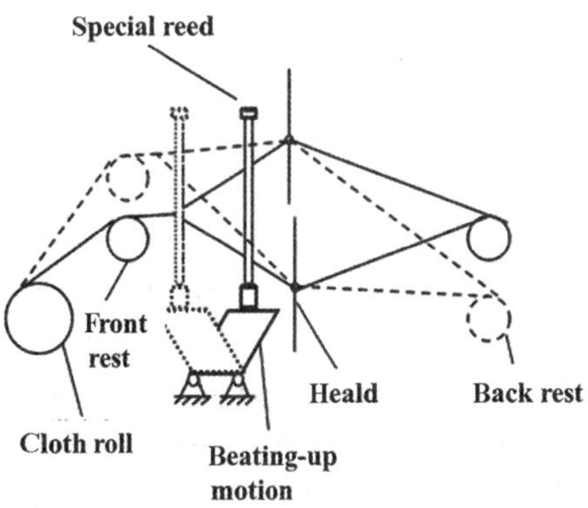

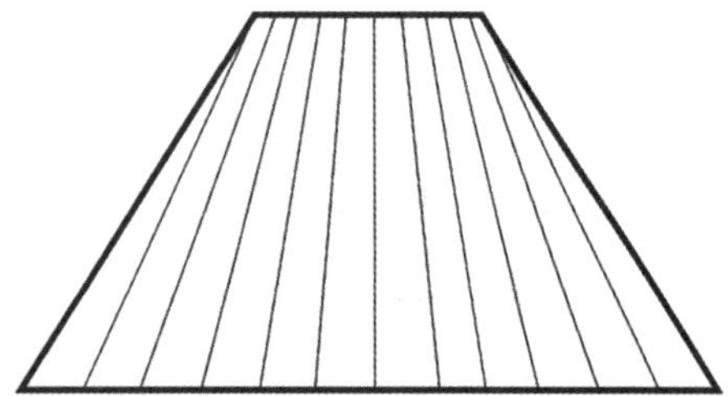

Fig. 4 (a) Fell moving; (b) special reed; (c) special reed model

2.2.2 large folding diameter ratio

It can be seen from Fig. 4 (a) and (b) that the change of the *FD* from large to small corresponds to the reed density from the bottom to the top end of the special reed. Due to the manufacturing technology of special reed, the reed density ratio can only reach 2, which is equivalent to the *FD* ratio cannot be larger than 2. For this reason, this paper develops a new weaving process to solve the necessary problem of weaving the large *FD* ratio for balloon woven fabrics.

The idea of this paper is to design a variable reed-in weaving process based on the existing special reed in accordance with the weaving requirements of variable diameter tubular fabrics with large *FD* ratio.

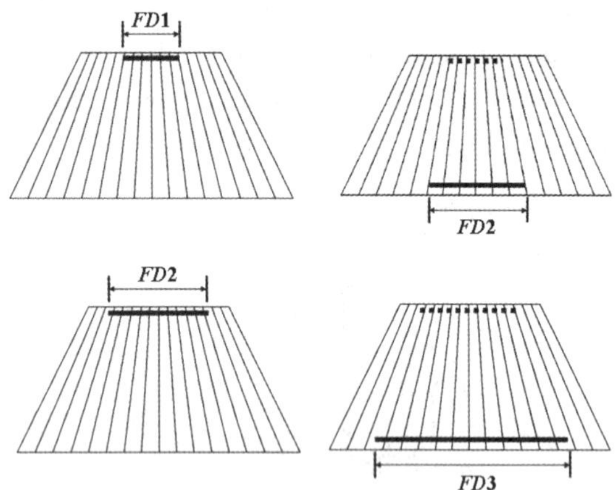

Fig. 5 (a) Reed-in; (b) fell move; (c) new reed-in; (d) fell move (solid line-position before fell movement and dotted line-position before fell movement)

Specifically, at this point, *FD* 1 completes an extension change to *FD* 2 as shown in Fig. 5 (a) to Fig. 5 (b). The warp yarn is removed from the front of the loom, the number of reed-in is changed and the yarn is re-threaded in Fig. 6. Then adjust the fell to the top of special reed (Fig. 5(c)), so that the fabric *FD* corresponding to the new fell position is consistent with the *FD* 2. Then start a new weaving like the above step to complete the large proportional change of 1 to 3. When decreasing the fabric width, the opposite of the above process.

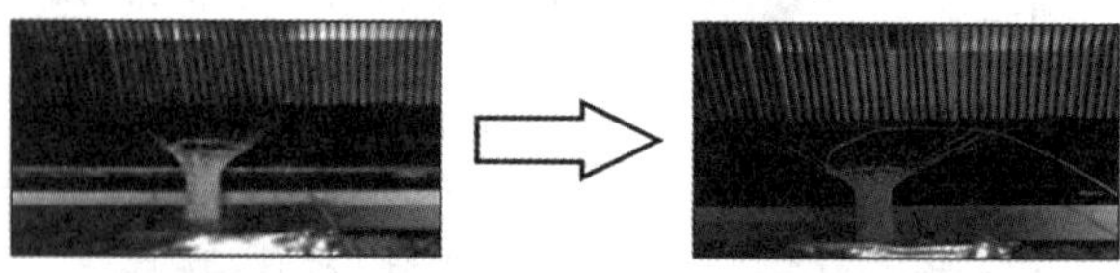

Fig. 6 Reed-in change

2.3 Weaving parameter design

First, according to the requirements of the balloon and the weaving difficulty, medical UHMWPE yarn and polyester monofilament were chosen as the warp and filling yarn. Secondly, in consideration of the strength needs the special structure of the fabric, the fabric weave is plain weave and the warp density will be determined from the small *FD* (left/right straight section). Finally, the filling density is chosen to be kept constant because of high bursting pressure[10].

In addition, it is known that the reed needs to be changed twice from $4<D_3/D_1\approx4.5<8$ according to the special reed specifications, and the calculation results are shown in Tab. 3.

Tab. 3 Fabric parameter

Yarn count (tex)	5.5	
Yarn count (tex)	2.5	
Fabric weaves	plain	
Draft plan	straight-over draft	
Warp density (piece/10 cm)	Left/Right straight section	1 050
	middle straight section	250
Filling density (piece/10 cm)	440	
Total yarns	64	
Reed (T/10 cm)	100–200	
Reed-in	8	
	4	
	2	

2.4 Weaving steps

Step 1: the 64 polyester monofilaments are warped, reeded and heddled according to the 8 yarns per reed and straight-over draft, and the fell is adjusted to the bottom of the special reed. Left straight section of the tubular billet with a length of 10 mm is woven according to the warp densities of 1050 piece/10 cm and the filling densities of 440 piece/10 cm.

Step 2: after the weaving of the left straight section is completed, the fell is gradually decreased from the top close to the bottom of the special reed with the weaving process.

Step 3: when fell moves to the bottom end, the warp yarn is removed from the reed at the front of the loom, the reed is threaded in reverse according to 4 yarns per reed, and the fell is adjusted to the top of special reed.

Step 4: start weaving again. As the fell gradually approaches the bottom of special reed, the fabric *FD* increases four times compared to the initial *FD*.

Step 5: repeat the content of step 3, in accordance with the 2 yarns per reed for reverse through the reed through the heald, until the fabric *FD* to D_3 to complete the left trapezoidal section weaving.

Step 6: the position of the fell remains unchanged, other basic parameters remain unchanged, and the middle straight section of the tubular billet with a length of 100 mm is woven.

Step 7: after completing the middle straight section of the tubular billet, start fell gradually close to the top end of special reed weaving, as weaving proceeds, the fabric *FD* gradually reduced.

Step 8: after the fell reaches the bottom end, the warp yarn is removed from the reed at the front of the loom, and the fell is adjusted to the bottom of reed in reverse according to 4 yarns per reed.

Step 9: start weaving again, wait until the fell is near the top end of reed, and repeat step 8, in accordance with the yarns per reed for reverse through the reed through the heald, until the fabric *FD* to D_5 to complete the right trapezoidal section weaving.

Step 10: keeping other basic parameters unchanged, weave right straight section of the tubular billet with a length of 10 mm.

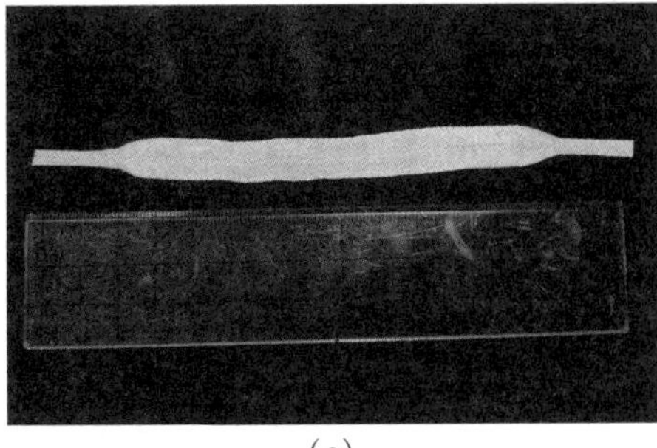

(a)

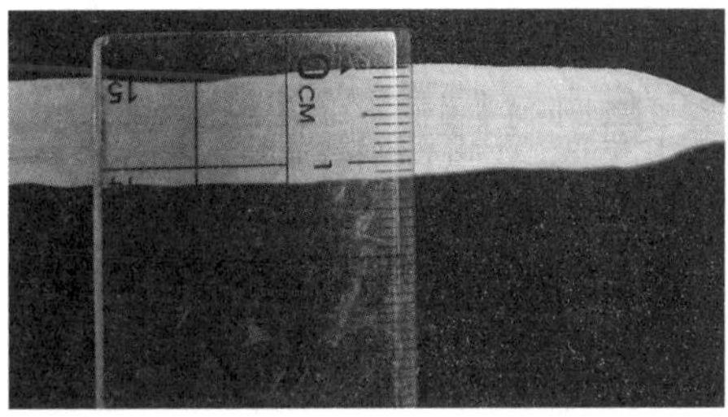

(b)

Fig. 7 (a) Simple length; (b) simple width

As shown in Fig. 7, the left trapezoidal section and the right trapezoidal section of the fabric each completed three *FD* changes, and the sample size basically matched Tab. 2 and cooperation's requirements.

3 Conclusions

In this study, the balloon model was converted into a tubular billet according to the "flattening-weaving-restoring" method, which shows that this geometry conversion idea is reasonable. And the weaving parameters of the tubular billet were designed based on the principle of equal filling density design, and the size of the samples shows that the design method of the weaving parameters is correct. In the process of looming, the preparation process of changing the position of fell and the number of reed-in of special reed can gradually realize the change of fabric *FD*. In summary, the geometric conversion and weaving parameters design method and preparation process of this research are completely reasonable and feasible.

References

[1] FU X L. Study on numerical simulation and process of balloon molding[D]. Beijing: Beijing University of Chemical Technology, 2017.

[2] YANG W X, QIAO S B. Current situation and prospect of drug-eluting balloon [J]. Chinese Circulation Journal, 2015, 30(5): 409-412.

[3] CUI K Y, LV S Z, LIU H, et al. Advancements for treating restenosis with drug-eluting balloon [J]. Journal of Cardiovascular and Pulmonary, 2016, 35(3): 246-250.

[4] MARCO R A, KUSHWAHA V P. Fractures treated with posterior decompression and pedicle screw instrumentation supplemented with balloon-assisted vertebroplasty and calcium phosphate reconstruction[J]. The Journal of Bone and Joint Surgery, 2009, 91(1): 20-28.

[5] YANG H L, et al. Clinical results of kyphoplasty in the treatment of osteoporotic vertebral compression fracture[J].

Suzhou University Journal of Medical Science, 2002(4): 406-409.

[6] GLAZIER J J, et al. , Laser balloon angioplasty combined with local intracoronary heparin therapy: immediate and short-term follow-up results[J]. American Heart Journal, 1997, 134(2).

[7] ZHU C Y, On the structures and weaving technology of 3D fabric preforms with various shapes[J]. Journal of Zhejiang Sci-Tech University, 2000(2): 3-7.

[8] DONG X Y. Research on the manufacture of tapered artificial vascular[D]. Shanghai: Donghua University, 2012.

[9] WANG Y Z. Research on auotomatic weaving of variable width fabric[D]. Shanghai: Donghua University, 2017.

[10] MENG F Y. Study on the design principle and flow behavior of conical woven tubular vascular graft[D]. Shanghai: Donghua University, 2021.

High-Performance Poly (Vinyl Alcohol) Composite Films Reinforced by Lignin-Graft-Poly (Vinyl Alcohol)

DUAN Lianjie[1,2], **LU Chunhong**[1,2*]

[1] *Key Laboratory of Textile Science & Technology, Ministry of Education, Donghua University, Shanghai 201620, China*

[2] *Department of Technical Textiles, College of Textiles, Donghua University, Shanghai 201620, China*

* *Corresponding author's email*: chlu@ dhu. edu. cn

Abstract: Polyvinyl alcohol (PVA) shows great prospects in packaging film due to its properties of non-toxic and non-hazardous. However, the water resistance is restricted as a result of abundant hydroxyl groups in the side chain of PVA, and inter- and intramolecular interactions strongly affect the mechanical properties. In this work, modified PVA (L-P) was prepared by grafting lignin onto PVA by acetal reaction. The graft-copolymerization process was carried out at 120 ℃ and 150 ℃ for 12 h under the nitrogen atmosphere, respectively. L-P was characterized by FTIR, DSC and water solubility. The increase of C—O—C bond in FTIR and the appearance of some lignin characteristic peaks confirmed the feasibility of the reaction. The DSC results revealed that the crystalline regions of the products were relatively diminished with an increase in temperature. Water solubility results showed that the intervention of lignin improves the water resistance of PVA. In addition, six groups of PVA composite films containing different contents of lignin and L-P were prepared via a solution casting method. Compared with pure PVA film, the composite films show better mechanical properties. The morphology of the films was characterized by SEM. Overall, this work enables the high-value utilization of modified polyvinyl alcohol fillers in composite films and potential applications in industrial fields.

Keywords: Polyvinyl alcohol; Lignin; Composite films; Mechanical properties

1 Introduction

Polyvinyl alcohol (PVA) holds an important position due to its excellent film-forming property, the barrier of oxygen, nontoxic and good chemical resistances, and is widely used as film. However, the water resistance and the mechanical property were affected by the abundant hydroxyl groups in the side chain of PVA which can form strong inter- and intramolecular interactions with water molecules or itself[1]. This will change the film structure, causing the film to expand and deteriorate its stability. Therefore, it is necessary to make proper modifications to the film structure to overcome the above-mentioned problems. Some different modification methods have been well developed in the past few decades, such as blending, grafting and crosslinking[2]. It is worth mentioning that molecular aggregates are formed between the polymer chains of the blends through the action of sub-valent bond force, which changes the distribution and structure of functional groups. In addition, some bio-based polymers are also added to polyvinyl alcohol, promoting the versatility of PVA, such as lignin[3], chitosan, nanocellulose, etc. Lignin is an aromatic polymer with many active functional groups such as hydroxy, methoxy, carboxyl and other groups which shows good reactivity[4]. Most notably, lignin also has hydrophobic, low-cost, bio-renewable and biodegradable advantages, making lignin a popular additional material for sustainable commodity plastics[5].

In this work, lignin was firstly grafted onto PVA by acetal reaction to obtain Lignin-graft-PVA (L-P) and then PVA, lignin and L-P were blended to prepare PVA composite (L-P/Lignin/PVA) film. The purpose of this study is to explore the effect of L-P on the mechanical properties of the film. L-P was characterized by FTIR, DSC and water solubility and the performance of the films was explored by tensile test and a scanning electron microscope (SEM).

2 Experiments

2.1 Materials

Organosolv lignin was purchased from Hubei Yunmei Technology Co., Ltd. China (90.0%), Dimethyl sulfoxide (DMSO) (99.0%) and acetone (99.5%), were purchased from Shanghai Titan Technology, PVA (Mw = 146 - 186 kDa, 99.0% hydrolysis) powders were purchased from Sigma-Aldrich Co., Ltd., para-toluene sulfonic acid (p-TSA), Water used in all experiments was deionized (DI).

2.2 Synthesis of L-P

The reaction was performed inside a 250 mL three-neck flask (Flask A) which was charged with 1.00 g polyvinyl alcohol and 30 ml DMSO. This flask was fitted with a bump trap and the other openings were respectively connected with a thermometer and a nitrogen gas pipe. Full dissolution was accomplished by heating to 90 ℃ for 4 hours with magnetic stirring. Meanwhile, 0.50 g Lignin and 0.077 g p-TSA were dissolved in 30 ml DMSO at 60 ℃ for 2 hours. Later, two solutions were mixed in flask A and then the system was equipped with flask A, condenser tube, horn tube, etc. The reaction was heated under nitrogen at 120 ℃ or 150 ℃ for 12 hours. The crude product was precipitated by adding the cooled reaction solution to a sufficient amount of acetone.

2.3 Preparation of L-P/Lignin/PVA composite film

The PVA and L-P and lignin solution was prepared by adding PVA (1 g), L-P and Lignin into DMSO successively, and then agitated at 85 ℃ for 2 h to ensure the full mixing. It was poured into the glass Petri dish (60 mm diameter) at once with a volume of 2.5 mL. The nanocomposite films were obtained by evaporating the moisture at room temperature.

2.4 Structural analysis

To verify the structural change of lignin after chemical modification, approximately 5 mg L-P was used to perform the structural analysis in the range of 4 000 - 500 cm^{-1} for 64 scans at 4 cm^{-1} spectral resolution by a Nicolet 6700 Fourier transform infrared spectrometer with attenuated total reflection-Fourier transform infrared (ATR-FTIR). DSC tests of L-P were carried out on a Modulated DSC 8500 in a nitrogen atmosphere at a heating and cooling rate of 10 ℃/min. The samples were dried at 100 ℃ for 12 h before testing, and then heated from 25 ℃ to 250 ℃ and held at that temperature for 5 min to eliminate the thermal history. The crystallization process was recorded from 250 ℃ to 25 ℃, where about 5 mg of each sample was measured and analyzed. The heating process is twice in total.

2.5 Water dissolution

5 mg L-P were placed in 20 mL of DI water and heated at 85 ℃ for 4 h on a hot plate. Solubility was calculated according to Equation (1):

$$S = \frac{m_b - m_a}{m_b} \times 100\% \quad (1)$$

where m_b is the mass before sample dissolution and m_a is the mass after sample dissolution.

2.6 Mechanical testing

The mechanical properties of all films were measured via the XQ-1C tensile testing system at a temperature of 20 ℃ and relative humidity of 65%. The sample size was 30 mm×5 mm with a gauge length of 20 mm and a loading rate of 50 mm/min according to ASTM D882 standard. Each sample was measured five times.

2.7 Image analysis

The morphology of the cross-section and surface of the film was observed using SU8010 (SEM) at an accelerating voltage of 5 kV. Before imaging, the samples were gold coated.

3 Results and Discussion

3.1 Synthesis of L-P

Graft copolymerization of lignin and PVA was performed with the p-TSA as a catalytic agent. The structural change of PVA caused by chemical modification is delineated by FTIR analysis in Fig. 1. For PVA, the weak peaks at 1221 and 1149 cm^{-1} are ascribed to the C—O—C stretching. L-P also contained the characteristic peaks of PVA aforementioned, yet it is much stronger, suggesting the increased content of C—O—C groups in the PVA structure after chemical modification. Comparing L-P at 120 ℃ and 150 ℃, it is found that the content of the C—O—C characteristic peak increases and —OH (peak at 3340 cm^{-1}) decreases. This strongly indicates that the higher temperature is conducive to chemical modification. Moreover, new peaks at 1 505 and 1 024 cm^{-1} in L-P were observed. These peaks are assigned to the aromatic skeletal vibration and C—H stretching of guaiacyl unit which belong to the characteristic peaks of lignin, respectively[6]. This observation partially affirms the successful chemical modification of L-P.

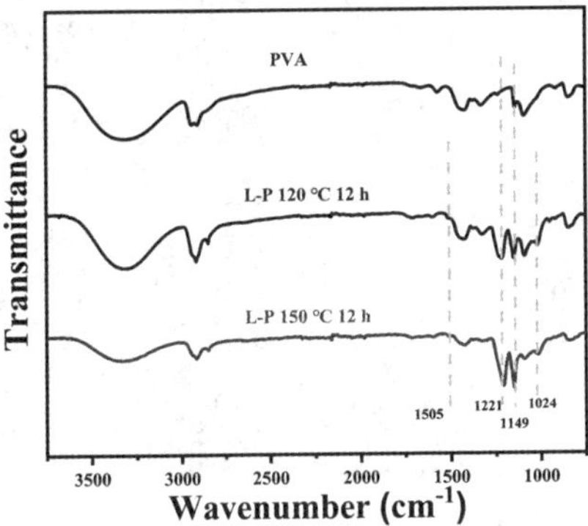

Fig. 1 FTIR spectra of PVA, L-P 120 ℃ 12 h, L-P 150 ℃ 12 h

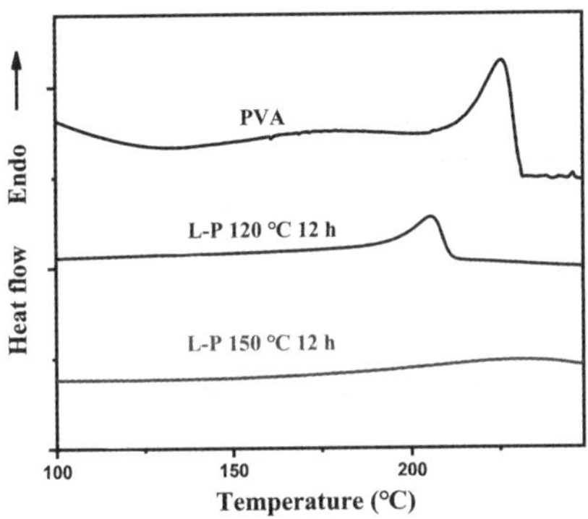

Fig. 2 DSC curves of PVA, L-P 120 ℃ 12 h, L-P 150 ℃ 12 h

Structural modification of PVA is revealed by Tm from DSC curves in Fig. 2. PVA showed a relatively large and sharp endothermic curve with a Tm range from 211 ℃ to 231 ℃. However, for L-P, the peak shifted toward the low temperature at 120 ℃ and the curve turn to wide and gentle at 150 ℃, respectively. This is because the introduction of amorphous lignin reduces the crystallinity of PVA and more lignin was grafted on PVA with the increase in temperature[7].

To indicate the structural difference of PVA before and after modification, the solubility of PVA in DI water has been measured in Fig. 3. In the water solvent, PVA is completely dissolved at 85 ℃. After structural modification, the solubility value is decreased to 72.11% and 22.19% for L-P at 120 ℃ and 150 ℃, respectively. This is related to the structure of grafted lignin which is approximately insoluble in water due to its abundant hydrophobic aromatic rings[8]. With a higher reaction temperature, modified PVA can have more hydrophobic groups for branched structures, therefore lower solubility in water.

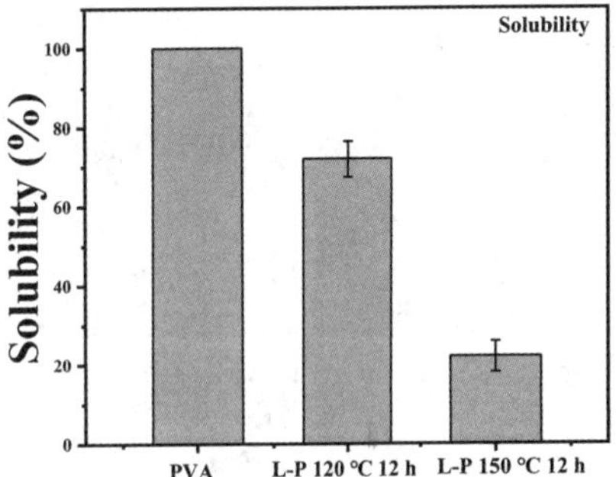

Fig. 3 Solubility of PVA, L-P 120 ℃ 12 h, L-P 150 ℃ 12 h

3.2 Effect of L-P on film mechanical properties

Fig. 4 (a) depicts the tensile strength of neat PVA film and other composite films. The mechanical strength of the composite films increases gradually with the increase of lignin or L-P content, and the maximum strength was nearly 3 times that of pure PVA film. The typical stress-strain curves are shown in Fig. 2 (b).

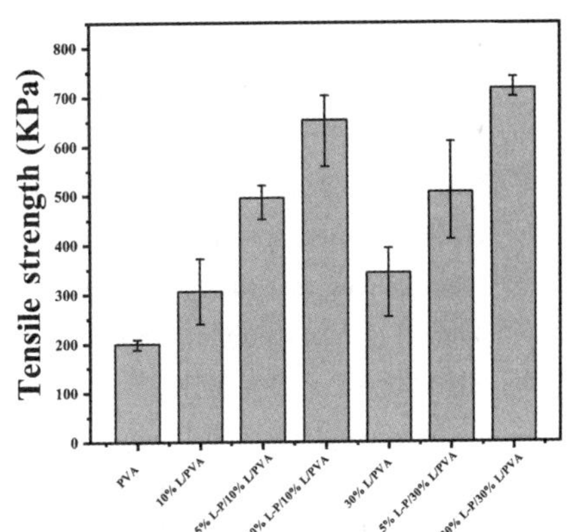

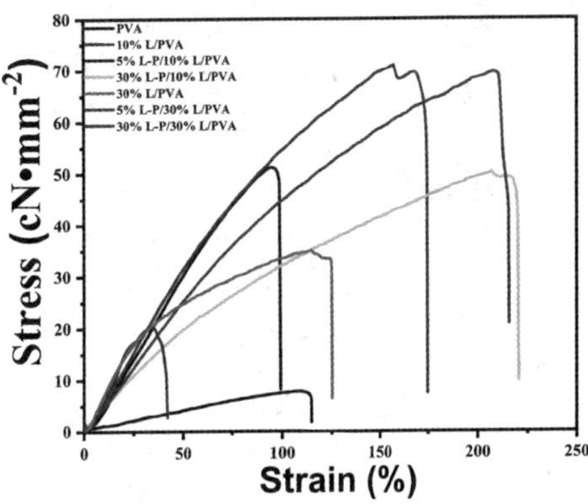

Fig. 4 (a) Tensile properties; (b) typical tensile stress-strain curves

We can find that the addition of lignin will improve the mechanical properties of the composite membrane, but it is not significant. By contrast, composite films containing L-P have larger tensile strength and strain. It is demonstrated that L-P can act as a bridge to improve the poor compatibility between PVA and lignin.

3.3 Effect of L-P on film macrostructure

To disclose the barrier mechanism of L-P/Lignin/PVA nanocomposite films, the dispersion of L-P in PVA matrix was studied by SEM, as shown in Fig. 5. The dispersion between the components of the composite films becomes better with the increase of L-P content. It is demonstrated with the conclusion of mechanical properties.

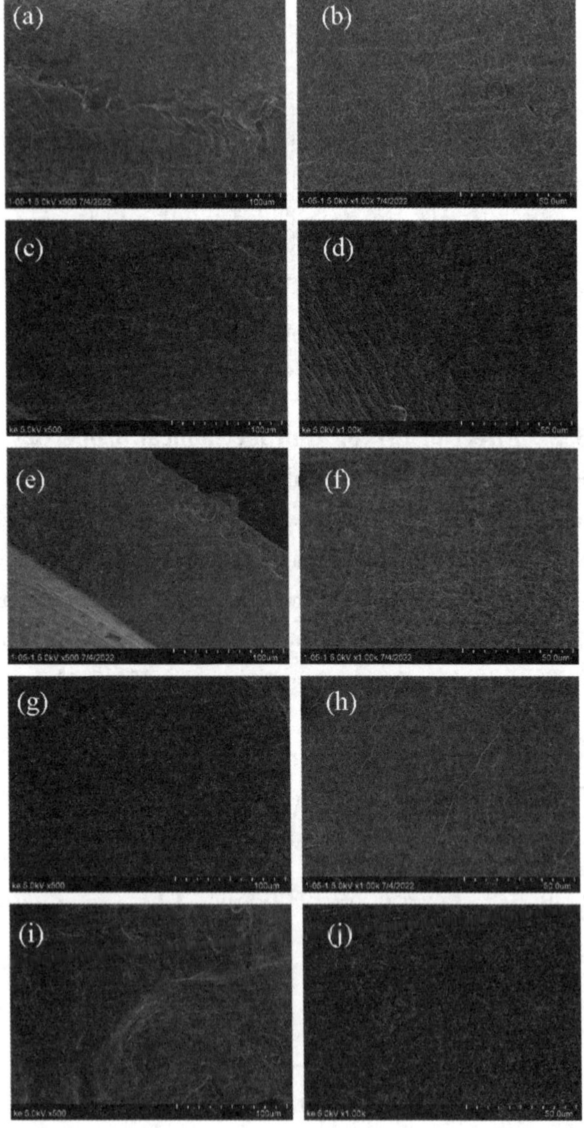

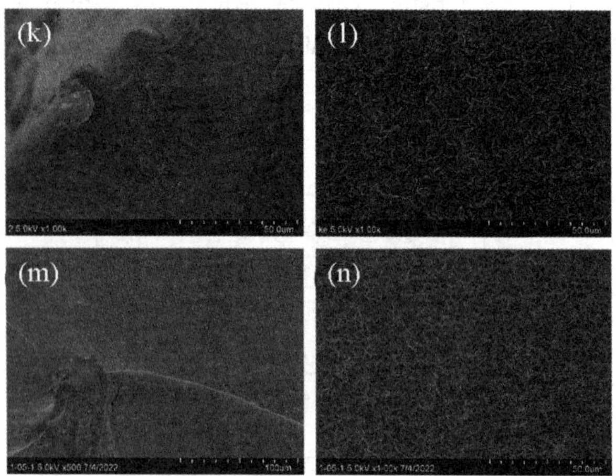

Fig. 5 SEM images of films of (a) and (b) pure PVA; (c) and (d) 10% Lignin/PVA; (e) and (f) 5% L-P/10% Lignin/PVA; (g) and (h) 30% L-P/10% Lignin/PVA; (i) and (j) 30% Lignin/PVA; (k) and (l) 5% L-P/30% Lignin/PVA; (m) and (n) 30% L-P/10% Lignin/PVA

4 Conclusions

In summary, L-P with excellent water resistance was successfully synthesized. L-P was introduced into the PVA film to construct a separate structure. The L-P/Lignin/PVA films exhibited much better mechanical performance than pure PVA, benefiting from the bridge function of L-P in the PVA matrix.

References

[1] RAO X, ZHOU Q, WEN Q, et al. High-performance and water resistant PVA-based films modified by air plasma treatment[J]. Membranes, 2022, 12(3): 551-558.

[2] GU J, LI J, SUN Y, et al. Progresses in the modification and application of poly (vinyl alcohol) membrane [J]. Chemical Industry and Engineering Progress, 2013, 32 (5): 1074-1080.

[3] SU L, XING Z H, WANG D, et al. Mechanical properties research and structural characterization of alkali lignin/poly(vinyl alcohol) reaction films[J]. Bioresources, 2013, 8(3): 3532-3543.

[4] HARARAK B, WINOTAPUN C, INYAI J, et al. Production of UV-shielded spherical lignin particles as multifunctional bio-additives for polyvinyl alcohol composite films [J]. Journal of Nanoparticle Research, 2021, 23(8): 1-15.

[5] ZHANG X, YANG M K, YUAN Q P, et al. Controlled preparation of corncob lignin nanoparticles and their size-dependent antioxidant properties: toward high value utilization of lignin[J]. ACS Sustainable Chemistry & Engineering, 2019, 7(20): 17166-17174.

[6] CHENG Y, LIN J, ZHENG Y, et al. High-performance

gel-spun poly(vinyl alcohol) fibers reinforced by organosolv lignin-graft-poly(acrylic acid)[J]. Industrial & Engineering Chemistry Research, 2022, 61(17): 6037-6051.

[7] PENG Z Q, CHEN D J. Study on the nonisothermal crystallization behavior of poly(vinyl alcohol)/attapulgite nanocomposites by DSC analysis[J]. Journal of Polymer Science Part B-Polymer Physics, 2006, 44(3): 534-540.

[8] LIN J, CHENG Y, LI Z, et al. Synthesis of modified lignin as an antiplasticizer for strengthening poly(vinyl alcohol)-lignin interactions toward quality gel-spun fibers[J]. ACS Applied Polymer Materials, 2022, 4(3): 1595-1607.

Study on Fatigue of Biceps Brachii in Movement Based on Characteristics of Muscle Thickness Detected by Fabric Sensor

YAO Fei[1], WANG Xi[1,2*], WANG Chuanling[1]

[1] *College of Information Science and Technology, Donghua University, Shanghai 201620, China*

[2] *Engineering Research Center of Digitized Textile & Apparel Technology, Ministry of Education, Donghua University, Shanghai 201620, China*

* *Corresponding author's email*: xiwang@ dhu. edu. cn

Abstract: With the rapid improvement of society and economics, health awareness has been increasingly focused on the whole world. Evaluating muscle fatigue and further preventing muscle damage have been arousing enormous interest from researchers. However, existing studies on muscle fatigue are mainly under static conditions and restrained in clinics or labs, other than in dynamic exercise. To address this problem, a flexible fabric sensor was utilized in this work to evaluate muscle fatigue by analyzing temporal muscle thickness. Contained inside elastic fabric and connected to a digital multimeter, variation of muscle thickness of biceps can be detected and recorded by the fabric sensors, along with the elbow joint angle and surface electromyography (EMG) on biceps by biometric goniometer and EMG sensor. Isotonic dumbbell curling test was designed for this work. Through data preprocess and combined analysis of temporal muscle thickness data in the time domain, it's found that unsupervised learning method such as hierarchical clustering can successfully separate fatigue and non-fatigue at an acceptable accuracy higher than 70%. Compared to conventional indicators collected by EMGs and angle signal indicators, the feasibility of this classification was verified. This work helps to open a new horizon of monitor muscle fatigue in dynamic motions, which is of huge practical significance for rehabilitation and simultaneously boosts the application of wearable technologies.

Keywords: Fabric sensor; Muscle fatigue; Classification; Unsupervised learning

1 Introduction

With the rapid expansion of society and the economy, health awareness has drawn increasing global interest. People are becoming increasingly concerned with their everyday health or exercise performance[1]. Nevertheless, excessive activity or exercise intensity will result in muscle fatigue and possibly tissue injury. Hence, evaluating muscle fatigue and further preventing muscle injury has been arousing enormous interest from researchers.

Muscle fatigue is the degenerating capacity of contraction or force production over time. Regarding studies on the evaluation of muscle fatigue, numerous domestic and abroad research teams have conducted research in this field. Since 1912, when Piper observed that the frequency of surface electromyography (EMG) decreased during fatiguing isometric contraction[2]. Cobb and Forbes then found that the frequency of EMG signals decreased and the amplitude increases for fatiguing muscles. For fatigue evaluation, EMG-related indicators, such as the mean frequency of the spectrum (MNF), have been utilized since the 1970s[3]. Recently, with the development of artificial intelligence technology and the improvement of algorithms, researchers have successfully developed an increasing number of fatigue identification systems. For one instance, Wang[4] uses the improved wavelet threshold and convolutional neural network-support vector machine (CNN-SVM) to categorize fatigue and non-fatigue EMGs.

However, existing monitoring of muscle fatigue is mainly under static conditions or mostly relies on EMG to evaluate muscle fatigue[2], restrained in well-conditioned labs only. Hence in this study, we tend to identify the muscle fatigue status using a novel perspective flexible fabric strain sensors, since it has been preliminary proved that muscle thickness can be an alternative index of muscle contraction[5]. To this end, this study employs a dynamic dumbbell curling test to gather data. And the method of hierarchical clustering is used to realize the evaluation of fatigue.

2 Experiments

In this paper, the fatigue research on the characteristics of bicep brachii thickness in dumb bell curling was studied, in which variation of muscle thickness of biceps, variation of muscle thickness of biceps and EMG on biceps were detected and recorded simultaneously. Details of experimental and data processing can be found in the following subsections.

2.1 Equipment

A fabric sensing belt was utilized, flexible fabric strain sensors (FSSs) was incorporated inside as sensing elements, which were carried and assembled on an elastic carrier made of highly-elastic cloth tape at a pre-tension of 5%[5]. The FSSs (purchased from AdvanPro limited, Hong Kong) were made by printing carbon nanocomposites on knitted fabrics. FSS is a type of soft, flexible, accurate, and deformable material with a wide working range and great durability. Silver-plated nylon wires were embedded as connection wires to the digital multimeter (DMM6500, Tektronix Inc., USA), forming a circuit to detect and store the continuous variations in biceps muscle thickness. Meanwhile, the elbow joint angle and surface EMG on biceps were recorded by wearable goniometer (BIOM-WS150, Biometric Ltd., USA), and biometric EMG sensor (BIOM-LE230, Biometric Ltd., USA) was also collected for Motion segmentation and reference.

2.2 Subjects

In this work, a total of 35 healthy male participants (age range: 20-22 years, height: 1.70-1.80 m, weight: 61-69 kg) were recruited. We opted to select male subjects with exercise habits, moderate arm girth (circumference>24 cm around biceps brachii belly), and comparable physical fitness in order to prevent the signal from being either too weak or exceed the range of FSS. All subjects were informed of the purpose, protocols methods as well as data collected of this test, and consents were obtained before test.

2.3 Experimental setup

Prior to the curling experiment, the participants were instructed to avoid strenuous training and ingestion of stimulant-containing foods, to avoid injury during the experiment or affect the experimental data. Proper warm up using small weights were performed by subjects before experiment to prevent damage, as well as help the subject to choose the proper load to achieve fatigue. Experimental setup can be seen in Fig. 1. Participants wore the test equipment after properly cleaning their skin with alcohol to reduce impedance and enhance signal strength. Finally, each participant was requested to select proper dumbbell weights according to their respective physical characteristics, to ensure the achievement of biceps fatigue after repetitions maximum (RM) of at least 8-12 flexion-extension cycles in the test. For an effect observation, the elbow joint angle shall cover a range wide than that between 10° and 120°.

In the work, the participants perform dumbbells curling with their left and right arms in groups ranging from 8 to 12 repetitions, while variation of muscle thickness of biceps brachii, the elbow joint angle and surface EMG on biceps belly were recorded simultaneously. The test ended when the participants subjectively felt tired and cannot perform one more flexion with a standard posture (means no change of posture or no enrollment of adjacent muscles).

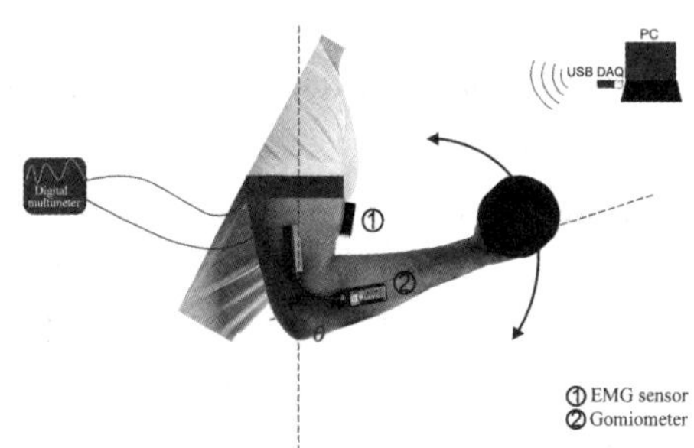

Fig. 1 Schematic diagram of dumbbell curling test

2.4 Data preprocessing

As the sensor's collected data cannot be utilized directly, it must be processed prior to further analysis.

To begin with, the elbow joint angle and surface Electromyography (EMG) on biceps were separated by DataLite software filter (Biometrics, Co., Ltd., USA). The collected data were then imported into MATLAB for preprocessing. The temporal muscle thickness signal was resampled by the interpolation function. This is because the time intervals of the muscle thickness measured by the digital multimeter were not identical and cannot be compared.

Then, a 10 Hz low pass filter was used to eliminate noises. The elbow joint angle signal and the muscle thickness signal were normalized, illustrated on a diagram, and then aligned. Then the two signals are segmented by the peak and valleys of curves. Due to the fact that the initial muscle was not immediately triggered or the participants immediately released the tension at the end of experiment, some of the data would be tailed and so on. In the same group of dumbbells curling tests, therefore, the angle range between 30° and 120° was selected, and the muscle thickness signal in this angle range was extracted as the data for fatigue recognition. Finally, the time span of each cycle of flexion-extension obtained was inconsistent, so the spline method in the interpolation function in MATLAB was used. Further, all rounds of observed samples were normalized by the maximum thickness (marked as 100%).

3 Characteristics of Muscle Thickness

After preprocessing, each cycle conveyed an observation of the whole data set, named one sample. We aimed to analyze the time domain characteristics of muscle thickness signal of biceps and explained it from a physiological perspective, trying to demonstrate that fatigue can be evaluated by muscle thickness signal.

3.1 Observation of time-domain characteristics of muscle thickness signal

If there is a significant difference between the first and last cycles, it was then reasonable to separate non-fatigue and fatigue cycles. Luckily as shown in Fig. 2, in the time domain, there was a clear distinction of muscle thickness signal characteristics between fatigue and non-fatigue. In the non-fatigue state, the rise and fall

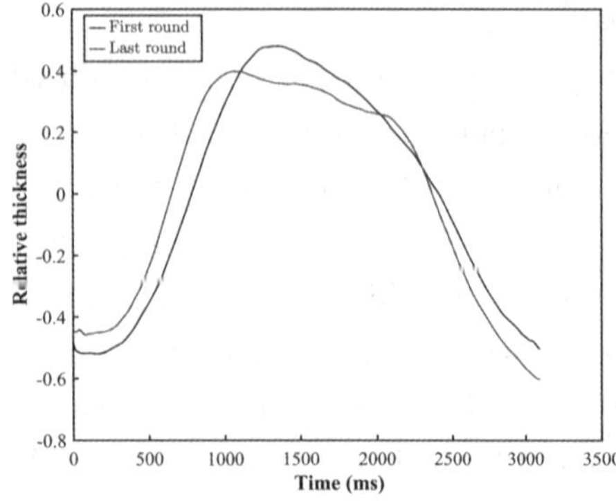

Fig. 2 Muscle thickness signal graphs of the first and last rounds

of muscle thickness signal was relatively regular, with obvious peak points. However, the rising and falling slope of the muscle thickness signal is visibly greater in the fatigue stage, showing that the rising and falling speed of the subjects is speeding up during fatigue. Simultaneously, we observed that in the fatigue state, the signal had no obvious peak point, i. e., a less-high plateau was observed for the time with maximum thickness.

3.2 Physiological mechanism

As a result of muscle tiredness, the nerve's inability to govern muscle contractions in a stable manner and the muscle's inability to get sufficient stimulation causes it to tremble[6]. From a physiological perspective, fatigued muscles cannot perform the same activity as non-fatigued muscles performing the same exercise.

Moreover, when the muscles are fatigued, the human body will recruit additional muscles to complete the action. Nevertheless, this will result in the instability of the action. It means that the subject will inadvertently speed up the movement to save energy, leading to accelerating of the rising phase. When the angle reached the top, it would not drop immediately, and the muscle signal would vibrate slightly. The reason is that the nerve was unable to not stably control the contraction of the fatiguing muscle, and the muscle would relax automatically. However, the nerve would constantly stimulate the muscle to continue contracting, and the muscle has to shift between the two states of contraction and relaxation, unable to guarantee that the movement can be maintained.

4 Fatigue Recognition

4.1 Fatigue state recognition

According to the above physiological mechanism, muscle thickness is promising to help classify non-fatigue and fatigue status during curling. In this paper, to preliminary dig into the difference in muscle thickness signals between non-fatigue and fatigue status, hierarchical clustering is used to recognize the muscle fatigue of all the collected rounds[7]. A total of 632 samples/observations of temporal muscle thickness data were collected, and the clustering was carried out according to the same group of dumbbells curling test. The average method in the hierarchical clustering function in MATLAB was used[8], due to the small amount of data, and

Euclidean distance is selected for distance calculation.

4.2 Recognition result

As can be confirmed from the dumbbells curling test, the first two rounds of the curling were generally in the state of non-fatigue, while the last two rounds are generally observed in a state of fatigue in the experiment. In fact, the testing protocols of using 8-12 RM ensure this observation. Hence, to evaluate the accuracy of hierarchical clustering for biceps fatigue, we compared the division of non-fatigue and fatigue by hierarchical clustering to the above observation during tests.

As per the selection of clustering categories for hierarchical clustering, in addition to being divided into fatigue state and non-fatigue state. Al-Mulla et al. [9] proposed an additional third kind of fatigue: transition-to-fatigue, which can be defined as the third state of transition from non-fatigue state to fatigue state. The third type of fatigue research highlights that it is insufficient to attempt to recognize fatigue level based on fatigue state and non-fatigue state, as the predictability of the measurement is lacking. As a result, the beginning of transitional fatigue is detected, followed by a gradual process until the beginning of fatigue. On the other hand, as shown in Fig. 3, fatigue status can be easily recognized because it was separated as one of the three distinct groups with a significant distance from each other.

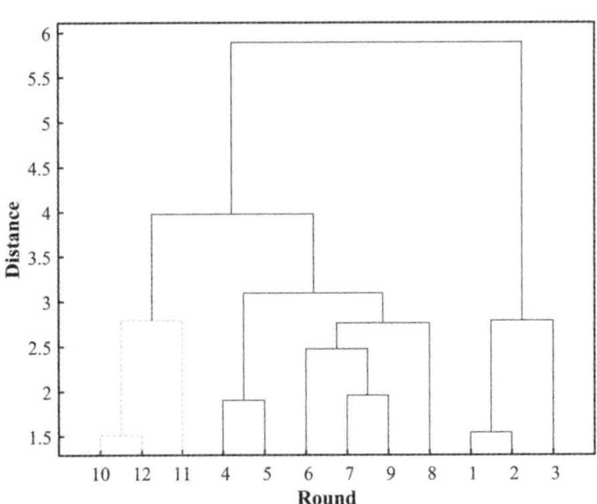

Fig. 3 Dendrograms of a group of dumbbells curling test

The classification results are shown in Fig. 4, fatigue and non-fatigue rounds are comparable, and Transition-to-Fatigue rounds make up the majority, which is consistent with physiological reality. The specific results are summarized in Table1. And hierarchical clustering separates fatigue state from a non-fatigue state with an accuracy of 78.69%.

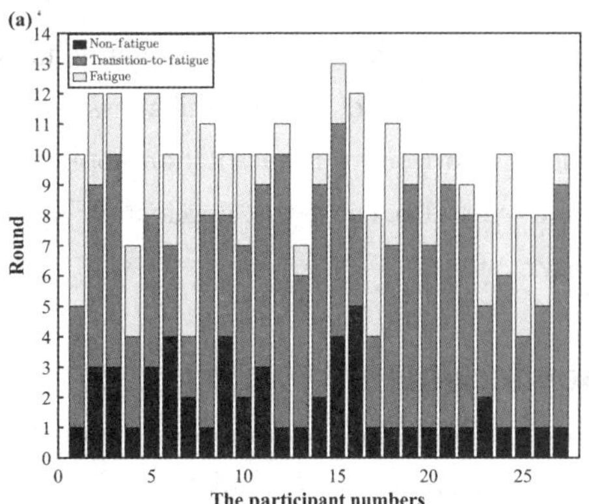

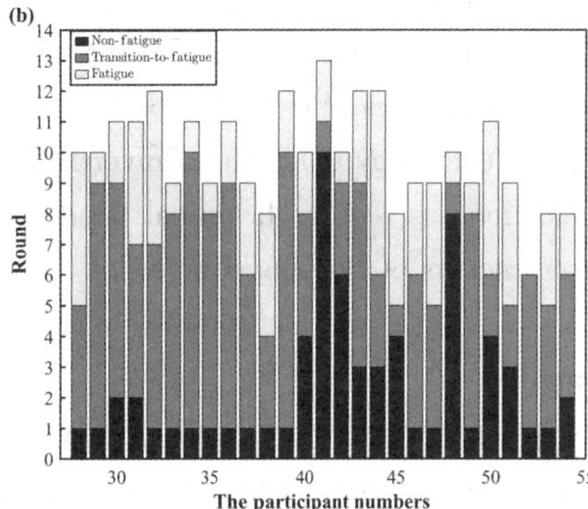

Fig. 4 Fatigue status classification diagram

Tab. 1 Fatigue status classification table

Muscle state	Non-fatigue	Transition-to-fatigue	Fatigue
Number of rounds	131	301	200

5 Conclusions

In this paper, a novel approach to identifying the fatigue status of biceps brachii is proposed based on the characteristics of muscle thickness. With a smart sensing belt with flexible fabric strain, it's feasible to accurately observe the variation of the thickness of the biceps, along with the elbow joint angle and surface EMG on biceps by biometric goniometer and EMG sensor. The observation of muscle thickness signal in time domain is realized through the data, from which the comparison of the characteristics between the signal of non-fatigue and fatigued were observed and understood with physiology. After that, hierarchical clustering was implemented and the fatigue and non-fatigue round of curling can be separated successfully with an accuracy of as high as 78.69%, when compared to conventional indicators by collected EMGs and angle signal indicators, generally verifying the feasibility of this method.

Hence, it's suggested that wearable devices powered by flexible sensors can be utilized to identify muscle fatigue through analysis of muscle thickness, which may open spacious horizon for rehabilitation applications and fitness and can also boost the application of wearable technology.

Acknowledgments

This research was funded by the National Natural Science Foundation of China (Nos. 12002085, 51603039), sponsored by Shanghai Pujiang Program, and supported by the Fundamental Research Funds for the Central Universities, the Key Laboratory of Textile Science and Technology (Donghua University), Ministry of Education, as well as the Initial Research Funds for Young Teachers of Donghua University.

References

[1] DEVRIES H A, MORITANI T, NAGATA A, et al. The relation between critical power and neuromuscular fatigue as estimated from electromyographic data[J]. Ergonomics, 1982, 25(9): 783-791.

[2] CIFREK M, MEDVED V, TONKOVIC S, et al. Surface emg based muscle fatigue evaluation in biomechanics[J]. Clin Biomech (Bristol, Avon), 2009, 24(4): 327-340.

[3] KWATNY E, THOMAS D H, KWATNY H G. An application of signal processing techniques to the study of myoelectric signals[J]. IEEE Transactions on Biomedical Engineering, 1970(4): 303-313.

[4] WANG J, SUN Y, SUN S. Recognition of muscle fatigue status based on improved wavelet threshold and cnn-svm[J]. IEEE Access, 2020, 8: 207914-207922.

[5] WANG X, TAO X M, SO R C, et al. Monitoring elbow isometric contraction by novel wearable fabric sensing device[J]. Smart Materials and Structures, 2016, 25(12): 125022.

[6] GORDON G, HOLBOURN A. The sounds from single motor units in a contracting muscle[J]. The Journal of Physiology, 1948, 107(4): 456.

[7] MADHULATHA T S. An overview on clustering methods[J]. arXiv preprint arXiv: 12051117, 2012.

[8] SOKAL R R. A statistical method for evaluating systematic relationships[J]. Univ Kansas, Sci Bull, 1958, 38: 1409-1438.

[9] AL-MULLA M R, SEPULVEDA F, COLLEY M, et al. Statistical class separation using semg features towards automated muscle fatigue detection and prediction[C]. Tianjin: 2009 2nd International Congress on Image and Signal Processing: IEEE, 2009-10-17.

Facile Fabrication of PP/PLA Composite Aerogel from Waste PP Melt-Blown Materials for Oil Absorption and Thermal Insulation Application

SU Yazhou[1,2], ZHAO Huawang[1,2], SUN Xiaoxia[1,2*]

[1] *Key Laboratory of Science & Technology of Eco-Textile, Ministry of Education, Donghua University, Shanghai 201620, China*

[2] *Department of Technical Textiles, College of Textiles, Donghua University, Shanghai 201620, China*

* *Corresponding author's email*: xxsun@ dhu. edu. cn

Abstract: The demand for polypropylene (PP) melt-blown materials has dramatically increased due to pandemic. It has caused serious environmental problems because of the lack of effective treatment for the waste PP melt-blown materials. In this work, we present an easy, practical, and expandable method to manufacture hierarchical porous PP/PLA composite aerogel for oil spill cleaning applications and thermal insulation applications. The composite aerogel with a three-dimensional porous structure is developed by combining recycled PP microfibers with polylactic acid (PLA) aerogel matrix by physical blending and thermally induced phase separation with the assistance of water. The morphology structure, oil absorption properties, and thermal insulation properties are tested and characterized. The composite aerogel exhibits a high porosity of 96.0%–97.1%, which leads to a good absorption capacity of <15.1 g/g for different oil and solvents. Moreover, the composite aerogels show an exceedingly low thermal conductivity of 0.036–0.039 W/(m · K), indicating good insulation properties. In particular, the addition of recycled PP microfibers enhances the toughness of the composite aerogel. The strategy developed in this study demonstrates the huge potential for applications in the field of recycling for the waste PP melt-blown materials.

Keywords: Waste PP melt-blown materials; Recycled PP microfibers; Porous materials; Oil absorption; Thermal insulation

1 Introduction

Except for the discarded masks, the production process of masks also produces a large amount of waste PP melt-blown materials. Usually, due to the uneven fiber distribution, both sides of the melt-blown materials must be chopped off during the fabrication process, which is usually discarded as trash and tends to be another waste resource[1].

At the present, the treatment of waste PP melt-blown materials, includes landfill, incineration, recycling, and other methods. However, the aforementioned solutions continue to have several drawbacks[2]. Thus, developing ecological and economical recycling processes for waste PP melt-blown materials is critical.

Nowadays, oily wastewater discharge and oil spills pose an immediate hazard to the natural environment and public health today. As a result, dealing with oily wastewater is critical for environmental preservation and resource conservation[3]. For its accessibility, high efficiency, and low cost, absorbent for physical absorption is now regarded as one of the most viable technologies for oil-water separation[4]. In addition, rapid industrial development and population growth have caused energy wastage all over the world. In particular, effective insulation is essential to maximize the energy efficiency of a building[5]. Three-dimensional porous materials have great potential for insulation applications.

In this work, the PP/PLA composite aerogels for oil absorption and thermal insulation applications were developed from the recycled PP melt-blown material and PLA using freeze-drying processes. The morphology, thermal properties, and absorption properties of composite aerogels were investigated comprehensively. This method provides the materials more porous structure. This study not only revealed a potential utilization of waste PP melt-blown materials with high added value but also partially solved the problem of environmental pollution and waste of resources.

2 Experiments

2.1 Materials

Waste PP melt-blown materials were received from Tianfeng Technology Co., Ltd. (Shandong, China). PLA particle (Mw = 80 000) was purchased from Wendon (Shanghai) Chemical Co., Ltd. (Shanghai, China). 1,4-Dioxane was purchased from Shanghai Sinopharm Chemical Reagent Co., Ltd. (Shanghai, China). Sudan I, Methylene blue hydrate and hexane were purchased from Shanghai Titan Technology Co., Ltd. (Shanghai, China). Sunflower oil and peanut oil were purchased from the market. Other reagents were used as received.

2.2 Fabrication of the PP/PLA composite aerogel

Firstly, 0.8 mL of deionized water was added to a mixture of PLA and 1,4-dioxane (0.3 g/10 mL dioxane). Then the mixture was heated at 60 ℃ for 1 h under constant magnetic stirring to obtain a homogeneous solution and then cooled to room temperature. Waste PP melt-blown materials were cut into small pieces and then dispersed into recycled PP microfibers by a blender. Afterward, the different quantities of recycled PP microfibers (0%, 5%, 10%, 15%, 20%, and 25% of PLA weight) were dispersed into the above solution under magnetic stirring to obtain a uniform composite dispersion. The obtained mixture was added into a mold, frozen at −16 ℃ for 12 h, and then transferred to a freeze-drying vessel at −56 ℃ for 24 h. For simplicity, the specimens containing 0%, 5%, 10%, 15%, 20%, and 25% recycled PP microfibers were indicated as F0, F5, F10, F15, F20, and F25. To further improve the mechanical properties of the material, a layer of melt-blown fabric is laminated on each of the upper and lower surfaces of the material when the mixture is poured into the mold. For example, a sample of F20 laminated with melt-blown fabric is called F20/PP_2.

2.3 The density and porosity of PP/PLA composite aerogel

The bulk density of aerogel was calculated by Eq. (1):

$$\rho = \frac{m}{V} \tag{1}$$

where m was the mass (g) and V was the volume (cm^3) of the aerogel. The porosity φ (%) of the aerogel could be calculated according to Eqs. (2) and (3):

$$\rho = \left(1 - \frac{\rho_1}{\rho_2}\right) \times 100\% \tag{2}$$

$$\rho_2 = \frac{m_{PP} + m_{PLA}}{\left(\frac{m_{PP}}{\rho_{PP}} + \frac{m_{PLA}}{\rho_{PLA}}\right)} \tag{3}$$

where ρ_1 and ρ_2 were the density of aerogel and the polymer scaffold respectively. The m_{PP} and m_{PLA} were the mass (g) of recycled PP fiber and PLA in the aerogel, respectively. ρ_{PP} and ρ_{PLA} were the density of recycled PP fiber and PLA respectively.

2.4 The morphology of PP/PLA composite aerogel

The morphology of PP/PLA composite aerogel was observed by the field emission spectroscopy transmission electron microscope (FESEM, S-4800, Hitachi, Japan). A thin layer of gold was sputtered on the surface of the samples before the FESEM procedure was performed.

2.5 The thermal stability analysis of PP/PLA composite aerogel

The thermal stability of the PP/PLA composite aerogel was demonstrated using a thermogravimetric analyzer (TGA 4000). The test was performed by progressively increasing the temperature in the air from 30 ℃ to 800 ℃ at a rate of 15 ℃ per minute and recording the mass of the sample at each temperature.

2.6 The liquid absorption capacity of PP/PLA composite aerogel

The liquid absorption capacity of the PP/PLA composite aerogel was examined by performing an absorption test at room temperature. The weighted samples were firstly immersed in oil or organic solvent for 5 min, then removed and dripped for 60 s, and weighted again. Each sample was tested at least three times and averaged. η (g/g) was the absorption capacity of the composite aerogels, which was calculated as:

$$\eta = \frac{m_1 - m_0}{m_0} \tag{4}$$

where m_0 (g) and m_1 (g) were the weights of the samples before and after the oil absorption test, respectively.

2.7 The thermal properties of PP/PLA composite aerogel

A TCI thermal conductivity analyzer was used to provide the thermal characterization of the thermal conductivities of the composite aerogels. Thermal images were taken by a thermal imager to demonstrate the thermal insulation effect of composite aerogels.

3 Results and Discussion

The preparation process of the composite aerogel is shown in Fig. 1 (a). Making use of physical blending and thermally induced phase separation process[6], the composite aerogels are formed without extensive shrinkage and structural damage (Fig. 1 (b)). Moreover, as can be seen in Fig. 1 (d), the composite aerogel shows a highly porous structure, which is critical for its applications. Fig. 1 (c) shows the porosity and density of the composite aerogels. Overall, the obtained composite aerogels show an extremely low density of 0.039–0.046 g/cm^3 and a high porosity level from 96.0% to 97.1%, predicting their potential for oily wastewater treatment and thermal insulation.

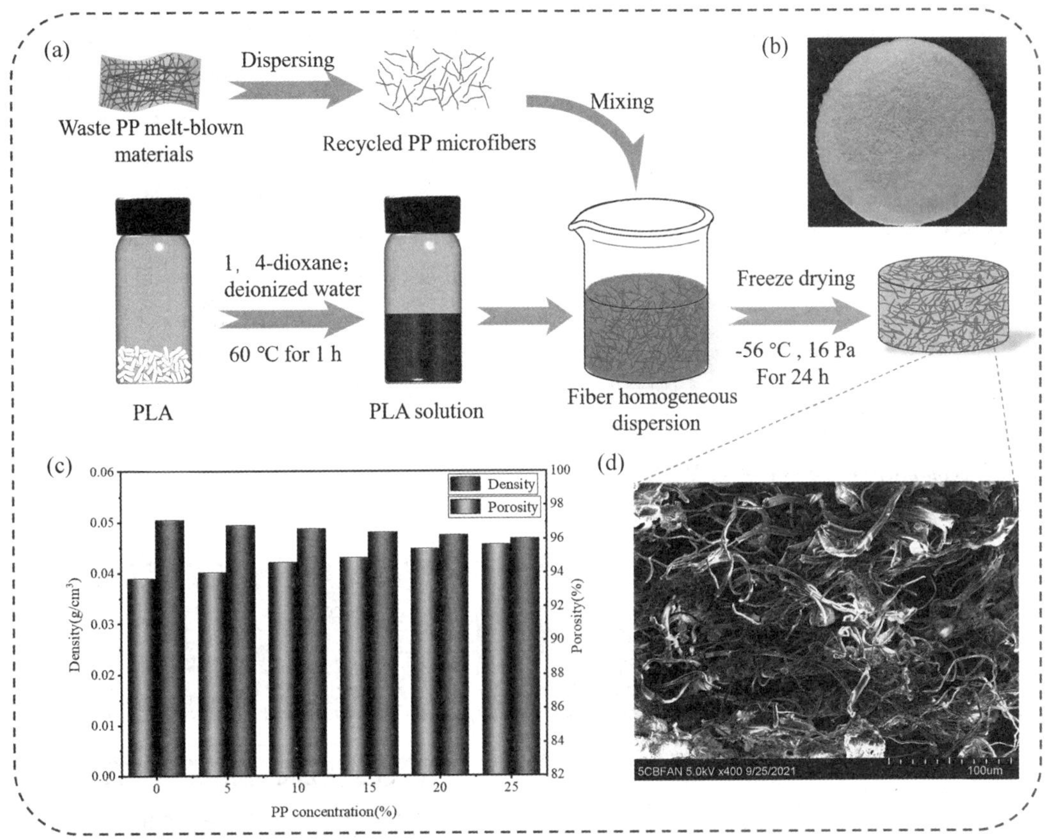

Fig. 1 (a) The preparation process of the PP/PLA composite aerogel; (b) photograph of the composite aerogel; (c) density and porosity of the PP/PLA composite aerogels in different PP concentrations; (d) SEM of the PP/PLA composite aerogel

3.1 Morphology of the PP/PLA composite aerogel

The morphology of aerogels is shown in Fig. 2. Therein, the recycled PP microfibers are arranged without apparent organization and are firmly bonded by PLA to form a porous structure. When PP fiber concentration increases from 0 wt. % to 25%, the aerogels become denser because of more and more PP fibers in the materials.

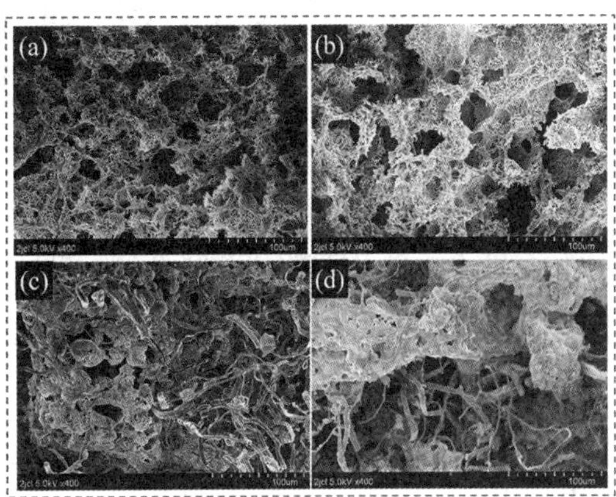

Fig. 2 SEM images of the aerogels with different recycled PP microfiber concentrations of (a) 0%; (b) 5%; (c) 15%; (d) 25%

3.2 Thermal stability of the PP/PLA composite aerogel

The thermal stabilities of the composite aerogels are shown in Fig. 3. The mass of the composite aerogels remained stable within a temperature range of 30 - 350 ℃. At the next temperature step of 350-500 ℃, all samples experienced thermal deterioration. Furthermore, the curve for sample F20 reveals the presence of two phases of thermal deterioration at 350 ℃ and 425 ℃, which correspond to the PP and PLA thermal degradation features, respectively, indicating the combination of two kinds of materials.

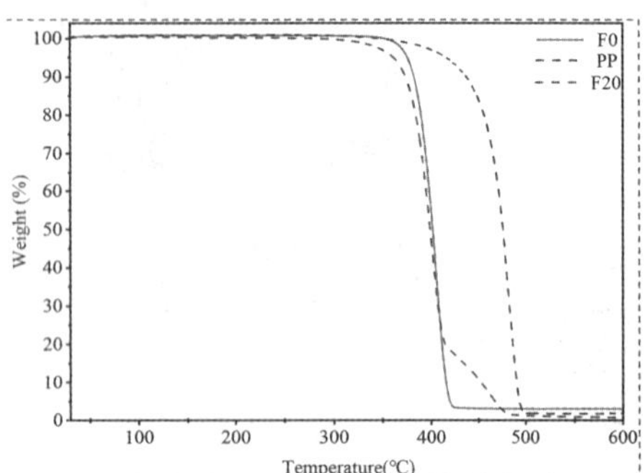

Fig. 3 Thermogravimetric curves of PP fiber and the composite aerogels

3.3 The oil absorption performance of the PP/PLA composite aerogel

Fig. 4 (a) shows the process of the oil slick separation from the water. After immersing the composite aerogel into the water/oil mixture, the oil (dyed with Sudan I) on the water surface was quickly and completely absorbed, while the water was not absorbed by the aerogel. In addition, the composite aerogel can be used to collect the oil underwater, due to its good hydrophobicity, the composite aerogel can selectively absorb the organic solvent like trichloromethane (dyed with Sudan I) from the bottom of water without absorbing any water (Fig. 4 (b)), thus realize the underwater oil-water separation. Fig. 4 (c) illustrates the oil absorption of composite aerogels with different fiber concentrations. It can be seen that with the increase of recycled PP microfiber concentration, the oil absorption capacity showed a downward trend. It is because of the increase of the recycled PP microfibers reduces the porosity of the composite aerogel. The absorption capacity of composite aerogels for different oils and organic solvents is shown in Fig. 4 (d).

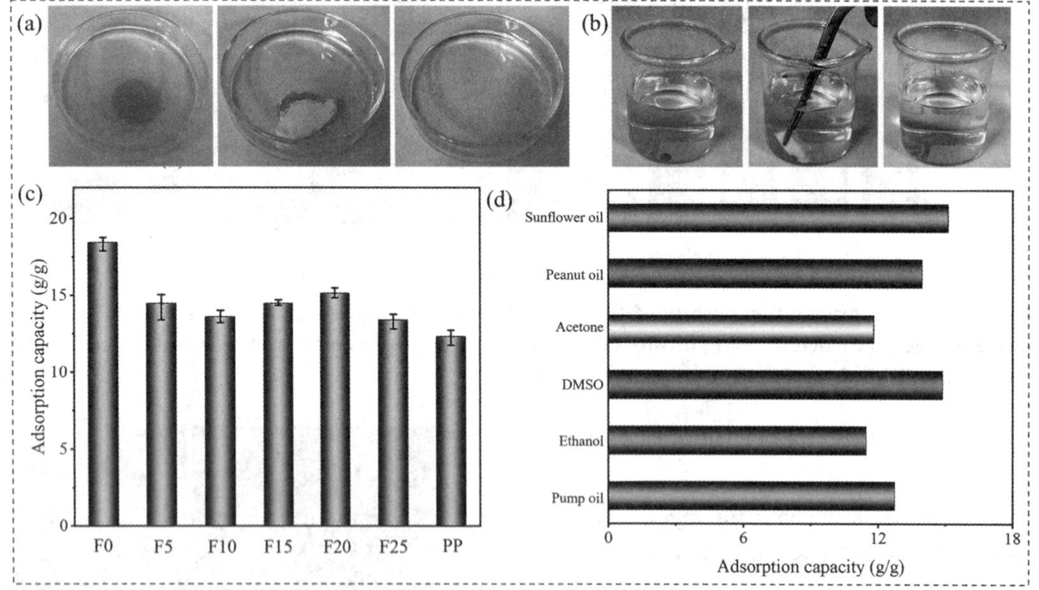

Fig. 4 The removal process of (a) sunflower oil and (b) trichloromethane with composite aerogels; (c) the absorption capacity of composite aerogels with different fiber concentrations; (d) the absorption capacity of composite aerogels for different oils and organic solvents

3.4 Thermal properties of the PP/PLA composite aerogel

The composite aerogel exhibits an ultra-low thermal conductivity of 0.036-0.039 W/(m · K), which is lower than single-layer melt-blown fabric (PP * 1) and the eight-layer melt-blown fabric (PP * 8). Fig. 5 (b) shows the infrared image of composite aerogel at different heat source temperatures. A temperature difference of 3-5 ℃ between the aerogel surface and the heat source, demonstrates its excellent thermal insulation effect.

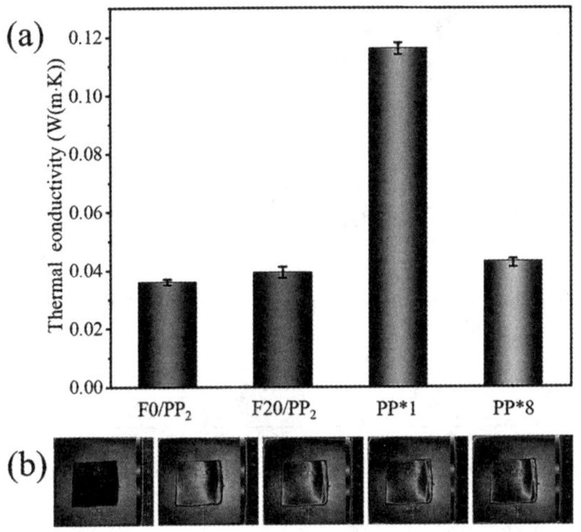

Fig. 5 (a) The thermal conductivities of different samples; (b) infrared images of composite aerogel

4 Conclusions

The PP/PLA composite aerogels have successfully developed from the waste PP melt-blown materials and PLA via the freeze-drying process. Due to their hydrophobicity, the composite aerogels could selectively adsorb various oils and organic solvents from water-oil mixtures. In addition, the composite aerogels exhibited a three-dimensional porous network structure, leading to a good potential application in the field of oil spill cleaning. In particular, the composite aerogels show an exceedingly low thermal conductivity of 0.036 – 0.039 W/(m · K), which indicates good insulation properties. The addition of recycled PP microfibers enhances the toughness of the composite aerogel. The strategy developed in this study is simple and available for large-scale manufacture, which could partially address the problems of environmental pollution and resource waste as well.

References

[1] YAN H. Development and application of melt blown fabric rim leftover[J]. Technical Textiles, 1998(2): 18–21.

[2] CHEN H M, DONG X, ZHAO Y, et al. Recycling and chemical upcycling of waste disposable medical masks[J]. Acta Polymerica Sinica, 2020, 51(12): 1295–1306.

[3] ADETUNJI A I, OLANIRAN A O. Treatment of industrial oily wastewater by advanced technologies: a review[J]. Applied Water Science, 2021, 11(6): 1–19.

[4] LU P, WANG X, PENG S, et al. Research progress in oil-absorption materials[J]. Modern Chemical Industry, 2019, 39(4): 22–26.

[5] PEREZ-LOMBARD L, ORTIZ J, POUT C. A review on buildings energy consumption information[J]. Energy and Buildings, 2008, 40(3): 394–398.

[6] WANG X, PAN Y, LIU X, et al. Facile fabrication of superhydrophobic and eco-Friendly poly(lactic acid) foam for oil-water separation via skin peeling[J]. ACS Applied Materials & Interfaces, 2019, 11(15): 14362–14367.

Shape Memory Properties of N-Octylamine/M-Xylylenediamine Epoxy Resin Toughening System

TANG Jia[1], JIANG Jinhua[1,2*], CHEN Nanliang[1,2]

[1] *College of Textiles, Donghua University, Shanghai 201620, China*

[2] *Engineering Research Center of Technical Textiles, Ministry of Education, Donghua University, Shanghai 201620, China*

[*] *Corresponding author's email*: jiangjinhua@ dhu. edu. cn

Abstract: The new resin-based shape memory composite material is regarded as an ideal material for the spatial expansion structure. This kind of epoxy resin-based shape memory composite has excellent properties such as high strength, high modulus, low specific density, and good weather resistance. However, the shortcoming of epoxy resin is brittle and easy to crack, which affects its wide application prospects to a large extent. This study prepared n-octylamine/m-xylylenediamine epoxy resin toughening system with shape memory properties, and explored the effects of different ratios of curing agent to crosslinking agent at the glass transition temperature and shape memory properties of the epoxy resin system. The results show that with the increase of m-xylylenediamine content, the crosslinking density of epoxy resin toughened system increases, the flexibility of the molecular chain becomes worse, and the glass transition temperature increases. With the decrease of n-octylamine content, the shape fixity ratio and shape recovery time of the epoxy resin toughening system increase, and the effect on the shape recovery ratio is not obvious, which can reach more than 98%.

Keywords: Shape memory; Epoxy resin; Glass transition temperature

1 Introduction

With the current rapid development of science and technology, the competition between countries is becoming more and more fierce. Due to the limited resources on the earth, the big countries are now focusing more on the competition for space resources. Therefore, the construction of large-scale space facilities required for the exploration of space resources becomes very important for a strong country, and the space deployment structure technology provides a strong support platform for the application of these large-scale space facilities[1-2]. At the end of the 20th century, scientists discovered a new type of resin-based thermosetting shape memory composite material with excellent properties such as high strength, high modulus, light specific density and good weather resistance. This new material is regarded as an ideal material for deployable space structures. Mainly because it is prone to mechanical deformation when it is subjected to external force, and at the same time, it can store strain during the mechanical deformation process, and these stored strains will recover their shape when subjected to certain external stimuli, so this material can almost meet the requirements of deployable structural materials for aerospace devices[3-4]. This new material is mainly composed of shape memory polymer and high-strength and high modulus fiber fabric. Therefore, some high-strength and light-weight deployable structural parts can be prepared by using this new material. These deployable structural parts can not only be combined with the space inflatable deployable structure as a rigid inflatable deployable structure, but also be used alone in the space deployable structure.

In recent years, the application of these new smart materials has attracted extensive attention of scholars, especially in the fields of aerospace and aviation, becoming a research hotspot of national defense departments. As a research institute affiliated to NASA, Composite Technology Development, Inc. has developed TEMBO® series epoxy resin-based thermosetting shape memory composite materials and successfully used them in space deployment structural components. However, there is still a lack of research in this field in China[5-7].

Although epoxy resin-based shape memory composites have excellent properties such as high strength, high modulus, low specific density and good weather resistance, their wide application prospects are greatly affected by the brittleness and easy cracking of epoxy resin materials. Therefore, it is extremely important to study the toughening and modification methods of epoxy resin. At present, there are many methods to reduce the brittleness of epoxy resin materials, but the more effective method is to blend them with elastomer or use a flexible curing agent for toughening, which plays an important role in the application and development of epoxy resin matrix shape memory composites in space deployable structures[8-10]. This study prepared n-octylamine/m-xylylenediamine epoxy resin (EP-OA/MXDA) toughening system with shape memory properties, and explored the effects of different ratios of OA and MXDA on the glass transition temperature and shape memory properties of the epoxy resin system.

2 Experiments

2.1 Materials

Epoxy E51 was provided by Qicai New Material Technology Company Limited (Jiangsu, China); OA and MXDA were provided by Shanghai Macklin Biochemical Company Limited (Shanghai, China).

2.2 Material ratio

According to the theoretical dosage formula of epoxy resin curing agent (1), the theoretical dosages of OA and MXDA required by 6 different proportions of the EP-OA/MXDA toughening system were calculated respectively. The formulas are shown in Tab. 1.

$$m = A \times \left(\frac{M}{n_H}\right) \times K \tag{1}$$

In Eq. (1), m is the mass of the curing agent required for 100 g epoxy resin (g); A is the correction coefficient; M is the molecular weight of the amine; n_H is the number of active hydrogens of the amine curing agent; K is the epoxy value.

Tab. 1 Formula of EP-OA/MXDA toughening system

System	EP(g)	OA(g)	MXDA(g)
EP-100OA	10	3.290	0
EP-80OA/20MXDA	10	2.632	0.346
EP-60OA/40MXDA	10	1.974	0.694
EP-40OA/60MXDA	10	1.316	1.040
EP-20OA/80MXDA	10	0.658	1.388
EP-100MXDA	10	0	1.734

2.3 Composite manufacture

The epoxy resin E51 was weighed into the beaker, and the beaker was placed in an oven for constant heating at 80 ℃ for 20 min. Then OA and MXDA were added to the beaker and stirred well. The mixture was evacuated in a vacuum oven for 5 min to remove air bubbles. Finally, the mixture was poured into the mold coated with mold release agent, and then placed in an oven at 80 ℃ for heating and curing for 3 h. After the sample was fully cured, mold release treatment was performed.

2.4 Differential scanning calorimetry

The glass transition temperature (T_g) of EP-OA/MXDA toughening system was measured by differential scanning calorimetry (DSC). In this paper, DSC8500 produced by PerkinElmer company is used to test the prepared samples, and the corresponding glass transition temperature is obtained. The test condition is that the heating rate is 10 ℃/min, the heating range is 0–140 ℃, and the atmosphere condition is nitrogen.

2.5 Shape memory performance test

The shape memory experiment was performed on the prepared sample at 20 ℃ higher than the glass transition temperature of the sample. The size of the sample was 80 mm×12 mm×2 mm. Firstly, the sample was heated at the corresponding temperature. When the sample became soft, it was folded and bent into a U-shape. Then the sample was cooled and fixed into a folded shape. Finally, the sample was heated to the corresponding temperature for recovery experiment.

Shape fixity ratio: in a high temperature environment, the ability of the shape memory polymer is deformed by force and then cooled to below T_g to fix the shape. The calculation formula is shown in Eq. (2).

$$R_f = \frac{\theta_f}{180°} \times 100\% \tag{2}$$

In Eq. (2), R_f is shape fixity ratio(%); θ_f is angle of the sample after cooling and fixing (°).

Shape recovery ratio: the ability of a polymer with a fixed shape after cooling to return to its original shape after being heated again in a high temperature environment (above T_g). The calculation formula is shown in Eq. (3).

$$R_r = \frac{\theta_f - \theta_r}{\theta_f} \times 100\% \tag{3}$$

In Eq. (3), R_r is shape recovery ratio(%); θ_f is angle after cooling and fixing of sample (°); θ_r is remaining angle after sample recovery (°).

Shape recovery rate: record the time required for the sample to recover from the U-shape to the original shape, so as to compare the recovery rate of the system.

3 Results and Discussion

3.1 DSC

In order to test the shape memory performance of the toughening system, the glass transition temperature of the system must be predicted. Therefore, differential scanning calorimetry tests were carried out on six different ratios of EP-OA/MXDA toughening systems. Fig. 1 shows the six different DSC curve of the ratio of EP-OA/MXDA toughening system. It can be seen from Fig. 1 that with the increase of MXDA content, the glass transition temperature of the system shows a gradually increasing trend. This is mainly because with the increase of MXDA content, the crosslinking density of the system increases, the molecular chains are packed more tightly, and the movement of molecular segments is hindered. At the same time, due to the decrease of OA content, the flexible segment in the system decreases, and the flexibility of the molecular segment becomes poor. Thus the glass transition temperature increases.

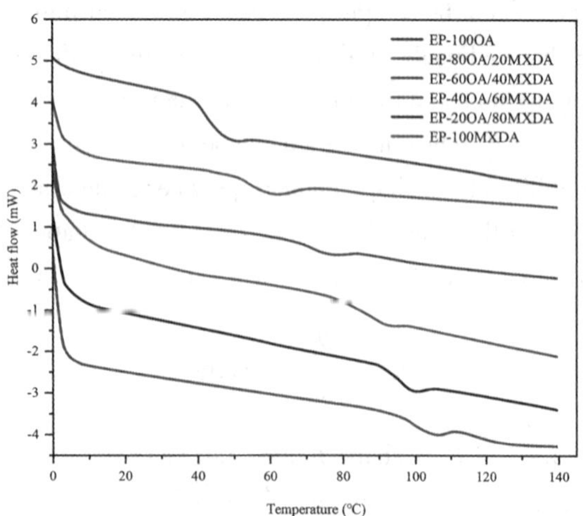

Fig. 1 DSC curves of EP-OA/MXDA toughening systems

Tab. 2 shows the approximate glass transition temperatures of 6 different ratios of EP-OA/MXDA toughening systems, and the glass transition temperatures of the systems range from 42.3 ℃ to 100 ℃.

Tab. 2 Glass transition temperatures of EP-OA/MXDA toughening systems

System	EP-100OA	EP-80OA/20MXDA	EP-60OA/40MXDA	EP-40OA/60MXDA	EP-20OA/80MXDA	EP-100 MXDA
T_g (℃)	42.3	55.4	70.8	83.5	93.8	100.0

3.2 Shape fixity ratio

For better shape memory materials, the shape fixity ratio must reach more than 98%, that is, the spring-back angle after the material is folded and bent into a U shape, cooled and fixed must be less than 4°. It can be seen from Fig. 2 that with the decrease of OA content and the increase of MXDA content, the shape fixity ratio of the system gradually increases.

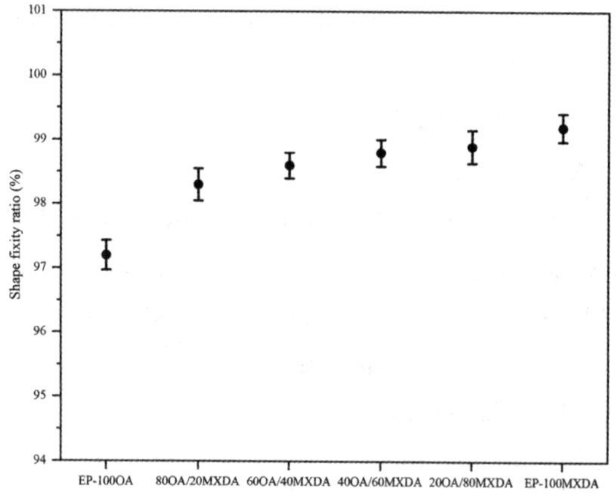

Fig. 2 Shape fixity ratio of EP-OA/MXDA toughening systems

It can be seen from Tab. 3 that the shape fixity ratio of EP-100OA toughening system is 97.2%, which does not reach more than 98%, so it cannot be used as a better shape memory material. This is mainly because OA is a flexible curing agent. Therefore, with the increase of OA content, the crosslinking degree of the system is low, so the content of fixed phase in the system is small, and the content of reversible phase is large. At the same time, the amount of flexible chain segment introduced increases, and the movement ability of chain segment is strengthened, Therefore, the shape fixity ratio of this system is relatively poor. The shape fixity ratio of the other EP-OA/MXDA toughening systems is more than 98%, and the difference between different proportions is not obvious, showing good shape fixation performance.

Tab. 3 Rebound angle and shape fixity ratio of EP-OA/MXDA toughening systems

System	EP-100OA	EP-80OA/20MXDA	EP-60OA/40MXDA	EP-40OA/60MXDA	EP-20OA/80MXDA	EP-100MXDA
Rebound angle(°)	5.0	3.0	2.5	2.2	2.0	1.5
Shape fixity ratio(%)	97.2	98.3	98.6	98.8	98.9	99.2

3.3 Shape recovery ratio

To get a better shape memory material, both high shape fixity ratio and high shape recovery ratio is necessary. The shape recovery ratio is generally required to be at least 98%, that is, the residual angle should not be higher than 4°, otherwise the use of functional material components will be affected. It can be seen from Fig. 3 that changing the proportion of OA and MXDA has no obvious effect on the shape recovery ratio of the six toughened systems.

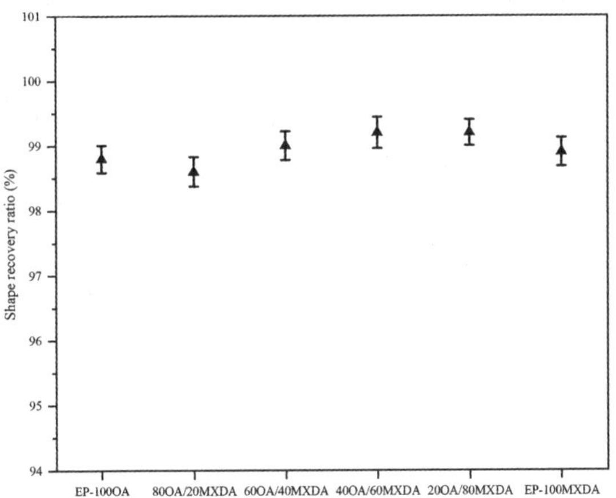

Fig. 3 Shape recovery ratio of EP-OA/MXDA toughening systems

From Tab. 4, it can be concluded that the residual angles of six toughening systems are less than 4°, that is, the shape recovery ratio reaches 98% or more.

Tab. 4 Residual angle and shape recovery ratio of EP-OA/MXDA toughening systems

System	EP-100OA	EP-80OA/20MXDA	EP-60OA/40MXDA	EP-40OA/60MXDA	EP-20OA/80MXDA	EP-100MXDA
Residual angle(°)	2.1	2.5	1.8	1.5	1.5	2.0
Shape recovery ratio(%)	98.8	98.6	99.0	99.2	99.2	98.9

Therefore, only in terms of the performance of shape recovery rate, the six systems can be used as better shape memory materials.

3.4 Shape recovery rate

Fig. 4 is a graph of the shape memory recovery time of six different ratios of EP-OA/MXDA toughened systems. It can be seen that with the decrease of OA content and the increase of MXDA content, the shape memory recovery time of the system increases, that is, the recovery rate decreases. The shape recovery time of the EP-100OA system is the shortest, which is mainly due to the fact that the system is completely cured by epoxy resin and OA. The glass transition temperature of this system is relatively low, the molecular chain flexibility of the system is the best and it is more sensitive to temperature. Therefore, when tested at a high temperature of 20 ℃ above T_g, the thermal equilibrium process is completed faster, so the total time required to recover to the initial shape driven by internal stress release is the shortest, and the recovery rate is the fastest. The content of MXDA in the other five toughening systems gradually increased, the chemical crosslinking degree in the system increased, and the thermal motion of molecular chains was hindered. The glass transition temperature of the system increases gradually, so the thermal equilibrium process time of the system increases in a high temperature environment. With the gradual decrease of the OA content in the system, the flexibility of the molecular chain also gradually deteriorated. Therefore, the total time taken to return to the initial shape driven by internal stress release increases, the recovery rate decreases.

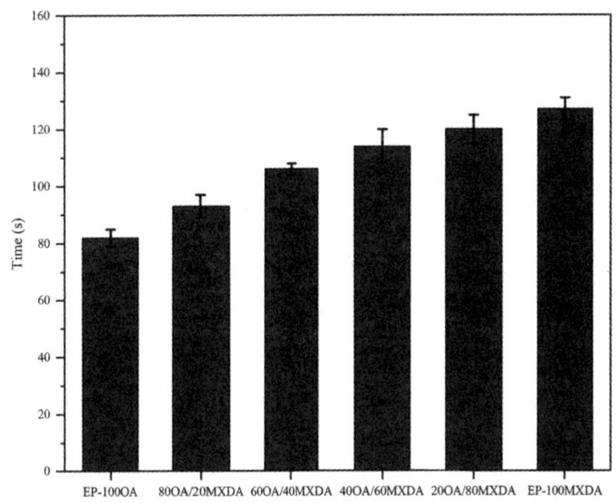

Fig. 4 Shape memory recovery time of EP-OA/MXDA toughening system

4 Conclusions

This study prepared an EP-OA/MXDA toughening system with shape memory properties, and explored the effects of different ratios of OA and MXDA on the glass transition temperature and shape memory properties of the epoxy resin system. The results show that with the increase of MXDA content, the crosslinking density of the epoxy resin toughened system increases, the flexibility of the molecular chain becomes worse, and the glass transition temperature of the epoxy resin toughened system increases gradually, ranging from 42.3 ℃ to 100 ℃. The shape memory performance test shows that with the decrease of OA content, the shape fixity ratio of the six systems increases; the shape recovery ratio has no obvious effect and can reach more than 98%; the recovery time increases and the recovery rate decreases.

References

[1] REN T, ZHU G, NIE J. Research progress on deployable structures of shape memory polymer composites[J]. Journal of Aeronautical Materials, 2018, 38(4): 47-55.

[2] WEI W, LIU J, HUANG J, et al. Recent advances and perspectives of shape memory polymer fibers[J]. European Polymer Journal, 2022, 175: 111385.

[3] LIU Z, LAN X, BIAN W, et al. Design, material properties and performances of a smart hinge based on shape memory polymer composites[J]. Composites Part B: Engineering, 2020, 193: 108056.

[4] PILATE F, TONCHEVA A, DUBOIS P, et al. Shape-memory polymers for multiple applications in the materials world[J]. European Polymer Journal, 2016, 80: 268-294.

[5] HU J, CHEN W, FAN P, et al. Epoxy shape memory polymer (SMP): material preparation, uniaxial tensile tests and dynamic mechanical analysis[J]. Polymer Testing, 2017, 62: 335-341.

[6] LIU T, LIU L, YU M, et al. Integrative hinge based on shape memory polymer composites: material, design, properties and application [J]. Composite Structures, 2018, 206: 164-176.

[7] LIU T, ZHU G, WEI K, et al. Advances in shape memory polymer composites[J]. Polymer Materials Science & Engineering, 2013, 29(11): 183-186,190.

[8] LUO L, ZHANG F, LENG J. Multi-performance shape memory epoxy resins and their composites with narrow transition temperature range[J]. Composites Science and Technology, 2021, 213.

[9] SHEN Y, WANG B, LI D, et al. Toughening shape-memory epoxy resins via sacrificial hydrogen bonds[J]. Polymer Chemistry, 2022, 13(8): 1130-1139.

[10] ZHAO S, LIU Q, YUAN A, et al. Intrinsic toughened conductive thermosetting epoxy resins: Utilizing dynamic bond and electrical conductivity to access electric and thermal dual-driven shape memory[J]. Materials Chemistry Frontiers, 2022, 6(14): 1989-1999.

Comparison of Tensile Properties and Breaking Morphology of Calotropis Gigantea, Kapok and Cotton Fibers

JIN Meiqi[1], SHEN Hua[1*], MA Yanxue[1], LI Yuling[1], XUE Wenliang[1], DING Yi[1], WANG Keyi[1], XU jianchu[2], CHEN Peng[3]

[1] *College of Textiles, Donghua University, Shanghai 201620, China*

[2] *Center for Future Studies of Mountain Areas, Kunming Institute of Botany, Chinese Academy of Sciences, Kunming 650201, China*

[3] *Yixing City Zhongchang Colorweaving Co., Ltd., Wuxi 214211, China*

* *Corresponding author's email:* shenhua@dhu.edu.cn

***Abstract*:** Calotropis Gigantea (C. Gigantea), kapok and cotton fibers are natural cellulose fibers. This paper chose African C. Gigantea fibers, Hainan kapok fibers, and China (Xinjiang) long-staple cotton fibers as representative samples, and compared their tensile properties and breaking morphology. The testing standard GB/T 14337—2008 was used to test the tensile properties of C. Gigantea and cotton fibers. According to the unique structure of C. Gigantea fibers, the test method was optimized. The macro-structure of the three kinds of fibers was observed under SEM. The results indicated that in the C. Gigantea fiber tensile property testing, the fiber's tip should be kept at the upper clamping end, and the distance between the position of the fiber's end and the lower edge of the lower clamping end is about 1 mm. The breaking load of C. Gigantea fiber is about 3.3 times than that of kapok fiber, and the breaking elongation is slightly less than that of kapok fiber, which is only 1/4 of that of cotton fiber. According to the breaking morphology of these three kinds of fibers, the breaking of C. Gigantea fiber is the smoothest, belonging to brittle breaking. Kapok fiber break belongs to intermediate breaking and cotton fiber break is serrated and belongs to ductile breaking.

***Keywords*:** Calotropis Gigantea fiber; Kapok; Cotton; Tensile properties; breaking load; breaking morphology

1 Introduction

Calotropis Gigantea (C. Gigantea) fiber is a new kind of natural cellulose fiber, which has a smooth and shiny surface. Its unique large hollow structure improves warmth retention, sound insulation, and moisture conductivity of products, so it can be used in the development of functional textiles[1].

The mechanical property of the fiber is an important indicator to measure the fiber's properties. And the tensile property is one of the important mechanical properties of C. Gigantea fibers, which is very important to the quality of yarns and the value of products. Many scholars[2-5] have compared the tensile properties of C. Gigantea, kapok and cotton fibers, and have made the following conclusions: the C. Gigantea fiber was a kind of high-load and low-elongation fiber; its breaking load was slightly lower than that of cotton fiber, but is significantly higher than kapok fiber; its breaking elongation was smaller than that of cotton fiber. Among them, Cui[5] compared the breaking morphology of C. Gigantea and cotton fibers. However, as C. Gigantea fiber has a hollow structure and the diameter of a single fiber is distributed from fine to coarse[3], it can be found that the relevant data in different literature[1-5] are very different, and there are obvious differences between the testing methods of tensile properties. Additionally, the morphology of three kinds of fibers after the tensile break has not been compared in details. Therefore, this paper optimized the original test method based on the unique structure of C. Gigantea fibers, and then compared the tensile properties and breaking morphology of C. Gigantea, kapok and cotton fibers. Findings in this paper are essential for the application of C. Gigantea fibers in the textile field, and also provide a theoretical basis for the further research and development of C. Gigantea fibers.

2 Experiment Methods

2.1 Materials

Samples: African C. Gigantea fibers, Hainan kapok fibers, Xinjiang long-staple cotton fibers. Before the test, all samples should be placed in the standard atmospheric temperature (20±2) ℃, humidity (65%±2%) constant temperature, and humidity room for 24 h humidity adjustment.

Sampling and clamping tools: a soft-headed silicone tweezer to avoid damaging the hollow structure of fibers, a black velvet board, and a 0.1 cN tension clip.

Sample preparation tools (for kapok fibers): double-sided adhesives, and 10 mm×10 mm hollow papers.

2.2 Test methods

The tensile properties of C. Gigantea, kapok and cotton fibers were measured by the LLY-06E electronic fiber strength tester. According to the test method in GB/T 14337—2008, the distance length was 10 mm, the number of test roots was 50, and the stretching speed was 5 mm/min.

One end of C. Gigantea and cotton fibers were held with a tension clamp and then another end with a tweezer and put directly into the air pump gripper for stretching. As C. Gigantea fiber has a hollow structure, and the diameter of a single fiber is distributed from fine to coarse[3], probably causing extrusion damage at the breaking site of the fiber during stretching, and it is difficult to obtain effective representative data. Therefore, this study optimized the testing method to avoid the damage of C. Gigantea fibers' unique hollow structure.

According to the "paper clip method"[6], kapok fibers were placed vertically on hollow papers, and then one end of the paper was cut off first. After gently placing the paper on the air pump holder with a tweezer, the other end was cut off before stretching.

TM 3000 tabletop scanning electron microscope (SEM) was used to observe the tensile breaking morphology of C. Gigantea, kapok and cotton fibers. The samples were sprayed with gold before the test.

2.3 Drawing method of stretching curve

There are two methods to make a representative curve from the measured tensile curve with large dispersion: one is the single fiber curve method, and the other is the average method of 5 curves. As the internal structure and external dimension of natural fibers are affected by the growth environment and other factors, the tensile curves of each fiber are quite different. Therefore, the average method of 5 curves was chosen in this paper to draw the representative tensile curves of three kinds of fibers.

The specific method was[7]: according to the average breaking load and elongation obtained from 50 fibers, 5 curves closest to the average breaking load and elongation were selected from the 50 tensile curves, and each curve was divided into several parts by equal breaking elongation, and the corresponding average breaking load was calculated. The average breaking load at the corresponding elongation point was obtained based on the measured average breaking elongation. Then the curve obtained by connecting each point was the representative curve of the fiber.

2.4 Optimization of the tensile properties test method for C. Gigantea fiber

Single C. Gigantea fibers' diameters from the tip to the end are varied from fine to coarse. In order to avoid extrusion damage at the breaking site of the fiber during stretching, the tip was unified on the upper gripper without stretching, and the end was placed on the lower gripper. After testing, it was found that the breaking positions of C. Gigantea fibers were all near the lower edge of the upper gripper, and the breaking surfaces were flat without extrusion.

At the same time, in order to obtain effective representative data, this paper tried to make the breaking position of C. Gigantea fiber at the upper clamping end in a relatively stable and thick part. After testing, it was found that the closer the clamping site was to the fiber end, the more stable the data was. For the convenience of operation, it was uniformly stipulated that when testing the tensile properties of C. Gigantea fiber, the tension clamp should be clamped at the bottom of the curved hook at the end of C. Gigantea fiber, and the fiber should be clamped in a way that the clamp mouth is perpendicular to the fiber. The distance between the tension clamp mouth and the lower edge of the lower clamping end should be about 1 mm.

3 Results and Discussion

3.1 Comparison of the tensile properties

The test method described in section 2.4 was used to measure the tensile properties of C. Gigantea fibers, and then test methods described in section 2. 2 were used to obtain the tensile properties of kapok and cotton fibers, as shown in Tab. 1.

Tab. 1 Comparison of the tensile properties of C. Gigantea, kapok and cotton fibers

Indicators		C. Gigantea	Kapok	Cotton
Breaking load/cN	Average	6. 19	1. 88	6. 37
	Range	4. 85–8. 62	0. 89–3. 60	4. 37–9. 72
	CV/%	11. 40	34. 42	22. 64
Breaking elongation/%	Average	2. 61	3. 08	12. 16
	Range	1. 50–3. 40	1. 10–6. 20	5. 10–18. 9
	CV/%	14. 43	36. 22	32. 37

As can be seen from Tab. 1, compared with kapok fiber, the breaking load of C. Gigantea fiber was about 3. 3 times than that of kapok fiber, and its breaking elongation was slightly smaller than that of kapok fiber. However, the *CV%* of C. Gigantea fiber was smaller, about 1/3 that of kapok fiber, and the numerical value was more concentrated. Compared with cotton fiber, the breaking load of C. Gigantea fiber was slightly smaller, but the breaking elongation was only 1/4 of cotton fiber, and the toughness was obviously worse.

The average method of 5 curves described in section 2. 3 was used to obtain representative tensile curves of three kinds of fibers, as shown in Fig. 1.

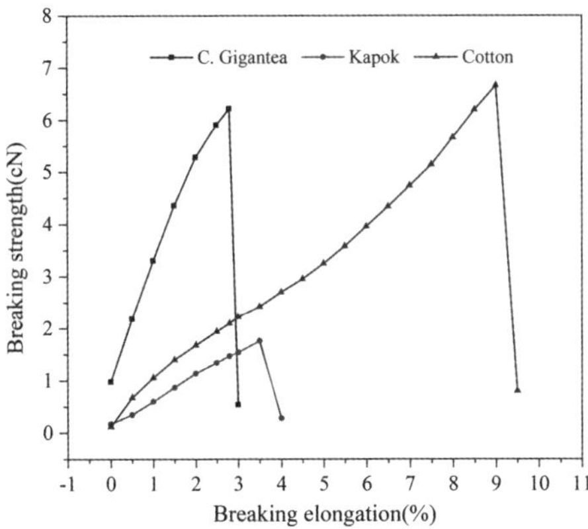

Fig. 1 Representative tensile curves

It can be seen from Fig. 1 that C. Gigantea fiber has a high load and low elongation. The tensile curve of C. Gigantea fiber was approximately a straight line, and the elongation was proportional to the breaking load, that is, the relationship between breaking load and elongation was linear. Compared with the tensile curves of kapok and cotton fibers, it can be found that the breaking load of C. Gigantea fiber was between them, while the elongation was the smallest.

3. 2 Comparison of the tensile breaking morphology of fibers

The tensile breaking morphology of C. Gigantea, kapok and cotton fibers is shown in Fig. 2.

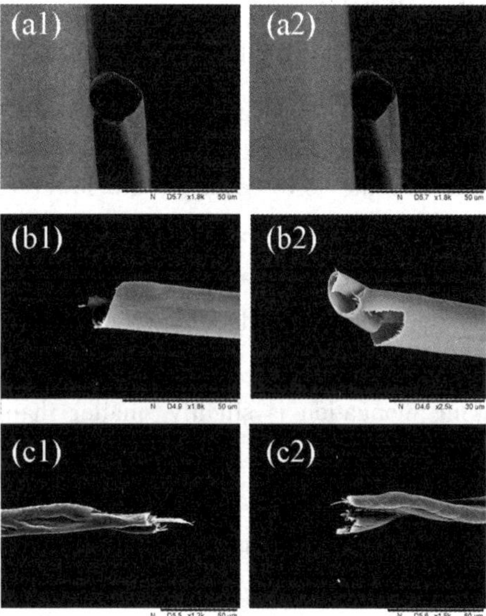

Fig. 2 The tensile breaking morphology (footnote 1, 2 represent the left and right breaks of a fiber): (a) C. Gigantea fiber; (b) kapok fiber; (c) cotton fiber

The breaking of polymer materials can be divided into brittle breaking, ductile breaking and intermediate breaking. The cross-section morphology of brittle breaking is obviously different from that of ductile breaking. The brittle breaking surface is smooth, while the ductile breaking surface is rough and granular, and the breaking surface in the intermediate state is in between, with both smooth and rough parts[7].

(1) Compared with kapok and cotton fibers, the tensile breaking morphology of C. Gigantea fiber was the smoothest, so the tensile breaking of C. Gigantea fiber is a typical brittle breaking. The tensile brittle breaking can replace the liquid nitrogen frozen brittle breaking to shoot the cross-section of fiber.

(2) The breaking of kapok fiber showed shedding traces along the relative transverse direction. This is because the spiral joint of the kapok fiber epidermal surface is the weak link between fiber load and stiffness, so it is easy to bend and breaking at the joint[8]. So the breaking of kapok fiber belongs to the intermediate state of breaking.

(3) The breaking of cotton fiber was serrated, and the

breaking surface was along the helix direction of the fibrils, which was torn in a long tail shape to form a small fibrous whisker plexus[7]. So the breaking of cotton fiber belongs to ductile breaking.

4 Conclusions

(1) When testing the tensile properties of C. Gigantea fiber, the fiber's tip should be kept at the upper clamping end, and the distance between the position of the fiber's end and the lower edge of the lower clamping end is about 1 mm.

(2) C. Gigantea fiber is a high-load and low-elongation fiber. Its breaking load is between kapok and cotton fibers, about 3.3 times of that of kapok fiber. And its breaking elongation is slightly smaller than that of kapok fiber, which is only 1/4 of that of cotton fiber, so its toughness is worse than that of cotton fiber.

(3) The tensile breaking morphology of C. Gigantea fiber is the smoothest, belonging to brittle breaking. Because of the spiral joint of the epidermis, the breaking of kapok fiber shows the trace of shedding along the relative transverse direction, which belongs to intermediate breaking. The breaking morphology of cotton fiber is serrated and belongs to ductile breaking.

References

[1] FANG G P, HU H M. Akund fiber and its application in knitting[J]. Knitting Industry, 2014(5): 26-30.

[2] GAO J, ZHAO T, CHEN J B. Composition, structure and property analysis of calotropis gigantea, kapok and cotton fibers[J]. Journal of Donghua Uuiversity (Natural Science Edition), 2012, 38(2): 151-155.

[3] QI Y H. Analysis of the structure and physicochemical performance of kenya's calotropis gigantea fiber [D]. Shanghai: Donghua University, 2016: 11-32.

[4] YANG X. Study on the correlation between the fiber's properties and its growing conditions and the structure and properties of akund fiber[D]. Shanghai: Donghua University, 2013: 34-36.

[5] CUI Y M. Calotropis gigantea fiber and textile technology and product design[D]. Shanghai: Donghua University, 2017: 30-88.

[6] XU G B, LIU W, LOU Y, et al. Analysis of the tensile property of kapok fiber[J]. Journal of Donghua Uuiversity (Natural Science Edition), 2009(5): 525-530.

[7] YU W D, CHU C Y. Textile materials[M]. Shanghai: Donghua University Press, 2009: 76-120.

[8] YI X J. Study on the microstructure and breaking mechanism of kapok fiber[D]. Shanghai: Donghua University, 2015: 11-32.

The Impact of COVID-19 Epidemic on Apparel Consumption Demand: A Survey from 2021 to 2022

ZHAO Junli[1*], **ZHAO Wei**[1], **ZHAN Yijia**[1]

[1] *Glorious Sun School of Business and Management, Donghua University, Shanghai 200051, China*

[*] *Corresponding author's email*: zjl@ dhu. edu. cn

Abstract: Since the outbreak of the epidemic in 2020, the demand for apparel consumption has changed a lot. But little is known about what dominates the change in apparel demand. The paper proposes a four-dimensional perceived value model to analyze the new characteristics of demand for apparel based on consumer value perception theory (CPV) and the technology acceptance model (TAM). The study collects a total of 3 576 valid sample data from the year 2021 to 2022. The paper finds that style and price are the primary factors influencing consumers' apparel purchases. Consumers spend less on clothing and intend to pay less than they did before the epidemic. Online and live webcast shopping increases. At the same time, the epidemic has influenced the demand structure for textiles and clothing. Consumers prefer to spend more on activewear, pajamas, or loungewear sports than before. About 80% of consumers state that they will not conduct retaliatorily purchasing in apparel after the epidemic which indicates that consumers are relatively rational. In addition, the increasing recognition of domestic apparel brands shows that domestic brand fashion become more favored by domestic consumers.

Keywords: Apparel; Consumers' Behavior; Demand; Covid-19

1 Introduction

Apparel is one of the basic necessities of life. When consumers perceive the benefits compared to the costs, they would like to pay for the product[1]. Perceived value refers to consumers' preference and evaluation of product attributes and performance[2]. For apparel consumption, each person has his own perceived value even for the same apparel. Zeithaml[3] first proposed the CPV theory, in which consumers believe that it is worthwhile to consume a product/service when they perceive it to be valuable. Researchers extend consumer perceived value from three-dimensional value, i. e. functional, social and emotional[4-5] to five-dimensional value, i. e. functional, social, emotional, cognitive and conditional[6]. Besides, some scholars have further investigated economic value[7].

However, the information technology improvement and the outbreak of Covid-19 have much impact on perceived value and demand for apparel. The TAM considers the impact of information technology on consumption. Put forward by Davis[8], the TAM attributes whether a new information technology will be used to six variables: perceived usefulness, perceived ease of use, user attitude, intention to use, actual usage behavior and external variables. Subsequently, Davis[9], Venkatesh & Davis[10], Venkatesh & Bala[11] and other scholars have improved TAM, whose main purpose is to explore the key factors and causal analysis of consumers' acceptance of information technology[12].

This paper combines consumer perceived value theory with the technology acceptance model to analyze the demand for apparel in four dimensions: functional, economic, emotional and digital. The functional value consists of the positive benefits perceived from the features of the product/service that satisfy an individual's needs[6]. Economic value refers to the economic income perceived when using the goods/services[7]. Emotional value is the value perceived as a result of satisfying an emotional need[13]. Digitization is the degree of consumer acceptance and use of new technologies. Based on CPV theory and TAM, this paper establishes a research model in Fig. 1.

The marginal contributions of this paper are twofold. First, we supplement the limited literature on apparel

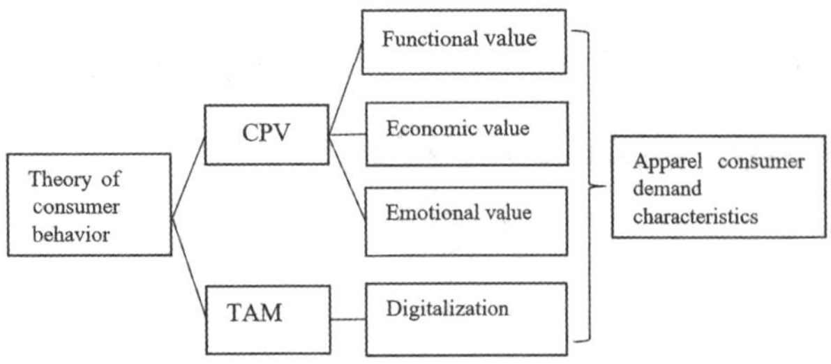

Fig. 1 Research model

demand in the context of the epidemic and advance our understanding of apparel consumption by combining CPV theory and TAM. In particular, this paper proposes a four-dimensional perceived value model: functional, economic, emotional and digital, which explored the consumer perceived value theory. Second, little literature studies the change in demand for apparel for two consecutive years after the outbreak of the epidemic. This paper compared the changes in apparel domestic demand based on 3576 valid questionnaires in the years 2021 and 2022.

2 Methodology

2.1 Questionnaire content design

Donghua University and China National Textile and Apparel Council (CNTAC) have conducted a survey on the demand for apparel in Shanghai and China since 2012. Since the outbreak of the COVID-19 pandemic in 2020, we have revised the content of the consumer questionnaire and added content related to the impact of the epidemic on apparel demand.

With the upgrading of consumer demand, the diverse and elastic demand characteristics of apparel demand such as functionality, fashion and personalization are becoming more prominent. The questionnaire mainly includes the following content. First, the basic information of consumers is investigated, including the demographic characteristics of the respondents, such as gender, age, occupation, city, education, monthly household income, and other basic information. Second, the survey focuses on three aspects: consumption channels, consumption ability and consumption habits based on the CPV theory and TAM. Third, the impact of the epidemic on respondents' consumption was investigated and studied mainly from consumers' income, consumer behavior and consumption psychology.

2.2 Questionnaire distribution design

We have an annual Summer Social Research Project. About 50 students participate in this project each year. We give them related survey training before the delivery of questionnaires. The questionnaire was designed to fill in only once per mobile phone and per computer terminal. Students participating in the project were asked to distribute no more than three student questionnaires to ensure the diversity of targeted respondents. In order to protect the privacy of individuals, the questionnaire is anonymous, making it easier for respondents to fill in their true willingness.

A total of 4075 questionnaires were distributed, of which 1504 questionnaires were distributed in 2021 and 2571 in 2022. A total of 3576 valid samples were obtained after dealing with missing and incorrect values of the variables, of which 1422 were valid in 2021 and 2154 in 2022, giving a validity rate of 87.8%.

3 Results and Discussion

3.1 Basic information about the respondents

The proportion of women in the respondents is slightly higher in 2022 than that in 2021. Most respondents age locates between 18-50 years old. The respondents are mainly from enterprises, institutions, and professional occupations. Respondents from Jiangsu province, Zhejiang province, Shanghai and Anhui province account for over 40%. 70% of respondents' monthly household income is below RMB 15 000. 90% of respondents spend less than RMB 1 000 on a single price of clothing. The proportion of the monthly household income of less than RMB 5 000 in 2022 increases compared with 2021. Individual income in 2022 decreases compared with that in 2021.

3.2 Consumer survey results

After the statistical analysis of 3576 questionnaires, the demand change in apparel focuses on five respects: functional value, economic value, emotional value, digitalization and pandemic impact.

3.2.1 Functional value

Style and price are the primary factors influencing consumers' apparel purchases.

Consumers are often perceived to make purchase decisions based on perceived acquisition value and perceived transaction value in respect of functional value[14]. The survey data shows that: style and price are two important factors to purchase apparel in both 2021 and 2022. 51.49% of people believe that the style of clothing is the primary factor influencing their purchase of clothing, followed by 33.57% believing that price is the most criticized factor. In addition, the two issues most criticized by consumers are the high price of clothing and the lack of individual design, with 49.16% and 42.20% respectively.

Fig. 2 Word cloud picture of top factors influencing consumers' apparel purchases in 2022

3.2.2 Economic value: lower budget

Consumers purchased clothing less frequently and spent less in 2022 than those in 2021. The budget for apparel purchases decreases in the latter half year. The highest frequency in 2021 was "once a month" shopping at 38.33%, which decreased to 29.20% in 2022 (Fig. 3(a)). The highest frequency in 2022 was "3 months and over" at 38.07% which shows that the frequency of purchase decreases (Fig. 3(a)). The overall spending on clothing in 2022 was significantly less than in the same period in 2021. 52.28% of respondents in 2022 reported spending less on clothing this year, an increase of 17.96 percentage points compared to 2021 (Fig. 3(c)). 41.65% of respondents will reduce their budget in the latter half year in 2022, which increases by 14.86% to 2021 (Fig. 3(b)).

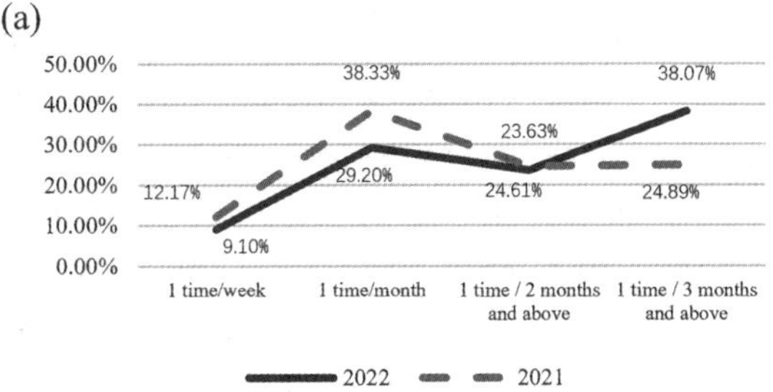

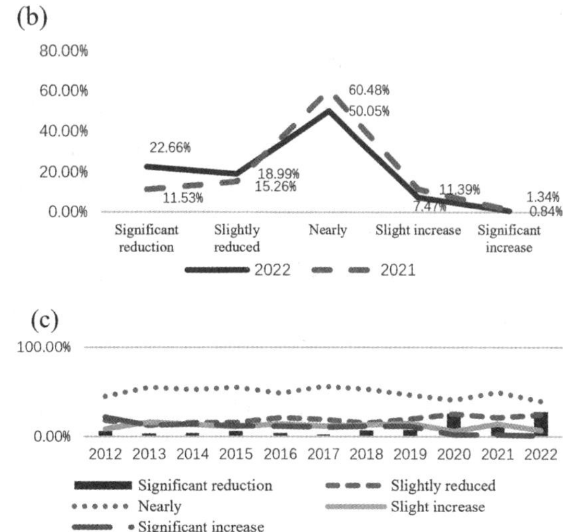

Fig. 3 (a) Comparison of consumer frequency of clothing purchases (2021-2022); (b) comparison of consumers' clothing budgets in the second half of the year (2021-2022); (c) overall spending on clothing purchases compared to the same period in the previous year (2012-2022)

The income deduction may account for above change. In 2022 survey, 23.68% of people have family members unemployed due to the epidemic, and the proportion of respondents whose individual income decreases by 17.34% compared to last year.

3.2.3 Emotional value: rising recognition for domestic brand

Consumer recognition of domestic apparel brands is on the rise from 2020 to 2022 (Fig. 4). 77.58% of respondents said they approve of domestic brands in the survey in 2022. In the survey, 68.48% of respondents believe that the biggest advantage of domestic brands is price, followed by 36.35% of respondents who believe that the biggest advantage of domestic brands is quality. The rise in consumer recognition of domestic brands is a reflection of rising emotional values based on patriotic sentiments and national cultural identity.

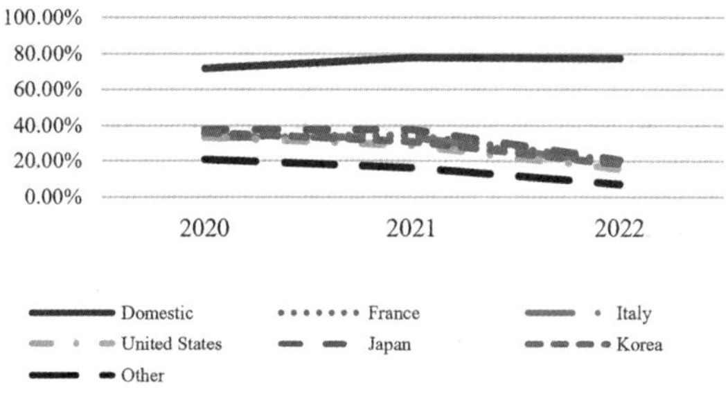

Fig. 4 Source of consumer recognize of apparel brands (2020-2022)

3.2.4 Digital: increasing online shopping

the proportion of online shopping is increasing from 2012 to 2022(Fig. 5(a)). 49.47% of respondents in 2022 said that their main source of clothing purchase is online shopping, which is similar to 2021. Moreover, the 2022 survey found that among the middle-aged and elderly (>50 years old) group, the proportion choosing to shop online also increased significantly from 31.72% in 2021 to 37.91% in 2022(Fig. 5(b)). Among the male group, the proportion choosing online shopping increased by 3.46% compared to last year. The perceived entertainment interactivity in live-streaming shopping has a significant positive impact on willingness to pay. In this survey, 94.34% of people said they knew or understood the live-stream shopping model. 17.55% of them said they were concerned about it, and 13.43% of them said they had already purchased through live-stream shopping.

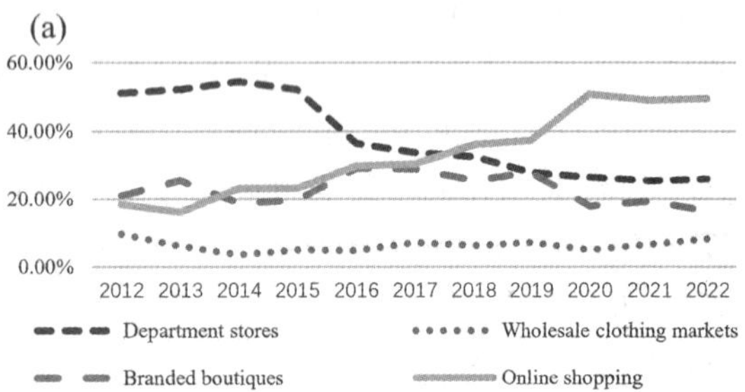

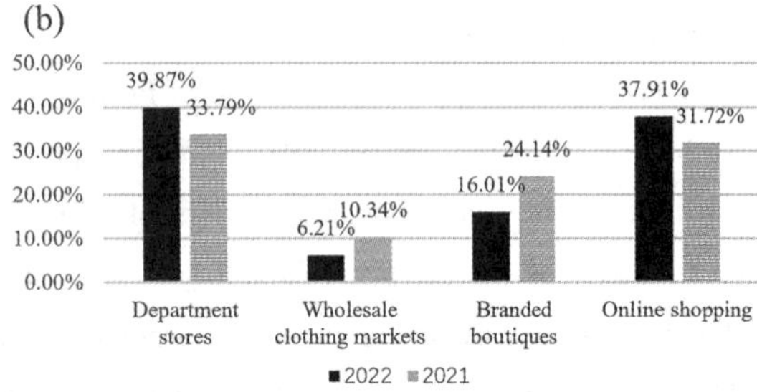

Fig. 5 (a) Access to clothing purchases (calendar year); (b) comparison of purchase channels for middle-aged and older people, 2021-2022

3.2.5 Impact of the epidemic

In the survey, even though the whole expenditure on apparel decreased (Fig. 6(b)), the proportion of casual sportswear and home wear purchases increased. Casual sportswear and loungewear dominated consumers' choice of clothing during the epidemic. 66.91% of respondents had purchased casual sportswear during the epidemic and 44.18% of respondents had purchased loungewear (Fig. 6(c)). In addition, 79% of respondents will have no retaliatory consumption after the epidemic which implies the majority of respondents are more rational (Fig. 6(a)).

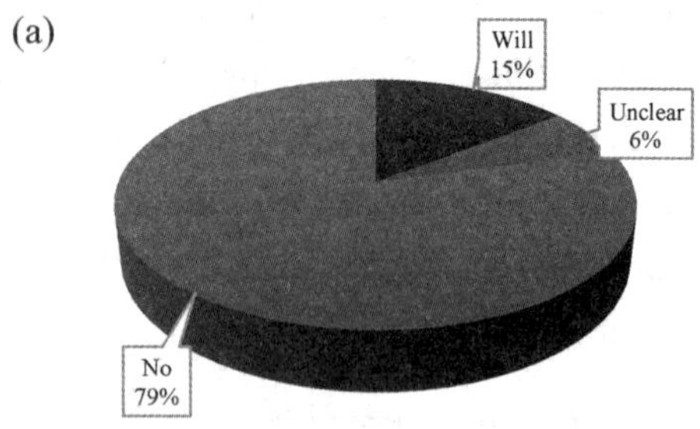

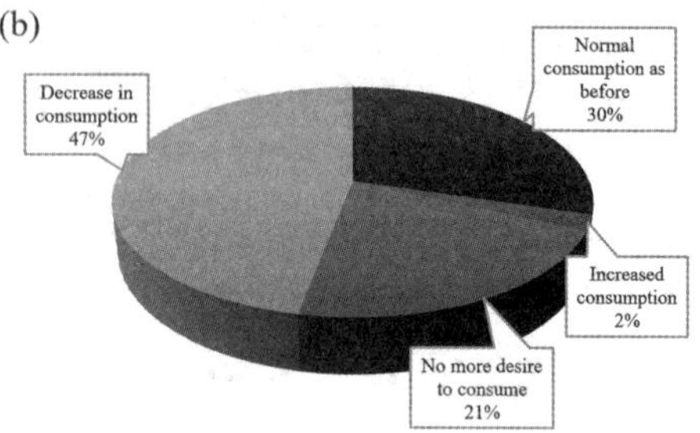

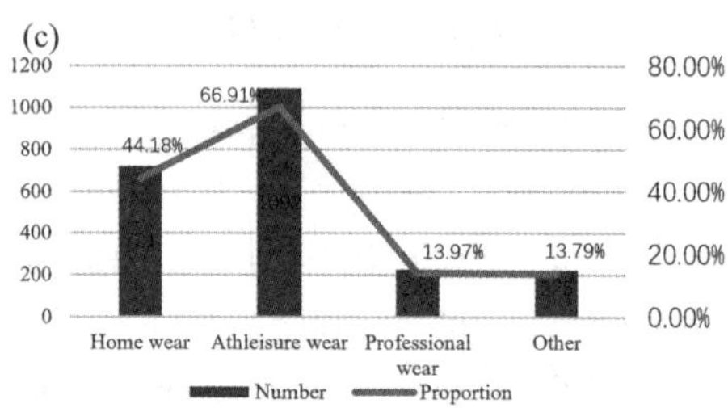

Fig. 6 (a) Retaliatory consumption after the outbreak; (b) change in apparel consumption desires during the outbreak; (c) types of clothing purchased during the outbreak

4 Conclusions

Based on consumer perceived value theory and technology acceptance model, this paper analyses apparel consumption demand from five aspects: functionality, economy, emotion, digitalization and the impact of the epidemic, using questionnaire survey data for two consecutive years in 2021 and 2022.

The study found that price and style were the primary factors influencing consumers' apparel purchases. Casual sportswear and home wear accounted for an increased share of purchases during the epidemic. With the double decline in the actual level of domestic residents' income as well as the expected level of income under the epidemic, the consumption expenditure, the frequency of purchases and expected expenditure on apparel decline. Consumers have increasing recognition of domestic apparel brands. The share of online and live-stream shopping increases. if the epidemic ends, most consumers will not adopt retaliatory consumption.

References

[1] YANG Z L, PETERSON R T. Customer perceived value, satisfaction, and loyalty: the role of switching costs[J]. Psychology & Marketing, 2004, 21(10): 799-822.

[2] WOODRUFF R B. Customer value: the next source for competitive advantage[J]. Journal of the Academy of Marketing Science, 1997, 25(2): 139.

[3] ZEITHAML V A. Consumer perceptions of price, quality, and value: a means-end model and synthesis of evidence[J]. Journal of Marketing, 1988, 52(3): 2-22.

[4] ROIG J C F, GARCIA J S, TENA M A M, et al. Customer perceived value in banking services[J]. International Journal of Bank Marketing, 2006, 24(5): 266-283.

[5] SANCHEZ J, CALLARISA L, RODRIGUEZ R M, et al. Perceived value of the purchase of a tourism product[J]. Tourism Management, 2006, 27(3): 394.

[6] SHETH J N, NEWMAN B I, GROSS B L. Why we buy what we buy: a theory of consumption values[J]. Journal of Business Research, 1991, 22(2): 159-170.

[7] REN J F, YANG J L, ZHU M Y, et al. Relationship between consumer participation behaviors and consumer stickiness on mobile short video social platform under the development of ICT: based on value co-creation theory perspective[J]. Information Technology for Development, 2021, 27(4): 697-717.

[8] FRED D, DAVIS J R. A technology acceptance model for empirically testing new end-user information systems: theory and results[D]. Massachusetts: Sloan School of Management, 1985.

[9] DAVIS F D. User acceptance of information technology: system characteristics, user perceptions and behavioral impacts[J]. International Journal of Man-machine Studies, 1993, 38(3): 475-487.

[10] VENKATESH V, DAVIS F D. A theoretical extension of the technology acceptance model: four longitudinal field studies[J]. Management Science, 2000, 46(2): 186-204.

[11] VENKATESH V, BALA H. Technology acceptance model 3 and a research agenda on interventions[J]. Decision Sciences, 2008, 39(2): 273-315.

[12] PAUL J, MODI A, PATEL J. Predicting green product consumption using theory of planned behavior and reasoned action[J]. Retailing and Consumer Services, 2016, 29: 123-134.

[13] JIANG K, YANG Z W, FENG Z X, et al. Mobile phone use while cycling: a study based on the theory of planned behavior[J]. Transportation Research Part F, 2019, 64: 388-400.

[14] SONI N, DAWRA J. Judgments of acquisition value and transaction value[J]. Indian Business Research, 2019, 12 (3): 389-410.

Establishment of Failure Analysis Process for Automotive Lightweight CFRP

WANG Han[1], ZHANG Yao[1], YANG Yuqiu[1]*

[1] *College of Textiles, Donghua University, Shanghai 201620, China*

* *Corresponding author's email*: amy_yuqiu_yang@ dhu. edu. cn

Abstract: Carbon fiber-reinforced resin matrix composite (CFRP) has been paid more attention in the field of automotive lightweight because its specific strength and design-ability are higher while density is lower than those of traditional metal for vehicle. An evaluation system which can predict the failure mode of CFRP for vehicle is needed. It includes such elements as the mechanical properties of raw materials, design method, simulation and so on. Tension and interlaminar-shearing performance of CFRP laminates were tested to get the basic mechanical properties of raw materials (carbon fiber prepregs). Three types of laminates with ply sequences including $[0°]_{16}$ (unidirectional laminate), $[0_2/45_2/90_2/-45_2]_s$ (isotropic laminate) and $[0/45_2/90/-45_2/0_2]_s$ (anisotropic laminate) were chosen to get the failure modes of laminates designed in different methods. Tension, compression and interlaminar-shearing simulation experiments were carried out and the failure modes were analyzed by Tsai-Wu failure criterion by ANSYS Workbench. As the result of simulation, in the tension simulation experiment, the failure of laminate with $[0°]_{16}$ firstly appears in the outermost layer, and gradually extends to the inner layers until complete failure. In $[0_2/45_2/90_2/-45_2]_s$ and $[0/45_2/90/-45_2/0_2]_s$, the failure occurs at the direction of 90° layer, followed by 45° and 0° layers. On the other hand, according to compression testing results, similar to the tensile mode, the failure of $[0°]_{16}$ shows the fracture occurs from the outside to inside by degrees. For $[0_2/45_2/90_2/-45_2]_s$ and $[0/45_2/90/-45_2/0_2]_s$, fractures firstly occur at the direction of 0° layer, followed by 45° layer and 90° layer. In the interlaminar-shearing simulation experiment, the failure firstly appears at where the sample contacts with the upper pressure head and the lower support position, and gradually extends to the inner layers until complete failure.

Keywords: Carbon fiber; Laminate; Mechanical properties; Failure; Finite element

1 Introduction

The fuel consumption is directly proportional to the weight of the vehicle[1]. According to data from the U. S. Department of Energy and other relevant departments, if the total vehicle weight would be reduced by 10%, the fuel consumption was reduced by 6% to 8% and exhaust emissions was reduced by 5% to 6%[2-3]. Automobile lightweight technology can save resources, reduce pollution, and play a role in protecting the environment.

Carbon fiber reinforced resin composite (CFRP) material is used in the field of automotive lightweight due to its high specific strength, good corrosion resistance and strong designability[2, 4-5].

Testing the mechanical properties of CFRP will cost a lot of material and financial resources. An evaluation system which can predict the failure mode of CFRP for vehicle is needed, including such elements as the mechanical properties of raw materials, design method, simulation and so on.

In this study, tension and interlaminar-shearing performance of CFRP laminates were tested to get the basic mechanical properties of raw materials (carbon fiber prepregs). Three types of laminates with ply sequences including $[0°]_{16}$ (unidirectional laminate), $[0_2/45_2/90_2/-45_2]_s$ (isotropic laminate) and $[0/45_2/90/-45_2/0_2]_s$ (anisotropic laminate) were chosen to get the failure modes of laminates designed in different method. Tension, compression and interlaminar-shearing simulation experiments were carried out and the failure modes were analyzed by Tsai-Wu failure criterion by ANSYS Workbench.

2 Experiments

2.1 Materials

Carbon fiber prepregs (JGC4225 carbon fiber/Jinfei 3725 epoxy resin), with resin content of 33% and 0.15 mm thick.

Three types of laying sequence of laminates including $[0°]_{16}$, $[0_2/45_2/90_2/-45_2]_s$ and $[0/45_2/90/-45_2/0_2]_s$ are designed respectively.

2.2 Basic mechanical properties test of prepregs

Tensile tests were carried out according to ASTM D_3039/D_3039M-17 in the 0° and 90° directions. The sample size of 0° was 250×15×1.14 mm with 4 tabs of 56×15×1.5 mm at both ends of the sample. Another (90°) was 175×25×2.06 mm with 4 tabs of 25×25×1.5 mm. Extensometer gage with a length of 50 mm was used in quasistatic test at a loading rate of 5 mm/min.

Interlaminar-shearing test was carried out (ASTM D_2344/D_2344M-16) with the sample size of 40×20×3.95 mm.

All composite samples were conditioned at 20 ℃ and 65% relatively humidity for 24 h before testing.

2.3 Finite element failure analysis process

2.3.1 Geometrical model of the structure (Design Modeler)

Three types of laminates models are established separately by Design Modeler of ANSYS according to ASTM D_3039/D_3039M-17, ASTM D_6641/D_6641-16 and ASTM D_2344/D_2344M-16 Standard Test Method.

The specimen geometry of finite element analysis is given in Tab. 1.

Tab. 1 Specimen geometry of finite element analysis

Type of sample	Overall length (mm)	Width (mm)	Thickness (mm)	Tab length (mm)	Tab thickness (mm)
Tensile	250	15	2.4	56	1.5
Compression	140	13	2.4	62.5	2
Shearing	40	20	2.4	—	—

2.3.2 Definition of material properties (engineering data)

According to the material characteristics and analysis needs, Epoxy Carbon UD (230 GPa)

Prepreg in the system is selected as the basic material, and some important performance parameters (shown in Tab. 2) are modified according to the basic mechanical properties of prepregs. In Tab. 2, E is the elastic modulus, μ is the Poisson ratio, and G is the shear modulus.

Tab. 2 Basic property parameters of sample material

E_x(GPa)	E_y(GPa)	E_z(GPa)	μ_x	μ_y	μ_z	G_{xy}(GPa)	G_{yz}(GPa)	G_{xz}(GPa)
121.92	7.00	7.00	0.27	0.40	0.27	4.7	3.1	4.7

2.3.3 Meshing

0.5 mm is selected to mesh the sample models to get more

accurate results, while 1 mm is chosen to mesh the tab models in tension and compression simulations and 0.5 mm with tetrahedral mesh is chosen to mesh the load nose and supports models in shearing simulation.

2.3.4 Defining the load and boundary conditions (static structural)

In the tension and compression simulations, the contact mode of sample and tabs is set to "bonded", and in the shear simulation, the contact mode of sample and load nose and supports is set to "No separation".

2.3.5 Post-treatment of the results (ACP Post)

The Tsai-Wu failure criterion is selected to determine the failure form of the laminates. The analysis results can be viewed in ACP-Post module, and the failure cloud map can be output.

3 Results and Discussion

3.1 Basic mechanical properties and failure process of the unidirectional laminate

In the 0° tensile experiment, the curves are linear and it is found that the fiber plays the main bearing role[6]. The slope of the front of the curve is large. Since the extensometer gage is added to obtain more accurate material modulus (121.92 GPa). As the displacement increases, the load reaches the peak with the tensile strength of 1877.79 MPa averagely, the curve drops sharply subsequently, while the sample is damaged, mainly due to the fiber fracture[5-6].

In the 90° tensile experiment, the curves are also linear. However, it is found that the epoxy resin matrix plays the main bearing role. The slope of the front of the curve is large, and more accurate material modulus is worked out (7.00 GPa). As the displacement increases, the load reaches the peak with the tensile strength of 31.45 MPa averagely, the curve drops sharply subsequently, while the sample is broken. Since the gaps first appear in the interface of carbon fiber and resin, and gradually expand, eventually leading to sample fracture.

The overall deformation of the sample is less under the shear force. The slope of the curve gradually decreases after the load grows to 50% of the limit load (7 538.2 N, with the shear strength of 71.55 MPa) and the yield phenomenon like the plastic material[6-7]. At the end of the curve, fluctuation can be found, which is due to the crack which first appears at the weak area of carbon fiber and resin matrix connection and gradually extends to the interface direction with the displacement increasing. When stress reaches the interface strength limit, interface debonds, while matrix is damaged by shearing force, followed by the adjacent fiber fracture, which lead to material damage[8].

3.2 Finite element simulation failure analysis

3.2.1 Tensile simulation

For the $[0°]_{16}$ laminate stretching simulation, because of the stress concentration, the fracture first appears near the edge of tabs at the 1st layer (shown in Fig. 1). As the displacement increases, the crack gradually expands to the center within the 1st plane (shown in Fig. 2) and then to the inner layer until the overall failure of the sample occurs.

For the $[0_2/45_2/90_2/-45_2]_s$ and $[0/45_2/90/-45_2/0_2]_s$,

Fig. 1 The tensile failure cloud map of the 1st layer of $[0°]_{16}$ with the displacement of 0.7 mm

Fig. 2 The tensile failure cloud map of the 1st layer of $[0°]_{16}$ with the displacement of 2.65 mm

the failure first appears at the edge of 90° layer (shown in Fig. 3), and then the edge positions of the 45° and −45° layer begin to fail (shown in Fig. 4), followed by the connection position of 0° layer and tabs.

Fig. 3 The tensile failure cloud map of the 5th (90°) layer of $[0_2/45_2/90_2/-45_2]_s$ with the displacement of 0.45 mm

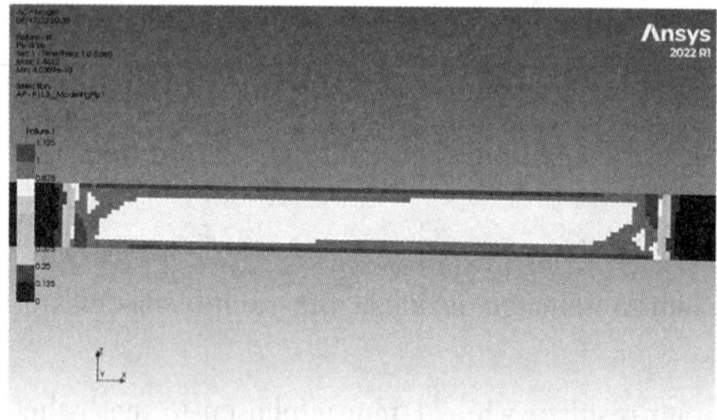

Fig. 4 The tensile failure cloud map of the 3rd (45°) layer of $[0_2/45_2/90_2/-45_2]_s$ with the displacement of 0.75 mm

3.2.2 Compression simulation

For the $[0°]_{16}$ laminate compression simulation, the fracture first appears at the outermost layer near the edge of tabs. As the load increases, the gap gradually extends to the center within the outermost layer (shown in Fig. 5), and then to the inner layer until complete failure.

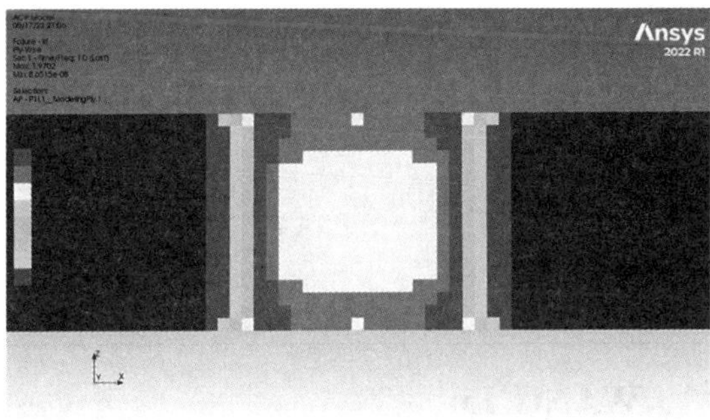

Fig. 5 The compression failure cloud map of the 1st layer of $[0°]_{16}$ with the displacement of 0.15 mm

For the $[0_2/45_2/90_2/-45_2]_s$ and $[0/45_2/90/-45_2/0_2]_s$, as the load increases, the failure first occurs at the outermost layer, near the tabs. Then the 45° and −45° layers fails, which are followed by the 90° layer. However, for the $[0/45_2/90/-45_2/0_2]_s$, it is noteworthy that after the 1st layer failed, gap occurs at the center of the 7th (0°) layer, and gradually expands (shown in Fig. 6).

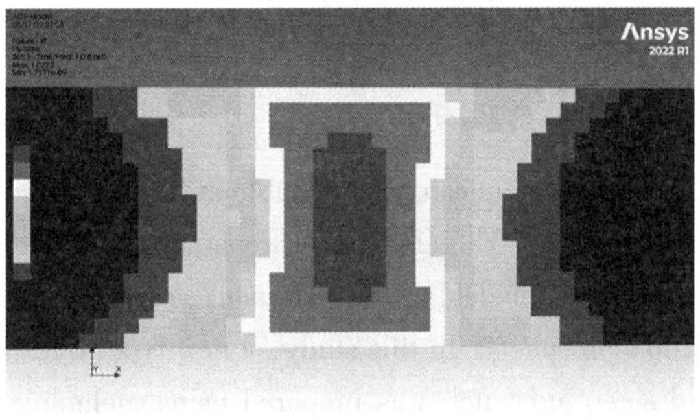

Fig. 6 The compression failure cloud map of the 1st layer of $[0/45_2/90/-45_2/0_2]_s$ with the displacement of 0.12 mm

3.2.3 Shearing simulation

For the $[0°]_{16}$, $[0_2/45_2/90_2/-45_2]_s$ and $[0/45_2/90/-45_2/0_2]_s$ shearing simulation, because of the

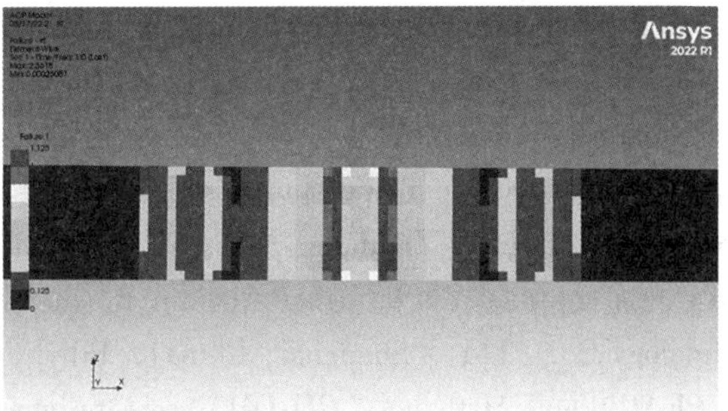

Fig. 7 The shearing failure cloud map of $[0°]_{16}$ laminate with the displacement of 0.12 mm

stress concentration, crack always firstly occurs at where the sample contacts with the upper pressure head and the lower support position (as the red parts shown in Fig. 7), and gradually extends to the inner layers until complete failure.

4 Conclusions

In this study, three laminates under tension, compression and shearing are selected to simulate to get the failure process. It is found that the crack always occurs first at the position with stress concentration, such as the sample edge, the connected place of sample and other parts, and the outermost layer. At the same time, the gap also easily occurs at the layer with weak resistance, such as 90° layers of the tensile sample, and 0° layers of the compression sample. Crack seems to extend within the failed planes and followed by other layers. The results obtained in this study can provide help for future experimental validation.

References

[1] FENG F. Application analysis of carbon fiber composite materials for automobile lightweight[J]. Synthetic Materials Aging and Application, 2022, 51(1): 123-125.

[2] LI L. The research and application of fiber reinforced plastics material in lightweight automobile field[J]. Intelligent City, 2021, 7(11): 167-168.

[3] YANG G Y, ZHAO R, XIAO B, et al. Application of CFRP in automobile lightweighting[J]. Petroleum & Petrochemical Today, 2020, 28(10): 24-28.

[4] PUTTARAJU D G, HANUMANTHARAJU H G. Finite element analysis and validation of tensile properties of carbon fiber reinforced polymer matrix composites[J]. Materials Today: Proceedings, 2022, 62(6): 2800-2807.

[5] FAN Z Y. Prediction and analysis of mechanical properties of carbon fiber reinforced composite laminates [D]. Dalian: Dalian University of Technology, 2018.

[6] ZHONG Z. Mechanical behavior and fracture failure mode simulation of CFRP [D]. Dalian: Dalian University of Technology, 2021.

[7] MENG X M, ZHONG Z, CHENG C Q, et al. Finite element simulation of in-plane shear of carbon fiber reinforced plastic laminates with two material models of LS-DYNA software[J]. Materials for Mechanical Engineering, 2020, 44(12): 85-90+96.

[8] HU X L. Study on the mechanical properties of phenolic/glass fiber composite[D]. Xi'an: Xi'an Technology University, 2014.

Construction of Textile-Based Transdermal Delivery System for Sleeping Aid

LIN Jiahan[1,2], **GAO Jing**[1,2*], **WANG Lu**[1,2]

[1] *Key Laboratory of Biomedical Textile Materials & Technology in Textile Industry, Shanghai 201620, China*

[2] *College of Textiles, Donghua University, Shanghai 201620, China*

* *Corresponding author's email*: gao2001jing@ dhu. edu. cn

Abstract: Transdermal delivery system (TDS) is a non-invasive drug delivery system that can avoid gastrointestinal "first-pass effect" and has the advantages of high patient compliance and low side effects. However, the type and thickness of the backing layer and the formula of the drug carrier layer affect the drug delivery effect and using comfort of the patch. This study develops a new textile-based transdermal patch, using Melatonin as a sleeping aid. Firstly, the interaction between the two types of backing layers and Melatonin was investigated by infrared spectral image and experiment of distribution coefficient, and polyester/viscose spunlaced non-woven fabric was selected as the backing layer of the patch. Based on DIA transdermal patch, the initial adhesion, holding adhesion and air permeability of the patch with different grammages of backing layer and thicknesses of drug carrier layer were investigated, respectively. The grammage of non-woven back layer of the patch was determined to be 60 g, and the thickness of wet film of drug carrier layer was 0. 500 mm.

Keywords: Transdermal delivery system; Textile-based; Pressure sensitive adhesive

1 Introduction

With the accelerating pace of people's life, more and more pressure from work and life, results in more and more patients with sleep disorders. Insomnia will not only interfere nighttime sleep to cause circadian rhythm disorder and affect daily life, but also cause cognitive decline, physical discomfort, emotional abnormalities and other problems, which are usually treated by oral melatonin[1-2]. However, melatonin passes through the digestive tract before being absorbed into the body, and the "first-pass effect" of the gastrointestinal tract not only reduces drug utilization, but also irritates the gastrointestinal tract. Transdermal delivery system (TDS) is the most widely studied non-invasive drug delivery system, which has the characteristics of small side effects and high patient compliance[3-5]. Moreover, TDS can achieve the effect of continuous and stable administration, especially suitable for kids[6]. Textile-based transdermal drug delivery system is a DIA type TDS composed of backing layer, drug carrier layer and release liner[7], in which the backing layer is composed of textiles. The transdermal drug delivery system has great advantages in combination with the comfortable, breathable, moisture permeable and soft characteristics of textiles. Therefore, the preparation of transdermal patch with melatonin for external use has broad application prospects. In this study, a new type of textile-based sleep aid patch was prepared using melatonin as therapeutic drug, and its formulation process was optimized.

2 Experiments

2.1 Materials

Polyacrylic acid pressure sensitive adhesive (PSA, DURO-TAK 1093) was provided by Henkel (Germany). Polypropylene spunlace non-woven fabrics (60 g/m^2) and polyester/viscose spunlace nonwoven fabrics (60 g/m^2) were obtained from TTXS Co., Ltd. (Jiangsu, China). Classine paper was purchased from Yishun Paper Co., Ltd. (Zhejiang, China). Melatonin (MT) was supplied by Shanghai Aladdin Biochemical Technology Co., Ltd. (Shanghai, China). Ethyl acetate (EAC) and Methanol (CH_3OH) were purchased from Collins (China).

2.2 Preparation of TDS

The patch formula was determined as PSA (g) : MT

(g) : EAC (mL) = 47 : 1 : 4.7. Weigh MT and EAC according to the recipe and put them into a beaker. Stirring them with a glass stick until they were well mixed. The solution was then added into the solvent glue and mixed evenly with magnetic stirring for 1h at room temperature at a speed of 200 r/min, and defoamed in ultrasonic for 30 min.

Pour the PSA onto the glassine paper and MS-ZN320B coating machine was used to prepare patches with wet film thickness of 0.2 mm, 0.3 mm, 0.4 mm, 0.5 mm and 0.6 mm, respectively. The content of MT in each patch (3 cm×3 cm) was about 5 mg. The samples were then left standing at room temperature for 15 min, vacuum-dried at 37 ℃ for 20 min, and stored at room temperature.

2.3 Morphology observation

Scanning electron microscopy (DXS - 10ACKT, China) was employed to observe the cross section of the TDS. The prepared sample was brittle broken in liquid nitrogen, and was coated with gold-palladium.

2.4 Adhesion testing of TDS

In order to clarify the adhesive force of the patch to the skin, the adhesion measurement method (0952) of the patch in Pharmacopoeia of the People's Republic of China was used for testing, and the device for testing initial adhesion and holding adhesion was set. Before the test, the prepared transdermal patch was packed and placed in a laboratory with the temperature of 18–25 ℃ and relative humidity of 40%–70% for more than 2 h. Each sample was tested three times.

2.5 Air permeability testing of TDS

Before the test, each sample was balanced at a temperature of (20±2) ℃ and relative humidity of 65%±2%. Fabric breathability meter was used to calculate the breathability of the sample by measuring the air flow rate per unit area within a certain time at a constant pressure difference on both sides of the sample.

2.6 Standard curve

10, 20, 30, 40, 50, 60, 70, 80, 90, and 100 μg/mL MT solution were prepared with methanol as solvent, respectively. The absorbance of solutions with different concentrations was measured and the standard curve was obtained.

2.7 Distribution coefficient

The different backing layers were cut into 1 cm×1 cm pieces and placed in a centrifuge tube, and 4 mL of 20 μg/mL MT solution was added into 1 : 49 (v : v) methanol and water. The centrifuge tube was placed in a shaker and oscillated at 32 ℃ for 24 h. The drug content was analyzed by ultraviolet spectrophotometry. According to the concentration of melatonin in solution before and after equilibrium, the distribution coefficient was calculated using eqn (1):

$$K = (C_0 - C)/C \quad (1)$$

where C_0 and C represent drug concentrations in MT solution before and after distribution, respectively.

2.8 Infrared spectral testing

The backing layer was cut into 1 cm×1 cm pieces and randomly divided into two groups. One group was added 4 mL MT solution of 20 μg/mL, the other group was added 4 mL blank solution. After shaking for 24 h, the backing layer was vacuum-dried at 50 ℃ for 2 h. Infrared spectral images of two groups of backlining layers were obtained by infrared spectrometer.

3 Results and Discussion

3.1 Morphology of TDS

The section morphology is shown in Fig. 1. It can be seen that the patch was well formed. The backing layer and the drug carrier layer were closely fitted, and the structure was clear without gaps and bubbles.

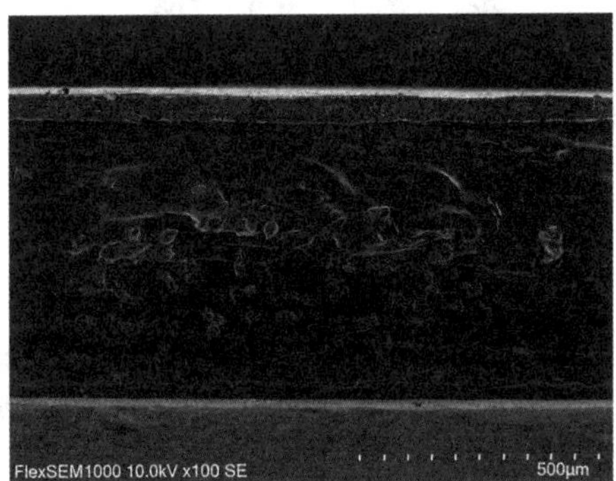

Fig. 1 SEM image of section of TDS

3.2 Adhesion analysis

The initial adhesion and holding adhesion of the patch with different thickness of drug carrier layer were measured. Initial adhesion represents the instantaneous adhesion of the transdermal delivery system, which is expressed by the maximum number of steel balls that the patch can adhere to. Holding adhesion refers to the ability of the patch to resist slippage and deformation caused by persistent external force, which is expressed by the length of time the patch falls off from the test plate. The experimental results are shown in Fig. 2. When the wet film thickness of the drug carrier layer increased from 0.2 mm to 0.6 mm, the maximum ball size of the drug carrier layer increased from 13 to 23, and the shedding time increased from 0.78 h to 2.50 h. The initial adhesion and holding adhesion were positively correlated with the wet film thickness. This phenomenon may be caused by the increase of the thickness of the drug carrier layer and the increase in the cohesion of the polyacrylate, thus increasing the shear strength, initial adhesion and adhesion.

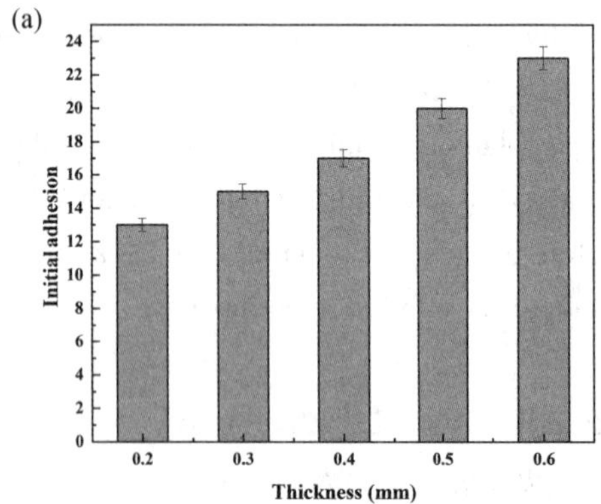

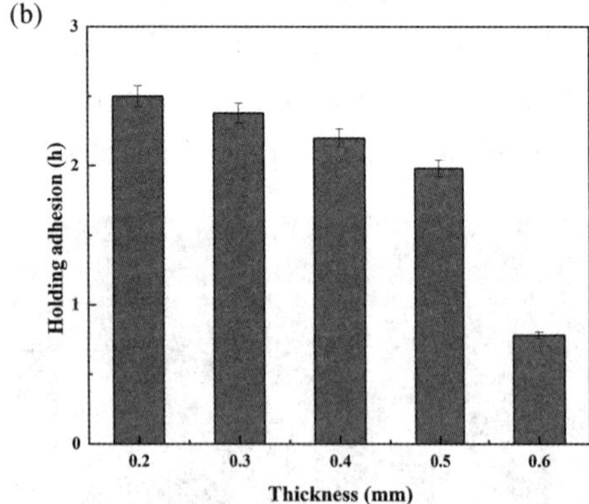

Fig. 2 (a) The curves of initial adhesion with different thicknesses of drug carrier layer; (b) the curves of holding adhesion with different thicknesses of drug carrier layer

3.3 Air permeability analysis

The permeability of the patch with different thicknesses was measured by fabric permeability meter. The experimental results are shown in Fig. 3. It can be seen that the air permeability decreased with the increase of the layer thickness. The possible reason is that the thicker the wet film thickness, the tighter the PSA structure, the longer the gas path through the pores, so the less gas passed through per unit time, and the lower the permeability. Poor air permeability may cause sweat on the skin to fail to expel in time and skin cells to overhydrate, resulting in impaired skin barrier function and allergic reactions. Therefore, while ensuring good adhesion of the patch, the air permeability of the patch should be increased as much as possible. Due to the opposite trend of adhesion and air permeability as the thickness of drug carrier layer changes, 0.5 mm was selected as the wet film thickness of drug carrier layer.

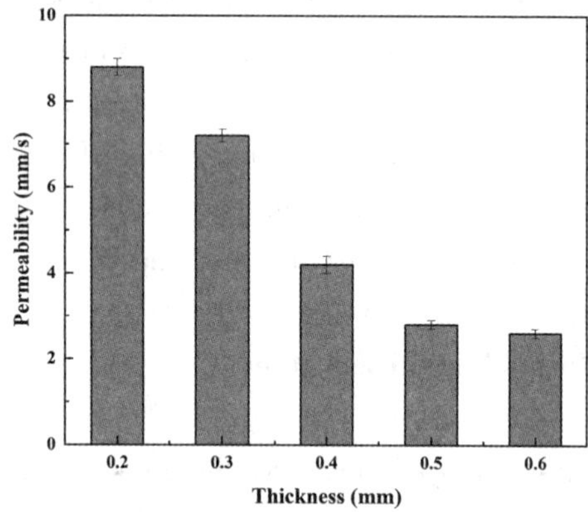

Fig. 3 The curves of air permeability with different thicknesses of drug carrier layer

3.4 Distribution coefficient analysis

Continuous scanning of 50 μg/mL solution in the wavelength range of 190–900 nm showed that MT had large absorption at both 229 nm and 277 nm. Considering the large interference of terminal absorption, 277 nm was selected as the measurement wavelength in this study. The absorbance of different concentrations of melatonin solutions at 277 nm was measured. The standard curve of MT is shown as Fig. 4.

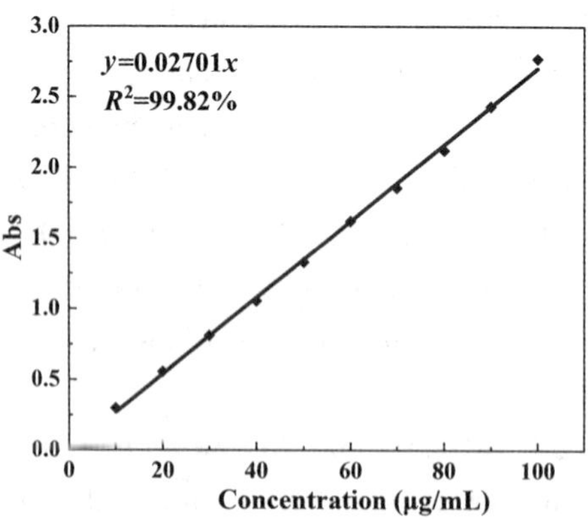

Fig. 4 Ultraviolet standard melatonin curve of MT

In the determination experiment of the distribution coefficient, the concentration of melatonin in the solution before and after equilibrium could be calculated according to the standard curve. The results shows that distribution coefficient of polyester/viscose spunlace nonwoven fabrics was 0.009±0.041, and distribution coefficient of PP spunlace nonwoven fabrics was 0.021±0.036.

Distribution coefficient can be used to determine the degree of the of MT adsorption by the backing layer, which can further explain the stability of the drug on the drug carrier layer. The higher the distribution coefficient, the stronger the adsorption between the backing layer and the drug. The experimental results shows that the distribution coefficients of the two different backing layers were small, indicating that the adsorption effect of the two backing layers on drugs was relatively small, and its influence could be ignored.

3.5 Infrared spectral analysis

Infrared spectrometer was used to test the backing layers before and after the balance to determine whether there were new chemical bonds between the drug and the backing layer, as shown in Fig. 5 and Fig. 6.

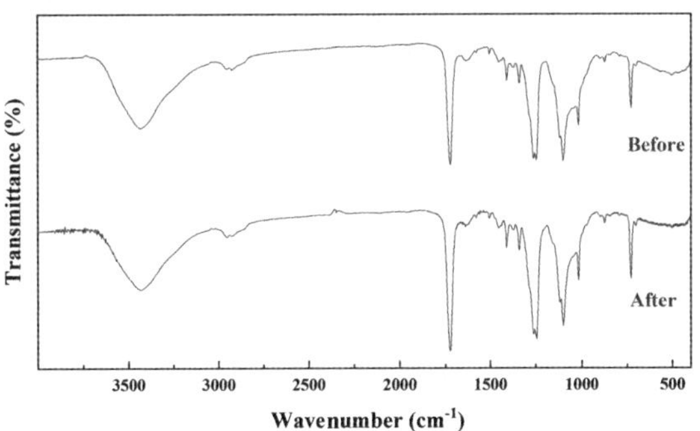

Fig. 5 Infrared spectral of polyester/viscose spunlace nonwoven fabrics

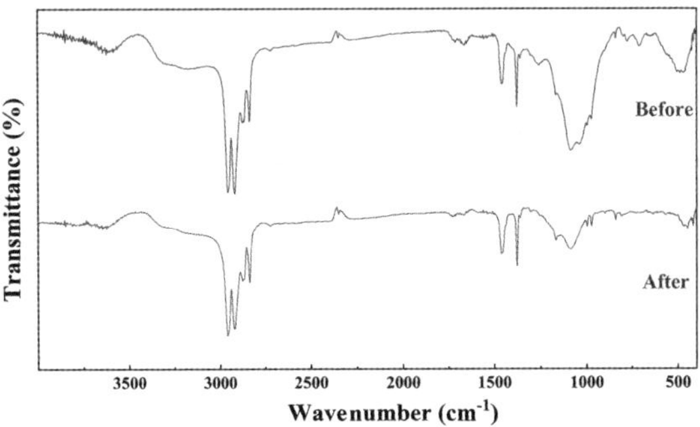

Fig. 6 Infrared spectral of PP spunlace nonwoven fabrics

It can be seen from Fig. 5 that the characteristic peaks of the polyester/viscose non-woven fabrics before and after the balance were completely consistent, which proved that no new chemical bond was formed. This was because the polyester/viscose spunbonded fabrics had excellent chemical inertness for drugs and would not react with drugs to cause drug loss.

As seen from Fig. 6, the functional groups in the sample are consistent with the standard infrared spectrum of polypropylene, because there was no reaction site between MT and polypropylene. Therefore, there was no reaction between the polypropylene non-woven fabric and MT. According to the infrared spectrum analysis of alcohols, C—O characteristic peak shift appeared in 1 300–1 000 cm^{-1}, which was presumed to be the primary alcohol absorption peak generated by methanol in blank solvent. Because the content was very low, the effect on the backing layer could be ignored.

4 Conclusions

In this study, we prepared a kind of melatonin transdermal patch which can be administered for a long time. It is well formed, breathable and comfortable. The preparation technology and optimum preparation parameters of the patch were determined as follows: PSA (g) : MT (g) = 47 : 1, PSA (g) : EAC (mL) = 10 : 1, wet film thickness of drug carrier layer was 0.5 mm, and the patch was well formed. SEM observation shows that the drug carrier layer is properly bonded with the backing layer, and tightly bonded with the release liner without clearance and bubbles. Polyester/viscose spunbonded fabric was selected as backing layer by distribution experiment and infrared spectrum analysis. The effect of polyester/viscose spunbonded fabrics on drug release was minimal, which proved that it had no adsorption and no reaction to melatonin and had excellent chemical inertia.

References

[1] KANIKKANNAN N, ANDEGA S, BURTON S, et al. Formulation and in vitro evaluation of transdermal patches of melatonin[J]. Drug Development and Industrial Pharmacy, 2004, 30(2): 205–212.

[2] SHI H, REITER R J, TAN D X, et al. Indole-3-acetic acid inducible 17 positively modulates natural leaf senescence through melatonin-mediated pathway in arabidopsis[J]. Journal of Pineal Research, 2015, 58(1): 26–33.

[3] JEONG W Y, KWON M, CHOI H E, et al. Recent advances in transdermal drug delivery systems: a review[J]. Biomaterials Research, 2021, 25(1): 24.

[4] PASTORE M N, KALIA Y N, HORSTMANN M, et al. Transdermal patches: history, development and pharma-

cology [J]. British Journal of Pharmacology, 2015, 172(9): 2179-2209.
[5] NG L C, GUPTA M. Transdermal drug delivery systems in diabetes management: a review [J]. Asian Journal of Pharmaceutical Sciences, 2020, 15(1): 13-25.
[6] DELGADO C M B, GUY R H. Effective use of transdermal drug delivery in children[J]. Advanced Drug Delivery Reviews, 2014, 73: 63-82.
[7] FYs 2013-2017 GDUFA Science and research report: Transdermal drug products[EB/OL]. https://www.fda.gov/industry/generic-drug-user-fee-amendments/fys-2013-2017-gdufa-science-and-research-report-transdermal-drug-products, 2021.

Application of Camellia Seed Husk Extraction on Silk Fabrics

CHEN Jie[1], **JIANG Huiyu**[1], **QI Huan**[1,2*]

[1] *College of Chemistry and Chemical Engineering, Wuhan Textile University, Wuhan 430200, China*

[2] *College of Textiles and Apparel, Quanzhou Normal University, Quanzhou 362002, China*

* *Corresponding author's email*: qhqh123@126.com

Abstract: Natural dyes have a series of problems such as low dyeing efficiency, single color, low dyeing fastness, etc. Therefore, the application of natural dyes in textile dyeing is greatly limited. To broaden the uses for Camellia seed husks and increase the variety of natural dyestuff. In this study, silk fabric dyeing with an extract from camellia seed husks was optimized. Six different metal mordants (ferrous sulfate, copper sulfate, potassium aluminum sulfate, chromium trichloride, neodymium chloride, and ferric chloride) were examined, and the differences in their effects and potential mechanisms were analyzed together with the color fastness of silk materials. Silk fabrics can benefit from a variety of dyeing processes, including antioxidant activity and UV resistance, both of which were examined in this study.

Keywords: Camellia seed husk; Dyeing process; Mordant dyeing; UV resistance; Antioxidant

1 Introduction

Natural dyes have gained attention as a result of the textile industry's quick development and the world's pressing environmental issues[1]. Natural dyes are said to be green dye with good biocompatibility and quick decomposition in the environment[2]. Numerous natural dyes also contain companion substances that provide textiles with additional specific qualities like UV resistance, antibacterial, antiviral, antioxidant, and flame retardant capabilities. Natural dyeing encounters issues like single color and difficulties in color matching because to the numerous and diverse sources of natural dyes. Therefore, it is crucial to develop new natural dyestuffs to expand the scope of their applications[3].

A common tree with large, evergreen leaves in certain alpine regions of Fujian Province, China, is the Camellia sinensis. Oil may be derived from Camellia sinensis fruit, which is a special type of premium natural oil in China. Before extracting the oil, the camellia husk, which contains the seed's outer husk and its saponins, proteins, and polysaccharides, was removed. It can be utilized again as raw materials for producing food, feed products, cosmetics, and other products[4]. The development of high value-added camellia seed products is urgently needed to achieve the green, holistic exploitation of resources since the effective use of camellia seed husk has historically been relatively low[5]. Silk is a naturally occurring protein fiber that has exceptional affinity with natural colors, making it appropriate for natural dyeing. In human society, silk has a long history of development. It is renowned for its outstanding taking capabilities, polished and attractive clothing qualities, and good mechanical properties[6]. Rarely was it reported that camellia seed husk extract was used to dye silk and that its qualities were studied after dyeing.

The qualities of the camellia seed husk extract used in this work to color silk were investigated. In order to optimize the dyeing procedure, a single-factor experiment was chosen. On the dyeing effect, the impact of various metal salt mordanting conditions was compared. Performances for color fastness, UV protection, and antioxidants were assessed. These findings served as a benchmark for future applications of camellia husk and range expansion of natural colors.

2 Experiments

2.1 Materials

Camellia seed husk were purchased from market (Quanzhou, China). Silk fabric (20 g/m^2) was purchased from the local market (Wuhan, China). Acetic acid, sodium chloride, sodium hydroxide, anhydrous ethanol, ferrous sulphate heptahydrate, copper sulphate pentahydrate, aluminium sulphate octadecahydrate were supplied by Xilong Science Co. Ltd. (Shantou, China). Methanol was purchased from Sinopharm Group Chemical Reagent Co. Ltd. (Wuhan, China). Chromium trichloride and neodymium chloride were purchased from Aladdin Biochemical Technology Co. Ltd. (Shanghai, China). Ferric chloride was supplied by Fuchen Chemical Reagent Co. Ltd. (Tianjin, China). 1,1-Diphenyl-2-trinitrophenylhydrazine (DPPH) was purchased from BOSF Biotechnology Co. Ltd. (Hefei, China). All reagents were used as received without any further purification.

2.2 Preparation of dyeing solution

Crushed camellia seed husks were put in a flask, collected, and then 50% ethanol was added at the ratio of 1 : 25. The solution was then heated and refluxed for an hour at 90 ℃. The suspension was run through a sieve with a mesh size of 100, and the ethanol in the filtrate was removed using a rotary evaporator. The necessary dye solution concentration for further dyeing processes can be produced through dilution or concentration.

2.3 Single-factor experiment

For the control group, the dyeing conditions were as follows: dyestuff mass of 30 g/L, holding time of 45 min, bath pH of 5.0, holding temperature of 80 ℃, and bath ratio of 1 : 30.

By using the single-factor experiment approach, the effects of dyestuff mass, holding temperature, holding time, and bath pH on the dyeing effect were examined. The holding temperature ranged from 50-130 ℃, the holding period ranged from 15-90 min, and the bath pH ranged from 2.5-10.0. The dyestuff mass ranged from 5-80 g/L. The silk fabric was dyed, then washed in tap water, dried, and tested.

2.4 Mordant dyeing

The silk fabric was mordant dyed with Nd^{3+}, Fe^{2+}, Cu^{2+}, Al^{3+}, Cr^{3+}, Fe^{3+} six kinds of metal ions respectively in three mordant dyeing methods: pre-mordant, meta-mordant and post-mordant, all with mordant concentration of 2% (o.m.f.).

When pre-mordant dyeing, the temperature was increased to 80 ℃ with a bath ratio of 1 : 30, and the fabric was removed after 45 min of heat preservation treatment, washed, and dyed according to the optimized dyeing process in 2.3, and the fabric was removed after dyeing, washed and dried.

When meta-mordant dyeing, adding mordant to the dyeing bath, dyeing is carried out according to the optimized dyeing process in 2.3, and the fabric is removed after dyeing, washed and dried.

When post-mordant dyeing, dyeing is carried out according to the optimized dyeing process in 2.3, and the fabric is taken out after dyeing, washed, and then heated up to 80 ℃ with bath ratio 1 : 30, and the fabric is taken out after holding treatment for 45 min, washed and dried.

2.5 Color characteristic

The color properties of dyed fabrics were assessed using the reflection method on a Datecolor 800 colorimeter. CIE laboratory color coordinates L^*, a^*, and b^* represent brightness, redness-greenness colors, and yellowness-blueness colors, respectively. The color characteristic (K/S) value is calculated from the Kubelka-Munk formula.

$$K/S = (1 - R)^2/2R \quad (1)$$

where R is the reflectivity of the fabric, K is the absorption coefficient, and S is the scattering coefficient.

2.6 Color fastness

Refer to GB/T 3920—2008 for rubbing color fastness test, GB/T 3921—2008 for washing color fastness, GB/T 250—2008 for discoloration degree, and GB/T 251—2008 for staining degree.

2.7 UV resistance

The colored silk fabric was folded into three layers, and its UV protection performance was measured on the UV protection performance tester (YG(B)912, China). The test standard was determined according to GB/T 18830—2009, and the values of UVA, UVB and UPF were recorded.

2.8 Antioxidant activity

Weigh 2 mg of DPPH in methanol solution and capacity it to 50 mL. The absorbance of DPPH solution at 517 nm was measured and recorded as A_1. 10 mL DPPH solution was added to 30 mg of the fabric to be tested. After 30 min, the absorbance at 517 nm was measured and recorded as A_2. Antioxidant activity was calculated according to Eq. (2).

$$Activity = (A_1 - A_2)/A_1 \times 100\% \quad (2)$$

3 Results and Discussion

3.1 Optimization of dyeing process

The optimized dyeing process conditions are shown in Fig. 1. As can be seen from Fig. 1 (a), when the dye mass reaches 50 g/L, the rising trend of color depth slows down. Fig. 1 (b) shows that it is still possible to improve the color depth when the insulation temperature reaches 130 ℃, but the silk fabric will be damaged if the insulation temperature is too high. After the holding time reaches 45 min, the color depth tends to be balanced as shown in Fig. 1 (c). Fig. 1 (d) shows that bath pH = 3.0 reaches the appropriate bath pH for dyeing camellia seed husk silk fabric. Therefore, the optimal dyeing conditions of silk fabric dyed with camellia seed husk can be obtained as follows: dyestuff mass 50 g/L, holding time 45 min, bath pH = 3.0, holding temperature 130 ℃.

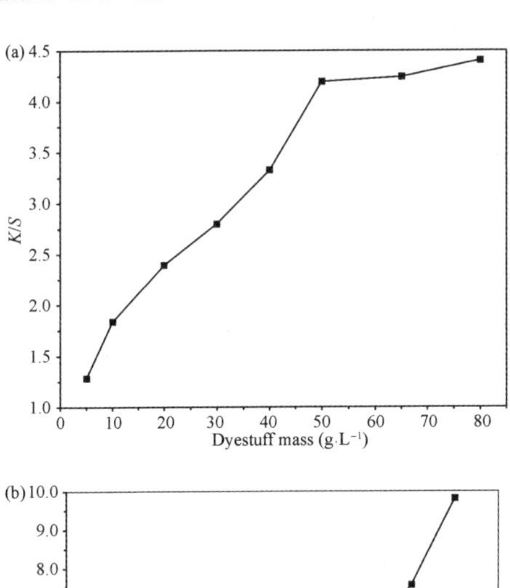

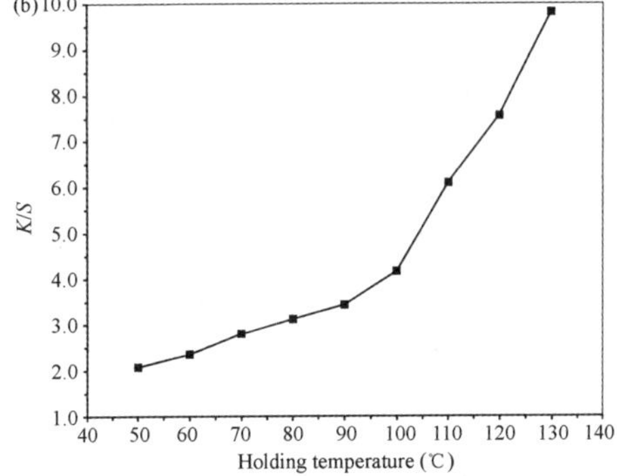

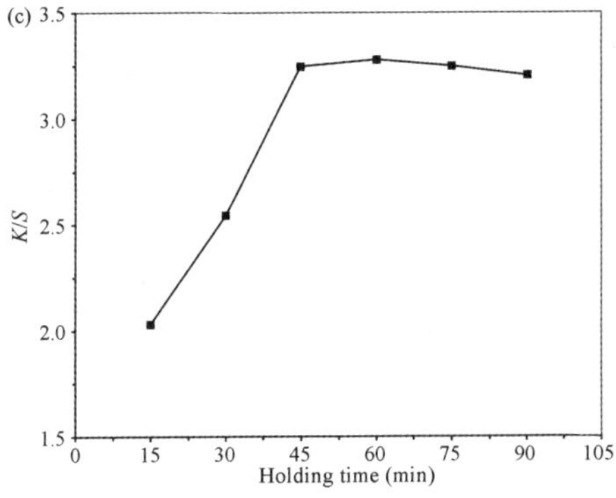

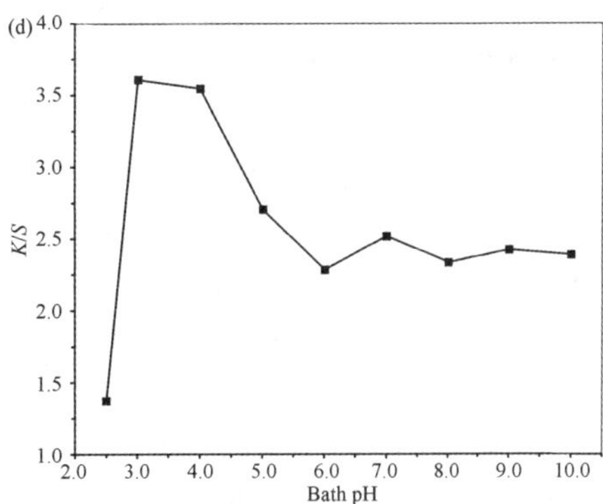

Fig. 1 Effect of each factor on *K/S* value: (a) dyestuff mass; (b) holding temperature; (c) holding time; (d) bath pH

3.2 Effect of mordanting dyeing

Six metal ions Nd^{3+}, Fe^{2+}, Cu^{2+}, Al^{3+}, Cr^{3+} and Fe^{3+} were used with a mordant concentration of 2.0% (o.m.f.) and three mordant dyeing methods were used: pre-mordant, meta-mordant and post-mordant[7]. The influence of different mordants and different mordant dyeing methods on color parameters was studied, and the results were summarized in Tab. 1.

Tab. 1 Effect of different mordants and mordanting methods on the color parameters of dyed silk fabrics

Mordants	Mordant dyeing methods	Color parameter fabric				Images of dyed
		L^*	a^*	b^*	*K/S*	
Blank	No mordanting	56.79	10.69	29.39	5.4900	
Neodymium trichloride	Pre-mordanting	55.41	10.16	29.20	6.1409	
	Meta-mordanting	56.16	10.21	29.43	5.9070	
	Post-mordanting	55.66	9.47	27.47	5.3333	
Ferrous sulphate	Pre-mordanting	31.22	2.96	4.76	9.4415	
	Meta-mordanting	31.54	2.72	3.47	8.5371	
	Post-mordanting	33.22	1.84	5.51	8.3764	
Copper sulphate	Pre-mordanting	52.65	10.30	27.49	6.5955	
	Meta-mordanting	51.87	10.64	27.73	6.9909	
	Post-mordanting	51.74	9.46	25.51	6.2201	
Aluminium sulphate	Pre-mordanting	55.35	10.74	29.45	6.4303	
	Meta-mordanting	55.72	10.86	29.47	6.1576	
	Post-mordanting	55.29	10.33	27.65	5.6071	
Chromium trichloride	Pre-mordanting	53.11	8.97	28.21	7.2498	
	Meta-mordanting	52.93	8.72	27.49	7.1846	
	Post-mordanting	52.11	8.56	25.58	6.3990	
Ferric chloride	Pre-mordanting	31.03	2.77	4.93	9.6213	
	Meta-mordanting	29.94	2.22	2.54	9.0888	
	Post-mordanting	34.23	1.96	6.58	8.2699	

The results in Tab. 1 show that dyed silk fabrics show the same K/S value order under both pre-mordant dyeing and metamotous dyeing conditions, that is, ferric chloride>ferrous sulfate>chromium trichloride>copper sulfate>aluminum sulfate>neodymium trichloride>non-mordant dyed silk fabrics, but under post-mordant dyeing conditions, they show another order. Ferrous sulfate>ferric chloride>chromium trichloride>copper sulfate>non-mordant dyed silk fabric>neodymium trichloride. In most cases, silk fabrics with mordant showed higher K/S values than those without mordant, and only neodymium trichloride showed lower K/S values than silk fabrics without mordant after mordant dyeing. In addition, the use of mordant dyeing will lead to changes in the color of the dyed silk fabric (redness-greenness value a^*, yellowness-blueness value b^*). Compared with the three mordant dyeing methods, the K/S value of pre-mordant dyeing is generally higher than that of primary mordant dyeing and post-mordant dyeing, except that the K/S value of pre-mordant dyeing is higher than that of post-mordant dyeing.

3.3 Color fastness property

The color fastness to rubbing and soaping of silk fabrics dyed with six metal ions was measured. The results were shown in Tab. 2.

As can be seen from Tab. 2, the color fastness to rubbing and soaping of silk fabrics after mordant dyeing has been improved, reaching grade 4.0 or above, which meets the requirements of daily clothing. The addition of metal salts leads to the improvement of fastness, which can be analyzed as follows: the addition of metal ions expands the structural system of pigment and enhances the force between pigment molecules and fibers. In addition, the expanded structural system reduces the water-solubility of the pigment, both of which make the dye less likely to peel off the fabric and maintain good durability.

Tab. 2 Effect of different mordants and mordanting methods on the color fastness of dyed silk fabrics

Mordant	Methods	Rubbing			Washing					
		Dry	Wet	Discoloration	Fiber staining					
					A_1	C	N	P	A_2	W
Blank		4	3-4	4-5	4-5	5	4	4-5	4	4
Nd^{3+}	①	4-5	4	5	5	4-5	4-5	5	4-5	4-5
	②	4-5	4	4-5	5	4-5	4-5	5	4-5	4-5
	③	4-5	4	5	5	4-5	4-5	5	4-5	4-5
Fe^{2+}	①	4-5	4	5	5	5	4-5	5	4-5	4-5
	②	4-5	4	4-5	5	5	4-5	5	4-5	4-5
	③	4-5	4	5	5	5	4-5	5	4-5	4-5
Cu^{2+}	①	4-5	4	5	5	5	4-5	5	4-5	4-5
	②	4-5	4	4-5	5	5	4-5	5	4-5	4-5
	③	4-5	4	5	5	5	4-5	5	4-5	4-5
Al^{3+}	①	4-5	4	5	5	5	4-5	5	4-5	4-5
	②	4-5	4	4-5	5	5	4-5	5	4-5	4-5
	③	4-5	4	5	5	5	4-5	5	4-5	4-5
Cr^{3+}	①	4-5	4	5	5	4-5	4-5	5	4-5	4-5
	②	4-5	4	4-5	5	4-5	4-5	5	4-5	4-5
	③	4-5	4	5	5	4-5	4-5	5	4-5	4-5
Fe^{3+}	①	4-5	4	5	5	5	4-5	5	4-5	4-5
	②	4-5	4	4-5	5	5	4-5	5	4-5	4-5
	③	4-5	4	5	5	5	4-5	5	4-5	4-5

Notes: ①—pre-mordant; ②—meta-mordant; ③—post-mordant;
A_1—acetate; C—cotton; N—nylon; P—polyester; A_2—acrylics; W—wool.

3.4 UV resistance property

Then the UV resistance performance was tested, and the results are shown in Tab. 3.

Tab. 3 UV resistance property before and after dyeing

	Undyed silk	80 ℃ dyed silk	130 ℃ dyed silk
UVA	16.15	2.25	1.46
UVB	10.15	1.94	1.46
UPF	8.24	49.91	67.8

It can be seen from Tab. 3 that the ultraviolet transmittance of dyed silk fabrics decreases and the UPF value increases. Due to the different amounts of dye on the silk, the silk dyed at 130 ℃ showed better UV resistance than the silk dyed at 80 ℃. The results showed that the silk fabric dyed with camellia seed husk extract had the UV resistance function, and there was a positive correlation between the amount of dye and UV resistance function.

3.5 Antioxidant activity

The antioxidant activity of dyed silk was shown in Fig. 2. As can be seen from the figure, the absorbance at wavelength 517 nm decreased after staining, indicating that DPPH free radical was cleared. The reason for the antioxidant activity of dyed silk fabrics is that there are more phenolic hydroxyl groups in the dye structure, which can transfer hydrogen atoms and stop the chain reaction of free radicals[8-9]. Dyed silk fabrics at 130 ℃ showed significantly stronger antioxidant activity than dyed silk fabrics at 80 ℃, which may be due to the difference in dye content on dyed silk fabrics.

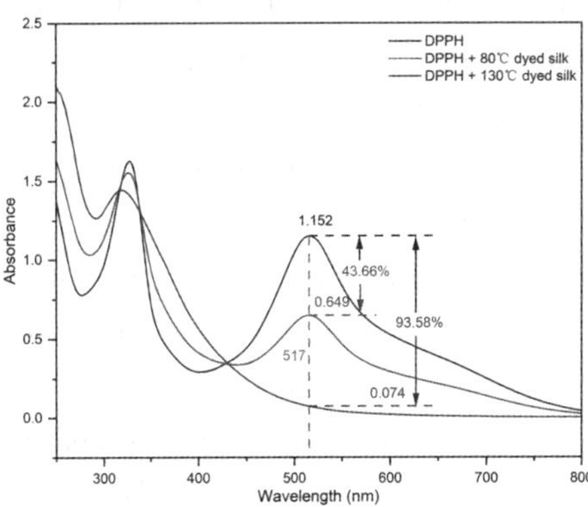

Fig. 2 Antioxidant activities of silk fabrics after dyeing

4 Conclusions

In this study, a method of extracting brown natural dye from camellia seed husk was proposed, and its application in silk fabric was discussed. Satisfactory dyeing results were achieved by optimizing dyeing process conditions. The six metal mordants have good dyeing effect on silk fabrics under pre-mordant dyeing condition, and the color fastness of the dyed silk fabrics is also good. The dyed silk fabrics showed better uv protection and antioxidant activity, and the difference in function was mainly related to the amount of dye on the silk fabrics.

References

[1] YANG T T, et al. Instrumental characterization and functional assessment of the two-color silk fabric coated by the extract from Dioscorea cirrhosa tuber and mordanted by iron salt-containing mud[J]. Industrial Crops and Products, 2018, 111: 117-125.

[2] JIANG H, et al. Extraction of natural dyes from the stem of Caulis spatholobi and their application on wool[J]. Textile Research Journal, 2019, 89(23, 24): 5209-5217.

[3] DEVI D, et al. Study of the structure of degummed Antheraea assamensis(muga) silk fibre[J]. The Journal of the Textile Institute, 2011, 102(6): 527-533.

[4] XIE Y, et al. Study on biomolecules in extractives of Camellia oleifera fruit shell by GC-MS[J]. Saudi J Biol Sci, 2018, 25(2): 234-236.

[5] ZHANG S, et al. Application of steam explosion in oil extraction of camellia seed (Camellia oleifera Abel.) and evaluation of its physicochemical properties, fatty acid, and antioxidant activities[J]. Food Sci Nutr, 2019, 7(3): 1004-1016.

[6] LI G, et al. Surface modification and functionalization of silk fibroin fibers/fabric toward high performance applications[J]. Materials Science and Engineering: C, 2012, 32(4): 627-636.

[7] EL-ZAHER N A, et al. Amendments of the structural and physical properties of cotton fabrics dyed with natural dye and treated with different mordants[J]. Journal of Natural Fibers, 2019, 18(9): 1247-1260.

[8] SUN S S, et al. Simultaneous coloration and functionalization of wool, silk, and nylon with the tyrosinase-catalyzed oxidation products of caffeic acid[J]. Industrial & Engineering Chemistry Research, 2013, 52(26): 8953-8961.

[9] LUO F, FEI X. Distribution and antioxidant activities of free, conjugated, and insoluble-bound phenolics from seven species of the genus camellia[J]. Journal of the American Oil Chemists' Society, 2019, 96(2): 159-170.

Study on Preparation and Properties of Polyacrylic Acid-Based Hydrogel Modified by Zwitterion/Cotton Fabric Composite Dressings

CHEN Hao[1], **KANG Gaowei**[1], **WU Dequn**[1,2*]

[1] *Key Laboratory of Textile Science & Technology, Ministry of Education, Donghua University, Shanghai 201620, China*

[2] *Innovation Center for Textile Science and Technology, Donghua University, Shanghai 201620, China*

* *Corresponding author's email*: dqwu@ dhu. edu. cn

Abstract: Hydrogel played an increasingly important role in biomedical fields such as tissue engineering, drug delivery and wound dressings. Polyacrylic acid-based hydrogel was widely used in transdermal drug delivery system, sweaty skin sensors and surgical dressings. However, with relatively low strength and poor toughness, the application of pure polyacrylic hydrogel was limited in treating some diseases. In this study, polyacrylic hydrogel was modified by incorporating into SBMA zwitterion. With increasing SBMA zwitterion, the fracture strength and elongation at break of the hydrogel increased, which indicated that the content of SBMA zwitterion within the gel affected its tensile strength and toughness. The cotton fabric was functionalized by grafting KH-570, strengthening the interfacial connection between hydrogel and fabric. Then, the hydrogel/cotton fabric composite dressings were prepared under radical polymerization reaction among functionalized cotton fabric, acrylic acid and SBMA monomers. The mechanical properties had also been improved. The resulting composite dressings were confirmed suitable for antibacterial and transdermal applications in the next further study.

Keywords: Hydrogel; Modified polyacrylic acid hydrogel; Functionalized cotton fabric; Composite dressing; Mechanical properties

1 Introduction

Hydrogels are a unique class of polymeric materials that possess an interconnected porous network and exhibit remarkable properties including high surface area, an accommodating matrix, flexibility, controllable mechanical strength, and excellent biocompatibility[1]. Hydrogels have widely broad applications in tissue adhesives, hemostatic agents, biomedical sensors, wound dressings, soft electronics and actuators[2-3]. Since hydrogels were firstly created, owing to its structure similar to tissue, it had been used as biological materials such as filling after enucleation of the eye[4]. Because of good compatibility, softness, the ability to load medicine drugs, hydrogels were also chosen as transdermal drug delivery system to treat many diseases, changing traditional delivery method. Methotrexate-loaded nanomicelles were carried by polyacrylic acid-based hydrogel for the purpose of treating Rheumatoid Arthritis, resulting higher encapsulation and expected therapeutic effect[5].

With a lot of hydroxyl groups on the fiber's surface, cotton fabrics were liable to functionalized in physical or chemical approaches, which makes they connect with hydrogel better[6-7]. The resulting composites composed of activated cotton fabrics and hydrogel were suitable for antibacterial dressings and transdermal drug delivery patches in the market.

To improve hydrogel's properties, adding moderate physical or chemical interactions including electrostatic, hydrophobic, host-guest interactions, $\pi-\pi$ conjugation, hydrogen bonding and other crosslinked networks was the general methods[8].

In this study, polyacrylic hydrogel was modified by incorporating into SBMA zwitterion. Acrylic monomers and SBMA monomers were copolymerized by UV light through radical polymerization. Afterwards, the tensile properties of the modified hydrogel were investigated. The cotton fabric was functionalized by grafting KH-570. Then, the hydrogel/cotton fabric composite dressings were prepared under radical polymerization reaction among functionalized cotton fabric, acrylic acid and SBMA monomers. The tensile properties of hydrogel/cotton composites were also investigated.

2 Experiments

2.1 Materials

Acrylic acid (AA), N, N'-Methylenebis (acrylamide) (MBA), 3-(Trimethoxysilyl) propyl methacrylate (KH-570) and Ethanol absolute (EtOH) were purchased from Sinopharm Chemical Reagent Co., Ltd. (Shanghai, China). [2-(Methacryloyloxy) ethyl] dimethyl-(3-sulfopropyl) ammonium hydroxide (SBMA) was purchased from Wendong Chemical Co., Ltd. (Shanghai, China). 2-hydroxy - 2-methylpropiophenone (HMP) was purchased from Aladdin (Shanghai, China). Cotton fabric (245 g/m^2) was provided by Yongyue Textile Technology Co., Ltd. (Huaifang, China).

2.2 Preparation of hydrogels

The hydrogels were synthesized utilizing a facile one-pot method. Briefly, AA, SBMA, MBA, and HMP were dissolved in deionized water and homogeneously stirred, followed by degassing with bubbling nitrogen for 10 min. Subsequently, the mixture was poured into customized molds and then transferred under a UV lamp (365 nm) with an intensity of 1 200 mW/cm^2 for photo polymerization within 20 min to form hydrogels. The recipe of the modified hydrogel is presented in Tab. 1.

Tab. 1 The recipe of modified hydrogel

Sample	AA (g)	SBMA (g)	MBA (g)	HMP (g)	H_2O (mL)
PAA	1.50	0	0.0075	0.015	4.5
PASA1	1.47	0.03	0.0075	0.015	4.5
PASA2	1.44	0.06	0.0075	0.015	4.5
PASA3	1.41	0.09	0.0075	0.015	4.5
PASA4	1.38	0.12	0.0075	0.015	4.5
PASA5	1.35	0.15	0.0075	0.015	4.5
PASA6	1.32	0.18	0.0075	0.015	4.5
PASA7	1.29	0.21	0.0075	0.015	4.5

2.3 Preparation of functionalized cotton fabric and hydrogel/cotton fabric composites

0.5 mL KH-570 was added into 20 mL water/ethanol (1 : 1) mixed solution and stirred a certain time, followed by immersing cotton fabrics (10 cm × 10 cm, 10 g) into it for 12 h under the dark condition. Then the cotton fabrics were washed with a lot of deionized water and dried at 60 ℃. Finally, the modified cotton fabric samples were prepared and denoted as Ac-o. To prepare the hydrogel/cotton composites, Co and Ac-Co were put into a customized sample mold, followed by injecting the hydrogel precursor solution into them, then the samples were cured under UV light for 20 min. The obtained samples were named as Co-Gel and Ac-Co-Gel.

2.4 Spectroscopic analyses

Fourier transform infrared (FTIR) spectra over the wavenumber ranging from 4 000 to 400 cm^{-1} were achieved at room temperature with PerkinElmer FTIR Spectra 1 000.

2.5 Mechanical tests of hydrogels

To research the effect of SBMA concentration in polyacrylic acid-based hydrogel, tensile tests of hydrogel were carried through computerized tension and pressure testing machine at a tensile rate of 50 mm/min with a gauge length of 30 mm. The hydrogel sample size was 50 mm×5 mm×2 mm.

2.6 Mechanical tests of hydrogel/cotton fabric composites

Tensile tests of hydrogel/cotton fabric composites were conducted on an electronic universal tensile testing machine at a tensile rate of 100 mm/min with a gauge length of 200 mm, to clarify the effect of treatment and connected hydrogel on the cotton fabric. The cotton fabric and composite size was 250 mm×50 mm.

3 Results and Discussion

3.1 Fabrication of hydrogel/cotton fabric composites

Fig. 1 showed the preparation process of hydrogel, modified cotton fabric and hydrogel/cotton fabric composite. As shown in Fig. 1 (a), the hydrogels were fabricated by the facile one-step radical copolymerization of SBMA and AA units by UV irradiation with MBA as the covalent cross-linkers and HMP as the initiators. Fig. 1 (b) demonstrated the one simple step of the modification of cotton fabric, aiming at making cotton fabrics connect with hydrogel better. Fig. 1 (c) illustrated the hydrogel/cotton fabric composite dressings were prepared under radical polymerization reaction among functionalized cotton fabric, acrylic acid and SBMA monomers.

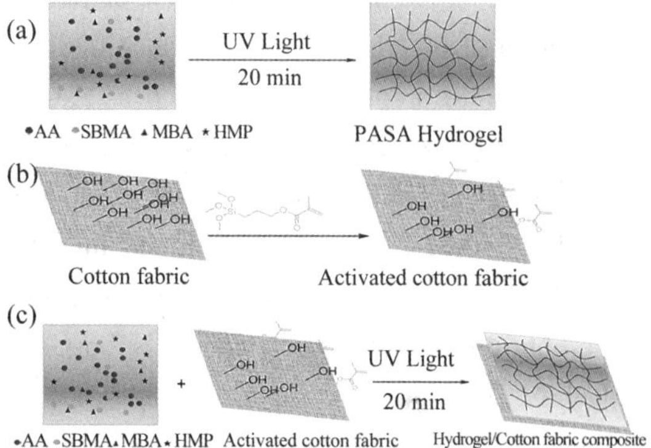

Fig. 1 (a) Preparation process of PASA hydrogel; (b) the functionalized route for the cotton fabric; (c) schematic diagram of the hydrogel/cotton fabric composite dressing

3.2 Synthesis and performance of hydrogels

FTIR spectra of PAA, PASA, SBMA were shown in the Fig. 2 (a). In the FTIR spectra of PASA, a band scope from 3 500 cm^{-1} to 2 500 cm^{-1} was due to the stretching of the —OH groups. Peak at 1 711 cm^{-1} was due to the stretching of the C =O groups. Peak at 1 035 cm^{-1} was due to the stretching of the S =O groups. Peak at 961 cm^{-1} and 924 cm^{-1} was due to the stretching of the S—O groups. These characteristic peaks showed that the PASA hydrogel was successfully synthesized. Hydrogels' optical images were shown in the Fig. 2 (b). All hydrogels were transparent and performed a little yellow color.

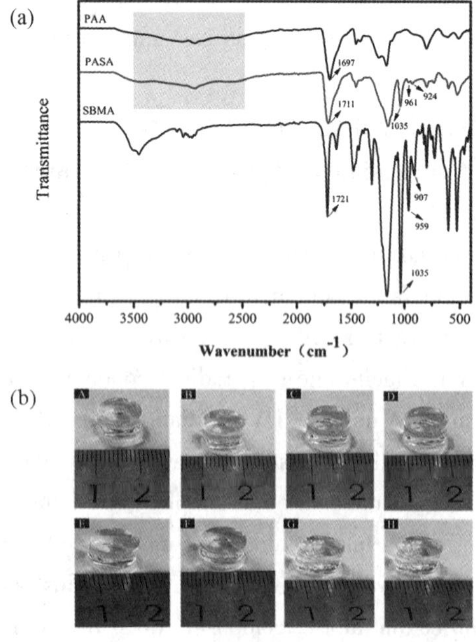

Fig. 2 (a) The FTIR spectra of PAA, PASA, SBMA; (b) hydrogels' optical images

3.3 Mechanical properties of modified hydrogels

Hydrogel needs to maintain certain mechanical properties during use. Modified polyacrylic acid-based hydrogels' stress-strain curve was illustrated in Fig. 3 (a). It was found that with the SBMA concentration increasing gradually, hydrogel's stress increasing performed a little change from Fig. 3 (b). However, the elongation at break of the hydrogel improved obviously in the Fig. 3 (c). Through analyzing stress and strain change, it was reflected that the intensity increased by 13.9% relatively and the elongation at break improved by 85.8% relatively on PASA7 hydrogel sample. On the one hand, with the increasing of zwitterion, strong dipole-dipole interactions could improve hydrogel system's strength properties. On the other hand, the positive and negative charges of zwitterion maybe attract other charged

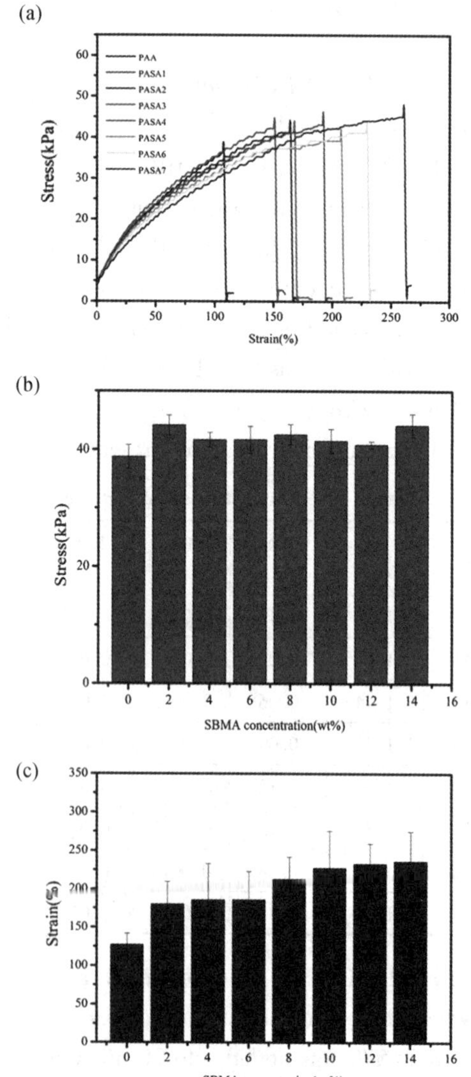

Fig. 3 (a) Modified polyacrylic acid-based hydrogels' stress-strain curve; (b) stress-SBMA concentration histogram; (c) stress-SBMA concentration histogram

functional groups, destroying internal structure in the hydrogel, thus lowering its strength properties. It was the positive and negative effect of zwitterion that SBMA mainly increased polyacrylic acid-based hydrogel's toughness.

3.4 Mechanical properties of hydrogel/cotton fabric composite dressing

It was conformed that cotton fabrics were functionalized successfully from Fig. 4 (a). In the FTIR spectrum, a new peak was found at 1 718 cm^{-1}, which is attributable to the stretching vibrations of C=O groups. Peak at 815 cm^{-1} was due to the stretching of the Si—O groups.

(a)

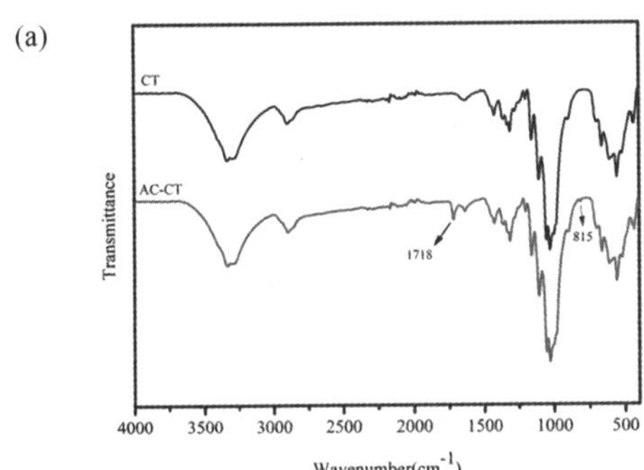

(b)

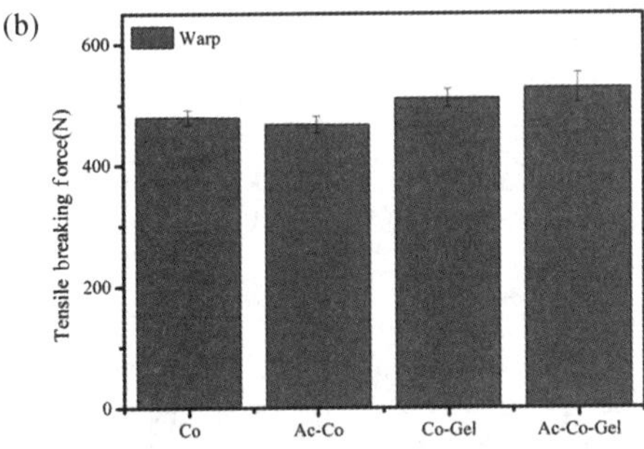

(c)

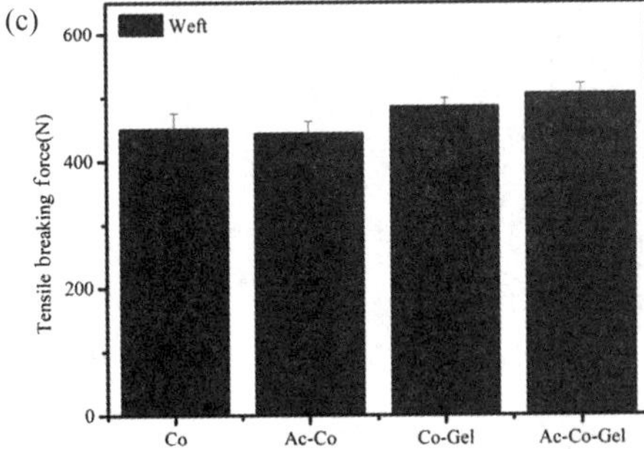

Fig. 4 (a) FTIR of cotton fabric and functionalized cotton fabric; (b) composites' tensile properties in the warp direction. (c) composites' tensile properties in the weft direction

Composites' tensile properties were presented at Figs. 4 (b) and 4(c). the tensile breaking force of the fabric before and after the modification decreased within a certain range in both warp and weft directions, indicating that the functionalization had little effect on the strength of the fabric. The mechanical tensile strength had been improved after connecting with hydrogel. The double bonds of the fabric functionalized can participate in the free radical copolymerization of hydrogels, and there is not only physical interaction but also chemical action between hydrogels and fabrics.

4 Conclusions

In summary, a kind of hydrogel/cotton fabric composite was designed that integrated the polyacrylic acid-based hydrogel modified by SBMA zwitterion and cotton fabric functionalized by grafting KH570. Because zwitterion had positive and negative influence on polyacrylic acid-based hydrogel system, incorporating into SBMA created a little bit of mechanical strength enhancement and mainly approved the toughness of the hydrogel. Owing to the existence of double bonds on the surface of functionalized cotton fabric, cotton fabric could connect with hydrogel network better, and the tensile properties of the hydrogel/cotton fabric composite dressing had been improved. The resulting composite dressings were confirmed suitable for antibacterial and transdermal applications in the next further study.

References

[1] ZHANG W, WANG R X, SUN Z M, et al. Catechol-functionalized hydrogels: biomimetic design, adhesion mechanism, and biomedical applications [J]. Chemical Society Reviews, 2020, 49(2): 433-464.

[2] PEI X, WANG J, CONG Y, et al. Recent progress in polymer hydrogel bioadhesives [J]. Journal of Polymer Science, 2021, 59(13): 1312-1337.

[3] ZHANG Y S, KHADEMHOSSEINI A. Advances in engineering hydrogels [J]. Science, 2017, 356 (6337): eaaf3627.

[4] WICHTERLE O, LIM D. Hydrophilic gels for biological use[J]. Nature, 1960, 185: 4706.

[5] QINDEEL M, KHAN D, AHMED N, et al. Surfactant-free, self-assembled nanomicelles-based transdermal hydrogel for safe and targeted delivery of methotrexate against Rheumatoid Arthritis [J]. ACS Nano, 2020, 14 (4): 4662-4681.

[6] BENLTOUFA S, MILED W, TRAD M, et al. Chitosan hydrogel-coated cellulosic fabric for medical end-use: antibacterial properties, basic mechanical and comfort properties[J]. Carbohydrate Polymers, 2020, 227: 115352.

[7] CHATTERJEE S, HUI P C L, WAT E, et al. Drug deliv-

ery system of dual-responsive PF127 hydrogel with polysaccharide-based nano-conjugate for textile-based transdermal therapy [J]. Carbohydrate Polymers, 2020, 236: 116074.

[8] SHAO Z, CHENG W, HU X, et al. An anti-pressure, fatigue-resistant and rapid self-healing hydrogel based on a nano-micelle assembly11Electronic supplementary information (ESI) available: materials and experimental procedures, FT-IR, 1H NMR, TEM, SEM, swelling tests, rheological analysis or additional details [J]. Polymer Chemistry, 2020, 11(13): 2300-2304.

Preparation and Performance of Electrospinning Nanofiber Membrane Carrying Menthol

YIN Haofei[1], ZHAO Yi[1*]

[1] *Engineering Research Center of Technical Textile, Ministry of Education, Shanghai 201620, China*

[*] *Corresponding author's email*: zhaoyi@ dhu. edu. cn

Abstract: Menthol, the main component in essential balm, has the potential function of antibacterial, antiseptic, and antioxidant properties. However, its application has been greatly limited due to the high volatility and poor water solubility of menthol itself. To overcome these issues, polyethylene pyrrolidone (PVP)/essential balm nanofibrous membranes are proposed as a potential way of utilizing menthol effectively through the electrospinning process. Infrared spectrometer analysis shows that menthol is successfully loaded into the fiber. When the concentration of the loaded substance is 12 wt%, the average diameter of the fibers is 303 nm. The contact angle test analysis investigates that the addition of menthol reduces the hydrophilicity and enhances the stability of the nanofiber membrane. Furthermore, the release characteristics of menthol are also investigated, showing that the release time of menthol has been effectively increased from 2 h to 5 h after electrospinning.

Keywords: Electrospinning; Drug loading; Fiber membrane; Slow release

1 Introduction

As the first line of defense of the human body, the skin participates in the functional activities of the whole body to resist all aspects of external attacks[1]. The essential balm can treat mosquito bites, burns, and scalds. After being applied to the skin, there is an obvious sense of coolness. It can eliminate skin inflammation and reduce the symptoms of redness and swelling. Its main active ingredient is menthol. Menthol has unique characteristics of aromatic and flavor compounds and exists widely in nature. Menthol has been widely used in cosmetics and over-the-counter drugs for more than a century because of its excellent functions and effects such as itching, antibacterial, and antioxidant property[2].

However, in practical application, the application of essential balm is real-time, so the drug can be wrapped in fiber to achieve a sustainable drug delivery process. In recent years, electrospinning fiber membranes have attracted a great quantity of attention in skin healing and care because of their high specific surface area, high porosity and controllable morphology, small fineness, and high uniformity[3]. The electrospinning fiber membrane not only can physically isolate viruses and bacteria but also is breathable and moisturizing, creating a good healing environment for the skin. The plant extract has antibacterial activity and natural environmental protection, so it is of great significance to prepare a functional fiber membrane.

The electrospinning technique is used because of the high volatility and poor water solubility of menthol, which limits its development and application. Compared with common polymers, the macromolecular carriers used for electrospinning are more environmentally friendly[4]. Polyvinyl pyrrolidone (PVP) is a synthetic polymer material with high hydrophilicity, good compatibility, non-toxicity, and biodegradability. It is currently widely used in the preparation of electrospinning medical antibacterial and fresh-keeping materials[5].

In this study, the composite nanofiber membrane was prepared by electrospinning technology using PVP as a carrier to load essential balm containing menthol with natural antibacterial activity. The membrane has excellent sustained-release performance, which maximizes the application potential of menthol. The effects of drug loading in electrospinning of PVP solution on fiber formation and its microscopic morphology and hydrophilicity properties were investigated.

2 Experiments

2.1 Materials

Polyvinylpyrrolidone (PVP, 250 kDa) and alcohol were obtained from Sinopharm Chemical Reagent Co., Ltd (Shanghai, China). The balm containing menthol was provided by Shanghai Zhonghua Pharma Co., Ltd (Shanghai, China).

2.2 Preparation of nanofibers

PVP was dissolved in absolute alcohol at 13% (wt/wt). Menthol extract was added into the PVP/alcohol solutions at 8%, 10%, and 12% (wt/wt) and then agitated at room temperature under constant stirring for at least 24 h before electrospinning.

Using a laboratory-made electrospinning device as shown in Fig. 1, under the environment of temperature (25±3) ℃ and relative humidity (40±5)%, the drum was covered with aluminum foil to receive nanofibers. The voltage was set at 15 kV. The bolus injection speed was set at 2 mL/h. The spinning distance was 15 cm. The drum rotation speed was 400 r/min.

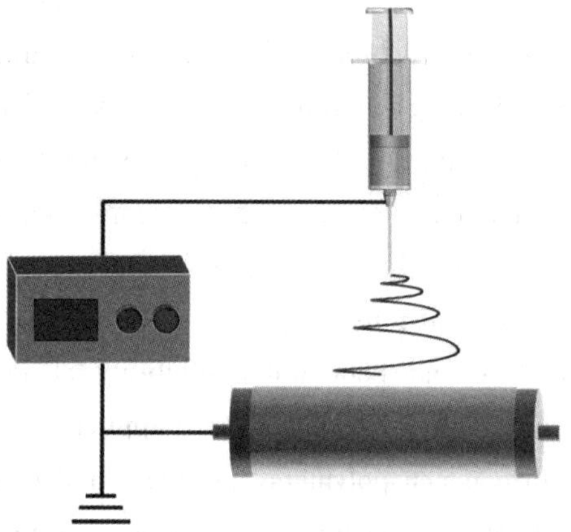

Fig. 1 Electrospinning device

2.3 Morphological characterization

The morphology of the fibers was examined by scanning electron microscopy (SEM, Thermo Fisher Scientific). The diameter of the fibers was measured by using the software called Nano Measurer 1.2.

2.4 Chemical structure characterization

Fourier transform infrared (FTIR) spectra were obtained from a Thermo Scientific Nicolet iS10 spectrometer.

2.5 Hydrophilic test

The contact angle was conducted using simulated body fluid (SBF) at a drop volume of 2 μL and an optical tensiometer (Attension Theta, Biolin Scientific, Germany) after ten seconds. Every type of sample is measured three times to get better experimental results.

2.6 Release test

A certain amount of menthol fiber membrane was taken and placed in an air blast oven for release, with the temperature set at 30 ℃. According to the settled time interval, use a balance to weigh the mass of the fiber membrane, record the change of the fiber membrane mass, and make the change curve of the weight loss of the fiber membrane with the release time.

3 Results and Discussion

3.1 Morphological analysis

The experimental results show that the concentration of the supporting material has a decisive effect on fiber formation and morphology. When the spinning solution was prepared with the essential balm in different ratios, the average fiber diameter increased slowly with the increase of drug loading. At the critical concentration of fiber formation corresponding to Fig. 2, if the solution concentration is too low, the solvent in the droplet in the electric field cannot volatilize completely, resulting in the coexistence of a large number of bead structures and fibers. When the concentration reaches 10%, the amount of coarse fiber decreases, and the fiber shows a trend to be uniform and smooth. When the concentration reaches the critical concentration above, the electric field force makes the fiber split, forming coarse fi-

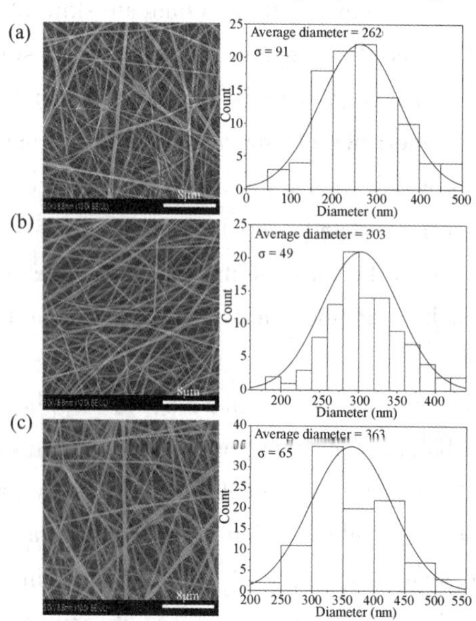

Fig. 2 The fibers morphology and the fibers diameter distribution histogram: (a) PVP/8% menthol; (b) PVP/10% menthol; (c) PVP/12% menthol

ber with a small specific surface area. Part of the solvent is difficult to volatilize completely, resulting in fiber adhesion. The optimal concentration of the menthol was determined to be 10%.

3.2 Chemical structure analysis

As shown in Fig. 3, the characteristic functional groups of menthol are —CH_3 and —OH, so the infrared spectra of essential balm have characteristic absorption peaks near 2 921 cm^{-1} and 1 442 cm^{-1}, respectively. Compared with pure PVP membranes, with the increasing mass fraction of essential balm embedded in nanofibers, the absorption peak at 1 657 cm^{-1} and 2 926 cm^{-1} become more and more obvious and shift to a long wavelength, which corresponds to menthol. These results indicate that menthol is loaded into PVP nanofibers and interacts with PVP molecules.

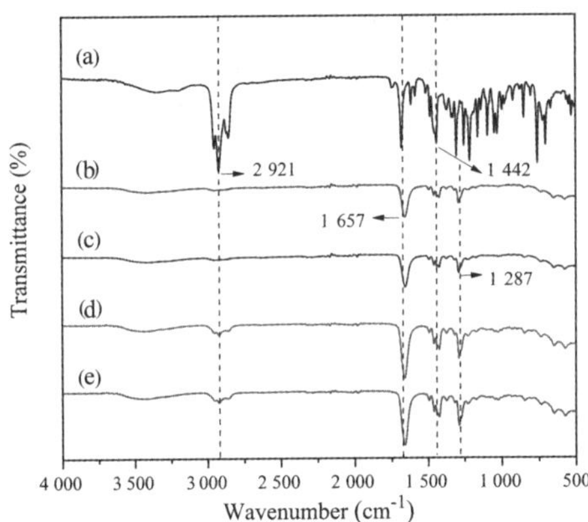

Fig. 3 Infrared spectrum analysis of fiber membrane: (a) pure balm; (b) PVP; (c) PVP/menthol 8%; (d) PVP/menthol 10%; (e) PVP/menthol 12%

3.3 Hydrophilic analysis

To determine hydrophilicity, the surface wettability was examined by using the SBF. As shown in Fig. 4(a), it can be observed that the contact angle of PVP fiber membrane is 5.8°, reflecting the relatively high hydrophilicity of 100%. However, the contact Angle of PVP/balm nanofiber membrane gradually increased from 30.5° to 51.5° as shown in Figs. 4(b)-(d). The hydrophilicity of the nanofiber membrane decreases while its stability increases. The contact angle of PVP/balm nanofiber membrane increases with the increase of balm content since menthol is hydrophobic. The addition of menthol reduces the hydrophilicity of PVP and enhances the stability of the fiber membrane.

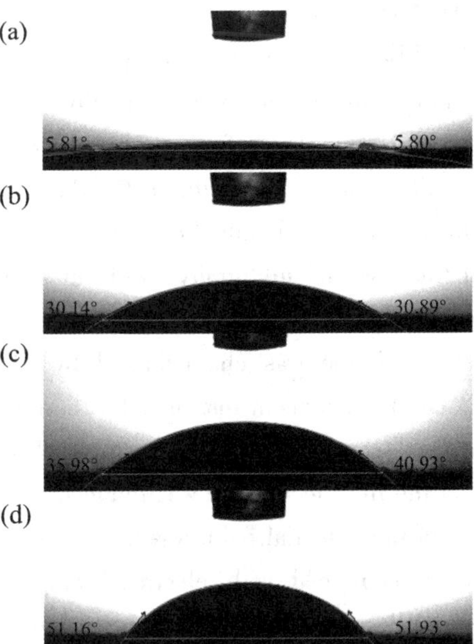

Fig. 4 Contact angle of SBF on electrospinning fibers with menthol concentrations of (a) 0%, (b) 8%, (c) 10%, and (d) 12%

3.4 Release analysis

The release of menthol was measured by the menthol-loaded fiber membrane at 30 ℃. As can be seen from Fig. 5, the sudden release behavior of balm was completed in 2 h. Menthol was gradually released from the electrospinning fiber membrane, and menthol was released at 5 h by PVP/15% menthol fiber membrane, showing a sustained-release effect. The release rate curves of the three different loading concentrations were basically the same, the release time was 5 h, and the cumulative release rate had little difference. The results showed that the drug loading had little effect on the release rate of menthol after electrospinning. It may be related to the high porosity of fiber membrane and the volatile nature of menthol.

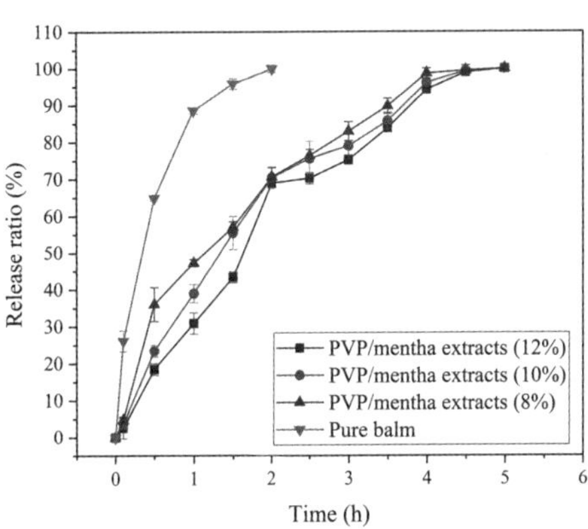

Fig. 5 The release curve of the membrane

4 Conclusions

The essential balm can be successfully incorporated into PVP spinning solutions to provide nanofiber membrane by electrostatic spinning. The results show that the composite fiber membrane with 13% PVP content, 10% balm content, and anhydrous ethanol as solvent has good fineness and uniformity, with an average diameter of 303 nm.

The fiber membrane was characterized by FITR and contact Angle measurement instrument. The release behavior of the essential balm was measured. The hydrophilicity of the fiber membrane was reduced and the release time of the essential balm was prolonged. The essential balm combined with electrospinning solutions based on PVP provides a promising strategy for skin management of the essential balm.

References

[1] GLIK J, ABU W, KITALA D, et al. A 2000 patient retrospective assessment of a new strategy for burn wound management in view of infection prevention and treatment [J]. International Wound Journal, 2017, 15(3): 344-349.

[2] REZAEINIA H; GHORANI B, EMADZADEH B, et al. Prolonged-release of menthol through a superhydrophilic multilayered structure of balangu (Lallemantia royleana)-gelatin nanofibers [J]. Materials Science & Engineering C, 2020, 115: 111115.

[3] QI R L, TIAN X J, GUO R, et al. Controlled release of doxorubicin from electrospun MWCNTs/PLGA hybrid nanofibers [J]. Chinese Journal of Polymer Science, 2016, 34(9): 1047-1059.

[4] SAJEEV U S, ANAND K A, MENON D, et al. Control of nanostructures in PVA, PVA/chitosan blends and PCL through electrospinning[J]. Bulletin of Materials Science, 2008, 31(3): 343-351.

[5] HUANG L B, CAO K, HU P, et al. Orthogonal experimental preparation of Sanguis Draconis- polyvinylpyrrolidone microfibers by electrospinning[J]. Journal of Biomaterials Science Polymer Edition, 2019, 30(4): 308-321.

Fabrication of Self-Healing Hydrogel/Cotton Fabric Wound Dressings with Antibacterial Properties

NIU Liting[1], LI Mengjuan[1], WU Dequn[1,2*]

[1] *Key Laboratory of Textile Science & Technology, Ministry of Education, Donghua University, Shanghai 201620, China*

[2] *Innovation Center for Textile Science and Technology, Donghua University, Shanghai 201620, China*

* *Corresponding author's email*: dqwu@ dhu. edu. cn

Abstract: Development of wound dressing with good antibacterial, self-healing and hygroscopic abilities is of great significance for management of wounds. Here, we prepared a hydrogel/cotton fabric wound dressing with self-healing ability and good antibacterial properties. Quaternary ammonium salt modified cotton fabric provides certain mechanical strength and good antibacterial effect. Hydrogels were synthesized by Schiff base reaction using carboxymethyl chitosan (CMC) and oxidized konjac glucomannan (OKGM) in aqueous solution. The hydrogel can keep the wound moist and can self-heal within 10 min for some external injuries. The hydrogel/cotton fabric wound dressing has good absorbent capacity and exhibits broad antibacterial activity against S. aureus and E. coli. Taken together, the dressing using self-healing hydrogels and cotton fabric with good antibacterial property are very promising self-adapting biomaterials for repairing irregular wounds.

Keywords: Hydrogel; Antibacterial cotton fabric; Self-healing; Wound dressing

1 Introduction

Wound healing is one of the important issues in the field of biomedicine, because skin damages have widespread existence. However, wound healing is a complexly physiological process including hemostasis, inflammation, proliferation, and tissue remodeling, and there will be overlapping and mutual influence between each stage. And the process of wound healing is often accompanied by the wound infection resulting from bacterial invasion. At present, traditional cotton dressings are still widely used in wound care due to their cheapness and convenience[1]. And many new wound dressings have been developed, such as semipermeable membranes, semipermeable membrane foams, hydrocolloids and hydrogels.

Although traditional dry cotton dressings can cover and protect the wound from infection and promote scabbing, they can adhere to the wound surface and cause secondary mechanical damage. Furthermore, dry cotton dressings are not antibacterial and do not inhibit bacteria, nor do they create a moist environment that is conducive to wound healing. As a result, these dressings cause wound dehydration and loss of biologically active substances, which further extends the wound healing time[2]. Hydrogels have become the most competitive candidates for wound dressings due to their good hydrophilicity, biocompatibility, and three-dimensional porous structure like extracellular matrix (ECM)[3-5]. Hydrogels can provide a moist environment at the wound interface, absorb excess tissue exudates and allow oxygen permeation[6]. In particular, self-healing hydrogels that can adapt to different wound shapes and can withstand a certain external force are considered to be promising skin wound healing materials[7-8]. Therefore, it can be considered to combine the hydrogel with cotton fabric to prepare a double-layer wound dressing. Both Supplement each other and play the advantages together.

Here, we prepared a hydrogel/cotton fabric wound dressing with self-healing ability and good antibacterial properties. For self-healing hydrogels, hydrogels were synthesized by Schiff base reaction using carboxymethyl chitosan (CMC) and oxidized konjac glucomannan (OKGM) in aqueous solution. Due to the dynamic covalent Schiff bases, the hydrogels exhibited rapid self-healing properties without any external stimuli. Cotton fabrics are modified with quaternary ammonium salts, so that the cotton fabrics have good antibacterial properties. The hydrogel precursor solution was then in situ gelled on cotton fabric to prepare a hydrogel/cotton fabric wound dressing.

2 Experiments

2.1 Materials

Carboxymethyl chitosan (CMC), konjac glucomannan (KGM), ethylene glycol ($C_2H_6O_2$) and sodium periodate ($NaIO_4$) were purchased from Shanghai Yuanye Bio-Technology Co., Ltd. (Shanghai, China). 3-(Trimethoxysilyl) propyl dimethyl octadecyl ammonium chloride (QAC) and Ethanol absolute (EtOH) were purchased from Sinopharm Chemical Reagent Co., Ltd. (Shanghai, China). Cotton fabric (245 g/m^2) was provided by Xiajin Renhe Textile Technology Co., Ltd. (Dezhou, China).

2.2 Synthesis of oxidized konjac glucomannan (OKGM)

OKGM is obtained by oxidizing KGM with $NaIO_4$. Briefly, dissolve 4 g of KGM in 400 mL of distilled water. Then 4 g of $NaIO_4$ was added and the mixture was stirred at room temperature in the dark for 8 hours. Finally, 3 mL of $C_2H_6O_2$ was added to the mixture and stirred for 1 hour to react the unreacted $NaIO_4$. The resulting product was dialyzed against distilled water (MWCO 14 000 Da) for 3 days and freeze-dried to obtain OKGM.

2.3 Synthesis and characterizations of hydrogels

5 wt% CMC, 5 wt% OKGM solutions were separately prepared in distilled water at room temperature. The hydrogels were prepared by mixing the CMC solution and the OKGM solution at 1 : 1 and 1 : 2 volume ratios, respectively, at 37 ℃, named CMC-OK-1 and CMC-OK-2.

Scanning electron microscopy (SEM), the Fourier transform infrared spectroscopy (FT-IR), the ^1H nuclear magnetic resonance (^1H NMR) were used to investigate the chemical and physical characterizations of copolymer samples and hydrogels.

2.4 Preparation of modified cotton fabric and hydrogel wound dressings

To prepare antibacterial cotton fabric (ACT), A certain mass of QAC was added to 20 mL of water/ethanol (1 : 1) mixed solution, and then a cotton fabric sample (5 g) was added to the above solution and stirred for 8 h. Afterwards, the cotton fabric was washed several times with deionized water and dried. The formula of antibacterial cotton fabric is detailed in Tab. 1. To prepare the hydrogel/cotton bilayer wound dressings, ACT was put into a custom-made sample mold, and then the hydrogel precursor solution was injected into it, and the obtained sample was named ACT-gel.

Tab. 1 The recipe for the antibacterial cotton fabrics

Sample	QAC (g)	Cotton fabrics (g)
CT	0	5
ACT-1	0.1	5
ACT-2	0.3	5
ACT-3	0.5	5

2.5 Rheological and self-healing property of hydrogels

To confirm the stability of the hydrogel, the storage moduli (G') and loss moduli (G'') of hydrogels were measured through a rheometer with a scan frequency of 1 Hz.

Macroscopic self-healing experiments and rheological recovery tests were chosen to evaluate the self-healing property of hydrogels. The hydrogel were made into two discs, which were stained with acid yellow and methyl red. The two discs were cut into two semicircles respectively, and the two semicircles were put together for 10 min. Then the hydrogel was self-healed to a disc. Then using rheological recovery tests to test quantitative self-healing property of hydrogels. A strain amplitude sweep (γ from 1% to 1 000%, with fixed angular frequency 1 rad/s) was taken to determine the critical phase transition stress point. Then an alternate step strain sweep with fixed angular frequency 1 rad/s was taken to assess the shear thinning and self-healing behavior of the hydrogels. Amplitude oscillatory strains were switched from $\gamma = 1\%$ to $\gamma = 400\%$, and 5 cycles were carried out.

2.6 Swelling test of hydrogels

CMC-OK hydrogels were immersed in distilled water until reaching the equilibrated state (these weight did not increase). Then freeze-dried the hydrogel, weighed (W_0) and immersed in distilled water again. After every 30 s, the hydrogel was taken out, wiped surface moisture and weighed (W_t). The test would not stop until the W_t kept constant. The swelling ratio (ESR) was calculated using the following formulas:

$$ESR = \frac{W_t - W_0}{W_0} \times 100\% \qquad (1)$$

2.7 Antimicrobial activity property

The antibacterial zone and colony count method were used to determine antimicrobial activity of the treated fabric against the bacterium Escherichia coli (E. coli) and Staphylococcus aureus (S. aureus) according to GB/T 20944.3—2008. The antibacterial rate was calculated by the following equation:

$$F = \frac{W_t - Q_t}{W_t} \qquad (2)$$

where W_t is the number of bacterial colonies in the control sample; and Q_t is the number of colonies of bacteria treated with antibacterial cotton fabrics.

3 Results and Discussion

3.1 Formation and characterizations of self-healing hydrogels

As shown in Fig. 1 (a), the OKGM was obtained by a ring-opening reaction using sodium periodate, which resulted in the introduction of the dialdehyde group into the OKGM dimer unit. The CMC-OK hydrogels occurred at 37 ℃ between amino groups from CMC and aldehyde groups from OK by dynamic Schiff base cross-linking.

The chemical structure of OKGM was characterized by FT-IR and ^{1}H NMR spectroscopy. As shown in Fig. 1 (b), there is a weaker C=O bending vibration peak at 1730 cm^{-1}, while the hemiacetal peak appears at 890 cm^{-1}. The ^{1}H NMR spectra of OKGM and KGM are shown in Fig. 1 (c). A new peak at 9.18 ppm was observed in the spectrum of OKGM. This corresponds to the aldehyde group. These results demonstrated the successful synthesis of OKGM.

3.2 Structure and performance of hydrogels

The SEM micrographs of the cross-section of hydrogel were presented in Fig. 1 (d). It was clearly observed that the developed hydrogel displayed a highly porous structure, which allowed the dressing to be permeable and was beneficial to wound healing. Fig. 2 (a) shows the swelling properties of the hydrogel, and the swelling ratio of the hydrogel reaches 13-18 times within 3 min, which is beneficial for absorbing wound exudate at the wound.

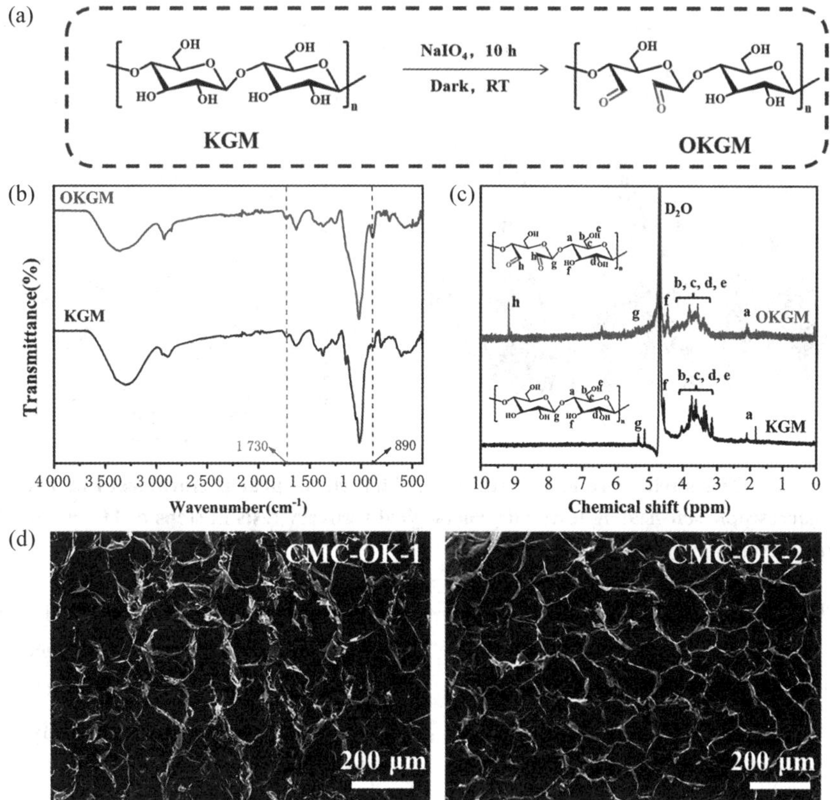

Fig. 1 (a) Route of synthesis of OKGM; (b) FTIR spectra of KGM and OKGM; (c) ^{1}H-NMR spectra of KGM and OKGM; (d) SEM images of the hydrogel

3.3 Rheological and self-healing property of hydrogels

The rheological properties of the hydrogels are shown in Fig. 2 (b). The frequency sweep profiles showed that the hydrogels of the indicated compositions had a stable G' approximately 10–20 times larger than the G'' in the range of 0.1–10 Hz.

Hydrogels used for wound dressing may be damaged by external force after being applied onto the wound site, and their self-healing property can prolong the lifespan. In Fig. 2 (c), the circular hydrogels were cut into half and the two different colors of fragments immediately healed into one piece when brought together in 10 min. In Fig. 2 (d), the rheological recovery performance of the hydrogels was measured by the successive step strain test. The G' value sharply decreased and was immediately lower than that of G'', when strain was 400% surpassing critical strain. While the strain returned to 1%, the G' and G'' values would recover or even surpassed the original values. The result demonstrated that a better hydrogel network was rebuilt in the renewed group after eliminating internal stress during the previous cycles. The collapse and renewal of the hydrogel network could repeat for several times, and the results further demonstrated that the hydrogels had a rapid and highly efficient self-healing capacity. The speedy and autogenic self-healing behavior was owing to the dynamic cross-linking with aldehyde groups from OKGM and amino groups from CMC. Thus, in this dynamic balance, the imine bonds would break and regenerate frequently to induce the self-healing capability of the hydrogels.

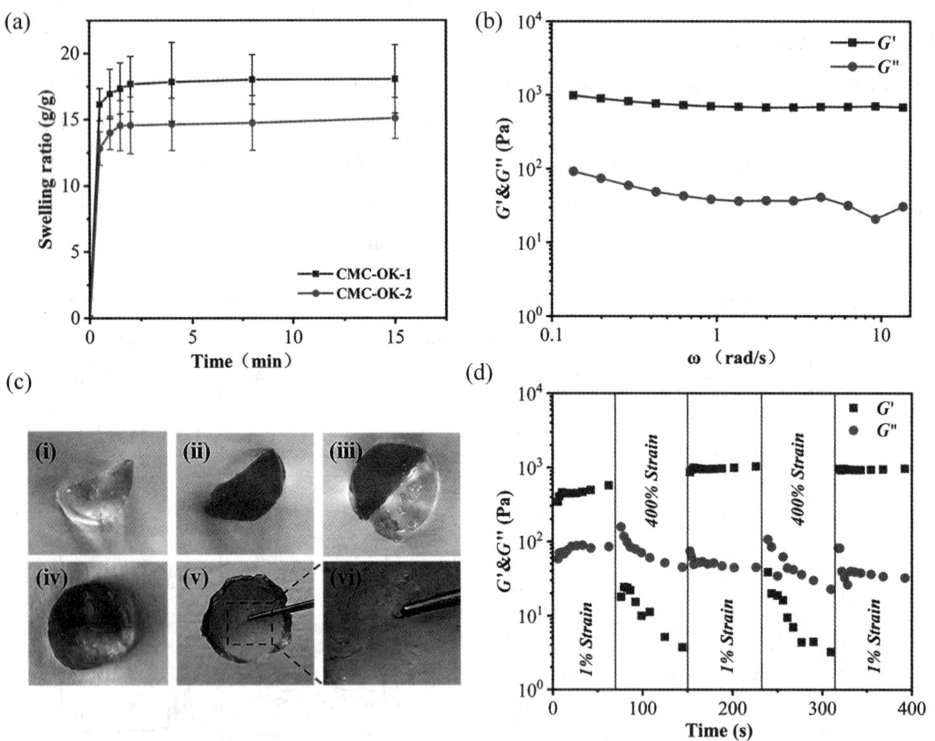

Fig. 2 (a) The swelling ratio of hydrogels; (b) rheological performances of hydrogels; (c) macroscopic self-healing test; (d) rheological recovery tests in steps of 1% and 400% oscillatory strain for cycles

3.4 Modification and characterization of cotton fabrics

The FTIR spectra, tensile properties and morphological observations of the antibacterial cotton fabrics are shown in Fig. 3. As shown in Fig. 3 (a), the absorption peaks 2 918 cm^{-1} and 2 851 cm^{-1} are assigned to the symmetrical stretching vibrations of the C—H peak, and the peak at 1 467 cm^{-1} is attributed to the bending vibrations of the C—H group from $—CH_3$ and $—CH_2—$. Furthermore, the peak appearing at 802 cm^{-1} is attributed to the Si—O—Si group, indicating that QAC has been successfully grafted on the cotton fabric surface.

The mechanical properties of modified cotton fabrics are crucial. As shown in Fig. 3 (b), the tensile breaking force of the fabric before and after modification decreased to a certain extent in both warp and weft directions, but after the hydrogel was combined with the fabric, the mechanical properties of the fabric were significantly improved. This can be seen in Fig. 3 (c).

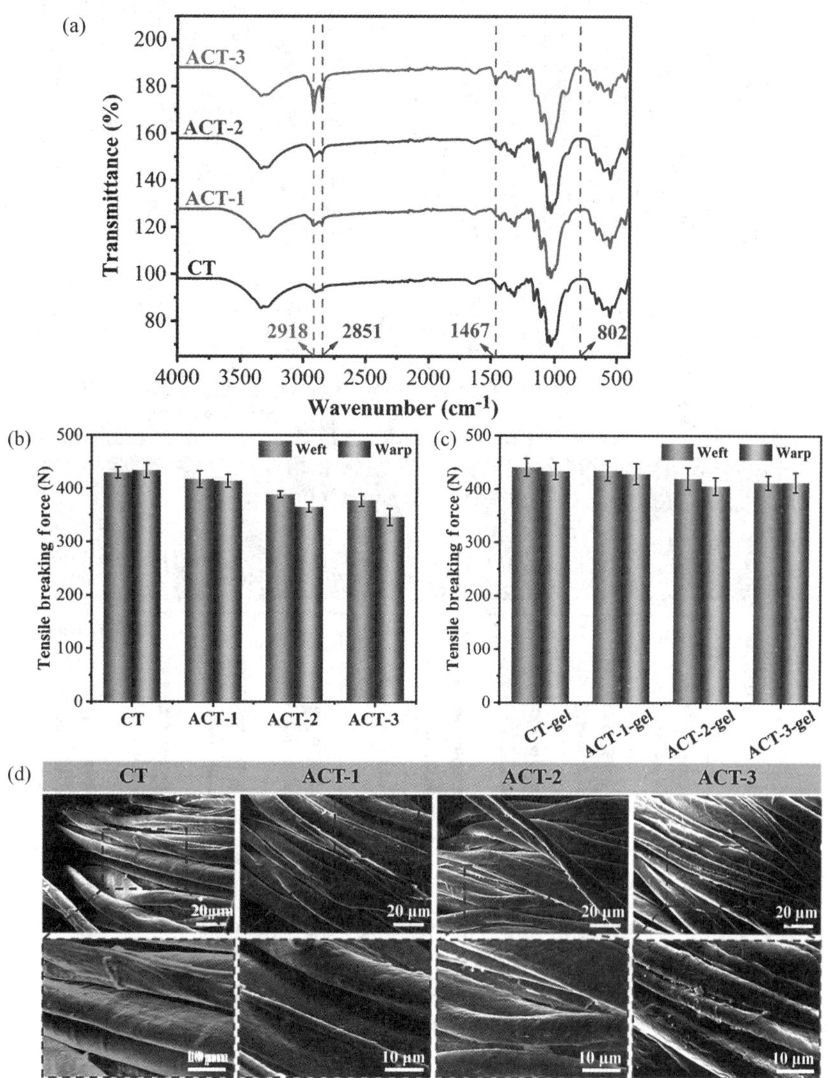

Fig. 3 (a) FTIR spectra of CT, ACT-1, ACT-2 and ACT-3; (b) tensile properties of cotton fabric; (c) tensile property of hydrogel/cotton wound dressing; (d) SEM micrograph of cotton fabric

SEM analysis to study the surface morphology of modified cotton fabrics. In Fig. 3 (d), it can be clearly seen that CT has a smooth longitudinal fibril structure, and after modification with antibacterial agent, the surface of the fabric is rough with trace particles. Fig. 4 (a) is a cross-sectional view of the hydrogel and cotton fabric composite dressing. It can be clearly seen that the hydrogel with large pores is on the left, and the dense fabric structure is on the right.

3.5 Antibacterial properties of modified cotton fabrics

Using E. coli and S. aureus as test strains, the antibacterial properties of antibacterial cotton and hydrogel/cotton fabric wound dressings were detected by colony counting method. As shown in Fig. 4 (b), there was little colony growth on the agar plates after treatment with antibacterial cotton fabric compared with the control. In Fig. 4 (c), the antibacterial rates of ACT-1, ACT-2, ACT-3, and ACT-3-gel were 92.24%, 97.75%, 99.14%, and 99.48% against E. coli and 93.65%, 96.51%, 98.73%, 99.36% against S. aureus. The results show that the modified cotton fabric has excellent antibacterial properties and meets the performance requirements of medical wound dressings.

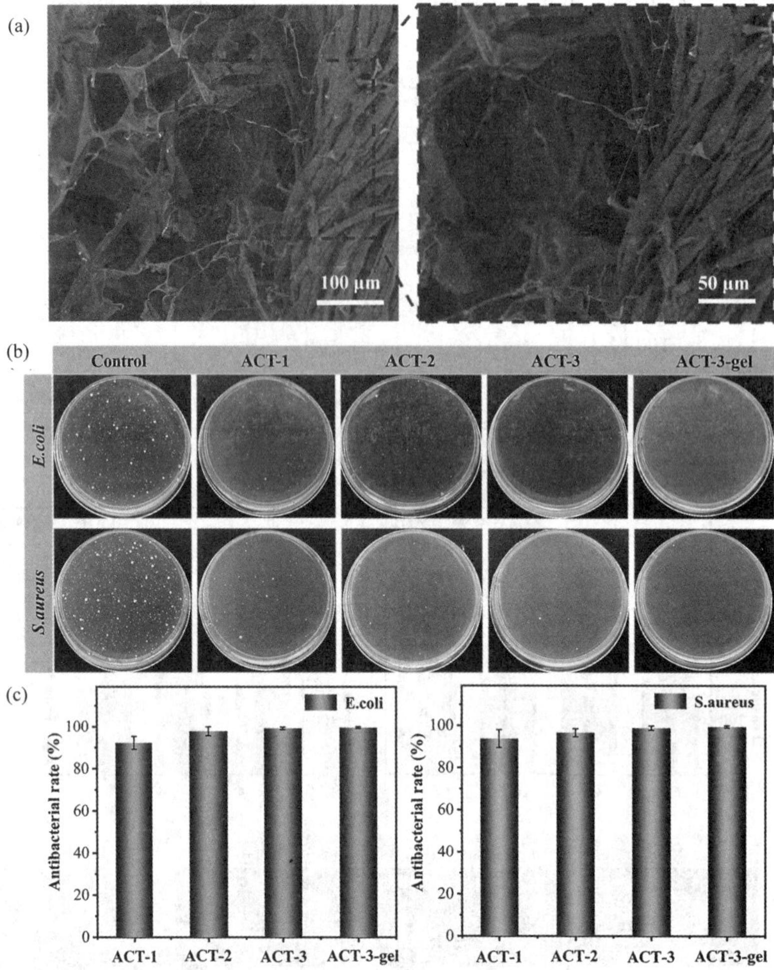

Fig. 4 (a) Cross-sectional SEM image of the hydrogel/cotton fabric wound dressing; (b) pictures of agar plates treated with antibacterial cotton fabric and hydrogel/cotton fabric wound dressing samples; (c) antibacterial rates of antibacterial cotton fabric and hydrogel/cotton fabric wound dressing samples

4 Conclusions

In this study, a hydrogel/cotton fabric wound dressing with self-healing ability and good antibacterial properties was designed. Among them, cotton fabric provides certain mechanical strength and good antibacterial effect. The hydrogel provides a moist environment and can achieve self-healing within 10 min for a certain external force damage. The swelling performance, tensile performance and antibacterial performance tests show that the composite dressing has good liquid absorption capacity, mechanical performance and antibacterial performance, and the bacteriostatic rate against E. coli and S. aureus are more than 99%, which can reduce wound infection chance. These properties endow this hydrogel/cotton fabric with the potential for use as wound dressings in moist and dynamic physiological environments.

References

[1] ZENG Q K, QI X L, ZHANG M, et al. Wound dressing: from nanomaterials to diagnostic dressings and healing evaluations[J]. ACS Nano, 2022, 16(2): 1708-1733.

[2] YANG C, LIU G F, CAI X J, et al. Chitosan and polyhexamethylene guanidine dual-functionalized cotton gauze as a versatile bandage for the management of chronic wounds. [J]. Carbohydrate Polymers, 2022, 282: 119130.

[3] YANG Z F, HUANG R K, WANG H, et al. Highly stretchable, adhesive, biocompatible, and antibacterial hydrogel dressings for wound healing[J]. Advanced Science, 2021, 8(8): 2003627.

[4] LIANG Y P, HE J H, GUO B L. Functional hydrogels as wound dressing to enhance wound healing[J]. ACS Nano, 2021, 15(4): 7078-7093.

[5] ZHU X, GUAN B G, LI X M, et al. Fabrication of an injectable hydrogel with inherent photothermal effects from tannic acid for synergistic photothermal-chemotherapy[J].

Journal of Materials Chemistry B, 2021, 9(30): 6084-6091.

[6] QIU W W, HAN H, WU D Q, et al. Nanofibers reinforced injectable hydrogel with self-healing, antibacterial, and hemostatic properties for chronic wound healing[J]. Journal of Colloid and Interface Science, 2021, 596: 312-323.

[7] QU J, ZHAO X, GUO B L, et al. Degradable conductive injectable hydrogels as novel antibacterial, anti-oxidant wound dressings for wound healing[J]. Chemical Engineering Journal, 2019, 362: 548-560.

[8] WANG Y X, XIE R Q, LU F, et al. A self-adapting hydrogel based on chitosan/oxidized konjac glucomannan/Ag NPs for repairing irregular wounds[J]. Biomaterials Science, 2020, 8(7): 1910-1922.

A Novel Epoxy Vitrimer for Pultrusion Molding of Carbon Fiber Composites

ZHANG Jian[1,2], ZHOU Gang[1,2], NING Na[1,2], WEI Yi[1*]

[1] *Center for Civil Aviation Composites, Donghua University, Shanghai 201620, China*

[2] *College of Textiles, Donghua University, Shanghai 201620, China*

[*] *Corresponding author's email*: Weiy@ dhu. edu. cn

Abstract: Based on the widely used pultrusion resin system in the market, we developed an epoxy vitrimer with triethanolamine as the bond exchange reaction catalyst, which meets the requirements of pultrusion process and is re-processable, degradable and recyclable. The rheological properties and curing behavior of the vitrimer were studied by a rheometer and a differential scanning calorimeter (DSC). The pultrusion process parameters were determined according to the simulation of the curing process by non-isothermal DSC. The thermal and mechanical properties were characterized by dynamic mechanical analysis (DMA) and universal testing machine. The results indicated that the thermal and mechanical properties of epoxy vitrimer were comparable to those of commercially available resins. Finally, the degradation of epoxy vitrimer and recycling of carbon fibers were also demonstrated.

Keywords: Epoxy resin; Vitrimer; Pultrusion process; Recyclable; Carbon fiber composite

1 Introduction

With the rapid development of composite, wind power industry has become the field with the largest consumption carbon fiber[1]. Pultrusion is commonly used to fabricate wind power blade carbon beams, and epoxy resin is commonly used as the matrix[2]. However, the non-melting and insoluble cross-linked structure makes them non-recyclable, resulting in huge resource waste and environmental pollution. Various existing recycling methods for thermoset composites, such as mechanical recycling methods[3] (mainly based on milling, crushing and grinding), thermal treatment[4] (pyrolysis and fluidized bed processes), chemical treatment[5] (hydrolysis, glycolysis and dissolution) and hybrid methods[6] (microwave-assisted chemical treatment) have been studied in depth for a series of years, but the overall effectiveness and benefits are not outstanding. The emergence of vitrimer is expected to solve this problem. By introducing dynamic covalent bonds into the cross-kinked structure, the reprocessing, self-healing, welding, shape memory and green recycling of epoxy resin can be achieved[7]. Currently, in addition to ester bonds, imine bonds[8], siloxane bonds[9], and disulfide bonds[10] can also exhibit reversible disconnection and reconfiguration equilibria under specific conditions. Vitrimers continue to develop abundantly and increase in variety, but their industrialization has not been studied enough[11]. There are no mature vitrimer products on the market. Based on the current carbon fiber composite market and the current research status of vitrimer materials, OLIN's pultruded epoxy anhydride resin system for wind turbine blades was used as a reference sample. The TEOA-mediated valence-adaptive network system was obtained by adding the hydroxylamine compound triethanolamine (TEOA) as a dynamic exchangeable catalyst to introduce ester bonds while keeping the original epoxy anhydride ratio unchanged. The comprehensive performance was determined to be best when the incorporation mass ratio was 2% by screening the hot pressing repeatability of the samples with different incorporation mass ratios. Since TEOA acted as a dynamic exchangeable catalyst and co-curing agent, three hydroxyl groups were involved in the curing, which ensured sufficient hydroxyl concentration in the cross-linked network and the curing degree of the resin system making the overall performance comparable to that of the reference sample, while exhibiting faster stress relaxation, good self-healing and recyclable properties.

2 Experiment

2.1 Materials

Epoxy resin (550E) with epoxy equivalent of 180–190, anhydride curing agent (555H) with anhydride equivalent of 190–200, and accelerator (555C) were provided by OLIN Co., Ltd., China. Triethanolamine (TEOA) and ethylene glycol (MEG) were purchased from Aladdin. Plain weave carbon fiber fabric, woven from SYT 45–12K carbon fiber, was supplied by Zhongfu Shenying Carbon Fiber Co., Ltd., China.

2.2 Resin sample preparation

At room temperature, 550E, 555H, 555C and TEOA were mixed with the weight ratio of 100 : 105 : 2.5. After degassing under a homogenizer (ZYMC-200V) at room temperature for 2 min, the obtained blend was poured into steel molds with specific dimensions. Finally, the sample was cured at 90 ℃ for 1 h and 130 ℃ for 2 h. After cooling to room temperature, the epoxy samples were demolded and designated as VR. As a control, the epoxy mixture without TEOA was also cured under the same condition and designated as RR.

2.3 Preparation of carbon fiber reinforced composite samples

The carbon fiber reinforced plastics (CFRP) were prepared by hand lay-up molding process. First, the epoxy system with and without TEOA was formulated in a flask and degassed in vacuum oven. Then, the formulated epoxy was brushed on a piece of carbon fiber fabric (15 cm×15 cm). After laying up four layers of carbon fibric with the warp direction, the wet composite was cured at 90 ℃ for 1 h and 130 ℃ for 2 h under 2 MPa using a hot press.

2.3 Reprocessing of cured resins

The cured VR resin was ground into fine powders by a ball mill at 800 r/min with a grinding time of 120 s. The obtained epoxy powders were tiled into steel mold and hot pressed at 240 ℃ and 2 MPa for 3 h. After cooling and demolding, a repeat molding sample was obtained, which was designated as VR-r.

2.4 Degradation of cured resins and carbon fiber composites

The 400 mg of VR and RR resins were placed in 10 mL of MEG and heated to 180 ℃ to observe whether the resins degraded or not, and the time was recorded. The CFRP sheets were cut into 60 mm×15 mm strips and heated to 180 ℃ in 10 mL of MEG.

2.5 Characterization

Differential scanning calorimetry (DSC, Netzsch DSC 214) was used to characterize the resin curing behavior and the determination of the pultrusion process parameters. The viscosity change of the resin under different shear rates at 45 ℃ was investigated on a rheometer (TA, New Castle, DE, USA). Dynamic thermomechanical analysis (DMA Q800, TA, USA) characterized the dynamic thermo-mechanical properties of the cured resin and the stress relaxation of the VR. A universal testing machine (Wance, ETM203B-EX, China) was used to investigate the tensile and bending properties of the cured resin. Each reported value for tensile and bending test was the average of five valid test specimens.

3 Results and Discussion

3.1 Curing kinetics and viscosity of resins

DSC is the most commonly used method to study the reaction kinetics of resin curing, and it is also an effective way to study the pultrusion process. Fig. 1 (a) shows the DSC exotherm diagrams of VR and RR resins cured at a ramp rate of 10 ℃/min. A large exothermic peak is observed in the curing process of both RR and VR, and the enthalpy are 310.7 J/g and 299.6 J/g, respectively. Although the peak and termination temperatures of VR and RR are basically keep the same, the initial temperature of VR is 6 ℃ lower than RR. This is mainly because TEOA, as a curing agent, can partially react with epoxy resin at a lower temperature, while acid anhydride reacts with epoxy at higher initial temperature. Fig. 1 (b) shows the DSC exothermic diagram of VR curing at four different ramp rates. with the increase of ramp rate, the exothermic peak of VR curing reaction gradually moved to a higher temperature. Fig. 1 (c) shows the curve of the peak temperature (T) of VR with the ramp rate (β). Using the T-β extrapolation method, the characteristic temperature value of the exothermic peak of the DSC curve corresponding to the ramp rate of 0 ℃/min is obtained, that is, the initial temperature is 97.5 ℃, the peak temperature is 118 ℃, and the terminal temperature is

138 ℃. According to the principle of selecting the three temperature zones of the pultrusion process, the three temperature zones can be roughly determined as 98-108 ℃, 120-130 ℃ and 140-165 ℃. Fig. 1 (d) shows the viscosity of VR and RR at 45 ℃. The viscosity of both resins are low and meet the requirements of the pultrusion process resin viscosity. Pultrusion process resin viscosity of 0.4-5.0 Pa·s is more reasonable to ensure good fiber output and avoid insufficient resin content of carbon fiber when entering the mold, which ensures the quality of CFRP.

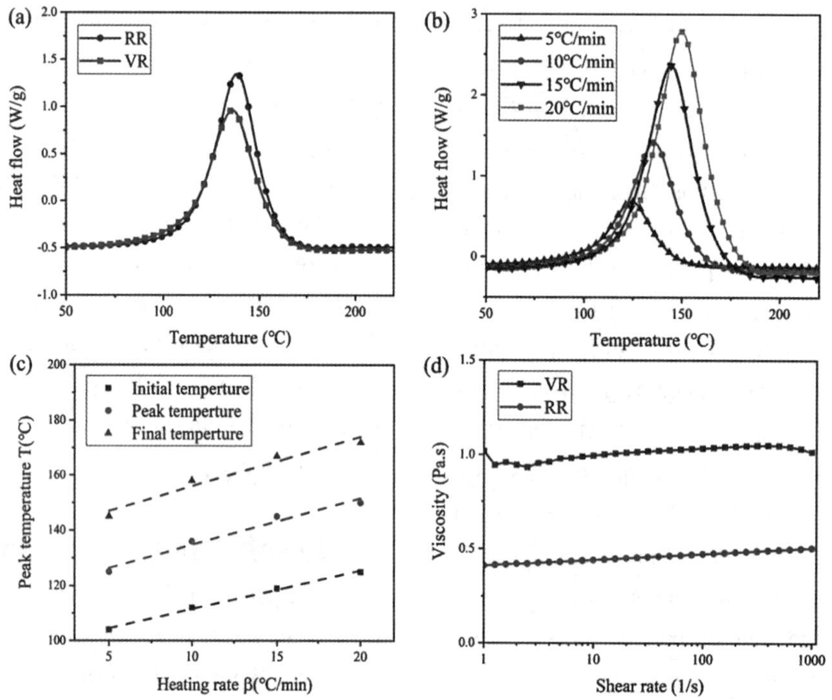

Fig. 1 (a) DSC thermograms of RR and VR at a ramp rate of 10 ℃/min; (b) DSC thermograms of VR at four different ramp rates. DSC thermogram at the rate; (c) linearity of peak temperature (T) versus heating rate (β) for VR at different heating rates Fitting; (d) viscosity of RR and VR resin at 45 ℃

3.2 Dynamic mechanical properties

Fig. 2 shows the DMA curves of RR, VR and VR-r. It can be seen from the Fig. 2 that the T_g value of VR and RR almost keep the same, indicating that the addition of 2% mass ratio TEOA does not cause any significant effect on the T_g of RR, which is in line with the wind turbine blade use conditions. However, after reprocessing and remolding, the T_g of VR-r of decrease by 13 ℃ compared to that of VR. The main reason for this phenomenon is due to TEOA as a catalytic co-curing agent, its addition helps to increase the conversion of reactants and cross-linking reactions, while the introduction of aliphatic TEOA into the cross-linking network structure also brings a flexible effect on vitrimer. In the process of VR repetitive molding, factors such as particle size, processing temperature, time and pressure during thermal processing will have an impact on the quality and performance of the reprocessed samples. Meanwhile, due to the higher processing temperature, some of the catalysts for ester exchange will be consumed, so that the exchange reactions cannot occur completely, so the T_g of VR-r will be reduced. It is worth noting that although the T_g of the VR decreases slightly after reprocessing, the value is high enough to satisfy most applications.

3.3 Stress relaxation of dynamic resin

Vitrimers can relax stress and flow, which is a typical behaviour of polymers with dynamic covalent bonds. Based on maxwell relation, the relaxation time (t^*) of vitrimer is customarily defined as the time when the modulus relaxes to $1/e$ (i.e., 36.7%) of the initial modulus. Fig. 3 (a) shows the normalized stress relaxation curves at 260 ℃ for RR and at 220, 240, 260 ℃ and 280 ℃ for VR. As shown in Fig. 3 (a), the VR exhibits clear stress relaxation upon heating. The relaxation time and corresponding normalized relaxation modulus decrease with increasing temperature. The relaxation time for VR was only 8.4 min at 280 ℃. It is worth noting that due to the DMP−30 catalyst contained in RR, which has the effect f catalyzing dynamic exchange bonds, RR exhibits a certain degree of relaxation at 260 ℃. Stress relaxation time versus temperature can be well fitted by Arrhenius equation. As shown in Fig. 3 (b), an excellent linear fitting curve was obtained. Meanwhile, the activation energy calculated from the curve was 140 kJ/mol.

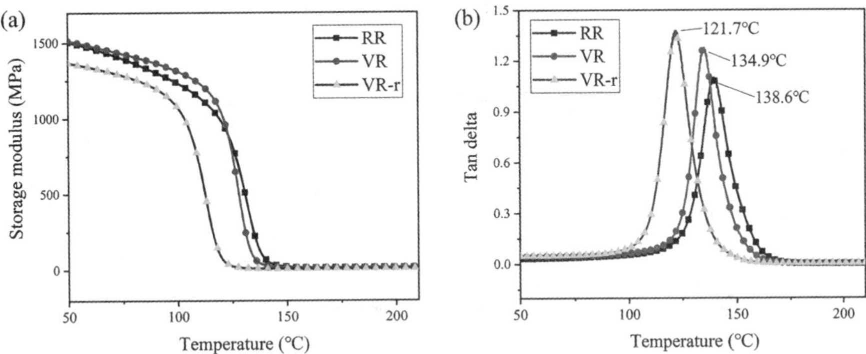

Fig. 2 The DMA test for the epoxy resin: (a) storage modulus; (b) tan delta

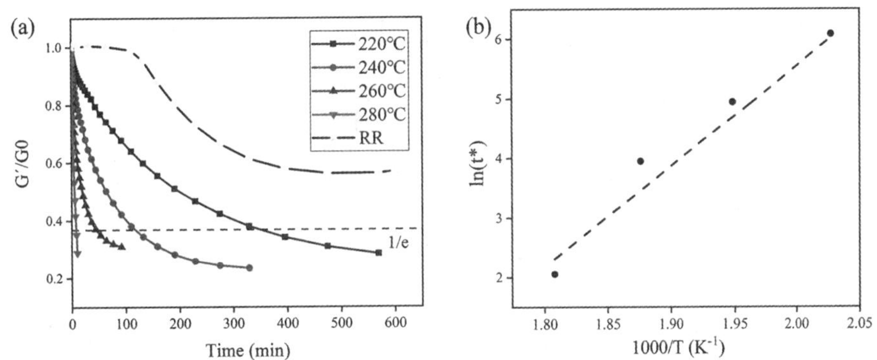

Fig. 3 (a) Normalized stress relaxation curves of RR at 260 ℃ and VR at different temperatures; (b) fitting of the relaxation times of VR to the Arrhenius equation

3.4 Mechanical properties of cured resins

In Fig. 4, flexural and tensile properties of RR and VR are depicted. As can be seen from Fig. 4 (a) that the flexural modulus of RR and VR are basically the same, while the flexural strength of VR is slightly higher than that of RR. This benefits from the addition of the aliphatic TEOA, which increases the degree of cure of the VR. The result of tensile test shows that the tensile strength and modulus almost keep the same. Generally, mechanical properties of the conventional resin RR and the dynamic resin VR exhibit similar level of flexural and tensile properties.

3.5 Degradation and recycling

Fig. 5 (a) shows the reprocessing and degradation process of VR resin. It can be seen that VR-r has a uniform appearance with certain mechanical properties after hot pressing. Combined with the previous DMA test, the thermomechanical properties of VR-r are maintained, and the resin after reprocessing can meet most application scenarios, such as automotive and construction. VR also has good degradation properties. The cured VR resin (400 mg) can be completely degraded in 10 ml MEG solution through transesterification at 180 ℃ for 5 hours, and the degradation can be accelerated by adding a small amount of TBD. Fig. 5 (b) shows the recycling process of carbon fiber, from which it can be seen that the resin on the surface of CFRP submerged in MEG solution can be effectively degraded. In the pultrusion process, carbon fibers are arranged axially, so the carbon fibers in carbon fiber pultrusion composites can be basically 100% recycled without damage, which greatly reduces the cost and protects the environment.

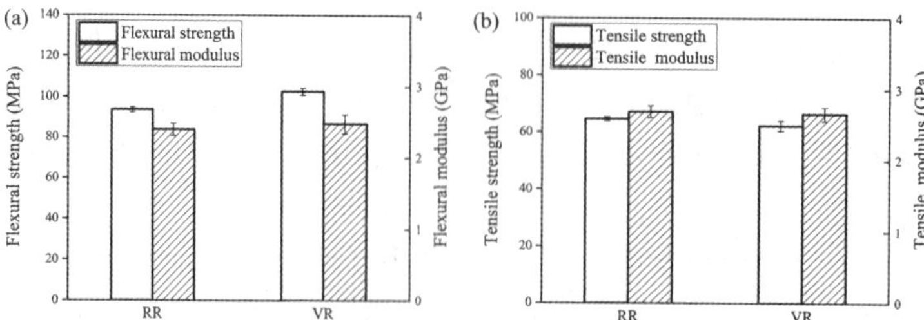

Fig. 4 Mechanical properties of the resin: (a) flexural properties; (b) tensile properties

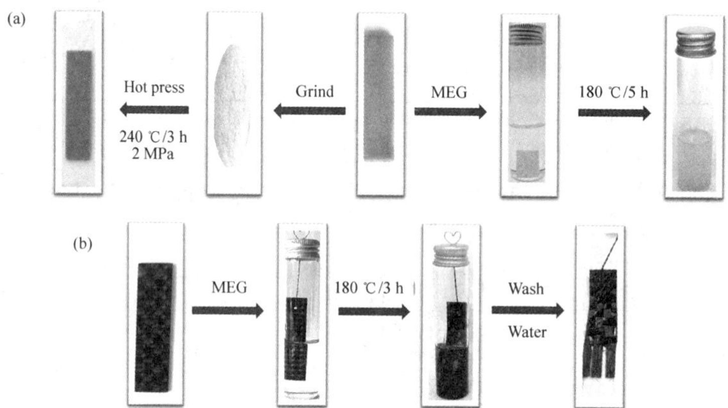

Fig. 5 (a) Remodeling and degradation of epoxy vitrimer; (b) recycling of carbon fiber

4 Conclusions

In summary, we have formulated an epoxy vitrimer with triethanolamine as the catalyst for the bond exchange reaction and to meet the requirements of the pultrusion process. The vitrimer has the same thermomechanical properties and excellent mechanical properties compared with the conventional epoxy anhydride resin system for pultrusion. The vitrimer has good reprocessability and degradability, and the carbon fiber composites made from it can also effectively recycle the carbon fibers in it, which has good market potential and industrial application value.

References

[1] LIN G, CINTE. Building "hard technology" advantage — global carbon fiber composites market report 2021[J]. Textile Science Research, 2022(Z1): 46-66.

[2] TARIQ M, NISAR S, SHAH A, et al. Effect of carbon fiber winding layer on torsional characteristics of filament wound composite shafts[J]. Journal of the Brazilian Society of Mechanical Sciences & Engineering, 2018, 40(4): 198.

[3] RAJAK D K, PAGAR D D, MENEZES P L, LINUL E. Fiber-reinforced polymer composites: manufacturing, properties, and applications[J]. Polymers, 2019, 11: 1-37.

[4] RANI M. A review on recycling and reuse methods for carbon fiber/glass fiber composites waste from wind turbine blades[J]. Composites Part B: Engineering, 2021, 215.

[5] ZHAO X, WANG X L, TIAN F, et al. A fast and mild closed-loop recycling of anhydride-cured epoxy through microwave-assisted catalytic degradation by trifunctional amine and subsequent reuse without separation[J]. Green Chem, 2019, 21: 2487-2493.

[6] MONTARNAL D, CAPELOT M. Silica-Like malleable materials from permanent organic networks[J]. Science 2011, 334, 965.

[7] JIARUI H. Hyperbranched polymer assisted curing and repairing of an epoxy coating[J]. Industrial & Engineering Chemistry Research, 2019, 115(8): 13-22.

[8] ZHENG P W, MCCARTHY T J. A surprise from 1954: siloxane equilibration is a simple, robust, and obvious polymer self-healing mechanism[J]. Journal of the American Chemical Society, 2012, 134(4): 2024-2027.

[9] LI W B. Self-healing silicon-containing eugenol-based epoxy resin based on disulfide bond exchange: synthesis and structure- property relationships [J]. Polymer, 2021, 229,345.
[10] WANG Y P, LIU X Q, JING X L, et al. Advances in the research of epoxy resin-based glass polymers[J]. Chemical Bulletin, 2021, 84(4): 313-321.
[11] BUDIYANTORO C, ROCARDJO H, NUGROHO G. Effects of processing variables of extrusion-pultrusion method on the impregnation quality of thermoplastic composite filaments[J]. Polymers, 2020, 12(12): 2833.

Optimization of Electromagnetic Properties of 3D Woven Spacer Composites by Carbon Nanotube-Doped Epoxy Resin Foaming Process

LIU Xueqiang[1], LI Wuzhou[1], XU Fujun[1,2*]

[1] *College of Textiles, Donghua University, Shanghai 201620, China*

[2] *Shanghai Key Laboratory of Advanced Micro & Nano Textile Materials, Shanghai 201620, China*

* *Corresponding authors email*: fjxu@ dhu. edu. cn

Abstract: In order to investigate the regulation of electromagnetic properties of 3D woven spacer composites, carbon nanotubes were used as the regulating factor of electromagnetic properties of 3D woven spacer composites, and 3D woven spacer composites with different CNT contents were prepared by epoxy resin foaming process. The 3D woven spacer fabric was microscopically modeled by 3DMax software, and the electromagnetic parameters of the composite were simulated and analyzed by CST Studio Suite simulation software, and the effect of carbon nanotube content on the electromagnetic properties of the 3D woven spacer composite was calculated by the finite element integration method. The results show that doped carbon nanotubes laminated with 3D woven spacer fabric using epoxy resin foaming process is an effective method to improve their electromagnetic properties, which can provide a reference for the design of lightweight and high strength electromagnetic wave absorbing materials.

Keywords: 3D woven spacer composites; Electromagnetic properties; CNTs; Epoxy resin foaming; Simulation

1 Introduction

The development of modern electromagnetic wave-absorbing materials is moving forward in the direction of good performance such as thin, light, wide and strong. Structural wave-absorbing composite material is to ensure the load-bearing capacity of the material at the same time, through a certain structural design, to give the material good wave-absorbing performance[1]. 3D woven spacer composite material is a new sandwich structure composite material prepared by 3D woven spacer fabric composed of upper and lower layers of fabric and middle column yarn as reinforcement system, which has the characteristics of high strength, impact resistance, delamination resistance and shear resistance, etc., and its unique structure of high and medium hollow degree is highly designable[2-4]. Therefore, 3D woven spacer composites are receiving increasing attention for research in developing lightweight and high-strength electromagnetic wave-absorbing materials.

The complex dielectric constant of a material is one of the most important characteristic parameters to describe the electromagnetic properties of a material and characterizes the interaction between the material and the electromagnetic field[5]. In the research and application of wave absorbing materials, the complex permittivity is an important factor affecting the good or bad electromagnetic absorption performance of wave absorbing materials[6]. Since there is still a gap in the research on the regulation of electromagnetic parameters of 3D woven spacer composites, this greatly hinders the development of 3D woven spacer composites in radar stealth materials, electromagnetic shielding materials and other applications. It is necessary to conduct systematic research on the regulation of electromagnetic parameters of 3D woven spacer composites.

This thesis investigates the effect of carbon nanotube powder on the modulation of electromagnetic properties of 3D woven spacer composites. The 3D woven spacer composites with different CNT contents were prepared by using epoxy resin foam foaming process, so as to regulate the electromagnetic properties of the 3D spacer composites[7, 8]. And 3DMax software was used to micro-model the 3D woven spacer fabric, and CST Studio

Suite simulation software was used to simulate and analyze the electromagnetic parameters of the composites[9].

2 Experiments

2.1 Materials

E-glass fiber (EDR13-300-386) was provided by China JUSHI Co. Ltd., Zhejiang, China. Microsphere foaming agent (particle size 20-25μm, expansion rate 4-5 times) was provided by Crerax Technology Co. Ltd., Shanghai, China. Dimethicone, Foam stabilizer, was provided by Sinopharm Chemical Reagent Co. A multiwalled carbon nanotube was provided by Chengdu Organic Chemical Co. Ltd., Chinese Academy of Sciences.

2.2 Design and weaving of fabric structure

The 3D woven spacer fabric in the experiment was woven by the three-dimensional loom integration, and its organization is shown in Fig. 1. The entire fabric is composed of the upper and lower surface fabrics and the middle column yarn[10]. The upper and lower surface fabrics are interwoven by two groups of warp and weft yarns, and the middle column yarn is interwoven with the weft yarns along the warp direction between the upper and lower surface fabrics. The column yarns are covered by the surface warp yarns and do not appear on the surface of the fabric, and their arrangement density is the same as that of the warp yarns on the upper and lower layers. The column yarn pattern is uniform and symmetrical, and the column yarns between each fabric structural unit support each other to form a stable structure.

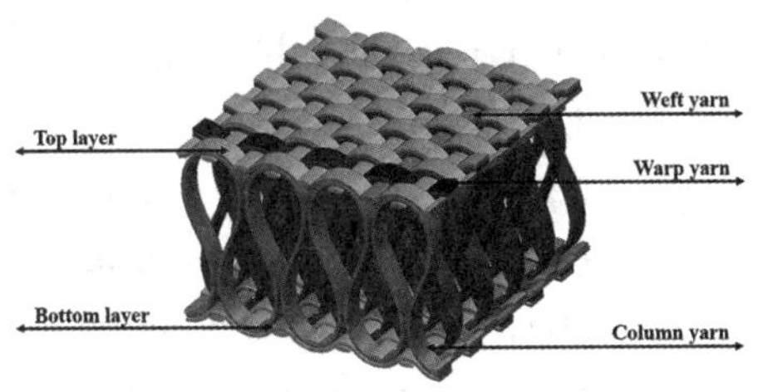

Fig. 1 The structure of 3D woven spacer fabric

2.3 Manufacture of composites with tunable dielectric properties

The preparation of composites with tunable dielectric properties was done in two main steps. First step, the carbon nanotube-doped epoxy resin foaming system was formulated. Bisphenol A epoxy resin and curing agent were mixed in the mass ratio of 100:33, and carbon nanotube powder, microsphere foaming agent and foam stabilizer were added into the blended epoxy resin in the ratio of 1%, 3% and 2%, respectively, and then evacuated and defoamed for 30 min after full mixing. The resin system to be foamed in the groove mold, retaining a certain amount of foaming space, heating to 95 ℃ free foaming 1h after cooling to 70 ℃ curing 7 h. The epoxy foam without doped carbon nanotubes is a common epoxy foam material (shown in Fig. 2(a)), and the epoxy foam composite with tunable dielectric properties obtained by doping 1% carbon nanotubes (shown in Fig. 2(b)).

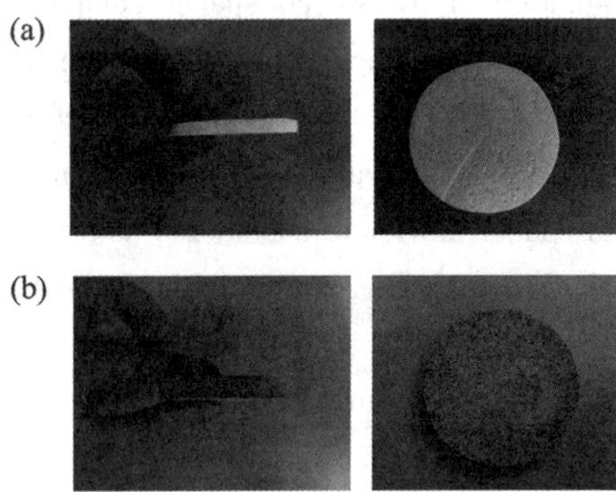

Fig. 2 Physical diagram of epoxy resin foam composites: (a) general type; (b) adjustable dielectric properties type

Second step, 3D woven spacer composites with tunable electromagnetic properties were prepared. The blended foaming system is dropped into the 3D spacer fabric via the cross-sectional direction and placed into the fluted mold to ensure that the resin completely and uniformly infiltrates the composite material, and the foaming is carried out according to the above foaming process to prepare a 3D woven spacer composite material with adjustable electromagnetic properties.

2.4 Dielectric properties testing of composites

To clarify the effect of doped carbon nanotube powder on the dielectric properties of 3D spacer composites, the dielectric constants testing of 3D woven spacer composites was conducted on a PNA3766 vector network analyzer (Nanjing Puna Technology Equipment Co. Ltd., Nanjing, China). The measurement method is the double-sided laminate method with the frequency set at 2.0 GHz, and the sample is a composite cylinder with a diameter of 70 mm.

2.5 Electromagnetic simulation of composites

In order to further investigate the effect of carbon nanotubes on the electromagnetic properties of 3D woven spacer composites, this study firstly used 3D Studio Max software to model the structure of 3D woven spacer fabrics. 3D Studio Max software is a 3D drawing and modeling software developed by Autodesk to visualize the microstructure of fabrics. 3D woven interval fabric using the yarn for the twist-free silk bundle, its upper and lower layers by the warp and weft yarn woven into the plain tissue structure, and the middle by the interval yarn up and down back and forth woven to connect the upper and lower layers to form a whole, its 3D fabric structure model as Fig. 3 (a). CST Microwave Studio (MWS) software was used to simulate electromagnetic simulation of 3D woven spacer composites, and the 3D woven composite structure model can be simplified in order to improve the computational efficiency as follows. (1) the upper and lower layers of the three-dimensional spacer fabric are considered as a regular rectangular body; (2) the conductivity of the individual yarns in the fabric is isotropic; (3) it is assumed that the yarns in the fabric are of the same size and there is no deformation. The simplified structural model of the CST MWS studio fabric unit is shown in Fig. 3 (b). The background is set as a vacuum, the boundary conditions in the x- and y-axis directions are periodic boundaries, and the electromagnetic waves propagate along the negative z-axis direction to investigate the effect of carbon nanotubes on the electromagnetic properties of 3D woven spacer composites in the frequency range from 2 to 18 GHz.

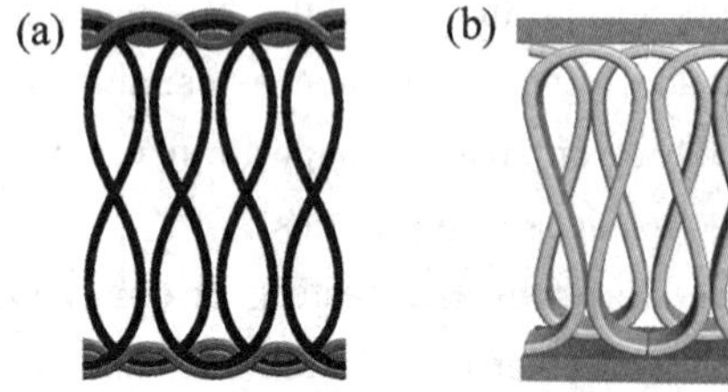

Fig. 3 (a) Structural model of 3D woven spacer fabric and (b) the simplified structural model of the CST MWS studio fabric unit

3 Results and Discussion

3.1 Dielectric properties of composites

The dielectric properties of plain epoxy resin foam and CNT-doped powder epoxy resin foam composites were shown in Fig. 4. The results showed that the dielectric constant of 1% CNT-doped powder epoxy resin foam composites increased from 1.25 to 2.37 by 89.6% and the dielectric loss increased from 0.06 to 0.27 by 350% compared with that of the common epoxy resin foam without CNT doping. It was shown that doping CNT into the epoxy foaming system can effectively improve the dielectric properties of the composites, and the dielectric loss of the material is enhanced faster than the dielectric constant, which is due to the structural defects of carbon nanotubes that enhance the polarization effect and thus increase the dielectric loss.

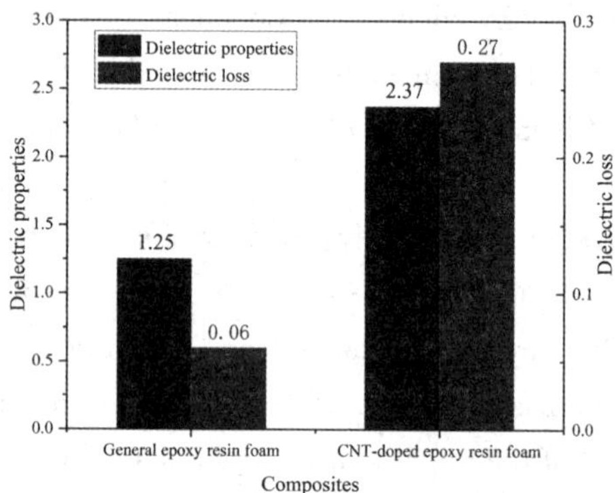

Fig. 4 Dielectric properties test results of plain epoxy resin foam and CNT-doped powder epoxy resin foam composites

3.2 Electromagnetic simulation results for composites

The electromagnetic simulation results of the 3D spacer composite in CST were shown in Figs. 5 (a) and 5(b). The glass fiber 3D woven spacer composite has low dielectric loss, and has over 90% transmission of microwaves in the frequency range of 2-12 GHz, with excellent wave transmission performance. This directly demonstrated that radar waves can pass through glass fiber panels very well, making them ideal for the development of structural wave-absorbing materials. When doped with 1% CNT epoxy resin foam process is used for the composite, the 3D woven spacer composite has >26 dB reflectivity for microwaves in the frequency range of 2-18 GHz, which increases the absorption of electromagnetic waves as shown in Fig. 5 (c) (shown in Fig. 5 (c)). As CNT is a non-magnetic material, the addition of CNT mainly changes the complex dielectric constant in the electromagnetic parameters of the composite, and its absorption mechanism for microwaves mainly comes from the loss of polarization and conductivity, as well as the multiple scattering due to its high specific surface area. Moreover, the electromagnetic wave absorption performance of 3D woven spacer composites can be easily controlled by the content of CNT, which makes 3D spacer composites show great superiority in the development of broadband and high strength electromagnetic wave absorbing materials.

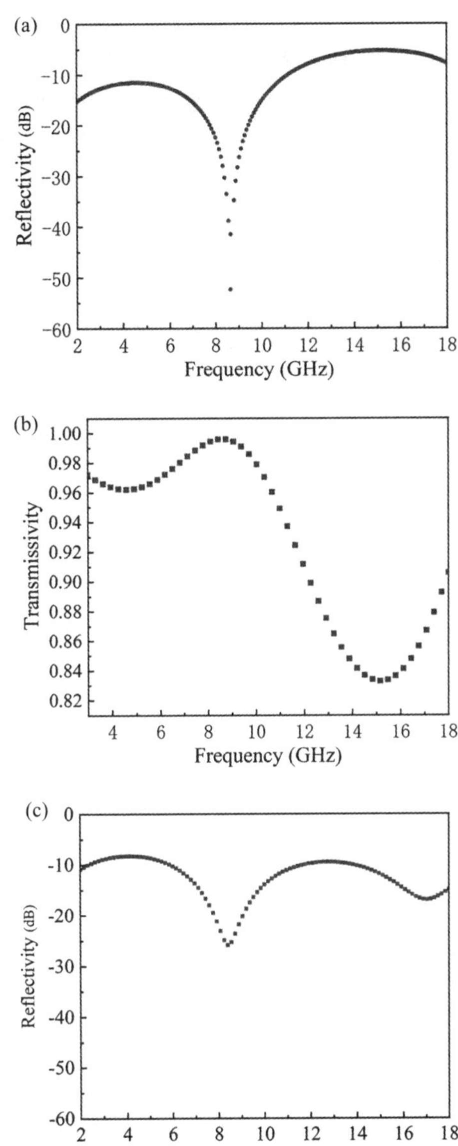

Fig. 5 Simulation results of electromagnetic simulation of 3D spacer composites in CST: (a) reflectivity of ordinary 3D woven spacer composites; (b) transmissivity of ordinary 3D woven spacer composites and (c) reflectivity of 3D woven spacer material doped with carbon nanotubes

4 Conclusions

In this paper, we studied the effect of doped CNT on the electromagnetic properties of 3D woven spacer composites. The epoxy foam composites with tunable dielectric properties were prepared by the doped carbon nanotube epoxy foaming process, and the 3D woven spacer fabrics were microscopically modeled and the electromagnetic properties of the composites were simulated and analyzed by using 3DMax and CST Studio Suite.

The results showed that CNT as a modulating factor for the electromagnetic properties of 3D woven spacer composites improves the dielectric properties of the composites, and the easy control of the electromagnetic properties of 3D spacer composites is expected to be achieved by varying the content of CNT. Compared with ordinary epoxy resin foam, the dielectric constant of doped 0.1% CNT powder epoxy resin foam composites was increased by 89.6%; the dielectric loss was increased by 350%. The main reason was that CNT is a non-magnetic material, and the addition of CNT greatly improves the complex dielectric constant in the electromagnetic parameters of the composite material, and its absorption mechanism for microwaves mainly comes from the loss of polarization and conductivity, as well as multiple scattering due to its high specific surface area. The doping of CNT is undoubtedly an effective way to modulate the electromagnetic properties of 3D woven spacer composites, and the use of epoxy foaming process provides ideas for the development of 3D spacer composites in broadband, high strength and light weight high performance electromagnetic wave absorbing materials.

References

[1] SHI K Z, LI J, WU Y, et al. Lightweight composite microwave absorbing materials based on graphene aerogels with honeycomb structure[J]. Physica Status Solidi-Rapid Research Letters, 2019, 13(8).

[2] ZHAO C Q, LI D S, GE T Q, et al. Experimental study on the compression properties and failure mechanism of 3D integrated woven spacer composites[J]. Materials & Design, 2014, 56: 50-59.

[3] ZHANG K, ZHENG L G, PEI R, et al. Light-weight, high-gain antenna with broad temperature adaptability based on multifunctional 3D woven spacer kevlar/polyimide composites[J]. Composites Communications, 2022, 30.

[4] WANG L, ZHANG K, FARHA F I, et al. Compressive strength and thermal insulation properties of the 3D woven spacer composites with connected spacer yarn structure[J]. Journal of Materials Science, 2020, 55(6): 2380-2388.

[5] YU Y, WANG C, LIU Y, et al. Research progress of

wave absorbing composites[J]. Journal of Silk, 2019, 56(12): 50-58.

[6] QUAN B, LIANG X H, JI G B, et al. Dielectric polarization in electromagnetic wave absorption: Review and perspective[J]. Journal of Alloys And Compounds, 2017, 728: 1065-1075.

[7] ZHONG F, HE J, WANG X, et al. Structure and properties of epoxy foams prepared by microwave irradiation[J]. Materials Review, 2006, 20(8): 149-151.

[8] KUS A, DURGUN I, ERTAN R. Experimental study on the flexural properties of 3D integrated woven spacer composites at room and subzero temperatures[J]. Journal of Sandwich Structures & Materials, 2018, 20(5): 517-530.

[9] LI X P, LV X L, WANG J. A broadband circuit analog absorber based on printed dipole antenna[Z]. 2013 International Workshop on Microwave And Millimeter Wave Circuits and System Technology (Mmwcst). 2013: 47-9.

[10] UMAIR M, HAMDANI S T A, NAWAB Y, et al. Effect of pile height on the mechanical properties of 3D woven spacer composites[J]. Fibers and Polymers, 2019, 20(6): 1258-1265.

Effect of Epoxide Structure on the Properties of Zein/Epoxide Ultrafine Fibrous Sensor Scaffolds

HUANG Lingxin[1], **ZHANG Tong**[1], **LIU Lu**[1], **LIU Fei**[1], **LIU Wanshuang**[2], **JIANG Qiuran**[1,3,4*]

[1] *Department of Technical Textiles, College of Textiles, Donghua University, Shanghai 201620, China*

[2] *Donghua University Center for Civil Aviation Composites, Donghua University, Shanghai 201620, China*

[3] *Key Laboratory of Textile Science & Technology, Ministry of Education, College of Textiles, Donghua University, Shanghai 201620, China*

[4] *Engineering Research Center of Technical Textiles, College of Textiles, Donghua University, Shanghai 201620, China*

* *Corresponding author's email*: jj@ dhu. edu. cn

Abstract: Flexible sensors have good application prospects in the treatment of orthopedic diseases, but existing flexible sensors cannot provide real-time monitoring of damaged tissues *in situ*. In this study, epoxides (EP) with high reactivity to active hydrogen structures were incorporated into protein zein ultrafine fibers and established sensing scaffolds with elastic base substrate, polydopamine bonding layer and silver conductive sensing layer. The effect of the epoxide structures on the zein/epoxide ultrafine fibrous scaffolds was investigated. It was proved that the epoxide with more epoxy groups could enhance the elasticity of scaffolds with the recovery rate increasing from 70. 4% to 93. 3%, stabilize the scaffold morphology with the deformation rate reducing from 52% to 8. 2% in 7 d, and adjust the sensing performance from a highly sensitive mode in a limited sensing range (118%) to a moderate sensitive mode in a wide sensing range (247%). The developed sensing scaffolds in this work exhibited promising potential for in-situ monitoring of the tissue repair process and fundamentally improving the therapeutic efficacy.

Keywords: Zein; Epoxide; Ultrafine fiber; Sensing tissue engineering scaffold; Elasticity

1 Introduction

Flexible strain sensors can monitor real-time tissue deformation during tissue repairing and have great potential for the treatment of orthopedic injuries to prevent malunion, nonunion, and secondary fracture complications[1-3]. However, most of the existing flexible sensors are non-degradable encapsulated films, which are mostly used for body surface and *in vitro* health detection[4-5]. *In situ* monitoring of injuries and the healing status *in vivo* are of great challenge. Current developed degradable *in vivo* sensing devices only possess sensing functions, and need to be attached to or near the affected area. Thus, they might cause secondary injuries to the defected tissues. If the flexible strain sensor is applied *in vivo*, it should be endowed with good biocompatibility and degradability. Therefore, the tissue engineering scaffold can be integrated with the strain sensing function to fulfill the *in situ* real-time monitoring task.

In this work, zein was used to prepare the base fibrous scaffold by electrospinning. The silver (Ag) was used as the sensing layer and a polydopamine (PDA) layer served as the bonding layer between the protein substrate and the Ag conductive layer. Elasticity and tensile properties were important for the zein fibrous scaffolds to serve as strain sensors. However, the electrospun zein mat has poor water stability and low wet strength with no elasticity[6]. Therefore, in our previous study, the epoxy, ethylene glycol diglycidyl ether (EGDE), with high reactivity to active hydrogens was introduced to the zein protein system which possesses abundant active hydrogen structures, and provided the fibers with high water stability, strength and elasticity[7]. However, the effect of the epoxide structures on the modification efficacy has not been revealed. In this current work, the key structural feature, the amount of epoxy groups of epoxides, was investigated as to the effects on the mechanical properties, water stability of the protein fibrous substrates and the signal output of the developed sensing scaffolds.

2 Experiments

2.1 Materials

Zein powders were purchased from the Hefei Bomei Biotechnology Co. , Ltd. (Anhui, China). EGDE, glycerol triglycidyl ether (GTE) and pentaerythritol glycidyl ether (PGE) were selected as the model epoxides with two to four epoxy groups and bought from the Guangdong Wengjiang Chemical Reagent Co. , Ltd. (Guangdong, China). Anhydrous ethanol, sodium chloride (NaCl), potassium chloride (KCl), disodium hydrogen phosphate dodecahydrate ($Na_2HPO_4 \cdot 12H_2O$), potassium dihydrogen phosphate (KH_2PO_4), glucose, silver nitrate and ammonia were provided by the Shanghai Aladdin Biochemical Technology Co. , Ltd. (Shanghai, China). Polyvinylpyrrolidone (PVP) was supplied by the Tokyo Chemical Industry Co. , Ltd. (Tokyo, Japan). Dopamine was provided by Shanghai McLean Biochemical Technology Co. , Ltd. (Shanghai, China).

2.2 Preparation of zein ultrafine fibrous scaffolds

The epoxides, EGDE, GTE and PGE, were individually diluted by 75%(V/V) ethanol with 5% PVP based on the weight of zein. Then, the zein powders were dissolved in these epoxide solutions at 16%, 17% and 18% with stirring and aged for 24 h. The solutions were electrospun at a voltage of 15 kV and a feed rate of 2 mL/h in a spinning chamber at 40% relative humidity. The spinning gap between the needle tip and the collection drum was kept at 18 cm and the collection drum rotation rate was set at 60 rpm. The obtained fibrous scaffolds were cross-linked at 120 ℃ for 1 h, and noted as ZE/EGDE-S, ZE/GTE-S, and ZE/PGE-S, respectively.

2.3 Preparation of zein fibrous sensing scaffolds

The bonding layer of PDA was formed by immersing the zein fiber mats (3 mm×25 mm) into the dopamine aqueous solution (2 g/L, pH 8.5) at 30 ℃ for 12 h with stirring at 180 rpm. After washing, the PDA coated mats were treated in the silver ammonia solution prepared with ammonia water and silver nitrate (Ag^+ 0.01 g/mL) for 0.5 h and then the glucose solution (0.16 g/mL) were added into the solution dropwise for silver ion reducing with stirring at 300 r/min for 1 h. The Ag plated PDA coated zein mats were rinsed in water. The obtained sensing scaffolds were noted as ZE/EGDE-SS, ZE/GTE-SS and ZE/PGE-SS, respectively.

2.4 Characterization of the base scaffolds

The appearances of the base zein ultrafine fibrous scaffolds were recorded by a digital camera (D7200, Nikon, Japan). After soaking the scaffolds in water for 30 min, the wet-state elasticity of the base zein ultrafine fibrous scaffolds (3 mm×28 mm) was evaluated on an electronic single fiber strength tester (LLY-06E, Laizhou Electronic Instrument Co. , Ltd. , China) at a test speed of 18 mm/min, a test gap of 18 mm and a test elongation of 18 mm (100%). The water stability of the scaffolds was characterized by measuring the deformations of the samples immersed in PBS at 37 ℃ for 15 min, 4 h, 2 d and 7 d and calculating the deformation ratios according to the following equation:

$$\eta = \frac{S_0 - S}{S_0} \tag{1}$$

where η is the deformation ratio, S_0 is the area of the sample at a dry state, and S is the area of the sample at wet state. To evaluate the strain sensing performance, the sensing scaffolds were mounted on a customized scaffold deformation controller moving at 5 mm/min and the surface resistance was monitored by a digital multimeter (34465A, Keysight, USA) which connected on the two ends of scaffolds.

3 Results and Discussion

3.1 Appearance of the base scaffolds

As shown in Fig. 1, the appearances of the zein ultrafine fibrous scaffolds modified by different epoxides were similar with a color of light yellow and a flat surface indicating the low fineness of fibers.

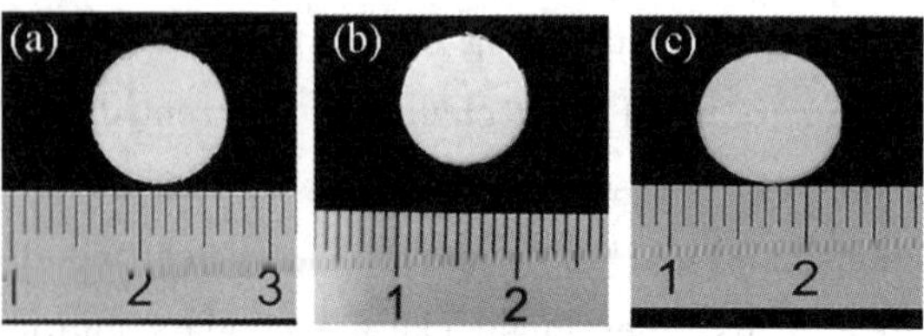

Fig. 1 The appearances of the zein ultrafine fibrous scaffolds modified by different epoxides: (a) ZE/EGDE-S; (b) ZE/GTE-S; (c) ZE/PGE-S

3.2 Wet-state elasticity of the base scaffolds

Fig. 2 shows the immediate and delayed elastic recoveries of the zein ultrafine fibrous scaffolds modified by different epoxides at a wet state. The total elastic recovery of the EGDE modified zein scaffolds was 70. 4% composed with 67. 1% of delayed elastic recovery and merely 3. 2% immediate elastic recovery. As the epoxy groups of the epoxide increased to three, the total elastic recovery elevated to 88. 2%, and the immediate elastic recovery was raised to 33. 2%. Further increase in the epoxy group amount endowed the scaffolds with a total elastic recovery to 93. 3% and an immediate elastic recovery of 43. 2%. More epoxy groups could provide epoxides with higher reactivity to proteins and resulted in more abundant cross-linkages in the protein structure. With a better polymer network, the deformation of fibers under tension was mainly due to the unfolding of the curly zein polymer chains in the amorphous phase, the changes in the angle and length of cross-linkages, which could be immediately recovered after removing the external tension. Permanent deformation was caused by the molecular slippage and reduced when the epoxide with more groups was used.

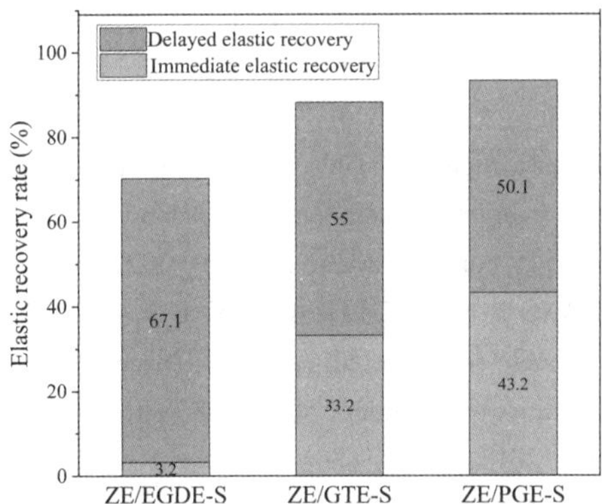

Fig. 2 Wet-state elastic recovery at 100% elongation of the zein ultrafine fibrous scaffolds cross-linked by different epoxides

3. 3 Wet stability of the base scaffolds

After the incubation in PBS, the zein scaffolds modified by the three epoxides displayed shrinkages in different extents. The deformation ratios of the scaffolds are listed in Tab. 1. Within the soaking duration for up to 7 d, the deformation rates of all samples increased gradually with reducing rates. When soaked for 15 min, the sample ZE/EGDE-S shrank rapidly with a deformation rate of 43. 3%, while the samples of ZE/GTE-S and ZE/PGE-S exhibited limited deformation with low rates of 2. 5% and 2. 3%. After incubation of 7 d, the deformation rate of the sample ZE/EGDE-S was elevated to 51. 5%, but the rates of the scaffolds modified by GTE and PGE were merely near 8. 2%.

Tab. 1 Deformation rates of the zein scaffolds modified by different epoxides

Incubation duration	ZE/EGDE-S	ZE/GTE-S	ZE/PGE-S
15 min	43. 3%	2. 5%	2. 3%
4 h	44. 3%	5. 9%	5. 2%
2 d	48. 1%	6. 7%	5. 7%
7 d	51. 5%	8. 2%	8. 2%

The epoxides were stably connected to the protein molecular chain through covalent bonds, which could not be broken by water molecules and greatly reduced the mobility of the molecular chain. Thereby, the area and pores of the scaffolds were reduced. The more epoxy groups on the epoxides, the denser the covalent bonded polymer networks formed, and the better the wet-state stability.

3. 4 Strain sensing performance of zein ultrafine fibrous sensing scaffolds

Fig. 3 presents relative resistance changes ($\Delta R/R_0$) of the sensing scaffolds modified by the three epoxides. The curves displayed exponential increase trends. When the strains of scaffolds were below 25%, the initial $\Delta R/R_0$ values of all scaffolds were similar and at low values (below 0. 4). With higher extent of stretching, the $\Delta R/R_0$ values were elevated substantially. The slope of the $\Delta R/R_0$ curve of the sample ZE/EGDE-SS was the highest in the elongation range above 25% and the maximum sensing elongation of the scaffold was 118%. The samples of ZE/GTE-SS and ZE/PGE-SS, though displayed lower $\Delta R/R_0$ values, could serve in wider deformation ranges with the maximum sensing elongations of 196% and 247%, respectively. In addition, the sample ZE/PGE-SS possessed the highest signal stability, while the sample ZE/EGDE-SS showed the lowest signal stability. With the increase in the epoxy quantity, the mechanical properties are enhanced. When subjected to external force, the samples with better mechanical properties could distribute the deformation evenly in the whole sample. Hence, the cracks of the sensing layer are evenly distributed on the surface of the sample. Uniform cracks increased the maximum sensing elongation, while locally concentrated, non-uniform cracks in the sensing layer resulted in an increase in signal output intensity, which in turn produced an increase in sensitivity. Thereby, the signal output of the sample with a higher cross-linking degree was more stable and uniform.

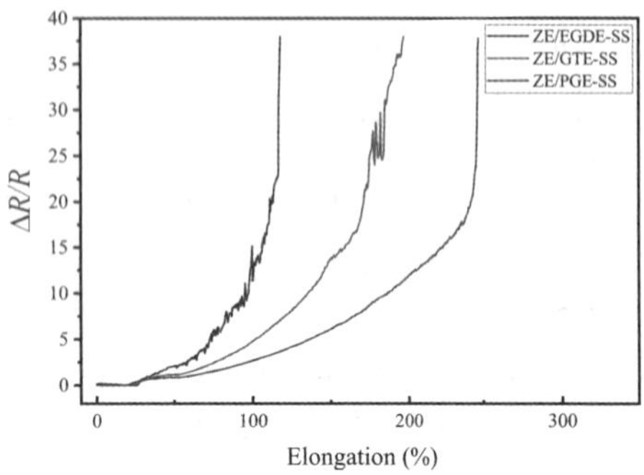

Fig. 3 Relative resistance changes of the sensing scaffolds composited with EGDE (black), GTE (red) and PGE (blue) with strain in PBS

4 Conclusions

This work revealed the influence of the key structural feature of epoxides on the morphological property, elasticity, water stability and sensing performance of the zein ultrafine fibrous sensing scaffolds constructed by the epoxide modified zein ultrafine fibers, the PDA bonding layer and the Ag conductive sensing layer. It was proved that more epoxy groups on the epoxides endowed the scaffolds with higher total elastic recovery, immediate elastic recovery, water stability as well as wider strain sensing range, but with the sacrifice of the sensitivity. Since the epoxide structure could control the stability, elasticity and sensing performances, the customized sensing scaffolds could be developed according to the requirements of specific applications for tissue repair. Further development in the tissue engineering scaffolds with the functions of in-situ postoperative healing monitoring could be promoted and provide patients with early-stage diagnosis and effectively avoid the complications of various injuries.

Acknowledgement

This work was financially supported by the Fundamental Research Funds of Shanghai Collaborative Innovation Center of High Performance Fibers and Composites (Province-Ministry Joint) (No. X12812101/015) and the Fundamental Research Funds for the Central Universities (No. 2232022D-13).

References

[1] D'LIMA D D, FREGLY B J, COLWELL C W. Implantable sensor technology: measuring bone and joint biomechanics of daily life in vivo [J]. Arthritis Research & Therapy, 2013, 15(1): 203.

[2] HAK D J, FITZPATRICK D, BISHOPJ A, et al. Delayed union and nonunions: epidemiology, clinical issues, and financial aspects [J]. Injury-International Journal of the Care of the Injured, 2014, 45: S3-S7.

[3] KLOTZBUECHER C M, ROSS P D, LANDSMAN P B, et al. Patients with prior fractures have an increased risk of future fractures: a summary of the literature and statistical synthesis [J]. Journal of Bone and Mineral Research, 2000, 15(4): 721-739.

[4] HEO J S, HOSSAIN M F, KIM I. Challenges in design and fabrication of flexible/stretchable carbon- and textile-based wearable sensors for health monitoring: a critical review[J]. Sensors, 2020, 20(14): 3927.

[5] HWANG B U, LEE J H, TRUNG T Q, et al. Transparent stretchable self-powered patchable sensor platform with ultrasensitive recognition of human activities [J]. ACS Nano, 2015, 9(9): 8801-8810.

[6] XU W J, KARST D, YANG W, et al. Novel zein-based electrospun fibers with the water stability and strength necessary for various applications[J]. Polymer International, 2008, 57(10): 1110-1117.

[7] LIU L, JIANG J C, JIN X Y, et al. Epoxide cross-linked and lysine-blocked zein ultrafine fibrous scaffolds with prominent wet stability and cytocompatibility [J]. ACS Applied Polymer Materials, 2021, 3(8): 3855-3866.

Aqueous Stable Gelatin/EGDE 3D Scaffolds Prepared via Ultralow Concentration Phase Separation with Ultraloose Fibrous Structure

SUN Yue[1], LI Shiyuan[1], LIU Fei[1], LIU Wanshuang[2], JIANG Qiuran[1,3,4*]

[1] *Department of Technical Textiles, College of Textiles, Donghua University, Shanghai 201620, China*

[2] *Donghua University Center for Civil Aviation Composites, Donghua University, Shanghai 201620, China*

[3] *Key Laboratory of Textile Science & Technology, Ministry of Education, College of Textiles, Donghua University, Shanghai 201620, China*

[4] *Engineering Research Center of Technical Textiles, College of Textiles, Donghua University, Shanghai 201620, China*

* *Corresponding author's email*: jj@ dhu. edu. cn

Abstract: Protein-based ultrafine fibrous tissue engineering scaffolds (PUFTESs) are favored due to their similar architecture and chemical composition to the natural extracellular matrices (ECMs). However, the compact structure and high deformation in aqueous environment of the conventional PUFTES limit their applications due to the difficulty in cell infiltration and the poor morphological stability, leading to the fail in three-dimensional (3D) tissue reconstruction. This work developed 3D ultraloose gelatin/epoxides scaffolds via using the ultra-low concentration phase separation (ULCPS) technique and introducing epoxides into the protein system. By adjusting the concentration of the phase separation solutions from 0. 01 to 0. 5 wt. %, the structures of the scaffolds shifted from ultrafine fibers to the fibers/sheets blended structure, the specific pore volume reduced from 366. 06 to 45. 96 cm^3/g, about 33. 3 to 4. 2 times of the values of the conventional electrospun scaffolds. The wet-state morphological stability was enhanced with merely 10% of the deformation rate, indicating the ability to maintain the highly porous structure for cell infiltration when served as tissue engineering scaffolds.

Keywords: 3D ultrafine fibrous scaffold; Gelatin; Ultra-low concentration phase separation; Water stability; Specific pore volume

1 Introduction

The extracellular matrices (ECMs) of tissues with unique ultrafine fibrous structures and protein-based chemical compositions serve as important scaffolds to provide mechanical support for cells[1]. The ideal tissue engineering scaffolds should perfectly mimic the ECMs as to structure and chemical composition in order to provide an appropriate bionic environment for complex tissue recovery[2]. Thus, two dimensional (2D) or three dimensional (3D) protein-based ultrafine fibrous scaffolds are preferred[3]. Currently, the 2D ultrafine fibrous membrane can be readily prepared by electrospinning to imitate the 2D ECMs[4], but it is difficult to establish the 3D randomly oriented ultrafine fibrous bulk that can be infiltrated by cells for 3D ECMs simulation[5]. The current techniques to form 3D ultrafine fibers include thermally induced phase separation[6], molecular self-assembly[7], 3D electrospinning[8]. Yet, the produced fiber bulks display either compact structure with small fiber gaps or unstable reserved pores, leading to the difficulty in cell infiltration[9]. Besides, the protein-based ultrafine fibers exhibit poor water stability which induces low mechanical properties and obvious deformations. Although various modifications have been applied, to maintain the sizes, pore volumes and fibrous structures of scaffolds is still a challenge due to the inherent hydrophilicity of proteins. Previously, we have successfully developed 3D randomly orientated ultrafine fibrous scaffolds from gelatin by an approach of ultra-low concentration phase separation (ULCPS)[10]. The specific pore volume could achieve (259. 5 ± 4. 7) cm^3/mg at dry state. Whereas, the scaffolds still displayed a high deformation in an aqueous environment, though the citric acid (CA) cross-linking was applied. In addition, the epoxide, ethylene glycol diglycidyl ether (EGDE), has been used instead of CA to cross-link protein zein electrospun fibers.

Since the epoxy groups have high reactivity to active hydrogen structures in proteins, the stabilization efficacy was substantial proven by a small deformation of the obtained scaffolds[11]. However, when the electrospinning approach is employed, the epoxide content is limited by the spinnability of the protein/epoxide spinning solution. It is difficult to acquire further improvement in the scaffold properties by elevating the epoxide content. Without a spinning process, the ULCPS method might be able to break through the limitation in the ratio of components. Thus, the combination of the epoxide modification and the ULCPS technology might provide the feasible approach to develop water stable and ultraloose 3D fibrous scaffolds. However, the solution composition and the processing parameters might affect the architectures and properties of the obtained protein scaffolds.

In this study, we employed gelatin and EGDE as the protein and epoxy models and systematically investigated the effects of the polymer content, gelatin/epoxy mass ratio, types of crosslinking agents on the morphological properties and aqueous stability. The possibility of the combination of the epoxy modification and the ULCPS fiber fabrication method for 3D fibrous scaffold fabrication was proven for potential application in cartilage repair.

2 Experiments

2.1 Preparation of scaffolds

The stock solution of gelatin and EGDE were prepared by dissolving the gelatin powders and EGDE in deionized water at 1%. The phase separation solutions were prepared by mixing the stock solutions of gelatin and EGDE at a mass ratio of 1 : 1 with designed total polymer contents (0.01% to 0.5%) and then loaded in aluminum cups (30 ml/cup) for pre-frozing at −80 ℃ for 1 h. The frozen solutions were lyophilized at −40 ℃ for 48 h to obtain 3D ultrafine fibers which were modified at 120 ℃ for 1 h. The CA cross linked sample was also fabricated as control according to our previous work[12].

2.2 Characterization of scaffolds

The macroscopic and microscopic architectures of scaffolds were recorded by a digital camera (D7200, Nikon, Japan) and a scanning electron microscopic (SEM) (S-4800, Hitachi, Shanghai, China). Instead of the porosity with limited differences in values due to the ultraloose structures of scaffolds, the specific pore volumes were calculated for porous structure description using the following equation:

$$V_{SP} = \frac{V_P}{m} = \frac{V_t}{m} - \frac{1}{\rho} \tag{1}$$

where V_{SP}, V_P and V_t are the specific pore volume of the scaffold (cm^3/g), the pore volume and the total volume of the scaffold (cm^3). m is the dry weight of the scaffold (g), and ρ is the gelatin density (g/cm^3). The dry weights of the scaffolds were recorded after cross-linking. The total volumes of the scaffolds were measured and calculated based on the digital images of the scaffolds taken from three different views (one top and two side views) using the software of Image-J (1.48 version). The density of gelatin was calculated from the dry weight and volume of the casted gelatin films. The water stabilities of scaffolds were evaluated based on the changes in the scaffold volumes before and after immersing in the phosphate buffered saline (PBS) at 37 ℃ for different durations (15 min, 1 h, 2 h, 3 h, and 4 h). The diameter (d) and height (h) were measured from the digital photos of each sample. The volume of scaffolds was calculated according to the following equation:

$$V = S \times h = \pi r^2 h = \pi\left(\frac{d}{2}\right)^2 h \tag{2}$$

The deformation rate was calculated by:

$$W = (V - V_0) \div V_0 \times 100\% \tag{3}$$

where W is the deformation rate of the scaffold(%), V is the volume of the scaffold after soaking in PBS solution (cm^3), and V_0 is the original volume of the scaffold at dry state (cm^3).

3 Results and Discussion

3.1 Effect of solution solid content on the water stability of scaffolds

Figs. 1 (a) and 1 (b) show the macro- and micro-morphologies of the 3D gelatin/EGDE ultrafine fibrous scaffolds prepared from the solutions with different polymer contents. The scaffolds made from the solutions at low polymer contents (0.01 and 0.02 wt. %) displayed a small "cotton ball" shape composed mainly with fine fibers and little chips. As the polymer content achieved 0. 1%, the shape of scaffolds shifted from "cotton ball" to thick "dish". In the "dish", thicker fibers and larger pieces of chips could be observed. Along with the elevation in the polymer content from 0. 1% to 0. 5%, the morphology of samples became closer to the shape of the mold cup with increasing thickness and the color of samples turned from milk white to light yellow indicating a more condensed structure. The SEM images revealed that the fibers became thicker and shorter, and the chips turned to sheets, resulting in a blended structure with fibers and sheets.

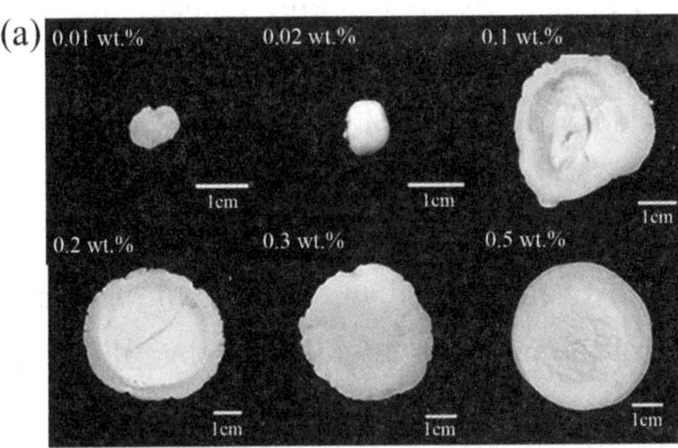

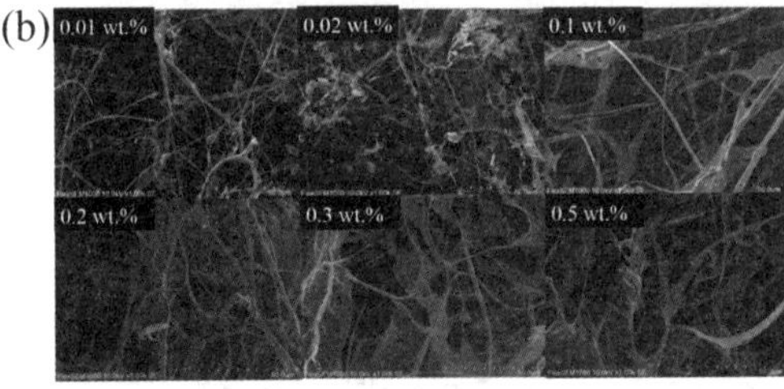

Fig. 1 Macro- (a) and micro-morphologies (b) of the 3D gelatin/EGDE scaffolds prepared from the solutions with different polymer contents via ULCPS

Fig. 2 exhibits the effect of the polymer content of the phase separation solutions on the specific pore volume of scaffolds. An ultrahigh specific pore volume of 366. 06 cm^3/g could be achieved when the polymer content of the phase separation solution was at a low value of 0. 01%. Although the scaffold showed a smaller volume (Fig. 1 (a)), the weight of the scaffold was also low with a small consumption of polymers. Hence, the volume of pores was large and the structure was loose. Similar to the shifting in macro-morphology, the specific pore volume dropped from 366. 06 to 181. 12 cm^3/g also at the polymer content of 0. 1%. The reason might be that the mainly fibrous structure turned to a mixed architecture of fibers and sheets at this polymer content, which substantially elevated the specific density of polymer in the scaffolds. Along with the further increase in the polymer content from 0. 1% to 0. 5%, the specific pore volume was then reduced in an almost linear trend from 181. 12 to 45. 96 cm^3/g. Although the specific pore volume of the 3D scaffolds was shifted in a large range from 366. 06 to 45. 96 cm^3/g, all the values were much higher than the values of the 2D gelatin ultrafine fibrous scaffolds (around 10 cm^3/g).

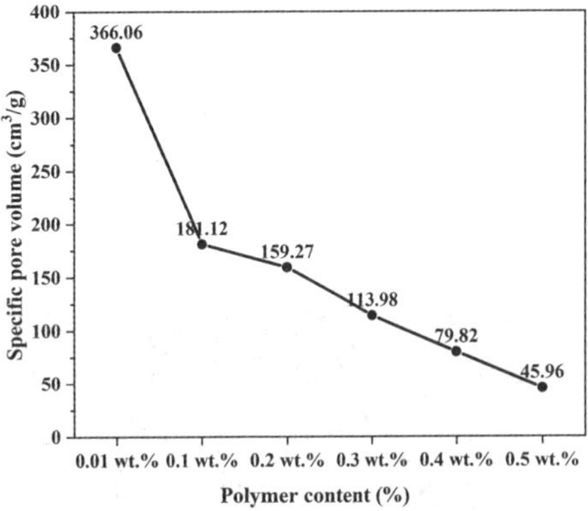

Fig. 2 Specific pore volumes of the 3D gelatin/EGDE scaffolds prepared from the solutions with different polymer contents via ULCPS

The wet-state morphological stabilities of the gelatin scaffolds were observed. Once immersed in PBS, the unmodified control scaffold was dissolved in a second. The CA modified scaffolds displayed an obvious shrinkage with a high deformation rate of 59% in the first 15 min. After 4 h, the shrinkage deformation rate increased to 67%. Since the measurement of the "cotton ball" volumes was not readily and the mode conformability of the samples prepared at low polymer content was poor, the scaffolds with the "dish" shapes were used for evaluation. The EGDE modified scaffold made from the solution with 0. 1% polymer content could maintain the "dish" shape after immersing in PBS but showed obvious swelling (Figs. 3 (d)-3 (f)). The initial swelling ratio was 36% at the first 15 min, and then reduced to around 22% in the following 4 h (Fig. 4). As the polymer content increased to 0. 2% and 0. 5%, the initial swelling ratios were reduced to 19% and 12% and the ratios slightly shifted to around 17% and 11% within 4 h, indicating the acquirement of higher wet stability. The possible reason might be that the ultrafine fibrous structure is more vulnerable compared to the gelatin sheets, leading to a higher stability of the fibrous scaffolds blended with sheets. Different from the CA cross-linked scaffolds, the deformation rates of the EGDE modified scaffolds were higher once immersed in PBS but reduced for different extents in the following 4 h. The possible reason might be that the deformation of the gelatin fibers was at both of the axial and radial directions resulting in the swelling of scaffolds. However, due the polymer orientation along the axial direction, the following expanding degree of fiber was higher at the radial direction, leading to the shrinkage of fibers in length and, hence, the reduction in the deformation of scaffolds. The higher the cross-linking extent, the weaker the swelling in the radial direction and thus the smaller the different between the initial and the following deformation rates.

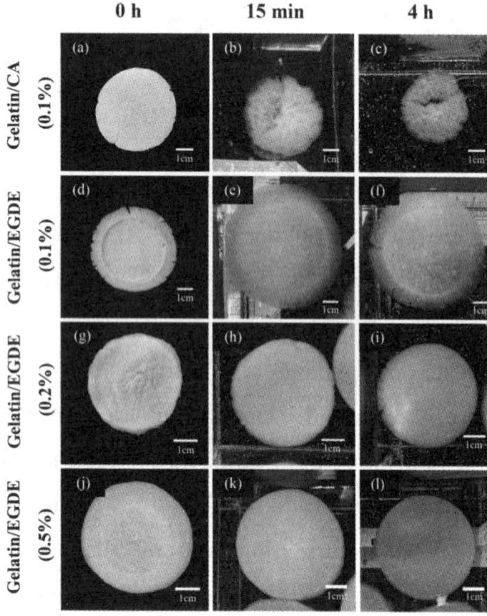

Fig. 3 The morphological deformation of the CA and EGDE modified 3D gelatin scaffolds prepared from the phase separation solutions with different polymer contents at dry state (a), (d), (g), and (j) or incubated in PBS for 15 min (b), (e), (h), and (k) and 4 h (c), (f), (i), and (l).

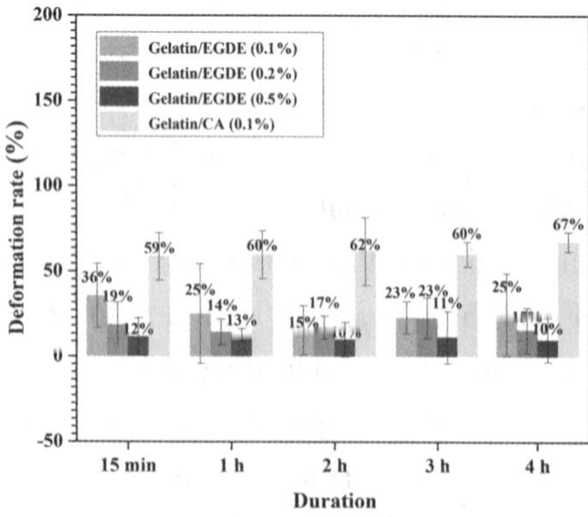

Fig. 4 The deformation rates of the 3D gelatin/CA and gelatin/EGDE fibrous scaffolds prepared from the phase separation solutions with different polymer contents incubated in PBS solution for up to 4 h

3.2 Effect of gelatin/EGDE ratio on the water stability of scaffolds

Fig. 5 shows the effect of the gelatin/EGDE ratio on the stability of the gelatin scaffolds (0.1% polymer content) in PBS. Once the gelatin/EGDE ratio was set at 2 : 1, the scaffolds displayed substantial swelling with an initial deformation rate of 48%. The rate was diminished to 38% in 4 h. The gelatin fibers with the gelatin/EGDE ratio of 1 : 1 possessed higher water stability proven by the lower deformation rates from 36% to 25%. With more EGDE content, the reactivity of the protein and epoxide system was more efficient, and produced an enhanced cross-linking efficacy and a denser polymer network. Further elevation in the EGDE content yielded melting of scaffolds and the collapse of the loose porous structure. Because the excessive EGDE molecules also served as plasticizer and the mobility of the protein chains raised introducing the softening and collapse of the porous scaffold. Thereby, the proper gelatin/EGDE ratio was set at 1 : 1.

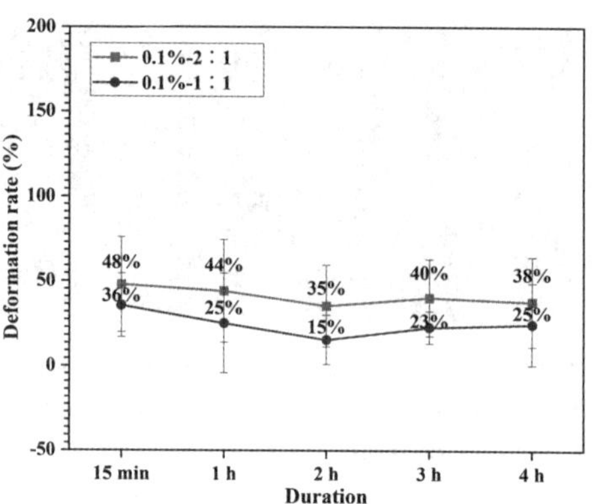

Fig. 5 The deformation rates of the 3D gelatin/EGDE fibrous scaffolds prepared from the phase separation solutions with 0.1 wt. % polymer content and different gelatin/EGDE ratio incubated in PBS solution for up to 4 h

4 Conclusions

In this work, ultraloose 3D gelatin/EGDE scaffolds with ultrafine fibrous or fiber/sheets mixed structures were formed via ULCPS. The morphologies were proven to be mainly controlled by the polymer content of the solutions for phase separation. The fiber/sheets mixed scaffolds showed better mode conformability. The specific pore volumes of the scaffolds could achieve a high value of 366.06 cm^3/g, which was about 33.3 times of the electrospun 2D gelatin scaffolds. With the assistance of the highly reactive epoxy, EGDE, the morphological stability at a wet state of the gelatin scaffolds could be enhanced and the deformation rate of the scaffolds was able to be reduced from 67% to 10%. This work paved the way for the construction of cell infiltrative water stable fibrous scaffolds from proteins for 3D tissue repair.

Acknowledgement

This work was financially supported by the Fundamental Research Funds for the Central Universities (No. 2232022D-13) and the Fundamental Research Funds of Shanghai Collaborative Innovation Center of High Performance Fibers and Composites (Province-Ministry Joint) (No. X12812101/015).

References

[1] GETOVA V E, DONGEN J A V, BROUWER L A, et al. Adipose tissue-derived ECM hydrogels and their use as 3D culture scaffold[J]. Artificial Cells Nanomedicine and Biotechnology, 2019, 47(1): 1693-1701.

[2] DAVARI N, BAKHTIARY N, KHAJEMOHAMMADI M, et al. Protein-based hydrogels: promising materials for tissue engineering[J]. Polymers, 2022, 14(5): 986.

[3] VOCETKOVA K, BUZGO M, SOVKOVA V, et al. A comparison of high throughput core-shell 2D electrospinning and 3D centrifugal spinning techniques to produce platelet lyophilisate-loaded fibrous scaffolds and their effects on skin cells[J]. RSC Advances, 2017, 7(85): 53706-53719.

[4] CHEN Y, JIA Z, SHAFIQ M, et al. Gas foaming of electrospun poly(L-lactide-co-caprolactone)/silk fibroin nanofiber scaffolds to promote cellular infiltration and tissue regeneration[J]. Colloids and Surfaces B: Biointerfaces, 2021, 201: 111637.

[5] XU H, CAI S, XU L, et al. Water-stable three-dimensional ultrafine fibrous scaffolds from keratin for cartilage tissue engineering[J]. Langmuir, 2014, 30(28): 8461-8470.

[6] UZ M, BUYUKOZ M, SHARMA A D, et al. Gelatin-based 3D conduits for transdifferentiation of mesenchymal stem cells into Schwann cell-like phenotypes[J]. Acta Biomaterialia, 2017, 53: 293-306.

[7] GELAIN F, LUO Z, RIOULT M, et al. Self-assembling peptide scaffolds in the clinic[J]. NPJ Regenerative Medicine, 2021, 6(1): 1-8.

[8] PUPPI D, DINUCCI D, BARTOLI C, et al. Development of 3D wet-spun polymeric scaffolds loaded with antimicrobial agents for bone engineering[J]. Journal of Bioactive and Compatible Polymers, 2011, 26(5): 478-492.

[9] BRENNAN M A, RENAUD A, GAMBLIN A-L, et al. 3D cell culture and osteogenic differentiation of human bone marrow stromal cells plated onto jet-sprayed or electrospun micro-fiber scaffolds [J]. Biomedical Materials, 2015, 10(4): 045019.

[10] JIANG Q, XU H, CAI S, et al. Ultrafine fibrous gelatin scaffolds with deep cell infiltration mimicking 3D ECMs for soft tissue repair[J]. Journal of Materials Science: Materials in Medicine, 2014, 25(7): 1789-1800.

[11] LIU L, JIANG J, JIN X, et al. Epoxide cross-linked and lysine-blocked zein ultrafine fibrous scaffolds with prominent wet stability and cytocompatibility[J]. ACS Applied Polymer Materials, 2021, 3(8): 3855-3866.

[12] JIANG Q, REDDY N, YANG Y. Cytocompatible cross-linking of electrospun zein fibers for the development of water-stable tissue engineering scaffolds[J]. Acta Biomaterialia, 2010, 6(10): 4042-4051.

Toughening of Epoxy Resins and Carbon Fibre Reinforced Composites by Core-Shell Rubber Nanoparticles

SHI Wenshuo[1,2], CAI Weidong[1,2], LIU Wanshuang[1,2*]

[1] *Center for Civil Aviation Composites, Donghua University, Shanghai 201620, China*

[2] *College of Textiles, Donghua University, Shanghai 201620, China*

* *Corresponding author's email*: wsliu@ dhu. edu. cn

Abstract: In this study, epoxy resins were toughened by core-shell rubber nanoparticles (CSRNPs) and used as the matrix for carbon fiber reinforced composites (CFRPs). The results showed that the addition of CSRNPs has a slight effect on the curing behaviors of epoxy resin. With the increase of CSRNPs loading, the glass transition temperature (T_g) of resulting epoxy composites decreased accordingly. The CSRNPs exhibited an excellent toughening effect for the epoxy resin, and maximum increase (200.0%) in critical stress intensity factor (K_{IC}) was achieved with the 10% CSRNP loading. The CFRP based on the toughened epoxy resin showed a 64.4% increase in mode I interlaminar fracture toughness (G_{IC}) and a 7.9% increase in flexural strength, respectively.

Keywords: Epoxy resins; Carbon fiber reinforced composites; Core-shell rubber nanoparticles; toughness

1 Introduction

Epoxy resins are one of the most widely used thermosetting polymer matrices for carbon fiber reinforced composites (CFRPs), because of their excellent stability, mechanical properties, and easy processability.[1-2] However, the inherent brittleness and poor fracture resistance of epoxy resins limit the application of CFRPs, because their highly crosslinked network structure cannot absorb and dissipate energy when an impact load is applied.[3] It is an effective approach to improving the interlaminar mechanical properties of CFRPs by toughening the epoxy matrix.[4] The most common method to improve the toughness of epoxy resins is to add a second phase modifier. For example, the incorporation of rubber particles, inorganic nanoparticles, and thermoplastic polymers can improve the fracture toughness of epoxy resins.[5-8]

As one of the most commonly toughening materials of epoxy resin, core-shell rubber nanoparticles (CSRNPs) consist of a glassy shell and a rubbery core. The glassy shell such as poly(methyl acrylate), gives good compatibility with epoxy resins, and the rubber core, such as polybutadiene or styrene-butadiene rubber, provides impact resistance to epoxy resins.[9-10] It has been reported that CSRNPs could improve the toughness of both epoxy resins and corresponding CFRP significantly while maintaining the thermal properties of epoxy resins.[11] The CSRNP would not be filtered by fibers during resin infusion due to their nanoscale dimensions.

In this study, we used a commercialized CSRNP to modify epoxy resins. The effects of CSRNP on the curing behaviors, thermal and mechanical properties of resulting epoxy composites were investigated. Emphasis was placed on the study of toughening effect of CSRNPs towards epoxy resins. Finally, the optimized toughened epoxy resin was used as the matrix for CFRPs. The flexural properties and interlaminar fracture toughness of CFRP were investigated. The toughening mechanism was revealed by microscopic observation of the cracked surfaces.

2 Experiments

2.1 Materials

The epoxy monomer (ML-5417A) and curing agent (ML-5417B) used in the work were purchased from Wells Advanced Materials (Shanghai) Co. The rubber core-shell particles were purchased from Wuhan Longtu New Material Co. Carbon fiber fabrics (T-700, 12K, 480 g/m^2) were purchased from Jiangsu Hengshen Co.

2.2 Preparation of CSRNP/epoxy composites

Different content of CSRNP (2%, 5% and 10%) were sufficiently mixed with epoxy mononer and curing agent. The weight ratio of epoxy monomer and curing agent was 100 : 30. The mixture was poured into an aluminum mold coated with the release agent and degassed for 30 min. The epoxy samples were cured at 50 ℃ for 6 h and 100 ℃ for 5 h. The pure epoxy resin was prepared using the same method.

2.3 Fabrication of CFRP

CFRPs based on the pure epoxy and CSRNP/epoxy composite with 10% CSRNP were fabricated by vacuum infusion process. Firstly, the release agent was evenly coated on a clean s glass mold, and 8-ply carbon fiber, suitable size peel-ply, infusion mesh, and vacuum bag are laid in sequence, sealed, and left the resin perfusion channel and the exhaust port in advance. Then, the resin perfusion channel was closed, the system was pumped vacuum, and the resin was transferred into the closed cavity of the carbon fiber layers at room temperature. Finally, the CFRPs ware obtained after the resin was cured at room temperature for 12 h and 80 ℃ for 4 h.

2.4 Characterization

Differential scanning calorimetry (DSC, NETZSCH, Germany) was utilized to study the curing behaviors at a heating rate of 10 ℃ min^{-1}. Dynamic mechanical analysis (DMA) was performed on a TA Instruments Q800 using double cantilever mode. The tensile properties and fracture toughness of all the epoxy samples were carried out according to ASTM-D 638 and ASTM-5015 standards respectively. Flexural properties and mode I interlaminar fracture toughness CFRP were measured according to ASTM-D790 and ASTM-D5528, respectively. Scanning electron microscope (SEM) was observed on a JCM-6000 plus NeoScope.

3 Results and Discussion

3.1 Curing behaviors

Fig. 1 shows the DSC scanning curves of pure epoxy and CERNP/epoxy composites. The peak temperatures of all epoxy systems were in the range of 100-105 ℃. The addition of CSRNP slightly increases the peak temperatures of the curing reaction due to the blunting effect. On the whole, CSRNP/epoxy composites showed similar curing behaviors compared with the pure epoxy system. Therefore, they could be cured using the same conditions

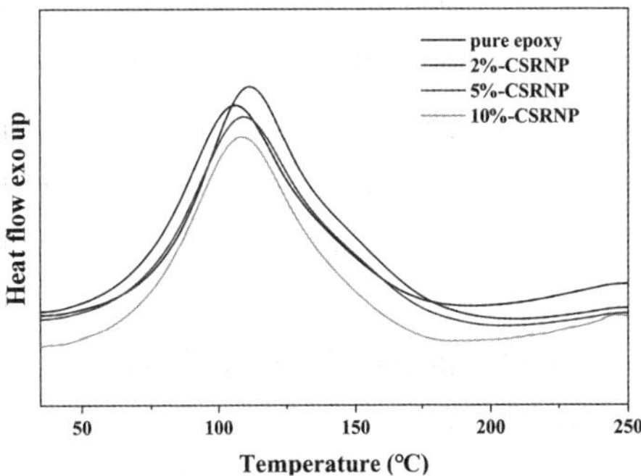

Fig. 1 DSC thermograms for non-isothermal curing behaviors of pure and CSRNP/epoxy systems

3.2 Thermal properties

The thermomechanical properties of the pure epoxy and CSRNP/epoxy composites were investigated by DMA. Fig. 2 displays the temperature dependence of the storage modulus (E') and the loss factor (tan δ) of all the samples. The storage modulus can reflect the elastic properties of polymers. As can be seen, all the epoxy samples also show the same E' values of-2 500 MPa at 25 ℃ (Fig. 2(a)). The temperature at the maximum tan δ value is often taken as the glass transition temperature (T_g) for polymers. As shown in Fig. 2(b), the T_g values of CSRNP/epoxy composites gradually decrease with increasing the CSRNP content. Typically,

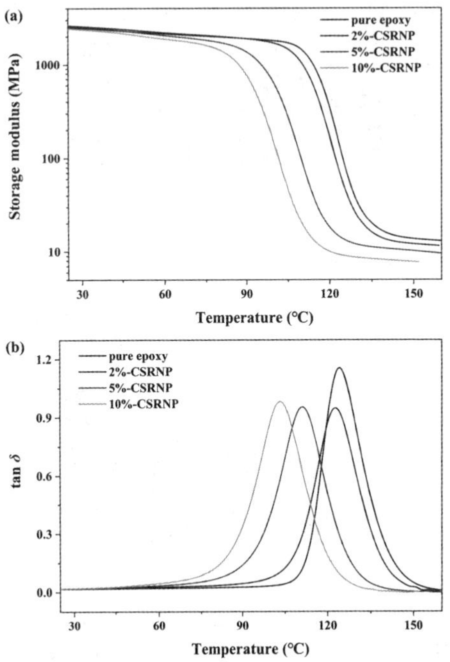

Fig. 2 Storage modulus (a) and tan δ (b) versus temperature curves of the pure epoxy and CSRNP/epoxy composites

when the content of the CSRNP reaches 10%, the T_g value of CSRNP/epoxy composite decreases from 124 ℃ to 111 ℃ compared with the pure epoxy resin. The decreased T_g of CSRNP/epoxy composites is because the T_g of CSRNP is much lower than that of the epoxy resin. High content of continuous phase with low T_g would result in the decrease in T_g for the resulting epoxy composites.

3.3 Mechanical properties of polymer

Fig. 3 shows the test results of the average tensile strength, tensile modulus, critical stress intensity factor (K_{IC}) and critical strain energy release rate (G_{IC}) for the pure epoxy and CSRNP/epoxy composites. The tensile strength of the pure epoxy resin was 82.4 MPa. The tensile strengths of CSRNP/epoxy composites decreased with the increase in the content of CSRNP. The tensile strength was 67.8 MPa with the addition of 10% CSRNP, which was 17.7% lower compared to the pure epoxy resin. The tensile modulus also decreased with the increase of CSRNP content. Compared with the pure epoxy resin (3455 MPa), the tensile modulus of epoxy composite containing 10% CSRNP decreased to 2825 MPa, owing to the lower modulus of CSRNP.

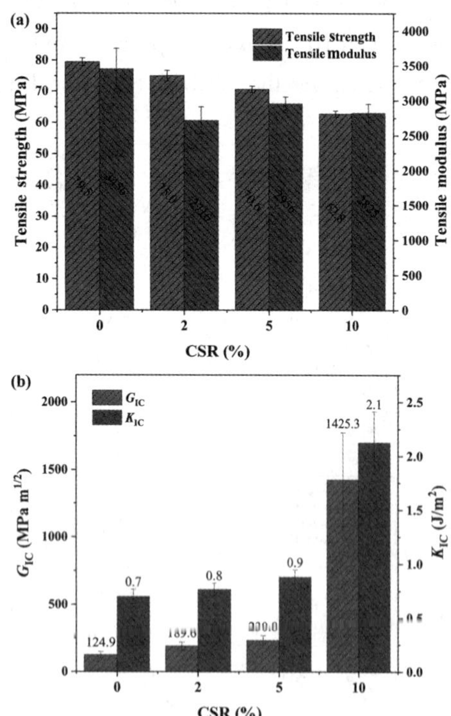

Fig. 3 Mechanical properties of the pure epoxy and CSRNP/EP epoxy composites: (a) tensile strength and tensile modulus; (b) fracture toughness (K_{IC}) and fracture energy (G_{IC})

The toughness improved positively with the content of CSRNP. With the addition of 5% CSRNP, K_{IC} increased from 0.7 J/m^2 to 0.9 J/m^2, and G_{IC} increased from 124.9 MPa · m$^{1/2}$ to 230.8 MPa · m$^{1/2}$, which increased by 28.6% and 92.8% compared to the pure epoxy, respectively. When 10% CSR was added, K_{IC} and C_{IC} reached 2.1 J/m^2 and 1425.3 MPa · m$^{1/2}$, respectively, which increased 200% and 1040% relative to the pure epoxy. This shows that the CSRNP can improve the fracture toughness of epoxy resin, but also hurt the tensile properties of epoxy resin.

3.4 Mechanical properties of CFRP

Fig. 4 shows the detail for all studied polymeric composite systems during the bending and fracture toughness tests.

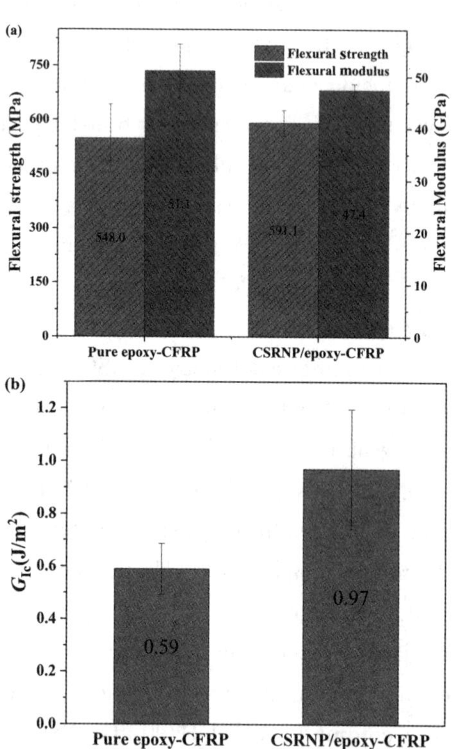

Fig. 4 (a) Flexural strength and modulus of CFRPs; (b) toughness of CFRP

The flexural strength of the CSRNP/epoxy-CFRP is 591.1 MPa, an increase of 7.9% relative to the pure EP-CFRP. The flexural modulus is 47.4 GPa, a decrease of 7.2%. Significantly improved toughness of CFRP based on modified EP, from 0.59 J/m^2 to 0.97 J/m^2, an increase of 64.4%. Carbon fiber has properties such as high strength and high modulus. The CFRP has excellent in-plane properties and weak interlaminar mechanical properties. The modified epoxy resin improves the interfacial properties between resin and fiber, and the increased toughness of the resin helps to improve the interlaminar properties of the CFRP laminate, therefore, the modified epoxy resin improves the flexural strength of the composite and fracture toughness.

3.5 Microstructure of CFRP

Figs. 5 (a), (c) and (e) show microstructure of pure epoxy-CFRP, there are cavities between the fibers because the resin fails to fully fill the inter-fiber voids. In addition, the surface of the carbon fiber is also relatively smooth, indicating that it has poor interfacial compatibility where between the resin matrix and the fibers. Compared to pure epoxy, the CSRNP/epoxy composites could fully fill the inter-fiber cavities, and a large amount of resin matrix adhered to the fiber surface. The addition of CSRNP not only improved the toughness of the epoxy resin, but also enhanced the adhesion between fibers and resin, and improved the interfacial properties of both.

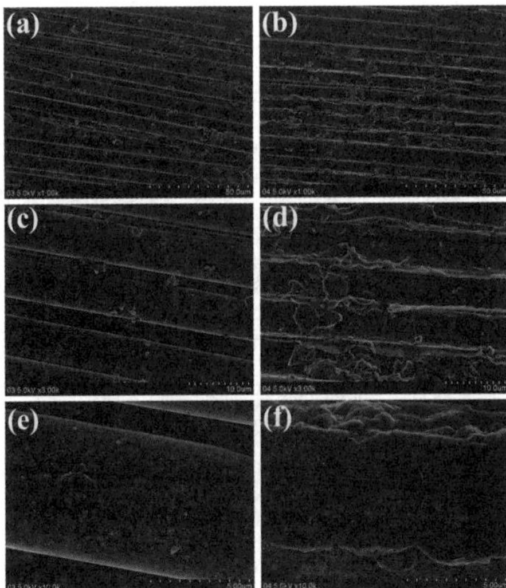

Fig. 5 (a), (c) and (e) SEM micrographs of the pure epoxy-CFRP at different magnifications; (b), (d) and (f) SEM micrographs of the CSRNP/epoxy-CFRP at different magnifications

4 Conclusions

In summer, CSRNP was utilized to toughen the epoxy resin. The higher-loading CSRNP decreases the T_g of pure EP. CSRNP dramatically improves the fracture toughness of resin and its CFRPs. In addition, CSRNP also improves the flexural strength of CFRPs when only adding 10% CSRNP, the K_{IC} and G_{IC} of the pure EP increase by 200% and 1040%, respectively. The flexural strength and toughness of CFRP increase by 7.9% and 64.4%, respectively.

References

[1] IMANAKA M, NARITA I, NAKAMURA Y, et al. Fracture properties of epoxy polymers modified with cross-linked and core-shell rubber particles[J]. Journal of Materials Science, 2020, 56(2): 1842-1854.

[2] SU X, ZHOU Z, LIU J, et al. A recyclable vanillin-based epoxy resin with high-performance that can compete with DGEBA[J]. European Polymer Journal, 2020, 140.

[3] TUO H, LU Z, MA X, et al. Damage and failure mechanism of thin composite laminates under low-velocity impact and compression-after-impact loading conditions[J]. Composites Part B: Engineering, 2019, 163: 642-654.

[4] CAROLAN D, IVANKOVIC A, KINLOCH A J, et al. Toughened carbon fibre-reinforced polymer composites with nanoparticle-modified epoxy matrices[J]. Journal of Materials Science, 2016, 52(3): 1767-1788.

[5] MOUSAVI S R, ESTAJI S, RAOUF JAVIDI M, et al. Toughening of epoxy resin systems using core-shell rubber particles: a literature review[J]. Journal of Materials Science, 2021, 56(33): 18345-18367.

[6] WANG J, XUE Z, LI Y, et al. Synergistically effects of copolymer and core-shell particles for toughening epoxy[J]. Polymer, 2018, 140: 39-46.

[7] CHEN D, LI J, YUAN Y, et al. A new strategy to improve the toughness of epoxy thermosets by introducing the thermoplastic epoxy[J]. Polymer, 2022, 240.

[8] AKBOLAT M Ç, KATNAM K B, SOUTIS C, et al. On mode-I and mode-II interlaminar crack migration and R-curves in carbon/epoxy laminates with hybrid toughening via core-shell rubber particles and thermoplastic micro-fibre veils[J]. Composites Part B: Engineering, 2022, 238.

[9] MOUSAVI S R, AMRAEI I A. Toughening of dicyandiamide-cured DGEBA-based epoxy resin using MBS core-shell rubber particles[J]. 2015, 49(19): 2357-2363.

[10] KLINGLER A, BAJPAI A, WETZEL B. The effect of block copolymer and core-shell rubber hybrid toughening on morphology and fracture of epoxy-based fibre reinforced composites[J]. Engineering Fracture Mechanics, 2018, 203: 81-101.

[11] QUAN D, IVANKOVIC A. Effect of core-shell rubber (CSR) nano-particles on mechanical properties and fracture toughness of an epoxy polymer[J]. Polymer, 2015, 66: 16-28.

Measuring the Frequency-Dependent Complex Permittivity of Fabrics

SU Qincheng[1*], **SHI Meiwu**[1,2]

[1] *College of Textiles, Donghua University, Shanghai 201620, China*

[2] *Institute of Quartermaster Engineering & Technology, Institute of System Engineering, Academy of Military Science, PLA, Beijing 100082, China*

[*] *Corresponding author's email*: su_qincheng@ 163. com

Abstract: The study of electromagnetic waves propagating in fabric in a complex problem because it's porous and heterogeneous, which demands a detailed understanding of its electromagnetic properties. In this study, the complex permittivity measurement principle of fabric was analyzed , the complex permittivity in 1 Hz−10 MHz of different textile materials and weave was measured respectively. The results showed that, fabric complex permittivity decreases with increased frequency, and with the rapid decline in frequency between 1−100 Hz continue to increase gradually stabilized; textile material has a significant effect on the fabric of complex permittivity, the complex permittivity of cotton much larger than polyester and nylon fabrics, and polyester complex permittivity is slightly greater than that of nylon; complex permittivity of plain, twill and satin fabrics decreases with increasing frequency, and rapidly declines between 1−100 Hz, wherein the complex permittivity of plain fabric is the largest, twill and satin close.

Keywords: Dielectric properties; Complex permittivity; Fabric

1 Introduction

Textiles as an electromagnetic wave propagation medium have not been widely studied. Although they are soft, low price and good wearability, they don't have excellent electromagnetic properties. In recent years, as the rapid development of conductive polymer and metalized fabric processing technology, conductive properties of textiles are improved, and forming a series of electromagnetic textiles. Metal fabric such as stainless steel fabric[1-2], copper mesh fabric[3-4], etc; metalized fabric such as chemical plating and magnetron sputtering[5] have been applied. The conductive mechanism for conductive polymers have been established, a variety of polymerization technologies for conductive polymers such as polypyrrole (PPy)[6-7], polythiophene (PTh)[8], and polyaniline (PANI)[9] were finishing to ordinary fiber or fabric surface have been widely studied, which also with light weight, large area deposition, and adjustable conductivity, have broad application prospects.

Permittivity is an important parameter influence the electromagnetic properties of textile, which characterization of the polarization properties of macroscopic quantities. The complex permittivity of fabric has a fundamental role to inhibit the propagation of electromagnetic waves. The permittivity has direct relevance to such applications as the electromagnetic shielding and radar absorbing fabrics, and the shielding effectiveness or return loss of fabric is directly related to the material chemistry and thickness[10]. More precisely, the real part of the complex permittivity shows the fabric's ability to store charge, and the imaginary part exhibits the energy loss in the dielectric relaxation[11]. In electromagnetic propagation calculation of a fabric structure over a wide frequency range, modelling the electrical properties of fabric closer to the experimental measurement is deemed advisable.

To describe the frequency dependence of the permittivity of fabric, it would be interesting to be to measure it on the material. However, it is not easy to determine the permittivity directly, computations to extract the value of the permittivity versus frequency from measurements is necessary. In this paper, the effective model to describe the frequency dependence of fabric and a possible relation between electromagnetic propagation and fabric's permittivity is investigated.

2 Experiments

2.1 Sample preparation

The polyester spun yarn (purchased from Lilaiya Textile Co., Ltd., China) was woven plain, twill and satin by rapier loom (developed by TJPU College of Textiles, China), which has the same warp and weft density. There are three specifications of plain fabric (purchased from Lilaiya Textile Corp., China) with different materials. The total parameters of samples are listed in Tab. 1.

Tab. 1 Fabric specification parameters

Sample No.	Material	Weave	Linear density (tex)		Density (threads/10 cm)		Thickness (mm)
			Warp	Weft	Warp	Weft	
1	Cotton	Plain	62.4	62.4	240	220	0.22
2	Polyester	Plain	55.8	55.8	250	200	0.16
3	Polyamide	Plain	57.5	57.5	450	400	0.12
4	Polyester	Plain	59.1	59.1	200	140	0.56
5	Polyester	3/1 Twill	59.1	59.1	200	140	0.70
6	Polyester	5/3 Satin	59.1	59.1	200	140	0.98

2.2 Experimental procedure

The permittivity measurements were taken two times on each samples with two directions according to fabric direction, taking the whole number of tests to 49 observations, which following a logarithmic law, between 1 Hz and 10 MHz. The sample holder is shown in Fig. 1, and measurements were taken at (20±1)℃, humidity (65±2)%.

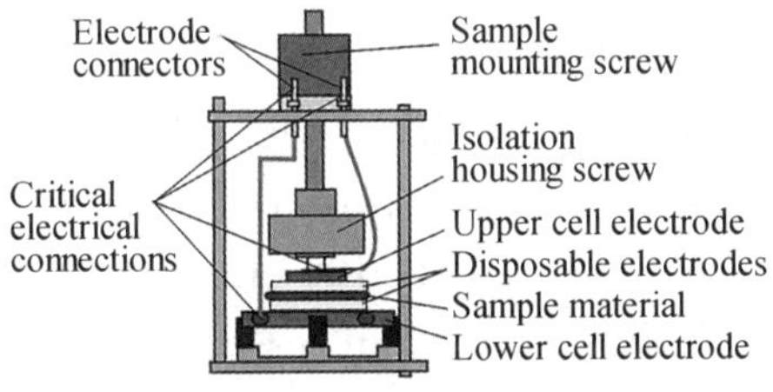

Fig. 1 Samples fixture: (a) schematic diagram; (b) test drawing

3 Results and Discussion

3.1 The influence of textile materials

The dielectric properties of the material depend on the polarization of the process, usually fiber molecular polarity is larger, the greater the dipole moment, therefore, the higher the degree of polarization, so the greater the permittivity[12]. The measurement results of complex permittivity of cotton, polyester and nylon plain weave fabrics are shown in Fig. 2.

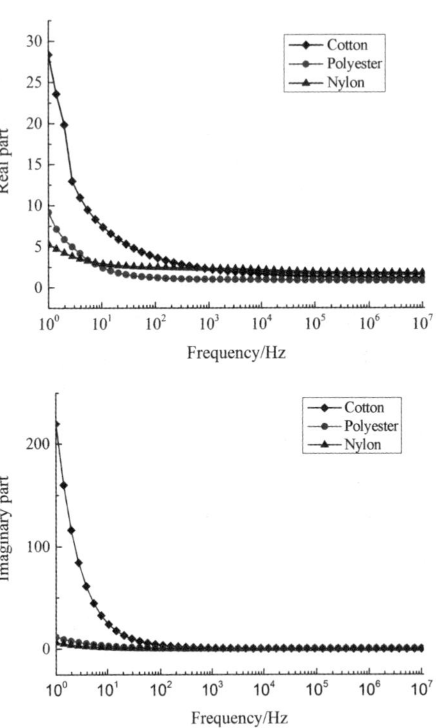

Fig. 2 Complex permittivity of different textile materials: (a) real part; (b) imaginary part

A number of observations can be made from the Fig. 2 presented in this section, namely the following. Fabric complex permittivity decreases with increased frequency, and with the rapid decline in frequency between 1–100 Hz continue to increase gradually stabilized. This is because with the increase of frequency, bound charge within the fiber molecules under the influence of an applied electric field produces an electric field polarization lags behind the speed of change of the applied electric field, resulting in decreasing the complex permittivity of the fabric.

Textile material has a significant effect on the fabric of complex permittivity, in cotton, polyester and polyamide fabrics, the complex permittivity of cotton much larger than polyester and nylon fabrics, and polyester complex permittivity is slightly greater than nylon. Because the molecular structure of cotton in more polar groups, while weakly polar polyester and nylon, moreover, water is a polar molecule, a permittivity of 81[12], cotton fabric moisture regain of 8.5% in the standard condition, polyester and polyamide fiber moisture regain of 0.4% and 4.5%, respectively. Thus increasing the moisture in the fabric so that the complex permittivity significantly increased.

3.2 The influence of weave

In the same warp and weft density polyester fabric, different weave reflects the change of the fabric structure to its permittivity. Polyester fabric of plain, twill and satin of measured values shown in Fig. 3. Complex permittivity of plain, twill and satin fabric decreases with increasing frequency, and rapidly decline between 1–100 Hz, wherein the complex permittivity of plain is the largest, twill and satin close. This is the reason for at the same latitude conditions, there is more interwoven points, less internal porosity, while the larger air content in 3/1 twill and 5/3 satin fabrics. Known dielectric constant of air is 1, which is smaller than the fabrics, therefore the complex permittivity of fabric decrease with the increase of the internal air volume.

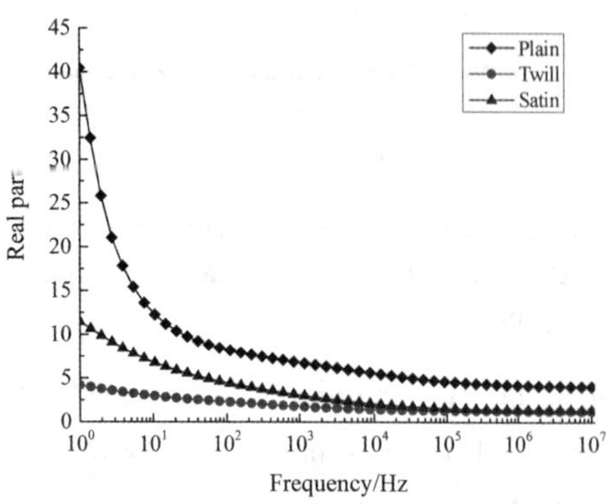

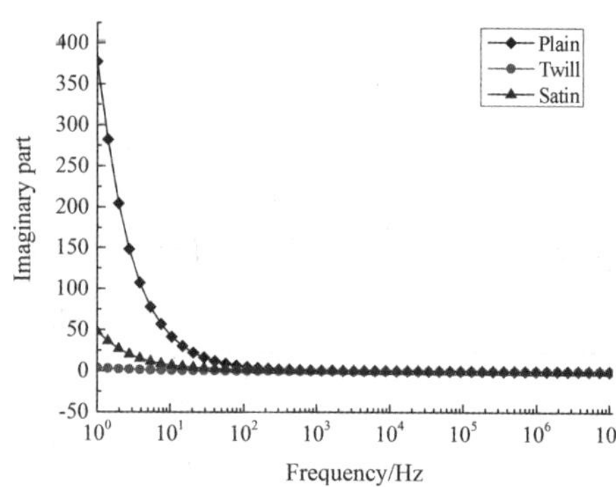

Fig. 3 Complex permittivity of different textile weave: (a) real part; (b) imaginary part

4 Conclusions

This study highlights the frequency-dependent complex permittivity of fabric, measured the complex permittivity in 1 Hz–10 MHz of different textile materials and weave respectively. The results showed that, fabric complex permittivity decreases with increased frequency, and with the rapid decline in frequency between 1–100 Hz continue to increase gradually stabilized; textile material has a significant effect on the fabric of complex permittivity, the complex permittivity of cotton much larger than polyester and nylon fabrics, and polyester complex permittivity is slightly greater than nylon; The complex permittivity of plain, twill and satin weave fabrics decreases with the increase of frequency, and decreases rapidly between 1–100 Hz, among which the complex of plain weave fabric is largest, and that of twill weave and satin weave is close.

References

[1] PERUMALRAJ R, DASARADHAN B, NALANKILLI G. Copper, stainless steel, glass core yarn, and ply yarn woven fabric composite materials properties[J]. Journal of Reinforced Plastics and Composites, 2010, 29(20): 3074-3082.

[2] SHYR T, SHIE J. Electromagnetic shielding mechanisms using soft magnetic stainless steel fiber enabled polyester textiles[J]. Journal of Magnetism and Magnetic Materials, 2012, 324(23): 4127-4132.

[3] OZDEMIR H, U URLU E S, OZKURT A. The electromagnetic shielding of textured steel yarn based woven fabrics used for clothing[J]. Journal of Industrial Textiles, 2015, 0(00): 1-21.

[4] XIAO H, SHI M W, WANG Q, et al. The electromagnetic shielding and reflective properties of electromagnetic

textiles with pores, planar periodic units and space structures[J]. Textile Research Journal, 2014, 84(16): 1679-1691.

[5] JIANG S, MIAO D, YANG G, et al. Fabrication of Ag thin film on polyester fabric by roll to roll magnetron sputtering system[J]. Journal of Materials Science: Materials in Electronics, 2015, 26(5): 3364-3369.

[6] QIAO Y, SHEN L, DOU T, et al. Polymerization and characterization of high conductivity and good adhension polypyrrole films for electromagnetic interference shielding[J]. Chinese Journal of Polymer Science, 2010, 28(6): 923-930.

[7] HONG Y K, BYUN S W, LEE J Y, et al. Electromagnetic interference shielding characteristics of fabric complexes coated with conductive polypyrrole and thermally evaporated Ag[J]. Current Applied Physics, 2001, 1(6): 439-442.

[8] HÅKANSSON E, AMIET A, KAYNAK A. Electromagnetic shielding properties of polypyrrole/polyester composites in the 1-18 GHz frequency range[J]. Synthetic Metals, 2006, 156(14/15): 917-925.

[9] KIM B R, LEE H K, PARK S H, et al. Electromagnetic interference shielding characteristics and shielding effectiveness of polyaniline-coated films[J]. Thin Solid Films, 2011, 519(11): 3492-3496.

[10] CHEN Z, XU C, MA C, et al. Lightweight and flexible graphene foam composites for high-performance electromagnetic interference shielding[J]. Advanced Materials, 2013, 25(9): 1296-1300.

[11] SANO E, AKIBA E. Electromagnetic absorbing materials using nonwoven fabrics coated with multi-walled carbon nanotubes[J]. Carbon, 2014, 78: 463-468.

[12] C C Y, YU W D. Textile Physics[M]. Shanghai: Donghua University Press, 2009.

Preparation and Evaluation of Textile-Based Silver Circuits for Flexible Joule Heaters by Electrohydrodynamic Printing Technology

LI Jiawei[1], GUO Wenjing[1], HU Jiyong[1*]

[1] *Key Laboratory of Textile Science & Technology, Ministry of Education, Donghua University, Shanghai 201620, China*

* *Corresponding author's email*: hujy@ dhu. edu. cn

Abstract: In the context of the wearable electronics era, smart heating textiles are developing rapidly and are widely used for cold protection and thermal therapy. In this study, the electrohydrodynamic (EHD) printing technology, a digitized printing process, is used to deposit conductive materials onto textile substrates and prepare flexible joule heaters with no contact, high precision and fast production. Particularly, a well-formed and conductively stable EHD spliced printing method with UV-curable silver nano-inks was chosen to prepare flexible joule heaters with circuits distributed in parallel. The results show that the heater has a fast response under the application of human adaptable low voltage, and achieve an excellent heating effect of 35 ℃ in 8 s during the high-speed heating phase. The final maximum equilibrium temperature fully meets the human heating requirements. These results are meaningful for the design and application of flexible heating electronics.

Keywords: Joule Heaters; Flexible electronics; Electrohydrodynamic (EHD) printing

1 Introduction

Low temperatures cause a heat imbalance between the body and the atmosphere, which can affect human physiological health, social productivity and energy storage[1]. Personal thermal management (PTM) technologies can create a local thermal envelope around the human body, effectively improving comfort and mitigating the negative effects of low temperatures on humans. Compared with traditional proximity heat delivery devices, smart heating textiles can further combine or even integrate temperature control elements with clothing and have demonstrated unique advantages in PTM heating technologies, which is widely used for cold prevention and heat therapy[2].

Two-dimensional fabrics are regarded as the preferred substrate for the preparation of wearable electronic devices due to their softness and breathability. One of the important ways to give the textile active heating function is to add conductive materials to the surface of the fabric to prepare a flexible joule heater (FJH) to make up for the heat gap on both sides through the joule heat effect. The key for the preparation of textile-based FJHs is to obtain a stable and reliable conductive pathway on the surface while retaining the flexibility of the fabric. Printing processes allow functionalized conductive substances to be deposited on fabric surfaces as nano-ink dispersions, thus promising flexible, large-scale preparation of flexible electronic devices[3]. Among the various printing methods, electrohydrodynamic (EHD) printing is one of the most promising non-contact printing technologies, which pulls ink from the printhead by an applied electrostatic field force to form a jet, requiring less ink printability than traditional "push-out" inkjet printing[4]. It has been demonstrated that the EHD printing process can be used to obtain high-resolution conductive patterns on the surface of complex porous fabrics[5] and discussed the impact of relevant factors on the forming quality of circuit, which has some insight into the use of EHD printing technology for the preparation of flexible FJHs.

Nano-silver is often doped in polymer substrates for the configuration of EHD printing inks due to its good electrical conductivity[6]. In this study, a UV-curable silver ink developed specifically for EHD printing was used, and the silver lines formed on the woven fabric surface showed good formability and resistance up to 20 Ω/cm. Considering the application requirements of the FJHs, a spliced printing was used to increase the width of the silver lines, effectively improving the conductivity and stability of the silver conductor. Afterwards, the heat generating circuit was arranged in a certain area and the FJH was printed in one-off by EHD printing.

2 Experiments

2.1 Materials

The UV curable nano-sliver conductive ink is mainly composed of the conductive filler. polymer, solvent, and Additives. The nano-silver flakes (≥99.9%) with high conductivity were purchased from Suzhou Tanfeng Graphene Technology Co. Ltd. (Jiangsu, China). Polyurethane acrylate (PUA, ≥99.8%) with features of high flexibility and low requirement on curing temperature was selected as prepolymer, which was produced by Chengdu Guangju Technology Co. Ltd. (Sichuan, China). TMPTA (≥ 99.0%), TPGDA, (≥ 99.0%) was employed as diluents to adjust the viscosity of conductive inks to be suitable for EHD printing. 1-(2-hydroxy-5-methylphenyl) 1-propanone (1 173, ≥ 99.0%) was employed as the photo-initiator to achieve inks curing. To improve the bond between the resin and the filler, γ-(2, 3epoxy-propoxy) propyltrimethoxysilane (KH－560, ≥ 99.0%) was used. Defoamers (BYK-A555) were purchased from BYK additives &instrument. All chemicals were analytical grade and used without further purification. Nylon woven fabric commonly used as a lining were chosen as the printing substrate. The number of warp and weft yarns per square inch of fabric is 330.

2.2 Fabric treatment

The woven fabric is first soaked in deionized water for 3 hours to relieve stress, followed by soaking in anhydrous ethanol to remove debris. The fabric is dried in an oven at 50 ℃ for one hour, and pressure is applied to the fabric to ensure that the fabric is flat. Soak the woven fabric in 1% antistatic agent water solution for 60 s, remove it, fold it neatly and dry it in a 50 ℃ oven for 1 h. [7]

2.3 Ink preparation

The nano-silver flakes, polymer, solvent, photo-initiator and additives were mixed to form UV-curable nano-silver ink according to the weight ratio in Tab. 1, during which they were stirred for 6 h and sonicated for 30 min to achieve uniform dispersion. The ink is usually left to stand for a period of time to reach a stable state as well as to remove surface blistering before use.

Tab. 1 Composition of UV-curable nano-silver ink

Ink components	Content(%)
Ag	20
PUA	32
TMPTA	19.2
TPGDA	19.2
1173	8
KH560	1.2
A555	0.4

2.4 Spliced printing of the heat generating silver circuits

The schematic diagram for electrostatic direct printing is shown in Fig. 1. The M08－001 electrohydrodynamic printer consists of a high voltage power supply, a three-axis platform motion control system (2D substrate motion control, height lift control) and an air driven micro syringe pump system. The high voltage power supply is set up to pull the liquid ink out of the nozzle head and create a conductive path on the fabric. The PC side is connected to the individual control systems of the printer via programmable input programs. Initially, the print height and position can be adjusted via the three-axis platform motion control system, the ink flow rate pulled out can be precisely adjusted by the micro syringe pump system and the final amount of ink per unit length of the silver line is directly related to the set flow rate and printing speed. The trajectory of the substrate during the entire printing process is determined by the CAD drawing. The printing process parameters for this experiment are recorded in Tab. 2, which has been proven to be appropriate in previous studies[8].

Tab. 2 Parameter setting of the EHD-printer

Parameter	Value
Voltage (kV)	2.5
Direct-writing height (mm)	6.4
Flow rate (mL h^{-1})	0.8
Relative humidity (%)	60±5
Temperature (℃)	25±2
Needle diameter (mm)	2

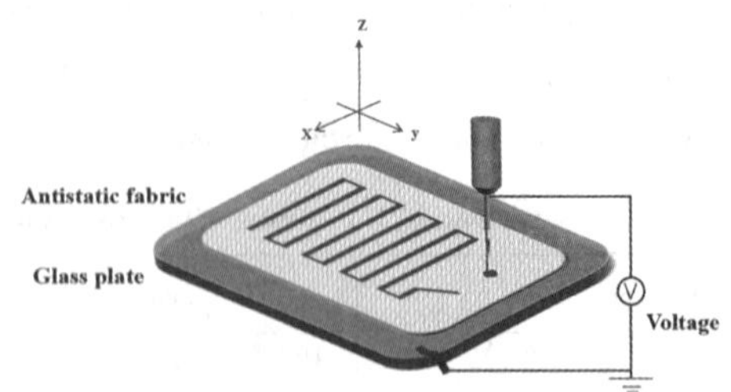

Fig. 1 Schematic diagram of the EHD-printing device

The aim of this experiment was to increase the line width by spliced printing of "line by line" to improve the aggregation of silver nanoparticles and make the printed conductive patterns more suitable for the preparation of FJHs. To determine the appropriate splice specifications, trajectories were designed and entered via CAD including different splice distance and number of splices to form. To avoid errors due to batches, comparison samples were designed to print to the same batch by head-to-tail splicing. Changes were also made to match out the appropriate print speed. Specific factor-level information is given in Tab. 3.

Tab. 3 Print track parameters for samples

Level	Factors		
	Splicing distance (mm)	No. of splices	Print speed (mm/s)
1	0.3	1	8
2	0.35	2	10
3	0.4	3	12
4	0.45	4	14
5	–	5	16

2.5 One-off printing of FJHs and evaluation of thermal properties

After a molding comparison of the printed samples in section 2.4, the printing design parameters for the preparation of the fabric-based FJHs were determined. The parallel distribution circuit shown in Fig. 2 (a) was designed and printed according to the print process parameters in Tab. 2. The entire trajectory was formed in one-off and the soft and flexible fabric base allowed the entire circuit to be cut at will. In the manner shown in Fig. 2 (b), this experiment used a HT-301 infrared thermometer to observe the infrared thermogram of the heat generator and record the temperature change of the heat generator with the applied voltage of 1 V to 5 V.

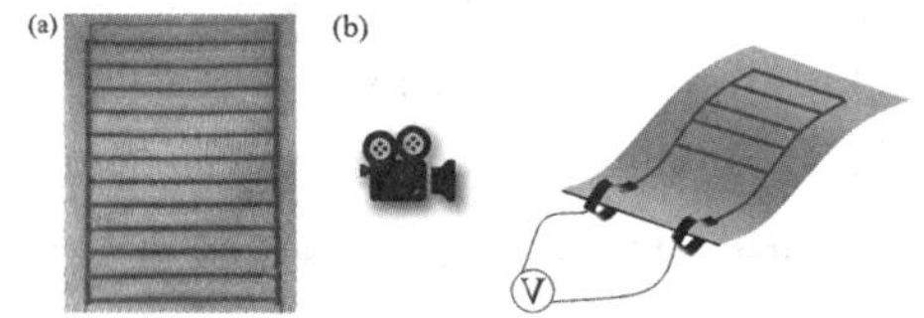

Fig. 2 (a) Printed trace diagram of a parallel circuit; (b) test method for thermal performance

3 Results and Discussion

3.1 Formability comparison and conductivity evaluation of splice printing

This set of experiments was carried out with three printed silver lines spliced together and printed with the process parameters shown in Tab. 2, aiming to determine the appropriate print separation distance and matching the corresponding print speed. The vertical coordinates in Fig. 3 (a) indicate the resistance values of the samples printed at different separation distances. It can be seen that the resistance value of the silver conductor increases with increasing separation distance, due to the reduced aggregation and more dispersion of the nano-silver on the fabric. At lower printing speeds, the silver ink loading per unit length is high enough to make this difference insignificant. As the printing speed increases, the effect of splice on the resistance values gradually becomes more pronounced and by the time the printing speed is increased to 14 mm/s, the

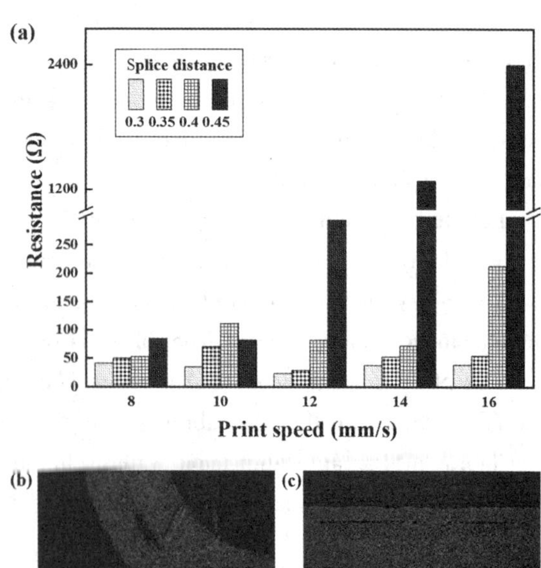

Fig. 3 (a) Comparison of resistance values for different splice distance at different print speeds; (b) and (c) printing defects caused by inappropriate spacing and printing speed

resistance values have reached the level of kΩ, which is completely inadequate for the application of FJHs. The surface morphology is also clearly visible through the electron microscope, with printing defects such as missed marks.

Based on the results of the previous set of experiments, a print spacing of 0. 35 mm and a print speed of 10 mm/s was set. This group of experiments was carried out to investigate the effect of the number of splice roots on the splice print quality, and the results are shown in Fig. 4. Combined with the left-hand vertical coordinate, it can be found that the conductive line width increases in an orderly manner as the number of splices increases. Combined with the right-hand vertical coordinate, it can be found that the resistance gradually decreases and stabilizes as the number of printing passes and line width increases. As the number of splices increases from 3 to 4, the resistance decreases slower and even rises when the number of splices reaches 5. Therefore, it is concluded that the resistance of the silver line basically reaches a stable value when the number of splices reaches 4.

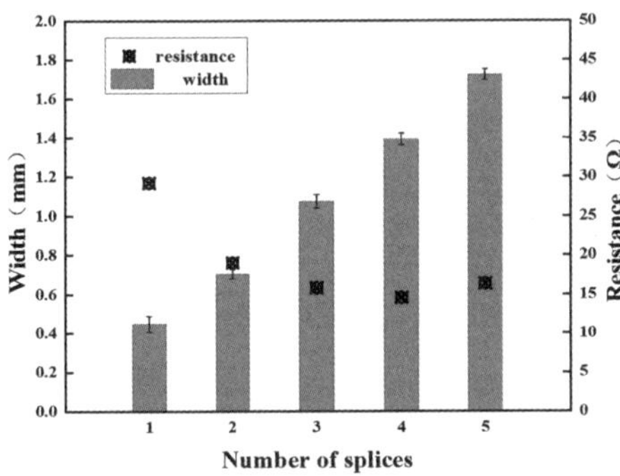

Fig. 4 Variation of line width and resistance values for different numbers of splices

3.2 Evaluation of the thermal properties of one-off molded FJHs

This section analyses the electro-thermal performance test results of the primary formed FJHs. Fig. 5 represents the temperature-time curve after applying 1-5 V voltage to the FRH. It can be seen that the FJH obtained from printing have a good temperature rise performance and can basically achieve a response within 2 s after applying the voltage, followed by a period of faster temperature rise region and basically reach the equilibrium temperature after 15 s. When a 5 V voltage is applied, the entire heat generating circuit can rise from room temperature to 60. 5 ℃ in 10 s. The entire heat generation circuit was analyzed and due to the limitations of conductivity and circuit design, the conventional style parallel heat generation circuit, which was printed and formed in one-off, inevitably produced heat generation differences. Theoretically, the printed circuit is transformed into an ideal linear heat generating resistor, and the values of the resistance of the shunts in parallel are marked in Fig. 5 (b). According to the principle of thermal equilibrium and the joule heat theory equation, the difference in heat generation between the shunts can be up to 3. 52 times greater, which has a significant impact on the uniformity of the circuit. This is expected to be further optimized by changing the applied voltage location and improving the circuit distribution design[9].

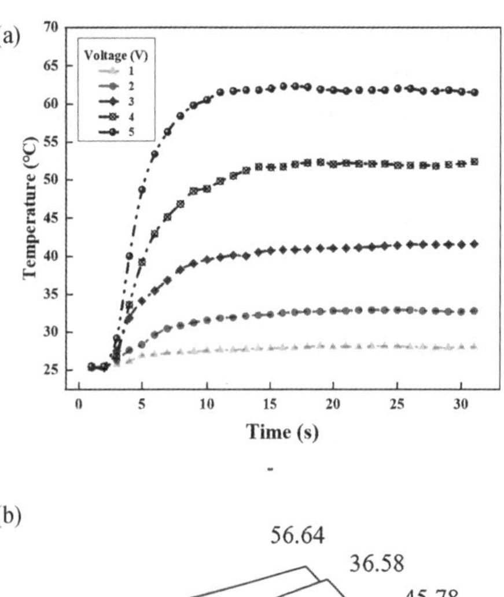

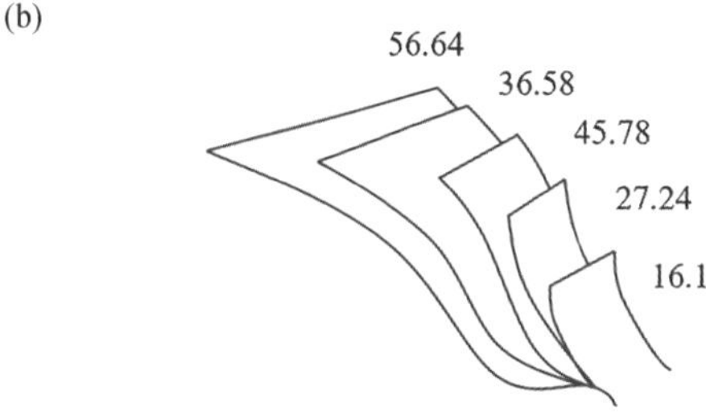

Fig. 5 (a) Temperature rise curve of the FRH after applying 1-5 voltage; (b) distribution of resistance values of heat generating circuits under the assumption of ideal linear resistance

4 Conclusions

In this study, a UV-curable nano-silver ink developed specifically for EHD printing was used to form conductive pathways on woven fabrics using spliced printing and to compare and analyze the conductivity and formability of different sizes of spliced conductive silver lines. The flexible resistive heater printed in this way is soft, portable, flexible in design and can be prepared quickly, while achieving a fast response in terms of heating and achieving sufficient heating equilibrium temperature. In the future, the heat distribution and embedding of the heater can be further optimized to improve its applicability in the field of cold protection and thermal therapy.

References

[1] PENG Y, CUI Y. Advanced textiles for personal thermal management and energy[J]. Joule, 2020, 4(4): 724-742.

[2] MA Z, ZHAO D, SHE C, et al. Personal thermal management techniques for thermal comfort and building energy saving[J]. Materials Today Physics, 2021, 20.

[3] LIU Q, TIAN B, LIANG J, et al. Recent advances in printed flexible heaters for portable and wearable thermal management[J]. Mater Horiz, 2021, 8(6): 1634-1656.

[4] GUO H, HE Y, YUE W, et al. Research on the conductivity of circuit on fabrics based on inkjet printing and electroless deposition technology[J]. IOP Conference Series: Earth and Environmental Science, 2021, 632(5).

[5] CUI Z, HAN Y, HUANG Q, et al. Electrohydrodynamic printing of silver nanowires for flexible and stretchable electronics[J]. Nanoscale, 2018, 10(15): 6806-6811.

[6] WU X, ZHOU Z, WANG Y, et al. Syntheses of silver nanowires ink and printable flexible transparent conductive film: a review[J]. Coatings, 2020, 10(9).

[7] CUI G, YOO J H, LEE J S, et al. Effect of pre-treatment on the surface and electrochemical properties of screen-printed carbon paste electrodes[J]. Analyst, 2001, 126(8): 1399-403.

[8] GUO W, HU J, YAN X. Effect of the solvent evaporation rate of silver ink on the electrohydrodynamic-printing formability of textile-based printing electronics[J]. Textile Research Journal, 2021, 92(5/6): 886-896.

[9] FANG S, WANG R, NI H, et al. Thermal field distribution investigation and simulation of silver paste heating fabric by screen printing based on Joule heating effect[J]. Journal of Materials Science: Materials in Electronics, 2021, 32(23): 27762-27776.

Evaluating the Properties of Cotton Fabrics Dyed with Extracts of *Lantana Camara* Leaves

MPOFU Nonsikelelo Sheron[1*], ODHIAMBO Derek[1], OKOTH Mickey[1], SOLACE Randa[1], MWASIAGI Josphat Igadwa[1]

[1] *Department of Manufacturing, Industrial and Textile Engineering, Moi University, Eldoret 3900-30100, Kenya*

* *Corresponding author's email*: nonsiempofu@ gmail. com

Abstract: The non-carcinogenic nature, lower toxicity and biodegradability of natural dyes have made them more preferred than synthetic dyes in the colouration of textile substrates. Natural dyes have been extracted from several plants, producing dyes with different properties. In this study, the leaves of *lantana camara* were exploited for their potential use as natural dyes. *Lantana Camara* is a perennial flowering plant in the verbena family. The plant has been used mainly as herbal medicine, source of microbicides, fungicides, insecticides and has also been used for treating malaria, ulcer, chickenpox, asthma, high blood pressure, measles, fevers, colds and tumours. The aim of this study was to extract natural dyes from the *Lantana Camara* leaves for use in the dyeing of cotton fabric. The efficiency of the dye on the cotton fabric was also done by performing colour fastness. The *Lantana Camara* leaves were collected from Moi University in Eldoret, Kenya. They were cleaned, dried and then grounded to powder. Dye extraction was done using the reflux method as ethanol and water being used with a material ratio of 10 : 1. Dyeing onto the cotton substrate was done using the heating oil bath method and techniques of pre mordanting, post mordanting and simultaneous mordanting with mango bark mordant. *Lantana Camara* leaves produce a green dye extract. Dyeing onto cotton fabric by different mordanting techniques resulted in different shades. Premordanted substrate had a darker shade followed by simultaneous substrates, and then finally the post mordanted ones. The colour fastness values varied from substrates of post-mordanted with those of pre and simultaneous ones. For better dyeing of cotton substrates using dye extracts from *Lantana Camara*, the pre-mordanting technique is recommended because it results in better dye uptake.

Keywords: Lantana camara; Natural dyes; Mango bark; Mordanting; Mechanical properties

1 Introduction

An increasing amount of people have gained interest in the use of natural dyes on natural fibres as a result of the recent increase in environmental consciousness around the world. Since natural dyes are made from renewable resources as opposed to synthetic dyes, which are made from non-renewable petroleum resources, they are seen to be more environmentally friendly because they are non-allergic, have better biodegradability, are non-carcinogenic and have lower toxicity than the synthetic dyes[1]. Synthetic dyes are extensively and often used, and while they produce vibrant colour, they are also carcinogenic and inhibit benthic photosynthesis[2]. Although natural dyes are more eco-friendly, they pose a challenge in that there is poor bonding between the dye molecules and the active sites in the fibre and they therefore require bridging chemicals. Mordanting compounds are useful for that purpose but the effectiveness of natural dyes is decreased by the use of synthetic mordants; some of which are toxic and not very environmentally friendly, raising questions about their use at times. The use of these synthetic mordants in the application of natural dyes makes the process equally as harmful and toxic.

To promote the use of renewable and fully eco-friendly dyeing, natural dyes and natural mordants are best used. In this study, *Lantana Camara* leaves were used as a source of natural dyes. *Lantana Camara* is a perennial flowering plant in the verbena family, Verbenaceae. It is native to tropical regions of the Americas and Africa but exists as an introduced species in numerous areas. The genus includes both herbaceous plants and shrubs growing to 0. 5-2 metres tall. The plant has sometimes been grown as an ornamental and at times it grows naturally as a weed[4]. Therefore, using it as a natural dye will not lead to deterioration of natural resources but will create value for this plant which grows as a weed. Mango bark contains tannins which are useful in the dyeing process. An earlier study effectively explored the use of mango bark extract as a mordant for bitter leave dye. The study demonstrated that mango extract may even be a more effective mordant for bitter leave dye than the standard metal salt; for instance, copper II sulphate which was used for comparison[5]. Other researchers have also effectively used mango bark as a mordant in the natural dyeing of cotton[6], leather[7] and wool[8]. This study therefore focused on the extraction of natural dyes from the leaves of the *lantana camara* plant and dyeing cotton fabric using mango bark as a mordant.

2 Experiments

2.1 Materials

The reagents used for the extractions and tests done in this research were of analytical grades and were purified. Plain woven cotton fabric was obtained from Rivatex East Africa Limited, Eldoret, Kenya. *Lantana camara* leaves were collected by hand pruning from Moi University main campus, Kesses, Eldoret, Kenya. Pruning was done at the base of the stem branches. The mango bark was collected from Siaya county in Kenya.

2.2 Sample preparation and extraction

The collected plant parts, that is, mango barks and lantana leaves were washed using clean running tap water and the leaves were stripped from the branches. After cleaning, the plant materials were dried both at room temperature and under the sun, ground into fine particles using an electric blender, sieved to get a fine powder and then stored in air tight containers. The Lantana camara leaf powder and Mango bark powder are shown in Fig. 1.

Fig. 1 (a) Lantana camara leaf powder and (b) Mango bark powder

Dye extraction was done using the reflux method using a solvent containing 70% ethanol and 30% water with material ratio of 10 : 1 (mL/g). The menstruum was then filtered into a conical flask using a filter paper and a funnel and the flask was then sealed and stored in a refrigerator for use within a few days.

2.3 Fabric pre-treatment

For optimum dye exhaustion, the cotton fabric was pre-treated to remove impurities and to increase its whiteness. This was done through desizing, scouring and bleaching. These pre-treatment processes were done at the Rivatex East Africa Limited. The fabrics were then boiled in washing soap for 1 h and then rinsed.

2.4 Fabric dyeing

The sample fabric was dyed by exhaust method using pre-mordanting, simultaneous and post-mordanting techniques using mango bark mordant[9]. The first set of pre-mordanted and simultaneously mordanted substrates were dyed without stirring and the second set was then done with continuous stirring at intervals of 10 min. The cotton sample size was 12 cm × 8 cm, the liquor amount was 40 mL, the dyeing time was 60 min and the dyeing was done at a temperature of 90 ℃ in an oil bath. After dyeing the substrate was air-dried.

2.5 Fastness testing of dyed fabrics

The following tests were then done on the samples; wash fastness test according to ISO 105 C06, rub fastness test according to ISO-105-X12 and light fastness test according to ISO 105-B06.

2.5.1 Wash fastness

The wash fastness test was performed on the wash master washing fastness machine. Each sample was cut in size of 10 cm by 4 cm and a plain white fabric of equal dimensions was cut and stapled on both sides of the fabric. Washing was done for 20 min and the samples were then allowed to cool and dry.

2.5.2 Rubbing fastness

The rubbing test was done on a Crock Master rubbing colour fastness tester according to ISO-105-X12. Both the dry rubbing tests and wet rubbing tests were carried out on the pre-mordanted, simultaneous and post-mordanted fabrics.

2.5.3 Light fastness

The light fastness test was done on a light fastness tester (STL ATLAS) for the pre-mordanted, simultaneously mordanted and post-mordanted fabrics using the mango bark mordant. Test specimens were cut in the dimension of 5 cm by 4 cm.

Grayscale analysis was used to analyse the test specimen, and sample findings from the pre-mordanted, simultaneously mordanted and post-mordanted samples were compared.

3 Results and Discussion

3.1 Effect of mordanting and stirring on dyeing

As shown in Fig. 2, the presence of a green colour on the cotton fabrics after dyeing confirmed that lantana camara leaf extracts can be used as a natural dye for cotton fabrics. There was a difference however in the shades of green for the pre-mordanted, simultaneously mordanted and the post-mordanted substrates. The pre-mordanted substrates gave a darker shade of green, followed by the simultaneously mordanted substrates. The post mordanted substrates however gave a very light shade of green. Therefore, pre-mordanted fabrics have a better dye uptake while post mordanted fabrics have the least.

A difference was also observed between fabrics dyed under continuous stirring and those dyed without stirring. The substrates dyed without stirring gave patches of uneven dyeing while those that were dyed under continuous stirring gave better and more even dyeing results.

Fig. 2 Dyed cotton fabric samples

3.2 Fastness properties of the dyed substrates

The results for the fastness properties were assessed against the grayscale ranging from 1-5 where 1 represents poor fastness properties with a large visual change on the textile substrate and 5 is represents excellent fastness properties with no visual change on the textile fabric. The rubbing fastness properties results shown in Tab. 1, reveal that all samples showed fairly good (2/3) to excellent (5) results with better dry rub fastness properties than wet rub fastness properties.

Tab. 1 Grayscale readings for rub fastness properties

Substrate	Dry	Wet
Pre-mordanted (without stirring)	3/4	4/5
Pre-mordanted (with stirring)	5	2/3
Post mordanted	4	5
Simultaneously mordanted (without stirring)	5	3/4
Simultaneously mordanted (with stirring)	5	3

The wash fastness results in Tab. 2, show that most samples had very good (4/5) to excellent (5) results except for the post mordanted samples that displayed poor wash fastness properties.

Tab. 2 Grayscale readings for wash fastness properties

Substrate	Gray Scale reading
Pre-mordanted (without stirring)	5
Pre-mordanted (with stirring)	5
Post mordanted	1
Simultaneously mordanted (without stirring)	4/5
Simultaneously mordanted (with stirring)	4/5

The light fastness properties in Tab. 3, reveal that all samples showed fairly good (3/4) to very good (4) results with post mordanted and simultaneously mordanted samples displaying better light fastness properties than the pre-mordanted samples.

Tab. 3 Grayscale readings for light fastness properties

Substrate	Gray Scale reading
Pre-mordanted (without stirring)	3/4
Pre-mordanted (with stirring)	3/4
Post mordanted	4
Simultaneously mordanted (without stirring)	—
Simultaneously mordanted (with stirring)	4

4 Conclusions

This study investigated the potential of using lantana camara leaf extracts for dyeing cotton fabrics using mango bark as the mordant. The green colour of the samples after dyeing showed that the *lantana camara* leaf extracts can be used for eco-friendly dyeing. For effective dyeing of cotton using *Lantana camara* dye and mango bark mordant, the dyeing techniques arranged in the order of their efficiency are pre-mordanting, simultaneous mordanting then post-mordanting. The results also demonstrated that both the fabric and the dyestuff need to be stirred continuously during dyeing to achieve even dyeing of the fabric. Most of the samples displayed good fastness properties to rubbing, washing and light with the exception of the post-mordanted samples that displayed poor wash fastness properties. The dye can therefore be considered as a substantive dye with good colour fastness properties. Future studies will focus on the efficacy of the dyed fabrics against bacteria and fungi for possible application in the medical field.

References

[1] CHOUDHURY A K R. Eco-friendly dyes and dyeing[J]. Advanced Materials and Technologies for Environmental, 2018, 2: 145-76.

[2] SINGH R, SRIVASTAVA S. Application of natural dye on silk fabric[J]. International Journal of Applied Home Science, 2017, 4(9): 805-809.

[4] GUNASEKARA C J, RANWALA S M. Growth responses of lantana (Lantana camara L.) varieties to varying water availability and light conditions[J]. Journal of the National Science Foundation of Sri Lanka, 2018, 46: 69-79.

[5] WANGATIA L M, TADESSE K, MOYO S. Mango bark mordant for dyeing cotton with natural dye: fully eco-friendly natural dyeing[J]. International Journal of Textile Science, 2015, 4(2): 36-41.

[6] AGULEI K D, MWASIAGI J I, GITHAIGA J T, et al. Use of plants as dyes: a case study of the use of pre-mordanting method to dye cotton fabric with extracts from Allium burdickii[J]. African Journal of Rural Development, 2017, 2 (2): 313-318.

[7] BERHANU T, RATNAPANDIAN S. Extraction and optimization of natural dye from hambo (Cassia singueana) plant ysed for coloration of tanned leather materials[J]. Advances in Materials Science and Engineering, 2017: 1-5.

[8] TEKLEMEDHIN T B, Dyeing of wool fabric using natural dye and natural mordant extracts[J]. Trends in Textile Engineering and Fashion Technology, 2018, 4(4): 538-542.

[10] JANANI L, HILLARY L, PHILLIPS K. Mordanting methods for dyeing cotton fabrics with dye from Albizia coriaria plant species[J]. International Journal of Scientific and Research Publications, 2014, 4(10): 1-6.

Epoxy Foam Filled 3D Woven Spacer Composites with Light-Weight and Excellent Mechanical Properties

ZHENG Liangang[1,2], **XU Fujun**[1,2*]

[1] *Shanghai Frontier Science Research Center for Modern Textiles, Donghua University, Shanghai 201620, China*

[2] *College of Textiles, Donghua University, Shanghai 201620, China*

[*] *Corresponding author's email*: fjxu@dhu.edu.cn

Abstract: In this work, epoxy foam-filled 3D woven spacer composites (FF−3DSCs) were obtained by a two-step method. The results show that the as-prepared FF−3DSC is lightweight, with a volume density of about 0.45 g/cm^3. In addition, based on the synergistic effect of the skeleton of 3D spacer composites and the epoxy foam, the eventual composites exhibit excellent compressive properties, of which value can reach 14.69 MPa. The results show that the composites prepared in this work have both light weight and high strength. This work can provide a reference for the design and preparation of composite materials with excellent comprehensive properties.

Keywords: Epoxy foam; 3D fabric; Lightweight composites; Mechanical properties

1 Introduction

With the development of aviation, aerospace and other industries, lightweight and high-performance composite materials have attracted more and more attention. Fiber reinforced composites play an important role in these fields. Among them, three-dimensional woven spacer composites have great application potential in aviation, aerospace, military, transportation, wind power and other fields because of its special three-dimensional structure, which brings light weight, high strength, high specific strength and specific modulus[1-2]. However, the hollow structure with high porosity also limits the strength of the three-dimensional woven spacer composite itself.

In order to improve the mechanical properties of three-dimensional woven spacer fabrics, researchers have proposed many methods, including designing thicker surface layer structure and using the core yarn with a larger diameter[3-4]. Although the above two methods can improve the mechanical properties of materials to a certain extent, the degree of improvement is small. In recent years, foam filled sandwich structure composites have attracted extensive attention, including polyurethane foam, phenolic foam, epoxy foam and so on[5]. However, sandwich composites based on traditional polyurethane, phenolic and other foam as core materials are mostly used for interior and exterior decoration structures of vehicles, and their mechanical properties cannot meet the service requirements of high-speed trains and medium and high-end vehicles. Compared with the former two, epoxy resin foam has excellent mechanical properties and is increasingly used as the core layer reinforcement structure[6].

In this work, a new type of epoxy foam reinforced three-dimensional woven spacer fabric composite was successfully prepared by two steps. Firstly, three-dimensional woven spacer Kevlar fabric/polyimide composites (3DSC) were prepared by self-made three-dimensional looms combined with thermal iridization. Then the prepared epoxy foam system and 3DSC were heat cured in the mold to obtain the eventual three-dimensional woven spacer composite filled with epoxy foam (FF−3DSC). Subsequently, the volume density of the as-prepared composites was tested and characterized. Finally, the flatwise compressive performance of FF−3DSC was tested by the universal tester. The results show that the composite not only has the characteristics of light weight, but also has excellent compressive properties. This work can provide reference for the design and preparation of high-performance composite materials with excellent comprehensive properties applied in aerospace, military, transportation and other high-end industrial fields.

2 Experiments

2.1 Materials

Kevlar 129 (type 964c) with low density (1.44 g/cm^3) and high strength (tensile strength 3.4 GPa) used to make three-dimensional fabrics was supplied by DuPont Co., Ltd. Polyamide acid (PAA) used to prepare polyimide was provided by the Institute of chemistry, Chinese Academy of Sciences. Epoxy resin (JL-235) and the curing agent (JH-242) were supplied by Changshu Jia Fa Chemical Co., Ltd. (Zhejiang, China). The foam stabilizer and microsphere foaming agent (Dimethicone, particle size 20-25 μm, expansion ratio 4-5 times) were purchased from Sinopharm Group Chemical Reagent Co., Ltd. and Crerax Science & Technology Co., Ltd., respectively.

2.2 3D weaving

Three-dimensional woven spacer Kevlar fabric has three sets of yarns, namely warp, weft and core (pile) yarns. The warp and weft yarns interweave to form the plain weave on the upper and lower surfaces. The pile yarns interweave with the upper and lower weft yarns to connect the upper and lower layers in a vertical direction. And the pile yarn is evenly distributed in positive and negative S-shapes along the warp direction. Therefore, a void structure is formed between two adjacent 'S', which will create the conditions for the realization of lightweight composite materials.

2.3 Manufacture of 3DSC

The three-dimensional woven spacer composites (3DSCs), which are composed of three-dimensional woven spacer Kevlar fabric and polyimide resin, are manufactured by the typical hand lay-up impregnation process. Specifically speaking, at first, the 3D Kevlar fabric was put on the glass plane. Secondly, PAA solution, the precursor of polyimide, was evenly transferred on the surface of the fabric by hand lay-up impregnation. After the solution penetrated completely into the fabric by wicking, a heat treatment was performed to facilitate the conversion of PAA to polyimide by thermal imidization. Finally, 3DKPC were obtained after the temperature increasing programmatically from 25 ℃ to 320 ℃.

2.4 Preparation of FF-3DSC

The detailed preparation process of FF-3DSC is described as follow. Firstly, the microsphere foaming agent and foam stabilizer were added to the prepared epoxy resin in the proportions of 3% and 2%, respectively. After being fully stirred, degassing in a vacuum was conducted for 30 min to eliminate air bubbles in the resin system. Subsequently, the pre-foaming resin was injected into the above-manufactured 3DSC with a dropper along the longitudinal section, and then the 3DSC filled with pre-foaming resin was transferred into the mold with the size of 20 mm×20 mm×12 mm. Finally, the foaming process was completed under the condition of 95 ℃ for 1 h and then 70 ℃ for 7 h. The schematical diagram of the FF-3DSC was shown in Fig. 1 (a).

2.5 Mechanical testing of composites

The mechanical property of the corresponding composites was measured through flatwise compression test. Specimens of size 20 mm×20 mm were made according to GB/T 1453—2005 to carry out compression by an Instron universal testing instrument (model 5967, Instron Instruments), with the loading rate of 2 mm/min at room temperature.

3 Results and Discussion

3.1 Characterization

A sample of the as-prepared composite material is shown in Fig. 1 (b), which is almost consistent with the structural schematic diagram of Fig. 1 (a). As can be seen from the Fig. 1 (b), along the warp direction, two adjacent piles form a 'goblet' shape, and the

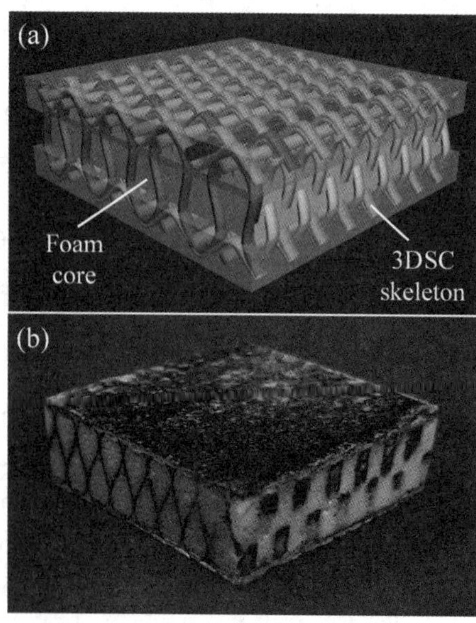

Fig. 1 Structural diagram (a) and digital picture (b) of foam-filled 3D woven spacer composites sample

space between the two pile yarns is completely filled with foam. And along the weft direction of the sample, the pile yarns are in the shape of Arabic numerals '1' arranged in parallel, and are completely embedded in the foam. This special structure formed by the combination of 3DSC skeleton and epoxy foam in favor of helping the composite obtain excellent comprehensive properties.

3.2 Composition of FF-3DSC

Fig. 2 is the mass percentage of each component of the composite material. It can be found from the figure that the mass ratio of 3D woven spacer aramid fabric and PI resin is similar, 25.5% and 23.02%, respectively, and both are approximately half of the foam mass in the composite material (51.48%). This is mainly attributable to the higher void content inside the 3DSC, which provides a lot of space for foam formation. Based on the fact that the epoxy foam itself contains a lot of air, the final average bulk density of the prepared composite material is about 0.45 g/cm^3, which shows the characteristics of light weight.

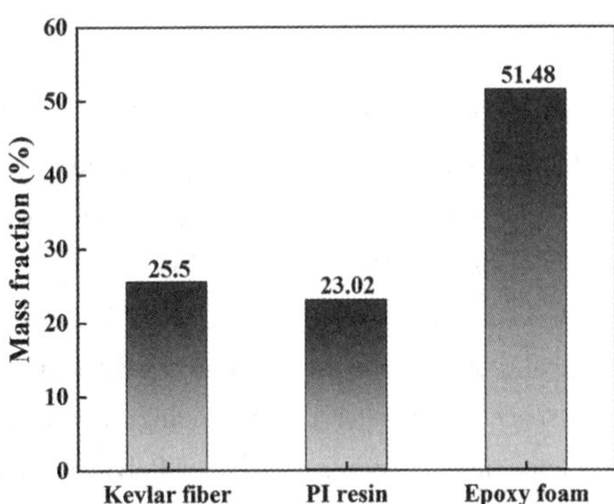

Fig. 2 Mass fraction of Kevlar fabric, PI resin, and epoxy foam in the as-prepared composites

3.3 Flatwise compression property of FF-3DSC

In order to study the mechanical properties of three-dimensional woven spacer fabric composites filled with foam, flatwise compression tests were performed. The typical stress-strain curves of the composites are shown in Fig. 3. It can be found that the curve mainly goes through three stages. The first is the linear elastic stage, followed by the near-linear elastic rising stage and finally the densification stage. It can also be found that FF-3DSC has a linear elastic stage and a densification stage similar to the three-dimensional woven spacer Kevlar/polyimide composites in previous work[7]. However, for FF-3DSC, the compressive strength value can continue to increase after a slight decrease due to the first failure of the sample. This can be explained as follows. First of all, the pile yarn and the foam bear the load at the same time in the linear elastic stage, and the stress is uniformly transmitted between the spool and the foam cell wall, so the force value increases linearly. With the increase of strain, when reaching the initial failure strength of the pile yarn and the foam, the pile yarn was broken by bending, accompanied by the rupture of part of the foam cells. The force value first decreased slightly and then continued to increase.

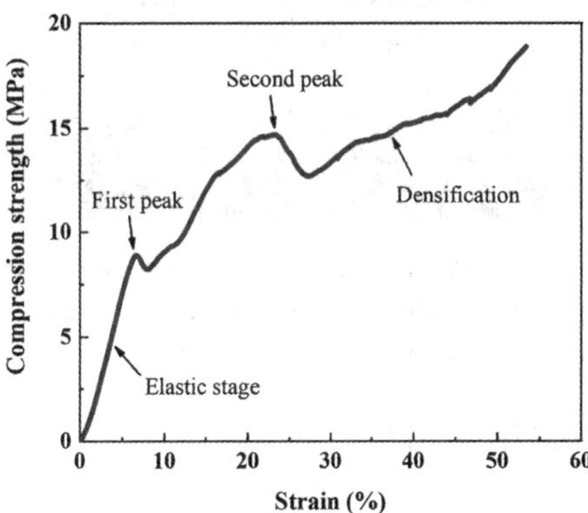

Fig. 3 Typical flatwise compressive stress-strain curve of epoxy foam-filled 3D woven spacer composites

During this process, the pile yarn tilted and further bent, and then broke and squeezed the foam. While the foam can restrain the tilt of the pile yarn in this process, so that the pile yarn continues to support the upper and lower panels. However, due to the destruction of the piles and the rupture of some foam cells, the slope of the curve is smaller than that in the first stage, showing an upward trend of near-linear elasticity until it reaches the maximum compressive strength value of the system. Afterwards, as the compression progresses, when the pile yarn is greatly inclined and the foam cells are completely ruptured and lose the effective fixation of the plies, the load begins to decrease. In the third stage, it can be found that the core layer space is compressed due to the closer distance between the upper and lower panels, and the space between the pile yarns and epoxy foam is compressed. Thus, the curve shows a significant rise again owing to the densification.

The typical digital picture of the specimen after failure is shown in Fig. 4. It can be seen from the weft view of Fig. 4(a) that the 'goblet'-shaped structure formed by two adjacent pile yarns along the warp direction is destroyed. And it can also be found that the single 'S'-shaped pile yarn was severely deformed. In addition, the extrusion failure of the foam can be seen in the picture. In the warp view, in addition to the bending failure of the '1'-shaped pile yarn and the extrusion failure of the foam, some fiber fracture can also be found at the bending point of the pile yarns.

Fig. 4 Digital pictures of the pile yarns of FF-3DSC after compression failure: (a) along weft view and (b) warp view

4 Conclusions

This study has focused on the manufacturing of the epoxy foam filled three-dimensional woven spacer composites (FF-3DSC). The final composites were obtained by a two-step method. The results show that the as-prepared FF-3DSC are lightweight, with a volume density of about 0.45 g/cm^3. In addition, based on the synergistic effect of the skeleton of the three-dimensional spacer fabric composites and the epoxy foam, the final composites exhibit excellent compressive properties, of which value can reach 14.69 MPa. The above results show that the composites prepared in this work have both light weight and high strength performances. This work can provide a reference for the design and preparation of composite materials with excellent comprehensive properties.

Acknowledgment

This work was financially supported by the Fundamental Research Funds for the Central Universities and Graduate Student Innovation Fund of Donghua University (CUSF-DH-D-2021017).

References

[1] WANGLY, ZHANG K, FARHA F I, et al. Compressive strength and thermal insulation properties of the 3D woven spacer composites with connected spacer yarn structure[J]. Journal of Material Science, 2020, 55: 2380-2388.

[2] ZHU L T, BIN RAHMAN M, WANG Z X. Effect of structural differences on the mechanical properties of 3D integrated woven spacer sandwich composites[J]. Materials, 2021, 14(15): 1-17.

[3] LI D, ZHAO C, JIANG N, et al. Fabrication, properties and failure of 3D integrated woven spacer composites with thickened face sheets[J]. Materials Letters, 2015, 148: 103-105.

[4] WANG L Y, CHEN L, LI W Z, et al. Ultra-high compressive strength of 3D woven spacer composites with bulked glass fiber and saturated resin absorption [J]. Composite Structures, 2022, 290: 1-7.

[5] SAFARI H, KARVEN M, NAHVI H. Mechanical characterization of natural nano-structured zeolite/polyurethane filled 3D woven glass fiber composite sandwich panels[J]. Polymer Testing, 2018, 67: 284-294.

[6] ZHENG L G, XIAO Y X, LIU L, et al. Experimental and numerical study of the behavior of epoxy foam-filled 3D woven spacer composites under bending load woven spacer composites under bending load[J]. Polymer Composites, 2022, 43: 3057-3067.

[7] ZHENG L G, ZHANG K, LIU L, et al. Biomimetic architectured Kevlar/polyimide composites with ultra-light, superior anti-compressive and flame-retardant properties[J]. Composites Part B: Engineering, 2022, 230: 1-9.

Enhanced Dyeing Properties of Gardenia Yellow on Cotton with Siloxane-Based Medium

LI Jingzhi[1], XIE Chendi[1], CHEN Yangyi[1,2], WU qingnong[1], CAI Xiuying[1], YUAN Xinyue[1], LIU jinglin[1], QIU Yiping[1,3*]

[1] *College of Textiles and Apparel, Quanzhou Normal University, Quanzhou 362000, China*

[2] *Shishi Institute of Smart & Ecological Textile, Quanzhou Normal University, Quanzhou 362000, China*

[3] *College of Textiles, Donghua University, Shanghai 201620, China*

* *Corresponding author's email*: ypqiu@ dhu. edu. cn

Abstract: Decamethyl cyclopentasiloxane (D5) was devoted to the medium in the dyeing process of gardenia yellow on cotton fabric to improve the dyeing properties. The effects of D5 percentages and mordants, dyes dosage as well as dyeing temperature and time on dyeing properties were investigated. The results show that optimal dyeing processes of gardenia yellow on cotton in siloxane-based medium are as follows: D5 percentage 94%vol, dye dosage 2% (o. w. f), alum as mordant dosage 3% (o. w. f), dyeing temperature 80 ℃, dyeing time 40 min. Compared with water bath dyeing, cotton fabrics dyeing with D5 present enhanced color strength (*K/S* value) and fine levelness. The color fastness of cotton fabrics dyeing with D5 is equal to that dyeing with water. This dyeing technique that immensely reduces dyeing water wasteness has potential application in natural dyestuffs.

Keywords: Gardenia yellow; D5; Dyeing kinetics, Color strength; Color fastness

1 Introduction

With the increasing awareness of environmental protection and health issues, natural dyes due to its eco-friendly, non-toxic characteristics recaptures humans' interests[1-2]. As a common natural dye, gardenia yellow processes the properties of bright color and color stability, contributing to its wide-spread applications[3-6]. Gardenia yellow can be used as a food additive and exhibits fine luster. It can also be applied in traditional Chinese medicine and presents good pharmacological activity. In textile, gardenia yellow is employed in dyeing of protein fibers such as silk and wool[7]. However, gardenia yellow dyeing on cellulose suffers from light color and unsatisfactory color fastness, which limits its application in cellulosic-textiles.

Recently, D5 as a novel dyeing medium has drawn much attentions[8-10]. It has been successfully applied and popularized in reactive dyes, VAT dyes (indigo), acid dyes and disperse dyes, which achieved good results, especially shown remarkable advantages in water, dyes and agents saving and pollution reduction.

Based on discussions above, in this paper, siloxane-based solvent D5 as dyeing medium was applied in gardenia yellow dyeing on cotton. The effects of D5 percentages, dyes and mordant dosages, dyeing temperature and time on dyeing properties were investigated. Besides, the color and light fastness of cotton fabrics were probed.

2 Experiments

2.1 Materials

The blenched cotton fabric (40×40, 180 g/m^2) was purchased from Shishi 668 Textile technology Co., Ltd, China. Gardenia yellow was obtained from Zhejiang Nuoyi Biotechnology Co., Ltd, China and purified before use. D5 was provided by Haining Lvyu Textile Technology Co., Ltd, China. The salts (Na_2SO_4, purity >99%) and alums ($KAl(SO4)_2 \cdot 12H_2O$, purity >99%) were purchased from Shanghai Aladdin, China.

2.2 Methods

2.2.1 Dyeing processes

Typical dyeing processes with water and D5/water bath are presented in Fig. 1.

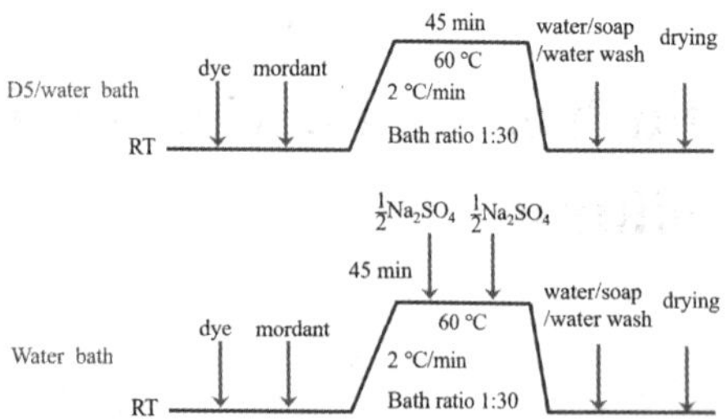

Fig. 1 Typical curves of dyeing process with water and D5/water bath

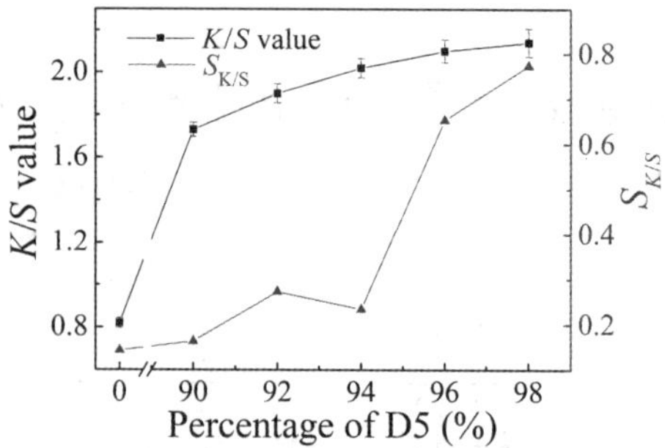

Fig. 2 Color strength and levelness of cotton fabrics dyeing with water and D5

2.2.2 *K/S* value and color levelness

The *K/S* values of cotton fabric under D65 illuminant was measured by a Datacolor 800 spectrophotometer (Datacolor, Co. USA). The color levelness of fabric was characterized by the sample standard deviation of *K/S* values ($S_{K/S}$) those were picked out from 12 test points of a sample. $S_{K/S}$ were calculated by Eqs. 1.

$$S_{K/S} = \sqrt{\frac{\sum_{i=1}^{n}[(K/S)_i - \bar{K/S}]^2}{n-1}} \tag{1}$$

2.2.3 Colorfastness

Laundering colorfastness of cotton fabrics was evaluated by a SW-12B washing fastness tester (Quanzhou Meibang Instrument Co., Ltd.) according to ISO 105-C09: 2001. Both dye and wet rubbing colorfastness were characterized by a YG571-11 rubbing fastness tester (Quanzhou Meibang Instrument Co., Ltd.) referring to ISO 105-X12: 2001. Light fastness was measure by a Q-sun B02 Xenon lamp color fastness tester (Q-Lab, Co., USA) according to ISO 105-B02: 1994.

3 Results and Discussion

3.1 Effect of D5 on dyeing property

The *K/S* value of cotton fabrics with varied D5 percentages were investigated (Fig. 2). To the fabric dying with water, the *K/S* value is 0.82. When the volume percentage of D5 rises to 90%, the *K/S* value remarkably reaches to 1.78 and continuously rises with the increasing percentage of D5. Finally, the *K/S* value is almost 3 time of that in water medium. The results demonstrates that the color strength of cotton fabrics dyeing with gardenia yellow in siloxane-based medium can be obviously enhanced comparing with water bath. Nevertheless, the high dye uptake, as a consequence, usually leads to the uneven dyeing. Hence, the color levelness of corresponding fabrics was characterized by calculating sample standard deviation of *K/S* values. As is showed in Fig. 2, $S_{K/S}$ of fabrics dyeing with 0-94%vol D5 are relatively low, implying acceptable color levelness. Nevertheless, it rises to 0.65 and 0.77 when dyeing with 96% and 98% D5 respectively. Meanwhile, the visual color spots can be observed on the fabrics (Fig. 3). Since gardenia yellow is insoluble in D5, when the amount of D5 in the dyeing bath reaches 90%, it is equivalent to a minimal bath ratio dyeing. In the dyeing process, abundant hydrophilic dyes and mordants will be divorced from D5 and aggregate on the cotton spontaneously. So compared with water bath dyeing, more dyes and mordants are immobilized on the fabric, resulting in a higher *K/S* value. Moreover, as a dyeing medium, D5 can also play the role of conveying material and transferring energy, effectively transferring the heat released by the hydrogen bond and van der Waals force formed by the reaction of dyes on the fabric, contributing to a good color levelness. However, further increasing D5, the water in dyeing bath is probability not sufficient to make the fiber swell evenly, which eventually leads to the color spots. Considering both *K/S* value and dyeing levelness, the optimum percentage of D5 is 94%vol. It is fixed to 94%vol in the following dyeing processes in this study.

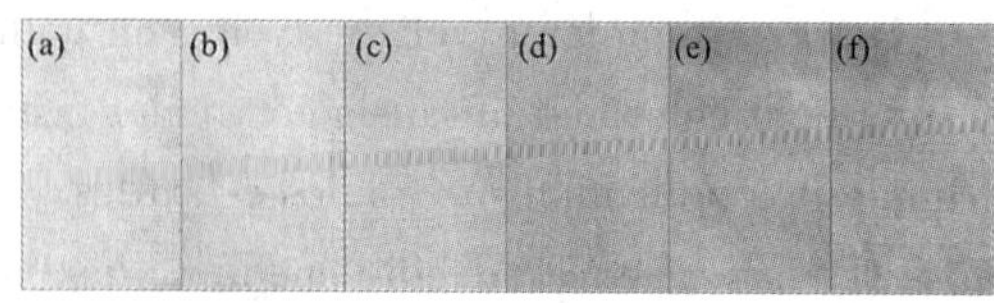

Fig. 3 Photographs of cotton fabrics dyeing with water and D5. a-f represent fabrics dyeing with 0%, 90%, 92%, 94%, 96%, 98% D5 in sequence

3.2 Effect of dyeing parameters on color strength

In order to investigate the effects of dyeing conditions on color strength, some dyeing experiments with varied parameters were designed and carried out. The results are shown in Fig. 4. With the dosage of dyes increasing from 0. 5% to 2% (o. w. f), the *K/S* value rises fast (Fig. 4(a)). When adding dyes dosage above 2%, it rises slightly to 1. 68 (3%). Further adding dyes, the *K/S* value nearly remains unchanged, which implies that the *K/S* value will be no more improved above 2% dyes dosages. This is owing to the limited combining capacity between dyes and cotton.

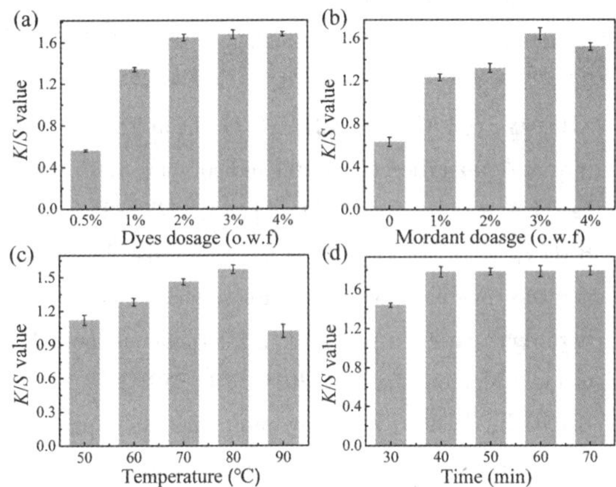

Fig. 4 Color strength (*K/S* value) of cotton fabrics as a function of (a) dye dosage, (b) mordant dosage, (c) temperature, (d) time

Fig. 4(b) illustrates the effect of mordant dosage on *K/S* value. At low dosage of mordant (1%–3%), the *K/S* value of the fabric continues to increase from 0. 63 to 1. 64. Further increasing mordant dosage, *K/S* value drops unexpectedly. The reason is that when the mordant is excessive, more dye molecules participate in the complexation reaction, resulting in molecular agglomeration, which induces a weaker affinity with the cotton. As a result, the *K/S* value decreases in turn.

Except for the dosages of dye and mordant, dyeing time and temperature also influence the *K/S* value. As seen from Fig. 4(c), The *K/S* value increased from 1. 12 to 1. 57 when the dyeing temperature rises from 50 ℃ to 80 ℃. When the temperature reaches to 90 ℃, the *K/S* value decreases significantly, which is even below the *K/S* value of 50 ℃. This is because that gardenia yellow as a natural dye tends to be oxidized at high temperature, resulting in the destruction of the conjugated structures of dye molecular. Hence, the color is faded, following by the decrease of *K/S* value of fabric. Therefore, the dyeing temperature of gardenia yellow should be controlled below 90 ℃.

To explore the improvement of gardenia yellow dyeing time on *K/S* value, a dyeing experiment from 30 – 70 min was carried out. As seen from Fig. 4(d), *K/S* value stays relatively constant after 40 min dyeing, then the improvement of *K/S* value is less prominent above 40 min. In initial dyeing process, the alum acts as a mordant, promoting the reaction between dye molecules and the fabric. With the extension of time, the more dye molecules are fixed on the fabric, and the *K/S* value gradually increases. However, after a certain period of time, the binding sites on the fiber, namely dye site, are occupied and exhaust, limiting the further chemical combinations. Based on the discussion above, the appropriate time for cotton fabric dyeing with gardenia yellow in siloxane-based medium is 40 min. Nevertheless, this time is obviously less than the optimal time of water bath dyeing of gardenia yellow in relevant literature, the reasons will be further explained below.

3.3 Colorfastness

Colorfastness plays an important role in the application of textiles. Therefore, the rubbing, laundering and light fastness of gardenia yellow dyed cotton fabric were assessed. As is shown in Tab. 1, The dry and wet rubbing fastness of both fabrics dyeing with water and D5 could reach grade 4 or above, this is mainly due to the mordant. The Al^{3+} of mordant acts as the center metal cation, and forms coordination ionic bond with negatively charged dye and cellulose. On the one hand, the coordination ionic bond reduces the solubility of gardenia yellow dye. On the other hand, compared with the hydrogen bond and van der Waals force forming between dyes and cellulose in the present of mordant, it possesses higher fastness. Therefore, both fabrics dying with water and D5 have good colorfastness to rubbing. Besides, the laundering color fastness of fabrics is above grade 3. However, due to photosensitivity of gardenia yellow, even after mordant dyeing, the light

Tab. 1 Colorfastness of gardenia yellow on cotton fabric in varied medium

Dyeing medium	T/℃	Colorfastness			
		dry rubbing	wet rubbing	laundering	light
100%water	40	4	3	3	2
	60	4-5	3-4	3	2-3
6%water and 94%D5	40	4-5	4	3-4	2-3
	60	4	3-4	3-4	2

colorfastness of dyed fabrics remains low (below 3), indicating that gardenia yellow dyed fabrics cannot be sun-cured or exposed outdoors for a long time. In general, gardenia yellow dyed textiles can be used for household apparels.

4 Conclusions

An eco-friendly dyeing process of gardenia yellow on cotton fabric with siloxane-based medium was carried out. When the percentage of D5 is 94%, the dyes dosages is 2% (o. w. f), alum as mordant dosage is 3% (o. w. f), dyeing temperature is 80 ℃, dyeing time is 40 min, the dyed cotton fabric exhibits excellent color strength and levelness.

The dry and wet rubbing fastness of both fabrics dyeing with water and D5 could reach grade 4 or above, but the light colorfastness of both fabrics remains low. This dyeing technique could reduce abundant of water and avoids the usage of salts, which is worth of popularization.

Acknowledgments

The authors are grateful to financial support provided by Fujian General project of Natural Science Foundation (No. 2019J01741), Fujian Young and Middle-aged teachers Education research project (No. JT180376).

References

[1] ZHANG H J, LIAN Y S, YU J, et al. Eco-friendly ph indicator based on natural anthocyanins from lycium ruthenicum[J]. Journal of Donghua University (English Edition), 2022, 39(2): 110-114.

[2] BECHTOLD T, TURCANU A, GANGLBERGER E, et al. Natural dyes in modern textile dyehouses-how to combine experiences of two centuries to meet the demands of the future[J]. Journal of Cleaner Production. 2003, 11(5): 499-509.

[3] ZHAO Y. Preparation of gardenia yellow pigment by membrane processes[J]. Journal of Nanjing University of Chemical Technology, 1997, 4(5): 159-172.

[4] LEI T F, PAN Y, ZHANG B, et al. Optimization of ultrasonic-assisted extraction of natural dyes from pomegranate rind using response surface methodology and its characterization[J]. Coloration Technology, 2021, 137(3): 259-271.

[5] ZHENG J, WANG Y, ZHANG Q, et al. Study on structure and properties of natural indigo spun-dyed viscose fiber[J]. e-Polymers, 2021, 21(1): 327-335.

[6] BAO L, ZHU X, DAI H, et al. Synthesis of porous starch xerogels modified with mercaptosuccinic acid to remove hazardous gardenia yellow[J]. International Journal of Biological Macromolecules, 2016, 89: 389-395.

[7] ZHOU L, SHAO J Z. Study on application of natural plant dye gardenia on cotton coloration and the dyeing mechanism[J]. Journal of Donghua University (English Edition), 2006, 23(5): 122-126.

[8] PEI L, LIU J, CAI G, et al. Study of hydrolytic kinetics of vinyl sulfone reactive dye in siloxane reverse micro-emulsion[J]. Textile Research Journal, 2016, 87(19): 2368-2378.

[9] SALEEM M A, PEI L, SALEEM M F, et al. Sustainable dyeing of nylon fabric with acid dyes in Decamethylcyclopentasiloxane (D5) solvent for improving dye uptake and reducing raw material consumption[J]. Journal of Cleaner Production, 2020, 279(5): 123480.

[10] AN Y, MIAO J, FAN J, et al. High-efficiency dispersant-free polyester dyeing using D5 non-aqueous medium[J]. Dyes and Pigments, 2021, 190: 10930.

Preparation and Properties of High Temperature Resistant Polyimide Nanofiber Membranes by Electrospinning

LI Ke[1,2], ZHANG Hongnan[1,2], QIN Xiaohong[1,2*]

[1] *College of Textiles, Donghua University, Shanghai 201620, China*

[2] *Key Laboratory of Textile Science & Technology, Ministry of Education, Donghua University, Shanghai 201620, China*

[*] *Corresponding author's email*: xhqin@ dhu. edu. cn

Abstract: Polyimide (PI) has been widely studied because of its excellent thermal stability and high mechanical properties. In order to study the structure and properties of PI nanofibers prepared by the two-step method, polyamide acid (PAA) and *N*, *N*-dimethylformamide (DMF) were used as raw materials to prepare PAA solutions with different solid contents. PAA nanofiber membranes were obtained by electrospinning technology, and then PI nanofiber membranes were obtained by gradient heating heat treatment by thermal amination technology. The performance was tested to explore the effects of PAA solid content on fiber morphology, chemical structure, mechanics, filtration performance and thermal stability of fiber membranes. The results showed that when the solid content was 18%, the morphology of PI nanofibers was the best, the average fiber diameter was 182.45 nm, the breaking strength and elongation at break were 10.533 MPa and 22.7% respectively, the filtration efficiency was 99.784%, and PI nanofiber membranes has excellent thermal stability below 500 ℃, in order to provide some useful help for the further research and application of PI in the field of high temperature resistant air filter materials.

Keywords: Polyimide; Electrospinning; Filter material; Thermal stability

1 Introduction

Polyimide (PI) is a kind of polymer containing imide rings in the main chain of macromolecules[1]. Due to its stable closed-loop structure, PI has many advantages, such as excellent high temperature resistance, excellent mechanical properties, chemical corrosion resistance and so on. It is widely used in special protection, military industry, high temperature dust removal and other fields[2]. Electrospinning is an important technology to prepare nanofiber materials. Continuous fibers are obtained by stretching and refining the sprayed polymer solution in high voltage electric field[3]. The nanofiber membranes prepared by electrospinning technology has the advantages of large specific surface area, small pore size and high porosity. It has great application value in the field of high efficiency and low resistance air filter materials[4].

In this experiment, PI nanofiber membranes was prepared by electrospinning combined with thermal imidization process, and the effect of PAA solid content on the morphology, structure, mechanics and filtration performance of PI nanofiber membranes was explored. The chemical structure and thermal stability of PI nanofiber membranes were tested and analyzed by infrared spectroscopy and thermogravimatry.

2 Experiments

2.1 Materials

Polyamide acid (PAA, 20%) were obtained from Teyue plastic (Jiangsu, China). *N*, *N*-dimethylformamide (DMF, analytically pure) were provided by Lingfeng Chemical Reagent Co. Ltd. (Shanghai, China).

2.2 Preparation of electrospun PAA nanofiber membranes

Accurately weighed the quantitative PAA and added it to the beaker containing DMF, used a magnetic stirrer to stir it magnetically, and reacted at room temperature for 4 h. Finally, a clear and transparent PAA solution without delamination was obtained. According to this method, PAA spinning solutions with solid contents of 16%, 18% and 20% were prepared, and they were defoaming for use.

Put the prepared PAA solution into a 10 mL syringe

and fixed it on the propulsion pump of the electrospinning device. Fig. 1 is the schematic diagram of the electrospinning experimental device. Covered the cylinder receiver with aluminum foil, adjusted the receiving distance and propulsion speed, set the spinning voltage as 16 kV, the pushing speed as 0.2 mL/h, the receiving distance as 20 cm, the cylinder speed as 100 r/min, the temperature as 20 ℃, the humidity as 30%, and the spinning time as 3 h. The electrospun PAA nanofiber membranes was prepared by putting the spun PAA nanofiber membranes into a vacuum drying oven for drying for 8 h to remove the solvent.

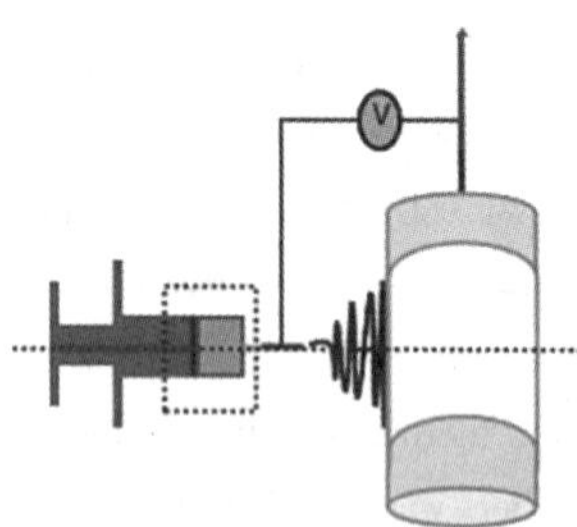

Fig. 1 Schematic diagram of single needle electrospinning experimental device

2.3 Preparation of PI nanofiber membranes

PI nanofiber membranes was prepared by thermal imidization process, and the membranes was transferred to muffle furnace. The thermal imidization process of PAA nanofiber membranes is mainly divided into three stages[3]:

(1) 150 ℃, 5 ℃/min, holding for 30 min to remove residual solvent.

(2) 250 ℃, 5 ℃/min, holding for 30 min, PAA molecule dehydration cyclization to PI.

(3) 300 ℃, 5 ℃/min, holding for 40 min, thermal imidization is completed.

2.4 Characterization

The micro morphology of PI nanofiber film was observed by scanning electron microscope (SEM, Ri Li Co. Ltd., China). Because PI nanofiber film is an insulating material, it needs to be sprayed with gold for 70 s before testing.

Fourier transform infrared spectroscopy (FTIR, Antaris 2, Somerfisher Instrument Co. Ltd., America) was used to characterize the chemical structure of the nanofiber film before and after imidization. The positions of the absorption peaks of each group in the sample were tested, and the changes of the chemical structure of the sample before and after imidization were analyzed.

Electronic single fiber strength meter (XQ-1C, Shanghai Xin Xian Instrument and Equipment Co. Ltd., China) was used to test the mechanical properties of PI nanofiber film. The clamping distance was 10 mm, and the tensile rate was 20 mm/min. the breaking strength of PI nanofiber film[5] was calculated as Equations (1).

$$\delta = \frac{F}{A} \tag{1}$$

where δ indicates the breaking strength; F refers to breaking strength; A refers to the original cross-sectional area.

Automatic filter material tester (TSI 8130, America) is used to test the filtration performance of PI nanofiber membranes. The dust source is NaCl aerosol, and the air flow rate is 32 L/min. The quality factor QF of PI nanofiber membranes was defined as Equations (2).

$$QF = \frac{-\ln(1-\eta)}{\Delta p} \tag{2}$$

where η is the filtration efficiency; ΔP refers to pressure drop; QF is the quality factor.

The thermal stability of PI nanofiber membranes was characterized by thermogravimetric analyzer (TGA 4000, Shanghai Perkin Elmer Instrument Co. Ltd., China). The protective gas was nitrogen. The gas flow rate was 40-50 mL/min, the heating range was 50-800 ℃, and the heating rate was 10 ℃/min.

3 Results and Discussion

3.1 Morphologies of PI nanofiber membranes

PAA with different solid content greatly affected the morphology, structure and fiber diameter of PI nanofiber film. Fig. 2 and Tab. 1 showed the scanning electron microscope photos and average fiber diameter of PI nanofiber film with different solid content. It could be seen from Fig. 2 and Tab. 1 that when the solid content was low, there were more broken fibers and round beads in the fiber membranes. With the increase of the solid content of the solution, the beads gradually disappeared, and the fiber diameter gradually increased. This was because when the solid content is low, the entanglement between polymer molecular chains was not enough, and the jet was easy to shrink and fracture under the action of electric field[6]. With the increase of the solid content, the viscosity of the solution increased, and the entanglement between molecular chains increased, the spinning process was gradually stable and the spun fibers were evenly distributed. However, when the solid content increased to 20%, the entanglement between molecular chains was too large, the electric field force at this time couldn't realize the uniform stretching and refinement of the fiber, the diameter of the nanofiber increased, and the solution was easy to block the needle, so the spinning process becomes obviously difficult.

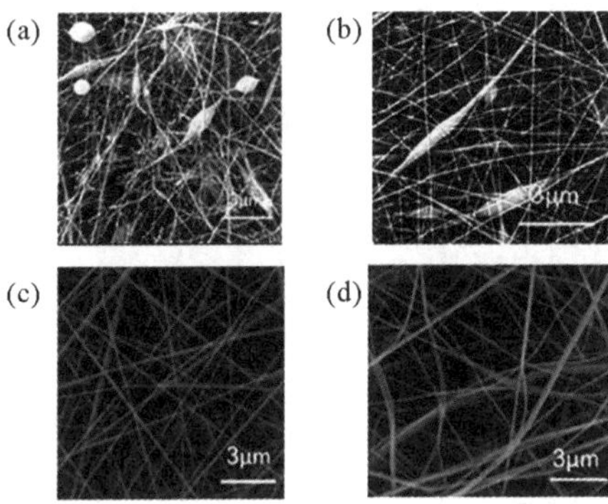

Fig. 2 SEM images of PI nanofiber membranes with different concentrations: (a) 14% PI; (b) 16% PI; (c) 18% PI; (d) 20% PI

Tab. 1 Average fiber diameter of PI nanofiber membranes with different concentrations

Concentration of PAA/%	14	16	18	20
Average fiber diameter/nm	129.49	155.95	182.45	288.75

3.2 Chemical structure of PAA and PI nanofiber membranes

Fig. 3 showed the FTIR of PAA and PI nanofiber membranes. It could be seen from the figure that in the infrared spectrum of PAA nanofiber film, there were obvious absorption peaks at 1 360 cm^{-1}, 1 550 cm^{-1}, 1 650 cm^{-1} and 1 715 cm^{-1}, corresponding to CO—NH symmetric stretching vibration, CO—NH asymmetric stretching vibration, C=O symmetric stretching vibration and C=O asymmetric stretching vibration absorption peaks respectively, which correspond to the characteristic peaks of PAA[7]. In the infrared spectrum of PI nanofiber membranes, the absorption peak C—N characteristic peak (1 373 cm^{-1}) of imine group at 720 cm^{-1}, 1 373 cm^{-1}, 1 720 cm^{-1} and 1 775 cm^{-1} indicated that after heating treatment, PAA nanofiber membranes had been completely dehydrated and cyclized into PI nanofiber membranes[8].

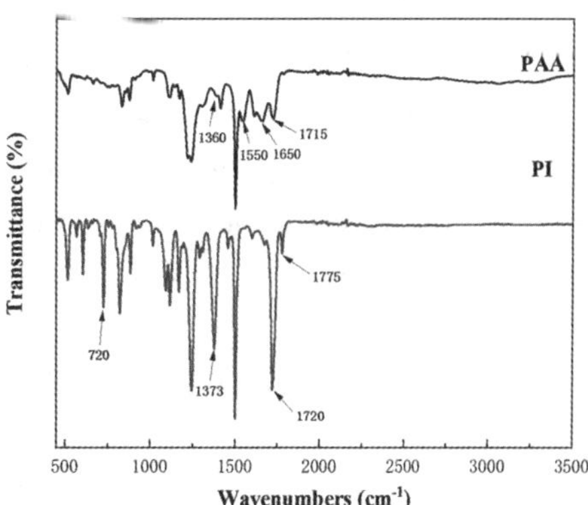

Fig. 3 FTIR spectrum of PAA and PI nanofiber membranes

3.3 Mechanical property of PI nanofiber membranes

Fig. 4 showed the stress-strain curve of PI nanofiber membranes with different concentrations. It could be seen from the figure that with the increase of PAA solid content, the elongation at break of PI nanofiber film increased, the breaking strength first increased and then decreased. This was because when the solid content was low, there were a large number of broken fibers and round beads in the fiber film, and the adhesion in the fiber film was small, so the breaking strength and elongation at break were small, which were 9.234 MPa and 11.9% respectively. With the increase of solid content, the beads disappeared, the fiber surface was smooth and the diameter was uniform. When subjected to the tensile force, the fibers would be arranged along the same direction from the original irregular arrangement. With the further increase of the tensile force, the fibers would gradually elongate to fracture, and finally lead to the fracture of the film[9]. Therefore, the breaking strength and elongation at break were greatly im-

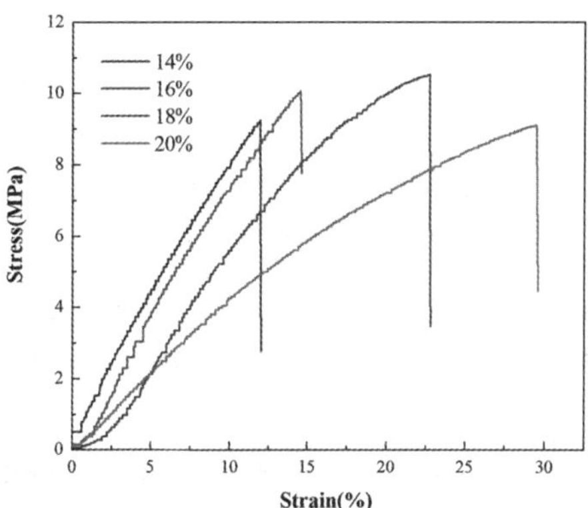

Fig. 4 Stress strain curves of PI nanofiber membranes with different concentrations

proved, as high as 10.533 MPa and 22.7%. However, with the further increase of PAA solid content, the diameter of PI nanofibers increased, and the fiber diameter was uneven, resulting in an increase in the thickness of PI film, a significant increase in the timing of fiber fracture, and thus the fracture strength decreased.

3.4 Filtration performance of PI nanofiber membranes

PAA with different solid content constructed PI nanofiber membranes with different morphologies and structures, which then affected the filtration performance of PI nanofiber membranes. As shown in Fig. 5, it was obvious that with the increase of solid content, the filtration efficiency and pressure drop of PI nanofiber membranes first increased and then decreased. The maximum filtration efficiency reached 99.784%, and the quality factor was 0.029. This was because when the solid content was low, the broken fiber and round beads made the fiber surface have more defects, the pore size was larger, and the filtration efficiency and pressure drop of the membranes were low[10]. With the increase of solid content, the fiber diameter was uniform, the beads disappeared, and the pore size decreased. When the solid content was 18%, the filtration efficiency and QF value of PI membranes reached the highest. With the further increase of solid content, the fiber diameter became thicker, the stacking density decreased, the PI nanofiber film was fluffy, and the pressure drop slightly decreased.

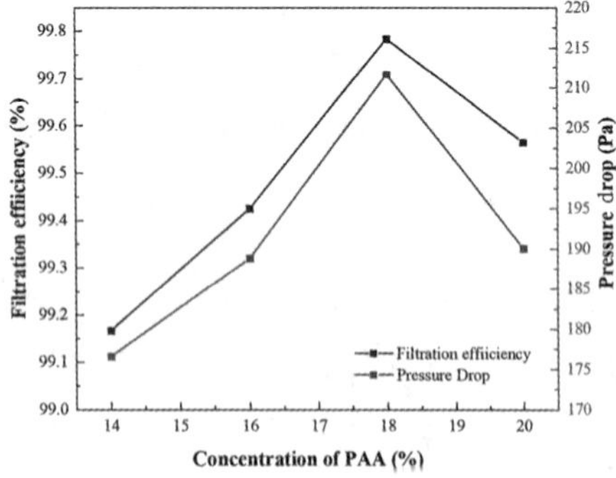

Fig. 5 Filtration performance of PI nanofiber membranes with different concentrations

3.5 Thermal stability of PI nanofiber membranes

TGA curve of PI nanofiber film was shown in Fig. 6. It could be seen from the figure that in N_2 atmosphere, the PI nanofiber film had only a small mass loss in the range of 100 ℃, which may be caused by the thermal weight loss of the residual solvent and water in the film. At 200 ℃, the weight loss rate of PI nanofiber membranes was 3%, and at 250 ℃, the weight loss rate was only 5%. At about 500 ℃, the thermal decomposition reaction of PI nanofiber membranes occurred. It could be concluded that the PI nanofiber membranes prepared by electrospinning in this experiment had excellent thermal stability and it had a wide application prospect in some high temperature environments.

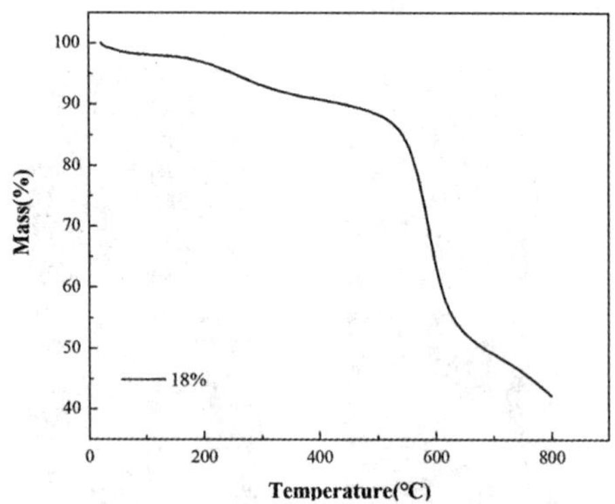

Fig. 6 TGA curve of PI nanofiber membranes

4 Conclusions

(1) The concentration of PAA had a great influence on the fiber morphology and fiber diameter of PI nanofiber membranes: with the increase of solid content, the beads in the fiber membranes disappeared, the broken fibers decreased, and finally the uniform and smooth fiber distribution was completely presented. The fiber diameter increased with the increase of solid content, and high solid content is not conducive to the experimental operation of electrospinning.

(2) The infrared spectrum test showed that with the thermal imination of gradient heating, the characteristic peaks of PAA disappeared completely, and the absorption peak of imine group appeared at the same time. The imination of PAA is full and the transformation is complete.

(3) When the solid content of PAA was 18%, the breaking strength and elongation at break of PI nanofiber membranes were 10.533 MPa and 22.7% respectively, and the filtration efficiency was as high as 99.784%. The comprehensive filtration performance was the best.

(4) TGA test showed that PI nanofiber membranes had excellent heat resistance, and the thermal weight loss rate at 500 ℃ was about 5%, which is promising to use this material as air filter material in high temperature field.

References

[1] DONG H, ZHENG S S, GUO T, et al. Preparation and properties of high heat resistant polyimide fiber[J] Journal of Textile Industry, 2022, 43(2): 19-23, 43.

[2] CHEN D, WANG R, TJIU W W, et al. High performance polyimide composite membranes prepared by homogeneity reinforcement of electrospun nanofibers[J]. Composites Science and Technology, 2011, 71(13): 1556-1562.

[3] LI L, LIU P, FU Q S, et al. Study on preparation of polyacrylonitrile/polyimide composite lithium-ion battery separator by electrospinning[J]. Journal of Materials Research, 2019, 34(4): 642-651.

[4] LI D W, SHEN Y, WANG L L, et al. Hierarchical structured polyimide-silica hybrid nano/microfiber filters welded by solvent vapor for air filtration[J]. Polymers, 2020, 12(11): 2494.

[5] JIANG S H, DUAN G G, CHEN L L, et al. Mechanical performance of aligned electrospun polyimide nanofiber belt at high temperature[J]. Materials Letters, 2015, 140: 12-15.

[6] NAH C, HAN S H, LEE M H, et al. Characteristics of polyimide ultrafine fibers prepared through electrospinning[J]. Polymer International, 2003, 52(3): 429-432.

[7] WANG Y, YANG Y, JIA Z X, et al. Effect of the amidation degree of precursor on the aggregated structure and properties of polyimide films[J]. Polymer Materials Science and Engineering, 2012, 28(10): 43-46.

[8] JIANG S, DUAN G, SCHOEBEL J, et al. Short electrospun polymeric nanofibers reinforced polyimide nanocomposites[J]. Composites Science and Technology, 2013, 88: 57-61.

[9] PARK S H, LEE H B, YEON S M, et al. Flexible and stretchable piezoelectric sensor with thickness-tunable configuration of electrospun nanofiber mat and elastomeric substrates[J]. ACS Applied Materials & Interfaces, 2016, 8(37): 24773-24781.

[10] WANG N, SI Y, WANG N, et al. Multilevel structured polyacrylonitrile/silica nanofibrous membranes for high-performance air filtration[J]. Separation and Purification Technology, 2014, 126: 44-51.

Low-Temperature Effect on Mechanical Properties of Cotton Fiber

CAO Jiqiang[1], XU Hong[1*]

[1] *College of Textile and Fashion, Xinjiang University, Urumqi 830017, China*

[*] *Corresponding author's email*: xjdxxuhong@ 163. com

Abstract: Cotton is an important raw material for the textile industry, and the processing quality of its initial product has an important impact on the properties of the final yarn and fabric. However, the effects of low temperature and humidity during ginning on its mechanical properties have not received extensive attention. In this paper, the mechanical properties of three groups of cotton fibers pretreated at different moisture regains and temperatures were analyzed. Furthermore, by observing the fiber fracture interface of different samples, the mechanical damage mechanism of cotton fibers under the action of humidity and temperature during initial processing was discussed. The results show that the mechanical properties of cotton fibers are negatively correlated with the ginning temperature. As the temperatures were set to −15 ℃, −10 ℃, −5 ℃, and 0 ℃, seed cotton moisture and temperature had a significant influence on fiber quality, which showed a reduction in mechanical properties. Especially, the mechanical property damage of low temperature high moisture regain(L-HMR) was 27.6% lower than the value at normal temperature. Finally, the cotton fiber in low moisture regains showed a brittle fracture. However, the high moisture regains had significant fibrils. This study provides theoretical support for the optimization of the ginning process in cotton initial processing enterprises in terms of moisture regain and low temperature.

Keywords: Low-temperature environment; Cotton fibers; Tensile mechanics; Fracture morphology; Fiber damage; Cotton ginning

1 Introduction

Cotton is an essential product of our daily life. The initial processing of cotton is an indispensable part of textile enterprises, and the ginning processing can directly reflect the initial quality of cotton fiber. Although the initial processing technology and equipment have been relatively advanced, the cotton fiber defect caused by the external environment and equipment strike is still worthy to attend. It has been widely studied that the initial quality of the fiber can be guaranteed by the initial processing of cotton in a normal environment. However, the mechanical damage of cotton fiber in the process of processing under an environment of low temperature and high moisture regain still needs further research. Temperature is an important environmental factor affecting cotton fiber properties, research shows that decreased night temperatures slowed both rate and amount of cellulose synthesis, which in turn resulted in underdeveloped fiber[1], characterized by poor physical and chemical properties[2-3], and paid more attention to the primary ecological factor affecting cotton yield and quality[4-5]. We all know the difference in performance between dry and continuous dry cotton, therefore, the initial processing temperature of cotton in enterprises should be controlled at 15−35 ℃ to ensure the quality of fiber. In cotton initial processing enterprises, attention should be paid to the change of moisture regain in the production process of process equipment. Changes in moisture and density can lead to changes in the structure of cellulose[6], while the normal moisture regains range is 6%−8%, but the actual production will be low or high, which will attract the attention of processing enterprises. The quantification of cotton fiber damage and the characterization of fracture interface may be helpful to ensure the quality of fiber. By comprehensively understanding the influence of the environment on the mechanical properties of cotton fiber, it provides new data reference for the practice of cotton primary processing enterprises in the world, which is of great significance to the processing process and environmental impact of cotton primary processing enterprises.

2 Experiments

2.1 Experimental materials

The cotton fiber samples are taken from cotton processing plants, cotton fibers are 20 kg, which were divided into five groups with 15 samples in each group. The cotton variety was xinluzao 13 and the cotton length was 28.9 mm. The fiber density is 1.55 g/cm^3, the linear density is 2.16 dtex, the tensile fracture strength of a single fiber is 2.52 cN/dtex, the elongation at break is 8.0%, and the initial modulus is 0.99 cN/dtex. The same cotton sample was used to observe the SEM fracture surface.

2.2 Experimental method

In this paper, the sampling and testing are completed by the experimental steps like Fig. 1 Firstly, three sample numbers were given, namely low-temperature high moisture regain (L-HMR), low-temperature low moisture regain (L-LMR), and normal control sample, respectively, 15 samples for each sample with 30 mg. Secondly, the conditioning pre-treatment. According to the experimental requirements, the samples are divided into two groups: low moisture regain and high moisture regain. Therefore, the low moisture regain was set to 3.5% in the oven. According to the above, the actual moisture regains of cotton fiber can reach 20% in the winter with snow. So, the high moisture regains of cotton fiber samples are adjusted to 20%. All the experimental samples are placed in sample seal bags for 48 h in the laboratory for moisture absorption balance. The third step was low-temperature treatment. The L-HMR fiber and L-LMR fiber samples in the second step were placed in refrigerators with different low temperatures for 24 h, and the temperatures were set to −20 ℃, −15 ℃, −10 ℃, −5 ℃, and 0 ℃, respectively. Finally, different samples were hooked, twisted, and bent using tweezers under the corresponding environmental conditions, including in normal temperature N-HMR fiber and N-LMR fiber samples, and then the samples were taken out to test the mechanical tensile properties.

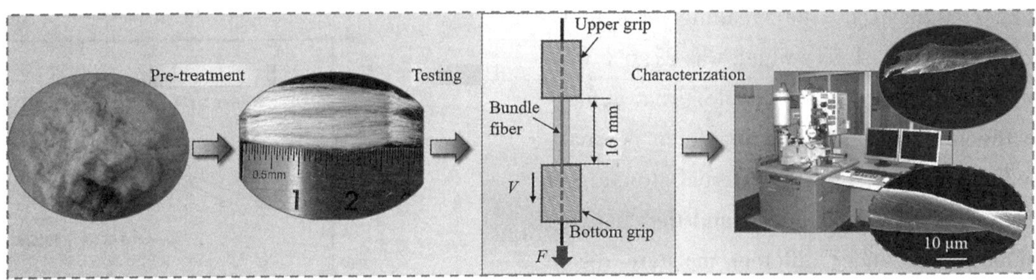

Fig. 1 Schematic representation for the steps of experimental

2.3 Testing and characterization

In this study, the mechanical properties of cotton fibers were mainly tested and the morphology of the fracture interface fibers was characterized. All fiber samples were tested in a standard laboratory environment, where the relative humidity and temperatures were 63% and 22 ℃, respectively. Using the calibration YG (B) 162D electron beam fiber strength meter, the mechanical properties of the cotton bundle fibers are measured, and the distance and the stretching rate are 10 mm and 10 mm/min, respectively. Bundle fiber test reference to GB 1103.1—2012 Cotton Part One-Sawtooth Processing of Fine Staple Cotton, each sample was tested 15 times, and the results were averaged. The fracture interface morphology analysis of fibers was initially observed using XSP−3CA optical microscopy. Specifically, the fracture morphology of the as-prepared normal control sample, L-HMR, and L-LMR samples was characterized by scanning electron microscopy (SEM; TM3000, Hitachi Group, Japan).

3 Results and Discussion

3.1 Results of mechanical properties of cotton fibers

The mechanical properties of the fiber were obtained by using the above research method. Tab. 1 shows the mechanical properties of the fiber under the conditions of normal moisture regain (6.5%) at normal temperature, the breaking force, fracture strength, and elongation at break were 4326.85 cN, 33.28 cN · tex^{-1}, and 7.7%, respectively. Tab. 1 shows the tensile mechanical properties resulting in low-temperature.

Tab. 1 Tensile mechanical properties result in low-temperature

Average of 15 data [standard variations]		Temperature(℃)				
		T_L=-20 ℃	T_L=-15 ℃	T_L=-10 ℃	T_L=-5 ℃	T_L=0 ℃
L-HMR	Breaking force(cN)	3 131.4[10.63]	3 287.31[10.22]	3 424.7[10.65]	3 886.5[9.98]	4 185.65[9.87]
L-LMR		3 355.0[11.22]	3 411.6[10.68]	3 713.37[10.02]	3 872.42[10.13]	3 906.85[9.92]
L-HMR	Fracture strength($cN \cdot tex^{-1}$)	24.09[10.63]	25.29[10.16]	26.34[10.65]	29.00[10.03]	32.20[9.87]
L-LMR		25.81[10.34]	26.24[10.68]	28.7[10.92]	29.79[10.13]	30.05[9.62]
L-HMR	Elongation at break(%)	6[10.37]	5.3[9.58]	6[11.34]	5.5[9.69]	7.2[9.12]
L-LMR		6.1[10.31]	6[10.11]	6.2[11.78]	6.6[9.80]	6.8[9.66]

3.1.1 Analysis of the breaking force of bundle fibers

From the results in Tab. 1, it is shown the breaking force index values of cotton fibers at low temperatures. Under the same other conditions, with the temperature from −20 ℃ to 0 ℃, the breaking force of L-LMR and L-HMR cotton samples gradually increases, and the breaking capacity of cotton fibers also increases continuously. It can also be seen from the results in Fig. 2 that the breaking forces of L-LMR and L-HMR are increasing continuously. The breaking force of L-HMR at −20 ℃ is 3 131.4 cN, which was 27.6% lower than that of the normal control sample. That is maybe because the intermolecular bonding force is weakened at low temperatures, and the internal structure of fibrils is destroyed. To further understand the changing trend of cotton fibers under different moisture regain and low-temperature conditions, combined with the data in Tab. 1, the results of different moisture regain and low-temperature conditions are simulated by a second-order polynomial fitting cooperation graph. The results are shown in Fig. 2, and the fitted prediction curves have high R^2 values.

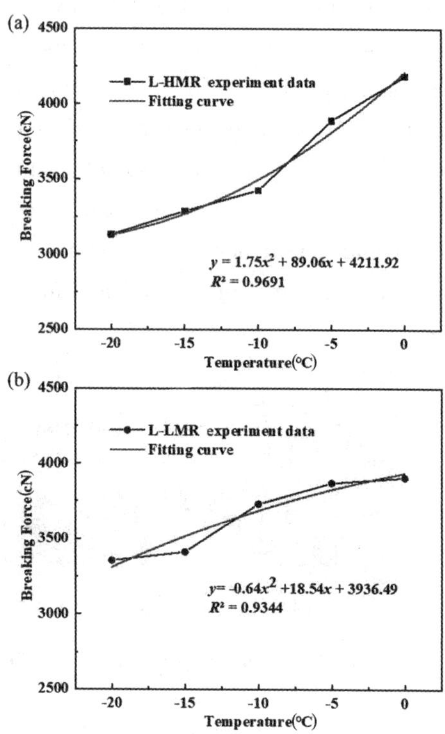

Fig. 2 Fiber breaking force curve of cotton bundle in low-temperature environment: (a) L-HMR fiber experimental curve and fitting curve; (b) L-LMR fiber experimental curve, and fitting curve

3.1.2 Analysis of the fracture strength and elongation at break of bundle fibers

The relationship between the bundle fiber strength and the average bundle fiber strength measured by the fiber strength tester is as follows[7]:

$$\bar{P} = n\frac{\bar{f}}{k} = \frac{N_{tB}}{N_{tf}}\frac{\bar{f}}{k} \quad (1)$$

In Eq. (1): $\bar{P}$ the average force of the fibers bundle (cN); $\bar{f}$ is the average strength of the single fibers (cN); n is the number of fibers in the fiber bundle; N_{tB} and N_{tf} are the linear density of digital and single fibers, respectively; k is the coefficient. Under the standard conditions of China, the cotton fiber is 1.412−1.481.

According to Eq. 1 and the fracture strength data in Tab. 1, and Fig. 3 (a), with the temperature rises (−20 ℃ to 0 ℃), the fracture strength of L-HMR and L-LMR bundle fibers has increased, especially at 0 ℃, and the fracture strength value of L-HMR has reached 32.20 cN/tex, and the value of L-HMR is 7.14% higher than that of L-LMR (30.05 cN/tex). At −5 ℃, the fracture strength values of L-HMR are much strong than that of L-LMR. These could explain the mechanical damage treatment at normal temperature had little damage to the fibers, while the ice crystals sandwiched between macromolecules destroyed the basic fibrils in low-temperature environment, especially in the case of high moisture regain.

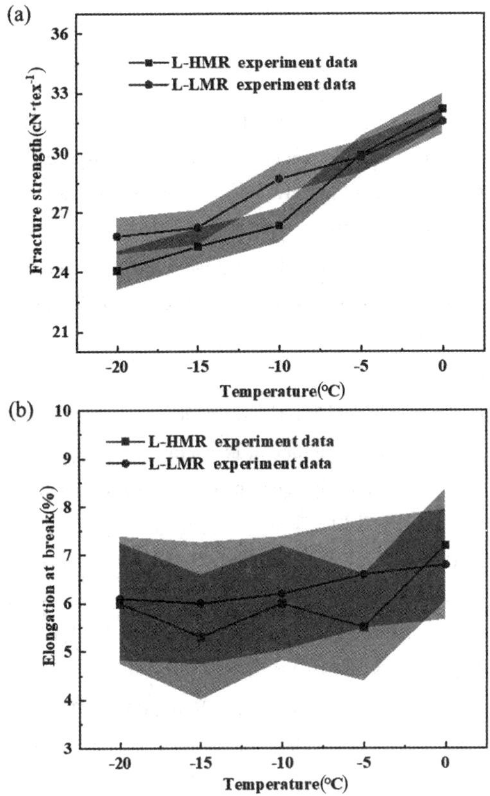

Fig. 3 The trend of cotton bundles in different low-temperature environments: (a) fracture strength; (b) elongation at break

The changing trend of elongation at the break of L-HMR and L-LMR cotton fiber can be observed in Fig. 3(b): With the increase in ambient temperature, the elongation at the break of L-HMR and L-LMR cotton fiber increases with the increase in temperature from −20 ℃ to 0 ℃. At the same temperature, the greater the moisture regain of the fiber, the higher the intermolecular secondary valence. the more bonds are disintegrated, the easier it is to loosen and slip between molecular chains, and the elongation at break increases[8]. The overall discreteness of elongation at the break of L-HMR and L-LMR cotton bundle fibers is relatively large in low-temperature environment, especially L-HMR performance is unstable. It can be observed that the elongation at the break of cotton bundle is the critical point at −10 ℃, there is little difference in the elongation at the break of cotton fiber. The analysis believes that the reason for the transformation of the macromolecular structure inside the fiber at this temperature state is also the reason for mechanical damage.

3.2 Analysis of Fracture interface of bundle fibers

As the well-known fact that cotton fiber is easy to exchange water with the surrounding environment, during the process of water exchange, the physical properties of the fiber have significant indigenous changes. The same relative moisture regain has a great influence on the fracture morphology of the cotton fiber bundle[9]. Especially in the case of extremely low temperature and high-water content, water molecules are prone to produce ice crystals that are easy to cut fibrils and are subjected to mechanical force during the ginning process to destroy the internal structure of cotton fibers.

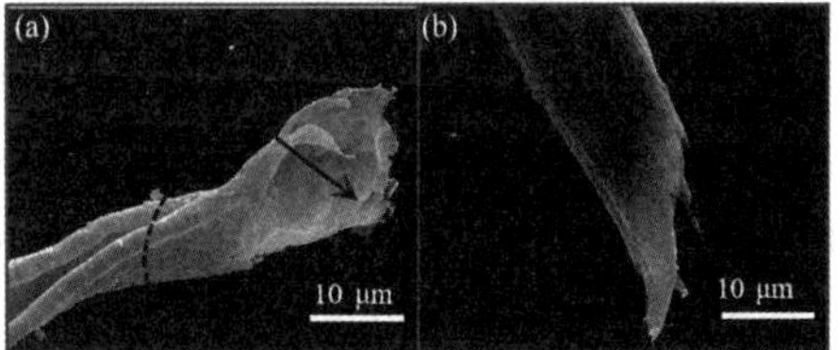

Fig. 4 Fracture morphologies of cotton fibers in different environments: (a) −20 ℃ L-HMR fracture interface morphology; (b) −20 ℃ L-LMR fracture interface morphology

Fig. 4 shows the fracture interface morphology of cotton fibers, which show at −20 ℃ temperature, as well as under high and low moisture regain. The fracture interface morphology of L-HMR cotton fibers at −20 ℃ was shown in Fig. 4(a). The circle shows the morphology of the fiber fracture interface, which can be seen in the fibrils and the damaged fiber cavity, the fibers are twisted and fractured and the same conclusion was drawn by Professor Hearle[10]. In the position indicated by the arrow in the area, there are many scratches, and tears on the interface of the fiber, the fibrils are dislocated, which is caused by the formation of ice crystals by water molecules in low-temperature environment and scratches the internal fibrils of the fibers. The force of the fiber fracture is relatively uniform, and the fracture interface is not as sharp as the fiber fracture of L-LMR. Fig. 4(b) is the topography of the fracture interface of the L-LMR fiber at −20 ℃. It can be observed that the fracture surface of the cotton fiber presents a sharp and brittle shape. The fractured form is a typical brittle fracture, and there is no obvious fibril exposed in the main body of the fiber. In the low-temperature environment, there are fewer effective macromolecular chains inside the fiber, which will be described in detail in the following articles for molecular conformation and structural adjustment.

4 Conclusions

Through the difference in temperature and moisture regain in actual production, the mechanical properties and structural damage of cotton fibers caused by different moisture regain at different low temperatures were studied, and the fracture interface morphology of a single fiber was characterized by SEM. The experimental results show that the mechanical properties at normal temperatures are higher than those at low temperatures. The breaking force of L-LMR at −20 ℃ was 27.6% lower than that of the normal control sample The mechanical properties of cotton fibers have changed, especially the difference in fracture strength value of L-HMR is 7.14% higher than that of L-LMR. The trend of mechanical and mechanical properties with the increase in temperature indicates that the conformation of macromolecules adjusts the structure with the increase in temperature, which increases the effective load-bearing macromolecular chains and leads to the improvement of uniformity. Finally, the fiber fracture interface was analyzed from the structural morphology. At −20 ℃, the fibrils were not significantly separated from the main fiber. Under high regain conditions, L-HMR has obvious fibrils, which corresponds to the high value of fracture strength. The fracture interface morphology of L-LMR fracture interface morphology is a typical brittle fracture. Through the test characterization and analysis of mechanical properties of cotton fibers. It provides theoretical support and effective pre-control for cotton primary processing enterprises in the low-temperature ginning processing environment.

Acknowledgements

We gratefully acknowledgements the financial support of the National Natural Science Foundation of China (No. 51963020). The authors also thank the College of Textile and Fashion for providing cotton samples, and fiber testing.

References

[1] GIPSON J, JOHAM H. Influence of night temperature on growth and development of cotton (Gossypium hirsutum L.) IV. seed quality 1[J]. Agronomy Journal, 1969, 61(3): 365-367.

[2] GIPSON J, JOHAM H. Influence of night temperature on growth and development of cotton (Gossypium birsutum L.). II Fiber properties 1[J]. Agronomy Journal, 1968, 60(3): 296-298.

[3] GIPSON J, JOHAM H. Influence of night temperature on growth and development of cotton (Gossypium hirsutum L.). III. fiber elongation 1[J]. Crop Science, 1969, 9(2): 127-129.

[4] MARTIN L K, HAIGLER C H. Cool temperature hinders flux from glucose to sucrose during cellulose synthesis in secondary wall stage cotton fibers[J]. Cellulose, 2004, 11(3): 339-349.

[5] HAIGLER C H, RAO N R, ROBERTS E M, et al. Cultured ovules as models for cotton fiber development under low temperatures[J]. Plant Physiology, 1991, 95(1): 88-96.

[6] ANDREWS F R, HONOLD E, GRANT J N. Heating, cleaning, and mechanical processing effects on cotton: part II: equilibrium moisture and density[J]. Textile Research Journal, 1963, 33(8): 609-616.

[7] YU W D. Textiles Materials[M]. 2nd Edition. Beijing: China Textile & Apparel Press, 2018.

[8] LAWSON R, RAMEY H H, KROWICKI R S. Cotton fiber tenacity and elongation in rapidly changing relative humidity[J]. Textile Research Journal, 1976, 46(10): 715-719.

[9] CINTRóN M S, INGBER B F. Preliminary examination of the effects of relative humidity on the fracture morphology of cotton flat bundles [J]. Textile Research Journal, 2013, 83(10): 1044-1054.

[10] HEARLE J W S, SPARROW J T. The fractography of cotton fibers[J]. Textile Research Journal, 1971, 41(9): 736-749.

Correlation Analysis of Pore Morphology and Liquid Diffusion Behavior of Fabric Structure

YIN Jiong[1,2], **ZHU Ying**[1], **SHAN Xiya**[1], **JI Feng**[1,2*]

[1] *Key Laboratory of Textile Science & Technology, Ministry of Education, Donghua University, Shanghai 201620, China*

[2] *College of Textiles, Donghua University, Shanghai 201620, China*

[*] *Corresponding author's email*: jifeng@ dhu. edu. cn

Abstract: In this paper, the effect of pore morphology on liquid diffusion behavior of fabric was investigated. Pore orientation angle (POA) and pore ellipticity (PE) were used to characterize the morphology of micropores. Accordingly, liquid diffusion angle (LDA) and liquid diffusion ellipticity (LDE) were used to characterize liquid diffusion on macroscopic fabric surface. Three kinds of knitted, woven and nonwoven fabric samples were selected. Firstly, the pore morphology index data of samples were measured by using image processing technology and equivalent elliptic fitting technology. Then, the liquid diffusion index data of each sample were obtained by using drop method and image technology. The results showed that both knitted and woven fabrics had obvious first and second level pores. LDA had a strong correlation with the orientation angle of first and second level pores. LDE was affected by the first level PE and the second level pore morphology. LDA on the nonwoven surface was consistent with the average of POA and LDE was consistent with the average of PE.

Keywords: Pore orientation; Pore morphology; Liquid diffusion angle (LDA); Liquid diffusion behavior

1 Introduction

All kinds of textile materials prepared with fiber as raw material have porous characteristics. Most of pores are small in scale and large in quantity, and pores are the key to affect moisture conductivity of textile materials, and play an important role in the comfort performance of textile fabrics. In different fabric structures, the pore law between fibers and yarns is different, which affects the transfer effect of liquid on the surface and inside of fabric materials. Singhal et al. [1] established a mathematical model of yarn wicking and proposed that the fiber-fiber contact and pore surface area jointly determine the degree of liquid rise in yarn. Monica et al. [2] proposed the natural fiber wicking model, connecting the modification of pore radius on micro scale of fiber with modification of porosity on micro scale of fabric. Researchers explored the influence of fiber type, cross section and porosity on liquid transport[3-5]. Zhang et al. [6] showed that fabric compactness and porosity were negatively correlated with wicking rate. Researchers established capillary model based on yarn structure parameters, and numerically simulated the movement trend of liquid in yarn and fabric in different pore structures[7-9]. Most of the researches on liquid transfer performance of fabric materials were based on porosity and pore size, and few of them were based on the actual multilevel pore structure.

In knitted and woven fabrics, there are multi-level pore structures, such as large pores between yarns and microscopic pores between fibers. In nonwovens, there are only single-level pore structures between fibers. Therefore, it is of great significance to study the influence of multi-level pore structure characteristics on liquid transfer behavior of textile.

In this paper, the effect of pore morphology on liquid transfer behavior of knitted, woven and nonwoven fabrics was studied. Firstly, the pore morphology in fabric was characterized, and pore orientation angle (POA) and pore ellipticity (PE) were proposed to characterize pore structure. Liquid diffusion angle (LDA) and liquid diffusion ellipticity (LDE) were used to characterize the diffusion of liquid in fabric. Then, the pore morphology index data of three kinds of samples and the diffusion behavior parameters of liquid in fabric were obtained through experiments. On this basis, the influence of pore morphology on liquid transfer behavior of fabric was analyzed.

2 Experimental Designs

2.1 Characterization of pore morphology

Woven and knitted fabrics were made of yarns and had two level pore structures. The first level pore was the pore between yarns. The pore was regular and slightly different in size, usually greater than or equal to 0.1 mm^2. The second level pore was the pore inside the yarn and between fibers, generally less than 0.1 mm^2. Nonwovens were directly made of fiber nets, so the internal pores had only second level pore structure, no first level pore structure. Internal pores of nonwovens (Fig. 1 (a)) were taken as an example to quantitatively characterize pore morphology. First, the pore was treated into plane geometry, and then the plane geometry selection was used to fit into the equivalent ellipse with the same area, centroid and orientation. The polar coordinate system was established with the center of the equivalent ellipse as the polar origin, the vertical laying-up direction of nonwovens (woven weft/knitted fabric transverse) as 0° direction and the counterclockwise direction as positive direction. The schematic diagram of equivalent ellipse of the pore selection is shown in Fig. 1 (b).

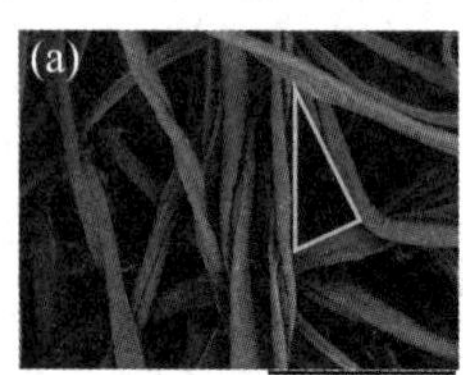

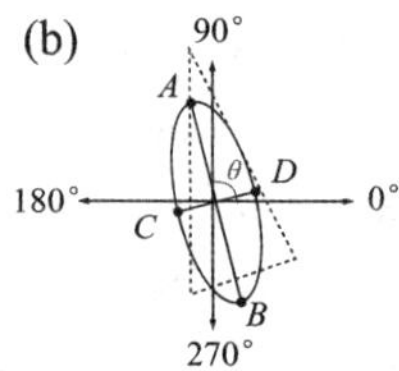

Fig. 1 (a) Actual pore; (b) schematic diagram of equivalent ellipse of pores

(1) In the range of polar coordinate plane 0°–180°, the angle formed between the long axis of the equivalent ellipse and the zero polar axis was defined as pore orientation angle θ, (°). POA reflected the directivity of a single pore in fabric.

(2) The ratio of the long axis length $2a$ to the short axis length $2b$ of equivalent ellipse was defined as porosity ellipticity ρ, as shown in Eq. (1), which was a dimensionless value.

$$\rho = \frac{2a}{2b} \tag{1}$$

where ρ is PE; $2a$ is the long axis length of equivalent ellipse; $2b$ is the short axis length of equivalent ellipse.

POA and PE were used to describe pore shape characteristics in fabric structure from the direction of pore and the degree of pore shape deviation from circle respectively. The second level pores of knitted and woven fabrics existed between fibers in yarn. In order to simplify the problem, it could be assumed that the second level POA was consistent with the yarn axis, and the direction of second level pores could be approximately characterized by the yarn direction.

2.2 Characterization of liquid diffusion

The droplet diffusion test was carried out to observe and analyze the diffusion behavior of liquid on fabric surface. In order to clearly characterize the liquid diffusion morphology on the fabric surface, image binarization and equivalent ellipse fitting were carried out by using ImageJ software. On this basis, a plane polar coordinate system was established with the starting point of liquid diffusion as the pole. The index of liquid diffusion angle δ was proposed: in the polar coordinate plane, the angle formed between the long axis of equivalent ellipse of liquid diffusion contour and the zero polar axis in the range of 0°–180° was defined as liquid diffusion angle δ, in the unit of (°). LDA was used to characterize the direction of liquid diffusion on fabric surface.

Taking nonwoven as an example, Fig. 2 shows schematic diagram of liquid diffusion morphology characterization.

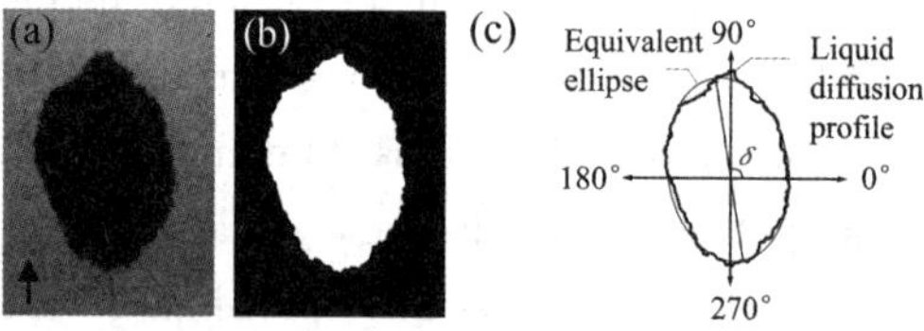

Fig. 2 (a) Actual diffusion; (b) binary image; (c) equivalent ellipse diagram

The ratio of length of the long axis to length of the short axis of equivalent ellipse of liquid diffusion profile was defined as liquid diffusion ellipticity, which was a dimensionless value greater than 1. The closer the value of LDE was to 1, the closer the liquid diffusion profile on the material surface was to circular. LDE is 1.52 as shown in Fig. 2.

3 Experiments

3.1 Materials

A total of 7 knitted, woven and nonwoven samples were

selected. The samples were provided by Nan Liu Enterprises Co. Ltd. Samples A1 and A2 were plain knitted fabric, samples B1 and B2 were plain weave fabric, and samples C1 – C3 were spunlaced nonwovens. All samples were made of cotton.

3.2 Determination of pore morphology

AD206 digital microscope was used to collect surface pore images of knitted and woven samples with magnification of 40 times. Nonwoven pore images were obtained by TM3000 benchtop scanning electron microscope. Six different positions of each sample were taken for image collection. A total of 42 images were obtained from 7 samples, and 100 pore data were extracted from each sample. The images were analyzed by ImageJ software and POA and PE were calculated.

3.3 Determination of liquid diffusion morphology

The samples were humidified at standard conditions (20 ℃ ± 1 ℃, relative humidity 65% ± 2%) for 24 h. The liquid diffusion behavior of quantitative liquid on fabric surface was observed by drip method. Conventional ink (produced in Ningbo Deli Co. Ltd., Zhejiang province) was selected as the titration liquid, and its viscosity value was 1.08 cP. After 20 μL ink was absorbed with a pipette, the fabric anisotropy characterization instrument[10] developed by the research group was used for droplet test.

4 Results and Discussion

According to the characterization and measurement methods described above, the test data obtained are shown in Tab. 1.

Tab. 1 Measurement data of pore morphology and liquid diffusion morphology of samples

Samples	First level POA (°)	First level PE	Second level POA (°)	Second level PE	LDA (°)	LDE
A1	91.87	1.92	61.16(right) 123.99(left)	8.69	96.92	1.26
A2	92.93	1.78	59.01(right) 116.8(left)	9.12	68.27	1.36
B1	85.19	1.56	90(warp) 0(weft)	7.83	91.86	1.17
B2	87.23	4.65	90(warp) 0(weft)	9.64	87.33	1.26
C1	—	—	99.07	2.23	134.56	1.40
C2	—	—	96.34	2.44	96.16	1.71
C3	—	—	94.03	3.03	88.47	1.82

4.1 Influence of knitted pore morphology on liquid diffusion morphology

For knitted fabrics, Origin software was used to draw the distribution map of POA. Taking sample A1 as an example, See Fig. 3.

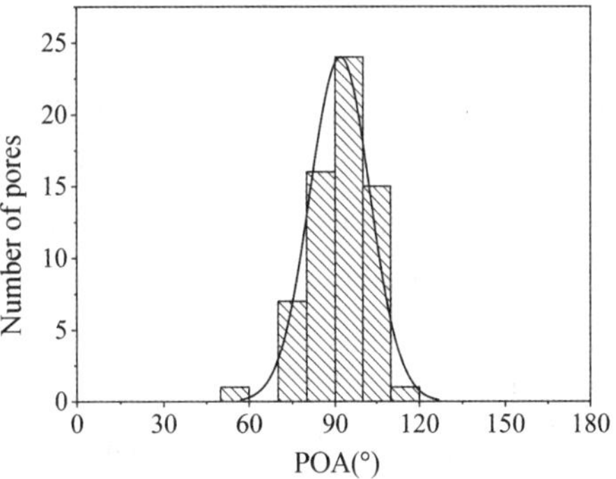

Fig. 3 Distribution map of first level POA of sample A1

According to the above statistics, the first level POA of sample A1 was mostly distributed between 80° and 110°, which accounted for 85.94% of the total. In the range of 80° – 110°, POA concentrated in 90° – 100°. The average of first level POA of both samples was close to the fabric longitudinal (90°). As can be seen from Tab. 1, the average of first level PE of samples A1 and A2 is 1.92 and 1.78 respectively, indicating that the pore form of sample A1 is narrower. Taking sample A1 as an example, the histogram of second level POA of knitted fabric is shown in Fig. 4.

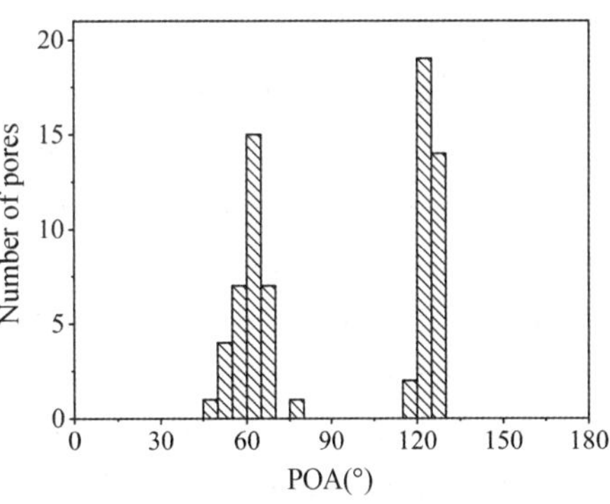

Fig. 4 Distribution map of second level POA of sample A1

As can be seen from Fig. 4, there are two peaks in the distribution of second level POA of the sample, corresponding to the inclination angles of the left and right coil columns respectively. The average of A1 and A2 second level POA were 92.58° and 87.92° respectively. The second level pore was mostly elongated and PE was higher than 5.

To sum up, the difference between the average of first level POA of sample A1 and sample A2 was only 1.06°, but the difference of LDA was 28.65°, indicating that the droplet diffusion direction was not only affected by first level pores. LDA of sample A1 was larger than the average of first level POA, and the average of second level POA was larger than the average of first level POA. LDA of sample A2 was smaller than the average of first level POA, and the average of second level POA was smaller than the average of first level POA. It could be preliminarily speculated that there was a certain correlation between second level pores and liquid diffusion direction.

Comparing LDE with PE, LDE of two samples was less than the average of first level PE, which indicated that LDE was not only related to first level PE. In second level pores, PE was very high, indicating that second level pores were narrow. The equivalent radius of the capillary along this direction was small and wicking was strong, so the capillary force along the left and right directions of the coil column was large. At the same time, due to differences in left and right coil column, the second level pore was an obvious tendency in two directions, and with the first level pore in a different direction, so that liquid transmission capability had multi-direction. The difference of liquid diffusion in all directions decreased, so it could be preliminarily speculated that second level pores had a certain influence on liquid diffusion.

According to the above data analysis, LDA was close to the average of first level POA, and the direction of its deviation from first level pores was consistent with that of second level pores. LDE was less than the average of first level PE, so it was speculated that second level pores had a certain influence on the multidirectional diffusion of liquid, which made LDE smaller.

4.2 Influence of woven pore morphology on liquid diffusion morphology

For woven fabrics, Origin software was used to draw the distribution map of POA. Taking sample B1 as an example, as shown in Fig. 5.

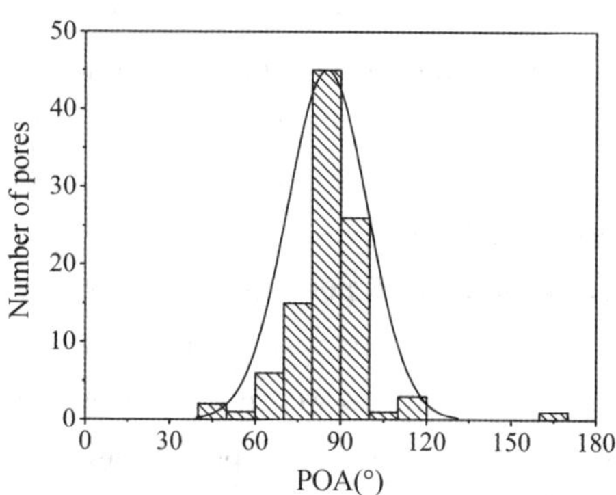

Fig. 5 Distribution map of first level POA of sample B1

It could be seen from Fig. 5 that first level POA of sample B1 was roughly between 85° and 95°. The average of first level POA of samples B1 and B2 were 85.19° and 87.23° respectively. The second level pores existed between fibers, so POA could be assumed to be consistent with the axial direction of yarns, so second level POAs were 90° (warp direction) and 0° (weft direction).

LDAs of samples B1 and B2 were 91.86° and 87.33° respectively, corresponding to the range of most first level POAs. This indicated that in samples B1 and B2, the pores with relatively consistent droplet diffusion direction accounted for a large proportion. It could be seen that droplet diffusion direction had a certain correlation with first level POA. LDAs of samples B1 and B2 were both greater than first level POAs, and second level POAs (warp) were both greater than their first level POAs. It could be analyzed that the direction of LDA deviating from the first level pore was the same as the direction of second level pore deviating from the first order pore, so the direction of liquid diffusion was related to the direction of first and second level pores.

It could be seen from Tab. 1 that LDE of both samples was less than the average of first level PE, that is, the short axis of fitting ellipse of liquid diffusion morphology was larger than that of first level pores, which indicated that transverse diffusion of liquid was not only related to first level pores, but also had a certain influence on second level pores in weft direction. In second level weft pores, PE was large, pores were narrow and wicking was strong. The liquid had a tendency to spread toward weft direction, so the difference of liquid transmission ability in the warp and weft direction of woven fabric was reduced, and LDE was small.

4.3 Influence of nonwoven pore morphology on liquid diffusion morphology

For nonwovens, Origin software was used to draw the distribution map of POA. Taking sample C1 as an example, as shown in Fig. 6.

As can be seen from Fig. 6, the number of pores with POA in the range of 100°–120° was the largest in sample C1. There was a peak value in the distribution of POA in all three samples. It could be seen form Tab. 1 that LDA of samples C1–C3 were consistent with the average POA. The larger the average POA was, the larger the LDA was. The relationship between LDA and average POA indicated that the direction with the strongest capillary force was closely related to the directivity of pore direction, so the average POA could reflect the direction with the strongest capillary force to a certain extent.

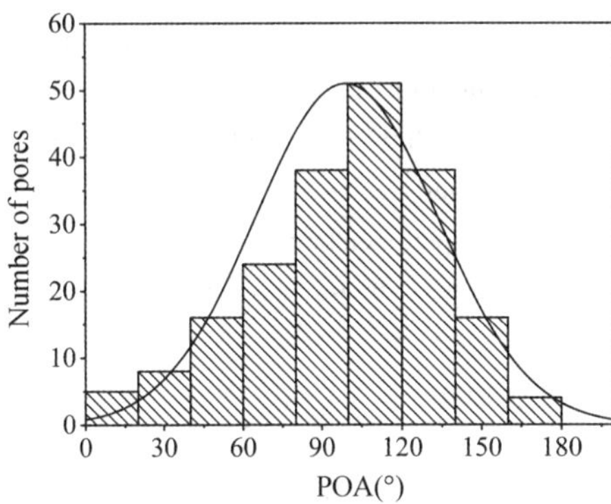

Fig. 6 Distribution histogram of POA of samples C1

Furthermore, PE of samples was calculated. As can be seen from Tab. 1, the average PE of three samples differed greatly from their respective LDE, and were all larger than LDE. Further analysis showed that LDE of samples C1–C3 was consistent with the average PE. The larger the average PE, the larger the LDE. The direction and shape of pores between fibers had a direct influence on the macroscopic liquid diffusion, and the capillary force of liquid on the fabric was closely related to the pore structure.

5 Conclusions

In summary, the indexes of POA, PE, LDA and LDE in fabric were proposed, and the characterization system of pore morphology and liquid diffusion morphology was established. Thus, the pore structure and liquid diffusion characteristics of fabric could be quantified. The diffusion direction and morphology of liquid in knitted and woven fabric samples were affected by both first and second level pores. Both knitted and woven fabrics had obvious first and second level pores. LDA had a strong correlation with first and second level pores. LDE was affected by both the first level PE and the morphology of second level pore. For nonwovens, LDA was close to the average POA. The larger the average POA was, the larger the LDA was. LDE was smaller than the average PE. The larger the average PE was, the larger the LDE was.

References

[1] SINGHAL K, DAS D. Modeling of liquid rise in twisted yarns[J]. The Journal of The Textile Institute, 2021, 112(12): 1896–1903.

[2] MONICA F P, PIERRE J L, SYLVAIN D. Capillary wicking in flax fabrics — effects of swelling in water[J]. Colloids and Surfaces A: Physicochemical and Engineering Aspects, 2016, 498: 176–184.

[3] NOVAKOVIC M S, MILANOVIC J, GRUJIC D, et al. Liquid transfer properties of textile fabrics as a function of moisture content[J]. Hemijska Industrija, 2020, 74(2): 119–132.

[4] VARSHNEY R K, KOTHARI V K, DHAMIJA S. A study on thermophysiological comfort properties of fabrics in relation to constituent fibre fineness and cross-sectional shapes [J]. The Journal of the Textile Institute, 2010, 101(6): 495–505.

[5] CHEN Q, TANG K M, MA P B, et al. Thermophysiological comfort properties of polyester weft-knitted fabrics for sports T-shirt[J]. The Journal of the Textile Institute, 2016, 108(8): 1421–1429.

[6] ZHANG H, ZHANG J C. Influence of cotton texture on wicking[J]. Cotton Textile Technology, 2003(11): 12–15.

[7] WANG L L. Study on the influence of structural parameters on the absorbability of polypropylene yarn and fabric wicking[D]. Tianjin: Tianjin Polytechnic University, 2017.

[8] FAN J, QI Y Y, LI C X, et al. Liquid transport mechanism in yarn based on configuration of non-uniform capillary[J]. Journal of Tianjin Polytechnic University, 2018, 37(5): 30–35.

[9] ZHANG M Q. A study on moisture transport of non-uniform fabric[D]. Shanghai: Donghua University, 2021.

[10] JI F, CHU X Y, ZHOU Q Q, et al. A fabric material anisotropy characterisation method and measurement device [P]. China: 201910417312.5, 2019-05-20.

Fabrication of 3D Gelatin Nanofibrous Scaffolds Using Self-Assemble One-Step Electrospinning

DU Man[1,2], ZHANG Kun[1,2*]

[1] Key Laboratory of Textile Science & Technology, Ministry of Education, Donghua University, Shanghai 201620, China

[2] College of Textiles, Donghua University, Shanghai 201620, China

* Corresponding author's email: kun. zhang@ dhu. edu. cn

Abstract: Recently, natural protein-based fibrous three-dimensional (3D) scaffolds with tunable physical properties have received more and more attention. Gelatin is one of the promising candidate materials due to its temperature sensitivity and excellent biocompatibility. Here, we propose a simple and efficient technique to fabricate fluffy and evenly 3D fibrous gelatin scaffolds through self-assemble electrospinning. This unique nanostructure can be instantly collected in the thickness direction owing to electrostatic repulsion, which is different from the conventional 2D electrospun membranes. In addition, we find that the physical properties of chemically cross-linkable scaffolds, including formability, stiffness, strength, and elasticity, can be further improved by adding glutaraldehyde.

Keywords: 3D scaffolds; Gelatin; Self-assemble electrospinning; Mechanical properties

1 Introduction

To date, a wide variety of naturally derived materials have been proven to be highly in demand in numerous areas such as tissue engineering[1], drug delivery[2], and soft robots[3], because of their intrinsic biocompatibility and excellent mechanical properties, as well as the unique structure, which provides more possibilities for their future development.

Gelatin is an attractive and promising natural protein-based hydrogel directly derived from the skin and shows superior biocompatibility and low Young's modulus, which can contribute to a good bio-mechanical match between material and human tissue[4]. The inexpensive and broad availability of gelatin allows its commercialization in various fields more feasible.

In the last few years, advances in material processing technology have promoted the development of various forms of micro and nanoscale scaffolds, such as freeze-drying[5], electrospinning[6], and 3D printing[7]. Electrospinning is a straightforward, efficient, and cheap way to fabricate nanofibers. However, the crucial restriction with electrospun nanofibrous is their inherently dense fibrous structure and the final fabrications mostly assembling into close-packed films. To overcome this limitation, researchers have come up with various methods for preparing 3D electrospun nanofibrous scaffolds, such as multilayer electrospinning[8], template-assisted electrospinning[9], and post-treated electrospinning[10]. The scaffolds often require complicated collector design, and thus their fabrication requires multiple steps, and special working conditions, and is hard to mass-produce.

In this study, we propose a simple and effective technique to fabricate an ultralight gelatin scaffold with fibers oriented randomly and evenly to form a 3D interconnected pores structure based on the principle of electrostatic repulsion. Based on the mechanism that enhancing the conductivity of the solution could contribute to degrading the surface resistivity of fibers, the 3D fibrous and fluffy scaffolds can be one step generated between the non-assisted collector and the charged jet. This unique nanofiber network, porous structure, and high surface area of gelatin scaffolds enable its great biocompatibility and satisfactory mechanical properties.

2 Experiments

2.1 Materials

The type A gelatin derived from porcine skin and HFIP was obtained from Signa-Aldrich. DPBS, sodium dodecyl sulfate (SDS), and glutaraldehyde was purchased from Thermo Fisher Scientific. All water used was deionized.

2.2 Fabrication of nanofiber scaffolds

The nanofiber scaffolds were fabricated by homemade electrospinning apparatus. 1.0 g gelatin was completely dissolved in 5 mL hexafluoroisopropanol (HFIP) and 5 mL DPBS to prepare the electrospinning solution for 2D gelatin scaffolds. The solution was prepared following a procedure previously reported[4, 11]. Three-dimensional gelatin scaffolds were prepared by electrospinning the above solution with the addition of 1.0 g SDS and 20 μL glutaraldehyde. Then the solution was placed in a 5 mL syringe mounted on the microinjection pump. The voltage was set to 23 kV, a flat collector was set 20 cm from the injector nozzle, and the solution was pumped out at a rate of 2 mL/h.

2.3 Characterization of 2D membranes and 3D scaffolds

Characterization of 2D membranes and 3D scaffolds including surface and inner morphological, average diameter, porosity, and bulk density were performed. The scaffolds were observed using a field emission scanning electron microscope (SU8010, Hitachi Inc, Schaumburg, IL). The distribution of fiber diameters was analyzed using ImageJ software (National Institutes of Health, USA). For the bulk density test, the sample was cut into a regular shape to measure its volume (V), and then accurately weighed (m) to calculate the density of the sample ($\rho = m/V$) ($n = 5$ for each group). For the porosity test, the hydrometer method was applied[4]. Dry samples were placed in a pycnometer filled with alcohol, used a vacuum to make the alcohol completely enter the samples, and then fill it with alcohol to weigh the pycnometer containing alcohol and sample. Then take out samples and weigh the total mass of the remaining alcohol and pycnometer. The porosity of the gelatin nanofiber membranes and scaffolds was calculated according to the following equation:

$$Porosity(\%) = \frac{W_2 - W_1 - W_s}{W_1 - W_3} \quad (1)$$

where W_1 is the weight of a pycnometer filled with alcohol, W_2 is the weight of a vacuum pycnometer with only sample and alcohol, W_3 is the weight of the remaining alcohol and pycnometer without sample, and W_s is the weight of dry samples.

3 Results and Discussion

3.1 Structure Characterization

Fig. 1 (a) shows the self-assembly of fibers at different points. The fibers were instantly deposited on the collector as the high voltage was supplied, and subsequent fibers were stacked layer by layer on the formerly deposited fibers loosely, which resulted in a build-up of fibers in the thickness direction. In the electrospinning process, the liquid droplet that acquired negative charges was elongated into nanofibers vertically gathered together and then lay on collector when the weight at the

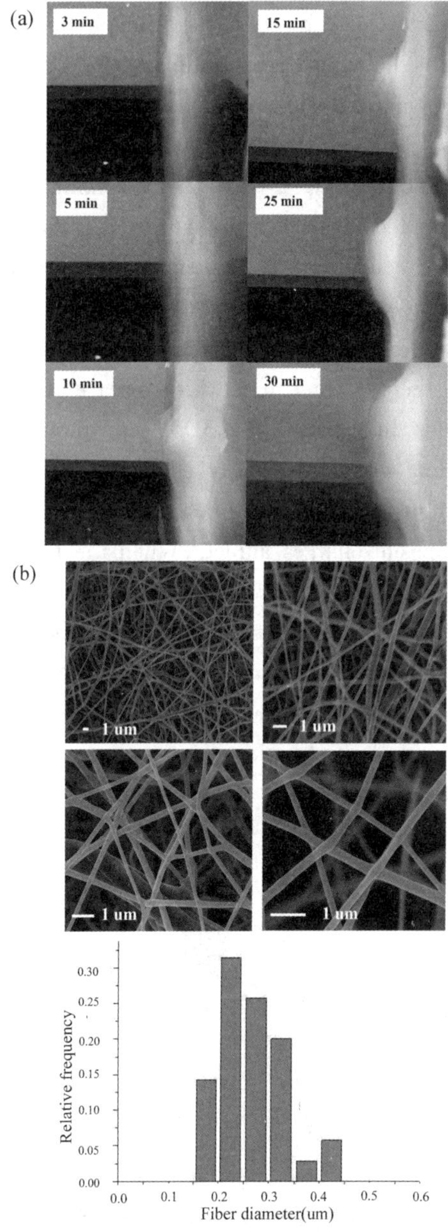

Fig. 1 (a) Electrospinning process of 3D gelatin nanofiber stacks; (b) SEM images and the average diameter of nanofibers in 3D gelatin scaffolds

top layer becomes heavy. Subsequently, the newly formed nanofibers repeated the same process and gradually formed fluffy three-dimensional nanofiber scaffolds. The scaffolds were formed by electrospinning the prepared gelatin solution for 30 min. Fig. 1 (b) shows SEM images of the porous structure of 3D gelatin scaffolds and the average diameter of the fibers which is in a wide range of 0. 15–0. 45 μm. We can see that the fibers in stacks are uniform, similar to those mats fabricated by conventional electrospinning.

3.2 Formation mechanism of the 3D scaffolds

Many investigations have indicated that the different transient electrical forces at the moment fibers hit the collector is the main reason leading to the difference between 2D and 3D electrospinning structures[4, 6, 12-14]. As shown in Fig. 2, during both conventional (Fig. 2 (a)) and 3D (Fig. 2 (b)) electrospinning, the positively charged fibers were first attracted to the negatively charged collector. For conventional electrospinning, the positive charges in fibers were tightly attached to the conductive collector but no electrons were transferred between them, owing to the high surface resistivity of the fibers. In contrast, for 3D electrospinning, sufficiently low surface resistivity can contribute to easier charge transformation. Like charges repel each other while opposite charges attract. When the fibers stroke the collector, the electrons were transferred to fibers rapidly and decreased attraction between fibers and collector. In some cases, the near portions of the fibers could even carry the same charges with the collector, thus forming multiple orientations and resulting in loose and fluffy 3D scaffolds.

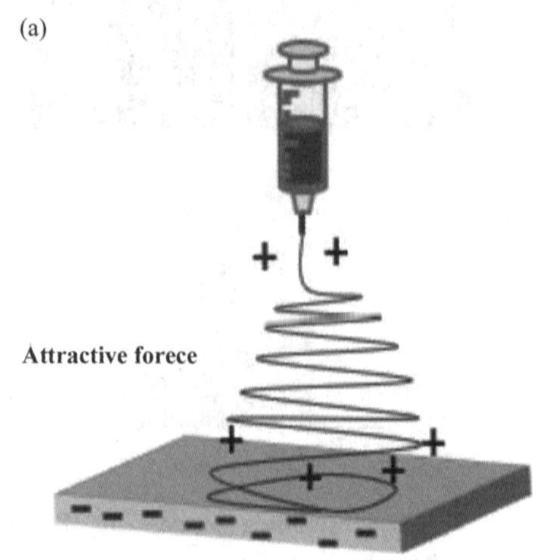

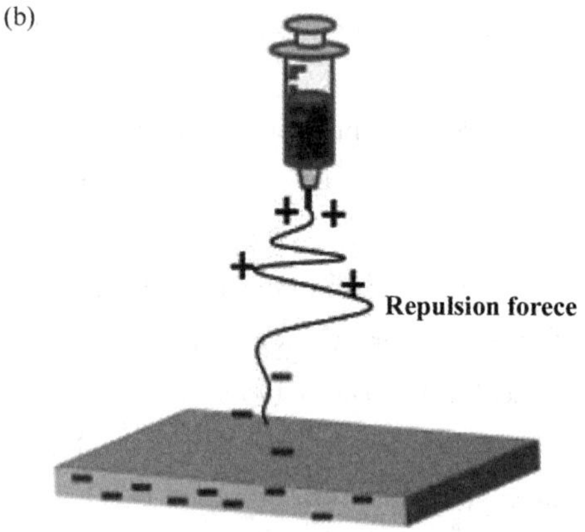

Fig. 2 Schematic diagram of deposition of fibers in 2D (a) and 3D (b) electrospinning processes. The needle was positively charged, and the collector was negatively charged. Fibers near the collector were attracted to form 2D membranes. In contrast, fibers near the collector were repulsed to form 3D scaffolds

In this study, it was found that 3D gelatin scaffolds can be fabricated by adding SDS to the solution and the surface resistivity of polymer decreased with the increasing SDS proportion. The sulfate groups of SDS that concentrates on the surface of fibers oriented toward the outside, and could induce the formation of a surface water layer on the fibers. On the surface of gelatin fibers, free movement of dissociable sodium ions from SDS effectively decreases the surface resistivity of the gelatin electrospun fibers. When using the pure 10% (w/v) gelatin solution, we can only observe a thin mat with a thickness of less than 2 mm appearing on the collector no matter how long it takes to spin. However, the 10% (w/v) gelatin solution with 10% (w/v) SDS electrospun onto the collector could form a thick 3D scaffold immediately.

3.3 Structure characterization

As above mentioned, gelatin is a macromolecular hydrophilic colloid and is easily soluble in water. The uncross-linked gelatin nanofibers would collapse into a 2D membrane after the electric field disappeared. Here, we add 20 μL of glutaraldehyde into the gelatin solution to chemically crosslink the gelatin nanofibers, maintaining the 3D structure without collapse even if the high voltage is removed. The highly fluffy 3D gelatin scaffolds show loose nanofibers with the random arrangement and exhibited porous, interconnected pores, which were totally different from 2D membranes composed of tightly packed nanofibers with micropores. The scaffold had a porosity of 94.5% ± 1.3%, which was significantly greater than that of the membrane's 74.9% ± 2.1% (Fig. 3 (a)). Fig. 3 (b) demonstrates the ultralight nature of the scaffold with a low density of (0.005 2± 0.000 64) g · cm^{-3} compared with the ultralight nature of the scaffold with a low density of (0.005 2 ± 0.000 64) g · cm^{-3} compared with the membrane (0.095 1 g · cm^{-3}±0.007 5 g · cm^{-3}).

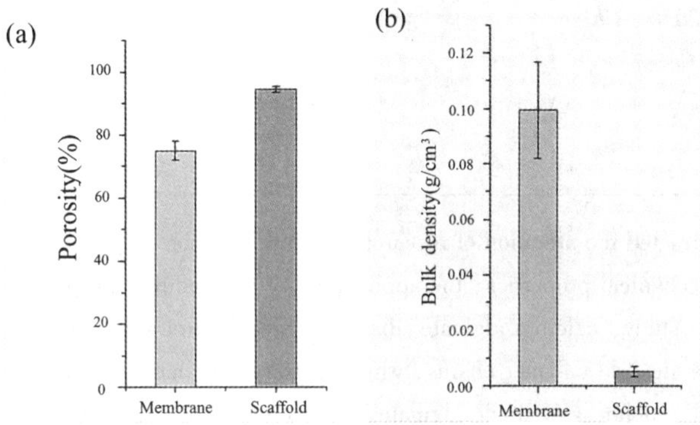

Fig. 3 (a) The porosity of the membranes and the scaffolds; (b) bulk density of the membrane and the scaffolds

4 Conclusions

In summary, we report a quick approach to self-assemble 3D fluffy scaffolds via electrospinning. By adding SDS and controlling the experimental conditions, the loose and randomly oriented gelatin fibers bundles can be self-assembled on aluminum foil within 30 min. A mechanism that increased in surface conductivity of material during electrospinning and could induce the formation of 3D structures was validated in the natural protein-based gelatin. With low bulk densities and high surface areas, gelatin scaffolds could be favorable for various applications.

References

[1] FANG X, XIE J, ZHONG L, et al. Biomimetic gelatin methacrylamide hydrogel scaffolds for bone tissue engineering[J]. Journal of Materials Chemistry B, Materials for Biology and Medicine, 2016, 4(6): 17-18.

[2] CHOY Y B, CHENG F, CHOI H, et al. Monodisperse gelatin microspheres as a drug delivery vehicle: release profile and effect of crosslinking density[J]. Macromolecular Bioscience, 2008, (8): 758-765.

[3] LI Z, ZHANG S, CHEN Y, et al. Gelatin methacryloyl-based tactile sensors for medical wearables[J]. Advanced Functional Materials, 2020, 30(49): 2003601.

[4] XIE X, LI D, CHEN Y, et al. Conjugate electrospun 3D gelatin nanofiber sponge for rapid hemostasis[J]. Advanced Healthcare Materials, 2021, 10(20): 2100918.

[5] CHEN W, CHEN S, MORSI Y, et al. Superabsorbent 3D scaffold based on electrospun nanofibers for cartilage tissue engineering[J]. ACS Applied Materials & Interfaces, 2016, 8(37): 24415-24425.

[6] CAI S, XU H, JIANG Q, et al. Novel 3D electrospun scaffolds with fibers oriented randomly and evenly in three dimensions to closely mimic the unique architectures of extracellular matrices in soft tissues: Fabrication and mechanism study[J]. Langmuir, 2013, 29(7): 2311-2318.

[7] DONG Z, CUI H, ZHANG H, et al. 3D printing of inherently nanoporous polymers via polymerization-induced phase separation[J]. Nature Communications, 2021, 12(1): 1-12.

[8] LEE J, JANG J, OH H, et al. Fabrication of a three-dimensional nanofibrous scaffold with lattice pores using direct-write electrospinning[J]. Materials Letters, 2013, 93: 397-400.

[9] ARRAS M M L, GRASL C, BERGMEISTER H, et al. Electrospinning of aligned fibers with adjustable orientation using auxiliary electrodes[J]. Science and Technology of Advanced Materials, 2012, 13(3): 35008.

[10] FADIL F, AFFANDI N D N, MISNON M I, et al. Review on electrospun nanofiber-applied products[J]. Polymers, 2021, 13(13): 2087.

[11] CHEN C, TANG J, GU Y, et al. Bioinspired hydrogel electrospun fibers for spinal cord regeneration[J]. Advanced Functional Materials, 2019, 29(4): 1806899.

[12] BONINO C A, EFIMENKO K, JEONG S I, et al. Three-dimensional electrospun alginate nanofiber mats via tailored charge repulsions[J]. Small, 2012, 8(12): 1928-1936.

[13] MI H, JING X, HUANG H, et al. Instantaneous self-assembly of three-dimensional silica fibers in electrospinning: Insights into fiber deposition behavior[J]. Materials Letters, 2017, 204: 45-48.

[14] SUN B, LONG Y, YU F, et al. Self-assembly of a three-dimensional fibrous polymer sponge by electrospinning[J]. Nanoscale, 2012, 4(6): 2134-2137.

Strong and Tough Alginate Helical Fibers via Mechanical Twisting

HAN Dongxu[1], DENG Linfeng[1], SHEN Hua[1], XU Fujun[1], NI Jiaqian[2], WANG Jilong[3*]

[1] *Key Laboratory of Textile Science & Technology, Ministry of Education, College of Textiles, Donghua University, Shanghai 201620, China*

[2] *Zhejiang Zhongda Testing Technology Service Co., Ltd., Shaoxing 312030, China*

[3] *College Textile and Garment, Shaoxing University, Shaoxing 312000, China*

* *Corresponding author's email*: 2022000018@ usx. edu. cn

Abstract: As a green natural polymer material, alginate has long attracted the attention of researchers, and it can be prepared into films, microspheres, fibers, etc. However, due to the poor mechanical properties, the applications of the traditional wet spinning alginate fibers have been limited in various fields. In this study, calcium alginate fibers prepared by wet spinning were twisted to produce helical structure and stretched to orient the alginate polymer chains, which improved the mechanical properties of alginate fibers. After twisting, the tensile strength and toughness of single alginate fiber increased by 112.83% and 155.38%, respectively. The tensile strength and toughness of twisted bundle fibers increased by 185.04% and 215.35%, respectively, compared with those of the untwisted bundle fibers. The dried fibers were then immersed in different metal ion solutions (Ca^{2+}, Zn^{2+}, Ba^{2+}, and Fe^{3+}). The results show that the ferric alginate fibers obtain the maximum tensile strength, and the calcium alginate fibers obtain the maximum toughness.

Keywords: Sodium alginate; Wet spinning; Helical structure; Mechanical properties

1 Introduction

Due to the non-renewability and environmental pollution of traditional resources, people began to devote themselves to the research of renewable and green resources. Some natural polymer materials (cotton, hemp, etc.) and biological macromolecular materials (alginate, cellulose, etc.) are biodegradable and eco-friendly, through the research and use of these materials, today's resource and environmental problems can be effectively alleviated[1]. Alginate is a random block polymer extracted from brown algae in the ocean, because of its good biocompatibility, biodegradability, high adsorption, gas permeability, and high flame retardancy, it is widely used in biomedicine, textile, food and wastewater treatment[2]. Alginate can prepare membranes, microspheres, fibers. At present, there are three spinning methods for preparing alginate fiber, namely microfluidic spinning[3], wet spinning[4] and electrospinning[5]. The spinning method commonly used by researchers is wet spinning, and the coagulation baths are mainly low-cost calcium chloride solutions. The G unit in the alginate macromolecule forms an "egg box" structure with calcium ions, so can obtain the solid calcium alginate fiber with stable structure[6].

However, the alginate fiber obtained by wet spinning are hard, brittle, and low in strength. Hence, how to improving the mechanical properties of alginate fiber has attracted many researchers. At present, the most commonly used method to improve the mechanical properties of sodium alginate is to blend it with other substances, such as chitosan[7], and graphene[8]. By adjusting the mixing ratio, fibers with better mechanical properties can be prepared.

Although alginate fiber can easily be prepared by wet spinning, its application is greatly limited due to the low strength. At present, most methods for improving the strength of sodium alginate fiber are blending with other substances. Although its strength has been increased, but its biocompatibility has been affected too. Therefore, in this study, we introduce a helical structure by twisting alginate fibers, the alginate polymer chains would be stretched and aligned, so that the originally loose and disordered molecular network is oriented along the helical direction, and the mechanical properties of alginate fibers are improved. Then the fiber was soaked in metal ions solutions to explore the effect of metal ions on the properties of the fiber.

2 Experiments

2.1 Materials

Sodium alginate (SA), medium viscosity, was obtained from Sigma-aldrich. Calcium chloride AR, 98% was obtained from General-reagent. Zinc chloride AR, 98% was obtained from Ron Reagent. Barium chloride dihydrate, analytically pure AR was obtained from Sinopharm Chemical Reagent Co., Ltd. Ferric chloride chemically pure, ≥97% was obtained from Sinopharm Chemical Reagent Co., Ltd.

2.2 Wet spinning calcium alginate fiber

Calcium alginate fibers were prepared by wet spinning. At first, 2.5% sodium alginate solution was prepared as the spinning solution, and a 4 wt.% calcium chloride solution was used as a coagulation bath. The prepared spinning solution was sucked into a syringe, and then it was fixed on the extrusion pump, and the spinning solution was extruded from the syringe at a certain spinning speed into a rectangular tank with coagulation bath. The fiber can be immersed in the coagulation bath for about one minute, and finally the fibers are collected by a winding device. Afterwards, fibers were placed at room temperature to dry naturally for 2 days to prepare for subsequent fiber twisting.

2.3 Preparation of helical alginate hydrogel fibers by twisting

Cut the previously dried fibers into short fibers of about 30 cm, and the diameter of the single fiber is about 150 μm, then the single fiber and bundle fibers (composed of 5 untwisted single fibers) are twisted on the yarn twister($Y331N^+$). The single fiber and bundle fibers are twisted in different twists (single fiber: 20, 40, 60 twists; bundle fiber: 40, 80, 120 twists), and then fix both ends to obtain dry helical calcium alginate single fiber and bundle fibers with stable twisting.

2.4 Treatment of different ions on fibers

The twisted bundle fibers were soaked in calcium chloride ($CaCl_2$), zinc chloride ($ZnCl_2$), barium chloride ($BaCl_2$) and ferric chloride ($FeCl_3$) solutions, respectively, the ion concentration in each solution remains the same. Finally, calcium alginate fiber (CA-Ca fiber), zinc alginate fiber (CA-Zn fiber), barium alginate fiber (CA-Ba fiber) and ferric alginate fiber (CA-Fe fiber) were obtained.

2.5 Mechanical testing of the fibers

Mechanical property of the fibers was measured through tensile test by using a single fiber strength tester (LLY-05E-500). The tensile tests were done according to GB/T 14344—2008 Standard Test Method. Test conditions: the pre-tension of single fiber is 0.1 cN, the pre-tension of bundle fibers is 0.5 cN, the test length is 10 mm, the stretching speed is 5 mm/min, and the return speed is 100 mm/min.

3 Results and Discussion

3.1 Effect of twisting on single fibers properties

Stress-strain curves of single fiber with different twists are given in Fig. 1 (a). Apparently, the strength of twisted single fibers is significantly higher than that of untwisted fibers. After twisting, the average tensile strength of the fiber increased by 112.83%. And the Fig. 1 (b) shows the toughness of single fiber with different twists, the average toughness of the twisted fibers increased by 155.38%. When twisting single fibers, the fibers are stretched to a certain extent in the length direction. At the same time, as the degree of twisting increases, internal stress is generated inside the fiber,

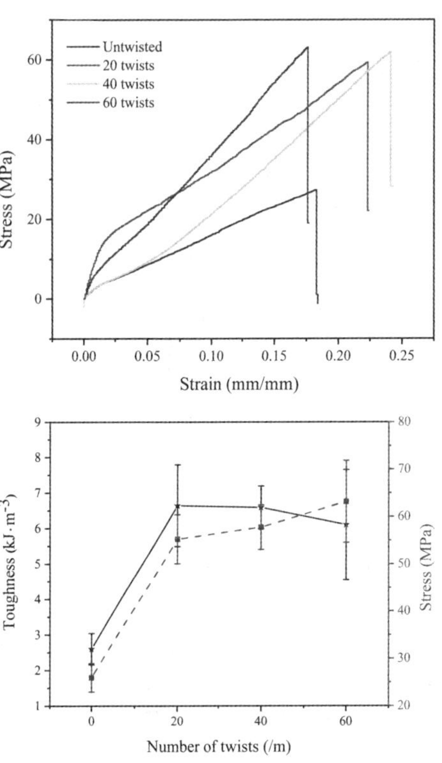

Fig. 1 (a) Stress-strain curves of single fiber with different twists; (b) toughness and maximum stress of single fiber with different twists

resulting in an increase in the rigidity of the molecular chain. Therefore, twisting can effectively improve the mechanical properties of single fibers.

3.2 Effect of twisting on bundle fibers properties

The stress-strain curves of the bundle fiber are shown in Fig. 2 (a). As the same of single fibers, the strength of bundle fibers has also been significantly improved after twisting. The average tensile strength of the twisted bundle fibers increased by 185.04% than the untwist fibers. And the Fig. 2 (b) shows the toughness of bundle fibers with different twists, the average toughness of the twisted fibers increased by 215.35%. Compared with untwisted bundle fibers, twisting can form a good twist between single fibers and stabilize the shape, thereby improving the strength of the twisted bundle fibers. In addition, like single fiber twisting, bundle fiber twisting will also stretch the fiber in the length direction. At the same time, the helical structure generated by twisting produces internal stress inside the fiber, causing the internal polymer chain to generate orientation, thereby enhancing its strength.

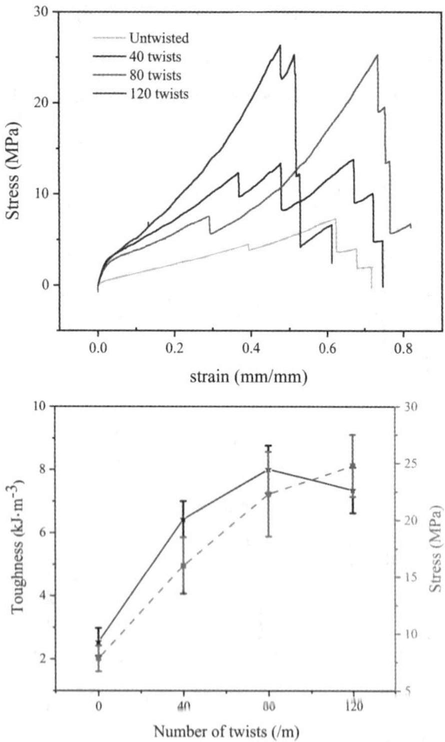

Fig. 2 (a) Stress-strain curves of bundle fiber with different twists; (b) toughness and maximum stress of bundle fiber with different twists

3.3 Effects of metal ions on the properties of bundle fibers

Stress-strain curves of bundle fibers (80 twists) soaked with different ions are given in Fig. 3 (a). It can be seen from the figure, tensile strength of CA-Fe fiber is the largest of the four fibers, elongation of the CA-Ca fiber is the largest of the four fibers. And from the Fig. 3 (b), we can see the toughness of the CA-Ca fiber is the largest, and the CA-Fe fiber is the smallest. This is because Fe^{3+} can interact with the three carboxyl groups of different alginate chains at the same time, so it will form a larger coordination number, the molecular formula is $(COO)_3M$, and build a three-dimensional valence bond structure. Then a more compact network is formed, and the ionic bonds are tightly bound, so it is not easy to break during stretching, and the tensile strength is large. However, because Fe^{3+} combined with more alginate macromolecules during the soaking process, the overall rigidity increases, so elongation of the CA-Fe fiber is low[9].

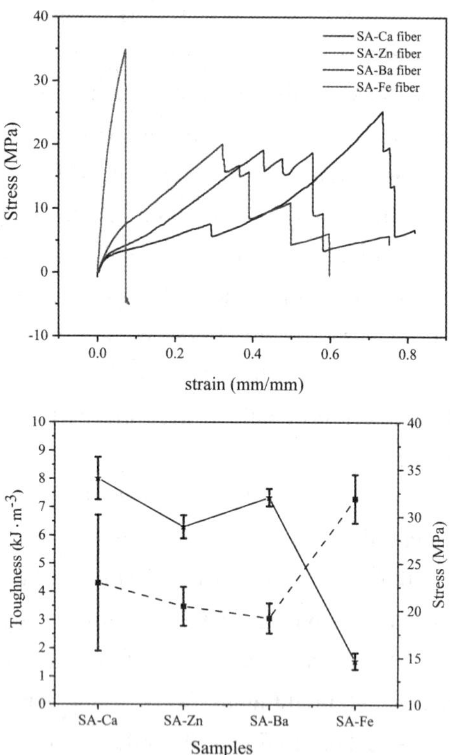

Fig. 3 (a) Stress-strain curves of bundle fibers (80 twists) soaked with different ions; (b) toughness and maximum stress of bundle fibers (80 twists) soaked with different ions

4 Conclusions

In this study, alginate fibers with helical structure were prepared by twisting after wet spinning. The effects of different twisting degrees on the properties of single fibers and bundled fibers were investigated. After twisting to form a helical structure, the strength and toughness of single fibers and bundled fibers were improved. After twisting, the strength and toughness of single fibers increased by 112.83% and 155.38%, respectively. And the strength and toughness of bundle fibers increased by 185.04% and 215.35%, respectively. Then the twisted bundle fibers are soaked in different metal ion solutions. The helical calcium alginate fiber obtained by soaking in Ca^{2+} has the maximum toughness, which is (8.01±0.75) $kJ \cdot m^{-3}$; The helical iron alginate fiber soaked in Fe^{3+} has the largest tensile breaking strength, which is (31.89±2.57) MPa. The results show that the mechanical properties of wet-spun alginate fibers can be effectively improved by twisting and soaking in metal ion solutions.

References

[1] ZHANG X, XIA Y, YAN X, et al. Efficient suppression of flammability in flame retardant viscose fiber through incorporating with alginate fiber[J]. Materials Letters, 2018, 215: 106-109.

[2] DAVARCI F, TURAN D, OZCELIK B, et al. The influence of solution viscosities and surface tension on calcium-alginate microbead formation using dripping technique[J]. Food Hydrocolloids, 2017, 62: 119-127.

[3] LIN S, LIU J, LIU X, et al. Muscle-like fatigue-resistant hydrogels by mechanical training[J]. Proceedings of the National Academy of Sciences, 2019, 116(21): 10244-10249.

[4] YANG Y, SUN J, LIU X, et al. Wet-spinning fabrication of shear-patterned alginate hydrogel microfibers and the guidance of cell alignment[J]. Regenerative Biomaterials, 2017, 4(5): 299-307.

[5] KIM Y J, YOON K J, KO S W. Preparation and properties of alginate superabsorbent filament fibers crosslinked with glutaraldehyde[J]. Journal of Applied Polymer Science, 2000, 78(10): 1797-1804.

[6] BRACCINI I, PÉREZ S. Molecular basis of Ca^{2+}-induced gelation in alginates and pectins: the egg-box model revisited[J]. Biomacromolecules, 2001, 2(4): 1089-1096.

[7] FAN L, DU Y, ZHANG B, et al. Preparation and properties of alginate/carboxymethyl chitosan blend fibers[J]. Carbohydrate Polymers, 2006, 65(4): 447-452.

[8] AN Y, GAO L, WANG T. Graphene oxide/alginate hydrogel fibers with hierarchically arranged helical structures for soft actuator application[J]. ACS Applied Nano Materials, 2020, 3(6): 5079-5087.

[9] HOU L, WU P. Exploring the hydrogen-bond structures in sodium alginate through two-dimensional correlation infrared spectroscopy[J]. Carbohydrate Polymers, 2019, 205: 420-426.

Superhydrophobic Nona-ZnO Crystal Coating for Steel Mesh

ZHAO Xin[1], LIU Yang[2], ZHOU Wenjun[3], WANG Yiru[1*]

[1] *School of mechanical engineering, Chengdu University, Chengdu 610100, China*

[2] *Officers College of PAP, Chengdu 610213, China*

[3] *School of Chemistry and Chemical Engineering, Neijiang Normal University, Neijiang 641100, China*

[*] *Corresponding author's email*: wangyiru@ cdu. edu. cn

Abstract: In order to avoid the harm of water to metal textiles, the hydrophobic modification of metal textiles has attracted more and more attention. Hydrophobic modification mostly depends on complicated method or expensive materials, which hinders the wide application of metal textiles. In this paper, a method of preparing superhydrophobic coating on steel textile is proposed, and related properties of modified steel mesh are characterized. Firstly, the surface of dry steel textile was covered with nano-ZnO crystal by a liquid deposition method at room temperature. Then the ZnO crystal was hydrophobically modified by stearic acid, and the ZnO superhydrophobic coating was obtained. The effects of ZnO concentration and deposition time on the surface crystal structure, surface hydrophobicity, and abrasion resistance were explored. The surface groups before and after hydrophobic modification were analyzed by XRD diffraction, and the hydrophobic properties of the material surface were analyzed through static contact angles. The result showed that after grinding with 1200 mesh sandpaper, superhydrophobic steel mesh with typical ZnO concentration was only reduced by 4%. This simple, cheap, effective and excellent preparation method is expected to play a positive role in related fields.

Keywords: Superhydrophobic; ZnO; Steel mesh; Nano crystal

1 Introduction

In some cases, the existence of water will cause metal corrosion, wire icing, bacterial breeding and other hazards, causing significant economic and property losses[1-2]. Therefore, how to avoid the harm of water in related fields is related to all aspects of human life[3-4]. Superhydrophobic materials have been developed by researchers. It also has very important research value in material science and interface structure design.

At present, the preparation of superhydrophobic materials mostly depends on expensive materials and equipment[5]. How to produce materials with more suitable price and better performance while playing the role of hydrophobicity has become one of the basic trends of current research[6-8]. In this paper, ZnO crystal structure was formed by the reaction of zinc salt and potassium hydroxide solution at room temperature, which was deposited on the steel mesh to form a dense film with a certain rough structure, and then a layer of ZnO superhydrophobic coating was obtained by stearic acid ethanol modification. This is a simple, cheap, effective and excellent method for the preparation of superhydrophobic surfaces.

2 Experiments

2.1 Materials

Steel fabrics with 800 mesh and 0. 05 mm thick were obtained from Qinxiang industrial factory in Changzhou, China. Potassium hydroxide solid, stearic acid and absolute ethanol are provided by Titan Technology Co., Ltd. (Guangdong, China).

2.2 Steel mesh treatment

Rinse the steel mesh with deionized water, put it into absolute ethanol after drying, wash it with ultrasonic for 40 min, and dry it for standby after washing.

2.3 Preparation of ZnO superhydrophobic coating

Firstly, the steel wire mesh used in the experiment was cleaned with clean water, put it into absolute ethanol, clean it with ultrasonic for 40 min, and then dry it for standby. Secondly, a certain mass of $ZnCl_2$ and KOH solids are weighed, pour them into a certain aqueous solution, $ZnCl_2$ solution and KOH solution with corresponding concentration are grouped and numbered in pairs. Slowly drop the same group of $ZnCl_2$ solution into

KOH solution and stir it fully to obtain a colorless and transparent mixed solution. The dried spare steel wire mesh is put into different components of zinc oxide aqueous solution and left standing at room temperature. Then, take out the steel mesh and dry it at 60 ℃. Superhydrophobic surfaces require low surface energy and high surface roughness. Finally, the dried steel mesh was soaked in a stearic acid ethanol solution with a concentration of 0. 1 mol/L, and then left for 1 h. After washing and drying with absolute ethanol, ZnO coating modified with stearic acid ethanol was obtained.

Tab. 1 Sample Composition

Sample	$ZnCl_2$ (g/50 mL)	KOH (g/50 mL)	Standing time (h)
1	1. 36	4. 48	12
2	0. 68	2. 24	12
3	1. 36	4. 48	4

2. 4 Characterization of zinc oxide superhydrophobic coating

The zinc oxide hydrophobic coating was characterized, including Contact angle measurement, chemical durability test, hydrophobic coating wear resistance test, oil-water separation test and environmental durability test. The surface wettability test is carried out by measuring the contact angle between water droplets and each sample and observing the penetration of water droplets on the surface. Put the experimental sample into the configured 0. 1 mol/L NaCl solution, and observe the change of contact angle after standing for 12 h to test its chemical durability. Place 1200 mesh sandpaper on the surface of the hydrophobic coating, place 500 g weight on the sample, and slowly pull the sandpaper back and forth to test the wear resistance of the hydrophobic coating. Put the experimental sample into deionized water with oil floating on the surface, observe whether it absorbs the oil on the surface and the integrity rate of oil absorption, and test its oil-water separation characteristics. The crystal structure of the sample surface was analyzed by X-ray diffractometer. The basic parameters are as follows: tube voltage 40 kV, tube current 30 mA, step angle 0. 06°, scanning range 5°- 80°, and sampling time 0. 5 s.

3 Results and Discussion

3. 1 Surface crystal structure

Through the XRD patterns of the samples (Fig. 1), it can be observed that their diffraction peaks are very sharp, indicating that they have good crystallinity. At the same time, in order to prove that stearic acid ethanol successfully modified the surface of zinc oxide, the corresponding angle should be found in the diffraction peak. After research, when 2θ is equal to 33. 40°, 34. 95°, 36. 79°, 46. 78°, 58. 40°, 63. 38°, and 69. 33°, it corresponds to ZnO crystal, indicating that the preparation method successfully generates ZnO. 2θ at 20. 03°, 20. 95°, and 27. 47° corresponds to stearic acid, which provides the successful modification of stearic acid to zinc oxide surface.

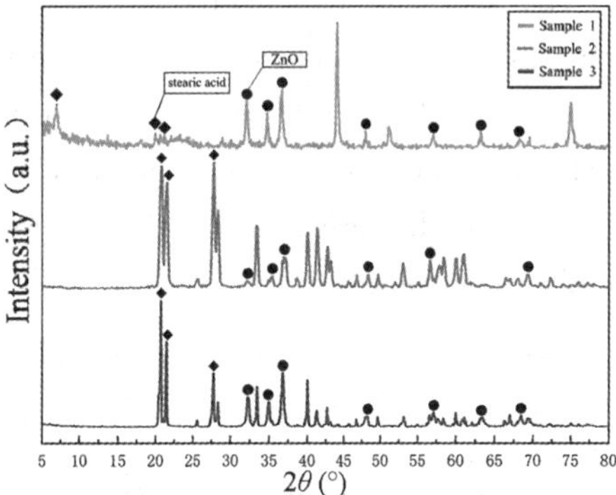

Fig. 1 XRD pattern of samples

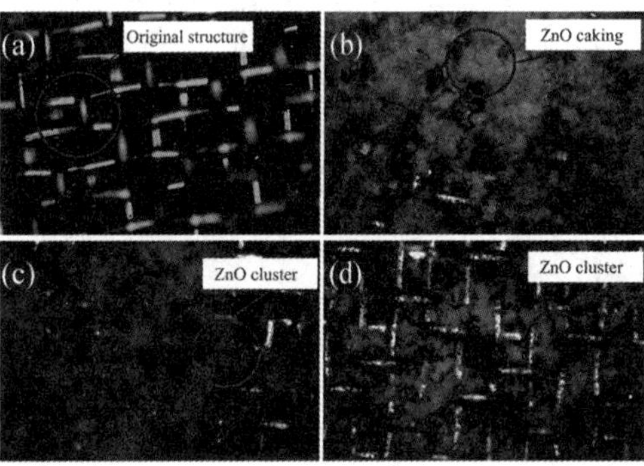

Fig. 2 MO image: (a) original mesh surface; (b) sample 1 surface; (c) sample 2 surface; (d) sample 3 surface

Based on the above data, ZnO crystals were successfully prepared from three groups of samples (Fig. 2), and the surface was successfully modified with stearic acid ethanol. However, the surface of sample 3 is not completely uniform, and the degree of crystal growth is weak because of its short deposition time. The concentration of zinc oxide on the surface is relatively low, so when stearic acid ethanol is used for modification, its concentration is also low. After measuring the contact angle, the pattern obtained by XRD diffraction reflects the residual water molecules in some areas that have not been successfully deposited to a certain extent.

3.2 Abrasion resistance test

After the wear resistance test of the three groups of samples, the contact angles were 101°, 118° and 146° respectively (Fig. 3), and the hydrophobicity was affected to varying degrees. However, compared with samples 2 and 3, the contact angle of sample 1 is as high as 50°. According to the surface morphology after wear, it can be concluded that although the surface roughness of sample 1 is high, its wear resistance test performance is greatly reduced, a large number of zinc oxide surfaces fall off, a large number of crystal structures fall off, and the steel matrix is clearly visible. There are still a large number of ZnO clusters on the surface of sample 2, the change is not obvious, and there is still a certain surface roughness. There is no obvious change on the surface of sample 3, and the surface is still hydrophobic.

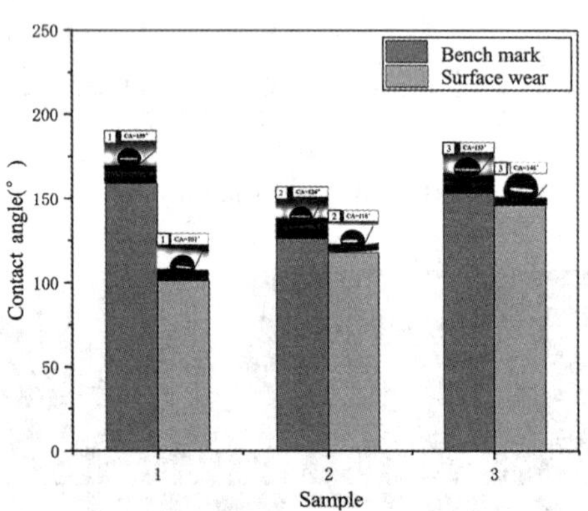

Fig. 3 Contact angle of sample after abrasion resistance test

4 Conclusions

In this paper, nano-ZnO crystal superhydrophobic coatings were prepared by liquid deposition method. The results show that the concentration of electrolyte and deposition time have important effects on the surface wettability of zinc oxide superhydrophobic coating. The higher the concentration of electrolyte, the higher the hydrophobicity of the sample surface, and the contact angle can reach 159°. But the higher the surface roughness greatly reduces its wear resistance test performance; The longer the deposition time of ZnO, the better its crystallinity and the more uniform the hydrophobic region of the sample. ZnO superhydrophobic coating has good chemical durability, environmental durability and oil-water separation, and is expected to play an active role in related fields.

References

[1] SHEN D, REN X G, MING W Y, et al. Progress in non-traditional processing for fabricating superhydrophobic Surfaces[J]. Micromachines, 2021, 12(9): 1003.

[2] REN J P, LU X F, TAO F R, et al. Biomass-based superhydrophobic coating with tunable colors and excellent robustness [J]. Carbohydrate Polymers, 2021, 270: 118401.

[3] SARAN R, ZHAI L, DAVID F, et al. Organic non-wettable superhydrophobic fullerite films[J]. Advanced Materials, 2021, 33(32): e2102108.

[4] SAJI V. Electrophoretic-deposited superhydrophobic coatings[J]. Chemistry, An Asian Journal, 2021, 16(5): 474-491.

[5] ZHENG L Z, HUANG W, LI H Q, et al. Light stimuli-responsive superhydrophobic films for electric switches and water-Droplet manipulation [J]. ACS Applied Polymer Materials, 2021, 13(30): 36621-36631.

[6] PATHREEKER S, CHEN F H, CHANDO P, et al. Superhydrophobic polymer composite surfaces developed via photopolymerization[J]. ACS Applied Polymer Materials, 2021, 3(9): 4661-4672.

[7] MEHANNA Y A, UPTON R L, SANDLE E, et al. The challenges, achievements and applications of submersible superhydrophobic materials [J]. Chemical Society Reviews, 2021, 50(11): 6569-661222.

[8] YU Z, YANG J L, ZHANG Z T, et al. A review of recent advances in superhydrophobic surfaces and their applications in drag reduction and heat transfer[J]. Nanomaterials, 2021, 12(1): 44.

Influencing Factors and Model Construction of Running Clothing Wearing Pressure and Comfort

GUO Ning[1,2], LI Zhijiang[1,2], ZHANG Peihua[1,2*], CHEN Nanliang[1,2]

[1] *Key Laboratory of Textile Science & Technology, Ministry of Education, Donghua University, Shanghai 201620, China*

[2] *Department of Technical Textiles, College of Textiles, Donghua University, Shanghai 201620, China*

* *Corresponding author's email*: phzh@ dhu. edu. cn

Abstract: In order to explore the relationship between the factors affecting the pressure and comfort of running clothes and the wearing pressure, the KES and the three-dimensional pressure test simulation system for fabrics developed by the research group were used. The same weave structures for 7 kinds of weft knitting. The elastic knitted fabrics with different interweaving ratios of raw materials were tested for mechanical properties such as tensile, shearing, and compression, as well as dress pressure tests. Based on the multiple linear regression analysis method, a prediction model of human dress pressure was established. The research results showed that the warp tensile work (*JWT*), the indexes of warp tensile recovery (*JRT*), weft tensile recovery (*WRT*), weft tensile linearity (*WLT*) and surface roughness (*SMD*) were significantly related to dressing pressure, among which *JWT* and *SMD* were significantly related to dressing pressure. Dressing pressure was positively correlated, and JRT, WRT and WLT were negatively correlated with dressing pressure. Through the linear regression analysis, it can be seen that there is a linear correlation between the mechanical properties of knitted fabrics and the dressing pressure, and the prediction model formula of weft knitted fabrics is $Y = 47.023 + 0.349 \times JWT + 0.066 \times JRT - 0.646 \times WRT - 20.838 \times WLT + 0.311 \times RC + 8.755 \times SMD$.

Keywords: Running clothing; Knitted fabric; Mechanical properties; Pressure comfort; Predictive model

1 Introduction

As a common physical property, pressure exists between tight-fitting clothing and the human body. The human body is an elastic body, and its ability to withstand the mechanical stimulation of clothing pressure is limited. Under ideal conditions, when the clothing moves 0. 001 mm, the human body will Pressure will be generated[1]. Appropriate clothing pressure has a positive effect on human body shaping, sports protection, and wound healing and recovery of burn patients. At the same time, it can also promote blood circulation, relieve and treat venous diseases, and improve motor function[2]. It has a certain influence on the psychological and physical comfort of the wearer, and is related to human health[3]. It can be seen that the research on the pressure comfort of clothing is very necessary.

In recent years, garment pressure is often used as an important indicator for evaluating the comfort and functionality of tight-fitting garments. Horiba et al. [4] obtained the 3D shape of the clothing by combining the CAD program with computer-aided design, and then used it as the initial shape to perform finite element analysis with the stress and strain generated when the clothing was worn as the load conditions, and calculated the contact pressure between the person and the clothing. Brubacher et al. [5] used a virtual fit technique to predict garment pressure, and compared the actual measured pressure with the predicted virtual pressure to evaluate the accuracy and practicality of the prediction. Although these prediction methods can predict the distribution law of clothing pressure, the process is cumbersome and the calculation is complicated, and the analysis of multi-factor and multi-indicators in these studies is usually independent, ignoring the comprehensiveness and balance between multi-indicators.

This paper selects two kinds of knitted fabrics commonly used in elastic knitted garments to study the mechanical properties of N/OP, T/OP interwoven weft knitted fabrics and the test and characterization of simulated dressing pressure, and find out through correlation analysis. The relationship between the mechanical properties of weft knitted fabrics and wearing pressure, and based on the multiple linear regression analysis method, a prediction model of human wearing pressure is established, so as to provide a certain reference for the development of tight-fitting knitted clothing products.

2 Experiments

2.1 Materials

Fabric information are shown in Tab. 1. In the table, N is nylon, OP is spandex, and T is polyester.

Tab. 1 Fabric information used in the test

Sample number	Organizational structure	Fabric components (%)	Thickness(mm)	Area density (g/m^2)	Transverse density (coil/(5 cm))	Longitudinal density (coil/(5 cm))
1#	Interlock	N 86/OP 14	0.74	170	63	102
2#		N 80/OP 20	0.69	280	47	86
3#		N 76/OP 24	0.70	250	56	83
4#		N 75/OP 25	0.58	210	65	80
5#		N 74/OP 26	0.58	200	71	106
6#		N 72/OP 28	0.48	180	71	103
7#		N 65/OP 35	0.47	140	63	120

2.2 Testing and characterization

The KES-FB1 tensile and shear tester was used to test the tensile and shear properties of fabrics. and the sample size is 200 mm×200 mm, take the average value of three test results. The KES-FB2 bending tester was used to test the bending properties of fabrics. and the sample size is 100 mm×100 mm, take the average value of three test results. The KES-FB3 compression tester was used to test the compressibility of fabrics. and the sample size is 200 mm×200 mm, take the average value of the three test results. The KES-FB4 surface friction tester was used to test the surface friction properties of fabrics. and the sample size is 200 mm × 200 mm, take the average of three test results. Refers to the standard of FZ/T 70006—2004 "Test Method for Tensile Elasticity Recovery Rate of Knitted Fabrics", and the LLY - 56B knitted fabric three-dimensional pressure simulation test system developed by the research group is used to simulate the dressing pressure of the human body in the process of wearing clothing[6]. The diameter of the sample is 35 mm.

3 Results and Analysis

3.1 Analysis of mechanical properties of knitted fabrics

As shown in Figs. 1 (a)-1(d), with the increase of the spandex ratio, the overall WT is larger when the warp is stretched, and the resistance to tensile deformation is strong. Weak, of which 1# and 7# have relatively strong resistance to deformation. When stretched in weft direction, WT and RT decreased obviously, and the resistance to deformation of the fabric gradually increased and the resilience decreased. The difference between the 1# and 7# longitude and weft directions is not large, and the performance is stable.

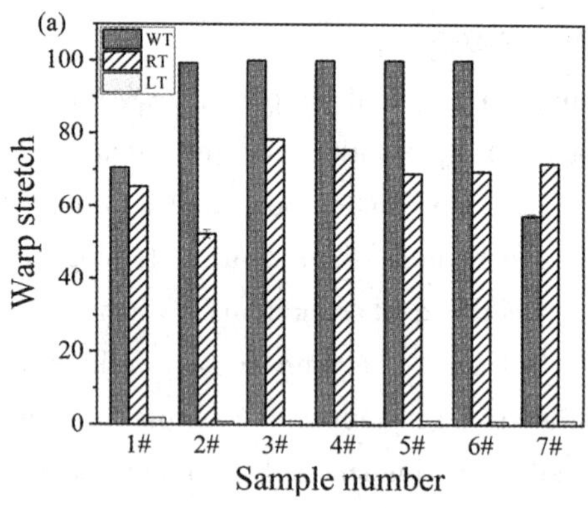

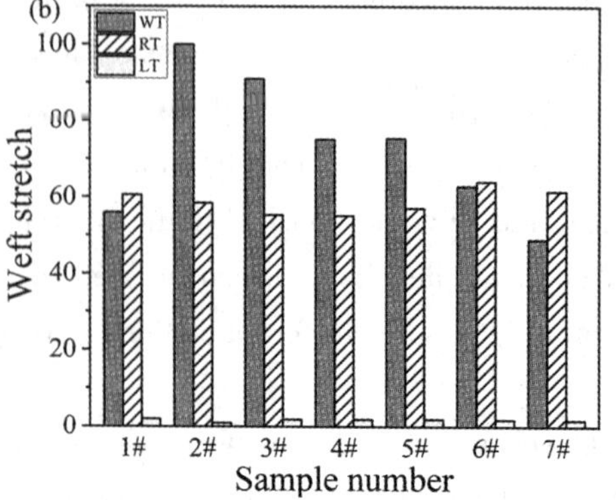

Fig. 1 (a) Warp stretch; (b) weft stretch

As shown in Figs. 2 (a)-2 (b), with the increase of the spandex ratio, the G, 2HG and 2HG5 indexes are significantly different when sheared in the warp direction. 1#, 3# and 5# are the largest, and have the worst resistance to shear deformation and recovery after shearing. During latitudinal shearing, 2HG and 2HG5 gradually decreased, and the recovery ability increased after shearing, but the overall difference trend was basically similar to that during warp direction shearing.

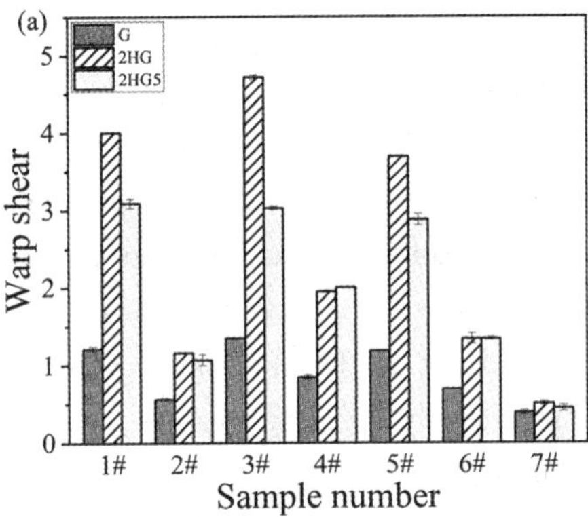

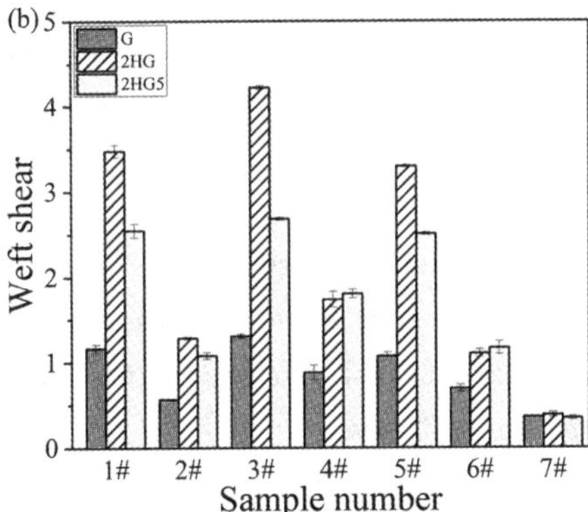

Fig. 2 (a) Warp shear; (b) weft shear

As shown in Figs. 3 (a)-3 (b), with the increase of the spandex ratio, B and 2HB basically increase gradually when the warp is bent, among which the B of 2#, 4# and 7# is relatively large, and the resistance to bending deformation is the worst, and the recovery after bending is poor. When zonal bending, comparing the

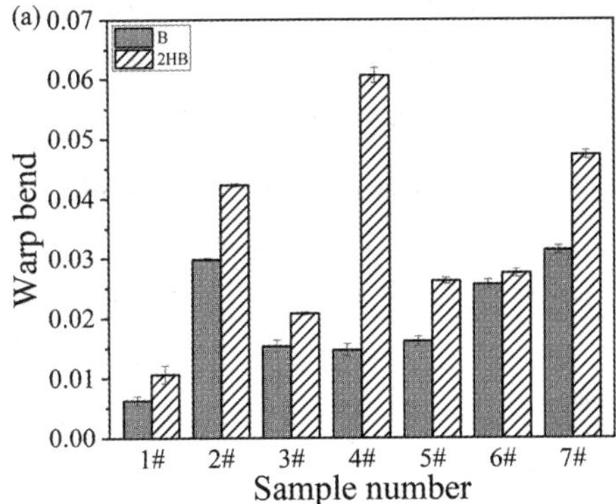

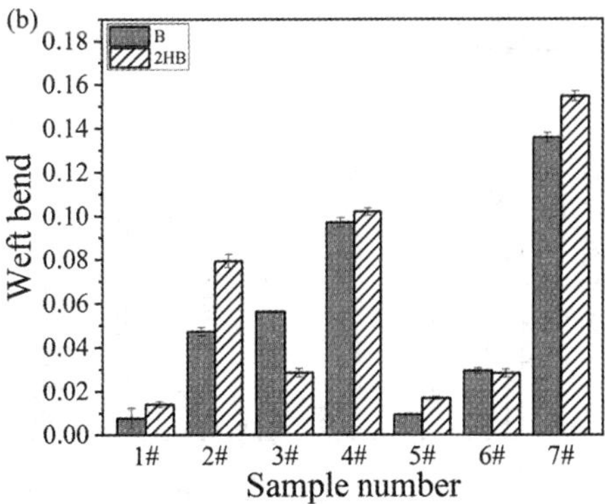

Fig. 3 (a) Warp bend; (b) weft bend

coordinate values, it is found that the overall difference trend of B and 2HB is basically similar to that of warp shearing, but B and 2HB of 2#, 3#, 4# and 7# increase, and the zonal bending performance is worse than that of warp shear.

As shown in Fig. 4, with the increase of the spandex ratio, the overall RC decreases and there is a significant difference, but it still has a good compression ratio. Responsiveness. Among them, 3# and 5# RC are smaller, but the recovery after compression is also above 44.2%; 6# has the best recovery ability, and the recovery after compression reaches 55.5%.

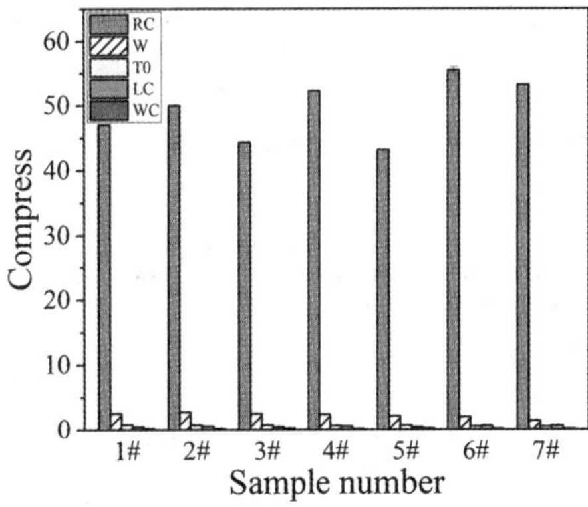

Fig. 4 Compress

As shown in Fig. 5, with the increase of the spandex ratio, the SMD is larger as a whole, but it basically shows a gradually decreasing trend. The flatness gradually increases, among which 7# SMD is the smallest, and the surface roughness of the fabric is low; while MIU and MMD are generally small, and the difference is not large, and the surface of the fabric has a certain degree of smoothness and uniformity.

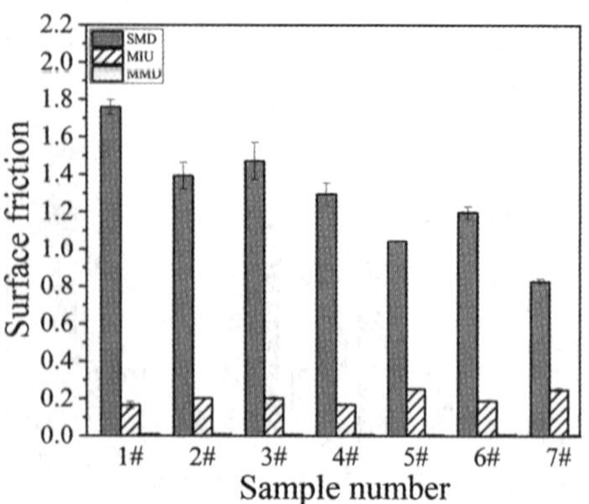

Fig. 5 Surface friction

3. 2 Simulation of human dressing pressure analysis

As shown in Fig. 6, with the increase of the spandex ratio, the pressure of the fabric gradually decreases, and the pressure of the 2# double rib fabric larger (56. 18 cN), the pressure of 7# double rib fabric is the smallest (21. 42 cN), and the reason why 2# double rib fabric has the highest pressure is its resistance to tensile deformation, poor recovery ability after bending, a certain surface, and the influence of mechanical indicators such as roughness.

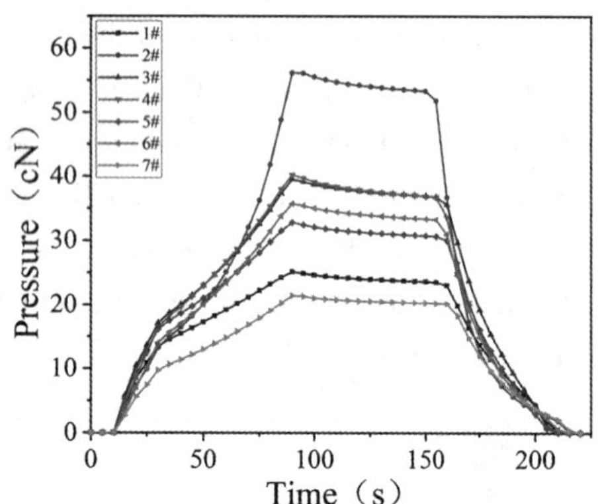

Fig. 6 Simulate dressing pressure

4 Construction of the Relationship Model Between Mechanical Properties and Dressing Pressure

4. 1 Correlation analysis

In order to study the correlation between mechanical performance indicators and dressing pressure, SPSS 23. 0 statistical software was used to analyze 16 of 7 weft knitted fabrics. A bivariate pearson significance test was carried out on the mechanical performance index and human body dressing pressure, and the results are shown in Tab. 2.

It can be seen from Tab. 2: *JWT*, *JRT*, *WRT*, *WLT* and *SMD* are significantly correlated with dressing pressure, among which *JWT* and *SMD* are positively correlated with dressing pressure, and *JRT*, *WRT* and *WLT* are negatively correlated with dressing pressure. The higher the smoothness, the smaller the pressure acting on the human body; the worse the tensile recovery, the weaker the ability to resist strain, the greater the pressure on the human body.

Tab. 2 Correlation analysis between mechanical performance indicators and dressing pressure

Performance parameter	Correlation coefficient type	Mechanical property index							
Contact pressure	Significance of pearson correlation coefficient	*JWT*	*WWT*	*JRT*	*WRT*	*JLT*	*WLT*	*JG*	*WG*
		0. 749**	0. 912	−0. 482*	−0. 396*	−0. 708	−0. 719*	−0. 087	−0. 038
		J2HG	*W2HG*	*J2HG5*	*W2HG5*	*JB*	*WB*	*J2HB*	*W2HB*
		−0. 110	−0. 039	−0. 096	−0. 008	0. 240	−0. 151	0. 271	−0. 076
		WC	*RC*	*LC*	*TO*	*W*	*MIU*	*MMD*	*SMD*
		0. 058	−0. 023	−0. 001	0. 282	0. 684	−0. 235	0. 539	0. 253**

4. 2 The multiple linear regression prediction model is constructed

On the basis of the correlation analysis, and the multiple linear regression prediction is carried out, and the validity of the established prediction formula is judge. The forecast results are shown in Tabs. 3−4.

Tab. 3 Prediction analysis and variance analysis results based on stepwise method between mechanical properties and dressing pressure

Model name	Forecast analysis			Project category	Variance analysis	
	R	R^2	*DW*		Sum of squares	Mean square
Pressure model of weft knitted fabric	1. 000[a]	1. 000	2. 083	Total regression residuals	777. 760	129. 627
					0. 000	
					777. 760	

Tab. 4 **Coefficient matrix of multiple liner regression model between mechanical properties and dressing pressure**

Index	Unstanda-rdized coefficients	Standardiz-ation coefficients	Collinea-rity statistics
	B	Beta	VIF
Constant	47. 023		
JWT	0. 349	0. 549	1. 328
JRT	0. 066	0. 049	5. 695
WRT	-0. 646	-0. 189	3. 342
WLT	-20. 838	-0. 670	5. 281
RC	0. 311	0. 128	2. 511
SMD	8. 755	0. 232	1. 521

It can be seen from Tab. 3 that R^2 is 1, and the fitting degree of the weft knitting models and the data has reached a perfect fit. The analysis shows that WT, RT, RT, LT, RC and SMD indexes of weft knitted fabric fit the dress pressure to a better linear model. Therefore, the dressing pressure can be predicted according to the mechanical properties of the fabric, and the formula of weft knitted fabric prediction model is $Y = 47.023 + 0.349 \times JWT + 0.066 \times JRT - 0.646 \times WRT - 20.838 \times WLT + 0.311 \times RC + 8.755 \times SMD$.

5 Conclusions

The results show that the warp knitted fabric has better tensile resilience in the warp direction than in the weft direction, but the weft direction has stronger resistance to tensile deformation and recovery after shearing, and the surface of the fabric is slightly rougher than the warp knitted fabric. Dressing pressure analysis results show that the increase of the spandex ratio, the pressure of the fabric gradually decreases, and the mechanical properties of fabrics are related to pressure to a certain extent.

The correlation analysis method is used to determine the correlation between each index and dressing pressure, and a multiple linear regression prediction model is established. The formula of weft knitted fabric prediction model is $Y = 47.023 + 0.349 \times JWT + 0.066 \times JRT - 0.646 \times WRT - 20.838 \times WLT + 0.311 \times RC + 8.755 \times SMD$.

References

[1] SUN Y B. Analysis of wearing pressure of knitted clothing based on finite element simulation[D]. Tianjin: Tiangong University, 2021.

[2] CHEN L H. Relation between fabric force at specified elongation and fitness of elastic clothing [J]. Journal of Textile Research, 2016, 37(7): 115-120.

[3] ZHOU J, ZHANG Y L, MAO Q. Research progress on the comfort of shaping clothing[J]. Journal of Xi'an Poly-Technic University, 2021, 35(4): 13-19.

[4] HORIBA Y, AMANO T, INUI S, et al. Proposal of method for estimating clothing pressure of tight-fitting garment made from highly elaatic materials: hybrid method using apparel CAD and finite element analysis software [J]. Journal of Fiber Science and Technology, 2021, 77(2): 76-87.

[5] BRUBACHER K, TYLER D, APEAGYEI P, et al. Evaluation of the accuracy and practicability of predicting compression garment presssure using virtual fit technology[J]. Clothing and Textiles Research Journal, 2021: 1940 - 2473.

[6] WANG Y R. Three-dimensional pressure simulation test system for knitting [P]. China: CN200810041348 X, 2010-09-8.

Objective Test Method of Hindrance to Body Movement by Sweaty Sportswear Fabrics

ZHU Ying[1], YIN Jiong[1], LIU Ziyv[1], HUANG Xinlin[2], JI Feng[1]*

[1] *Key Laboratory of Textile Science & Technology, Ministry of Education, Donghua University, Shanghai 201620, China*

[2] *Shanghai New Fiber Instrument Co., Ltd., Shanghai 200051, China*

* *Corresponding author's email*: jifeng@ dhu. edu. cn

Abstract: During exercise, the body often sweats much and clothes will get wet and stick to the skin surface. Thus, our skin cannot stretch freely and we feel blocked. At present, there is no proper method that could objectively and quantitatively characterize the skin sensation of being blocked by wet sticky fabrics. Hence the development and evaluation of anti-sticky fabrics were hindered due to lack of objective evidences. In this paper, a new method was proposed to objectively evaluate the blocking sensation of the skin by wet sticky fabrics during body movement. At first, the wet sample was sticked to the artificial skin. And the artificial skin began to stretch just like what it was as the body moved. Then the fabric was made to slip on the skin surface along tangential direction, meanwhile the tangential adhesion force (that was) equal to the stretch force of the fabric was measured. At last, the displacement-tangential adhesion force curve was drawn, and indices such as the average tangential adhesion force and the wave periods of the adhesion force were obtained from the curve. In this way, the skin sensation by sweaty sticky fabrics was objectively characterized. Ten different types of commonly used fabrics of sportswear were collected and the tangential adhesion forces of each wet samples were tested. The result of the study showed that the method could effectively evaluate the characteristics of the wet sticky sensation of different types of sportswear.

Keywords: Tangential adhesion force; Blocking sensation; Objective test; Fabrics of sportswear

1 Introduction

The human body sweats much during exercise, making the sportswear get wet and adhere to the skin surface. Then the skin can't stretch freely and feel the sensation of being hindered[1]. This sensation can be quantified by curves. And it can also be expressed by the mechanical characteristic values of sticky objects as they detach from the skin[2]. The comfort of sweaty fabrics can be divided into two directions: the normal and the tangential direction.

As for the wet sticky sensation in the normal direction, it was first measured by the XAD-1 instrument, which was developed by our team[3]. The instrument was added a device simulated the skin which could control the amount of sweat[4]. By simulating the normal separation of the skin and the fabric, the adhesion curve was measured. Thus, the results visually presented the adhesion and separation between wet fabrics and the skin along the normal direction. What's more, this method was also used to study the effect of tensile and bending performances on wet adhesion[5]. However, the link between the adhesion curve and the blocking sensation hasn't been systematically proposed yet.

In fact, the adhesion between sweaty fabrics and the skin was mainly measured in tangential direction. Therefore, in the development and evaluation of fabrics, the study of the tangential wet adhesion is necessary. Wang et al. [6] developed a device to detect the frictional performance between fabrics and the skin. A friction roller was used to simulate the human arm, but the results were significantly affected by the arm's moment of rotational inertia. Zhou et al. [7] installed a power device to make the wet fabric slide relatively along the artificial skin surface. So, the fabric tension in the direction of its slippage was measured as the wet adhesion force. But its accuracy was restricted to the unadvanced device at that time, making it hard to obtain the accurate data. In order to precisely analyze the skin sensation of being hindered, our team developed the XWF-1 instrument. It could objectively and accurately measure the tangential adhesion force between sweaty fabrics and the skin in real time[8].

In this paper, the XWF-1 was used to measure the wet adhesion and the blocking sensation of different types of fabrics of sportswear. 10 common summer sportswear fabrics were collected to be tested and analyzed.

2 Experiments

2.1 Principles and instruments

The wet fabric adhered to the skin surface and followed its movement, hindering the skin from stretching. And the tangential adhesion force was overcome during its movement. The quantity and duration of the force were the main sources of the skin sensation of being blocked.

The XWF-1 was selected in this research. During the experiment, the wet sample were clamped by clips and adhered to the artificial skin surface. Then, the instrument made the skin move under the sample, simulating the skin stretch during the body movement. So, the wet fabric slipped relatively to the skin along the tangential direction. The motion path was shown in Fig. 1.

The dynamic load was detected by the sensor in real time, and the tangential adhesion force during the movement was measured. Therefore, the adhesion curve was generated, and it could be divided into two periods: rise and slip period. So, indices could be obtained. The adhesion curve was shown in Fig. 2.

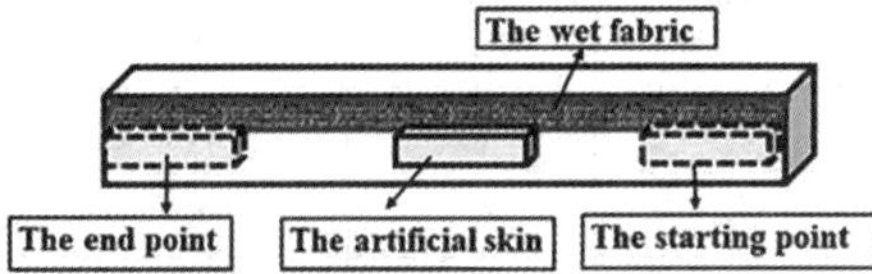

Fig. 1 Relative movement between the wet fabric and the artificial skin

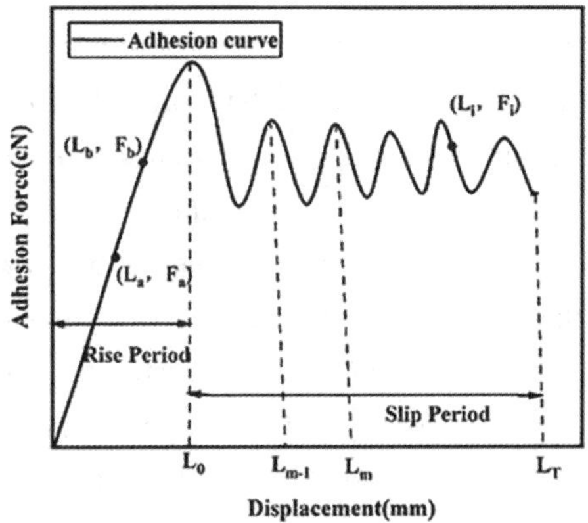

Fig. 2 Adhesion force-displacement curve

(1) Rise Period: The wet fabric was stretched by the skin movement, and it wasn't fully into a state of relative motion yet.

L_0(mm) was the distance the fabric had moved, which represented the fabric ability to stretch with the skin movement. And it symbolized the duration of skin sensation of being hindered.

K (mm, cN) was the increasing rate of fabric tension, which represented the rate of the blocking sensation of the skin. The larger the K was, the faster the blocking sensation increased while the body moved after sweating. The calculation formula was shown in Eq. (1).

$$K = \frac{F_a - F_b}{L_a - L_b} \tag{1}$$

(2) Slip Period: When the fabric tension increased enough to compete with the resistance, the curve reached to its first peak, and the adhesion between the fabric and the skin were continuously formed and unraveled.

T(mm) was the fluctuating cycle of the fabric movement, representing the duration of the blocking sensation. The calculation formulas were shown in Eq. (2)-(3). L_m and L_{m-1} were the displacement of adjacent wave peaks.

F(cN) was the average value of the adhesion force between the wet fabric and the skin, symbolizing the intensity of the skin sensation of being hindered. (L_i, F_i) (mm, cN) was any point in slip period, and L_T was the adhesion distance at the end of slip period. The calculation formula was shown in Eq. (4).

$$\Delta L_m = L_m - L_{m-1} \tag{2}$$

$$T = \frac{1}{n}\sum_{m=1}^{n} \Delta L_m \tag{3}$$

$$F = \frac{1}{L_T - L_0}\int F_i \mathrm{d}l \tag{4}$$

2.2 Materials

In the experiment, 10 common summer sportswear knitted fabrics were collected. The structure, material, grammage, thickness, course and wale of the fabrics were recorded in Tab. 1 and Tab. 2.

Tab. 1 Structures and materials of fabrics

Name	Structure	Material
K1	Weft plain	Wool
K2	Rib	Wool

(Continued)

Name	Structure	Material
K3	Purl	Wool
K4	Weft plain	Wool
K5	Purl	Wool
K6	Weft plain	Wool
K7	Weft plain	Nylon
K8	Weft plain	Spandex
K9	Weft plain	Polyester
K10	Weft plain	Cotton/Spandex

Tab. 2 Structure parameters of fabrics

Name	Grammage (g/m^2)	Thickness (mm)	Course (/5 cm)	Wale (/5 cm)
K1	115.46	1.78	100	70
K2	135.66	2.33	140	90
K3	71.31	0.84	80	60
K4	101.88	1.81	110	90
K5	129.04	1.86	110	80
K6	145.84	2.53	150	140
K7	104.10	1.11	85	69
K8	141.10	1.92	150	130
K9	96.10	0.93	80	60
K10	143.00	2.15	130	100

Pictures of fabrics of sportwear are shown in Fig. 3.

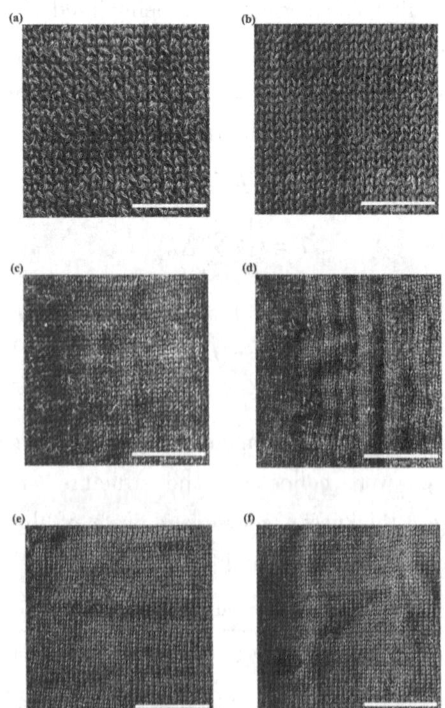

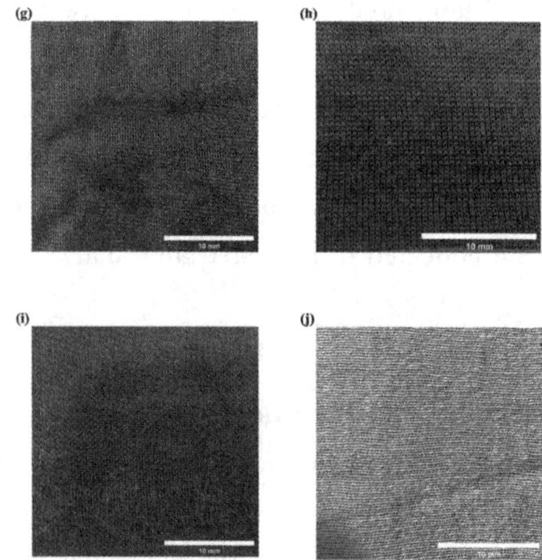

Fig. 3 Pictures of fabrics: (a) K1; (b) K2; (c) K3; (d) K4; (e) K5; (f) K6; (g) K7; (h) K8; (i) K9; (j) K10

3 Results and Discussion

According to method above, the indices of adhesion curves of 10 selected fabrics were obtained. Therefore, indices could be used to analyze the rate, duration and intensity of the skin sensation of being blocked. Indices were given in Tab. 3.

Tab. 3 Indices of adhesion curves of sweaty sportswear fabrics

Name	L_0(mm)	K(cN/mm)	T(cN)	F(cN)
K1	2.40	85.42	6.01	293.61
K2	2.29	86.90	5.07	297.92
K3	2.61	70.98	6.60	294.34
K4	2.70	81.31	8.18	300.47
K5	2.57	88.49	5.21	313.54
K6	3.00	58.78	9.80	301.78
K7	2.38	87.08	6.66	256.76
K8	2.29	77.12	4.62	267.09
K9	2.19	62.85	5.74	301.62
K10	2.08	60.16	5.25	276.81

K6 was a wool weft plain knitted fabric with high grammage, thickness and density. Its values of L_0, T, and F were larger, while K was smaller. Thus, it could stretch easily with the skin movement, and would adhere to the skin tightly and persistently. While wearing wet K6 and slightly moved, the sensation of being hindered increased slowly, but the fabric would stick to the skin surface for long and brought a strong blocking sensation.

K7 was a nylon weft plain knitted fabric with low grammage, thickness and density. Its values of L_0, T, and F were smaller, while K was larger. Therefore, it couldn't keep up with the skin stretch, and quickly separated from the skin. When wearing wet K7 and moved slightly, the feeling of hindrance would rise rapidly. The wet fabric sticked to the skin surface, bringing a soft and short blocking sensation.

According to the analytical method above, wearing wet K1, K2, K3 and K5 fabrics would bring the skin a short but strong feeling of hindrance, and this sensation would be quickly enhanced by slight movements. When wearing wet K4, it would give the skin a long-lasting and strong blocking sensation. K8 and K10 fabrics would bring a brief and gentle feeling of being hindered to the skin, while K9 would bring a brief but strong blocking sensation.

4 Conclusions

The research focused on the phenomenon of the adhesion between the wet fabric and the skin. It also proposed an objective and effective method for measurement.

The XWF-1 was used to test fabrics and draw their adhesion curves, and curves were divided into two periods. Thus, indices obtained from the curves were correlated with duration and intensity of the skin sensation of being hindered. So, the adhesion between different types of sportswear fabrics and the skin could be quantitatively characterized and analyzed.

References

[1] JIANG M F, WANG Y Yi. Research progress of adhesive sensation in human clothing[J]. Journal of Textile Research, 2019(05 VO 40): 177-184.

[2] SONG M G. Investigation of the contribution of garment design to thermal protection. Part 2: Instrumented female mannequin flash-fire evaluation system[J]. Textile Research Journal, 2010, 80(14): 1473-1487.

[3] XIE J J, JI F, LUO J, et al. Research on development and test method of fabric wet fitting comfort test instrument[J]. China Cilium Inspection, 2018(2): 92-93.

[4] WANG Y, JI F, HUANG X L. Development and research of fabric wet adhesion tester with controllable sweat volume[D]. Shanghai: Donghua University, 2014.

[5] SUN S Y, JI F, WANG Y, et al. Small stress under the fabric tensile and bending properties of the impact of the wet stick[J]. Journal of Textile Science and Technology, 2014, 35(8): 27-31.

[6] WANG F M, WANG X. A device and method for detecting frictional properties between fabric and skin[P]. China: CN102175600A, 2011-09-07.

[7] ZHOU G C, GAO R H, SU Y J, et al. A method and device for measuring the viscosity between fabric and skin[P]. China: TW565697, 2003-12-11.

[8] JI F, ZHANG S Y, HUANG X L, et al. A method and device for characterizing deformation behavior of wet moving viscose of textile fabric[P]. China: 2021-04-26.

Electrospun PVDF/Zn Core-Spun Yarn for Smart Textiles

WANG Ci[1], JI Dongxiao[1], ZHANG Hongnan[1*], QIN Xiaohong[1]

[1] *Department of Technical Textiles, College of Textiles, Donghua University, Shanghai 201620, China*

* *Corresponding author's email*: hnzhang@ dhu. edu. cn

Abstract: The piezoelectric sensor is currently a key research area for real-time health monitoring. Most of the piezoelectric sensors are made into thin films because of the simple preparation process, but they show poor air permeability. In this study, the piezoelectric core yarns were made by conjugate electrospinning. The metal zinc wire was used as the core electrode, and polyvinylidene difluoride (PVDF) nanofibers were electrospun onto the core electrode as the piezoelectric layer. The core-shell yarns were woven to prepare the piezoelectric sensor fabric. The X-ray diffraction (XRD), Fourier transform infrared (FTIR) spectroscopy, and differential scanning calorimetry (DSC) verify the superiority of piezoelectric performance in PVDF after electrospinning. The fabric with a size of 4 cm×1. 5 cm could generate a voltage of 7. 2 V and a current of 500 nA at 3 Hz vibration frequency, which had a good piezoelectric monitoring effect.

Keywords: PVDF; Electrospinning; Core yarn; Piezoelectric effect; Smart fabric

1 Introduction

Under the background of national sports and fitness boom, people pay more attention to physical health, and the demand for mobile and intelligent sports monitoring equipment is also increasing. Among them, intelligent devices used to monitor basic body health indicators such as heart rate, blood pressure and sweat during exercise are the most common in daily life. However, in the process of human movement, force will be attached to the human joints as the main form of expression[1]. Due to its unique electromechanical coupling effect, piezoelectric sensors play an important role in human movement monitoring[2-3]. At present, there are some wearable piezoelectric devices for movement monitoring, but the vast majority of piezoelectric equipment is film-type, which is difficult to compatible with complex changes in human activities. With the continuous development of the intelligent wearable device industry and the emergence of various new preparation technologies, various flexible non-film composites have been studied, laying a solid foundation for the innovative development of intelligent equipment for human movement monitoring.

Among various piezoelectric materials, piezoelectric ceramics have the best piezoelectric properties, such as lead zirconte titanate (PZT)[4-5], barium-titanate ($BaTiO_3$)[6-7], zinc oxide (ZnO)[8-9], etc. However, most of them are rigid and brittle and partially toxic. They have certain limitations in the application of flexible piezoelectric equipment, so they are often used as auxiliary additives. Compared with piezoelectric ceramics, organic piezoelectric polymers have good processing properties that can form yarns or fabrics, in addition to biocompatibility and non-toxicity. Such as polyvinylidene fluoride (PVDF) and its copolymers[10-11], polyacrylonitrile (PAN)[12], polylactic acid (PLA)[13], etc. Among them, because of its unique piezoelectric crystal phase (β-phase), the piezoelectric response of PVDF and its copolymers is the highest among polymers[14]. In addition, PVDF is easier to spin and has lower costs, so it is widely used.

In this study, we investigated the effect of electrospinning on the crystalline phase of PVDF. And a single electrode piezoelectric yarn (SEPY) was prepared by conjugate electrospinning. The SEPY consisting of PVDF micro-nano fibers (MNFs) as the shell coating material and conductive metal zinc wire as the core. Finally, the SEPYs-based piezoelectric devices into textiles was designed and investigated.

2 Experiments

2.1 Materials

PVDF powder (M_W = 11 000 000, HSV900, Arkema, France) and zinc wire (Daoguan Co. Ltd., Shanghai, China) with diameters of 0.3 mm were purchased from Taobao market. *N*, *N*-dimethylformamide (DMF, AR) and Acetone (AR) were purchased from Sinopharm Chemical Reagent Co. Ltd., China. All chemicals were used without further purification.

2.2 Preparation of the electrospinning PVDF/Zn yarn and fabric

First, PVDF powder was dissolved into a mixture solution of DMF/acetone (v : v=3 : 2) under a 50 ℃ water bath and stirred for 4 h to obtain a 10% uniform spinning solution. Then, two plastic syringes filled with PVDF precursor solution were fitted with 18G metal nozzles, and they were fixed on the two symmetrical pumps, and the distance between the two metal nozzles was 18 cm. The metal zinc wire was extended and fixed to the winding winder through the small hole of twist device, and the distance between the winding winder and the needle was 22.5 cm. After that, the propulsion pump rate was set as 1.4 mL/h, the speed of the twist device was set as 600 r/min, and the speed of winding winder was set as 700 r/min. Next, the positive and negative high voltages of ±7.8 kV were connected to two metal needles respectively. Under the condition of a relative humidity of 45%±5% and a temperature of (22±3) ℃, the pump pumped the solution out of the metal needle. With the action of positive and negative voltage, the solution jet with the opposite charges formed a Taylor triangle cone at the needle and sprayed into filaments. Under the rotation of the twist drum, it gathers and covers it on the zinc wire to obtain continuous core yarn with uniform stucture. Repeat the above electrospinning process seven times to obtain stable core-spun yarns. Finally, through a simple homemade weaving device, the SEPYs were divided into warp and weft yarns, and they were woven into a fabric with a size of 4 cm×1.5 cm, and the relevant properties of the fabric were measured and characterized.

2.3 Characterization

After gold spraying of the SEPYs, the longitudinal and transverse morphological structures of the SEPYs were observed by field emission scanning electron microscopy (FE-SEM, HITACHI SU8010, Japan). Before SEM imaging, samples were coated with ~15 nm Pd/Au (Leica EM ACE 600). The diameter distribution of submicron fibers was calculated by measuring the diameters of 100 fibers with Nano Measurer software (Laboratory of Surface Chemistry and Catalysis, Department of Chemistry, Fudan University, China). The infrared absorption spectra of PVDF before and after electrospinning and their β-phase content were determined by Fourier transform infrared spectroscopy (FTIR, Nicolet 6700, America). The diffraction patterns of PVDF before and after electrospinning were obtained by X-ray diffractometer (XRD, Bruker D8 ADVANCE, Germany) at a scan rate of 5° per minute in the scan interval of 5°–90°, and their grain sizes were calculated and compared. The crystallinity of PVDF before and after electrospinning were characterized by Differential scanning calorimetry (DSC, DSC8500, Perkinelmer, America) analysis. The magnitudes of piezoelectric constant of PVDF solution films and PVDF electrospun MNFs films were analyzed and compared by a piezoelectric coefficient (d_{33}) measuring instrument (ZJ-3AN, Institute of Acoustics, CAS, China). The output of open-circuit voltage and SC of the SEPYs fabric was recorded by an electrometer (kethley 6514).

3 Results and Discussion

3.1 Morphological structure of PVDF MNFs

As shown in Fig. 1 (a), the PVDF MNFs after electrospinning without beads, which had uniform morphology, and the average diameter of the MNFs was 0.81 μm obtained by statistics. The longitudinal and transverse electron micrographs of the SEPY were shown in Fig. 1 (b) and Fig. 1 (c). The SEPY had a uniform structure with an average diameter of 0.5 mm, and the shell MNFs layer had good encapsulation and a certain orientation structure along the zinc wire.

3.2 Crystallinity of materials

It had been indicated that the electrospun PVDF might form more β crystalline phases to enhance its piezoelectricity[15]. Thus, FTIR, DSC and XRD were used to investigate the crystal structures of PVDF before and after electrospinning. Fig. 2 (a) show the FTIR spectra of PVDF before and after electrospinning, and the figure had two typical peaks, each corresponding to the bonding vibration of α or β phase: 840β and 763α. Each peak corresponds to a single phase, so their

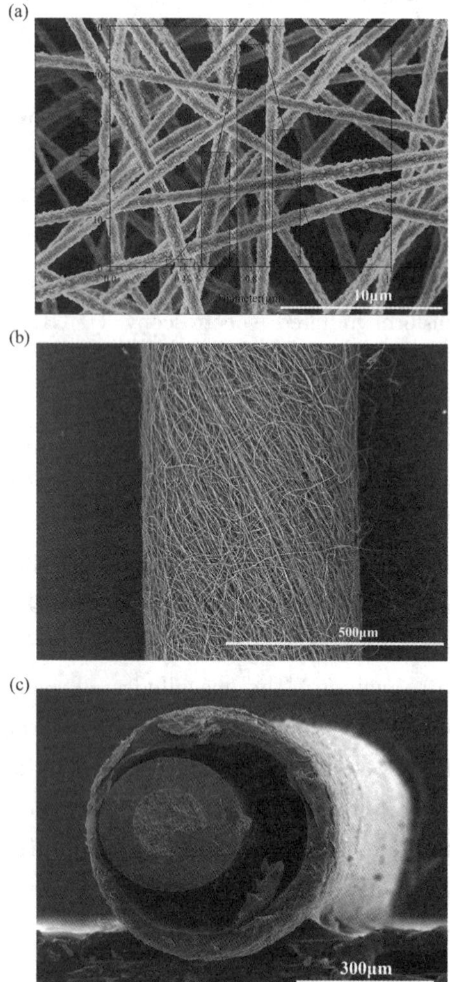

Fig. 1 (a) SEM image of PVDF MNFs; (b) SEM image of the single SEPY; (c) cross-sectional image of the SEPY

relative intensities can be used to quantify the mass fraction of each phase. So 840β and 763α were often used to quantify the β-phase content with the follow formula[16]:

$$F(\beta) = \frac{A_\beta}{(K_\beta / K_\alpha) A_\alpha + A_\beta} \quad (1)$$

where $F(\beta)$ is the β phase fraction in the material, A is the absorption band intensity, and K is the absorption coefficient of each phase, K_α and K_β are the absorption coefficients of 6.1×10^4 cm^2mol^{-1} and 7.7×10^4 cm^2mol^{-1} respectively. After calculating, the $F(\beta)$ of PVDF powder and PVDF MNFs were 55.03% and 93.65% respectively. It shown that the mass fraction of $F(\beta)$ in PVDF is nearly doubled after electrospinning, indicating that the high-pressure traction effect in the electrospinning process can play a positive role in the generation of polar β-phase of PVDF.

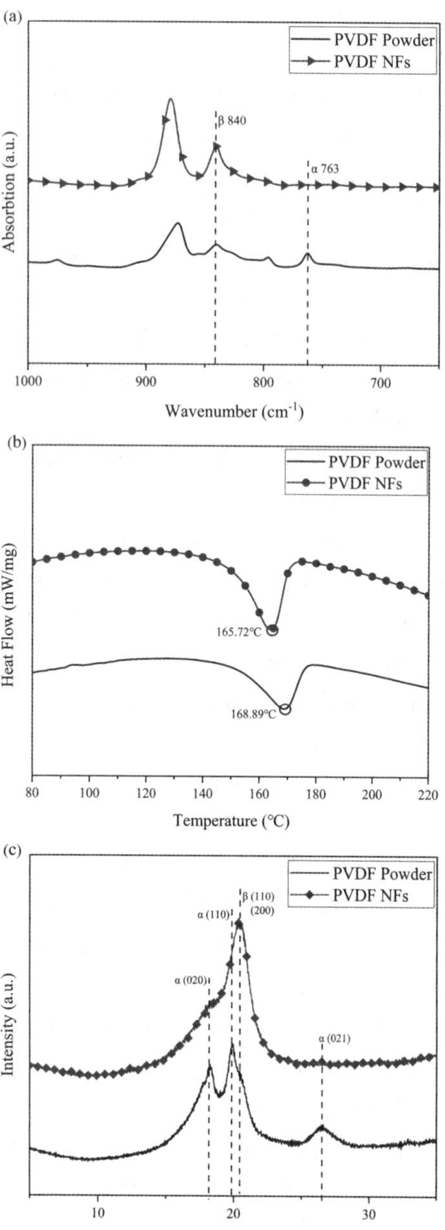

Fig. 2 (a) FTIR spectra of the PVDF powders and PVDF MNFs; (b) DSC traces of the PVDF powders and PVDF MNFs; (c) XRD patterns of the PVDF powders and PVDF MNFs

As α-phase and β-phase are the main crystalline phases in PVDF, the crystallinity of PVDF before and after electro spinning can be quantitatively analyzed by DSC, which can reflect the piezoelectric properties of PVDF sideways. The degree of crystallinity (X_c) of samples were calculated by the follow formula[17]:

$$X_c = \frac{\Delta H_f}{\Delta H_f^*} \times 100\% \quad (2)$$

where ΔH_f is the melting enthalpy of PVDF specimen

and ΔH_f^* (104.6 J/g) is the melting enthalpy of 100% crystalline PVDF. The DSC melting curves of PVDF before and after electrospinning are shown in Fig. 2 (b). The crystallinity of the PVDF powder was calculated to be 19.36% and that of the PVDF MNFs was 37.81%, which verified the FTIR experimental results. Also, because the heat absorption peaks of PVDF powder and PVDF MNFs were 168.89 ℃ and 165.72 ℃, respectively, indicating that each sample contained a major crystalline phase. It indicates that the β-phase is the main crystalline phase in PVDF after electrospinning, and the melting point of β-phase is lower than that of α-phase.

Moreover, XRD was measured in order to get a more intuitive feeling of the direct changes in the crystal structure of PVDF before and after electrospinning. As shown in Fig. 2 (c), there were large differences between the peaks of diffraction curves of PVDF powder and PVDF MNFs. Among them, 2θ were most significant at 18.24° for α (020) and 20.39° for β (110/200). It can be seen that the absorption intensity of α phase of PVDF MNFs was greatly reduced after electrospinning, and the absorption intensity of β-phase diffraction peak of PVDF MNFs was much larger than that of PVDF powder, which again verifies that electrostatic spinning technology improves the β phase in PVDF MNFs.

3.3 Piezoelectric properties of the SEPYs fabric

According to the above analysis, that electrospinning had significant advantages in enhancing the dominant crystalline phase (β-phase) of PVDF piezoelectric properties. Next, the d_{33} of PVDF solution film and PVDF MNFs film were measured and compared to further verify that. The test results were shown in Fig. 3, the d_{33} of PVDF solution film and PVDF MNFs film were 0.8 pC/N and 2.4 pC/N respectively. The d_{33} of PVDF was enhanced to three times after electrospinning, which can provides better piezoelectricity. The results demonstrate that electrospinning is an important technology for enhancing the piezoelectric properties of PVDF and that electrospun PVDF MNFs have some potential for piezoelectric textile applications.

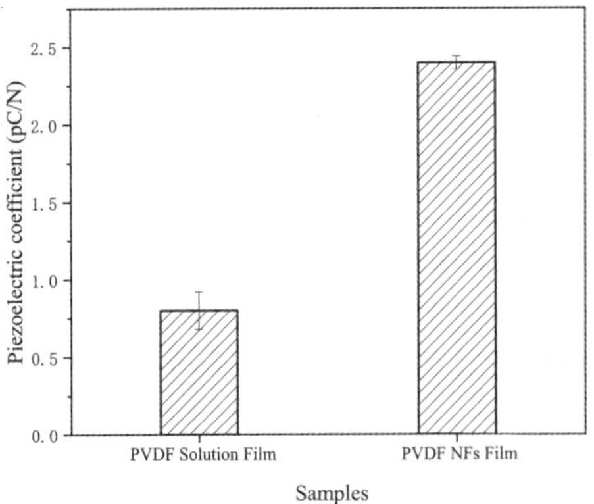

Fig. 3 The piezoelectric constants of the PVDF samples in different states

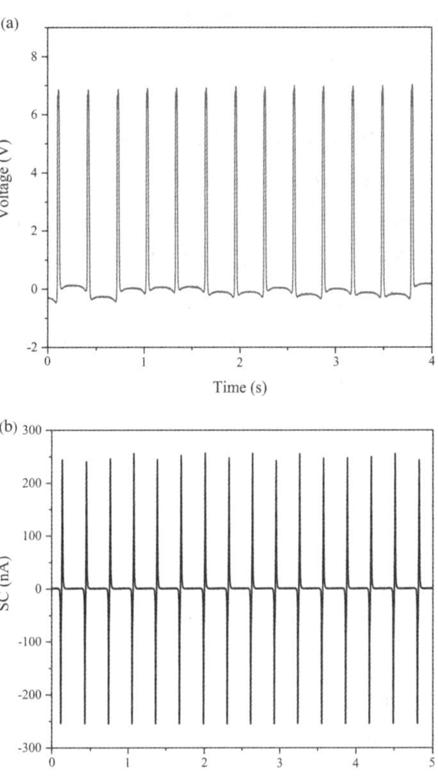

Fig. 4 Electrical performances of the prepared 4 cm × 1.5 cm SEPYs fabric under 3 Hz frequency: (a) open-circuit voltage; (b) SC

To investigate the piezoelectric properties of SEPYs for the smart textile applications, the SEPYs were woven to obtain a fabric sample with interlaced warp and weft. And the fabric generated an open-circuit voltage of 7.2 V (Fig. 4 (a)) and an SC of 500 nA (Fig. 4 (b)) under the beat with a frequency of 3 Hz, which indicating a good piezoelectric response to external signals. Compared to the lower piezoelectric sensing effect of solution piezoelectric films, it can be seen that SEPYs are capable of sensing external stimuli. Therefore, they have great potential for piezoelectric sensing smart wearable textile applications.

4 Conclusions

This study has focused on the effect of electrospinning on piezoelectric properties of PVDF MNFs. Obviously, in the FTIR, DSC, and XRD graphic curves, the β crystalline phases absorption intensity of PVDF MNFS had significantly enhanced after electrospinning. As the results, the crystallinity content of PVDF MNFs had increased from 19.36% to 37.81%, and the percentage of β-phase had increased from 55.03% to 93.65% after electrospinning. These results all verified the electrospinning technology had played a positive role in the enhancement of PVDF piezoelectric properties. In addition, the SEPYs prepared by conjugate electrospinning were able to be woven into flexible fabrics with piezoelectric voltage and can generated an open-circuit voltage of 7.2 V and an SC of 500 nA. This method of core-spun piezoelectric yarns preparation will provide a new strategy for designing and improving the performance of wearable human movement monitoring textiles in the future.

References

[1] PAUL J P. Force actions transmitted by joints in the human body[J]. Proceedings of the Royal Society B: Biological Sciences, 1976, 192(1107): 163-172.

[2] ZHU M L, SHI Q F, HE T Y Y, et al. Self-powered and self-functional cotton sock using piezoelectric and triboelectric hybrid mechanism for healthcare and sports monitoring[J]. ACS Nano, 2019, 13(2): 1940-1952.

[3] CHEN Z F, WANG Z, LI X M, et al. Flexible piezoelectric-induced pressure sensors for static measurements based on nanowires/graphene heterostructures[J]. ACS Nano, 2017, 11(5): 4507-4513.

[4] SAPPATI K K, BHADRA S. Flexible piezoelectric 0-3 PZT-PDMS thin film for tactile sensing[J]. IEEE Sensors Journal, 2020, 20(9): 4610-4617.

[5] HOU X J, ZHANG S N, YU J B, et al. Flexible piezoelectric nanofibers/polydimethylsiloxane-based pressure sensor for self-powered human motion monitoring[J]. Energy Technology, 2020, 8(3): 1901242.

[6] YUAN L J, FAN W, YANG X, et al. Piezoelectric PAN/$BaTiO_3$ nanofiber membranes sensor for structural health monitoring of real-time damage detection in composite[J]. Composites Communications, 2021, 25: 100680.

[7] MU J L, XIAN S, YU J B, et al. Flexible and wearable $BaTiO_3$/polyacrylonitrile-based piezoelectric sensor for human posture monitoring[J]. Science China Technological Sciences, 2022, 65(4): 858-869.

[8] TAN Y S, YANG K, WANG B, et al. High-performance textile piezoelectric pressure sensor with novel structural hierarchy based on ZnO nanorods array for wearable application[J]. Nano Research, 2021, 14(11): 3969-3976.

[9] SHIRLEY J A, FLORENCE S E, SREEJA B S, et al. Zinc oxide nanostructure-based textile pressure sensor for wearable applications[J]. Journal of Materials Science: Materials in Electronics, 2020, 31(19): 16519-16530.

[10] ZHANG C, FAN W, WANG S, et al. Recent progress of wearable piezoelectric nanogenerators[J]. ACS Applied Electronic Materials, 2021, 3(6): 2449-2467.

[11] UDDIN M M, BLEVINS B, et al. Highly flexible and conductive stainless-steel thread based piezoelectric coaxial yarn nanogenerators via solution coating and touch-spun nanofibers coating methods[J]. Smart Materials and Structures, 2022, 31(3): 035028.

[12] MA L Y, ZHOU M J, WU R H, et al. Continuous and scalable manufacture of hybridized nano-micro triboelectric yarns for energy harvesting and signal sensing[J]. ACS Nano, 2020, 14(4): 4716-4726.

[13] ZHENG T, YUE Z L, WALLACE G G, et al. Nanoscale piezoelectric effect of biodegradable PLA-based composite fibers by piezoresponse force microscopy[J]. Nanotechnology, 2020, 31(37): 375708.

[14] YANG E L, XU Z, CHUR L K, et al. Nanofibrous smart fabrics from twisted yarns of electrospun piezopolymer[J]. ACS Applied Materials & Interfaces, 2017, 9(28): 24220-24229.

[15] DAI Z, WANG N, YU Y, et al. One-step preparation of a core-spun Cu/P(VDF-TrFE) nanofibrous yarn for wearable smart textile to monitor human movement[J]. ACS Applied Materials & Interfaces, 2021, 13(37): 44234-44242.

[16] ZHAO X J, CHEN S J, ZHANG J, et al. Crystallization of PVDF in the PVDF/PMMA blends precipitated from their non-solvents: special "orientation" behavior, morphology, and thermal properties[J]. Journal of Crystal Growth, 2011, 328(1): 74-80.

[17] PARK S R, KWON Y B, SUNG M C, et al. Poling-free spinning process of manufacturing piezoelectric yarns for textile applications[J]. Materials & Design, 2019, 179: 107889.

Compressive Behavior of Braided Tubular Joint Composite Using Digital Image Correlation (DIC) Method

GAO Xinyu[1], **WANG Qihang**[1], **GU Bohong**[1]*

[1] *College of Textiles, Donghua University, Shanghai 201620, China*

* *Corresponding author's email*: gubh@ dhu. edu. cn

Abstract: Tubular joints are commonly used in architecture and aeronautics. In this study, a new braiding method was proposed to perform the tubular joint based on a 4-step 1×1 braiding process. The compressive behavior of the three-way tube was analyzed in longitudinal and axial directions respectively. And digital image correlation (DIC) was used to capture the 3-D deformation of the connection region and measure its full-field strain. The results show that the connection region experienced bulging deformation under compression and failed due to the shear failure of the intersection. It is expected that such a study could shed light on the development of composite joints.

Keywords: Tubular joint composite; Mechanical testing; Digital image correlation; Compressive behavior

1 Introduction

Tubular joints are important components in the engineering industry such as gas, fluid transportation and truss[1-4]. The main way to produce tubular joints is by adhesive or welding, which features corrosion and stress concentration in the welded joints. Fu[5] fabricated the tubular joint strengthened by carbon fiber-reinforced polymer (CFRP) and analyzed its static performance, and found that after CFRP application, ultimate load-bearing capacity, initial stiffness of tubular joint was enhanced a lot. Despite recent progress in the application of fiber-reinforced polymer (FRP) to strengthen the steel joint[6-7], it remains challenging to produce joints with smooth weld and seamless bonding. Nowadays, with the development of textile, it is worth noting that tubular joints have made progress in knitting but the knitting joint shows relatively poor strength[8]. Knitted preforms are relatively loose structure in which curled loops can deform easily under low carrying load levels. Braiding methods are expected to perform joints with high strength/weight ratio and design flexibility. Yang[9] has proved that 3-D braided composite truss joints had excellent loading capacities through experimental and numerical methods.

Here, we report an integral braiding method to achieve the braided three-way tube with uniform structure and good yarn cover ratio of the connection region. Compressive strength of the three-way tube was compared in axial and longitudinal directions. Strain field and 3-D deformation of the connection region were obtained by digital image correlation system. The purpose of this study is to prepare for the future development of braided composite joints.

2 Materials and Method

2.1 Specimen preparation

Carbon fiber (SYT49S-12K Zhongfu Shenying Carbon Fiber Co. LTD., China) was selected for three-way tube braiding. The braiding method was consistent with the tube braiding, 4-step 1×1 braiding process based on row-and-column actions in a braiding sequence which forms an interlacing pattern. 2×16 braiding pattern was used in which 2 represents the number of layers and 16 for the number of yarns in circular direction. In addition, self-designed braiding mold was used due to its assemble and disassemble function, as shown in Fig. 1.

The braiding process of the three-way tube is schematically shown in Fig. 1. Once the first tube was performed, there's a need for adding yarns before braiding the second branch tube. Yarns in Part 1, 3 were removed to the non-working area so that the ends of adding yarns could attach to the empty carriers in Part 1, 3. Besides, yarns in Part 4 were all moved to the

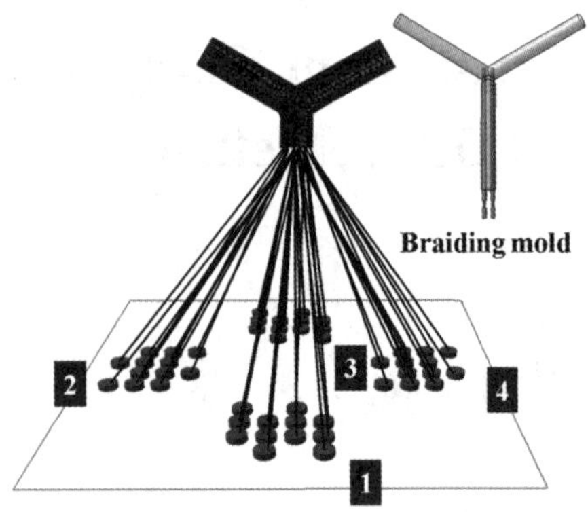

Fig. 1 Braiding process of the three-way tube

non-working area and the method of yarns added was similar to that in Part 1, 3. The reserved yarns in the non-working area were moved back to the working area when the second tube was performed. Therefore, three branch tubes have equal yarn distribution and the yarns in the connection region were orderly interlaced to form a uniform structure.

The preparation of impregnation is that the braiding mold was replaced by curing mold comprising three separated silica gel tubes and then the braiding preforms with curing mold were placed in the metal mold with a slightly larger diameter (see in Fig. 2). The epoxy resin (JC-02A/JC-02B, Changshu Jiafa Chemical Co. Ltd., China) was injected into preforms with two tubes (resin inlet) and one tube (vacuum vent) with vacuum-assisted resin transfer molding (VARTM) technique. The curing process was done in stages of 90 ℃ for 2 h, 110 ℃ for 1 h, and 130 ℃ for 4 h respectively. The size of specimens was 22 mm×2 mm×1.69 mm.

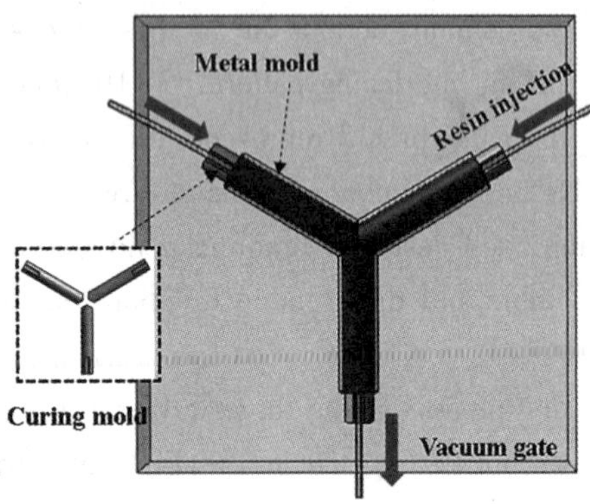

Fig. 2 VARTM technique curing

2.2 Compression test

The compression test was conducted on the Instron 5982 at a compression rate of 2 mm/min. The size of the load cell is 100 kN. Specimens were fixed in two self-designed round clamps clenched to the metal plate with screws shown in Fig. 3.

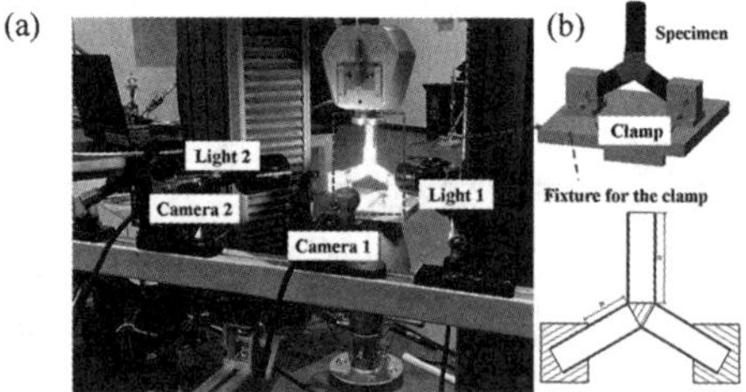

Fig. 3 (a) Compression test system (MTS 810) with digital image correlation (DIC: VIC-3D, corrected solution USA) system; (b) self-designed clamp and fixture

The digital image correlation (DIC, VIC-3D, Correlated Solutions, USA) system is a non-contact optical measurement method to find the field of displacement and strain on the surface of the three-way tubular joint. Paint is applied on the surface of a specimen to obtain a random speckle pattern by which appropriate subsets were formed and their deformations were tracked by Vic-3D software. Here, DIC was used for deformation analysis of the three-way composite tubes under compression. Two cameras were set in front of the specimen, as well as two lights worked for uniform illumination, as shown in Fig. 3. 3-D surface coordinates of the specimens were computed by calibration images with 14×10 dots through stereo-triangulation in Vic-3D software. Successive speckle images were later analyzed in Vic-3D software in which the speckle image was divided into subsets with 35 pixels, and the movement of the subsets was tracked from one image to the next. The deformation process of the specimens was recorded at a rate of 50 Hz during the test.

3 Results and Discussion

3.1 Load-displacement curve

Fig. 4 (a) shows the load-displacement curves obtained from longitudinal and axial directions respectively. The load in the longitudinal direction is constantly stable after peak load, and this peak load is much greater than that in the axial direction. This is because three branch tubes bore most load after the connection region was broken at the earlier stage. However, the load in the axial direction sharply declined after peak load due to the broken connection region of the upper branch tube.

Here, we pay more attention to the failure process of the three-way tube under axial compression load, as shown in Fig. 4 (b). Resin cracks firstly occurred on the connection region as a result of bulging movement due to the compression of three branch tubes. Then the damage escalated to the shear failure and stretched to the intersection region, which caused a total breakage of the intersection.

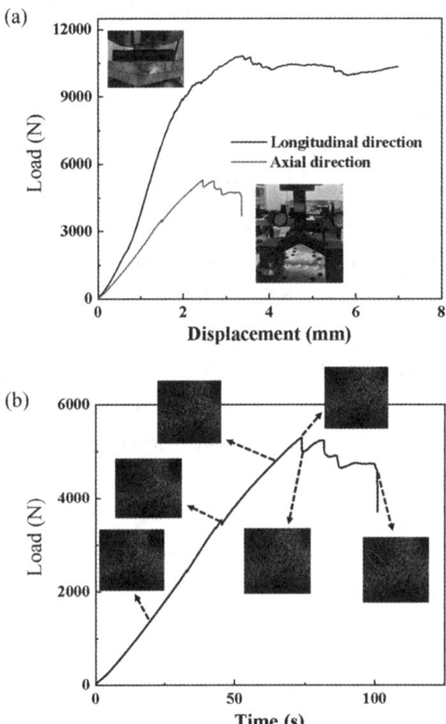

Fig. 4 (a) Load-displacement curves in longitudinal and axial directions; (b) damage evolution in axial direction

3.2 Surface strain distribution on the connection region

Fig. 5 (a) shows the out-of-plane displacement and strain field of the connection region at the elastic stage where e1 represents the major principal strain. The positive color showing on the connection region of the upper branch tube indicates a bulging movement due to the compression of the three branch tubes with the increasing axial load. However, the lower surface presents a compression strain region resulting from the squeeze of the upper. At peak load, the shear failure located on the right intersection region initiated by the resin cracks on the bulging surface, as shown in Fig. 5 (b). Fig. 5 (c) demonstrates the final failure of the three-way tube with the fracture of the left and right intersection. The breakage expanded across the bulging area.

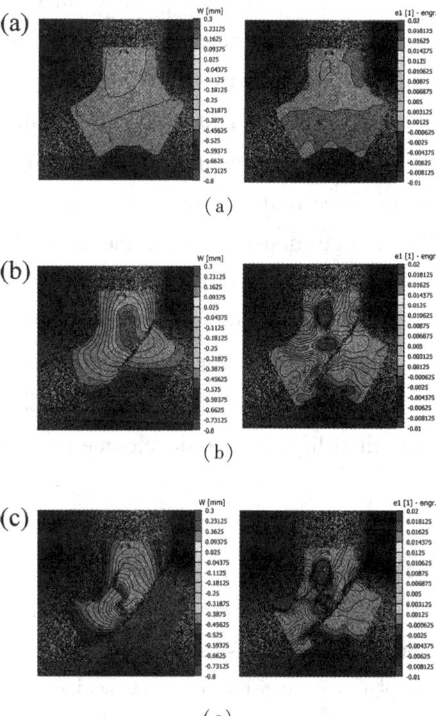

Fig. 5 DIC results at (a) elastic stage; (b) peak load; (c) final failure

Furthermore, Fig. 6 shows the extracted out-of-plane displacement of the selected line (yellow line), in which three colored lines were in accord with the three specified stages. A progressively bulging movement occurred on the selected line resulting from the compression of two side branch tubes. At final failure stage, the damage developed through the connection region, which led to the separate curve (blue curve).

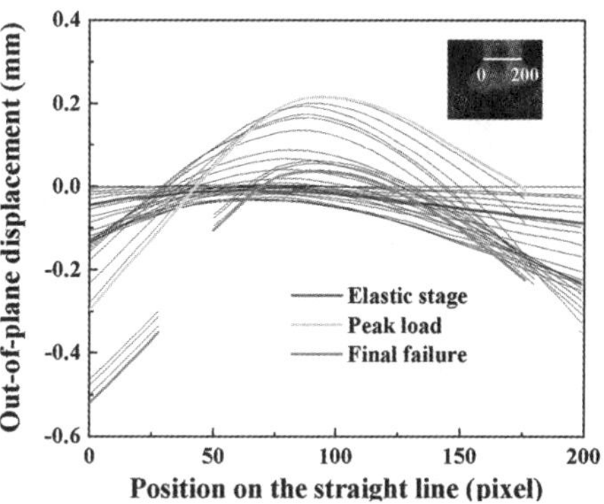

Fig. 6 Evolution of out-of-plane displacement on the selected path (the yellow solid line) under compression

4 Conclusions

A novel integrated braiding method was proposed to perform three-way tubes. The compressive behavior of this three-way composite tube was investigated. Progressive failure and deformation of the connection region were obtained by digital image correlation system.

We found that braiding yarns in the connection region were uniformly interlaced and evenly distributed. Compressive strength in the longitudinal direction is mainly related to the three branch tubes, while in the axial direction the strength depends on the mechanical property of the intersection parts. The failure mode of the three-way tube is the shear failure of the intersection region initiated by the bulging damage due to the compression of the three branch tubes. It is expected that such a study could shed light on the development of composite joints.

References

[1] CHOO Y, VAN DER VEGTE G, ZETTLEMOYER N, et al. Static strength of T-joints reinforced with doubler or collar plates. I: experimental investigations[J]. Journal of Structural Engineering, 2005, 131(1): 119-28.

[2] HOON K H, WONG L K, SOH A K. Experimental investigation of a doubler-plate reinforced tubular T-joint subjected to combined loadings[J]. Journal of Constructional Steel Research, 2001, 57(9): 1015-39.

[3] BANEA M D, DA SILVA L F. Adhesively bonded joints in composite materials: an overview[J]. Proceedings of the Institution of Mechanical Engineers, Part L: Journal of Materials: Design and Applications, 2009, 223(1): 1-18.

[4] LI Z, JIANG X, LODEWIJKS G. The latest development of reinforcement techniques on tubular joints[M]//Progress in the Analysis and Design of Marine Structures. Florida: CRC Press. 2017: 783-90.

[5] FU Y, TONG L, HE L, et al. Experimental and numerical investigation on behavior of CFRP-strengthened circular hollow section gap K-joints[J]. Thin-Walled Structures, 2016, 102: 80-97.

[6] MOHAMED H, ZHANG L, SHAO Y, et al. Stress concentration factors of CFRP-reinforced tubular K-joints via zero point structural stress approach[J]. Marine Structures, 2022, 84: 103239.

[7] HOSSEINI A S, BAHAARI M R, LESANI M. SCF distribution in FRP-strengthened tubular T-joints under brace axial loading[J]. Scientia Iranica, 2020, 27(3): 1113-29.

[8] LEONG K H, RAMAKRISHNA S, HUANG Z, et al. The potential of knitting for engineering composites — a review[J]. Composites Part A: Applied Science and Manufacturing, 2000, 31(3): 197-220.

[9] HONG Y, YAN Y, TIAN Z, et al. Mechanical behavior analysis of 3D braided composite joint via experiment and multiscale finite element method[J]. Composite Structures, 2019, 208: 200-12.

Structure Formation and Coverage Rate of t-SCSY Based on Modified Ring Spinning

CAO Jiqiang[1,2], **DU Zhaoqun**[1], **YU Weidong**[1*]

[1] *College of Textiles, Donghua University, Shanghai 201620, China*

[2] *College of Textile and Fashion, Xinjiang University, Urumqi 830017, China*

[*] *Corresponding author's email*: wdyu@dhu.edu.cn

Abstract: In order to improve the structural stability and wrapping property of the helical wrapping structure yarn, a new helical wrapping structure tri-component spreading composite structure yarn (t-SCSY) with three-axis yarn spreading mechanism and its forming method are designed. The coverage rate was evaluated. By improving the spinning spreading device, the cladding layer is expanded into a quasi-single layer to achieve a "hard structure" of the surface layer and the whole of t-SCSY with elastic performance. Through the geometric relationship between fiber transfer and yarn structure, a geometric relationship model for calculating coverage rate was constructed, and the axial coverage rate of t-SCSY was calculated. The experimental results show that the t-SCSY obtained by the spreading device has good structural stability in the strain range of 30%, and the coverage rate is effectively improved. Combined with the calculation of the coverage theory model, the coverage rate reaches 79.6%. The effective combination of the yarn spreading device and the ring spinning machine to spinning t-SCSY can realize the multi-functionalization and intelligence of the yarn so that different types of elastic-covered composite yarns can be obtained by changing the structure, and make application possibilities for the smart wearable fashion design aspects.

Keywords: Spreading device; Composite structure; Forming control; t-SCSY; Coverage model

1 Introduction

Multifunctional textiles are more and more popular, and covered yarns with high elastic cores and structural stability have attracted much attention. The methods and equipment for preparing functional yarns are continuously improved and developed. At present, most of the covered yarns use hollow spindles[1], friction spinning[2], rotor spinning, and ring spinning methods[3-4]. It is an important development direction of the yarn to effectively combine the yarn spreading device and the ring spinning to spin the elastic-covered composite yarn, so as to realize the multi-functionalization and intelligence of the yarn. In the past few years, textile fibers are often used as basic materials in various fields due to their high mechanical freedom and softness. At the same time, the elasticity and stiffness of textiles and their deformation have become the focus of research. Elastic conductive materials are used in wearable electronic devices, health care, etc.[5-6]. With textile industry has grown rapidly, converting from 3-dimensional fabrics to two-dimensional fabric substrates or structures to one-dimensional yarn fiber structures[7-9], and also used in various types of strain sensors, such as one-dimensional composite yarns[10]. Spinning of CNT/cotton/spandex composite yarns with sirofil by incorporating carbon nanotubes (CNTs) into fiber strands, and yarn structure of elastic conductive composite yarn is reported[11]; there are also high elastic conductive yarns prepared by improved ring spinning to achieve protective equipment applications[8, 12]. Those research focuses on the application of yarn and does not theoretically analyze the coverage rate of yarn. This paper focuses on the structural design and forming technology of tri-component spreading composite structure yarn (t-SCSY) that give the elastic and axial tensile deformation a faster response under small deformation. The coverage rate of t-SCSY is studied, and the composite yarn is spun by a modified spreading device on the ring spinning, which can change the yarn coverage structure, which is expected to provide a reference for the preparation of functional composite structural yarn.

2 Experiments

2.1 Manufacturing of composite structural yarn

The Textile Materials and Technology & Fibrous Soft Matter team has improved the equipment on the existing ring spinning machine, in which the multifilament spreading device is installed on the cradle, and the fiber bundles are dispersed and thinned in the lateral direction so that the filaments change from bundles to flat quasi-single layers[13]. In this paper the staple fiber roving with a quantitative value of 3.2 g/(10 m), elastane filament (EF) is 140 D, polyester multifilament (PET-MF) is 150D/36F, a total draft ratio of 22.22, twist with 750 tpm, and the spindle speed is 6 000 rpm. The coverage structure of the yarn surface was observed by digital microscope.

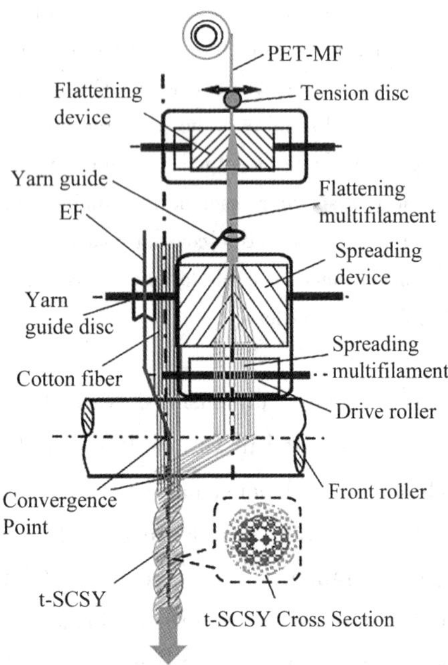

Fig. 1 Schematic diagram of spinning t-SCSY and the cross-sectional yarn

Fig. 1 is a schematic diagram of the t-SCSY process prepared by the yarn spreading device. The cotton roving in the way of conventional ring spinning, which output from the back roller-middle roller — front roller in sequence; the EF is fed from the back seat of the ring spinning through the yarn guide plate installed on the cradle, that is located just above cotton fibers, output from the jaws of the front roller, and the cotton fiber/EF yarn is located at the center line. The untwisted PET-MF is output from the tension disc of the ring spinning frame, which is flattened by the herringbone-shaped roller in the flatting device, then further expanded into a quasi-single layer, which is fed into the front roller nip by the driving roller, and the output quasi-single layers is covered with cotton fiber/EF yarn. Through the twisting action, the short fiber/elastic yarn and quasi-single layers wrapping t-SCSY are formed at the convergence point.

2.2 Helical structure of the composite yarn

The schematic diagram of the t-SCSY is shown in Fig. 2. The EF is located in the center of the t-SCSY, the cotton fiber layer is wrapped around the EF, and the outermost layer is the quasi-single covering layer. When the t-SCSY is stretched tightly and elastically, the core yarn is drawn and thinned, the pitch of the outer covering layer becomes larger, and a certain rigidity is maintained so that the strength of the composite yarn with the helical structure is effectively guaranteed.

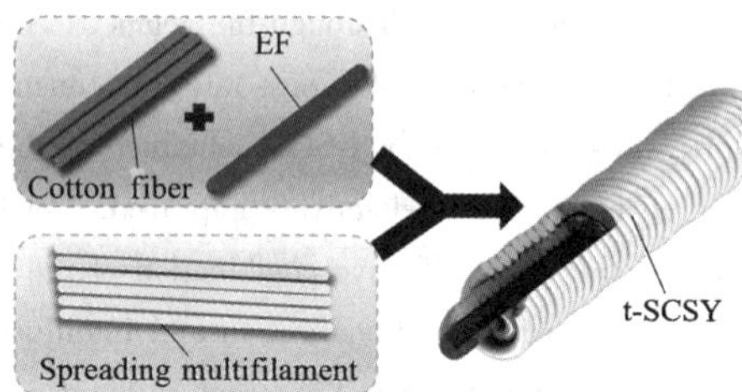

Fig. 2 Schematic of two-step yarn production artificially

2.3 Theoretical evaluation of tunable coverage rate

The structural morphology of t-SCSY is made of core yarn and the helical rigidity formed by the quasi-single layers or soft matter fibrous aggregates twist. During the process, not only the yarn is rotated by the twist, but every single fiber is rotated relative to the axial direction, and there is bending and torsional deformation. According to the basic principles of geometry, the t-SCSY coverage model was established by using the helix equation, the relationship between the cladding layer and the core layer, and the relative turning angle to analyze the fiber morphology. Taking the winding of the monofilament in the multifilament as the research object, according to the helical Eqs. (5)-(7), all the fibers are spirally wound along the length of the composite yarn, as shown in Fig. 3. The geometry of the helical yarn[14] as shown.

According to the simplified helical geometry, if the cladding length L is used as the parameter, then:

Core length

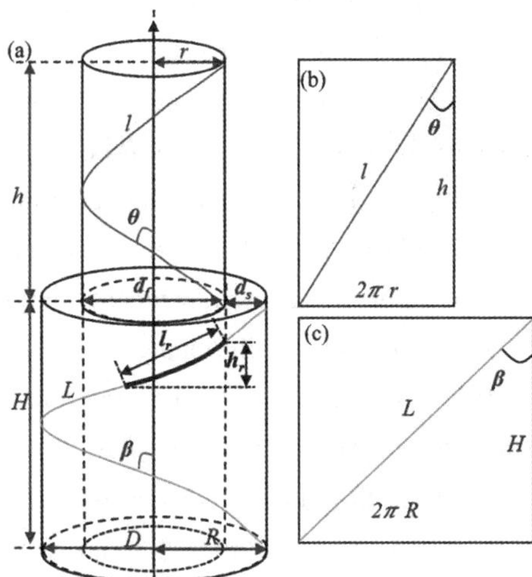

Fig. 3 Cylindrical helical yarn geometry and yarn spreading diagram. (a) Helix geometry; (b) core layer spreading diagram; (c) outer covering layer spreading diagram

$$l^2 = h^2 + (2\pi r)^2 \tag{1}$$

Coverage length

$$L^2 = H^2 + (2\pi R)^2 \tag{2}$$

Core helix angle

$$\tan \theta = \frac{2\pi r}{h} = 2\pi r T_0 \tag{3}$$

$$\tan \beta = \frac{2\pi R}{H} = 2\pi R T \tag{4}$$

where l and L are the fiber lengths (mm) of a twist at the radii r and R, respectively, θ and β are the helix angles (°) at the rad r and R, respectively, β is also called the twist angle, h and H respectively the pitch of the core layer and the cladding layer (mm), T_0 is the twist of the core yarn, T is the twist of the covering layer, twist/100 mm.

The helix equation of the composite yarn covering layer in combination with the spatial geometric relationship is shown in (5), (6), and (7):

$$x = R\cos \alpha = R\cos\left(\frac{\pi}{2} - \beta\right) = R\sin \beta \tag{5}$$

$$y = R\sin \alpha = R\sin\left(\frac{\pi}{2} - \beta\right) = R\cos \beta \tag{6}$$

$$z = \frac{H}{2\pi}\alpha = \frac{H}{2\pi}\left(\frac{\pi}{2} - \beta\right) \tag{7}$$

where α is the rise angle (°) of the helix.

It can be seen from the above equation that the basic structural parameters of the t-SCSY are mainly determined by the helix angle β, the rise angle α, and the pitch H. Those can be used to reflect the covering of the composite yarn during stretching, and can also reflect the changes in the geometry of composite yarn structures.

As shown in Fig. 3(a), it is assumed that the pitch length of the polyester-covered filaments in one spiral of the triaxial t-SCSY is H, and the cross-sectional diameter of the central axis core yarn is d_f, the thickness of the covering fiber is d_s, the twist number of the cladding layer is T, the projected length of the exposed core yarn is l_r, and the angle between the core yarn axis and the yarn axis at the exposed part is β_r. Then a twist can expression:

Pitch height

$$H = \frac{100}{T} \tag{8}$$

Covered helix diameter

$$D = d_f + d_s \tag{9}$$

Total length of covered

$$L = \sqrt{(\pi D)^2 + H^2} = \sqrt{(\pi D)^2 + \left(\frac{100}{T}\right)^2} \tag{10}$$

The axial length of exposed

$$h_r = l_r \cos \beta_r \tag{11}$$

Core yarn exposed length

$$l_r = \frac{h_r}{H}\sqrt{(\pi D)^2 + \left(\frac{100}{T}\right)^2} \tag{12}$$

The helix angle of the yarn in the exposed area is equal to $\beta_r{=}\beta$, so the coverage rate is

$$c\% = \frac{H - h_r}{H} \times 100\% = \left(1 - \frac{l_r \cos \beta}{H}\right) \times 100\% \tag{13}$$

where: D is the diameter (mm) of the covering helix in a twist; d_f is the cross-sectional diameter of the central axis core yarn (mm); d_s is the thickness of the covered layer (mm); L is the total length of the covered in one twist (mm); h_r is the axial length of the exposed part of a twisted (mm); l_r is the actual length of the exposed (mm); β_r is the angle between the exposed core and the axis; c is the coverage rate of the t-SCSY.

From Eq. (13), we verified that the pitch and helix angle mentioned above are the main parameters of the structure and formation of the yarn, where l_r the actual length of the exposed core is an indirect parameter

3 Results and Discussion

Theoretically, in the process of applying an external force to the t-SCSY, the balance of the covering layer's inward stress points is the same as the outward stress of the core yarn. When the pitch and helix angle of the covering layer is changed, the core structure and covering the structure of the composite yarn have obvious helical mandrel changes during the stretching process.

3.1 Results of structure and morphology

The cotton fiber/EF yarn in the t-SCSY is relatively soft to form a "soft structure", and the rigid covered quasi-single layer forms a "hard structure", so the yarn body presents the effect of being rigid on the outside and soft on the inside. The three-component fibers have different contributions to the yarn performance. The cotton fiber/EF core yarn provides high elasticity to the composite yarn, and the cotton fiber can fix and strengthen the core yarn. The covered quasi-single layer gives the composite yarn enhanced protection.

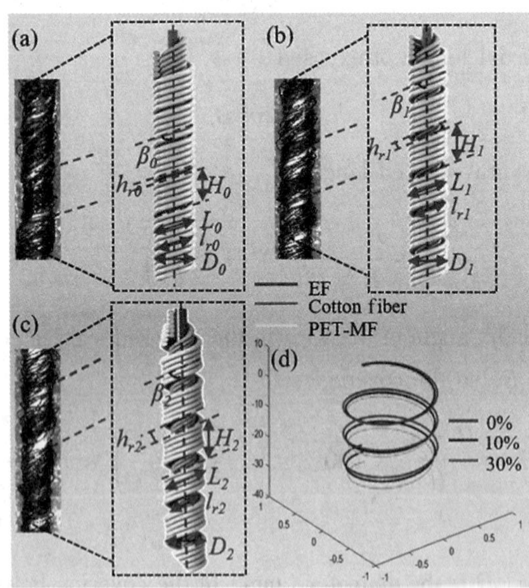

Fig. 4 Axial strain and helical diagram of composite yarns with different helical structures. (a) 0%; (b) 10%; (c) 30%; (d) helical changes with different strains

In order to further observe the local structure and morphology, fiber orientation distribution, and coverage rate of the yarn, different small strain stretches (10%, 30%) were used to observe the yarn coverage rate by using a USB digital microscope. In the case of axial covering, the obtained 0%, 10%, and 30% axial strain forming appearance effect and the corresponding helical are shown in Fig. 4. In Fig. 4(a), the spreading black PET-MF is covered on the outer layer of the core yarn, and the cotton fiber is white translucent and covered white EF. Fig. 4(b) shows the change of the coverage rate after 10% axial strain, the covered quasi-single layer is subjected to stress, from Fig. 4(a) to the shape of Fig. 4(b) where the pitch H_1 increases, and the helix angle β_1 decreases. and then through 30% axial strain, as shown in Fig. 4(c), the helix pitch H_2 increases and the helix angle β_2 decreases, the t-SCSY shrinks as a whole, the orientation degree of the covered monofilament increases, the diameter of the t-SCSY decreases, the coverage rate decreases, and the wrapping rate decreases. The coverage rate becomes worse, but the support and protection effect of the rigid structure on the covered layer is gradually manifested.

3.2 Calculation of the coverage rate of t-SCSY

The coverage rate of the composite yarn can be seen in the actual picture in Fig. 4. By measuring and analyzing the length and angle, we obtained the initial pitch of 0.9 mm, the apparent diameter of 0.55 mm, and the initial helix angle of 68.5°; the pitch when stretched to 10% is 0.967 mm, and the apparent diameter is 0.52 mm, and the helix angle after stretching is 64.61°; the pitch, when stretched to 30%, is 1.067 mm, the apparent diameter is 0.5 mm, and the helix angle after stretching is 58.14°. According to the helical Eqs. (5)-(7), under the condition of small deformation, 10% and 30% axile strains are applied to the t-SCSY in the axial direction, respectively. The change in the helix angle in the line is β_1, β_2, and the change in tensile pitch is H_1, H_2, based on Eq. (13) and the coverage rate c, the pitch H and the helix angle β, It can be obtained that the coverage rate of the t-SCSY in the initial state, 10% strain state and 30% strain state can be as 79.6%, 74.1%, and 68.75%, respectively, the corresponding spiral changes are shown in Fig. 4(d). Under the action of different axial strains, the coverage rate of the t-SCSY showed a downward trend, and the structure and morphology of the yarn also changed.

4 Conclusions

By retrofitting the ring spinning frame with a yarn spreading device, the yarn spreading method is used to expand the multifilament and then spinning t-SCSY, which has a good covering structure. We get the t-SC-SY has a structure of "external rigidity and internal softness", that the covering structure is composed of the self-generated torsion caused by the helical coil and twist of the covering quasi-single to form a rigid outer layer, and has a "hard structure", then the core layer has a "soft structure". According to the basic principle of the geometry of spiral line equation to study the spiral torsional deformation of t-SCSY, the composite yarn has better structural stability and covering effect, the coverage rate reaches 79.6%. The size of the structural parameters such as the diameter of the fiber, the pitch of the helix, and the helix angle determine the change in the composite yarn structure. An in-depth analysis of the structure of such yarns will promote the development of composite yarns in the direction of intelligence, which is meaningful for further research on the performance of structural composite yarns and provides possibilities for applications in smart wearable fashion design.

Acknowledgements

We gratefully acknowledgements the financial support of the National Key R&D Program of China (No. 2016YFC0802802), the authors also thank the College of Textile and Fashion for providing cotton samples, and fiber testing.

References

[1] FCSA J. Electrically conductive, elastic composite yarn, corresponding device, and manufacturing method: Europe, 2236654[P/OL]. 2010-2-08.

[2] VOIGT A. Electrically conductive, elastically stretchable hybrid yarn, method for manufacture thereof and textile product with a hybrid yarn of this kind: U. S, 511, 814[P/OL]. 2007-3-15.

[3] LIU S, DU Z Q, XIE K F, et al. A novel interlaced-helical wrapping yarn with negative Poisson's ratio[J]. Fibers and Polymers, 2018, 19(11): 2411-7.

[4] CHEN J, WEN X, LIU X, et al. Flexible hierarchical helical yarn with broad strain range for self-powered motion signal monitoring and human-machine interactive [J]. Nano Energy, 2021, 80(6):1-11.

[5] SHI J, LIU S, ZHANG L, et al. Smart textile-integrated microelectronic systems for wearable applications[J]. Advanced Materials, 2020, 32(5): 1-37.

[6] SCHWARTZ G, TEE B C K, MEI J G, et al. Flexible polymer transistors with high pressure sensitivity for application in electronic skin and health monitoring[J]. Nature Communications, 2013, 4(1): 1-8.

[7] JEONG C K, LEE J, HAN S, et al. A hyper-stretchable elastic-composite energy harvester[J]. Advanced Materials, 2015, 27(18): 2866-2875.

[8] WANG Y, GORDON S, YU W, et al. Structural architecture of wearable materials based on tri-component elastic-conductive composite yarn: toward a Joule heating application[J]. Textile Research Journal, 2019, 89(16): 3303-11.

[9] LEE J, ZAMBRANO B L, WOO J, et al. Recent advances in 1D stretchable electrodes and devices for textile and wearable electronics: materials, fabrications, and applications[J]. Advanced Materials, 2020, 32(5): 1-28.

[10] CAI G, YANG M, PAN J, et al. Large-scale production of highly stretchable CNT/cotton/spandex composite yarn for wearable applications[J]. ACS Applied Materials & Interfaces, 2018, 10(38): 32726-35.

[11] SPELEERS P. Electrically conductive elastic composite yarn: U. S, 11/915, 771[P/OL]. 2008-11-20.

[12] WANG Y, CHU C, CAO J, et al. Mechanically robust and highly stretchable woven fabric containing metal wire for personal protective clothing[J]. Journal of Industrial Textiles, 2020, 51(4): 5938-5951.

[13] CAO J Q, YU W D, DU Z Q. Progress of composite structural spinning technology based on filament spreading and splitting[J]. Wool Textile Journal, 2020, 48(6): 1-7.

[14] HEARLE J W, GROSBERG P, BACKER S. Structural mechanics of fibers, yarns, and fabrics[M]. New York; Wiley Interscience, 1969: 175-212.

Preparation and Piezoelectric Properties of Carbon Nanotube Coated Nanofiber Core-Spun Yarn

JIN Jingye[1], WU Tianmei[1], LANG Chenhong[1*], YANG Weibin[1], QIU Yiping[1,2], ZHANG Chuyang[1]

[1] *College of Textiles and Apparel, Quanzhou Normal University, Quanzhou 362000, China*

[2] *College of Textiles, Donghua University, Shanghai 201620, China*

* *Corresponding author's email*: jin_jingye@ 163. com

Abstract: Flexible pressure sensors are important components of flexible wearable electronic textiles. In this study, a yarn-based pressure sensor was prepared with silver-coated nylon filament yarn as the core layer, electrospinning nanofibers with piezoelectric properties as the skin layer, and carbon nanotubes coated on the surface layer. The sensor uses silver-plated nylon filaments and carbon nanotube coatings as inner and outer electrodes, and uses electrospinning nanofibers as insulating piezoelectric layers, which can output electrical signals when subjected to external low-frequency pressure. At different pressures and pressure frequencies, the effects of process factors and piezoelectric properties of yarn-based pressure sensors were studied. It is found that the double-layer skin structure can effectively prevent the carbon nanotube solution from infiltrating into the core layer to form a short circuit; the increase of the concentration of carbon nanotube solution and the number of coating times can effectively improve the electrical conductivity of the surface electrode and improve the piezoelectric signal. At the pressure of 5. 56–22. 22 kPa and the pressure frequency of 1–3 Hz, the yarn-based pressure sensor with a length of 3 cm can stably output an alternating current signal of 0. 33–1. 21 V, showing its great potential in building smart wearable electronic textiles.

Keywords: Electrospinning nanofibers; Carbon nanotube; Piezoelectric properties; Core-spun yarn

1 Introduction

Electrospinning piezoelectric nanofibers is a kind of micro-nano fiber materials made of polymers by electrospinning. During the curing process, the polymers are subjected to high-speed drafting and polarization by an electric field, resulting in a specific crystal form with piezoelectric effect. After years of development, electrospun piezoelectric nanofiber aggregates have become a representative form of flexible piezoelectric materials, and have accumulated rich application cases in emerging fields such as smart wearable textiles[1-3].

Clothing is the second skin of human beings, and yarn, as the basic material of clothing fabrics, has a range and duration of contact with the human body that is unmatched by any other material. The yarn based on electrospun nanofibers not only has the typical anisotropic structural characteristics of traditional yarns, effectively overcomes the disadvantages of low mechanical properties, poor delamination resistance, and unstable structure of membrane-structured nanofiber aggregates. but also greatly expand the application scope of nanomaterials, which have broad application prospects in functional textiles, drug carriers, microelectronic devices and other fields[4-7].

In this paper, the silver-coated nylon yarn is used as the core layer and the inner electrode, the electrospun piezoelectric nanofiber is used as the skin layer and the piezoelectric insulating layer, the carbon nanotube serosa is the outer electrode. By studying the effects of spinning solution concentration, cortex structure, carbon nanotube concentration and coating times on the serosa resistance, the preparation process was optimized, and the piezoelectric response of carbon nanotube-coating piezoelectric nanofiber core-spun yarn was investigated.

2 Experiments

2. 1 Materials

PVDF-HFP (average relative molecular weight

400 000) was obtained from Sigma-Aldrich Trading Co., Ltd. (Shanghai, China). Silver-plated nylon filament yarn (fineness 350 D) was supplied by Shandong Boyin Surface Functional Materials Co., Ltd. (Shandong, China). Carbon nanotube dispersion (0.4% by mass) was obtained from Shenyang Huijing Nanotechnology Co., Ltd. (Liaoning, China). N, N-Dimethylformamide (analytical grade) was supplied by Shanghai McLean Biochemical Technology Co., Ltd. (Shanghai, China).

2.2 Sample preparation

A certain amount of PVDF-HFP powder was added to a certain volume of DMF solvent, stirred with a magnetic stirrer at 70 ℃ for 2 h, then added a certain volume of acetone and continued to stir at 30 ℃ for 8 h, finally the mass fractions of 14%, 16%, 18%, 20%, 22% and 24% of PVDF-HFP electrospinning solution were prepared. The volume ratio of DMF to acetone in the solution was 2 : 3.

In the preparation process, the spinning voltage was 10 kV, the spinning solution advancing speed was 1mL/h, the winding speed was 12 cm/min; the twisting speed was 300 r/min.

The carbon nanotube pre-dispersion with N-methylpyrrolidone as the carrier was diluted with deionized water to prepare carbon nanotube slurry with mass fractions of 0.4%, 0.3% and 0.2%. The spun nanofiber core-spun yarn was immersed in the carbon nanotube slurry for 10 s and taken out, and the excess slurry on the surface of the yarn was removed with a soft brush, and then placed in a drying oven for constant temperature drying for 10 min.

2.3 Testing and characterization

The microstructure and morphology of the yarn were observed by scanning electron microscope, and the diameter of nanofibers was measured and characterized by Nano Measurer software. A digital multimeter was used to measure the resistance per unit length of the carbon nanotube serous film on the yarn surface along the yarn axis to measure its conductivity as an outer electrode. In addition, one probe of the digital multimeter was touched to the outer electrode and the other to one end of the conductive core yarn (the inner electrode) to check the continuity of the inner and outer electrodes.

A certain length of carbon nanotube coated nanofiber core-spun yarn was fixed on the detection head of the dynamometer, and the linear motor was controlled to apply a periodic pressure of 1 Hz, 2 Hz or 3 Hz to the yarn. The contact area between the detection head of the dynamometer and the head end of the linear motor was a square of 3 cm×3 cm, and a silicone rubber layer was glued on each side to protect the core-spun yarn. The length of the core-spun yarn was 3 cm under force. The inner and outer electrodes of the core-spun yarn were connected to the two probes of the oscilloscope with wires to collect piezoelectric signals for further filtering and analysis.

3 Results and Discussion

3.1 Effect of spinning solution concentration on nanofiber diameter

A scanning electron microscope photo of the core-spun yarn was taken to observe the morphology and actual coating of the cortical nanofibers, as shown in Fig. 1. It can be seen that under the winding and twisting action of the spinning device, the nanofibers of the cortex layer were locally disordered and ordered in an overall state, and the core layer was well covered with a certain twist without core leakage. Further magnification revealed that the nanofibers had smooth surfaces and u-

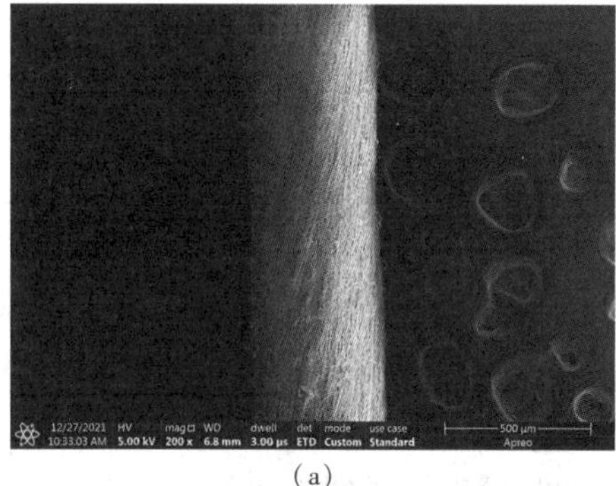

(a)

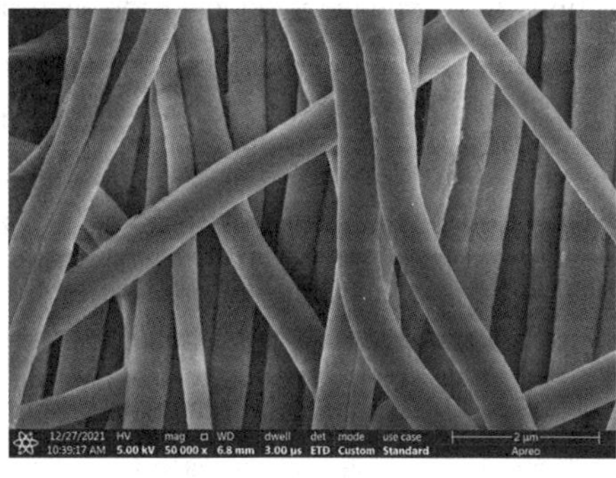

(b)

Fig. 1 SEM images of cortical nanofibers: (a) ×200; (b) ×50 000

niform diameters. The measurement showed that as the concentration of the spinning solution increases from 14% to 24%, the average diameter of the cortical nanofibers gradually increased from 454 nm to 711 nm, and the diameter distribution of the nanofibers with the same concentration was more concentrated, as shown in Fig. 2.

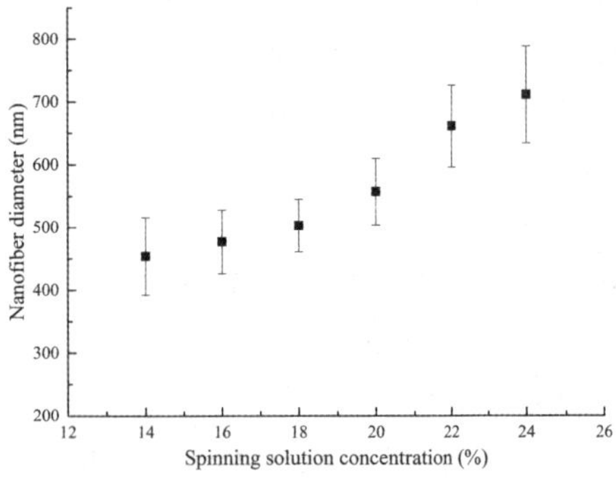

Fig. 2 Nanofiber diameters with different spinning solution concentrations

3.2 Influence of nanofiber diameter on the effect of carbon nanotube slurry coating

In order to further improve the stability of the nanofiber insulating layer, a core-spun yarn covered with double-layer nanofibers was prepared in this paper. That is, spinning with 24% concentration spinning solution first, and the core layer of the conductive yarn was covered to form a single-layer core-spun yarn; then spin the spinning solution with 14%, 16%, 18%, 20% and 22% concentration respectively to cover the single-layer core-spun yarn to obtain a double-layer core-spun yarn. The double-layer core-spun yarn not only had a thicker nanofiber insulating layer, but also the outer fiber diameter was smaller than that of the inner fiber, which had a certain reverse wicking effect.

Forty positions were randomly selected for each concentration of spinning solution samples, and the resistance per unit length (per 1 cm) was tested. The statistical results are shown in Fig. 3. When the spinning solution concentration of the outermost nanofibers was 18%, the surface electrode resistance of the core-spun yarn after coating with carbon nanotubes was the smallest, which was 307.41 $\Omega \cdot cm^{-1}$; when the spinning solution concentration further decreased or increased, the resistance values all showed an increasing trend. The three-dimensional void structure constructed by the outermost nanofibers had a wicking effect on the liquid carbon nanotube slurry. However, due to the large specific surface area of carbon nanotubes, the viscosity of the slurry was relatively large, which hinders the penetration of the slurry in the pores of small size. These two factors work together to make the surface electrode resistance of the core-spun yarn show the characteristics of first decreasing and then increasing with the increase of spinning solution concentration.

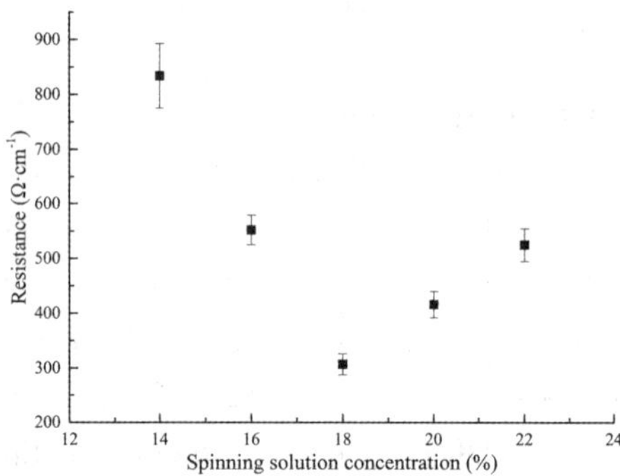

Fig. 3 Surface resistance of core-spun yarn with different concentrations of spinning solution

3.3 Influence of coating times on coating effect of carbon nanotube slurry

Observation on the surface of the core-spun yarn after coating once found that the carbon nanofibers were evenly dispersed and attached to the surface of the yarn, and were closely combined with the nanofiber skin, but the formed carbon nanotube serosa was full of holes, the yarn surface was not completely covered, as shown in Figs. 4 (a) and 4 (b).

Repeated coating experiments were carried out on the double-layer core-spun yarn prepared by the same process. Electron microscope observation found that a relatively complete carbon nanotube sizing film had been formed on the surface of the core-spun yarn after coating twice. The tube serosa basically formed a continuous coverage of the core-spun yarn; the carbon nanotubes had a higher filling degree in the nanofiber skin layer, and even form a three-dimensional film-like network in a certain depth below the surface layer, as shown in Figs. 4 (c) and 4 (d). When the core-spun yarn was repeatedly coated for 3 times, a dense and smooth continuous serous film would be formed on the surface of the yarn with a large thickness, so that the nanofiber texture almost disappeared, and carbon nanotubes fully filled the nanofiber skin within a certain thickness, as shown in the Figs. 4 (e) and 4(f). Further increasing the number of coatings, the surface micro-morphology of the core-spun yarn no longer changes significantly.

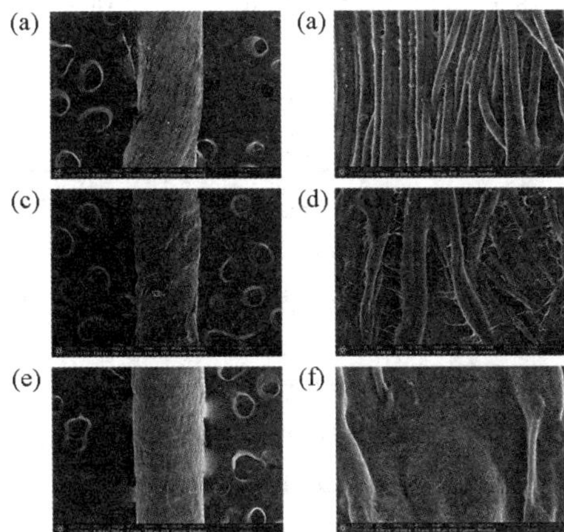

Fig. 4 SEM images of the serosal film on the surface of the core-spun yarn with different times of coating: (a) once, ×200; (b) once, ×20 000; (c) twice, ×200; (d) twice, × 20 000; (e) 3 times, × 200; (f) 3 times, ×20 000

As shown in Fig. 5, with the increase of the number of coatings, the surface resistance of the core-spun yarn decreased rapidly and then tended to be stable. It had dropped to 65.34 $\Omega \cdot cm^{-1}$ after 3 coating times, and if the number of coatings continued to increase, the resistance only fluctuated between 60.77 $\Omega \cdot cm^{-1}$ and 46.94 $\Omega \cdot cm^{-1}$ without significant change.

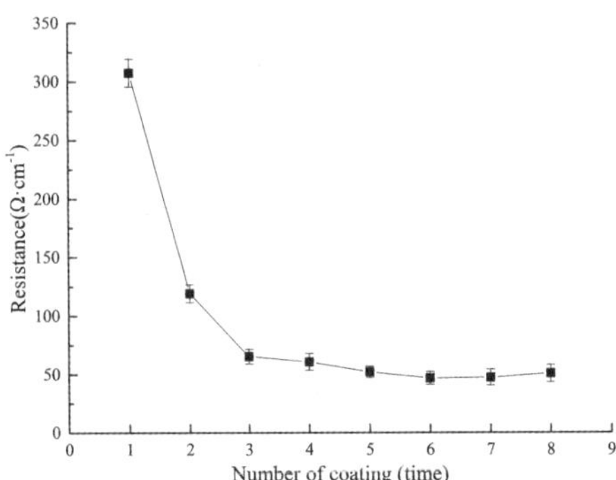

Fig. 5 Surface resistance of core-spun yarn after different times of coating

3.4 Piezoelectric response of carbon nanotube coated nanofiber core-spun yarn

As the pressures used in this paper were 5 N, 10 N and 20 N, the corresponding pressures were 5.56 kPa, 11.11 kPa and 22.22 kPa, which were between the pressure levels of daily elastic knitted clothing and sports shoe insoles on the human body[8].

As shown in Fig. 6. In the low-frequency pressure modes of 1 Hz, 2 Hz and 3 Hz, when the force frequency was constant, the output piezoelectric voltage of the core-spun yarn increased steadily with the gradual increase of the pressure. When the pressure remained unchanged, with the increase of the force frequency, the output voltage of the core-spun yarn also showed a significant upward trend. When subjected to a pressure of 5 N at 1 Hz, the output voltage was relatively low at 0.33 V, and when subjected to a pressure of 20 N at 3 Hz, the output voltage was relatively high, reaching 1.21 V.

When subjected to pressure, the core-spun yarn first underwent a deformation dominated by structural deformation between fibers. At this time, the external pressure overcame the surface friction of the skin nanofibers and the core conductive yarn fibers, so that the position and orientation of the fibers changed until the reconstruction of the position was completed and a new equilibrium state was reached. Then, under the further action of external force, the fiber itself underwent elastic deformation, the orientation of the internal dipole changed, and charges were generated on the surface of the fiber. The greater the external pressure, the more sufficient the elastic deformation of the fibers, the more fibers that underwent elastic deformation, and the stronger the piezoelectric signal. In the case of low frequency, increasing the pressure frequency, the pressure acceleration was higher, and the fiber discharge time was shorter, which was beneficial to generate a higher voltage. In addition, the interval between two adjacent pressures was shortened, and the structural deformation inside the yarn was not fully recovered, resulting in the elastic deformation of the fiber itself to bear more pressure, which also helped to increase the piezoelectric voltage.

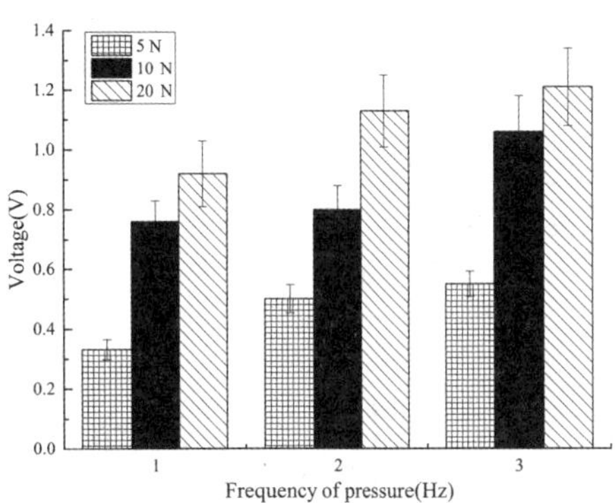

Fig. 6 Piezoelectric response of carbon nanotube coated nanofiber core-spun yarn

4 Conclusions

This paper introduced the preparation and piezoelectric properties of carbon nanotube coated nanofiber core-spun yarn, which can be used as a flexible sensor in a single form, or can be made into a fabric as a flexible fabric power source. The spinning solution concentration had a significant impact on the penetration and spreading effect of carbon nanotube slurry on the surface of the core-spun yarn by adjusting the diameter of the nanofibers. The double-layer nanofiber core-spun structure with inner thickness and outer fineness could effectively prevent the carbon nanotube slurry penetrating into the conductive yarn core layer to form a short circuit. When the pressure frequency and pressure value increased, the output voltage of the carbon nanotube coated nanofiber core-spun yarn increased accordingly, and the output voltage could reach 1.21 V when the frequency was 3 Hz and the pressure was 22.22 kPa. For a single core-spun yarn, this level of output voltage was sufficient to meet the signal strength requirements of general wearable devices for flexible pressure sensors. If multiple core-spun yarns were connected in parallel to make a flexible fabric power supply, the electronic components could be driven after rectification.

Acknowledgments

This work was supported by National Natural Science Foundation of China (No. 51803108), Fujian Provincial Natural Science Foundation of China (Nos. 2021J01977 and 2019J05106), Quanzhou City Science & Technology Program of China (Nos. 2018G011 and 2019G028), and Quanzhou Home-bay Recruitment Program of Global Talents (No. 2017ZT002).

References

[1] ANTON S R, SODANO H A. A review of power harvesting using piezoelectric materials[J]. Smart Materials and Structures, 2007, 16(3): 1-21

[2] AI Y, LOU Z, CHEN S, et al. All rGO-on-PVDF-nanofibers based self-powered electric skins[J]. Nano Energy, 2017, (35): 121-127.

[3] LEE J, YOON H, KIM T, et al. Micropatterned P(VDF-TrFE) film-based piezoelectric nanogenerators for highly sensitive self-powered pressure sensors [J]. Advanced Functional Materials, 2015, 25(21): 3203-3209.

[4] WU S, ZHANG C, QIN X, et al. Progress in electrospun aligned nanofiber bundles and yarns[J]. Polymer Materials Science and Engineering, 2014, 30(6): 182-186.

[5] YANG Y, QIN X, YU J. Research progress of transforming electrospun nanofibers into functional yarns [J]. Journal of Textile Research, 2021, 42(1): 1-9.

[6] LIU C, HE H, SUN R, et al. Influence of spinning process on covering property of core-spun nanofiber yarn[J]. Polymer Materials Science and Engineering, 2016, 32(12): 82-86.

[7] YU Y. Preparation and properties of PVDF piezoelectric yarn and flexible wearable fabric[D]. Xi'an: Xi'an Polytechnic University, 2021.

[8] LIU J, GAO Z. Development of clothing pressure detector based on flexible film sensor[J]. Journal of Textile Research, 2021, 42(12): 159-165.

Characterization of Ethiopian Cotton Fiber Properties for Its Suitability in Producing Quality Ring Spun Yarn

TILIKSEW Addis[1,2*], **ABERA Kachi**[1], **NICHOLUS Tayari Akankwasa**[2], **WANG Jun**[2]

[1] *Ethiopian Institute of Textile and Fashion Technology, Bahir Dar University, Bahir Dar 6000, Ethiopia*

[2] *The State Key Laboratory of Textiles, Donghua University, 201620 Shanghai, China*

* *Corresponding author's email*: hulieaddis27@gmail.com

Abstract: Yarn property is highly dependent on fiber properties, the arrangement of fibers within the yarn, and the mass distribution of fibers along its length. The research was focused to analyze the effect of the constituent cotton fibers on the yarn properties of Ethiopian cotton. Commercially cultivated cotton (DP 90, Stam 59 A), and one variety which is cultivated by small-scale farmers were considered for this study. Fiber properties were tested using USTER® HVI 1000. The results were analyzed and compared for the three varieties of cotton fibers. Of the three varieties of cotton, the local cotton variant has better fiber quality for the main cotton fiber properties that determine yarn properties. The comparison of the effect of different cotton types on yarn properties was assessed by producing 30 tex count yarn which is commonly produced in Ethiopian textile factories. Yarn samples were tested and analyzed for mass variation, strength, elongation, and hairiness. The analysis shows that the yarn produced from the local cotton variant exhibits better quality than the yarn samples produced from the other two variants of cotton fibers in all yarn properties except for yarn elongation and hairiness property.

Keywords: DP 90; Hairiness; Stam 59; Local; Yarn evenness; Yarn

1 Introduction

The quality of fiber used for yarn production predominantly determines the yarn quality and property. The process of yarn production can be influenced by several factors such as properties of raw material, type and level of technology used for production and skill of workers[1-2]. The demand for raw-materials for textile and cloth manufacturing become increasing since the recent years due to the expansion of textile and clothing industries in Ethiopia. Hence, the country is focusing and doing effort to have persistent supply of quality raw-materials mostly cotton for its booming textile and clothing industries[3]. Then the two commercially producing cotton varieties of DP 90, Stam 59A and the Local variant of cotton samples were used for producing ring spun yarn. The growing area for the considered cotton variants are the three regional states of Amhara, Afar and Benishangul Gumuz, situated in North-West, North-East and West Ethiopia. The main objective of this work is to test the important cotton fiber properties required for yarn production and testing the quality of yarn produced from the considered varieties of cotton samples growing in Ethiopia and show which variety of cotton can produce better quality yarn.

2 Materials and Methods

2.1 Materials

Three cotton varieties of DP 90, Stam 59A and the Local ones were collected with sufficient amount for testing and production of yarn samples, from the three regional states of Amhara, Benishangul Gumuz, and Afar Regional States of Ethiopia. The samples were collected in specific places of Metema and Quara (Amhara), Worer Agricultural Research Center (Afar) and Asosa Agricultural Research Center (Benishangul Gumuz). The cotton sample varieties were collected from the 2019 and 2020 years of crop.

2.2 Sampling, production and testing

2.2.1 Fiber quality testing

The experimental work includes fiber property testing, yarn production, and the produced yarn quality testing. Before conducting testing the samples were conditioned using standard conditioning chambers at Ethiopian Institute of Textile and Fashion Technology (EiTEX),

Bahir Dar University Laboratory, Ethiopia and at State Key Laboratory of Textiles, Donghua University Shanghai, China. The three variants of the main cotton fiber quality determining, properties of fiber length, length uniformity, short fiber content, microniare value, elongation, and strength were tested using USTER® HVI 1 000 instrument and ASTMD4605 - 86 test method. The obtained test results were analyzed using software application.

2.2.2 Yarn production and quality testing

After testing and evaluation of the physical properties the three variants of cotton fiber samples of DP 90, Stam 59A, and Local, the preparation for yarn production carried out. Accordingly, the cotton fiber samples were manually opened and processed using carding and drawing units with appropriate machine settings Tab. 1 Yarn production was carried out using Laboratory-Scale of Chinese Mini Mill Spinning Machine (Huarui Jiahe Textile Machines). Yarn count of (30 tex) with twist of 750 turns per meter was produced from the three cotton sample variants. For the yarn produced from the three variants of cotton fiber samples, comparison for mass variation, thick place, thin place, neps, strength, elongation and hairiness were assessed.

Tab. 1 The machine used for production of intermediate products and yarn

No.	Product	Count (tex)	Machine type/model
1	Sliver	4 921	Micro-blowing combined with card: FA158-50H
2	Sliver	4 921	Draw frame: CNST-B01
3	Roving	421	Roving frame: CNST-C02
4	Yarn	30	Spinning frame: CNST-X01

The comparison of yarn quality determining properties was carried out in order to study, the variety of cotton, yield yarn with acceptable quality parameters. After production of yarn, the quality of yarn were tested per the standards method described in Tab. 2

Tab. 2 Yarn quality standard test methods

No.	Type of test	Instrument	Test method
1	Twist	Y331N	GB/T 2543—2011
2	Strength	YG061—1500	ISO 2062:2009
3	Evenness	CT 3000	ASTM D1425

2.2.3 Statistical analysis

The three variants of cotton fiber samples were tested and analyzed for the suitability of ring-spinning process to produce yarn with required quality parametrs. Yarn production comparative analysis was carried out to establish a quantitative relationship between produced yarn and cotton fiber sample variants used for yarn production. The input parameters used for testing and analysis are cotton varieties with their corresponding quality determining fiber properties, while the output parameters are yarn mass variation (CvM), tensile strength and hairiness. The analysis was made by using Minitab® 17 statistical software (MINITAB®, 2006). Several methods of mathematical analysis could be used to study the relation between fiber and yarn properties namely, fiber quality index (FQI), the spinning consistency index (SCI), the premium-discount index (PDI), and multiple-criteria decision-making (MCDM) are the main ones[4]. The feasibility of utilizing HVI, fiber micronaire and strength property data, as a semi-quantitative and fast tool, to compare the yarn tenacity performance can be used for yarn to fiber property relations analysis[5].

3 Results and Discussion

3.1 The three variant of cotton fiber quality test result and discussion

The cotton fiber quality test results for the three variants of cotton fiber samples are depicted in Tab. 1. The cotton quality parameters were tested using USTER® HVI 1 000. The tested cotton quality parameters are microniare, maturity, length, length uniformity (UL), short fiber content (SFC), strength, elongation and spinning consistency index (SCI).

The microniare value for the three variants of cotton fibers samples were measured and recorded in increasing order of 3.74, 3.92, and 4.12, for Local, DP 90 and Stam 59 A respectively. According to the microniare value range classification, the microniare value from 3.1 to 3.9 is fine cotton fiber category, and hence the DP 90 and Local cotton fiber variants are fine cotton fibers. While the microniare value of Stam 59 A cotton is relatively coarser than the two variants of cotton fiber samples. The microniare value of 4.0 to 4.9 leveled as medium fine and hence Stam 59 A cotton fiber variant is medium fine according to USDA cotton classification system.

The maturity for the three variants of cotton fiber samples were measured and recorded as 0. 86 for Stam 59 A, 0. 85 for DP 90 and 0. 84 for Local cotton as depicted in Tab. 3. As per the maturity level designation, the maturity ratio of all the fibers is categorized above average maturity level. Yarn strength and fineness are dependent on fiber maturity[6].

As shown in Tab. 3, the length uniformity index (UI) for the three variants of cotton fiber samples were measured as 81. 53% for Stam 59A, 83. 28% Dp 90 and 86. 02% for Local cotton variants.

Tab. 3 Fiber property statistics for DP 90, Stam59A and Local cotton Variant

	Variable	No	Mean	St. Dev	Coef. Var
Dp 90	Microniare	50	3. 92	0. 41	10. 41
	Maturity	50	0. 85	0. 01	1. 31
	UHML	50	28. 62	1. 52	5. 30
	UI	50	83. 28	2. 53	3. 03
	SFC	50	12. 5	1. 17	9. 36
	Strength	50	28. 04	2. 29	8. 18
	Elongation.	50	6. 37	0. 76	11. 93
	SCI	50	118	5. 8	4. 9
Stam 59 A	Microniare	52	4. 12	0. 18	4. 49
	Maturity	52	0. 86	0. 01	0. 72
	UHML	52	27. 77	0. 79	2. 84
	UI	52	81. 53	1. 23	1. 51
	SFC	52	11. 40	1. 87	16. 35
	Strength	52	27. 33	1. 64	5. 99
	Elongation.	52	6. 04	0. 52	8. 64
	SCI	52	121	10. 8	8. 88
Local	Microniare	52	3. 74	0. 47	12. 48
	Maturity	52	0. 84	0. 01	1. 31
	UHML	52	30. 09	1. 38	4. 59
	UI	52	86. 02	1. 48	1. 72
	SFC	52	6. 51	0. 91	13. 89
	Strength	52	29. 65	2. 86	9. 65
	Elongation	52	6. 89	0. 69	10. 02
	SCI	52	159	16. 6	10. 44

Therefore, according to HVI, classification the UI for Local cotton fiber variant is very high, while for DP 90 is high and for Stam 59 A is intermediate.

The strength of cotton fibers under investigation were measured in (cN/tex) as 27. 33 for Stam 59 A, 28. 04 for DP 90, and 29. 65 for Local cotton variants Tab. 3. The high strength value implies the strong the fiber and correspondingly strong yarn will be produced from it. As per HVI fiber strength classification the Local cotton fiber sample (29. 65 cN/tex) variant is under strong fiber classification, and while the strength values of DP 90 (28. 04 cN/tex) and Stam 59 A (27. 33 cN/tex) are at average strong fiber classification.

The average increase in length of the sample at its break (rupture) point is explained as elongation. Cotton fiber usually elongates from 4% to 8% before breaking. As shown in Tab. 3, the elongation values of all the three cotton fiber variants fell in this elongation range with measured value of 6. 04% for Stam 59A, 6. 37% for DP 90 and 6. 89% Local cotton.

As shown in Tab. 3 UHML of 28. 62mm for DP 90, 27. 77 mm for Stam 59 A, and 30. 09 mm for Local cotton fiber variants have been measured. Hence, the Local cotton fiber variant is longer than the two variants of cotton fibers.

3. 2 Yarn production, quality test result and discussion

The produced yarn quality was tested for yarn count, twist, evenness, tensile strength, elongation, and hairiness using the laboratory instruments and standards described in Tab. 1 and Tab. 2. The yarns produced from the three cotton fiber variant samples Yarn1, Yarn 2, Yarn 3 were produced from variants Dp 90, Stam 59A and Local cotton respectively.

3. 2. 1 Yarn count and twist

To study the effect of fiber properties on ring spun-yarn quality, yarn with 750 turns per meter average twist with nominal count of 30 tex were produced. As illustrated in Tab. 2.

3. 2. 2 Yarn evenness

Yarn evenness, was measured using CT3000 evenness tester. The CT3000 machine adopts advanced charge-coupled device (CCD) digital imaging technology to test the appearance of the quality of yarns[7]. From the graphs, it is observed that Yarn 3 spun from Local variant of cotton has a lower coefficient of variation (CV) as described in Fig. 1.

3. 2. 3 Yarn strength and elongation

Accordingly, strength test is carried out for the produced sample Yarn 1, Yarn 2 and Yarn 3 from the three variants of cotton fibers. The tensile properties of yarns were determined by using YG 061—1500 yarn

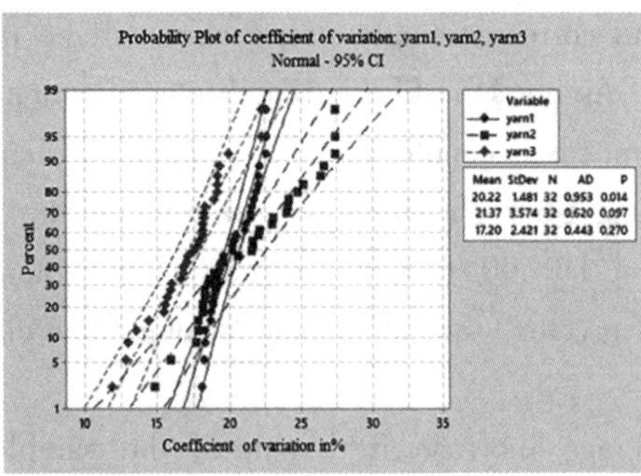

Fig. 1 Probability plot of mass variation CV% for Yarn 1, Yarn 2 and Yarn 3 spun from three different cotton varieties: local, Stam59A and DP 90

tensile tester at room temperature with applying ISO 2062: 2009 FZ/T01034 standard test methods.

As per the assessed result Y3, spun from Local cotton variant was exhibited high tensile strength (cN/tex) value as shown in Fig. 2. Therefore, as per measured tensile strength value, the three yarn samples preference for their strength ordered as Yarn 3>Yarn 2> Yarn 1.

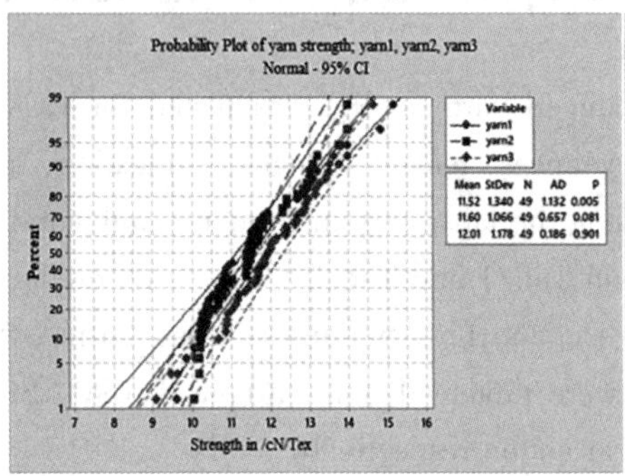

Fig. 2 Probability plot of the Tensile strength of strength in cN/tex of Yarn 1, Yarn 2 and Yarn 3 spun from three different cotton varieties (local, Stam59A and DP 90)

The elongation of yarn plays a very important role during weaving. According to test result of elongation percentage for the three yarn samples, their preference for elongation property is ordered as Yarn 1> Yarn 3> Yarn 2, Fig. 3

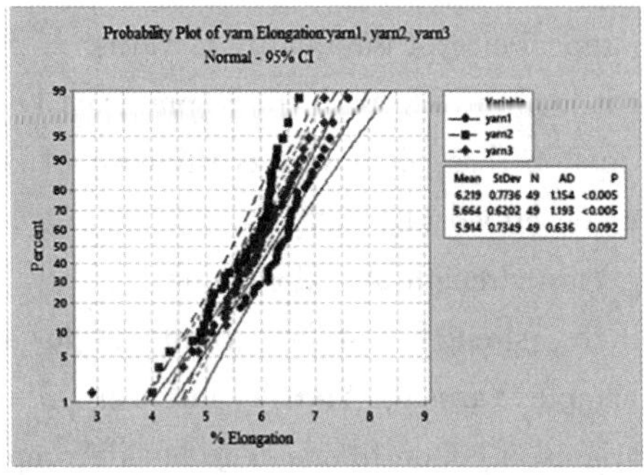

Fig. 3 Probability plot of percentage elongation of Yarn 1, Yarn 2 and Yarn 3

Yarn hairiness was tested using YG172A yarn hairiness tester and the instrument measures hairiness length, hairiness number, hairiness distribution and hairiness index automatically. Testing speed of 100 m/min and the indices of hairiness length 3 mm were considered for measuring and analyzing hairiness as described in Fig. 4. The degree of hairiness is relatively less for Yarn 2 sample, produced from Stam 59 A cotton variant. As per degree of hairiness of the three yarn samples produced, the yarn with less degree of hairiness is preferable and their preference ordered as Yarn 2 > Yarn 3 > Yarn 1.

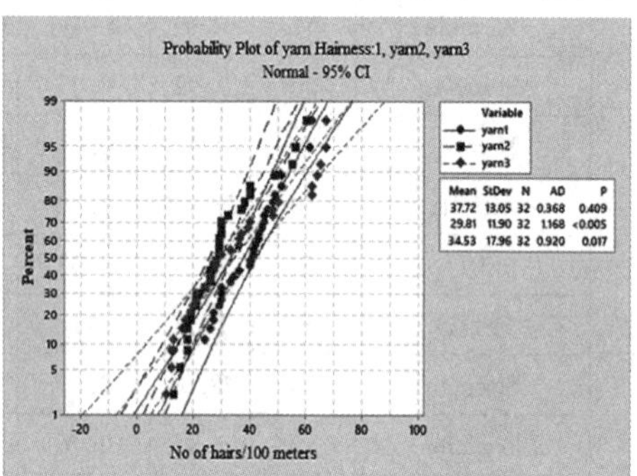

Fig. 4 Probability plot of the number of hairs of yarn in 100 meters of yarn samples spun from three different cotton varieties: local, Stam59A and DP 90

4 Conclusions

All the three cotton variants considered under the study in general have good spinnable properties. From all the three cotton fiber variants, Local cotton has superior properties microniare, strength, length, and length uniformity than the two commercially cultivated cotton variants in Ethiopia. Accordingly, except elongation and hairiness, the yarn spun from Local cotton variant is more preferable with its quality particularly in evenness and strength. Therefore, research and development has to be intensively carried out further to improve the local cotton types which is handled by small holder farmers.

References

[1] KOTB N A. Predicting yarn quality performance based on fibers types and yarn structure[J]. Life Science Journal, 2012. 5(28 (3)): 13-15.

[2] FAULKNER W B, HEQUET E F, WANJURA J. Boman-relationships of cotton fiber properties to ring-spun yarn quality on selected high plains cottons[J]. Textile Re-

search Journal, 2012, 82(4): 400-414.

[3] ADDIS T, KACHI A, WANG J. A review of current state and future directions of cotton production in Ethiopia[J]. Cogent Food & Agriculture, 2021, 7(1): 0-12.

[4] THIBODEAUX D, SENTER H, KNOWLTON J L, et al. The impact of short fiber content on the quality of cotton ring spun yarn[J]. Journal of Cotton Science, 2008, 12(4): 368-377.

[5] MOHAMMAD A M, HABIB M M. Supply chain management of textile industry: a case study on Bangladesh[J]. Supply Chain Management-an International Journal, 2012, 1(2): 35-40

[6] HARIJAN Y. Comparative study for seed cotton yield and fibre quality parameters in cotton (Gossypium sp. L.)[J]. International Journal of Pure & Applied Bioscience, 2017, 5(2): 519-525.

[7] ZHANG D, CHENG L. Comparison of two different yarn evenness test methods[J]. Modern Applied Science, 2010, 6(3): 71-76.

Tactile Comfort Evaluation of Knitted Automobile Seat Fabric

LI Ling[1], ZHANG Peihua[1*]

[1] *Department of Technical Textiles, College of Textiles, Donghua University, Shanghai 201620, China*

[*] *Corresponding author's email*: phzh@ dhu. edu. cn

Abstract: As an important part of the car's interior, the car seat fabric is also the most direct part of the ride. This article will select 50 kinds of different structural parameters of knitting structure car seat fabrics, and comprehensively evaluate the touch comfort of car seat fabrics through the combination of subjective evaluation and objective testing and evaluation. Firstly, the subjective evaluation of knitted car seat fabric touch comfort is completed through two methods: questionnaire investigation and fabric feel evaluation. Then, determine the objective mechanical performance index of fabric physics based on subjective evaluation indicators, including fabric stretching and break, stretching 17 indicators such as reply, bending, and compression. Finally, subjective and objective evaluation are linked with objective evaluation through data analysis.

Keywords: Car seat fabric; Tactile comfort; Knitted fabric; Fabric handle evaluation

1 Introduction

In the context of the new economy, my country's automobile industry has developed rapidly, the per capita vehicle ownership rate continues to increase, and consumers' awareness of car comfort is getting stronger and stronger. Car seats are the core part of the car interior. It is composed of the seat skeleton, filling layer and epidermal surface. As an important part of the car seat, it not only plays a protective and decorative role in the seat, but also the first interface between passengers and cars. The seat fabric has received special attention.

2 Experiments

2.1 Materials

In order to make the research conclusions of this article applied to most conventional knitted car seat fabrics, 44 knitted car seat fabrics with different linear densities, thicknesses, area densities, fabric structures and appearance effects were selected as the research objects.

2.2 Experimental equipment

This experiment mainly uses KES-FB－3 compression tester, KES-FB－2 pure bending tester and YG026MB multifunctional electronic fabric strength machine.

2.3 Experimental method

2.3.1 Subjective evaluation

In this paper, the subjective evaluation of tactile comfort of knitted car seat fabric is completed by questionnaire survey and fabric hand feeling evaluation. Firstly, through questionnaire survey, the evaluation indicators that affect the tactile comfort of knitted car seat fabric are selected: roughness, softness, spring back and comprehensive comfort. After that, the five grades of the above indexes are measured by the way of hand feeling evaluation, so as to complete the subjective evaluation of knitted car seat fabrics.

A total of 268 valid questionnaires were collected on the question "what properties should automobile seat fabrics have". The questionnaire results are shown in Tab. 1.

Tab. 1 Questionnaire survey options and results

Option	Statistical results
Soft	81.34%
Stiff	18.28%
Rough	10.82%
Smooth	32.84%
Elastic force	58.21%
Gloss	18.66%
Ventilation	76.12%
Moisture absorption and conduction	48.13%
No static electricity	46.27%
Warm in winter and cool in summer	66.79%

According to the survey results of 268 questionnaires, we get three indicators of subjective evaluation: soft and hard, rough, rebound and comprehensive comfort. In this test, a five-level scale was used to evaluate rough — smooth, stiff — soft, no rebound — rebound, and uncomfortable — comfort. The subjective rating scale is shown in Tab. 2.

Tab. 2 Subjective rating scale

Index	Score				
	1	2	3	4	5
Rough	Rough	Coarser	Commonly	Smoother	Smooth
Soft and hard	Stiff	Relatively stiff	Commonly	Softer	Soft
Rebound	No rebound	Less elastic	Commonly	More elastic	Rebound
Comfort	Uncomfortable	Less comfortable	Commonly	More comfortable	Comfortable

2.3.2 Objective test

In this article, objective experiments will use the KES-FB fabric style instrument to test the physical mechanical properties of bending and compressing. The stretch and fracture performance of the fabric is tested with multi-functional electronic fabric strong machine. And thickness instrument will be used to test the thickness of the fabric. Precision weighing instrument will be used to test the area density of the fabric.

3 Results and Discussion

Comparative analysis of 11 kinds of fabrics of objective index of measured data and subjective experiments measured data, such as the objective measurement of the fabric thickness and roughness index comparison, compared with hard and tough, bending performance response performance test was compared with the rebound index, to evaluate the car seat fabric touch comfort perception test consistency between subject and object.

3.1 Contrastive analysis of subjective roughness and fabric area density

It can be seen from Fig. 1 (a) that the apparent density is closely related to the subjective roughness, and the correlation between the structural area density of different fabrics and the subjective roughness is different. For plain fabrics, the greater the area density, the finer and smoother the fabric, and the smaller the subjective roughness.

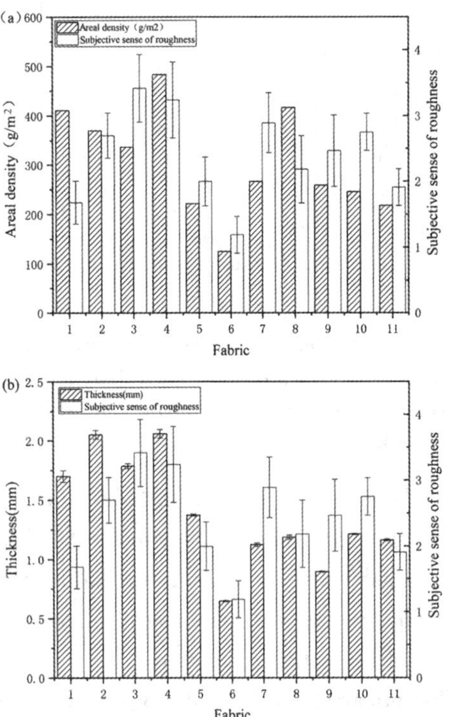

Fig. 1 (a) Contrastive analysis of subjective roughness and fabric area density; (b) contrastive analysis of subjective roughness and fabric thickness

For mesh fabric, the greater the area density, the stronger the subjective roughness of the fabric.

It can be seen from Figs. 1 (a) and 1 (b) that the thickness of the fabric of the car seat fabric can reflect the roughness of the fabric in the subjective evaluation. The thicker the fabric, the thicker the yarn, and the relatively rough surface. The larger the rough indicators in the subjective evaluation, the rough the fabric.

3.2 Comparative analysis of subjective softness and stiffness and fabric bending lag

It can be seen from Figs. 2 (a) and 2 (b) that the bending hysteresis moment 2hb of knitted car seat fabric can reflect the softness and hardness of the fabric. The larger the value of bending lag moment is, the

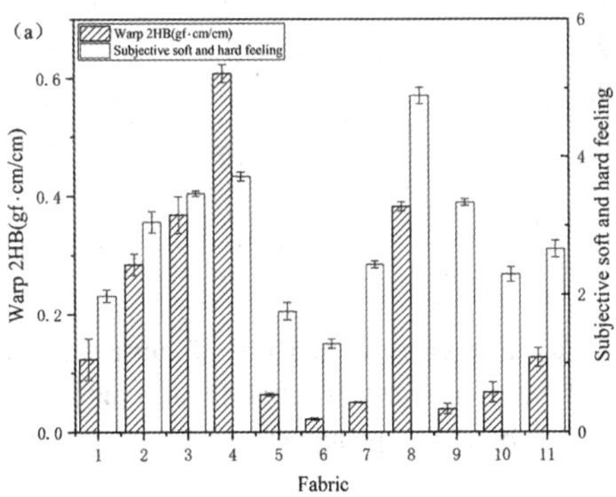

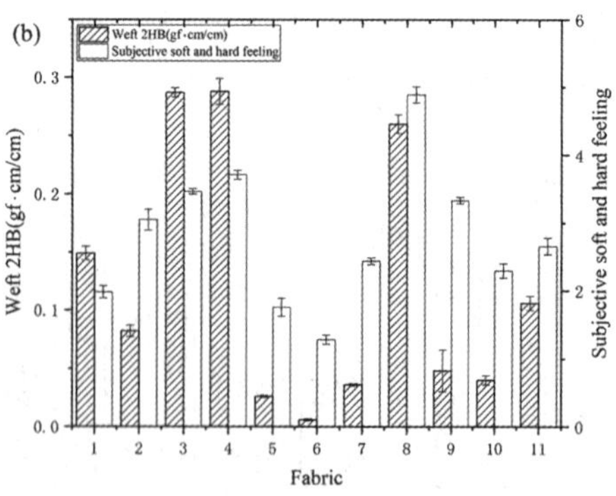

Fig. 2 (a) Comparative analysis of subjective softness and stiffness and warp bending lag of fabric; (b) comparative analysis of subjective softness and stiffness and weft bending lag of fabric

worse the bending resilience of the fabric is, and the fabric is more rigid; The less the score of softness and hardness in subjective evaluation, the harder the fabric is. The objective results are basically consistent with the results of subjective evaluation.

3.3 Comparative analysis of subjective springback and fabric recovery

It can be seen from Figs. 3 (a) and 3(b) that the recovery rate of knitted car seat fabric reflects the resilience of the fabric. The larger the recovery rate, the better the bending recovery of the fabric, that is, the better the resilience of the fabric. The larger the value of resilience in subjective evaluation, the better the resilience of the fabric. This objective result is basically consistent with the result of subjective evaluation.

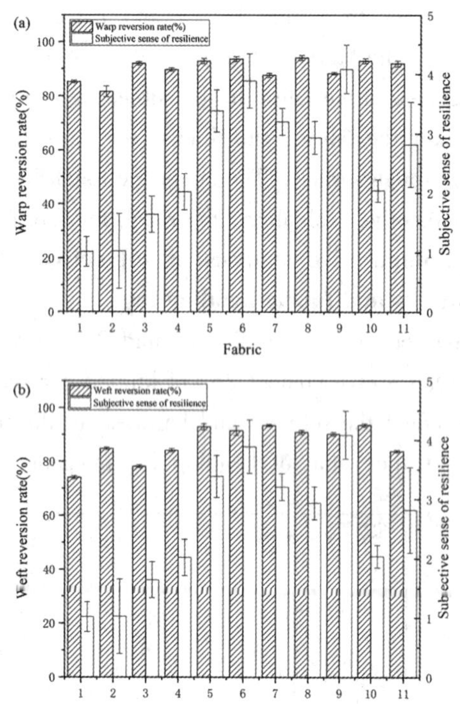

Fig. 3 (a) Comparative analysis of subjective springback and warp recovery of fabric; (b) comparative analysis of subjective springback and weft recovery of fabric

4 Comprehensive evaluation

Definition: $z1$ = thickness, $z2$ = areal density, $z3$ = warp breaking strength, $z4$ = warp elongation, $z5$ = weft breaking strength, $z6$ = weft elongation, $z7$ = warp test elongation, $z8$ = warp recovery, $z9$ = warp deformation rate, $z10$ = weft test elongation, $z11$ = weft recovery, $z12$ = weft deformation rate, $z13$ = warp bending stiffness Ba, $z14$ = warp bending lag 2hba, $z15$ = weft bending stiffness be, $z16$ = weft bending lag 2hbe, $z17$ = compression linearity LC, $z18$ = compression work WC, $z19$ = compression rebound rate RC. Establish the original data 44×19 matrix, then import the matrix data into the SPSS software, and $Z_{ij}(i=1, 2, \cdots, 44; j=1, 2, \cdots, 19)$ is standardized. The transformed index quality is $X_{ij}(i=1, 2, \cdots, 44; j=1, 2, \cdots, 19)$. SPSS software itself does not provide the function of principal component analysis, so it needs to use some functions of factor analysis to complete principal component analysis. Factor analysis is carried out on the standardized indicators of car seat fabrics, and the results of factor analysis are shown in Tab. 3.

Tab. 3 Partial table of correlation coefficient of component matrix

Index	Component				
	1	2	3	4	5
Z02	0.873	0.156	-0.244	-0.214	0.231
Z16	0.867	0.022	0.043	0.279	0.184
Z14	0.848	0.032	0.052	0.016	0.323
Z01	0.807	-0.141	0.299	-0.102	0.153
Z05	0.710	0.050	-0.403	-0.271	-0.230
Z08	0.135	0.722	0.598	-0.049	-0.210
Z09	-0.105	-0.718	-0.600	0.044	0.205
Z18	0.120	-0.245	0.757	-0.027	0.323
Z13	0.086	0.221	-0.223	0.846	0.031
Z06	-0.290	0.226	0.039	0.047	0.860

It can be seen from Tab. 4 that the characteristic values of five principal components extracted are $b1 = 5.930$, $b2 = 3.358$, $b3 = 2.335$, $b4 = 2.018$, $b5 = 1.699$, which are all greater than 1, and their cumulative variance contribution rate is 80.736%, which basically reflects the overall situation of the original data. On the whole, the result of factor analysis is ideal.

Tab. 4 Partial table of total variance of Index Interpretation

Component	Initial eigenvalue		
	Total	Variance Proportion %	Cumulative total %
1	5.930	31.212	31.212
2	3.358	17.673	48.885
3	2.335	12.288	61.173
4	2.018	10.623	71.796
5	1.699	8.940	80.736

Calculate the eigenvector and get the corresponding coefficient of each index in the five principal components. The characteristic vector is multiplied by the value of each index after standardization to obtain the score of the main component. The 19 indexes are converted into five indexes F_1, F_2, F_3, F_4 and F_5 to reflect the tactile comfort of the 44 knitted structure automobile seat fabrics tested. The linear equation is: ($i = 1, 2, \cdots, 19$)

$$F_{i1} = 0.331X_{i1} + 0.358X_{i2} + 0.143X_{i3} + 0.144X_{i4} + 0.292X_{i5} - 0.119X_{i6} - 0.246X_{i7} + 0.055X_{i8} - 0.043X_{i9} - 0.238X_{i10} - 0.223X_{i11} + 0.241X_{i12} + 0.035X_{i13} + 0.348X_{i14} + 0.282X_{i15} + 0.356X_{i16} - 0.083X_{i17} + 0.049X_{i18} - 0.231X_{i19} \quad (1)$$

$$F_{i2} = -0.077X_{i1} + 0.085X_{i2} + 0.117X_{i3} + 0.283X_{i4} + 0.027X_{i5} + 0.123X_{i6} - 0.273X_{i7} + 0.394X_{i8} - 0.392X_{i9} - 0.200X_{i10} - 0.371X_{i11} - 0.355X_{i12} + 0.121X_{i13} + 0.017X_{i14} + 0.069X_{i15} + 0.012X_{i16} + 0.353X_{i17} - 0.134X_{i18} + 0.169X_{i19} \quad (2)$$

$$F_{i3} = 0.196X_{i1} - 0.160X_{i2} - 0.298X_{i3} + 0.271X_{i4} - 0.264X_{i5} + 0.026X_{i6} + 0.099X_{i7} + 0.391X_{i8} - 0.393X_{i9} + 0.209X_{i10} - 0.154X_{i11} + 0.141X_{i12} - 0.146X_{i13} + 0.034X_{i14} - 0.105X_{i15} + 0.028X_{i16} - 0.095X_{i17} + 0.495X_{i18} - 0.102X_{i19} \quad (3)$$

$$F_{i4} = -0.072X_{i1} - 0.151X_{i2} - 0.434X_{i3} - 0.113X_{i4} - 0.191X_{i5} + 0.033X_{i6} + 0.009X_{i7} - 0.034X_{i8} + 0.031X_{i9} - 0.031X_{i10} - 0.173X_{i11} + 0.173X_{i12} + 0.596X_{i13} + 0.011X_{i14} + 0.391X_{i15} + 0.196X_{i16} + 0.286X_{i17} - 0.019X_{i18} + 0.206X_{i19} \quad (4)$$

$$F_{i5} = 0.117X_{i1} + 0.117X_{i2} + 0.222X_{i3} - 0.094X_{i4} - 0.176X_{i5} + 0.660X_{i6} + 0.069X_{i7} - 0.161X_{i8} + 0.157X_{i9} + 0.345X_{i10} + 0.113X_{i11} - 0.133X_{i12} + 0.024X_{i13} + 0.248X_{i14} + 0.071X_{i15} + 0.141X_{i16} + 0.239X_{i17} + 0.248X_{i18} - 0.138X_{i19} \quad (5)$$

The proportion of the eigenvalue corresponding to each principal component to the sum of the total eigenvalues of the extracted principal components is used as the weight to calculate the main feature index F of the comprehensive evaluation of the principal components. The comprehensive evaluation and score of the main ingredients are shown in Tab. 5.

Tab. 5 Comparison of subjective evaluation and objective evaluation

Fabric	Objective		Subjective	
	F score	Rank	Comprehensive comfort	Rank
1#	0.975	10	2.542	16
2#	0.507	11	1.780	29
3#	2.306	1	1.565	35
4#	1.274	8	1.363	37
5#	1.680	5	1.594	32
6#	-0.433	28	2.669	13
7#	-0.616	30	3.052	8
8#	0.029	20	3.197	6
9#	0.056	18	3.236	4
10#	-0.047	22	3.203	5
11#	1.472	7	0.959	41
12#	0.506	12	1.879	26
13#	0.125	17	1.325	38
14#	-2.016	44	2.427	19
15#	-1.161	41	3.762	1
16#	0.048	19	2.598	15
17#	-1.984	43	3.465	3
18#	-0.877	34	3.674	2
19#	1.854	3	1.554	36
20#	-0.970	35	1.845	27
21#	-0.198	24	3.025	9
22#	-1.123	40	2.854	11
23#	0.027	21	1.807	28
24#	0.483	13	1.923	25
25#	-0.703	32	1.070	39
26#	-0.174	23	2.393	20
27#	-0.757	33	2.117	22
28#	1.046	9	1.020	40
29#	0.455	14	1.615	31

(Continued)

Fabric	Objective		Subjective	
	F score	Rank	Comprehensive comfort	Rank
30#	-0.276	26	2.172	21
31#	-0.418	27	1.580	33
32#	-0.275	25	2.520	18
33#	0.157	15	2.643	14
34#	-1.012	36	2.542	17
35#	-1.104	39	2.872	10
36#	-0.515	29	2.706	12
37#	-1.064	38	1.571	34
38#	-1.207	42	3.086	7
39#	2.189	2	0.667	44
40#	1.502	6	0.828	43
41#	1.834	4	0.856	42
42#	0.142	16	1.756	30
43#	-0.687	31	2.115	23
44#	-1.049	37	2.113	24

The analysis and calculation of the dual variable correlation with SPSS can be used to obtain the correlation between objective evaluation and comfort indicators in subjective evaluation. According to analysis and calculation, the comprehensive evaluation results of objective touch comfort are significantly negatively related to the evaluation results of subjective fabric touch and comfort. The more obvious the objective mechanical performance of the seat fabric, the worse the contact comfort of the fabric.

5 Conclusions

The evaluation and analysis of the objective physical properties such as compression, bending and stretching of 44 kinds of knitted structure automobile seat surface show that the fabric parameters that affect the contact comfort are mainly fabric structure and yarn count. Generally, the coarser the yarn is, the greater the flexion required for interweaving, and the stiffer the fabric is formed. Mesh fabrics and double-sided jacquard fabrics are thicker than other fabric structures. The thickness durability of mesh fabric, warp flat fabric and single-sided fabric is relatively strong. Warp knitted fabrics have smaller bending stiffness and bending lag moment, and the fabric is softer, while mesh fabrics, pile fabrics and double-sided jacquard fabrics are thicker and stiffer, and it is more difficult to recover after bending. The thicker the yarn, the greater the breaking strength of the fabric, and the fabric with high breaking strength is more difficult to deform in the process of use.

The subjective evaluation and analysis of the roughness, softness, resilience and comfort of 44 kinds of knitted structure automobile seat fabrics show that the subjective evaluation results are close to the objective analysis results. Mainly reflected in the following aspects: fabric with small bending stiffness and bending lag moment is considered to be softer in subjective analysis. The better the fabric's bending recovery, the better the fabric's elasticity in subjective analysis. Through the comparison and analysis between the subjective and objective index test results, the tactile comfort of automotive seat fabrics has been more comprehensively evaluated.

References

[1] AHIRWAR M, BEHERA B K. A mathematical model for objective hand evaluation of automotive seat fabrics[J]. Industrial Textiles, 2020, 0(0): 1-21.

[2] CALIFANO R, NADDEO A, COCCARO C. Comfort seat design: thermal sensitivity of human back and buttock[J]. International Journal of Industrial Ergonomics, 2020, 78: 19-35.

[3] WEGNER M, MARTIC R, FRANZ M, VINK P. A system to measure seat-human interaction parameters which might be comfort relevant [J]. Applied Ergonomics, 2020, 84(C): 56-75.

[4] RUTH P, MIRJAM G, BRENDA H G, PRUYN AD T H. Comfortable seating: the influence of seating comfort and acoustic comfort on customers' experience of hospitality in a self-service restaurant[J]. Applied Ergonomics, 2019, 81: 41-61.

[5] SCHEFFELMEIER M, CLASSEN E. Development of a method for rating climate seat comfort[J]. Iop Conference Series: Materials Science and Engineering, 2017, 254 (18): 47-70.

[6] HATOUM O, GHADDAR N, GHALI K, ISMAIL N. Experimental and numerical study of back-cooling car-seat system using embedded heat pipes to improve passenger's comfort[J]. Energy Conversion and Management, 2017, 144: 28-39.

[7] MIHAELA S, LAVINIA S, PAULA. Unguresan factors which influence the thermal comfort inside of vehicles[J]. Energy Procedia, 2016, 85: 36-51.

[8] JOOYONG K, BAEKSOUNG P, CHAEJUNG L. A product-focused process design system (pfpds) for high comforts artificial leather fabrics [J]. Textile Coloration and Finishing, 2008, 20(6): 57–76.
[9] KOLICH M. Predicting automobile seat comfort using a neural network[J]. International Journal of Industrial Ergonomics, 2003, 33(4): 44–61.
[10] MINORU T, TOYONORI N, EIJI T. Relation between seat shape and sitting comfort evaluation on car driver's seat[J]. Sen'i Gakkaishi, 1999, 55(9): 9–28.

The Effect of Financial Constraints on OFDI of China's Textile Firm

ZHAO Junli[1*], **GAO Yujun**[1], **DING Jieli**[1]

[1] *Glorious Sun School of Business and Management, Donghua University, Shanghai 200051, China*

[*] *Corresponding author's email*: zjl @ dhu. edu. cn

Abstract: Based on the financial constraint theory and new-new trade theory, this paper studies the impact of financial constraints on outward foreign direct investment (OFDI) of China's textile listed companies using the data of textile listed firms from 2011 to 2019. First, we construct a theoretical model and develop a series of hypotheses to analyze how financial constraints influence textile firms' OFDI decisions. Second, we construct indexes ScoreA and SA to measure financial constraints. We find that the financial constraints have a negative effect on the OFDI decision of textile enterprises. Internal financial constraints, commercial credit constraints, and external financial constraints have different effects on the OFDI of textile enterprises, among which external financial constraints have the greatest influence. In addition, the moderating test shows that higher productivity will reduce the negative effect of the financial constraints on OFDI choice of textile enterprises.

Keywords: Financial constraints; OFDI; Listed textile companies; Productivity heterogeneity

1 Introduction

The prior studies on outward foreign direct investment (OFDI) are mainly conducted in the context of developed countries. The choice of outward foreign direct investment from developing countries is under-researched[1]. China has become the largest in production and export of textile and clothing products in the world since 1994. The textile industry is traditionally important and plays a critical role in terms of output, exports, and employees in China's manufacturing sector. With internal and external pressures such as the rising cost of raw material and labor, environmental regulation, carbon reduction regulation, and increasing trade barriers in the textile industry, some Chinese textile firms began to transfer their production overseas. From 2003 to 2018, the amount of OFDI of China's textile industry reached \$ 9.796 billion, with an average annual growth rate of 15.6%.

While as the private and small-scale enterprises dominate the textile and apparel industry, China's textile firms face credit discrimination and high financial cost. Compare with export, OFDI has a larger fixed cost and needs much more financial support. Thus, the paper addresses the following issues. How will OFDI be affected by the financing constraints in textile firms? Will financing constraints have the same impact on the heterogeneous firms with different productivity in the textile firm?

Financial constraint was first put forward by Fazzari & Poterba (1988) who believed that financial constraint was the financial dilemma encountered by enterprises when they were faced with the lack of internal capital, high financial costs, and cannot obtain external fund under an incomplete capital market[2]. Some literature analyzes the impact of financing constraints on the exporting decision of heterogeneous firms[3-4]. Limit research has studied whether financial constraints have varied impacts on OFDI in different sources and different productivity firms. Therefore, the contribution of this paper is two-fold. First, we incorporate productivity heterogeneity and financial constraints into a unified analytical framework, to examine the heterogeneous effect of the financial constraints on OFDI. Second, we explore the effects of different financial constraints on the OFDI of textile enterprises.

2 Theory and Hypothesis

The theory of firm heterogeneity was first discussed from the perspective of firm productivity heterogeneity[5]. However, researchers found that firm productivity is not the only factor that influence OFDI. Some scholars propose other influencing factors, including firm size, firm ownership, human capital level, and financial heterogeneity. The latter study argues that firms, when affected by financing constraints, often have no choice but to give up foreign direct investment. On the one hand, it is due to the inability to pay for the fixed costs of OFDI. On the other hand, it is due to the lack of strong liquidity and financing capacity under the larger uncertain risks of overseas investment. Therefore, hypothesis 1 is proposed.

Hypothesis 1: The financing constraints on textile firms will inhibit their behavioral choices for OFDI.

The theory of firm heterogeneity in inter-international trade and investment argues that firm productivity has an important impact on OFDI[56]. The firm with highest productivity will choose OFDI, while the next higher productivity firm will choose outsourcing and export, and the low productivity firm will produce in domestic market[6-7]. Thus, we put forward hypothesis 2.

Hypothesis 2: Textile firms with higher productivity levels will somewhat reduce the deterrent effect of financing constraints on their OFDI.

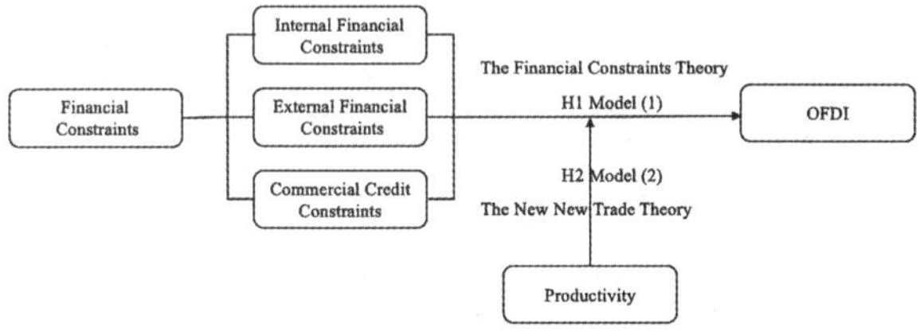

Fig. 1 Theoretical models

3 Variable and Model

3.1 Data and variables

3.1.1 Dependent and independent variables

Using the methods adopted by Wang et al. (2015)[8], the paper constructs a comprehensive index ScoreA to measure financing constraints from internal financial constraints, commercial credit constraints, and external financing constraints. The comprehensive index of financial constraints SA was built to test its robustness[9]. The internal financing mainly comes from their funds and profits. Commercial credit comes from the amount of credit granted by other enterprises. External financing is mainly divided into issuing stocks or bonds, equity financing, and loans to banks and other financial institutions.

In this paper, the data of listed textile companies are selected and the variables are chosen by constructing a dichotomous value for OFDI in listed textile companies, in which listed textile companies are assigned a value of 1 for *OFDI* and 0 otherwise. Data of listed textile companies were from CSMAR and Wind databases.

3.1.2 Control variables

(1) Firm productivity ($\ln P$). In this paper, the Solow residual value method is used to measure firm productivity by deflating the base year of 2010 and setting the capital depreciation rate at 5%.

(2) Enterprise size ($\ln S$). According to the existing literature on enterprise size indicators and the availability of data, we use the logarithm of the number of employees to measure the enterprise size[8].

(3) Human capital ($\ln H$). The investment in human capital is an important reflection of the cost of innovation for enterprises and plays an important role in the future development of textile enterprises in the international market in the long term and to increase their market share.

(4) Enterprise technology level ($\ln LA$). This paper selects the enterprise intangible assets to measure the technological level of an enterprise.

(5) The listed age of enterprises ($\ln Age$). We use the logarithm of the difference between the year of observation and the year the firm went public to measure the age of the listed firm. Generally speaking, the older an enterprise is listed, the richer its accumulated resources, business management experience.

(6) Enterprise nature (D). This paper divides the enterprise ownership according to the Wind database, where state-owned enterprises ($D=1$) and non-state-owned enterprises ($D=0$).

All the variables and data sources can be found in Tab. 1.

Tab. 1 Variables and data sources

Variables	Symbol	Name	Definition
Dependent variables	OFDI	Enterprise OFDI	OFDI or not (0, 1)

(**Continued**)

Variables	Symbol	Name	Definition
Independent variables	Score	Financial Constraints Composite Indicator	ScoreA or SA
	lnProfit	Internal financial constraints	Log of total profit
	lnRA	Business Credit Constraints	Log of accounts receivable
	lnIE	External Financing Constraints	Log of interest expense
Control variables	lnP	Business productivity	Log of productivity
	lnS	Enterprise size	Log of the number of people employed
	lnH	Human Capital	Log of (total wages/number of employees)
	lnIA	Enterprise technology	Log of intangible assets
	lnAge	Age of listed companies	Log of (observation year — listing year)
	D	Characteristic of enterprise	State owned or not (0, 1)

3.2 Model

In this paper, we constructed the following logit model (1) to test hypothesis 1. The dependent variable (*OFDI*) is a binary choice variable. The independent variable *Score* is represented by *ScoreA* or *SA*, the composite indicator of financial constraints constructed in this paper. *Control* is the control variable, including log firm productivity (ln *P*), log firm size (ln *S*), log human capital (ln *H*), log firm technology level (ln *LA*), firm age (ln *Age*), and the dummy variable firm nature (*D*).

$$P(OFDI_{it} = 1 \mid x_{it}) = \beta_0 + \beta_1 Score_{it} + \beta_2 Control_{it} + D_{it} + \varepsilon_{it} \quad (1)$$

Hypothesis 2 is moderating test about textile firms with higher productivity levels that will somewhat reduce the deterrent effect of financing constraints on their OFDI. To distinguish the productivity of textile companies, the productivity of textile listed companies in the sample is ranked from high to low, a dummy variable of whether they are high productivity (*Top*ln *P*) is constructed, and the top 25% of productivity are taken as high productivity companies and their *Top*ln *P* is defined to take the while the *Top*ln *P* of the remaining 75% of companies takes the value of 0. To investigate productivity heterogeneity, the interaction term between the dummy variable *Top*ln *P* and *ScoreA* (*ScoreA* ∗ ln *P*) is introduced to construct model (2), and the meanings of other variables are the same as the model (1).

$$P(OFDI_{it} = 1 \mid x_{it}) = \beta_0 + \beta_1 ScoreA_{it} + \beta_2 ScoreA_{it} * Topln\ P_{it} + \beta_3 Control_{it} + D_{it} + \varepsilon_{it} \quad (2)$$

3.3 Results

According to the regression results of Probit and Logit of the model (1), it can be seen in Tab. 2 that the regression coefficients of the financial constraint indicator (*ScoreA*) are both significantly negative, thus verifying hypothesis 1: the financial constraint suffered by textile firms inhibits their behavioral choice of OFDI.

Tab. 2 Results of impact of financial constraints on OFDI

	Model (1)		Model (2)	
Variables	Probit (1)	Logit (1)	Probit (2)	Logit (2)
ScoreA	−0.220*** (−3.66)	−0.369*** (−3.29)	−0.283*** (−4.48)	−0.476*** (−3.99)
ScoreA ∗ ToplnP	—	—	0.255*** (4.08)	0.466*** (3.95)
lnP	1.685*** (6.04)	3.409*** (5.77)	1.034*** (3.31)	2.117*** (3.36)
lnS	0.903*** (3.03)	1.476*** (2.69)	0.829*** (2.84)	1.403*** (2.57)
lnH	1.669*** (3.60)	2.845*** (3.23)	1.503*** (3.26)	2.670*** (3.07)
lnIA	0.333* (1.86)	0.681** (2.03)	0.321* (1.75)	0.654* (1.91)
lnAge	0.534 (0.90)	1.176 (1.06)	0.661 (1.13)	1.380 (1.26)
D	0.465 (0.76)	0.840 (0.67)	0.177 (0.30)	0.153 (0.13)
Constant	−16.460*** (−6.25)	−29.845*** (−5.96)	−15.515*** (−5.98)	−28.696*** (−5.74)
N	857	857	857	857
Pseudo R^2	0.331	0.339	0.360	0.364

Note: The t-values are in brackets, where *, ** and *** indicate significant at 10%, 5%, 1% significant level respectively.

In addition, the regression results of the remaining explanatory variables of the Logit and Probit models are largely consistent. It can be seen that the regression coefficient of firm productivity (ln *P*) is significantly positive, which is consistent with the findings of the impact of productivity on firms' internationalization choices highlighted by the theory of firm heterogeneity, i. e. , firms with high productivity are more inclined to engage in OFDI. The regression coefficient of the firm size variable (ln *S*) is significantly positive, the larger the size of listed textile firms, the easier it is to generate economies of scale to alleviate the impact of overseas investment costs. The regression coefficient of human capital (ln *H*) is significantly positive, which indicates that and the higher the level of human capital, the larger probility enterprises choose OFDI. The regression coefficient of the enterprise's technology level (ln *LA*) is significantly positive at a 10% level, the higher the level of ln *IA*, the more the enterprises tend to choose OFDI. The effect of enterprise nature (*D*) and enterprise listing age (ln *Age*) on the OFDI of China's listed textile companies is not significant.

Furthermore, the paper tests the impact of different financing constraints, including internal financial constraints, commercial credit constraints and external financial constraints on OFDI of textile enterprises. The results show that different financing constraints have different influences on the OFDI of textile enterprises, among which external financing constraints have the greatest influence.

In the regression results of model (2), the regression coefficient of the interaction term *ScoreA* * *Top*ln *P* in Probit (2) is significantly positive with a value of 0.255, and the regression coefficient of *ScoreA* index is significantly negative with a value of −0.283, which is opposite to the positive and negative regression coefficient of the interaction term *ScoreA* * *Top*ln *P*. Therefore, the regression coefficient of the interaction term *ScoreA* * *Top*ln *P* in Logit (2) is significantly positive, with a value of 0.466, and the regression coefficient of the *ScoreA* index is significantly negative, with a value of −0.476, which is also opposite to the regression coefficient of the interaction term *ScoreA* * *Top*ln *P*. Thus, *ScoreA* in Probit (2) is −0.028 (−0.283+0.255). The coefficient of the combined effect of *ScoreA* in Logit (2) is −0.01 (−0.476+0.466). Thus, hypothesis 2 is verified: textile firms with higher productivity levels will reduce to some extent the hindering effect of financing constraints on their OFDI.

4 Conclusions and Implications

Based on the financing constraint theory and the new trade theory, this paper examines the impact of financing constraints of textile listed companies on their OFDI by using the data of textile listed companies in China from 2011−2019.

We find that financial constraints will inhibit their OFDI behavior choice. Additionally, internal financial constraints, commercial credit constraints, and external financial constraints have different effects on the OFDI of textile enterprises, among which external financing constraints have the greatest influence.

Textile firms with higher productivity levels will reduce the hindering effect of financing constraints on their OFDI to some extent. It further validates the theory of firm heterogeneity that firm productivity has an important impact on OFDI from perspective of financial constraints.

Based on the findings of this paper, we propose the following suggestions. Textile enterprises should pay more attention to external financial constraints. At the same time, textile enterprises should increase the investment in human capital and technology costs to promote their productivity. The investment in human capital can obtain innovative and high-quality talents, which is beneficial to textile enterprises to improve product design capabilities and brand awareness. The investment in technology and research is an important measure to improve enterprise productivity, which is important for textile enterprises to alleviate the financial constraints.

References

[1] LI M, LI D, MARJORIE L, et al. Chinese MNEs' outward FDI and home country productivity: the moderating effect of technology gap [J]. Global Strategy Journal, 2016(6): 289−308.

[2] FAZZARI S M, POTERBA J M. Financing constraints and corporate investment [J]. NBER working papers, 1988, 33(4): 657−657.

[3] MANOVA K. Credit constraints, equity market liberalizations and international trade [J]. Journal of International Economics, 2008, 76(1): 33−47.

[4] CHANEY T. Liquidity constrained exporters [J]. Journal of Economic Dynamics and Control, 2016, 72: 141−154.

[5] MELITZ M J. The impact of trade on intra-industry reallocations and aggregate industry productivity[J]. Econometrica, 2003, 71(6): 1695-725.

[6] HELPMAN E, MELITZ M J, YEAPLE S R. Export versus FDI with heterogeneous firms[J]. American Economic Review, 2004, 94(1): 300-316.

[7] GROSSMAN G M, HELPMAN E. Outsourcing in a global economy[J]. Review of Economic Studies, 2005, 72(1): 135-159.

[8] WANG B J, TAN Y Y, YU M J, et al. Do financing constraints inhibit outward foreign direct investment by Chinese private firms? [J]. The Journal of World Economy, 2015, 38(12): 54-78.

[9] YANG J Y. Financing constraints and firms' export behavior: An empirical study based on industrial firms' data [J]. China Economic Quarterly, 2012, 11(4): 1503-24.

[10] MU X R, SUN L Y. The impact of productivity and financing heterogeneity on firms' exports — an analysis based on threshold effects[J]. Journal of International Trade, 2017, 10: 61-73.

[11] LIU L Y, HE Y L, WANG Z F, et al. Do financing constraints affect Chinese firms' outward FDI? — A theoretical and empirical analysis based on a micro perspective. [J]. Journal of Financial Research, 2015, 8: 124-40.

Study on Preparation of Hydrophilic Polypropylene Nonwovens by In-Situ Photoinitiated Grafting of Zwitterionic Polymer with Urushiol

XU Danyao[1,2], **LI Yan**[1,2*], **WANG Lu**[1,2]

[1] *Key Laboratory of Textile Science & Technology, Ministry of Education, Donghua University, Shanghai 201620, China*

[2] *Department of Technical Textiles, College of Textiles, Donghua University, Shanghai 201620, China*

[*] *Corresponding author's email*: yanli@ dhu. edu. cn

Abstract: Polypropylene (PP) nonwovens have many excellent properties, but their poor hydrophilicity greatly limits their application as functional materials in many fields. To develop a more facile and mild method for hydrophilic modification of PP nonwovens, we devised a method for the preparation of hydrophilic coatings using urushiol (UR) in situ photoinitiated grafting of zwitterionic polymers on PP nonwovens. The adhesion stability and thermal stability of the UR photoinitiated layer were explored, then the coating preparation process was studied in detail, and the prepared hydrophilic modified PP nonwoven (PSU-PP) was evaluated. The results show that the UR photoinitiated layer has good stability, the PSU coating can cover the fiber surface uniformly, and the surface hydrophilicity of PSU-PP is significantly improved. In addition, UR could still play an initiator effect at a low concentration of 0. 5 mg/mL. In conclusion, this research can provide a useful reference for a new hydrophilic modification method for PP nonwovens.

Keywords: Polypropylene nonwovens; Zwitterionic polymer; Urushiol; Hydrophilic modification

1 Introduction

Polypropylene (PP) nonwovens have a simple manufacturing process, low cost and high output, excellent mechanical properties, good air permeability, abrasion resistance, acid and alkali resistance, and other excellent properties. However, the molecular structure of PP consists with hydrophobic methyl groups, with high crystallinity and circular cross-section, which give the fiber a tight structure as well as the lack micropores and gaps, resulting in poor hydrophilicity and the extremely inert surface of PP nonwovens[1]. The poor wettability of PP greatly limits its application as a functional material in the fields of medical and sanitary materials, geotechnical construction materials, clothing materials, wiping materials, and automotive interiors [2]. Therefore, improving the wettability of PP nonwovens has theoretical and practical significance.

At present, the surface hydrophilic modification methods mainly include bulk modification, blend modification, surface coating, graft modification, corona treatment, and ozone treatment[3]. Although these methods can improve the surface hydrophilicity of PP nonwovens to a certain extent, there are still some shortcomings. For example, the coating formed by surface coating or adsorption method on PP nonwovens has poor stability and is easy to fall off; surface chemical treatment could result in pollution, and the cost of surface oxidation treatment is high[4]; plasma treatment would cause material damage and is difficult to control[5]. Therefore, there is an actual need to develop a more facile method which is capable of surface hydrophilic modification of PP nonwovens under mild conditions.

Surface photochemical initiation can form a uniform polymer layer on a solid surface with good stability, fast initiation speed, low toxicity, good biocompatibility, and mild conditions. The zwitterionic polymer contains both positive and negative charges, which can form a tight and stable super hydrophilic hydration layer through ionic interaction. The polymer also has thermal stability, chemical stability, and biocompatibility[6].

Urushiol (UR) is a natural catechol derivative derived from raw lacquer, which is environmentally friendly, inexpensive, and has long unsaturated hydrocarbon side chains and catechol end groups. In this study, the free radicals generated during the polymerization of UR are used as the photo initiator for the polymerization of zwitterionic monomers. The method of *in-situ* photoinitiated grafting of UR to the zwitterionic polymer to prepare hydrophilic modified PP nonwoven was explored. The stability of UR and the preparation process of hydrophilic coatings were explored, and the hydrophilicity of the prepared coatings was characterized. The obtained results can provide a useful reference for new hydrophilic modification methods of PP nonwovens.

2 Experiments

2.1 Materials

Raw lacquer was purchased from Xi'an Raw Lacquer Research Institute. Anhydrous ethanol, analytically pure, purchased from Shanghai Titan Technology Co., Ltd. Triethanolamine and ferric chloride, analytically pure, purchased from Sinopharm Chemical Reagent Co., Ltd. [2-(methacryloyloxy) ethyl] dimethyl-(3-sulfonic acid propyl) ammonium hydroxide (SBMA), purity ≥97%, purchased from Shanghai Aladdin Biochemical Technology Co., Ltd.

2.2 Stability test of UR initiator layer

UR elution experiment. To explore whether UR would be eluted by the solution during the preparation process, the PP nonwoven samples were soaked in the UR solution and then air-dried. Then the samples were respectively soaked in absolute ethanol, deionized water, and SBMA monomer solution for 2 h. The other sample was prepared according to the preparation process, and the residual solution was taken for testing during the process to determine whether UR would be eluted under the condition of ultraviolet irradiation. Using ferric chloride solution for color development test.

Research on the heating oxidation of UR. The UR anhydrous ethanol solution obtained by extraction was heated in a water bath at 80 ℃ to evaporate the solvent, and then dissolved in anhydrous ethanol again. The UV-Vis spectrophotometer was used to detect the change of UR content after repeated heating.

2.3 Preparation of hydrophilic coatings

The PP nonwovens were soaked in UR solutions with different concentrations The resultant samples were named UR-PP. Then soak it in SBMA monomer solutions with different concentrations, and irradiate it under the condition of 365 nm ultraviolet light. A composite coating was formed and the coating sample was named PSU-PP. After the reaction was completed, the residual unreacted UR and SBMA were removed by soaking and washing with deionized water and absolute ethanol for 24 h.

2.4 UR content test

UR and ferric chloride can form complexes with characteristic UV absorption. Prepare 1.6% triethanolamine and 0.1% ferric chloride ethanol solution respectively, take 25 mL of each of the above solutions in a 100 mL volumetric flask, and mix them with anhydrous ethanol to obtain a mixed solution. Take 1 mL of UR anhydrous ethanol solutions of different concentrations into a 10 mL volumetric flask, and dilute to the volume with the mixture. After mixing, the absorbance at 625 nm of the complex solution was measured using a UV-Vis spectrophotometer.

2.5 Characterization

The samples were characterized by using a SU 8010 field emission scanning electron microscope (FE-SEM), and the microscopic morphology of fibers in the untreated samples and the samples after coating modification was observed. The chemical groups on the surface of the different modified samples were confirmed using a Nicolet 6700 Fourier transform infrared spectrometer. The static water contact angle of the sample surface was determined by the suspension method using an OCA15EC contact angle meter.

3 Results and Discussion

3.1 Stability of UR initiating layer

Regardless of whether the UR content in the solution was high (25 mg/mL) or low (0.5 mg/mL), a blue-green complex was always formed when the solution containing UR was added to the ferric chloride solution. Use ferric chloride solution to test whether the immersion solution contains eluted UR. As shown in Fig. 1, neither the immersion solution directly immersing the sample nor the residual liquid in the preparation process produces complexes. It is proved that the UR attached to the PP nonwovens will not be eluted during the whole preparation process, and the effect of UR elution does not need to be considered in the subsequent exploration.

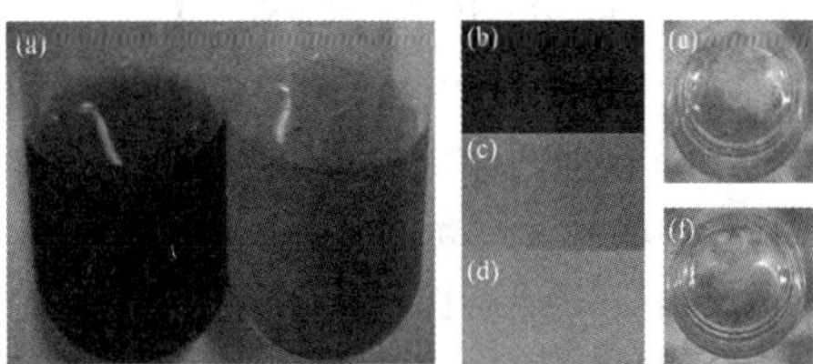

Fig. 1 (a) UR with high (left) or low (right) concentration can form a blue-green complex with ferric chloride; using absolute ethanol (b), deionized water (c), and SBMA solution (d); after soaking UR-PP for two hours, add ferric chloride solution to the soaking solution; take the residual solution (e) during sample preparation, and add ferric chloride solution (f) to the soaking solution

After the extracted UR solution was evaporated to remove the solvent, it was dissolved in anhydrous ethanol again for testing. The complex formed by UR and ferric chloride had characteristic absorption at ultraviolet 625 nm. The change in UR content after heating was detected. As shown in Fig. 2, the content of the extracted UR did not change significantly after repeated heating, which confirmed that UR was not affected by heating under aerobic conditions, and the effect of UR on heat loss did not need to be considered in the subsequent exploration.

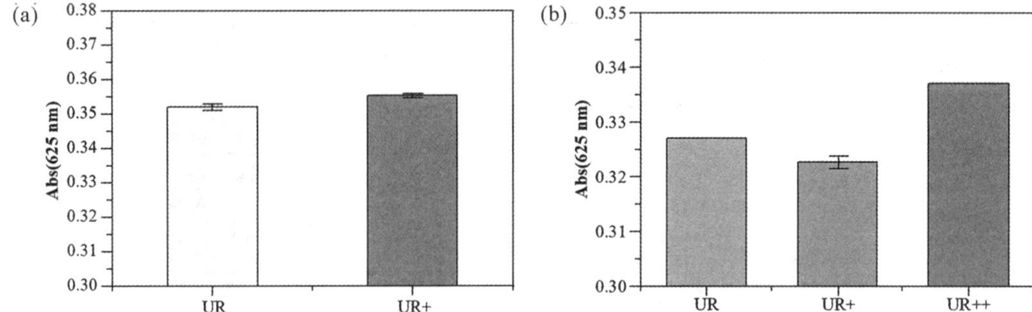

Fig. 2 Absorbance at 625 nm of the complex solution of UR and ferric chloride: (a) heated once; (b) heated twice (UR: UR anhydrous ethanol solution. UR+: the UR product is obtained by heating the water bath at 80 ℃ to evaporate the solvent. UR++: evaporate the solvent twice.)

3.2 Preliminary study on coating preparation

The concentration of UR solution was 25 mg/mL, and the concentration of SBMA solution was 140 mg/mL. Two groups of samples were set up to compare whether the drip infiltration method or the immersion method was more favorable for grafting when adding the SBMA solution. As shown in Fig. 3 (a), due to the strong hydration ability of PSBMA, which can absorb moisture inthe air, a strong and broad characteristic peak of phenolic hydroxyl groups is formed at 3 200-3 500 cm^{-1}. The characteristic peaks of carbonyl group at 1 650-1 690 cm^{-1}, 1 710-1 725 cm^{-1}, ether bond at 1 215 cm^{-1}, and sulfo group at 1 040 cm^{-1} can also prove that PSBMA was successfully grafted on the sample,

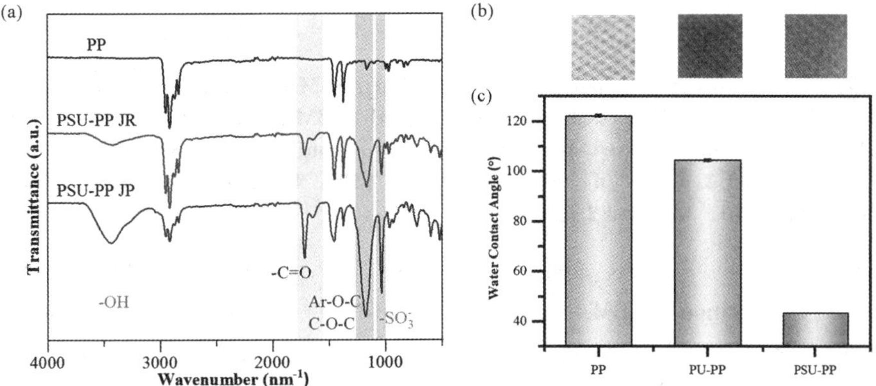

Fig. 3 (a) FTIR results of PP nonwovens, coating samples prepared by drop-wetting method (PSU-PP JR), and coating samples prepared by soaking method (PSU-PP JP); (b) color change of PP nonwovens, PU-PP and PSU-PP; (c) water contact angles of PP nonwovens, PU-PP and PSU-PP

and also confirmed the practical feasibility of this preparation process. In addition, by comparing the peak intensities of the characteristic peaks, we can see that the grafting effect of the soaking method is better than that of the infiltrating method. As can be seen from Fig. 3 (c), comparing the PSU-PP prepared by the soaking method, and the pure polyurushiol coated sample (PU-PP) without adding SBMA solution, the PP nonwovens are very hydrophobic. And the contact angle of PU-PP decreases slightly, because UR contains hydrophilic phenolic hydroxyl groups, while the surface hydrophilicity of PSU-PP is significantly improved.

In addition, the PP nonwovens are white, and the modified samples have different shades of gray-black or brown-black. Due to the formation of UR quinones, polymers, and high polymers during the polymerization of UR, the color changes from yellow to reddish-brown, to dark brown, and finally to black. At this time, most of the PU-PP samples are gray-black, or brown-black. While the lighter color of PSU-PP is due to the reduction of UR polymers s. The free radicals generated during the polymerization of UR itself can initiate the polymerization of SBMA. Therefore, we can speculate that the UR monomer and SBMA monomer will compete for free radicals produced by UR when the two polymerized reactions occur simultaneously.

3.3 Coating preparation parameters

Through preliminary experiments, it was found that the coating formed by higher concentrations of UR and SBMA would damage the structure of non-woven samples. Therefore, the concentration of UR was reduced in subsequent experiments, and an attempt was made to explore the concentration limit of UR-induced grafting.

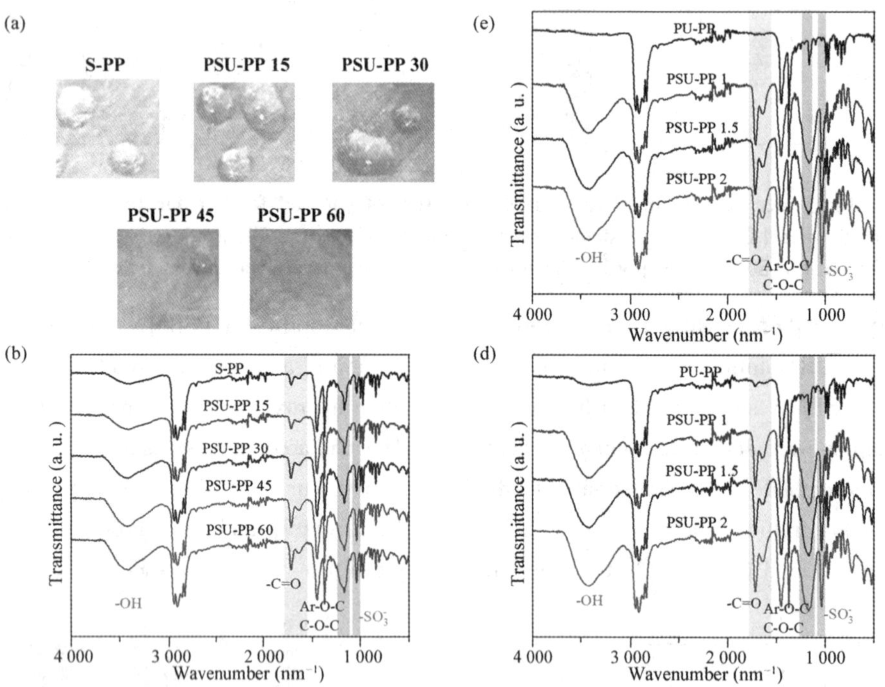

Fig. 4 (a) Surface wettability (UR: 0.5 mg/mL, SBMA: 50 mg/mL, time: 15, 30, 45, 60 min); (b) FTIR results (UR: 0.5 mg/mL, SBMA: 50 mg/mL, time: 15, 30, 45, 60 min); (c) infrared spectrum (UR: 0.5 mg/mL, SBMA: 50 mg/mL, time: 1.0, 1.5, and 2.0 h); (d) infrared spectrum (UR: 0.5 mg/mL, SBMA: 50 mg/mL, time: 1.0, 1.5, and 2.0 h)

The minimum concentration of UR was adjusted to 0.5 mg/mL, and the concentration of SBMA was fixed at 50 mg/mL. At the same time, the reaction time of UV irradiation was explored. Use UR concentration of 0.5 mg/mL and SBMA concentration of 50 mg/mL, irradiated for 15, 30, 45, and 60 min. Four groups of samples were named PSU-PP 15, PSU-PP 30, and PSU-PP 45 and PSU-PP 60. A group of samples (S-PP) in which the PP nonwovens which was not soaked in the UR solution but put into the SBMA solution were prepared at the same time and irradiated for 60 mins. As shown in Fig. 4 (b), the grafting starts after the 30 minutes point of reaction time, and stays relatively unchanged after 45 minutes. S-PP also showed a small amount of zwitterion grafting, which may be caused by the zwitterion itself having a certain viscosity. As can be seen from Fig. 5 (a), due to the extremely low concentration of UR, the color of the PSU-PP sample is white. The results of hydrophilic are consistent with the FTIR results. The surface wettability of the samples was significantly improved when the irradiation time was more than 45 min.

Using the same concentration, irradiated for 1.0, 1.5, and 2.0 h, a group of polyurushiol coating samples was prepared at the same time, and 4 groups of samples were prepared, named PSU-PP 1, PSU-PP 1.5, PSU-PP 2, and PU-PP. As shown in Fig. 4 (c), due to the extremely low UR concentration, even after normalization, the characteristic peaks of the UR polymer coating are very weak. The three groups of PSU-PP samples still showed obvious PSBMA peaks, which proved that UR still played a role in initiating grafting. The infrared characteristic peak intensities of these three groups of samples are the same, so it is speculated that the concentration of UR may restrict the grafting of SBMA. Therefore, increasing the UR concentration to 1 mg/mL, four groups of samples were prepared, named PSU-PP 1, PSU-PP 1.5, PSU-PP 2, and PU-PP. As shown in Fig. 5 (d), the characteristic peaks of the UR polymer coating can be reflected after normalization, while the infrared characteristic peaks of the three groups of PSU-PP samples still have no significant difference. Therefore, 2 h can be temporarily used as the time for complete reaction.

Through the analysis of the above experimental results, the concentration of SBMA monomer was further regulated, the concentration of UR was 1 mg/mL, and the concentration of SBMA was adjusted to 30, 40, and 50 mg/mL. Four kinds of samples were prepared, named PSU-PP 30, PSU-PP 40, PSU-PP 50, and PU-PP. As shown in Fig. 5 (a), from the infrared results, PSBMA was successfully grafted in all three groups of samples, but the effect of changing the SBMA monomer concentration could not be judged from the infrared results.

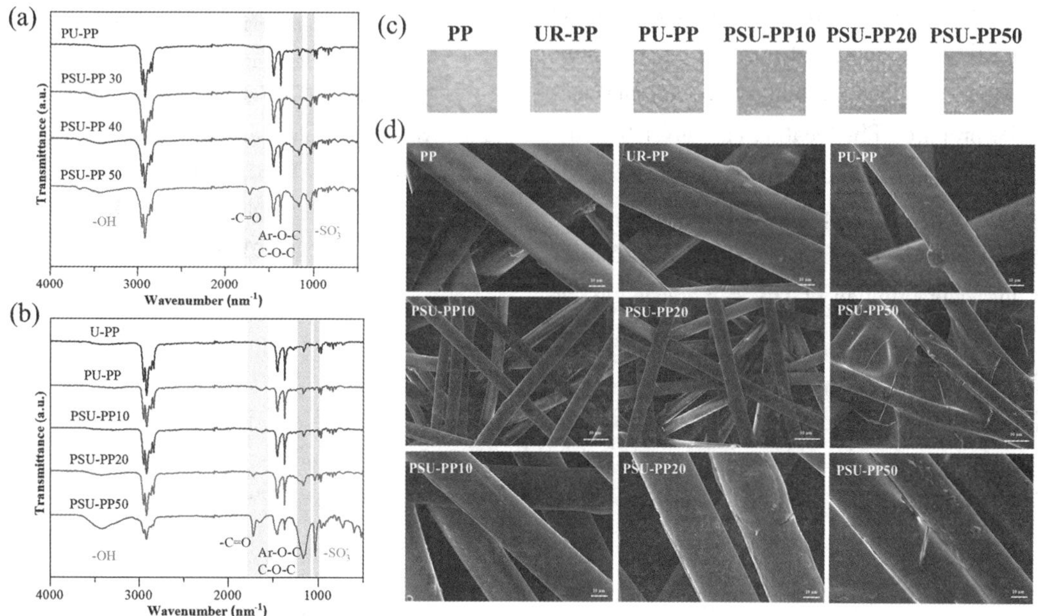

Fig. 5 (a) FTIR results (UR: 1 mg/mL, SBMA: 30, 40, 50 mg/mL, time: 2 h); FTIR results (b), macroscopic morphology (c) and electron microscope (d) of the sample (UR: 2.5 mg/mL, SBMA: 10, 20, 50 mg/mL, time: 2 h)

Considering that the UR concentration may still limit the amount of PSBMA grafting, the UR concentration was further increased to 2.5 mg/mL, the SBMA concentration was adjusted to 10, 20, and 50 mg/mL. 4 groups of samples were prepared and named PSU-PP10, PSU-PP20, PSU-PP50, and PU-PP. As shown in Fig. 5 (c), the color of this group of samples is light yellow. From Fig. 5 (b), there is no characteristic peak of PSBMA in FTIR results, indicating that PSU-PP10 did not successfully graft with PSBMA. And the characteristic peak of PSU-PP20 was weaker than that of PSU-PP50. As shown in Fig. 5 (d), it can be observed from the SEM that the surface of the PP fiber was originally smooth, and the surface of the UR-PP was covered with a non-granular uniform and smooth coating. The morphology of PU coating is similar to that of UR coating. However, the surface morphology of PSU-PP10 has not changed, which again confirms that there is no grafting of PSBMA, and it can be observed that the fibers of PSU-PP20 become thicker, while the coating formed on the surface of PSU-PP50 is very obvious, and even the gap between the PP fibers has been blocked. Therefore, it can be speculated that after increasing the concentration of UR, if the concentration of SBMA is lower than a certain limit, it will not be able to gain an advantage in the competitive reaction, which will lead to the failure of the preparation of hydrophilic coatings. When the SBMA concentration is also high, which will affect the fiber structure of the nonwovens.

4 Conclusions

In summary, this paper prepares a hydrophilic coating on PP nonwovens by *in-situ* photoinitiated grafting of PSBMA with UR. It is confirmed that the UR photoinitiated layer has good adhesion stability and thermal stability, and the prepared PSU coating can cover the surface of PP nonwovens uniformly and improve the its surface hydrophilicity significantly. UR can still show an initiating effect at a low concentration of 0. 5 mg/mL, and a low concentration of UR is more conducive to preparing hydrophilic coatings that do not affect the structure of PP nonwovens. This new method of hydrophilic modification of PP nonwovens has good feasibility and potential application prospects.

References

[1] ABEER A, XU G B. Comparative study of polypropylene nonwoven on structure and wetting characteristics[J]. The Journal of the Textile Institute, 2021, 112(7): 1100-1107.

[2] FAEZEH K, PARHAM S, MOHAMMAD Z, et al. Acoustic and thermal performance of polypropylene nonwoven fabrics for insulation in buildings[J]. Journal of Building Engineering, 2022, 50: 104125.

[3] WEI P, LOU H J, XU X M, et al. Preparation of PP non-woven fabric with good heavy metal adsorption performance via plasma modification and graft polymerization[J]. Applied Surface Science, 2021, 539: 148195.

[4] YANG X H, LI S P, YAO Y Q, et al. Preparation and characterization of polypropylene non-woven fabric/ZIF-8 composite film for efficient oil/water separation[J]. Polymer Testing, 2021, 100: 107263.

[5] YANISA L, THITIRAT R, PATTRA L, et al. Adsorption of Cu(II) ions from aqueous solution using PE/PP non-woven fabric grafted with poly(bis[2-(methacryloyloxy) ethyl] phosphate)[J]. Journal of Environmental Chemical Engineering, 2021, 9(6): 106440.

[6] CHEN T Y, OU S, CHIEN H. Biomimetic mineralization of tannic acid-supplemented HEMA/SBMA nanocomposite hydrogels[J]. Polymers, 2021, 13(11): 1697.

Modeling of a Weft Knitted Loop

ZHANG Fengli[1], WANG Xi[2], ZHANG Tianshu[3], LI Qiao[1*]

[1] *Key Laboratory of Textile Science & Technology, Ministry of Education, College of Textiles, Donghua University, Shanghai 201620, China*

[2] *College of Information Science and Technology, Donghua University, Shanghai 201620, China*

[3] *College of Science, Donghua University, Shanghai 201620, China*

* *Corresponding author's email*: qiaoli@ dhu. edu. cn

Abstract: The loop structure of the knitted fabrics is closely related to the physical properties of the yarn and fabrics, and it is very difficult to establish the loop geometries with a high accuracy By comparing the advantages and disadvantages of different approaches to the geometrical modeling of the knitted loops, the NURBS spline curve is finally selected to establish the geometry of the weft knitted loop. Based on the fabric specification parameters, the coordinates of several selected points of the unit loop are determined, and the control points are inversely calculated to obtain the geometry of the loop.

Keywords: Weft knitted fabric; Loop configuration; NUBRS spline curve; Geometrical modeling

1 Introduction

The weft knitted structure is diversiform and the loop structure of the knitted fabrics is closely related to the physical properties of the yarn and fabrics, for example, decoherence, extensibility, strength and elasticity, etc. The analysis of weft knitted structure can correctly predict the appearance of the weft knitted fabric[1], improving the efficiency of weaving and the quality of the fabric. It has been 80 years since the research on the loop model of weft knitted fabric began in the 1940 s, and many models have been established, but there are some defects. To analyze the characteristics of the deformation of traditional models and knitted fabrics, different models with wider practicability are adopted for model establishment.

2 Pure Geometric Model

2.1 Chamberlain model

In this model, a coil is assumed to be composed of a half-arc coil and a straight line, and they are not tangent at the connection point. The loop distance A and the line of the center of the settling arc of adjacent arcs are also equal to the diameter d of the four yarns, this model lays a foundation for the subsequent model study. However, there is a great difference between the simulated effect and the real fabric[2].

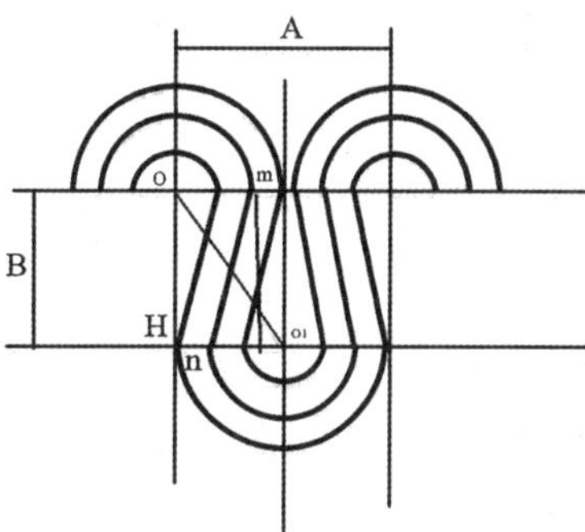

Fig. 1 Chamberlain model

2.2 Pierce model

This model is constructed based on the Chamberlain model[3]. The loop is regarded as the connection of the arc and straight line. But they are detailed at the connecting point. It presents an arc shape in the thickness direction and is located on the cylindrical surface with

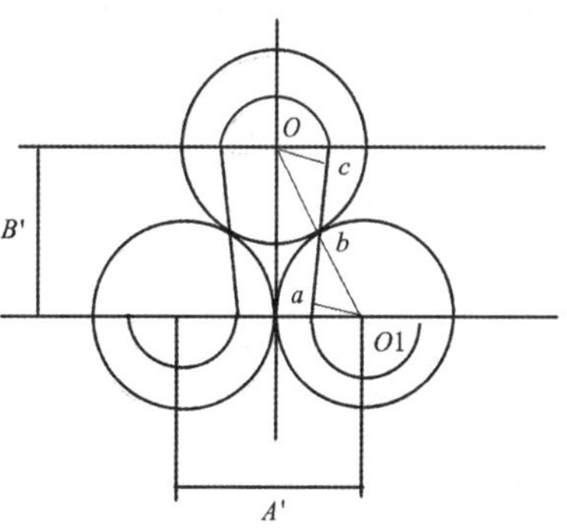

Fig. 2 Pierce model

radius R in the transverse direction. However, under this model, the loop torsion is discontinuous, so the loop provided under this model cannot form a stable fabric.

2.3 Leaf and Glaskin model

The purpose of this model is to find a model suitable for both wet-relaxed and dry-relaxed fabrics to solve the problem of loop torsion discontinuity[4]. In this model, the loop is composed of circular arcs of radius R that are tangent to each other. The center of the circular arc is used as the origin of coordinates. So as to ensure the continuity of torsion.

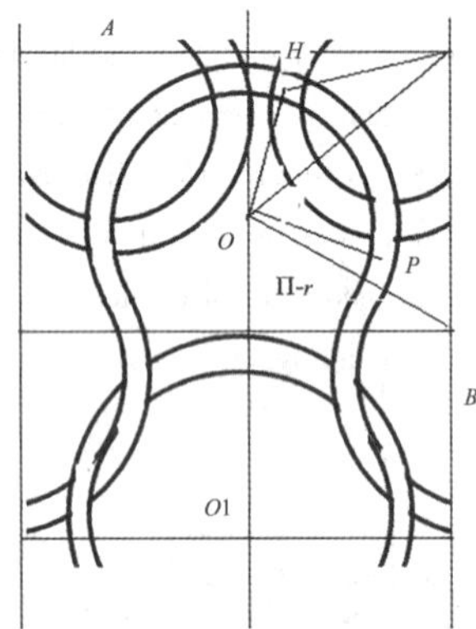

Fig. 3 Leaf and Glaskin model

The above pure geometric models are all formed by connecting circular arcs with straight lines or completely circular arcs when the fabric is completely relaxed. Although they are too ideal, so they are limited by the essence of two-dimensional, so the three-dimensional sense is not strong, which leads to a certain gap between the model and the actual fabric loop, but it also lays a foundation for subsequent research.

3 Segmented Function Coil Model

Kurbak[5-6] model divides the loop into eight parts, and each part has a relative function, so some loop models of weft knitting organization are established. However, due to a large amount of calculation and very complex characteristics of the segmented function, there are still certain restrictions on the flexibility.

4 Spline Curve Coil Modeling

In order to study the three-dimensional structure of weft knitting and the authenticity of the coil, a spline curve is used to simulate the coil. Bezier is the first computer tool applied in industrial design[7]. However, when changing one of the value points on the coil, the whole curve will change, and some defects will appear when multiple curve segments are simulated[8]. Non uniform rational B-spline curve (NURBS) has the advantage of changing any control point without affecting the shape of other curves, and it is easier to be close to the actual shape of fabric, so the NURBS curve is used for simulation. This paper mainly introduces the geometric modeling of weft knitting coil based on NURBS spline curve.

5 Modeling

5.1 Characteristics of NURBS curve

NURBS curve is a kind of B-spline curve, which is determined by control vertices and spline basis function. Let $n+1$ control points $P_i(I=0, 1, 2, \cdots, n)$, also known as the vertex of the characteristic polygon, then the expression of $k+1$ order (k degree) uniform rational B-spline curve is[9]:

$$Q(U)=\frac{\sum_{i=0}^{n} W_i P_i N_{i,k}(u)}{\sum_{i=0}^{n} W_i N_{i,k}(u)}$$

where k is the power of the NURBS spline curve, W_i is the weight factor corresponding to the control vertex P_i, N_i, $k(u)$ is the basis function of the spline of degree k. Among them, the weight factors W_0, $W_n>0$, and other weight factors $W_i \geqslant 0$, and the sequential k weight factors are zero at different times. The spline curve $Q(u)$ is determined by the values of the control point P_i, the weight factor W_i and function N_i, $k(u)$[10].

A point $P(u)$ on the NURBS curve with parameter $u \in [ui, ui+1]$ is at most $k+1$ and the control vertex dj, $j=i-k$, $i-k+1$, $\cdots$, i, And moving the ith control point di on the curve is at most related to part of the curve on the defined interval (UI, UI $+k+1$), but independent of the rest[11].

5.2 Establishment of loop unit model

Before the coil model is established to a set of modeling conditions, assuming that:

For the yarn, the cross-section is uniform and after deformation the yarn is smooth and continuous, the geometric center of the yarn passes through the value point.

For the coils, they are symmetrical and does not skew and the structure of each coil is uniform, so that the centerline of the yarn can be guaranteed to be a second-order continuous space curve.

According to the research of U. Güdükbay[12] and others, in a complete loop, 12 type value points are selected according to the process characteristics of weft knitting coil, which are t1-t12 respectively. As shown in Fig. 4, the basic parameters of weft flat needle loop structure are: coil height h, coil spacing ω, fabric thickness t, black points in the figure are particle points, and gray points are type value points, ω is the lap distance, $h1+h2$ is the lap height.

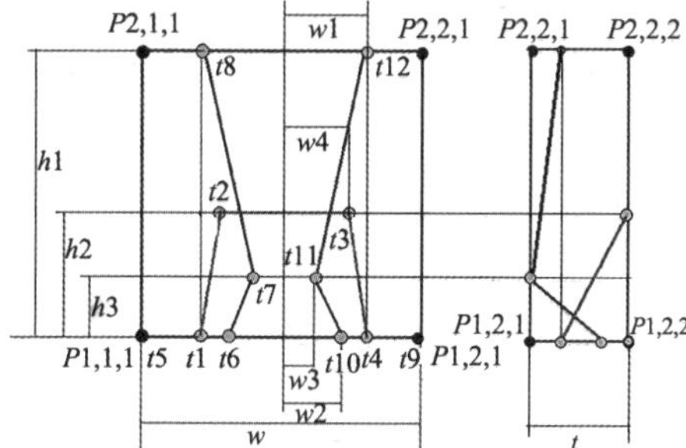

Fig. 4 Coil model[5]

Then the coordinates of the eight particles can be obtained as follows:

$P1, 1, 1 = (0, 0, 0)$; $P1, 2, 1 = (\omega, 0, 0)$;
$P2, 1, 1 = (0, h1, 0)$; $P2, 2, 1 = (\omega, h1, 0)$;
$P1, 1, 1 = (0, 0, t)$; $P1, 2, 1 = (\omega, 0, t)$;
$P2, 2, 2 = (\omega, h1, t)$; $P1, 2, 1 = (0, h1, t)$

Where Pi, j, k ( ), i is the row where the point is located, j is the column where the point is located, and k is the layer where the point is located.

Taking the first complete loop as an example, according to the statistical law, that is:

$$w1 = \frac{13}{40}w$$

$$w2 = \frac{11}{50}w$$

$$w3 = \frac{7.1}{50}w$$

$$w4 = \frac{10.6}{40}w$$

$$h2 = \frac{17.8}{50}h1$$

$$h3 = \frac{11}{40}h1$$

The coordinates of each type value point can be obtained, for example, the coordinates of $t1$ are:

$$t1 = \frac{3.5}{20}P_{(1,2,1)}[n] + \frac{16.5}{20}P_{(1,1,1)}[n]$$

5.3 Determination of control points

Cubic B-spline curve is adopted, and each type value point is a node, which determines that the curve passes through the control point. A cubic curve needs at least four control points, so that the shape of the curve can change when a point is displaced, to fit the real curve more closely. Two adjacent nodes can be used as both ends of a curve, Then the other two control points are calculated by interpolation method and obtained by de-bour algorithm, namely:

$$P(u) = \sum_{j=i-k+1}^{i} d_j^l N_{j,k-1}(u) = \cdots = d_i^l$$

$$u_i \leqslant u < u_{i+1}$$

$$d_j^l = d_j \quad l = 0$$

$$d_j^1 = (1 - \alpha_j^l) d_{j-1}^{l-1} + \alpha_j^{l-1} \quad i = 1, 2, \cdots, k$$

$$j = i - k + 1, \cdots, i$$

$$\alpha_j^l = \frac{u - u_i}{u_{j+k+1-1} - u_j}$$

Where: u is the control factor ($u_0 = u_1 = u_{2=} u_3 = 0$, $u_{i+3} = u_{i+2} + |p_{i-1}|$) (,$k$ is the number, i is the node position ($i=1, 2, \cdots, 12$), and $P(u)$ is the interpolation point coordinate[13].

6 Conclusions

Any curve and surface can be described theoretically by NURBS, and complex curves and surfaces can be described based on their spline basis functions. At the same time, the curve can be locally adjusted by the control points. It has good morphological characteristics, flexible control by the weight factor curve, and stable and rapid calculation. According to each section of the curve in turn adjustment can better fit the real curve.

References

[1] XIN R U, PENG L, MINGLAI L, et al. Modelling and algorithm of weft knitted fabric[J]. Journal of Textile Research, 2010, 39(09): 44-49.

[2] CHAMBERLAIN J. Hosiery yarns and fabrics[J]. City of Leicester Technical College. 1949, 107.

[3] PEIRCE, FREDERICK T. 5-The geometry of cloth structure[J]. Journal of the Textile Institute Transactions, 1937, 28(3): 45-96.

[4] LEAF G A V. 4-Models of the plain knitted loop[J]. Journal of the Textile Institute Transactions, 1960, 51.

[5] SHA S, JIANG G M, ZHANG A J, et al. 3-D simulation of weft knitted fabric based on improved mass-spring model[J]. Journal of Textile Research, 2015(2): 111-115.

[6] KURBAK A. Plain knitted fabric dimensions. II. [J]. Textile Asia, 1998(4): 36-44.

[7] KURBAK A. Plain knitted fabric dimensions. I. [J]. Textile Asia, 1998(3): 41-44.

[8] SURHONE L M, TIMPLEDON M T, MARSEKEN S F, et al. Non-uniform rational B-spline[J]. Betascript Publishing, 2010. 76.

[9] SHA S, JIANG G M, ZHANG A J, et al. Three-dimensional modeling and deformation for weft knitted fabric loops[J]. Journal of Textile Research, 2017, 38(2): 177-183.

[10] YUAN F, JU T T, XIA F T. Experimental analysis and modeling research of the morphological mtructure of the weft-nnitted Loop[J]. Journal of Fiber Bioengineering & Informatics, 2012, 5(3): 299-307.

[11] SHI F Z. Cagd & nurbs[M]. Beijing: Higher Education Press, 2001: 265-269.

[12] LI Y, YANG L, CHEN S, et al. Three dimensional simulation of weft knitted fabric based on surface model[J]. Computer Modelling & New Technologies, 2014, 18(3): 57.

[13] UǦUR GÜDÜKBAY, SERKAN B, ÇETIN K, et al. Particle-based simulation of the interaction between fluid and knitwear[J]. Signal, Image and Video Processing, 2014, 8(3): 415-422.

Influence of Fabric Structure on Patterned Nanofiber Membranes

WANG Xuening[1,2], XIE Xiaojing[1,2], WANG Fujun[1,2], GUAN Guoping[1,2], LI Chaojing[1,2*], LIN Jing[1,2*], WANG Lu[1,2]

[1] *Key Laboratory of Textile Fabric Technology, Ministry of Education, Donghua University, Shanghai 201620, China*

[2] *College of Textiles, Donghua University, Shanghai 201620, China*

* *Corresponding author's email*: lcj@dhu.edu.cn; jlin@dhu.edu.cn

Abstract: Electrospun nanofibers are widely used in tissue engineering and regenerative medicine for their property of mimicking the structure of natural extracellular matrix. Normal electrospun nanofiber membranes are usually two-dimensional and dense, which is not conducive to cell infiltration. Thus, the preparation of patterned nanofiber membranes has become a hot research topic. Metal template method is a common method to prepare patterned nanofiber membranes. However, although this method can achieve controlled deposition of nanofibers and solve the difficulties caused by the over-density of normal nanofiber membranes, it cannot produce multi-scale scaffolds with three-dimensional structures, which are necessary for tissue engineering scaffolds. In this study, we first wove two fabrics in different structures, T2/2 and T4/2, which were then used as receiver templates for electrospinning to produce patterned three-dimensional multiscale scaffolds. The effects of fabric structure (T2/2 and T4/2) and spinning time (0.5 h, 1.0 h, and 1.5 h) on the surface morphology of nanofiber films were investigated. The results show that the patterned nanofiber film can be achieved by using fabric as a template. The prepared nanofiber films showed a diagonal lattice pattern, which replicated the surface structure of the fabric. The T4/2 fabric had a more distinct patterning effect than the T2/2 fabric. Optimal patterning was achieved with a spinning time of 0.5 h. The fiber diameters of the T2/2 and T4/2 groups at 0.5 h were (703.78±171.44) nm, and (646.37±88.76) nm. The probability of fiber orientation in the T2/2 and T4/2 groups was 77.21% and 71.03%, and the average pore diameters were 19.54 μm and 12.81 μm.

Keywords: Nanofiber membranes; Patterning; Fabric template; Electrospinning

1 Introduction

Electrospinning is a technique to produce nanofibers economically and conveniently. The nanofibers made by electrospinning are widely used in the preparation of various tissue engineering scaffolds due to their large specific surface area, high porosity, and ability to mimic the physiological structure of natural extracellular matrix[1-3]. Despite the great potential of electrospinning, the nanofiber membranes produced by conventional electrospinning are usually dense structures in 2D, difficult for cell infiltration[4].

Controlled deposition and patterning of nanofibers is a novel way to address the obstacles caused by the dense arrangement of fibers in conventional electrospinning processes[5]. To achieve controlled deposition and patterning of nanofibers, researchers often use customized metal templates as a receiver device for electrospinning, and the use of metal templates of different shapes allows the fabrication of nanofibers with different topologies[6-7]. However, the patterned nanofiber membranes obtained by using metal templates alone are mostly two-dimensional planes, which are not conducive to cell growth and nutrient exchange, and are still far from adequate for a tissue engineering scaffold. Scaffolds with different multi-scale nano/micro fibers and topological morphologies can provide structural and mechanical support for potential tissue engineering applications[8-9]. The preparation of patterned nanofiber films using fabric as a template is a promising method, which can produce three-dimensional multiscale scaffolds. But few people have studied the feasibility of using fabric as a receiving template for electrospinning.

In this study, we investigated the feasibility of fabrics as templates to prepare patterning. Fabrics with different microstructures were fabricated and used as a receiving device for electrospinning fibers. The deposition of nanofibers on fabrics of different microstructures were evaluated, as well as the effect of different spinning times on the patterning effect of nanofibers. Specifically, the diameter and orientation of the nanofibers were analyzed, and the pore size of the composites formed by the fabric and the nanofibers was performed.

2 Experiments

2.1 Materials

Silk yarn (diameter 0.26 mm±0.02 mm) from Yangzhou Jinhuan Medical Device Factory (Jiangsu, China). PGA yarn (diameter 0.19 mm±0.01 mm) from Suzhou Maticon Medical Technology Co., Ltd. (Jiangsu, China). Polycaprolactone (PCL, relative molecular mass 80 000), from Sigma-Aldrich (USA). Dichloromethane (DCM) and *N*, *N*-dimethylformamide (DMF) are both from Sinopharm Chemical Reagent Co., Ltd. (China).

2.2 Weaving of fabrics

Two types of fabrics, 2/2 right twill (noted as T2/2) and 4/2 right twill (noted as T4/2), are woven with silk yarn as the warp yarn and silk yarn and PGA yarn in the ratio of 1 : 3 as the weft yarn using a four shuttle box split tube special loom (YASL2300-12-24, China).

2.3 Electrospinning

First, PCL particles were added to DMF and DCM (3 : 7 volume ratio of the two) to prepare a PCL solution with a mass concentration of 0.185 g/mL. To ensure adequate dissolution of the polymer, the prepared solution was stirred on a magnetic stirrer for 12 h. Then the above spinning solution was transferred into a 10 mL syringe equipped with a 23 G needle and spun on an electrospinning machine (TL-Pro-BM, Shenzhen Tongli Micro-Nano Technology Co., Ltd., China). The fabric was fixed on the aluminum foil and the aluminum foil with the fabric was fixed on the receiver roller to receive the nanofibers. The spinning time was set to 0.5 h, 1.0 h and 1.5 h, the spinning voltage was 16 kV, the pushing speed was 0.6 mL/h, the spinning distance was 20 cm, and the roller speed was 1 600 r/min. The obtained complexes formed by electrostatic spun fiber film and fabric were placed in a vacuum drying oven at 37 ℃ for 12 h to fully remove the residual organic solvent.

2.4 Surface morphology observation of nanofiber membrane

The complexes formed by nanofibers and fabrics were sprayed with gold after vacuum drying, and then their apparent morphology was observed using a field emission scanning electron microscope (SU 8010, Hitachi, Japan). The diameter and orientation of the nanofibers were measured using Image J software. The probability of 90°±15° under the fitted curve obtained using the orientation distribution data was used as an indicator of the degree of fiber orientation.

2.5 Pore size distribution test of composite materials

The complex formed by nanofiber and fabric was cut to the size of 30 mm×30 mm, and then its pore size was analyzed using a porous material pore size analyzer (CFP-1100AI, Porous Materials, Inc., USA).

Statistical analysis:

Data on nanofiber diameters were tested for significance using One-way ANOVA test. A value of $p<0.05$ was considered significant and indicated with (*), $p<0.01$ was indicated with (**), and $p<0.001$ was indicated with (***).

3 Results and Discussion

3.1 Effect of different spinning time and different structure of the receiving fabric on the surface morphology of nanofiber film

Tab. 1 shows the basic structural parameters of both T2/2 and T4/2 fabrics. The effects of different spinning times and different receiving substrates on the surface topography of nanofiber films are shown in Fig. 1. Obviously, it can be clearly seen that compared to the control group (aluminum foil as the receiving template), using fabric as the receiving template enables the controlled deposition of nanofibers, so that patterned nanofiber films can be formed. Patterned nanofiber films can be achieved for both T2/2 fabric and T4/2 fabric. The nanofibers show a twill-like grid pattern, which replicates the surface topology of the fabric. Comparing the images of different spinning times, the patterning effect of the nanofiber film gradually decreases with the increase of the spinning time, and the best patterning effect of the nanofiber film is found at 0.5 h. Fig. 2 shows SEM images of the surface morphology of nanofiber films on T2/2 fabric and T4/2 fabric at 0.5 h spinning time. Nanofibers are deposited more on the twill bulge and the warp float length line of the twill fabric, while there are fewer nanofibers deposited in the middle of the two adjacent warp float length lines. The area where the nanofibers are deposited on the longitudinal floating line can be called "fiber-dense zone", and the area where the nanofibers are deposited in the middle of the two floating lines can be called "fiber-sparse zone". Nanofibers in the "fiber-dense zone" are more numerous and more uniformly oriented, while the nanofibers in the "fiber-sparse zone" are sparse and haphazardly arranged. Comparing the T2/2 group with the T4/2 group, the T4/2 group seems to have a longer "fiber-dense zone" and "fiber-sparse zone" than the T2/2 group. This may be related to the fabric structure, the T4/2 fabric has a longer floating length line than the T2/2 fabric.

Tab. 1 Basic structural parameters of T2/2 fabric and T4/2 fabric

	T2/2	T4/2
Warp density (cpi/(10 cm))	385.53±4.79	384.89±15.74
Weft density (cpi/(10 cm))	404.72±23.09	512.89±37.62
Tightness (%)	100.04	99.99
Thickness (mm)	0.32±0.01	0.38±0.01

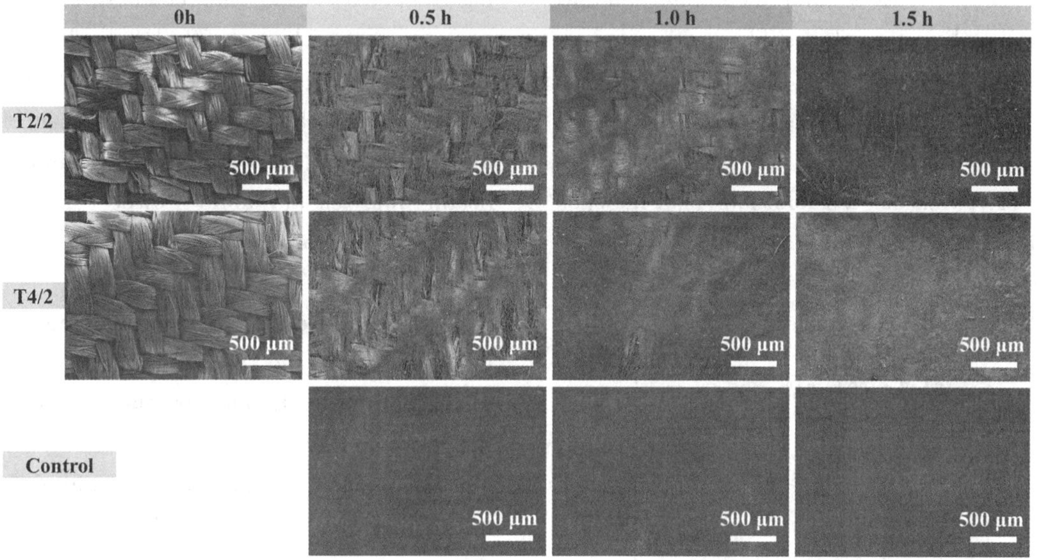

Fig. 1 SEM images of composites with different spinning times and different receiving templates (Control group for aluminum foil receiving)

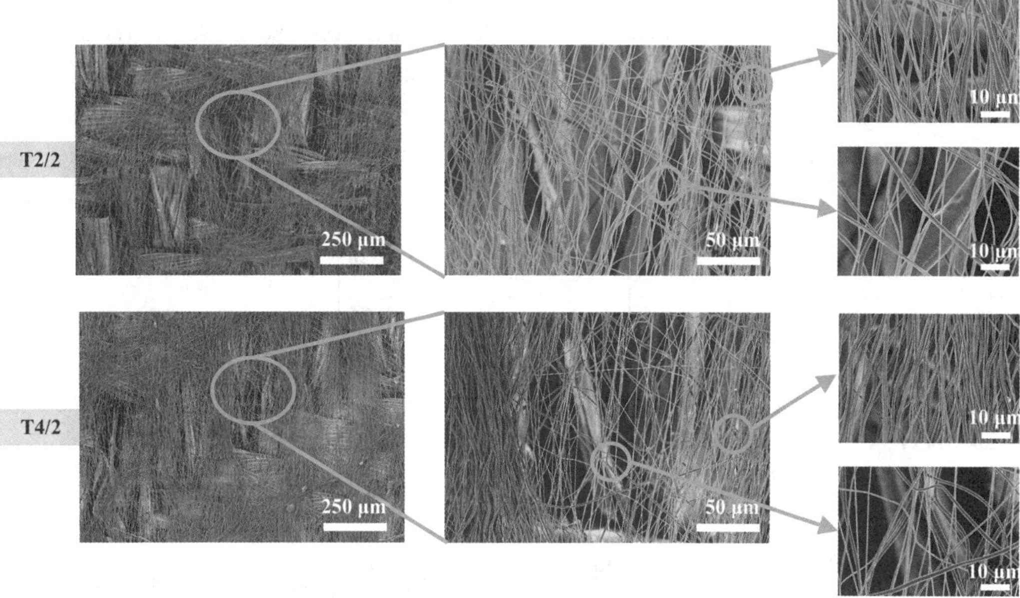

Fig. 2 SEM images of T2/2 and T4/2 groups at a spinning time of 0.5 h

3.2 Effect of different spinning time and different structure of the receiving fabric on fiber diameter and fiber orientation of nanofiber film

The diameters of nanofibers on different spinning times and different receiving templates are shown in Fig. 3 (a). In the T2/2 group, the fiber diameter was (703.78±171.44) nm at 0.5 h, while it decreased to (629.76±76.27) nm at 1 h. A similar situation was observed in the T4/2 group, where the fiber diameter was (646.37±88.76) nm at 0.5 h, and decreased to (619.70±67.92) nm at 1 h. As can be seen from the figure, the diameter of nanofibers on each receiving template basically decreases with increasing spinning time, while the diameters of nanofibers deposited on T2/2 fabric and T4/2 fabric and aluminum foil do not differ much from each other. This may be since at the beginning of electrospinning, the generated charged jet stream has not yet reached a stable state, and the spun nanofibers are affected by its coarse diameter, while after a period, the charged jet stream starts to become stable and the fiber diameter decreases.

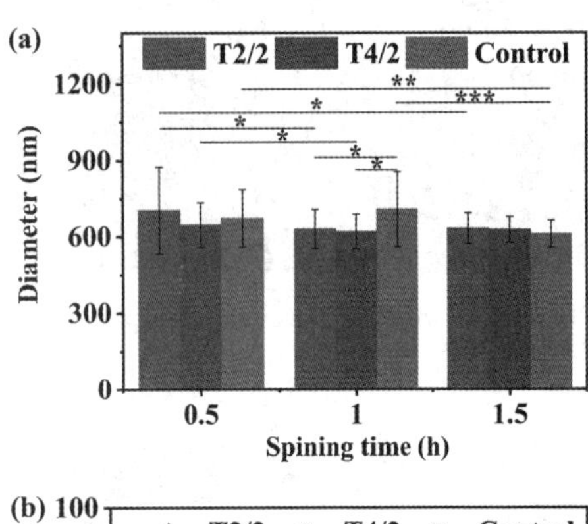

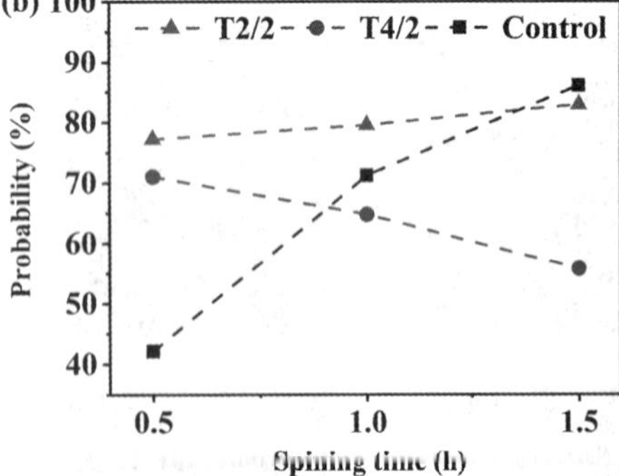

Fig. 3 (a) Average diameter of nanofibers on different spinning times and different receiving substrates; (b) probability of nanofiber orientation angle distribution at 90°±15° on different spinning times and different receiving substrates

Fig. 3 (b) shows the probability of nanofiber orientation angle distribution at 90° ± 15° on different spinning times and different receiving templates. Generally, researchers believe that a fiber orientation angle of 90° is considered to have a good degree of orientation, so the higher the probability of fiber orientation angle distribution at 90°±15°, the higher the degree of orientation. As can be seen from the figure, the degree of fiber orientation increases with increasing spinning time when T2/2 fabric and aluminum foil are used as receiving templates, while the opposite is true for T4/2 fabric. In addition, it can be seen from the data in the figure that at 0.5 h spinning time, the fabric receiving group (including T2/2 group and T4/2 group) has a higher degree of fiber orientation than the aluminum foil receiving group. At 0.5 h, the probability of the degree of orientation was as high as 77.21% and 71.03% in the T2/2 and T4/2 groups, yet only 42.15% in the control group. This may be since at the early stage of spinning, the voltage is not yet stable, and the raised longitudinal floating lines on the fabric provide certain topographic cues for the deposition of nanofibers, resulting in a higher degree of fiber orientation.

3.3 Effect of different spinning times and different structure of the receiving fabric on the pore size of the complexes formed by the nanofiber film and the fabric

The average pore size of two fabrics with different structures (including T2/2 and T4/2 groups) at different spinning times is shown in Fig. 4. Pore size of the complexes formed by the fabrics and nanofiber membranes of both structures decreased significantly with the increase of spinning time. The average pore size of the T2/2 group was about 19.54 μm at 0.5 h spinning time, which was only a decrease of 8.95% compared to the average pore size of the original fabric, while by 1 h and 1.5 h there was a significant decrease of about

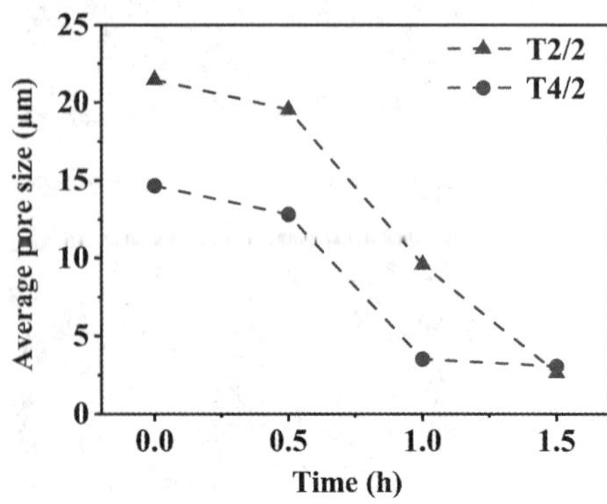

Fig. 4 Average pore size of T2/2 and T4/2 fabrics at different spinning times

55. 36% and 87. 74% each. The average pore size of the T4/2 group was about 12. 81 μm at 0. 5 h, which was only a decrease of 12. 62%, but at 1 h decreased by 75. 99% and at 1. 5h the average pore size decreased by 79. 20%. The pore size of the T2/2 group is almost always larger than that of the T4/2 group at all spinning times, because the original T2/2 fabric has a larger pore size than the T4/2 fabric.

4 Conclusions

This study focused on the feasibility of using fabric as a template to prepare patterned nanofiber membranes. The diameter and degree of orientation of nanofibers on fabric were analyzed. The average pore size of the composite formed by nanofiber membranes and fabric was evaluated. The preparation of patterned nanofiber films can be achieved by using fabric as a receiving template for electrospinning. Three spinning times of 0. 5 h, 1. 0 h and 1. 5 h were compared, and it was found that the best patterning of the nanofiber film on the fabric at 0. 5 h showed a clear twill lattice pattern, which replicated the structure of the fabric. The different structures of fabrics lead to various lattice patterns with different sizes. The patterned nanofiber membrane fabricated from the T4/2 fabric as a template has a larger lattice size, compared to the T2/2 group. The fiber diameters of the T2/2 and T4/2 groups at 0. 5 h were (703. 78±171. 44) nm and (646. 37±88. 76) nm, respectively. Using the fabric as the receiving template facilitates the nanofibers to maintain a high degree of orientation at the early stage of electrospinning. At 0. 5 h, the probability of the degree of orientation was as high as 77. 21% and 71. 03% in the T2/2 and T4/2 groups, respectively. At 0. 5 h spinning time, the complexes had larger average pore size, 19. 54 μm and 12. 81 μm for the T2/2 and T4/2 groups, respectively. This study provides a new idea for the patterned preparation of nanofiber membranes. The production of patterned nanofiber membranes using fabric as a template allows the production of patterned nanofiber membranes while harvesting three-dimensional multiscale scaffolds.

References

[1] ALI M G, MOUSA H M, BLAUDEZ F, et al. Dual nanofiber scaffolds composed of polyurethane-gelatin/nylon 6-gelatin for bone tissue engineering[J]. Colloids and Surfaces A: physicochemical and Engineering Aspects, 2020, 597: 124817.

[2] STOCCO T D, BASSOUS N J, ZHAO S, et al. Nanofibrous scaffolds for biomedical applications [J]. Nanoscale, 2018, 10(26): 12228-12255.

[3] ZHU S, HE Z, JI L, et al. Advanced nanofiber-based scaffolds for achilles tendon regenerative engineering[J]. Frontiers in Bioengineering and Biotechnology, 2022, 10: 897010.

[4] SEMITELA Â, GIRÃO A F, FERNANDES C, et al. Electrospinning of bioactive polycaprolactone-gelatin nanofibres with increased pore size for cartilage tissue engineering applications[J]. Journal of Biomaterials Applications, 2020, 35(4/5): 471-484.

[5] LIU H, MUKHERJEE S, LIU Y, et al. Recent studies on electrospinning preparation of patterned, core-shell, and aligned scaffolds[J]. Journal of Applied Polymer Science, 2018, 135(31): 46570.

[6] KHALILI S, NOURI KHORASANI S, RAZAVI M, et al. Nanofibrous scaffolds with biomimetic structure[J]. Journal of Biomedical Materials Research Part A, 2018, 106(2): 370-376.

[7] PALOMARES D, AMMANN K R, SALDANA PEREZ J J, et al. Patterned electrospinning: a method of generating defined fibrous constructs influencing cell adhesion and Retention[J]. ACS Applied Bio Materials, 2021, 4(5): 4084-4093.

[8] FENG J, XU R, ZHAO J, et al. Multi-scale nano/micro fiber scaffolds with different topological morphologies[J]. Fibers and Polymers, 2022, 23(4): 935-943.

[9] SHALUMON K T, CHENNAZHI K P, TAMURA H, et al. Fabrication of three-dimensional nano, micro and micro/nano scaffolds of porous poly(lactic acid) by electrospinning and comparison of cell infiltration by Z-stacking/three-dimensional projection technique[J]. Iet Nanobiotechnology, 2012, 6(1): 16-25.

Study on the Influence of Structural Parameters on the Mechanical Properties of Warp-Knitted Spacer Fabrics

ZENG Pei[1], **LI Tiantian**[1], **LIU Qi**[1], **XIANG Xihao**[1], **FU Shaoju**[1*]

[1] *Key Laboratory of Textile Science and Technology, Ministry of Education, College of Textiles, Donghua University, Shanghai 201620, China*

* *Corresponding author's email*: sjfu@ dhu. edu. cn

Abstract: Warp-knitted spacer fabrics play a great role in clothing, decoration, medical care, architecture and other fields. However, the design and mechanical properties of warp-knitted spacer fabrics are still unclear. In this study, five types of warp-knitted spacer fabrics with different structures are designed and knitted from the polyester filament, and the relationships between the fabric structure and the mechanical, abrasion resistance properties of warp-knitted spacer fabrics are investigated. The results show that the fabric strength is enhanced with the increase of warp and weft densities and bulk density. The tighter the loop structure, the smaller the mesh hole, and the higher the strength. The strength of warp-knitted spacer fabric can greatly increase by adding weft insertion structure to the surface of warp-knitted spacer fabric. However, the strength enhancement effect is not obvious when the filling yarn is added only on one surface. The warp-knitted spacer fabrics made of polyester, coupled with their smooth and thick appearance, have good abrasion resistance property, so it has a wide range of applications.

Keywords: Warp-knitted spacer fabrics; Mechanical properties; Weft insertion structure

1 Introduction

Three-dimensional fabrics play a great role in clothing, decoration, medical care, architecture and other fields[1]. Three-dimensional knitted fabrics can be divided into warp knitted and weft knitted spacer fabrics according to their structure and knitting methods. Compared with weft-knitted spacer fabric, warp-knitted spacer fabric has more complete technology and more flexible product design, so it has a huge market prospect[2]. Warp-knitted spacer fabric is composed of the surface layer knitted by two systems and the spacer layer knitted by one system[3-4]. As a three-dimensional porous material, warp-knitted spacer fabric has good bending resistance, sound absorption, compression resistance and other properties[5], which makes warp-knitted spacer fabric widely used in clothing, decoration, medical care, construction and other fields[6-7]. However, the design and mechanical properties of warp-knitted spacer fabrics are still unclear. In this study, we would like to design 5 kinds of warp-knitted spacer fabrics, and study the effect of structural parameters on fabric tensile property and abrasion resistance property. Polyester (PET), the most common synthetic fiber, was selected to knit spacer fabrics.

2 Experiments

2.1 Materials

The yarns (64 tex polyester filament or 42 tex polyester filament) for the surface structure of warp-knitted spacer fabrics and the spacer yarns (42 tex polyester monofilament) were obtained from NTEC Monofilament Technology Co. Ltd., Nantong, China.

2.2 Warp-knitted spacer fabrics design

The spacer fabrics were designed with polyester monofilament, and knitted by Raschel double needle bed warp knitting machine with gauge E18.

The types of spacer yarns are shown in Tab. 1. The structures of the surface layers were selected as hexagonal net, rhombic mesh and pillar stitch-weft insertion structures, which are shown in Tab. 2, and the detailed structural parameters, such as stitch notations, pictures of the spacer fabrics are shown in Tab. 3.

Tab. 1 Types of spacer yarns

Types	Material	Lapping movement and threading method

I	42 tex PET	GB3:(1-0)-(3-4)//
		1 threading 1 empty
		GB4:(1-0)-(3-4)//
		1 threading 1 empty

II	42 tex PET	GB3:(1-0)-(1-0)//
		1 threading 1 empty
		GB4:(1-0)-(1-0)//
		1 empty 1 threading

Tab. 2 Types of warp-knitted spacer fabrics

Sample No.	Material	Surface structure
A	64 tex PET	Hexagonal net
B	64 tex PET	Pillar Stitch + weft insertion
C	64 tex PET	Rhombus mesh
D	64 tex PET	Rhombus mesh Pillar Stitch + weft insertion
E	42 tex PET	Rhombus mesh

Tab. 3 Detailed structural parameters of the fabrics

No.	Stitch notions		Surface structure	Cross section
A	GB1, GB5: (2-1/2-3)x2/2-1/(3-4/3-2) X2/3-4// 1 threading 1 empty (2-3/2-1)X2/2-3/(1-0/1-2) X2/1-0// 1 empty 1 threading Typers of spacer yarns: I	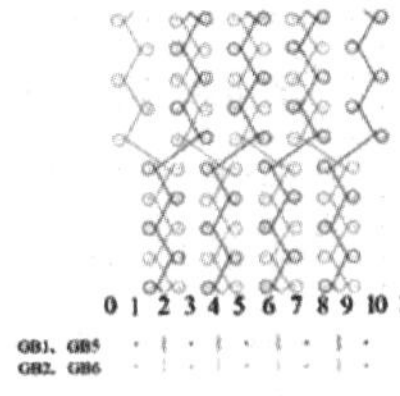		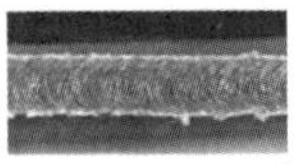
B	GB1:1-0// full-set threading GB5:1-0// full-set threading double side full weft insertion Types of spacer yarns: II	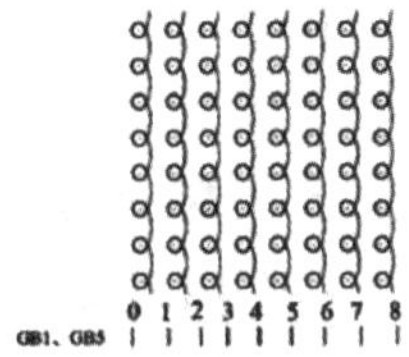		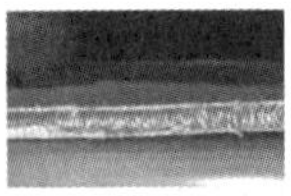
C	GB1, GB5: (3-2/4-5)X3/ (6-7/5-4)X3// 2 threading 2 empty GB2, GB6: (4-5/3-2)X3/ (10-/2-3)X3// 2 empty 2 threading Types of spacer yarns: II	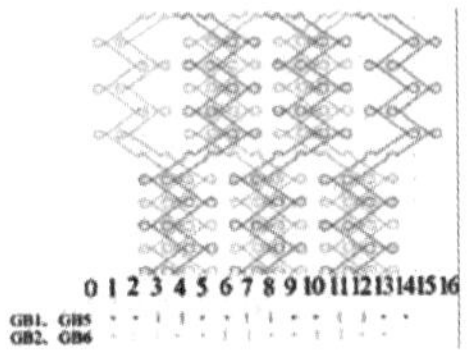		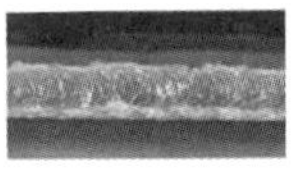
D	GB1:1-1// full-set threading	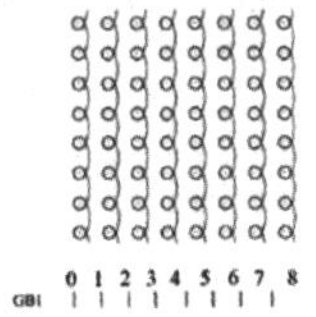		
	GB5: (3-2/4-5)X4/(6-7/5-4) X4//2 threading 2 empty GB6: (4-5/3-2)X4/(1-0/2-3)X 4//2 empty 2 threading single side full weft insertion Types of spacer yarns: II			
E	GB1, GB5: (3-2/4-5)X3/(6-7/5-4)X3 //2 threading 2 empty GB2, GB6: (4-5/3-2)X3/(1-0/2-3)X3 //2 empty 2 threading Types of spacer yarns: II	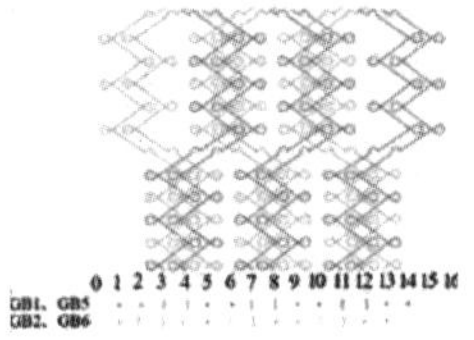		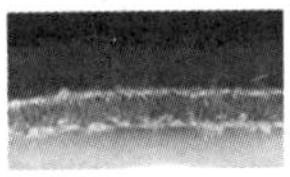

2.3 Basic performance testing of fabrics

Courses per cm (CPC) and wales per cm (WPC) were calculated using a digital optical microscope according to ASTMD 3887.

According to ASTM D1059, the areal density of warp-knitted spacer fabrics was measured. To this end, five $10 \times 10\ cm^2$ square fabric samples were cut randomly from different locations and weighed by an electronic balance.

According to GB/T 1380—1997, the thickness of warp-knitted spacer fabrics is measured, and YG141N digital fabric thickness gauge is used in this study. The experiments are repeated 5 times at different locations of each sample.

Bulk density of warp-knitted spacer fabrics was measured by dividing the surface density of the fabric by thickness.

Spacer filament arrangement density of warp-knitted spacer fabrics was measured by means of visual count, and its value is related to the mechanical properties of warp-knitted spacer fabric.

2.4 Tensile testing of fabrics

To investigate the effect of different structural parameters on the tensile properties of warp-knitted spacer fabrics, the tensile test of fabrics was conducted on a YG026 Electronic Strength Tester. The test is based on GB/T 3923.1—2013, using the strip method to do the tensile test, each indicator measured five times. Before testing, warp-knitted spacer fabrics were conditioned at 20 ℃ and 65% relative humidity for 24 h.

2.5 Abrasion resistance testing of fabrics

Through the YG502 fabric pilling tester, in accordance with the standard GB/T 4802.3—2008 to measure the fabric abrasion resistance, each piece of fabric grinding 150 times. The abrasion resistance of the fabric was evaluated by the appearance change of the fabric after pilling under the same conditions. Anti-pilling performance is usually divided into five levels, the fifth level is the best, and the first level is the worst. Specific standards are, the fifth level: no change on the fabric surface; fourth level: the fabric surface has a light surface lint, but not formed fluffy balls; third level: the fabric surface produces a medium degree of lint or fluffy balls; second level: the fabric surface produces obvious lint or fluffy balls; first level: the fabric surface is covered with lint or fluffy balls. Before testing, warp-knitted spacer fabrics were conditioned at 20 ℃ and 65% relative humidity for 24 h.

3 Results and Discussion

3.1 Effect of structural parameters on fabric tensile properties

The basic performance results of designed warp-knitted spacer fabrics are given in Tab. 4. The experimental results of the tensile mechanical properties of the spacer fabrics are shown in Tab. 5. The strength-elongation curves obtained during the tensile test of warp-knitted spacer fabrics are shown in Fig. 1.

Tab. 4 Basic performance of warp-knitted spacer fabric

Sample No.	Thickness (mm)	CPC (courses/(5 cm))	WPC (wales/(5 cm))	Areal density (g/m^2)	Bulk density (kg/m^2)	Spacer filament density ($yarn/cm^2$)
A	18.10	46	8	652.24	36.04	33
B	7.12	30	36	861.10	120.94	86
C	11.13	34	33	764.62	68.70	71
D	7.77	38	32	688.52	88.61	72
E	6.36	46	32	759.85	119.47	65

Tab. 5 Tensile performance results of warp-knitted spacer fabric

Sample No.	Breaking strength (N)		Breaking elongation (cm)		Breaking elongation (%)		Breaking strength (N/mm)	
	Warp-wise	Weft-wise	Warp-wise	Weft-wise	Warp-wise	Weft-wise	Warp-wise	Weft-wise
A	380.4	176.7	6.83	10.96	68.3	109.6	7.61	3.53
B	785.9	1141.6	4.31	11.97	43.1	119.7	15.72	22.83
C	820.7	493.5	6.92	14.21	69.2	142.1	16.42	9.87
D	741.9	545.6	5.69	10.76	56.9	107.6	14.84	10.91
E	795.8	447.2	6.52	13.93	65.2	139.3	15.88	10.27

From Fig. 1, it can be concluded that the warp spacer fabric A has the lowest warp strength, while the warp spacer fabrics B, C, D and E have the same range of warp strength, with C having the highest strength and D being slightly weaker.

This is mainly because the coursewise density of fabric A is small, and the coursewise density of fabric B, C, D, E is at the same level. When the same tensile force is applied to the fabric of the same size and the same raw material, fabric A has less stretched yarn, so its warp strength is the lowest. For fabric C, D, E, because the structures of the two sides of D are different, there is force inhomogeneity, so it has the lowest warp strength. Fabric E uses a finer yarn than fabric C, so the strength of fabric E is lower than fabric C. And the surface structure of fabric B is pillar stitch, which has less longitudinal stretchable space than the mesh structure, so its strength is lower than fabric C and E.

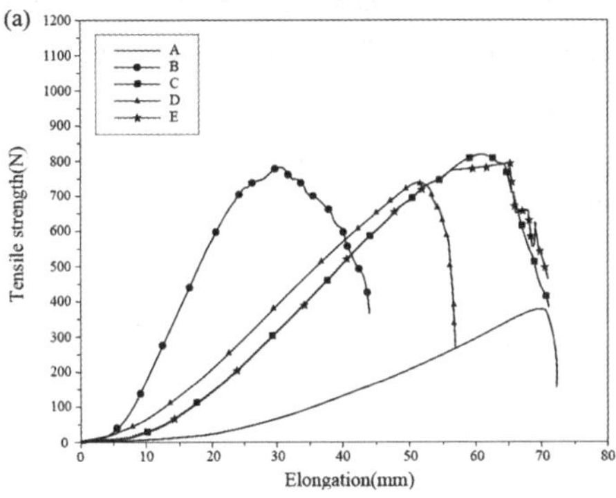

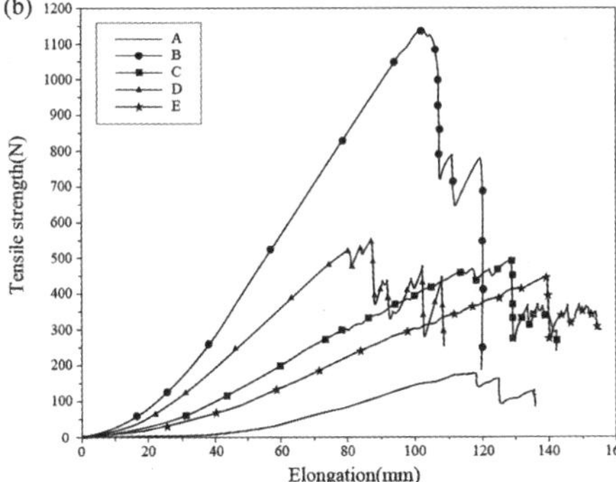

Fig. 1 (a) The strength-elongation curves of warp-wise tensile testing; (b) the strength-elongation curves of weft-wise tensile testing

Generally speaking, the warp strength of fabrics is higher than the weft strength, such as sample A, C, D and E. This is because when the warp-knitted spacer fabrics are stretched in the warp direction, the underlap and loop leg of knitted loops provide a large space for extension. while when the warp-knitted spacer fabrics are stretched in the weft direction, the needle loop and sinker loop can only provide a small extension space, resulting in the weft strength being lower than the warp strength. However, the surface structures of fabric B and D have the addition of weft insertion, especially the fabric B has a structure of weft insertion and pillar stitch on both sides, so its weft strength significantly exceeds its warp strength.

3.2 Effect of structural parameters on fabric abrasion resistance properties

The experimental results of the abrasion resistance of the spacer fabrics are shown in Tab. 6.

After 150 times of grinding, warp-knitted spacer fabrics B and D have obvious lint and fluffy balls, and the corresponding abrasion resistance property is poor. However, the surfaces of warp knitted spacer fabric A, C, and E have no visible lint and fluffy balls, corresponding to excellent abrasion resistance property.

The main reason for the poor abrasion resistance of warp-knitted spacer fabric B and D is that their surfaces are weft-lined combined with pillar stitch, which is similar to the structure of woven fabric. Therefore, when grinding, the weft-lined yarn is easy to be picked out, resulting in the pilling phenomenon.

The reason of warp knitted-spacer fabrics A, C and E have better abrasion resistance properties is that the mesh structure of these warp knitted spacer fabrics is tight, the fabric surface is smooth, and the fabric has a relatively larger thickness, which has a certain relief and dispersion effect on the external grinding[8].

Tab. 6 Abrasion resistance results of warp-knitted spacer fabric

Sample No.	Fuzz resistance	Pill resistance
A	Level 5	Level 5
B	Level 3	Level 3
C	Level 5	Level 5
D	Level 3	Level 3
E	Level 5	Level 5

4 Conclusions

By designing 5 kinds of warp-knitted spacer fabrics with different structures and testing the mechanical properties and abrasion resistance properties of the fabrics, the main conclusions are drawn as follows. First, with the increase of warp and weft density and bulk density, the fabric strength of warp-knitted spacer fabrics generally shows an upward trend. The tighter the loop structure, the smaller the mesh hole, and the higher the strength. Second, the strength of warp-knitted spacer fabric can be greatly increased by adding weft insertion structure to the surface of warp-knitted spacer fabric. However, the strength enhancement effect is not obvious when the filling yarn is added only on one surface. Third, the warp-knitted spacer fabric made of polyester, coupled with its smooth and thick appearance, has good abrasion resistance property, so it has a wide range of applications.

Acknowledgement

This work was supported by the Graduate Student Innovation Fund of Donghua University (No. GSIF-DH-M-2022001).

References

[1] DONG K, PENG X, CHENG R, et al. Advances in high-performance autonomous energy and self-powered sensing textiles with novel 3D fabric structures[J]. Advanced Materials (Deerfield Beach, Fla.), 2022, 34 (21): 2109355.

[2] CHANG Y P, LIU Y P, ZHAO S Q, et al. Design and manufacture of three-dimensional auxetic warp-knitted spacer fabrics based on re-entrant and rotating geometries [J]. Textile Research Journal, 2022, 92(3/4): 467-478.

[3] RAJAN T. P, KANDHAVADIVU, P, PERIYASAMY A P. Influence of porosity and yarn linear density on the thermal behaviour of polyester warp knitted spacer fabrics[J]. Fibers and Polymers, 2021, 22(11): 3212-3221.

[4] ZHI C, LONG H. Sound absorption properties of syntactic foam reinforced by warp-knitted spacer fabric[J]. Cellular Polymers, 2016, 35(5): 271-286.

[5] LIU L, LIN H, MA P. Influences of spacer monofilament distributions on mechanical properties of warp-knitted spacer fabric/resin composites[J]. Textile Research Journal, 2022, 17(4): 292-295.

[6] YU S, DONG M C, JIANG G M, MA P B. Compressive characteristics of warp-knitted spacer fabrics with multi-layers[J]. Composite Structures, 2021, 256: 113016.

[7] YAN G, et al. Comparing properties of the warp-knitted spacer fabric instead of sponge for automobile seat fabric[J]. Journal of Physics: Conference Series, 2021, 1948(1): 012196.

[8] YU W D. Textiles Materials[M]. 2nd Edition. Beijing: China Textile & Apparel Press, 2018.

Mechanical Properties Evaluation of Suede Artificial Leather Based Cloth for Car Seats

ZHU Hui[1], WANG Ruihua[2], XIA Rong[2], DU Lijie[1], MA Yao[1], GAO Jing[2*]

[1] *Benecke Changshun Auto Trim (Zhangjiagang) Co., Ltd, Zhangjiagang 215600, China*

[2] *College of Textiles, Donghua University, Shanghai 201620, China*

[*] *Corresponding author's email*: gao2001jing@dhu.edu.cn

Abstract: Taking the suede artificial leather base cloth applied to the car seat surface as the research object, and the COPET/PET sea-island filament, and DTY as the fabric warp and weft yarn respectively, the main factors affecting the mechanical properties of the suede artificial leather base cloth in the process of molding and post-processing were discussed. While the process parameters were optimized with the tear resistance as the main criterion. Experimental results show that when light net DTY or heavy net DTY is used as fabric weft yarns, and the impact on the tear properties of the base cloth is not significant. However, the fiber splitting process to get the micro-fiber affects the tear strength of the base cloth obviously, as well as the drying and shaping process. When the fabric is treated with 18 g/L of lye, and the initial and second setting temperature is 160 ℃ and 200 ℃ respectively, the fibers in the base cloth are fully split to obtain the suede artificial leather base cloth plump and soft surface. Thus, the base cloth has the excellent mechanical strength, and can meet the needs of the car seat cover.

Keywords: Car seat cover; Suede artificial leather; Base cloth; Splitting process; Mechanical properties

1 Introduction

The car seat cover is an important part of the car interior, which connects the seat foam and skeleton mechanism. It provides comfort for passengers and improves the beauty of the seat. According to the different materials of car seat cover, it is generally divided into leather cover, artificial leather cover and fabric cover. Among them, the microfiber leather[1] in the artificial leather cover is loved by consumers because of its leather texture, good physical properties and eco-friendly features[2]. Suede artificial leather is precisely one of the microfiber leather.

The most commonly used weaving methods of suede fabrics are woven suede and knitted suede. The knitted suede has good air and moisture permeability, excellent resilience, so is widely used in the clothing industry[3]. The woven suede is compact and durable, and is mostly used in the field of automotive interior. In order to improve the competitiveness of suede artificial leather products for seat covers, relevant enterprises began to pay attention to research and develop ultra-high strength suede base cloth, to improve the overall safety performance of the automobile interior in case of emergencies. Therefore, the mechanical properties of suede base cloth have become the focus of attention. Among them, the tearing performance has become the most important index for the mechanical property evaluation of base cloth by relevant enterprises.

In this study, taking the suede artificial leather base cloth applied to the car seat surface as the research object, and the COPET/PET sea-island filament as warp yarn, and light net and heavy net DTY as weft yarn respectively to suede fabric preparation, then through the fiber splitting process to get the sea-island micro-fiber from transitional composite state, so as to form a suede fabric style with fine, smooth and soft surface[4]. Finally, the suede fabric is dried and shaped. By optimizing the relevant preparation and post-processing parameters, the study obtains suede base cloth with excellent mechanical properties for seat cover.

2 Materials and Methods

2.1 Materials and Instruments

COPET/PET sea-island filament (220D/72F, Jiangsu

Beltford New Materials Co. , Ltd.), light mesh and heavy mesh DTY (300D/96F, Jiangsu Beltford New Materials Co. , Ltd.), liquid alkali sodium hydroxide (Zhongyan Changzhou Chemical Co. , Ltd.), neutralizing acid (Changzhou Bangyu Chemical Co. , Ltd.), hydrophilic softener DS (9%-11%, Changzhou Jichi New Material Co. , Ltd.).

The equipment and instruments required for the experiment include: a warping machine (Karmaier (China) Co. , Ltd.), JAT810 air jet loom (Toyota Automatic Loom Company of Japan), HJFA-2 overflow cylinder (Liaoning Donglin Hydraulic Technology Co. , Ltd.), custom setting machine (Lixinmen Fuji Textile Machinery (Zhongshan) Co. , Ltd), custom inspection machine (Lixinmen Fuji Textile Machinery (Zhongshan) Co. , Ltd.), TP-114 precision electronic balance (Denver Instruments (Beijing) Co. , Ltd.), YG141N fabric thickness instrument (Nantong Hongda Experimental Instrument Co. , Ltd.), TM-3 000 scanning electron microscope (Hitachi, Japan), YG026MB-250 electronic fabric strength machine (Changzhou China Fiber Inspection Instrument Equipment Co. , Ltd.), HD026N+ electronic fabric strength instrument (Nantong Hongda Experimental Instrument Co. , Ltd.).

2.2 Experiments

The preparation process of suede base cloth for automobile synthetic leather is as follows: warping → weaving → fiber splitting → drying → shaping.

Firstly, the warping machine is used to warp the warp yarn COPET/PET sea-island filament, light net DTY and heavy net DTY is as weft yarn respectively, The woven texture is 5/2 warp satin. The grey cloth with uniform density is prepared by air-jet loom.

Then, the grey fabric is treated by the fiber splitting process in the sample cylinder[5]. Using liquid alkali NaOH as the fiber splitting agent, set the concentration of NaOH at (8. 0±0. 5) g/L, (13. 0±0. 5) g/L, (18. 0±0. 5) g/L and (23. 0±0. 5) g/L, respectively[6-7].

Then use the setting machine to shape the base cloth after fiber splitting process. With reference to the setting temperature required in the actual industrial production of suede, two drying and setting schemes are designed. The initial and second setting temperature is 160 ℃ and 200 ℃ respectively to get the base fabric sample A, and the initial and second setting temperature is 200 ℃ and 180 ℃ respectively to get the base fabric sample B. The hydrophilic softener DS is used for surface softening treatment. The molding speed of the molding machine is 16 m/min.

2.3 Performance tests

According to the enterprise's standard for evaluating the mechanical strength of automotive seat cover base cloth, the tearing strength is selected to evaluate the mechanical properties of suede base cloth. According to GB/T 3917. 2—2009, the experiment was repeated five times for all suede base cloth samples, and the tear strength in the warp and weft directions was measured and characterized.

Using SEM, different magnification was set to observe the surface morphology of the base fabric and its warp and weft yarns. The diameter of the warp and weft yarns was measured by Nano Measurer, and the diameter data were statistically analyzed by Origin.

3 Results and Discussion

3.1 Effect of light and heavy mesh weft yarn on the mechanical properties of the base cloth

Polyester stretch textured yarn is processed by slight and short stretching under the of compressed air. The number of its network can be controlled by adjusting the compressed air. Too many nets will cause network spots on the fabric surface, which will affect the dye uptake of the fabric during dyeing. Too few nets will make the fabric loose and fluff under tension, thus affecting the overall fabric strength. In this study, the light mesh DTY with a network number of 40-50 knots/m and the heavy mesh DTY with a network number of 100-120 knots/m were respectively selected as weft raw materials of suede fabrics.

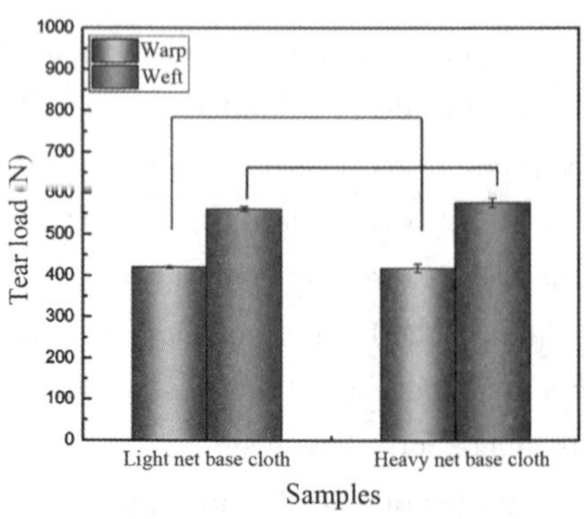

Fig. 1 Histogram of tearing strength of base fabric sample

Fig. 1 shows the histogram of the longitudinal and latitudinal tear strength of suede base fabric. The tear strength of the light net base fabric sample in warp and weft direction is (420.18±3.84) N and (560.69±5.98) N respectively, the tear strength of the heavy net base fabric sample in warp and weft direction is (418.23±10.42) N and (566.57±11.34) N respectively. SPSS software is used to analyze the difference significance between the tear strength in warp and weft direction of the light and heavy net base fabrics. Within the 95% confidence interval of T-test, the significance values of warp and weft are 0.066 and 1.000 respectively, both of which are greater than 0.05. That is, light net DTY and heavy net DTY as weft yarn to weave suede base cloth have no significant effect on its tearing performance.

3.2 Optimization of fiber splitting lye concentration

Fig. 2 shows the SEM of fiber-splitting fabrics treated with different concentrations of NaOH. The warp yarn of the base cloth after 8 g/L and 13 g/L fiber splitting process do not form superfine fibers due to the low alkali concentration and strong adhesion between the sea component and the island component. When the alkali concentration increased to 18 g/L and 23 g/L, the strong concentration alkali liquid can effectively reduce the adhesion between the sea component and the island component[8-9]. The difference between the shrinkage property of the sea component and the island component can be used to produce cracking. Meanwhile, the alkali liquid can etch fibers to form ultra-fine fibers.

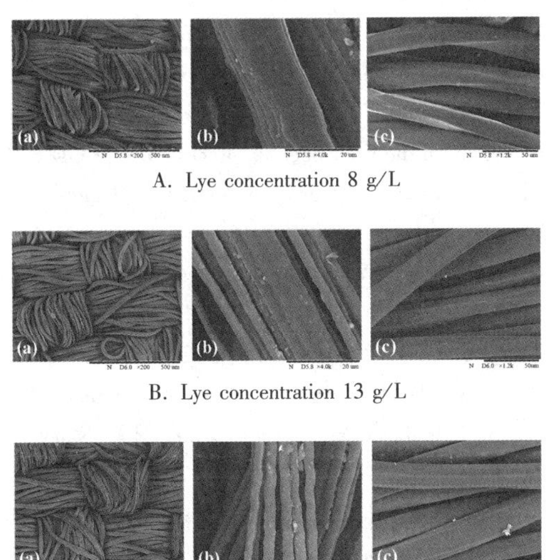

A. Lye concentration 8 g/L

B. Lye concentration 13 g/L

C. Lye concentration 18g/L

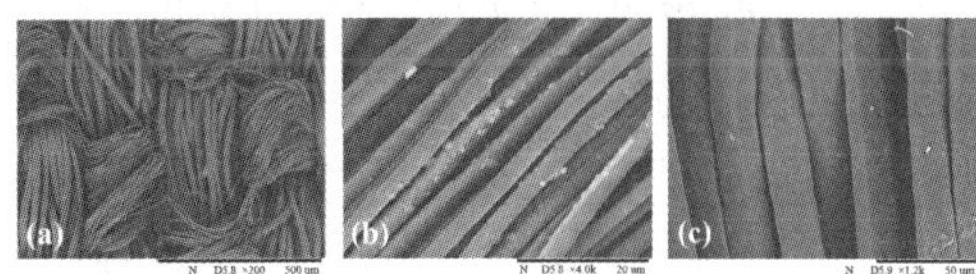

D. Lye concentration 23g/L

Fig. 2 SEM of fabrics treated with different concentration of lye (a. base cloth; b. warp; c. weft)

Observe and characterize the actual suede effect of the base cloth, as shown in Fig. 3. With the increase of the lye concentration, the surface color of the base fabric gradually becomes lighter, the gloss gradually increases, and the suede effect is enhanced. However, when the alkali concentration is lower than 18 g/L, the warp and weft interweaving structure of the base cloth is clear and the suede effect is not obvious, which cannot meet the requirements of the plump suede base cloth.

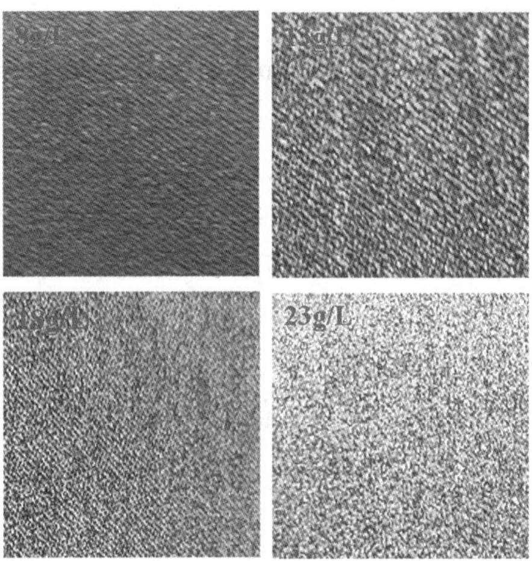

Fig. 3 Suede effect comparison of base fabric under different concentration of lye

The tearing performance of fabric samples under four different liquid alkali concentrations was characterized, as shown in Fig. 4. It can be observed that the alkali concentration has a great impact on the strength of the base cloth. The tearing strength of the base cloth in warp and weft direction gradually decreases with the increase of the liquid alkali concentration. That is mainly because the increase of the alkali concentration improves the etching degree of the yarn, and reduce the overall strength of the fabric. The decline rate of tearing strength in weft direction shows the law of slow first and then fast, which is mainly due to the different hydrolysis rates of COPET in different stages of fiber splitting process[10]. At the initial stage of hydrolysis, the solvent molecules penetrate the polyester to expand their volume, and then gradually form a completely dissolved molecular dispersed homogeneous phase system.

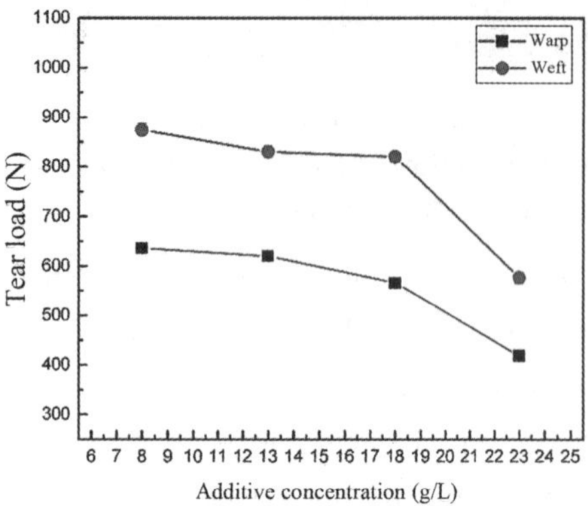

Fig. 4 The tearing strength in warp and weft direction of the base cloth

According to the suede effect of different concentrations of lye, the lye concentration of 18 g/L was selected for fiber splitting treatment.

3.3 Effect of drying temperature on mechanical properties of base cloth

The tearing performance evaluation and difference significance analysis were carried out on the base fabric sample A and B, respectively. The results are shown in Fig. 5. The tear strength in warp direction of sample A is (418.23 ± 10.42) N and in weft direction is (566.57±11.34) N. The tearing strength in warp direction of sample B is (380.20±6.58) N and in weft direction is (511.35±8.3) N. The tear strength of sample A is significantly higher than that of the base cloth sample B. Under T-test, there is a significant difference between tear strength in the warp and weft of them ($p<0.05$), thus the drying temperature has a significant impact on the tear performance of base cloth. The suede base cloth set and shaped by first 160 ℃ and then 200 ℃ has higher mechanical strength.

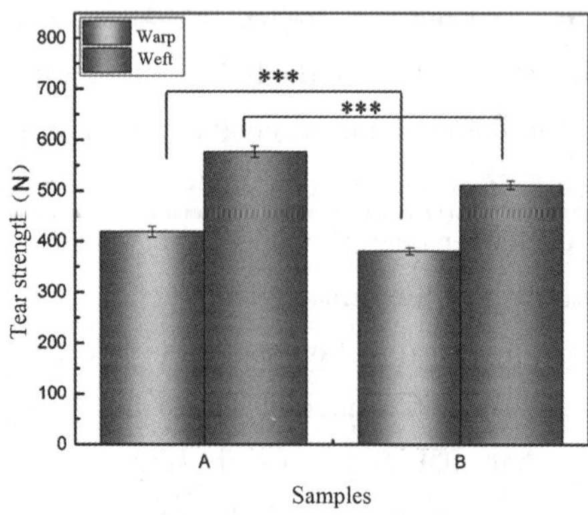

Fig. 5 Histogram of tearing strength of base fabric sample

4 Conclusions

(1) The suede base fabric is produced with COPET/PET sea-island yarn as warp yarn and light net and heavy net polyester stretch textured yarn as weft yarn respectively. The influence of light and heavy net DTY on the mechanical properties of suede base fabric is not significant. Considering the flatness of the base fabric, the heavy net DTY is selected as weft yarn for interior base fabric weaving.

(2) The fiber splitting process has a significant influence on both the suede effect and the mechanical properties. When the concentration of NaOH is lower than 18 g/L, the sea-island filament is not fully split, which cannot achieve the full plump suede effect. When the alkali concentration is set at 18 g/L, the suede base cloth has both the plumper suede surface and higher mechanical strength.

(3) The drying and molding temperature also has a great impact on the mechanical properties of the suede base cloth. When the initial and second setting temperature is 160 ℃ and 200 ℃ respectively, it can effectively improve the mechanical strength of suede base cloth.

References

[1] SANDIP V, PURANE, NARSINGH R, et al. Microfibers, microfilaments & their applications[J]. AUTEX Research Journal, 2007, 7(3): 148-158.

[2] FERNANDES I P, AMARAL J S, PINTO V, et al. Development of chitosan-based antimicrobial leather coatings[J]. Carbohydrate Polymers, 2013, 98(1): 1229-1235.

[3] XV H B. Research on structural properties and style evaluation of suede fabrics[D]. Suzhou: Suzhou University, 2008.

[4] Research on the fiber opening process of COPET/PA6 sea-island microfiber leather base cloth[D]. Hangzhou: Zhejiang Sci-tech University, 2015.

[5] MA X Y, LV L Y, LI X. Enzyme hydrolyzation of polyamide hyperfine fiber synthetic leather base[J]. Carbohydrate Polymers, 2013, 98(1): 1229-1235.

[6] RENEKER D H, CHUN I. Nanometre diameter fibers of polymer, produced by electrospin-ning[J]. Nanotechnology, 1996, (7): 216.

[7] BRESEE R R, KO W C. Fiber formation during melt blowing[J]. International Nonwovens Journal, 2003, 12(2): 21-28.

[8] LIU H, WU Q, ZHANG Q. Preparation and properties of

banana fiber-reinforced composites based on high density polyethylene (HDPE)/nylon-6 blends[J]. Bioresource Technology, 2009, (60): 6088-6097.

[9] HUANG J X, ZHANG L. Production methods and research progress of island fiber[J]. Chemical Fiber and Textile Technology, 2019, 48(1): 34-38.

[10] SUN Q J, YANG D L. Fiber opening treatment process of suede fabric [J]. Printing and Dyeing, 2005 (7): 34-36.

Kaolin/$CaCl_2$ Modified Chitosan Nonwoven Fabrics and the Rapid Hemostatic Properties

LIU Jian[1,2], LEI Zhen[1,2], GUAN Guoping[1,2*], WANG Lu[1,2]

[1] Key Laboratory of Textile Science & Technology, Ministry of Education, Donghua University, Shanghai 201620, China

[2] College of Textiles, Donghua University, Shanghai 201620, China

* Corresponding author's email: ggp@ dhu. edu. cn

Abstract: In order to obtain products with excellent hemostatic properties, chitosan nonwoven fabrics were prepared by nonwoven technology using chitosan fibers as raw materials. To enhance the hemostatic properties of the nonwoven fabrics, different concentrations of kaolin and Ca^{2+} solutions were used for modification, and the blood clotting time (BCT), blood coagulation index (BCI), activated partial thromboplastin time (APTT), and prothrombin time (PT) were characterized. The test results show that kaolin and Ca^{2+} can be well attached to the chitosan nonwoven fabrics, and the chitosan nonwoven fabric treated with 1% kaolin and 0.3% Ca^{2+} has the best hemostatic properties. It is in vitro BCT was (55±15) s, which was 161 s shorter than that of commercial hemostatic sponges. Compared with the blank control, BCI was shortened by 91.1%, APTT was shortened by 54.9%, and PT was shortened by 28.4%. Combined with previous research, it is implied that both endogenous and exogenous coagulation pathways are involved to promote coagulation in the present study.

Keywords: Chitosan; Nonwoven fabrics; Kaolin; Hemostatic property

1 Introduction

The uncontrolled bleeding caused by various accidents and natural disasters is the primary cause of human trauma death. According to the root survey, about 60% of the casualties in various accidents were due to excessive blood loss[1]. Excessive bleeding in a short time will lead to severe symptoms such as hypothermia, acidosis, coagulation disorders, and multiple organ failure. When the blood loss exceeds 30% of the total human blood, there is a risk of death[2]. Therefore, effective hemostasis in the early stages of bleeding is very important.

Global sales of hemostatic materials reached $2.761 billion in 2020, but domestic products for rapid hemostasis were scarce. Generally, hemostasis sponge, Yunnan Baiyao and tourniquet are used to stop bleeding. Biological products are often used to stop bleeding in moderate and severe bleeding. However, there are still problems such as high price, poor environmental adaptability, inconvenient to carry and store, and unable to meet the needs of military and civilian applications[3]. Therefore, it is very meaningful to study a hemostatic material with low cost, safety, high efficiency, easy to use, and easy to carry and store. Chitosan is a linear polysaccharide obtained by deacetylation of natural macromolecular chitin[4]. The cations on the molecular chain can generate electrostatic adsorption with the anions on the surface of red blood cells, and promote the aggregation of red blood cells to stop bleeding[5-7]. However, the hemostatic effect of single chitosan material is limited. Kaolin is a kind of aluminum silicate minerals, which can contact with blood to promote endogenous coagulation. Moreover, SiO_2 in kaolin can adsorb water in blood and concentrate coagulation factors to achieve the purpose of hemostasis[8-10]. Ca^{2+} promotes coagulation in both endogenous and exogenous coagulation pathways[11]. Therefore, in this study, chitosan fiber materials were used as raw materials, kaolin and Ca^{2+} were used to modify chitosan fiber materials, hoping to obtain materials with rapid hemostasis function.

2 Experiments

2.1 Materials

Chitosan fiber materials were purchased from Qingdao Jifa New Materials Co., Ltd. Hemostatic sponge was purchased from Guilin Fukangsen Medical Device Co. , Ltd. Kaolin, calcium chloride, APTT kit and PT kit were purchased from Shanghai Yuanye Biotechnology Co. , Ltd. Pluronic® F127 and anhydrous ethanol were purchased from Shanghai McLin Biochemical Technology Co. , Ltd. Anticoagulated rabbit blood was purchased from Hongquan biology.

2.2 Modification of chitosan fiber materials

Firstly, chitosan nonwovens (CS) were obtained by the nonwoven process. Pluronic® F127 solution with a mass fraction of 3% was prepared as the surfactant, and then kaolin was added to prepare 0.5%, 1% and 3% kaolin solutions, respectively. The solutions were heated and stirred in a magnetic stirrer for 2 h. After the heating was stopped, CS was added to the solution. After dipping for 1 h, it was taken out and dried in an oven at 60 ℃ for 30 min. The samples were named CS1, CS2 and CS3. The CS treated with 0.5%, 1%, 3% $CaCl_2$ solution was named CS4, CS5, CS6. CS treated with 1% kaolin and 0.5% $CaCl_2$ solution was named CS7.

2.3 Scanning electron microscopy

The prepared samples were sprayed with gold and the surface morphology of the nonwovens was characterized by scanning electron microscopy (SEM). At the same time, the elemental analysis of different samples was carried out by an elemental analyzer to observe the distribution of elements.

2.4 Characterization of hemostatic properties in vitro

Determination of BCT: Cut 5 mg sample and commercial hemostatic sponge into centrifuge tube. 300 μL anticoagulant rabbit blood was incubated with the sample at 37 ℃ for 1 min. Then, 30 μL $CaCl_2$ solution (0.2 M) was added to induce coagulation reaction. The centrifuge tube was tilted 90° every 5 s to observe whether the blood flowed, and the coagulation time was recorded when the blood was coagulated. Six parallel samples were set for each group with no hemostatic material as the blank control, and the experimental results were averaged.

Determination of BCI : 5 mg of samples to be tested and commercial hemostatic sponge were cut and placed in a centrifuge tube, and incubated in a constant temperature incubator at 37 ℃ for 5 min. 200 μL anticoagulant rabbit blood was dropped on the surface of each sample, and then 20 μL $CaCl_2$ solution (0.2 M) was added to the surface to neutralize the anticoagulant to induce a coagulation reaction. After the centrifuge tube was incubated at 37 ℃ for 3 min, 10 mL of deionized water was added to it for 10 min, and the blood cells that were not involved in coagulation burst. 200 μL anticoagulant rabbit blood was added to 10 mL deionized water and 10 mL PBS respectively as a positive control group and negative control group. After incubation, 200 μL supernatant was transferred to 96-well plate, and the absorbance of hemoglobin in the supernatant was measured at 540 nm. Calculate BCI according to Eq. (1).

$$BCI = \frac{OD_t - OD_n}{OD_p - OD_n} \times 100\% \tag{1}$$

Where OD_t is the absorbance value of the experimental group, OD_p is the absorbance value of the positive control group, and OD_n is the absorbance value of the negative control group.

2.4 Whole blood coagulation kinetics

Several surface dishes were prepared, and the sample to be tested was made into a 2 cm × 2 cm size and placed in a surface dish. Five surface dishes were used for each sample. The whole blood which neutralized the anticoagulant was dropped on the surface of each material in order, 200 μL for each sample. After the drop was completed, the time was counted, and 20 mL of deionized water was slowly added to a surface dish every 2 min. Then 200 μL supernatant was transferred into a 96 — well plate, and the absorbance of hemoglobin was measured at 540 nm using a microplate reader. The obtained data were plotted as curves.

2.5 APTT and PT characterization

First, 10 mL anticoagulant rabbit blood was centrifuged (3 000 r/min, 10 min) to collect the upper liquid-poor platelet plasma (PPP). Test according to APTT kit and PT kit instructions.

3 Results and Discussion

3.1 Morphology of the samples

As shown in Fig. 1 (a), the chitosan nonwoven fabric obtained by the acupuncture process has good apparent morphology, which is composed of fibers with regular morphology and diameter. The weight of the chitosan nonwoven fabric is (43.4–49.9) g/m^2, the pore size is (55 ± 43) μm, and the porosity is as high as 89.7%. Kaolin adhesion can be clearly observed on the surface of CS1, CS2 and CS3 samples treated with kaolin solution. Solid particles can also be clearly seen on the surface of CS4, CS5 and CS6 samples treated with calcium chloride solution. As shown in Fig. 1 (b), a large number of Si, Al and Ca were detected in CS7, which proved that kaolin and Ca^{2+} were successfully attached to chitosan nonwovens.

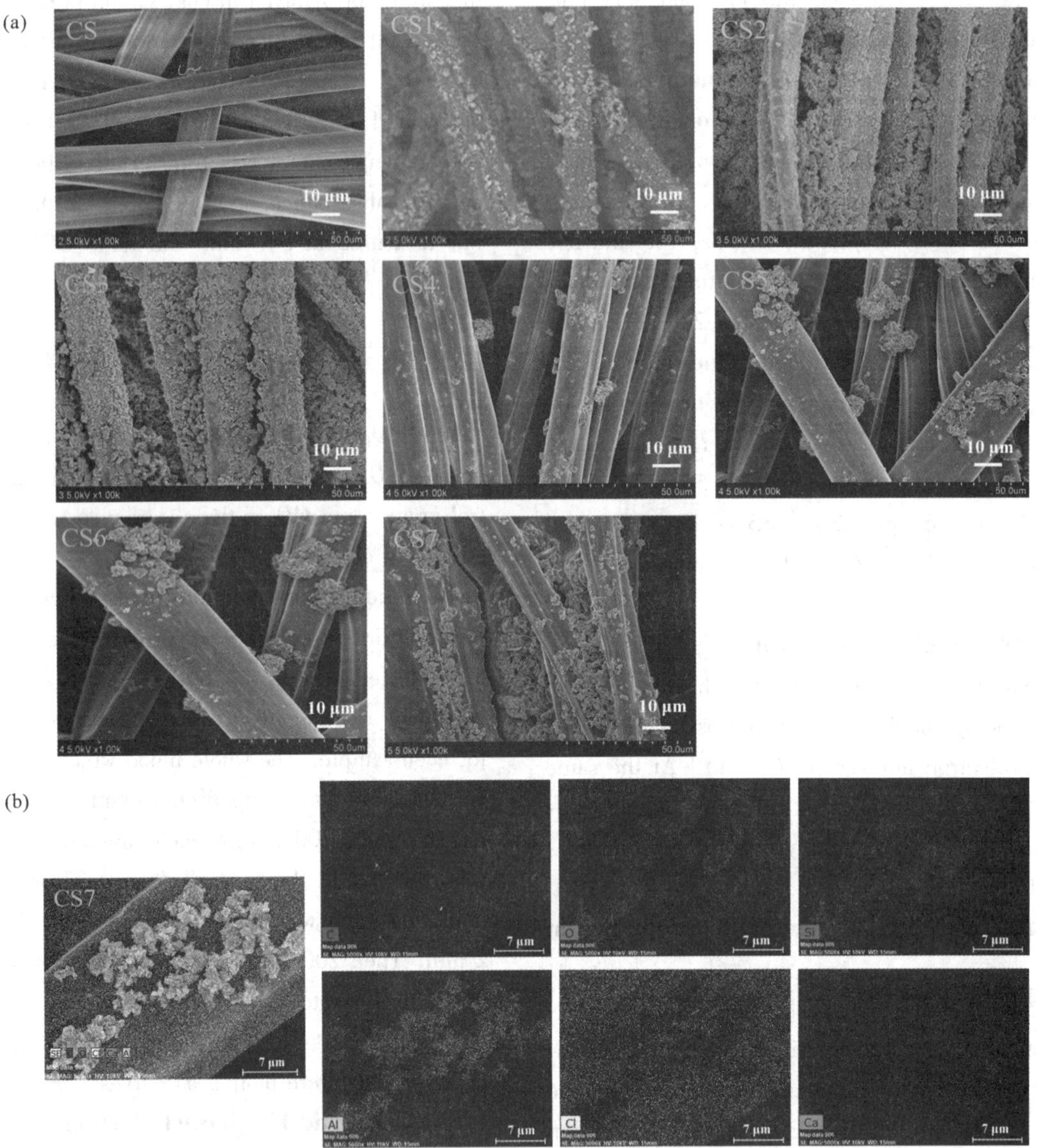

Fig. 1 (a) SEM images of chitosan nonwovens modified by different methods; (b) element distribution map of modified chitosan nonwoven fabric

3.2 Hemostatic properties in vitro

The results of the material mixed with fresh anticoagulant rabbit blood are shown in Fig. 2 (a) Under normal conditions, the blood coagulation time in vitro was 450 s-550 s. CS1, CS2, CS3, and CS7 can coagulate blood in a short time after mixing with blood. The shortest coagulation time of CS7 (55 s±15 s) was 503 s shorter than that of blank control and 161 s shorter than that of the hemostatic sponge. By comparison, it can be seen that kaolin has an excellent hemostatic effect. When the concentration of kaolin is greater than 1%, the hemostatic effect has little change. The coagulation time of CS, CS4 and hemostatic sponge is between 100 s and 250 s, and the coagulation time of CS6 and CS7 is longer than that of blank control, indicating that a low concentration of Ca^{2+} will promote the hemostasis process, and when the concentration of Ca^{2+} is too high, it will inhibit coagulation[12].

BCI index is an index reflecting the coagulation effect, which represents the content of hemoglobin released by incompletely solidified blood cells broken by deionized water after full action of the blood and hemostatic materials. The smaller BCI indicates a better coagulation effect of the material. Fig. 2 (b) shows that the BCI values of CS1, CS2, CS3 and CS7 are lower, the BCI value of CS7 is the smallest, and the hemostatic effect is the best. BCI test results were consistent with BCT.

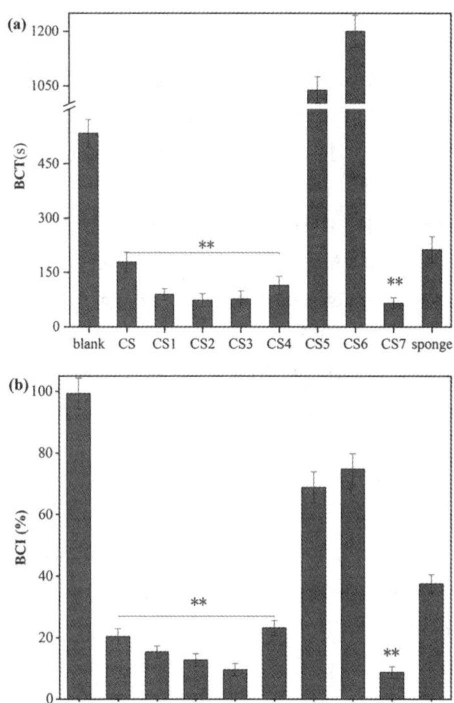

Fig. 2 (a) BCT results of different samples; (b) BCI results of different samples. ** indicates statistical significance ($p<0.01$)

3.3 Whole blood coagulation kinetics

The dynamic coagulation test is the time and stability of the formation of a blood clot after the contact of biological materials with blood. The higher the slope of the curve, the faster the coagulation of the material. In Fig. 3, it can be seen that the dynamic coagulation curve of CS1, CS2, CS3, CS7 appear at steep drop points at the same time, between 0–2 min, indicating that these four kinds of hemostatic gauze samples can absorb blood coagulation into clots within 2 min. The coagulation curve of CS, CS5 and hemostatic sponge changed greatly within 0–5 min, indicating that the hemostatic time was about 5 min. The dynamic curves of CS6 and CS7 changed slowly, and the absorbance value was significantly greater than that of other samples, indicating that the hemostatic effect of the two samples was poor. After 8 min, the dynamic coagulation curve maintained a relatively stable state, because the formation of blood clots, most of the red blood cells in the solution were included in the blood clots. With the change of time, the absorbance value of the solution changed little.

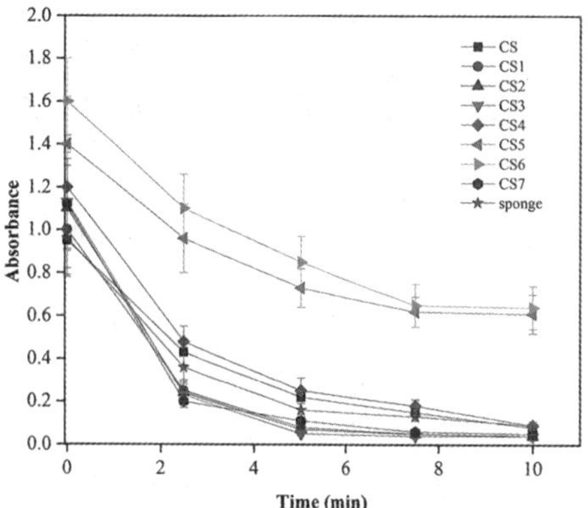

Fig. 3 Dynamic coagulation curves of different samples

3.4 APTT and PT

As shown in Fig. 4 (a), the APTT time of CS1, CS2, CS3 and CS7 were significantly lower than that of the blank control group, and the result of CS7 was the best, which was 54.9% shorter than that of the blank control group, indicating that kaolin could promote the endogenous coagulation pathway and accelerate hemostasis, and the promoting effect increased with the increase of kaolin concentration[13]. From the experimental results of CS4, CS5 and CS6, low concentration of Ca^{2+} also promoted the endogenous coagulation pathway, and the promoting effect gradually disappeared with the increase of Ca^{2+} concentration. Hemostatic sponge had little effect on the activation of partial thromboplastin time.

As shown in Fig. 4 (b), there was no significant difference between the PP results of CS, CS1, CS2, CS3,

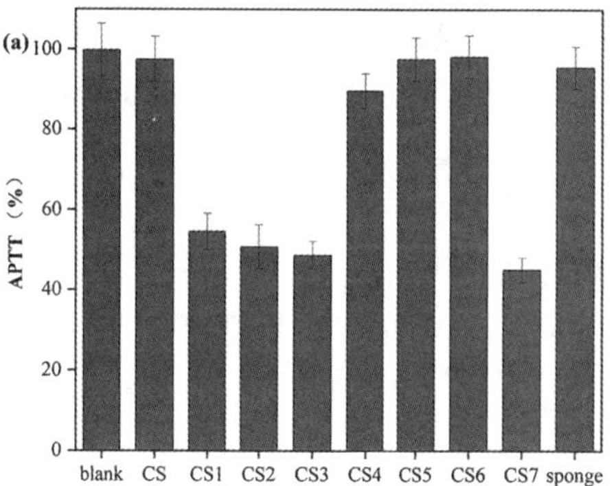

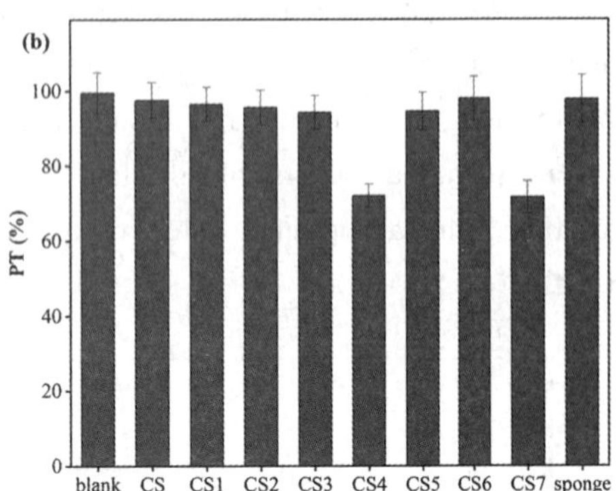

Fig. 4 (a) APTT results of different samples; (b) PT results of different samples

CS5, CS6 and hemostatic sponge and the blank control. Only the PP results of CS4 and CS7 were 28.4% shorter than that of the blank control. The experimental results showed that chitosan and kaolin did not significantly affect the prothrombin time. Only a low concentration of Ca^{2+} had a promoting effect on the exogenous coagulation pathway. With the increase of Ca^{2+} concentration, the promoting effect gradually disappeared.

4 Conclusions

In the present work, chitosan nonwovens were prepared by a nonwoven process using chitosan fiber as raw materials. The nonwoven fabrics were modified by kaolin and Ca^{2+} solutions with different concentrations. The results show that kaolin could promote the acceleration of hemostasis by endogenous coagulation pathway, and a low concentration of Ca^{2+} could promote the acceleration of hemostasis by endogenous coagulation pathway and exogenous coagulation pathway. The chitosan nonwoven fabric treated with 1% kaolin and 0.3% Ca^{2+} shows the best hemostatic performance. The BCT was (55±15) s. Compared with the blank control, BCI was shortened by 91.1%, APTT was shortened by 54.9%, and PT was shortened by 28.4%. The sample can promote red blood cells aggregation, and favorable water absorption, and is a safe and efficient hemostatic material.

References

[1] WELCH M, BARRATT J, PETERS A, et al. Systematic review of prehospital haemostatic dressings[J]. BMJ Military Health, 2020, 166(3): 194-200.

[2] STANWORTH S J, DAVENPORT R, CURRY N, et al. Mortality from trauma haemorrhage and opportunities for improvement in transfusion practice[J]. British Journal of Surgery, 2016, 103(4): 357-365.

[3] KHOSHMOHABAT H, PAYDAR S, KAZEMI H M, et al. Overview of agents used for emergency hemostasis[J]. Trauma Mon, 2016, 21(1): e26023.

[4] HU Z, ZHANG D Y, LU S T, et al. Chitosan-based composite materials for prospective hemostatic applications[J]. Marine Drugs, 2018, 16(8): 25-30.

[5] CHATURVEDI A, DOWLING M B, GUSTIN J P, et al. Hydrophobically modified chitosan gauze: a novel topical hemostat[J]. Journal of Surgical Research, 2017, 207: 45-52.

[6] HATTORI H, ISHIHARA M. Changes in blood aggregation with differences in molecular weight and degree of deacetylation of chitosan [J]. Biomedical Materials, 2015, 10(1): 015014.

[7] LEE G D, DOH S J, KIM Y, et al. Preparation and characterization of carboxymethyl chitosan nonwovens for wound dressing[J]. Fibers and Polymers, 2020, 21(9): 1894-1905.

[8] ELSABAHY M, HAMAD M A. Design and preclinical evaluation of chitosan/kaolin nanocomposites with enhanced hemostatic efficiency[J]. Marine Drugs, 2021, 19(2): 134-123.

[9] CHEN S, YANG Z, WANG F. Preparation and characterization of polyimide/kaolinite nanocomposite films based on functionalized kaolinite [J]. Polymer Engineering & Science, 2019, 45(3): 375-369.

[10] SUN X, TANG Z, PAN M, et al. Chitosan/kaolin composite porous microspheres with high hemostatic efficacy[J]. Carbohydrate Polymers, 2017, 177: 135-143.

[11] ZHU J, SUN Y, SUN W, et al. Calcium ion-exchange cross-linked porous starch microparticles with improved hemostatic properties[J]. International Journal of Biological Macromolecules, 2019, 134: 435-444.

[12] APRYATINA K V, KORYAGIN A S, SMIRNOVA O N, et al. Hemostatic compositions based on chitosan complexes with calcium ions[J]. Pharmaceutical Chemistry Journal, 2021, 54(12): 1269-1273.

[13] LUNDIN J G, MCGANN C L, DANIELS G C, et al. Hemostatic kaolin-polyurethane foam composites for multifunctional wound dressing applications[J]. Materials Science and Engineering C, 2017, 79: 702-709.

Multi-Layered Micro/Nano Filters with Excellent Filtration Performance

MA Wenhua[1], **QI Huan**[2], **ZHANG Chuyang**[1,2*]

[1] *College of Textile and Clothing, Xinjiang University, Urumqi 830046, China*

[2] *College of Textile and Clothing, Quanzhou Normal University, Quanzhou 362000, China*

[*] *Corresponding author's email*: chuyangzhang@126.com

Abstract: Multi-layered micro/nano filters were prepared via meltblown and electrospun technology for effectively capturing particulate matter (PM). The total areal density of micro/nano filter was kept the same (about 18.5 g/m^2) by controlling the electrospinning time and the areal density of meltblown. Polyvinylidene fluoride (PVDF) nanofibers with an average diameter of 348 nm were evenly distributed in the multilayer structure of the micro/nano filters. For comparison, the multi-layered meltblown filters were also prepared. The filtration performance of these two kinds of filters was studied. The surface morphology, pore structure and porosity were investigated to analyze the filtration mechanism of micro/nano filters. The results showed that introducing nanofibers could significantly improve the filtration performance of meltblown. The quality factor of the micro/nano filter increased to 0.073 Pa^{-1} when nanofibers existed in the form of five ultrathin layers among the meltblown, which was 40.4% higher than stacked on a single layer of meltblown. At the same time, the filtration efficiency for 0.3 μm NaCl aerosol particles was 97.94% and the pressure drop was only 53.2 Pa. The multi-layered micro/nano filters developed here would be promising in the field of air filtration.

Keywords: Multi-layered; Electrospinning; Meltblown; Air filtration; Particulate matter

1 Introduction

With the development of industrialization and urbanization, air pollution is becoming more and more serious[1-2]. It has become the biggest environment problem, seriously damage human health and ecosystems[3-4]. Particulate matter (PM) is one of the most important sources of air pollution. Actually, PM can load many harmful components such as toxic liquids and gases which further increase the harm to human health[5]. Especially, PM 2.5 whose aerodynamic diameter less than 2.5 μm is more harmful. It can penetrate into people's lungs and bronchi more easily because of its small particle size and thus trigger various respiratory diseases such as asthma, lung cancer, cardiovascular disease and so on[6-7]. In addition, The COVID-19 outbreak in 2019 posed a serious threat to human health and life and has a great impact on the global economy. In view of this situation, the development of air filters with high efficiency and low resistance to prevent the inhalation of PM 2.5 and COVID-19 is urgent. Actually, air filters such as masks, air purifiers and window screen have been used for outdoor personal protection or indoor air-quality improvement[8-10]. The filtration mechanism of air filters mainly includes the following five: interception, inertial impaction, gravity settling, Brownian diffusion and electrostatic adsorption[11-13].

Nanofiber-based electrospinning filters have been used for air filtration in recent years because of its small diameter, relatively small pore size, high specific surface area and porosity[14]. However, pure nanofiber membranes are not suitable for air filtration because of its high pressure drop and poor mechanical properties which leads to low quality factor and unstable filtration property. Many scholars have tried a variety of means to fabricate nanofiber-based electrospinning air filters with excellent filtration performances. Generally speaking, there are two major ways to fabricate excellent nanofiber-based electrospinning filters: Introducing inorganic particles and preparing micro/nano filters. For example, Yao et al.[15] introduced flower-like ZnO into the PVA-co-PE nanofiber-interconnected networks, which significantly reduced the pressure drop and improved the quality factor of the pure PVA-co-PE nanofiber air filter. The flower-like ZnO served as scaffolds and increased the porosity, specific surface area, and surface roughness. Li et al. [16] added SiO_2 nano-particles (NPS) into the PI spinning solution to fabricate SiO_2@PI Hybrid membranes. The SiO_2 NPs distributed in the surfaces of PI nanofibers could generate hierarchical structures, which enhanced the mechanical properties and filtration performance of composite filters. Compared with introducing inorganic particles, it seems that preparing micro/nano filters attracted more attention. The lamination of microfibers and nanofibers is a simple and effective way to prepare high performance micro/nano filters[17-21]. To the best of our knowledge, there was no report about the preparation of multi-layered micro/nano filters with the same total areal density by using ultra-thin meltblown and electrospinning nanofibers. In this work, we prepared multi-layered micro/nano filters via meltblown and electrospun technology. Nanofibers were evenly distributed in the multilayer structure of the micro/nano filters. For comparison, the multi-layered meltblown filters were also prepared. The filtration performance of these two kinds of filters was studied. The results showed that introducing nanofibers could significantly improve the filtration performance of meltblown. The quality factor of micro/nano filters increased with the increase of layers. This work provided a versatile strategy for further design and development of air filters with excellent filtration performance.

2 Experiments

2.1 Materials

Analytical-grade Polyvinylidene fluoride (PDVF, Mw = 1 000 000) and N,N-dimethylformamide (DMF) were purchased from Sinopharm Chemical Reagent Company Limited (Shanghai, China). Polypropylene pellets (Y1500) with a melt flow index (MFI) of 1 500 g 10 min^{-1} was purchased from China-Base Petrochemical Corporation (Ningbo, China). All chemicals were used without further purification.

2.2 Preparation of meltblown

PP meltblown was fabricated by the spunbond meltblown spunbond (SMS) pilot line at Quanzhou Normal University (Shishi) Ecological Intelligent Fabric Engineering Technology Research Institute. Three kinds of meltblown (M1, M3, M5) with different areal densities were fabricated by regulating the speed of collector belt. These meltblown was cut into dimensions of 31 cm×20 cm to match the surface area of the receiving roller of electrospinning device. Detailed process parameters are listed in Tab. 1.

Tab. 1 Major processing parameters of meltblown

Spinning Temperature (℃)	Drawing air temperature (℃)	Receive distance (mm)	Drawing air pressure (MPa)	Metering pump supply (cc/min)	Airgap (mm)
260	270	300	0.1	45	0.89

2.3 Preparation of multi-layered micro/nano filters

PVDF powder was dissolved into DMF by continuously stirring for 10 h at 25 ℃ with a magnetic stirring apparatus. The prepared PVDF solution (18%) was loaded into a 5 ml syringe placed on an injection pump. The prepared meltblown was wound on the receiving roller as the substrate to receive nanofibers. After spinning for a certain amount of time, a new layer of meltblown with the same areal density was overlaid on the top of the nanofibers. And then nannofibers were deposited on the new meltblown for the equal time. In this way, we can obtain the desired multi-layered micro/nano filters with homogeneous structure on the receiving roller with the assistance of electrostatic force and adhesion. All fabricated micro/nano filters were vacuum-dried at 80 ℃ for 10 h to remove residual solvent. The rate of the receiving roller is 800 r/min. Other process parameters are listed in Tab. 2.

Tab. 2 Major processing parameters of electrospinning

Injection speed (mL/h)	Applied voltage (kV)	Receive distance (cm)	Temperature (℃)	Relative Humidity (%)	Gauge of needle (G)
1.0	25	15	22±2	48±2	18

2.4 Characterization of multi-layered micro/nano filters

The surface morphology of micro/nano filters was investigated by scanning electron microscope (SEM, TESAN MIRA LMS, Tescan China Ltd., China) after coating with gold. The diameter and its distribution of the PVDF nonofibers and PP microfibers were measured by an analysis software (nano measure 1.2). 150 fibers were chosen randomly in different areas to ensure the validity of the data. The filtration performance of fabricated filters was measured by an automated filter tester (DR251XL, Wenzhou Darong Textile Instrument Co., Ltd., China). Aerosol particles passed through the testing area area of 100 cm^2 at the air flow rate of 32 L/min. The quality factor (QF) is used to evaluate the filtration performance of a filter.

3 Results and Discussion

3.1 Structure design of micro/nano filters

In order to study the influence of fiber distribution of micro and nano fibers on filtration performance, three kinds of multi-layered micro/nano filters were fabricated as described above. These three kinds of samples contained the equal amount of microfibers and nanofibers were named M1-E1 * 1, M3-E3 * 3, M5-E5 * 5. Where M1, M3, and M5 represent three kinds of meltblown whose areal density are 16.50, 5.43, 3.35 g/m^2 respectively. The speed of collector belt were 10, 30 and 50 m/min during fabrication of meltblown and other parameters were not been adjusted. In theory, the areal density of M1 is five times of the M5. E1, E3 and E5 represent three layers of nanofibers with different spinning time. M5-E5 * 5 represents five layers of M5-E5. That is to say, five layers of meltblown and five layers of nanofibers were arranged alternately. Detailed parameters of these three kinds of samples are listed in Tab. 3.

Tab. 3 Detailed parameters of three different kinds of samples

Sample	Structure parameter	Total areal density of meltblown (g/m^2)	Cumulative spinning time (min)	Speed of collector belt of meltblown (m/min)
M1-E1 * 1	M1+30 min E	16.50	30	10
M3-E3 * 3	(M3+10 min E) * 3	16.29	30	30
M5-E5 * 5	(M5+6 min E) * 5	16.75	30	50

3.2 Morphology

The surface morphology of micro/nano filters after NaCl aerosol filtration test is shown in Fig. 1 (a)-(c). It was obvious that micro/nano filters exhibited the hierarchical structure consisting of random microfibers and nanofibers. In the multilayer structure, microfibers acted as scaffold and coarse filter and nanofibers acted as fine filter. The prepared micro/nano filters integrated the advantages of microfibers and nanofibers and thus enhanced filtration performance as a result of their synergistic effect. Besides, it was not difficult to find that NaCl aerosols were mainly captured by nanofibers with relatively fine diameters. The nanofibers of M3-E3 * 3 and M5-E5 * 5 captured significantly fewer aerosol particles than M1-E1 * 1. That was because the microfibers acted as surface layer of micro/nano filters which was the first layer to encounter the airflow carrying particles during filtration tests. Before the particles reached the innermost nanofiber layer, they had encountered several layers of nanofibers and most of them had been captured. On the contrary, the particles only passed through one layer of microfibers before they reached the innermost nanofiber layer of M1-E1 * 1. The microfibers had a weak ability to capture particles especially ultrafine particles and thus a large number of ultrafine particles penetrated microfibers and eventually were captured by the nanofibers of M1-E1 * 1.

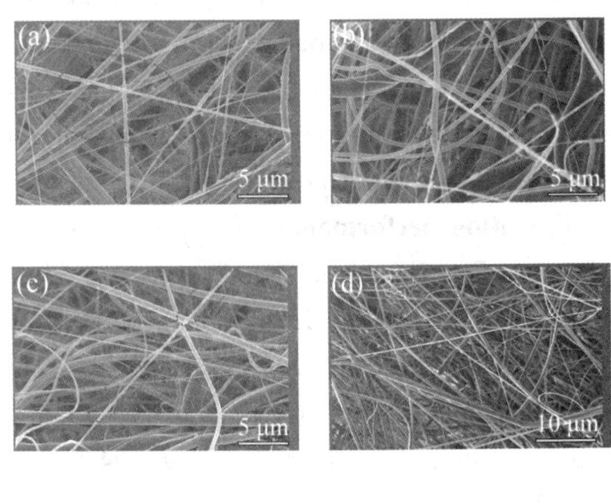

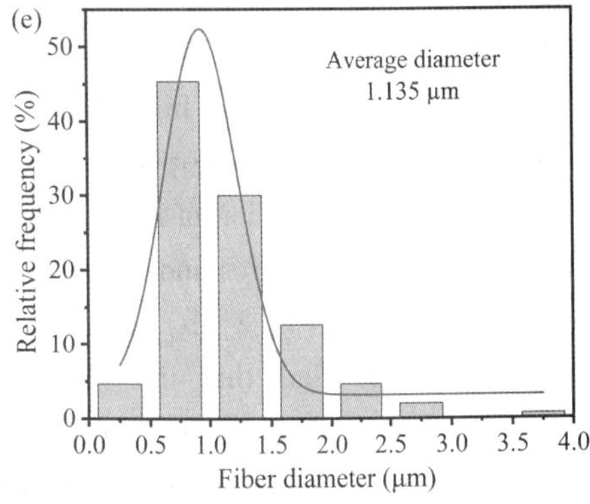

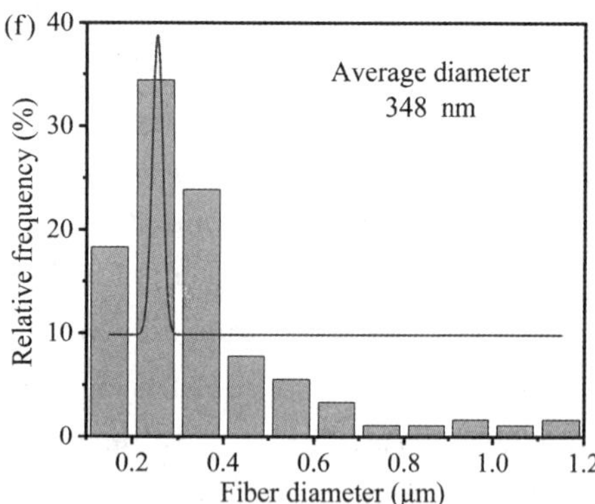

Fig. 1 The surface SEM images of the multi-layered micro/nano filters: (a) M1-E1 * 1; (b) M3-E3 * 3; (c) M5-E5 * 5 and pure meltblown filter; (d) M1 Fiber diameter distribution of (e) melt-blown microfibers and (f) electrospinning nanofibers

Most of the fiber diameters of M1 were less than 1.5 μm and the average fiber diameter was 1.135 μm. Interestingly, about 45% of the fibers were less than 1μm in diameter which reached nanoscale (Fig. 1 (e)). That was why the difference in fiber size between the two types was not as obvious as we expected in the SEM images of the micro/nano filters. Most of the diameters of electrospinning nanofibers were less than 400 nm and the average fiber diameter was 348 nm (Fig. 1 (f)). The prepared micro/nano filters had versatile fiber diameters, which was conducive to the formation of complex and variable pore structure and increased the tortuosity of the pore and thus making us speculate that the multi-layered micro/nano filters may possess satisfactory filtration efficiency

3.3 Filtration performance

As shown in Fig. 2 (a) and Fig. 2 (c), the filtration efficiencies of M1 * 1, M3 * 3 and M5 * 5 were 64.88%, 67.43% and 69.79% respectively. The filtration efficiencies of M1-E1 * 1, M3-E3 * 3 and M5-E5 * 5 were 95.36%, 97.99% and 97.94% respectively. The introduction of nanofibers significantly improved filtration efficiency. On the other hand, the pressure drops of them changed from 44.8, 44.4, 34.1 to 58.6, 53.9, 53.2 Pa respectively. The reason might be that the introduction of nanofibers blocked the original relatively larger pores and reduced the average pore size. Comparing Fig. 2 (b) with Fig. 2 (d), it was not difficult to find that the quality factors (QFs) of micro/nano filters were 0.052, 0.072, 0.073 Pa^{-1} respectively, which were significantly high-

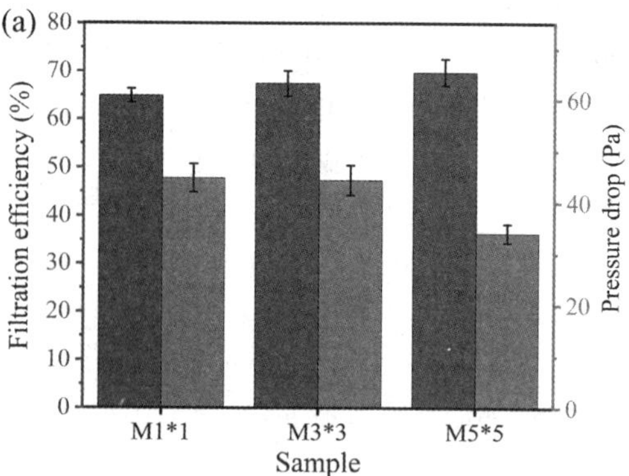

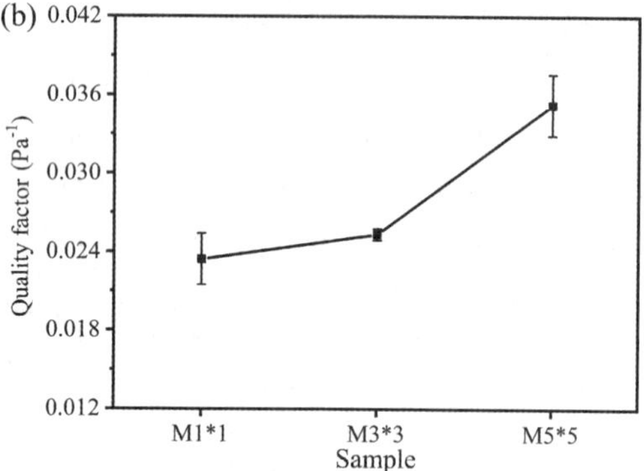

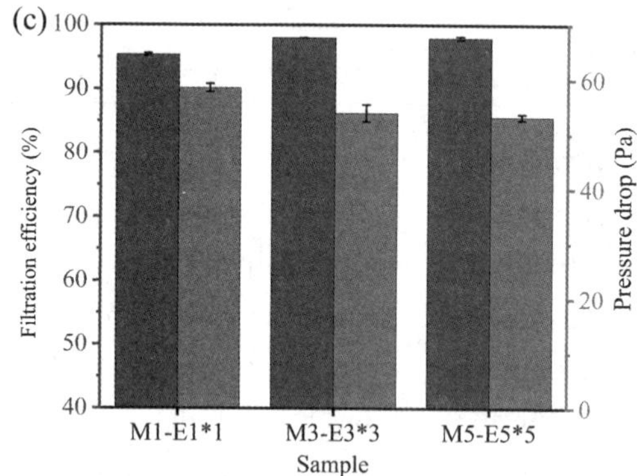

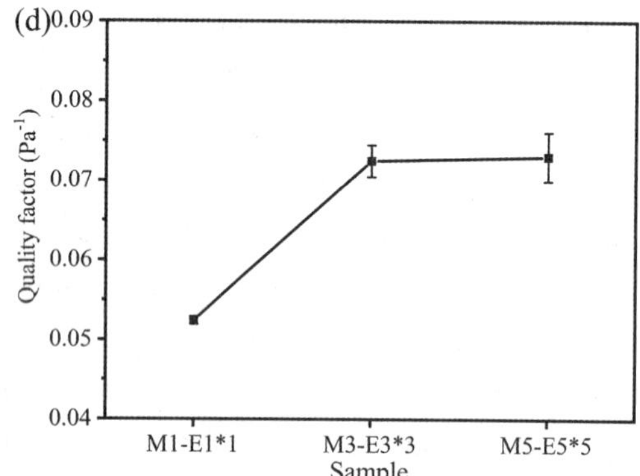

Fig 2 (a) Filtration efficiency and pressure drop and (b) quality factors of different pure meltblown filters; (c) filtration efficiency and pressure drop; (d) quality factors of micro/nano filters

er than pure meltblown filters M1 * 1, M3 * 3 and M5 * 5 (0.023, 0.025, 0.035 Pa^{-1}). Besides, the quality factor of the micro/nano filter increased from

0. 052 to 0. 073 Pa^{-1} when nanofibers existed in the form of fivo ultrathin layers among the meltblown, which was 40. 4% higher than stacked on a single layer of meltblown (0. 052 Pa^{-1}). We suspected that the multilayered structure may enhance the the tortuosity of the pores and thus improved the ability to capture particles. In addition, the relatively larger pore size and porosity of the M5 − E5 * 5 endowed it lower resistance. These two factors made M5 − E5 * 5 have satisfactory quality factor.

4 Conclusions

In summary, a multi-layered micro/nano filters via meltblown and electrospun technology was fabricated. Nanofibers were evenly distributed in the multilayer structure of the micro/nano filters. The prepared micro/nano filters integrated the advantages of microfibers and nanofibers and thus enhanced filtration performance as a result of their synergistic effect. The quality factor of the micro/nano filter increased to 0. 073 Pa^{-1} when nanofibers existed in the form of five ultrathin layers among the meltblown, which was 40. 4% higher than stacked on a single layer of meltblown.

The results demonstrate that fabricating micro/nano filters with multilayered structure is an efficient mean to obtain high performance air filtration composite filter. The micro/nano filters prepared by this method can meet most of the air purification requirements

References

[1] ZHANG R, JING J, TAO J, et al. Chemical characterization and source apportionment of $PM_{2.5}$ in Beijing: seasonal perspective[J]. Atmospheric Chemistry and Physics, 2013, 13: 7053.

[2] OBERDORSTER G, UTELL M J. Ultrafine particles in the urban air: to the respiratory tract and beyond? [J]. Environmental Health Perspectives, 2002, 110 (8): A440−A441.

[3] NEL A. Air pollution-related illness: effects of particles[J]. Science, 2005, 5739(309): 1326−1326.

[4] LIU C, HSU P, LEE H, et al. Transparent air filter for high-efficiency $PM_{2.5}$ capture [J]. Nature Communications, 2015, 6: 6205.

[5] SONG J, LIU Z, LI Z, et al. Continuous production and properties of mutil-level nanofiber air filters by blow spinning[J]. RSC Advances, 2020, 10: 19615.

[6] TIMONEN K L, VANNINEN E, HARTOG J D, et al. Effects of ultrafine and fine particulate and gaseous air pollution on cardiac autonomic control in subjects with coronary artery disease: the ultra study[J]. Journal of Exposure Science & Environmental Epidemiology, 2006, 16: 332−341.

[7] PETERS A D, DOCKERY D W S, MULLER J, et al. Increased particulate air pollution and the triggering of myocardial infarction [J]. Circulation, 2001, 103 (23): 2810−2815.

[8] ZUO F, ZHANG S, HUI L, et al. Free-standing polyurethane nanofiber/nets air filters for effective pm capture[J]. Small, 2017, 46(13): 1702139.

[9] ULLAH S, ULLAH A, LEE J, et al. Supporting information reusability comparison of melt-blown vs. nanofiber face mask filters for use in the coronavirus pandemic[J]. ACS Applied Nano Materials, 2020, 7(3): 7231−7241.

[10] CAMPOS R K, JIN J, RAFAEL G H, et al. Decontamination of SARS-COV − 2 and other RNA viruses from N95 level meltblown polypropylene fabric using heat under different humidities [J]. ACS Nano, 2020, 14 (10): 14017−14025.

[11] LU T, CUI J, QU Q, et al. Multistructured electrospun nanofibers for air filtration: a review[J]. ACS Applied materials & Interfaces, 2021, 13(20): 23293−23313.

[12] LI P, WANG C, ZHANG Y, et al. Air filtration in the free molecular flow regime: a review of high-efficiency particulate air filters based on carbon nanotubes[J]. Small, 2014, 10(22): 4543−4561.

[13] KIM J, HONG S C, BAE G, et al. Electrospun magnetic nanoparticle-decorated nanofiber filter and its applications to high-efficiency air filtration[J]. Environmental Science & Technology, 2017, 51(20): 11967−11975.

[14] LIU H, LIU L, YU J, et al. High-efficiency and superbreathable air filters based on biomimetic ultrathin nanofiber networks [J]. Composites Communications, 2020, 22: 100493.

[15] YAO Z, XIA M, XIONG Z, et al. A hierarchical structure of flower-like zinc oxide and poly (vinyl alcohol-coethylene) nanofiber hybrid membranes for high-performance air filters[J]. ACS Omega, 2022, 7(3): 3030−3036.

[16] LI D, SHEN Y, WANG L, et al. Hierarchical structured polyimide-silica hybrid nano/microfiber filters welded by solvent vapor for air filtration[J]. Polymers, 2020, 12 (11): 2494.

[17] CHOI Y O. Effect of process control parameters on the filtration performance of pan-ctab nanofiber/nanonet web combined with meltblown nonwoven[J]. Polymers, 2021, 20(13): 3591.

[18] LOU L H, QIN X H, ZHANG H. Preparation and study of low-resistance polyacrylonitrile nano membranes for gas filtration[J]. Textile Research Journal, 2017, 87(2):

208–215.

[19] WANG N, SI Y, WNAG N, et al. Multilevel structured polyacrylonitrile/silica nanofibrous membranes for high-performance air filtration[J]. Separation and Purification Technology, 2014, 126(15): 44–51.

[20] HUNG C H, LEUNG W F. Filtration of nano-aerosol using nanofiber filter under low peclet number and transitional flow regime[J]. Separation and Purification Technology, 2011, 79(1): 34–42.

[21] OMOLLO E, ZHANG C, MADSIAGI J I, et al. Electrospinning cellulose acetate nanofibers and a study of their possible use in high-efficiency filtration[J]. Journal of Industrial Textiles, 2016, 0(00), 1–14.

Application of Triple Repellent and Antistatic Agent on Polylactic Acid SMS Nonwovens

ZHANG Yongmeng[1], **QI Huan**[2], **ZHANG Chuyang**[1,2*]

[1] *College of Textile and Apparel, Xinjiang University, Urumqi 830000, China*

[2] *Institute of Smart & Ecological Textile, Quanzhou Normal University, Quanzhou 362002, China*

[3] *College of Textile and Apparel, Quanzhou Normal University, Quanzhou 362002, China*

[*] *Corresponding author's email*: chuyangzhang@ 126. com

Abstract: In this work, NW-PA C6 and NW−16 were applied on PLA (polylactic acid) spunbond meltblown spunbond (SMS) for triple-repellent and antistatic finishing. Single factor test and orthogonal experiment were used to select and optimize the process parameters. The primary and secondary relationship for the process is explored. The results show that when the concentration of the triple-reject finishing agent is 40 g/L, antistatic agent is 10 g/L, drying at 60 ℃ for 240 s and baking at 120 ℃ for 15 s, the finished polylactic acid composite nonwovens have excellent performance. At the same time, the mechanical and comfort properties of the treated nonwoven fabric are not significantly reduced and can meet the needs of medical surgical gown materials.

Keywords: Polylactic acid; SMS nonwovens; triple-repellent, antistatic finishing; Process optimization; Performance testing

1 Introduction

The development of textile materials with excellent comprehensive performance for medical protection is always an important research area. Surgical gowns are used by doctors and nurses as protective clothing for the operating room, protecting healthcare workers from contamination by various blood-borne pathogens due to occupational characteristics in contact with a patient's blood and body fluids, thereby reducing the incidence of hospital-acquired infections[1]. Meanwhile, the surgical gown also acts as a sterile barrier to protect patients, preventing the transfer of microorganisms, particles and liquids from the operator to the patient[2]. Therefore, the surgical gown should have the dual function of protecting both the patient and the doctor[3].

In some articles, the effects of structural parameters, finishing processes on the barrier performance and wearing comfort of surgical gowns were studied. Seong et al. [4] used a fluorocarbon siloxane finishing agent to sort out the polyester/lyocell fabric by pad-dry-cure and obtained a breathable and antistatic superhydrophobic fabric. Krishnasamy et al. [5] used nano-silver and fluorocarbon thought particle pad-dry-cure methods to organize the woven surgical gown, which has antibacterial, anti-liquid and breathable functions at the same time. Montazer et al. [6] use antimicrobial agents and fluorinated chemical waterproofing agents to treat different fabrics. Zhan et al. [7] report on PU/triclosan composites for versatile protective coating materials for PLA/LMPLA nonwovens through a simple needle-reinforced scraping and impregnation process technology. The introduction of PU and triclosan has resulted in improved hydrophobic, abrasion-resistant, mechanical and antimicrobial properties of the resulting nonwovens. In this work, a triple-repellent and antistatic finishing agent is used to treat SMS composite nonwoven fabrics. Univariate experiments and orthogonal tests were conducted to optimize each process condition, and the final protective performance is used as the evaluation index to screen out the best process and obtain protective nonwoven materials with good waterproof, anti-fouling and antistatic properties.

2 Experiments

2.1 Materials

Experimental materials and instruments: polylactic acid meltblown nonwovens (M), weight 30 g/m^2, Polylactic acid spunbonded nonwovens (S), weight 20 g/m^2,

(Quanzhou simartin import and export trade Co., Ltd.); NW-PA Non-woven triple repellent finishing agent, NW-16 antistatic agent, (Guangzhou Lianzhuang Technology Co., Ltd.); R-3 automatic stereotyping dryer (Foshan Yuanmao Precision Machinery Manufacturing Co., Ltd.); P-AO Pneumatic Electric Small Rolling Mill (Foshan Yuanmao Precision Machinery Manufacturing Co., Ltd.);

2.2 Sample preparation

SMS three-layer composite nonwovens prepared by fusing machine with the temperature of 65 ℃, wherein S layer is 20 g/cm^2, M layer is 30 g/cm^2. The prepared SMS nonwovens sample is treated according to the "pad-dry-cure" process, and the wet pick-up is maintained at 130%-140%.

2.3 Testing and characterization

Water repellency: According to the AATCC 42-2017 standard, the water impermeability spray impact test of the sample is carried out.

Alcohol repellency: Taking GB/T 24120—2009 "Textiles — Determination of resistance to water/alcohol solution" as the reference standard to estimate the alcohol repellency of fabrics.

Contact angle test: With reference to the standard DB44/T 1872—2016, the contact angle tester is used to determine the contact angle of the sample before and after finishing.

Comfort performance: The comfort performance of nonwovens is evaluated by testing the breaking strength, elongation at break, breathability and other properties of the fabric.

3 Results and Discussion

3.1 Single-factor experiments

Finishing liquid concentration: 10 g of antistatic agent is added to the three-reject finishing solution with a concentration of 50 g/L to make a mixed finishing solution;

Pad process: One dip and one padding process; padding pressure is 0.3 MPa;

Drying and curing process: Drying at 60 ℃ for 240 s, while baking at 110 ℃ for 60 s shaping;

3.1.1 Effect of finishing agent concentration on the performance of nonwovens

The triple-repellent and antistatic agents are added to 1 L distilled water in the ratio of 4 : 1 and prepared into a mixed finishing solution, as shown in Tab. 1.

Tab. 1 Preparation of triple-repellent and antistatic finishing solution

No.	Triple repellent agent	Antistatic agent
0#	0 g/L	0 g/L
1#	20 g/L	5 g/L
2#	30 g/L	7.5 g/L
3#	40 g/L	10 g/L

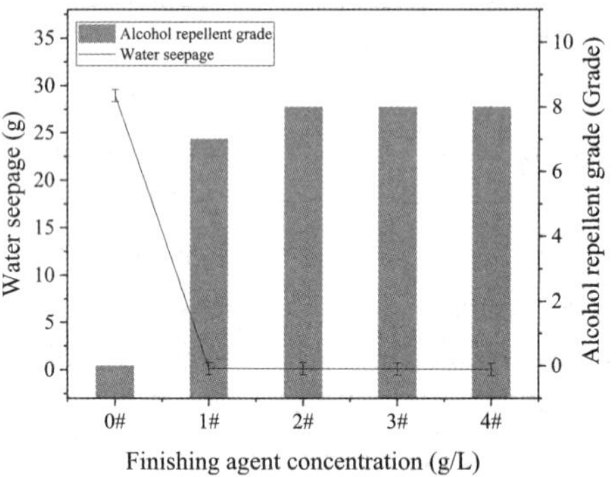

Fig. 1 Effect of different concentrations on the water/alcohol repellency of fabrics

Fig. 1 shows the alcohol repellency rating and water permeation resistance of SMS nonwovens under different finishing agent concentrations. The water seepage amount of the untreated SMS is 28.95 g, which decreased sharply after treatment, and the anti-seepage effect gradually increased with the increase of the finishing agent concentration. The alcohol resistance rating of the nonwovens also increased with finish concentration. When the concentration of the finishing agent is above 30 g/L, the alcohol repellency level of the fabric can meet the requirements of the isolation gown.

3.1.2 Effect of baking temperature on properties of nonwovens

In order to ensure the finishing effect, the sample is finished with the No. 4# finishing solution. The samples are dipped and rolled with a rolling residual rate of 140% and dried at 60 ℃ for 240 s. Then bake at different temperatures for 60 s to explore the effect of baking temporature on the properties of nonwovens.

It can be seen from Fig. 2 that the alcohol resistance grade of the sample increases with the rise of the baking temperature, from grade 0 untreated to grade 8 after treatment. However, when the baking temperature is 130 ℃, the water permeability of the sample increases sharply to 0.15 g, resulting in a decrease in water re-

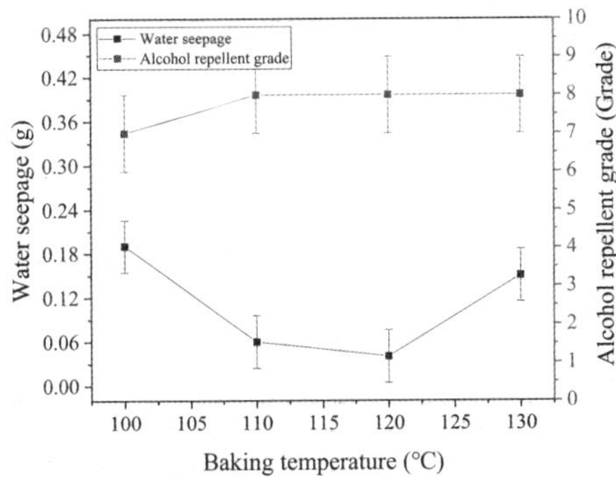

Fig. 2 Effect of different temperatures on the water/alcohol repellency of fabrics

sistance. This is because the higher temperature changes the internal structure of the PLA melt blown nonwovens. It increases the pores between the fibers to reduces the impermeability of the sample.

3.1.3 Effect of baking time on the properties of nonwovens

The sample is dipped and rolled in No. 4# finishing solution, and the rolling residue rate is 140%. The sample is dried at 60 ℃ and baked at 110 ℃, and the water and alcohol repellency under different baking time is investigated. The results are shown in Fig. 3.

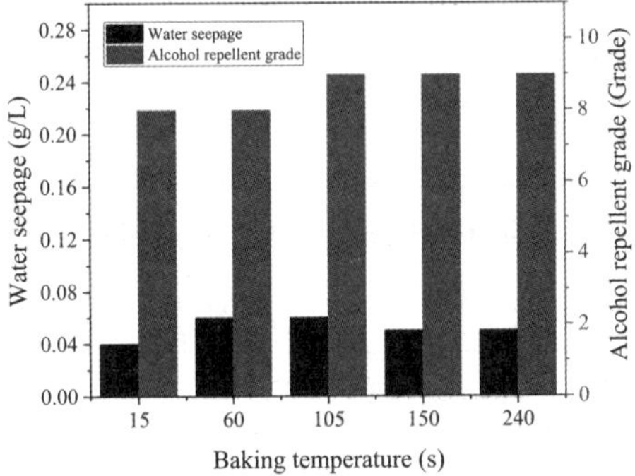

Fig. 3 Effect of different baking time on water/alcohol repellency of nonwovens

As can be seen from Fig. 3, at 110 ℃, the water permeability of the sample does not change obviously with the increase of baking time. The amount of water seepage is kept between 0.4–0.6 g and there is little difference.

3.2 Process optimization

3.2.1 Orthogonal experimental design

In this paper, the effects of the factors on the properties of nonwovens are discussed. Three kinds of influencing factors are set up with four gradients respectively. The experimental scheme is shown in Tab. 2. At the same time, in order to increase the reliability of the experiment, two blank columns were added and $L16(4^5)$ is used to carry out the experiment. Water/alcohol repellent tests were carried out on 16 kinds of samples, and the sum of square of data deviation is analyzed.

Tab. 2 Orthogonal test factor-level

Level	A Concentration (g/L)	B Temperature (℃)	C Time (s)
1	1#	100	15
2	2#	110	60
3	3#	120	105
4	4#	130	150

3.2.2 Analysis of orthogonal test results

3.2.2.1 Analysis of variance

Tab. 3 Analysis of variance of impervious water performance

Factor	Sum of squares of deviation	Freedom	F Ratio	F Critical value
A	0.002	3	0.750	2.810
B	0.004	3	1.500	2.810
C	0.002	3	0.750	2.810
D	0.007	3	2.625	2.810
E	0.001	3	0.375	2.810
Error	0.01	9	—	—

As can be seen from Tab. 3, the effect of the factors on water seepage resistance of SMS composite nonwovens is not obvious. The reason may be that the samples treated with triple-repellent and antistatic agent all have good water permeability resistance, and the difference is very small.

Tab. 4 Analysis of variance of alcohol grade

Factor	SSD	DF	FR	FC
A	1.500	3	2.250	2.810
B	0.000	3	0.000	2.810
C	0.500	3	0.750	2.810
D	0.500	3	0.750	2.810
E	0.500	3	0.750	2.810
Error	2.00	9	—	—

SSD: Sum of squares of deviation; DF: Degree of freedom; RF: F Ratio; FCV: F Critical value

Tab. 4 shows that the concentration of finishing agent has a great influence on the alcohol repellent grade, while the baking temperature and baking time have no significant effect on the alcohol repellent performance of the samples.

3.2.2.2 Statistical analysis of experimental results

Through the analysis of the above orthogonal experimental results, the best finishing process of polylactic acid SMS composite nonwovens can be determined.

(1) Water repellency

Finishing agent concentration A: $I_A > II_A > IV_A > III_A$; Baking temperature B: $I_B > II_B > IV_B > III_B$; Baking time C: $II_C = III_C > IV_C > I_C$;

(2) Alcohol repellency

Finishing agent concentration A: $II_A = III_A > IV_A > I_A$; Baking temperature B: $I_B = II_B = III_B = IV_B$; Baking time C: $I_C > II_C = III_C = IV_C$;

The higher the alcohol repellent grade, the better the alcohol repellent performance of the surface fabric. When the concentration of finishing agent is II_A and III_A, the baking temperature is I_B, II_B, III_B, IV_B and the baking time is I_C, the alcohol repellent property of the fabric is better.

3.3 Blood repellency

The composite nonwovens finished with No. 3 finishing solution were cut to the size of 175 mm×325 mm for blood rejection test. The results show that when the synthetic blood is sprayed from the height of 610 cm on a 45° tilted sample, the liquid beads and rolls down rapidly. At the same time, no synthetic blood is observed on the other side of the fabric, as shown in Fig. 4. This shows that the finished material has a certain degree of blood repellency.

At the same time, the anti-synthetic blood contact angle of the fabric is tested. the results showed that the contact angle of SMS nonwovens is increased from 110° to 134°, as shown in Fig. 5.

Fig. 4 Fabric contamination after spraying: (a) front; (b) reverse

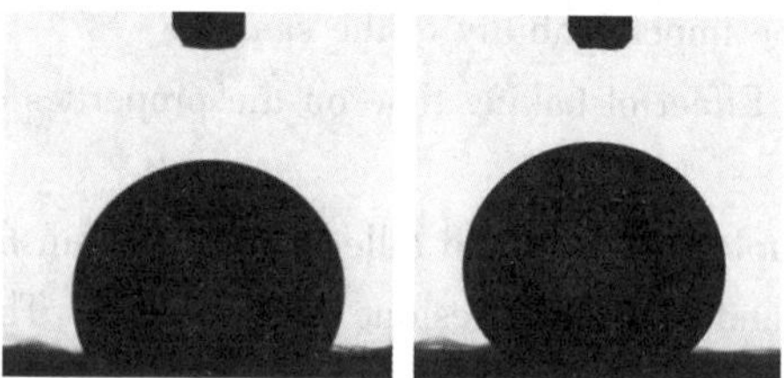

Fig. 5 Anti-blood contact angle before and after finishing: (a) before; (b) after

3.4 Test of other properties of polylactic acid SMS nonwovens

After triple-repellent and antistatic finishing, the test results of mechanical properties and comfort properties of polylactic acid SMS nonwovens are shown in Tab. 5.

Tab. 5 Other properties of SMS nonwovens

SMS nonwovens	Breaking strength(N)		Elongation at break(%)	Air permeability ($cm^3/(cm^2 \cdot s)$)	Moisture permeability ($g \cdot m^{-2} \cdot d^{-1}$)	Static contact angle(°)
Before	vertical	81.45	13.23	11.74	6 254.417	100.42
	Horizontal	37.94	16.97			
After	vertical	83.98	6.74	10.14	6 212.014	127.47
	Horizontal	38.39	5.66			

It can be seen from Tab. 5 that the breaking strength of SMS composite nonwovens increases slightly, while the elongation at break decreases after triple-repellent and antistatic finishing. There is no significant difference in moisture permeability and air permeability of nonwovens before and after finishing, because the finishing agent only covers the surface of the fiber and does not plug the pores between the nonwovens. So, the triple-repellent and antistatic finishing basically has no negative effect on the air permeability and moisture permeability of the products, and can meet the needs of medical protective products.

4 Conclusions

(1) From the single factor and orthogonal experiment, it can be concluded that the finishing concentration has a great influence on the alcohol repellency of the material, followed by baking temperature and time, but the three main factors have little influence on the impermeability of the fabric. The best finishing process is as fol-

lows: concentration of finishing agent is 3#, baking temperature 120 ℃ and baking time 15 s. At this time, the water leakage of the sample is 0.05 g and the alcohol repellent grade is 8.

(2) The sample treated with the selected optimum process conditions has good properties, and the air and moisture permeability of the treated fabric are basically the same as those before treatment.

References

[1] UNSAL E, DANE J H, SCHWARTZ P. Effect of liquid characteristics on the wetting, capillary migration, and retention properties of fibrous polymer networks[J]. Journal of Applied Polymer science, 2010, 97(1): 282-292.

[2] BELKIN N L. The new standard for barrier surgical gowns and drapes: what it means to the infection control practitioner[J]. Infection Control Today, 2006.

[3] ERYURUK S H, BK K, KALAOGLU F. A study on ultrasonic welding of nonwovens used for surgical gowns[J]. International Journal of Clothing Science and Technology, 2017, 29(1): 00.

[4] SEONG O K, CHUNG H P, JOOYOUN K. Breathable, antistatic and superhydrophobic PET/Lyocell Fabric[J]. Journal of Engineered Fibers and Fabrics, 2015, 10, 3.

[5] KRISHNASAMY J, SENTHILKUMAR T, NEELAKANDAN R. Development of breathable and liquid/microbes barrier woven surgical gowns for hospital usage[J]. Indian Journal of Fibre and Textile Research, 2017, 42(4): 453-464.

[6] MONTAZER M, RANGCHI F. Simultaneous antimicrobial, water and blood repellent finishing of disposal nonwovens using ctab and fluorochemical[J]. Tekstil ve Konfeksiyon, 2009, 19(2): 128-132.

[7] ZHANG Y, LI T T, SHIU B C, et al. Eco-friendly versatile protective polyurethane/triclosan coated polylactic acid nonwovens for medical covers application[J]. Journal of Cleaner Production, 2020, 282(1): 124455.

Extremely Tough Glass Fiber Fabric/Hydrogel Composites Based on Interfacial Modification and Physicochemical Dual Cross-Linking Strategy

FU Yingping[1,2], HU Xuefeng[1], NI Jiaqian[4], WANG Jilong[3*], SHEN Hua[1], XU Fujun[1,2]

[1] *Key Laboratory of Textile Science & Technology, Ministry of Education, Donghua University, Shanghai 201620, China*

[2] *Department of Textile Engineering, College of Textiles, Donghua University, Shanghai 201620, China*

[3] *College Textile and Garment, Shaoxing University, Shaoxing 312000, China*

[4] *Zhejiang Zhongda Testing Technology Services Co., Ltd., Shaoxing 312030, China*

[*] *Corresponding author's email*: 2022000018@usx.edu.cn

Abstract: Hydrogels hold promising applications, but the lack of toughness greatly limits their functional applications. The introduction of rigid fiber/fabric reinforcement into soft hydrogel matrix to toughen hydrogel is an emerging hotspot in recent years. However, the problem of easy debonding of the interface makes the enhancement effect of fiber/fabric usually unable to give full play. In this work, silane molecular bridges were introduced into the interface between poly (acrylamide-acrylic acid)-Fe^{3+} chemical-ionic dual cross-linked hydrogel matrix and rigid glass fiber fabric reinforcement. Finally, a silane interface chemical modified glass fiber fabric/dual cross-linked hydrogel composite with strong interfacial adhesion, which was extremely tough, soft and thin, was successfully obtained. Its tearing toughness was as high as 280 000 J/m^2, which was about 4.83 times higher than that of unmodified composite hydrogel, while maintained a high water content of 51%. This idea provides a new strategy for producing tough hydrogel composites with high interfacial strength, and is of great significance for promoting the application of hydrogel in complex fields such as soft robots.

Keywords: Hydrogel composites; Glass fiber fabric; Dual cross-linking; Interfacial strength; Tearing toughness

1 Introduction

Hydrogels are soft and wet polymer network materials, which have high water content (>50%) as well as good bio-compatibility and show promising applications[1]. However, the insufficient toughness (~100 J/m^2) greatly limits their functional applications[3]. In recent years, the introduction of rigid fiber or fabric reinforcement into soft hydrogel matrix to further toughen hydrogel is an emerging hotspot[2]. However, the key lies both in the strong interfacial bond and the high toughness of matrix to effectively transfer stress. But most researchers tend to develop fabric reinforced composite hydrogels with different fiber types or woven structures, while lacking systematic study on the interfacial effects, which makes the strengthening effect of rigid fibers cannot be fully brought into play[5].

In this work, we adopt poly (acrylamide-acrylic acid) chemical-ionic double cross-linked tough hydrogel as matrix[4], glass fiber fabric as rigid fiber reinforcement. And the fabric was further modified with silane (TMSPMA) to introduce vinyl silane macromolecular bridges at the interface. Finally, the interfacial enhancement effect of silane modification was characterized and evaluated by adjusting the concentration of silane and combining with Fourier transform infrared spectroscopy (FTIR), contact angle (CA) and trouser tearing test.

2 Experiments

2.1 Materials

Acrylamide (≥98%, AAm) (Sigma Aldrich); Acrylic acid (≥99%, AA) (Sinopharm reagent Shanghai Co., Ltd.); Ammonium persulfate (≥98%, APS) (Honeywell); N, N′-Methylenebis (acrylamide) (99%, MBAA) (Sigma Aldrich Shanghai Co., Ltd.); Tetramethylenediamine (99%, TEMED) (Aladdin reagent); Absolute ethanol (95%), Acetone (AR), Acetic acid (AR), Anhydrous calcium chloride (AR), Anhydrous ferric chloride (AR) (Sinopharm reagent); γ-(Methacryloxy)-propyltrimethoxy-silane (TMSPMA, ≥98%, 1.045 g/mL) (Aladdin reagent). Alkali-free plain weave glass fiber fabric (thickness 0.1 mm, areal density 10×10 roots/cm) (Hebei Xinsheng Fire Insulation Material Co., Ltd).

2.2 Preparation of silane hydrolysis solution

The silane hydrolysis solution consisted of absolute ethanol and deionized water (volume ratio 1 : 10) was added acetic acid to adjust pH (−4.5), then the mixture was hydrolyzed under stirring for 5 h. Five solutions (0.5%, 1.0%, 2.0%, 3.0%, 5.0%) were prepared, respectively.

2.3 Preparation of silane modified glass fiber fabric reinforcement (GF)

The fabric rinsed with acetone and air-dried was immersed in silanol hydrolysis solution for sealing reaction for 6 h, and finally rinsed with acetone again and air-dried for later use.

2.4 Preparation of dual crosslinked hydrogel matrix (DG)

The hydrogel precursor solution was transferred into the self-made glass mold by a syringe, and then was heated in oven at 55 ℃ for 2.5 h. Then the as-prepared hydrogel after demoulding and simply washing the surface, was immersed in 0.1 M $FeCl_3$ or $CaCl_2$ solution for secondary crosslinking for 1 h or 1 min, and then was immersed in deionized water dialysising for 12 h ($CaCl_2$ crosslinked due to swelling, not for dialysis). The AAm and AA were 1 200 mg/10 mL and 200 μL/10 mL[4]. Initiator (APS) and crosslinking agent (MBAA) were 1% and 0.25% of the total monomer, respectively. And crosslinking accelerator (TEMED) was 10 μL/10 mL.

2.5 Preparation of glass fiber fabric composite hydrogel (DG/GF)

The glass fiber fabric fully infiltrated by hydrogel precursor solution, was transferred to a self-made glass mold by using a syringe and a scraper to form a sandwich structure. And the remaining steps were all same as the preparation of hydrogel matrix.

2.6 Characterization of glass fiber fabric reinforcement

Infrared (FTIR) test: the thoroughly dried fabric was characterized by the ATR attachment of Fourier transform infrared spectrometer, which scanned each sample 16 times at a resolution of 4 cm^{-1} in the range of 4 000–400 cm^{-1}. Contact angle (CA) test: the wettability of ordinary and silane modified glass fabric was measured by contact angle goniometer. Bubble capture and seat drop methods were used for ordinary fabric and silane modified fabric, respectively.

2.7 Characterization of glass fiber fabric composite hydrogel

Swelling test: sample dimensions immersed in 0.1 M ion solution (Fe^{3+} or Ca^{2+}) were measured by a digital vernier caliper. Each sample was measured at 3 locations and averaged. Water content test: first wiped off the water on sample surface and weighed. And then dried it at 85 ℃ to a constant weight, weighed again. Trouser tearing test: the sample sizes were 45 mm in width and 65 mm in length, and a 37.5 mm notch was set. The tensile device of INSTRON universal material testing machine was adopted with a displacement speed of 50 mm/min[5]. The tearing toughness (T, J/m^2) was obtained by dividing the integral of the tearing curve by the cross-sectional area of tearing path of the sample.

3 Results and Discussion

3.1 FTIR of silane modification glass fiber fabric

The ordinary glass fiber fabric has many hydroxyls (—OH) on surface, thus the silicon hydroxyls (Si—OH) at the end of the TMSPMA macromolecule, which also contains carbon-carbon bond (C═C) and carbonyl group (C═O), could undergo dehydration condensation reaction with them. As shown in Fig. 1 (a), FT-IR results show that there were strong overlapping silicon-oxygen-silicon bond (Si—O—Si) absorption peaks at 460 cm^{-1}, 760 cm^{-1} and 910 cm^{-1}. In addition, silane modified fabrics exhibited significant characteristic peaks of C═O and C═C at 1 717 cm^{-1} and

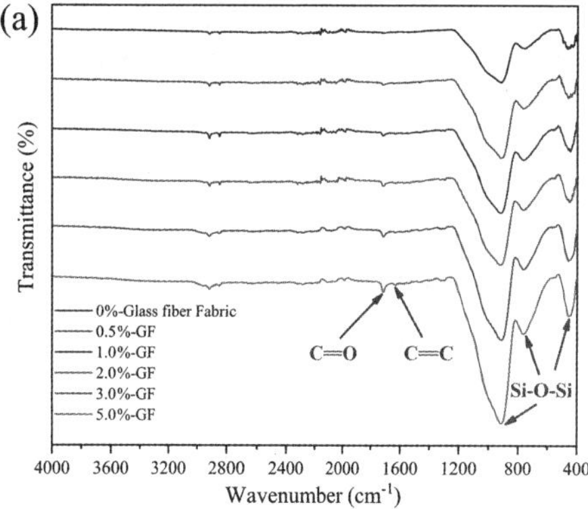

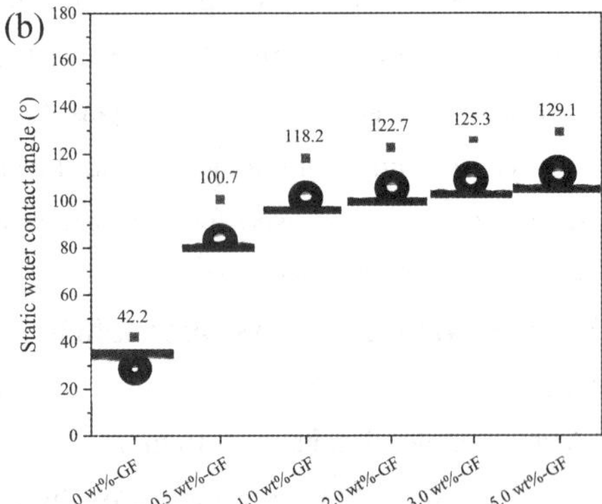

Fig. 1 FTIR (a) and contact angle (b) of modified glass fiber fabric (GF)

1 660 cm^{-1}, respectively. Thus, it can be speculated that silane TMSPMA may be successfully grafted onto the surface of glass fibers.

3.2 Contact angle of silane modification glass fiber fabric

The ordinary glass fiber fabric is highly hydrophilic, but silane is hydrophobic substance. Therefore, the hydrophobicity will increase if successfully grafting. As shown in Fig. 1 (b), the water contact angle of silane modified glass fiber fabric gradually increased with the increase of silane concentration, and reached a maximum of 129.1°. These results also indicate that silane TMSPMA may be successfully grafted from the side.

3.3 Swelling property and water content of composite hydrogel

Ions have different interaction strength with carboxyl (COO^-), thus affecting the swelling property of dual crosslinked hydrogel[4]. Therefore, the effects of ion-crosslinking with $CaCl_2$ and $FeCl_3$ solution (both 0.1 M) on the ordinary and silane modified (2%) composite hydrogels (initial thickness (0.30 ± 0.01) mm) were investigated. As shown in Fig. 2 (a), it can be seen that the thickness changes of the two were basically the same, all reached the extreme value in 1 h, but two opposite trends of strong swelling (200%) and small shrinkage (33%) occured for Ca^{2+} and Fe^{3+}-DG/GF. Thus, for Ca^{2+}, only 1 min was crosslinked to reduce the effect of swelling, and Fe^{3+} was crosslinked for 1 h. Besides, from Fig. 2 (b), we can see that the Fe^{3+}-DG and Fe^{3+}-DG/GF all kept high water content of about 60% and 51%, respectively.

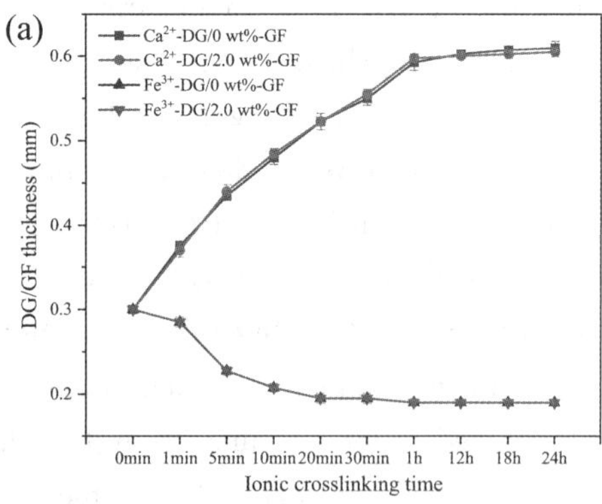

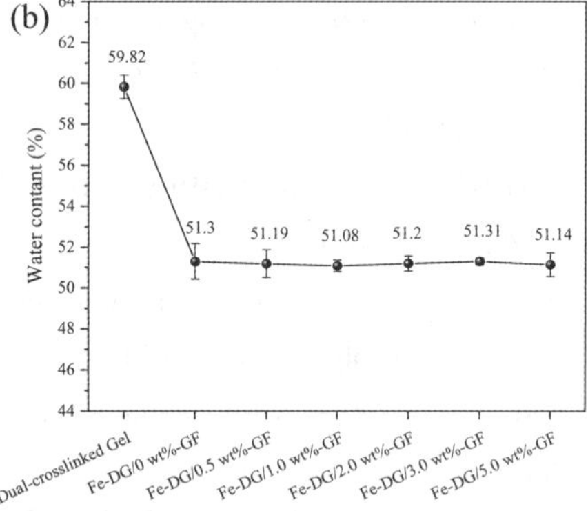

Fig. 2 Thickness change (a) and water content (b) of composite hydrogels (DG/GF)

3.4 Tearing toughness of composite hydrogel

The thickness of GF, Ca^{2+}-DG/GF and Fe^{3+}-DG/GF were 0.10 mm, (0.33 ± 0.02) mm and (0.25 ± 0.01) mm.

Firstly, effects of different cross-linking ions. From Figs. 3 (a)–3(c), the tearing curve of the Fe^{3+}-DG/GF (serrated) was much wider than Ca^{2+}-DG/GF (hilly), which indicates that former formed a larger energy dissipation zone at the tearing triangle zone, resulting in simultaneously breakage of the glass fiber bundles and matrix, while only slippage of fiber bundles occurred for Ca^{2+}-DG/GF. Besides, for ordinary DG/GF, the tearing toughness (58.08 kJ/m^2) of Fe^{3+} was 1.2 times that of Ca^{2+}(48.10 kJ/m^2); and for silane modified DG/GF, the tearing toughness of Fe^{3+} was nearly 4 times that of Ca^{2+}, which indicates that the stronger Fe^{3+}-COO^- bond significantly increased the interfacial strength.

Secondly, different silane concentrations. From Figs. 3 (b)–3(d), the tearing performance of silane modified DG/GF was significantly higher than ordinary one as a whole, and reached the maximum value at 2% silane, which may be attributed to the competitive effects of the decrease of the infiltration of precursor solution

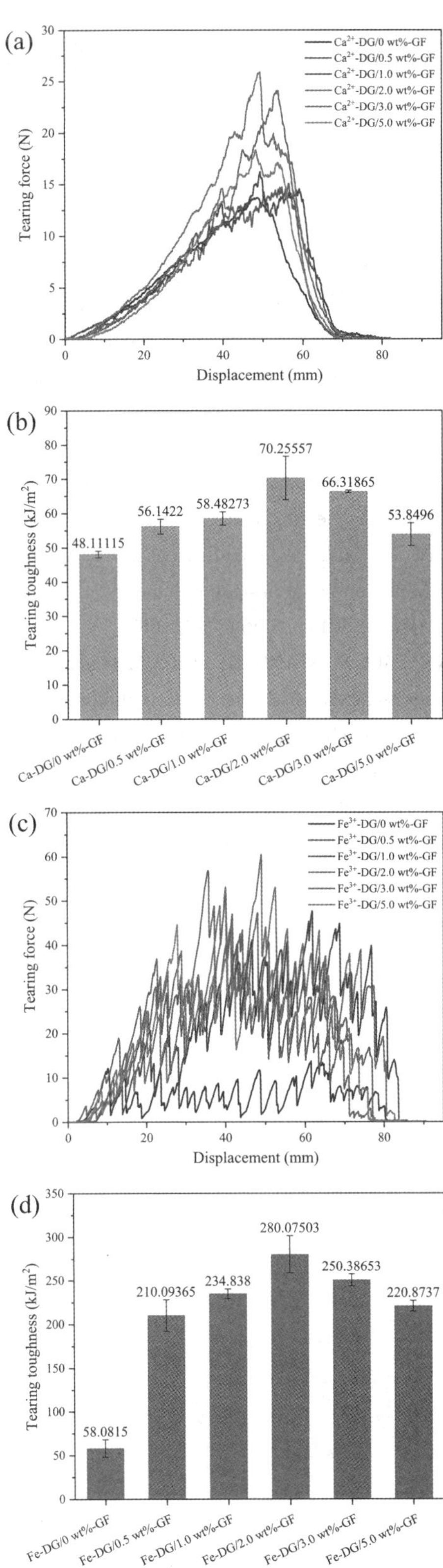

Fig. 3　Tearing curve and tearing toughness of Ca^{2+}(a)-(b) and Fe^{3+}(c)-(d) dual crosslinked composite hydrogel (DG/GF)

and silane reinforcement reaching a balance state. For Fe^{3+}, the tearing toughness (280 kJ/m^2) of silane (2%) modified DG/GF was about 4.83 times higher than ordinary one (58.08 kJ/m^2); while Ca^{2+} increased by about 1.46 times. It indicates that Fe^{3+} and silane achieved synergy to improve the interface strength.

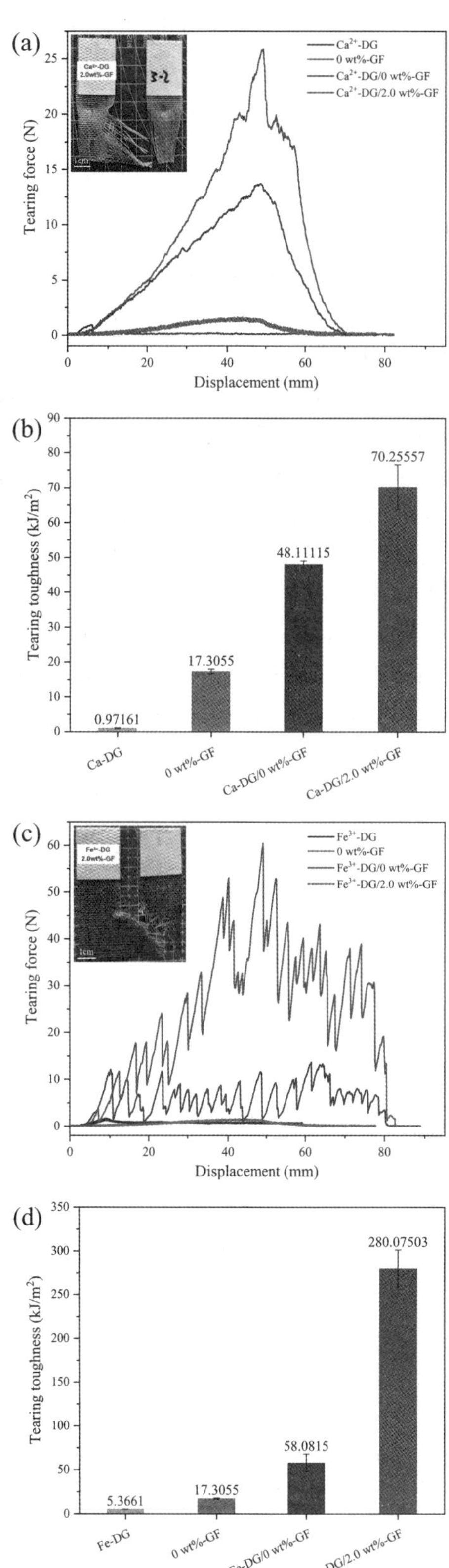

Fig. 4　Tearing curve and tearing toughness of hydrogel, fabric, and composite hydrogels for Ca^{2+}(a)-(b) and Fe^{3+}(c)-(d)

To visualize the synergistic effect, from Fig. 4, we can see that silane molecular bridges significantly improved the tearing performance. Besides, the tearing toughness of Fe^{3+}-DG (5.37 kJ/m^2) was much higher than that of Ca^{2+}(0.97 kJ/m^2), which indicates that the effective synergy between the tougher matrix and the optimal silane (2%) can effectively improve interfacial adhesion strength.

4 Conclusions

To sum up, we reported a simple and effective method for preparing extremely tough, soft and thin composite hydrogel with strong interfacial adhesion. Wherein, the thin glass fiber fabric (0.1 mm thick) provided a very high anisotropic tearing stiffness; the high fluidity of poly (acrylamide-acrylic acid) precursor solution realized the full infiltration and filling of the glass fiber tow; and the high strength Fe^{3+}-COO^- coordination bond further enhanced the physical entanglement of the interface; while the introduction of silane molecular bridges between glass fiber and hydrogel significantly enhanced the interfacial strength. Finally, an extremely tough ((280.00±21.35) kJ/m^2) composite hydrogel (0.25 mm thick) was obtained, which also maintained a high water content of 51%. This idea provides a new strategy for producing tough hydrogel composites with high interfacial strength, and is of great significance for promoting the application of hydrogels in complex fields such as soft robots.

References

[1] AHMAD Z, AMIN A, ALI S, et al. Versatility of hydrogels: from synthetic strategies, classification, and properties to biomedical applications[J]. Gels, 2022, 8(3): 167.

[2] HE Y, ZHAO X Y, RAO P, et al. Saline tolerant tough-yet-strong fiber-reinforced gel-nacre for soft actuator[J]. Chemical Engineering Journal, 2022: 137091.

[3] ZHAO X. Multi-scale multi-mechanism design of tough hydrogels: building dissipation into stretchy networks[J]. Soft Matter, 2014, 10(5): 672-687.

[4] LIN P, MA S, WANG X, et al. Molecularly engineered dual-crosslinked hydrogel with ultrahigh mechanical strength, toughness, and good self-recovery[J]. Advanced Materials, 2015, 27(12): 2054-2059.

[5] KING D R, SUN T L, HUANG Y, et al. Extremely tough composites from fabric reinforced polyampholyte hydrogels[J]. Materials Horizons, 2015, 2(6): 584-591.

The Preparation of Thermoregulatory Kapok Fiber Containing Microencapsulated Phase Change Material

CHNE Yu[1,2], DING Yi[1,2*], SHEN Hua[1,2], LIU Yunying[1,2]

[1] *Key Laboratory of Textile Science & Technology, Ministry of Education, Donghua University, Shanghai 201620, China*

[2] *Department of Technical Textiles, College of Textiles, Donghua University, Shanghai 201620, China*

[*] *Corresponding author's email*: dingyidhu@ dhu. edu. cn

Abstract: As a kind of natural fiber, kapok fiber (KF) has the characteristics of large hollow structure, high cellulose content and large specific surface area, which is conducive to the physical deposition and chemical crosslinking on kapok fiber. For the perspective of improving the use function and added value of KF, a facile vacuum impregnation method is used to fill the microencapsulated phase change material (MPCM) into the hollow structure of KF, and obtain the natural textile fiber with the function of heat storage and temperature regulation. The MPCM was confined inside the lumen of KF, leaving the interfiber space and surface free of MPCM. The measured peak melting temperature was 29 ℃ and crystallization temperature was 26 ℃. The results show that the finished solution consisting of 25% MPCM and 7% binder endows the KF with the melting phase change enthalpy of 11. 202 9 J/g and the crystallization phase change enthalpy of 7. 913 7 J/g. The suitable phase-change temperature, the high latent heat, the excellent moisture absorption, and air permeability provide the potential for the novel application of KF/MPCM in intelligent thermoregulated textiles.

Keywords: Kapok fiber; Microencapsulated phase change material; Thermal regulation; Vacuum impregnation

1 Introduction

Natural fiber has always been widely used in the textile industry as its thermal-wet comfort. Kapok fiber (KF) is a natural seed fiber extracted from the kapok tree fruit[1], with a light brown color and light weight, fluffy texture, known as the "cashmere growing on the tree". KF is a cellulose fiber composed of cellulose, hemicellulose and lignin, similar to cotton and hemp fiber. A thin layer of wax on the surface presents lipophilic and hydrophobic properties. KF has a smooth surface, a thin head end and a thick tail end, and is generally in a closed state. It is a unique thin-walled large hollow structure, the hollow degree is up to 80-90%, and far beyond the hollow degree is only about 40% of the chemical fiber[2]. Although KF was used as textile filler for a very early time, it is difficult to spin alone because of its short length, poor cohesive force and low strength, so it has been limited in textile and clothing. However, its unique advantages of high hollow structure, large specific surface area and light weight make it be excellent thermal material.

Smart passive thermoregulatory textiles form a highly fluffy three-dimensional porous structure by adding fiber-filling materials and storing a large amount of static air to reduce the heat transfer rate, thus achieving the effect of thermal insulation. It has the disadvantage of bulky size and heavy weight. Smart active thermoregulatory textiles are the current development hotspot. By introducing some elements that can produce heat independently into the fabric, such as moisture absorption heating, phase change heating, electric energy and solar energy conversion heating, it can ensure thermal comfort and meet the needs of clothing aesthetics[3].

Microencapsulated phase change material (MPCM) can absorb and store heat, delaying the increase of microclimate temperature in clothing. When the ambient temperature decreases, the MPCM releases heat to delay the decrease of the microclimate temperature in clothing, resulting in a relatively minor temperature variation of the microclimate environment in clothing.

Currently, there are mainly three types of processing methods for thermoregulation textiles using MPCM: spinning method[4], finishing method[5] and hollow fiber filling method. The spinning method is to prepare synthetic fiber by adding MPCM. The advantage of the spinning method is that the MPCM is fixed inside the fiber and shows excellent washing fastness. However, the process of this method is complex, the latent heat value of the prepared thermoregulatory fiber is low, and the spinning process probably causes certain damage to the MPCM. The finishing method is that MPCM is finished on the surface of the fabric by coating or padding. It's simple but provides poor washing fastness. The hollow fiber filling method is to use the hollow structure of the fiber to store MPCM. It provides a stable form during phase change and is suitable for large-scale production. In the mid-1980s, Vigo and Frost[6-7] proposed filling hydrated inorganic salts and PEG into hollow rayon and polypropylene (PP) fibers to achieve temperature control textiles; they found that the heat capacity of PCM-filled fibers decreased with repeating heating-cooling cycles. Song et al.[8] filled lauric acid in KF microtubules by the vacuum impregnation method, and the obtained materials had high thermal energy storage capacity, good cycle stability and reusability.

As a kind of natural fiber, KF has the characteristics of large hollow structure, high cellulose content and large specific surface area. In order to improve the filling percentage of KF, this paper use a facile vacuum impregnation method to fill MPCM into the hollow structure and to achieve the heat storage and temperature regulation function. The composite energy storage material can not only meet people's increasingly functional demand for thermoregulatory textiles but also meet people's demand for moisture-absorbing and breathable textiles.

2 Experiments

2.1 Materials

MPCM (core material is octadecane, wall material is polyurea, diameter 2–3 μm, latent heat value is 180 J/g) was provided by Shanghai Xinwu Textile Technology Co. Ltd. Polyacrylate binder was provided by Shanghai Xinwu Textile Technology Co. Ltd.

2.2 Fiber treatment

Since the two ends of KF are closed, it is not conducive to the adhesion of MPCM inside the fiber hollow structure. Therefore, before the sample preparation, the KF was simply manually combed, and then cut fibers to about 5 mm.

An appropriate concentration of alkali treatment solution can remove wax, pectin and impurities on the fiber surface, which is conducive to improving the core absorption effect of fiber and increasing its hydrophilicity. The reagents and mass concentrations used in the alkali treatment solution were NaOH 6.0 g/L, JFC 0.6 g/L, Na_2SO_3 1.5 g/L, Na_2SiO_3 1.5 g/L, Na_3PO_4 0.5 g/L. The bath ratio is 100 : 1. The fiber was treated at 85 ℃ for 30 min, then washed with deionized water and dried in a vacuum dryer at 60 ℃.

2.3 Preparation of KF/MPCM

The thermoregulated KF was prepared by vacuum impregnation method in the phase-change finished solution composed of 25% MPCM and 7% binder. In short, the phase-change finished solution and KF were mixed uniformly in the beaker (bath ratio 200 : 1), and then the beaker was transferred to the vacuum drying oven at 30 ℃ for 100 min. At this temperature, the phase-change finished solution was absorbed into the lumen of KF under capillary force. Afterward, the beaker was vacuumed to 0.1 MPa in a vacuum drying oven to remove the air in the lumen, which allowed for the diffusion of the phase-change finished solution into the KF. And then the vacuum pump was turned off to enable the entering air to further induce infiltration of the phase-change finished solution. The above operation steps were repeated at least three times to ensure that the lumen of KF was fully filled with the phase-change finished solution.

2.4 Characterization

The surface and cross section morphology of the sample was observed by Scanning electron microscopy (SEM, TM3000), all samples were sputter-coated with gold to avoid charge accumulation. The chemical structure evolution of samples was examined by Fourier transform infrared spectrophotometry (FTIR, Spectrum Two) in a wavenumber range of 400–4 000 cm^{-1} using KBr pellets. The phase-change temperatures and latent heat were measured by Differential scanning calorimetry (DSC, DSC8500, PerkinElmer, USA) measurement. Sample analysis was conducted on 3–5 mg of material each time with a heating and cooling rate of 5 ℃/min for 15–45 ℃ under nitrogen. The latent heat was calculated based on the integrated area under the peak.

3 Results and Discussion

3.1 Fibers morphology

The surface morphology and cross-section of the KF and obtained KF/MPCM were examined by SEM, and representative results are shown in Fig. 1. It is evident that KF presents a cylindrical structure with a smooth and ripple-free surface. A large hollow interior was clearly observed at the cross-section of the KF cut through the vertical axis. As shown in Fig. 1 (d), The MPCM was confined inside the lumen of KF, leaving the interfiber space and surface free of MPCM. The hollow structure of KF can play an important role as a microcontainer for MPCM, and the binder also can increase the fastness of the MPCM during water washing and phase change.

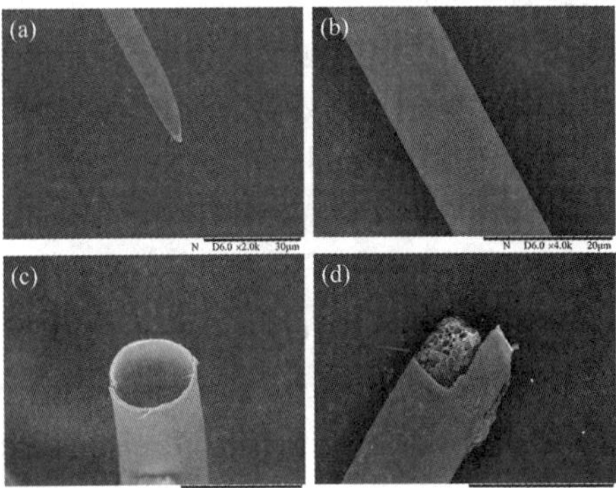

Fig. 1 SEM images of (a), (b), and (c) kapok; (d) KF/MPCM

3.2 Chemical structure characterization

FTIR analysis was used to monitor the chemical structure evolution of intermediates. Fig. 2 compares the FT-IR spectra of KF, binder, MPCM and KF/MPCM. As shown in Fig. 2, KF is a cellulose fiber, and its characteristic absorption peaks are O—H stretching vibration absorption band at 3 365 cm^{-1}, C—O—H asymmetric stretching vibration absorption band near 1 030 cm^{-1} in the fingerprint area and C—H asymmetric stretching vibration absorption band near 2 921 cm^{-1}. In the FTIR spectrum of binder (polyacrylate), the characteristic absorption peak at 1 735 cm^{-1} arises from C=O, and at 2 919 cm^{-1} and 2 851 cm^{-1} arise from —CH_2—. The infrared spectra of MPCM showed strong absorption peaks at 2 919 cm^{-1} and 2 851 cm^{-1}, which were caused by the C—H stretching vibration of CH_2 and CH_3 groups. And a moderate absorption peak near 1 473 cm^{-1} due to the shear bending vibration of CH_2 and a strong absorption peak near 720 cm^{-1} due to the in-plane rolling vibration of —CH—, which were exactly the absorption peaks of n-octadecane (the core material of MPCM).

KF showed C=O non-conjugated carbonyl vibration near 1 735 cm^{-1} due to the small part of lignin, while KF/MPCM showed significantly enhanced C=O vibration at 1 735 cm^{-1} due to the adhesion of polyacrylate binder on the basis of the original cellulose characteristic peak. At the same time, the infrared spectrum of KF/MPCM also showed the C—H stretching vibration near 2 919 cm^{-1} and 2 851 cm^{-1}, the CH_2 shear bending vibration near 1 473 cm^{-1}, and the-plane rocking vibration near 720 cm^{-1}. Therefore, the FTIR results show that effective attachment of MPCM and binder to KF can be achieved by the vacuum impregnation method.

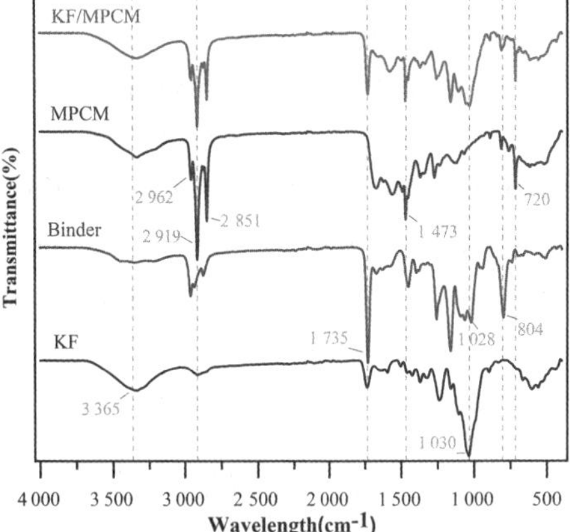

Fig. 2 FTIR spectra of KF, binder, MPCM and KF/MPCM

3.3 Thermophysical properties of KF/MPCM

Six groups of different treatment processes were used to prepare thermoregulatory KF as shown in Tab. 1. The weight gain rate of KF before and after treatment indicates the attachment amount of phase change finishing solution, calculated as shown in Eq. (1).

$$w = \frac{w_1 - w_0}{w_0} \times 100\% \tag{1}$$

where w is the weight gain rate of KF, w_1 is the weight before treatment and w_2 is the weight after treatment.

Melting phase change enthalpy and crystallization phase change enthalpy were used to indicate the temperature regulating capacity of the treated fibers.

As shown in Fig. 3, from the 1# and 2# samples, it can be seen that the weight gain rate and phase change enthalpy of the fibers obtained by the vacuum impregnation method were better than those obtained by the water bath impregnation method. This was because the vacuum impregnation method could use the pressure difference to extract the air in the lumen of KF, so as to assist the MPCM in fully entering the lumen of KF. The 3# sample increased the bath ratio, and the fiber temperature regulating capacity was improved again. The 4 # sample increased the high-temperature baking process so that the weight gain rate of KF was significantly improved, but the phase change enthalpy did not increase, indicating that it is mainly caused by the binder film-forming effect, which will improve the washing fastness of the KF/MPCM. The 5# sample was pretreated by shearing before impregnation so that the ends of kapok fiber changed from the closed state to the open state, which was more conducive to the filling of MPCM. The 6# sample was treated with 6 g/L alkali solution before impregnation, and the impregnation effect of KF was significantly decreased, which may be due to the high concentration of alkali treatment destroying the hollow structure of KF, thus affecting the adhesion of MPCM in the interior. Subsequently, the concentration of alkali treatment can be reduced or other more moderate modification technologies can be found to improve the impregnation effect.

Tab. 1 The treatment process scheme

Number	Method	Bath ratio	Pretreatment	Baking
1	Water bath immersion	100	Without	Without
2	Vacuum impregnation	100	Without	Without
3	Vacuum impregnation	200	Without	Without
4	Vacuum impregnation	200	Without	With
5	Vacuum impregnation	200	Cut	With
6	Vacuum impregnation	200	Alkali treatment	With

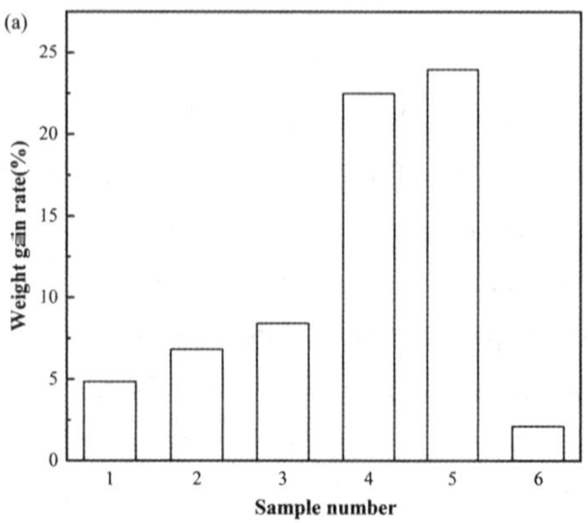

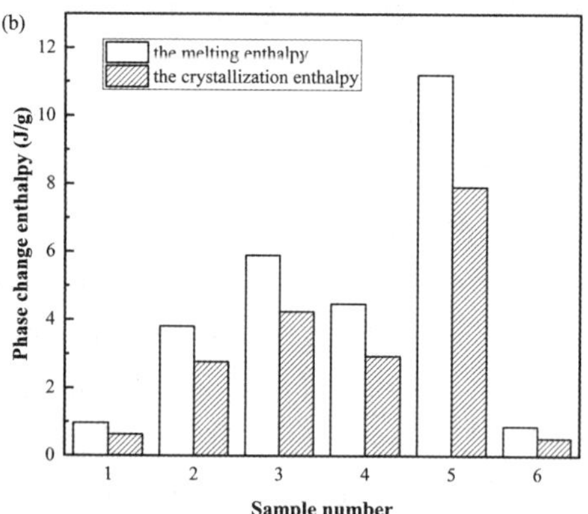

Fig. 3 (a) Weight gain rate and (b) DSC of KF/MPCM under different treatment processes

4 Conclusions

In summary, KF with temperature regulating capacity was successfully prepared via a facile vacuum impregnation method. MPCM and binder were filled into the lumen of KF. The obtained KF/MPCM melted at 29 ℃ with melting phase change enthalpy of 11. 2029 J/g, and freezes at 26 ℃ with crystallization phase change enthalpy of 7. 9137 J/g, achieving an unprecedentedly natural fiber with not only temperature regulating capacity, but also the excellent moisture absorption and air permeability.

References

[1] SANGALANG R H. Kapok fiber-structure, characteristics and applications: a review[J]. Oriental Journal of Chemistry, 2021, 37(3): 513-523.

[2] DING Y, CAI Z, WANG L, et al. The adsorption character of kapok fiber and reactive dyeing technology on modified kapok fiber[J]. Journal of Engineered Fibers and Fabrics, 2014, 9(3): 57-64.

[3] FANG Y, CHEN G, BICK M, et al. Smart textiles for personalized thermoregulation[J]. Chemical Society Reviews, 2021, 50: 9357-9375.

[4] IQBAL K, SUN D M. Development of thermal stable multifilament yarn containing micro-encapsulated phase change materials[J]. Fibers and Polymers, 2015, 16(5): 1156 1162.

[5] SHAHID A, MIAH S, RAHIM A. Thermal and breathability management of microencapsulated phase change material (MPCM) incorporated jute fabric[J]. Journal of Engineered Fibers and Fabrics, 2021, 16: 1-7.

[6] VIGO T L, FROST C M. Temperature adaptable hollow fibers containing polyethylene glycols[J]. Journal of Coated

Fabrics, 1983, 12(4): 243-254.

[7] VIGO T L, FROST C. Temperature-sensitive hollow fibers containing phase change salts[J]. Textile Research Journal, 1982, 52(10): 633-637.

[8] SONG S, ZHAO T, QIU F, et al. Natural microtubule encapsulated phase change material with high thermal energy storage capacity[J]. Energy, 2019, 172: 1144-1150.

Study on Electrical Properties of Screen Printing Fabric-Based Microstrip Line

TU Huating[1] *, HONG Hong[2], ZHANG Yaya[3]

[1] *College of Medical Instruments, Shanghai University of Medicine & Health Sciences, Shanghai 201318, China*

[2] *School of Materials Science and Engineering, Shanghai Institute of Technology, Shanghai 201418, China*

[3] *College of Fashion Design, Jiaxing Nanhu University, Jiaxing 314001, China*

* *Corresponding author's email*: tuht@ sumhs. edu. cn

Abstract: As one of the basic units of wearable electronics, the electrical properties of screen printing fabric-based microstrip line is of great importance. However, usually only the line resistance or impedance of the printed film was discussed and less attention was paid to the cause of signal attenuation at high frequencies. In this paper, the theoretical analysis, simulation, and experimental verification methods were adopted in this research to systematically study the electrical properties of screen printing fabric-based microstrip line. The results show that the dielectric loss tangent of the substrate material is the main factor, and the smaller the loss tangent, the better the electrical properties of the microstrip line. The conductivity of the signal line is another factor which affects by the printing thickness, uniformity, and conductivity of the conductive ink. In all, it makes a foundation for printing wearable electronics, especially in high frequency applications.

Keywords: Electrical property; Resonant frequency; Attenuation; Screen printing; Microstrip line

1 Introduction

Nowadays, conductive ink has been widely used in screen printing fabric-based wearable electronics[1-3]. As the basic unit of the above applications, the electrical properties of the microstrip line were widely discussed. Up to now, most of the works focus on the direct current (DC) properties by measuring the sheet resistance of the film[4-5] or resistivity of the printing trace[6-7]. They cannot reflect all the electrical properties of the microstrip line, especially at high frequency. Therefore, the performance of radio frequency (RF) response characteristics is also important for the electrical properties of the microstrip line.

In this paper, the theoretical analysis, simulation, and experimental verification methods were adopted in this research to systematically study the electrical properties of screen printing fabric-based microstrip line. Firstly, the microstrip line fabricated on a commercial FR-4 printed circuit board (PCB) was analyzed to demonstrate the accuracy of our simulation. Then, a group of fabric-based microstrip line with different conductive materials were fabricated for analysis of the influence of substrate and conductor materials on the electrical properties.

After all, the influence of dielectric loss and conductivity on the frequency response characteristics of fabric-based microstrip line were investigated in this research. It will offer guidance for designing wearable RF devices with lossy flexible materials by utilizing printing technology.

2 Fabrication

A microstrip line consists of a conductive signal line and a grounded plane fabricated on the two sides of the dielectric substrate[8]. The geometry of the microstrip line was illustrated as the model in Fig. 1. The width (w) of the signal line is one of the critical parameters and must be designed according to the dielectric properties of its substrate. Usually, it will be determined to achieve 50 impedance matching. The thickness (t) and conductivity (σ) of the signal line will affect the skin effect[9] and thus change the electrical properties. Additionally, the dielectric properties of the substrate will affect the resonant frequency and attenuation.

In order to study the electrical properties of screen

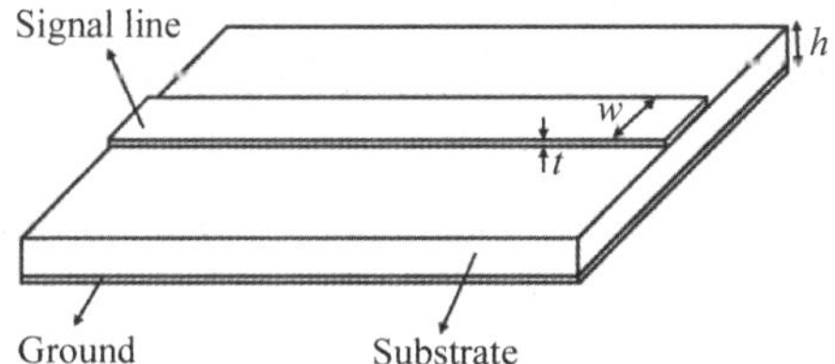

Fig. 1 The geometry of microstrip line

printing fabric-based microstrip line, a group of microstrip line with different materials were fabricated for comparison. Firstly, the microstrip line fabricated on the substrate of a commercial FR-4 printed circuit board (PCB) was measured and experimented to verify the accuracy of this research method (shown in Fig. 2). The dielectric constant and loss tangent of FR-4 are 5.2 and 0.015, and the thickness is 1.5 mm. The signal line was made of copper, which was fabricated by pasting copper tape on the upside of the FR-4. By calculation, the width of the microstrip line is 2.5 mm to feed 50 Ω port impedance.

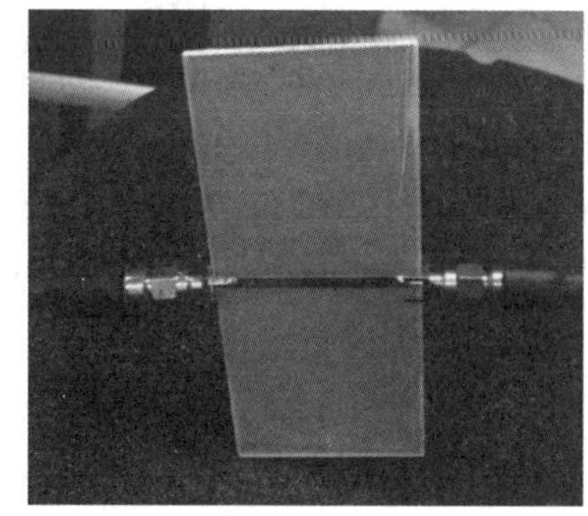

Fig. 2 Microstrip line fabricated with copper on FR-4 PCB

Then, a group of fabric-based microstrip line made up of copper and conductive ink was studied (shown in Figs. 3 (b) and 3(c)). The conductive ink is ET-4F flexible silver conductive ink (Guangzhou UV New Material Co., Ltd.), with a viscosity of 20 000 cps, and 65% silver content. The fabricated process is screen printing, and the screen mesh is nylon filament with 300 counts. The curing condition was 120 ℃ for 30 min in the oven to make the solvent volatilization.

Tab. 1 The properties of fabric substrate

Fabric	Material	Coating	Thickness (mm)	Weight (g/m^2)	Thread count (/10 cm)
Woven	Nylon	Polyamide	0.172±0.002	1.46±0.06	193±2

Tab. 2 Design parameters of the fabric-based microstrip line

	Microstrip line #1	Microstrip line #2
Substrate	Nylon fabric	Nylon fabric
Upper conductor	Copper tape	Conductive ink
Height of substrate (mm)	0.172	0.172
Length of substrate (cm)	50	50
Width of conductor (mm)	1	1
Thickness of conductor (μm)	10	15

Here, the above fabric was purchased by Pingzheng Bar Code Co., Ltd. (Shanghai, China), which has excellent washing durability and printability. The permittivity of this fabric is 2.4, the dielectric loss tangent is 0.05, and the other properties are listed in Tab. 1. The design parameters of the fabric-based microstrip lines were listed in Tab. 2.

3 Measurement

In order to characterize the electrical properties at high frequency, the S-parameter of microstrip lines with different printing layers was studied

with Agilent E5071C ENA. Before measurement, the full two-port calibration was done with a Calibration Kit (Agilent 85052E). Then the samples were measured. The measurement process was shown in Fig. 3(a).

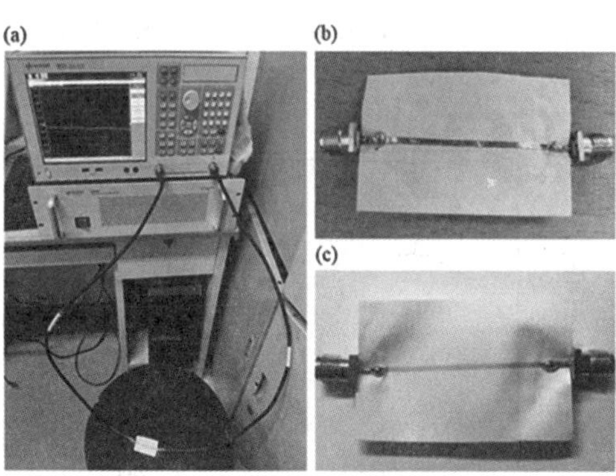

Fig. 3 The experiment of fabric-based microstrip line: (a) the measured process; (b) fabric-based microstrip line fabricated with copper; (c) fabric-based microstrip line fabricated with screen printing

As a comparison, a group of microstrip line fabricated with copper tape and conductive ink with the same dimension was measured (shown in Fig. 3(b) and Fig. 3(c)). The measurement results of screen printing conductive ink, as well as copper tape microstrip line, were compared with simulation in Ansoft HFSS.

4 Results and Discussion

The first step of this research was to demonstrate the accuracy of our simulation. The resonant model of the microstrip line was established in Ansoft HFSS simulation software. Parameters for all material properties as well as structure dimensions of the microstrip line were set identically to the actual sample.

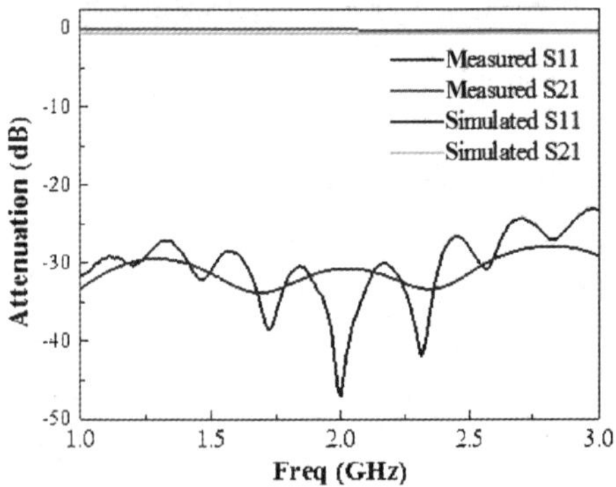

Fig. 4 The frequency response characteristics of simulated and measured results for FR-4-based microstrip line

From Fig. 4, the frequency response characteristics of simulated results are in line with the measured results. As the S21 is close to 0 dB, and S11 is much lower than −20 dB, which means that the signal loss is very small during the transmission. Furthermore, most of the energy is transmitted to the other port through the microstrip line, and the energy loss can almost be ignored.

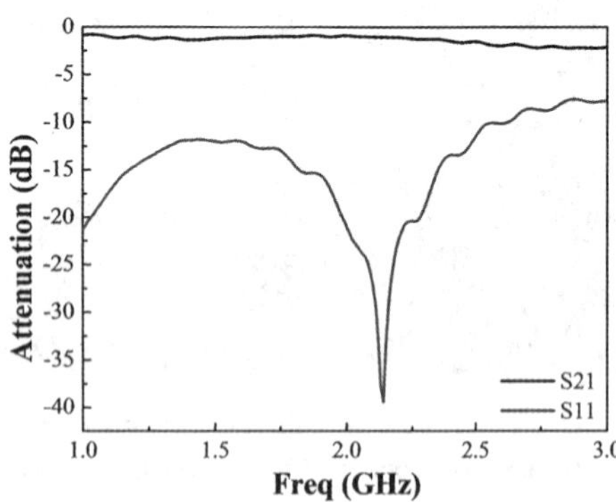

Fig. 5 The frequency response characteristics of measured results for fabric-based microstrip line (copper)

The attenuation curves in Fig. 4 have shown the high transmission effect for the FR-4-based microstrip line. Then the microstrip lines fabricated with copper and conductive ink on the fabric were investigated to study the influence of dielectric properties and conductivity of the conductor on the electrical properties. The measured results for the fabric-based microstrip line were shown in Fig. 5 and Fig. 6. The curves of S21 in Fig. 5 are away from 0 dB, and the values of S11 are larger than −20 dB, with means a big loss produced during the transmission.

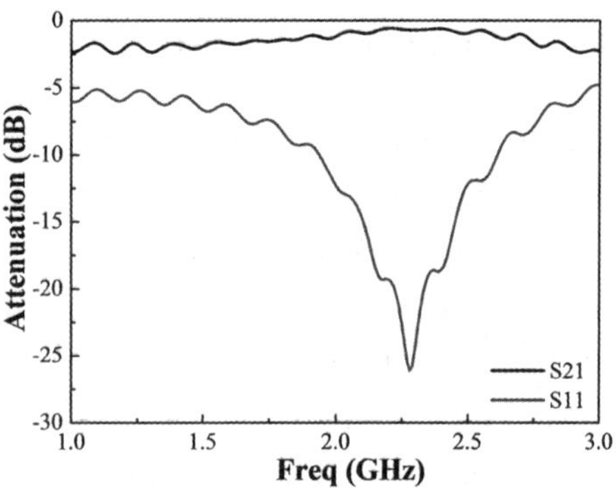

Fig. 6 The frequency response characteristics of measured results for fabric-based microstrip line (conductive ink)

Compared with Fig. 4 and Fig. 5, the signal lines are fabricated with copper and the variable is only the substrate material. The main reason is that the dielectric loss tangent of FR-4 substrate is very small compared to fabric substrate, resulting in a small dielectric loss. That's to say, In the process of signal transmission in microstrip line, compared with PCB, the loss of electromagnetic energy in the fabric is larger, resulting in low transmission efficiency[10-11].

Compared with Fig. 5 and Fig. 6, the microstrip line was fabricated on the same fabric, and the variable is only the conductor of the signal line. Due to the excellent conductivity (near perfect conductor) and flat surface of copper, the signal transmission performance is better than that of the printed sample. At the same time, the printed transmission line has a relatively large resistance (ranging from several ohms to several tens of ohms)[12] due to the incomplete continuity of the conductive layer, resulting in serious signal attenuation.

In all, from Fig. 4, Fig. 5, and Fig. 6, it can be deduced that the RF properties of microstrip lines are affected by the dielectric properties of substrate materials and the electrical properties of conductors. Screen printing is a good way to prepare flexible RF devices. The conductivity of the ink will affect the signal transmission performance of the device. However, compared with the influence of dielectric properties of the substrate, the impact of conductive ink is small.

5 Conclusions

In this research, contrast tests were performed to study the influence of electrical properties of screen printing fabric-based microstrip line. The results show that the dielectric loss tangent of the substrate material is the main factor affecting the electromagnetic loss during signal transmission. The smaller the loss tangent, the better the electrical properties for the microstrip line, resulting in good transmission efficiency. The conductivity of the ink is another factor to affect the electrical properties of the fabric-based microstrip line. But there is no obvious difference between the performance of conductive ink and pure copper for the material of the signal line.

Acknowledgements

This work was financially supported by the Foundation of National Key R&D Program of China (No. 2020YFC2008700) and National Key Clinical Specialty Discipline Construction Program of China (No. GWV-10.1-XK05), and Shanghai Science and Technology Plan Project (No. 22010502400).

References

[1] HASN U, PPPER M E, LUNDQIST J, et al. Screen-printed fabric antennas for wearable applications[J]. IEEE Open Journal of Antennas and Propagation, 2021, 2: 591-598.

[2] WAGIH M, KOMOLAFE A, TORAH R, et al. Screen printing reliable wearable microstrip antennas on rough textile substrates[C]//2021 IEEE International Symposium on Antennas and Propagation and USNC-URSI Radio Science Meeting, 2021, 2021-11-4.

[3] BLACHOWICZ T, EHRMANN G, EHRMANN A. Textile-based sensors for biosignal detection and monitoring[J]. Sensors, 2021, 21(18): 6042.

[4] MERILAMPI S, LAINE M T, RUUSKANEN P. The characterization of electrically conductive silver ink patterns on flexible substrates[J]. Microelectronics Reliability, 2009, 49(7): 782-790.

[5] KAZANI I, HERTLEER C, MEY G D, et al. Electrical conductive textiles obtained by screen printing[J]. Fibres and Textiles in Eastern Europe, 2012, 20(1(90)): 57-63.

[6] XU H, TANG X, SUN H, et al. Conductivity of silver and copper film printed by particle-free reactive inks [C]//18th International Conference on Electronic Packaging Technology, 2017-8-16.

[7] NIE X, WANG H, ZOU J. Inkjet printing of silver citrate conductive ink on PET substrate[J]. Applied Surface Science, 2012, 261: 554-560.

[8] GARG R, BHARTIA P. Microstrip Antenna Design Handbook[M]. Artech House, 2001.

[9] HWANG L, TURLIK I. A review of the skin effect as applied to thin film interconnections[J]. IEEE Transactions on Components, Hybrids, and Manufacturing Technology, 1992, 15(1): 43-55.

[10] SANKARALINGAM S, GUPTA B. Determination of dielectric constant of fabric materials and their use as substrates for design and development of antennas for wearable applications[J]. IEEE Transactions on Instrumentation and Measurement, 2010, 59(12): 3122-3130.

[11] DENLINGER E J. Losses of microstrip lines[J]. IEEE Transactions on Microwave Theory and Techniques, 1980, 28(6): 513-522.

[12] KAZANI I, HERTLEER C, MEY G D, et al. Electrical conductive textiles obtained by screen printing[J]. Fibres & Textiles in Eastern Europe, 2012, 20(1): 57-63.

Structure and Properties of Cryogenic Treated Aramid III Fiber

JIN Qianbo[1], ZHANG Yinnan[2], QIU Yiping[1], XU Fujun[1]*

[1] *College of Textiles, Donghua University, Shanghai 201620, China*

[2] *College of Fashion Technology, Shanghai University of Engineering Science, Shanghai 201620, China*

* *E-mail addresses*: fjxu@ dhu. edu. cn

Abstract: The influence of cryogenic treatment (−196 ℃) on structure and properties of aramid III fiber are investigated. Scanning electron microscope (SEM) results show that the surface of aramid III fiber is uneven, grooves increase and surface roughness increases after cryogenic treatment, which is beneficial to improve the interfacial properties of the fiber. After cryogenic treatment, the tensile strength of aramid III fiber increases by 24%, and the wear fracture time of the fiber increases by 54. 2%.

Keywords: Cryogenic treatment; Aramid III fiber; Tensile strength; SEM

1 Introduction

Aramid III fiber is a new ternary copolymer polyamide fiber developed independently in China[1]. Compared with common para-aramid and interpara-aramid, the third monomer is added in the polycondensation process of aramid III fiber, and N, O, S and other elements are introduced into its macromolecular chain segment, which makes the molecular chain with a certain heterocyclic structure. The chemical structure is shown in Fig. 1[2].

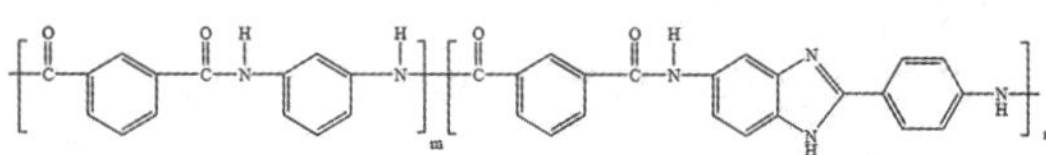

Fig. 1 Chemical structure of aramid III fiber

This special molecular structure endows aramid III fiber with high crystallinity and orientation[3], which makes it exhibit excellent mechanical properties. The tensile strength of aramid III fiber is 1. 5 times that of Kevlar49, and the elongation at break is 45%–65% higher than Kevlar49. In addition, aramid III fiber has the advantages of low density, high modulus, high temperature resistance, corrosion resistance and good dimensional stability[4-5]. It can be used as reinforcement material or functional material, which makes it has great application potential in aerospace, military protection, automobile transportation and other fields[6-7].

With the in-depth study of aramid III fiber, it is found that the fiber has a "skin-core" structure and poor radial

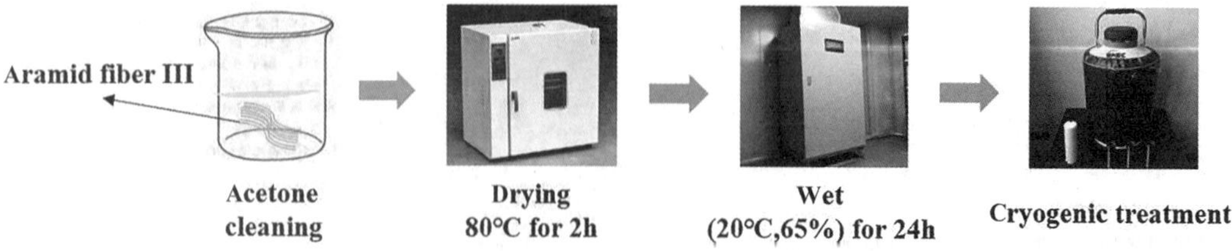

Fig. 2 Schematic diagram of cryogenic treatment process

mechanical properties. When subjected to external friction, the fiber is prone to fibrillation, which is manifested as increased hairs on the fiber surface. In addition, the surface of the fiber lacks active groups, and the adhesion to the resin matrix is poor when forming a composite material, which affects the mechanical properties of the composite material.

Cryogenic treatment is a simple, efficient and environmentally friendly modification method. It takes liquid nitrogen as the refrigerant and places the material under −130 ℃−−196 ℃ environment for cold treatment, so as to achieve the purpose of material modification[8]. Cryogenic treatment technology is initially applied to the modification of metal materials and achieves great success. Later, researchers tried to apply it to the modification of plastics and polymers. Indumathi J conducted cryocooling treatment on PI, PEI and their composites, and the results showed that the wear resistance of the materials was improved[9]. After the cryogenic treatment of PTFE and PA by K. N. Pande, the wear resistance and thermal properties of the polymers were significantly improved[10-11].

In this study, aramid III fiber independently developed in China was used as raw materials to test their mechanical properties and friction properties after cryogenic treatment and used SEM to observe the surface microstructure of the fiber, tried to discuss the influence of cryogenic treatment on the structure and properties of aramid III fiber.

2 Experiments

2.1 Material selection and cryogenic treatment

The selected aramid III fiber was provided by Zhonglan Chenguang Chemical Co. LTD. The material should be pretreated before cryogenic treatment. As shown in Fig. 2, firstly, the aramid III fiber was immersed in acetone solvent for 30min to clean impurities and oil, and then washed with distilled water, and finally the fiber was dried and moistened. The fiber was cryogenic by quenching method (QM), that is, the fiber was directly placed in liquid nitrogen environment (−196 ℃) for 12 h, and then fiber was removed and returned to room temperature naturally.

2.2 Fiber surface morphology detection

Aramid III fiber was first sprayed with gold to enhance its electrical conductivity, and then the surface morphology of the fibers was studied by scanning electron microscopy (SEM) at a voltage of 5 kV.

2.3 Single fiber performance test

The tensile properties of aramid III single fibers were tested by single-fiber strength tester (XS(08)XD−3). The parameters of the instrument were set as the test distance of 20 mm, the tensile speed of 20 mm/min, and 50 fibers were tested in each group. The average strength of fiber is calculated according to Formula 2−1:

$$\sigma_f = \frac{\sum_{1}^{n} \frac{4F_i}{\pi d^2}}{n} \times 10 \qquad (2-1)$$

Where σ_f is the average breaking strength of fiber, F_i is the breaking strength of single fiber, d is fiber diameter, n is the number of fiber roots.

A self-made fiber abrasion tester was used to characterize the abrasion resistance of single fiber through the friction fracture time of fiber, and 30 fibers were tested in each group.

3 Results and Discussion

3.1 Fiber surface morphology detection

The SEM images of the aramid III fibers before and after cryogenic treatment are shown in Fig. 3. As shown in Fig. 3 (a), there are some fiber fragments or small impurities on the surface of the fiber without cryogenic treatment, and the fiber is relatively flat as a whole. However, for the cryogenically treated fibers, as shown in Fig. 3 (b), the surface of the fibers is uneven and some longitudinal grooves appear, which increases the surface roughness of the fibers.

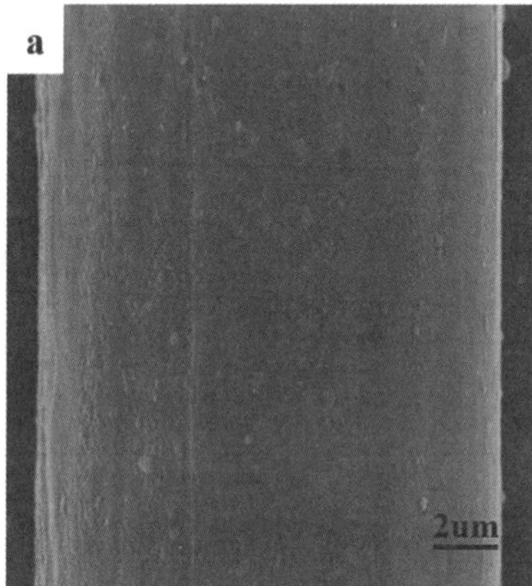

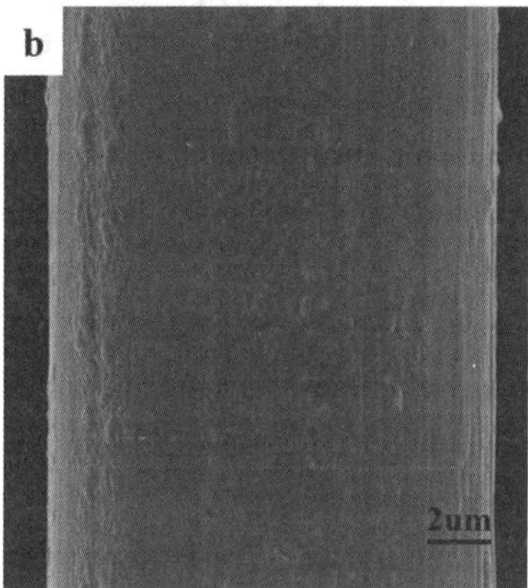

Fig. 3 SEM images of the (a) untreated aramid III fiber; (b) QM treated aramid III fiber

This may be due to the inconsistent shrinkage of the internal structure of the fiber in a low temperature environment. The increased surface roughness of the fibers is beneficial to the interface bonding of the aramid III fibers with the resin matrix when making composite materials, thereby improving the properties of the composite materials.

3.2 Single fiber tensile property

It can be seen from the test results in Tab. 1 and Fig. 4, the breaking strength and breaking elongation of aramid III fiber are enhanced, and the tensile strength is increased by 24% after cryogenic treatment. The results indicate that cryogenic treatment can improve the surface roughness of aramid III fiber without damaging the tensile properties of the fiber body.

Tab. 1 Tensile properties of Aramid III fibers before and after cryogenic treatment

	Control	QM
Sample size	50	50
Breaking strength(cN)	109.34	129.72
Breaking elongation(%)	3.53	4.09

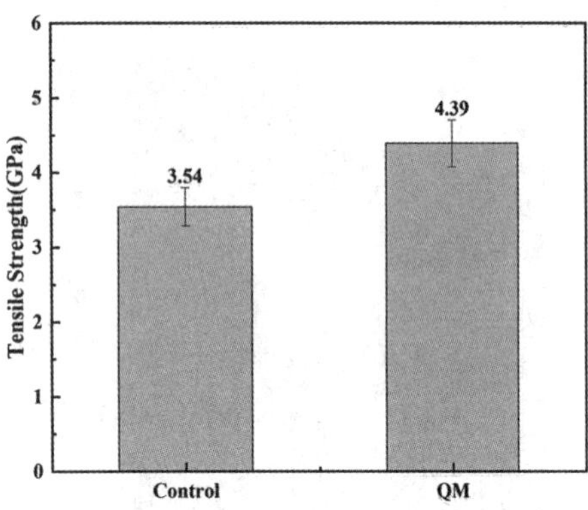

Fig. 4 Effect of cryogenic treatment on tensile strength of aramid III single fiber

The reasons for the improvement of fiber tensile properties after cryogenic treatment may be as follows: 1) Cryogenic treatment shrinks the bond length and angle of molecular bonds and increases the binding force of molecular bonds; 2) The compressive stress generated during cryogenic treatment reduces the spacing of molecular chains and the arrangement of molecular chains more compact, thereby increasing the interaction force between molecular chains.

3.3 Single fiber abrasion resistance property

As shown in Fig. 5 (a), a fiber abrasion tester was made to test the wear resistance of aramid III single fiber. One end of the aramid III single fiber was fixed, and the other end was bound with a 10 g hammer. After that, the aramid III single fiber was suspended on the friction roller, and the number of turns of the friction roller was recorded when the single fiber was broken, so as to calculate the friction fracture time.

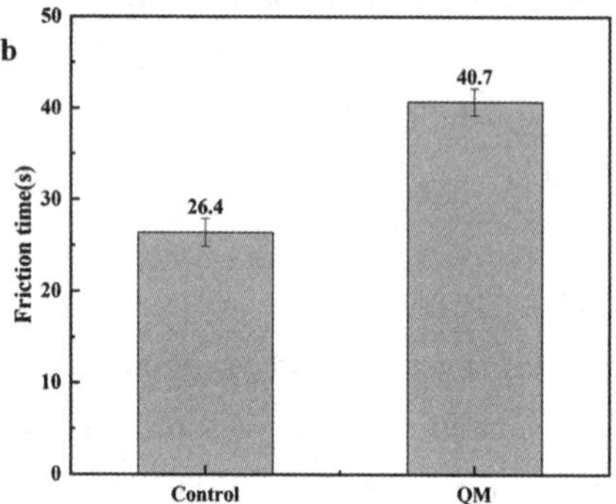

Fig. 5 (a) A self-made fiber abrasion tester; (b) friction breaking time of aramid III single fiber

It can be seen from Fig. 5 (b) that after cryogenic treatment (QM), the wear resistance time of aramid III single fiber increases from 26.4 s to 40.7 s, which is an increase of 54.2%. This is because after cryogenic treatment, the distance between the fiber macromolecular chains is reduced, the molecular chains are arranged more closely, and the interaction force between the molecular chains is increased. When subjected to external friction, the aramid III fiber is not easy to "fibrillate", which is beneficial to improve the wear resistance of the fiber.

4 Conclusions

In this paper, we studied the effect of cryogenic treatment on the structure and properties of aramid III fibers. The results show that after cryogenic treatment, the wear resistance and tensile properties of aramid III fiber are improved, the tensile strength of aramid III single fiber is increased by 24%, and the wear fracture time of single fiber is increased by 54.2%. In addition, the surface roughness of the fiber increases after cryogenic treatment, which is beneficial to improve the interfacial properties of the fiber.

References

[1] LIU K J, YANY Q, ZHU L H, et al. Introduction of organic fiber[J]. Synthetic Fiber, 2013, 42(1): 25.

[2] MACHALABA N N, KURYLEVA N N. Tver' fibers of the armos type: manufacture and properties[J]. Fibre Chemistry, 2000, 33: 319-324.

[3] YEUNG K K H, RAO K P. Mechanical properties of kevlar-49 fibre reinforced thermoplastic composites[J]. Polymers & Polymer Composites, 2012, 20(5): 411-424.

[4] CHOI I, KIM J G, LEE D G, et al. Effects of a damaged composite face to the electromagnetic wave transmission characteristics of low-observable radomes[J]. Composite Structures, 2011, 93(11): 2740-2747.

[5] PEREPELKIN K E. Properties of armos para-aramid fibres in conditions of use. Comparison with other para-aramids[J]. Fibre Chemistry, 2001, 33(2): 105-114.

[6] MM A, CKD A, APK B, et al. Properties of syndiotactic polystyrene composites with surface modified short Kevlar fiber[J]. Materials Science and Engineering: A, 2006, 441(1/2): 206-214.

[7] DAGHER H J, IQBAL A, BOGNER B. Durability of isophthalic polyester composites used in civil engineering applications[J]. Polymers and Polymer Composites, 2004, 12(3): 169-182.

[8] PODGORNIK B, PAULIN I, ZAJEC B, et al. Deep cryogenic treatment of tool steels[J]. Journal of Materials Processing Tech, 2015, 229.

[9] INDUMATHI J, BIJWE J, GHOSH A K, et al. Wear of cryo-treated engineering polymers and composites[J]. Wear, 1999, 225.

[10] PANDE K N, PESHWE D R, KUMAR A. Optimization of cryo-treatment parameters for PTFE by quantum-chemical approach and its evaluation through mechanical, thermal and structural characterization[J]. Transactions of the Indian Institute of Metals, 2012. 65(4): 365-373.

[11] PANDE K N, PESHWE D R, KUMAR A. Effect of the cryogenic treatment on polyamide and optimization of its parameters for the enhancement of wear performance[J]. Transactions of the Indian Institute of Metals, 2012, 65(3): 313-331.

Low-Cost and Customizable Weft Knitted Fabric as E-Skin for Power and Sensing

XU Shuai[1], **SUN Jiangdong**[1], **SUN Pengfei**[1], **SUN Shuang**[1], **NIE Wenqi**[1,2*]

[1] *School of Textile and Garment, Anhui Polytechnic University, Wuhu 241000, China*

[2] *Key Laboratory of Bio-Fibers and Eco-Textiles, Qingdao University, Qingdao 266071, China*

* *Corresponding author's email*: wenqinie@ 163. com

Abstract: Electronic skins (E-skins) play an important role in areas such as health monitoring and human interaction. As an epidermal sensor device, the realization of self-powered sensing without the energy supply of a rigid battery makes the E-skin even more practical. This study investigates a method of preparing low-cost and customizable E-skin by the design of weft knitting and triboelectric nanogenerator (TENG). We have selected a suitable strategy for the preparation of PET knitted fabrics from PET monofilaments using machine tests on various weft knitting structures, and on this basis, an E-skin has been prepared. Tests have shown that the E-skin has an open-circuit voltage of 31 V under 1 Hz mechanical frequency. This approach is an inspiration for the future development of epidermal electronic devices.

Keywords: E-skin; Triboelectric; PDMS; Weft knitting; PET

1 Introduction

In recent years, epidermal electronics have attracted worldwide attention for their flexibility and applications in the sensing field[1-4]. The electronic skin (E-skin) is a design that mimics human skin and is endowed with functions such as measuring strain, pressure, and temperature. Although, most of the E-skin prepared in previous studies have high flexibility, thin thickness, and are suitable for interface with human skin[5]. However, they still rely on rigid batteries for their energy supply, resulting in poor portability or causing discomfort to the human skin. In contrast, triboelectric nanogenerator (TENG) has tremendous advantages to harvest energy around the environment. Besides, TENGs offer superior power generation properties and simple processing routes for flexible devices[6]. Consequently, the introduction of TENGs into electronic skins can be used not only as sensors but also as power sources. This strategy is the preferred choice for designing self-powered E-skin.

For the materials of E-skin, there is various kind of functional materials. Among them, polydimethylsiloxane (PDMS) is widely used in many areas of epidermal electronics research due to its accessibility, high flexibility, and thermal stability[5-7]. At the same time, PDMS is also a common material for TENG device design, especially for its strong electronegativity. However, the TENG-based e-skin has more than just a PDMS layer; a conductive layer for electrostatic induction is indispensable. Most methods aim to coat PDMS with a conductive layer, but PDMS itself is hydrophobic and conductive substances do not bond well to it[8]. In addition, the E-skin is subjected to a lot of mechanical deformation, which can lead to an extremely unstable conductive layer. Inspired by the fact that textile constructions act as reinforcing elements in composite materials, the use of textile structures to assume the conductive layer in electronic skins will result in strong mechanical properties and robustness[9].

Herein, we report a low-cost and customizable method based on weft knitting technology to prepare E-skin. PET monofilaments are designed into various knitting patterns by combining flat knitting and fleximesh structure. After heat-setting the PET fabric has stable mechanical properties and is a good substrate for building electronic skins. The prepared E-skin exhibits a stable voltage output at different mechanical frequencies with an open circuit voltage of 31 V. And, the voltage output can be used both as an energy supply source and as a sensing signal for a self-powered energy system, providing an alternative route to delicate wearable epidermal electronics.

2 Experiments

2.1 Materials

PET monofilament (0.1 mm) was obtained from Huatong Color Spinning. PDMS (Sylgard 184) was provided by Dow Corning Co., Ltd, which contains component A and component B. Liquid bandage (LB) was supplied by 3M Nexcare (Shanghai, China). Conductive silver pulp was obtained from Jingjie Technology (Shenzhen, China). Its curing temperature was 50 ℃.

2.2 Preparation of PET monofilament weft knitted fabric

Design of weft knitting structures from PET on computerized flat knitting machines (LXA-252S, needle size 12), including flat knitting, 1+1 ribbing, loop transfer, jacquard, and fleximesh. Before the formal knitting of the PET monofilaments, the parameters of the computerized flat knitting machine were adjusted. The bottom part of the fabric was chosen from a four-flat structure with alternating color codes 8 and 9. Before the machine worked, the yarn was threaded into the nozzle. No. 1 nozzle was used to weave the waste yarn and No. 3 nozzle was used to weave the PET monofilament. After the preparation work is completed, the knitting machine was carried out, and the waste was fed into the needle bed to knit the substrate. When the fabric roll at the bottom of the machine pulls the woven waste yarn, the machine was put on hold and the head was reset. Then the PET monofilament was fed into the needle bed to weave formal patterns. And all the sample size was 7 cm×8 cm.

2.3 Preparation of E-skin

First, the weft knitted fabric was heated to 200 ℃ in a drying oven to eliminate the internal stress. Then, conductive silver pulp fully coated on the PET surface prepared the Ag/PET electrode at 60 ℃ thermally solidify temperature. Second, the PDMS solution was configured with a mass ratio of 10 : 1 (component A and component B, respectively). Third, the heat-setting fabric was sandwiched between two acrylic plates and evenly coated with PDSM solution. Then the structure was taken into a hot press machine with a pressure of 1 MPa and a temperature of 60 ℃. Next, removed the fabric (PDMS/Ag/PET) after 2 hours and washed the surface with deionized water to gain the e-skin. Before the pattern was placed on the human skin, the targeted area was sprayed with an ultrathin LB layer. With the high stickiness of the LB, the E-skin could be easily fitted on human skin.

2.4 Electrical performance of E-skin

The electrical output performance is measured by simulating the various designs of E-skin with a linear motion motor (LinMot P01-23). In addition, the striking force is 20 N, the maximum separation distance is approximately 30 mm, and the applied frequency is fixed at 1-3 Hz. V_{OC} (open-circuit voltage), I_{SC} (short-circuit current), and Q_{SC} (short-circuit charge transfer) were gauged by an electrometer (Keithley 6514) and recorded by a custom LabVIEW program.

3 Results and Discussion

3.1 Structure design of weft knitted fabric

There are many different tissue designs for knitted weft knitted fabrics, but the properties of PET monofilaments differ significantly from those of conventional knitting yarns. Therefore, we tried out various structures and tested them on the machine (fabric density: transverse density 25, longitudinal density 28), and their simulation previews are shown in Fig 1. It is found that 1+1 ribbing, loop transfer, and jacquard structure cannot be prepared. For 1+1 ribbing, the structure is tighter, while PET monofilaments are more rigid and cannot be bent several times in succession in one area. In the case of the loop transfer, the rigidity of the PET monofilament is too limited for the moving loop to be stretched, resulting in easy breakage. The back side of

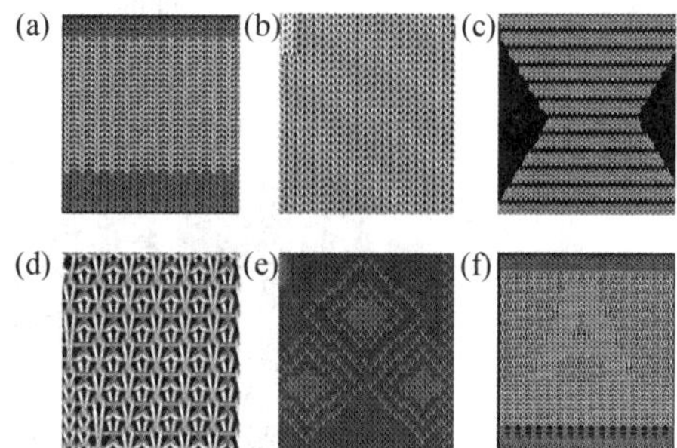

Fig. 1 Preview of the simulation of the structure, including (a) flat knitting, (b) 1+1 ribbing, (c) loop transfer, (d) fleximesh, and (e) jacquard; (f) combination flat knitting and fleximesh structure to design pattern of letter A

the jacquard tissue is also difficult to knit due to the presence of dense sesame dots. Fortunately, flat knitting and fleximesh structures allow for the structural forming of PET monofilaments, as shown in Fig. 1 (a) and (d). So, we have combined the advantages of both structures and designed the overview diagram in Fig. 1 (f), which has a concave and convex structural design and an ability to personalize the design.

3.2 Preparation of E-skin and working mechanism

We have successfully prepared PET fabrics utilizing computerized flat knitting machines and tissue chart design. Thanks to the superior shape recovery of PET monofilaments after heat setting, PET fabrics can be used as a base material for E-skin that can withstand complex deformation situations after fitting to the skin. Fig. 2 (a) shows the schematic illusion of the E-skin of a sandwich structure. Conductive Ag coated on PET fabric to form electrodes for electronic skin (Ag/PET). And, PDMS is selected as the dielectric material due to its high electronegativity triboelectric sequence. At the same time, PDMS has outstanding skin-friendly and mechanical properties and is considered a potential material for skin replacement. In the end, we chose LB as the binder to bond the E-skin to the real skin for the sensing function application.

The working principle of E-skin is illustrated in Fig. 2(b), which is interpreted by the TENG single-electrode mode. When materials such as clothing are in contact with PDMS, there is no potential difference between the conductive layer and external circuit. Once the separation progress happened, the Ag/PET gains an additional charge. This is because PDMS is stronger in gaining charge compared to clothing. And after separation, the charge is generated at the Ag/PET because of electrostatic induction. Meanwhile, the Ag/PET creates a potential difference with the external circuit, resulting in the current generation. Conversely, the cloth is approaching the PDMS layer again, the reduction of the distance will make the cloth obtain a higher electric potential than the Ag/PET electrode, which generates a reversed current signal to the separation process.

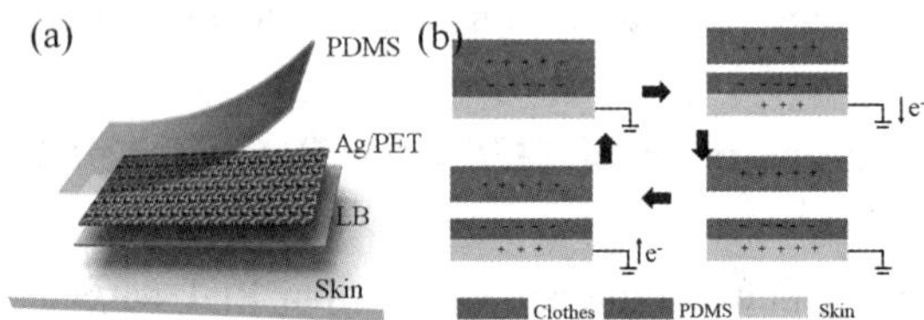

Fig. 2 (a) The schematic illusion of e-skin; (b) the working mechanism of e-skin

3.3 Electrical performance of E-skin

The electrical performance of E-skin as Figs. 3 (a)-3(c) contains V_{OC}, I_{SC}, and Q_{SC}. Take the design of the letter A as an example, the V_{OC} and Q_{SC} almost remain constant 31 V and 16 nC, respectively. And an increasing I_{SC} (from 0.122 μA to 0.338 μA) is observed when the frequency rises from 1 to 3 Hz. As the typical TENG device, the output of E-skin can be explained as follows by the Gauss theorem.

$$V = -\frac{Q}{S\varepsilon_0}\left(\frac{d}{\varepsilon} + x(t)\right) + \frac{\sigma x(t)}{\varepsilon_0}$$

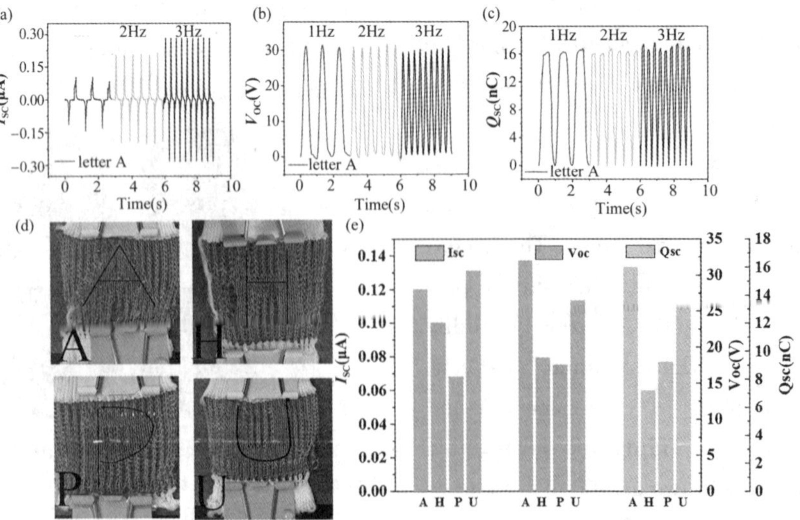

Fig. 3 (a)-(c) Electrical output performance of the F-TENG under different loading frequencies (1–3 Hz), including (a) I_{SC}, (b) V_{OC}, and (c) Q_{SC}; (d) the photographs of E-skin with different letters structure; (e) the output performance comparison of different letters structure

where Q is the transferred charge and x is the separation distance between layers. And d, S, and ε are the thickness, electrode area, and relative permittivity of the triboelectric materials, respectively. ε_0 is the permittivity of the vacuum and σ is the electrostatic charge density.

The prepared E-skin substrate is shown in Fig. 3 (d) and the letter shapes can be clearly distinguished, demonstrating the feasibility of personalizing the E-skin through the knitting process. As depicted in Fig. 3 (e), the different letters show different outputs at the same mechanical frequency. Normally, we use the voltage signal as sensing information for characterization. It is found that the difference in voltage output between different letters can be used as an E-skin sensing signal. This feature indicates that E-skin designed in knitting has great potential for wearable applications.

4 Conclusions

In summary, a low-cost and customizable method based on weft knitting technology has been developed to prepare E-skin. It allows the conversion of mechanical energy into electrical energy within potential applications for flexible sensing and human-machine interface. From a structural point of view, the combination of flat knitting and fleximesh structure provides the opportunity to personalize PET monofilaments into various skin patterns. The E-skin showed an open-circuit voltage of 31 V under 1 Hz mechanical frequency. This voltage output can be used both as a source of energy supply and as a sensing signal, which acts as a self-powered energy system. Meanwhile, it is necessary to inspire the preparation of high-performance E-skin at a low cost in the future.

References

[1] HAN X, JIANG D, QU X, et al. A stretchable, self-healable triboelectric nanogenerator as electronic skin for energy harvesting and tactile sensing[J]. Materials, 2021, 14(7): 1689.

[2] WEN J, NIU C, HE H, et al. A load-dependent model of triboelectric nanogenerators for surface roughness sensing[J]. IEEE Sensors Journal, 2021, 21(18): 20220-20228.

[3] ZHANG W, LIU Q, CHAO S, et al. Ultrathin stretchable triboelectric nanogenerators improved by postcharging electrode material[J]. ACS Applied Materials & Interfaces, 2021, 13(36): 42966-42976.

[4] ZHAO J, WANG Y, SONG X, et al. Flexible triboelectric nanogenerator based on polyester conductive cloth for biomechanical energy harvesting and self-powered sensors[J]. Nanoscale, 2021, 13(43): 18363-18373.

[5] WONG T H, LIU Y, LI J, et al. Triboelectric nanogenerator tattoos enabled by epidermal electronic technologies[J]. Advanced Functional Materials, 2021, 32(15): 2111269.

[6] CAI Y W, ZHANG X N, WANG G G, et al. A flexible ultra-sensitive triboelectric tactile sensor of wrinkled PDMS/MXene composite films for E-skin[J]. Nano Energy, 2021, 81: 105663.

[7] HE W, SOHN M, MA R, et al. Flexible single-electrode triboelectric nanogenerators with MXene/PDMS composite film for biomechanical motion sensors[J]. Nano Energy, 2020, 78: 105383.

[8] YU C, YU C, CUI L, et al. Facile preparation of the porous PDMS oil-absorbent for oil/water separation[J]. Advanced Materials Interfaces, 2017, 4(3): 1600862.

[9] WANG Z, MA Z, SUN J, et al. Recent advances in natural functional biopolymers and their applications of electronic skins and flexible strain sensors[J]. Polymers (Basel), 2021, 13(5): 813.

Double-Layer Nanofiber Membrane Composite Hydrogel Dressing with High Moisture Retention and Mechanical Enhancement

LI Mengya[1], LI Xiaoran[1*], DING Bin[1]

[1] *Innovation Center for Textile Science and Technology, College of Textiles, Donghua University, Shanghai 201620, China*

* *Corresponding author's email*: xiaoranli@ dhu. edu. cn

Abstract: Designing bilayer nanofiber composite hydrogel dressings with good water retention, mechanical properties, self-healing properties, and biocompatibility to promote full-thickness skin wound regeneration is highly desired in clinical applications. Here, a series of bilayer hydrogels based on chitosan-grafted MA and self-healing hydrogels using $NaIO_4$-oxidized chitosan combined with highly elastic hydrophobic nanofibrous membranes were used for wound dressings. These composite hydrogels exhibited similar or superior stretchability, ductility, and fatigue resistance to human skin, and still exhibited similar stress to original hydrogels after 30 load-unload compression cycles at 90% stress-strain curve. In addition, the composite bilayer hydrogel has more excellent water retention performance, and it can still retain 50% of the water of the original hydrogel at room temperature for 7 d.

Keywords: Double-layer nanofiber composite hydrogel; Water retention; Self-healing; Mechanical properties

1 Introduction

The skin is the first barrier for the human body to protect from external damage. At the same time, skin wounds and defects are also the most common problems in current clinical medicine. Therefore, the research on functional wound healing dressings is particularly important[1-2]. Studies have shown that a variety of new dressings have been applied to full-thickness wound healing and tissue repair, among which natural hydrogel dressings are widely used due to their good biocompatibility[3-4]. Various functional hydrogels based on wound healing stages have been developed, which can help skin wounds to quickly stop bleeding, avoid inflammation, and promote cell proliferation[5-6]. However, in actual clinical use, hydrogel dressings still have problems such as poor mechanical properties, insufficient moisturizing properties, and insignificant healing effects[7]. In addition, the self-healing hydrogel can heal itself after repeated injuries in the human body, which not only improves the service life of the hydrogel but also effectively protects the wound and promotes repair[8]. The research on self-healing hydrogels in the field of medical dressings has caused a great sensation, but the water retention and mechanical properties of self-healing hydrogels have yet to be explored.

Hydrogels are 3D network materials containing a large amount of water or biological fluids with soft 3D cross-linked structures that can mimic the natural extracellular matrix (ECM)[9-10]. In particular, wounds like skin joints and cavities are difficult to repair well, so self-healing hydrogels show great advantages, not only can fit irregular wounds perfectly, but also self-heal broken networks. However, the hydrogels based on natural polymers reported so far still have a series of problems, either poor mechanical properties or low water retention, which lead to the inability of the hydrogels to be compact in practical applications. After fitting the wound, the hydrogel will also deform, tear and fall off, which greatly shortens the service life of the hydrogel, resulting in wound infection or inflammatory response[11-12]. Therefore, it is crucial to develop natural hydrogels with good self-healing and mechanical properties, which can allow surgeons to use convenient and effective hydrogels dressings replace traditional sutures and gauze.

Inspired by the above facts, this work designs a multifunctional bilayer nanofiber hydrogel for wound healing.

The upper layer is a highly elastic hydrophobic nanofiber membrane, which is mainly blended with polydimethylsiloxane (PDMS) and thermoplastic polyurethane (TPU) in a certain proportion, which has excellent strength, elasticity and hydrophobicity. The lower layer is a chitosan-based hydrogel. Our results show that the bilayer composite hydrogel has good elastic mechanical properties and fatigue resistance. The composite hydrogel can recover within 5–10 min after damage or fracture, and still exhibits excellent ductility and compressibility under multiple cyclic deformations. At the same time, the composite hydrogel also has good moisturizing performance, and the biological environment of continuous moisturizing is conducive to the rapid repair of wounds.

2 Experiments

2.1 Materials

Chitosan (CS, degree of deacetylation ≥95%, viscosity: 100–200 mPa · s) was purchased from Aladdin (Shanghai, China). Methacrylic anhydride (MA, 94%) and sodium periodate ($NaIO_4$, 99.5%) were purchased from Bentong Chemical (Shanghai, China).

2.2 Synthesis of methacryloyl chitosan (CS-MA) and aldehyde-chitosan (CS-CHO)

CS-MA and CS-CHO was obtained by modification according to the previous synthetic method. Dissolve 1 g chitosan powder (about 5.45 mmol NH_2) in 50 mL 1% acetic acid solution, and then add 50 mL ethanol solution for dilution. After complete dissolution, 0.42 g MA (about 5.45 mmol carbon-carbon double bond) was added dropwise to the CS solution. The resulting solution was reacted at room temperature for 12 h, then the mixed solution was neutralized and dialyzed against deionized water for 5 d (dialysis bag specification: MWCO = 14 kDa), water was changed 3 times a day, and finally obtained by freeze-dried and stored at 20 ℃ until use. In addition, CS-CHO was also synthesized by a similar method. Dissolve 2 g chitosan powder in 160 mL 1% acetic acid solution, add 1 g $NaIO_4$ powder with stirring under the dark condition, react at room temperature for 24 h in the dark, then add 2 mL EG (ethylene glycol), and let stand for 0.5 h stop the reaction. The resulting solution was neutralized and dialyzed against deionized water for 7 d with daily water changes, and finally was obtained by freeze-dried and stored at −20 ℃ until use.

2.3 Fabrication of nanofiber membrane

Thermoplastic polyurethane (TPU) was dissolved in N, N-Dimethylformamide (DMF) solvent to prepare a 14% solution, and DOW SYLGARD 184 polydimethylsiloxane (PDMS) main agent was dissolved in Tetrahydrofuran (THF) solvent to prepare 14% solution, add curing agent (main agent mass: curing agent mass = 10 : 1), mix the above two solutions in a mass ratio of 2 : 1, and stir for 30 min to obtain a uniform and stable precursor spinning solution. The precursor spinning solution was placed in a 10 mL syringe, and an electrospinning device was used for spinning. The spinning voltage was 30 kV, the receiving distance was 25 cm, the perfusion speed was 1.6 mL/h, the temperature was (5±2) ℃, and the relative humidity was 50%±5%. The obtained fiber membrane was placed in an oven for heat treatment at a temperature of 60 ℃, and a time of 120 min, and finally a PDMS/TPU nanofiber membrane was prepared.

2.4 Preparation of bilayer composite nanofibrous hydrogels

CS-MA was dissolved in deionized water to make a 4% CS-MA solution. CS-CHO was dissolved in deionized water to prepare CS-CHO solutions of 0%, 8%, 10%, and 12%, respectively. The two-layer composite self-healing hydrogel can be obtained by mixing the above two solutions in a volume ratio of 1 : 1 and placing them on the PDMS/TPU nanofiber membrane.

2.5 Water retention tests

The prepared hydrogels were placed in a room temperature environment of 25 ℃ and 54% RH for 7 d. The water volatilization of the composite hydrogels was observed, photographed and weighed every day. The water retention of the composite hydrogel at different times was calculated by the following formula:

$$\text{Water rentention}(\%) = \frac{m_t}{m_0} \times 100\% \qquad (1)$$

where m_0 and m_t are the hydrogel weights on day 0 and day t, respectively.

2.6 Mechanical performance tests

Mechanical tensile stress-strain properties were tested with a universal material testing machine using a 50 N tension transducer. The composite hydrogel samples were prepared into strips (length 20 mm×width 5 mm×

thickness 1 mm), and the tensile strength and elongation at break were obtained at a tensile rate of 10 mm/min. The compressive stress-strain performance test is to prepare the composite hydrogel sample into a cylindrical shape (height 10 mm× diameter 15 mm), study the test at 90% strain at a speed of 5 mm/min, and then return to 0% strain at a speed of 5 mm/min, the cycle was repeated 50 times to study compression and recovery properties.

2.7 Self-healing property of hydrogels

In the macroscopic self-healing experiment, two cylindrical hydrogels stained with rhodamine B and methylene blue, represented by CS-MA/CHO10-PT hydrogels, were cut into two halves. At room temperature and in the absence of any external stimulation, the cutting interface of each piece was immediately put together and healed after 5 minutes. The self-healing process of the hydrogel was recorded by taking a video with a mobile phone and observing the changes of the cutting interface under a microscope.

2.8 Cytocompatibility evaluation of hydrogels

The cytotoxicity of the hydrogel was assessed by direct contact test between the hydrogel and fibroblasts (L929). L929 cells were seeded on hydrogel groups with different ratios at a density of 2×10^4 cells/well, and three parallel samples per group were used for testing. Cell viability was further tested by the LIVE/DEAD® Viability/Cytotoxicity Kit method. The hydrogel obtained by live and dead staining was carefully taken out and scanned under a fluorescence microscope to obtain the cell growth on the hydrogel.

Statistical analysis:

All test data at least three parallel samples were used for statistical analysis, and the data results were expressed as mean±standard deviation (SD).

3 Results and Discussion

3.1 Characterization of hydrogels

CS-CHO and CS-MA were formed by $NaIO_4$ oxidation and MA grafting, respectively. The structural characterizations of CS-CHO and CS-MA were obtained by FT-IR and ^{1}HNMR in Fig. 1. The peaks of CS at 3 350 cm^{-1}, 3 289 cm^{-1}, 2 867 cm^{-1}, 1 646 cm^{-1} and 1 065 cm^{-1} indicated O—H stretching, N—H stretching of $—NH_2$, C—H stretching, $—NH_2$ deformation in plan (usually called NH_2 band), and C—O—C bridge symmetric stretching, respectively. For CS-CHO, a hemiacetal proton 895 cm^{-1} was formed between the aldehyde group and the hydroxyl group. Furthermore, for CS-MA, several new peaks contributed by newly formed amides were observed at 1 638 cm^{-1}, 1 535 cm^{-1} and 1 354 cm^{-1}, corresponding to the amide I band (C=O stretching), amide II Band (N—H deformation and C—N stretching) and amide III band. In addition, a new peak at 8.12×10^{-6} corresponded to the aldehyde group on the CS-CHO chain and the corresponding hemiacetal proton at 5.32 ppm, as confirmed by the ^{1}HNMR spectrum. The NMR spectrum of CS-MA clearly showed vinyl proton signals at 5.44 and 5.66×10^{-6}.

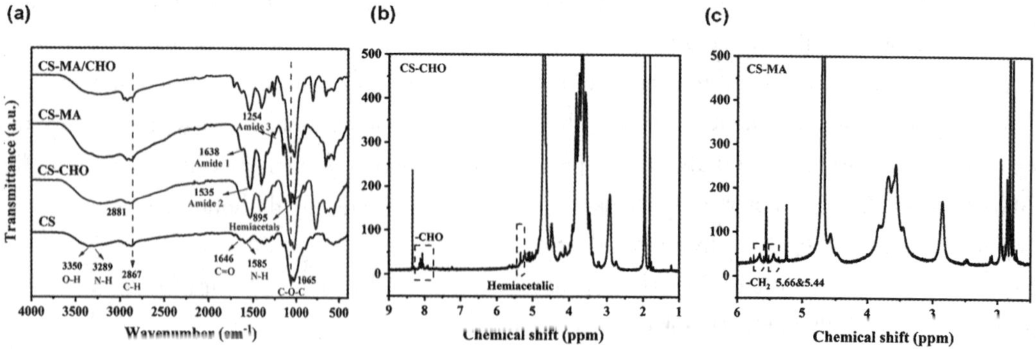

Fig. 1 (a) FT-IR spectra of CS, CS-CHO, CS-MA, CS-MA/CHO; (b) ^{1}H NMR spectra of CS-CHO; (c) ^{1}H NMR spectra of CS-MA

3.2 Water retention of hydrogels

Due to the evaporation of water at the interface, hydrogel materials generally have poor water retention capacity, which greatly limits their lifetime in practical applications. It could be seen in Fig. 2 that under the protection of the hydrophobic nanofiber membrane PDMS/TPU, the water dispersion of the double-layer nanofiber composite hydrogel is significantly slower, and the water retention rate can be increased by more than half. It is found that the water retention performance of CS-MA/CHO10-PT is compared. Optimally, leave it at room temperature for 7 d and still retain 50% of moisture.

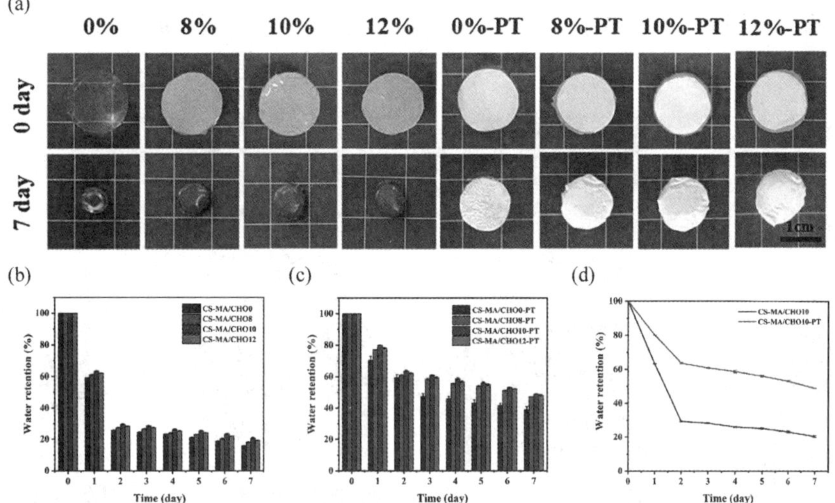

Fig. 2 (a) Images of hydrogels stored at 25 ℃ and 54% RH for 0 and 7 d; (b) comparison of water retention of hydrogels for 0–7 d; (c) comparison of water retention of bilayer composite hydrogels for 0–7 d; (d) comparison of water retention between CS-MA/CHO10 and CS-MA/CHO10-PT in 0–7 d

3.3 Mechanical performance of hydrogels

Mechanical properties similar to skin are beneficial to skin wound repair materials, not only to maintain the integrity of the material, but also to facilitate good adhesion when the skin tissue is deformed by external forces. The tensile and compressive test results of the hydrogel are shown in Fig. 3. These hydrogels show good stretchability, among which the tensile strength and elongation of the composite hydrogels are greater than those of the original hydrogels, and the CS-MA/CHO10-PT composite hydrogel has the best tensile mechanical properties. glue. In addition, after 30 loading-unloading compression experiments with 90% strain, it can be seen that the hydrogel also has good fatigue resistance.

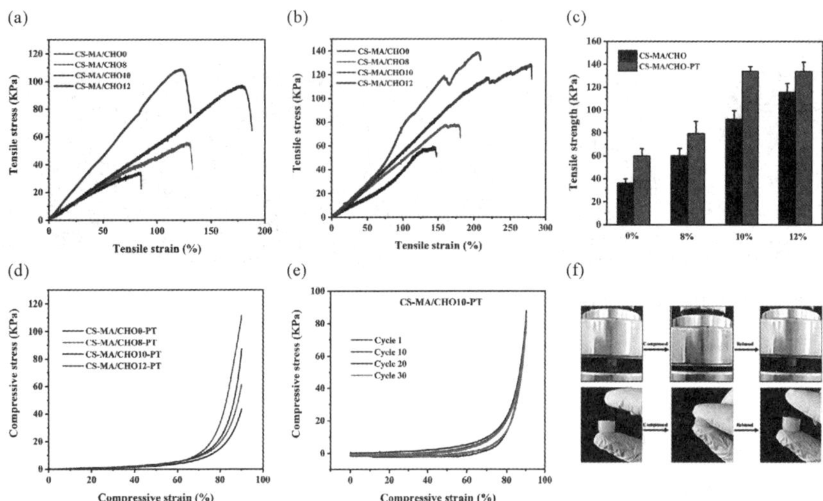

Fig. 3 (a) Tensile stress-strain curves of original hydrogels; (b) tensile stress-strain curves of composite hydrogels; (c) tensile strength comparison of hydrogels; (d) compressive stress-strain curves of composite hydrogels; (e) stress-strain cyclic curve with 90% compressive strain; (f) compressive schematic

3.4 Self-healing property of hydrogels

The self-healing properties of hydrogels can be assessed by intuitive methods. As shown in Fig. 4, the circular hydrogels dyed with different colors were cut into two pieces, and they healed within 5 min without any stimulation by physical contact. Meanwhile, the self-healing process of the hydrogel was further visualized using the bright-field mode of the fluorescence microscope. The self-healing hydrogel still has good tensile mechanical properties. In particular, the tensile mechanical properties of the composite hydrogel after healing are better than the original hydrogel.

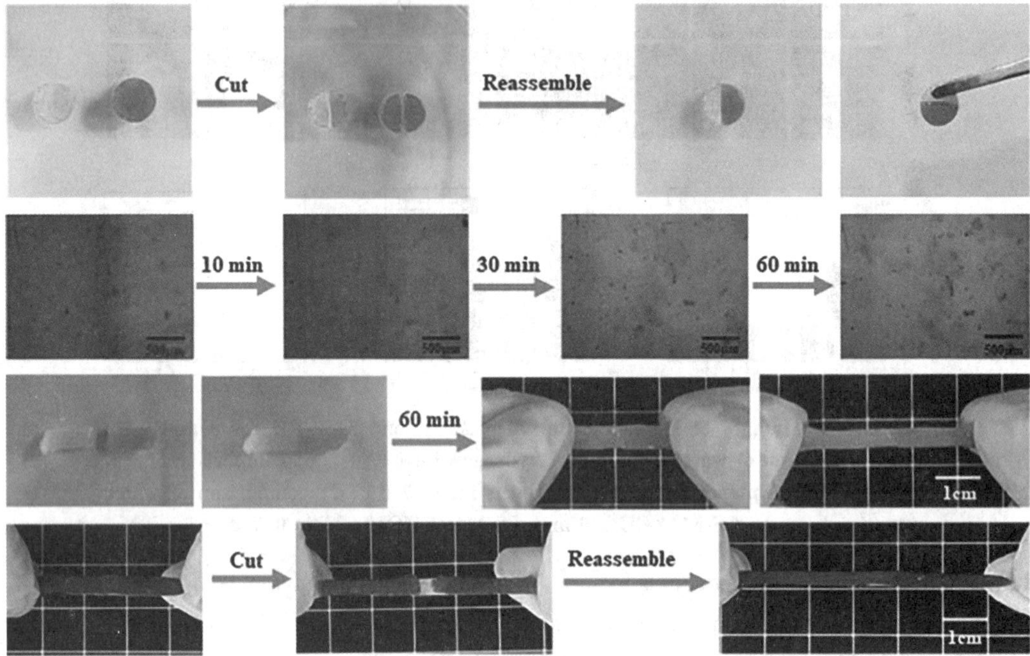

Fig. 4 Macroscopic self-healing experiment

3.5 Cytocompatibility evaluation of hydrogels

Good biocompatibility is an important factor for biomaterials. Fig. 5 shows the live/dead staining photos of L929 cells cultured on the hydrogel material for the third day. It can be seen that the cells grow well, and the cell viability is the highest on the CS-MA/CHO10 hydrogel, which was enough to prove its excellent biocompatibility.

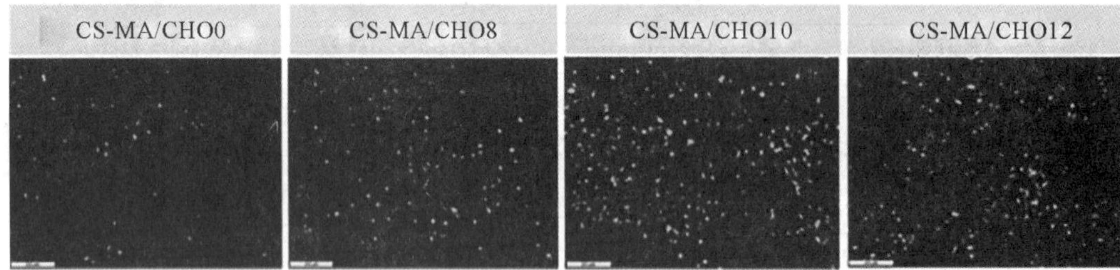

Fig. 5 Live/dead staining of cells after 3 d of incubation in hydrogels

4 Conclusions

By introducing two chitosan polymers (CS-MA and CS-CHO) and nanofiber membrane PDMS/TPU, a bilayer composite hydrogel with enhanced water retention, mechanical properties and self-healing properties was prepared. The hydrogel has excellent water retention properties, and it can maintain 50% of the original hydrogel's moisture for 7 d at room temperature. At the same time, it has good tensile properties (tensile stress can reach 132.3 kPa, tensile strain can reach 276.5%), compression recovery properties (90% strain can still be well recovered after 30 cycles of compression) and self-healing properties. In addition, these hydrogels also exhibited good cytocompatibility, which is beneficial to promote wound healing.

References

[1] QU J, ZHAO X, LIANG Y, et al. Antibacterial adhesive injectable hydrogels with rapid self-healing, extensibility and compressibility as wound dressing for joints skin wound healing[J]. Biomaterials, 2018, 183: 185-199.

[2] LIANG Y, ZHAO X, HU T, et al. Adhesive hemostatic conducting injectable composite hydrogels with sustained drug release and photothermal antibacterial activity to promote full-thickness skin regeneration during wound healing[J]. Small, 2019, 15 (12): e1900046.

[3] WU S, ZHANG Z, XU R, et al. A spray-filming, tissue-adhesive, and bioactive polysaccharide self-healing hydrogel for skin regeneration[J]. Materials & Design, 2022, 217.

[4] ZHANG K, LV H, ZHENG Y, et al. Nanofibrous hydrogels embedded with phase-change materials: temperature-responsive dressings for accelerating skin wound healing[J]. Composites Communications, 2021, 25.

[5] LIANG Y, LI Z, HUANG Y, et al. Dual-dynamic-bond cross-linked antibacterial adhesive hydrogel sealants with on-demand removability for post-wound-closure and infected wound healing[J]. ACS Nano, 2021, 15 (4): 7078-7093.

[6] SHEN Y, XU G, HUANG H, et al. Sequential release of small extracellular vesicles from bilayered thiolated alginate/polyethylene glycol diacrylate hydrogels for scarless wound healing[J]. ACS Nano, 2021, 15 (4): 6352-6368.

[7] ZHOU L, DAI C, FAN L, et al. Injectable self-healing natural biopolymer-based hydrogel adhesive with thermoresponsive reversible adhesion for minimally invasive surgery[J]. Advanced Functional Materials, 2021, 31 (14).

[8] YANG Y, XU L, WANG J, et al. Recent advances in polysaccharide-based self-healing hydrogels for biomedical applications[J]. Carbohydrate Polymers, 2022, 283: 119161.

[9] LU X, CAO L, YIN X, et al. Stretchable, tough and elastic nanofibrous hydrogels with dermis-mimicking network structure[J]. Journal of Colloid and Interface Science, 2021, 582: 387-395.

[10] DU J, YAO Y, WANG M, et al. Programmable building of radially gradient nanofibrous patches enables deployment, bursting bearing capability, and stem cell recruitment[J]. Advanced Functional Materials, 2021, 32 (8).

[11] WANG L, ZHANG X, YANG K, et al. A novel double-crosslinking-double-network design for injectable hydrogels with enhanced tissue adhesion and antibacterial capability for wound treatment[J]. Advanced Functional Materials, 2019, 30 (1).

[12] HE X, LIU X, YANG J, et al. Tannic acid-reinforced methacrylated chitosan/methacrylated silk fibroin hydrogels with multifunctionality for accelerating wound healing[J]. Carbohydrate Polymers, 2020, 247: 116689.

Preparation of Soy Protein Nonwoven Materials for Air Filtration and Formaldehyde Purification

CHEN Yuansen[1,2], **HU Yong**[1,2], **WU Haibo**[1,2*]

[1] *Engineering Research Center of Technical Textiles, Ministry of Education, Donghua University, Shanghai 201620, China*

[2] *Department of Nonwoven Materials and Engineering, College of Textiles, Donghua University, Shanghai 201620, China*

* *Corresponding author's email*: hbwu @ dhu. edu. cn

Abstract: Conventional air filters can only filter solid particulate matter (PM) but cannot chemically capture and remove toxic gaseous molecules. Soy protein nonwoven filtration materials are prepared by the impregnation method. To select the best impregnation process conditions, we tested the PM filtration efficiency of the filtration materials by regulating the concentration of soy protein solution and impregnation time. In addition, we conducted formaldehyde purification test to verify the formaldehyde purification effect. The results show that the purification efficiency of formaldehyde can reach 45% in 0.2 mg/m^3 formaldehyde environment, thus achieving initial purification for indoor formaldehyde. The soy protein nonwoven air filtration materials in this research are biodegradable and environmentally friendly. It can not only achieve the filtration of solid PM, but also purify the formaldehyde in the air.

Keywords: Air filtration; Soy protein; Nonwoven material; Impregnation method; Formaldehyde purification

1 Introduction

With the rapid development of economy, indoor air pollution has caused a widespread concern, which can lead to an increase in mortality for various diseases[1]. According to the physical form, indoor air pollutants can be divided into solid particulate pollutants and gas pollutants[2]. The most common indoor gaseous pollutant is formaldehyde. Excessive amount of formaldehyde inhaled by human body will cause various health problems such as chronic poisoning, dermatitis symptoms, and even endanger life in severe cases[3]. In order to assure human health, national standard (GB/T 18883—2002) stipulates that the volume concentration of formaldehyde in the indoor air environment should not exceed 0.08 mg/m^3.

Nonwoven filtration material is one of the most widely used air filtration materials today, which can effectively remove various pollutants. Basically, non-woven materials can capture micro-sized particles based on five different mechanisms including interception, diffusion, inertial impaction, gravitational settlement and electrostatic attraction[4]. However, conventional nonwoven air filters can filter tiny PM effectively but cannot chemically capture and remove toxic gaseous molecules. Although activated carbon fiber filter media enables the filtration of solid particles and polluting gases in the air, the adsorption capacity for activated carbon is prone to saturation, even releasing adsorbed harmful gases[5]. Therefore, developing a filtration material capable of purifying indoor toxic gases has attracted attention[6].

Soy protein is a kind of low-cost plant protein in abundant supply. It contains plenty of functional groups (e.g. —OH, —COOH, $—NH_2$, $—NH_3^+$ and —SH) and include 18 types of amino groups. Notably, these functional groups and amino groups can be considered for effectively capturing the pollutants at the molecular level[7]. These reactive functional groups endow soybean protein excellent ability to interact with various particles or chemicals.

In this research, we prepared soy protein nonwoven filtration materials by impregnation method. Cellulose fibers were used due to short process, low cost and environmental protection in the preparation process. As a result, filtration materials we prepared are biodegradable and environmentally friendly, which can not only achieve the filtration for solid particles, but also have a certain purification effect on formaldehyde in the air.

2 Experiments

2.1 Materials

Loycell fiber (5.62 dtex×64 mm) was obtained from Sinopec Yizheng Chemical Fiber Co. Ltd. (Jiangsu, China). Soy protein isolate (SPI, 98%) was obtained from DuPont China Group Co. Ltd. (Shanghai, China). Alkali, analytical grade, was supplied by Maclean Technology Co. Ltd. (Shanghai, China). Glycerol, was obtained from Shanghai Lingfeng Chemical Reagent Co. Ltd. (Shanghai, China).

2.2 Preparation of soy protein nonwoven materials

The Loycell non-woven base fabrics were prepared by opening, de-mixing and needle-punching process. The net weight of base fabric was 200 g/m^3, needling depth was 6 mm, feed speed was 1 m/min, needling frequency was 300 times per minute.

According to the gradient and dichotomous method to configure the solution, three soybean isolate protein solution concentration gradients were set: 1%, 3%, 5% and 7%. During the configuration of soy protein impregnation solution, adding 0.1% propanetriol to promote protein dissolution and appropriate amount of NaOH to adjust the solution pH. The solution was prepared in water bath while heating and stirring (Water bath temperature was 50 ℃, heating time was 0.5 h, rotor speed was 400 r/min), until the soy protein was completely dissolved.

The non-woven base fabric was firstly impregnated in a glass dish with protein impregnation solution. After impregnation, it was taken out and de-liquefied on a pneumatic trolley (the pressure of the clamping stick was set to 0.5 MPa). Finally, drying it in an oven.

2.3 Filtration performance testing

In order to analyze the influence of impregnation conditions on the filtration performance of soybean isolate protein nonwoven air filtration materials, the filtration efficiency of each group of samples was tested separately by using a TSI 8130 automated filter tester. The filtration performance was tested by using sodium chloride aerosol with a median mass diameter of 0.25 um, geometric standard deviation less than 1.83, and gas flow rate of 16 L/min. After obtaining the filtration efficiency η and filtration resistance ΔP (The unit is Pa), the filter material quality factor Q_F was calculated by Eq. (1).

$$Q_F = \frac{-\ln(1-\eta)}{\Delta P} \qquad (1)$$

2.4 Formaldehyde purification performance testing

In order to investigate the formaldehyde purification effect of soy protein nonwoven filter material, UV spectrometer u-4100 was used to test formaldehyde concentration before and after filtration. Formaldehyde purification efficiency was measured through a device shown in Fig. 1. A 38% formaldehyde solution was used as the volatile source of formaldehyde gas, and formaldehyde degradation efficiency was characterized by measuring the difference in concentration of formaldehyde gas in the air on both sides.

Placed the sample in the circular hole between the two sealed boxes A and B, then used the sealing baffle on the A side of the nonwoven to prevent the air circulation between the two sides of the A and B sealed boxes. Then, measured 2 mL of 38% formaldehyde solution in the storage bottle, placed it on the side of the sealed box A and opened the storage bottle. When the formaldehyde tester reading in the sealed box A reached about 0.2 mg/m^3, covered the lid of the storage bottle to remove the formaldehyde solution. When the index of formaldehyde tester in A sealed box was stable, the reading was recorded as C_0(the unit is mg/m^3), then the baffle in front of the plant protein nonwoven fabric in A sealed box was removed and a small fan was turned on to blow the gas in A sealed box to B side, and the readings of C_1 and C_2(the unit is mg/m^3) of formaldehyde tester in A and B sealed boxes were recorded per 30 s. The final formaldehyde degradation efficiency was calculated as follows.

$$E = \left(1 - \frac{C_1 + C_2}{C_0}\right) \times 100\% \qquad (2)$$

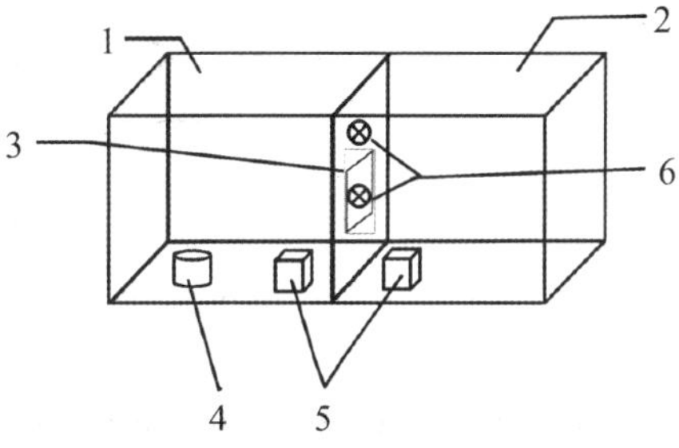

1—seal box A; 2—seal box B; 3—soy protein nonwoven filter material; 4—formaldehyde solution; 5—formaldehyde detector; 6—small fan

Fig. 1 Formaldehyde purification efficiency test device

3 Results and Discussion

3.1 Effect of soy protein solution concentration on filtration performance

Filtration efficiency and filtration resistance under impregnation treatment with different soy protein solution concentrations are given in Fig. 2 (a). Apparently, with the increase of the concentration of soybean protein infusion solution, the filtration efficiency and filtration resistance of the non-woven filter material increased gradually. This can be attributed to that increased concentration resulted in more soybean protein loading on the non-woven base fabric, which has a certain blocking effect on the pore size of the base fabric and reduces the pore size, thus improving the filtration efficiency and filtration resistance. Fig. 2 (b) shows the quality factors of nonwoven filtration materials under different soy protein impregnation treatments. Compared to the untreated base fabric, the quality factors of the treated materials slightly decreased. The quality factors of the filtration materials under the 3% soy protein impregnation treatment is relatively better. Therefore, 3% soy protein maceration solution was used in all subsequent analyses.

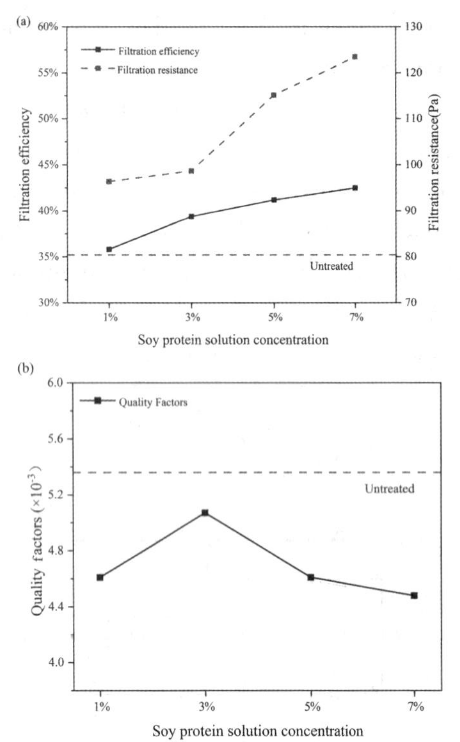

Fig. 2 (a) Filtration efficiency and filtration resistance after impregnation treatment with different soy protein solution concentrations. The impregnation time was 1 h and the drying temperature was 40 ℃; (b) quality factors of filtration materials after impregnation treatments with different soy protein solution concentration

3.2 Effect of impregnation time on filtration performance

The effect of impregnation time on the soy protein nonwoven air filtration material is shown in Fig. 3 (a). As the impregnation time increases, the loading of soy protein on the nonwoven substrate increases, resulting in a relative decrease in pore size. Compared with the untreated nonwoven substrate, the filtration efficiency and filtration resistance of the impregnated nonwoven material continuously increased. Interestingly, the filtration efficiency increased little when the impregnation time was 1.5 h, but the filtration resistance increased sharply. Fig. 3 (b) presents the quality factor of the nonwoven filtration materials obtained at different impregnation times. Compared to the untreated nonwoven base fabric, there was a slight decrease in the quality factor. The longer the impregnation time, the more quality factor decreased. Therefore, the nonwoven filtration material obtained under the treatment of soy protein solution with a concentration of 3% and impregnation time of 1 h was used for formaldehyde purification testing.

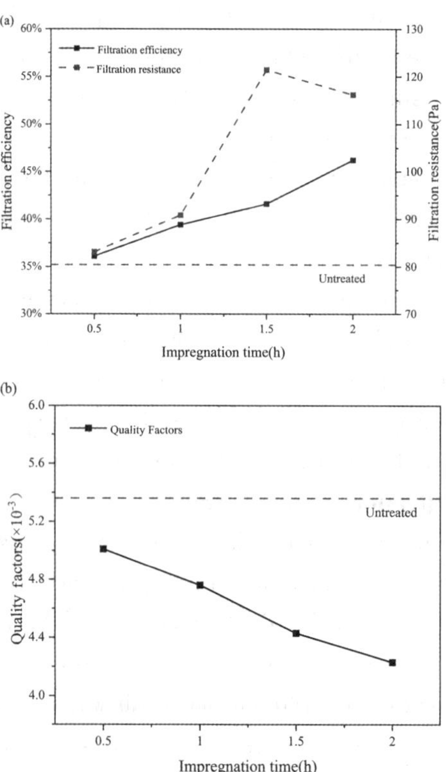

Fig. 3 (a) Filtration efficiency and filtration resistance after impregnation treatment with different impregnation time. The soy protein solution concentration was 3% and the drying temperature was 40 ℃; (b) quality factor of filtration materials under different impregnation time

3.3 Formaldehyde purification performance

The UV spectra recorded were compared to those published in the literature. The absorption spectrum of formaldehyde is found at wavelengths from 300 nm to 340 nm[8]. UV spectra of the formaldehyde before and after filtration are shown in Fig. 4. The figure shows that formaldehyde is significantly purified by soy protein nonwoven air filtration material, confirming the purification effect of soy protein nonwoven material on formaldehyde.

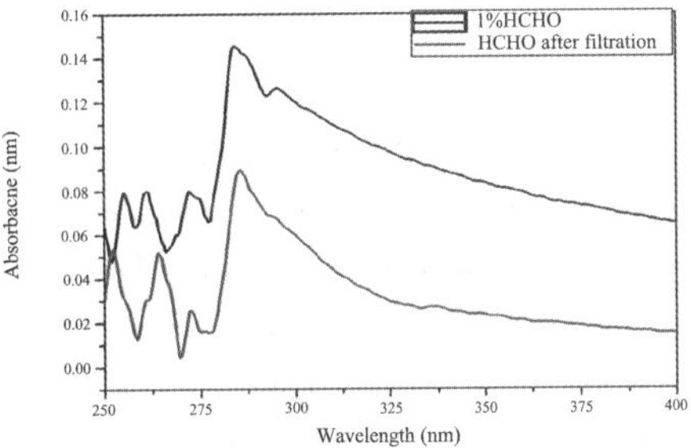

Fig. 4 UV spectra of the HCHO before and after Filtration

The formaldehyde purification test results are shown in Fig. 5, the purification proportion of formaldehyde by soy protein nonwoven filter material gradually increases with the increment of filtration time. Meanwhile, the purification efficiency of formaldehyde can reach up to 45%, which preliminarily achieves the purification of formaldehyde.

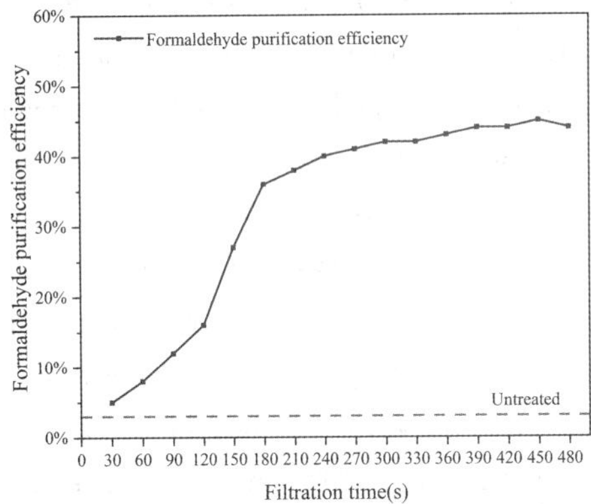

Fig. 5 Formaldehyde purification efficiency diagram of soy protein non-woven air filtration material

4 Conclusions

This research prepared soy protein nonwoven filtration materials by impregnation method. The effect of impregnation process on filtration efficiency was compared to select the best process condition by regulating the concentration of soy protein solution and impregnation time. The final formaldehyde purification test verifies the good formaldehyde purification effect. The results show that the soy protein nonwoven filtration materials can achieve filtration for both solid particles and formaldehyde in the air. The purification efficiency of the filtration materials for formaldehyde can reach up to 45%, which initially realizes the purification of formaldehyde gas.

References

[1] CAREY I M, Atkinson. Mortality associations with long-term exposure to outdoor air pollution in a national English cohort[J]. American Journal of Respiratory and Critical Care Medicine, 2013, 187 (11): 1226-1233.

[2] SUO H. The hazards and treatment of indoor environmental air pollution[J]. Chemical Design News, 2021, 47 (2): 179-180.

[3] YANG J B, CHEN J. Discussion on indoor air pollutants and pollution control measures[J]. Resource Conservation and Environmental Protection, 2019, 12(10): 100-103.

[4] YOON K, HSIAO B S, et al. Functional nanofibers for environmental applications [J]. Journal of Materials Chemistry, 2008, 44: 5326-5334.

[5] YANG J B, CHEN J. Exploration of indoor air pollutants and pollution control measures[J]. Resource Conservation and Environmental, 2019, 12(10): 10-13.

[6] ZHAO C B, CHU Z J. Ultrasound heat treatment effects on structure and acid-induced cold set gel properties of soybean protein isolate[J]. Food Bioscience, 2021, 39.

[7] VEGALUGO A C, LIM L T. Controlled release of allyl isothiocyanate using soy protein and poly (lactic acid) electrospun fibers[J]. Food Res. Int., 2009, 42 (8): 930-940

[8] DAVENPORT J J, HODGKINSON J, et al. A measurement strategy for non-dispersive ultra-violet detection of formaldehyde in indoor air: spectral analysis and interferent gases[J]. Measurement Science and Technology, 2016, 27 (1): 11.

A New Method for Virtual Try-On Based on Multi-Head Channel Attention

SHI Yi[1], WAN Xianfu[1*], WANG Jun[2]

[1] *College of Textiles, Donghua University, Shanghai 201620, China*

[2] *Key Laboratory of Textile Science & Technology, Ministry of Education, Donghua University, Shanghai 201620, China*

[*] *Corresponding author's email*: wanxianfu@dhu.edu.cn

Abstract: Try-on cannot be conducted for online clothing shopping, resulting in high refund rate. Virtual try-on can simulate the visual effects of try-on results through computer simulation technology to provide suggestions for online shopping. Virtual try-on has progressed from the traditional database matching-based virtual try-on to the emerging deep learning-based virtual try-on, and the recently introduced cascade dual attention mechanism further improves the try-on effects. However, there is still a gap between the effect of virtual try-on and the actual try-on in the case of clothing covered by limbs. In this paper, we proposed a new method through development of a multi-head channel attention mechanism for virtual try-on model to improve the effect of virtual try-on. Compared with the cascaded dual attention mechanism, it achieves a better virtual try-on effect, and the FID is decreased by 4%.

Keywords: Virtual try-on; Attention mechanism; Deep learning; Multi-head channel attention

1 Introduction

E-commerce has grown rapidly in recent years. Online clothing shopping is becoming more common. However, unlike offline clothing shopping, try-on can not be conducted for online clothing shopping. Virtual try-on simulates the visual effects of try-on results through computer simulation technology to provide suggestions for online shopping, which has attracted researcher's attention.

Virtual try-on takes an image of target clothes and an image of person as inputs and outputs an image of the person wearing the target clothes. Traditional virtual try-on was based on database matching. Ehara[1] and Tanaka[2] first build a database, then compare the image of person with the images in the database to find the most similar image in the database to guide the virtual try-on.

Emerging virtual try-on was based on deep learning. Han[3] introduced deep learning into virtual try-on, and divided the virtual try-on task into two steps. The first step deforms the input clothes to fit the posture and body shape of the input person. The second step synthesizes the try-on result with deformed clothes and image of person. TPS is used to deform clothes and a neural network with an autoencoder structure is used to generate try-on result. CPVTON[4] used neural network to calculate deformation parameters, which made the deformation more accurate, and became the basis of many subsequent works.

The synthesis of try-on result requires neural networks to extract appropriate features from the images. The channel attention mechanism and the spatial attention mechanism explicitly model the weights of channels and spaces of image, respectively, which are more effective than the implicit modeling of convolutional neural networks. Hu[5] introduced cascaded dual attention mechanism into virtual try-on model, which improved the feature extraction ability of the neural network and improved the effect of virtual try-on, however in the case of clothes that is covered by limbs, there is still a gap between the effect of virtual try-on and the actual try-on.

To address this problem, we propose four attention mechanisms and introduce them into the try-on result synthesis network of virtual try-on. The proposed multi-head channel attention mechanism improves the model's ability to deal with the covered clothes.

2 Experiments

2.1 Dataset

The dataset used in the experiments is the VITON dataset[3], which has a total of 16,253 pairs of images. Each pair of images includes an image of clothes and an image of person wearing the clothes. The dataset is divided into two parts, in which 14,221 pairs of images are used as the training set and the remaining 2,032 pairs are used as the testing set.

2.2 Model

Based on Hu's model[5], we obtain new models by proposing new attention mechanisms. The location of the attention mechanism in the model is shown in Fig. 1. Different attention mechanisms correspond to different models.

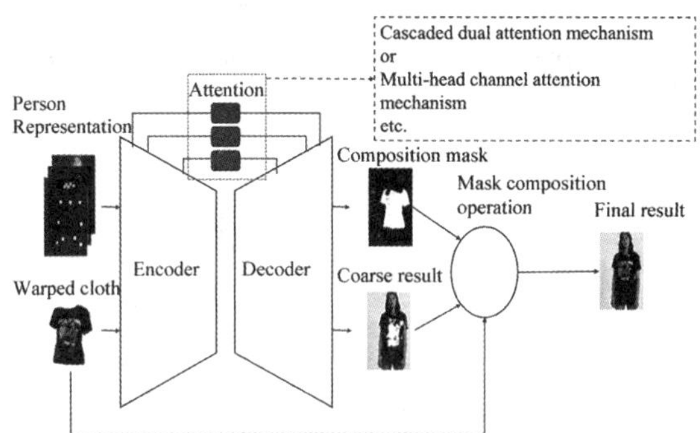

Fig. 1 Location of the attention mechanism in the model

2.2.1 Multi-head channel attention

The attention mechanisms usually applied to images include channel attention mechanism and spatial attention mechanism. Channel attention can generate different weights for each channel of the feature maps so that each channel has different effects on the generation of features. Spatial attention can assign different weights to each pixel of the feature maps so that the network can focus on more important spatial positions when generating new features. The cascade dual attention mechanism introduced by Hu[5] is composed of channel attention and spatial attention in series. The channel attention mechanism uses the compression and expansion module of three linear layers to model the channel attention. The model is simple and the modeling ability may be insufficient. Inspired by the multi-head attention[6], we propose a multi-head channel attention mechanism to improve the model's ability to model channel attention. The multi-head channel attention mechanism is shown in Fig. 2. The channel attention of the cascaded dual attention mechanism is shown in Fig. 3.

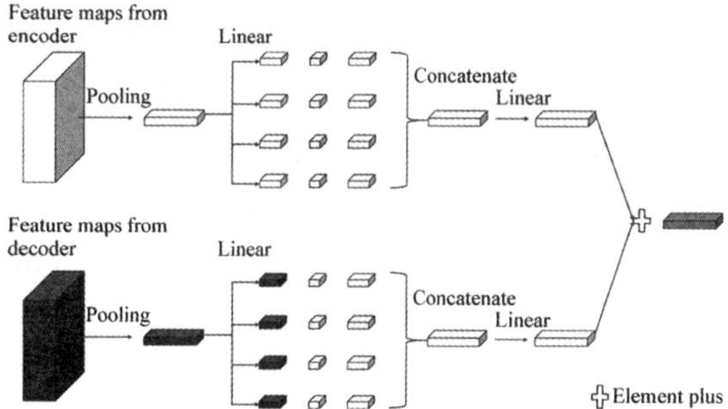

Fig. 2 Multi-head channel attention mechanism

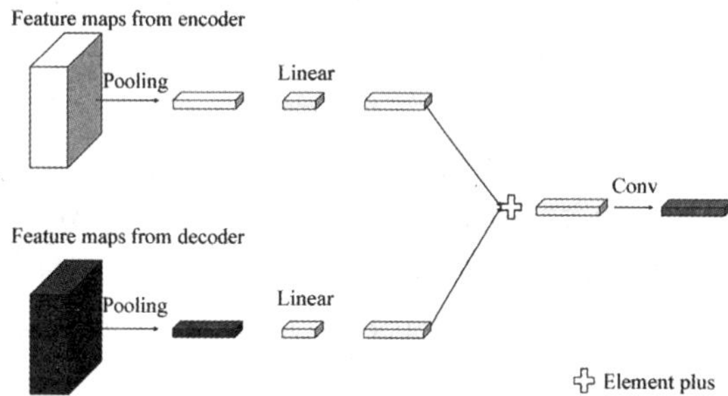

Fig. 3 Channel attention of the cascaded dual attention mechanism

The multi-head channel attention mechanism first uses a pooling layer to pool the feature maps from the encoder and decoder into a vector, and then uses the multi-head compression and expansion module to process the two vectors separately, and then add the two vectors together and activate it by sigmoid function to obtain the attention of each channel, finally the feature maps from the encoder is weighted by the channel attention. The multi-head compression and expansion module uses h linear layers to map the input vector to h different vector spaces, and then uses compression and expansion modules to process these vectors separately, and then combines these vectors and feeds them to the linear layer for fusion. The dimension of the mapped space is $1/h$ of the dimension of the original vector space, so the total computation is comparable to the original channel attention. In this paper h was set to 4.

2.2.2 Spatial attention design

The spatial attention mechanism in the cascade dual attention mechanism introduced by Hu[5] uses 1×1 convolution to model attention. Due to the limitation of the size of the convolution kernel, the generation of the attention of each pixel is only related to the pixels in the area covered by the convolution kernel. Intuitively, after seeing all the pixels, it can be determined whether a pixel deserves attention. This spatial attention mechanism lacks consideration of pixels outside the convolution kernel region. Therefore, we propose three spatial attention mechanisms that determine whether a part deserves attention after seeing the global information.

(2.1) The first spatial attention mechanism is shown in Fig. 4 marked (b). It first uses two 1×1 convolutional layers to compress the channels of the feature maps from the encoder and decoder to 1 channel, and then use the pooling layer to reduce the resolution of the two feature maps, then the two feature maps are added and flattened into a vector, the vector is processed by the compression and expansion module and activated by sigmoid function, and the processed vector is restored to feature maps and upsampled to the original size as spatial attention to weight the feature maps from encoder. Because the number of image pixels is generally much more than the number of channels, it takes a lot of computation to directly flatten the image into a vector and use the compression and expansion module to calculate the attention. Therefore, we pool the image into a low-resolution image before applying compression and expansion module.

(2.2) The second spatial attention mechanism is shown in Fig. 4 marked (c). This attention can be considered as a variant of the attention described in (2.1). The difference is that the attention in (2.1) first fuses the feature maps from the encoder and decoder and then calculates the attention, while this attention calculates the attention separately and then fuses them.

(2.3) The third spatial attention mechanism is shown in Fig. 5. It first uses two 1×1 convolutional layers to compress the channels of the feature maps from the encoder and the decoder to 1 channel respectively, and then compress the two feature maps in the height and width directions respectively to obtain two vectors. Then the compression and expansion module are used to process these two vectors respectively to obtain horizontal attention and vertical attention. These two attentions are respectively expanded to the size of the feature maps and added to obtain the spatial attention. The final spatial attention is obtained by summing the two spatial attentions of the encoder and decoder, and used to weight the feature maps from the encoder. Regarding the feature maps as a cube, the channel is regarded as the growth direction, and the space is left with the height and width directions. This design refers to the channel attention calculation method, and calculates the attention separately in the two directions of the space to synthesize the total spatial attention.

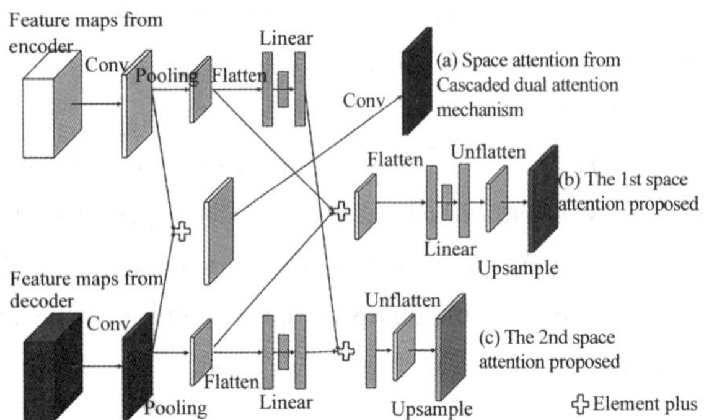

Fig. 4 Three different spatial attention mechanism

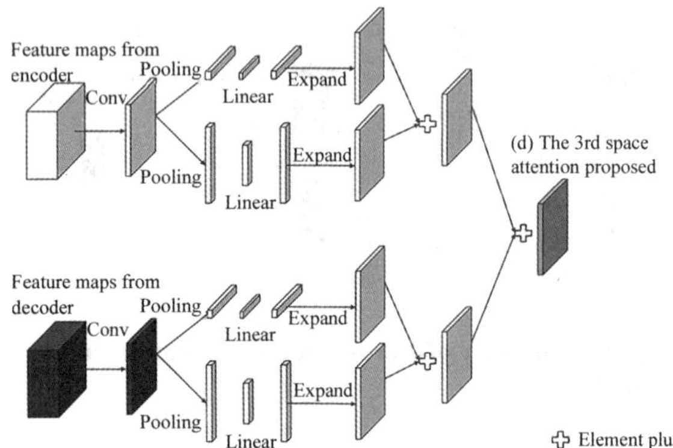

Fig. 5 The Proposed third spatial attention mechanism

2.2.3 Model generation

The following three models are obtained by replacing the spatial attention in the cascaded dual attention mechanism with the three spatial attention mechanisms described in (2).

Model ATT-S1, replacing the spatial attention in the cascaded dual attention mechanism with the spatial attention mechanism described in (2.1).

Model ATT-S2, replacing the spatial attention in the cascaded dual attention mechanism with the spatial attention mechanism described in (2.2).

Model ATT-S3, replacing the spatial attention in the cascaded dual attention mechanism with the spatial attention mechanism described in (2.3).

The model ATT-C1 is obtained by replacing the channel attention in the cascaded dual attention mechanism with the multi-head attention mechanism described in (1).

Channel attention and spatial attention are applied to the channel and spatial direction of the feature maps successively, so the attention that works later may affect the attention that works before. To explore if channel attention and spatial attention cooperate well, we propose the following models.

The model ATT-NS removes the spatial attention in the cascaded dual attention mechanism.

The model ATT-NC removes the channel attention in the cascaded dual attention mechanism.

The model ATT-C1-NC removes the spatial attention in ATT-C1.

2.3 Model training

For each model, the number of training iterations was set to 200, 000. Adam optimizer with parameters β_1 = 0.5 and β_2 = 0.999 was used for training. For the first 100, 000 iterations, the learning rate η was set to 0.0001 and fixed. For the last 100, 000 iterations, η was decreasing linearly from 0.0001 to 0. The image size used for training was 256×192.

2.4 Evaluation metrics

Three metrics, FID, SSIM, and PSNR were used for evaluation. FID was used to measure the similarity between two sets of images. The lower the FID value, the better. SSIM calculates the similarity of brightness, structure and contrast between the synthetic image and the label image with the value range from 0 to 1. The higher the SSIM value, the better. PSNR stands for peak signal-to-noise ratio, the bigger the better.

Testing the trained model on testing set would yield 2, 032 try-on result images. When the input person wears the same clothes as the input clothes, the test is a paired test, otherwise the test is an unpaired test. The FID is calculated using unpaired test results. The remaining two metrics are calculated using paired test results.

3 Results and Discussion

3.1 Spatial attention effect on virtual try-on

The experimental results of the effect of different spatial attention on virtual try-on were show in Tab. 1. Hu's model[5] has achieved the best results on the three metrics of FID, SSIM and PSNR, indicating that the spatial attention used in the cascaded dual attention mechanism is better than the three spatial attention mechanisms proposed in this paper.

Tab. 1 Spatial attention effect on virtual try-on

model	FID ↓	SSIM ↑	PSNR ↑
Hu's model[5]	18.6595	0.8233	23.5786
ATT-S1	19.3494	0.8218	23.4407
ATT-S2	19.6538	0.8231	23.5440
ATT-S3	18.9569	0.8222	23.5512

ATT-S1 and ATT-S2 did not achieve ideal results. The reason may be that when calculating the attention, the image was pooled into a low-resolution image first, then the attention calculated on the low-resolution image was upsampled to the original resolution, resulting some rectangular attention areas, which is not fine enough and was not suitable for high-resolution images. The horizontal attention and vertical attention in ATT-S3 are superimposed, and the attention also formed some rectangular areas, which is not fine enough, so the effect was not good.

3.2 Channel attention effect on virtual try-on

The experimental results for channel attention were shown in Tab. 2. According to the experimental results of ATT-ROM, ATT-NC and ATT-NS, the metrics results obtained by ATT-NC and Hu's model[5] were comparable, while the FID of ATT-NS was worse than the former two. The reason may be that the effect of the channel attention is overwritten by the stronger spatial attention. It seems like that channel attention and spatial attention may not be suitable for coexistence in the try-on module. Strong attention may cover weak attention and weak attention may affect strong attention leading to effect reduction.

Tab. 2 Channel attention effect on virtual try-on

Model	FID ↓	SSIM ↑	PSNR ↑
CPVTON[4]	27.1484	0.7871	21.3103
Hu's model[5]	18.6595	0.8233	23.5786
ATT-C1	18.3682	0.8237	23.6217
ATT-NS	19.5363	0.8229	23.5417
ATT-NC	18.6429	0.8223	23.4853
ATT-C1-NS	17.8969	0.8238	23.6381

The ATT-C1 model achieved better results than Hu's model in FID, SSIM, and PSNR, indicating that the multi-head channel attention mechanism proposed in this paper was better than the original channel attention. The multi-head channel attention maps the original features to multiple spaces, which is more comprehensive than the channel attention mapped to a single space, so the expressive ability is stronger, and the obtained attention is more accurate so that generating a better try-on result.

The ATT-C1-NS model achieved the best results on the three metrics, the FID was decreased by 4% compared to Hu' model, surpassing the ATT-C1 model, indicating that the multi-head channel attention mechanism does not cooperate well with spatial attention. Appending the spatial attention after the multi-head attention will reduce the model effect. Different layers of the feature maps are generated by different convolution kernels, representing different features. Channel attention gives different weights to these features, allowing them to be combined into more effective new features, while spatial attention weights feature maps in the spatial direction, which may change these original features, so that the weight generated by channel attention is no longer suitable, resulting in a decrease in the effect.

3.3 Qualitative results

Some target images, try-on results synthesized by Hu's model[5], and try-on results synthesized by the ATT-C1-NS model are shown in Fig. 6. The first column is the target image, the second column is the try-on results synthesized by Hu's model, and the third column is the try-on results synthesized by the ATT-C1-NS model. The try-on results synthesized by the try-on module using multi-head attention is better than the try-on module using the cascade dual attention in the overlapping areas of the clothes and the human body.

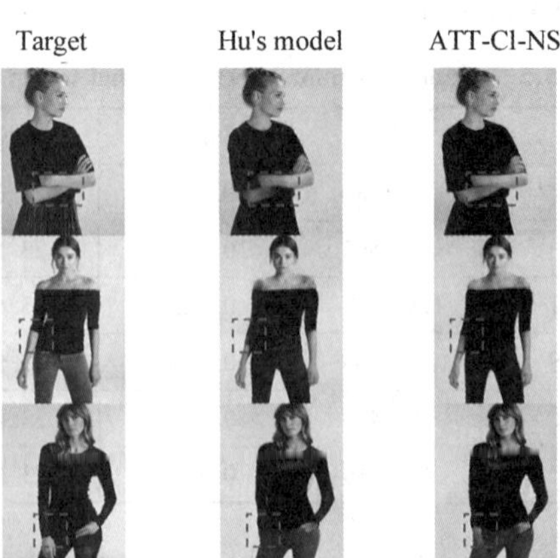

Fig. 6 Some try-on results

4 Conclusions

In this paper, three spatial attention mechanisms and one channel attention mechanism were proposed and introduced into the try-on result synthesis network of CPVTON, and five new try-on models are obtained. Compared with Hu's model, proposed ATT-C1-NS performs better on the three metrics of FID, SSIM and PSNR. When synthesizing try-on images with overlapping areas between clothing and human body, ATT-C1-NS performs better.

References

[1] EHARA J, SAITO H. Texture overlay for virtual clothing based on PCA of silhouettes[C]//IEEE/ACM International Symposium on Mixed and Augmented Reality, 2006.

[2] TANAKA H, SAITO H. Texture overlay onto flexible object with PCA of silhouettes and *k*-means method for search into database[C]//Proceedings of the 11th IAPR Conference on Machine Vision Applications, 2009.

[3] HAN X, WU Z, WU Z, et al. VITON: an image-based virtual try-on network[C]//IEEE/CVF Conference on Computer Vision and Pattern Recognition, 2018.

[4] WANG B, ZHENG H, LIANG X, CHEN Y, et al. Toward characteristic-preserving image-based virtual try-on network[M]. Cham: Computer Vision — ECCV 2018, 2018.

[5] HU X, ZHANG J, PENG T, et al. Cascaded cross-domain feature fusion for virtual try-on[J]. Journal of Computer Applications, 2022, 42(4): 1269-1274.

[6] VASWANI A, SHAZEER N, PARMAR N, et al. Attention is all you need[C]//Advances in Neural Information Processing Systems, 2017.

Design Management Process Within Micro and Small Textile Enterprises in Kenya

OMBURA Adonijah[1*], **VIKIRU George**[1], **ANDANGO Elizabeth**[1], **KINUTHIA Lydia**[2]

[1] *School of Creative Arts, Film and Media Studies, Kenyatta University, 43844-00100 Nairobi, Kenya*

[2] *Faculty of Education and Community Studies, Egerton University, 536-20115 Egerton, Kenya*

[*] *Lead and Corresponding author*: ombura. adonijah@ ku. ac. ke

Abstract: Design management is the link between design and business. The process covers the complex interaction of design and management factors. Design factors include "the Planning, the improvement of a product or creation of a new product while considering materials and technology for aesthetic and utilitarian product value". Management factors include "the process of planning, organization, and control of scarce resources to reach pre-established enterprise objectives". The design management process considers both internal and external parameters in the decision-making process. These parameters work together and can be explained by product design strategy framework and theories like the CToC and CK theory. A study of the Textile MSEs in Nairobi found that management strategies namely recognition, motivation, and teamwork were most preferred in product design and development, with respondents' top preferences of 78.4%. 77.8% and 77.8% for the respective variables. The decision on the product design implementation is left with the customers; this was mentioned by forty-one percent (41%) of those who responded. Findings should help Tex MSEs review their priorities in design management.

Keywords: Micro and small textile enterprises; Product design & development; Design management; Product innovation

1 Introduction

1.1 Background

Design management is the process that covers the complex interaction of design and business management factors. Design factors include the planning, improvement, or creation of a new product with materials and technology taken into account for aesthetic and utilitarian product value. Management factors include the process of planning, organising, and controlling scarce resources to achieve the enterprises pre-determined goals. The focus on design management efficacy in micro and small textile enterprises to develop successful product innovations is pertinent. An enterprise able to design its own product will have an advantage in a competitive market. In order to compete in the global marketplace, textile enterprises must be able to design and develop innovative textile goods using design, marketing, materials, and technology interfaces[1]. According to the management approach of product design, as well as the existing support structure and capacity, are critical for innovation. Considerations for management support include linkages, R&D investments, and internal capability-building[2].

1.2 Design management and product design process

The design management process takes into account both internal and external environmental enablers in order to complete a given design task and target. The acronym "PRODUCTS" represents the eight (8) product design and development work environment constructs, which is an interesting way of presenting these factors. P_ for people, R_ for resources, O_ for opposition (competitor), D_ for design, U_ for use, C_ for construction, T_ for technology, and S_ for society are the factors that Garrat (2008)[3] model identifies when designing a product.

2 Theoretic Frameworks

Product design and development consider the above mentioned internal and external environmental factors and are also guided by design and management theories like the Componential Theory of Creativity (CToC) and

Concept and Knowledge (CK) theory.

2.1 Componential theory of creativity

The Componential Theory of Creativity (CToC) was founded by Teresa Amabile to serve as a realistic illustration of the psychological requirements for producing creative work[4]. Four factors are required to boost enterprise innovation: creativity, motivation, expertise, and the social environment. The components are illustrated in Fig. 1.

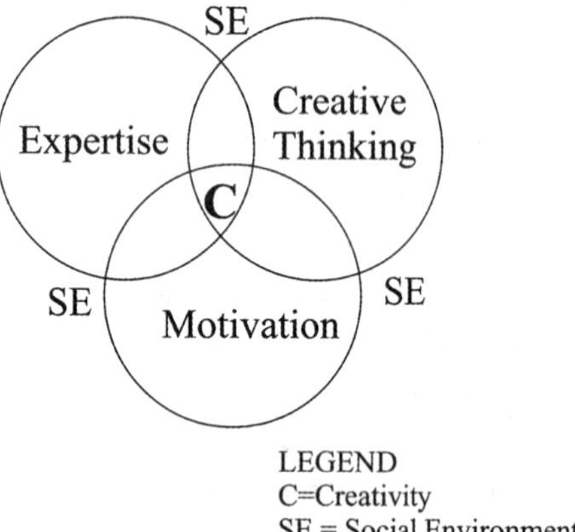

Fig. 1 Component model of creativity
Source: Amabile (1998)

The CToC stipulates that "creativity should be at its highest when an intrinsically motivated person with high domain expertise and high skills in creative thinking works in an environment high in support for creativity"[4]. It is a widely used theory to describe both the process of individual creativity and the process of organisational innovation, as well as the ways in which the two are connected through mutual influence[5]. To clarify the symbiotic relationship of the individual creativity aspect and the enterprise innovation in product design, the Amabile's component model of 1988 was further enhanced to show the interaction between creativity as spelt out in the CToC and innovation in a firm. Figure 2 depicts the symbiotic relationship between individual/group/team creativity and organizational/enterprise innovation, while the organisational work environment also influences individual/group/team creativity.

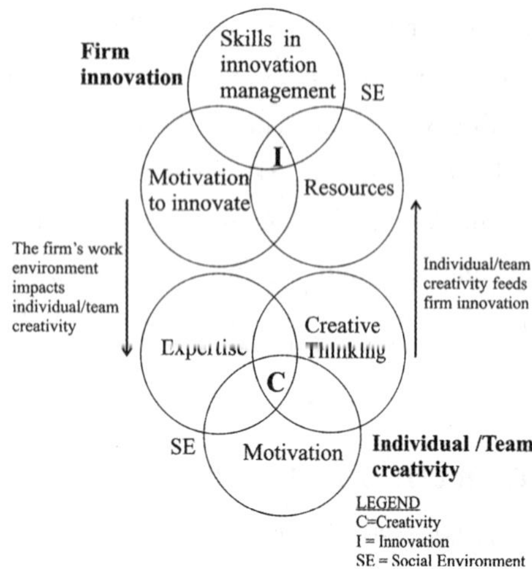

Fig. 2 Interaction between creativity and innovation
Source: Adapted from Amabile & Pratt (2016)

2.2 Concept-Knowledge theory

The C-K theory provides a formal basis for understanding existing design concepts and places creative thinking and innovation at the centre of the product design process[6]. The theory addresses two main spaces in the innovation process: the concept space (C) and the knowledge space (K). A 'design square' diagram depicts the interaction between the operators; these operators record the versatility of the innovation process as well as the range of design scenarios. It models the design dynamics as a joint expansion of a space of concepts (C) and a space of knowledge (K), employing four operators (C→K), (K→C), (C→C), (K→K)[7]. Concepts and knowledge interact for effective creativity, with concepts relying on aspects from the nominal world and knowledge relying on the logical/factual world[8] as shown in Fig. 3.

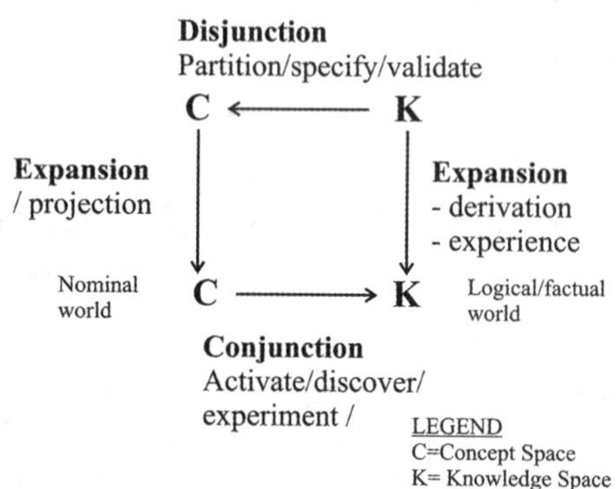

Fig. 3 The design square model of the C-K theory
Source: Shai, Reich, Hatchuel, & Subrahmanian, 2012

The (C→K) operator seeks to broaden a company's current available knowledge (K) through concepts (C) formed in the organization's external environment. Bringing together (conjunction) and activating existing concepts originating outside the company to generate new knowledge for product design is the operation. The (K→C) operator expands a firm's concept space (C) by bringing in external knowledge (K) elements. The operation includes knowledge disjunction and validation to be applied to concept development. If not done carefully (K→C), it may result in 'imitation' since it may involve the use of previously proven market/product concepts (designs); in other words, fixation on current product ideas. The internal operator (C →C) implies the partitioning or inclusion of existing internal concepts in the development of new enterprise design concepts. This could be a common approach for companies to develop new and distinct designs from internal concepts. The (K→K) operator enables the knowledge space to self-expand. Here internal enterprise knowledge is created through deduction and experimentation with existing knowledge. Knowledge stimulates innovation and is at the heart of the 'core competency' argument, which asserts that businesses must grow and manage their knowledge resources[9]. Enterprises that create and apply their knowledge and expertise have a better chance of establishing and maintaining a competitive advantage in their industry[9]. This suggests that textile MSEs must update their knowledge and skills in order to add value to existing products and develop new ones.

3 Results and Discussion

This section discusses some design management parameters based on data from a study titled "Product Design Strategies and their Influence on the Performance of Micro and Small Textile Enterprises in Nairobi City County, Kenya"[10].

3.1 Innovation orientation

In small businesses, the leader usually makes the majority of the key decisions in the innovation process. In the organization's approach to innovation, the leader can be very direct and pervasive[11]. In addition, the manager or owner is critical in creating an enabling environment for the product design process. The study uncovered the product design process status of textile MSEs (N = 153). The results revealed a high level of customer participation in the design process. The majority of key decisions in the innovation process are made by the leaders of many small businesses. According to a study on textile MSEs in Nairobi, the enterprise's product design decisions are made by the customer (41%), the owner (35%), the designers (17%), the supervisor (5%), and the artisan/employee (2%).

3.2 Knowledge management

A company can gain a competitive advantage by implementing sound knowledge management, which includes information acquisition, dissemination, and reaction, as well as a market orientation that prioritizes the customer and competitors[12]. Greater ability to integrate external knowledge and apply it for commercial purposes strengthens the relationship between customer participation and new product performance[13]. Both managers and employees of the textile MSEs approved efforts to improve design knowledge in their enterprise.

3.3 Skill/Expertise

According to the findings of the Textile MSE study in Nairobi, hiring competent individuals and enhancing in-house capabilities was a top priority for Textile MSEs. Improving design and production skills is important; however, due to human resource and financial constraints, many small businesses are unable to benefit from thorough training. The results revealed a high score of 94% who encouraged design skill upgrade in the enterprise.

3.4 Motivation

Amabile (2012) recognized that motivation is a crucial factor for creativity as stipulated in the Componential Theory of Creativity. Apart from Teresa Amabile's theory, several theories support the need for motivation in a firm, including the following two: Herzberg's Two-Factor Theory states that intrinsic motivators tend to create motivation when they are present, whereas the absence of extrinsic motivators tends to reduce motivation in the firm[14]. The other theory, the Equity Theory of Motivation, created by J. Stacey Adams in the early 1960s, recognizes that motivation can be influenced by an individual's view of fair treatment in an environment.

3.5 Social Environment

To produce a suitable product for the client/market, the product design and innovation process begins with a situation/market analysis. The market study includes a self-evaluation comparison with outstanding performers in the same trade (benchmarking). This market orientation requires awareness to competitor and customer dynamics, which should generate extrinsic motivation to drive a firm's innovation agenda. The textile product designer analyses the needs of the consumer in a subtle way, establishing a harmony valued by the customer while also establishing unique and functional product attributes. Association with outsiders or external environment participants is also a useful method for MSEs to gain necessary resources. Creativity and innovation procedure should be ethical, follow all regulatory rules of the industry and country, and also be attentive to societal trends. The customer was the main source of the design idea (85.6%) and it was noted that the textile MSEs had more information about their customers than they did about their competitors.

3.6 Internal work environment

The work environment is another pertinent factor considered in design management. To achieve their goals, enterprises must have appropriate resources. Most of the time older businesses can innovate more effectively than newer ones because they have had time to establish the knowledge, resources, and relationships that favourably impact product innovation[15]. The study done on the textile MSEs in Nairobi checked the work environment variables of formality of operations; communication; teamwork; delegation of authority; motivation; and recognition and revealed that of these variables, recognition, motivation, and teamwork were the most prevalent. The three variables had relatively similar scores of 78.4%. 77.8% and 77.8%, supported by 120, 119, and 119 textile MSEs, respectively.

4 Conclusions

This study has focused on the important component of design management for effective textile MSE performance. This paper illustrated with the guidance of the CToC and C-K theories on how internal and external environmental variables are critical for creativity and innovation in a textile enterprise. A recipe for successful textile product design and development must include the essential variables of innovation orientation, knowledge management, skill and expertise upgrade, motivation, and social environment aspects. All of which need to work together in a favourable internal work environment. It is reasonable to conclude that creativity and innovation are critical processes in product design and design management that demand the coordination of personality, knowledge, thinking styles, intellectual capacities, and motivation in a conducive environment.[16].

References

[1] LOTTERSBERGER A. Design, Innovation and Competitiveness in the Textile Industry [D]. Unpublished PhD Thesis, Politecnico Di Milano, Italy.

[2] LAWAL F A, WORLU R E, AYOADE O E. Critical success factors for sustainable entrepreneurship in SMEs: Nigerian perspective [J]. Mediterranean Journal of Social Sciences, 2016, 7(3 S1): 338.

[3] GARRATT J. Design and technology [M]. Cambridge: Cambridge University Press, 2008.

[4] AMABILE T M. Componential theory of creativity [Harvard Business School Working Paper No. 12-096] [J]. Encyclopedia of management theory, Cambridge, MA: Sage, doi, 10(9781452276090), n42.

[5] AMABILE T M, PRATT M G. The dynamic componential model of creativity and innovation in organizations: making progress, making meaning[J]. Research in Organizational Behavior, 2016,36, 157-183.

[6] HATCHUEL A, MASSON P L, WEIL B. CK theory in practice: lessons from industrial applications [C]. Paper presented at the 8th International Design Conference, Dubrovnik, Croatia, 2004: 245-258.

[7] HATCHUEL A, WEIL B. A new approach of innovative design: an introduction to C-K theory[C]. Paper presented at the 14th International Conference on Engineering Design, Design Society. Stockholm.

[8] SHAI O, REICH Y, HATCHUEL A, et al. Creativity and scientific discovery with infused design and its analysis with C-K theory [J]. Research in Engineering Design, 24(2), 201-214. https://doi.org/10.1007/s00163-012-0137-x

[9] BESSANT J R. High-involvement innovation: building and sustaining competitive advantage through continuous change[J]. John Wiley & Sons Incorporated.

[10] OMBURA A. Product design strategies and their influence on the performance of micro and small textile enterprises in nairobi city county [D]. Kenya: Kenyatta University, 2022.

[11] VAGNANI G, VOLPE L. Innovation attributes and managers' decisions about the adoption of innovations in organizations: a meta-analytical review [J]. International Journal of Innovation Studies, 2017, 1(2), 107-133.

[12] KAMYASUP M T, NTAYISUP J M, AHIAUZUSUP A. Knowledge management and competitive advantage: the interaction effect of market orientation[J]. African Journal of Business Management, 2010, 4(14), 2971-2980.

[13] MORGAN T, OBAL M, ANOKHIN S. Customer participation and new product performance: towards the understanding of the mechanisms and key contingencies [J]. Research Policy, 2010, 47(2), 498-510.

[14] GODSPOWER D. Theories of motivation[J]. Academia. edu — Share research, 2015, https://www.academia.edu/13445161/Theories_of_Motivation

[15] JAJJA M S S, KANNAN V R, BRAH S A, et al. Linkages between firm innovation strategy, suppliers, product innovation, and business performance: Insights from resource dependence theory[J]. International Journal of Operations & Production Management, 2017.

[16] STERNBERG R J. The assessment of creativity: An investment-based approach[J]. Creativity Research Journal, 24(1): 3-12. https://doi.org/10.1080/10400419.2012.652925.

Application of Dyed Valerian Apocynum in Cultural and Creative Products from the Perspective of the Silk Road

NI Yu[1], LI Bowen[1], QI Huan[1,2*]

[1] *College of Textiles and Apparel, Quanzhou Normal University, Quanzhou 362000, China*

[2] *Ecological Inelligent Fabric Engineering Pechnology Research Institute of Shishi, Quanzhou 362000, China*

* *Corresponding author's email*: qhqh123@126.com

Abstract: As a traditional intangible cultural heritage craft loved by local people, Anxi indigo dyeing is a testimony to the existence of cultures along the Silk Road. Apocynum is a local characteristic crop with health benefits in Xinjiang. Based on literature research and other materials, the article analyzes the performance of apocynum, and sorts out the historical changes of Anxi indigo dyeing, aiming to better understand the essence of dyeing techniques, so as to provide dyeing apocynum products and modern cultural and creative products. Combine to provide a source of inspiration.

Keywords: Silk Road; Anxi indigo dyeing; Apocynum; Cultural and creative design

1 Introduction

Since its opening in Qin and Han dynasties, the "Silk Road" has been a bridge for economic and cultural exchanges between ancient China and the outside world. It has promoted the collision and integration of technology, culture and art among countries along the route, and promoted the progress of world civilization.

Apocynum is a perennial wild plant along the Silk Road. Apocynum has abundant health care functions, and the fabric made of apocynum fiber has the functions of breathable antibacterial, far-infrared, and anti-ultraviolet. During the market research, it is found that the overall style of hemp cultural and creative products on the market is single, and the phenomenon of homogeneity in style is serious. The author believes that combining the product with other intangible cultural heritage can make the product more attractive and connotative. Anxi indigo dyeing is also an endangered traditional handicraft along the Silk Road. The combination of apocynum and Anxi indigo can not only increase the cultural attributes of the product, but also expand its functionality to achieve mutual benefit and win-win results.

2 Apocynum

Apocynum(Fig. 1) is a high-grade fiber plant native to Bazhou, Xinjiang. The fabric made of apocynum fiber has the characteristics of moisture absorption, breathability, antibacterial, far-infrared, anti-ultraviolet, etc. It is suitable for long-term wear. However, there are few apocynum fabrics on the market at present, mainly due to the complex degumming process of apocynum, which easily leads to unclean degumming, and the lack of innovation in related product design and low cultural value.

Fig. 1 Xinjiang Apocynum plants

Nowadays, with the improvement of living standards, people's demand for environmentally friendly and healthy fabrics has further increased, and various properties of apocynum just meet people's expectations, which also provides a broader market foundation for the development of apocynum. In order to lay stress on in the torrent of the times, apocynum must continue to innovate itself to create apocynum products with better performance, practicality and beauty.

China has a tradition of dyeing fabrics with natural dyes since ancient times. Many plant dyes themselves are Chinese herbal medicines, which have health care and medicinal functions. During the dyeing process, the active ingredients of natural dyes and natural pigments are absorbed by the fabric together, which makes the dyed fabric have a certain antibacterial effect[1]. The author's experimental conclusion in the article "Nectarine Dyeing and Properties of Apocynum/Cotton Knitted Fabrics" shows that the apocynum fabrics dyed with nectarine have better far-infrared properties and antibacterial properties. Among them, Candida albicans and Escherichia coli The bacteriostatic rate of bacteria increased from 38% and 36% before staining to 84% and 78% after staining, respectively. Apocynum fabrics dyed with Chinese herbal medicines are expected to achieve full-spectrum antibacterials and contribute to the development of textiles in the medical industries[2].

3 Anxi Blue Dye

3.1 Overview of Anxi indigo dyeing techniques

Anxi blue dye, also known as Fujian Qing, is a printing and dyeing product made by extracting dyes from natural plants and using traditional anti-dyeing techniques. During Song and Yuan Dynasties, cotton in southern China was spread through the Maritime Silk Road, which truly realized the world of clothing and quilts, and successfully changed the clothing of the bottom people from hemp to cotton[3]. The carrier of Anxi blue calico is mainly linen and cotton. The "green" of Fujian green refers to blue with green components. Although "green" is one of the five major colors, it was not considered a noble color in all dynasties.

Zhu Xi, built and deepened most of his theoretical achievements in the Quanzhou area. Popularity also affected the aesthetics of people at that time. In addition, the imperial court's stricter control over the "ban on clothing color" made indigo dyed fabrics be loved by the local people.

3.2 Analysis of current situation of Anxi indigo dyeing

Anxi was called Blue Stream in ancient times, because there was a river in Anxi that rinses and dyes indigo cloth all the year round, thus presenting a spectacular scene of "the stream is extremely blue". In the early Qing Dynasty, the printing and dyeing skills of Anxi blue calico had become mature[4]. The above historical data show that Anxi indigo dyeing has reached its heyday in Ming and Qing Dynasties.

In modern times, Anxi indigo dyeing(Fig. 2), like the hand-printing and dyeing techniques in other regions, began to be impacted by modern Western machine production and new chemical dyes. Anxi indigo dyeing entered a stagnant stage of development. In 1995, Mr. Huang Jiongran, who worked in the Anxi Museum, carried out rescue protection for Anxi indigo dyeing, and restored the ancient Anxi indigo dyeing process. In 2005, Anxi blue calico was listed as the first batch of intangible cultural heritage in Fujian Province. In 2010, Mr. Huang passed away, and Anxi Indigo has no successor and is on the verge of being lost again. In 2016, Mr. Chen Guorui, who was teaching in Anxi Overseas Chinese Vocational School, conducted field inspections on blue calico in various places, and established a blue calico studio to guide students to inherit and protect traditional skills. In general, the status of Anxi Indigo's inheritance is not good. It takes the joint efforts of generations to make Anxi Blue enter the lives of more people.

Fig. 2 Anxi indigo dye

3.3 Analysis of reasons for decline of Anxi blue dye

The main reasons for the stagnation of the development of Anxi blue dyeing in Fujian are as follows. Firstly, the local government does not pay enough attention to protection. As a provincial-level intangible cultural heritage project, Anxi indigo dyeing is very rare. In addition, the blue-dyed patterns of Anxi are relatively simple[5]. Today, some blue calico patterns in Anxi are mainly traditional auspicious patterns in ancient times. The color and composition of the patterns are too simple and cannot meet the diverse aesthetics and needs of current consumers. The third is failure to integrate with local characteristic industries. In 2021, Quanzhou successfully applied for the World Cultural Heritage, and the development of Quanzhou's tourism industry in the next few years will attract much attention. Traditional Anxi indigo dyeing is mainly used in clothes, sheets, mattresses, door curtains, and other items. The people who buy such items are relatively limited, and most of them are local residents. If Anxi Indigo wants to develop further, it must seek new breakthroughs. The first solution is to solve the problem that the target consumer group is relatively limited. Therefore, how to combine the traditional craftsmanship of Anxi indigo dyeing with tourism products to make creative products and generate economic benefits is a problem that we must consider.

3.4 Analysis of traditional craftsmanship of Anxi indigo dyeing

In Anxi indigo dyeing, the selection of plant dyes is relatively simple, generally using indigo, and the main techniques are wax, clip and twist.

3.4.1 Wax valerian

The wax valerian in Anxi indigo dyeing refers to a traditional printing and dyeing process that uses wax as a resist (Fig. 3). Common types of wax are beeswax, wood wax and paraffin. A mixture of beeswax and paraffin is more commonly used in batik today. Batik mainly uses the waterproof of wax. The main operation method is to first draw a draft on the fabric with a removable pen, then use wax to draw a specific pattern on the fabric, and immerse it in the dyeing solution. The dye solution cannot be immersed in the wax-coated part, but can only produce a dyeing effect on the unwaxed part. After the wax is removed by boiling at high temperature, the undyed ground color pattern will appear on the fabric.

Fig. 3 Utensils needed for batik: including wax knife and beeswax

3.4.2 Clipping

The principle of clipping in Anxi indigo dyeing is to use anti-staining agent to make pre-designed patterns appear on the fabric. The main method is to sandwich the fabric between two flower plates with the same pattern that has been coated with anti-staining agent, and then immerse it in the vegetable dye solution, so that the concave part of the flower plate can fully contact the dye and dye it successful(Fig. 4). However, the protruding part will be hindered by the pattern, and the undyed fabric will appear on the fabric.

Fig. 4 Utensils needed for clipping: flower plates

3.4.3 Twisted valerian

In Anxi indigo dyeing, the twisted valerian is tie-dyeing. There are three main methods of twisting valerian: bundling, sewing and interleaving. The beauty of tie-dyeing is reflected in its randomness. The technological process of tie-dyeing is mainly divided into three types according to the different operation methods. One is to wrap hard fillers such as walnuts in the fabric with the fabric to be dyed, and then sew with good tenacity cotton thread, and then immerse it in the vegetable dye solution. Remove the stitches after dyeing, so that the tie will form a gradient of color halo. The second is to dye the fabric to be dyed by twisting it into "horns". The dyeing principle is the same, mainly using the dye resistance of the ligation part to the fabric. The third is to use wooden clips, bamboó chopsticks and other small tools to assist the dyeing of the fabric to be dyed, and the patterns dyed with the help of Zsmall tools are generally more regular(Fig. 5).

Fig. 5 Utensils needed for twisted valerian: wrap the walnut with fabric and wrap it with thin thread to achieve the effect of anti-staining

3.5 Anxi indigo dyeing innovative process exploration

Compared with wax valerian and clip valerian, the beauty of twisted valerian is that it is formed naturally without deliberate design. The part of traditional dyeing that needs to be inherited and carried forward. As for the traditional wax valerian and twisted valerian, the extraction and processing of its composition and image are more meaningful. Therefore, the author believes that the use of digital printing to simulate the more complex wax and valerian patterns in traditional dyeing in design works, combined with the hand-made twisted valerian patterns that cannot be reproduced by modern machines, can alleviate the industrialization crisis faced by traditional printing and dyeing to a certain extent. On the other hand, the essence of traditional craftsmanship can be preserved.

4 Innovative Application of Dyed Valerian Hemp in Cultural and Creative Products

As a special handicraft in Fujian Province, Anxi indigo dyeing has strong regional characteristics. The single tone and simple pattern of Anxi indigo dye can no longer meet people's diverse aesthetic needs. With the success of Quanzhou's application for a world cultural heritage, the market's demand for cultural and creative products has reached an unprecedented new height. However, the tourism products on the market are mixed. Through the investigation of cultural and creative stores in Quanzhou, the author found that the phenomenon of commodity homogeneity is very serious. Therefore, the author launched a redesign of cultural and creative products around the Silk Road culture.

4.1 Design inspiration and sources

The author extracted the local bamboo shoot jelly and batter from Quanzhou, egg and waffle ice cream from Kuala Lumpur, etc., and redesigned them in the way of watercolor hand-painting. The fabric is made of apocynum fabric with antibacterial far-infrared function, and designed a series of cultural and creative products. The author hopes that more people will pay attention to the intangible cultural heritage, pay attention to the Silk Road culture, and be able to participate in the innovation of the Silk Road culture. (Fig. 6).

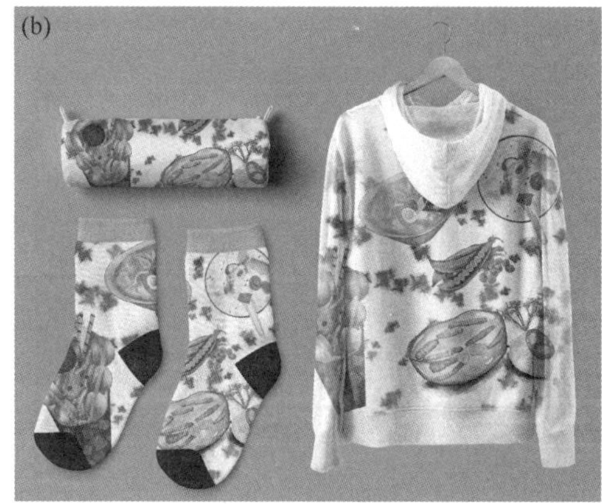

Fig. 6 (a) Innovative indigo pattern design; (b) innovative indigo products(author's own painting)

5 Conclusions

Anxi indigo dyeing is an important treasure in Silk Road culture, and it also embodies the wisdom of the vast number of craftsmen at that time. Apocynum is a characteristic plant from Lop Nur, the key node of the Silk Road. Its significance in Silk Road culture is not only that it can be made into an antibacterial far-infrared health care fabric, but also a great fabric those who have experienced the Silk Road culture. As a designer, paying attention to the use of Anxi indigo dye and apocynum can better preserve the Silk Road culture in the contemporary era, and further trigger people's thinking on the innovative application of intangible cultural heritage, interpret national culture, and enhance people's nationality.

References

[1] HAN J, ZHENG L, SUN Y. Application progress of antibacterial natural dyes in textile dyeing[J]. Textile Industry and Technology, 2020, 49(12): 19-21,24.

[2] NI Y, LI X Q, QI H. Dyeing and properties of nectarine red pigment of apocynum/cotton knitted fabrics[J]. Printing and Dyeing, 2021, 47(10): 25-29.

[3] ZHAO H Y, HU H J L, GUO K W. Research on cotton propagation in South China from the perspective of the Maritime Silk Road[J]. Silk, 2019, 56(8): 99-105.

[4] ZHUANG C. Anxi county chronicle[M]. Xiamen: Xiamen University Press, 2012.

[5] LEI Y, WANG Y L, TANG G. Deconstruction and remodeling of village cultural space: taking Anxi Lanyan as an example[J]. Urban Architecture, 2020.

Study on the Method of Fabric Damage Measurement

GU Xianyu[1,2], CUI Lina[1,2], QIU Yiping[1,2,3*]

[1] *College of Textiles and Apparel, Quanzhou Normal University, Quanzhou 362000, China*

[2] *Ecological Intelligent Weaving Engineering Research Institute of Quanzhou Normal University (Shishi), Quanzhou 362000, China*

[3] *College of Textiles, Donghua University, Shanghai 201620, China*

* *Corresponding author's email*: ypqiu@ dhu. edu. cn

Abstract: Sewing damage and fabric damage are unacceptable in the garment industry. In order to detect the damage of fabrics, a testing device for fabric damage is built in this study. The device can measure the luminous flux of fabrics to determine the damage of fabrics. Copper ammonia fabric is a kind of filament fabric with obvious sewing damage. In order to verify the accuracy of equipment detection, this study designed a group of sewing experiments in combination with the garment manufacturing process parameters, and detected the change of light transmission area of sewing damage before and after plasma treatment of copper ammonia fabric, and quantitatively analyzed the improvement of damage. Through the comparison of experimental tests and calculation analysis, it is proved that the experimental device can detect more accurate sewing damage.

Keywords: Fabric; Sewing damage; Measurement method

1 Introduction

The sewing damage of fabric refers to the damage caused by the sewing machine needle to the fabric during operation. The sewing damage of woven fabric is still a common problem in the factory at present. Improving the sewing damage can improve the quality of products. How to evaluate whether the sewing damage of fabrics has been improved needs to develop a suitable test method. The former is the test method used by Wu Jiaju, that is, to determine the relationship between the strength of needle puncture and the number of needle holes in the sewing track during sewing, so as to obtain the correlation between the puncture strength and sewing damage[1]. Shi Jianguo, the former, pointed out that the best way to improve the sewing damage is to increase the sliding performance of the yarn. The stitchability of the fabric after sandolube finishing has been greatly improved[2].

Currently, a fabric defect picture marking auxiliary system is installed on the cloth inspection machine. The cloth on the cloth inspection machine moves horizontally, and the camera directly above the cloth inspection surface can stably and quickly take photos of defect points for storage or uploading[3]. A fabric defect detection table, which places the fabric on the detection board, can increase the brightness through the light strips on both sides of the table during the detection, so that the detection workers can better complete the fabric detection[4]. Based on the improved cascade r-cnn fabric defect detection method, according to the uneven distribution of defect types, some defects have extreme aspect ratio and many small targets, the prediction model is improved to make the detection effect better[5-6]. There are also invention patents that can detect defects on the fabric, and use the image acquisition method to collect defects on the fabric and form detection results to improve the detection efficiency[7-8].

2 Device Design and Construction

The device design mainly uses the transmission principle of light to test the sewing damage of fabrics. In a closed photographic frame, a diffuser is placed between the uniform LED light panel and the sun visor to make the light more evenly distributed. The white light can pass through the diffuser and then pass through the fabric. When the light passes through the damaged part, it can completely penetrate, so the damaged part can be clearly seen. Since the damage of sewing is basically around the thread track, the research is mainly focused on the damaged part of the fabric. Therefore, a light transmission test window with adjustable size is set, as shown in Fig. 1, which is the schematic diagram of light transmission.

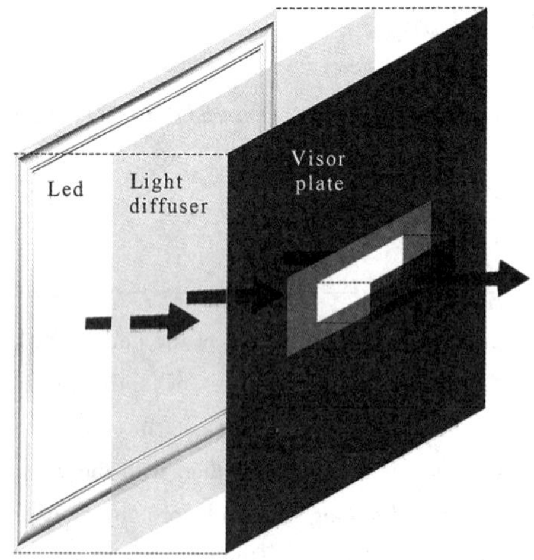

Fig. 1 Diagram of light transmission

The camera frame of 60×60×60 size is connected to the power supply. The back of the camera frame is close to the LED lamp, the LED lamp is close to the diffuser, and the sunshade is placed 10 cm away from the diffuser. Place the fabric to be captured on the test window, close the surrounding to prevent reflection, and use Canon es-68 camera to sample the fabric respectively. Then the image is input into the computer to calculate the area of the fabric surface damage, to judge whether the fabric sewing damage has been improved. The equipment structure is shown in Fig. 2.

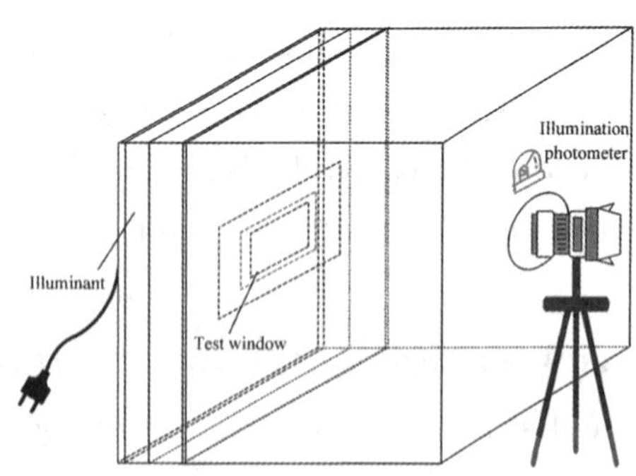

Fig. 2 Equipment structure

3 Sample Preparation

In order to accurately test the damage of fabrics after sewing, the sewing effects of untreated fabrics and a series of plasma treated fabrics were designed to test and compare the sewing damage of the two groups. Since most of the damage lengths are no more than 2 cm, the designed actual test window size is 8 cm×2 cm.

Prepare standard samples (4×9) and warp cut 18 pieces of fabrics, ja001, ja002 and ja003. The needle pitch is 1, 2, and 3 mm, respectively. By analogy, the results after plasma treatment at different times are shown in Tab. 1.

Tab. 1 Fabric samples

Needle pitch	1 mm	2 mm	3 mm
Untreated	Ja001	Ja002	Ja003
Plasma	Jb151	Jb152	Jb153
—	Jb301	Jb302	Jb303
—	Jb451	Jb452	Jb453
—	Jb601	Jb602	Jb603
—	Jb751	Jb752	Jb753

4 Test

4.1 Image acquisition and analysis

When collecting images, place the camera at the center in front of the equipment, keep the distance unchanged, adjust the pixel and focal length, and keep it unchanged. Stick the fabric on the panel at a certain distance from the lamp while keeping the fabric fit. The camera box can realize a light environment with uniform light and stable color temperature. The camera used is Canon es-68 camera, which is equipped with 50 mm lens to ensure high-quality images. The fabrics before and after plasma treatment were placed in the self-made photographic frame. Import the collected images into imageproplus to measure the damage area.

4.2 Damage area ratio

The light transmittance without plasma treatment is S_0, the light transmittance after plasma treatment is S_P, and then the calculation formula of improvement degree is

$$\frac{S_0 - S_P}{S_0} \times 100\%$$

5 Results and Discussion

As shown in the Fig. 3, compare the damage areas of ja001, ja002, and ja003 with those after plasma treatment, and discuss the sewing damage after plasma treatment of different groups. As shown in the figure, compare the sample without plasma treatment, such as the sample without plasma treatment in Figs. 3 (a)-(c) with the sample after plasma treatment in Figs.

3 (d)-(f), and calculate the damage area and damage degree.

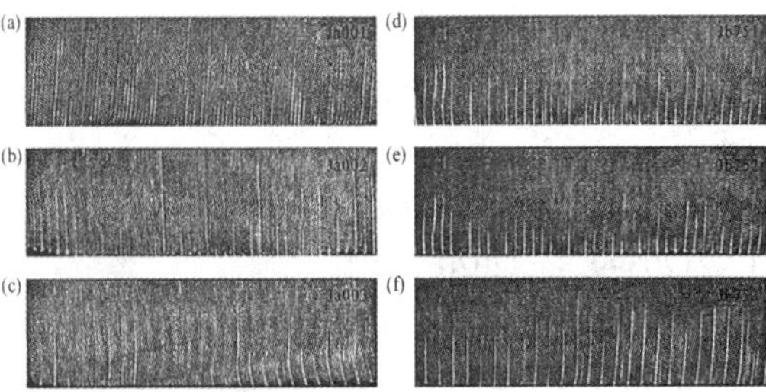

Fig. 3 Comparison between untreated and plasma treated samples

5.1 Analysis of damage area data results

See Tab. 2 for damage area and improvement degree of fabric samples before and after plasma treatment. ja001, ja002 and ja003 represent the damage areas without plasma treatment, which are compared with other damage areas after plasma treatment at different times. The difference of comparison is divided by the improvement degree of the damage areas of ja001, ja002 and ja003.

Tab. 2 Damage area and improvement of fabric samples

Number	Fabric damage area(cm^2)	Improvement(%)
Ja001	1.08	—
Ja002	0.72	—
Ja003	0.90	—
Jb151	0.88	1.85
Jb152	0.40	44.44
Jb153	0.35	61.11
Jb301	0.66	38.89
Jb302	0.46	36.11
Jb303	0.48	46.66
Jb451	0.53	50.92
Jb452	0.26	63.88
Jb453	0.24	73.33
Jb601	0.25	76.85
Jb602	0.22	69.44
Jb603	0.22	75.55
Jb751	0.31	71.29
Jb752	0.22	69.44
Jb753	0.40	55.55

Tab. 2 shows the comparison and improvement degree of the damage area between the non-plasma treated group and the plasma treated group after sewing for different times. The damage area of fabrics without plasma treatment is large. When the treatment time increases to 30 s and 45 s, the damage area is significantly reduced. When the treatment time reaches 60 s, the damage area reaches the minimum. When the processing time reaches 75 s, the damage area starts to increase due to excessive etching, as shown in Fig. 4.

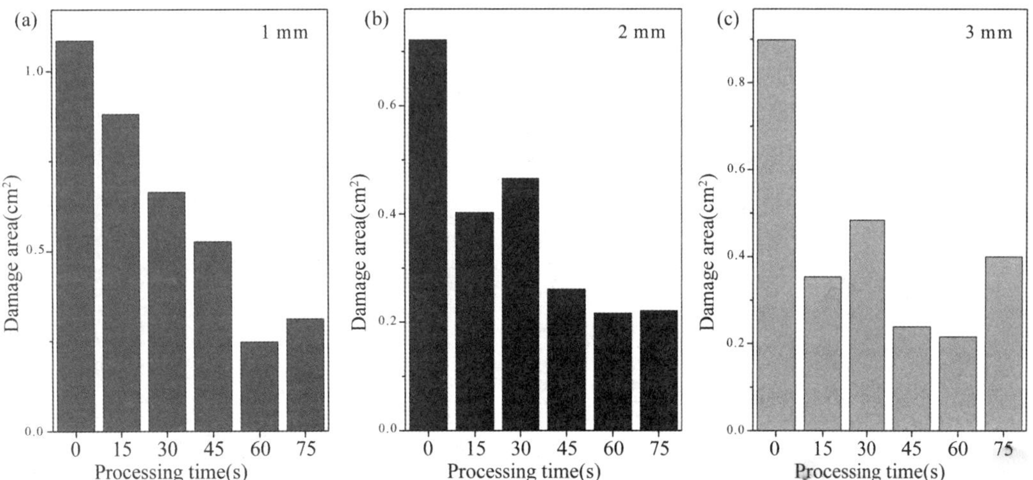

Fig. 4 Damage area distribution at different plasma treatment times: (a) 1 mm needle pitch; (b) 2 mm needle pitch; (c) 3 mm needle pitch

The etching phenomenon on the fiber surface produced by the plasma treatment process increases the roughness of the fabric and increases the friction force between the fabrics, to reduce the sewing damage. The damage improvement effect is the best when the treatment time is 60 s. When the sewing needle distance is larger, the damage area is almost smaller.

6 Conclusions

In conclusion, the damage area of fabric can be accurately calculated by using the image method. Therefore, this measuring device can accurately judge the degree of fabric damage and measure the damage improvement effect. The method of image measurement is effective and feasible. After plasma treatment, it has obvious improvement effect. Therefore, it can be used in industrial production to accurately measure the degree of fabric damage. From the data results of the damage area of the fabric tested by the above equipment, it can be seen that the accuracy of the measuring equipment is high, and the method of borrowing image test is effective and feasible.

References

[1] WU J J. Sewing damage and sewing property test of knitted fabrics[J]. Shanghai Textile Science and Technology, 1983 (10): 50-53.

[2] SHI J G. Sewing injury and its prevention[J]. Foreign Textile Technology (Knitting and Clothing), 1988 (15): 41-43.

[3] SUN L, GU M F, WU C R, Fabric defect picture marking auxiliary system[P]. China: CN208224110u, 2018-12-11.

[4] YU Z X. A fabric defect detection table[P] China: CN212059953u, 2020-12-1.

[5] XU S B, ZHENG Y M, YUAN D C. Fabric defect detection method based on improved cascade r-cnn[J/OL]. Modern Textile Technology: 1-9[2022-01-11].

[6] AN M. Design and implementation of fabric intelligent detection vision system for garment flexible production line[D]. Shenyang: University of Chinese Academy of Sciences, 2021.

[7] QI W B, ZHENG D Q, LI Y. A fabric defect detection system convenient for rapid positioning[P]. China: CN113218964a, 2021-8-6.

[8] QI W B, ZHENG D Q, LI G. Control system and method for fabric defect detection[P]. China: CN113325003a, 2021-8-31.

Development of Medical Textile Products Focused on the Ugandan Market

KIIZA Solomon[1*], ASHABA Osbert[1], MOUREEN Namiiro[1]

[1] *Product Development Directorate, Department of Textile Materials and Polymers, Uganda Industrial Research Institute (UIRI); Nakawa Industrial Area, P. O Box, 7086, Kampala.*

* *Corresponding author's email*: solomonkiizabakora19@gmail.com.

Abstract: This study offers a way forward for the utilization of the resources available in Uganda to fabricate personal protective equipment (PPE); a class of garments or equipment that reduces the risk of the wearer from getting injured or contracting an infection.

Uganda has already experienced bouts of infectious outbreaks, the most recent of which have been Ebola and COVID-19. During such outbreaks, the nation has always struggled to protect its people and especially the frontline health workers. The project targets to fulfill the PPE needs of the Ugandan health sector, schools, institutions, the general public and the armed forces. Implementation of the study's findings will help the country prepare for similar outbreaks in the future.

Accordingly, a variety of products key to the health workers attending to their work at the frontline have been designed and fabricated at the UIRI Textile Workshop and their properties are characterized at the institute's textile materials and polymer testing laboratory. The products focused on were disposable coverall (Hazmat suit), disposable shoe covers, and disposable head covers.

During the analysis, due attention was focused on compliance with the international standards. Accordingly, standards relevant to the given products were considered. These standards included ASTM D5034 (tensile strength), ASTM D4966 (abrasion resistance), ASTM D3786 (bursting strength), ASTM D1388 (stiffness), ISO 16603 (synthetic blood penetration of the fabric material and seams), and EN-ISO 7854 (30% Sulphuric acid and 10% Sodium Hydroxide penetration).

Keywords: Medical textiles; Ugandan medical textile market

1 Introduction

Uganda has already experienced spells of infectious outbreaks and is currently battling a global infectious pandemic. Now is the time to ensure that most of the vital PPE used in safeguarding those at the forefront of responding to such outbreaks are produced within the country.

In a move to develop local capacity, a team of experts from Uganda industrial research institute department of textile materials and polymers researched and developed a number of PPEs using available technology and materials. The developed PPEs include disposable coverall (Hazmat suit), disposable shoe covers, and disposable head covers. This was aimed at enabling the nation set-up self-reliant PPE production facilities, ensuring that the Ugandan population can have access to affordable, yet good quality PPE in times of need, and minimize healthcare workers' risk of exposure during infectious outbreaks. An overall goal was to develop our technological capabilities by utilizing the locally available raw materials.

The study focused on innovating PPEs that are needed by Ugandan health sector, schools and institutions, the general public and the armed forces.

In the wake of pandemic ravaging the whole globe, the initial cases of this disease in Uganda were imported into the country by returnees from abroad. Prior to this, the country had experienced Ebola and Marburg hemorrhagic fever outbreaks[1, 2]. This reminds us of the vulnerability we face as a country to the outbreaks of the infectious diseases. These are often very deadly. The Zaire *ebolavirus* had fatality of 90%, the Sudan ebolavirus with fatality of 50%-55% and the Bundibugyo ebolavirus at 34%[3]. In addition, they tend to spread fast and through channels that are very hard to control, putting the lives of medical workers, care takers and the general public in danger. For the pandemic that the country is currently facing, it is transmitted by skin-to-skin contact with an infected person, touching contaminated items, or by droplets of a patient's cough, sneeze and saliva. Despite these very challenging transmission routes, when new cases are uncovered, family members and doctors have to take care of the victims. In the course of their work, PPE become their last line of defense, making them very essential for survival. Regarding COVID-19's risk of contraction and its prevention, PPE constitute respirators, gloves, face shields, shoe covers, head covers and gowns. In addition to protecting handlers from contracting the infection, some of the PPE such as the masks go an extra notch, reducing the transmissibility from patients to healthcare workers. All these factors make PPE such a treasured asset in the fight against pandemic and other similar infections.

In times of outbreaks, healthcare workers (HCWs) work around the clock to save our lives and limit casualties. Thus, an infection to just one HCW would have a dramatic effect on the whole health care system. However, the likelihood of a HCW becoming infected has been reported to be more than three times that of the general population[4]. If not well equipped with the right PPE, work becomes too risky.

Their selfless commitments can only be rewarded by maximally providing for their safety during work. This is an ethical commitment both in terms of addressing the occupational risks they face and providing for the continuity of patient care and services needed to lower the infections. Additionally, if a pandemic is not appropriately managed, it plunges the country into economic problems. This is manifested by the effects of the lockdown on Uganda's economy[5].

When the frontline workers are fully equipped with PPE, they are confident about their personal security that boosts their morale enabling them to work around the clock to save lives. However, it has been revealed that HCWs in the country are poorly equipped with PPE that is frightening amidst the rising cases of pandemic in Uganda, and with the infections beginning to spread among the frontline workers. Medical textiles have been singled out as the most essential tool in efficiently dealing with the infections. Despite their significance, we source most of them from outside the country which exposes us to the inefficiencies of the international inventories deliveries.

This shortage has been partly attributed to the fact that most of these materials are imported and as the major PPE manufacturers and exporters have been dealing with pandemic in their homelands. This has been further exacerbated by the disruption in the international travels and logistics flow.

Having local capacity to produce PPE, would remove the lapses and inefficiencies in delivery that result from sourcing our products from outside the country. This would in turn put us in an optimal position to thoroughly deal with similar infectious outbreaks.

2 Methods and Materials

Potential fabrics were sourced from the local market, analyzed at the UIRI textile materials and polymer testing laboratory. Spun bond polypropylene coated with micro-porous polyethylene of 75GSM was zeroed down onto for the manufacture of disposable coverall (hazardous materials suit) and fabric of similar composition with 26GSM was zeroed down onto for the manufacture of head and shoe covers. Working prototypes were then fabricated at the UIRI garment manufacturing facility.

2.1 Disposable coverall/Hazmat suit

A polypropylene spun bond fabric coated with micro-porous polyethylene film that had been sourced from the local fabric market in town was used in making the coverall. Such a fabric construction is chosen for it is optimal for protecting the user from the hazards of the workers' environment (as it can offer a high barrier protection to the penetration of fluids).

Product details are shown in Fig. 1.

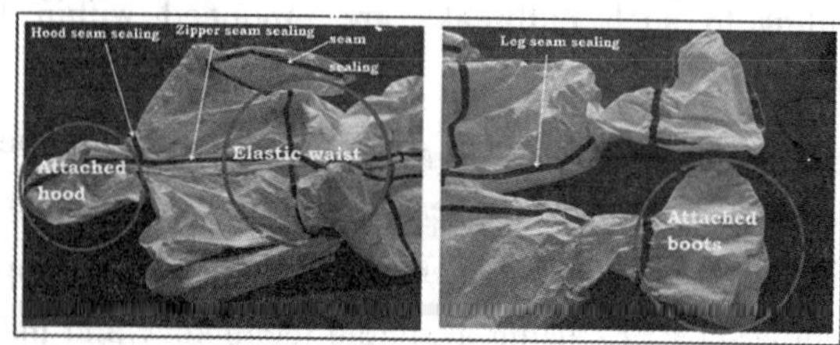

Fig. 1 UIRI designed coverall (hazmat suit), with an attached hood, elastic waist, attached boots with all the seams sealed to ensure that there are no avenues for contamination during usage

Product's attributes list in Tab. 1.

Tab. 1 Product's attributes

In line with the technical specifications of personal protective equipment for pandemic, the coverall was produced with sealed seams, an attached hood, attached boots, elastic wrists, elastic waist, could be closed at chin with a zipper front and had a storm-fly; **Fig. 1.**	
Materials Physical Properties	
Material	Tyvek (Polyethylene laminate material)
Basis weight	75 GSM
Thickness	0.23mm
Basis weight	26 GSM
Fashion	Covering the shoes and elasticated at the ankle

2.2 Disposable shoe & head covers

Disposable shoe and head covers shown in Fig. 2 provide a barrier against possible exposure within a contaminated environment.

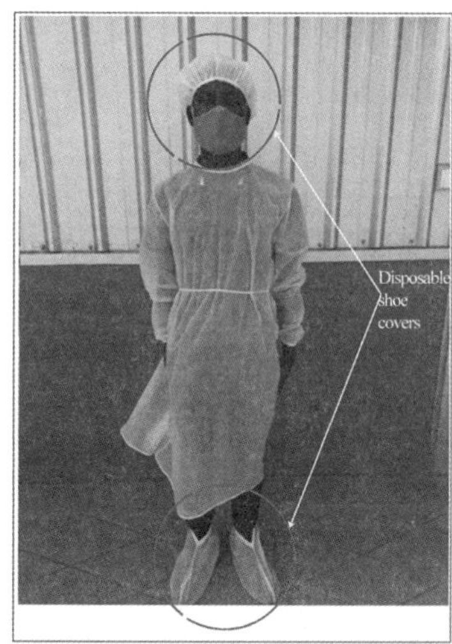

Fig. 2 Head covers and shoe covers

3 Results and Discussion

3.1 Conformity assessment test

Accordingly, the developed samples were subjected to lab analysis at the textile materials and polymer testing laboratory (Fig. 3).

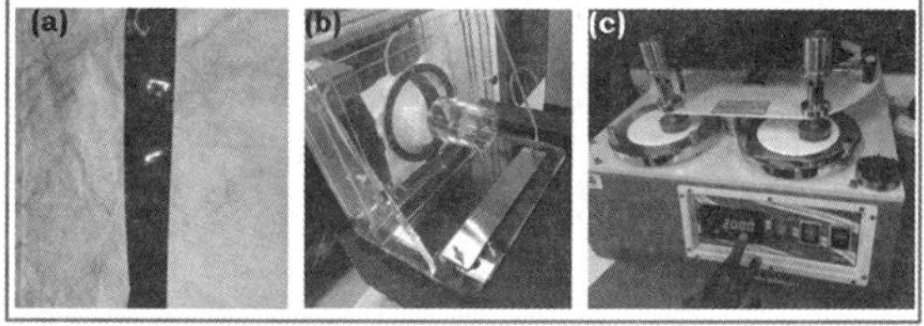

Fig. 3 (a) A sample of a sealed coverall seam, used in the synthetic blood penetration test; (b) the seam area of the coverall mounted on the synthetic blood tester in a testing procedure and (c) testing for the fabric used in making the coverall for abrasion resistance

3.2 Hazmat suit characterization

The characters of Hazmat suit list in Tab. 2.

Tab. 2 The characters of hazmat suit

Fabric Tensile	Grab — Lengthwise	ASTM D5034	200.84 N
	Grab — Transverse	ASTM D5034	169.16 N
Abrasion resistance		ASTM D4966	No wear change at 2 000 rubs
Bursting Strength		ASTM D3786	4.26 kg/cm^2
Stiffness		ASTM D1388	2.613 kg
Fabric and taped seams resistance to permeation by Liquids			
Test		**Test Method**	**Result**
Sulphuric acid (30%) penetration of the main fabric		EN-ISO 7854	No penetration after 480 minutes
Sulphuric acid (30%) penetration of the seams		EN-ISO 7854	No penetration after 480 minutes
Sodium hydroxide (10%)		EN-ISO 7854	No penetration after 480 minutes
Fabric and taped seams resistance to permeation by fluids			
Test		**Test Method**	**Result**
Synthetic blood penetration of the fabric material		ISO 16603	Does not permit the penetration of synthetic blood
Synthetic blood penetration of the seams		ISO 16603	Does not permit the penetration of synthetic blood

4 Conclusions

It can be concluded that the designed coverall is suitable for use in emergency medical response. It can find application in laboratories, animal research, sanitation, crime scene clean-up, forensics and embalming. The shoe and head covers if worn properly, protect the hair, feet and scalp from possible contamination when sprays or airborne exposure is anticipated.

The study shows that PPEs can be produced locally. Having local capacity to produce PPE, would remove the lapses and inefficiencies in delivery that result from sourcing our products from outside the country. This would in turn put us in an optimal position to thoroughly deal with similar infectious outbreaks.

References

[1] OKWARE S I, OMASWA F G, ZARAMBA S, et al. An outbreak of Ebola in Uganda[J]. Tropical Medicine & In-

ternational Health, 2002, 7: 1068–1075.

[2] MBONYE A, WAMALA J, KABOYO W, et al. Repeated outbreaks of viral hemorrhagic fevers in Uganda[J]. African Health Sciences, 2012, 12: 579–589.

[3] OKWARE S. Managing Ebola in low-resource settings: experiences from Uganda. Ebola[C]: Intech Open; Ebola, 2016.

[4] HIRSCHMANN M T, HART A, HENCKEL J, et al. COVID–19 coronavirus: recommended personal protective equipment for the orthopaedic and trauma surgeon[J]. Knee Surgery, Sports Traumatology, Arthroscopy, 2020, 28: 1690–1698.

[5] MADINAH N. The COVID – 19 pandemic: Economic effects and government measures in Uganda[J]. SARJANA, 2020, 35: 10–22.

Strain Rate and Gauge Length Effects on Tensile Properties of Polymeric Monofilaments

ALRAYAH. H. D. Yousif[1,3*], **SYED Rashedul Islam**[2], **LIU Yanping**[2*], **ZHANG Yumei**[1]

[1] *State Key Laboratory for Modification of Chemical Fibers and Polymer Materials, College of materials Science and Engineering, Donghua University, Shanghai 201620, China*

[2] *Key Laboratory of Textile Science & Technology of Ministry of Education, College of Textiles, Donghua University, Shanghai 201620, China*

[3] *Industries Engineering and Technology, University of Gezira, Sudan*

* *Corresponding author's email*: liuyp@ dhu. edu. cn; alrayahhassan31@ gmail. com

Abstract: Polymeric monofilaments are ideal materials as spacer yarns in the production of 3D knitted spacer fabrics due to their excellent mechanical properties. The compression and impact properties of spacer fabrics are vital and mainly dependent on the tensile properties of the monofilaments. Thus, it is crucial to define their mechanical properties under extreme loadings, for example varying strain rates. In this paper, three polymeric monofilaments with the same diameter of 0. 12 mm, namely PE, PB, and LMPET were tested on the universe mechanical tester at three different strain rates of 0. 0003, 0. 0070, and 0. 0300/s and five gauge lengths of 50, 100, 150, 200, and 250 mm to assess their tensile properties sensitivity. In addition, the fractured ends of the monofilaments after failure were recorded by means of SEM. Based on the measured stress-strain curves, there was an obvious transition in the monofilament's behavior as the strain rate or gauge length increased. This transition in the monofilament behavior on tensile loading was attributed to their transition temperatures T_g. The experimental results showed that the strain rate and the gauge length have significant effects on the tensile strength, modulus, elongation at break, and toughness of the monofilaments. Although the PE monofilament's modulus was massively increased with increasing the strain rate and gauge length, it is not recommended to be spacer yarns for spacer fabrics under high impact and compression loads. LMPET monofilament is an ideal material to be used as spacer yarns for spacer fabrics, particularly under high impact loading, because its modulus is higher than PE and PBT. The increase in the gauge length was found to have negligible effects on the tensile strength of the monofilaments. Different types of fracture faces at high strain rates were observed.

Keywords: Strain rate; Guage length; Mechanical properties; Polymeric monofilaments; Spacer fabrics

1 Introduction

Polymeric monofilaments (PMs) are currently used as spacer yarns in the production of 3D knitted spacer fabrics due to their excellent mechanical properties and good knit-ability[1-2]. They are knitted together with two flexible outer layers of spacer fabrics using different knitting technologies. Due to their unique properties, these monofilaments serve as the main-load carrier for the spacer fabric against compression or impact loadings. Also, spacer fabrics are produced in different thicknesses and shapes and used in human body protection to protect them from injuries during impacts. However, the spacer fabric's compression and impact properties are hugely related to the monofilaments' mechanical properties. Therefore, it is crucial to determine monofilament mechanical properties at extreme loadings to develop high-quality spacer fabrics for impact protector applications.

Experimental investigations have already been conducted to determine the mechanical properties of spacer fabrics and single polymeric fibres or bundles. Zhao et al. [3] conducted compression tests on weft-knitted spacer fabrics using a universal mechanical testing machine. Results showed that spacer fabrics knitted with a shorter spacer yarn span distance, coarser monofilaments, and higher spacer yarn density had better compression resistance and absorption energy. Liu et al. [4] showed that warp spacer fabrics knitted with large-size mesh structures had higher peak contact force and peak transmitted force than spacer fabrics knitted with small-size mesh and close structures when subjected to impact force. Stamoulis et al. [5] performed tensile experiments on a single PA6. 6 monofilament under six different speeds varying from 5 to 250 mm/min. The experimental results showed that the only quantity that changed was elongation at break, and the mean Young's modulus remained constant (2. 5 GPa) as a function of increasing the speed. Basu et al. [6] performed quasi-static tensile tests on commercially PET fibers using the Keysight T150 technologies. Results showed no significant strain-rate sensitivity on the tensile strength in PET fibers. However, the yield strength showed significant strain-rate sensitivity similar to other semi-crystalline polymers. More recently, Bai et al. [7] studied the dynamic tensile properties of the large-rupture-strain PET fiber bundle using an MTS machine and an Instron drop-weight impact system. The experimental results showed that the tensile strength, failure strain, elastic modulus, and toughness of the PET fiber bundle were sensitive to the strain rate. Unfortunately, their study did not mention the effect of gauge length and high strain rate. Wang et al. [8] also studied the tensile properties of UHMWPE yarns under high strain rate loading using a universal testing machine and drop-weight impact apparatus. The results showed that the tensile behavior of PE yarns was highly strain-rate dependent over the range from 0. 000 033 to 0. 33/s. However, the measured tensile properties were independent at a rate greater than 0. 33/s and tested gauge lengths of 25 mm and 50 mm. Jenket et al. [9] investigated the effects of temperature and strain rate on strain-to-failure of PE fiber using a Split-Hopkinson Tension Bar tester. The results showed that tensile strength increased as the strain rate increased and decreased as the temperature increased; unfortunately, they only considered a short gauge length in their study, which was 10 mm.

Moreover, many efforts have already been made to study the effects of strain rate on tensile properties of conventional and synthetic fibers using the quasi-static and impact tests[10-11]. Ou et al. [10] studied the strain rate effect on the dynamic tensile mechanical properties of the glass fiber bundle. The experimental results showed that the tensile strength increased with the increasing strain rate from 1/600 to 160/s. Zhu et al. [11] investigated the effect of the strain rate (20 to 100/s) on the dynamic tensile mechanical properties of the Kevlar 49 fiber bundle. The results showed that the tensile strength, the elastic modulus, and the toughness increased with the increasing strain rate.

These previous studies have provided abundant and valuable knowledge about the mechanical properties and the effects of strain rate on the mechanical properties of multifilament fibers and yarns of small diameters. However, the effects of different strain rates and gauge lengths on the mechanical properties of different polymeric monofilaments were not found in the literature.

This paper reports the effects of strain rate sensitivity on the mechanical properties of three polymeric monofilaments used as spacer yarns for impact protective applications. The tensile tester was employed to test these monofilaments upon break, employing three initial velocities (5, 100, and 500 mm/min) and 5 gauge lengths (50, 100, 150, 200, and 250 mm). Their stress-strain curves obtained the tensile strength, Young's modulus, elongation at break, and toughness of monofilaments. The effects of strain rate and gauge length on these properties were evaluated. After breakage, the fractured surface ends of the samples were analyzed using SEM. It is expected that this study's findings could help develop high-quality spacer fabrics for impact protective applications.

2 Experiments

2. 1 Materials

Three monofilaments, namely PE, PBT, and LMPET, were prepared with different polymers by Jiangsu DOWAY New Material Science and Technology CO., Ltd., China. Their specifications from the original material are listed in Tab. 1. LMPET monofilament was spun with low viscosity, meaning. That its macromolecular chains are shorter with a lower molecular weight.

Tab. 1 Specifications of the polymeric monofilaments

Sample codes	Diameter (mm)	Linear density (dtex)	Intrinsic viscosities ($g \cdot dL^{-1}$)
PE	0. 120	122. 0	—
PBT	0. 120	177. 0	—
LMPET	0. 120	179. 0	0. 67

2. 2 Tensile tests

Tensile tests were performed on a universe mechanical tester (YG082) manufactured by Ningbo Dahe Instrument Co., Ltd., Zhejiang, China. Two tensile test methods are carried out in this study. (1) Monofilament with a gauge length of 250 mm was loaded upon break at three loading speeds of 5, 100, and 500 mm/min, denoted as strain rates of 0.000 3, 0.007 0, and 0.030 0/s, respectively. (2) Monofilament with five gauge lengths, namely 50, 100, 150, 200, and 250 mm, was loaded upon break at a constant strain of 500 mm/min. Both methods adopted a pretension of (0.05±0.005) cN/dtex and five tests for each sample.

Here, tensile tests were carried out to evaluate the tensile properties of the samples at different strain rates and gauge lengths. The samples' tensile strength, Young's modulus, elongation at break, and toughness were obtained based on their stress-strain curves and calculated by averaging the five tests.

2.3 SEM tests

SEM (FlexSEM 1 000, Hitachi, Japan) was used to detect the fractured ends of the monofilaments after breakage. Samples were first coated with gold to obtain better conductivity at imaging time and then placed on the SEM machine. Fractured samples were observed using a voltage of 10 kV. The SEM images of broken samples at the high strain rate of 0.030 0/s were taken in this study.

3 Results and Discussion

3.1 Stress-strain curves behavior of the monofilaments

The tensile stress-strain curves of three monofilaments at strain rates of 0.000 3, 0.007 0, and 0.030 0/s are chosen here and shown in Fig. 1, and the tensile strength, modulus, elongation at break, and toughness are presented in Fig. 2. The results indicate that monofilaments' tensile stress-strain curves are changed differently after strain rate increases. In the stress-strain curves shown in Fig. 1, the monofilaments had similar characteristics in the linear elastic phase of the curves regardless of increasing the strain rate. However, in the plastic deformation phase, the response of the monofilaments to tensile load varied with increasing the strain rate. It was found that the PE stress-strain curve was first hardened in the elastic phase and then softened at a lower strain rate of 0.000 3/s and exhibited a brittle manner at higher strain rates of 0.007 0/s and 0.030 0/s in the plastic phase, as illustrated in Fig. 1 (a). This finding is in line with the results of

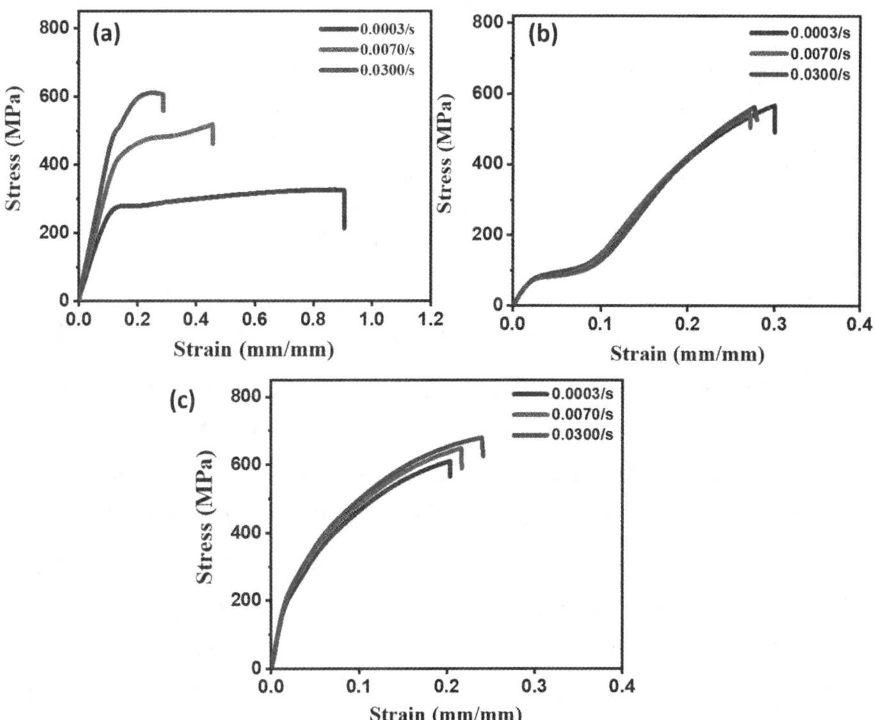

Fig. 1 Tensile stress-strain curves of the monofilaments at strain rates of 0.000 3, 0.007 0, and 0.030 0/s: (a) PE; (b) PBT; (c) LMPET

Schwartz and Cansfield et al. [12-13] who found that the PE transited or failed from ductile to brittle at about 0. 0017, 0. 0033, and 0. 33/s, respectively. Unlike PE, the stress-strain curves of LMPET showed an opposite response with increasing the strain rate. It is found that a lower strain rate results in lower tensile strength, elongation at break, and toughness, while a higher strain rate produces a much larger tensile strength, elongation at break, and toughness, demonstrating that the strain rate also has a significant influence on the LMPET stress-strain curves. In the case of the stress-strain curves of the PBT monofilament, the response of the curves was different from that of PE and LMPET as increasing the strain rate. The stress-strain curves were not changed in the elastic and plastic deformation phases, and the tensile properties only possessed slight changes as the strain rate increased to a maximum strain rate of 0. 0300/s compared to PE and LMPET monofilaments. The results mentioned above demonstrated that the strain rate has significant effects on the tensile behavior of the PE and LMPET monofilaments while having negligible effects on PBT tensile behavior.

3.2 Strain rate effects

In order to study the strain-rate effects on the tensile properties of PE, PBT, and LMPET monofilaments, we performed tensile tests at three strain rates: 0. 000 3, 0. 007 0, and 0. 030 0/s. Fig. 2 shows the tensile strength, modulus, elongation at break, and toughness of monofilaments as a function of strain rates. As shown in Fig. 2, the experimental results of PBT monofilament do not show significant strain-rate effects on tensile properties within the strain-rate range from 0. 000 3/s to 0. 030 0/s. The tensile strength, modulus, and toughness of PBT monofilament increase only by 6. 3%, 4. 4%, and 6. 7%, respectively, as the strain rate increases from 0. 000 3/s to 0. 030 0/s. The elongation at break of PBT monofilament decreased by 2. 9% over the same strain ranges.

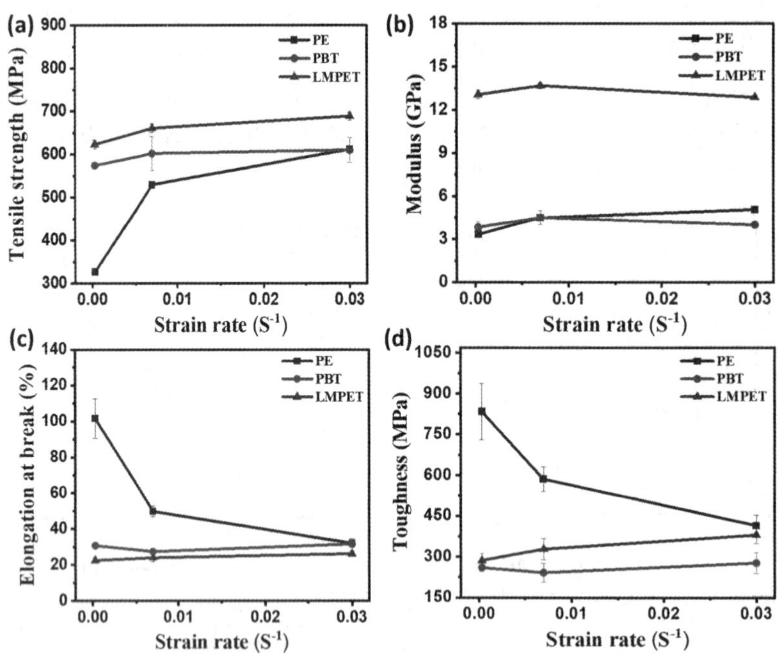

Fig. 2 Tensile properties of PE, PBT, and LMPET monofilaments at a strain rate of 0. 000 3, 0. 007 0, and 0. 030 0/s: (a) tensile strength; (b) modulus; (c) elongation at break; (d) toughness

Interestingly, PE and LMPET monofilaments show significant strain-rate effects on the tensile properties over 0. 000 3/s to 0. 030 0/s. The tensile strength and modulus of PE monofilament massively increase by 87. 3% and 52. 7%, and the elongation at break and toughness decrease by 101. 6% and 50. 3%. Also, the tensile strength, modulus, elongation at break, and toughness of LMPET monofilament increase by 10. 5%, 4. 6%, 16. 6%, and 32. 7%, respectively, when the strain rate increases from 0. 000 3/s to 0. 030 0/s. Overall, the rapid transition from ductile to brittle in PE with increasing strain rate was associated with its glass transition temperature T_g. The glass transition temperature of PE (−78 ℃) is much lower than PBT and LMPET monofilaments (79 ℃ and 55 ℃). Therefore, at room temperature, the monofilaments behave differently on tensile loading as the strain rate increases.

3.3 Gauge length effects

The tensile properties of PE, PBT, and LMPET monofilaments at different gauge lengths are shown in Figs. 3 and 4. These results are obtained using five gauge lengths of 50, 100, 150, 200, and 250mm at a constant strain rate of 500 mm/min. Results show that the increase in the gauge length has insignificant effects on the tensile strength of the monofilaments. The tensile strength of the PBT and LMPET monofilaments decreased only by 5.7% and 2.6% as the gauge length increased from 50 mm to 250 mm, respectively, while the tensile strength of PE exhibited uneven values at the same gauge length ranges as shown in Fig. 3. As a comparison, previous literature demonstrates that the tensile strengths of some high-performance fibres (PBO, PPTA, and PIPD) increase with decreasing gauge length. Therefore, our results indicate that the tensile strength-limiting flaws are numerous and evenly distributed along the length of PE, PBT, and LMPET monofilaments. In other words, the significant defect along the length limited the tensile strength of the monofilaments. However, the experimental results presented in Figs. 3 and 4. showed that the gauge length significantly affected the PE, PBT, and LMPET monofilaments' modulus, elongation at break, and toughness. The modulus and toughness massively increase while the elongation at break decreases with increasing the gauge length from 50 mm to 250 mm, respectively. Thus, it can be concluded that monofilament with longer length under tensile loading has higher modulus and toughness but lower elongation at break, while monofilament with shorter length under tensile loading has lower modulus and toughness but higher elongation at break.

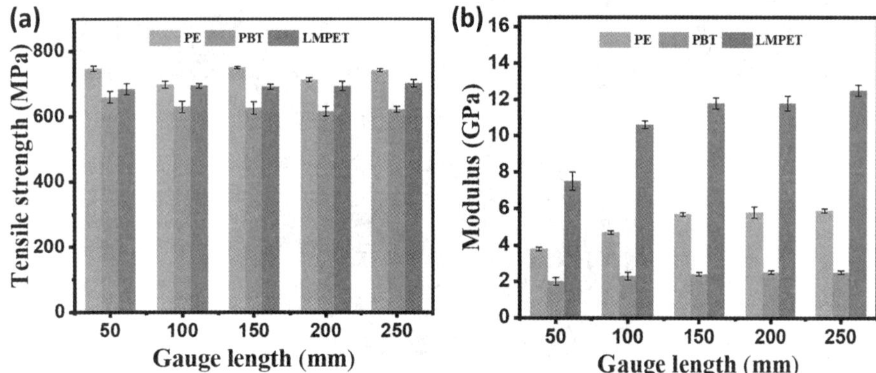

Fig. 3 Tensile strength and modulus of PE, PBT, and LMPET monofilaments with different gauge lengths

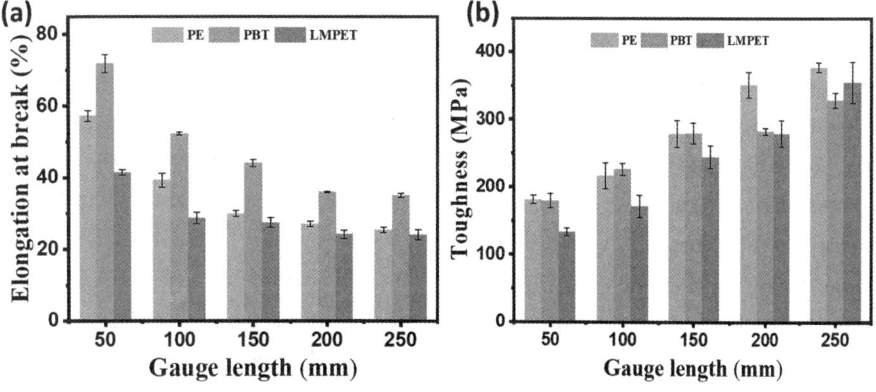

Fig. 4 Elongation at break and toughness of PE, PBT, and LMPET monofilaments with different gauge lengths

3.4 Fracture surfaces of monofilaments

For example, the fractured ends of monofilaments tested under a high strain rate of 0.030/s are shown in Fig. 5. Unlike PBT and PET monofilaments, the fractured ends of PE at the strain rate of 0.030/s are markedly different because its Tg is far away than PBT and LMPET monofilaments and crystalline structure can be modified in the initial cold stretching processes[1]. The fractured ends of PE are illustrated in Fig. 5(a). The sample split axially, thinner towards the tips, and the fibrils become longer at the fractured ends. This result is in line with the report by Wang et al.[8] who have found the same fractured end appearances and transition from ductile to brittle failure in UHMWPE fiber as a function of increasing the strain rate from 0.000 033/s to 400/s. Figure 5(b) shows the fractured ends of PBT monofilament tested at the strain rate of 0.030/s. It was found that the sample showed irregular faces with shorter fibrillation near the surface. This result indicated that the fibrils within the PBT structure are still bonded together and broken at different times at high strain rates resulting in greater tensile strength. From Fig. 5(c), it can be seen that the LMPET spun with the low molecular weight material and short molecular chains broken with a regular diameter and its fracture surface appeared smooth and flat with no trace near the surface when the strain rate was high. This result indicated that the molecular chains within the LMPET structure deformed simultaneously and acted as bearing for the tensile loading resulting in high stress compared with samples at lower strain rates.

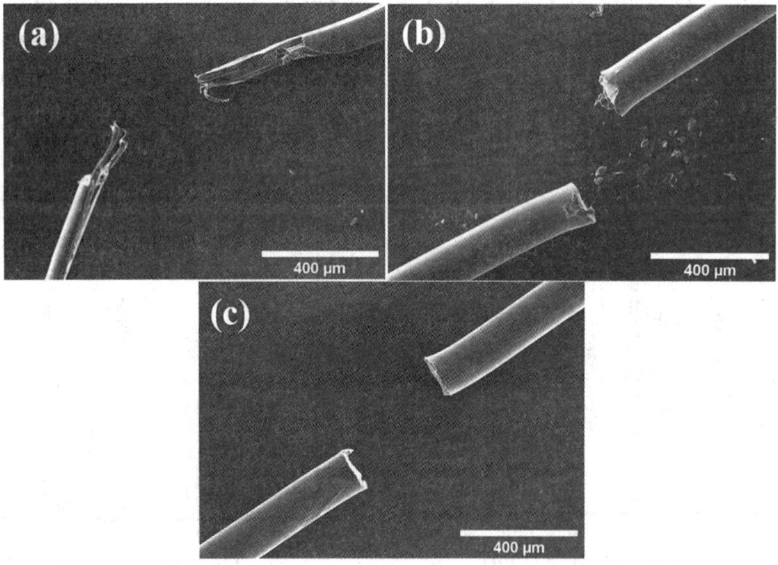

Fig. 5 Fractured ends images of PE monofilament tested at the strain rate of 0.030/s: (a) PE; (b) PBT; (c) LMPET

4 Conclusions

Three polymeric monofilaments with the same gauge length used as spacer yarns for spacer fabrics were tested on a universe mechanical tester using six speeds. In addition, the same monofilaments with five gauge lengths were also tested under a universe mechanical tester at a constant strain of 500 mm/min. The effects of strain rate and gauge length on the mechanical properties of monofilaments, including tensile strength, modulus elongation at break, toughness, and fractured ends, were investigated. The study findings are recorded as follows.

(1) The mechanical properties of PE monofilaments are more insensitive to the strain rate than the other monofilaments. The PE monofilament possessed a higher strain rate sensitively in terms of tensile strength, elongation at break, and toughness than PBT and LMPET. Since the PE exhibites a higher strain rate sensitively in terms of the mechanical properties, it is not recommended to be used as spacer yarn for spacer fabrics, particularly under high impact loading, because its modulus is much lower than LMPET.

(2) The gauge length has significant effects on modulus, elongation at break, and toughness of the monofilaments. But it has insignificant effects on the tensile strength due to the tensile strength-limiting defects distributed along the length of the monofilaments.

(3) Experimental results also showed considerable differences in the tensile strength and failure of the monofilaments with increasing the strain rate and gauge length, indicating the strong sensitivity of these monofilaments to tensile and impact loads.

References

[1] GUO H, DAFAALAH A H, LIU Y, et al. A comparative study on the mechanical properties of polymeric monofilaments[J]. Journal of Engineered Fibers and Fabrics, 2020, 15: 1558925020962790.

[2] BAI Y L, YAN Z W, OZBAKKALOGLU T, et al. Dynamic behavior of PET FRP and its preliminary application in impact strengthening of concrete columns[J]. Applied Sciences, 2019, 9(23): 4987.

[3] ZHAO T, LONG H, YANG T, et al. Cushioning properties of weft-knitted spacer fabrics[J]. Textile Research Journal, 2018, 88(14): 1628-1640.

[4] LIU Y, HU H, LONG H, et al. Impact compressive behavior of warp-knitted spacer fabrics for protective applications[J]. Textile Research Journal, 2012, 82(8): 773-788.

[5] OU Y, ZHU D, ZHANG H, et al. Mechanical characterization of the tensile properties of glass fiber and its reinforced polymer (GFRP) composite under varying strain rates and temperatures[J]. Polymers, 2016, 8(5): 196.

[6] BASU S. Statistical variation and strain-rate sensitivity of the mechanical properties of individual pet fibers[J]. Agilent Technologies Application Note, 2012.

[7] BAI Y L, YAN Z W, OZBAKKALOGLU T, et al. Quasi-static and dynamic tensile properties of large-rupture-strain (LRS) polyethylene terephthalate fiber bundle[J]. Construction and Building Materials, 2020, 232: 117241.

[8] WANG H, HAZELL P J, SHANKAR K, et al. Tensile properties of ultra-high-molecular-weight polyethylene single yarns at different strain rates[J]. Journal of Composite Materials, 2020, 54(11): 1453-1466.

[9] JENKET D R, FORSTER A M, PAULTER N G, et al. An investigation of the temperature and strain-rate effects on strain-to-failure of UHMWPE fibers[B]. Challenges in Mechanics of Time Dependent Materials, 2017, Volume 2. Conference Proceedings of the Society for Experimental Mechanics Series. Springer, Cham. https://doi.org/10.1007/978-3-319-41543-7_4

[10] OU Y, ZHU D, ZHANG H, et al. Mechanical characterization of the tensile properties of glass fiber and its reinforced polymer (GFRP) composite under varying strain rates and temperatures[J]. Polymers, 2016, 8(5): 196.

[11] ZHU D, MOBASHER B, ERNI J, et al. Strain rate and gage length effects on tensile behavior of Kevlar 49 single yarn[J]. Composites Part A: Applied Science and Manufacturing, 2014, 43(11): 2021-2029.

[12] SCHWARTZ P, NETRAVALI A, SEMBACH S. Effects of strain rate and gauge length on the failure of ultra-high strength polyethylene fibers[J]. Textile Research Journal, 1986, 56(8): 502-508.

[13] CANSFIELD D L M, et al. Tensile strength of ultra high modulus linear polyethylene filaments[J]. Polymer Communications, 1983, 24: 130-131.

Thermal and Tensile Properties of 3D Weft-Knitted Spacer Fabrics Coated by SiO_2 Aerogels with Two Different Concentrations

SYED Rashedul Islam[1*], ALRAYAH H. D. Yousif[1], SYED Amdad Hosen[2], TAOSIF Ahmed[3], SAGAR Sarker[1], WENDI Wang[1], JIANG Jinhua[1*]

[1] *Shanghai Frontier Science Research Center for Modern Textiles, College of Textiles, Donghua University, Shanghai 201620, China*

[2] *School of Economics, Shanghai University of Political Science and Law, Shanghai 201701, China*

[3] *Faculty of Wood Engineering and Creative Industries, University of Sopron, Sopron, 9400, Hungary*

* *Corresponding author's email*: jiangjinhua@ dhu. edu. cn, sri060791@ gmail. com

Abstract: The work studied the surface, thermal, and tensile properties of untreated and treated 3D weft-knitted spacer fabrics. In order to improve the thermal and tensile properties of weft-knitted spacer fabrics, silica aerogels coating was applied. Scanning electron microscopy (SEM) and Fourier transform infrared spectroscopy (FTIR) studies of untreated and treated spacer fabrics were considered to confirm the existence of silica aerogels and understand the molecular structure of the coated spacer fabrics. The elementary properties of untreated and treated 3D weft-knitted spacer fabrics with silica aerogels were investigated and compared. The experimental results revealed that silica aerogels coating had a substantial effect on the thermal and tensile properties of 3D weft-knitted spacer fabrics. Hence, these 3D fabrics could be appropriate for industrial applications as thermal insulating and enduring materials.

Keywords: 3D spacer fabrics; Silica aerogels; Sol-gel method; Thermal properties; Tensile properties

1 Introduction

3D weft-knitted spacer fabrics are made on a double jersey circular knitting machine with a movable needle cylinder and dial[1]. These fabrics typically involve one inner layer of spacer yarns and two outside layers of elastic yarns, i. e., an upper layer and a bottom layer[2]. They possess strong planar flexibility and transverse deformability[3]. These 3D fabrics can be a fantastic option if durability, thermal insulator, and cost are the key concerns[4-5]. Furthermore, they can be reused, and they are deliberated as environmentally textile materials[6-8]. Nano-porous silica aerogels (SAs) are highly thermal insulating materials. The retardation of textile fabrics has been improved by the incorporation of various compounds[9-10].

Small pore size, high pore volume, low density, and large internal surface area are all characteristics of SAs[4, 11]. The lowering in conduction, convection, and radiation offers them low thermal conductivity, good thermal insulation qualities, and high strength[12-13]. As a result, SAs are now widely employed to provide the required alterations in commercial textiles, such as weft-knitted spacer fabric surfaces used mostly for biomedical, hydrophobic, maritime, oleophilic, aviation applications, as well as winter clothing and other apparel[13-17]. The foundation for this research is the dearth of extensive investigation on the aspects, particularly the thermal and tensile attributes of 3D weft-knitted spacer fabrics. The associations among the 3D weft-knitted spacer fabrics, SAs, thermal, and tensile properties are indeed relatively limited.

The authors still have not found significant research work on the 3D weft-knitted spacer fabrics with the silica aerogel coating for studying its thermal and tensile properties. In this study, the effect of silica aerogels on thermal conductivity, thermal resistance, tensile strength, initial modulus, and strain properties of weft-knitted spacer fabrics was evaluated systematically by scanning electron microscopy (SEM), Fourier transforms infrared spectroscopy (FTIR), thermal insulation and tensile tests. Thus, 3D weft-knitted spacer fabrics with SAs coating were discovered as an alternating source for thermal insulating and long-lasting materials.

2 Experimental

2.1 Materials

3D weft-knitted spacer fabric sample (92% polyester and 8% spandex) was bought from Tianbin Textile Co. Ltd., Changshu, China. The weft-knitted tuck spacer fabric was made on a circular knitting machine of gauge 28. Tetraethyl-orthosilicate (>99%, TEOS) and N-hexane (>99%, n-C_6H_{14}) were purchased from National Drug Group Chemical Reagent Co. Ltd., Shanghai, China. All details of fabrics and other chemicals used in this study were described in our earlier published paper[15]. The chemicals and reagents were applied without any further rectification.

2.2 Silica aerogels preparation

The silica sol-gel process was carried out in two stages, i.e., initially acid-catalyzed TEOS hydrolysis and finally base-catalyzed gelation[1]. The entire process of silica aerogel preparation was clearly explained in our former work[10]. Two different concentrations of silica aerogels were prepared to study the thermal and tensile properties of 3D weft-knitted spacer fabrics, as listed in Tab. 1.

Tab. 1 Two different concentrations of silica aerogels (SAs)

Chemicals	SAs (Con. 1)	SAs (Con. 2)
TEOS	80.5	90.5
EtOH	698	688
HCl	28.5	28.5
H_2O	95.5	95.5
NH_4OH	32.5	32.5
DMF	32.5	32.5
HMDS	32.5	32.5

2.3 Fabrication of samples with silica aerogels

3D weft-knitted spacer fabric samples were cut into 35 cm×35 cm pieces and dipped into the silica sol for 15 min. The production process of coated weft-knitted spacer fabrics with silica aerogels was intensely described in our earlier published work[3].

2.4 SEM analysis

The scanning electron microscopy (SEM) was completed with a Flex-SEM 1000 (SU1000, Hitachi Ltd. Japan) to study the surface morphology of 3D spacer fabrics. Prior to analysis, a vacuum sputter coater was used to provide a gold layer to all weft-knitted spacer fabric samples.

2.5 FTIR test

The Fourier transform infrared spectroscopy (FTIR) was completed with an American Nicolet TM 5700 FT-IR spectrometer. The spectra were produced in the variety of 500–4 000 cm^{-1} founded on the attenuated total reflection technique (ATR). The test was done with 64 scans and set the resolution at 8 cm^{-1}.

2.6 Thermal insulation test

The thermal resistance tester (YG606E) from China was used to endorse the thermal insulation properties of spacer fabrics. The testing plate was maintained at a temperature of 35 ℃. Three readings of each sample were taken, and the average results were computed. The thickness of the 3D spacer fabric and thermal conductivity were related to the thermal resistance. Equation 1 determined the correlation between thermal conductivity and thermal resistance.

$$R = \frac{T}{\lambda} \tag{1}$$

where R is thermal resistance in ($m^2 \cdot K \cdot W^{-1}$), λ is thermal conductivity ($W \cdot m^{-1} \cdot K^{-1}$), and T is fabric thickness (mm).

2.7 Tensile test

Tensile test was performed in accordance with GB/T 3923 test standards on testing machine YG026MB – 250 from China. The samples were divided into 13 cm× 2 cm pieces. The usual test conditions were kept at 20 ℃ and 65% relative humidity. The computer calculated the relevant elongation in mm based on the highest force in Newtons used to break the samples. The stretching speed was 100 mm/min. The gauge length was 70 mm. Every sample was examined three times in different locations for this test, and the average results were taken into account.

3 Results and Discussion

3.1 Surface morphology of the untreated and treated weft-knitted spacer fabrics

The surface morphology and microstructure of the untreated and treated samples with silica aerogels were observed by the SEM test, as shown in Fig. 1. Figure 1(a) showed the untreated spacer fabric surface whereas Figs. 1 (b) and 1 (c) exhibited the treated samples with compacted surface and ceaseless structure due to silica aerogel coatings. The silica aerogel-treated surfaces of spacer fabrics were observed to be broader than those of untreated fabrics.

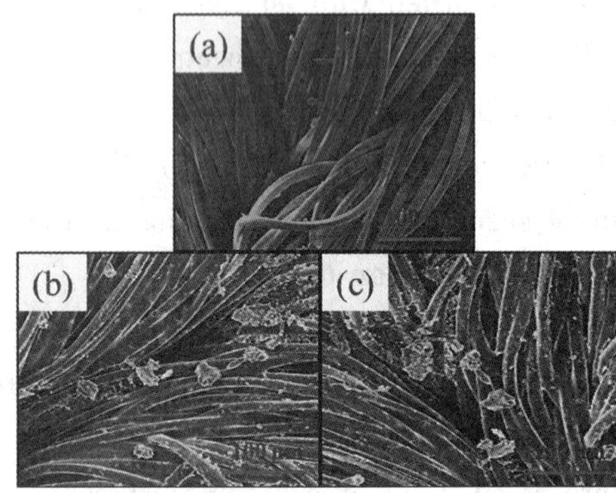

Fig. 1 SEM images: (a) untreated sample; (b) and (c) silica aerogel treated samples

In the gaps between the fibers of the treated fabrics, numerous significant interphase crusts and bridges were seen. Untreated spacer fabric samples had a smooth surface; however, treated spacer fabric surfaces had burls and stubs, which gave them a severe and harsh appearance. Moreover, the SiO_2 network was present in the structures of the treated fabrics. TS 2 had the most silica aerogel as compared to TS 1 on the surface with fewer friezes and more porosity ratio. Thus, it caused better thermal resistance and tensile properties as compared to TS 1.

3.2 FTIR analysis

Figure 2 showed the results of the FTIR analysis. The appearance of the peak position at around 1 050 cm^{-1} was due to the bending and stretching of C—O bonds in both untreated and treated samples. The uneven stretching and vibration of Si—O—Si bonds caused the overlapping absorbance peak at 1 050 cm^{-1}, which was dissimilar from the untreated sample. Thus, it confirmed the construction of a nanostructured silica aerogel network on the treated samples[15].

Due to the stretching and bending of C—H bonds, absorption bands at approximately 1 350 cm^{-1} and 2 950 cm^{-1} were observed. Both samples possessed C = O groups in their compositions, according to the prominent peak at 1 700 cm^{-1}. The asymmetric stretching of —C≡C— bonds caused the peak to appear at a wavelength of 2 200 cm^{-1} in the treated samples. Therefore, the presence of Si—O—Si bonds in the spectra established the development of the thermal insulating and tensile properties of the spacer fabrics.

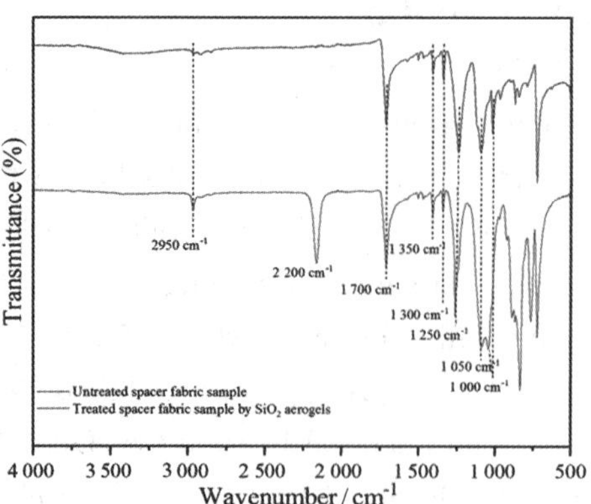

Fig. 2 The FTIR spectra of untreated and treated samples with silica aerogels

3.3 Thermal insulation test

The thickness, thermal conductivity, and thermal resistance values of 3D weft-knitted spacer fabrics were listed in Tab. 2. The thermal resistance increased from 0.062 5 $m^2 \cdot K \cdot W^{-1}$ to 0.073 9 $m^2 \cdot K \cdot W^{-1}$ and 0.081 3 $m^2 \cdot K \cdot W^{-1}$ for TS 1 and TS 2, respectively, and the thermal conductivity decreased from 0.035 19 $W \cdot m^{-1} \cdot K^{-1}$ to 0.027 06 $W \cdot m^{-1} \cdot K^{-1}$ and 0.032 86 $W \cdot m^{-1} \cdot K^{-1}$ for TS 1 and TS 2, respectively, after coating with silica aerogels. This was due to higher add-on percent of silica aerogels[3].

Tab. 2 Thermal properties of untreated and treated samples with silica aerogels

Samples	Thickness/ T (mm)	Thermal conductivity/ λ ($W \cdot m^{-1} \cdot K^{-1}$)	Thermal resistance/ R ($m^2 \cdot K \cdot W^{-1}$)
UTS	3	0.0480	0.0625
TS 1	3	0.0405	0.0739
TS 2	3	0.0369	0.0813

The samples that were exposed to silica aerogels had less thermal conductivity because there was less conduction, convection, and radiation. Lower densities and the treated consistent nanoporous structure's convoluted route for thermal transmission reduced conduction. The convection was minimized by the silica aerogels' smaller pore sizes. The effect of the radiating thermal conductivity with dispersing dopants caused the overall heat transfer to decrease. The greater thermal resistance outcomes and lesser thermal conductivity outcomes indicated that there was less heat transfer from the body to the spacer fabric interface. This was interrelated to a warmer sensation and thus, made an interesting thermal insulating material.

3.4 Tensile test

The tensile strength, initial modulus, and strain of untreated and treated samples were listed in Tab. 3. The tensile strength of treated weft-knitted spacer fabrics was higher compared to their corresponding untreated weft-knitted spacer fabrics, as shown in Fig. 3. The tensile strength increased from 3.39 MPa to 3.61 MPa and 3.75 MPa for TS 1 and TS 2, respectively. This could be the silica aerogels' adherence to the spacer fabrics. The discharge of the ethanol-containing fluid might just have caused the treated samples to possess a larger surface area than that of the untreated samples, which in turn created more opportunity for the generation of van der Waals interactions between the silica aerogels and 3D spacer fabrics.

The fibers of the spacer fabrics were continuously linked to the Si—OH in the silica aerogel nanoparticles by hydrogen and dipolar-dipolar bonds with thermodynamic interaction. This could also be verified by the fact that treated spacer fabric samples showed lower break elongation than untreated spacer fabric samples in selected samples. It disclosed how strain and stress were related to each other in a material. When a particular force was attached to strong material, it was used to flex. The strain of treated spacer fabrics was perceived as lower than that of their involving untreated spacer fabrics. The strain decreased from 1.38 to 1.07 and 1.02 for TS 1 and TS 2, respectively.

Tab. 3 The tensile properties of untreated and treated samples spacer fabrics

Sample	Strength or Stress/σ (MPa)	Strain/ϵ	Modulus/ E (MPa)
UTS	3.39	1.38	2.46
TS 1	3.61	1.07	3.37
TS 2	3.75	1.02	3.68

Both weft-knitted samples treated with silica aerogel showed a notable enhancement in initial modulus. The initial modulus increased from 2.46 MPa to 3.37 MPa and 3.68 MPa for TS 1 and TS 2, respectively. The key part of the stress-strain curve's higher stress and lower strain resulted in treated samples with a higher initial modulus. The treated weft-knitted spacer fabrics' inflexibility was represented by the initial modulus. The overall fluctuations in the initial modulus of the treated fabric samples and their rigidity may be affected in a few ways by the distribution of the silica aerogel nanoparticles.

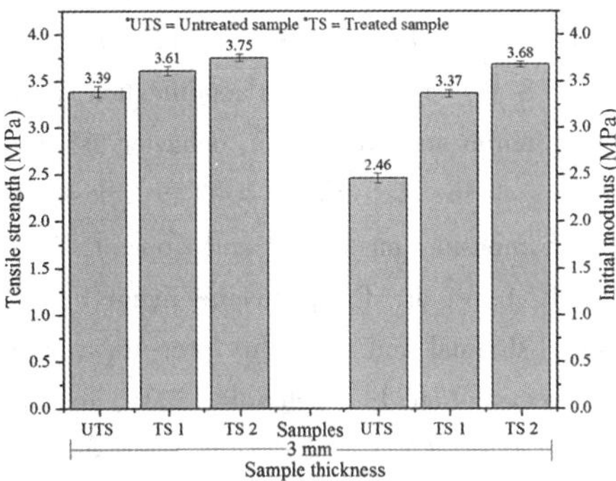

Fig. 3 Tensile strength and initial modulus of untreated and treated samples with silica aerogels

4 Conclusions

This research work effectively characterized the thermal conductivity, thermal resistance, tensile strength, initial modulus, and strain features utilizing silica aerogel coating to study the thermal and tensile properties. In order to meet the requirements for solidity, softness, and insulating sensitive qualities in diverse garments, extraordinary formation with the silica aerogel coating could be designed to provide better thermal insulating properties with improved tensile properties of 3D weft-knitted spacer fabrics.

Acknowledgment

This research work was financially supported by Donghua University China.

References

[1] ISLAM S R, et al. 3D weft-knitted spacer fabrics (WKSFs) coated with silica aerogels as oil intercepting sorbents for use in static and dynamic water tests[J]. Industrial Crops and Products, 2022, 186: 115169.

[2] YIP J, et al. Study of three-dimensional spacer fabrics: Physical and mechanical properties[J]. Journal of Materials Processing Technology, 2008, 206(1/2/3): 359-364.

[3] ISLAM S R, et al. Influence of silica aerogels on fabric structural feature for thermal isolation properties of weft-knitted spacer fabrics[J]. Journal of Engineered Fibers and Fabrics, 2019, 1558925019866446(14): 1558-9250.

[4] ZHENG H, et al. Assembly of silica aerogels within silica nanofibers: towards a super-insulating flexible hybrid aerogel membrane[J]. RSC Advances, 2015, 5(111):

91813-91820.

[5] MAO N, et al. The thermal insulation properties of spacer fabrics with a mechanically integrated wool fiber surface[J]. Textile Research Journal, 2007, 77(12): 914-922.

[6] ISLAM S R, et al. Surface modification of weft-knitted spacer fabrics for hydrophobic behavior using silica aerogels dip-coating[C]. Shanghai: 2019 Sino-Africa International Symposium on Textiles and Apparel, 2019-11-8.

[7] ABEER A, et al. Polypropylene/lignin/POSS nanocomposites: thermal and wettability properties, application in water remediation[J]. Materials, 2021, 14(14): 3950.

[8] XIU Y, et al. UV and thermally stable superhydrophobic coatings from sol-gel processing[J]. Journal of Colloid and Interface Science, 2008, 326(2): 465-470.

[9] ARUMUGAM V, et al. Investigation on thermo-physiological and compression characteristics of weft-knitted 3D spacer fabrics[J]. The Journal of the Textile Institute, 2017, 108(7): 1095-1105.

[10] ISLAM S R, et al. Using various concentrations of SiO_2 aerogel for oil wicking, spreading, and interception tests of 3D weft-knitted spacer fabrics[J]. The Journal of the Textile Institute, 2022, 10: 1754-2340.

[11] ZHU Q, et al. Modified silica sol coatings for highly hydrophobic cotton and polyester fabrics using a one-step procedure[J]. Industrial and Engineering Chemistry Research, 2011, 50(10): 5881-5888.

[12] ISLAM S R, et al. Hydrophobic and oleophilic 3D weft-knitted spacer fabrics coated by silica aerogels with five different concentrations[J]. Journal of Industrial Textiles, 2022, 15280837211029048 (52): 1-25.

[13] ISLAM S R, et al. Mechanical properties of weft-knitted spacer fabrics integrated with silica aerogels[J]. Journal of Donghua University (English Edition), 2019, 36 (6): 559-563.

[14] ZIAEI M, et al. Thermal insulation property of spacer fabrics integrated by ceramic powder impregnated fabrics[J]. Journal of Indian Textile, 2013, 43(1): 20-33.

[15] ISLAM S R, et al. The study of hydrophobicity and oleophilicity of 3D weft-knitted spacer fabrics integrated with silica aerogels[J]. Journal of Industrial Textiles, 2022, 15280837211029048 (51): 8804S-8825S.

[16] ISLAM S R, et al. Oil/water separation of weft-knitted spacer fabrics integrated with silica aerogels by sol-gel method[C]. Shanghai: 2021 Sino-Africa International Symposium on Textiles and Apparel, 2021-11-5.

[17] STARK C, FRICKE J. Improved heat-transfer models for fibrous insulations[J]. International Journal of Heat and Mass Transfer, 1993, 36(3): 617-625.

Synthesis and Characterization of Polyborosilazane (PBSZ) for SiBNC Ceramic

MALIK Hamza[1], SHEN Jian[1], TANG Zicheng[1], LIU Yong[1,2*]

[1] *State Key Laboratory for Modification of Chemical Fibers and Polymer Materials, College of Materials Science and Engineering, Donghua University, Shanghai 201620, China*

[2] *The Key Laboratory of High-Performance Fiber and Product, Ministry of Education, Donghua University, Shanghai 201620, China*

[*] *Correspondence author's email*: liuyong@ dhu. edu. cn

Abstract: A SiBNC ceramic precursor, polyborosilazane (PBSZ), was successfully synthesized using boron trichloride (BCl_3), trichlorosilane ($HSiCl_3$), and hexamethyldisilazane (Me_6Si_2NH) as raw materials. Fourier transform infrared (FTIR) spectroscopy, nuclear magnetic resonance (NMR) analysis, X-ray photoelectron spectroscopy (XPS), thermogravimetric analysis (TGA), and density test (DSC) were used to analyze the structure and the high-temperature performance of the obtained PBSZ. Results showed that the network of Si-N-B and six-membered B-N rings were presented in the PBSZ structure. The ceramic yield of synthesized PBSZ at 800 ℃ in an N_2 atmosphere was 53.9%.

Keywords: SiBNC ceramic; Ceramic precursor; Polyborosilazane; Polymer-derived method

1 Introduction

SiBNC ceramics are extremely promising materials for applications in extreme environments due to their excellent thermal properties, high mechanical properties, and excellent oxidation and creep resistance characteristics[1-2]. Different studies proved that the SiBNC quaternary system exhibits excellent high-temperature performance as compared to binary and ternary ceramics. As Muller shows, the decomposition temperature was about 2 000 ℃ for the 5.9% boron content of SiBNC ceramics, while for SiCN ceramics it was only about 1 500 ℃[3]. The low density and excellent thermal properties, mainly because of the presence of boron content in SiBNC ceramics, are also highly suitable for applications in the aerospace and aviation industry[3-4]. However, because of the very low diffusion coefficients of Si and B, the polymer-derived ceramics (PDCs) method is the effective route to synthesize SiBNC ceramics[3, 5-7]. Hence, the preparation of high-performance PBSZ is important to obtain the high-temperature performance of SiBNC ceramics[6, 8-9].

With the aim of improving the performance of PBSZ, many researchers have synthesized many different kinds of PBSZ. Markus and coworkers[3] synthesized high-yield polyborosilazane precursors by using different polymer routes. Soluble and meltable hyperbranched polyborosilazanes (PBSZ) with Si—H and CH =CH_2 were synthesized by Kong et al.[10]. Bernard et al.[6] synthesized [B($C_2H_4SiCH_3NCH_3$)$_3$] by aminolysis of [B($C_2H_4SiCH_3Cl_2$)$_3$] (tris (dichlorosilylethyl) boranes). At ambient conditions, through the reaction of [(trichlorosilyl) dichloroboryl] ethane (TSDE) and hexamethyldisilazane borazine precursors were synthesized by Jaschke. Borazine precursors were also synthesized by Jansen et al. via cross-linking of Cl_3Si—CH_2—BCl_2, (CH_3)Cl_2Si—CH_2—BCl_2 and methylamine. The obtained SiBNC ceramics exhibit excellent thermal properties and oxidation resistance at temperatures up to 1 300 ℃[6]. However, it is difficult to avoid moisture and keep the complete process in an inert atmosphere during multiple-step processes.

Herein, we synthesized a PBSZ for SiBNC ceramic by using boron trichloride (BCl_3), trichlorosilane ($HSiCl_3$), and hexamethyldisilazane (Me_6Si_2NH) as starting materials via a mono-step method. In addition, the structure and properties of synthesized PBSZ were also analyzed.

2 Experiments

2.1 Materials

N_2 with 99.99% purity was purchased from Shanghai Shenzhong Industrial Gas Co. Ltd., China, BCl_3 with 99.5% purity was purchased from Beijing Multi-Technology Co. Ltd., China. $HSiCl_3$ with 99.7% purity, and Me_6Si_2NH with 99% purity were purchased from Bailingwei Chemical Technology Co. Ltd., China.

2.2 Polymer precursor synthesis

PBSZ precursors were synthesized from BCl_3, $HSiCl_3$, and Me_6Si_2NH, and the substance ratio of $HSiCl_3$: BCl_3 : Me_6Si_2NH is 3 : 2 : 15. The raw materials were added to the flask at −30 ℃ slowly and simultaneously, since this reaction was a violent process accompanied by much heat release. The byproducts and n-hexane were evaporated for 2 h at 80 ℃. To obtain PBSZ with appropriate viscosity, the flask was subsequently heated at a rate of 20 ℃/h up to 200–250 ℃, where it was maintained for 2 h.

2.3 Characterization

FTIR spectroscopy of PBSZ was recorded on Nicolet 8700 by the KBr method. The scanning wave number range is 4 000–500 cm^{-1}, and the number of scans are 32. The NMR spectra of samples, ^{1}H, ^{11}B, ^{13}C, and ^{29}Si were analyzed by an Avance 400/600 spectrometer. The frequencies of ^{29}Si spectrum, ^{11}B spectrum, and ^{13}C spectrum are 400.13 MHz, 79.5 MHz, 128.37 MHz, and 100.6 MHz, respectively, under MAS conditions. XPS measurements were done on the Escalab250 instrument. The degree of vacuum is 2×10^{-7} Pa, and the sample surface was sputtered for 10 min at a resolution of 0.9 eV/104 counts per second (CPS) and a signal-to-noise ratio of 30 : 1. TGA of the precursor was analyzed by the TG209 F1 Tris TGA from Germany. The precursor was heated to 800 ℃ at a constant temperature of 10 ℃/min under the condition of 1–5 mg, nitrogen protection and flow rate of 20 mL/min. The density of the precursor was measured by the 1000UPY–20T automatic true density analyzer produced by the American Quanta Instrument Company, under a nitrogen atmosphere, the gas flow rate was 50 mL/min, the heating rate was 10 ℃/min, the temperature range was 50–250 ℃, and the sample amount was 5–10 mg.

3 Results and Discussion

FTIR and NMR spectroscopy were used to analyze the structures of PBSZ. As shown in Fig. 1, the peaks at about 3 440 cm^{-1} and 1 230 cm^{-1} were caused by the N—H vibration. A similar phenomenon was also observed for the peak at 2 190 cm^{-1}, which was assigned to the Si—H bond, and the signal observed at 1 430 cm^{-1} was attributed to the Si—C bond. While the stretching vibration peaks at 2 930 cm^{-1} and 780 cm^{-1} were assigned to the C—H bond. The Si-CH_3 group was assigned to the peaks, which were at 880 cm^{-1} and 710 cm^{-1}. While the peak observed at 1 020–1 070 cm^{1} was assigned to the B—C bond in accordance with the chemical construction of PBSZ.

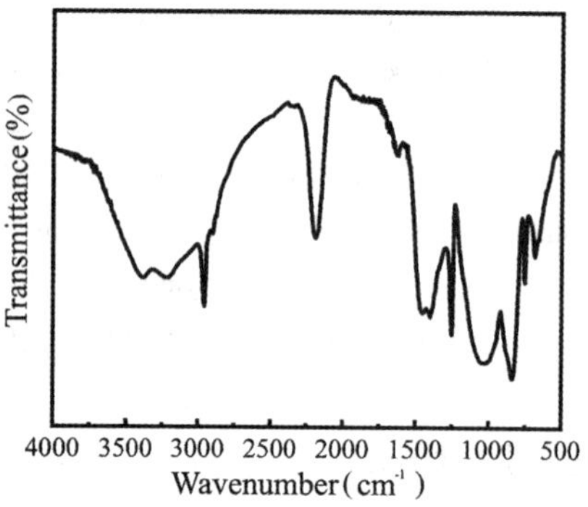

Fig. 1 FTIR spectra of PBSZ

Figure 2 represents the FT-IR spectroscopy of the PBSZ synthesized at different temperatures. Figure 2 shows that the stretching vibration peak of Si—H becomes weak at 230 ℃ and almost vanishes at 250 ℃ as compared to that at 200 ℃, which indicates the synthesis of the PBSZ at 250 ℃ via dehydrogenation coupling between Si—H or Si—H and N—H. The sharpness of the characteristic absorption peaks of C—H and —Si$(CH_3)_3$ is also reduced at 250 ℃ as compared to that

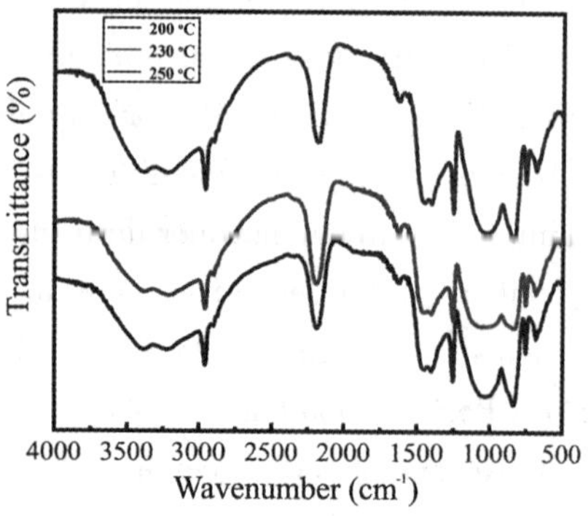

Fig. 2 FT-IR spectra of PBSZ polymerized at 200 ℃, 230 ℃, and 250 ℃

at 200 ℃, which is also confirmed by the TGA curve of PBSZ. The FTIR spectra show an amorphous precursor with Si—N bonds at 918 cm^{-1}, B—N bonds at 1 438 cm^{-1}, Si—H bonds at 2 150 and 1 205 cm^{-1}, and an N—H bond with a strong and broad peak at 3 383 cm^{-1}. The composition of the PBSZ at different polymerization temperatures is shown in Tab. 1.

Tab. 1 Composition of the PBSZ at different polymerization temperatures

Temperature (℃)	Compositions(%)				
	Si	B	N	C	H
200	42.3	4.1	17.7	27.1	8.6
230	41.8	4.2	18.5	26.8	8.5
250	41.5	4.2	18.9	26.7	8.4

In addition, the NMR spectra of the precursor exhibit typical PBSZ characteristics. Figure 3 shows the results of ^{1}H-NMR, ^{13}C-NMR, ^{11}B-NMR, and ^{29}Si-NMR. Figure 3 (a) represents the particular spectra of ^{1}H-NMR. Peaks at 4.5–5.0 parts per million (ppm) and 0.6–2.2 ppm were attributed to the bonds of Si—H and N—H, while the Si—CH_3 group belonged to the strong peak at 0–0.45 ppm. The broad peaks at 0–10 ppm were related to the Si—CH_3 group. Si—CH_3 is the only form in which the C element exists (Fig. 3 (b)), which be released while polymerizing at a high temperature. In ^{11}B-NMR spectra, the BN_3 site was attributed to the signal at 27.2 ppm, while SiC_3N was attributed to the signal at 3.4 ppm (Fig. 3 (c)). The signal at 3.5 ppm in the ^{29}Si-NMR spectra of the cross-linked single-source precursor was attributed to the $SiC_2^{(sp3)}N_2$ sites[11], while the signal at −38 ppm was assigned to $SiC_2^{(sp3)}N_3$ sites. Thus, the ^{29}Si-NMR spectra revealed that the hydroboration and vinyl polymerization reactions that occurred under certain conditions resulted in the transformation of the sp^2 hybridized carbons of the vinyl groups into sp^3 carbon sites (Fig. 3 (d)).

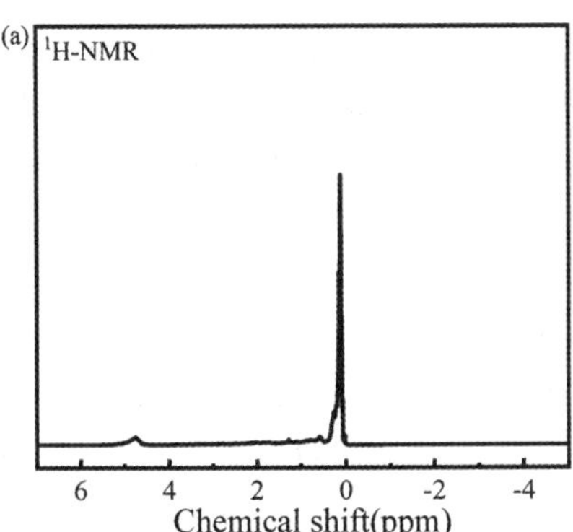

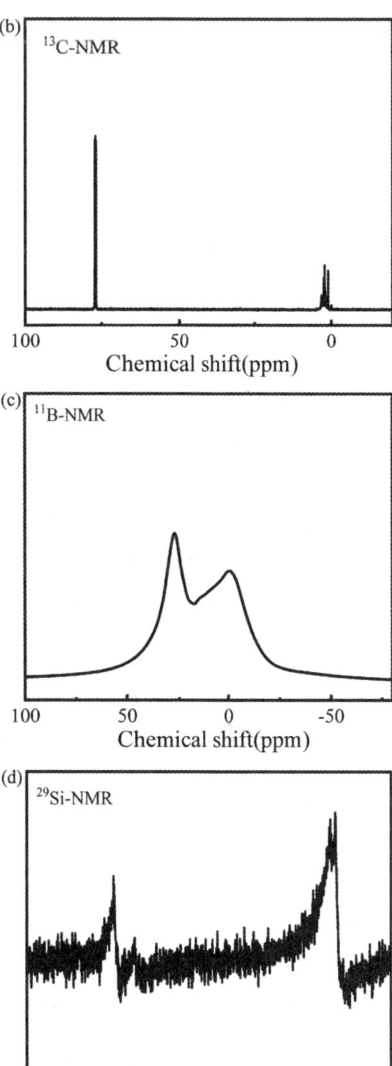

Fig. 3 NMR spectra of PBSZ: (a) ^{1}H-NMR; (b) ^{13}C-NMR; (c) ^{11}B-NMR; (d) ^{29}Si-NMR

In XPS analysis (Fig. 4) Si, N, and B elements were fitted according to the Gaussian function to explore the existing form and the binding state of each element in the precursors. Figure 4 (a) represents that the Si element mainly exists in three forms, Si—C (101.1 eV), Si—N (101.9 eV), and Si—O (103.0 eV) in the precursor. During the experiment, PBSZ is partially oxidized, resulting in the formation of Si—O. Similar to Si pattern, N1s pattern has three peaks: N—H (399.0 eV), N—Si (398.1 eV), and N—B (398.3 eV) (Fig. 4 (b)). In the B1s spectra, the only peak at 191.1 eV belongs to the B—N bond, which represents its presence in PBSZ. This shows that these atoms were incorporated into the ceramics successfully (Fig. 4 (c)). XPS shows that multi-phase

PBSZ were formed (Fig. 4 (d)). Spectra show that B1s occurs at 190.8 eV, C1s at 280 eV, N1s at 395.4 eV, Si-2p at 99.0 eV, Si-2s at 180.4 eV, and O1s at 545.4 eV.

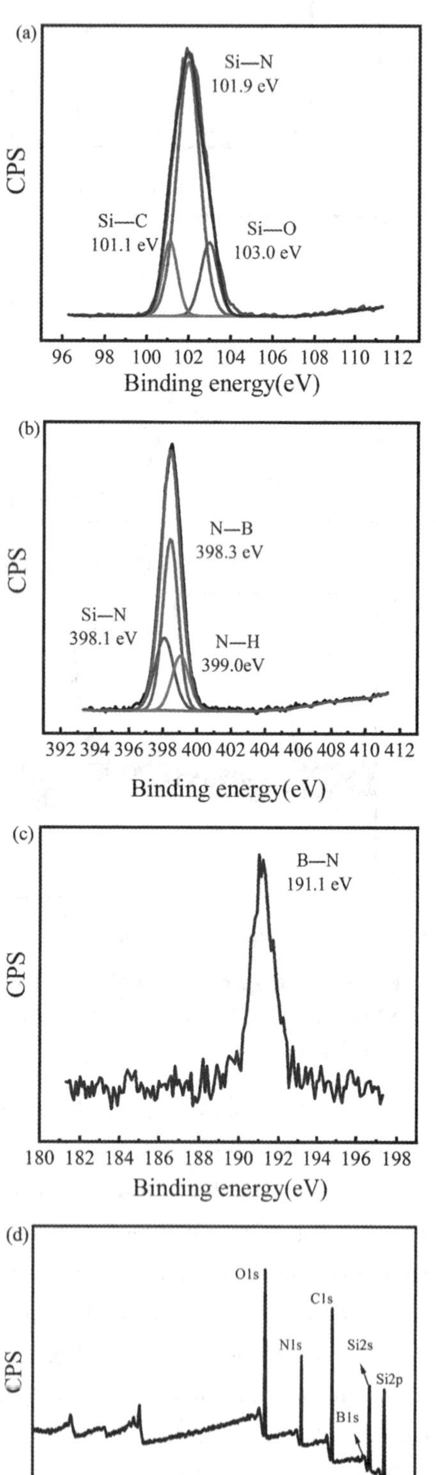

Fig. 4 XPS: (a) Si; (b) N1s; (c) B1s; (d) PBSZ

Figure 5 shows TGA of PBSZ for analysis of the weight loss of PBSZ[12]. The ceramics yield at 800 ℃ under an N_2 atmosphere of PBSZ was 53.9%. The thermal weight loss of the precursor PBSZ was mainly divided into two stages. The first one was from room temperature up to 200 ℃, and the weight loss was about 6%. The main reason for the weight loss in this process was the removal of unreacted precursor small molecules, such as Me_6Si_2NH and Me_3SiCl. The second stage was from 200-550 ℃, and the weight loss was about 36%. The weight loss in the first part of this stage was relatively fast, mainly due to the transamination reaction and cross-linking reaction of the precursor polymer, reaching a highly cross-linked state. The second half of the weight loss was mainly due to the organic-to-inorganic conversion reaction. A small number of methyl radicals and hydrogen radicals combine with each other to generate methane and hydrogen. After 700 ℃, there was basically no weight loss, indicating that the polymer to ceramic transformation has been basically completed.

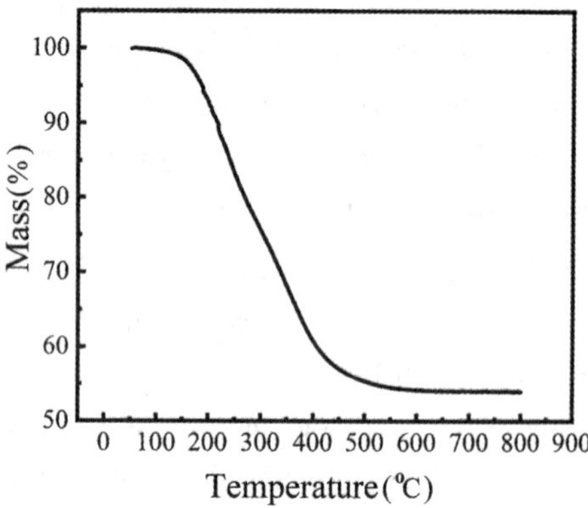

Fig. 5 TGA of PBSZ from room temperature to 800 ℃

Tab. 2 is the density test result of the PBSZ precursor obtained by polymerizing for 3 h at 250 ℃. As can be seen from Table 2, the average density of the precursor at room temperature was 1.037 8 g · cm^{-3}. Figure 6 shows the DSC heating curve of PBSZ prepared at a polymerization temperature of 250 ℃ when the polymerization time is 3 h. It can be seen from the curve that with the increase in the polymerization temperature, the vitrification of PBSZ increases. As the transition temperature gradually increased, the glass transition temperature of the precursor prepared under the condition of 250 ℃ was about 79 ℃. In addition, with the increase in temperature, there was no obvious melting peak but a line extending in the endothermic direction. The curve shows that the polymer is amorphous.

Tab. 2 The test results of PBSZ density

Times/n	Volume /V(cm^{-3})	Density/ρ($g \cdot cm^{-3}$)
1	1.93	1.03
2	1.92	1.03
3	1.92	1.04
Average value	1.92	1.03

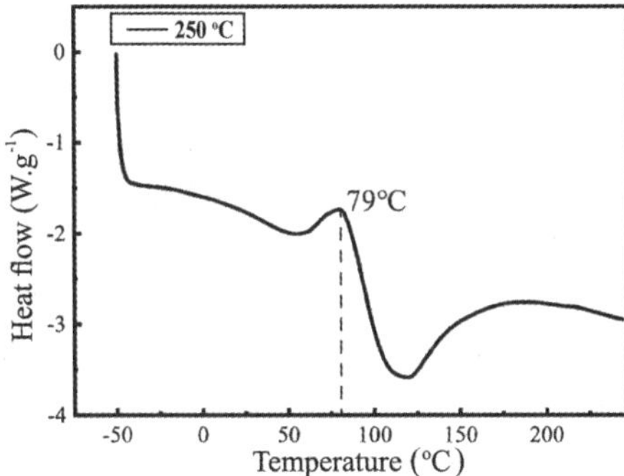

Fig. 6 DSC curve of PBSZ synthesized at 250 ℃

4 Conclusions

A PBSZ precursor was successfully synthesized by using BCl_3, $HSiCl_3$, and Me_6Si_2NH as raw materials via a mono-step method. FTIR and NMR spectroscopy show Si—N, N—H, and B—N bonds are in the structure. TGA indicates that PBSZ contains a ceramic yield of about 53.9% at 800 ℃ in an N_2 atmosphere. In addition, it shows that the precursor is amorphous. Thus, the study indicates that the PBSZ precursor is suitable for producing high-performance SiBNC ceramics.

References

[1] KAMAL A, RAJASEKHAR B V, PAINULY A, et al. A novel precursor for the synthesis of mixed non-oxide ultra high temperature ceramics[J]. Journal of Inorganic and Organometallic Polymers and Materials, 2020, 30(5): 1578-1588.

[2] JI X, SHAO C, WANG H, et al. A simple and efficient method for the synthesis of SiBNC ceramics with different Si/B atomic ratios[J]. Ceramics International, 2017, 43(10): 7469-7476.

[3] ZHAO H, CHEN L, LUAN X, et al. Synthesis, pyrolysis of a novel liquid SiBCN ceramic precursor and its application in ceramic matrix composites[J]. Journal of the European Ceramic Society, 2017, 37(4): 1321-1329.

[4] PUERTA A R, REMSEN E E, BRADLEY M G, et al. Synthesis and ceramic conversion reactions of 9-BBN-modified allylhydridopolycarbosilane: a new single-source precursor to boron-modified silicon carbide[J]. Chemistry of Materials, 2003, 15(2): 478-485.

[5] GE Y, SUN S, WANG Q, et al. Effect of Fe-contained species on the preparation of α-Si_3N_4 fibers in combustion synthesis[J]. Journal of the American Ceramic Society, 2016, 99(4): 1464-1471.

[6] ZHANG C Y, LIU Y, HAN K Q, et al. Synthesis and characterization of a novel polyborosilazane for SiBNC ceramic[C]. Kitakyushu: 2018 5th Global Conference on Polymers and Composite Materials, 2018-04-10.

[7] COLOMBO P, MERA G, RIEDEL R, et al. Polymer-derived ceramics: 40 years of research and innovation in advanced ceramics[J]. Ceramics Science and Technology, 2013, 245-320.

[8] WANG Y, CHEN L, XU T, et al. High char yield novolac modified by Si-BNC precursor: thermal stability and structural evolution[J]. Polymer Degradation and Stability, 2017, 137: 184-96.

[9] ZHANG Q, YANG Z, JIA D, et al. Synthesis and structural evolution of dual-boron-source-modified polysilazane derived SiBCN ceramics[J]. New Journal of Chemistry, 2016, 40(8): 7034-7042.

[10] KONG J, WANG M, ZOU J, et al. Soluble and meltable hyperbranched polyborosilazanes toward high-temperature stable SiBCN ceramics[J]. ACS Applied Materials & Interfaces, 2015, 7(12): 6733-6744.

[11] YUAN J, HAPIS S, BREITZKE H, et al. Single-source-precursor synthesis of hafnium-containing ultrahigh-temperature ceramic nanocomposites (UHTC-NCs)[J]. Inorganic Chemistry, 2014, 53(19): 10443-10455.

[12] GU J, LIANG C, DANG J, et al. Ideal dielectric thermally conductive bismaleimide nanocomposites filled with polyhedral oligomeric silsesquioxane functionalized nanosized boron nitride[J]. RSC Advances, 2016, 6(42): 35809-35814.

Tensile Properties of Injection Molded Recycled Carbon Fiber Reinforced Polycarbonate Bolts

YANG Yang[1], WANG Xianchen[1], YU Ying[2], YANG Yuqiu[1*]

[1] *College of Textiles, Donghua University, Shanghai 201620, China*

[2] *Huzhou College, Huzhou 313002, China*

[*] *Corresponding author's email*: amy_yuqiu_yang@dhu.edu.cn

Abstract: When traditional metal bolt fasteners connect resin-based composite materials, the connection performance of the joint is weakened due to the difference in thermal expansion coefficient between the metal and the composite material. In addition, metal bolt fasteners also have shortcomings such as easy corrosion and high density. Fiber-reinforced resin matrix composite bolts not only overcome the inherent defects of metal bolts, but also provide a useful exploration for the recycling of carbon fibers. In this paper, by blending recycled carbon fiber (rCF) with polycarbonate resin (PC), the dumbbell splines and bolt fasteners of polycarbonate composite reinforced with chopped carbon fiber were prepared by injection molding process. The uniaxial tensile tests were carried out on the dumbbell-shaped splines, bolt heads, screws and other regions respectively, and the samples were observed by an scanning electron microscope (SEM) to explore their uniaxial tensile mechanical properties and fracture failure mechanism. According to the standard of strength test for metal fasteners, this paper presents a strength test method for composite bolts. The uniaxial tensile strength and failure mode of rCF/PC bolts were also studied. The results show that the tensile strength of rCF/PC bolt is lower than that of rCF/PC dumbbell spline bolt. In composite bolts, the tensile strength of the screw area is greater than that of the bolt head. This is because the jet phenomenon occurs in the injection molding process of rCF/PC bolts, which makes the tensile strength of raw materials higher, but the tensile strength of bolts decreases significantly after preparation. The failure mode of rCF/PC bolts is different from that of metal bolts, where the screw is partially broken.

Keywords: Carbon fiber reinforced composite; Composite bolt; Fiber orientation; Recycled carbon fiber; Injection molding

1 Introduction

With the application of resin-based composites in various fields ranging from hardware appliances, electronic products, and sporting goods to aviation, automobile engineering, shipbuilding, and wind power generation, the application has become more and more extensive[1-3]. In the connection technology of composites, glue, riveting, bolt connection, and welding are widely used. Traditional bolting technology uses mainly metal materials. Although it has the advantages of high strength, high hardness and strong bearing capacity, it has inherent defects such as high density, poor corrosion resistance, and different linear thermal expansion coefficients when connecting with composites, which will reduce the quality of joints. Resin-based composite bolts not only have the advantages of light weight, but also have a similar linear thermal expansion coefficient with composite parts, and are more corrosion-resistant. They have broad development prospects in the emerging field of resin-based composite material connection technology[4].

The rising use of carbon fiber composites leads to an increase in waste containing carbon fibers. Wastes containing carbon fibers mainly include discarded carbon fibers, scraps from the production of composite components, scraps with substandard quality, and carbon fiber composite products that have reached their service life (such as aircraft, automobiles, and fan blades that contain carbon fiber composite components). These waste carbon fiber composites contain a large amount of carbon fibers with recycling value. Markets and Markets™ has analyzed and forecasted that the global recycled carbon fiber market will grow from $109 million in 2020 to $193 million in 2025[5-7].

This paper uses recycled carbon fiber (rCF) reinforced polycarbonate resin (PC) thermoplastic resin co-mixing and granulating, using injection molding technology, obtains injection molding recycled short-cut carbon fiber reinforced resin matrix composite dumbbell-shaped splines and bolt fasteners. The uniaxial tensile tests were carried out on the dumbbell-shaped splines, bolt heads, screws and other regions respectively, and the samples were observed by an scanning electron microscope (SEM) to explore their uniaxial tensile mechanical properties and fracture failure mechanism[8].

2 Experiments

2.1 Raw materials

The raw material used in this experiment is to blend the recycled short-cut carbon fiber with polycarbonate (PC), and adopt the screw extrusion mechanism to prepare the pellet. The length of the recovered short-cut carbon fiber is 0.5 mm, the diameter is 8 μm, and the mass fraction in the pellet is 25%. The raw material was supplied by Nantong Fuyuan New Material Co. Ltd., (Jiangsu, China).

2.2 Sample preparation

The samples used in this experiment are prepared by injection molding process with the rCF/PC pellets mentioned above. One is a standard dumbbell-shaped stretch spline in accordance with GB/T 1447—2005 (shown in Fig. 1) and the other is a rough-toothed composite bolt of M8×60 specification (shown in Fig. 2). The injection molding process of dumbbell spline is set as follows. The drying temperature is 100 ℃, the drying time is 4 h, the injection molding temperature from the first section to the nozzle is 265 ℃, 275 ℃, 285 ℃, 285 ℃, 285 ℃, and 285 ℃, the holding pressure is 50MPa, the pressure holding time is 10 s, the mold temperature is 70 ℃, the screw back pressure is 3 bar (1 bar = 0.1 MPa), and the cooling time is 20 s. The injection molding process of composite bolts is set as follows. The temperature of drying material is 100 ℃, the drying time is 4 h, the injection molding temperature from the first section to the nozzle is 220 ℃, 220 ℃, 245 ℃, 230 ℃, and 230 ℃, the holding pressure is 50 MPa, the holding time is 10 s, the mold temperature is 70 ℃, the screw back pressure is 3 bar, and the cooling time is 25 s. The density of rCF/PC sample is 1.23 g/cm^3, and the fiber volume fraction is 33.33%.

Fig. 1 Dumbbell spline

Fig. 2 Composite bolts

2.3 Methods

2.3.1 Dumbbell spline tensile test

According to the relevant provisions of GB/T 1447—2005, the QJ-212C electronic universal testing machine was used to carry out the tensile test on the dumbbell-shaped sample of rCF/PC. The tensile loading speed is set as 2 mm/min, and the test is repeated for 3 times.

2.3.2 Composite bolt tensile test

Due to the particularity of the shape of the composite bolt, when subjected to tensile stress, there is a stress concentration phenomenon at the connection between the bolt and the screw. Therefore, when the loading method specified in the National Military Standard of the People's Republic of China GJB 715.23A—2015 is used for the test, and the sample always shows the fracture failure at the junction of the bolt head and the screw. The tensile strength of the screw could not be obtained.

In order to fully reflect the tensile mechanical properties of injection-molded composite bolts. This article divides the tensile strength test of composite bolts into two parts: the tensile strength of the bolt head and the tensile strength of the screw.

The first part is the bolt head tensile strength test, which aims to test the connection strength of the transition zone between the bolt head and the screw. The clamping method is shown in Fig. 3, using a metal nut with sufficient strength to fasten the composite material bolts on the fixture from below the fixture, and the tensile loading speed is set to be 10 mm/min.

The second part is the screw tensile strength test, which aims to test the tensile strength of the screw part. The clamping method is shown in Fig. 4. The screw part of the composite bolt is fixed on the clamp from above and below the clamp by two metal nuts with sufficient strength. The joint between the bolt and the screw is much higher than the metal nut. The tensile loading speed is set to be 10 mm/min.

Fig. 3 Loading diagram for uniaxial tensile strength test of bolt head

Fig. 4 Load diagram of screw uniaxial tensile strength test

2. 3. 3 Fiber orientation research

To observe the fiber orientation of composite bolts, the sample must be embedded in thermosetting resin for curing. In this test, epoxy resin and a curing agent with an amount of 80% by weight of epoxy resin were selected, and after stirring, they were put into a vacuum pump for degassing for 1 h. Put the sample to be observed into a square silicone mold, pour the defoamed resin, and put it in an oven preheated to 90 ℃. The specific curing parameters are 2 h at 90 ℃, 1 h at 110 ℃, and 4 h at 130 ℃.

The cured sample was demolded, then cut, ground and polished. The samples were magnified 50 times, 100 times and 200 times respectively under the super depth of the field microscope for observation, then photographed and recorded.

3 Results and Discussion

3. 1 Raw material tensile properties

The results show that the tensile strength of the raw material is 148. 98 MPa, the tensile strength of the bolt head is 23. 99 MPa, and the tensile strength of the screw is 43. 66 MPa.

Use an SEM to observe the fractured section. The fractured section of the rCF/PC splines after the tensile test is shown in Fig. 5.

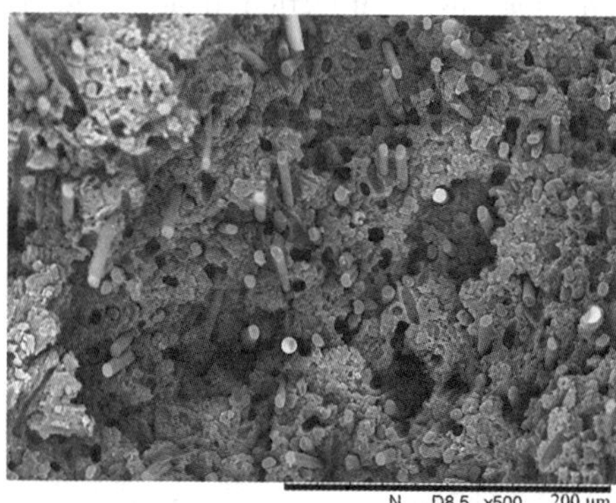

Fig. 5 Electron microscopic view of broken section of rCF/PC dumbbell spline at 500× magnification

3. 2 Tensile properties of bolt head

Photographs of the rCF/PC bolt after fracture are shown in Fig. 6. It can be seen that the rCF/PC bolt breaks at the junction of the bolt head and the screw. And the rCF/PC bolt has a protruding cone on the bolt head, and the screw head has a corresponding cone hole (shown in Fig. 7). The main failure modes of rCF/PC bolt epidermis are resin matrix fracture and fiber fracture and pull out. The resin matrix in the lumbar region of the cone presents a smooth surface, and a small amount of fibers are damaged by tension, leaving some holes. The tensile failure of the resin matrix at the cross section is caused by the poor fusion between the cone and other areas caused by the spray grain phenomenon.

Fig. 6 Picture of rCF/PC bolt that completed uniaxial tensile strength test of bolt head

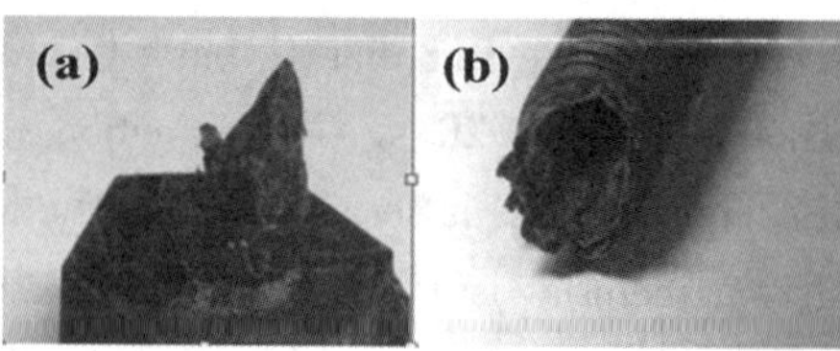

Fig. 7 Partial enlarged view of rCF/PC bolt: (a) bolt head taper; (b) screw head taper hole

Due to the poor fluidity of PC resin, it is sensitive to temperature. After the rCF/PC melt enters the mold with a lower temperature, the fluidity of the melt will be further reduced, and the melt will become a strip. The strip-shaped melt will be folded in the mold, making it difficult for the carbon fibers at the melt fold to pass through the melt interface and unable to connect the melt at the fold.

3.3 Screw tensile properties

In order to comprehensively study the tensile mechanical properties of composite bolts, this paper designs a test method for screw strength with reference to relevant standards.

After completing the uniaxial tensile test, the fracture section of the screw is shown in Fig. 8. The failure mode of rCF/PC bolt is the fracture of screw layer.

Fig. 8 Picture of the rCF/PC bolt that completed the screw uniaxial tensile test

During the test, under the action of uniaxial tensile force, slip occurs between the inner thread layer of the metal nut and the outer thread layer of the bolt, and the metal nut begins to move upward. At the same time, the threaded layer on the upper half of the bolt began to be squeezed by the metal nut, causing squeeze damage. After that, the head of the bolt is directly subjected to the pressure given by the metal nut, and the load increases rapidly until brittle fracture occurs.

After the rCF/PC bolt is damaged (shown in Fig. 9), the thread layer above the metal nut shows a flower-like damage due to the extrusion of the metal nut. The thread of the part engaged with the metal nut can still be rotated out of the nut without obvious deformation and damage. The junction of the bolt head and the screw is broken, the thread layer and the screw are peeled off, and the smooth round rod is exposed (as shown in Fig. 9(b). Enlarge the fracture of the thread layer (shown in Fig. 10). The thread layer and the round rod area have obvious delamination phenomenon. Enlarge the round bar area (shown in Fig. 11). There are undamaged smooth surfaces and a small amount of damaged matrix on the surface of the round rod area. This is due to the formation of spray lines due to poor fusion between the resin substrates during the injection molding process. The poor welding between the resins makes the thread layer and the round rod area prone to delamination. When subjected to a uniaxial tensile force, it is difficult to fully exert the strength of the screw.

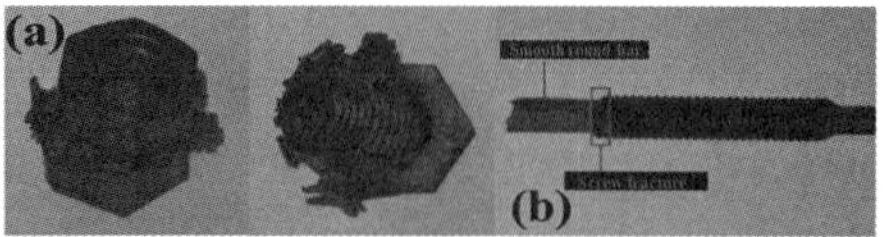

Fig. 9 Pictures of rCF/PC bolts that have completed uniaxial tensile strength testing of bolts: (a) after bolt head damage; (b) after screw damage

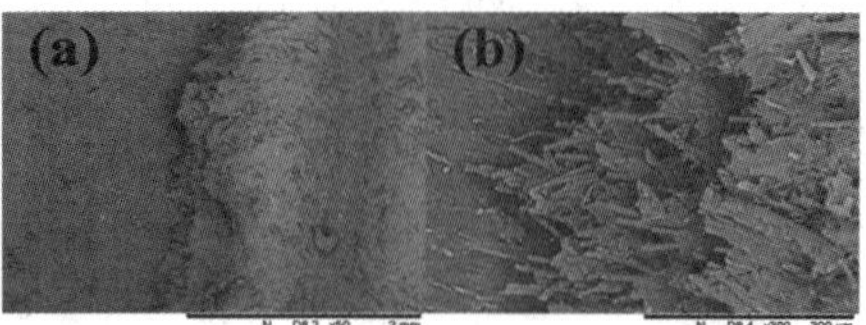

Fig. 10 Enlarged view of different multiples of threaded fracture: (a) 50× magnification; (b) 300× magnification

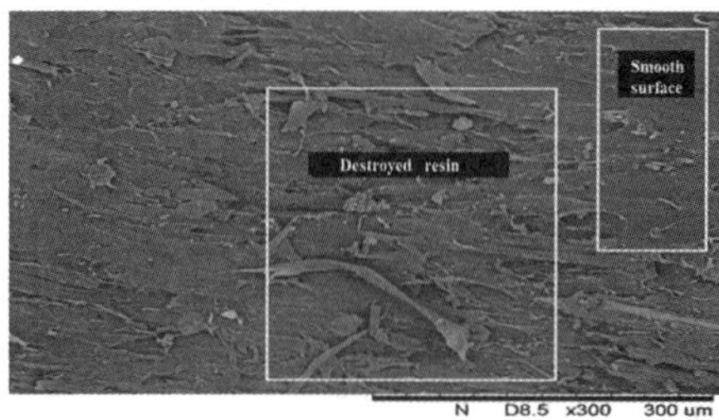

Fig. 11 Partial enlarged view of the round rod area

4 Conclusions

(1) Through the uniaxial tensile test of the bolt head, it is found that the failure mechanism of rCF/PC bolt is: the fracture of resin matrix, the fracture and extraction of carbon fibers in the epidermis layer, and the failure and fracture of resin matrix due to poor welding in the core layer.

(2) The uniaxial tensile test of the screw part shows that the failure mode of the screw part of the rCF/PC bolt is the fracture of the thread layer, and the failure mechanism is the fracture of the resin matrix, the fracture of the fiber and the fracture of the cortex caused by the pulling out.

(3) Compared with the products made of rCF/PC, the tensile strength of dumbbell spline > screw > bolt head. More specifically, due to the injection molding process of rCF/PC bolts will appear jet grain phenomenon, which makes the tensile strength of the raw material it-

self is very large, but the tensile strength of the bolt is significantly decreased after it is made.

References

[1] OLHAN S, KHATKAR V, BEHERA B K. Review: textile-based natural fibre-reinforced polymeric composites in automotive lightweighting [J]. Journal of Materials Science, 2021, 56: 18867-18910.

[2] SEO J, KIM D Y, KIM D C, et al. Recent developments and challenges on machining of carbon fiber reinforced polymer composite laminates [J]. International Journal of Precision Engineering and Manufacturing, 2021, 22: 2027-2044.

[3] SHEHRYAR A, SANAN A, et al. A comparative study on the effect of carbon-based and ceramic additives on the properties of fiber reinforced polymer matrix composites for high temperature applications[J]. Ceramics International, 2021, 24(47): 33956-33971.

[4] ZHOU L, QIN Z W, LIU S, et al. Progress on joining technology of thermoplastic resin matrix composites [J]. Materials Reports, 2019, 33(19): 3177-3183.

[5] SHEN C. Application of recycled carbon fiber in automotive industry[J]. Auto Time, 2019(21): 10-12.

[6] DU X Y, CHEN X Q, WANG Z Y, et al. Recycling and reusing of carbon fiber reinforced polymer: a current technical progres [J]. Polymer Materials Science and Engineering. 2020, 36(8): 182-190.

[7] LI L Y, YIN X P, WANG D, et al. Research progress on recovery technology of carbon fiber reinforced plastic [J/OL]. New Chemical Materials: 1-6[2021-10-07]. http://kns. cnki. net/kcms/detail/11. 2357. TQ. 20210511. 1455. 302. html.

[8] CUI S S, WANG S, ZHANG T. Design and experiment of m5 composite material bolts [J]. Aerospace Materials & Technology, 2017, 47(4): 23-26.